T0391227

The Palgrave Handbook of the History of Human Sciences

David McCallum
Editor

The Palgrave Handbook of the History of Human Sciences

Volume 1

With 45 Figures and 7 Tables

palgrave
macmillan

Editor
David McCallum
Victoria University
Melbourne, VIC, Australia

ISBN 978-981-16-7254-5   ISBN 978-981-16-7255-2 (eBook)
https://doi.org/10.1007/978-981-16-7255-2

© Springer Nature Singapore Pte Ltd. 2022
This work is subject to copyright. All rights are reserved by the Publisher, whether the whole or part of the material is concerned, specifically the rights of translation, reprinting, reuse of illustrations, recitation, broadcasting, reproduction on microfilms or in any other physical way, and transmission or information storage and retrieval, electronic adaptation, computer software, or by similar or dissimilar methodology now known or hereafter developed.
The use of general descriptive names, registered names, trademarks, service marks, etc. in this publication does not imply, even in the absence of a specific statement, that such names are exempt from the relevant protective laws and regulations and therefore free for general use.
The publisher, the authors and the editors are safe to assume that the advice and information in this book are believed to be true and accurate at the date of publication. Neither the publisher nor the authors or the editors give a warranty, expressed or implied, with respect to the material contained herein or for any errors or omissions that may have been made. The publisher remains neutral with regard to jurisdictional claims in published maps and institutional affiliations.

This Palgrave Macmillan imprint is published by the registered company Springer Nature Singapore Pte Ltd.
The registered company address is: 152 Beach Road, #21-01/04 Gateway East, Singapore 189721, Singapore

Contents

Volume 1

Part I What Is the History of Human Sciences **1**

1 What Is the History of the Human Sciences? 3
   Roger Smith

2 Kant After Kant: Towards a History of the Human Sciences
   from a Cosmopolitan Standpoint 29
   Steve Fuller

3 History of the Human Sciences in France: From Science
   de l'homme to Sciences Humaines et sociales 57
   Nathalie Richard

4 Narrative and the Human Sciences 79
   Kim M. Hajek

5 Diltheyan Understanding and Contextual Orientation in the
   Human Sciences 109
   Rudolf Makkreel

6 Durkheimian Revolution in Understanding Morality: Socially
   Created, Scientifically Grasped 133
   Raquel Weiss

7 Problematizing Societal Practice: Histories of the Present and
   Their Genealogies 159
   Ulrich Koch

8 Human Sciences and Theories of Religion 181
   Robert A. Segal

9 Historical Studies in Nineteenth-Century Germany: The Case
   of Hartwig Floto 207
   Herman Paul

v

Part II Visualizations 227

10 Anatomy: Representations of the Body in Two and Three Dimensions 229
Anna Maerker

11 History of Embryology: Visualizations Through Series and Animation 259
Janina Wellmann

12 Visualizations in the Sciences of Human Origins and Evolution 291
Marianne Sommer

Part III Self and Personhood 321

13 Made-Up People: Conceptualizing Histories of the Self and the Human Sciences 323
Elwin Hofman

14 Inner Lives and the Human Sciences from the Eighteenth Century to the Present 349
Kirsi Tuohela

15 Michel Foucault and the Practices of "Spirituality": Self-Transformation in the History of the Human Sciences 375
Nima Bassiri

16 Human Sciences and Technologies of the Self Since the Nineteenth Century 401
Fenneke Sysling

17 The Sex of the Self and Its Ambiguities, 1899–1964 423
Geertje Mak

Part IV Anthropology 455

18 Economic Anthropology in View of the Global Financial Crisis 457
Timothy Heffernan

19 Anthropology, the Environment, and Environmental Crisis 483
María A. Guzmán-Gallegos and Esben Leifsen

20 On the Commonness of Skin: An Anthropology of Being in a More Than Human World 505
Simone Dennis

21	Indigeneity: An Historical Reflection on a Very European Idea Judith Friedlander	533

Part V Historical Sociology **559**

22	The Past and the Future of Historical Sociology: An Introduction Marta Bucholc and Stephen Mennell	561
23	Power and Politics: State Formation in Historical Sociology Helmut Kuzmics	583
24	Organized Violence and Historical Sociology Christian Olsson and Siniša Malešević	625
25	Historical Sociology of Law Marta Bucholc	651
26	Sport and Leisure: A Historical Sociological Study Dominic Malcolm	677
27	Norbert Elias and Psychoanalysis: The Historical Sociology of Emotions Robert van Krieken	699
28	Identity, Identification, Habitus: A Process Sociology Approach Florence Delmotte	725
29	Hidden Gender Orders: Socio-historical Dynamics of Power and Inequality Between the Sexes Stefanie Ernst	749
30	Collective Memory and Historical Sociology Joanna Wawrzyniak	775

Part VI History of Sociology **805**

31	Historiography and National Histories of Sociology: Methods and Methodologies Fran Collyer	807
32	The History of Sociology as Disciplinary Self-Reflexivity George Steinmetz	833
33	Locating the History of Sociology: Inequality, Exclusion, and Diversity Wiebke Keim	865

34	**Colonialism and Its Knowledges**	893
	Sujata Patel	
35	**Knowledge Boundaries and the History of Sociology**	917
	Per Wisselgren	
36	**Social Theory and the History of Sociology**	935
	Hon-Fai Chen	
37	**José Carlos Mariátegui and the Origins of the Latin American Sociology**	961
	Fernanda Beigel	

Volume 2

Part VII History of Psychology **975**

38	**Psychologies: Their Diverse Histories**	977
	Roger Smith	
39	**Social Psychology: Exemplary Interdiscipline or Subdiscipline**	1005
	James M. M. Good	
40	**Of Power and Problems: Gender in Psychology's Past**	1035
	Elissa N. Rodkey and Krista L. Rodkey	
41	**Indigenous Psychologies: Resources for Future Histories**	1065
	Wade E. Pickren and Gülşah Taşçı	
42	**Children as Psychological Objects: A History of Psychological Research of Child Development in Hungary**	1087
	Zsuzsanna Vajda	
43	**A History of Self-Esteem: From a Just Honoring to a Social Vaccine**	1117
	Alan F. Collins, George Turner, and Susan Condor	
44	**Vygotsky, Luria, and Cross-Cultural Research in the Soviet Union**	1145
	René van der Veer	
45	**Values and Persons: The Persistent Problem of Values in Science and Psychology**	1167
	Lisa M. Osbeck	
46	**Politics and Ideology in the History of Psychology: Stratification Theory in Germany**	1195
	Martin Wieser	

| Part VIII | History of Psychiatry | 1221 |

47 The Mental Patient in History 1223
Peter Barham

48 Asylums and Alienists: The Institutional Foundations of
Psychiatry, 1760–1914 1253
David Wright

49 Forensic Psychiatry: Human Science in the Borderlands
Between Crime and Madness 1273
Eric J. Engstrom

50 Psychiatry and Society 1305
Petteri Pietikäinen

51 Early Child Psychiatry in Britain 1331
Nicola Sugden

52 Human Experimentation and Clinical Trials in Psychiatry 1357
Erika Dyck and Emmanuel Delille

53 Colonial and Transcultural Psychiatries: What We Learn From
History 1379
Sloan Mahone

54 Geriatric Psychiatry and Its Development in History 1403
Jesse F. Ballenger

55 Antipsychiatry: The Mid-Twentieth Century Era
(1960–1980) 1419
Allan Beveridge

| Part IX | History of Economics | 1451 |

56 History of Thought of Economics as a Guide for the Future 1453
Dieter Bögenhold

57 Classical Political Economy 1473
Heinz D. Kurz

58 Neoclassical Economics: Origins, Evolution, and Critique 1501
Reinhard Neck

59 The Austrian School and the Theory of Markets 1541
David Emanuel Andersson and Marek Hudik

60 Joseph Schumpeter and the Origins of His Thought 1563
Panayotis G. Michaelides and Theofanis Papageorgiou

61	Learning from Intellectual History: Reflection on Sen's Capabilities Approach and Human Development Farah Naz	1585

Part X History of Ethnography and Ethnology **1611**

62	History of Ethnography and Ethnology: Section Introduction Bican Polat	1613
63	Before Fieldwork: Textual and Visual Stereotypes of Indigenous Peoples and the Emergence of World Ethnography in Hungary in the Seventeenth to Nineteenth Centuries Ildikó S. Kristóf	1621
64	Scientists and Specimens: Early Anthropology Networks in and Between Nations and the Natural and Human Sciences Brooke Penaloza-Patzak	1651
65	Center and Periphery: Anthropological Theory and National Identity in Portuguese Ethnography João Leal	1679
66	Making and Unmaking of Ethnos Theory in Twentieth-Century Russia Sergei Alymov	1699
67	Assessing Ethnographic Representations of Micronesia Under the Japanese Administration Shingo Iitaka	1729
68	Boasian Cultural Anthropologists, Interdisciplinary Initiatives, and the Making of Personality and Culture during the Interwar Years Dennis Bryson	1757

Part XI Gender and Health in the Social Sciences **1783**

69	Gender and Health in the Social Sciences: Section Introduction Meagan Tyler and Natalie Jovanovski	1785
70	Our Bodies, Ourselves: 50 Years of Education and Activism Amy Agigian and Wendy Sanford	1791
71	Reclaiming Indigenous Health Research and Knowledges As Self-Determination in Canada Carrie Bourassa, Danette Starblanket, Mikayla Hagel, Marlin Legare, Miranda Keewatin, Nathan Oakes, Sebastien Lefebvre, Betty McKenna, Margaret Kîsikâw Piyêsîs, and Gail Boehme	1805

72	**The Limits of Public Health Approaches and Discourses of Masculinities in Violence Against Women Prevention** .... Bob Pease	1831
73	**The Madness of Women: Myth and Experience** .... Jane M. Ussher	1853
74	**Feminine Hunger: A Brief History of Women's Food Restriction Practices in the West** .... Natalie Jovanovski	1877
75	**Systems of Prostitution and Pornography: Harm, Health, and Gendered Inequalities** .... Meagan Tyler and Maddy Coy	1897
Index	....	1921

About the Editor

David McCallum is Emeritus Professor of Sociology at Victoria University, Melbourne, Australia.

He publishes work in the fields of historical sociology, law and society, history of psychiatry, Indigenous studies, and sociology of education, and has most recently published a book titled *Criminalizing Children: Welfare and the State in Australia*, published in the Cambridge Studies in Law and Society series (Cambridge University Press, 2017).

Victoria University (VU) acknowledges, honors, recognizes, and respects the Ancestors, Elders, and families of the Boonwurrung (Bunurong), Woiwurrung (Wurundjeri), and Wadawurrung (Wathaurung) people of the Kulin Nation on the Melbourne campuses, and the Gadigal and Guring-gai people of the Eora Nation on our Sydney Campus. These groups are the custodians of university land and have been for many centuries.

Section Editors

Elwin Hofman
Cultural History Research Group
KU Leuven
Leuven, Belgium

Fran Collyer
Department of Social and Psychological Studies
University of Karlstad
Karlstad, Sweden

Dieter Bögenhold
Department of Sociology
Klagenfurt University
Klagenfurt, Austria

Stephen Mennell
School of Sociology
University College Dublin
Dublin, Ireland

Roger Smith
Reader Emeritus, History of Science
Lancaster University
Lancaster, UK

Bican Polat
Department of Philosophy
Boğaziçi University
Bebek, İstanbul, Turkey

Marianne Sommer
Department for Cultural and Science Studies
University of Lucerne
Luzern, Switzerland

Eric J. Engstrom
Department of History
Humboldt University
Berlin, Germany

Natalie Jovanovski
Centre for Health Equity | Melbourne School of
Population and Global Health
University of Melbourne
Melbourne, VIC, Australia

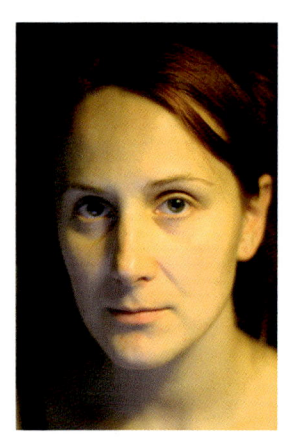

Marta Bucholc
Faculty of Sociology
University of Warsaw
Warsaw, Poland

Centre de recherche en science politique
Université Saint-Louis Bruxelles
Brussels, Belgium

Meagan Tyler
Centre for People, Organisation and Work (CPOW)
RMIT University
Melbourne, VIC, Australia

Muller-Wille Department of History and Philosophy of Science, University of Cambridge, Cambridge, UK

Contributors

Amy Agigian Department of Sociology and Criminal Justice, Suffolk University, Boston, MA, USA

Sergei Alymov Institute of Ethnology and Anthropology, Russian Academy of Sciences, Moscow, Russia

David Emanuel Andersson IBMBA Program, College of Management, National Sun Yat-sen University, Kaohsiung City, Taiwan

Jesse F. Ballenger College of Nursing and Health Professions, Drexel University, Philadelphia, PA, USA

Peter Barham London, UK

Nima Bassiri Duke University, Durham, NC, USA

Fernanda Beigel Consejo Nacional de Investigaciones Científicas y Técnicas (CONICET), Mendoza, Argentina

CECIC- Universidad Nacional de Cuyo, Mendoza, Argentina

Allan Beveridge Royal College of Physicians of Edinburgh, Edinburgh, UK

Gail Boehme File Hills Qu'Appelle Tribal Council, Fort Qu'Appelle, SK, Canada

Dieter Bögenhold Department of Sociology, Universitat Klagenfurt, Klagenfurt am Wörthersee, Austria

Carrie Bourassa University of Saskatchewan, Regina, SK, Canada

Dennis Bryson Department of American Culture and Literature, Faculty of Humanities and Letters, İhsan Doğramacı Bilkent University, Ankara, Turkey

Marta Bucholc University of Warsaw, Warsaw, Poland

Hon-Fai Chen Lingnan University, New Territories, Hong Kong

Alan F. Collins Lancaster, UK

Fran Collyer Sociology and Social Policy, University of Sydney, Sydney, NSW, Australia

Susan Condor Emeritus Professor of Social Psychology, Loughborough University, Loughborough, UK

Maddy Coy Center for Gender, Sexualities and Women's Studies Research, University of Florida, Gainesville, FL, USA

Emmanuel Delille Department of Contemporary History, University Johannes Gutenberg, Mainz, Germany

Centre Marc Bloch, Berlin, Germany

Florence Delmotte Fund for Scientific Research (FNRS) / Université Saint-Louis – Bruxelles, Bruxelles, Belgium

Simone Dennis School of Archaeology and Anthropology, College of Arts and Social Sciences, Australian National University, Canberra, ACT, Australia

Erika Dyck Department of History, University of Saskatchewan, Saskatoon, Saskatchewan, Canada

Eric J. Engstrom Department of History, Humboldt University, Berlin, Germany

Stefanie Ernst Institut für Soziologie, University of Münster, Münster, Germany

Judith Friedlander Hunter College of the City University of New York, New York, NY, USA

Steve Fuller Department of Sociology, University of Warwick, Coventry, UK

James M. M. Good Department of Psychology, Durham University, Durham, UK

María A. Guzmán-Gallegos VID Specialized University, Oslo, Norway

Mikayla Hagel University of Saskatchewan, Regina, SK, Canada

Kim M. Hajek Institute for History, Leiden University, Leiden, The Netherlands

Timothy Heffernan School of Built Environment, Faculty of Arts, Design and Architecture, University of New South Wales, Sydney, NSW, Australia

Elwin Hofman Cultural History Research Group, KU Leuven, Leuven, Belgium

Marek Hudik Faculty of Business Administration, Prague University of Economics and Business, Prague, Czechia

Center for Theoretical Study, Charles University, Prague, Czechia

Shingo Iitaka Faculty of Cultural Studies, University of Kochi, Kochi, Japan

Natalie Jovanovski University of Melbourne, Melbourne, VIC, Australia

Miranda Keewatin University of Saskatchewan, Regina, SK, Canada

All Nations Hope Network, Regina, SK, Canada

Wiebke Keim CNRS/SAGE (University of Strasbourg), Strasbourg, France

Margaret Kîsikâw Piyêsîs All Nations Hope Network, Regina, SK, Canada

Ulrich Koch George Washington University, Washington, DC, USA

Ildikó S. Kristóf Institute of Ethnology, Hungarian Academy of Sciences, Budapest, Hungary

Heinz D. Kurz Graz Schumpeter Centre, University of Graz, Graz, Austria

Helmut Kuzmics University of Graz, Graz, Österreich

João Leal CRIA, Universidade Nova de Lisboa, Lisboa, Portugal

Sebastien Lefebvre Laurentian University, Sudbury, ON, Canada

Marlin Legare University of Saskatchewan, Regina, SK, Canada

Esben Leifsen Department of International Environment and Development Studies, Norwegian University of Life Sciences (NMBU), Aas, Norway

Anna Maerker History, King's College London, London, UK

Sloan Mahone University of Oxford, Oxford, UK

Geertje Mak NL-lab, KNAW Humanities Cluster, Amsterdam, The Netherlands

Amsterdam School of Historical Studies, University of Amsterdam, Amsterdam, The Netherlands

Rudolf Makkreel Atlanta, GA, USA

Dominic Malcolm School of Sport, Exercise and Health Sciences, Loughborough University, Loughborough, UK

Siniša Malešević School of Sociology, University College, Dublin, Ireland

Betty McKenna University of Regina, Regina, SK, Canada

Stephen Mennell School of Sociology, University College Dublin, Dublin, Ireland

Panayotis G. Michaelides Laboratory of Theoretical and Applied Economics and Law, National Technical University of Athens, Athens, Greece

Farah Naz Department of Sociology and Criminology, University of Sargodha, Sargodha, Pakistan

Reinhard Neck Department of Economics, Alpen-Adria-Universität Klagenfurt; Karl Popper Foundation Klagenfurt; Kärntner Institut für Höhere Studien; and CESifo, Klagenfurt, Austria

Nathan Oakes University of Saskatchewan, Regina, SK, Canada

Christian Olsson Université libre de Bruxelles, Bruxelles, Belgium

Lisa M. Osbeck Department of Anthropology, Psychology, and Sociology, University of West Georgia, Carrollton, GA, USA

Theofanis Papageorgiou University of Patras, Patras, Greece

Sujata Patel Umea University, Umea, Sweden

Herman Paul Institute for History, Leiden University, Leiden, The Netherlands

Bob Pease Institute for Social Change, University of Tasmania, Hobart, TAS, Australia

School of Humanities and Social Sciences, Deakin University, Geelong, VIC, Australia

Brooke Penaloza-Patzak Department of History and Sociology of Science, University of Pennsylvania, Philadelphia, PA, USA

Wade E. Pickren Dryden, NY, USA

Petteri Pietikäinen History of Sciences and Ideas, University of Oulu, Oulu, Finland

Bican Polat Tsinghua University, Beijing, China

Nathalie Richard Department of History, Le Mans Université, TEMOS CNRS UMR 9016, Le Mans, France

Elissa N. Rodkey Crandall University, Moncton, NB, Canada

Krista L. Rodkey Santa Barbara, CA, USA

Wendy Sanford Our Bodies, Ourselves, Newton, MA, USA

Robert A. Segal School of Divinity, History and Philosophy, King's College, University of Aberdeen, Aberdeen, UK

Roger Smith Institute of Philosophy, Russian Academy of Sciences, Moscow, Russian Federation

Marianne Sommer Department for Cultural and Sciences Studies, University of Lucerne, Luzern, Switzerland

Danette Starblanket University of Saskatchewan, Regina, SK, Canada

George Steinmetz Department of Sociology, University of Michigan, Ann Arbor, MI, USA

Nicola Sugden Centre for the History of Science, Technology and Medicine, University of Manchester, Manchester, UK

Fenneke Sysling Leiden University, Leiden, The Netherlands

Gülşah Taşçı Istanbul 29 Mayıs University, Istanbul, Turkey

Kirsi Tuohela University of Turku, Turku, Finland

George Turner Preston, UK

Meagan Tyler RMIT, Melbourne, VIC, Australia

Jane M. Ussher Translational Health Research Institute, Western Sydney University, Penrith, NSW, Australia

Zsuzsanna Vajda Department of Psychology, Károli Gáspár University of the Reformed Church in Hungary, Budapest, Hungary

Robert van Krieken University of Sydney, Sydney, NSW, Australia

School of Sociology, University College Dublin, Dublin, Ireland

René van der Veer University of Leiden, Leiden, The Netherlands

Joanna Wawrzyniak University of Warsaw, Warsaw, Poland

Raquel Weiss Department of Sociology, Federal University of Rio Grande do Sul (UFRGS), Porto Alegre, Brazil

Janina Wellmann Institute for Advanced Study on Media Cultures of Computer Simulation (MECS), Leuphana Universität Lüneburg, Lüneburg, Germany

Martin Wieser Faculty of Psychology, Sigmund Freud Private University Berlin, Berlin, Germany

Per Wisselgren Department of History of Science and Ideas, Uppsala University, Uppsala, Sweden

David Wright Department of History and Classical Studies, McGill University, Montreal, QC, Canada

Part I

What Is the History of Human Sciences

What Is the History of the Human Sciences?

Roger Smith

Contents

Introduction: The Question of Human Identity	4
Philosophy and the History of the Human Sciences	10
History and the History of the Human Sciences	16
Conclusion: "History of the Present"	22
References	26

Abstract

The history of the human sciences, like references to the *human sciences*, denotes a domain of inquiry not a discipline with definable subject matter and established institutional structures. The institutional arrangements supporting the domain vary with local circumstances. The terms have come into widespread use over the last half century or so, often as terms of convenience bringing into relation different disciplines of the psychological and social sciences. The general term encompasses these disciplines insofar as they are not straightforwardly natural sciences but have affinities with the humanities. This often supports interdisciplinary or transdisciplinary projects. This chapter explains these usages. More substantially, it argues that the many dimensions of the human subject give value to a heading under which to debate the relation of different approaches to knowledge and practice associated with differences between the natural science, social sciences, and humanities. Work in the history of the human sciences, it is argued, necessarily raises deep-lying, wide-ranging questions about the very conception of *the human*.

Providing an overview of these matters is unusual, is difficult, and cannot be comprehensive. Presenting the variety of philosophical and disciplinary orientations at stake, the chapter therefore presents an open-ended survey. At the same

---

R. Smith (✉)
Institute of Philosophy, Russian Academy of Sciences, Moscow, Russian Federation
e-mail: rogersmith1945@gmail.com

© The Author(s), under exclusive licence to Springer Nature Singapore Pte Ltd. 2022
D. McCallum (ed.), *The Palgrave Handbook of the History of Human Sciences*,
https://doi.org/10.1007/978-981-16-7255-2_83

time, it attempts rational advocacy of a view that orientates the human sciences in relation to their history and to historical knowledge.

The sections that follow the opening discussion of the question of identity introduce the relations of the domain of the history of the human sciences to philosophical questions and to history. The reference to *history* raises distinctive issues, since many researchers working in the domain do so from a position within psychological and social science disciplines and take an interest in history to the degree that it contributes to knowledge, practice, and critique in these disciplines. This orients history of the human sciences toward policy matters and politics in the broad sense, hence the interest in "the history of the present," which is discussed in the concluding section. Counterbalancing this, the chapter makes a substantial argument for the epistemological necessity of historical knowledge in the human sciences, pointing to the ways historical forms of knowledge are built into human self-understanding, given that humans are reflective, or reflexive, subjects.

Keywords

Human science · History · Human nature · Historicism · Reflection · Reflexivity · History of the present · Humanities

Introduction: The Question of Human Identity

The publication of a *Handbook of the History of the Human Sciences* confirms the existence of the domain. But what is it? It is a complex, intellectually challenging, and far from trivial question. This introductory chapter therefore reflects in general terms on the domain's identity.

There is no socially well-defined entity or institution that answers to the title, history of the human sciences (henceforth HHS). Rather, the phrase points to potentialities for raising large questions about the appropriate methods and content for studies of "the human," past, present, and future. Indeed, there neither is nor could be a well-defined domain, for the profound reason that the category, "the human," is not empirically delimited, not unambiguously "given," but intrinsically self-questioning and self-defining: "The difficulty of providing precise definition comes from the subject. 'Man' is not an empirically delimited subject" (Smith 2001: 7027). The subject matter and forms of knowledge in the domain are in their nature various and open-ended in scope.

The question of identity is therefore a philosophical matter. At the same time, it is ethical and political, since it concerns the organization of knowledge and practice about the nature and lives of different people. What people generally, in groups, or individually, hold to be true about themselves is not separate from evaluative judgments about ways of life. This is a pressing contemporary issue. This is most obviously so, in the sense that scholars working in the human sciences are conscious of the large obligation to orient the domain toward the interests of a global humanity, indigenous intellectual life, and postcolonial circumstances. The *Handbook* as a whole, edited in Australia, addresses this. The roots and hence identity of the notion of the human sciences, however, have a European formation (the modern forms of which will be loosely characterized as "western"); this chapter focuses on this. What the future holds is another matter.

It is a social fact that references to the human sciences are loose and variable. For some academics, the title is simply a matter of utility; others, however, think there is more to usage than this, though they understand this "more" in various ways. The usefulness of the heading follows from the fact that there is no agreed list of subject areas, commonly called academic disciplines, which address "the human" as a subject. There is psychology, social science, anthropology, political science, linguistics, human geography, human biology, jurisprudence, economics, etc. An open-ended term, "the human sciences," is therefore needed. This is all the more helpful when scholars wish to draw in other subject areas, such as management science, or medicine and the health sciences, or large parts of the humanities (e.g., incorporating presuppositions about "the human" in literary theory). It is pragmatic to have a general term open to being given different denotations in different institutional settings. As a result, the term is at the very least a term of convenience, serving, in particular, local settings as an organizational umbrella under which to coordinate different activities sharing the human subject. If "the human sciences" is a term convenient for organizational purposes, a search for precise definition seems inappropriate. There are (to the author's knowledge) no academic departments of HHS, nor is it the purpose of the present chapter to argue that there should be. Rather, the phrase supports critique and fosters and coordinates debate in a large range of disciplinary and institutional settings. Reference to HHS helps *to hold in relation* the many dimensions of activity in intellectual, academic, and organizational settings concerned with the human subject.

Nevertheless, there is more than utility at stake. The arguments that follow explain why. The chapter deals first with the human science content, then with philosophy, and then with the reference to history.

In English-language institutional contexts, the human sciences center on a cluster of disciplines that do not appear to belong straightforwardly either to the natural sciences or to the humanities. In this respect, the human sciences are to a significant extent heir to what formerly were called, and often still are called, the psychological and social sciences – the domain that Wolf Lepenies once referred to as "between literature and science" (Lepenies 1988). Use of the word "sciences" places the human sciences in a central position in debate about the nature of knowledge in general and about the distinctiveness, or not, of knowledge of being human. The ramifications of this are enormous. It would appear in principle possible to use the term human "studies," rather than human "sciences," analogous to reference to an area like cultural studies. Yet, this is not the usage. The reasons may be found in the history of academic activity and the debt of contemporary disciplines to the German-language and continental European tradition of referring to all bodies of systematic knowledge, held to be true and grounded in rational principles, as sciences. Especially as a legacy of the work of the German philosopher Wilhelm Dilthey (discussed in ▶ Chap. 5, "Diltheyan Understanding and Contextual Orientation in the Human Sciences," by Rudolf Makkreel in this volume), it became common to distinguish the cluster of disciplines not belonging with the natural sciences (*Naturwissenschaften*) as *Geisteswissenschaften*. The latter term raises questions for translation, but the contemporary custom is to render *Wissenschaft* as science (not studies, as was done earlier) and *Geisteswissenschaft* sometimes but not always as human science, as at least until recently German universities in general included the social sciences *and* the humanities (philosophy, history, philology, literary and art studies, classics) under this term. (See Rudolf Makkreel and Frithjof Rodi, Editors' Preface to all volumes of Dilthey, *Selected Works*; referenced here to Dilthey 2002: xi–xii.) This combination of social sciences and humanities marks out English-language reference to the human sciences and the usage in this *Handbook*. It must also be acknowledged, however, that there are scholars who think that these sciences properly belong with biology.

The spread of reference to the human sciences in the last half century or so has been marked with deep ambivalence about the inclusion, or not, of the humanities. When Odo Marquard, philosopher and Rector of the German university of Giessen, argued for the position of the humanities in the modern university, he did so under the heading rendered in translation as "the human sciences" (Marquard 1995). He thereby grouped together the "interpretive" disciplines (including the humanities and the branches of the psychological and social sciences where knowledge has interpretive form). This introductory chapter is broadly sympathetic to this understanding of the human sciences. In institutional practice in English-language universities, however, the human sciences and the humanities are often separate, and there is a degree of innovation in relating the human sciences and the humanities in the English-speaking world.

Behind this discussion of titles and translation are substantial questions about the form of knowledge appropriate for human self-understanding. They concern the nature and comprehensiveness of the psychological and social sciences as headings under which to coordinate knowledge of the human subject. Discussion lies open to the idea that different disciplines may contribute in perhaps as yet undefined ways to study of "the human." It definitely rejects belief that the human sciences, to be sciences, necessarily belong with the natural sciences. Given the contemporary standing of biology, and the great emphasis on knowledge in evolutionary and neuronal terms, this is controversial, even provocative. The present discussion argues that the domain of the human sciences is and should be committed to *debate about forms of knowledge* and to *diversity* of approach. This openness, or receptivity, justifies the term. The chapters in this general section of the *Handbook* support this approach.

The issues are undoubtedly complex and conflicted. All the same, it is straightforward to observe that HHS is well placed to foster inquiry, as the domain as a whole does not presuppose any one fixed position as regards the foundations of knowledge. While there were once substantial hopes that Marxism would provide a unifying framework, and while there is support in some parts of the world for an overall, even authoritarian, religious framework, in western academic settings, it is not known where new insights about being human will emerge or in which fields there will be productive developments. Further, when one field does claim foundational status, as the neurosciences have sometimes done in recent decades, there is need for institutional and intellectual space in which to put the claim into critical perspective.

All the same, it must be acknowledged that numerous writers on the history of their disciplines in the social sciences, broadly understood, do not refer to the human sciences to describe these disciplines or the historiography (Backhouse and Fontaine 2014). It is unclear whether this is a matter of contingent institutional factors or a principled choice. Clearly, though, reference to the human sciences and to their history cannot be taken for granted as an unproblematic part of the intellectual and academic world.

References to the human sciences were uncommon in English half a century ago, and the phrase still sounds strange in some languages, e.g., in Russian. By contrast, both *sciences humaines* and *science de l'homme* have long been established and more commonly used, though with different meanings and associations, in the French-language world (see ▶ Chap. 3, "History of the Human Sciences in France: From Science de l'homme to Sciences Humaines et sociales," by Nathalie Richard's in this volume). *Science de l'homme* corresponds to the English-language "science of man," an expression widely used from the eighteenth century until the later part of the twentieth century and distinctively associated with the Enlightenment aspiration to know and thereby reform the human world within the scope of natural law (Fox et al. 1995). This Enlightenment inheritance persists in modern references to the human sciences, though the content and consequences of Enlightenment thought are subject to severe critique, commonly in the light of Adorno's and Horkheimer's much argued-over work (Adorno and Horkheimer 1986). Reference to "the science of man" has clearly become untenable in English-language academic settings. This is a philosophical matter, as well as a matter of the politics of gender, because the earlier presumption equated "certain men" (particular), "Man" (the human species in general), and the male sex with the normative form or type of the rational human being. Further, if there were once hopes to establish the science of man in the singular, recent references are to *sciences* in the plural, reflecting a troubled appreciation of the complexity of what is being called for. In the eighteenth and nineteenth centuries, the relevant areas of study were also called the moral sciences or *les sciences morales*. In these expressions, the word "moral" drew a contrast with "physical" and marked out the domain thought to be distinctively human, which included philosophy, ethics, the study of language, political economy, anthropology, and psychology. In large ways, the human sciences, as viewed in this chapter, are heir historically to the moral sciences (Makkreel 2012). Significantly, they have often shared in common a structural integration of descriptive and evaluative content, based on the presumption that knowledge of natural law provides the grounds for rational practice or, in other words, that the natural is normative.

There is no systematic empirical work that charts references to the human sciences in recent decades; description is therefore impressionistic. Beginning in the 1960s, references to the term became increasingly common in English, so that it became an unremarked part of many people's vocabulary. This acceptance may have been influenced to a degree by the use of "the human sciences" in the translation of Michel Foucault's *The Order of Things: An Archaeology of the Human Sciences* (Foucault 1970), the originality of which prompted huge commentary. Foucault, seemingly irreversibly, problematized and historicized the disciplinary content of the human sciences. What that means is for debate, and the debate gives the domain a significant part of its content. When Foucault was elected to the prestigious Collège de France in 1970, it was to the chair of the history of systems of thought (Eribon 1991). His claim to this chair was precisely that his writings (at this stage) had delineated the modern notion of "Man" as the subject, and conceivably transitory subject, of the human sciences.

It is easy to suppose that usage has spread in large part because of the appeal of a term that brings together, at least in name, disciplines otherwise working independently, though all claiming in a general way to address the human subject. The term is now fashionably interdisciplinary. The specialization of the pursuit of knowledge, so conspicuous since the creation and consolidation of professional academic disciplines in the nineteenth century, and though generally thought inevitable, indeed natural in the advance of science, has always had critics. As Marquard noted, when anthropology was institutionalized as a discipline in the nineteenth century, it was not as a universal field, a comprehensive science of "Man"; this has left space in the modern university where, instead of a universal science, there is interdisciplinarity. The way forward for integrated human self-knowledge is not, he proposed, through a unified science but through "interdisciplinary conversation" (Marquard 1995: 104). The twentieth century saw a number of large-scale projects sharing this goal, projects driven by the enlightened aspiration to unite knowledge of human beings and thereby provide the foundation of rational social policy. Such projects followed the example of business and medical organizations in subjecting complex practices to a rational plan. Examples include the Yale Institute of Social Relations in the 1930s, and the Ford Foundation funded project in the 1950s to establish "the behavioral sciences" and unite research in human physiology, psychology, political sciences, social science, and the organizational sciences like systems thinking (Morawski 1986; Solovey 2013: Chap. 3; Ward 1981). The project to establish Soviet Marxism as a unified body of knowledge at the foundation of all the sciences may also be thought of as a huge political experiment in this regard (Graham 1987; Todes and Krementsov 2010). In turn, it may be that general disillusionment with the possibility of a comprehensive Marxist philosophy and social science has greatly encouraged reference to a more diversified view of the possibilities under the heading of the human sciences.

Where there is a clearly defined purpose to integrate research effort to achieve social goals, it has become appropriate to assemble expertise and investment under the heading of the human sciences. This supports proposals for cross-, inter-, or transdisciplinary fields (the labels and purposes vary) – this is the intention – not dominated by any one specialized discipline. "Human science" has value to institutions because it does not specify which disciplines or specialties are to be coordinated, and an institution is therefore able under this heading to bring together whatever fields suit its local history and management policies. For the same reasons, the term is attractive to many individuals trained in a specialized field but seeking a heading under which to escape the constraint the disciplinary speciality imposes, in order to achieve a broader perspective and to contribute to the integration of knowledge thought necessary for practical goals.

Each of the established human science disciplines is extremely heterogeneous (e.g., though economics less so than psychology) and all of them, even if united socially by national and international professional bodies, are not easily definable by subject content. (The difficulties of defining psychology, for instance, are raised in the introductory chapter to the *Handbook* volume on psychology.) This is surely another reason why reference to the human sciences has gained favor: it escapes from the intellectually unrewarding business of policing discipline boundaries. It also explains why the study of discipline formation, and the role in it of professionalization, the laboratory, the archive, and methods (centrally statistical methods), is a core area of research in the domain. This research substantially contributes to explaining the social processes responsible for the particular institutional forms and classification of knowledge. The sheer diversity of what goes on under human science headings, the divergence of views about whether the fields should seek to follow the explanatory form of the natural sciences or whether there are explanatory forms proper to their own spheres of inquiry, and the constitutive activity of methods and techniques of inquiry have all attracted historical research (e.g., Danziger 1990; Porter 1986, 1995; Haskell 1977).

It is necessary to say something about medicine, which many people might informally think the oldest and most widely practiced area of the human sciences (Porter 1997). A parallel case might be made that the various branches of "economy," understood as right management of the individual person, the household, and the state, are also ancient. Indeed, it could be said that in the early years of the nineteenth century, political economy was the most developed of all the human sciences (Manicas 1987; Smith 1997). Interest in medicine, however, perhaps more than economics, has informed recent work under the HHS heading. The promotion of modern scientific medicine from the nineteenth century, with pathological anatomy, germ theory, genetics, evidence-based medicine, and so on, would seem to have brought medicine firmly into the domain of the natural sciences. Yet the demands of therapy, necessarily directed at individual people not bodies, the flourishing of alternative medicines, and the overwhelming importance to health of social conditions and modes of life, place medicine at the center of debates about the way the human sciences range beyond the purview of the natural sciences. Further, studies of particular areas of ill-health, including mental illnesses as well as hard-to-classify conditions like hysteria, autism, and post-traumatic stress, have a very prominent place in HHS. (This is evident in the *Handbook*.) HHS has proved to be a space in which to examine, both objectively and normatively, the institutional and intellectual formation of current practice in the area of health and well-being. In this space, history engages with contemporary social and medical policy issues.

The very fact of undertaking work in HHS, including work encompassing medicine, at least tacitly affirms the existence and necessity of forms of knowledge about "the human" that cannot be reduced to biological natural science knowledge. This irreducibility has rational content and implies that different forms of knowledge are just that – different. (Durkheim's defense of the irreducible content of sociological explanations is a classic case of the argument; see ▶ Chap. 6, "Durkheimian Revolution in Understanding Morality: Socially Created, Scientifically Grasped," by Raquel Weiss's in this volume.) The presumed irreducibility is also evaluative: it has been customary in modern humanistic culture to attach special value to being human and to require of knowledge that it do justice to the special, even unique, value of persons (see *Handbook* section on "the person"; for the special value of the person in medicine, Tauber 1999). The interpretation and history of this humanistic evaluation is very much an issue for HHS. As Steve Fuller draws attention to in his chapter in this volume (▶ Chap. 2, "Kant After Kant: Towards a History of the Human Sciences from a Cosmopolitan Standpoint"), there is much contemporary discussion about the manner in which technological innovation and transfiguration of previously taken-for-granted human identities is overtaking "the human."

Philosophy and the History of the Human Sciences

References to human sciences as a category draw in deep-lying philosophical questions. Discussion implicitly, and sometimes explicitly, encompasses debate about the very conception of what it is to be human, debate over the differentiation of "the human" as a subject area, and debate about the relations between knowledge of the physical world and knowledge of the human world of history, mind, society, language, religion, and culture.

Work under the heading of the human sciences in no way questions the importance of the specialized biological, sociological, economic, and other disciplines. But it does ask how different bodies of specialized knowledge relate to each other and to historical knowledge and philosophical inquiry. This necessarily encompasses systematic discussion of the nature and forms of discourse, theories of meaning and communication, hermeneutics, and the comprehension of rules, including the rules of language. All such discussion does not seek knowledge in the causal explanatory form of much natural science knowledge, and this seems to mark out the domain of the human sciences in contrast to the natural sciences.

To be sure, there have been and there are biologists and promoters of biology for whom any body of knowledge about humankind, if it is to be called a *science*, must ground its knowledge on biology. There have also been social scientists who have claimed that sociological explanations should be causal in form, just like natural science explanations. Marxists, for their part, anticipated the unification of all science in a theory of the material conditions (or "base") of its production. But inquiry under the heading of HHS makes the epistemic priority thus given to biological, sociological, or historical materialist argument itself the subject of study. It develops a space that welcomes different approaches to systematic knowledge of being human and to debate about the relations between them. It has intensive interests in nonnatural science forms of knowledge.

It is therefore a source of some confusion to say that the subject of the human sciences is "human nature." This term has biological connotations in scientific as well as public contexts, and reference to it commonly enough connotes belief in biological knowledge (with all its anthropological and medical dimensions) as the basic or foundational content of the human sciences. This is a position the interpretation of HHS offered in this chapter argues against. The leading point is simple but significant. The subject matter of the field of HHS involves the consciousness, or reflective awareness, of the knowing subject about itself. (How this relates to notions of the "self" or "I" is a substantial area of inquiry.) Thus, work in HHS, this chapter argues, includes debate about the forms of knowledge appropriate for understanding human nature or, to use a term without the same biological connotations, for understanding "being human." Going further, the implication, developed in the following section of the chapter, is that HHS conceives the forms of understanding as *historical formations*.

HHS advocates a mode of intellectual and scholarly practice rather than promotes a particular body of knowledge. If it engages very substantial philosophical debates, it does not pretend to resolve philosophical issues. As a consequence, its "liberal" or, better, open-minded practice tends to towards treating knowledge and debate as historically specific, not transhistorical, in form and content. This is of course, in practice, to take a philosophical position, and it hardly needs to be said that there is extensive argument among philosophers, and among those who write in the social sciences and humanities under the heading of "theory," about realism and anti-realism and the nature of transhistorical truth. Different positions are legion. HHS does not attempt to resolve such questions philosophically, though it surely needs a working knowledge of what is at stake. Practice, however, tends to put forward historically contextual discussion of the debates as a resource. This historical treatment raises the specter of epistemic relativism, bringing the field into relation with debates that have gone at least since the late nineteenth century over the "historicist" interpretation of human self-understanding. Whatever position is taken on these matters – and it varies – the contextually oriented practice of HHS is important in attracting scholars to the domain, as it fosters critique, disciplinary self-questioning, and perspective in relation to knowledge and practice in existing fields like sociology, psychology, and political science.

Work under the heading of HHS has to decide how far to engage the philosophical issues concerning the relations of different forms of knowledge and the potential relativity of knowledge, or how far to leave the issues unexamined and to get on with more defined, empirically oriented tasks in which closure, at least for the present, appears possible. Nevertheless, the commitment to HHS itself is a commitment to fostering debate and not to taking any one intellectual practice (or "paradigm") for granted. This commitment leads to considerations that there may be no one foundation for human self-knowledge, that different forms of knowledge serve different purposes, and hence that moral or evaluative debate is intrinsic to the undertaking.

Though the modern divergence between continental European and English-language usage of the word "science" is familiar enough, the implications are perhaps insufficiently acknowledged. Academicians of the Russian Academy of

Sciences, for instance, include philosophers and philologists as well as physicists and biologists, while in Britain the former belong in the British Academy and the latter in the Royal Society, entirely separate institutions. In the continental European tradition influenced by German-language scholarship, history and theology, for example, insofar as they are formally rational disciplines, may be called sciences. Naturally, there is still debate as to whether they really have this formal standing. But in English-language usage, such subject fields are not sciences, though there have been attempts to turn them into sciences, for example, to make history scientific through use of rigorous quantitative methods (as in the cliometrics practiced in the 1970s and, later, under labels such as quantitative history). Scholars have debated at length what constitutes a science, but in the continental European setting it has led to debate about the relations between the sciences in all their range, whereas in the English-speaking world it has led to debate about relations between the natural sciences, that is, "the sciences," and those disciplines that are not sciences, "the humanities." The human sciences have problematic standing in both settings, but in the continental European tradition, this derives from the question as to whether these sciences actually have the formal form constituting a science, while in the English-speaking world, it derives from the question whether these disciplines share the form of knowledge found in the natural sciences.

It is now widely agreed that rational attempts to construct a formal demarcation between science and nonscience, of the kind earlier philosophers of science (such as Karl Popper with the criterion of falsification) attempted, have not been successful. All the same, as a matter of practice, workers in the human sciences, faced as they often are by unqualified and self-styled experts, not to mention sheer prejudice and openly irrational argument, often feel called upon to divide science from pseudoscience. This brings to the fore questions about the sources of authority for knowledge as well as about the rational validity of knowledge. How has the authority claimed for knowledge and practice in the human sciences been established and how is it maintained? This is a large preoccupation. It is, for instance, a reason why Foucault's later body of work, centered on the interrogation of power/knowledge, has had such central importance (Foucault 1980). Historical case studies of the establishment of authoritative knowledge, or about the failure of this goal, may have great contemporary relevance (discussed further in the last section of this chapter). This is the case, for example, for studies of the inconclusive nature of arguments for and against spiritualist phenomena (in the nineteenth century) and ESP (in the twentieth) (Lamont 2013).

By training and practice, people who work in particular human sciences, as in other scholarly fields, know what constitutes disciplined knowledge in their particular area. English speakers conventionally refer to disciplined knowledge about politics, economics, language, human behavior, and so on, as science without more ado (as in references to "political science" or "behavioral science"). It is nevertheless always open to question the rational constitution of a field as science, or the adequacy of its empirically formulated statements about what is the case, or claims to truth; indeed argument about this appears endemic in the human sciences, which have even been dubbed "the uncertain sciences" (Mazlish 1998). There is, for instance, recurrent argument in psychology about the standing of qualitative methods as opposed to methods of participant observation, while in economics there is debate about the standing of knowledge (presupposed in rational-choice theory) based on supposed universals of individual human nature. Rather than arguing about such issues in abstract terms, much work in HHS engages debate by putting issues *in context*, that is, by specifying the historical particulars concerning who takes part in debate, in which institutions, when, for what reasons, and with which audience in mind. Such scholarship merges with *critique*, the analysis of the epistemic, economic, social, and political conditions that have fostered particular forms of knowledge and practice, analysis carried out with a view to removing any presumption that these forms are natural or unquestionable. There is also reference to "deconstruction," using the term originating in literary theory to describe demonstrations that texts do not refer to an objectively and independently verifiable world but to other texts purportedly about such a world. In the human sciences, the reference is most often informal, pointing to a demonstration that a claim that something is naturally or objectively the case originates in humanly constructed social processes.

The earlier extensive influence of Marxist arguments in the social sciences focused a great deal of critical work on the analysis of ideology. This showed, not least, that the concept of ideology itself has a history. It is now a question as to whether the concept retains analytic value and a political cutting edge in the face of usage equating it with any socially embedded belief system. But how belief systems relate to politics, and hence the question of the interpretation of ideology, is of central interest for HHS. There is, for example, a substantial body of work about post-1945 US social and psychological science and its role as a belief system supporting American claims to world leadership (e.g., Herman 1995; Cohen-Cole 2014). This has also been placed in international context (Kirtchik and Boldyrev 2016).

Work in this vein well illustrates the tendency of scholarship in HHS to shift questions about the nature and standing of the human sciences from abstract philosophical contexts to specific historical contexts. This said, it is nevertheless necessary to recognize that the separation of what belongs to philosophical discourse and what to empirical historical discourse is by no means clear or agreed and itself has a history. If the separation between rational (philosophical) statements and empirical (scientific) statements has its classic modern formulation in Kant, Kant provided the terms for what has turned out to be a continuing debate rather than a resolution. Later generations of philosophers, including Frege, Husserl, and Russell, rearticulated the distinction, and indeed for much of the twentieth-century philosophers and scientists agreed on a mutually convenient division of intellectual labor. But the distinction has always been much questioned in the human sciences: the very characterization of the human subject as a reasoning subject makes analytic knowledge of reason (philosophy) of inescapable interest even where the project is the empirical (scientific) knowledge of "the human." Among promoters of the psychological and social sciences at the beginning of the twentieth century, the argument was common that "at last" science was substituting progressive empirical inquiry for the unproductive circling back of philosophical questions on the same unresolved, and perhaps irresolvable, aporias or perplexing difficulties. Earlier generations of historians of the human sciences conceived of a history telling a story about these sciences distancing themselves from philosophy in the interests of empirical, scientific reason. But historians more recently have re-examined the evidence and reasoning and shown that key figures associated with the founding of modern scientific disciplines were markedly concerned with philosophy (for Wundt and psychology, Araujo 2016; for Durkheim and social science, Lukes 1973).

The debate shifted again in the last decades of the twentieth century, especially in response to the philosophy of mind rethought in the light of the advance of the neurosciences. Some philosophers and scientists argued strongly that philosophy needed scientific knowledge and scientific knowledge needed philosophy. The arguments markedly divided philosophers themselves, between those who, broadly following Wittgenstein (Bennett and Hacker 2003), argued for the necessity of distinguishing statements concerning reason and statements concerning neuronal facts and those committed to a naturalist approach to philosophy (e.g., Dennett 1993) who argued that new knowledge of the brain opened new answers to philosophical questions. In such circumstances, historical and social knowledge not party to the dispute, that is, knowledge providing perspective, has much to contribute, especially by showing how such divisions have come about and continue to structure intellectual activity. (As concerns the neurosciences, see Vidal and Ortega 2017; and for a study of the relations of neurological and narrative forms of knowledge in a particular biographical and social context, see Proctor 2020.) HHS gives perspective a voice. It is a voice with specificity, which speaks about actual practices, rather than, or at least in addition to, a voice of "high theory."

Still relevant terms for reflection on all this became current in German-language academic writing in the last decades of the nineteenth century. The terms grew out of a century of development of the sciences of all kinds (history as well as physiology) as specialized disciplines. The sciences, as opposed in this setting to philosophy, acquired new authority in the academic sphere and in educated culture generally, accompanied by academic professionalization (creating the categories of the specialist and the amateur). Provoked by the arguments of Auguste Comte, revised and transmitted as the logic of the moral sciences by J. S. Mill, German scholars debated whether the sciences concerned with psychology, history, society, language, and culture, and perhaps even philosophy itself, should follow the road to knowledge established in the natural sciences, as Comte and Mill had said they must. The outcome was the extended examination of the special conditions of knowledge formation in the human sciences, introduced above in connection with the distinction between *die Naturwissenschaften* (the natural sciences) and *die Geisteswissenschaften* (the sciences of the human spirit, or, in some versions, *die Kulturwissenschaften*, the sciences of culture), and also involving discussing of the relations of causal explanation and hermeneutic understanding, and of nomothetic knowledge (concerned with finding general laws) and idiographic knowledge (concerned with the historically specific understanding of particulars). These terms and arguments have become a standard part of the intellectual framework of HHS; However, much recent scholarship has also turned to other sources for philosophical inspiration. Thus, the philosopher Charles Taylor, in 1971, in an influential paper, restated the case for the distinctiveness of the human sciences (which, however, he called "the sciences of man"), because of the interpretive form of knowledge in their fields (Taylor 1985).

In the light of such work, it might be a temptation to characterize HHS as heir to *die Geisteswissenschaften*, a home for forms of knowledge of the human that are not, and cannot be, translated into biological subject matter, or reduced to causal explanation in material terms, or represented as a series of general laws. This characterization, however, is not viable: the modern human sciences are not the nineteenth-century *Geisteswissenschaften*. In the nineteenth century, and well into the twentieth, much work in *Geistes-* or *Kulturwissenschaft* rested on idealist forms of reason and perpetuated the project of philosophical anthropology devoted to interpretation of the workings of the universal human spirit. As Makkreel has argued, even Dilthey himself was not completely satisfied with the term *Geisteswissenschaften* because of idealistic connotations linking the human spirit with Hegel's notion of absolute spirit. Dilthey once referred to the human sciences as *Humanwissenschaften*, wanted a more empirical approach than Hegel had entertained, and rejected the idea of absolute spirit. He reinterpreted objective spirit as the medium of local commonalties like language and social customs that nurture people from birth. Nevertheless, in this like twentieth-century philosophical anthropologists, Dilthey continued to uphold language about the human spirit, assuming its universality, even if the forms of expression of the spirit, and hence the subject matters to be studied, were local. In the eyes of many western secular writers, this kind of basis for human self-knowledge, or anthropology, has not survived either skeptical and analytic critique in philosophy or the calamitous course of history in the twentieth century (e.g., Ringer 1990; Megill 1985).

Elements of philosophical anthropology nevertheless remain in the widely held, often taken for granted, but far from universal humanism lurking in the background of many people who work in the human sciences and have hopes, in some sense, for social improvement. (For a study of such hopes for the human sciences persisting from Tsarist Russia into the Bolshevik Soviet Union, see Beer 2008.) Humanist belief, in this like a religious faith, posits an absolute *value*, the value of being human, a value "in itself," a value *given* in the being of a living person as an individual. (For discussion of religion and the human sciences, see Roberts 2002; for sociological understanding of religion, see ▶ Chap. 8, "Human Sciences and Theories of Religion," by Robert Segal's in this volume.) Positing this belief is an ontology-forming practice, which renders the study of history, language, and culture different from the study of states of nature. (Some kinds of environmental philosophy posit the same kind of value, or "sacredness," for all existence, for physical nature, as well as for "the human," but this is beyond the scope of the *Handbook*.) Taking up discussion of human value, HHS takes on existential and moral significance: it interprets human self-understanding and its articulation as a religious or humanist form of life, not only as a body of knowledge. Yet there is also a large body of literature in or relevant to HHS, sometimes self-identified as "anti-humanist," committed to demonstrating the historical and theoretical contingency of the value posited for human subjects. The stances exist side by side in the human sciences. HHS, it must be conceded, is "a broad church."

History and the History of the Human Sciences

The "*history* of the human sciences" is surely not a phrase made attractive for reasons of institutional convenience. Indeed, it must be said clearly that there is little agreement about the import of the reference to history in HHS, as this *Handbook* amply shows. This is not necessarily a problem or a critical focus. But it is inquiry into the forms of knowledge judged appropriate for "the human," and particularly about the place of historical knowledge in this judgment, which gives HHS its particular intellectual purpose.

Once again, there is a nineteenth-century intellectual background to the issues. To clarify this requires comment on "historicism" as a term. The term (and variants of it), to be sure, has been used in a number of different ways and no longer attracts much attention. Nevertheless, the issues underlying earlier usages have not gone away. In the nineteenth century, the fields now forming the human sciences were part of and central to an educated culture that established the meaning of things and events by deriving meaning from the place of every aspect of human activity in history, history understood in the form of a story. Educated western culture looked to human reason to transcend historically specific circumstances and establish the unique, truthful nature of the historical stories that scientists (historians as well as biologists, philologists as well as physicists) were constructing. The European claim to take the lead in civilization appeared to depend on establishing this truth. The culture raised the political and cultural status of history as the discipline studying the unifying plot of European and national progress, and of progress in knowledge itself. The historicist writing that resulted treated each aspect of the human sphere as part of a historical process, and it found meaning and value (and indeed disruption and degeneration) in each aspect by identifying each aspect's particular, individual relation to the historical plot as a whole. There was precedent in Judeo-Christian belief in Providence. There was also the beginnings of a historiography for HHS that found in Vico, writing in the early eighteenth century, the "classic" presentation of the issues (the historiography introduced in Smith 2007). The historical way of thought also accommodated burgeoning evolutionary ideas in biology and ideals of national progress in the sociopolitical sphere. At the same time, there was deep disquiet among the academics whose vocation was to study these things that historical reason and absolute reason appeared to be in contradiction, or at least to push in different directions. The Hegelian philosophy that had claimed to effect a reconciliation of history and reason had been widely criticized since Hegel's death, and other attempts at a synthesis, like Dilthey's, remained incomplete. The language of "crisis" was common.

The whole plot, the whole way of thought, unraveled in the twentieth century. Belief in the possibility of telling one story about human progress, or belief in "universal history," almost totally disappeared from the academic world in the

West, though this was not the case in other settings, and notably not in Christian fundamentalist teaching or in Muslim universities. Notions of the human essence, however formulated, and Enlightenment and nineteenth-century ideals of "Man" or "the human spirit" – in most western university settings – were analyzed (or "deconstructed") as representations of culturally specific, historically local values, values of men not women, for example, or of white colonizers not black slaves, of one tradition of discourse not another. In the contemporary academic world, there is acute consciousness that the historian necessarily selects materials with which to create knowledge. History is rarely thought of as providing the unifying content of knowledge. It is no longer controversial to assert, as Claude Lévi-Strauss once did, that "history is therefore never history, but history – for" (Lévi-Strauss 1972: 257).

A significant contribution to rejection of historicist argument came from within the social science or human science disciplines themselves. The rejection was a mark of growing professionalization, and it took the form of critical self-consciousness about the distinction between speculation, so common in historical and evolutionary reconstructions, and empirically verified knowledge about the relations of structure and function. Thus, Durkheim substantially built his argument for sociology as a science on a critique of Herbert Spencer's "synthetic philosophy" of evolutionary progress, Bronisław Malinowski established a pattern for social anthropology as a science through fieldwork, and linguists such as Ferdinand de Saussure and Roman Jakobson turned to what they claimed was rigorous structuralist, not historical, analysis of language. The large exception to this intellectual trend was Marxism, which in the Soviet form of dialectical materialism recreated its own specific type of universal historical understanding as the basis of the human sciences.

Historicism, if understood as belief in progress and the interpretation of local events in terms of their place in universal history, is widely thought to have collapsed with the political and moral barbarism of the twentieth and twenty-first centuries. If understood, however, as a label for a positive emphasis on knowledge based on the explanation and evaluation of the particularity of events, on the necessity of knowledge of individual entities within particular contexts, it continues to flourish, not least in the activities of HHS. Just as biography gives meaning to an individual life, so it is possible to think of history giving meaning to the life of individual communities, nations, and social entities of all kinds, including bodies of knowledge and including the sciences (MacIntyre 1977). To provide a theory of meaning, and to decide whether meaning originates with historical knowledge, with empirical statements, with structural features of cognition, the unconscious, or language, or with some other factor, is a matter for philosophy. But historical understanding in many forms, including storytelling and autobiography as well academic writing, is still fundamental in many aspects of many cultures. Debate about the temporality of being human, that is, debate about knowledge with historical form, has existential relevance. Reference to the *history* of the human sciences alludes to all these issues.

If theoretical debate about the nature of the human sciences owes much to late nineteenth-century German sources, it is also substantially indebted to the completely different work of Foucault and to post-structuralist French thought.

Whether Foucault was able to sustain his large-scale claim about the coming into existence, around 1800, of discursive conditions for the human sciences, and more provocatively for "the human" as the content of these modern sciences, was doubtful. His statements proved questionable on empirical grounds and vulnerable to philosophical criticism aimed at structuralist forms of argument in general. Indeed, Foucault shifted his attention to historically localized "regimes of truth," which he interrogated in terms of the operations of power as knowledge-generating practices (spreading reference to "power/knowledge"). The study of the constitution of power/knowledge in areas of relatively delimited and historically specific practice, for instance, in the life of medical conditions like hysteria and autism, or in fields like child observation and social hygiene, has proved enormously fruitful. The resulting studies have in effect established one kind of paradigm, or type of "normal science," in HHS. There are many studies of the micropolitics of power and governance (especially bio-power, the power exercised in and through individual human bodies), as Foucault's approach to governance was specifically applicable in the liberal democracies in which many western scholars lived. It had enormous importance as a way of thinking about power, just at the time when Marxist and functionalist notions of "top-down" power were being judged inadequate (Rose 1998; Rose and Miller 2008).

Foucault's work found a large audience in the English-speaking world alongside, and sometimes in competition with, a radical Marxian interpretation of knowledge, including scientific knowledge, as the expression of the social relations of production (Young 1977, 1985), and a major reworking of the sociology of knowledge that argued for a fully fledged sociological approach to knowledge formation (Barnes 1974; Bloor 1976). The considerable expansion of science and technology studies (STS) in the 1970s and 1980s, incorporating arguments for social epistemology, deeply challenged the whole domain of the history and philosophy of science (summarized in Golinski 1998). While the great bulk of work in STS was not concerned with the human sciences, it nevertheless contributed to the academic conditions in which HHS was to flourish, since it established arguments and pioneered case studies in the social, and thereby historical, formation of knowledge. There was enthusiasm for empirical historical studies, focused as mainstream historical work focused, on the detail of particular events and their contexts, in knowledge-making practices and the achievement of legitimacy and authority. Indeed, a number of scholars, such as the philosopher of science Ian Hacking, explicitly invested intellectual work in detailed historical studies, as they considered this likely to be more fruitful in epistemology than yet further debate about abstract, general issues, such as realism versus relativism, which had become bogged down. Hacking wrote under the heading of "historical ontology" (Hacking 1995a, 2002); the historian of science Lorraine Daston developed the notion of "the biography of scientific objects" (Daston 2000); Simon Schaffer similarly argued that historical studies, rather than abstract argument, will prove fruitful in the theory of knowledge (Schaffer 2008). The social psychologist and historian of psychology, Kurt Danziger, made a major contribution with the historical study of psychological categories, including personality and memory (Danziger 1990, 2008). The attitude these positions express is widely shared in HHS: "small," that is, local, historical studies illuminate the "big" issues at stake in inquiry into being human. At the same time, however, HHS has continued to be home to an interest in more directly philosophical forms of argument about the nature of "the human" (e.g., Smith 2007; Tauber 2009, 2013).

The western philosophical tradition has assigned a central place to conscious reason's capacity to reflect on itself as conscious reason, making reflection a central topic in ontological and epistemic inquiry. Descartes predicated his philosophy on the existence of such reflection, while Locke attributed knowledge to both sensory experience and reflection on that experience. Kant's critical philosophy, followed in this respect by idealist philosophers, attempted systematically to delineate the rational conditions in which such reflection on the exercise of reason and judgment occurs and knowledge becomes logically possible. These philosophies posited the "I" as the ground of such reflection, rendering the notion of the ego or self the reflexive condition of cognition. Everyday contemporary language is indeed full of phrases (as in, "I know what I'm doing") denoting reflective awareness or consciousness of mental and bodily action. Twentieth-century phenomenology provided formal description of reflection in terms of the intentionality of mental acts, the property of awareness as awareness of something, including awareness of awareness itself. Questions about the content and nature of such reflective awareness, and about whether it is veridical (do I really know what I am doing, and how do I know this? – a question often formulated in Freudian terms) have been fundamental for HHS.

The literature refers to "reflexivity" as well as to "reflection," though it does so in a number of ways, so that the term may confuse more than it clarifies. In some usages, "reflexivity" names the process by which, as it is argued, any statement can be shown to be dependent on other statements, and every self-reflection dependent on other self-reflections, leading to statements that references to entities of world and of self with claimed absolutely objective standing are false: they reference reflexivity not entities (Lawson 1985). This was one foundation for debate around the culture of postmodernism. Of more specific concern to HHS, the term was also used to denote the process in which action and reflection on action – that is, knowledge – partner each other in a circle of mutual influence. (Hence it may be said that cognitive events are also actions.) It is widely argued that human cognition and conduct exist in a circle, a "reflexive" temporal, or historical, circle. Hacking studied specific examples of the processes, which he referred to as "looping" effects (Hacking 1995b, 2007; also Gigerenzer 1992). A reflecting subject changes the nature of the subject through reflection. This is exemplified in belief in the possibility of therapeutic change through talk, in which a patient's self-knowledge affects action, and the action in turn affects self-knowledge. The implications of this kind of reflexivity for belief about the "natural," as opposed to "constructed," standing of basic categories in the human sciences (the debate about the presence of "natural kinds" as opposed to "human kinds") continue to be argued over.

A number of earlier scholars, including the archaeologist and philosopher of history R. G. Collingwood, argued that the presence of reflexive processes differentiates the subject matter of the human sciences (for the purpose of this argument, including the humanities) from the natural sciences. It appears common sense to recognize that physical properties remain what they are while subject to observation, while the reflective human subject, through reflection, changes over time (Smith 2005, 2007). Being human has a deep temporality. The belief has supported a notion of "the human" as an ontologically distinct condition, as a self-creative or autopoietic nature, in which action (language, culture, social and political organization) brings into existence new forms of being human. The argument has appealed to philosophical idealists from the time of Herder, who declared, "we live in a world we ourselves have made" (quoted in Berlin 2000: 168), through Marxist theory characterizing people in terms of the way, by means of production, they address the demands of material existence. Collingwood deduced from the argument the rather idiosyncratic conclusion that historical knowledge is the reliving of the thoughts of historical actors in their reflective engagement with the world, making their self-understanding the core knowledge of the human sciences (Collingwood 1961).

This view of what is here being called reflexivity (even when detached from Collingwood's specific philosophy of history) elevates history to a key position as knowledge in the human sciences: history is a science in terms of which all other human sciences have meaning. Historical work, establishing knowledge of what being human has been thought to entail, records and interprets the enactment of that entailment, including the activity of generating scientific knowledge. This, in the most general sense, is the subject of HHS.

The strong version of the argument, which argues for the coming into being of new human states – not just ways of thinking about human states – over historical time through human action, finds support both in localized case studies (e.g., of new categories like PTSD; Hacking 2007 talked of kinds of people as "moving targets") and in studies of large-scale technological change. The latter has considerable contemporary resonance. The development of reproductive, virtual reality, prosthetic, robotic, neuropharmacological, and digital technologies is having a profound impact on scientific and public willingness to countenance temporal shifts in human nature (or, indeed, shifts out of human nature). The sociologist Nikolas Rose has taken a lead in calling on the social sciences to respond to these changes rather than, as they have been inclined to do, to remain fearful of the supposed naturalizing, sociologically blind discourse of biology (Rose 2013). "To recognize that subjectivity is itself a matter of the technologizing of humans is not to regard this process as amounting to some kind of crushing of the human spirit . . . Human capacities are . . . inevitably and inescapably technological. An analytics of technology has, therefore, to devote itself to the sober and painstaking task of describing the consequences, the possibilities invented as much as the limits imposed, of particular ways of subjectifying humans" (Barry et al. 2005: 278). References to statements about "the human" in the contemporary world have to face the dramatic ways in which science and technology are changing, or appear about to change, core dimensions of what, hitherto, were accepted to be universal biological or spiritual conditions of human nature. This is so; yet it can still be asked whether the circumstances are unprecedented. Earlier changes of belief and action – the domestication of the horse, monotheism, and printing, for instance – may also have changed the nature of being human. The issues would seem to increase the relevance and appeal of

HHS. The domain appears well equipped to contribute to assessing whether contemporary technological change really does mark a qualitatively new stage in human evolution.

There is no intention in these comments to reduce all kinds of knowledge to historical knowledge. Reductionist claims, in all versions, whether to a supposed ground in material nature, in human history, or in language, are equable vulnerable to argument that different forms of knowledge have different purposes. "History has no more any such final and simple word to divulge that would express its true meaning than nature does" (Dilthey 1988: 131).

Weaker versions of the argument about the occurrence of historical change in the nature of human nature, or in being human, merge with studies of the ways in which what is said to be known about the past has profound significance in the present. Few people, perhaps, would deny this in the realm of politics or social life; there is much interest in the way, for example, histories of nation states act in political imagination (not to mention feature in school textbooks). The argument is more contentious in the human sciences (and much more contentious in the so-called hard sciences), of course, because scientists generally believe that they legitimate knowledge claims by objective reference and not through a historical story. The issues are complex, however, since what counts as an objective reference does so by virtue of its place in a historical story, and what counts as a valid historical story does so by virtue of the objective procedures deployed to establish it in the first place. An appreciation of complexity in these issues is intrinsic to HHS. Empirical historical studies and theoretical analysis need each other in coping with complexity. As Makkreel said about Dilthey's advocacy of *die Geisteswissenschaften*, from this "perspective of understanding the real task of the human sciences lies at the intermediary level where the intersection of the particular and the universal occurs" (Makkreel 2012: 318).

Weaker versions of the argument that belief structures action and action structures belief also inform sympathy in the human sciences – where such sympathy exists – for historical work because it creates or sustains *perspective*. Perspective, in this context, refers to a capacity to see and analyze currently held knowledge and activity as the creation of specific circumstances, however rigorous and objective the observational and experimental methods of an expert community. Achieving perspective is an argument, or "defense," for doing HHS (or history of psychology, of social theory, of criminology, of ethnographic fieldwork, etc.) that historically oriented researchers commonly use to legitimate their speciality to colleagues. It is necessary to speak of "defense" because scholars in the human sciences who do historical work are nearly always outnumbered by those who do not, and they are vulnerable to the judgment that they would be better off "doing science." (An argument for the history of psychology based on the value of "perspective" is provided in the introductory chapter to the *Handbook* volume on psychology.)

Discussion of knowledge generally, and human self-knowledge in particular, starts from and builds on what has been stated and argued over before. If this appears a banal assertion, it is nevertheless a central reason for taking historical knowledge seriously, and it is thus far from banal in implication. There are, of course, philosophers, like Heidegger, who have made the claim to start afresh (or to start again from "the ancient wisdom" of the pre-Socratics), but even they have argued through criticism and example. As concerns the human sciences, training and education commonly include a substantial introduction to the interpretation of canonical authors and texts. This educational practice has point because the very complexity of what is at stake in formulations of human self-knowledge renders studies in terms of historical particularity a key means to make abstract discussion accessible. The texts of Adam Smith, Comte, Marx, Darwin, William James, Durkheim, Freud, Weber, Collingwood, Foucault, and many others have thus become part of the working resources of historians of the human sciences. It is also part of HHS to understand how and why such authors and texts acquired canonical status. This is an essential part of the reflective activity that HHS fosters in the human sciences, the activity intrinsic to achieving perspective.

Conclusion: "History of the Present"

It is the case, though figures are not available (and would be difficult to obtain because of the nebulous nature of the domain), that a majority, perhaps the great majority, of people who might describe themselves as doing work in HHS do not have training or positions in the history discipline. This obviously reflects employment opportunities, since human sciences claim to provide specific social benefits, and thus they attract funds in a way historical work does not. The reasons go beyond this, however. Many people working in the human sciences think that history is pertinent to the human sciences in general, or to particular disciplines, or to themselves, only insofar as it contributes to what psychologists, social scientists, economists, or other people in the contemporary human sciences do. The history of interest to these scholars, and some of them might say the only history that is of interest, is, to use a phrase thrown out by Foucault and taken up in English translation by Rose among others, "the history of the present" (Garland 2014; Barry et al. 2005; also the journal, *History of the Present*, founded 2011). The phrase accurately conveys the idea that the purpose of historical work is to expose to view the manner in which present theories, social practices, and policies (e.g., concerning the mentally ill, children, the indigent, or drug users) have come to be what they are. Such history may be directed at the micropolitics of some aspect of social management; or it may be directed at the shaping of the specific current content and practices of the human sciences. Its objective is to demonstrate the contingency of some aspect of the present and the workings of power in present practices. Thus, the turn to history is often coupled with a wish to open up intellectual imagination about the ways a particular current situation may be changed. In short, "history of the present" serves practice-oriented critique. Its aim is perspectivist, not relativist. The purpose is to understand the conditions of possibility of the present, not to create historical knowledge as such. In this context, it is popular to describe history as "genealogy," another term employed by Foucault, who was in turn indebted to Nietzsche.

The resulting historical practice is self-evidently attractive to people working in the human sciences because it prioritizes the current content of these sciences and the politics of the present. It finds value in serious historical work, but it does not waste precious intellectual and financial resources on research into historical knowledge "for its own sake." Scholars attracted to "the history of the present" are able to stay within their parent disciplines while undertaking circumscribed historical work in order to analyze and, where thought necessary, critique present practices and policy. Hence the popularity of Nietzsche's early essay, "On the uses and disadvantages of history for life," in which Nietzsche turned on German academic historians for creating a discipline impotent when it came to addressing questions pressing on individual lives (Nietzsche 1983).

As the majority position, and therefore easily taken for granted, "the history of the present" may even foster a certain antagonism to history as a disciplinary field in the humanities, an antagonism evident, for example, in statements that certain kinds of historical detail are "uninteresting," invoking an epistemic judgment and not just an individual emotional preference: "The 'history' *itself* is not so much of interest for us today; it matters first of all that we see how it relates to our present concerns and preoccupations" (Personal communication 2018). Natural scientists, faced by the expansion of the history of science as an autonomous discipline responsive to the standards of general historical scholarship rather than to the interests of natural scientists, also made this judgment. If meant as more than an expression of individual preference or group conventions (and it is necessary to ask, who exactly is "we" in the quoted statement), there is something misguided about such statements. All historical writing, after all, is written "in the present" inhabited by the authors, and all historical writing, at least tacitly, takes a stance on the epistemic relations between the writer's situation and what ordinary language calls "the past." There is, therefore, a sense in which all historical writing necessarily addresses the present. The question at issue is: *which present, which person's present*, the present that matters to the social or natural scientist, or the present that matters to the writer of history, or the present that matters to someone else? The assertion that one audience not another matters is an assertion of the value of one form of knowledge over another to one group of people rather than another. It is not self-evident that natural or social science knowledge is more significant, of more value in some general or universal way, than historical knowledge. Statements to the effect that an area is not "interesting" do not constitute rational argument but reflect the assumptions, and as it may be prejudices, of a working life within one community rather than another. It remains the case, however, for many people working in the human or natural sciences, for whom the history is interesting insofar as it is thought to contribute to advancing scientific work or political argument in the present, their interest appears "natural." This is because they think scientific knowledge fundamental, or they think scientific knowledge useful, in ways they think historical knowledge is not.

This, then, is the nub: on what grounds is it more basic or fundamental to pursue natural or social science knowledge as opposed to historical knowledge? Rather than answering this question, should it not be recognized instead that different forms of knowledge have different purposes, and that the presumption that one form of knowledge is more fundamental than another is misguided? The blunt fact is that historical knowledge does not have utility, measured in the terms that the natural or social sciences are represented as having and that attract funding sources. But this is a matter of politics and ways of life not reason.

Examination of the relationship between the writer's present and discursive practice as the historical subject is a well-established area of inquiry in the theory of historical knowledge. "The history of the present" is not distinctive as a philosophical project but, rather, distinctive through the priority it gives, for contingent social reasons, to the implications of history for a current policy issue or a scholar's disciplinary identity. It is important to recall, with Dilthey, that the human sciences "have arisen out of the practical activity of life itself and have been developed through the demands of vocational training ... Indeed, their first concepts and rules were discovered for the most part in the exercise of social functions themselves" (Dilthey 1988: 88). History has not arisen by some special practice, different from the practice of the other human sciences.

Moreover, if all knowledge is in some way a historical formation, the separation of science and history breaks down: everything said about the world has historical character. There are, of course, positivist historians, just as there are positivist natural and social scientists, who write as if the sole business to hand is the application of appropriate methods to uncover the facts, whether of history or of nature or of society, facts that are immutable and "outside history." The search for facts informally understood in this way, in history as a discipline as much as in physiology as a discipline, constitutes a large part of what T. S. Kuhn called "normal science," the practice of which imposes no pressing need to justify the content of knowledge shared in a discipline or to make it interesting to outsiders (Kuhn 1970). Positivist discourse comes into its own in this setting. It is perfectly legitimate and understandable to find scientists looking to history in a restricted way as a resource for their own work in normal science. This neither legitimates nor delegitimates history as a form of knowledge. What "history of the present" clearly does do, as a matter of practice, is put the political nature and consequences of human science knowledge at the center of the agenda, in a way academic history normally does not. This, however, is a matter of practice and purposes not principles of reasoning. It is also easy to think of examples of mainstream history writing that have had great political significance (on the historical facts concerning the founding of nation states, for instance).

If it is the case that few scholars trained in academic history work under the heading of HHS, this is a function of conventions within academia, in which historians have not trained to study formally constituted bodies of knowledge and practice (whether in philosophy or science), while scholars coming from the human science disciplines have. There is nothing immutable about this, and indeed in the history of science, medicine, and technology, there is constructive crossover between trained historians, historical studies of present knowledge and practice, and the human sciences. All the same, for most historians of science, the history of the human sciences remains an unfamiliar area, on the margins of historical studies of the physical and biological sciences that are said to have taken the lead in the establishment of modern science. It was, therefore, thought necessary to create a special forum for the promotion of the interests of historians of the human sciences within the History of Science Society (international, though based in the USA). A large-scale thesis, which proposes that modern science is indebted to the systematization of studies of "the human" in history, medicine, jurisprudence, anthropology, and political economy, rather than interpreting these studies as dependent offshoots of a model of science established in the sciences of nature, remains to be argued for (Gaukroger 2011, 2016; Smith 1997).

The conventional interest of historians of science in the natural sciences, if in recent decades very much broadened to include "public science" and "the public understanding of science," itself has a history. This is significant for understanding what goes on under the heading of HHS. The history of science as a discipline first developed in the context of a positivist understanding of science as the progressive achievement of civilization, with the physical sciences taking the lead and the psychological and social sciences following their example (Thackray 1984a, b). The plot of the history of the human sciences, as far as early historians of science were concerned, was the gradual incorporation of "the human" into nature, the demonstration of "man's place in nature," in T. H. Huxley's phrase (Huxley 1906), hence the overwhelming prominence given to Darwin. If, however, there are different forms of knowledge, and if knowledge in the human sciences is in certain large respects different from knowledge of nature, there must be other plots waiting to be written. HHS encompasses this large issue, in a way that histories of particular disciplines do not, thus leading the domain to engagement with philosophical questions of human self-understanding. It can be no surprise and no criticism to say that HHS as a domain has no predetermined boundaries. A picture of how scholars actually get on with questions that can be delimited for stated purposes stands in for a definition of the field. The *Handbook* attempts such a picture.

It is noteworthy that institutions or academic societies committed by name to fostering HHS have origins mainly in the interests of scholars coming from the psychological and social science disciplines as well as the history of science: the *Société Français pour l'Histoire des Sciences de l'Homme* (f.1986), the journal *History of the Human Sciences* (f.1988), the Forum for History of Human Science (f.1988) in the History of Science Society, and the European Society for the History of the Human Sciences (originating as Cheiron: The European Society for the History of the Behavioural and Social Sciences, f.1982, and adopting its new name in 1995).

The future is unclear and prediction a notoriously speculative branch of knowledge. Whatever the complexities of human self-knowledge, and indeed in response to the complexities, the history of the human sciences currently, and for the foreseeable future, offers an intellectual framework for discussion, and an umbrella term of convenience, within the terms of which constructively to integrate argument coming out of different disciplines. The *Handbook* seeks to be a key resource for this purpose.

References

Adorno T, Horkheimer M (1986) Dialectic of enlightenment (trans: Cumming J). Verso, London (first German publ. 1947)
Araujo S d F (2016) Wundt and the philosophical foundations of psychology: a reappraisal. Springer, Cham
Backhouse RE, Fontaine P (eds) (2014) A historiography of the modern social sciences. Cambridge University Press, New York
Barnes B (1974) Scientific knowledge and sociological theory. Routledge & Kegan Paul, London
Barry A, Osborne T, Rose N (2005) Writing the history of the present. In: Joseph J (ed) Social theory: a reader. Edinburgh University Press, Edinburgh, pp 269–289
Beer D (2008) Renovating Russia: the human sciences and the fate of liberal modernity, 1880–1930. University of Chicago Press, Chicago
Bennett MR, Hacker PMS (2003) Philosophical foundations of neuroscience. Blackwell, Malden/Oxford
Berlin I (2000) Vico and Herder. In: Three critics of the enlightenment: Vico, Hamann, Herder. Princeton University Press, Princeton, pp 1–242
Bloor D (1976) Knowledge and social imagery. Routledge & Kegan Paul, London
Cohen-Cole J (2014) The open mind: Cold War politics and the science of human nature. University of Chicago Press, Chicago
Collingwood RG (1961) The idea of history. Oxford University Press, Oxford (first publ. 1946)
Danziger K (1990) Constructing the subject: historical origins of psychological research. Cambridge University Press, Cambridge
Danziger K (2008) Marking the mind: a history of memory. Cambridge University Press, Cambridge
Daston L (ed) (2000) Biographies of scientific objects. University of Chicago Press, Chicago
Dennett DC (1993) Consciousness explained. Penguin Books, London
Dilthey W (1988) Introduction to the human sciences: an attempt to lay a foundation for the study of society and history (trans: Betanzos J). Harvester Weatsheaf, London (first German publ. 1883)
Dilthey W (2002) The formation of the historical world in the human sciences. In: Makkreel RA, Frithjof R (eds) Selected works (trans: Makkreel RA, Scanlon J, Oman WH), vol 3. Princeton University Press, Princeton
Eribon D (1991) Michel Foucault (trans: Wing B). Harvard University Press, Cambridge, MA (first French publ. 1989)
Foucault M (1970) The order of things: an archaeology of the human sciences (trans: Sheridan A). Tavistock, London (first French publ. 1966)
Foucault M (1980) In: Gordon C (ed) Power/knowledge: selected interviews and other writings by Michel Foucault, 1972–1977. Pantheon Books, New York
Fox C, Porter R, Wokler R (eds) (1995) Inventing human science: eighteenth-century domains. University of California Press, Berkeley
Garland D (2014) What is a "history of the present"? On Foucault's genealogies and their critical preconditions. Punishment Soc 16(4):365–384. https://doi.org/10.1177/1462474514541711
Gaukroger S (2011) The collapse of mechanism and the rise of sensibility: science and the shaping of modernity, 1680–1750. Oxford University Press, Oxford
Gaukroger S (2016) The natural and the human: science in the shaping of modernity, 1739–1841. Oxford University Press, Oxford
Gigerenzer G (1992) Discovery in cognitive psychology: new tools inspire new theories. Sci Context 5:329–350
Golinski J (1998) Making natural knowledge: constructivism and the history of science. Cambridge University Press, Cambridge
Graham LR (1987) Science, philosophy, and human behavior in the Soviet Union. Columbia University Press, New York

Hacking I (1995a) Rewriting the soul: multiple personality and the sciences of memory. Princeton University Press, Princeton

Hacking I (1995b) The looping effects of human kinds. In: Sperber D, Premack D, Premack AJ (eds) Causal cognition: a multidisciplinary debate. Clarendon Press, Oxford, pp 351–394

Hacking I (2002) Historical ontology. Harvard University Press, Cambridge, MA

Hacking I (2007) Kinds of people: moving targets. Proc Br Acad 151:285–318

Haskell TL (1977) The emergence of professional social science: the American Social Science Association and the nineteenth-century crisis of authority. University of Illinois Press, Urbana

Herman E (1995) The romance of American psychology: political culture in the age of experts. University of California Press, Berkeley

Huxley TH (1906) Man's place in nature and other essays. J M Dent, London (first publ. 1863)

Kirtchik O, Boldyrev I (eds) (2016) Special issue, the human sciences across the Iron Curtain. Hist Hum Sci 29(4–5)

Kuhn TS (1970) The structure of scientific revolutions, 2nd edn. University of Chicago Press, Chicago

Lamont P (2013) Extraordinary beliefs: a historical approach to a psychological problem. Cambridge University Press, Cambridge

Lawson H (1985) Reflexivity: the post-modern predicament. Hutchinson, London

Lepenies W (1988) Between literature and science: the rise of sociology (trans: Hollingdale RJ). Cambridge University Press, Cambridge (first German publ. 1985)

Lévi-Strauss C (1972) The savage mind (La pensée sauvage), (trans: anon). Weidenfeld and Nicolson, London (first French publ. 1962)

Lukes S (1973) Emile Durkheim: his life and work. A historical and critical study. Allen Lane, The Penguin Press, London

MacIntyre A (1977) Epistemological crises, dramatic narrative and the philosophy of science. Monist 60:453–472

Makkreel RA (2012) The emergence of the human sciences from the moral sciences. In: Wood AW, Hahn SS (eds) The Cambridge history of philosophy in the nineteenth century (1790–1870). Cambridge University Press, Cambridge, pp 293–322

Manicas PT (1987) A history and philosophy of the social sciences. Basil Blackwell, Oxford/New York

Marquard O (1995) On the unavoidability of the human sciences. In: The defence of the accidental: philosophical studies (pp 91–108) (trans: Wallace RW). Oxford University Press, New York (first German publ. 1985)

Mazlish B (1998) The uncertain sciences. Yale University Press, New Haven

Megill A (1985) Prophets of extremity: Nietzsche, Heidegger, Foucault, Derrida. University of California Press, Berkeley

Morawski JG (1986) Organizing knowledge and behavior at Yale's Institute of Human Relations. Isis 77:219–242

Nietzsche F (1983) On the uses and disadvantages of history for life. In: Stern JP (ed) Untimely meditations (trans: Hollingdale RJ). Cambridge University Press, Cambridge, pp 57–123 (first German publ. 1874)

Personal communication (2018) Anonymous referee's report on book ms

Porter T (1986) The rise of statistical thinking 1820–1900. Princeton University Press, Princeton

Porter T (1995) Trust in numbers: the pursuit of objectivity in science and public life. Princeton University Press, Princeton

Porter R (1997) The greatest benefit to mankind: a medical history of humanity from antiquity to the present. HarperCollins, London

Proctor H (2020) Psychologies in revolution: Alexander Luria's 'romantic science' and Soviet social history. Palgrave Macmillan, Cham

Ringer FK (1990) The decline of the German mandarins: the German academic community, 1890–1933. Wesleyan University Press/University Presses of New England, Hanover/London

Roberts RH (2002) Religion, theology and the human sciences. Cambridge University Press, Cambridge

Rose N (1998) Inventing our selves: psychology, power, and personhood. Cambridge University Press, Cambridge

Rose N (2013) The human sciences in a biological age. Theory Cult Soc 30(1):3–34. https://doi.org/10.1177/0263276412456569

Rose N, Miller P (2008) Governing the present: administering economic, social and personal life. Polity Press, Cambridge

Schaffer S (2008) Interview of Simon Schaffer. Video file. http://www.dspace.cam.ac.uk/handle/1810/205060

Smith R (1997) The Fontana history of the human sciences. Fontana Press, London. (Identical with The Norton history of the human sciences. W W Norton, New York)

Smith R (2001) Human sciences, history and sociology. In: Smelser NJ, Bates P (eds) International encyclopedia of the social and behavioral sciences. Pergamon, Oxford, pp 7027–7031

Smith R (2005) Does reflexivity separate the human sciences from the natural sciences? In: Smith R (ed) Special issue on reflexivity, Hist Hum Sci 18(4):1–25

Smith R (2007) Being human: historical knowledge and the creation of human nature. Manchester University Press/Columbia University Press, Manchester/New York

Solovey M (2013) Shaky foundation: the politics-patronage-social science nexus in Cold War America. Rutgers University Press, New Brunswick

Tauber AI (1999) Confessions of a medicine man: an essay in popular philosophy. MIT Press, Cambridge, MA

Tauber AI (2009) Science and the quest for meaning. Baylor University Press, Waco

Tauber AI (2013) Requiem for the ego: Freud and the origins of postmodernism. Stanford University Press, Stanford

Taylor C (1985) Interpretation and the sciences of man. In: Philosophy and the human sciences: philosophical papers, vol 2. Cambridge University Press, Cambridge, pp 15–57

Thackray A (1984a) Sarton, science and history. Isis 75:7–9

Thackray A (1984b) The pre-history of an academic discipline: the study of the history of science in the United States, 1891–1941. In: Mendelsohn E (ed) Transformation and tradition in the sciences: essays in honor of I. Bernard Cohen. Cambridge University Press, Cambridge, pp 395–420

Todes D, Krementsov N (2010) Dialectical materialism and Soviet science in the 1920s and 1930s. In: Leatherbarrow W, Offord D (eds) A history of Russian thought. Cambridge University Press, Cambridge, pp 340–367

Vidal F, Ortega F (2017) Being brains: making the cerebral subject. Fordham University Press, New York

Ward JF (1981) Arthur F. Bentley and the foundations of behavioral science. J Hist Behav Sci 17:222–231

Young RM (1977) Science *is* social relations. Radical Sci J 5:65–129

Young RM (1985) The historiographic and ideological contexts of the nineteenth-century debate on man's place in nature. In: Darwin's metaphor: nature's place in Victorian culture. Cambridge University Press, Cambridge, pp 164–247

Kant After Kant: Towards a History of the Human Sciences from a Cosmopolitan Standpoint

2

Steve Fuller

Contents

Introduction: Using Kant as a Lens to See into the Future of Humanity 30
Kant's Anthropology: The Ascent of the Ape in Human Self-Understanding 31
Remnants of the Divine Soul in the Naturalized Mind: Of Nerves and Aether 33
Between Mind and Matter: Life as Extended Self-Correction 36
Wrestling with the Rousseauian Legacy: The Eugenic Versus the Statistical State 40
Kant Between the Stoics and the Epicureans: The Rise of the Cosmopolitan Standpoint 42
Cosmopolitanism's Anti-sentimental Expansion of the Circle of Moral Concern 48
Conclusion: Testing Cosmopolitanism on Machines: Extending the Human Beyond the Animal 50
References 54

Abstract

This chapter considers the history of the human sciences as propaedeutic to humanity's future self-understanding. Immanuel Kant is pivotal in this context, not merely as someone whose views about the human have been influential, but more importantly as someone who deeply problematized what it means to be "human" in ways that remain relevant. In particular, Kant updated his understanding of the Judaeo-Christian and Greco-Roman traditions to project an indefinitely extendable vision of humanity, which is captured by the Stoic idea of *cosmopolitanism*. So, how would Kant define humanity today? The chapter explores the question largely by drawing on Kant's fertile appeal in his later "critical" writings to the distinction between the "Stoic" and "Epicurean" worldview, both of which acknowledge the centrality of chance to the cosmos, with the Stoic adopting the more hopeful and even risk-embracing approach to such existential uncertainty. The overall import of Kant's Stoic cosmopolitanism is to undermine the intuitiveness of the "sentimentalism" associated with

S. Fuller (✉)
Department of Sociology, University of Warwick, Coventry, UK
e-mail: s.w.fuller@warwick.ac.uk

© The Author(s), under exclusive licence to Springer Nature Singapore Pte Ltd. 2022
D. McCallum (ed.), *The Palgrave Handbook of the History of Human Sciences*,
https://doi.org/10.1007/978-981-16-7255-2_42

the animal-based conceptions of humanity favoured by the Epicurean approach. In this respect, Kant opens the door to what transhumanists call a "morphologically free" conception of humanity that is in principle open to membership by both extraterrestrials – a prospect Kant himself entertained – and artificially intelligent machines, a move with significant implications for what the history of the human sciences has been about and might be in the future.

Keywords

Kant · Stoic · Epicurean · Humanity · Cosmopolitanism

Introduction: Using Kant as a Lens to See into the Future of Humanity

This chapter considers the history of the human sciences as propaedeutic to humanity's future self-understanding. Immanuel Kant is pivotal in this context, not merely as someone whose views about the human have been influential, but more importantly as someone who deeply problematized what it means to be "human" in ways that remain relevant. In particular, Kant updated his understanding of the Judaeo-Christian and Greco-Roman traditions to project a future vision of humanity, which is captured by the Stoic idea of *cosmopolitanism*. In this respect, Michel Foucault (1970) short-changed Kant's significance by stressing his conversion of Linnaeus' *Homo sapiens* into the proper object of the human sciences (aka "anthropology"). Nevertheless, it is not so hard to reinvent Kant's problematic for the future that faces us today; hence the title "Kant after Kant." It amounts to asking, "How would Kant define humanity now?" The chapter explores the question by recalling several signature features of the Kantian approach, latter-day developments that support his approach, as well as challenges that it faces.

At the same time Linnaeus's coinage of *Homo sapiens* identified the human with a distinct sort of (apelike) body, some alternative and equally lasting moves were also being made to sustain the spiritual nature of humanity by broadly materialist means. These are surveyed as well, but in the end, Kant arrived at an indefinitely extendable conception of the human that characterizes his cosmopolitanism's sense of "universalism," one rooted more in the moral imagination than in empirical reality. Here the chapter draws on Kant's fertile appeal in his later "critical" writings to the distinction between the "Stoic" and "Epicurean" worldview, both of which acknowledge the centrality of chance to the cosmos, with the Stoic adopting the more hopeful and even risk-embracing approach to such existential uncertainty. The overall import of Kant's Stoic cosmopolitanism is to undermine the intuitiveness of the "sentimentalism" associated with the animal-based conceptions of humanity favored by the Epicurean approach. In this respect, Kant opens the door to what transhumanists call a "morphologically free" conception of humanity that is in principle open to membership by both extraterrestrials – a prospect Kant himself entertained – and artificially intelligent machines, a move with significant implications for what the history of the human sciences has been about and might be in the future.

Kant's Anthropology: The Ascent of the Ape in Human Self-Understanding

Immanuel Kant was the first philosopher who came fully to grips with the conception of the human that has been focus of the history of the human sciences. He used the term *anthropology* in the 1790s to name the line of empirico-normative inquiry surrounding this conception, of which Michel Foucault (2008) famously made a great deal (cf. Zammito 2002). Nowadays this moment is presented as the founding moment in "scientific racism," since Kant certainly ordered the various "races" known to him according to their degree of "humanity." However, "race" in Kant's day was completely innocent of anything resembling today's "genetics." Indeed, he wrote nearly a century before this science was implicated in any "racist" explanations or policies with regard to human differences. Instead, "race" was understood as the intergenerational consequence of "culture," understood in the same spirit as "agriculture." It was a sort of "blood and soil" amalgam that produced a distinctive collective way of being in the world. This perspective was explicitly developed by Kant's student Johann Gottfried Herder, based on which various "Romantic" views of culture, many associated with nationalism, were popularized in the nineteenth and twentieth centuries, often with drastic consequences for individual human lives.

Kant explicitly charged "anthropology" with inferring people's souls from their external appearances, with the aim of judging their fitness for "humanity." In his day, customs, artifacts, and physiognomy – later to include phrenology – were central to such judgements. While it is easy more than two centuries later to bemoan the shortsightedness of Kant's approach to the subject, we should not overlook the enduring explanatory appeal of such "morphological" accounts of life – across both the social and biological sciences – even after genetics began to present a more complex and what has come to be a profoundly alternative way of understanding the same phenomena. Walter Lippmann's critical coinage of "stereotype" in the 1920s reflected the persistence of this mindset, which arguably remains among biologists studying non-human animal and plant life, where the concept of "specimen" (a semantic derivative of "species") still retains a conceptual hold. More to the point, much of what passed for "philosophical anthropology" in the early twentieth century – in, say, Ernst Cassirer, Max Scheler, and Helmuth Plessner – can be easily understood within the ambit of Kant's original project, only now without the explicit desire to rank order the various specimens of humanity. As for the social science that now bears the name "anthropology," it has generally followed the lead of Franz Boas, who at the dawn of the twentieth century decided in favor of Darwin's over Lamarck's account of evolution – even while genetics was still in its infancy – precisely because the former did not presuppose an "orthogenesis" that invariably cast non-Europeans as "backward" peoples. The full moral import of this position is reduced by calling it "relativism."

Kant's own conception of the human was empirically stabilized in the upright ape, which Carolus Linnaeus only a generation earlier had identified with the human: *Homo sapiens*. Nevertheless, Kant's normative orientation remained anchored in an older sensibility that in principle allowed the predicate "human" to be extended across the animal kingdom (*à la* Aesop, Francis of Assisi) as well as to any extraterrestrial beings (*à la* Bruno, Leibniz). In this respect, Kantian "anthropology" was a creature of syncretism: a new paradigm case of the human being superimposed on an older understanding of humanity, to whose Stoic/Christian roots this chapter shall return.

Nevertheless, the influence of this image on human self-understanding was massive. In particular, by restricting the eligibility requirements for counting as human to apelike creatures, the calling card of "scientific racism" was introduced, namely, a sliding scale of peoples in which ape and human define polar opposite states of being human. Thomas Henry Huxley (1863) incorporated this vision into Darwin's theory of evolution via a drawing of ordered skeletons, from gibbon to "man" as the frontispiece of a book designed to provide evidence for the theory. It served as the prototype for natural history and ethnology exhibits from the late nineteenth century onwards. To be sure, Darwinists have endlessly repeated the explicitly non-teleological – indeed, "non-progressive" – cast of their version of evolution. Nevertheless, 1965 marked the publication of the famous Huxley-inspired "March of Progress" (real name: "The Road to *Homo sapiens*") illustration of an ascending array of apelike creatures in the "Early Man" volume of the *Life Nature Library*, a 25-volume mass circulation book series published in eight languages and designed to reconfigure popular understanding of all of nature from a broadly evolutionary perspective.

It would be difficult to overplay the role of this self-styled "progressive" image of the human as epitomizing the mindset of the enemy in both anti-racist and anti-speciesist campaigns over the past 150 years. In these contexts, Blacks and apes have been, respectively, the primary subjects for emancipation from the enemy, which in both cases is cast as the Caucasian variety of *Homo sapiens*. Yet at the same time, an increasing number of those who also self-identify as "anti-racist" and "anti-speciesist" have complained about the relative lack of attention given to comparably oppressed non-Blacks and non-apes. This suggests that, for better or worse, the Linnaean equation of "Human = *Homo sapiens*" has been the real enemy of those keen to free humanity from its stereotyped self-understanding. This is not to deny that various degrees of xenophobia have been displayed throughout history, be it in the pages of Herodotus' *Histories* or the *Travels of Marco Polo*. But such aversion to foreigners was normally more focused on their customs than their looks. However, the Linnaean equation led to a fixation on the morphology of the *body* of the "human" as criterial.

A good point of reference is so-called "Vitruvian Man," Leonardo da Vinci's 1490 sketch of the human body's mathematical proportions, which he thought – in the Neo-Platonic fashion of the day – represented a microcosm of the universe, a sign that we were a part specially created by God to understand the whole of divine creation. Unfortunately, once the Linnaean equation eclipsed Leonardo's own understanding of the human, "racism" and "sexism" of the sort that continues to blight the human condition came to the fore. That the image depicted in "Vitruvian Man" happened to be of a white male specimen of *Homo sapiens* thus proved damning.

Kant is complicit insofar as his fixation on the empirical character of our knowledge of the world resulted in a morphologically driven conception of the human that persisted long after genetics had begun to shatter those empirical stereotypes – in the case of "The March of Progress," even after the molecular revolution in biology had reasserted the fundamental material similarity of all life forms. The likely source is Kant's intellectual thralldom to David Hume, whose own blended view of "civil history" and "natural history" presaged many features of today's evolutionary psychology. Certainly T.H. Huxley's (1879) own elevation of Hume to the status of the "naturalistic" philosopher of choice for evolutionists set this train of thought in motion. In contrast, Linnaeus had been clear that *Homo sapiens* would be nothing but an ape, had God not injected it with a soul, as animal morphology alone would never be sufficient to redeem humans from their fallen nature. However, such creationist arguments for human exceptionalism did not move Kant. On the contrary, *à la* Hume, he regarded them as the sort of wishful thinking that Wittgenstein would later dub as "language on holiday," which regards ontology – in this case, of an autonomous soul or "mind" – as the product of our believing our own hype.

Remnants of the Divine Soul in the Naturalized Mind: Of Nerves and Aether

The demystification of human exceptionalism in the nineteenth century has continued to leave the science of psychology's exact object of inquiry under a metaphysical cloud. More generally, from Kant onwards, the naturalistic turn in the history of the human sciences has privileged nominalism over realism. Thus, "mental states," understood as the secular successor to spiritual states, have been defined as talk about them, behaviors associated with such talk, corresponding brain states, or some complex relationship among these three sorts of entities. However, the idea that mental states might be *sui generis* has met strong secular resistance, as the increasingly ghettoized history of "psychical" (now called "parapsychological") research illustrates. It would seem that such states are permitted in secular liberal societies only when policed by established religions that stay within their precincts. Otherwise, as the early twentieth century failure of introspection as a method in experimental psychology demonstrated, no clear agreement could be reached on a characterization of mental states, let alone their explanation.

The one significant exception to this sorry tale was a field-based conception of mind that by the end of the nineteenth century provided the theoretical underpinnings for psychiatry as a form of secular pastoral theology, as it came to be understood especially in Third Republic France. It travelled under such names as "intercerebral psychology" (Gabriel Tarde) and "genetic memory" (Theodule Ribot).

Here the mind was understood as the emergent collective effect of the interaction of brains and their artefacts distributed across potentially vast regions of space and time (Valsiner and van der Veer 2000: Chap. 2). Indeed, the sense of heritability implied in the concept of "genetic memory"was meant to carry not only Lamarckian overtones of intergenerational improvement but also a more atavistic, Darwinian sense of past generations as retarding or deforming the capacities of future ones (Fuller 2018). The material characterization of this conception of mind was often modelled on the aether of physics popularized by James Clerk Maxwell, which was increasingly understood in the statistical terms that he had imported from the human sciences (Porter 1986: Chap. 7). Maxwell was enamored of the idea that an elastic aether would permit micro-level indeterminacy to have a triggering effect, resulting in a stabilized field of action that amounted to a lawlike regime. In this way, the metaphysical problem of free will and determinism might be solved by science. Whatever one makes of this strategy as scientific metaphysics, Maxwell was arguably gesturing towards "chaos theory" (Hacking 1990: Chap. 18).

The overall effect of these developments was that memory lost the solidity it had enjoyed as the repository of tradition, on which even the disciples of "progress" had wished to build. Instead, memory became a biologized version of Marx's protean image of "capital" as a resource that might be multiply deployed to various effects in the present. Even the followers of Emile Durkheim contrasted the "objectivity" of historical scholarship to the unreliability of both individual and group memory, notwithstanding the presence of commemorative artefacts and rituals (Halbwachs 1992). The way had been semantically prepared in the late eighteenth century by the adaptation of *susceptibility* from the study of magnetism to the practice of mesmerism as "animal magnetism." A century later, the term of choice had become *suggestibility* as the mental disposition that made hypnosis possible. In both cases, individual memory appears as eminently malleable, its liabilities exploitable under the right conditions, ranging from deliberately induced semiconscious states in therapeutic sessions to spontaneously generated collective action occurring at mass encounters. Sigmund Freud, and especially his followers, not least the apostate Carl Jung, did the most in the twentieth century to elaborate and popularize this "deep plasticity" in the human condition.

Kant is relevant to this story because of the basis on which he upheld the autonomous moral agent in the modern era. He follows the Stoics in defining "autonomy" as the overcoming of *heteronomy,* a state of competing psychic forces that the autonomous agent disciplines through a weighted integration of their significance, issuing in authoritative action. We thus come to "own" our actions in a state of literal "self-possession." This is the source of the modern understanding of "decision," a generalized self-application of the experience had by the Greek citizen who is chosen by lot to decide a court case (*dikai*). Modern formal accounts of rationality (e.g., Bayes Theorem, expected utility maximization) are theories of "decision-making" in this sense. Kant imagined humans as, on the one hand, animals whose responses are primed by multiple conflicting forces in the external environment, and on the other hand, fallen deities who find it very hard to resolve those forces. Indeed, our fallen state pulls us in *more* directions than ordinary animals –

and with potentially worse consequences. For Kant, animals are simpler, insofar as the informativeness of their senses involve less cognitive mediation. Thus, they are more easily disciplined by nature. We shall later see that this understanding of animals also underwrites the normative stance adopted by the Epicureans, Kant's foil to the Stoics in his seminal "critical" period.

Noam Chomsky (1959) provides a relatively contemporary and complementary way into Kant's point. His revolutionary reassertion of humanity's unique linguistic capacity rested on arguing that nature's stimuli are "impoverished" when it comes to prompting a human response because nature speaks in many tongues at once, and thus presents a confused picture to the human receiver. Both Kant and Chomsky argue that an overdetermination of external forces on individuals underdetermines their potential response: They are forced to choose. In other words, to rise above the animal to the level of the human is to recognize this openness of response. For Kant, we are morally obliged to act in the face of this openness in order to earn our humanity.

The combination of naturalistic and theological issues that led Kant to contrast the complex and evil nature of humans with the relative simplicity and innocence of animals had been adumbrated in the mid-eighteenth century, which had witnessed the rise of a wide range of Christian dissenters influenced by John Locke, typically with a foot in theology and in medicine, who were beginning to distinguish humans from other animals in terms of relative neural complexity and brain size. These broadly defined "theological materialists" included such intensely controversial figures as John Wesley, Emanuel Swedenborg, and later Joseph Priestley, who popularized the most on point of all these thinkers, David Hartley. Hartley is now known primarily for his laws of sensory association, which were foundational in the development of empirical psychology in both its introspectionist and behaviorist phases, and then more recently in so-called "neural net" models of the brain (Cobb 2020: Chaps. 2, 10). Hartley speculated that the brain is a cosmic transducer in which the nerves vibrate synchronously, ideally in harmony with the world outside the brain. Anticipating Maxwell, he envisaged an elastic aether common to activity both inside and outside of the brain that served as the common medium of transmission. Some have even argued that Hartley's heroic conjecture marked the dawn of the "method of hypothesis" in modern science (Laudan 1981: Chap. 8). In any case, the conjecture continues to have its fans, now fortified with quantum mechanics (e.g., Penrose 1994).

We would now say that Hartley's (and by implication, Maxwell's) was a "soft-wired" holistic conception of the brain, in contrast to the more "hard-wired" version that emerged in the nineteenth century, courtesy of Hermann von Helmholtz, who looked to the rigidity of telegraph wires as the appropriate neural metaphor for their fidelity of signal transmission (Cobb 2020: Chap. 3). At stake here is a point about the academic contextualization of brain studies that gets to the metaphysical heart of humanity, which shall be explored in the next section. The implied difference is between *brain as transducer* and *brain as governor*: the former outward looking and soft-wired, the latter inward looking and hard-wired. The history of neuroscience has straddled these two images, which are basically how the brain looks to contemporary physics and biology, respectively (Fuller 2014). On the one hand, the brain might be seen as the best instrument currently available to register all that is happening in the cosmos – but which might have its functions improved or even replaced in the future. This favors a more substrate-neutral (carbon-neutral?) conception of humanity of the sort associated with a cosmopolitan extraterrestrialism. On the other hand, the brain might be regarded as the evolved product of the body of which it is a part, whose own survival depends on that of the body it regulates. Implied here is a more substrate-biased earthbound conception of humanity, as exemplified by the Linnaean equation of "Human $=$ *Homo sapiens*" (Fuller 2021). As we shall see, Kant's own conception of humanity straddled those two images, the former corresponding to the ethical and the latter to the epistemic sphere.

Between Mind and Matter: Life as Extended Self-Correction

In terms of the history of naturalism, Jean-Baptiste Lamarck's 1802 coinage of *biology* to designate the science of "life as such" aimed to capture the conceptual space that had opened up, once mind was seen as, in some sense, an emergent feature of matter. Lamarck's main strategy was to elevate *besoin*, by which the French physiocratic school of political economy had meant the natural basis of what economists today call "demand," to a metaphysical principle – life's own needs – that is then further differentiated as the wants (*désir*) entertained by individual organisms (Fuller 2013). Curiously, while Lamarck's needs/wants distinction has now been long intuitively clear to philosophers, psychologists, and marketing people (Abraham Maslow embodying all three in one person), it continues to bedevil discussions in both the field from which his ideas derived (political economy) and the field that he wished to found (biology). In that respect, Lamarck's legacy has remained close to the "philosophy" side of his magnum opus, *Zoological Philosophy*. Specifically, as Arthur Lovejoy (1936) famously suggested, Lamarck was the most diligent disciple of Count Buffon's counter-Linnaean attempt to render a materialist version of the Neo-Platonic "Great Chain of Being." Buffon replaced "mind" with "life" as the generative principle that is "communicated" across space and time, resulting in what we conventionally identify as "organisms," which are in turn organized into "species" for purposes of governing nature, perhaps by God but certainly by humans.

The atheism intimated here haunted both great nineteenth century masters of evolutionary theory, Lamarck and Darwin. It is worth contrasting the charges brought against them. Darwin's "atheism" was based on the lack of need for God either to explain or justify the outcomes that have unfolded on Earth because the entire process ultimately happened by chance. It need not have happened at all – which is to say, as Stephen Jay Gould (1989) memorably explained, had it happened again, it probably would not have happened the same way. In other words, Darwin does not beg the question in favor of God's existence, as Leibniz did when he first asked, "Why is there something rather than nothing?" In contrast, Lamarck's alleged atheism came not from fully dispensing with God but from depicting the deity as randomly tossing out spores that are only semi-formed yet equipped with self-reproductive and adaptive capacities. He was seen as having replaced the deity who intelligently designs the cosmos from outside of time (a priori) with one who does it through trial and error in real time (a posteriori). While it resonated well with the empiricist epistemology on the rise in the late eighteenth and early nineteenth centuries, it placed the divine and the human on a continuum that reversed the poles that St. Augustine had established for the Christian world (Toulmin and Goodfield 1965: Chap. 8). Instead of humans cast as imperfect images of God by virtue of their "fallen" status, God appeared to have become the grandest human experimenter, perhaps even a venture capitalist, whose life-forms are like start-ups, the more successful of which end up merging their capacities to become more complex organisms. Herbert Spencer and his intellectual progeny are usefully evaluated from this standpoint.

Lamarck's framing of the problem of life is nowadays associated with Stuart Kauffman (1993) and the Santa Fe Institute's attempts to simulate the world's full biological complexity based on minimal computational constraints. Moreover, even followers of Darwin, worried about his silence on the very origins of life, have proposed the crypto-Lamarckian idea of "panspermia," whereby cosmically dispersed seeds sowed the Earth's natural history, which then unfolded on Darwinian terms. Lamarck took his word for this evolutionary process from the physiocrat economist Turgot, which carried into Léon Walras' late nineteenth century "neo-classical" account of how prices are stabilized in a market: *tâtonnement* ("groping"). It is also the word that figures in Pierre Teilhard de Chardin's equally heretical account of "creative evolution," in which eugenics appears as the latest form of human participation in the cosmic trial and error process (Fuller 2011: Chap. 4). To understand the theological scandal that this vision originally caused, imagine that all the algorithms running our computer systems did not result from a set of benevolent programmers but have been casually sowed by various hackers as viruses, some of which over time develop lives of their own by virtue of adapting to each other and their common environment, resulting in ever more complex and powerful programs, which eventuate in an algorithmic "singularity." Such a Lamarckian singularity would leave even Ray Kurzweil's transhumanist followers with sleepless nights, since it might well not be the one that they either would have wanted or expected.

The peculiarly positive role assigned to the embodiment of error – the seemingly random sowing of seeds – in Lamarckian evolution has redounded across the human sciences. One major albeit subtle legacy over the past two centuries has been for a distinction to be drawn between extant forms of human life ("cultures" in a judgemental sense) that are *merely* surviving from an earlier era ("atavisms") and others that are ascendant. Thus, instead of a polygenic species, humanity is seen as monogenic but divided into "stages" of development. An updated version of this vision underwrites science fiction writer William Gibson's famous remark that the future is already here but unevenly distributed. It has animated thoughts and policies for "uplifting" the human condition, typically through the concerted alleviation of "poverty," a term for personal deprivation that in the modern era came to be understood negatively as arrested development on multiple levels and relative to presumed norms of "quality of life." Imperialism was the first global political standard-bearer of this idea, which Lenin then repurposed in the name of Marxism. In the twentieth century, these norms were increasingly standardized across the world and continue to be used as a pretext for action by both governments and nongovernmental agencies. This idea of "normalizing error," which has spawned such taken for granted tools of statistical reasoning as "standard deviation," is discussed below and taken up more explicitly in the next section.

Lamarck's view is usefully contrasted with that of his younger contemporary Xavier Bichat, whose pioneering medical research into organic tissues, combined with a distinctive approach to the nature of life, inspired a diverse range of nineteenth century thinkers, including Auguste Comte, Arthur Schopenhauer, and Claude Bernard. Bichat believed that life itself was a kind of error, a temporary resistance to the normalcy of death. It is from here that arguably comes the modern biomedical imperative to place the prevention of death even above the Hippocratic prevention of harm. Lamarck disagreed, not because he differed from Bichat over the statistically exotic character of life; rather, he believed that one of life's functions is to continually redraw the boundary between the organism and the environment in its attempt to establish a temporary global equilibrium in cosmic forces (Klarsfeld and Revah 2003: Chap. 1). From this standpoint, what Linnaeus might identify as a distinct species is really the outworking of *besoin* as a specific *désir* – and what Darwin might identify as the ecological replacement of one species by another is really the abandonment of an old *désir* for a new one. On this view, biological extinction looks like "planned obsolescence" of products in the dynamics of a market whose contours are always in flux. Clearly this feature of Lamarck's vision, which he derived from the great Enlightenment theorist of universal progress, the Marquis de Condorcet, resonates with today's transhumanist themes about the defeat of entropy ("extropianism") and humanity's "morphological freedom" (cf. Burckhardt 2013; Fuller 2019a: Chap. 2).

Additionally, in the second half of the twentieth century, Lamarck's vision exerted enormous influence over the "post-structuralist" French philosophers (Deleuze, Foucault, Derrida). Their common Sorbonne lecturer Georges Canguilhem used "error" as a red thread to weave a compelling history and philosophy of biomedicine that folded the Lamarckian story of the origins of life into a Neo-Darwinian one, in which mutations are understood as what the geneticist Richard Goldschmidt (1940) memorably called "hopeful monsters." It effectively renders error (aka "pathological") as a by-product of the reproduction of the truth (aka "normal"), sometimes resulting in a "new normal" (Talcott 2019: Chap. 3). Put another way, Lamarck's approach to the nature of life managed to simulate how an event in time and space might appear to a being (a Deist deity?) with total spatiotemporal comprehension – as Spinoza put it, *sub specie aeternitatis*. From that standpoint, any event – possibly every event – may constitute an "error" in some sense, if it has both advanced and impeded the overall direction of history. While this may sound like Hegel's philosophy of history, Hegel's history was panoramic: The "cunning of reason" unfolded on a scale and with a focus comparable to a theatre performance, which ultimately takes place primarily for the benefit of the spectators –

not the actors. There is relatively little of what Lamarck called *transformisme*. In contrast, the drama of life depicted by Lamarck was more like watching a soap opera unfolding simultaneously in a common space over an indefinitely long stretch of time (e.g., different families in a neighborhood). In that case, "errors" appear in the first instance as pretexts for further plot development.

It is worth contrasting Lamarck's attitude to error as a feature of history with two other conceptions that enjoyed a theological and philosophical following in Lamarck's day. One, exemplified by Leibniz's *Theodicy*, held that history amounts to God teaching directly to humans by example, which is the most efficient way for a perfect being to communicate with an imperfect one. After all, the Fall was supposedly the result of Adam's inability to distinguish good from evil. In that case, the succession of beneficial and harmful events in human history is designed to provide instruction on the path to salvation. On this view, evil is what evil does (which includes natural catastrophes), and the question then is how to learn from it. The second view came from Lamarck's contemporary, Thomas Malthus, who inspired Darwin's conception of natural selection. Malthus saw the evil of high infant mortality as an "error" that would be rectified over time as humanity learned to live "sustainably" (in today's terms) with the natural environment. This would require a measure of technological progress but mostly moral restraint. A Calvinist by disposition, Malthus saw God as teaching more indirectly, with fallen humans having to learn the deity's lessons in a painful and piecemeal fashion. Nevertheless, Malthus' clearly believed that humanity will have learned its lesson once it achieves what we would now call a "steady state" economy – that is, one *not* given to the sort of indefinite expansion that might be promoted by an activist state. Indeed, his famed *Essay on Population* had been triggered in response to Condorcet's vision of human progress engineered through the increased proliferation and integration of a state-steered industrious humanity (Fuller 2006: Chap. 13). Lamarck may be seen as having extended Condorcet's vision across all of life, just as Darwin extended Malthus'.

It would be difficult to exaggerate the long-term significance of the Lamarckian shift in horizons. It collapsed the distinction between life and mind, while rendering the differences between forms of life/mind matters of degree rather than of kind, such that greater "mentality" comes to be associated with a kind of increased material, ultimately neural, complexity. Together they define what we now call the "evolutionary" worldview, the main scientific beneficiaries of which have been Darwin and his followers. Sometimes this worldview has been called "vitalism," as opposed to "mechanism," understood as still drawing a clear ontological distinction between the creative mind (the "mechanic") and created life (the "machine"). By the second half of the nineteenth century, such "vitalism" had come to be generalized as the metaphysics of "panpsychism," in which humanity's distinctiveness rests simply on possessing greater "consciousness," understood as a continuum applying to all forms of life, if not matter more generally. In that case, humans are simply the "most" conscious beings. A century later, Gregory Bateson (1979) smartened Lamarck's image by portraying him as having anointed humanity the pilot (*cybernetes* in Greek) of life as a project potentially involving a great cosmic awakening.

Put in brute theological terms, for Lamarck the "human" named neither a uniquely embodied soul nor Linnaeus' upright ape but whatever happened to be the leading edge of the overall outworking of the life-mind process. From that standpoint, Teilhard de Chardin (1961) had tried (unsuccessfully, at least in his lifetime) to theologize the point by portraying the human form of God, Jesus, as the ultimate face of creation, or "Omega Man" – arguably a rather optimistic gloss on the idea of the "Anthropocene" (Fuller 2011: Chap. 4).

Wrestling with the Rousseauian Legacy: The Eugenic Versus the Statistical State

If Lamarck envisaged individuals as distributed moments in a spatiotemporally extended life process, Jean-Jacques Rousseau held nearly the diametrically opposed view, insofar as Rousseau regarded individuals as sources of hidden potential, and *education* as the environment where that potential might be elicited in acts of self-definition. Reading against Linnaeus' own grain, Rousseau and his fellow *philosophes* entertained the prospect that even apes might be eligible to undergo such an edifying regime. An important strand of the debate in Germany, associated with the concept of *Bildung*, was championed by the Kant-inspired Prussian education minister Wilhelm von Humboldt in the context of renovating the universities. The Humboldtian vision featured educators as exemplars of Enlightenment who brought their vanguard research and thinking alive in the lecture hall, followed by seminars in which teacher related to student in more medieval style as master to apprentice. At least that is how the vision unfolded in the nineteenth century and survived into the second half of the twentieth century. It is still what we mean by the "modern" university today. Its aim, however imperfectly executed, was to prepare individuals drawn from the entire population to be at once autonomous agents and specialist workers: equal builders of the whole of which they also constituted a proper part. Such people would be the standard bearers of republican democracy.

This broadly Rousseauian way of thinking about personal development also infiltrated thinking about more ordinary forms of education and even general child-rearing practices. Indeed, it provides the background for understanding the prima facie attractiveness of Francis Galton's coinage of the binary "nature/nurture," which resulted in twentieth century welfare state policies to provide "equal opportunity" for youth to mature into optimally functioning adults – at least from society's standpoint. *À la* Plato's *Republic*, but in the context of a liberal democratic regime, the task boiled down to establishing a "level playing field" to determine who belonged where in the social order. It resulted in many policies designed to uplift people from "backward" social backgrounds who might be suited for "better things." Its epistemic calling card – and political lightning rod – was the regular administration of tests of "intelligence" or "aptitude" in schools that aimed to channel young people so as to enable them to make the most use of their talents. Notwithstanding the criticism and even scorn heaped on the envisaged "meritocracy," the promise of such a social order was behind the UK Labour Party's electoral success in the first century of its history (Young 1958). This approach will be revisited in the final section in the context of a "Turing Test" for humanity.

More to the point, even Galton's "eugenic" update of Plato's republic of exemplary individuals, each well-adjusted to his or her place in society, was more in the spirit of Rousseau than the statistical image we saw emerging in the previous section – namely, of a population whose members vary around a norm that serves to stabilize their common environment. Behind this difference is a shift in the role of the state from executor of a plan that channels individuals into determinate roles to manager of diversity in a world of flux. This shift in perspective amounted to a return – albeit in a statistical key – to the medieval origins of the state in *status*, which Aquinas understood in terms of the restoration of a natural order but is now what we call "homoeostasis," a balancing out of differences around an assumed mean; hence *status quo* (Kantorowitz 1957: Chaps. 5 and 6). To be sure, these two conceptions of the state – the eugenic and the statistical – are not mutually exclusive (Porter 1986: Chap. 5). Indeed, they clearly coexisted in twentieth century welfare states. But that still leaves their relative priority open.

Instead of education, the statistical state stresses *socialization*, a process focused mainly on the "interiorization" of the norms. In the middle third of the twentieth century, state efforts in this regard spawned a strategy of "inoculation" that aimed to mentally safeguard people against largely subconscious forces capable of undermining belief in the facticity of the social norms. It was the ideational equivalent of the mass vaccinations that had begun a generation earlier, fuelled by the epidemiological imagination that had been ignited by the rival biomedical commanders-in-chief in the 1870–1871 Franco-Prussian War, Louis Pasteur and Robert Koch. They jointly established "microbes" as a fixture in the European imagination. Talk of "influence" (from "influenza") and, more recently, "memes" (modelled on viruses) remain the legacy, whose natural home in the twentieth and twenty-first centuries has been mass and social media, respectively (Fuller 2009: Chap. 3). Inoculation in turn has taken the form of "propaganda" campaigns, not least the introduction of "civics" as a school subject, specifically targeting immigrant and minority populations in the attempt to assimilate or "normalize" them. The overall effect shifted the function of education toward indoctrination, which by the late 1960s resulted in a "counterculture" backlash, most notably Paolo Freire's "pedagogy of the oppressed" and Ivan Illich's "de-schooling," both of which had more than a whiff of Rousseau about them.

In fact, Rousseau's hand can be found equally in approaches to education focused on individual self-realization and on collective self-maintenance. A legacy of Rousseau's Calvinist upbringing in Geneva was a strong sense of Original Sin, secular versions of which are shot through his writings. Of relevance here is the idea that children incur a debt by virtue of having been brought into the world in the first place. However, the debt is not to be paid to the parents specifically but to the society into which they were born. (Hence Auguste Comte's coinage of "altruism," which he turned into a universal, future-facing "pay it forward" ethic.) In that case, once children reach adulthood, they must take a decision: Either to affirm and work for the society's aims or freely depart but in the spirit of ostracism. As in the expulsion from the Garden of Eden, humans may lose formal recognition but never their freedom. This moment of decision has been enshrined in ideas of "national service," including compulsory voting for citizens, regardless of nation of origin. In Rousseau's case, it was underwritten by a seemingly "totalitarian" sense of a "general will" based on the society's agreed value system. To his credit, however, he understood the burden that such a conception of citizenship placed on individuals. Indeed, Rousseau's "anarcho-communist" followers insist that a just world order requires that no society exceed a certain scale and scope, precisely due to the need for individuals to freely agree on the society's basic aims and performance standards. Thus, his ideal world would probably be a patchwork of city-states like Geneva.

Kant complicated Rousseau's picture of humanity by shifting the sense of "universal" from the consensus-seeking general will to something closer to the Stoic conception of *cosmopolitanism* and the Christian conception of *agape*. Instead of stressing the literal merging, or "communion," of individual souls to form a self-governing corporate whole, Kant wanted to remove and transcend the differences between the souls that came from their individual embodiment, so as to bring about a second-order communion that Kant called the "kingdom of ends." In principle, this strategy allowed for an indefinite expansion in the scale and scope of the normative order, which Kant himself periodically countenanced as including extraterrestrial intelligences, all in search of ever higher forms of cosmopolitan unity – not least at the end of the regular set of student lectures that constituted *Anthropology*.

Kant Between the Stoics and the Epicureans: The Rise of the Cosmopolitan Standpoint

Before academic disciplines as we know them became philosophically salient in the second half of the nineteenth century (aka Neo-Kantianism), philosophers generally ranged quite widely in search of sources of inspiration and insight – and this included tapping into the mathematical imagination. Kant's own sense of "universalism" anticipates the modern mathematical idea of "universe" introduced in the mid-nineteenth century by George Boole and which logicians normally encounter in set theory. It is captured well in mathematician David Hilbert's defense of Georg Cantor, whose idea of "nested infinities" defied the early twentieth century imagination. Commonly known as the "Grand Hotel Paradox," Hilbert asks us to imagine a putatively full hotel of infinitely many rooms. Nevertheless, it can still accommodate more guests by having each current guest shift to the neighboring room to enable the new guest to occupy the first room. Hilbert's point is that the idea is coherent, albeit counterintuitive. The relevant implication for cosmopolitanism is that, at least in principle, membership in the set of humans can be expanded indefinitely without diluting the concept of humanity (Fuller 2019b). Of course, this does not obviate the sociologist Georg Simmel's shrewd observation about the disproportionate tendency for the poor and the weak to champion group identity: "What is common to all can only be truly possessed by the one who possesses the least of it" (Simmel 2008: 491). Not surprisingly, then, "human rights" tend to matter most to those who are not normally recognized in any way other than that they possess humanity.

This is the backdrop against which we should understand Kant's view that "humanity" ultimately names a normative standard toward which all beings might strive to meet. He conceived of the "rational" life as the application of the will to achieve this ideal. Plato had originally proposed this way of being in the world as the unique product of philosophical training, which then entitled one to rule others. The Stoics placed this training within everyone's reach by making it the essence of "virtue," such that self-control becomes the key to mastery over all things in one's life. Christianity arguably turned it into a divine obligation. In Kant's hands, it reappears as "practical reason," whose signature principle is the categorical imperative. As Charles Sanders Peirce realized at the end of the nineteenth century, Kant's general understanding is driven by a powerful mathematical intuition, which Peirce himself associated with the *asymptote*, namely, the endless approximation of a curve (one's life) to a straight line (the standard of one's life). For him and Karl Popper after him, it charted the trajectory of *scientific progress*, characterized in terms of increasing "verisimilitude" (Laudan 1981: Chap. 14). Nowadays this mathematical image is most closely associated with the transhumanist vision of the "singularity," driven by such exponential equations such as "Moore's Law," which governs computer processing power. But arguably it had also lay behind Engels' Hegel-inspired "First Law of Dialectics," whereby incremental change eventuates in a qualitative state shift, as when gas results from the gradual heating of water.

To be sure, the Stoics themselves had encouraged this line of thought with their understanding of energy (*ergon*) not as an object's intrinsic potential for motion (*à la* Aristotle's *dynamos*) but as sustained effort applied to an object toward an end, during which resistance – or "friction," as physicists would later say – is overcome. Once again, "end" isn't simply an anticipated concrete product but also the matching of an ideal through successive approximations that is captured in the sustained effort. The labor theory of value – both in David Ricardo and Karl Marx – trades on this Stoic imagery. In the second half of the nineteenth century, physicists and economists – using versions of the same equations – dubbed the application of effort to objects "work." Meanwhile theologians and psychologists (notably William James) spoke of "willpower" when the effort was applied to oneself. In the twentieth century, when the labor theory of value fell out of favor in capitalist economics, the requisite sense of "work" was transferred to the "entrepreneur" and the "manager," both armed with "innovation," who occupy a second-order position vis-à-vis the material economy. In this respect, Saint-Simon's "utopian socialism" has had the last laugh on Marx's more "scientific" version, which has resolutely stuck to the original first-order labor theory of value (Fuller 2020b).

Informing all this "economism" was a metaphysics based on *returns to scale*. So far, only the Stoic view of the matter has been discussed, but it also involved the Hellenistic school that Kant in his later "critical" phase counterposes to Stoicism: *Epicureanism*. The Stoic mentality focuses on *increasing* returns to scale, what is popularly known as "economies of scale." Benjamin Franklin epitomized this sensibility in the aphorism, "God helps those who help themselves," which Max

Weber made central to his account of capitalism's secularization of the Protestant ethic. It justified the accrual of interest on capital as a reward for investment, an activity that had been proscribed in the Middle Ages for treating the artificial as if it were natural (i.e., money making money), and hence left to the Jews, who were already regarded as infidels. The opposite turn of mind, represented by the Epicureans, focuses on *decreasing* returns to scale, or "diseconomies of scale." The intuition here is that of diminishing returns on investment and diminishing marginal utility, whereby at some point – if not always – more effort leads to proportionally less satisfaction.

While the Stoic is "magnanimously" open to the world in the hope – but not the expectation – of reciprocation, the Epicurean seeks to limit her encounters with the world out of fear of endless disappointment. Historically, this difference in attitude was tied to the relative material wealth of the Stoics, which they converted into a spiritual currency that gave them the courage to withstand material loss. In contrast, Epicureans would never place themselves in high-risk situations that might incur such loss – and the ensuing pain. Indeed, *pain* is a bellwether sensation for both the Epicurean and Stoic: Whereas the former tends to turn it into an object of fear and loathing, the latter tries to reduce it to a mere inconvenience that might even provide an opportunity for learning.

Kant's philosophy can be understood as a Herculean attempt to give both the Epicurean and the Stoic their due. At stake was the constitution of the human as a realizable being. His basic solution was a divided settlement, whereby the Epicurean wins in the *Critique of Pure Reason* and the Stoic wins in the *Critique of Practical Reason*. The key is the contrasting accounts of the mind provided by the two sides, both subsequently played out in the history of psychology, though typically without sufficient care to the real bone of contention between the Epicureans and Stoics. The labels "empiricism vs. rationalism" and "materialism vs. idealism" have been especially misleading in this regard. The Epicureans held that the powers of the mind are confined to the limits of the body bearing it, whereas the Stoics defined the mind as the power to transcend the body bearing it. Both are normative statements masquerading as empirical ones, projecting alternative ways of being in the world. Much closer to the spirit of their disagreement is the *precautionary* versus *proactionary* attitudes toward risk (Fuller and Lipinska 2014). At multiple levels, Epicureans are risk-avoiders, Stoics risk-seekers. Thus, medical practitioners have traditionally tended toward the Epicurean, medical researchers toward the Stoic. Sinclair Lewis' 1925 novel, *Arrowsmith*, captured this tension in the life of one person who has ostensibly dedicated his life to "medicine" in a sense that encompasses both research and practice.

The Epicureans stressed the wisdom of the senses in the way that evolutionary psychologists nowadays observe when they present pain as a danger sign to the organism: a feeling that already carries information for the body to respond appropriately. Higher-order cognitive mediation is unnecessary and may even cause confusion. Indeed, it is an open question whether the Epicurean sees higher-order cognition as anything other than delusional. Certainly Darwin entertained that prospect, when he traced the human propensity for violence to our large brains, which provided more space for the development of fixed ideas. The overall picture of the human that emerges from the Epicurean world view is that our senses are instruments of self-regulation in the context of a larger world of which we are a part but over which we ultimately have no long-term determinate control. Mental pathology arises from thinking otherwise.

To be sure, this did not stop Epicureans from speculating about the constitution of this world, the Roman poet Lucretius most enduringly. Here the image of endless random atomic recombination loomed large. However, it was meant in the spirit of consolation – to arrest the impulse to inquire further. But of course, the exact opposite happened. The last thing that the ancient Epicureans would have wished is for this therapeutic moment to open up into a theory-led, experiment-driven empirical research program, as happened in the early seventeenth century. This required a more Stoic spirit, which has arguably characterized the atom's progress toward full ontological recognition over the ensuing centuries. The history has been marked by intensely abstract proposals punctuated by hotly contested trials of faith that were temporarily resolved by experiment. Newton's inconclusive empirical inquiries into the nature of light set the precedent for this turn in the atom's fortunes (Dobbs 1991). Einstein's explanation of so-called "Brownian motion" generated in the laboratory seemed to settle the atom's existence in the early twentieth century. But soon thereafter the battle was resumed at the "sub-atomic" level that shapes research in quantum mechanics to this day. What makes this entire trajectory "Stoic" rather than "Epicurean"in spirit is that the cognitive is clearly separated from the sensory sphere with the two then placed in confrontation with each other, via hypothesis testing. Whereas the Epicureans had presumed that their story of atoms was simply a consoling myth in the sense that an atheist such as Richard Dawkins might regard Genesis, the Stoics approached the matter in a more "literal" spirit, *à la logos*, the self-creative word, resulting in such "theoretical" entities as atoms, genes, and microbes, the existence of which require that inquirers exercise their will to test the projective capacity of their imagination upon the world.

A linguistic benchmark for this privileging of the cognitive over the sensory is what J.L. Austin (1962) famously called the "performative" character of utterances. Nearly two decades before Austin, the logician (and practicing Presbyterian) Arthur Prior had shrewdly characterized the paradigmatic performative utterance – promising – as an act of "special creation" (Prior 1942: Chap. 5). From that standpoint, science looks like an endlessly agonistic enterprise about getting the world to live up to one's word, where the first impressions of the senses are faced as the enemy's front line – what Francis Bacon called the "idols of the mind." In the face of such empirical resistance, one remains cool, since in this sense a hypothesis is a promise waiting to be broken, in which case the theory-world (*Weltbild*) one had created would need to be aborted in at least some respects. This was the *modus operandi* of Bacon's *experimentum crucis*, the basis of Popper's falsificationist methodology of science. Indeed, had Bacon as a trained lawyer got his way, what we now call "hypotheses" would be a version of what we now call "contracts."

An illuminating way to capture the difference between the Epicurean and the Stoic mentalities is in terms of their respective attitudes towards *chance*, the reality of which both schools admit. Whereas the Epicurean accepts chance as "the other" beyond one's control but to which one must adapt, the Stoic incorporates chance as part of oneself – indeed, as constitutive of one's sense of agency. This conversion of chance from an "outside" (Epicurean) to an "inside" (Stoic) state of the world is tantamount to a Gestalt switch in one's experience of the world's uncertainty from objective indeterminacy to subjective freedom. It is a distinction that haunts the modern history of probability, which reached physics first through Maxwell, then Boltzmann and ultimately Heisenberg. Kant himself effectively split the difference between the two perspectives in his contrasting appeals to *noumenon*. On the one hand, he adopted an Epicurean version – as whatever always remains other to our cognitive horizons – to set the limits of pure reason: the "thing-in-itself." On the other hand, he adopted a Stoic version – as whatever always allows the extension of our cognitive horizons – to define the ideal of practical reason as the projection of one's will as universal law: the categorical imperative.

The perennial appeal of the Kantian settlement is epitomized in Antonio Gramsci's slogan for aspiring revolutionaries: "Pessimism of the intellect, optimism of the will." Nevertheless, the Kantian settlement also threw the nineteenth century mind into a bipolar disorder that resulted in the sort of "genius" displayed by Schopenhauer and Nietzsche. It was bounded by, on the one hand, the "*ignoramus ignorabimus*" attitude of positivist science and, on the other, the transcendental egoism of Romantic poetry. Alternative takes include Nietzsche's own Zarathustrian "What doesn't kill me makes me stronger" and its ironic complement, the Existentialist punchline: "Be careful what you wish for." All of these striking expressions attest to the fact that no matter the expansiveness of our will, we are ultimately constrained in our capacity to act. In the religious terms that had animated Kant, we may aim for divinity but we remain tainted by Original Sin. In the more prosaic terms favored by secular liberal theorists, we cannot anticipate all the consequences of our actions. In between the sacred and secular reading sat Bernard Mandeville's influential eighteenth-century satire *The Fable of the Bees*, which went so far as to imply that most of what passes for "public benefit" amounted to the externalities of private vices pursued in aggregate. Echoes of this sensibility remained in the centrality of uncertainty and risk to the founder of the Chicago School of Economics, Frank Knight's understanding of economic transactions, the source of which was the Calvinist theodicy surrounding Original Sin on which he had been raised (Nelson 2001).

In response to humanity's fundamental finitude, be it cast in sacred or secular terms, Karl Popper updated Kant by treating the laboratory as a kind of cradle of liberty, insofar as "our ideas die in our stead," with the lucky ideational survivors eligible to inform – with equal tentativeness – the organization of life outside the lab. Here Popper traded on the ambiguity surrounding the word "experiment," which simply meant "experience" in Bacon's day; hence Bacon's own qualification of *experimentum* with *crucis* to indicate that he did not mean the Epicurean's spontaneous experience but the Stoic's tested experience. For his own part, it is by no means obvious that Bacon would be so quick to generalize the ways of science to the ways of politics in the spirit of Popper's "open society." If the political sensibilities of the early Royal Society, not to mention that of his amanuensis Thomas Hobbes are any indication, he probably would not. But none of this is to deny that Popper picked up something deep about the Kantian settlement that would enable the "human sciences" to be "sciences" in the proper sense – that is, in the same sense as the natural sciences.

We can best get at this point in terms of the similarity between Newton's sense of "universal law" and the modern state's approach to legislation, which served to make Kant's elision of the two when formulating the categorical imperative seem so intuitively compelling. Underlying both is the liberal principle that whatever is not prohibited is permitted. Logically speaking, the laws of physics do not say what must happen but what cannot happen. To determine what *must* happen, *à la* Laplacian "scientific determinism," one needs to add "boundary conditions," which are basically ad hoc constraints on the possibility for action. Such epistemic "stage-setting" is much more easily achieved in a laboratory or a simulation than in the empirical world. In the law, similar boundary conditions are established in judicial interpretation, in the form of "precedents" or "exceptions" whereby a universal principle is significantly restricted in its application. As probability theory slowly crept into the mathematical imagination of physicists in the nineteenth century, this emerging image of the scientific attitude toward the world altered the meaning of "necessity," whereby it became the "taming of chance." Charles Sanders Peirce and Max Weber stood at the cusp of the relevant developments in probability theory across the natural and human sciences at the *fin de siècle* (Hacking 1990; Neumann 2006).

In true Stoic fashion, humanity had come to incorporate chance into itself, thereby bringing closer together the tasks that Kant had set for science and ethics. In practice, this meant that humans shifted from operating freely albeit fallibly within known constraints – the essence of the nineteenth century engineering mentality – to being implicated in the very determination of those constraints, at once self-legislators and world-legislators. Thus, we do not simply make the most out of the world, perhaps in some residual sacred sense that it was designed for us to know and use. Rather, we are literally responsible for the world's very existence and survival. Moreover, this shift significantly predated Heisenberg's Uncertainty Principle. It was already present in, say, the novels of Dostoevsky a half-century earlier, whereby the state of the world appears as the consequence of acts in which the agents remain in spiritual turmoil because they know they could have always acted otherwise. Indeed, the dramatic tension of such works rests in not quite understanding the true identities of the agents until all the action has been resolved and their fates are revealed – to the extent that the narrative permits. The twentieth century was very much about *decisionism* in just this sense, whether we mean Max Weber and Carl Schmitt, Niels Bohr and Erwin Schrödinger, Karl Popper and Jean-Paul Sartre – or John van Neumann and Alan Turing. It amounts to a triumph of "observation" as a motivated activity, for which we are then held responsible, whatever the outcome, which only in retrospect can be known to have been right or wrong.

Underlying this sensibility is the idea that "humanity" mainly refers to a process of collective self-discovery whereby the other comes to be recognized as either one's own or something else through an encounter that tests both parties. It is worth contrasting this view with Darwin's theory of evolution, which after all might be seen as treating the natural world as a proving ground for organisms, *à la* "struggle for survival," in which a variant generated in the process of species reproduction faces an alien environment that then determines its fate. However, as Darwin himself stressed, "natural selection" bears the Epicurean signature of being a superordinate force *blind* to human concerns. In contrast, "artificial selection," the stuff of husbandry and eugenics, happens with eyes wide open, even in the face of chance and the unforeseen consequences it brings. That is more the stuff of Stoics. What Epicureans see as our ultimate lack of control, Stoics regard as a surmountable obstacle, the course of which we might steer even if not completely control.

A measure of humanity's Stoic mentality is its instinctive repugnance to unconscious manipulation, be it done in the name of operant conditioning or revealed preferences. But this is not because the Stoic opposes manipulation per se. "Self-mastery" is quite obviously a form of manipulation – but one to which the self willingly submits. Indeed, that staple of consumer contracts and research ethics – the "informed consent" clause – would be regarded by Stoics in the spirit of caveat emptor as a legitimate field of play for both sides. It is only the *unconscious* aspect of manipulation to which the Stoic objects. For this reason, the Stoic tends to construe in epistemic terms what the Epicurean casts in ontic terms: simple ignorance instead of outright impotence. Moreover, because the Stoic shares the Epicurean's belief in the reality of chance, such ignorance is often relieved not by meeting it on its own terms but by displacing its significance. Even if something cannot be known, it does not follow that it must be worth knowing. There may be other ways around our ignorance that reduce it to a temporary barrier rather than remain a mark of permanent subordination, as, say, Kant's "thing-in-itself" interpretation of the *noumenon*. Of course, if this strategy fails, "sour grapes" may result.

Cosmopolitanism's Anti-sentimental Expansion of the Circle of Moral Concern

In effect, the Stoic proposes the epistemological equivalent of "beating the odds" in gambling: The will offers proxies to make up for what the intellect lacks. The result is an "overcoming" ("sublation" in the Hegelian jargon) of the original situation, so there is no longer a need to address the outstanding ignorance. You simply enter a new field of play, which brings its own opportunities – and obstacles. Freud's ironic take on this idea was "sublimation," whereby what modernity's Stoic heirs presented as "progress" is recast as a burden that is never quite relieved, a debt that is never quite repaid, the cumulative effect of which serves as a drag on the "march of civilization." The challenge that Freudian sublimation poses to Hegelian sublation – a latter-day version of the Epicurean challenge to the Stoic – is that, as economists say, the "externalities" can never be fully "internalized." More specifically, "we," understood as humanity in the making, will ultimately fail to bear the cost of incorporating the other in some endless quest to beat the odds against the prospect of a more expansive universe. The cosmic casino is ultimately stacked against us.

Nevertheless, this is the "cosmopolitan" standard to which the Kantian conception of humanity ultimately aspires. As the Epicurean sees it, what the Stoic regards as humanity's journey to discover its place in the cosmos – the topic of Max Scheler's (2009) unfinished *magnum opus* – has amounted to an abandonment of parts of our prior selves on behalf of future selves. Such a change in our self-understanding as material beings is not positive if it simply generates more problems down the line. While the Stoic is mentally prepared to face problems, even those of one's own making, the deeper challenge is whether in solving an indefinite sequence of such problems the aim of revealing humanity's nature remains in view. The philosophical discourse surrounding the temporal continuity of personal identity, starting with John Locke, may be understood as a metaphysical laboratory for exploring this challenge, the most recent distinguished contributor to which has been Derek Parfit (1984).

A good way to approach the challenge is through a favorite cosmopolitan image: *expanding the circle of moral concern*. Both Stoics and Christians (in the context of *agape*) characterize the affect associated with this process as a reduction in the difference in one's feeling toward the familiar and the alien: Both are accorded the same dignity, respect, etc. It is a decidedly "cool" emotional state, comparable to a gas diffused over a greater space. It is opposed to *compassion*, which implies a capacity to feel as the other does, if only because one happens to share a common world with them. In this respect, *sympathy* may be seen as the limiting – or "coolest" – case of compassion, but in both cases, one "feels the other's pain." Schopenhauer provocatively staked his own position against Kant's, whose metaphysics he otherwise shared, by accusing him of lacking compassion, the emotion that Schopenhauer believed should anchor ethics. However, he also seemed to believe that our capacity for compassion was limitless, or at least should operate without limits, even if it results in self-extinction. This might then be presented as "self-sacrifice," as in the spontaneous saving of any form of life in distress. The term *Weltschmerz* ("world-suffering") was coined – and ironized – in the nineteenth century to capture this sensibility.

While Kant would certainly not endorse Schopenhauer, neither would Kant's great contemporary Adam Smith, who had effectively reversed the polarities to regard compassion as an exaggerated version of sympathy – which most certainly has limits. Indeed, Smith regarded sympathy and self-interest as complementary "moral sentiments" that trade on the disposition and location of our bodies. In effect, for Smith, one's own aversion to pain is the benchmark of being concerned for the other's pain. However, the reach of such sentiments depends on the extent that the agents inhabit a common lifeworld. With regard to strangers and others with whom one has a more episodic and transactional relationship, it is more rational to adopt the sort of "self-interested" standpoint that has become the calling card of *Homo oeconomicus*. The significance of the correlation of emotional distance and moral distance implied in Smith's account of moral sentiments would be hard to overstate. It underwrote Ferdinand Toennies' *Gemeinschaft/Gesellschaft* distinction, on which most sociological accounts of the transition from "traditional" to "modern" societies have traded (Fuller 2006: Chap. 10).

Nevertheless, for Kant, Smith's entire approach confuses mores with genuine morality, which aspires to universal coverage, and hence an ever-expanding circle of concern. It follows that even strangers cannot be treated as merely contingencies in one's life who can be handled strategically. Rather, one needs to affirm that the other *ought* to be in this world. This in turn probably requires some redistribution of material resources from oneself to the other, so that everyone in this expanded circle has an equal opportunity to flourish and hence to live in accordance with the dignity that they deserve. Indeed, the early socialists acknowledged their debt to Kant's ethics, which was swept away only once Lenin's self-styled "materialism" positioned Kant's epistemology as the "idealist" enemy. Of course, this has not stopped Communists from trying to stick to the spirit of Kant's ethics without mentioning his name.

Kant and Smith's disagreement over the ethical standing of moral sentiments reflects a profound difference in metaphysical orientation towards humanity's "species unity." Smith's understanding of "species" in "species unity" was as close to the modern biological understanding as one could have before Darwin. It was thoroughly naturalistic. In contrast, while Linnaeus had distinguished one species from another by the mode of physical reproduction, he nevertheless formally named them with an eye to their imagined function in divine creation. Even a philosopher as avowedly naturalistic as Aristotle tended to discuss species in terms of what makes an organism what it is – that is, its "essence." He made the question turn on the organizational principle (or "soul") governing the matter constituting the organism, which left the door open for the same spiritual principle to possess alternative embodiments. Once the Aristotelian corpus became generally available to Christian theologians in the thirteenth century, this prospect came to the fore in debates over the interpretation of the Trinitarian nature of the deity: In what sense might there be one God who exists in three persons, only one of which (Jesus) is materially configured as human? A secular version of this question in which "human" replaces "God" has increasingly stalked the modern era, especially once the Linnaean equation came to stereotype our understanding of the human. Kant's "cosmopolitan" vision of the human as a being who might be found anywhere in the cosmos is the clearest expression of that sensibility, preparing us for any future recognition of the human.

Conclusion: Testing Cosmopolitanism on Machines: Extending the Human Beyond the Animal

At the dawn of our computer era, the late Harvard philosopher Hilary Putnam (1960) reworked "functionalism" to refer to the view that the same mental states can be multiply instantiated. After all, even materialists concede that two people might acquire the same beliefs by substantially different means, which in turn would reflect the individuals' particular life histories, as registered in both brain and behavior. Here Putnam was bearing witness to the birth pangs of "artificial intelligence" (AI), which aimed to produce machines capable of instantiating the same beliefs as humans, at least to the satisfaction of the humans involved. For Putnam, all such instantiations – both human and machine – counted as "functionally equivalent." Perhaps it was no accident that he came to philosophy by way of mathematics, in which "equals" establishes an identity between differences, to put it in the paradoxical language of German idealism. Indeed, this general approach to the problem of mental states had been first proposed a decade earlier by another mathematician, Alan Turing, who used it to develop a decision procedure for determining whether a machine can think. The basic idea, which continues to animate AI research, is for a human and a machine to answer a set of questions testing their "intelligence" sufficiently well that a tester ignorant of their respective identities can't infer which is which. Analytic philosophers follow Willard Quine (1953) in speaking of "referential opacity," whereby the same object – in this case, intelligence – can appear in radically different guises, or the same end can be accessed by mutually exclusive means.

When the medieval theologians spoke of the trinitarian nature of God as a "mystery," they presumably meant such opacity. By the end of the nineteenth century, the paradigm case of referential opacity had shifted from theology to science, courtesy of the German mathematician Gottlob Frege. Frege observed that the ancient Greeks had two names for the planet Venus, "Morning Star" and "Evening Star," which they thought referred to different celestial entities but which we now say were different ways ("senses") of referring to the same entity. Quine, Turing, and Putnam all proceed from Frege. Common to them is the idea that for something to exist, it must be both *predicable* and *identifiable* – and these are separate capacities: The former is the power to define what follows from being X and the latter is the power to decide that something is X. In the classical syllogism, the definition corresponds to the major premise and the decision to the minor premise, based on which the status of one's being in the world is then deduced.

Kant anticipated them all, returning to Plato's *Sophist* and holding, *contra* Aristotle, that existence can be exhaustively decomposed into predication and identification: There is no need for an additional property, "being." For example, in the Turing Test, a candidate entity is identified as "intelligent" in the same way as any other such entity, namely, in virtue of certain qualities that are predicated of it based on passing examination questions that set the criteria for intelligence. Thus, the relationship between predication and identification is somewhat like that of legislation and adjudication: The terms of engagement are first defined (e.g., the nature of intelligence), and then they are shown to be satisfied (or not) by particular entities. This "functionalist" approach to ontology effectively reduces existence to legitimate standing in a regime. When Hans Kelsen and other twentieth century legal positivists spoke of norms as enjoying the status of "facts," they had just this in mind: the standards that actions need to meet to count as permissible in a given regime. This formula provided concrete expression of Alfred Tarski's "semantic" theory of truth, which translates "true" as true in a specific language – one of the few points on which the original logical positivists and Popper's renegade followers agreed.

Of course, sociology and anthropology had developed their own versions of "functionalism" around the same time, but largely based on modelling societies on organisms. However, within these fields, the turn *against* their own sense of "functionalism" starting in the 1960s corresponds to the stance taken by Turing, Putnam and their intellectual progeny vis-à-vis AI. It resulted in a "constructivist" (later "postmodern") conception of the social order as an ongoing achievement of successful role-playing and "passing." What made such constructivism "anti-functionalist" in the sociological sense was its suggestion that the social order – including the social order of science itself – is more fragile than advertised: Things are not as regular as they appear; they are just presented that way by those in charge of presentation. Through the strategic deployment of ethnographic methods across a wide range of fields, the constructivists showed empirically that even the most seemingly mundane episodes of social life are as finely crafted and judged as theatre, the success in which requires not only a strong script but also a stage performance that is well-received by the audience (Gouldner 1970: Chap. 10).

Various normative conclusions were drawn at the time based on the demonstrated manipulability of situations and arbitrariness of outcomes. They tended to be focused on whether the avowed norms were enforced fairly – and even whether "fairness" was intelligible, especially given the general state of the world. Thus, psychiatry's professional control over the sane/insane binary became a prime target. At the same time, and in partial response, social democrats argued that the situated and arbitrary character of social judgements could be harnessed for the greater public good. They focused on the manufacture of situations, the uncertainties of which would encourage people to decide in ways that would be beneficial for all concerned. It effectively revived the Stoic idea of allowing self-mastery to overcome ignorance by binding the will to compensate for intellectual deficiency. This hopeful line of thought ran from game theory and John Rawls' theory of justice to legislation requiring that employment practices be "blind" (i.e., nondiscriminatory) vis-à-vis the religion, class, race, and/or gender origin of job candidates (Fuller 2020a: Chap. 11).

Might not the next step in this trajectory be a *substrate-blind* test of humanity, modelled on the Turing Test? That a Turing Test-like procedure might be used to qualify candidates not merely as intelligent but as human was first suggested – albeit backhandedly – after the surprising success of the elementary computer program ELIZA in addressing mental health issues: Patients were convinced that on the other side of the computer's interface was a human, not an algorithm. I say "backhandedly" because the program's designer, Joseph Weizenbaum (1976), believed that the patients should *not* have been so easily confused. The fact that they were revealed the extent to which technological mediation of the human lifeworld had dulled our understanding of what it means to be human. Of course, this gloomy prognosis went against the spirit of the anti-psychiatry movement, in terms of which ELIZA's success shed a harsh light on the hold that the medical profession continued to enjoy over people's lives. After all, if a computer doctor can treat patients as well as a human doctor, then what is the added value of employing the human – and does the therapeutic value of what either of them says amount to anything more than a placebo effect?

Moreover, this question quickly acquired added significance, as the nascent field of cognitive science began to show that in diagnostic settings involving the statistical weighting of symptoms, computers outperformed even seasoned professionals, who routinely fell afoul of poor memory, various biases and processing limitations. Indeed, by the 1980s these findings had spurred the development of computerized "expert systems," domain-specific forms of artificial intelligence designed to capture an idealized version of a human expert's decision-making practice. But the proliferation of such machines still left very open the question of whether they – and their promised superior successors – marked a moment in the expansion, dilution, or perhaps even falsification of what it means to be human.

In the early 1970s, Hubert Dreyfus (1972) had already stepped into the breach, fortifying the Linnaean equation of "Human = *Homo sapiens*" with Heideggerian phenomenology and later Wittgensteinian talk of "forms of life." He effectively tried to turn "being human" into a kind of meta-social expertise, captured by such omnibus phrases as "tacit knowledge" and "skilled practices." To be sure, this was not an unknown approach, though Dreyfus's use of it was ironic from the long view of history. After all, appeals to "human rights" have been invoked after the establishment of "civil rights" in the modern era because in practice full civil rights have been enjoyed only by a subset of all the humans formally entitled to those rights; hence the need for, say, the "blind" employment legislation mentioned above. In this respect, Dreyfus' meta-social expertise amounted to "insider knowledge," the possession of which presupposes a certain upbringing, training, credentialing, etc. Thus, in the Dreyfusian rendition of the human, a candidate needs to have already participated substantially in the human lifeworld, which probably means having been the biological product of a recognized human – and of course raised accordingly. To be sure, Dreyfus and his followers rarely talk in such "speciesist" terms, yet their dependency on fine points of acculturation for determining a candidate's standing as "human" de facto rules out non-*Homo sapiens*.

One telling development from Dreyfus' line of thought has been the rise of "expertise studies" in the sociology of science. Its champion, Harry Collins, originally detailed the differences in the activities surrounding the judgements reached by human experts and their computer counterparts (Collins 1990). But later he moved to study the differences between such experts and various groups of lay people who are defined by their distance from formal training and recognition in the expertise (Collins and Evans 2007). The red thread that runs through this line of research is that "expertise," be it defined ontologically (e.g., "being a human") or epistemologically (e.g., "being a physicist"), is *self-recognizing*: It takes one to know one. However, in violation of Kantian strictures, this approach effectively collapses predication and identification into one supposedly intuitive conception of "being." Missing is, so to speak, a metaphysical "separation of powers," whereby those who determine *what* something is (predication) are detached from those who determine *that* something is (identification). Instead, the Dreyfusian approach toward candidate humans produces a version of "confirmation bias" (aka "prejudice"), whereby one expects that a successful candidate will not only satisfy the official criteria but also fit in smoothly with the other already recognized humans. Indeed, to the unforgiving eye, "sharing a form of life" is simply a euphemism for "old boys' club." Such an attitude inhibits any process of self-discovery of what it means to be human by refusing to allow new cases to test the limits of the current criteria.

This chapter began by observing that the exclusive identification of the human with an upright ape was a mid-eighteenth century innovation. From a world-historic standpoint, it may turn out to have been an aberration, thereby vindicating Foucault (1970) on his major thesis, but perhaps not as he intended. The recent rise of "trans-" and "post-" human philosophies is returning in a new key to the materially more variable conception of the human that prevailed prior to the coinage of "*Homo sapiens*" (Fuller 2012: Chap. 3). It is fuelled by developments no less scientific than the ones that influenced Linnaeus, Kant, and their contemporaries. The key difference today is the presence of stronger scientific arguments for the superficiality of morphological differences, ranging from the massive genetic overlap of all life forms to the increasing difficulty people have had in telling difference between the carbon- and silicon-based intelligence. In most general terms, the paths to becoming human (or expert, *pace* Harry Collins) are multiple: They do *not* rely on a privileged biological (or disciplinary) genealogy. Indeed, the validity of the "human" as a concept depends on its independent corroboration; hence, the need for a separation of powers between the predicators and the identifiers of the human. Kant's cosmopolitan vision of humanity understood this point as requiring the ethical standpoint to transcend the animal basis for spontaneous moral sentiments, thereby expanding the circle of concern. But equally, it may require generating a different sort of legitimizing historical narrative for humanity, one in which the human is not an emergent feature of the ape, but rather the ape is a contingent expression of the human. At this point, we cross the threshold into "Humanity 2.0," which will define tomorrow's history of the "human sciences" (Fuller 2011, 2012).

References

Austin JL (1962) How to do things with words. Clarendon Press, Oxford. (Orig. 1955)
Bateson G (1979) Mind and nature: a necessary unity. E.P. Dutton, New York
Burckhardt R (2013) Lamarck, evolution and the inheritance of acquired traits. Genetics 194:793–805
Chomsky N (1959) A review of BF Skinner's *verbal behavior*. Language 35(1):26–58
Cobb M (2020) The idea of the brain: a history. Profile Books, London
Collins H (1990) Artificial experts. MIT Press, Cambridge, MA
Collins H, Evans R (2007) Rethinking expertise. University of Chicago Press, Chicago
Dobbs BJ (1991) Stoic and Epicurean doctrines in Newton's system of the world. In: Osler M (ed) Atoms, pneuma and tranquillity: Epicurean and stoic themes in European thought. Cambridge University Press, Cambridge, UK, pp 155–174
Dreyfus H (1972) What computers can't do. MIT Press, Cambridge, MA
Foucault M (1970) The order of things. Random House, New York. (Orig. 1966)
Foucault M (2008) Introduction to Kant's anthropology. MIT Press, Cambridge, MA. (Orig. 1961)
Fuller S (2006) The new sociological imagination. Sage, London
Fuller S (2009) The sociology of intellectual life. Sage, London
Fuller S (2011) Humanity 2.0: what it means to be human past, present and future. Palgrave, London
Fuller S (2012) Preparing for life in humanity 2.0. Palgrave, London
Fuller S (2013) Deviant interdisciplinarity as philosophical practice: prolegomena to deep intellectual history. Synthese 190:1899–1916

Fuller S (2014) Neuroscience, neurohistory and the history of science: a tale of two brain images. Isis 105:100–109
Fuller S (2018) The hour of political biology: Lamarck in a eugenic key? Hist Hum Sci 31:1–7
Fuller S (2019a) Nietzschean meditations: untimely thoughts at the dawn of the transhuman era. Schwabe Verlag, Basel SZ
Fuller S (2019b) The metaphysical standing of the human: a future for the history of the human sciences. Hist Hum Sci 32:23–40
Fuller S (2020a) A player's guide to the post-truth condition: the name of the game. Anthem, London
Fuller S (2020b) Knowledge socialism purged of Marx: the return of organized capitalism. In: Peters M et al (eds) Knowledge socialism. Springer, Berlin, pp 117–134
Fuller S (2021) The mind-technology problem. Postdigital Sci Educ. https://link.springer.com/article/10.1007/s42438-021-00226-8
Fuller S, Lipinska V (2014) The proactionary imperative: a foundation for transhumanism. Palgrave, London
Goldschmidt R (1940) The material basis of evolution. Yale University Press, New Haven
Gould SJ (1989) Wonderful life. Norton, New York
Gouldner A (1970) The coming crisis in western sociology. Basic Books, New York
Hacking I (1990) The taming of chance. Cambridge University Press, Cambridge, UK
Halbwachs M (1992) On collective memory. University of Chicago Press, Chicago. (Orig. 1925)
Huxley TH (1863) Evidence as to man's place in nature. Williams & Norgate, London
Huxley TH (1879) Hume. Macmillan, London
Kantorowitz E (1957) The king's two bodies: a study in medieval political philosophy. Princeton University Press, Princeton
Kauffman S (1993) The origins of order: self-organization and selection in evolution. Oxford University Press, Oxford
Klarsfeld A, Revah F (2003) The biology of death: origins of mortality. Cornell University Press, Ithaca
Laudan L (1981) Science and hypothesis. Kluwer, Dordrecht
Lovejoy A (1936) The great chain of being. Harvard University Press, Cambridge, MA
Nelson R (2001) Frank knight and original sin. Indep Rev 6(1):5–25
Neumann M (2006) A formal bridge between epistemic cultures: objective possibility in the time of the second empire. In: Löwe B et al (eds) Foundations of the formal sciences: the history of the concept of the formal sciences. Kings College Publications, London, pp 169–182
Parfit D (1984) Reasons and persons. Oxford University Press, Oxford
Penrose R (1994) Shadows of the mind. Oxford University Press, Oxford
Porter T (1986) The rise of statistical thinking, 1820–1900. Princeton University Press, Princeton
Prior A (1942) Logic and the basis of ethics. Clarendon Press, Oxford
Putnam H (1960) Minds and machines. In: Hook S (ed) Dimensions of mind. New York University Press, New York, pp 138–164
Quine WVO (1953) From a logical point of view. Harvard University Press, Cambridge, MA
Scheler M (2009) The human place in the cosmos. Northwestern University Press, Evanston. (Orig. 1928)
Simmel G (2008) Sociology: inquiry into the origins of social forms. Brill, Leiden. (Orig. 1908)
Talcott S (2019) Georges Canguilhem and the problem of error. Palgrave, London
Teilhard de Chardin P (1961) The phenomenon of man. Harper and Row, New York. (Orig. 1955)
Toulmin S, Goodfield J (1965) The discovery of time. Harper and Row, New York
Valsiner J, van der Veer R (2000) The social mind: the construction of an idea. Cambridge University Press, Cambridge, UK
Weizenbaum J (1976) Computer power and human reason. W.H. Freeman, San Francisco
Young M (1958) The rise of the meritocracy. Penguin, Harmondsworth
Zammito J (2002) Kant, Herder and the birth of anthropology. University of Chicago Press, Chicago

History of the Human Sciences in France: From Science de l'homme to Sciences Humaines et sociales

3

Nathalie Richard

Contents

Introduction	58
The Historical Roots	58
The Political Crucible of "la Science de l'homme"	60
Dismembering "Science de l'homme" in Nineteenth-Century France	62
"Science de l'homme," "Sciences Humaines," and "Sciences Sociales" in the Twentieth Century	66
Reflexivity	69
Conclusion	71
References	72

Abstract

Focused on the phrases used to name these sciences, this chapter gives an historical sketch of the development of modern human and social sciences in France. It highlights the importance of discipline formation and growing specialization, as well the recurrence of unifying projects counterbalancing this trend. Its central argument is that a moment of crystallization took place between 1770 and 1810, when the phrase "*science de l'homme*" took a new meaning. This founding moment for the human sciences coincided with the genesis of the political fundamentals of modern France. The convergence of these large-scale processes conferred on the project of building a body of scientific knowledge about man intellectual and political dimensions that have proved nondissociable. It bestowed specific characteristics on the "*science de l'homme*" or "*sciences humaines et sociales*," which marked its fate in France and have persisted to the present day.

N. Richard (✉)
Department of History, Le Mans Université, TEMOS CNRS UMR 9016, Le Mans, France
e-mail: nathalie.richard@univ-lemans.fr

© The Author(s), under exclusive licence to Springer Nature Singapore Pte Ltd. 2022
D. McCallum (ed.), *The Palgrave Handbook of the History of Human Sciences*,
https://doi.org/10.1007/978-981-16-7255-2_43

Keywords

"Science de l'homme" · Discipline formation · Specialization · Unity of the human and social sciences · Science/politics relation

Introduction

This chapter gives an historical sketch of the founding and development of modern human and social sciences in France. It is organized in chronological sections and focuses, from the eighteenth century onward, on the phrases which were chosen in order to designate these sciences, as a whole or in part. This history is determined by major trends, such as the demise of the medieval and early modern distinction between sacred and profane, and the rise of new dividing lines between nature and culture, "moral" (mind) and "physical" (body). This history is also strongly impacted by the formation of the modern disciplines and the specialization process going with it. These major evolutions created ongoing tensions, which still shape the field of the human and social sciences in the present days. One aim of this chapter is therefore to demonstrate that history is relevant to understand the present state of these sciences.

History is also relevant to point out and try to explain the differences of the human and social sciences in the Francophone and the Anglophone worlds. Many French phrases used to name and classify these sciences do not translate easily into English. A central argument in this chapter is that this difference is due to the special relation the human and social sciences have with politics, and to the specificity of French political history in which they developed. Indeed, modern human and social sciences emerged in France shortly before, during and after the 1789 Revolution, so that there is a strong convergence between the intellectual process of shaping a new field of knowledge, and the political process of forging a new nation-state.

The Historical Roots

Although it is usual to locate the intellectual origins of modern humanities and social sciences in the eighteenth century (Moravia 1970; Gusdorf 1973; Fox et al. 1995; see Christie 1993), the phrases "human sciences" and, less frequently, "science(s) of man" have a longer history. In France and in the French-speaking world as early as the sixteenth century, many humanists used the terms (Bruter 2006). Guillaume Budé mentioned the *"scientias humanas"* in *De studio literarum recte et commode instituendo* [*On the Proper Institution of the Study of Letters*], 1532), as did Jean Calvin in his *Ordonnances ecclésiastiques* [*Ecclesiastical Ordinances*] of 1541. The expression "science of man" was used by Polycarpe de la Rivière in his treatise on moral theology, *Angélique. Des excellences et perfections immortelles de l'âme* ([*Angelic: On the excellences and immortal perfections of the soul*], 1626), whose first discourse was entitled "Que l'homme est la vraie science de l'homme" (« That man is the true science of man »).

Yet, within the humanist culture, the opposition was not between the human and the natural sciences. It pointed to a partition of the sacred and the profane, and to the dichotomy between the human sciences (or the sciences of man) and the divine sciences (or the sciences of God). Man then was not the object of the human sciences, but the knowing subject, whose flawed reason and uncertain knowledge contrasted with the perfect wisdom of God and the certainties derived from faith. This primary significance of the phrase lasted at least until the eighteenth century, when it gradually faded away (Vidal 1999). While the Catholic abbot Nicolas-Sylvestre Bergier still used the phrase in its old meaning in the article "*sciences humaines*" of the *Encyclopédie méthodique* dictionary of *Théologie* (1788–1790; Masseau 2006), a new signification had already emerged in Nicolas Malebranche's work in the early years of the century. A frequently reproduced quotation indeed superimposed the two meanings. The Catholic philosopher wrote, "of all the human sciences, the science of man is the most worthy of man," thus defining the science of man as a science in which man was both the object to be known and the knowing subject. For this reason, Malebranche has been credited, along with Descartes, with having contributed to the secularization of the science of the human mind in the French tradition (Vidal 2006). Malebranche, endorsing the hierarchy of forms of knowledge inherited from Renaissance Humanism, nevertheless distinguished between human self-knowledge and true "revealed" knowledge. He specified that "science of man" was "only an experimental science, which result[ed] from the reflection that one makes on what happens in oneself," whereas "the knowledge of God [was] not experimental" (*La Recherche de la vérité* [*The Search for Truth*], "Foreword," 1712; Carbone 2007).

In this long-term process of secularization and redefinition of the limits of human rational knowledge, which is not specific to the French-speaking world, the Enlightenment was a key period. Historians have emphasized the diversification in topics of inquiry as well as the expansion of the fields of investigation (for instance: Auroux 1991, 2003; Besse et al. 2010; Blais and Laboulais 2006; Duchet 1995; Kaufmann and Guilhaumou 2003). In the works of well-known philosophers and natural scientists (such as Denis Diderot, Voltaire, Jean-Jacques Rousseau, Montesquieu, Maupertuis, Condillac, and Buffon) and of more specialized scholars such as the antiquarians Bernard de Montfaucon and Antoine Court de Gébelin (Mercier-Faivre 1999), human language, society, customs, religions, and their diversity in time and space became the focus of secular analysis. Part of this new knowledge was synthetized in Diderot and d'Alembert's *Encyclopédie ou Dictionnaire raisonné des sciences, des arts et des métiers* [*Encyclopedia, or a Systematic Dictionary of the Sciences, Arts, and Crafts*] (1751–1772) (Groult 2003, part III). Historians have also depicted the Enlightenment as a change in epistemological perspective, highlighting the deleting of the partition between the human and the divine sciences and the promotion of a new boundary between the human and the natural sciences. This new boundary was drawn on the base of the differing nature of the objects of the two kinds of sciences, although the former could partly borrow their methods from the latter, as suggested by the expression "natural history of man" adopted by the naturalist Buffon (Blanckaert 2006a, b). Indeed, regarding the proper method of the human sciences, two options emerged in this period. Following Buffon and the naturalists-philosophers, the first option emphasized observation and linked the study of man to the natural sciences and to medicine. The second option tied the science of man to philosophy by focusing on the specificity of self-knowledge. It emphasized the self-observation of the internal operations of body and mind. This supported the advocacy of philosophical introspection as a method, appearing, for example, in the work of Pierre-Jacques Brissot, leader of the *Girondin* faction during the French Revolution (*De la vérité, ou méditations sur les moyens de parvenir à la vérité dans toutes les connaissances humaines* [*On truth, or meditations on the means of arriving at the truth in all the forms of human understanding*], 1782; Coursin 2018).

In France, a moment of crystallization took place between 1770 and 1810. This founding moment for the human sciences coincided with the genesis of the political fundamentals of modern France. The coincidence of these large-scale processes conferred on the project of building a body of scientific knowledge about man intellectual and political dimensions that have proved nondissociable. It bestowed specific characteristics on the "*science de l'homme,*" which marked its fate in France and have persisted to the present day.

The Political Crucible of "la Science de l'homme"

A number of studies have pointed out the importance of the last decades of the eighteenth century in the constitution of the modern human sciences in France. They differ in the emphasis they place on the continuity of a long evolution or on the abrupt change from one epistemic regime to another, as canonically illustrated by the contrasting visions of Georges Gusdorf and Michel Foucault (Gusdorf 1960; Foucault 1966; see Blanckaert 1999).

The singularity of the French case is due to the political context in which these developments took place. Crucial political and ideological changes and major reconfigurations in the human sciences tightly intermingled in France. As several authors have pointed out, the discourses on the reform of science were also discourses on political reform (Chappey 2006; Gauchet and Swain 2007; Staum 1980). In 1773, Helvétius' *De l'homme* [*Treatise on Man*] marked an important step in the setting of this new agenda (Duchet 1995: 377–406). The science of man as he understood it would not be limited to the study of the human understanding, as Malebranche had suggested and as Hume had advocated (Jones 1989) but would place at the heart of its program the study of the relationships between the mental faculties and the physiological organization of man. The emergence of the expression "*science de l'homme,*" used in the singular, in the title of works published in French between 1770 and 1800 was indicative of this redefinition. The purpose of this epistemic operation was primarily methodological. Helvétius was a materialist, and he refocused his science on the relation between the physical and the moral. To study this new subject, he advocated refounding the knowledge of man on more reliable methods, borrowed from anatomists and physiologists, in order to establish a more valid science. However, the goals assigned by Helvétius to this new science were not merely intellectual. By providing a more solid knowledge of man, the *"science de l'homme"* was expected to improve humanity, and indeed, to be the very engine of progress. Thus, the science of man was also a "science of government," as Helvétius formulated it. In the years before 1789, this political-scientific project was at the core of the struggles between the promoters of the Enlightenment and the anti-Enlightenment. Faced with criticism by Catholic apologists, such as Bergier, the *"science de l'homme"* became a banner (Chappey 2006: 47), fostering the rise of the new figure of the physician-expert among its defenders, as illustrated by Paul-Joseph Barthez (*Nouveaux Eléments de la science de l'homme* [New Elements of the Science of Man] 1778; Carbone 2017; Williams 1994).

Challenged by other formulations such as "social art" and "social science" (Baker 1973; Audren 2006), the phrase somewhat faded away during the first years of the French Revolution, only to reemerge at the forefront of the intellectual and political stage later, when the revolutionary regime was facing internal and external threats and was radicalized under the Jacobins' rule. Several authors promoted the *"science de l'homme"* as a tool capable of reducing misconceptions in public opinion and in fighting the enemies of the Revolution (Chappey 2006). The *Idéologues* [Ideologues] represented the high point of this evolution. They were the artisans of the success of the *"science de l'homme"* after the fall of Robespierre in 1794 and the establishment of the Directoire (Moravia 1974). Represented in particular by Antoine Destutt de Tracy and by Georges Cabanis (Goetz 1993; Staum 1980), the Ideologues were the architects of the moderate and conservative republican turn of the French Revolution. They played an important role in drafting the new Constitution of Year III, and their influence was reflected in the creation of new scientific institutions, including the new Academy of "Moral and Political Sciences," actually a section of the *Institut de France*, in 1795 (Azouvi 1992). As a whole, the *Institut*, which took over from the *Académies royales* [Royal Academies] abolished after 1789, was designed as the main institutional concretization of the scholarly project of the Ideologues. Its promoters conceived the *Institut* as a "living encyclopedia" (Cabanis, *Rapports du physique et du moral de l'homme* [*On the relations between the physical and moral aspects of man*], 1802). Within this new institution, all sciences would join and contribute to a better understanding of the world and humanity. Based on this sounder knowledge, the *Institut*'s ultimate goal was to "civilize" men, in order to reduce the savagery that had manifested itself from within during the Terror, and which, according to the Ideologues prevailed outside France in all countries which had not endorsed the principles of the Revolution.

The 1800 "Ideological moment" (Blanckaert 2000; Citton and Dumasy 2013), with its intellectual content unfolding, for example, in the columns of the magazine *La Décade philosophique* (Boulad-Ayoub 2004), permanently tied the links between a scientific enterprise and a universalist and moderate republican project which was to become the destiny of the *"science de l'homme"* in France. Yet this golden moment was short-lived. Finding the Directoire ineffective in setting up the administrative structures necessary for reform, the Ideologists rallied to Napoleon Bonaparte and supported the coup of the 18th Brumaire (1799). The outcome, however, saw their expectations unfulfilled, and their ambition to bring about reform opposed by the Napoleonic regime, as manifested by the suppression of the Moral and Political Sciences section when the *Institut de France* was restructured in 1803. Thus, the 1800 moment, regarded as fundamental in the recomposition of the sciences in France, is of great complexity. On the one hand, it was indeed the time when the science of man "consecrated the alliance of the physical sciences and the arts of government" (Blanckaert 2000) and seemed triumphant. On the other hand, Bonaparte's rise to power marked an inflection from the revolutionary spirit. The intellectual and political reformist ambition of the Revolution had remained largely unfulfilled, and the new regime undertook the partial realization of some reform projects, albeit in a more conservative and technocratic way. As a result, the Consulate and the Empire started the demise of the unifying project projected by the Ideologues, and marked the end of the revolutionary reformist utopias with which this project was associated, ushering in a sense of loss and nostalgia constitutive of French political modernity (Knee 2014). From 1802 onward, there was a reorganization of the academic system under strict State control, and the promotion of an increasingly specialized and professional scientific expertise in the service of the state (Fox and Weisz 1982; Fox 2013).

The complexity of this inaugural period can be appreciated in the conflicting interpretations formulated about the *Société des Observateurs de l'Homme* [Society of Observers of Man] (1799–1804). Once considered as home of the Ideologues, this learned society proved to be, in reality, one of the places where the unified science of man was dismembered along several lines—between body and mind, reason and faith, science and literature, and professionals and amateur scientists—as well as according to the new relations between state and experts (Chappey 2002, 2019). Burdened by the constitutive ambiguities in which it had come into existence, the "*science de l'homme*" unfolded along dividing lines that both caused its fragmentation and ensured its success, since various groups could claim to be the upholders of its legacy in nineteenth-century France.

Dismembering "Science de l'homme" in Nineteenth-Century France

The breakdown of the "*science de l'homme*" in the nineteenth century is reflected in the disappearance of the expression and in the diversification of the terms used to designate it as a whole or in parts. The words anthropology, psychology, sociology, and history took precedence. This change in vocabulary revealed several lines of partition.

The first divide concerned the method of the human and social sciences and their position in the cartography of the sciences. After 1800, the two ways of conceiving the science of man which had emerged in the eighteenth century coexisted. On the one hand, the naturalistic and medical lineage gave birth to phrenology (Renneville 2020) and anthropology (Blanckaert 1989, 2009), which subordinated the knowledge of the mind to the study of the anatomical and physiological characteristics of the body. On the other hand, a tradition developed emphasizing self-observation and the proximity between the human and social sciences and philosophy. It sometimes focused on the relations between internal organic sensations and the mind, as in the work of Pierre Maine de Biran (Azouvi 2000; Vigarello 2016). It also refocused on the analysis of mental functions alone, giving rise to a strong French tradition of philosophical (introspective) psychology, which included Theodore Jouffroy and Victor Cousin among its founders (Antoine-Mahut 2020; Goldstein 2005). Leading to the assertion of transcendence, this dualistic and spiritualist tradition, called eclecticism, dominated the humanities in French University until the beginning of the twentieth century (Brooks III 1998).

By 1850, however, influential authors were criticizing philosophical introspection, and giving new orientations to the science of man. This evolution took place under the influence of positivism and the new methodological requirements that Auguste Comte (Bourdeau et al. 2018; Petit 2003) and John Stuart Mill (Guillin 2017; Guillin and Souafa 2010) set up for the sciences taking human behavior and society as their objects. The most notable of the scholars taking lessons from positivism but transforming it for the purposes of the human sciences was Ernest Renan. He reasserted dualism and the partition of the physical sciences and the sciences of the mind, which he named "*sciences de l'humanité*" [sciences of humankind]. While not giving up philosophical psychology, he proposed to refound the study of the human mind on the observation of historical and social (i.e., "positive") "facts," mainly languages and religions. Bringing the German philological tradition and biblical criticism to France, Renan gave an unprecedented influence to the new positive science of religions thanks to the huge scandal caused by his *Vie de Jésus* [*Life of Jesus*] in 1863 (Richard 2015). Other authors rejected Renan's dualism as well as philosophical introspection. They argued that introspection was a legitimate method for the human sciences, because it was not a mere philosophical tool and could achieve the same level of objectivity as observation within the natural sciences or medicine. It could have an experimental dimension, when self-observation focused on altered states of consciousness, in which drugs or sleep modified mental functions. Alfred Maury, who held the Chair in History and Morals at the *Collège de France* and was the author of one of the most famous books on dreams in the nineteenth century (*Le Sommeil et les rêves* [*On Sleep and Dreams*] 1861), was an eminent promoter of this new science of man (Carroy 2012; Carroy and Richard 2007). Yet its most influential advocate was Hippolyte Taine, whose 1870 book *De l'intelligence* [*On Intelligence*] stands out as seminal in the history of French scientific psychology. In this book, Taine described the human mind's relation to the brain. He argued that psychology, like physiology, should rely on positive facts, which were provided by self-observation in experimental conditions and by the alienists (psychiatrists). Psychologists could also gather positive facts regarding the functioning of the human psyche in literary sources and in history. Taine's entire work, from literary criticism to the history of the French Revolution, therefore built a science of man centered on scientific psychology, in which psychological factors explained all human behavior (Richard 2013). Renan, Maury, and Taine differed from eclecticism by referring to positivism. Yet they shared with

Victor Cousin and his disciples the conception of the science of man as part of philosophy. They identified themselves as philosophers and considered that the human sciences were the driving force in a global reform of philosophy. They did not mingle with the anthropologists gathered in the *Société d'anthropologie de Paris* [Society of Anthropology of Paris] created in 1859 by Paul Broca, and they did not endorse the scientific agenda centered on anatomy and physiology.

In nineteenth-century France, another major dividing line, grounded in differing political traditions, ran through the "*sciences de l'homme.*" Many works, to be sure, fell within the revolutionary and republican heritage, though with multiple nuances. Some authors identified themselves with the heritage of the Radical Enlightenment, as did some 1848 utopian socialists and, after 1860, anarchists, such as Elisée Reclus (Bourdeau and Macé 2018; Rignol 2014). But the majority of authors were more moderate and claimed a 1789 legacy after making an inventory compatible with the extension of the voting franchise and constitutional monarchy. Such was the case, in a paradigmatic way, for Victor Cousin, whose philosophy served as a quasi-official ideology during the July Monarchy (Goldstein 2005; Vermeren 1995). During the Second Empire, Taine and Renan joined the ranks of the opponents to Napoléon III's authoritarian regime, and their secular scientific worldview was perceived as being at the forefront of the battle against the union between the State and the Catholic Church. Although Renan and Taine criticized universal suffrage and distanced themselves from the new regime of the Third Republic, they received recognition and honors during its first decades. Ernest Renan, indeed, whose course on the history of Christianity at the *Collège de France* had been discontinued during the Second Empire, became the powerful administrator of that very institution. In 1903, the inauguration of his monument in his native town of Tréguier, in Brittany, provided an opportunity for the public authorities to stage the confrontation between the Republican state and the Catholic Church, at a time when the Separation Bill of 1905, which established the complete separation between the French State and religions, was planned (Priest 2015).

Most proponents of the human sciences in nineteenth-century France endorsed the principles inherited from 1789, but a counterrevolutionary science of man also existed. Opposed to belief in free will and individual liberties, it promoted belief that social bounds and tradition are primeval in human identity. This counterrevolutionary trend, which was reminiscent of German romanticism, could be found in Louis de Bonald's work and was perhaps as influential in shaping French sociology as its liberal republican counterpart (Heilbron 2015; Moreau 2020).

After 1870 and the establishment of the Third Republic, the full range of these political positions remained visible among the promoters of the human and social sciences. Within sociology, for example, Emile Durkheim and his disciples, promoters of a university-based sociology, were companions to the socialists and inspired solidarism, a more moderate vision of socialism, which became the central political ideology of the new regime (Blais 2007). The Durkheimians' theoretical and political ideas opposed those of Gabriel Tarde, whose sociological irrationalism was based on belief in the power of imitation and charisma, and inspired the antirepublican royalist extreme right. The Durkheimians also opposed the followers of Frédéric Le Play, who promoted a sociology committed to social reform led by a paternalist and Christian elite (Mucchielli 1998; Kalaora and Savoye 1989). Thus, the French "*science de l'homme*" saw the political project with which it was originally associated splintered into multiple, and often enough opposing, trends.

By 1900, the Ideologues' unified science of man was fragmented along several lines, which did not always overlap. In addition to the methodological and political oppositions mentioned above, there were also significant debates about the relative capacities of science and literature to provide legitimate knowledge about humankind. As early as 1800, men of letters advanced the claim for the distinctive power of literary creation against the claim of scientists and philosophers to be the sole provider of knowledge about man. These debates traversed the *Société des Observateurs de l'Homme* and were constitutive of the modern direction of literature in France. Germaine de Staël and François de Chateaubriand, Bonald, and Cabanis, among others, intervened (Citton and Dumasy 2013; Zékian 2013). Throughout the century, the territories of literature and human sciences morphed or merged according to institutional evolutions and individual trajectories. During the Second Empire, political adversity shut the path to academia for Taine and Renan. They were therefore forced to live by their pen, and they published both academic and literary works, novels, autobiographical texts, and plays, some of which, such as Renan's *Souvenirs d'enfance et de jeunesse* [*Recollections of my youth*] (1883), became part of the national literary canon. In this context, literary criticism gained a special importance, as it became the cradle where a discourse reunifying literature and human sciences could emerge (Lepenies 2002; Richard 2013). At the end of the century, the competition between science and literature was reenacted in an extremely confrontational political climate marked by the Dreyfus Affair. A number of authors, often connected to the antirepublican right, not accepted for university positions, and publishing in influential cultural magazines, claimed for literature the exclusive prerogative of telling the truth about human beings, opposing materialist academics who, they held, jeopardized the moral and Christian foundations of society. The novelist Paul Bourget, for example, called himself a psychologist. In *Le Disciple* [*The Disciple*] in 1889, he portrayed a scholar reminiscent of Taine and denounced the dangers of a deterministic psychology, capable of manipulating individuals into committing crimes. Other writers asserted their identity as sociologists, and made similar claims against Durkheimian sociology, while pointing out its compromise with the republican ideology (Mosbah-Natanson 2017; Sapiro 2004). These debates durably blurred the distinction between professionals and amateurs in the area of human studies, a distinction which does not exactly coincide with the academic/nonacademic separation.

Shortly before 1900, the largely dismembered project of the "*science de l'homme*" may appear to have met its end under the impact of the academic reform (1896) which created, in France, the modern universities and specialized curricula. Combining teaching and research more closely, these new curricula consolidated disciplinary identities, based on specific skills and practices. Thus, the 1800 moment contrasts with a second founding moment, the 1900 moment of the birth of the humanities and social sciences in the form of separate and professionalized disciplines, with the university as their institutional basis (*Les débuts...* 1992; Noiriel 1990; Mucchielli 1998).

"Science de l'homme," "Sciences Humaines," and "Sciences Sociales" in the Twentieth Century

Nevertheless, the ideal of a synthetic science of man inherited from the eighteenth century and later embodied in the works of Taine or Renan still shaped a horizon of expectation. It generated a series of unresolved tensions within the disciplines now called "*sciences humaines et sociales*" [the human and social sciences].

The first tension concerns the place of philosophy among these disciplines. If in France sociology and psychology usually assume closeness and critical, reciprocal links with philosophy, the relations between history and philosophy have been less peaceful. The two disciplines have recurrently competed for the domination of the academic hierarchy of the humanities and have produced rival claims about their ability to train republican citizens (Fabiani 1988). A second tension concerns differing views about the legitimate settings for the institution and dissemination of the human and social sciences, and about the social role of their practitioners. The institutional platform for professional historians, sociologists, anthropologists, geographers, and so forth certainly became the universities and academic journals. However, the figure of "the Author," a figure capable of borrowing from several disciplines, acting in the public arena, and publishing in the general press, as embodied at the time of the French Revolution by Cabanis or Destutt de Tracy, and then again later by Taine or Renan, continued to haunt twentieth-century human and social sciences. In the wake of the political controversies of the Dreyfus Affair, this character was renamed "*Intellectuel*" [Intellectual], a difficult to translate label that further underlines the distinctiveness of French humanities and social sciences (Charle 1990; Ory and Sirinelli 1986). A third tension concerns the balance and contrast between the technical and literary nature of styles of writing within the human and social sciences and, as a result, the closeness or distance of writing in these fields from literature. While university training promotes the ideal of a technical and stylistically neutral manner of writing, justified by the scientific objectivity at which these sciences aim, an injunction to achieve a more elegant style, which is the trademark of the "intellectual/author," remains. The writings of Foucault, whose style attracted many comments (for example, Favreau 2012), and the tradition of the "two books"—the scholarly monograph and the literary travelogue—in French ethnology, illustrate this tendency (Debaene 2010).

The very phrase "*science de l'homme*" did not disappear after 1900. It was used, for example, by the Parisian publisher Payot to name a new collection launched in 1963. Created within the famous series "Bibliothèque scientifique," which had existed since 1925, this collection was alive until the beginning of the 2000s. But its evolution reveals the uncertainties of the fate of the expression "*science de l'homme*" in the twentieth century, since the series was progressively narrowed down until only psychoanalysis remained. In the same years, other publishers chose other expressions, such as Gallimard's "Bibliothèque des sciences humaines" in 1965, which published Foucault's books (Chaubet 2012). After 1900, the phrase "*science de l'homme*" therefore competed with other terms such as "*sciences sociales*" (social sciences), "*sciences humaines*" (human sciences) (Reubi 2020), or "*synthèse*" (Biard et al. 1997). Each of these labels referred to the totality of disciplines that, in the English-speaking world, fall within the relatively distinct fields of the humanities, the social sciences, and the behavioral sciences, so that their translation into English is fraught with misinterpretation.

These proliferating expressions not only disclosed the presence of a still lively ambition to go beyond disciplinary specialization, but they also exposed conflicting agendas. The use of one of these phrases may have denoted the intention of ranking the sciences that have "man" as their object and of subordinating the various academic disciplines to one that would encompass the others by virtue of the reach of its more globalizing explanations. The expression "social sciences" thus sometimes conveyed the intention, made explicit by Emile Durkheim and his disciples at the beginning of the twentieth century, to place the other sciences of man under the heading of sociology, establishing a hierarchy of disciplines that institutionalized a conception that social determinants, above all, are capable of explaining humankind. On the other hand, some uses reflected the intention to strengthen interdisciplinary collaborations between disciplines conceived to be of equal standing, focused on an object ("man") whose specificity called for multiple viewpoints. Such was the "synthesis" agenda launched by Henri Berr from 1900 onward, through the *Revue de synthèse* (founded 1900) and the *Centre international de synthèse* [International Synthesis Center] (1929), an innovative forum for the confrontation of disciplinary points of view in the inter-war period (Biard et al. 1997). A further novelty of the interwar period was the arrival of American foundations (Tournès 2011). The Rockefeller Foundation promoted the "social sciences" in France and turned them into a tool of soft power, giving to the phrase a strong connotation of applied sciences, so that a new opposition between "*sciences sociales*" (applied) and "*sciences humaines*" (less strictly subordinated to political application) developed (Reubi 2020).

The political divisions present in the nineteenth century persisted. Many projects in the human sciences remained associated with republican agendas. Some of them were conservative in tendency, as with Charles Rist's *Institut de recherches économiques et sociales* [Institute for Economic and Social Research] (1933) (Tournès 2006), while others were more left-wing, as illustrated by the *Encyclopédie française* [The French Encyclopedia]. This publication, the first volume of which was published in 1935, was coordinated by the historian Lucien Febvre and by Anatole de Monzie, who was Minister of Education several times during the interwar period. Its editorial team brought together representatives of the scientific community and politicians and was a forum for discussion of the institutional reform of the sciences, including the humanities. The volume, *La civilisation écrite* [*The Writing Civilization*], which appeared in 1939, clearly established a connection between the promotion of the humanities, the republican ideal, and resistance to authoritarian regimes (Lucien Febvre et l'Encyclopédie française 2002; Reubi 2020). Other projects lent toward the extreme right, as the right became redefined by the rise of fascisms in the inter-war period. This was the case, for example, for George Montandon's "anthroposophy," which put physical and cultural anthropology at the core of the human sciences. Montandon was a physician from Geneva, who had studied ethnology and had joined the Swiss communist party before moving to Paris in the 1920s. In the early 1930s, he was granted the chair in ethnology at the *Ecole d'anthropologie de Paris* [Paris School of Anthropology], and his work then fell into the racialist tradition of French physical anthropology. At the end of the decade, he endorsed fierce anti-Semitism, and he contributed to the French anti-Semitic policy after 1940 (Reynaud-Paligot 2010).

From the 1930s onward, the French *"sciences de l'homme"* found new institutional foundations. There were new national institutions that defined the human sciences as a whole or divided them into several categories. The *Caisse nationale des sciences* [National Science Fund] was founded in 1930 with the purpose of awarding scholarships and grants to young scientists. It was divided into two sections, organizing the sciences into the "mathematical and experimental sciences" and the *"sciences humaines"* (Sonnet 2019). In 1933, the *Conseil supérieur de la recherche scientifique* [Higher Council for Scientific Research], a body designed to implement a proper policy of research, abandoned the expression and differentiated between the "historical and philological sciences," on the one hand, and the "philosophical and sociological sciences," on the other, embodying the competition mentioned above between *"sciences humaines"* and *"sciences sociales,"* and between history and philosophy. In 1935, the two agencies merged and the new *Caisse nationale de la recherche scientifique* [National Fund for Scientific Research] reverted to use of the label *"sciences humaines."* It not only supported existing research, but also helped found new laboratories in fields located at the interfaces of established disciplines, such as genetic editing (the critical edition of texts taking into account the different phases of writing and rewriting of the manuscript in order to understand the process of creation), for which the *Institut de recherche et d'histoire des textes* [Institute for Textual Research and History] was created in 1937.

The *Caisse nationale* became the *Centre national de la recherche scientifique* ([National Center for Scientific Research] CNRS) in 1939 (Guthleben 2009). It pioneered a research policy that is still partly relevant today, with the promotion of interdisciplinary collaboration and the provision of support to niche disciplines, whereas university curricula fostered disciplinary specialization. On the eve of the war, the CNRS divided its activities between "pure sciences" and "applied sciences," but after 1945, it reverted to the distinction between mathematical, physical-chemical, biological, and natural sciences, on the one hand, and *"sciences humaines,"* on the other. The *"sciences humaines"* thus became established on a long-term basis in France. They are now part of the *Institut des sciences humaines et sociales* [Institute of Humanities and Social Sciences] whose primary ambition is to "promote the emergence of new research topics, multi-disciplinarity within the humanities and social sciences, and inter-disciplinarity with other sciences" (https://inshs.cnrs.fr). Though sometimes more as a proclaimed ideal than as a reality, the label "*sciences humaines*" was also imposed on universities, when a 1958 decree renamed the old faculties of letters as "faculties of letters and human sciences." But the labels that were imposed with a strong orientation toward interdisciplinary collaboration still retained their ambiguity. This was well illustrated by the first *Maison des sciences de l'homme*, established in Paris in 1963. In the mind of its sponsors, this institution embraced a structuring interdisciplinary ideal (Bruhns et al. 2018; Tournès 2011), but, in reality, it was chaired by the historian Fernand Braudel for more than 20 years and served to effect the intellectual and institutional domination of history over the humanities and social sciences (Dosse 2010).

Although not easily translatable into English, the label "*sciences humaines*" was also adopted in postwar international institutions under the influence of French delegates. In 1945, the Preparatory Commission for the creation of UNESCO drafted a report that included a chapter devoted to "*sciences humaines*" ("humanistic studies" in the English version). In 1949, the International Council for Philosophy and "*sciences humaines*" (Humanistic Studies) was created, and in 1963, social sciences and humanities were combined into a single sector of "*sciences humaines et sociales*" ("social and human sciences") (Conil-Lacoste 1968).

Reflexivity

The success of the terms "*sciences humaines*" and "*sciences humaines et sociales*" in the 1960s did not eliminate scientific competition between disciplines and political tensions. The postwar history of these sciences was marked in France by two new factors, the academic institution of psychoanalysis and the rise of Marxism.

Psychoanalysis developed in France after the First World War, with well-known promoters such as Marie Bonaparte (Amouroux 2012) and as a clinical tool in child psychiatry. The first psychoanalytical associations were created during this period (*Société psychanalytique de Paris* [Paris Psychoanalytical Society], 1926). But a new step was taken after 1945, when psychoanalysis entered the universities. At the *Université de Paris* (the Sorbonne), Daniel Lagache made psychoanalysis important in the new psychology department created in 1947 (Ohayon 2006). After 1968 and the separation of the Sorbonne in many universities, a department of "*Sciences humaines cliniques*" [Clinical human sciences] was established at the *Université Paris 7* (founded 1969). Juliette Favez-Boutonnier strongly oriented it toward psychoanalysis (Ohayon 2006). At the same university, Jean Laplanche created the first diploma of advanced studies (DEA, *diplôme d'études approfondies*) in psychoanalysis in 1970, as well as an academic research laboratory, the *Centre de recherches en psychanalyse et pathopsychologie* [Research Center in Psychoanalysis and Pathopsychology]. Psychoanalytical tools became central in the training of most psychologists and were important in the theoretical research on pathopsychology. Jacques Lacan was a central figure in this postwar development of French psychoanalysis. His seminars, given between 1953 and 1979 at the Sainte-Anne Hospital in Paris, then at the *Ecole normale supérieure* and the Sorbonne, became a world-famous forum of discussion of Freudian concepts (Roudisnesco

2010). This academic platform gave psychoanalysis an unprecedented audience among students and the general public in France, as illustrated by the journal *Psyche* (1946–1963) founded by the journalist and feminist Maryse Choisy. It made psychoanalysis one of the most visible disciplines within the humanities, before the radical critique of the 2000s unsettled this prominent situation with the publication of *Le livre noir de la psychanalyse* [The black book of psychonalysis] (Catherine Meyer ed. 2005).

The postwar period also witnessed the rise of Marxism in France. Already between the two world wars, Marxism became a pole of interest for intellectuals of diverse background in search of a new unifying paradigm for humanities and social sciences. Such was the case, for example, of the historians Marc Bloch and Lucien Febvre, the founders of the journal *Les Annales* (Gouarné 2013). The cold war context polarized and radicalized the political tensions in the humanities and social sciences, as illustrated in the opposition between two leading "*intellectuels*" of the time Jean-Paul Sartre and Raymond Aron who became eponymous figures of the opposition between two ideological sides. No synthesis is available on this topic, and historians tend to speak of "marxisms" or "uses of Marx" in the plural form (see Barbe 2014). Marx' philosophy itself was subject to various interpretations and could even be split between the young and the older Marx' ideas, following Louis Althusser's analysis in his best-seller *Pour Marx* ([*For Marx*], 1965). Increasing references to Marx in all the human and social sciences could therefore point to different theoretical and political standpoints, depending on the emphasis put on a dialectical or on a materialist analysis of human and social phenomenon. Some scholars used Marx in a purely theoretical or philosophical manner. Others interpreted their endorsement of Marxism as the necessity to link intellectual and activist commitments. This trend however was divided into many groups, with various attitude toward Stalinism and the French Communist Party. The group and magazine *Socialisme ou barbarie* [Socialism or Barbary], for example, founded in 1948 (1949 for the magazine) by the philosophers and human scientists Cornelius Castoriadis and Claude Lefort, represented an anti-Stalinist Marxist trend. In the wake of the 1968 events in France, these theoretical and political tensions became more radical (Orain 2015).

The success of the terms "*sciences humaines*" and "*sciences humaines et sociales*" in the 1960s did not settle either the question of their unity. Is it a matter of similarity between the sciences themselves, united in the same category by the similarity of their methods, and thus appropriately classified together in an encyclopedic order of the sciences? Or does their unity lie in the unity of their object? Debating the balance between encyclopedism and humanism has marked the history of the human sciences since the end of the eighteenth century (Vidal 1999). The debate was reactivated in France in the critical, political, and philosophical context of the 1960s through confrontations that combined historiographical options and philosophical-political postures regarding the definition of man. It was most notably embodied in the conflicting points of view of Michel Foucault and Georges Gusdorf. The latter developed a long-term history that described the progressive establishment of the human sciences on the basis of the unity of their object, with modern humanism as their ideological core. In contrast, the former intentionally adopted a discontinuous perspective, which interpreted the discourses creating an object affirming the unity of "man"—humanism—and of the disciplines of knowledge about it, as an ideology in the Marxist sense of the term (Blanckaert 1999).

These debates, which were not and probably cannot be settled, have had the structuring effect of shedding light on reflexivity understood in a double sense. On the one hand, Foucault's perspective has underlined the structural relation between the scientific discourses and their objects. This represents the beginning of an epistemological reflection on what could create the unity of the humanities and social sciences: the existence of a reflexive circle between co-occurring discourses and objects. The contributions of Ian Hacking (1999) and Roger Smith (2007) are, in their respective ways, in line with these perspectives. On the other hand, competing historiographical options have emphasized the importance of studying the history in order to better understand the human and social sciences. Reflexivity in this case refers to the capacity of these disciplines to develop a critical view on themselves by intertwining historiographical, methodological, and epistemological considerations (Blondiaux and Richard 1999; Smith 1997).

In France, this line of inquiry has developed particularly as a result of the work carried out by the *Société française pour l'histoire des sciences de l'homme* [French Society for the History of Human Sciences]. Founded in 1986, this academic society has promoted collective research on topics bridging different disciplines, in a spirit of methodological open-mindedness. In the name of this association, the word "*pour*" had a double meaning. One aim of the *Société* was to resist overspecialization and to provide a platform for interdisciplinary cooperation on reflexive topics. Its other aim was to promote the history of human and social sciences as a professional activity, which could not be delegated solely to their practitioners, and had to be tackled with specific historical and epistemological tools (Blanckaert 1993; Blanckaert 2006a, b). Since 1986, following a similar agenda, there has been a seminar at the *Ecole des Hautes études en sciences sociales* (EHESS), and a scientific journal, *La Revue d'histoire des sciences humaines*, dedicated to the same topics (founded 1999, https://journals.openedition.org/rhsh/).

Conclusion

Gradually supplanted by the phrase "*sciences humaines et sociales,*" the expression "*science de l'homme*" is not very common in French today, but it has not totally disappeared. Its use is then a deliberate choice.

To refer to the "*science de l'homme*" entails to question the designations more frequently used nowadays and to emphasize that the current configuration of disciplines, as well as the institutions in which they are implemented, is a historically recent construction. At least until the interwar period, the disciplinary specialization that is now prevalent in French universities was not so fixed, and the boundaries between psychology and philosophy, sociology and ethnology, for example, remained porous. Choosing the expression "*science de l'homme*" thus makes it possible to look to the past while avoiding the "tunnel-vision"—i.e., the tendency to consider that the disciplines that exist today have always existed—denounced by Stefan Collini (Collini 1988).

The phrase, with its outdated character, thus underlines the historicity of the humanities and social sciences, and the importance of taking their history into account in order to understand their present state. This statement applies not only to France, but also to any other national context. However, the expression "*science de l'homme*" also points to the particularity of the history of disciplines that take man as their object in France. This chapter argues that the main reason for this particularity is political, an argument that, again, could be made about the history of the human sciences in other contexts.

References

Amouroux R (2012) Marie Bonaparte: entre Biologie et Freudisme [Marie Bonaparte. Between Biology and Freudism]. PUR, Rennes

Antoine-Mahut D (2020) The 'empowered king' of French spiritualism: Théodore Jouffroy. Br J Hist Philos 28(5):923–943

Audren F (2006) Naissances de la science sociale, 1750-1850 [The origins of social science]. Revue d'histoire des sciences humaines 15

Auroux S (1991) Histoire des idées linguistiques. II. Le développement de la grammaire occidentale [The history of linguistic ideas. II. The development of Western grammar]. Mardaga, Paris

Auroux S (2003) La naissance de la grammaire générale [The Birth of General grammar]. In: Groult M (ed) L'*Encyclopédie* ou la création des disciplines [the *Encyclopédie* or the creation of disciplines]. CNRS Editions, Paris, pp 213–223

Azouvi F (ed) (1992) L'Institution de la raison [The instituting of reason]. Vrin, Paris

Azouvi F (2000) Maine de Biran : la Science de l'homme [Maine de Biran. The Science of Man]. Vrin, Paris

Baker KM (1973) Politics and social science in eighteenth-century France: the society of 1789. In: Bosher JF (ed) French government and society, 1500–1850. The Athlone Press, London, pp 208–230

Barbe N (ed) (2014) « Sciences sociales et marxisme » [Social sciences and Marxism]. Le Portique 2014:32

Besse JM, Blais H, Surun I (eds) (2010) Naissances de la géographie moderne (1760–1860). Lieux, pratiques et formation des savoirs de l'espace [The beginnings of modern geography (1760–1860). Places, practices, and the making of spatial knowledge]. ENS, Paris

Biard A, Bourel D, Brian E (eds) (1997) Henri Berr et la culture du XXe siècle [Henri Berr and 20th Century Culture]. Albin Michel, Paris

Blais M-C (2007) La solidarité. Histoire d'une idée [Solidarity. The history of an idea]. Gallimard, Paris

Blais H, Laboulais I (2006) Géographies plurielles. Les sciences géographiques au moment de l'émergence des sciences humaines (1750–1850) [Plural Geographies. Geographical sciences at the time of the emergence of the human sciences]. L'Harmattan, Paris

Blanckaert C (1989) L'Anthropologie en France. Le mot et l'histoire [anthropology in France. The word and the story]. In: Blanckaert C, Ducros A, Hublin J-J (eds) Histoire de l'Anthropologie: hommes, Idées, moments [History of anthropology. People, ideas, moments], vol 3-4. Bulletins et Mémoires de la Société d'Anthropologie de Paris, pp 13–44

Blanckaert C (1993) La Société française pour l'histoire des sciences de l'homme. Bilan, enjeux et « questions vives » [The French Society for the History of Human Sciences. Overview, issues and "live questions"]. Genèses 10:124–135

Blanckaert C (1999) L'histoire générale des sciences de l'homme. Principes et périodisation [The general history of the human sciences. Principles and periodicization]. In: Blanckaert C, Blondiaux L, Loty L, Renneville M, Richard N (eds) L'histoire des sciences de l'homme. Trajectoire, enjeux et questions vives [The history of the human sciences. Trajectory, challenges and live questions]. L'Harmattan, Paris, pp 23–60

Blanckaert C (2000) 1800–Le moment « naturaliste » des sciences de l'homme [The "naturalist" moment of the sciences of man]. Revue d'Histoire des Sciences Humaines 2(3):117–160

Blanckaert C (2006a) "Notre immortel naturaliste": Buffon, la science de l'homme et l'écriture de l'histoire ["Our immortal naturalist". Buffon, the science of man and the writing of history]. In: Duchet M (ed) Buffon. De l'homme [on man]. L'Harmattan, Paris, pp 409–467

Blanckaert C (2006b) L'histoire des sciences de l'homme, Une culture au présent [The history of the human sciences, a present-oriented culture]. La revue pour l'histoire du CNRS 15. https://doi.org/10.4000/histoire-cnrs.529

Blanckaert C (2009) De la race à l'évolution. Paul Broca et l'anthropologie française (1850–1900) [From Race to Evolution. Paul Broca and French Anthropology]. L'Harmattan, Paris

Blondiaux L, Richard N (1999) A quoi sert l'histoire des sciences de l'homme ? [what is the purpose of the history of the human sciences?]. In: Blanckaert C, Blondiaux L, Loty L, Renneville M, Richard N (eds) L'histoire des sciences de l'homme. Trajectoire, enjeux et questions vives [The history of the human sciences. Trajectory, challenges and live questions]. L'Harmattan, Paris, pp 109–130

Boulad-Ayoub J (2004) La *Décade philosophique* comme système (1794–1807) [The *Décade philosophique* as a System], vol 9. PUR, Rennes

Bourdeau M, Pickering M, Schmaus W (eds) (2018) Love, order and Progress. The science, philosophy and politics of Auguste Comte. University of Pittsburgh Press

Bourdeau V, Macé A (2018) La nature du socialisme. Pensée sociale et conceptions de la nature au XIXe siècle [The nature of socialism. Social thought and conceptions of nature in the 19th century]. Presses universitaires de Franche-Comté, Besançon

Brooks J III (1998) The eclectic legacy: academic philosophy and the human sciences in nineteenth-century France. University of Delaware Press, Newark

Bruhns H, Nettelbeck J, Aymard M (eds) (2018) Clemens Heller, imprésario des sciences de l'homme [Clemens Heller, Impresario of the Human Sciences]. Éditions de la Maison des sciences de l'homme, Paris

Bruter A (2006) L'histoire enseignée et les sciences humaines au temps des humanités [Teaching history and the human sciences in the age of the humanities]. In: Pelus-Kaplan M-L (ed) Unité et globalité de l'homme [Unity and Globality of man]. Syllepse, Paris, pp 33–49

Carbone R (2007) Infini et science de l'homme: l'horizon et les paysages de l'anthropologie chez Malebranche [The infinite and the science of man. Horizon and landscapes of anthropology in Malebranche]. Vrin, Paris

Carbone R (2017) Medicina e scienza dell'uomo: Paul-Joseph Barthez e la Scuola di Montpellier [Medicine and the science of man. Paul-Joseph Barthez and the Montpellier school]. Università degli studi, Napoli Federico II, Napoli

Carroy J (2012) Nuits savantes. Une histoire des rêves (1800–1945) [Scholarly nights. A history of dreams]. EHESS, Paris

Carroy J, Richard N (eds) (2007) Alfred Maury, érudit et rêveur. Les sciences de l'homme au milieu du 19e siècle [Alfred Maury, scholar and dreamer. The sciences of man in mid-19th century]. Rennes, PUR

Chappey J-L (2002) La Société des Observateurs de l'Homme (1799–1804). Des anthropologues au temps de Bonaparte [The Society of the Observers of man (1799–1804). Anthropologists at the time of Bonaparte]. Société des études robespierristes, Paris

Chappey J-L (2006) De la science de l'homme aux sciences humaines: enjeux politiques d'une configuration de savoir (1770-1808) [From the science of man to the human sciences. Political issues in a knowledge configuration]. Revue d'Histoire des Sciences Humaines 2(15):43–68. https://doi.org/10.3917/rhsh.015.0043

Chappey J-L (2019) The *Société des Observateurs de l'homme* (1799–1804) and the circulation of state-related knowledge under Napoléon. In: Schilling L, Vogel J (eds) The transnational culture of expertise: circulating state related knowledge in the 18th and 19th centuries. DeGruyter, Berlin, pp 88–106

Chaubet F (2012) Gallimard et les sciences humaines: le tournant des années 1940 [Gallimard and the human sciences. The turn of the 1940's]. Histoire@Politique 2(2):112–129. https://doi.org/10.3917/hp.017.0112

Charle C (1990) Naissance des intellectuels (1880–1900) [The birth of the intellectual]. Editions de Minuit, Paris

Christie J (1993) The human sciences: origins and histories. Hist Hum Sci 6(1):1–12. https://doi.org/10.1177/095269519300600101

Citton Y, Dumasy L (eds) (2013) Le moment idéologique: Littérature et sciences de l'homme [The Ideologic Moment. Literature and the Sciences of Man]. ENS Éditions, Lyon. https://doi.org/10.4000/books.enseditions.2532

Collini S (1988) Discipline history and intellectual history, reflections on the history of social sciences in Britain and France. Rev Synth 109(3–4):387–399

Conil-Lacoste M (1968) Vingt ans d'activité de l'Unesco dans le domaine des sciences sociales [Twenty years of Unesco's activity in the field of the social sciences]. Rev Fr Sociol 9(3):390–404. https://doi.org/10.2307/3320564

Coursin R (2018) Brissot et la république en acte [Brissot and the republic in action]. La Révolution française 13. https://doi.org/10.4000/lrf.1894

Debaene V (2010) L'adieu au voyage. L'ethnologie française entre science et littérature [Farewell to travel. French Ethnology between Science and Literature]. Gallimard, Paris

Débuts des sciences de l'homme (Les) (1992) The beginnings of the sciences of man. Communication 54

Dosse F (2010 [1987]) L'Histoire en miettes. Des *Annales* à la « nouvelle histoire » [The shattered history. From the *Annales* to the « new history »]. La Découverte: Paris

Duchet M (1995) Anthropologie et histoire au siècle des Lumières [Anthropology and history in the enlightenment], 1st edn. Albin Michel, Paris, p 1971

Fabiani J-L (1988) Les philosophes de la République [The philosophers of the republic]. Editions de Minuit, Paris

Favreau J-F (2012) Vertige de l'écriture: Michel Foucault et la littérature (1954–1970) [The Vertigo of Writing. Michel Foucault and Literature]. ENS Éditions, Lyon

Fox C, Porter R, Wokler R (eds) (1995) Inventing human sciences. Eighteenth-century domain. University of California Press, Berkeley

Fox R, Weisz G (eds) (1982) The Organization of Science and Technology in France, 1808–1914. Cambridge University Press and Editions MSH, Paris

Fox R (2013) The savant and the state: science and cultural politics in nineteenth-century France. The Johns Hopkins University Press, Baltimore

Foucault M (1966) Les mots et les choses. Une archéologie des sciences humaines [The order of things. An archaeology of the human sciences]. Gallimard, Paris

Goetz R (1993) Destutt de Tracy. Philosophie du langage et science de l'homme [Destutt de Tracy. Philosophy of language and the science of man]. Droz, Genève

Goldstein J (2005) The post-revolutionary self. Politics and psyche in France. Harvard University Press, pp 1750–1850

Gouarné I (2013) L'introduction du marxisme en France. Philosoviétisme et sciences humaines (1920–1939) [The introduction of Marxism in France. Philosovietism and the Humanities (1920–1939)]. Presses universitaires de Rennes, Rennes.

Groult M (ed) (2003) L'*Encyclopédie* ou la création des disciplines [The *Encyclopédie* or the creation of disciplines]. CNRS Éditions, Paris

Guillin V (2017) The French influence. In: Macleod C, Miller DE (eds) A companion to mill. Wiley-Blackwell, Chichester, pp 126–141

Guillin V, Souafa D (2010) The reception of Stuart mill in France. *La Vie des idées*, https://booksandideas.net/The-Reception-of-John-Stuart-Mill-in-France.html

Heilbron J (2015) French sociology. Cornell University Press, Ithaca/London

Jones P (ed) (1989) The "science of man" in the Scottish enlightenment. Edinburgh University Press, Hume, Reid and their Contemporaries

Gauchet M, Swain G (2007) La pratique de l'esprit humain. L'Institution asilaire et la révolution démocratique [Madness and Democracy. The Modern Psychiatric Universe]. New Preface by M. Gauchet, Gallimard. Paris (1st ed 1980)

Gusdorf G (1960) Introduction aux sciences humaines. Essai critique Sur leurs origines et leur développement [Introduction to the human sciences. A critical essay on their origins and development]. Les Belles Lettres, Paris

Gusdorf G (1973) L'avènement des sciences humaines au siècle des Lumières [The advent of the human sciences in the age of enlightenment]. Payot, Paris (Les sciences humaines et la pensée occidentale [The Human Sciences and Western Thought], vol VI

Guthleben D (2009) Histoire du CNRS de 1939 à nos jours: Une ambition nationale pour la France [History of the CNRS from 1939 to the present. A National Ambition for France]. Armand Colin, Paris

Hacking I (1999) The social construction of what? Harvard University Press, Cambridge, MA

Kalaora B, Savoye A (1989) Les inventeurs oubliés. Le play et ses continuateurs aux origines des sciences sociales [The forgotten inventors. Le play and his followers at the origins of the social sciences]. Champ Vallon, Paris

Kaufmann L, Guilhaumou J (eds) (2003) L'Invention de la société. Nominalisme politique et science sociale au 18e siècle [The invention of society. Political nominalism and social science in the 18th century]. EHESS, Paris

Knee P (2014) L'expérience de la perte autour du moment 1800 [The experience of loss around the 1800 moment]. Voltaire Foundation, Oxford

Lepenies W (2002) Sainte-Beuve. Au seuil de la modernité [Sainte-Beuve. On the Threshold of Modernity]. Gallimard, Paris

Lucien Febvre et l'Encyclopédie française [Lucien Febvre and the Encyclopédie française] (2002) Cahiers Jaurès 163-164. www.cairn.info/revue-cahiers-jaures-2002-1.htm

Masseau D (2006) Un apologiste au service de l'*Encyclopédie méthodique*: Bergier et le dictionnaire de *Théologie* [an apologist at the service of the *Encyclopédie méthodique*. Bergier and the dictionary of *theology*]. In: Blanckaert C, Porret M (eds) *L'Encyclopédie méthodique* (1782–1832). Des Lumières au Positivisme [The *Encyclopédie méthodique*. From Enlightenments to Positivism]. Genève, Droz, pp 153–167

Mercier-Faivre A-M (1999) Un supplément à l'*Encyclopédie*: le monde primitif d'Antoine court de Gébelin [A supplement to the *Encyclopédie*. The primitive world of Antoine court de Gébelin]. H. Champion, Paris

Meyer C (ed) (2005) Le Livre noir de la psychanalyse. Vivre, penser et aller mieux sans Freud [The Black Book of Psychoanalysis. Living, thinking and feeling better without Freud]. Les Arènes, Paris.

Moravia S (1970) La scienza dell'uomo nel Settecento [The science of man in the 18th century]. Laterza, Bari

Moravia S (1974) Il pensiero degli Idéologues [The doctrine of the ideologues]. La nuova Italia, Firenze

Moreau E (2020) À propos de Louis de Bonald et de sa sociologie [On Louis de Bonald and his Sociology]. Sociétés 4(4):139–150. https://doi.org/10.3917/soc.150.0139

Mosbah-Natanson S (2017) Une "mode" de la sociologie Publications et vocations sociologiques en France en 1900 [The "Fashion" of Sociology. Sociological publications and vocations in France in 1900]. Classiques Garnier, Paris

Mucchielli L (1998) La découverte du social. Histoire de la sociologie en France (1870–1914) [The discovery of the social. The history of sociology in France]. La Découverte, Paris

Noiriel G (1990) Naissance du métier d'historien [The birth of the Historian's profession]. Genèses 1:58–85. https://doi.org/10.3406/genes.1990.1014

Ohayon A (2006 [1999]) Psychologie et psychanalyse en France. L'impossible rencontre (1919–1969)[Psychology and Psychoanalysis in France. The Impossible Encounter]. Paris: La Découverte

Orain O (ed) (2015) Les "années 68" des sciences humaines et sociales [The "68s" of the human and social sciences]. Revue d'histoire des sciences humaines 2015:26

Ory P, Sirinelli J-F (1986) Les intellectuels en France, de l'affaire Dreyfus à nos jours [Intellectuals in France, from the Dreyfus affair to the present day]. A Colin:Paris

Petit A (2003) Auguste Comte. Trajectoires positivistes (1798–1998) [Auguste Comte. Positivist pathways]. L'Harmattan, Paris

Priest RD (2015) The gospel according to Renan. Reading, writing, and religion in nineteenth-century France, Oxford U. P

Renneville M (2020) Le Langage des crânes: Une histoire de la phrénologie [The language of skulls. A history of phrenology], 1st edn. La Découverte, Paris

Reubi S (2020) À quoi sert l'organisation des sciences ? L'émergence de la catégorie "sciences humaines" en français (1880-1950) [What is the purpose of the Organization of Sciences? The emergence of the category "sciences humaines" in French]. Revue d'histoire des sciences humaines 37. https://doi.org/10.4000/rhsh.5286

Reynaud-Paligot C (2010) L'émergence de l'antisémitisme scientifique chez les anthropologues français [The Emergence of Scientific Anti-Semitism among French Anthropologists]. Archives Juives 1(1):66–76. https://doi.org/10.3917/aj.431.0066

Richard N (2013) Hippolyte Taine. Histoire, psychologie, littérature [Hippolyte Taine. History, psychology, literature]. Classiques Garnier, Paris

Richard N (2015) La *vie de Jésus* de Renan. La fabrique d'un best-seller [Renan's life of Jesus. The making of a bestseller]. Presses universitaires de Rennes, Rennes

Rignol L (2014) Les Hiéroglyphes de la nature. Le socialisme scientifique en France dans le premier XIXe siècle [The hieroglyphs of nature. Scientific socialism in France in the first 19th century]. Les presses du réel, Dijon

Roudinesco E (2010 [1982–1986]) Histoire de la psychanalyse en France [History of psychoanalysis in France]. Paris: Le livre de Poche

Sapiro G (2004) Défense et illustration de "l'honnête homme": les hommes de lettres contre la sociologie [Defense and Illustration of the Honest Man. Men of Letters against Sociology]. Actes de la recherche en sciences sociales 153:11–27

Smith R (1997) The Fontana history of the human sciences. Fontana Press, London

Smith R (2007) Being human: historical knowledge and the creation of human nature. Columbia University Press, New York

Sonnet M (2019) Faire de la recherche son métier? Les « sciences humaines » à la Caisse nationale des sciences (1930–1939) [Making Research one' s Profession? The "Human Sciences" at the Caisse nationale des sciences]. Revue d'histoire des sciences humaines 34:125–154. https://doi.org/10.4000/rhsh.3158

Staum MS (1980) Cabanis. Enlightenment and medical philosophy in the French revolution. Princeton University Press

Tournès L (2006) L'Institut scientifique de recherches économiques et sociales et les débuts de l'expertise économique en France (1933-1940) [The scientific Institute for Economic and Social Research and the beginnings of economic expertise in France]. Genèses 4(4):49–70. https://doi.org/10.3917/gen.065.0049

Tournès L (2011) Sciences de l'homme et politique. Les fondations philanthropiques américaines en France au XXe siècle [The human sciences and politics. American philanthropic foundations in France in the 20th century]. Garnier, Paris

Vermeren P (1995) Victor cousin. Le jeu de la Philosophie et de l'État [Victor cousin. The interplay of philosophy and the state]. L'Harmattan, Paris

Vidal F (1999) La science de l'homme: désirs d'unité et juxtapositions encyclopédiques [The science of man. Aspirations for Unity and Encyclopedic juxtapositions]. In: Blanckaert C, Blondiaux L, Loty L, Renneville M, Richard N (eds) L'Histoire des sciences de l'homme: Trajectoire, enjeux et questions vives [The history of the human sciences. Trajectory, challenges and live questions]. L'Harmattan, Paris, pp 61–78

Vidal F (2006) Les sciences de l'âme. XVIe-XVIIIe siècle [The sciences of the soul. 16th–18th century]. Honoré Champion, Paris

Vigarello G (2016) Le sentiment de soi. Histoire de la perception du corps [The sense of self. History of the perception of the body]. Seuil, Paris

Williams EA (1994) The physical and the moral. Anthropology, physiology, and philosophical medicine in France. Cambridge University Press, pp 1750–1850

Zékian S (2013) Les enjeux littéraires de la science de l'homme : Bonald et Cabanis dans la guerre des sciences et des lettres [The literary stakes of the science of man: Bonald and Cabanis in the war between sciences and letters]. In: Citton Y, Dumasy L (eds) Le moment idéologique: Littérature et sciences de l'homme [The Ideologic moment. Literature and the sciences of man]. ENS Éditions, Lyon, pp 47–67

Narrative and the Human Sciences

Kim M. Hajek

Contents

Introduction	80
Science, Narrative, and Historiography	82
Texts, Things, and Inclusiveness	86
Narrative, Knowledge, and Self	90
Narratology	93
Narrative in Practice	97
Conclusion	103
References	104

Abstract

The relevance of narrative to the history of the human sciences (HHS) is undeniable. It is equally significant, however, that scholars can mean a number of very different things when they write about narrative. As a consequence, the sense of "narrative" in HHS (and beyond) has expanded enormously to the point where it risks becoming a trivial or meaningless term of analysis. This chapter unpacks the various uses of "narrative" that appear in scholarly work, and it relates them to "everyday" notions of narrative as storytelling or as an overarching position on some topic (e.g., the narrative of climate change). The chapter provides a conceptual guide to ways narrative appears on three entangled levels of HHS enquiry: first, in reflections on methodology, both of historical work itself, as well as of the human sciences about which HHS researchers write; second, in theorizations of narrative as it has been disciplined as an object of study in several sciences, notably psychology and narratology; and third, in textual studies of where and how narrative is part of the practice of human-scientific activity, and how those textual forms have changed. The three levels are clearly interlinked: A historian of psychology might write a narrative about the use of narrative forms of therapy

K. M. Hajek (✉)
Institute for History, Leiden University, Leiden, The Netherlands
e-mail: k.m.hajek@hum.leidenuniv.nl

in the history of psychology, for instance, and make use of narratological concepts to analyze textual features of resulting case histories. Attention to narrative at each level can shed light on what it means to make knowledge about the human. This chapter argues that HHS work will be most productive when different meanings of narrative are distinguished with greater precision. Undertaking such analysis, finally, reveals the relative sparsity to date of careful textual analysis of what human scientists write.

Keywords

Narrative · Narratology · Anthropology · Hayden White · Psychoanalysis · Medical humanities · Storytelling · Case history

Introduction

Narrative has an important place in the history of the human sciences (HHS). Few scholars would disagree with this statement; some would call the place of narrative essential or inescapable, others might add that it is controversial. Where the disagreement begins is over what we mean by narrative, and, correspondingly, how we construe its place in the history of the human sciences. For just as the grouping of the human sciences is itself "loose and variable" (Smith), so too has narrative become a fluid and capacious term, particularly since the so-called "linguistic" and "narrative" turns. But where Roger Smith (Chap. 1, "What Is the History of the Human Sciences?" to this volume) explains persuasively how it can be productive for scholarship to leave membership of the human sciences somewhat open, the same is not the case for narrative. Unless we delineate clearly what we mean by "narrative" in different areas of enquiry, the term risks becoming so broad and diffuse as to lose its epistemic cogency. That would be a loss for historical study of attempts to understand humanity, as well as for ongoing scientific work in this area, and indeed it is nothing especially new to deplore "hyperextension" or "trivialisation" of "narrative" in scholarship (e.g., Herman 1998; Ryan 2007; Woods 2011; Carrard 2015; Beatty 2017, pp. 38–39). Narrative – taken variously as a subject of methodological reflection, an analytical category, and an object of study of the human sciences – still has a great deal to offer in elucidating what it means to generate knowledge of the human, and how specific cultural and historical formations play into such efforts. There is notably space to explore further how writing as a scientific activity – with frequent narrative elements – informs what human scientists know and how they know it. This chapter aims to provide the clarity needed to ensure that narrative can continue to have an important place in HHS scholarship, by disentangling how narrative has been construed in past human-scientific enquiry and historical study of these sciences.

Conceptual and terminological precision are the focus, therefore, since they can help us avoid conflating or confusing distinct ways that story-like representations inflect scientific knowing about "human nature." In other words, this chapter serves only secondarily as a bibliographical resource; its primary contribution is as a conceptual guide to the range of meanings and uses of narrative in the history of the human sciences. Certainly, space limitations prevent any truly comprehensive survey of narrative in HHS, even if that were the aim. But it follows that much exciting and relevant scholarship will necessarily be absent from these pages. The chapter's scope broadly follows the contours of HHS as an intellectual grouping, meaning that it is concerned, on the one hand, with ideas and activities from disciplines that aim to produce knowledge about the human, and on the other hand, with historical study practiced as a scholarly endeavor. Following the lead of Roger Smith's introduction, this chapter makes no attempt to delimit the range of the human sciences, but takes examples somewhat eclectically from disciplines including psychology, archeology, and anthropology. These examples tend to fall temporally in the modern period, especially dating from the second half of the nineteenth century or later. Thinkers and contexts are brought together here for the light they can shed on narrative in HHS; that should not be taken to imply there was any real historical engagement between them. And reciprocally, other lines of influence or dialogue are set aside in these pages. There is also no discussion of science popularization or "vulgarisation," neither to the broad nonspecialist public, nor to funders or policy makers, although it should be evident that storytelling plays an important role in these efforts (examined in several chapters in Fludernik and Ryan 2020; Dillon and Craig 2022). The emphasis lies rather on the roles of narrative in knowledge-making and scholarly reflections on those roles.

Everyday senses of "narrative" do, however, run through HHS, adding to the complexity of the subject. Most often, "narrative" enters everyday language as a high-register way to denote storytelling in all its forms, with the additional connotation that the story is likely somewhat fictional (cf Cohn 1999). A second common usage links narrative to an overarching position on some issue, often one that is contested – for example, news media might ask who controls the narrative on COVID-19 vaccinations in a given country. These two common senses of narrative will appear as recurring threads through this chapter. The chapter's structure, however, unpacks the intersection of narrative and HHS on three levels. The first level considers narrative as it enters into the methodological preoccupations of human sciences, beginning with history, because HHS scholarship is by definition historical, and indeed, it is often assumed that producing a history of something is equivalent to producing a narrative of or about it. Questions of narrative and method are primarily normative in nature, both when researchers view narrative as ideologically or epistemologically suspect, and when they valorize its inclusiveness. Taking historiography as the starting point in section "Science, Narrative, and Historiography," the chapter examines issues of narrative and methodology in sections "Science, Narrative, and Historiography" and "Texts, Things, and Inclusiveness," with the focus shifting in the latter section to disciplines of anthropology, archeology, and medicine. Next, several human-scientific disciplines have taken narrative as an object of study; this is the second level on which narrative and HHS intersect.

Notably, psychological enquiry has implicated narrative in the functioning of human cognition (section "Narrative, Knowledge, and Self"), while narratology has theorized the textual detail of narrative in technical terms (section "Narratology"). Tools provided by narratology, as well as by literary and rhetorical studies, play a role in the third level explored here (in section "Narrative in Practice"): narrative forms as they have participated in the practices of various human sciences through the past centuries. Research on this level has been concentrated especially around case-based modes of knowing, such as those in medicine, psychoanalysis, and psychology.

By disentangling these three levels, the chapter proposes to help orient fellow scholars within the complexity of multiple approaches to narrative, and to encourage greater conceptual and terminological clarity. For we can encounter unproductive and recursive loops of narrative, if we are not careful with definitions. A researcher in HHS might first consider whether what she writes – i.e., history of some kind – is or should be narrative in form. Her concerns might mirror those of the historical actors she considers, if history itself is studied as a human science, asking "what is the history of the way historians have used narrative"? Our issues loop around again if the human science studied takes narrative as an object of enquiry, such as in cognitive psychology or narratology: Can we do a narrative analysis of the way narratologists have used narrative in their writing? To talk unreflectively about narrative could be to evoke any one of these levels, either of the everyday senses, or a more technical conception emerging from a discipline like narratology. The potential for confusion or scholars "talking past" one another is vast, though the confusion can also be negated – that is the goal of this chapter.

What matters is to avoid conflating or confusing distinct ways that story-like representations inflect scientific knowing about "human nature." We should be very hesitant to sketch broad conclusions about narrative as some monolithic entity, just as we, as HHS scholars, privilege attention to perspective and context over characterizing the sciences we study in monolithic terms. That may mean making less frequent use of "narrative" in its everyday senses, and limiting recourse to the term to more technical contexts. It would also be beneficial to examine more carefully the ways narrative has entered in practice into human-scientific activity, not only for what this could tell us about intersections between writing and knowing, but also to enrich ongoing endeavors to discipline narrative under the aegis of factual narratology and cognitive psychology.

Science, Narrative, and Historiography

History-writing is a logical place to begin, since this is an activity common to all researchers in HHS, as well as being a preoccupation of those who study history as a human-scientific discipline. A common assumption – and one which still persists despite empirical evidence to the contrary (Carrard 2015, 2017/2013) – is that history and narrative are largely interchangeable terms; that is, historians are held primarily to produce narratives about the past. Lexical similarities between "history" and "story" reinforce this view and invoke conceptual associations between the two.

(The two terms are even more closely tied in French, where the difference is often marked only by a capital letter: *Histoire* against *histoire*.) Furthermore, stories are widely supposed to have a privileged place in human cognition (see section "Narrative, Knowledge, and Self"), especially when expressed in natural language (Wise 2022; Caracciolo 2020); that much historical scholarship is accessible to educated nonspecialists does nothing to trouble "narrativist" coupling of history and narrative. Although such "narrativism" has its own historicity (see Carrard 2015 for the Francophone case), what matter for our purposes are the ideological and epistemological implications for history as a field of systematic knowledge about the past (as a *science* in the French sense, or *Geisteswissenschaft* in German). We are dealing with two interlinked questions. Put most basically: Assuming historiography produces narratives, can narrative serve to generate properly scientific knowledge? Should, therefore, historians use narrative (or continue to do so) in their work?

Answers to these questions usually fall under the field of philosophy of history and, as such, might seem remote from the routine activities of HHS scholarship. Asking about the value of narrative for making systematic knowledge (i.e., for science), however, is clearly a matter which bears upon human sciences beyond history itself, even if it has chiefly been framed in those disciplinary terms. Two major debates from the second half of the twentieth century continue to inform how researchers think about narrative, science, and history (e.g., Morgan and Wise 2017). They thus merit examination here in their main lines, even if specialists might point out that philosophical enquiry has moved on (e.g., Carrard 2015).

At issue, first, is the epistemological validity of narrative – initially taken to mean historiographical narratives. Can narratives provide a properly scientific explanation for their subject matter, or do only law-like findings, such as emerged from the physical sciences, count as explanations for science? CG Hempel is one figure strongly associated with a view of narrative as scientifically inadequate; for him, narratives provide only "explanation sketches" rather than explanations properly constituted. Under this account, proper explanation proceeds by deductive-nomological methods, insofar as they produce generally applicable laws – the exemplar of explanation here is obviously the major laws of classical physics. Hempel and his like were strongly opposed by philosophers of history from the 1960s and 1970s; key thinkers include Arthur Danto, Louis Mink, William Gallie, and Paul Ricoeur (lucid overview in Carrard 2015; see also discussions in Beatty 2017; Wise 2017). These scholars argued variously for the autonomy of history as a mode of knowledge-making and, in the process, valorized narrative by claiming that certain of its characteristics are essential to explaining human events. Gallie advocated the importance of "followability," Mink of "synoptic judgement," and Ricoeur of human experiences of time (see Wise 2017). Their work might be said to have vanquished the Hempelian stance on narrative and explanation, or perhaps more precisely, engendered a move away from questioning what distinguishes natural sciences and "opposing" forms of enquiry like history.

Nevertheless, binaries opposing (natural) science and history, or laws and narrative, have persisted into the twenty-first century as something of a cultural commonplace, notably for professional scientists, but also for historians of science.

Scholarship in history and philosophy of science cannot ignore the influential views of certain scientists that perpetuate such binaries. One well-known figure is paleobiologist Stephen J. Gould. He invoked the science-narrative opposition to critique what he conceived as the speculative "storytelling" prevalent in scientific disciplines such as sociobiology. Gould's dismissal of narrative for the purposes of science is already evident in his terminology; he refers to "storytelling" and "just-so stories," rather than "narrative." Under this logic, narrative explanations or arguments are marked as scientifically invalid by association with the commonplace and fictional, by implicit contrast with technically rigorous scientific accounts. Paula Olmos' (2022) careful analysis reveals, however, that Gouldian "just-so" criticism is not necessarily directed to narrative as a discursive scientific form, but may be applied to many forms of reasoning present in science. Importantly, Olmos draws these conclusions by engaging in depth with precisely the philosophical premises mobilized by Gould. Her work can thus help decouple the twin assumptions underlying Hempel's and Gould's ideas: that history and narrative are broadly synonymous, and that both are less "scientific" than natural or social sciences. For as Roger Smith outlines in his introductory chapter, historical study seemingly continues to need such persuasive advocates, if HHS is to engage in meaningful dialogue with professional human scientists. Smith indeed serves as one such advocate in his introduction, from a broader perspective than that of narrative.

The second persistent debate about narrative and history has chiefly exercised historians, and additionally some scholars from "humanities" disciplines. Again, narrative is viewed with a certain suspicion, though here, ideological issues are linked to historians' epistemological concerns. The worry is that narrative is artificially coherent in a way that "real" events are not, such that to write in narrative discourse about the past is inevitably to encode some ideological stance on the events in question. In the words of Hayden White (1987), whose work inescapably dominates on this question, the narrative "form" implies a certain ideological and moral "content," usually one which reinforces the authority of a (nonprogressive) status quo. Whereas "real" events manifest discontinuously and heterogeneously, the narrative discursive form imposes an entirely constructed "coherence, integrity, fullness, and closure" (White 1987 p. 24) on whichever events a historian happens to select for inclusion in the narrative. Narrative is not necessarily problematic in itself, for White, although his views on the subject changed over time. In an earlier work (White 1973), he proposed that scholars of differing schools tend to "emplot" historical events under one of four esthetic modes: tragedy, comedy, romance, and satire. The "formal coherency" of narrative poses an epistemic risk when historians do not signal the moral implications of their narrative choices (or perhaps are unaware of them), but pretend to an objectivity which "feigns to make the world speak itself and speak itself as a story" (White 1987, p. 2). Underlying White's concerns about historiography is the narrativist position we met above, which assumes historical output to be predominantly narrative in discursive form and also fundamentally narrative in structure. And it was properly an assumption, or at least a premise on White's part; his analyses are primarily rhetorical rather than narrative, and limited to short excerpts of exemplary texts – the purpose is normative not descriptive.

At stake, under the logic of this suspicion, is what it means to represent the real, and what to make of ambitions to do so in a so-called scientific and objective way (Dean 2019). For all White's scorn for nineteenth-century realism (admittedly Germanic and rather naïve), these were also of concern for literary writers and critics of the period. In a letter of 1864, Emile Zola famously wrote on the "window" opened between art and the world, in which sat "a sort of transparent screen." Developing the metaphor, he outlined the predominant qualities of the "screen" employed by a number of esthetic movements (classicism, romanticism, and realism), all of which nonetheless modified reality as it passed (quoted in Becker 2005, my translation). White indeed conceived history and fiction as fundamentally of the same order; when we return in section "Narratology" to questions of factual versus fictional narrative, we might notice a number of points of connection, if not direct engagement, between White's thinking and the contemporaneous intellectual project of narratology (more discussion in Puckett 2016).

Strikingly, in HHS scholarship, epistemic unease around the supposed transparency and value-neutrality of extended prose is not always couched in terms of narrative. Historian of psychology Jill Morawski (2016) has critiqued selectivity and ontological presuppositions encoded into S Milgram's obedience studies. The problems Morawski raises share much with White's positions, including a concern for the political project of the human science in question (psychology). Morawski, however, writes in terms of "description," in contrast to White's "narrative," and draws upon notions of "thick" and "thin" description in science (as developed by Ted Porter (2012), following Clifford Geertz). One might wonder how much these respective arguments would change if "description" and "narrative" were swapped, as in Mat Paskins' (2022) study of "thin" chemical narratives. Does the (somewhat) more assured scientific status of psychology promote use of the more neutral "description"? This is not to disparage Morawski's pertinent analysis, which was no doubt framed by the extrascientific factor that her article appeared as part of a special issue on description in science. Rather, to think through what changes in Morawski's or White's positions by swapping "narrative" and "description" is further to see ways that references to "narrative" in HHS may be polyvalent or ambiguous, especially when bound up with assumptions about where narrative is to be found. Experiment reports in modern psychology journals would seem equally apt as historiography to appear in properly narrative form, to the extent that both give accounts of human actions in time. Similarly, both discourse types contain descriptive passages. (On modern scientific journal articles as narrative, see Meunier 2022; Jajdelska 2022; also Bazerman 1988.) A more technical textual study might distinguish between narrative and description in scientific documents, using tools from narratology. Where polyvalence in language remains, however, it leaves HHS work on narrative open to distortion or disparagement, especially coupled with perceptions of narrative as biased or less scientifically valid. Hence this chapter calls for greater precision and clarity when dealing with narrative in HHS, and for restraint in using the term, if "narrative" is meant more as a synonym for "prose," "history," or simply "position" (on some question).

It is worth noting at this point just how common it is for HHS scholars – like historians generally – to use "narrative" to refer to overarching accounts of change in some central condition or central subject over time. Such accounts count straightforwardly as narratives by everyday standards, although they have low narrativity in the technical sense employed by narratology, given their necessarily limited complexity (see below section "Narratology"). On the one hand, this usage of narrative relates to the grand narratives or metanarratives (in Lyotard's terms) of nineteenth-century philosophies of history, which broadly tell "the story of humanity as the greatest historical narrative of history" (Simon 2019; also Smith introduction). The subject of such grand narratives is not only broad, but also unitary, their temporal scope wide, if not ahistorical, while their bold claims tend to bear political or moral charge. A prime example would be Auguste Comte's three stages of human development, though of course Comte's scientific aspirations led him to label his account a "law." In other words, grand narratives are precisely the object of twentieth-century critical disillusionment and dismissal – of the kind enunciated by White and others (including Lyotard). Contemporary researchers sometimes employ "narrative" self-consciously in these terms, as BZ Simon (2019) does when interrogating how notions of posthumanity tend to universalize the "humanity" that is the subject of progressist histories. Other times, as in several chapters in the recent book *Analysing Historical Narratives* (Berger et al. 2021), the status of the "narratives" in question seems to hesitate between grand narrative and simply a "framing of the past" (which is perhaps not coincidentally the book's subtitle). This latter usage is, on the other hand, encountered frequently in scholarship, with "narrative" standing as shorthand for any large claims or received views within a given field. For instance, M Braat and coauthors (2020) contrast "the narrative" about behaviorism – its supposed dominance in interwar American psychology – with the results of a bibliometric analysis. Note the definite article here: *the* narrative, not *a* narrative. Often the point is to set up a (singular) position on some historiographical topic which the HHS scholar will oppose or, more usually, complicate. Against *the* narrative invoked, HHS researchers tend to call for a multiplicity of narratives and perspectives – plural instead of singular. Indeed, this is precisely what Braat et al. conclude: the need for "a more diverse and subtler story" (2020, p. 277). If the rise and fall of behaviorism in American psychology qualifies minimally as a narrative in technical terms, one could equally well refer to "the received view" or the "usual claim/s." Here most especially, taking pains to avoid "narrative" – or specify "grand/meta narrative," if appropriate – would not be a particularly demanding move for HHS researchers, yet it could reduce pernicious conceptual broadening of the notion of narrative.

Texts, Things, and Inclusiveness

Beyond history, with its metalevel interest for HHS scholars, other human-scientific domains have reflected on the methodological appropriateness of narrative. The stakes here, as for historiography, are primarily epistemological – is narrative a valid way to generate knowledge about a given disciplinary object? – but do also sometimes include explicit ethical aspects. In general, a positivist-inspired notion of science as experimental and nomological is set against the supposed imprecision, subjectivity, and artificiality of narrative. Alternatively, narrative is reclaimed for its capacity to weave together heterogeneous elements, to incorporate context and singularity (Morgan and Wise 2017), or for promoting inclusion and experientiality, in contrast to a scientism which portrays human subjects as mere objects or numbers. The remainder of this section explicates these broad themes within three distinct contexts of reflection on narrative: the explicit antinarrativity and antihistoricism of nineteenth-century German anthropology; twenty-first-century calls for greater balance in archeology between narrative and nonnarrative methods; and the contemporary movement to reintroduce narrative into clinical medicine.

As Smith evokes in his introduction, many of the core human sciences participated in a shift in accounts of human development around the end of the nineteenth and early twentieth centuries. In place of speculative grand narratives about humanity, based on canonical European texts, researchers advocated for empirical study of humans. Their key epistemic values included objectivity, experimentation, and (Bernardian) positivism. This was the era of what Daston and Galison (2007) have called "mechanical objectivity," and of expansion of experimental methods from the physical sciences into physiology, psychology, and beyond. Anthropology in the German-speaking sphere was one budding discipline to renounce narrative with particular thoroughness and zeal. Figures like Adolf Bastian, Rudolf Virchow, or Robert Hartmann posed the same kinds of questions about human nature and cultural development as the historico-philosophical work of Hegel or Ranke but aspired to do so using the methods of the natural sciences (Zimmerman 2001, esp. Chap. 2). As Andrew Zimmerman outlines in his excellent and still-topical study (on which this outline draws heavily), this version of anthropology offered "the possibility of a nonnarrative, natural science of humanity" (2001, 183).

Anthropologists enacted a double rejection of narrative texts, both as object of study and as form of communication. Instead of analyzing the written production of great European thinkers, they focused on collecting objects from so-called "natural peoples" (*Naturvölker*, i.e., usually colonized groups). Once again, the underlying motivation was to access reality – in this case of human nature – more directly and objectively. Since "natural peoples" were thought to have no culture or history, their bodies and artifacts were considered to reveal the human without any distorting mediation. Even the languages of non-Europeans were potentially suspect, insofar as language can be learned or taught. The tangible objects of "natural peoples," in contrast, were held to stand outside time, outside history, and as such, seemed a fixed point anchoring the "reality" of (European) human nature. With these views, German anthropology redefined "the relation of time to the human sciences" (Zimmerman 2001, p. 59). Unlike their counterparts in England or France, German anthropologists repudiated evolutionary perspectives on the relations between "primitive" and European peoples; change in time simply had no place in their aspiring science.

This concentration on objects situated outside time was doubled by attempts to devise antinarrative forms through which to disseminate anthropological findings. One such was the questionnaire, which served to collect together specific observations of heterogeneous aspects of natural peoples, ideally supplemented by a collection of physical objects. Published without further interpretation, the questionnaire had a form that explicitly blocked narrative, as the author was "continually stopping and starting with each new question ... rather than spinning a yarn to connect the details" (Zimmerman 2001, p. 55). In a similar vein, Bastian conceived his publications as equivalent to an encyclopedia, with elements bearing on a given topic only brought together in the index. He even experimented with sentence structures that resembled collections of disparate details far more than any conventional grammatical utterance – to the detriment of readers, even experts, who struggled to comprehend Bastian's ideas. Indeed, that was precisely the point: to present the artifact collected with as little interpretative distortion as possible on the part of the author.

Bastian's antinarrative writing, like German anthropology's peculiar relation to time, is no doubt an outlier in the history of attempts to purify knowledge of the human from the supposed pitfalls of narrative. Their motivating epistemic considerations run through human-scientific reflection on methods more generally, however: words versus things (or numbers); connecting versus fragmenting; inclusion versus exclusion; and author interpretation versus reader reconstruction. They were sufficiently present in cultivated discourse to form the subject of one of Anatole France's short stories, "Monsieur Pigeonneau" in 1887 (see Hajek 2017). The story functions as a commentary on the progress-oriented work of scientific purification in play in two areas of human-scientific research: hypnotism and archeology. Serious archeologist Pigeonneau is diverted from his efforts to study one single object and to avoid any generalization; once hypnotized, he "poetizes" and generalizes from his artifact and finishes by writing not just any narrative, not even derided history, but a fairy-tale (*conte*), surely the epitome of dangerous imaginative speculation. Pigeonneau might be the "dupe" of the duo who hypnotize him, but was he not already the "dupe" of sterile antinarrativity? Alongside HHS scholars, literary authors continue to interrogate relations between "novelistic fiction," "historical truth," and how each might be expressed (e.g., Binet 2009).

If archeology attempted to purge its activities of narrative in the nineteenth century, in the twenty-first, a number of thinkers have called for narrative to play a greater role in knowledge-making about prehistory. Time and connectivity are key epistemic factors here. Archeologist Anne Teather's (2022) concern is the way archeologists of the Neolithic employ sophisticated dating practices (radiocarbon dating, etc.). Often combined with Bayesian statistical methods, this kind of "absolute" or "direct" dating offers a precision and objectivity that resemble the methods of the natural sciences. The attraction of absolute dating methods can lead, however, to what Teather terms "reductive narratives": accounts which only situate the occurrence of a particular kind of Neolithic structure (e.g., a causewayed enclosure) in time. These function like Hayden White's "chronicles," contends Teather, and are therefore less productive of rich knowledge about prehistory than apparently less-scientific archeological narratives. The latter are "built from aggregations of evidence from many excavated sites"; relative analyses of each kind of object ("genealogies") are woven together with attention to context to produce what might properly be called a "narrative." Against White's argument, in effect, Teather revalorizes the coherence and fullness offered by narrative. Narrative opens up rich possibilities for research, in this view, whereas it is the apparently more objective dating projects which tend to achieve an artificial closure. Referring back to the categories at stake in German anthropology, the approach Teather espouses is one that privileges connecting objects, interpretation, and numbers, over potential fragmentation of knowledge into artifact-specific chronologies.

Another call to renarrativize archeology steps back from fieldwork to focus on archeologists' writing practices. Here, the issue is explicitly how to generate more ontologically inclusive knowledge of the distant past – i.e., knowledge that is less anthropocentric – without having to abandon the human subject as a key narrative actor (Ribeiro 2018). Using the philosophical notion of "intentionality," Artur Ribeiro (2018) essentially argues for a rehistoricization of archeological writing that figures human subjects, in parallel with objects, as active agents informed by the specificity of their temporal development. Such historicity (akin to a metanarrative) provides a "context of intelligibility" for human social action. The resultant knowledge claims – narratives – would then likely possess a greater degree of narrativity than the current practice of attributing temporal change to abstract "forces," although Ribeiro does not put it in these terms.

Including the human subject as a properly subjective agent in knowledge-making is also precisely the ambition of the movement known as "narrative medicine." Where twenty-first-century clinical medicine is usually considered more a natural or biological science than a human science, "narrative medicine" proposes quite properly to humanize it. The stakes here are more ethical than epistemic. Advocates like Rita Charon (2006) focus on improving the patient's experience of healthcare by training medical practitioners in "narrative competence." Practitioners are supposed to approach patient self-narratives (of complaints, fears, and experiences) with greater openness than the usual "evidence-based" perspective. Regardless of what narratives (if any) are actually exchanged, "narrative medicine" aims to foster qualities usually associated with (literary) narratives, such as inclusion of different perspectives, reflexivity, and experientiality (Charon 2006). Clearly, narrative medicine differs from other human-scientific endeavors in that it largely does not aim to generate knowledge about the human condition. Rather, we might class it among the human sciences insofar as it deals with particular knowledge produced and exchanged along encounters between patient, doctor, and healthcare system. We should not forget, however, that improving knowledge exchange is only an intermediary goal, because the primary goal is the ethical project of empowering the individual patient in what is perceived as a depersonalizing and technocratic medical system. Under this logic, "narrative is regarded as not merely expressive but as transformative and even therapeutic" (Woods 2011, p. 73).

The narrative medicine approach encodes a number of assumptions about human existence, as Angela Woods (2011) presciently identifies. Most notably, it treats a person's narrative and their humanity as "coextensive"; the self of narrative medicine is constituted autobiographically. Implicit is a normative sense that it is healthy or virtuous for humans confronted with illness to express themselves in narrative form (and the inverse). This narrative form, moreover, tacitly universalizes what is a "Western middle-class" discursive form, in Woods' view (2011, p. 76). One might also question the extent to which the breadth and imprecision of "narrative" in narrative medicine works against more systematic efforts to understand differences between the diverse verbal forms that comprise a clinical encounter (74). (We will return to scholars' analysis of specific human-scientific texts in section "Narrative in Practice.") Many of the assumptions Woods identifies have their origins in broadly psychological areas of enquiry, which brings us to the ways psychology and other human sciences have taken narrative as their object of study.

Narrative, Knowledge, and Self

We have already met two assumptions about narrative that are broadly psychological in nature: that humans tend to structure knowledge of history in narrative mode (in White's suspicion of historical narratives), and that human selfhood is constituted through narrative autobiography (in narrative medicine). Both connect to ways the human sciences have attempted to discipline knowledge about narrative and the mind. Psychology is perhaps the most evident of these – the mind is principally a psychological object of study – but since at least the 1960s, systematic study of relations between narrative and mind (or self) has been found in many disciplinary homes, including neuroscience, philosophy, linguistics, and narratology. The various theories and themes of enquiry would merit an entire book to themselves and, indeed, are also to be found explicated in other chapters in this handbook. What follows here thus provides an overview of some of those theories and enquiries, with an emphasis on those which connect to other thinking on narrative discussed in this chapter.

We can usefully separate theorizing about narrative and mind into three main strands (adapting the schema from Pléh 2018). First, researchers have asked how stories work: How do humans read and listen to narratives? How are stories constructed out of cultural resources? Why might stories be more easily remembered than other discursive forms? A second broad topic is investigation into how narrative can serve to organize knowledge, while the third area of enquiry asks whether narrative enters into humans' notions of the self. Research is ongoing in all three areas across various human-scientific disciplines, but broadly speaking, the three emerged in chronological sequence.

Already in fin de siècle France, philosophers interested in esthetics, including Henri Bergson, called upon ideas from newly scientific psychology to explain what it meant to contemplate a work of art or read a novel. Reading a novel was akin to receiving successive (hypnotic) suggestions, for French philosopher Paul Souriau, who held that a writer's language should not distract readers from the images or ideas generated by it. Nonetheless, it was for the reader to "fill in" the image from the writer's suggestive words (Souriau 1893). How stories were filled in by patterns of cultural experience was at issue for British psychologist Frederick Bartlett in the 1930s, as he explored connections between narrative, memory, and culture (Pléh 2018). The problem is one of the ambiguity of language: how does a reader or listener understand a sequence of verbal utterances as a story? From the 1960s, proposed answers have emerged from multiple disciplines, and engaged researchers across psychology, artificial intelligence (AI), linguistics, and anthropology. (The reciprocal problem of identifying a grammar for story sequences conventionally sits under narratology and is thus examined in section "Narratology.") Key names include Schank and Abelson, Colby, and Swinney, to name just a few – for more detail, see Pléh (2018), Jahn (2010), or Vincent Hevern's online guide (2013).

This was the time of the first so-called cognitive revolution, which conceived human cognition in terms of the overarching metaphor of the computer (Herman 2007). Under its logic, "narrative competence can be redescribed as a nested structure of processing strategies operating at different levels" (Herman 1997, p. 1053). On the small scale, readers are held to call upon "scripts" and "frames" to "supply the defaults that fill gaps ... and provide the presuppositions that enable one to understand what the text is about" (Jahn 2010). A "frame" might cover the situation of making a promise, while a "script" consists of a "stylised generalized episode" like going to a birthday party (see Herman 1997, Jahn 2010). A range of different scenarios can be slotted into a single script, and different scripts can be built up to form larger structures – the whole, of course, anchored in cultural patterns of experience. When human (or anthropomorphized) actors appear, readers "mobilize our naïve social psychology about the structure of human action and the usual motives for action" (Pléh 2018, p. 240). On a larger scale, schemata and templates provide "ready-made" forms of the kind of actions likely to be encountered in story, and the kinds of stories they belong to – e.g., the poor woodcutter who wants to marry a princess, in a fairy tale (Pléh 2018).

Where, then, do humans derive these cultural templates? A second strand of enquiry holds narration to have a fundamental role not only in familiarizing readers with story forms, but also more broadly in structuring how humans understand our world. Psychologist Jerome Bruner proposed dividing human cognition about the world into two principal modes of thought: the "paradigmatic" and the "narrative" (see useful overview in Pléh 2018). Narrative cognition is more personal and experiential, linked to local or contextualized situations. The paradigmatic mode, in contrast, is descriptive and impersonal and organizes the world into timeless categories. Insofar as the two modes are held to be incommensurable, Bruner's scheme evokes other binary oppositions between what may broadly be called science and narrative, as remarked by Cheyne and Tarulli (1998). This is the territory of opposition of nomothetic and idiographic knowledge (see Smith introduction) and returns us in some sense to CG Hempel, whose ideas we encountered in the first section. Bruner's account, however, tends to invert the hierarchization of science over narrative, by construing the narrative mode of thought as "always more primary, more elementaristic, and readily available" (Pléh 2018, p. 242). (Pléh also signals fascinating ways that Bruner's ideas were inflected in the Hungarian context by János László.) In other words, narrative is more natural – more basically human –

under this conception. Its written form would then be construed in natural language, as against the formal language of the paradigmatic mode, which strives for impersonal and transparent expression (on narrative and natural language in science, see Wise 2022).

Once narrative modes of thought and representation are viewed as primary and natural for humans, it is a small step to construing almost all scientific expression as tacitly narrative in nature. This is precisely what psychologists Cheyne and Tarulli (1998) maintain about the experimental report in twentieth-century psychological research. Borrowing from literary theory, they mobilize Bakhtinian notions of genre to argue "that the paradigmatic style is simply a specialized development of the narrative" (p. 3). The result would be that research articles in psychology, and indeed psychological theorizing, should be understood as having a narrative character, specifically that of the *Bildungsroman* (or narrative of personal development), rather than the paradigmatic form associated with efforts to assure a scientific status for psychology (Cheyne and Tarulli 1998). If science and narrative are opposed, this logic elevates narrative to the apex of human ways of knowing the world; it replaces "the myth of science as univocal rationality" with an "idyll of narrative," as narratologist David Herman (1998, pp. 384–385) remarks in a penetrating critique of Cheyne and Tarulli's argument. The problem, according to Herman, is precisely the one that this chapter also aims to address: a tendency to use the one term "narrative" to cover a range of often-disparate concepts. The result is an inflation both of the term, and of the place of narrative in scientific activity. Despite the other merits of Cheyne and Tarulli's piece, it can serve as a prescient warning of the risks of extending ideas about narrative-as-cognitive-mode to scholarly study of the human sciences, without careful attention to conceptual precision. If narrative is innate to and fundamental in human ways of knowing, then it no longer has useful analytical force when contrasted with other forms of knowing and writing – other forms will always already be derivatives of the narrative. Indeed, just as care should be taken in specifying the sense of "narrative" in play in a given HHS study, it can also be problematic to allow one theory or schema to stand in metonymically for "narrative" or "science" when the two are brought into engagement. As Herman (1998) notes, Bakhtin's ideas provide only one way of describing narrative genres. Correspondingly, Bruner's paradigmatic and narrative modes do not sum up human-scientific investigation of narrative and cognition.

This brings us to the third strand of research by which narrative has been linked not only to mental processes, but more particularly to humans' notions of selfhood. Here, philosophical scholarship, notably that of Paul Ricoeur, takes a larger place alongside psychology, cultural theory, and other human-scientific disciplines, to the point that conceptions of narrative identity are marked by their "interdisciplinary reach and connectivity" (Klepper 2013, p. 4 – Klepper cogently surveys the state of the field, in much greater depth than is possible here). Fundamentally, research in this third strand construes human selfhood as configured through narration. Selfhood, or identity, is neither fixed nor stable in this view, but rather prone to "crises" of dissolution or decomposition, of the kind explored in modernist literature during the early decades of the twentieth century (Pléh 2018; Ryan J 1991 studies this modernist "vanishing" self in detail). Paul Ricoeur drew directly upon this literature in formulating his complex and extensive theory of narrative identity. Modernist fiction notably points to a potential "crisis in identity" by articulating dissonances between two aspects of the self: identity as maintained in time, and an individual's sense of self as agent, distinct from other humans (Klepper 2013). (Literary autobiography also often plays with disjunctions between the two – semiologist Roland Barthes' (1975) autobiography is one example.) For Ricoeur and others in his vein, narrative works to hold together these two modalities of identity and to negotiate between them. Humans make retrospective sense of events they experience by configuring themselves as characters in a plot, within the narrative forms available in their historical and cultural context. The self, it is supposed, is always in a dynamic process of narration – always in the middle of the story, projecting forward and making sense retrospectively, to make intelligible one's whole life (Klepper 2013). The experience of time becomes inseparable from narrative in Ricoeur's thinking, which renders "human time [as] fundamentally narrative time" (Klepper 2013). It follows that "the experience of time and the story of the self in time is a recursive process" (Klepper 2013, p. 8). But importantly, it is a process informed by social and cultural institutions and what sociologists have termed their "identity protocols" (Klepper 2013). In the case of narrative psychotherapies, a psychologist or psychoanalyst might guide a patient's process of narrative construction and reconstruction explicitly (Pléh 2018). What is intentionally dialogical narrative work in psychotherapy takes a looser form in the narrative medicine approaches discussed above, but both bind together narrative with health. If narrative self-identification is dialogical – whether explicitly in spoken language, or in interaction with social contexts – it can also be construed as "socially distributed cognition"; such notions pertain to the domain of "discursive psychology." There, the emphasis is on storytelling in practice, and how cognitive processes can be shared across participants, or across time and place, as narratives are told and interpreted (Herman 2007).

Narratology

How narratives may be interpreted or understood as a specific discourse type is precisely the concern of narratology. The term "narratology" was coined in 1969 by Tzvetan Todorov to designate the project of analyzing "the structure and operation of the literary discourse" (Todorov 1969, p. 71). At that point, the field encompassed the work of primarily Russian and French scholars working along more-or-less structuralist lines; this theorizing (with some poststructuralist work) is now known as "classical narratology," in contrast to the self-conscious movement of "postclassical narratology," which dates from the 1990s and continues today.

It might initially seem surprising to count narratology among the human sciences; however, in the early ("classical") period the field had explicitly scientific ambitions. It was to be "a kind of propaedutic for a future science of literature," according to Todorov (1969, p. 71), insofar as its aim was to understand the structure of literary texts on an abstract level. That is, novels would be examined as exemplars for what they could reveal about the workings of literature in general, rather than being explicated as individual texts or read as revealing their authors' intentions. (Nonfictional texts are a more recent object of enquiry, to be discussed below under "factual narratology.") Todorov (1969), for example, analyzed Boccacio's *Decameron* in relation to the concept of plot and used it to set up a number of categories that could be used to describe all plots. We can understand this project by analogy with linguistics: Just as categories such as "noun" or "verb" allow the linguist to describe a sentence, the categories of narratology would provide for "a more extensive and precise description" of any (literary) narrative (Todorov 1969, p. 75). Indeed, linguistics served as more than just a useful analogy, for structuralist narratology was closely tied to semiotics, and was influenced by the work of linguists such as Roman Jakobson and Algirdas Greimas (see Puckett 2016). Anthropology was another human science to contribute to narrative analysis, notably through the ideas of Claude Lévi-Strauss. This multidisciplinary exchange continues today and further adds psychology and cognitive sciences to the mix.

Importantly for the subject of this chapter, the aim of narratology is to break down our broad term "narrative" into its various elements, so as to elucidate what it is and how it functions in comparison to other kinds of discourse. As Kent Puckett (2016, p. 8) puts it, "looking at and naming different aspects of [narrative] gives us the ability *to see* what is weird about almost any narrative." The fundamental distinction is between what is most commonly called *story* (*histoire* in French; *syuzhet* in Russian), and *discourse* (*récit*; *fabula*). (There are other terms in use, and French theorist Gérard Genette added a third element in the form of *narration*.) Basically, the "story" consists of the set of events recounted, while "discourse" encompasses the way those events are told: A man's day in Dublin might be told in flashback or in order from dawn to dusk, for instance. Any narrative therefore works as the dynamic relation between story and discourse, with the important corollary that we usually only have access to the discourse in the form of what is written down (or told aloud). From this fundamental distinction, narratology interrogates the workings of plot, genre, order, and perspective, to name just a few. The next paragraphs outline some key ideas and thinkers, with an emphasis on those which connect to the human sciences, or may be useful for HHS scholars. (More detailed overviews of narratology are found in Culler (2011), Herman et al. (2010), Puckett (2016), and the *Living Handbook of Narratology* maintained online out of the University of Hamburg. For more on how narratological concepts can be useful in HHS scholarship, see Hajek (2022).)

In the 1920s, Russian formalists such as Victor Shklovsky and Vladimir Propp worked to account for the "strangeness" and "literariness" of literature, and their work became an important source for structuralism in general, and narratology more particularly, once it began to be translated in the 1950s and 1960s (Puckett 2016). Propp is particularly known for systematizing the plot elements of Russian folktales into a limited set of actions (known as "functions") and characters (or "actants"). Like the chemical elements of the periodic table – an explicit analogy for Propp and later Greimas and Lévi-Strauss – these narrative elements could be arranged into various sequences to construct different folktales, although the system also imposed constraints on the kind of functions which could follow one another (Herman 1997; Puckett 2016, Chap. 5). Tzvetan Todorov also worked on plot, while semiologist Roland Barthes labeled some events as "hinges" around which a narrative might branch into realized and unrealized outcomes (see Beatty 2017 for an application of this).

Perhaps the most extensive system for understanding narrative was proposed by French scholar Gérard Genette (1972), using Marcel Proust's *A la recherche du temps perdu* as his specific object of enquiry (see also Puckett 2016, Chap. 6). Genette's ramifying system and careful technical vocabulary first covered questions of temporality (*"temps"*) between story and discourse, such as the relative ordering of events (flashbacks or flash-forwards, for instance – *"analepse"* and *"prolepse,"* for Genette), their relative duration (does the discourse skip over certain story periods?), and frequency (when a given event is revisited several times in the discourse). Under the heading "mood," Genette enunciated different possibilities for narrative distance (does the discourse pretend to "show" or does it "tell" events?) and narrative perspective (is the narrator a character, or an omniscient observer?). His notions of "focalization" and "narrative levels" are particularly useful for the kind of critical work done in HHS, as they let us scrutinize the perspective/s from which a narrative is recounted, and attend carefully to who speaks in each situation. A nineteenth-century psychological case history, for instance, might consist of several levels: the "observation" proper in which the events of diagnosis and treatment are supposed to be given transparently to the reader; and a "commentary" in which the psychologist explicitly addresses his peers and links the case to broader theories. Yet despite ambitions for transparency, we might find the observation to be focalized internally at key points, that is, with the narrator "sitting behind" the patient to portray her inner sentiments, something which would be strictly inaccessible were the narrator only to view the patient from the outside (see Hajek 2022).

That narratology is a properly human, or perhaps social, science becomes clear in Mikhail Bakhtin's proposition that novels have an intrinsic "heteroglossia" (or "polyphony"), in that they incorporate multiple points of view. Similarly, Julia Kristeva's notion of "intertextuality" has narratives enmeshed in a web of mutual references – a way of opening up narratives as writing, which Kent Puckett (2016) links to a broader political project of critique. We can indeed see broad shared ideas between Kristeva's efforts to open up texts as *writing*, opposed to closed and fixed *books*, and ways that Hayden White valorized narrative in historiography for its capacity to provide multiple (re-)tellings of the same events. Roland Barthes' studies also participated in the same kind of political project (as Kristeva, if not in any direct engagement with White), notably though the designation "*lisible*" (readerly) for texts that are closed, and simply present themselves to a passive reader. The opposite term is "*scriptible*" (writerly), for texts (re-)written in a process of active reading, and which allow for many (re-)interpretations (Puckett 2016, Chap. 6).

It is in enquiry into the work done by readers in (re)constructing a narrative that we can find the broad transition to "post-classical" narratology. Study of form gives way to ideas on cognition, with narratives no longer construed as "self-sufficient products," but rather "texts to be reconstructed in an ongoing and revisable ... process" (Jahn 2010). Studies of implicit readers, and their "horizon of expectations," slide into investigation of narrative understanding – bringing us back to issues where narratology intersects with psychological questions of the narrative mind and self. A good place to illustrate differences between the two approaches is in the vexed matter of how exactly to define a narrative. Marie-Laure Ryan outlines the problem, and some of the solutions, in a 2007 article: We see how definitional attempts move largely from concerns about (textual) representation of events, to the mental representation of what we call story. Ryan herself takes a helpfully pragmatic view, by reframing the question in terms of degrees of narrativity. Rather than ruling a particular text in or out of the category "narrative," she proposes that we consider "narrative texts as a fuzzy set allowing variable degrees of membership, but centered on prototypical cases that everybody recognizes as stories" (2007, p. 28). A given text thus would have higher or lower narrativity depending on whether it included intelligent agents with a mental life, for example. Under this criterion, a psychological case history has higher narrativity than an archeologist's account of changes in grave goods, even as both texts treat transformations over time and causal chains. Specifying degree of narrativity using Ryan's schema would help HHS scholars deal with the kind of texts that are routinely part of human-scientific activity (see also Hajek 2022). This would help us move away from assuming that such texts are (always) narratives (as we saw in the case of historiography), to closer scrutiny of different parts of a given text, or to more rigorous comparison between texts. Indeed, narratology in general has much to offer HHS scholarship in the way of careful terminology for describing and comparing what we mean by narrative.

Some of this terminology, however, is borrowed from, or shared with, other human sciences, notably when it comes to postclassical narratology. David Herman (1997), a key figure in the subfield of "cognitive narratology," has applied the AI concepts of scripts and sequences to the study of narrative. (We met these concepts in section "Narrative, Knowledge, and Self.") For the nonspecialist, cognitive narratology seems to differ little from studies of narrative and cognition in psychology, neuroscience, or AI (see Jahn 2010). It is not only a matter of narratology borrowing concepts from other human-scientific disciplines, however. Herman (2007) also argues for a reciprocal exchange of ideas and explores several areas where analytical categories from narratology (such as experientiality) might fruitfully be overlapped with those from other sciences of the mind (e.g., the notion of qualia), as a way to generate new insights for both fields. If Herman's earlier work on scripts links to computer-based models for human cognition, the notion of experientiality is better associated with work on narrative and selfhood. Monika Fludernik has indeed made experientiality a privileged marker of narrative itself, through the subfield known as "natural narratology" (Caracciolo 2020). Here, so-called "natural" modes of cognition are primary in the ways readers encounter a narrative. Conversational storytelling about an anthropomorphic protagonist, expressed in natural language, serves as the prototype of narrative. Narrative cognition, under this mode, is embodied and situated, as readers project their past experiences onto characters' intentions and emotions. As Marco Caracciolo (2020) enunciates, this is a space for interdisciplinary exchange and even disciplinary convergence. But therein also lies a risk of blurring between textual properties and structures and "the (more or less shared) cognitive make-up and presuppositions of the audience" (Caracciolo 2020). This is another version of our already-encountered problem of conceptual spread in narrative, insofar as narrative elements can lose their autonomy as "independently describable objects" (Caracciolo 2020).

In the twenty-first century, there has been a concerted move to interrogate the narrative structure and reader experience of texts beyond the fictional. Fludernik is again the key figure in what is known as "factual narratology." Here, scholars have tended to focus, on the one hand, on whether narratological categories developed for the analysis of fictional texts – especially Genette's – also apply to nonfictional forms such as journalism, legal narratives, and scientific output (Fludernik 2020). Interestingly, insofar as scientific narratives have been an object of enquiry in this domain, they have chiefly emerged from human and social sciences, rather than the natural sciences. On the other hand, the contours of "factuality," as contrasted with "fictionality," are of prime concern. Indeed, if chapters in the recent handbook of *Narrative Factuality* (Fludernik and Ryan 2020) are a guide, the second strand of enquiry seems to dominate, leading researchers to interrogate such thorny concepts as "truth," "fact," or "authenticity" as they apply to given nonfictional text-forms. This is rather to the detriment of fine-grained narratological analysis and theorizing about factual narratives in their particular historical and disciplinary contexts. Fludernik (2020) herself signals the tendency for scholars working in factual narratology to treat broad disciplinary narrative forms as monolithic, by positing universal characteristics for "scientific narratives" or "sociological narratives" based on examination of only a few individual texts (see also Hajek 2022). Interestingly, the self-consciously innovative factual narratology movement seems to have little recourse to more detailed textual analyses of scientific documents which emerged from studies on the rhetoric of science in the 1980s (e.g., Bazerman's 1988 work on experimental reports). Overall, there has been relatively little detailed narrative analysis of documents emerging from the past and present of the human sciences, with medicine and historiography as perhaps the exceptions. HHS as a field is well placed not only to provide further exploration of human-scientific texts as narratives, but furthermore to feed results of this empirical study back to factual narratology (or narratology in general).

Narrative in Practice

Although HHS scholars take scientific documents – research articles, manuscript notes, and monographs – as key sources in examining the workings of the human sciences, it remains somewhat rare for scholars to analyze these documents for their textual and narrative workings. As we have seen, one common assumption is that (most) human scientific output takes the form of narrative. When "narrative" is understood simply as "story," or identified with "fiction" (see Cohn 1999), associations with literature readily come to mind. As a result, scholarship in (history of) science and literature is often viewed unreflectively as synonymous with studying narrative. In 2010, it could thus seem straightforward to evoke a "common focus on narrative" as underlying the "close affinity" between literary studies and history of science (Otis 2010). Both literary and scientific texts may be narratives and may be analyzed for their narrative attributes, of course. However, much of what has emerged from the rich scholarship in science and literature (and the related field of literature and medicine) are investigations of *thematic* exchanges between the two kinds of knowledge. By the same token, attending to linguistic or rhetorical elements of scientific texts does not necessarily shed (much) light on their narrative structure. Charles Bazerman's (1988, Chap. 9) pioneering study of experimental reports in American psychology is predominantly rhetorical and sociological, for example, and would be complemented by narratological analysis of shifts in narrator focalization and narrative level. Similarly, most of the existing enquiry into metaphor in science says nothing about narrative in any strict sense, even though it is common to see the terms used somewhat interchangeably, especially in article titles and abstracts. This is not to dismiss the tremendously fruitful work in the rhetoric of science or in bringing human-scientific texts into dialogue with their fictional/literary counterparts. At their best, studies in this area offer careful readings of constellations of texts – scientific, fictional, and beyond – as they gathered around some culturally and historically specific topic pertinent to systematic enquiry about the human. Two examples (among many possibilities) are Jacqueline Carroy's (1991) exploration of the psychological subject between medicalized "object" and dramatized character in French cultures of *magnétisme* and hypnotism along the nineteenth century, and Anne Vila's (1998) masterful interweaving of multiple discursive threads to elucidate eighteenth-century French conceptions of sensibility.

At issue in such scholarship are questions that extend our understanding of the history of the human sciences: Under what conditions did the psychological subject emerge? What can this tell us about psychology as a scientific discipline? How were philosophy and medicine intertwined in Enlightenment conceptions of body and mind? Concerns emerging from HHS as a scholarly domain similarly motivate the majority of studies of narrative, in the stricter sense of a particular discourse type. Only rarely do researchers generalize from their analysis of narrative to propose narratological typologies of human-scientific writing in some discipline, or to develop formal equivalences between scientific and literary production in a given historical context. (One recent study of the latter kind is Devin Griffiths' (2016) argument for analogy as the form that structures both historical novels and natural history in nineteenth-century Britain – though topically Griffiths' work sits at the very boundary of what we might consider the human sciences.) Investigation of narrative in the human sciences properly speaking has clustered around two disciplines: historiography and medicine, with medicine extending via physiology into the "psy" sciences (psychiatry, psychology, and psychoanalysis). The remainder of this section sketches out some of the issues in play in such investigations, with an emphasis on twenty-first-century scholarship. It closes by considering several recent projects that have adopted a systematic and interdisciplinary approach to examining narrative knowing.

Just as there has been considerable (prescriptive) theoretical reflection on narrative in historiography, the narrative features of history writing have also been studied (descriptively) to a significant extent. Of particular interest for the purposes of this chapter is the use of narratological tools, combined with consideration of rhetoric, to analyze the textual features of histories produced by academic historians. Philippe Carrard (2015, 2017/2013) has argued persuasively that contemporary French historians mostly do not write their scholarly work – their journal articles and monographs – in narrative form, as against assumptions that history and narrative are interchangeable. Carrard's enquiry relates to the "macrotextual structure" of historians' texts and asks to what extent this structure is organized by time sequence. He recognizes that certain portions of the texts may constitute (mini-)narratives, but such interspersed use of narrative is not his primary interest. In place of an overarching temporal structure with its own plot, historians chiefly write in "synchronic cross-sections," according to Carrard, who further divides this textual form into the subforms of the "tableau" and the "analysis." When historians' writing does proceed as a narrative, such diachronic developments predominantly comprise a series of "tableaux," or "analyses," and result in a "stage narrative" tracing broad changes in attitudes. Beyond its intrinsic interest, Carrard's typology further establishes a schema against which textual output by historians in other national contexts or time-periods may be studied. It would also lend itself to further development at a microtextual level.

In addition, empirically detailed analyses of the kind Carrard proposes may be mobilized to explore the narratological characteristics of human-scientific (i.e., nonfictional) texts. Carrard (2017) extends Dorrit Cohn's (1999) seemingly underappreciated theorization of the uniqueness of fictional narrative (especially the historical novel), in comparison with its nonfictional counterpart (i.e., historiography), in order to engage directly with the project of factual narratology (see Fludernik and Ryan 2020). He notably nuances Cohn's stance (which itself follows Genette) on nonfiction, in considering that the author and narrator of contemporary scholarly histories are not always identical with one another. (Robert Meunier (2022) argues along the same lines regarding research articles in twentieth-century biology.) The context of utterance is of prime importance for historiographic texts, for Carrard, because it sets up the initial pact between text and reader; even if history could be "read" as fiction, it cannot (easily) be "taken" as fictional, given the presence of paratextual elements such as footnotes or reference to a scholarly book series. What Carrard does in terms of explicit engagement with contemporary narratology in his 2017 chapter, this chapter would encourage in other HHS areas of enquiry, especially "core" human sciences such as sociology and anthropology.

Alongside historiography, the other principal human science to have been the object of sustained and extensive textual analysis is medicine, especially in relation to medical observations (or case histories). The journal *Literature and Medicine* is a major site for scholarly exploration of medical case histories, both in the recent and more distant past. Researchers in this area "take seriously the case history as narrative and narrative as a legitimate, thoroughly medical way of knowing," as Kathryn Hunter enunciated in a 1992 (p. 169) special issue devoted to the "Art of the

Case History." Many contributions to the issue, including Hunter's, encode political aims, in addition to offering more or less broad accounts of particular narrative features. In support of the emerging narrative medicine movement, these scholars affirmed the epistemological legitimacy of the narrative case history in knowledge-making and, concomitantly, insisted on its continuing relevance in professional medical practice. Attention to medical case-writing in contemporary practice is indeed an important feature of work in literature and medicine. One research article in the 1992 special issue tracked the physician's narrative voice in medical students' case histories, with a focus on what we might call narrative distance between the physician-narrator and the patient (Flood and Soricelli 1992). Interestingly, research on medical case histories tends to make rather sparing use of narratological terms, even where we could readily connect scholars' analyses to concepts from narratology, such as "distance" in the previous example. This remains the case for detailed readings of the structure and functions of single texts, such as Brian Hurwitz's (2014) delightful explication of the cumulative narrative work done by presenting patient observations in series in Parkinson's essay on the shaking palsy. The 2020s, however, seem well placed to see the emergence of explicitly narratological studies of medical case histories, both historical and very contemporary, following from a 2019 workshop in Fribourg revisiting "Doctors' Stories," and associated research projects led by Martina King.

The medical case itself has a long history, which Gianna Pomata (2005, 2011) has traced through Renaissance and early modern genres of *historia*, *observationes*, and *curationes*. In a number of influential articles and book chapters, Pomata proposes that we conceive the medical observation as an "epistemic genre," a textual form read for its cognitive content by a community of scholars, and which lends itself to ready circulation among members of that "intellectual shared space" (Pomata 2011). Pomata's studies, and those following her lead, have prompted examination of how individual observations are arranged into larger collections, whether that be in chronological or alphabetical sequence, or for narrative effect (as Hurwitz (2014) demonstrates for Parkinson's text). Another strand of research explores how medical cases traveled both geographically between individual physicians, and discursively between intellectual and cultural contexts.

The case observation can also be framed a scholarly writing form that spans what are now disciplinary boundaries. Of notable relevance for HHS is the area where the medical subfields of physiology, neurology, and psychiatry shade into psychology and psychoanalysis. How narrative practices (and other textual features) are transformed between medico-surgical case-writing and 1870s psychology is the subject of a 2020 article by Kim Hajek, which takes the work of French surgeon-turned-double-personality-expert Eugène Azam as its exemplar. For the earlier part of the nineteenth-century, Juan Rigoli's (2001) lengthy monograph dissects French writing about madness – by *aliénistes* (psychiatrists) in case form, by patients as autobiography, and by *aliénistes* about patients' words – in extensive and comprehensive detail. Sabine Arnaud's 2015/2014 monograph interrogates even earlier French "psy" writing in the form of observations of hysteria, which she places in dialogue with their literary counterparts. Of the three, Arnaud draws most self-consciously on narratological concepts, such as when she elucidates shifts in the second part of the eighteenth century in the temporality of observations of hysteria toward retrospective modes of configuring a patient's (life) history (2015/2014, Chap.5). There are nonetheless other analytic lenses derived from narratology to be found through these examples and similar scholarship; Genette's work theorizing narratorial mood and voice – e.g., distance, perspective, and focalization – has proven especially productive for elucidating the interlinked textual and epistemic dynamics of "psy" writing. With such narratological tools, HHS scholars can explore the extent to which human scientists investigating the human mind share or pretend to share novelists' "entirely *un*natural power to penetrate the psyches of their protagonists" (to use Dorrit Cohn's (1999, p. 42) characterization of fiction). Reciprocally, these tools open up interrogation of instances when the mentally troubled subject turns narrator – or ostensible narrator, for subject-narrators are marked by paradox and ambiguity, as Rigoli (2001, Chap. 5) and others have noted, framed between madness (*folie*) and reason, between implausibility (*l'invraisemblable*) and truth, by the very act of offering their stories to an attentive *savant*.

Reference to implausible or ambiguous patient-narrators will bring the Freudian case history to mind for many HHS scholars, and it should be no surprise that Freud's narratives have received extensive attention from literary scholars and historians. Notably, they have not only served to elucidate the epistemic, institutional, or political dynamics of psychoanalysis as a field of study (e.g., Sealey 2011), but also informed explicitly narratological enquiry. The well-known comparisons between psychoanalytic case-writing and short stories that appear in several of Freud's texts might even seem to invite readings across the borderlines of fiction and science. Thus, although Cohn (1999) otherwise enlists historical (i.e., historiographical) narratives as exemplars of nonfiction, which she sets against the posited narratological distinctiveness of fiction, she nonetheless includes a chapter on Freud's three long case histories (Dora, the Wolf Man, and the Rat Man). Cohn (1999, Chap. 3) likely takes too strong a stance when she argues for the evident nonfictionality of Freud's cases on the basis of the narrator's (Freud's) expressions of doubt, the potential openness of the narratives, and the frequent verbs of telling used to signal information received from the patient. If these textual features are accompanied by a striking (near) absence of internal focalization (i.e., narration of the subject's inner mental states) in Freud's texts, the combination does not apply to the "psy" sciences more widely. To draw only on our previous examples, Arnaud (2015/2014, Chap. 5) argues that observations of hysteria achieve narrative *closure* (not openness) by means of their retrospective temporality. And Hajek (2020) finds self-conscious internal focalization alongside Azam's expressions of doubt and repeated reference to tellings and retellings in the case of Félida's double personality. A richer area of enquiry for current HHS scholarship would seem to lie in Cohn's (1999, Chap. 7) interrogation of the overlap (or rather the separation in fiction) between the author and narrator of a narrative, and the implicit norms of each one. Researchers might extend Bazerman's (1988, Chap. 9) study of behaviorist rhetoric in psychology journal articles to other fields and contexts, and examine the extent to which epistemic virtues espoused by human scientists as *authors* are inscribed into the implicit stance of scientist-*narrators* recounting their observations or experiments. (Bazerman does not use quite these terms.)

Just as Cohn enlists Freud's texts to explore narratological questions, so too (evidently from the title) does the recent strand of research known as factual narratology. Led by Monika Fludernik, with findings to date encapsulated in the 2020 handbook *Narrative factuality* (Fludernik and Ryan eds), this project is self-consciously systematizing, broad, and interdisciplinary. It brings together research spanning a range of disciplinary perspectives, with studies of text-types from advertising and journalism, via the literary, to the legal, the religious, and the scientific. Human sciences are well represented, in chapters examining issues of factuality and truth in sociology, anthropology, linguistics, and historiography. Principal preoccupations are questions of factuality (what is a fact?), authenticity, and truth, with contributors perhaps overinvested in these notably complex philosophical concepts at the expense of contextualized empirical studies. CE Scheidt and A Stukenbrock's (2020) chapter on the real in psychoanalysis is a notable exception, in that it relates narrative elements of clinical case histories to the need for analysts and patients to negotiate the "real" throughout the therapeutic process. Where *Narrative factuality* is a fertile resource for HHS is first in its disciplinary breadth, and furthermore in providing a framework against which to situate studies of narrative practices in human-scientific activity. The factual narratology project further offers a direct means to connect theories and categories from narratology with historically contextualized investigations of scientific narratives. Ideally, this would promote substantive dialogue between the two domains, not only respective borrowings.

How such a dialogue might occur in more concrete terms is precisely the subject of one introductory chapter in the forthcoming collection *Narrative science* (Morgan et al. eds). There, Hajek works from the output of the ERC-funded Narrative Science (NS) Project to argue for the benefit for history and philosophy of science (HPS) of using narratological tools to unpack narrative forms of systematic knowing, and signals potential areas of joint HPS and narratological interest. As a whole, *Narrative science* proposes an integrated, yet wide-ranging set of studies into how narrative has been used by scientists in doing science over the past two centuries. Led by historian of economics Mary Morgan, the project had self-consciously systematic ambitions and includes fine-grained explorations of narrative as it contributes to knowledge-making in domains of HHS interest including archeology, psychology, medicine, and economics (see Morgan et al. eds; Morgan and Wise 2017; Morgan 2017; www.narrative-science.org). Our concerns above with HHS, narratology, and the narrative self are notably intertwined in Elspeth Jajdelska's (2022) chapter, in which the literary scholar reads a contemporary journal article in cognitive psychology for its experiential and emotional narrativitity. Investigations of human-scientific knowledge-making such as Jajdelska's are brought into relation through the Narrative Science Project with narrative forms of knowing across the natural, historical, and social sciences, from mathematics, through geology, to political science – intersections between different case-studies are an explicit feature of Morgan et al. (2022). Despite the extensive scope of Morgan's project, there remains space for more extensive HHS scholarship along similar lines, especially concerning linguistics, sociology, and anthropology. More narrowly focused on narrative as a comparative device across the sciences and humanities is a recent (2021) volume edited by Martin Carrier and his colleagues. If no human sciences (strictly speaking) appear in the work, it joins the Narrative Science Project as a model for integrating scholarly reflection on scientific narratives, while still taking account of the particularities of different disciplinary and historical contexts. Of especial relevance for the purposes of this chapter is the way contributions to these collections mobilize differing conceptions of narrative – and in large part, take care to delimit their particular conceptual lens.

Conclusion

This chapter has drawn attention to the polyvalence of narrative as a term and concept as it enters into HHS scholarship. It unpacks the various senses of narrative by means of a tripartite framework, elucidating the major ways narrative poses normative issues of method for the human sciences, constitutes one of their objects of study, and has been studied in the textual practices of various human sciences. There are doubtless many other ways to construct such a schema, and the point here is not to advocate for this explication in preference to any other. Rather, what matters most for HHS is to attend carefully to distinctions between differing senses of narrative, so as to avoid confusion between them. Perhaps the simplest change would be to avoid using "narrative" to mean "bold (temporal) claim" or "received notion." Without attention to conceptual clarity, the risk is that narrative becomes so broad a notion as to lose meaning for HHS – that it no longer adds anything to our investigations of what it has meant to generate knowledge of the human.

As we have seen, narrative raises epistemic issues for human-scientific knowledge-making, sometimes supplemented by ethical or political concerns. Narrative, as an intertwined mode of writing and knowing, has all too often been set in opposition to positivist-inspired "scientific" methods, and epistemic values like objectivity. Everyday usage of "narrative" to mean "storytelling" supports such a perception, reinforcing connotations of narratives as artificial, even fictional. Where such views persist, so too does the assumption that history is (always) about writing narratives. It follows that historical study – common to all HHS scholars – often has to be defended as properly explanatory. Into the twenty-first century, human scientists must sometimes similarly defend their self-conscious recourse to narrative. This seems to hold even though a number of recent interdisciplinary projects have demonstrated that narratives are to be encountered across the full range of scientific disciplines, and that they serve important epistemic functions even in so-called hard sciences such as mathematics or chemistry (Morgan et al. 2022).

In twenty-first-century human sciences, however, many of the epistemic functions of narrative have been revalorized. Its capacity to weave together multiple perspectives and to configure heterogeneous elements are held to promote inclusion and coherence over fragmentation, comprehensibility over density. Some psychologists and narratologists contend that narrative's power derives from its close relation to natural language and the fundamental structures of human experientiality. Grouped under this view are theories that human beings experience temporally unfolding events as narrative, or construct their self-identity around narrative, or are hard-wired to organize knowledge in narrative form.

Yet narrative also clearly exists in text as a specific mode of discourse, with a formal structure – one notably formal*ized* by classical narratology. What is still relatively unexplored in HHS is how the narratives produced in the course of scientific knowledge-making might differ formally from their fictional counterparts. This matters for narratological theories. More acutely, it matters that HHS has rather limited understanding of when and whether human-scientific disciplines have employed narrative. To take just one example, historical writing might not take a narrative structure when considered on a macrotextual level (Carrard 2015, 2017/2013), but what kinds of events are represented as mini-narratives within larger historical texts? We also need further detailed narratological analysis to understand precisely what is going on in human-scientific writing, and what such textual practices mean for how these documents produce knowledge. Whose perspective is portrayed as different sections of case histories or experiment reports are narrated? With what authority does the narrator present different kinds of information? Such questions apply especially to routine textual production in the human sciences, not only to special or unusual documents like Freud's case histories. Fine-grained narrative analysis along these lines will complement existing studies of the rhetoric and metaphors in play in scientific writing. When it comes to human sciences other than medicine, history, and the "psy" disciplines, there is particularly wide scope for investigation – think sociology and anthropology. There is much that HHS can gain from further attention to narrative in its discursive particularities as a means of thinking and knowing the human.

Acknowledgments The Narrative Science Project at the London School of Economics and Political Science received funding from the European Research Council under the European Union's Horizon 2020 research and innovation program (grant agreement No. 694732).

References

Arnaud S (2015 [2014]) On hysteria. University of Chicago Press, Chicago
Barthes R (1975) Roland Barthes par Roland Barthes. Seuil, Paris
Bazerman C (1988) Shaping written knowledge: the genre and activity of the experimental article in science. University of Wisconsin Press, Madison
Beatty J (2017) Narrative possibility and narrative explanation. Stud Hist Phil Sci 62:31–41
Becker C (2005) Lire le réalisme et le naturalisme. Armand Colin, Paris
Berger S, Brauch N, Lorenz C (eds) (2021) Analysing historical narratives: on academic, popular and educational framings of the past. Berghahn Books, New York & Oxford
Binet L (2009) HHhH. Grasset (Poche), Paris
Braat M, Engelen J, van Gemert T, Verhaegh S (2020) The rise and fall of behaviorism: the narrative and the numbers. Hist Psychol 23(3):252–280

Caracciolo M (2020) Experientiality. In: Hühn P et al (eds) The living handbook of narratology. Hamburg University, Hamburg. http://www.lhn.uni-hamburg.de/article/experientiality
Carrard P (2017 [2013]) History as a kind of writing. University of Chicago Press, Chicago
Carrard P (2015) History and narrative: an overview. Narrative Works 5(1):174–196
Carrard P (2017) Historiographic discourse and narratology: a footnote to Fludernik's work on factual narrative. In: Alber J, Olson G (eds) How to do things with narrative: cognitive and diachronic perspectives. De Gruyter, Berlin, pp 125–140
Carrier M, Mertens R, Reinhardt C (eds) (2021) Narratives and comparisons: adversaries or allies in understanding science? Bielefeld University Press, Bielefeld
Carroy J (1991) Hypnose, suggestion et psychologie. L'invention des sujets. P.U.F., Paris
Charon R (2006) Narrative medicine: honoring the stories of illness. Oxford University Press, Oxford
Cheyne A, Tarulli D (1998) Paradigmatic psychology in narrative perspective: adventure, ordeal, and *Bildung*. Narrat Inq 8(1):1–25
Cohn D (1999) The distinction of fiction. The Johns Hopkins University Press, Baltimore & London
Culler J (2011) Literary theory: a very short introduction, 2nd edn. Oxford University Press, Oxford
Daston L, Galison P (2007) Objectivity. Zone Books, New York
Dean CJ (2019) Metahistory: the historical imagination in nineteenth-century Europe, by Hayden White. Am Hist Rev 124(4):1337–1350
Dillon S, Craig C (2022) Storylistening: narrative evidence and public reasoning. Routledge, London
Flood DH, Soricelli RL (1992) Development of the Physician's narrative voice in the medical case history. Lit Med 11(1):64–83
Fludernik M (2020) Factual narration in narratology. In: Fludernik M, Ryan M-L (eds) Narrative factuality: a handbook. De Gruyter, Berlin, pp 51–74
Fludernik M, Ryan M-L (eds) (2020) Narrative factuality: a handbook. De Gruyter, Berlin
Genette G (1972) Discours du récit. In: Figures III. Seuil, Paris, pp 65–278
Griffiths D (2016) The age of analogy: science and literature between the Darwins. Johns Hopkins University Press, Baltimore
Hajek KM (2017) 'A portion of truth': demarcating the boundaries of scientific hypnotism in late nineteenth-century France. Notes Rec 71(2):125–139
Hajek KM (2020) Periodical amnesia and *dédoublement* in case-reasoning: writing psychological cases in late nineteenth-century France. Hist Hum Sci 33(3–4):95–110
Hajek KM (2022) What is narrative in narrative science? The narrative science approach. In: Morgan MS, Hajek KM, Berry DJ (eds) Narrative science: reasoning, representing and knowing since 1800. Cambridge University Press, Cambridge
Herman D (1997) Scripts, sequences, and stories: elements of a postclassical narratology. PMLA 112(5):1046–1059
Herman D (1998) Narrative, science, and narrative science. Narrat Inq 8(2):379–390
Herman D (2007) Storytelling and the sciences of mind: cognitive narratology, discursive psychology, and narratives in face-to-face interaction. Narrative 15(3):306–334
Herman D, Jahn M, Ryan M-L (eds) (2010) Routledge encyclopedia of narrative theory. Routledge, London
Hevern VW (2013) Narrative psychology: internet and resource guide. Psychology Department, Le Moyne College, Syracuse. https://web.lemoyne.edu/~hevern/narpsych/narpsych.html
Hunter KM (1992) Remaking the case. Lit Med 11(1):163–197
Hurwitz B (2014) Urban observation and sentiment in James Parkinson's essay on the Shaking Palsy (1817). Lit Med 32(1):74–104
Jahn M (2010) Cognitive narratology. In: Herman D, Jahn M, Ryan M-L (eds) Routledge encyclopedia of narrative theory. Routledge, London.
Jajdelska E (2022) Narrative performance and the 'taboo on causal inference': a case study of conceptual remodelling and implicit causation. In: Morgan MS, Hajek KM, Berry DJ (eds)

Narrative science: reasoning, representing and knowing since 1800. Cambridge University Press, Cambridge

Klepper M (2013) Rethinking narrative identity: persona and perspective. In: Holler C, Klepper M (eds) Rethinking narrative identity: person and perspective. John Benjamins Publishing, Amsterdam

Meunier R (2022) Research articles as narratives: familiarizing communities with an approach. In: Morgan MS, Hajek KM, Berry DJ (eds) Narrative science: reasoning, representing and knowing since 1800. Cambridge University Press, Cambridge

Morawski J (2016) Description in the psychological sciences. Representations 135(1):119–139

Morgan MS (2017) Narrative ordering and explanation. Stud Hist Phil Sci Part A 62:86–97

Morgan MS, Wise MN (2017) Narrative science and narrative knowing: introduction to special issue on narrative science. Stud Hist Phil Sci Part A 62:1–5

Morgan MS, Hajek KM, Berry DJ (eds) (2022) Narrative science: reasoning, representing and knowing since 1800. Cambridge University Press, Cambridge

Olmos P (2022) Just-so what? In: Morgan MS, Hajek KM, Berry DJ (eds) Narrative science: reasoning, representing and knowing since 1800. Cambridge University Press, Cambridge

Otis L (2010) Science surveys and histories of literature: reflections on an uneasy kinship. Isis 101(3):570–577

Paskins M (2022) Thick and thin chemical narratives. In: Morgan MS, Hajek KM, Berry DJ (eds) Narrative science: reasoning, representing and knowing since 1800. Cambridge University Press, Cambridge

Pléh C (2018) Narrative psychology as cultural psychology. In: Jovanović G, Allolio-Näcke L, Ratner C (eds) The challenges of cultural psychology: historical legacies and future responsibilities. Routledge, London, pp 237–249

Pomata G (2005) Praxis historialis: the uses of historia in early modern medicine. In: Pomata G, Siraisi NG (eds) Historia: empiricism and erudition in early modern Europe. MIT Press, Cambridge, pp 105–146

Pomata G (2011) Observation rising: birth of an epistemic genre, 1500–1650. In: Daston L, Lunbeck E (eds) Histories of scientific observation. University of Chicago Press, Chicago, pp 45–80

Porter T (2012) Thin description: surface and depth in science and science studies. Osiris 27(1): 209–226

Puckett K (2016) Narrative theory: a critical introduction. Cambridge University Press, Cambridge

Ribeiro A (2018) Death of the passive subject: intentional action and narrative explanation in archaeological studies. Hist Hum Sci 31(3):105–121

Rigoli J (2001) Lire le délire. Aliénisme, rhétorique et littérature en France au XIXe siècle. Fayard, Paris

Ryan J (1991) The vanishing self. University of Chicago Press, Chicago

Ryan M-L (2007) Toward a definition of narrative. In: Herman D (ed) The Cambridge companion to narrative. Cambridge University Press, Cambridge, pp 22–35

Scheidt CE, Stukenbrock A (2020) Factual narratives and the real in therapy and psychoanalysis. In: Fludernik M, Ryan M-L (eds) Narrative factuality: a handbook. De Gruyter, Berlin, pp 297–312

Sealey A (2011) The strange case of the Freudian case history: the role of long case histories in the development of psychoanalysis. Hist Hum Sci 24:36–50

Simon ZB (2019) The story of humanity and the challenge of posthumanity. Hist Hum Sci 32(2): 101–120

Souriau P (1893) La suggestion dans l'art. Alcan, Paris

Teather A (2022) Stored and storied time in archaeology. In: Morgan MS, Hajek KM, Berry DJ (eds) Narrative science: reasoning, representing and knowing since 1800. Cambridge University Press, Cambridge

Todorov T (1969) Structural analysis of narrative (trans: Weinstein A). Novel 3(1):70–76

Vila AC (1998) Enlightenment and pathology: sensibility in the literature and medicine of eighteenth-century France. John Hopkins University Press, Baltimore

White HV (1973) Metahistory: the historical imagination in nineteenth-century Europe. Johns Hopkins University Press, Baltimore

White HV (1987) The content of the form: narrative discourse and historical representation. Johns Hopkins University Press, Baltimore

Wise MN (2017) On the narrative form of simulations. Stud Hist Phil Sci Part A 62:74–85

Wise MN (2022) Narrative and natural language. In: Morgan MS, Hajek KM, Berry DJ (eds) Narrative science: reasoning, representing and knowing since 1800. Cambridge University Press, Cambridge

Woods A (2011) The limits of narrative: provocations for the medical humanities. Med Humanit 37:73–78

Zimmerman A (2001) Anthropology and antihumanism in Imperial Germany. University of Chicago Press, Chicago

Diltheyan Understanding and Contextual Orientation in the Human Sciences

5

Rudolf Makkreel

Contents

Introduction	110
Understanding in the Human Sciences	111
Theoretical Cognition and Reflective Knowledge	115
Dilthey's Approach to History	116
Hermeneutical Objectivity	117
Three Kinds of Objectifications	118
Kant and the Importance of Reflective Judgment for Interpretation	121
Rethinking Historical Reason as Historical Judgment	123
Dilthey's Theory of Worldviews	124
Dilthey's Formative Social Ethics	127
Conclusion	129
References	130

Abstract

Wilhelm Dilthey proposed a theory of the human sciences that includes both the humanities and the social sciences. All these disciplines have the task of understanding human interaction and productivity by focusing on the relevant contextual conditions that are in play. Understanding is an orientational process that is not merely cognitive but also engages the life of our feelings and the interests of the will. Dilthey often contrasted understanding in the human sciences with explanation in the natural sciences, but that does not mean that the human sciences never need to explain. Since the natural sciences deal primarily with external experience for which things appear to be discrete, their connectedness is hypothetical from the start. However, a human science such as psychology starts

---

Rudolf Makkreel was deceased

R. Makkreel (✉)
Atlanta, GA, USA
e-mail: philrm@emory.edu

© The Author(s), under exclusive licence to Springer Nature Singapore Pte Ltd. 2022
D. McCallum (ed.), *The Palgrave Handbook of the History of Human Sciences*,
https://doi.org/10.1007/978-981-16-7255-2_45

with states of mind that merge with each other from the start. This connectedness may be indeterminate and needs further analysis to understand its structural organization, but it is not hypothetical. If explanations are needed, it is for specific contextual purposes. The relation of understanding to the interpretation of objectifications of human life such as expressions and actions becomes increasingly important as Dilthey develops his hermeneutics. The historical significance of these objectifications cannot be fully understood in universal cognitive terms without also relating them to the sociocultural systems and specific epochs that frame them. Together they produce a kind of reflective knowledge that is evaluative and practical. Dilthey always thought of philosophy as more than a systematic discipline and as integrally related to the contiguous fields of religion and poetry. They all contribute to our worldview.

Keywords

Dilthey · Kant · Human sciences · Explanation · Understanding · Contextualization · Worldview · Hermeneutics · Orientation

Introduction

In this chapter, Wilhelm Dilthey's contributions to the human sciences will be reexamined. Dilthey was very influential during his lifetime, but because much of his work had not yet been published when he died, many of his successors made premature conclusions about his final overall position. These partial assessments by Husserl, Heidegger, Gadamer, and others were taken at face value by their disciples. The slow rate at which Dilthey's writings were published both in the *Gesammelte Schriften* (Dilthey 1914–2005) and the *Selected Works* allowed many misleading claims about Dilthey to linger. In what follows, a comprehensive developmental account of Dilthey's philosophy will be provided.

Wilhelm Dilthey was a German philosopher born in 1833 in Biebrich am Rhein. He first not only studied theology at Heidelberg but also read the works of Hegel and Friedrich Schleiermacher there with Kuno Fischer. Later he studied with Friedrich Adolf Trendelenburg in Berlin and completed his dissertation on Schleiermacher's ethics in 1864. His first professorship was in Basel, but he soon moved to universities in Germany to be closer to Berlin where he felt most at home. When the Hegel chair in philosophy at the University of Berlin opened up with Hermann Lotze's unexpected death in 1881, Dilthey rushed the first volume of his *Introduction to the Human Sciences* into print. This made it possible for him to be appointed to this prestigious position where Dilthey remained for the rest of his life. He became one of the leading members of the Prussian Academy of the Sciences and presented many of his most important writings at their meetings. He also initiated the project of the immense Academy edition of Kant and appointed the other editors. As early as 1860, Dilthey wrote in his dairy that he intended to write a "new critique of pure reason" that would eventually turn into his life project of a Critique of Historical Reason. It would revise the idea of a Kantian critique in light of the historical philosophical turn that began with Hegel. What Kant had done, to ground the natural sciences, would now also be done for the human sciences and the understanding of history.

Dilthey's own philosophical thinking was informed by an incredible knowledge of its history. He also had a great interest in religion and the arts. He wrote innumerable essays about European literature and music. Dilthey died in 1911 in the Dolomites where he was writing an essay on the nature of religious feeling in light of the breakthroughs made by Schleiermacher and William James. Dilthey and James met once at a dinner at the home of Hermann Grimm in Berlin and James wrote about this meeting in a long letter to his sister: "he was the first man I have ever met ... to whom learning has become as natural as breathing." James later invited Dilthey to a psychology conference in Boston, but Dilthey never made the trip.

Understanding in the Human Sciences

Dilthey made important contributions to our modern conception of the human sciences and is probably best known for the way he made contextual understanding central to them and how that differs from explanation in the natural sciences (Makkreel 1992; Lessing et al. 2011; Nelson 2019; Rodi 2003). However, this distinction between explanation and understanding is provisional for Dilthey and is not meant to sharply set these two kinds of sciences apart. It points to two ways of responding to the initial conditions of empirical inquiry. Thus, in trying to reconceive psychology as a human science, Dilthey wrote that whereas the natural sciences have as their object facts that are observed as isolated external phenomena, the data of the human science of psychology are given directly from within as a real living continuum. A similar claim can be made for the study of social organizations to the extent that we think of ourselves as participant observers. This means that the natural sciences need to explain how discrete material phenomena are related and that at least some human sciences can start with an immediate sense of connectedness that is at least partly understandable. The causal relations that the natural sciences seek are hypothetical from the start and can only be ascertained when they are derivable from universally accepted laws or accord with statistical generalizations. The human sciences are less hypothetical, but they have their own challenges which at times also require explanations. The holistic understanding of psychic and social life that is more easily attained tends to be vague, and each of us adds our own perspective. So we do not necessarily understand human relations objectively. The initial context of understanding tends to be rather limited or provincial and often needs to be revised or expanded. For this reason, Dilthey argues that the continuum of understanding needs to be constantly reflectively reassessed.

For the natural sciences, the main challenge is to find ways to synthetically connect things that are external to each other. By contrast, the human sciences need to analyze processes and activities that are already indeterminately connected to find more determinate ways of organizing them as part of the world. That is why Dilthey called his psychology both descriptive and analytical. The general ways in which we feel our mental life to be internally connected must be analytically restructured to properly understand more specific functions that then become the concerns of other human and natural sciences.

In his *Introduction to the Human Sciences* of 1883, Dilthey designated his descriptive and analytical psychology as the first of the human sciences since the experienced nexus of consciousness is directly available to us. To those who shed doubt on our ability to introspect because attentive self-observation will interrupt the ordinary flow of experience, Dilthey responds by claiming that our basic access to our experience is a "reflexive awareness [*Innewerden*]" (Dilthey 2002: 48; 2010: 169). Reflexive awareness constitutes the felt self-givenness of our experience as it extends from within us to the world we inhabit. Reflexive awareness is prereflective. It is directly presentational and precedes the reflective distinctions between act and content, subject and object, associated with the representational states of traditional epistemology and psychology. Dilthey conceived his new psychology as the first of the human sciences, not in the sense that it grounds any others, but in that it provides a preliminary orientational framework for all of them. The other human sciences include not merely the humanities that deal with meaningful cultural subject matters such as art history, literary theory, philology, philosophy, and religion. The human sciences also encompass what we call social sciences such as anthropology, economics, history, and political theory and what we now call cognitive sciences such as linguistics.

Dilthey's term for the human sciences is *Geisteswissenschaften*, which literally means "sciences of spirit." The term is reminiscent of Hegel, but he had only used the singular expression *Wissenschaft vom Geist* to refer to philosophy as one comprehensive science. The first plural usage is generally attributed to the 1849 translation of that part of John Stuart Mill's *System of Logic* that deals with the moral sciences, but there were earlier uses (see Makkreel 2012: 295–296). Because Dilthey was not a Hegelian, he hesitated about invoking an idealistic and totalizing notion of spirit. In 1875, he chose a more mundane title when he published the essay "On the Study of the History of the Sciences of Human Beings (*Menschen*), Society, and the State." He also toyed with the terms *Humanwissenschaften* and *Kulturwissenschaften*. The latter term was chosen by the Neo-Kantian Wilhelm Windelband but does not include psychology. So in 1883, Dilthey ended up choosing the now standard German term *Geisteswissenschaften* as the most convenient shorthand with the stipulation that they only refer to the historical domains designated by Hegel as objective spirit. Whereas Hegel considered art, religion, and philosophy as merging into a comprehensive absolute spirit, for Dilthey these fields remain distinct parts of what continues to be historically handed down to us as a more regionalized objective spirit. There is no appeal to an overarching universal spirit, merely to a manifolded web of spirit. Spirit is now the local context of customs, norms, assumptions, and conventions that we find ourselves part of. Just as Dilthey's descriptive psychology oriented each of us to the world mentally, so objective spirit is a medium that orients each of us publicly and historically.

Each of us is contextually embedded in the world. In an 1890 essay entitled "The Origin of Our Belief in the Reality of the External World and Its

Justification," Dilthey argues that this belief is not an intellectual inference but an attitudinal response based on directly felt resistance (*Widerstand*) to our motor impulses. Here again, the distinction between the reflexive and the reflective is important. Felt resistance is reflexive and does not become the positing of an external reality until it is reflectively explicated as a restraint (*Hemmung*) of some willed intention (see Dilthey 2010: xvii, 21–72). The reflexively felt world is the human world that we are a part of; the external world of nature has become alien to us in modern times.

Windelband moved away from this attitudinal and content-focused approach to the human and natural worlds and attempted a cleaner way of distinguishing the cultural sciences from the natural sciences. He defined the cultural sciences as being idiographic in method and the natural sciences as nomothetic or law based. He thought that the psychology laboratories of Wilhelm Wundt would eventually come up with more precise natural laws of mental life than Mill's laws of association. Dilthey responded by claiming that these laws would be too hypothetical to be properly tested. But his main objection was to Windelband's notion of the idiographic. To conceive of history and cultural life in terms of singular events is reductionistic. The *Geisteswissenchaften* do not just describe particulars and leave explanation by means of universal laws to the natural sciences. There are human sciences such as linguistics and economics that can also come up with explanative laws of development. This means that the initial distinction between the natural sciences as explanative and the human sciences as concerned with understanding is only approximately right. It needs to be qualified as follows. It is not the case that causal explanations are impossible in the human sciences, but that we are not likely to find underlying pervasive laws like the law of gravitation. Nor can we have experimental human sciences such as chemistry where we can determine what happens when a finite number of elements or particles are made to interact in a test tube without outside interference. All the natural sciences can be successful to the extent that they relate a limited number of variables in abstract contexts. In the sociocultural world, those conditions are more difficult to produce because of human involvement. Nevertheless, within certain disciplinary domains, it is possible to find uniformities if only for limited periods. Thus, Dilthey thought that within capitalist economies like England in the nineteenth century, we could determine certain causal laws of development.

Thus, Dilthey's response to Windelband was that there are systematic human sciences that "connect the general and the individual" (Dilthey 2010: 227) and are not merely idiographic. Indeed historical singularity cannot be determined without placing it into a general context to give it meaning. The highest task of the human sciences is to understand what individuates or sets apart one singular event from another. All the sciences relate what is particular and what is general, but the human sciences do it differently than the natural sciences. To causally explain something like the fall of an object is to subsume a particular to an already accepted universal law. To understand the meaning of our experience is to place it into some general context that then needs to be reflectively analyzed and articulated. This is explicated most fully in Dilthey's "Ideas for a Descriptive and Analytic Psychology" of 1894 that regards the psychic nexus as a continuum that provides an initial sense of coherence. Instead of using the ordinary term *Erfahrung* that conceives of experience as an external aggregate, Dilthey introduces the term *Erlebnis* that connotes an experience that possesses the initial unity of something lived through from within. *Erlebnis* can thus be translated as lived experience. Dilthey seems to sometimes equate it with inner experience. But what he speaks of as inner experience is not just introspective. He gives the example of looking at a picture of Goethe in his study as an inner experience. It is perceptually outward directed, but because it reminds him that it was inherited from his father, it leads him to reminisce in an emotionally charged way. Lived experience is really both inner and outer because it is situationally oriented. Another characteristic of lived experience is that it builds on itself. We never experience something exactly the same way the second time we see it. Over time, the psychic nexus becomes an "acquired psychic nexus" that informs our ongoing lived experience of the world. This acquired nexus produces the structural articulation of our psychic life and becomes the basis for the development of individual character formation (see Dilthey 2010, 188–189). The acquired psychic nexus works in the background to automatically filter out those constituents of experience that are no longer of interest to us and intensifies constituents that are of interest. This means that the imagery of our mental life is habitually being transformed.

In an earlier work of 1887 entitled "The Imagination of the Poet," Dilthey delineated three processes of poetic metamorphosis that are relevant here. Poetic imagery is transformed beyond the bounds of reality under the influence of the life of the poet's feelings. Dilthey compares this to what happens when we dream, but he resists the Romantic penchant to stress the pathological tendencies associated with genius. The first stage of image transformation serves to exclude those recurring constituents of ordinary experiential imagery that are no longer of interest to the poet's feelings. The second process is to expand or contract what is left in the psychic nexus. The third transformation is the creative one that draws "actively" on "the whole acquired psychic nexus." Rather than passively filtering out what is no longer of interest, this process finds content in the acquired psychic nexus to "complete" the process of image transformation (see Dilthey 1985: 104). Thereby what is ordinarily peripheral in our consciousness enters more fully into the nuclear images in the poet's consciousness and enlivens them. Instead of responding directly to their worldly context, poets use their imagination to respond through the intermediary context of their acquired psychic nexus. This allows at least some circumstantial constituents to penetrate into the innermost core of what is attended to. Dilthey then goes on to reformulate this process of imaginative completion in language that recalls Schiller when he writes: "That process of completion by which something outer is enlivened by something inner or something inner is made visible and intuitable by something outer is especially important for poetry which proceeds from lived experience" (Dilthey 1985: 104).

The imagination is usually conceived of as synthesizing representations that are not ordinarily linked or as recombining them, but here the imagination also articulates representations by structurally recontextualizing them. Thus, the imagination serves not only to expand our horizon but also to individuate what is given there. For Dilthey, individuality is not to be defined by some qualitatively distinctive or unique features, but by the structural articulation of common elements for which the acquired psychic nexus is the model. This acquired structural nexus of a developed human being "encompasses imagery, concepts, value-determinations and ideals as much as well-established volitional aims" (Dilthey 2010, 195–196). This means that it encompasses all three of the main aspects of our life: the cognitive, affective, and volitional. Dilthey stresses their interconnectedness in all lived experience. Thus, any cognitive experience is also evaluative in terms of selecting what is of interest and volitional in that it demands being attentive. What individuates an experience is how these general constituents are proportioned, that is, which ones are predominant.

Theoretical Cognition and Reflective Knowledge

Early in his life, Dilthey spoke of the need for a new epistemology (*Erkenntnistheorie*) for the human sciences to do justice to the fullness of lived experience. But what he really did is to supplement the epistemology of the natural sciences that focuses on purely theoretical "cognition (*Erkenntnis*)" with a theory of self-reflection that also makes room for the more encompassing "knowledge (*Wissen*)" (Dilthey 2002: 25) that is relevant to the human sciences. "Self-reflection (*Selbstbesinnung*)" examines all three of the aspects of human experience that were just discussed and thus probes the conditions "for action as well as for thought" (Dilthey 1989: 268). Dilthey never fully formulated how cognition and knowledge are to be integrated, but if one reads between the lines of how he pursues his philosophical approach to both the natural and the human sciences, one can explicate it as follows. Much of our lived experience is reflexively possessed and provides an immediate kind of *knowing* that is prescientific. Both the natural and human sciences attempt to refine that knowing by means of conceptually mediated discursive *cognition* that can be publicly and methodically tested. Because the natural sciences abstract from human involvement, they can more readily produce measurable cognitive result than the human sciences. Also because there are limits to experimentation in human affairs, the human sciences must supplement what can be intellectually cognized to generate a more inclusive reflective knowledge to deal with normative issues about our affective and volitional life. Dilthey writes that it "must refer *to all classes of knowledge. It must extend to the conceptual cognition of reality*, to the *positing of values*, and to the *determination of purposes* and to *the establishment of rules*" (Dilthey 2002: 25). This reflective knowing aimed at by the human sciences will be comprehensive and use the cognition of universal truths to illuminate the specificity of human productivity. Here the human sciences become philosophical. Instead of seeing philosophy as the queen of the natural sciences, Dilthey asserts that only as a human science can philosophy properly survey the encyclopedia of all the sciences and assess worldviews about the mysteries and obscurities of existence (de Mul 2004).

Dilthey's Approach to History

Dilthey's concern to understand individuality as well as the fullness of life also extends to his approach to history. He often pointed to the value of autobiographies for how they illuminate historical life, and he wrote an important intellectual biography of Friedrich Schleiermacher whose work on hermeneutics interested him. But he also tried to understand this religious thinker as part of the surrounding context of the cultural life of Berlin. He made Schleiermacher a point of intersection of many intellectual currents. Dilthey objected to the way Hegel submerged individuals to more encompassing forces. This raises the question how general a context a human science like history should aim for. Traditional theorists of history such as Kant, Hegel, and Ranke focused on universal history and sought for laws of human development valid for all times. Dilthey was skeptical about arriving at laws that could dialectically deduce stages of historical progress as well as Comte's positivistic claim that all the sciences must pass through three stages of theological speculation, metaphysical abstraction, and factual exactitude. History can only be understood and explained by replacing traditional constructive synthesis (*Konstruktion*) with formative analysis (*Aufbau*). History is not just some grand narrative but a much more complex array of medium-range forces and activities whose effects can at least be partially traced through a series of sociocultural systems where individuals are also able to exert some influence. Dilthey's main work is entitled *The Formation (Aufbau) of the Historical World in the Human Sciences*, which amounts to the claim that the historical world is given its understandable form in the specific systems that the human sciences focus on. The *Aufbau* in the German title has sometimes been translated as "construction," but that ignores the way Dilthey distinguished between an analytical *Aufbau* and an artificial synthetic *Konstruktion*. To properly understand a momentous event like a revolution or a war is to focus on those sociocultural systems that intersect there. They provide the relevant contexts for historical understanding.

A sociocultural system could be a political, religious, or educational institution, a court of law, a parliament or stock exchange, a trade union, or musical academy. Dilthey initially called them all purposive systems that perform functions that require cooperation and organization. Each of them is characterized by what Kant had called the "immanent purposiveness" of systems such as organisms that preserve and advance themselves through change. But whereas organisms are primarily centered on themselves, sociocultural systems are more adaptive to their context and can undergo greater transformations over time. That is why Dilthey ended up calling them *Wirkungszusammenhänge* or productive contexts or systems because their purposes often change. These productive systems produce cooperative effects that are often unexpected. History has to be understood as the product of human agency working through productive sociocultural systems which themselves interact with each other either cooperatively or competitively. History is approached as a kind of cosmic weather where there is an intersection of various forces. This new approach ends up with what is now called *Wirkungsgeschichte* or a history of effects.

Hermeneutical Objectivity

So far, we have spoken of both private psychological and public spiritual meaning contexts. Dilthey originally assumed that the nexus of our own experience has its own immanent intelligibility that makes self-understanding the basis for understanding others. But starting with his 1900 essay "The Rise of Hermeneutics," he begins to question that assumption. It is not the case that hermeneutics was a new discovery for Dilthey. In 1860, he had already written a long prizewinning essay entitled "Schleiermacher's Hermeneutical System in Relation to Earlier Protestant Hermeneutics" (Dilthey 1996: 33–227). But he now saw that when hermeneutics is fully extended to deal with secular texts and other historical objectifications, it takes on a new philosophical significance for our self-understanding. Just because we give a certain sense to our own experience does not mean that we have properly understood it. He writes that "Even the apprehension of our own states can only be called understanding in a figurative sense.... Only in language does human inner life find its complete, exhaustive, and objectively understandable expression. That is why the art of understanding centers on the exegesis or interpretation of those remains of human reality preserved in written form" (Dilthey 1996: 236–237). Hermeneutically, the objective understanding of our own experience does not occur until we have expressed it and then interpret it. This means rethinking what is inner from the outside. And the fact that we grow up in the medium of objective spirit means that what is inner is not just psychological. The words we use to express what we experience are not just our own but the common possession of the language we inherit. Thus, as Dilthey makes clear in his most definitive hermeneutical essay that he presented as a talk at the Prussian Academy in 1910, "The Understanding of Other Persons and Their Manifestations of Life," "Before the child learns to speak, it is already wholly immersed in the medium of commonalities. The child only learns to understand the gestures and facial expressions, movements and exclamations, words and sentences, because it constantly encounters them as the same and in the same relation to what they mean and express. Thus the individual becomes oriented in the world of objective spirit" (Dilthey 2002: 229–230). This oriented mode of understanding is called elementary understanding. Here meaning is supported by a framework of commonality in which we partake.

These two quotations represent a shift in Dilthey's approach to understanding. Originally, the acquired psychic nexus provided the structural framework for finding meaning, but now objective spirit does. The first context gains us private certainty [*Gewißheit*], but the second provides the conditions for public reliability [*Sicherheit*]. Certainty is a subjective criterion, but hermeneutics requires an objective reliability that can be confirmed by others. What Dilthey comes to expect from the human sciences is reliable understanding, and this will require what he calls higher understanding.

The elementary understanding made possible by the commonality of our native language and the customs and conventions of our local community is important for orienting us in this world but only gets us so far. It relies on a sense of familiarity but does not prepare us for dealing with people from different regions who may speak another language or make use of different conventions. Even someone using our own language may say something puzzling or incongruent that will require reflection or higher understanding. Then we must adjust our contextual horizon by consulting a dictionary with a broader scope or by applying more general logical means to analyze what has been expressed. And if there is a complexity about the topic under discussion, then some of the disciplinary cognitive tools of the human sciences may be able to explicate what is meant. Higher understanding is not content with implicit elementary understanding and seeks to replace a sense of commonality with a broader consensus based on universal standards.

Dilthey adds that if there is some reason to suspect deceit or some practical interest that causes an obscurity of meaning, then the relation between expression and what is expressed passes over into that of productivity to end product. This introduces an undeclared intent or even unconscious concern into what is to be understood and requires us to search for an "overall connectedness or unity of a work or person – a life-relationship" (Dilthey 2002: 233). At this point, we approximate the highest task of understanding that Dilthey calls a transpositional understanding of otherness. He sometimes calls it a process of reexperiencing (*Nacherleben*) that is also important for historical understanding. It is necessary to point out that reexperiencing is not a literal reproducing or reliving. Just as re-cognition is not just cognizing something over again but introduces a retrospective meta-awareness, so reexperiencing something is to more deeply understand it. This is why the hermeneutic circle that proceeds from part to whole and back is not a vicious but a productive circle. Whereas originally understanding reverses the process that was experienced, reexperiencing goes forward with it again and can produce new insight. Thus, historians can sometimes understand what happened in the past better than those who experienced it. They may not be able to reproduce all the details of what happened, but they have a more extensive context in which to assess the significance of it.

Three Kinds of Objectifications

Not only are there different modes of understanding and interpretation, but there are also, as the title of his 1910 talk suggested, different kinds of manifestations of life to be understood that present distinctive challenges. There are those life-manifestations that we have called expressions intended to mean something as well as those manifestations such as actions that can mean something without any communicative intent. We will start with the theoretical class of life-manifestations that "consists of concepts, judgments, and larger thought-formations. As constituents of science, they have been detached from the lived experience in which they arose" (Dilthey 2002: 226). Here understanding is focused on "mere logical content which remains identical in every context" (Dilthey 2002: 226) and conforms to universal norms. Expressions such as "water boils at 212F" and "15+39=54" are intended to convey a determinate meaning that can be fully shared. But this an impersonal meaning and reveals nothing about the real state of mind of the speaker. Dilthey's second class of life-manifestations are actions that are intended to have an effect without necessarily wanting to communicate this to others. If we see someone lift a hammer, we expect that this is to pound a nail into something. But we cannot be sure that the hammer will not be used as a weapon even if we happen to know that the person holding it is a carpenter. Here the situational context is decisive in surmising the momentary purposive intent. However, this will not reveal any overall state of mind. Only the third class of life-manifestations are meant to do that. These are what Dilthey calls expressions of lived experience. One might think that here he is just thinking of expressions of our conscious states of mind such as joy and grief, exclamations of affection, admissions of regret, and declarations of intent to punish someone. Expressions of lived experience may seem easy to understand, but Dilthey notes that they are often subject to the power of hidden interests and may be deceitful. His main focus, however, is on artistic works that "contain more of the nexus of psychic life than any introspection can catch sight of" and draw "from depths not illuminated by consciousness" (Dilthey 2002: 227). Here understanding can never be complete, and psychology will prove to be an inadequate guide. Dilthey never rejected the importance of psychology, but his hermeneutic turn led him to no longer regard it as the first of the human sciences. Hermeneutically, any starting premises and assumptions are subject to revision. Great works of art are expression of lived experience not in the sense that they are true in a confirmable way to the artist's life but truthful in a deeper human sense. Kant had spoken of the need of the esthetic spectator to respond to beauty in nature and in art in a disinterested contemplative way to fully appreciate it. Dilthey supplements that with the claim that artists must remold their lived experience disinterestedly into something impersonal (see Dilthey 1985: 227). Thus, what we should look for as expressed in a poem "is not the inner processes of the poet; it is rather a nexus created in them but separable from them... here a spiritual nexus is realized that enters the world of the senses, and which we understand by a regress from that world" (Dilthey 2002: 107). Similarly, as readers of a novel, we must understand hermeneutically by proceeding from the outside in, and this inside is the structural meaning nexus of the novel whose constituents are its characters and plot.

These comments require us to revisit Dilthey's earlier claims about the poetic imagination where the acquired psychic nexus of the poet was made the point of reference for how the meaning of a literary work is created. To the extent that this suggested that the reader of a poem must reexperience the state of mind of the poet, this is now shown to be wrong. This is because there is no assurance that artists are fully conscious of what their acquired psychic nexus contains. We can still refer to the acquired psychic nexus of the poet as providing a partial explanation of how the work is generated without allowing it to define the full meaning of the work. And in so far as the acquired psychic nexus orients us to the world, it represents an outlook on life, but not yet an explicit worldview.

The way an artist's creative process surpasses mere psychological states is suggested most clearly when Dilthey characterizes a composer as not merely converting private feelings into sounds. The feelings of a Beethoven or a Schubert are musical from the start because they live in a tonal world that provides a medial context for them to work in. A medial context encompasses both the material medium used in a specific art and the conventional means that formally organize and convey its meaning (see Makkreel 2015: 212–214). For Beethoven and Schubert, this includes the well-tempered or chromatic scale of 12 semitones and the allegros, andantes, and rondos of the sonata form. Understanding the relation between the lived experience of these composers and their compositions does not require us to go back to their psychic life per se but to the way they responded to the medial tonal world they inhabited in Vienna. This is the most relevant context to be considered to assess their musical accomplishments. We only need to consider their private life to explain some possible oddities about their output as might be the case with some of Schumann's symphonies. Now psychology becomes a last resort rather than a point of departure.

From all this, we can conclude that the task of hermeneutics is not merely the exegesis and philological authentication of literary and theoretical texts, as well as of archival records. Nonlinguistic objectifications, such as paintings and musical scores, historical monuments, and industrial products, can be considered as textual in a figurative sense. But the overriding task of hermeneutics must be a consideration of the relevant contexts that orient the interpretation of these life-manifestations. As a philosopher of history and human sciences more generally, Dilthey focused on those relatively enduring sociocultural productive contexts (*Zusammenhänge*) where human action is concentrated. But he also spoke of the importance of tracing historical development in terms of more limited epochs. These temporal contexts are not to be equated with the more general notion of ages (*Zeitälter*). He wrote an important essay on "The Three Epochs of Modern Aesthetics and Its Present Task" that bridges the division between the ages of Enlightenment and Romanticism. The essay traces the development of the philosophy of art from the rationalist esthetics of Leibniz, Baumgarten, Mendelssohn, and Lessing, to the more experimental analysis of esthetic impressions of Hutcheson, Gerard, Burke, Kames, and Rameau, to the historical method of the nineteenth century that makes the creative capacity of artists central. The rationalist epoch upholds the perspective of art critics and their rules. It "conceives of beauty as a manifestation of the logical in the sensuous and of art as a visible presentation of the harmonious nexus of the world" (Dilthey 1985: 187). The analysis of esthetic impressions of the second epoch considers how certain qualities in objects can produce pleasurable feelings in the esthetic spectator. It turns away from intellectual and ideal considerations to the ways in which artists can produce their desired effects on the tastes of their public. This more empirical analytical approach was important, but it reduced a work of art to "a mere aggregate of properties capable of arousing us" (Dilthey 1985: 199). The third epoch attempts to understand the creative historical development of art as rooted in both the spontaneity of transcendental consciousness and the unconscious power of genius. The latter allowed estheticians to weave the arts "into the larger fabric of cultural and spiritual development" (Dilthey 1985: 203). The present task of esthetics as Dilthey saw it in 1892 was to recognize that all these epochs had run their course. He found himself in a transitional period where the naturalism of Dickens and Balzac was becoming dominant and their search for down to earth truths had left behind the ideals of rational order and creative profundity. Dilthey saw the need for a new theory of style that would give works of art structural coherence in a nonidealizing way by incorporating "precisely that which is most brittle, factual, and particular" (Dilthey 1985: 219) in the new medium of the novel. Style is defined as the inner medial form that gives works of art their contextual bearings. Now what Dilthey had described as a mental transformative completion of a nuclear image is formally redefined in terms of the style of a work of art that brings its overall features into focus in some dominant point of impression (see Dilthey 1985: 217).

Great works of art are able to imaginatively recontextualize the world to enhance the meaning of our lives whatever our circumstances. But when historians recontextualize their subject matter from their ever-changing perspectives, they must let what they reimagine intersect with the generally agreed-on reality. The basic parameters of the original contextual conditions must remain the point of departure. What is added with each reinterpretation must take note of previous accounts if only to revise them in light of new information about what happened or what resulted. The complexity of history is such that different contextual sociocultural systems intersect in some larger contextual nexus such as a war or revolution, the collapse of a regime, or the rise of a nation. Even if there is agreement about the relevant contributing factors, there may be disagreement about their relative importance. Here judgment becomes central for the reflective assessment of purposiveness in history.

Kant and the Importance of Reflective Judgment for Interpretation

Kant's *Critique of the Power of Judgment* of 1790 was focused on the assessment of esthetic and teleological purposiveness, but it also provides general guidelines for evaluating and interpreting all human affairs that have been overlooked (Kant 2002). In this work, Kant makes a distinction between the determinant judgments that apply the mechanical principles of most natural sciences, and the reflective judgments needed to make sense of organic creatures and to find meaning in human creativity. In a natural science such as physics, particular processes are determined by preexisting universal laws like gravity. But when we judge organisms and works of art, we are confronted with particulars for which there are often no available laws or rules to define how they function. Here we must search for new universal concepts to characterize them, and that requires reflection. When we exclaim that something is beautiful, this is just a placeholder for a better qualitative ideational understanding of its significance, which we can only comprehend if we engage with others in judging it. What this means is that determinant judgments explain by means of known universals and reflective judgments seek for what Dilthey called the contextual understanding of the human sciences. As we saw earlier, the idea of human sciences was articulated later in the nineteenth century, but Kant paved the way for what he called useful sciences when he lectured extensively about the need to replace traditional psychology that was introspective and posited a metaphysical soul with a new pragmatic anthropology that orients us to the larger context of the world of nations and human commerce. We can now expand on this generally overlooked overlap between Kant's theory of reflective judgment and Dilthey's search for holistic understanding in the human sciences.

This kinship between Kant and Dilthey may not be immediately apparent, for when Kant speaks of understanding, he means what can be theoretically determined or explained by the intellectual faculty of *Verstand*. Dilthey's understanding involves a more encompassing reflective process of *Verstehen* that is both theoretical and practical. What needs to be brought out is that for Kant, practical tasks require ideas of reason rather than concepts of understanding. Reason projects what our freedom makes possible, but it sometimes must also confront what is not possible, and then it must settle for what he called ideational comprehension [*Begreifen = comprehendere*] in his early lectures on logic. To comprehend is to have "sufficient insight, insofar as something serves a certain purpose" (Kant 1992: 300). Comprehension thus comes closer to Dilthey's *Verstehen,* for it is both theoretical and practical. And when the idea of comprehension reoccurs in the *Critique of Judgment,* it is linked to what Kant calls the purposiveness of reflective judgment.

When reflective judgment confronts an unfamiliar particular phenomenon or situation in an effort to comprehend it in more general terms, it must first properly contextualize it relative to the world. According to Kant, there are four kinds of judgmental worldly contexts. If something can be thought or imagined but has never been experienced, then we can merely place it in an open-ended field. If it is experienceable, then it belongs to a territory that we can share with others. Objects that are determinantly known and can be causally explained belong to a lawful domain. And things that are merely part of our acquaintance because they happen to be familiar to us may be considered as belonging to a local habitat. It turns out that we can assign a different modality to each of these judgmental contexts. A field contains mere possibilities, a territory is actual for us humans, a domain is marked by necessary order, and a habitat displays the contingency of what we happen to come across. We can now apply these contexts to what Kant writes about the esthetic experience. There is for Kant something unexpected about discovering a beautiful flower along a country path and instantly liking it. Esthetically, our task is to establish whether the pleasure we happen to feel in the habitat of our inner sense can be made communicable to the territorial *sensus communis* of all human beings. Because the pleasure we feel is only cognitive in a very general sense, it cannot be communicated to others in the determinate conceptual terms linked to a rule-bound domain. Esthetically, we can only resort to symbolical analogies that are indirect and rely on playful contextual comparisons. Esthetic pleasure requires us to symbolically transpose ourselves into the contextual standpoint of others. Kant has been accused of keeping esthetic pleasure too subjective, but what he really does is to make it intersubjective and world oriented. Similarly, judgments about human actions and products are normative and will only be valid if we find the right modal context for evaluating them (see Makkreel 2015: 100–108).

Rethinking Historical Reason as Historical Judgment

Relating reflective judgment in Kant and contextual understanding in Dilthey have some important consequences for Dilthey's original project of a Critique of Historical Reason. It shows that reflective contextualization provides an orientational strategy for comprehending matters that need to be both understood theoretically and judged normatively. It also provides the framework for a more diagnostic hermeneutical approach to the human sciences that can prioritize the most relevant factors in a situation. Although Kant presented his four judgmental contexts rather abstractly, their importance can be illustrated more concretely by imagining how they can be applied to the complexities of human history. Thus, in gathering the various accounts and documents about a past war, historians can contextually frame them in terms of the actual territorial disputes at stake, and the contingent local habitats that may have sparked the conflict. Local newspaper reports about battles and casualties will have a different status than the legal papers that document the outcome of the war. The latter will have to be made sense of in terms of the respective laws of the governmental domains involved. The various contextual discourses distinguished here intersect and are all relevant, but they do not have equal force or validity and must be weighed reflectively by means of historical judgment. Each sphere may demand its own norms of assessment. What makes history so fascinating is the possibility of also exploring the field of counterfactuals. They challenge our judgment. What would have happened if Hitler's scientists had developed the atom bomb before the USA did? Would that have made all the military advances of the Allied forces irrelevant?

We saw that Dilthey's approach to art history focused on epochal disciplinary developments that had an independent life from the more standard general division between Enlightenment and Romanticism. Another historical concept that Dilthey helped to define was that of a generation. In his massive *Life of Schleiermacher*, he showed how this religious thinker interchanged new ideas about the profundity of life with a whole group of literary and intellectual figures. Schleiermacher was born in 1768, but the generation that Dilthey delineates around his subject does not begin to crystallize until 1796 after he arrived in Berlin. It is Dilthey's judgment that this is the point in time when Schleiermacher and contemporaries like the Schlegels began to engage each other and define themselves against the earlier generation of Lessing, Kant, Fichte, and Goethe that had so powerfully influenced them. A historical generation is not a biological generative life span but an ideational formation that is delineated from within itself. Both involve an immanent productivity that is not reducible to the external conditioning of causality. Here again, we see the afterlife of Kant's notion of functioning systems that are immanently purposive. They are not closed off from what occurs outside of them but respond in a characteristic way.

What Dilthey tends to look for in history and life are typical patterns. When we are unable to find lawlike universals about human life, we should focus on the generic forms in which its history organizes itself. This interest in typicality was already exhibited in Dilthey's *Poetics* where the task of the imagination is said to be to bring out what is typical in human experience. The plots of literary works can create certain individuals to typify what happens around them. A literary type embodies something generic without being merely general. It is contextually significant but cannot be given a universal meaning or be scientifically defined. They are too localized to be universalizable but nevertheless provide insight into human existence.

Dilthey's Theory of Worldviews

When we turn to Dilthey's theory of worldviews, we find him distinguishing types of them that are morphological in that they can be clearly delineated from each other. These are not temporally and contextually individuated as in literature but more general and stable. Since worldviews deal with ultimate questions that have not been scientifically resolved, they are too broad in scope to be universalizable. Worldviews reflect on all three aspects of lived experience, but they differ on which of them is to be given priority. The worldview type that gives priority to the empirical and cognitive aspects of experience is called naturalism by Dilthey. The type that grants priority to our affective life and to what is valued is called objective idealism. It searches for the overall coherence of the universe and values harmony. The worldview type that prizes our volitional life above all else is called the idealism of freedom. Dilthey sees these three types recurring throughout the history of Western philosophy. The worldview of naturalism goes back to ancient Greek thinkers such as Democritus and Epicurus. Other important proponents of naturalism recognized by Dilthey are Hobbes and Hume, Mill and Comte, Feuerbach, Marx, and Nietzsche. Although naturalism reappears in slight variations, it is pluralistic in nature and open to contingency. The life of the feelings and the will is considered to be physically determined so that freedom is a mere illusion. The idealism of freedom is its extreme opposite. It conceives the world dualistically. Natural processes are determined, but human beings are also spiritual beings that can transcend their natural urges. Dilthey regards the idealism of freedom as first occurring in Attic philosophy "as a conception of formative reason that shapes matter into a world. The great discovery of conceptual thought and moral will and their connection with a spiritual order that is independent of the natural order of things derives from Plato and remains fundamental in Aristotle" (Dilthey 2019: 285). Many Christian thinkers adopted this division of a natural and spiritual order without accepting reason as shaping the world. In religious terms, the idealism of freedom merges with theism. The most prominent modern proponents of the idealism of freedom are Kant and Fichte who defined its ethical imperatives and consequences. In France, Maine de Biran and Bergson modified it into an idealism of personality. William James's pragmatism is judged to be a new version of the idealism of freedom.

Dilthey goes on to say that "the majority of philosophical systems cannot be classed with either naturalism or the idealism of freedom" (Dilthey 2019: 287). Here he points to thinkers as different as Heraclitus and Parmenides, the Stoics, Giordano Bruno, Spinoza, Leibniz, Schelling, Hegel, Schopenhauer, and Schleiermacher. He calls this third main worldview type "objective idealism" because it finds ideals embodied in the actual world. It is monistic and views the world contemplatively for meaningful coherence. "To conceive the universe as a single whole in which each part is determined by the ideal meaning nexus of the whole: that is the great new thought of Leibniz's system. It is wholly defined by the question of the sense and meaning of the world" (Dilthey 2019: 288). Although he does not say so, objective idealism offers a universal hermeneutics. It searches for the kind of truthful insight that Dilthey finds in great art, whereas the other two worldview types fight about ultimate truths.

Dilthey turned to Western metaphysics to formulate his three worldview types. But they have less explicit analogues in religion and the arts. Just as the idealism of freedom can be allied with Christian theism, so objective idealism has a pantheistic counterpart. The tension between them manifested itself in Germany when defenders of the Christian faith such as Friedrich Jacobi attacked figures such as Lessing for coming too close to Spinoza's pantheism. Artistically, Schiller was the dramatist of the idealism of freedom, Goethe the poet of objective idealism, and Balzac and Zola the novelists of naturalism.

Naturalism could be said to be the worldview that holds that everything can be explained. By contrast, the idealism of freedom places limits on our powers to explain. As Kant points out, human freedom cannot be explained but needs to be posited to render the idea of moral responsibility possible. Thus, Kant writes that "had not the moral law already been distinctly thought in our reason, we should never consider ourselves justified in assuming such a thing as freedom" (Kant 1997: 4n). We cannot explain or understand our freedom as such, but we can somehow practically comprehend it through the moral law. Objective idealism by contrast is the view that nothing can be adequately explained without an overall understanding of everything. This sets a very high standard on human inquiry and on what a worldview must provide. Of course one can make it a less demanding procedural principle that allows overall understanding to be indeterminate and leaves the determinacy to more detailed explanations.

Many of those who followed Dilthey in exploring the proper use of the explanation-understanding distinction have diluted the above procedural principle even more. Thus, Max Weber and Ernst Cassirer regard all understanding to be like what Dilthey called the elementary understanding of conventional common sense and allow explanation to replace the higher understanding that aims at universally valid cognition. Thereby they reduce understanding to a mere prelude to explanation. Some analytical philosophers of the Vienna circle such as Otto Neurath and their American allies went even further by reducing Dilthey's understanding to a mode of empathy that may be suggestive but unreliable for getting to the truth. But Dilthey himself had little use for empathy as it was being explored by his contemporaries like Theodor Lipps. They conceived empathy (*Einfühling*) as an immediate feeling of oneness with something appealing that leads one to project oneself into it. Nothing could be further from Dilthey's higher understanding which requires conceptual mediation on the way to a more comprehensive reflective knowing. Today, empathy is being used in a much broader sense that is not merely a one-way projection but makes room for reciprocal engagement and communication.

This mode of feeling-with others can contribute to the process of understanding but still falls short of it.

To take Dilthey's theory of *Verstehen* seriously is to allow for both determinate understanding and determinate explanations. Both elementary and higher understanding are highly descriptive in nature, but they also organize their subject matter contextually in ways that define the meaning of what is at stake. Explanations may further clarify aspects of what was described, but the contextual circumstances exposed by understanding remain the framework for testing the validity of these explanations. An explanation is only as good as the understanding that went into it, and the results, if valid, will in turn deepen our overall understanding. Thus, the relation between understanding and explanation should be reciprocal.

It has been shown that Dilthey's early descriptive psychology lost some of its importance in light of his increasing interest in hermeneutics. A psychology focused on the inner life of individuals can no longer ground the truth of our interpretations. However, when psychology is expanded into a broader anthropology, it can still provide clues to how we understand each other. We spoke about how traditional epistemology needs to be replaced by philosophical self-reflection that integrates our cognition with the evaluative and volitional dimensions of our life. Late in his life, Dilthey expanded his idea of self-reflection into that of anthropological reflection. This led him to consider whether the instinctive sense of solidarity that develops on the basis of family and other elemental kinship relations can be made use of by more advanced ethical systems. But before going into that problem, it should be noted that in his essay about our belief in the reality of the external world, Dilthey showed that the bodily felt resistance to our will that derives from the world around us also incorporates our relation to other human beings. Human differentiation is more complex in that it also can provide feelings of support. Thus, he wrote that "The emotional and volitional processes that give color and strength to the reality of other life-units consists of dominance, dependence, and community. Through them we gain the lived experience of a *thou*, which at the same time deepens the sense of the *I*. A constant slight change of pressure, resistance, and support causes us to feel that we are never alone" (Dilthey 2010: 29). Whether this inspired Martin Buber's work on the I-Thou relation is not known, but we do know that Buber studied with Dilthey and was close enough to the family to bring Dilthey's corpse back to Berlin after he died in the Dolomites in 1911.

The way that Dilthey defined the I-Thou relation in bodily terms was prescient. We think of the hermeneutics of bodily practices as a recent creation, yet Dilthey makes our embodiedness very much part of his thought. Whereas traditional epistemology made our mental life exclusively phenomenal, Dilthey insisted that our lived experience is always real and sometimes physical. In dreams and hallucinations, images are phenomenal, but my experience of the outside world can become real to the extent that I feel it resisting "the corporeality of my self" (Dilthey 2010: 56) and its motor impulses. The world is not immediately real, but once I recognize it and other human beings as restraining my will, I become part of it. This allows me to develop what Dilthey calls a fellow feeling (*Mifgefühl*).

Dilthey's Formative Social Ethics

In his *System of Ethics* based on a University of Berlin lecture course in 1890, Dilthey argued for a social ethics that is rooted in anthropological reflection. He criticized Mill's utilitarian moral theory for using an associationist psychology that links human feelings to sense-based impressions that come from without. What is lacking according to Dilthey is a real understanding of how feelings are rooted in our inner drives and desires. Mill cannot fully account for what motivates human agents. His social program placed too much emphasis on governmental legislation to motivate us from on high. A true ethics must be able to motivate individuals from within rather than through legislation from without. Dilthey defines his task in the lectures on ethics as developing a "psycho-ethical" approach that is rooted in "anthropological-historical analysis" (Dilthey 2019: 104) that can make use of a more primitive mode of kinship or group identity. Here individuals grow up with a sense of solidarity that stems from within – an inner sense of belonging to something larger than themselves. Dilthey contrasts this kind of fellow feeling or feeling-with (*Mitgefühl*) to the sympathy that Hume and Adam Smith had invoked to motivate people to help others. According to Dilthey, sympathy is a feeling "transferred from one living being to another" (Dilthey 2019: 89). Sympathy affects us from the outside and cannot motivate us from within the way that a shared feeling like solidarity can. The task of morality is to broaden that sense of a fellow feeling so as to treat all others equally. That is a challenge since we know that group solidarity can often produce an us-against-them mentality. The task is to transform this instinctive bond with some others to a willed bond with more others.

Dilthey differentiates three kinds of ethics in terms of how they appeal to the will. The first kind uses the will to *negate* our natural instincts and desires as in Buddhism and Christianity. Its culmination can be found in Schopenhauer's negation of the individual self. The second kind of ethics only *restricts* our natural impulses in so far as they are contrary to our reason. Here Dilthey points to the Stoics and Kant. The third or *formative* use of the will that Dilthey supports reshapes our natural impulses from within. The model for this is the moderating lifestyle of the ancient Greeks where self-control and resolve were cultivated (see Dilthey 2019: 81–82).

The formative stance that Dilthey wants to cultivate in his social ethics is to broaden solidarity based on kinship into a concern for all human beings. Together, solidarity and benevolence are to provide the background for the recognition that we have obligations to other human beings that can be institutionalized as a social system of justice. In the final analysis, individuals must become willing to interact with each other on the basis of a sense of duty and justice. "From a personal standpoint, this commitment involves a sense of what is right or just. It comes with its own feeling of duty to mutual order and possesses a moral value completely independent of any purposes" (Dilthey 2019: 107). The expression, "sense of what is right or just," was chosen by the editors of the *Selected Works* to convey what Dilthey means by *Rechtschaffenheit,* which is normally translated as "uprightness." But uprightness is usually thought of as a private virtue, which loses the social dimension that Dilthey has in mind.

This becomes more evident in the next section of the lectures where Dilthey delineates the three main volitional incentives that drive the evolution of ethical life. The first incentive of the will is the striving for personal excellence along the lines of the formative kind of ethics we saw him espouse. The second incentive centers again on benevolence as a social virtue. The third volitional incentive is "the consciousness of the commitment that inheres in the duty to do what is right" (Dilthey 2019: 128). At the heart of this sense of commitment is the respect for others as ends in themselves. The respect for others that was reflexive or implicit in instinctive solidarity and in benevolence is now recognized to be at the core of the reflective commitment to do what is right. This socially directed sense of rightness is independent of any external enforceability.

In the final lecture, Dilthey moves from these subjective volitional incentives to the level of objective ethical principles. He does so by drawing on an early essay where he had agreed with Kant that moral oughts are unconditional and may be considered as "synthetic-practical judgments *a priori*" (Dilthey 2019: 134). This is unexpected not only because Dilthey expressed his reservations about synthetic a priori theoretical judgments throughout his life, but also because the just discussed incentives were based on empirical instinctive relations such as solidarity. This requires Dilthey to argue that the ethical obligations we adopt as autonomous adults have a prescriptive and normative quality that is not empirically derivable. Now the commitment to what is right or just is given priority over the feeling of benevolence because it is our most fundamental obligation. Dilthey affirms that the moral commitment to do what is right or just is unconditional. However, it is based on respect for other human beings as ends in themselves rather than on Kant's respect for a higher law. This moral principle is recognized by Dilthey as a *synthetic principle of unity* in that it incorporates the obligation to identify with the rights of the other. The second ethical principle loosens this being bound by the other into the broader feeling of benevolence. Benevolence "does not place us into that rigid chain of mutual obligation through the will's sense of what is right, but rather in a free reciprocal relation of human sentiments that, without a feeling of compulsion, pervades the whole moral world" (Dilthey 2019: 135). The principle of benevolence transforms the respect for the rights of others into a caring for their well-being. It provides an open-ended *synthetic principle of multiplicity [Vielheit]* that incorporates both what unites and differentiates human beings. Although benevolence was also a Kantian virtue, Dilthey's affirmation of it is more in the spirit of Lessing who was known for his tolerance of human differences.

It is Dilthey's third ethical principle that first invokes the universal *validity* of an ideal. This is a projected ideal of perfection which is explicated as the striving for inner moral worth. Dilthey adds that this "urge toward perfection … involves a creative synthesis of our organization" which must be "obtained in combination with the theoretical content of the human spirit. Thus, there are as many different ways to understand the nature and basis of this urge for perfection and value as there are cultural stages" (Dilthey 2019: 136). Accordingly, Dilthey's third principle entails that a *synthetic plurality [Mehrheit]* of articulated cultural systems will be needed to fulfill the moral destiny of humanity. By replacing multiplicity with plurality, Dilthey departs from Kant, whose third formulation of the categorial imperative moved from multiplicity to the *allness [Allheit]* of the kingdom of ends. One can ask what sets multiplicity and plurality apart, and why there is no final communal allness? The answer is that multiplicity amounts to an undifferentiated aggregate, whereas a plurality is a more differentiated series. Dilthey does not believe that history can or should move us to some homogeneous all-inclusive communal state. Instead, he argues that human attempts to create a universally valid form of morality will produce culturally distinct ethical systems, each with its own inner worth. The striving for a universal ideal does not need to bring about one uniform moral system that wipes away the ethnic heritage of the objective spirit we inherit.

When Dilthey does discuss universal history, he does not project some final telos but examines the main systematic aspects of life that can be coordinated within a specific time span such as an epoch. Each human science can apply its universally accepted disciplinary conceptual tools to the sociocultural systems that factor into the historical events that need to be understood. Human sciences can at times even explain what happened whether through some developmental law or by contextual analysis. Thus, we can expand on this survey of Dilthey's work by adding that whereas "explanation tends to be exclusive and subordinative by ruling out factors not relevant to the search for generalizations, understanding should be inclusive and coordinative by diagnosing how various strands of influence interact in a specific situation. Explanations are primarily theoretical proposals that can be tested like specific hypotheses. Understanding, however, is not something that is proposed like a hypothesis. It is something we possess, either as an inherited assessment rooted in life or as a subsequent reflective achievement" (Makkreel 2015: 191).

The understanding we possess at any point may be provisional and in need of more refined explanations. But the realization that our understanding can be revised and improved by means of valid explanations should not stand in the way of the fact that it is understanding that puts explanations in context and must ultimately judge their validity and significance. This is when understanding becomes objectively interpretive and reflectively proceeds from the outside in. Understanding is cumulative and ongoing whereas explanations are piecemeal. Explanations produce connections at clearly defined levels whereas understanding develops interconnectedness at many levels. And that is why contextual understanding and hermeneutic orientation remain important.

Conclusion

The last two paragraphs have extended Dilthey's own stated positions to reinforce the current relevance of his theory of the human sciences. Some of his main contributions that have been covered include his ideas about the direct reflexive awareness that goes behind the phenomenal representational consciousness of

Kant's epistemology and Mill's psychological laws of association. Lived experience comes with a reflexive self-presence that is directly felt as real and ongoing. And the reflection that will be needed to be able to discourse about this with others requires what Dilthey called a mode of self-reflection that is both cognitive and practical. Whereas the cognition of the natural sciences is purely theoretical and highly hypothetical, the human sciences use conceptual cognitive tools to relate the direct knowing of lived experience and the commonalities we take for granted to a more reflective knowledge that assesses the human world at large. Here we move from everyday elementary understanding to the higher understanding made possible by hermeneutics to reach for what is ultimately an evaluative worldview. Dilthey's descriptive psychology was important for delineating the holistic structural features of our experience that orient us to the world. It provides us a self-certainty that builds the acquired nexus of individual character. But the higher understanding aimed at by the human sciences demands a more objective reliability based on how our experience is expressed. To test what seems self-evident (*selbstverständlich*) or obvious to us from within, we must step outside ourselves and interpret our own expressions in the same way that others do. Only then can we reach valid self-understanding (*Selbstverständnis*). The understanding of individuality is only possible if it is historically contextualized. It requires us to consider a person as a point of intersection of many more general forces and interacting sociocultural systems. Indeed the fruitfulness of Dilthey's approach to the human sciences is the way he makes use of contextual intersectionality to understand the richness of human life and its history.

References

de Mul J (2004) The tragedy of finitude: Dilthey's hermeneutics of life. Yale University Press, New Haven

Dilthey (1985) Poetry and experience, selected works, vol 5. Edited and translated, with an Introduction, by Makkreel RA, and Rodi F. Princeton University Press, Princeton

Dilthey (1989) Introduction to the human sciences, selected works, vol 1. Edited and translated, with an Introduction, by Makkreel RA, and Rodi F. Princeton University Press, Princeton

Dilthey (1996) Hermeneutics and the study of history, selected works, vol 4. Edited and translated, with an Introduction, by Makkreel RA, and Rodi F. Princeton University Press, Princeton

Dilthey (2002) The formation of the historical world in the human sciences, selected works, vol 3. Edited and translated, with an Introduction, by Makkreel RA, and Rodi F. Princeton University Press, Princeton

Dilthey (2010) Understanding the human world, selected works, vol 2. Edited and translated, with an Introduction, by Makkreel RA, and Rodi F. Princeton University Press, Princeton

Dilthey (2019) Ethical and world-view philosophy, selected works, vol 6. Edited and translated, with an Introduction, by Makkreel RA, and Rodi F. Princeton University Press, Princeton

Dilthey W (1914–2005) Gesammelte Schriften, 26 vols. Vandenhoeck & Ruprecht, Göttingen

Kant I (1992) Lectures on logic. Cambridge University Press, Cambridge

Kant I (2002) Critique of the power of judgment. Cambridge University Press, Cambridge

Lessing H-U, Makkreel R, Pozzo R (eds) (2011) Recent contributions to Dilthey's philosophy of the human sciences. Stuttgart-Bad Cannstatt: frommann-holzboog
Makkreel (2012) The emergence of the human sciences from the moral sciences. In: The Cambridge history of philosophy in the nineteenth century (1790—1870). Cambridge University Press, Cambridge, pp 293–322
Makkreel (2015) Orientation and judgment in hermeneutics. University of Chicago Press, Chicago
Makkreel R (1992) Dilthey: philosopher of the human studies, 2nd edn. Princeton University Press, Princeton
Nelson E (ed) (2019) Interpreting Dilthey. Cambridge University Press, Cambridge
Rodi F (2003) Das Strukturierte Ganze. Studien zum Werk von Wilhelm Dilthey. Göttingen, Velbrück Wissenschaft

Durkheimian Revolution in Understanding Morality: Socially Created, Scientifically Grasped

6

Raquel Weiss

Contents

Introduction	134
Toward a Sociology of Morality	135
Durkheim's Social and Intellectual Background	136
Moral Philosophy in France	138
A Sociological Theory of Morality	139
"The" Moral Does Not Exist	140
An Inquiry into the Soul of Society	145
Taking a Stand: Normative Boundaries	151
When a Moral Ideal Is Not Really "Normal"	151
Secular Morality and Moral Individualism	153
Conclusion	155
References	157

Abstract

The question of how we should act has, for many centuries, remained under the guard of philosophy. The many philosophical schools that have succeeded each other since the founding times of Plato and Aristotle have offered different answers to this question, always maintaining the concern to offer a well-founded ethical principle to guide human action. In the nineteenth century, however, the emergence of the human sciences produced important challenges to philosophical discourse, shifting the question of how things should be to the question of why things are what they are. Émile Durkheim was one of the important figures on this journey, with his efforts to found a new discipline, Sociology, guided by the central concern of unveiling the social origin of morality. This text presents this sociological project from four fundamental axes. First, it situates the author in relation to the social, political, and intellectual context of his time, so as to justify his methodological choices and his main interlocutors. In the second axes we

R. Weiss (✉)
Department of Sociology, Federal University of Rio Grande do Sul (UFRGS), Porto Alegre, Brazil
e-mail: weiss.raquel@gmail.com

© The Author(s), under exclusive licence to Springer Nature Singapore Pte Ltd. 2022
D. McCallum (ed.), *The Palgrave Handbook of the History of Human Sciences*,
https://doi.org/10.1007/978-981-16-7255-2_46

have a discussion of the main constitutive elements of morality according to this theoretical perspective. After that, fundamental concepts from his sociology of religion are mobilized, which are considered necessary to complete the framework of his moral theory, revealing the full potential of Durkheimian sociology for understanding contemporary society. Finally, there is an exposition of his defense of the ideal of the human person, accompanied by a discussion of how it is possible to take a position on moral phenomena without exceeding the limits of science.

Keywords

Émile Durkheim · Sociology of Morality · Secular Morality · Ethics

Introduction

How *should* we act? This apparently simple question is one of the most longstanding in the history of human thought. In Western tradition, it is in Ancient Greece that we find the main records of the first attempts to answer it. The pages of the great epics, such as those attributed to Homer and Hesiod, narrate the achievements, choices, and characteristics of the gods and heroes, who at the same time expressed and founded the central values of that culture. In that historical moment, the poets became the spokesperson of "right action," because they hold privileged access to the wisdom of the ancestors and it is in these that lies the answer on how to act.

Reiterated for centuries in tragedies and even in comedies, the values of the Greek people remained for a long time based on the premises of the identification between "good" and "ours" – the correct action was that in accordance with the precepts and demands of the ancestors and gods of the city. Antigone, although considered the first character to incarnate the ideal of autonomy, refuses the laws of the city and acts according to higher principles, risking her own life to guarantee her brother a *proper* burial, that is, as prescribed by the gods of the city.

However, at the time of Sophocles, the mythical narrative as moral foundation began to be disputed. Between the fifth and fourth centuries BC, there was a consolidation of democracy and the political and cultural center of Greece moved to Athens, establishing a context propitious for the flourishing of the sophistic school. Gorgias and Protagoras, well versed in history, harshly challenged the universalistic and nonrational character of mythology, implying a critique of tradition and law based on the authority of the ancestors.

Parallel to the skeptical criticism of the sophists, the guide to action of the poetic-mythological tradition was faced by another lineage of thought that was beginning to emerge, positioning its weapons against the relativism of the sophists. Even without having formulated a theoretical system and without the legacy of written texts, Socrates represents the inaugural milestone of a new way of dealing with the question of how we should act. Based on the assumption that truth is not *outside*, but *inside* us, he established morality as the central theme of philosophy.

Although the ideas formulated by Cyrenians and Epicurists differed widely from those advanced by skeptics and stoics, and all of them followed a very different direction from Plato and Aristotle, it is important to point out that a long journey began, in Western thought, whose task was to find a *solid foundation* for moral action. The question, "how should we act?" started to be treated beginning with the assumption that an exact, universal and rationally established answer would be possible. Even when religion occupied a preponderant place, in the context of a Christian hegemony that transformed philosophy into theology, reason continued as the common thread. In the doctrines of St Augustine and St Thomas Aquinas, the absolute character of morality is guaranteed by the idea of revelation as a direct expression of God's will, yet there was constant, refined effort to show reason as a divine instrument, as the way God manifests himself in the world.

When philosophy began a path of renunciation of strong theological and metaphysical foundations, as in Kant's philosophy and Utilitarianism, reason – whatever conception was held of it – continued to figure as a plumb line in order to guarantee the human being a sound parameter for her action. Alongside this, some philosophical currents, as exhibited in Shaftesbury, David Hume, and the British school of moral sense, attributed to the heart and emotions the key to finding the right path to follow.

In the nineteenth century a critique – internal to philosophy itself – began to address the whole of the post-Socratic tradition. The German philosopher Friedrich Nietzsche became the best known detractor of this long and diverse ethical tradition, introducing a genealogical perspective that aims to critically reconstitute the dominant morality, which he calls "herd morality." However, even if this marks a great distance from previous moral thought and introduces a historical perspective, Nietzsche continued to hold a fundamentally normative position, that is, one that is aimed at finding the best possible answer to the question, "how should we act?", based on a very distinctive conception of human perfection. It is also an investigation aimed at establishing what is the "good."

Against this backdrop, with long roots in the history of Western thought, the intellectual field saw the emergence of a new way of addressing properly "human" issues. This chapter will examine this movement of progressive detachment from moral philosophy following the trajectory of Émile Durkheim, one of the founders of one of the main new scientific disciplines to emerge in the nineteenth Century, sociology.

Toward a Sociology of Morality

Born in 1858 in the Alsace-Lorraine region, on the border of France and Germany, Émile Durkheim has entered the history of the human sciences as the founder of French sociology. Although he conceived sociology to be a broad undertaking, encompassing all objects that could be considered social facts, we will look more closely at his proposal for the constitution of a sociology of morality. It involved applying the fundamental principles of sociology to a domain that for centuries remained under the guard of philosophy, in its most varied traditions. Before presenting the moral theory that supports his project, however, it is important to understand his social and intellectual context, which justifies the first movement made by the author, namely, the establishment of a distance from the main schools of moral philosophy of his time, namely, Kantianism and Utilitarianism.

Durkheim's Social and Intellectual Background

Raised in a Jewish family with a long tradition of rabbis, Durkheim was pushed in this direction from a very young age. Still in his early youth, he decided his destiny would be different, and he began preparatory studies to become a teacher. This was the path that took him to Paris, where he had the chance to study at one of the most prestigious French educational institutions, the *École Normale Supérieure* (ENS). There he had access to a classical education in the field of humanities and became familiar with the main currents of thought that were circulating in his country at that time, among which two in particular took his attention.

The first of these was largely due to the readings he made of Charles Renouvier, a philosopher who considered his thought a continuation of the work of Kant, who founded a lineage called neocriticism or French neokantism. Besides his doctrine on the categories of understanding, which had great repercussions, especially in Durkheim's later work, this philosopher was a crucial influence by calling attention to the centrality of the phenomenon of moral action. For Renouvier, morality was the most important phenomenon of human life and, for this reason, the political changes that French society was going through – he wrote at the time of the consolidation of the Third Republic – demanded that science be at the service of the country's moral transformation (Logue 1993). As Durkheim himself acknowledged (Deploige and Durkheim 1907), Renouvier was his great intellectual master, and it was by reading his texts that he understood that morality should be the great theme of his career. Although Renouvier spoke of the importance of science for the moral transformation of the country and pointed to the centrality of the moral phenomenon (Renouvier 1848), it was another French author, Auguste Comte, who offered Durkheim a more tangible idea of which science he could follow, sociology. It was, however, at this time a discipline that was more a generic idea, still strongly anchored in philosophical precepts, than a field of knowledge. By deciding to follow this path, Durkheim took upon himself the task of paving it, that is, of laying the foundations for sociology to cease to be just an idea and become a real science.

To understand what was involved in this founding project of sociology, making morality one of its most important objects, we need to take into account some important factors. During this period, France was experiencing its third attempt to implement the Republic, a project that had suffered many setbacks since the overthrow of the monarchy during the French Revolution of 1789. The main focus of the academic and political community was on strengthening science and the process of secularization of education, from elementary school to universities.

In 1885–1886, the young high school teacher received an invitation from the French government to join a study mission to Germany in order to learn about university reform and advances in science in that country. During this period, he had contact with experimental research in the field of psychology, especially as carried out in Wilhelm Wundt's laboratory. Although he diverged in various aspects of his theory from other authors who sought to investigate morals scientifically, Durkheim returned enthusiastic about this trip, which consolidated his conviction in the possibility of founding a science of society (Durkheim 1993 [1897]; Jones 1994).

Soon after the publication of two articles on what he had learned in this mission, Durkheim was invited to take up a position as a substitute professor at the University of Bordeaux and, from that moment on, he was able to dedicate himself more systematically to establishing the foundations of the new science.

As far as the scientific field is concerned, it must be pointed out that the dominant epistemological paradigm in the French sciences was methodological naturalism. While in Germany the main difference instituted was between the *Naturwissenschaften* and the *Geistenwissenschaften*, in France there was an assumption that science should form a whole. Every scientific discipline should be guided by the same methods, whose paradigmatic references are those of the natural sciences. The assumption was that all realms of reality are to some extent natural and, therefore, can be explained by scientific rationality. One consequence of this was the idea that for a new science legitimately to establish itself, it would require an object of its own, an object not belonging to any other existing science.

Durkheim's first step was, therefore, to ensure the new science he intended to found satisfied all these parameters. The general principles of this new science were systematized in the book, *The Rules of Sociological Method*, which expressed its adherence to a "lighter" form of methodological naturalism (Massella 2006) and had as its main purpose the legitimization of sociology as a science. As far as our focus here is concerned, it is important to highlight its concept of social fact and its first rule, that social facts should be treated as "things."

The concept of "social fact" was forged to justify the claim that sociology had its own special subject matter, thus establishing a boundary line in relation to psychology, for example. Basically, social facts encompass everything that has been created by society, that is, is based on intersubjective processes, whose main characteristic are exteriority in relation to individuals, the ability to impose on them and to be shared by a large number of other individuals of the same group or territory. To treat these facts as "things" does not mean to say they are things, but to adopt toward them the same mental attitude we have when we investigate nature, that is, the awareness that nothing is known about them before we observe them. To defend the possibility of a sociology of morality was therefore to affirm that morality is a social fact, that is to say, it is not determined by God's will, nor by reason, nor is it an expression of innate feelings. In order to become the object of sociology, morality had to be rendered as something else. One of the greatest challenges was therefore to bring to the field of science a subject that had for centuries been under the aegis of philosophy.

Moral Philosophy in France

Despite the constant dispute with Germany, Immanuel Kant's philosophy remained highly influential in France throughout the nineteenth century, although his reception was mediated by native interpretations that gave rise to philosophical movements that we may consider properly French. The most influential movement in the university milieu during this period was the so-called "French spiritualism," whose main protagonist was Victor Cousin (Schmaus 2004). Cousin brought the Kantian philosophy to the center of the debate, presenting it as the most appropriate for the interests of the Republic – the most adequate regime to fit the ideals of French Revolution – since it allowed religion to be discarded without creating a moral vacuum. At the University, the curricula of the philosophy courses were all marked by this, and their institutional power was so great that it made life very difficult for those who did not support Cousin's idealism and were not under its tutelage.

When Cousin slowly began to lose influence, Kantianism continued to set in. With Charles Renouvier, Kantianism received its most important reworking on French soil, quickly supplanting Cousin's eclectic reading. The founder of the movement called "neo-criticism" was considered one of the most important thinkers of the Third Republic, having influenced an entire generation of intellectuals in their commitment to the republican ideology.

Reference to Kant was, therefore, unavoidable. The presence of the Königsberg philosopher crosses every moment of Durkheimian work, though always in an ambiguous way. Durkheim seems to reproduce the Kantian architecture, but replaces its foundation: the position Kant ascribes to transcendental reason is, in Durkheim, occupied by society. This inversion implied an ontological shift, with multiple consequences, allowing Durkheim to build a radical critique of Kantian moral philosophy and to forge the general concepts of his own theory.

The second lineage of moral philosophy that increasingly gained ground on French soil was the utilitarian. In particular, authors such as the English John Stuart Mill and Herbert Spencer were mobilized more and more frequently to try to explain and, above all, legitimize the social transformations produced by the division of labor in capitalist society. From the ethical point of view, the great appeal of utilitarian doctrine was to morally justify actions consistent with this new social model, sustaining the affirmation that the pursuit of happiness – or the maximization of utility – was not only something acceptable but the right thing to do. Although modern utilitarianism, founded by Jeremy Bentham and James Mill, introduced a concern with the social dimension of human life, they still remained heirs to the ancient Greek tradition of epicurists and cyrenaics, who identified the classic theme of happiness with the attainment of pleasure.

Although he aimed more directly at contemporary utilitarianism (especially Spencer), Durkheim sought to attack the more general premises of this philosophy, in particular the idea that human motivation is and should be individual self-interest. Even if modern utilitarians tried to think of a possible articulation between individual happiness/pleasure and collective happiness/pleasure, self-interest still remained the foundation of moral action. As with Kant, Durkheim's process of dismantling this philosophical system is directly aimed at the foundation: if morality is always a social construction, it makes no sense to conceive a universal idea of the human being and to justify moral conduct based on this characteristic.

There is a common point in Durkheim's criticism of these two lineages, which is the starting point for his detachment from all moral philosophy. In both cases, the initial object of his criticism is deduction as the privileged mode of access to knowledge. We cannot forget that we are still at the turn of the nineteenth Century and twentieth centuries, when biology began to have great cultural influence and legitimized the experimental method and induction as the scientific procedure par excellence. In the introduction to the first edition of his book *The Division of Social Labor*, suppressed in later editions, we find an excerpt that sums up well this refusal of deduction, particularly directed against Kantian philosophy:

> It may be that there is an everlasting morality, written in some transcendental spirit, or immanent to things and whose historical morals are only successive approximations: this is a metaphysical hypothesis that we do not need to discuss. But, in any case, this morality is relative to a certain state of humanity and, as long as this state is not realized, not only can it not be obligatory for normal consciousness, but it may also be our duty to fight it. (Durkheim 1975 [1873], p. 273 own translation)

In order for morality to be the object of science, Durkheim needed to demonstrate that it would be possible to investigate it empirically, and so it could not be a rational truth a priori, whose discovery would require merely the methodical use of thought. The moral principles of philosophers would always be just another moral system created by human beings, and could become moral "in fact" only to the extent its ideas were socially embraced, which would only happen in a partial way. In short, from the sociological point of view, "the" moral does not exist; all that exists are socially constituted moral systems, which vary according to each place and each historical moment.

A Sociological Theory of Morality

Let's go back to the question we started with: how should we act? While moral philosophy was driven by the expectation of finding a certain and universal answer, Durkheimian sociology displaces the terms of the question. A moral science should not scientifically find a formula to guide human action, as if it were an objective truth to be discovered. The fundamental postulate of this theory is that the parameters of human action, which we perceive as mandatory, are always socially constituted. It is the group to which we are linked that says how we should act. The task of the sociology of morality is, therefore, to understand how different groups produce their own moral rules and what the implications of different moral systems are for society – and also for individuals.

"The" Moral Does Not Exist

To understand what the sociology of morality proposed by Durkheim consists of implies certain challenges. This theme underlies almost all of his writings; at the same time, we do not have a systematized and definitive text on the subject. By the time of his death in 1917, Durkheim was 59 years old, well established in the university milieu, and leading a group of prominent researchers around the journal he founded, *L'Année Sociologique*. This solid academic structure offered the best prospects for a future with an extensive and long life to a group that constituted what we now call the "Durkheimian school" or "French school of sociology."

At the peak of this trajectory, when everything seemed to go as smoothly as possible, the First World War broke out. Many of Durkheim's main collaborators left their academic posts to serve on the front; many of them never returned, including his son, André, who was to have been the main continuer of his intellectual legacy. The founder of the sociology chair at the Sorbonne reduced his academic production and dedicated himself to producing political texts, severely condemning the horrors of the War.

As the conflict extended over time, he gradually lost his vitality, until he ended up in a period of seclusion. It was at this moment that he gathered all his remaining forces, in the expectation of writing what was to be his great work, a book entitled La Morale. This ambitious project did not advance beyond its first steps, being interrupted by the death of the author on November 15. All we have is a summary of the book, accompanied by some indications of the path that could be followed (Durkheim 1975 [1920]), with reference to texts already published and some manuscripts.

We cannot know, consequently, what the definitive format of his sociology of morality would have been. Nevertheless, from the index of the aforementioned book and, above all, from the methodical analysis of his writings, the expert literature has gathered elements solid enough to present the general outlines of this project. We shall see below what the hard core of his moral theory consists of, which gives support to the more general project of scientific investigation of morality.

The starting point of Durkheim's moral theory is the premise that, as a social fact, morality is always socially constructed. Nevertheless, although he constantly reaffirmed the limitations of philosophical speculation, it was by navigating the pages of philosophy books that he arrived at a central proposition, namely, that morality is always a *duty* and a *good*.

Kantian moral theory is the most conspicuous example of a deontological ethics; after all, for Kant, the truly moral action is the one performed exclusively out of duty. In this author's system, moral action is the one informed only by the principles of practical reason, and an action according to duty would be a kind of guarantee that the motivation for the action is exclusively a law of reason, without any "contamination" by our own desires linked to our sensitive existence.

Durkheim recognizes the appropriateness of the Kantian thesis that morality always assumes an imperative form, being, therefore, a duty. For him, this is a universal characteristic, because when we talk about morality, we are referring to a rule, a system of necessary rules, which regulate what can and cannot be done. He refuses to accept, however, that morality is determined by pure practical reason, and so its imperative character must be explained in another way.

Morality always appears to individuals as a duty because it is a social imperative. Since it is a collective creation, the values and rules constituted in the processes of social interaction never fully coincide with the will of singular individuals. Even when individuals take an active part in determining moral rules, there is always something beyond them. Once objectified in the "collective consciousness," morality is, at least in part, external to the individual, and there is pressure from the group for moral precepts to be followed. This is why morality always takes an imperative form, with duty being its external and formal aspect.

Durkheim's criticism extends to the Kantian conception of human nature, making room in his theory for the establishment of a second characteristic of morality, namely, the good or the desirability of something:

> "Kant's hypothesis, according to which the sentiment of obligation was due to the heterogeneity of reason and sensibility, is not easy to reconcile with the fact that moral ends are in one aspect objects of desire. If to a certain extent sensibility has the same end as reason, it cannot be humbled by submitting to the latter." (Durkheim 2010 [1906], p. 22)

For Kant, our existence as phenomenal beings would be a burden that submits us to natural causality, making us bound to the contingencies of the physical world and also of our own passions. In contrast, understood as beings with rational access to the numinous, we have the possibility of leading a rational existence, making us the legislators of our own will. To act only according to rational principles would therefore be the prerogative for our freedom. For Durkheim, however, this would be a mistaken moral conception: morality is not only an obligation, it is also an object of desire.

On this point, the author comes close to utilitarianism. The second characteristic of morality, the good, is based on the assumption that human beings need a sensible motive for their action, something that affects their sensibility. If morality were only pure duty, it would either be impossible – and have no effectiveness – or it would be founded on fear. But fear itself is a type of feeling, which is the foundation of a partial morality, something that is imposed in a unilateral and authoritarian way, without the sincere adherence of those who follow it. However, this proximity to utilitarianism is only partial. By affirming that morality is a good and, therefore, something desirable, Durkheim supposes the existence of a certain pleasure for group life and, by extension, for what is produced by the group. For the author, his position is very different from that sustained by the utilitarians because, for them, subjective motivation is anchored in self-interest linked to the selfish nature of the human being. Although Durkheim recognizes that there is a selfish dimension in us, morality is not based on it, but on altruism:

> We will say nothing about morality based on individual interest, because it can be seen as abandoned. There is nothing that comes from nothing; it would be a logical miracle if we

could deduce altruism from selfishness, love for society from self-love, the whole from the part. The best proof of this is the form that Mr. Spencer has recently given to this doctrine. He has only been able to remain consistent with its principle by putting the most generally accepted morality on trial, by treating as superstitious practices those duties which imply a true disinterestedness, a more or less complete oblivion of oneself. Thus he himself was able to say of his own conclusions that they would probably not gain much adherence, because 'they do not agree sufficiently either with current ideas or with the most widespread feelings'.... What would one say of a biologist who, instead of explaining biological phenomena, would contest their right to exist?. (Durkheim 1975 [1893], p. 263, *own translation*)

As with accepting one aspect of Kant, Durkheim seems to accept one aspect of utilitarian theory only to immediately depart from it. Once again, it is the premise that morality is a social fact that leads the author down another path. To get a more comprehensive view of how the author explains this double characteristic of morality, the best way is to follow the arguments presented in a set of lectures written for a course on "Moral Education," published posthumously in 1925 in a book of the same name (Durkheim 1925, Transl. 2002).

These lectures were aimed at future pedagogues and school teachers, and in them Durkheim seeks to answer three questions: what is morality in general, what kind of morality can replace the one taught in French schools, which is still religious in nature, and how it could be taught. Although the emphasis is on the "dispositions" that must be formed in the individual in order to become a moral subject, this is the most important resource we have to reconstruct his moral theory. The discussion about duty and good appears, therefore, linked to these two fundamental dispositions, namely, the "spirit of discipline" and the "spirit of attachment to the group."

Another interesting characteristic of this text is the use of a recurrent argumentative strategy, that is, to draw out the approximation of purpose between morality and religion, which allows a theoretical development based on a historical and genealogical reconstruction. To arrive at a general formulation of what morality is, says Durkheim, it is necessary to look at religion, because historically morality has always existed in a religious form. Morality and religion are practically overlapping phenomena.

Let us begin, then, with the idea of duty, which appears in the pages devoted to the disposition Durkheim calls the "spirit of discipline." Duty is the external aspect of morality, that is, the most easily identifiable sign; it refers to obligation, as a prescription to be observed. In order to know when we are facing a moral precept, it is enough to ask what would happen if this precept is not obeyed. When it comes to a moral precept, noncompliance always implies a sanction, that is, a reaction that does not immediately derive from the act, but from the reaction of the group.

Here we shall introduce an example which should help us to elucidate the general sense of the theoretical proposal formulated by Durkheim and, at the same time, will serve the purpose of indicating to what extent his intellectual project can be taken as an important tool for the understanding of contemporary moral phenomena. In normal times, if an individual disregards the rule of washing his hands before eating food, he may have some health problem. This is an immediate consequence of the action; there is no social mediation. No one would look at this individual with a condemning eye; at most, someone might think that he has poor hygiene habits, but no one would stop treating him for this reason. In a context such as the pandemic caused by the Coronavirus, this same act acquires a moral connotation. Of course, the question of direct consequence remains – I might get sick because of this very act. But in the context of a pandemic, individual behavior with regard to certain hygienic cares, such as washing one's hands, starts to have consequences that go far beyond the individual sphere – it can cause illness and even death for other people.

Following the logic of the Durkheimian theory, we can state that, in the historical moment in question, washing hands becomes a moral rule. It means that it is obligatory; it is perceived by individuals as a duty, it is not an optional action. This coercive character will be perceived at the very moment when this precept is publicly disobeyed: disapproving looks, warning glances, or even other more forceful reactions.

The individual who has neglected to wash his hands may have a variety of reactions. It may be that it was just distracted, and as soon as he realizes the disapproval, he becomes embarrassed or even feels guilty. But it could also be that this individual does not feel any embarrassment or even enjoys not following the rule because he does not consider it valid. In either case, we must ask the following question: What is the purpose of the moral rule? In Durkheim's theory, the objective of a moral rule is to standardize behavior in a given direction and, by doing so, lead society as much as possible in a defined direction, that is, to a desired future condition. It is about safeguarding the highest values of the group.

The rule is thus only the *shape* of morality, the way in which it appears. To grasp the *meaning* of a moral rule, one must ask what value it protects – this value corresponds to the dimension of desirability, that is, it refers to morality as a good. This means that, for there to be a moral fact in the full sense, it is not enough to have a rule whose noncompliance implies a sanction on the part of the group. It is also necessary that the members of the group consider it as something important, exactly because its coercive character serves to protect values. A moral value is considered good, fair, and desirable, and for this reason it is likely to attract the support of individuals.

To return to our example, a sociological investigation of this hypothetical scenario should ask the following questions: Why is washing hands no longer just a hygiene precept but a moral rule? To what moral value is this rule related? For which motives does someone deliberately decide not to comply with this rule? Is there disagreement about the value in question?

Before proceeding with this exercise of imagination, let us recall an important point. In Durkheimian theory, what is at stake is not a debate about which action is morally correct, or whether the rule we use as an example is really a moral rule. It is not a matter, as would be the case in Kantian philosophy, of asking whether the maxim guiding the individual's action (of washing or not washing one's hands) can be taken as a universal law – that is, taking the form of the categorical imperative –

and whether it was performed only by the duty stemming from this law. Nor is it a question, as in utilitarian logic, of assessing whether the subject correctly made the calculation of the difference between pleasure and pain, and acted in such a way as to maximize his utility. What matters is to identify which social forces are at work, which socially produced values are circulating and are linked to the moral rule under consideration. The evaluation of individuals' actions is always context-specific, that is, related to prevailing rules and values.

With this in mind, we can formulate as a hypothetical answer to the first question: washing hands has become a moral rule because scientific research has shown that this is one of the main ways to avoid contamination with the Coronavirus. Within this scenario, individual behavior can have a significant impact on the rate of spread of the virus, which, in turn, affects the number of illnesses and even deaths caused by COVID-19. The main value attached to this moral rule is, therefore, respect for human life: Handwashing serves to standardize behavior in the direction of reducing the transmission of the virus and, consequently, decreasing rates of illness and death. This – human life – is the fundamental value at stake. And what about the other two questions? The failure to comply with the rule may be due to a nonrecognition of the validity of scientific research supporting the necessity of hand washing or the lethal character of the virus; or it may be that human life is simply not a moral value for this particular person.

In this last case, it could be that this is just an individual "deviation" of someone who has not fully embraced values prevalent in "society." This is, however, a very simplistic way of understanding Durkheim's theory. Although one of his best known postulates is that society is more than the total sum of its individuals (Durkheim 2013a [1895]), this does not mean society is a homogeneous entity, in which unique and totalizing values exist. In fact, one of the most striking features of contemporary societies is precisely this axiological diversity, which results from the fact that society is not a large cohesive organism, but is formed by a plurality of interacting groups. One of Durkheim's fundamental arguments points out that the formation of the moral subject requires the development of a capacity for attachment to the group, i.e., the ability to feel pleasure and to bind oneself to a collectivity; if morality is always a collective creation, the feeling of belonging to the group is an important process for adhering to these moral values. Differences in behavior, therefore, should not be explained solely at an individual level, in terms of adherence to or refusal to accept a specific set of values, but rather require an understanding of how different groups produce different values.

As we had the opportunity to observe during the coronavirus pandemic, even in the face of massive scientific research demonstrating the lethality of COVID-19 and the recommended practices for preventing the transmission of the virus, several people and even political leaders ignored these recommendations. Behavior of this kind makes explicit the fact that there is no consensus even on such structural values of democratic societies as the dignity of human life. A crisis that initially seemed to be only a health crisis, with a single common enemy, soon revealed itself to be also a *moral crisis*, with radically different positions clashing, provoking serious divisions and even major conflicts.

An Inquiry into the Soul of Society

From the example mobilized so far, we tried to show how the sociology of morality proposed by Durkheim is guided by the purpose of elucidating which dimensions are at stake when considering morality as a social phenomenon, that is, as something that is socially constructed and that operate in everyday life. To the extent that morality is brought into the field of human sciences, it is no longer thought of in abstract terms, but as a real force that affects the lives of individuals, taking on an imperative form that expresses values considered good and fair.

However, up to this point it has not been possible to explain how these values are constituted and acquire such a significant regulatory force. Moreover, in the example we are following, we are faced with the fact that morality produces cohesion, but also produces tensions between irreconcilable points of view. If morality is a social construction, why is it so difficult that different values can be negotiated between different groups?

To understand the intensity of these ruptures and the difficulty in producing conciliatory solutions, it is necessary to appreciate an underlying aspect of both rules (duty) and moral ideals (good): their sacred character. For this, we must now examine one of the main characteristics of Durkheim's moral theory, namely, the proximity between morality and religion. After all, it is in his discussion on religion that we find the missing pieces to put this puzzle together, gathering all the necessary elements to understand how, for Durkheim, morality is a sui generis type of social fact, which occupies a central place in collective life. We need to examine his sociology of religion to find the key concepts for his sociology of morality: sacred, ritual, belief, and collective effervescence.

In the lessons on *Moral Education*, Durkheim departed from the thesis regarding the co-origination of morality and religion. According to the author, for a long time and in all societies, morality figured under the aegis of religion. The solution to the question of what morality is, therefore, involves an investigation into what religious morality has been. On this occasion, Durkheim explains the duty and desirability of morality from an allusion to divinity. God is presented as a power both feared and loved by his believers, a feature transferred to morality. These attributes of God are, in turn, derived from the fact that the very idea of God is a social creation, a kind of symbolic expression of collective life. This transfiguration, however, is only announced. We know very little about it and, above all, Durkheim lacks the empirical and conceptual tools to explain the relationship between god and society. It is still not known why, sociologically speaking, morality has a sacred character.

At the turn of the century, there was a new direction in Durkheim's work, culminating with the publication of *The Elementary Forms of Religious Life* (1912), when the new empirical findings received a thoroughly refined theoretical treatment. One of the most important events was the reading of the ethnographic records of Walter Baldwin Spencer and Francis James Gillen, English anthropologists who spent long periods among the Australian aborigines.

Nowadays, we are aware that, to a large extent, these ethnographic investigations were carried out in the broader context of colonial domination, both on the African continent and in Oceania, leaving an ethical stain which should be pointed out (Connell 1997; Glowczewski 2014; Mbembe 2017). In addition to contributing to a narrative that establishes a mythical view of non-European peoples, they also held very distorted views of their daily lives, values, and practices.

Despite all this, we can argue that the accounts of these authors served the theoretical imagination of Durkheim, who made a "creative" reading of the data reaching him at second hand. As William Watts Miller argues, *The Elementary Forms of Religious Life* cannot be considered a distortion of empirical data or a misreading of Spencer and Gillen's accounts, but "It is instead an imaginative re-construction, which involved its author in developing a whole new seminal theory of his own. The work is both a transfiguration of Spencer and Gillen's Australia and a transfiguration of the old Durkheimian Australia" (Miller 2012, p. 113).

In his book, Miller describes the long process followed by Durkheim's ideas until they reach the most definitive formulations in The Elementary Forms, which are of interest here. One of those ideas is the notion of the sacred. In a text of 1902, dedicated to a discussion of totemism, we already see a first formulation of the sacred as a force (Durkheim 1902). However, it was only in 1904, after an important reflection developed by Marcel Mauss, his nephew and one of his main collaborators, that a clearer notion of religion as a phenomenon linked to the idea of the existence of an "impersonal force" came into existence. For Durkheim, the absence of evidence for the existence of an impersonal force in Spencer and Gillen's accounts was a problem, but Marcel Mauss was able to find this evidence from the reference to the arungkilha, a magical force of evil. Mauss was able to link this belief with an already well-known and documented belief, namely, the idea of mana, present among the inhabitants of Melanesia.

Durkheim's whole endeavor goes in the direction of formulating an explanatory hypothesis of mana, this impersonal force central to his concept of the sacred and, therefore, to his own definition of religion and, by extension, of morality. *The Elementary Forms of Religious Life* became widely known as a treatise on religion, inaugurating an important tradition in this field of studies (Pickering 2009; Rosati 2009). In it, we also find an attempt to address the question of the origin of categories that organize human thought, through a debate with empiricism and idealism. This is the reason why it became an essential reference for thinking about the sociology of knowledge (Schmaus 1994) and even epistemology (Rawls 2004). At the same time, however, given the centrality of the religious phenomenon in Durkheim's theoretical perspective, it is possible to affirm that this quest for the most elementary forms, that is, the most essential forms of religious life, was at the same time an investigation into the basic elements of social life in general, in which morality plays a central role (Weiss 2012; Rosati and Weiss 2015).

In his introduction to the book, the author presents a general definition that we can take as the common thread of the concepts pivotal for our purposes:

> A religion is a unified system of beliefs and practices relative to sacred things, that is to say, things set apart and forbidden – beliefs and practices which unite into one single moral

community called a Church, all those who adhere to them. (...) In showing that the idea of religion is inseparable from the idea of a Church, it conveys the notion that religion must be an eminently collective thing (Durkheim, 1995, 44).

For the moment, it is important to emphasize that religion involves beliefs, which are special types of representation, that is, ideas, and also practices, which in the religious context are called rites. Those are beliefs and rites concerning sacred things – note well, the author does not bind religion to the existence of a deity, which enables us to greatly expand the idea of religion, applying it to phenomena we often do not consider as such. What matters is whether there is something "sacred" in relation to which these beliefs and practices refer. Finally, it is worth mentioning the idea of "moral community"; with this, the author assumes that a religion is always anchored in a collective dimension. A moral community presumes the presence of a set of values and rules that guide the conduct of its members in a given direction, circumscribing a set of actions as being obligatory, and another set of actions as being forbidden. What this means is that, in the context of a religion, certain actions and ideas are not optional: breaking these precepts always implies some kind of sanction.

As we have seen above, Durkheim claims that, historically, every moral existence has always been circumscribed by some religion, so that to understand morality "in itself" implies the comparative observation of different moral systems, which have always been under the auspices of religion. In the text we are now considering, the author reverses the terms of the equation: the notion of morality – of moral community – becomes part of the definition of religion. What, after all, comes first, morality or religion?

A closer look at Durkheimian theory reveals that this is a misleading question. These are largely overlapping and co-originating phenomena. Every moral community – that is, a community with strong shared notions of good and duty among its members – is a form of religion. At the same time, there is no religion that is not a moral community. Religion does not fully coincide with morality, for there are certain dimensions that go beyond the strictly moral dimension. Likewise, morality is not always confined to the context of a community and assumes more complex forms. However, both share a central characteristic, the sacred.

We come, once again, to the point where we concluded the previous section: the sacred character of morality, which we can now analyze based on the discussions Durkheim brought into play to explain religion. The first important aspect to be made explicit is that the sacred is not something in itself, it is not an entity or a substance, but an attribute of things, people, and even ideas. The second important dimension is that it is not an innate property, meaning it is a characteristic not derived from things/people/ideas in themselves, but attributed through a process of consecration. Finally, the constitutive feature of the sacred is its character of exceptionality, which leads everything that holds this attribute to be placed in a separate orders. The sacred cannot be challenged or touched, and the sacred is in a relationship of opposition to things that are profane, ordinary, mundane, day-to-day. One of the forms taken by the sacred is precisely that of an impersonal force, called mana by the Melanesians.

That is why the contact – even if indirect – that Durkheim had with these cultures, so distant from his reality, was so central to his theory.

One of Durkheim's greatest contributions to the field of social sciences is his theory of the sacred, which presupposes an appreciation of how this impersonal force is produced. The author's central hypothesis in this regard is linked to his concept of collective effervescence, a kind of trance induced not by the use of psychoactive substances, but as a result of collective encounters with a high degree of intensity, capable of producing an alteration in psychic life. Ritual occasions are the prototypical example of this type of association, but effervescence can be produced whenever individuals gather around a shared purpose, engendering interactions of different kinds: bodily movements, such as dancing and clapping, and mental ones, such as a heated debate or an assembly that collectively follows the speech of a leader. The intensity of the effervescence can vary due to different factors, such as the quality of the participants' involvement, the number of people involved, and the degree of psychic, physical, and emotional fusion. But whenever individuals find themselves gathered around a shared meaning there is, even if in a small degree, an alteration in perception, capable of transforming or reinforcing the beliefs of those who participate in this interaction process.

The sacred arises from the perception of the disparity between the psychic intensity experienced in moments of effervescence and that which occurs in ordinary life. There is a condensation of this experience, which is remembered as something of another nature, and which becomes even more poignant when it becomes embedded in a symbol, which remains as a receptacle of this energy, and serves as a reminder of there being something more important than the banality of everyday concerns. The things, ideas, and persons that hold an important place in the processes of effervescence are enveloped by this: "energy" – it is they, and not so much the moment itself that become invested with a sacred character. Collective effervescence engendering processes are, therefore, the central mechanism of consecration: what is touched by this "force" takes a different place in the order of affairs, and requires, for those who share this perception, a different kind of attitude.

A sacralized thing, idea, or person has the prerogative to evade criticism, to demand an attitude of respect and, above all, to be invested with immeasurable value. It is easy to see how this mechanism operates in the context of an instituted religion: beliefs demand from the faithful an uncontested adherence, sacred objects cannot be manipulated without care and demand a ritual action, sacralized persons (saints, prophets, etc.) possess a moral authority and their behavior cannot be questioned, nor can they be defamed. Any action that puts the virtues of the sacred into question, which makes them a cause for laughter or slander, is strongly disapproved of by the members of the community.

It is important to bear in mind that the sacred is ambiguous: it is the holy and the evil, the pure and the impure/source of pollution. When we think in moral terms, this means that the sacred circumscribes not only what is good, but also what is evil. This distinction appears when discussing the piacular rites, which can be of different types. The pure sacred is associated with positive feelings, of joy and celebration, since they aim to reiterate that which is good for the collectivity concerned, that which nourishes it and reinforces its fundamental values. The impure sacred, on the other hand, is associated with painful feelings, which threaten the meaning of things or the very physical or symbolic survival of the group. They are intended to reaffirm the repudiation of actions and events that must never happen again, and thus contribute to overcoming collective suffering and to strengthening the bonds of solidarity among group members.

The moral rule, that is, its imperative character, has the prerogative of regulating conduct in order to ensure that "evil" does not cross the group boundaries and that the purely sacred is protected not only from evil, but also from what is profane. But such a rule only makes sense insofar as it stands for what really matters, that is, the moral ideal.

The moral ideal is another word referring to the desirability of morality, to morality not so much as a rule, but as a value. It is a more precise concept within Durkheimian theory, in which it can be defined as a special kind of representation, an idea, invested with a sacred character. The moral ideal is the soul of a group, inasmuch as it condenses all those things the group considers most valuable, incommensurable with anything else that cannot be quantified or equated with profane demands. These ideals can take the form of deities, or remain as values in the strictest sense, such as the ideal of justice, human dignity, etc.

In either case, it is not physical characteristics or personal qualities that matter or that are worthy of this status. Moral ideals matter precisely to the extent they are collective creations, symbolic embodiments of the group itself. The famous Durkheimian statement that "God is society hypostatized" sums up well the mechanism by which the group itself is worshipped in the things it enshrines. This is the sense in which every moral ideal carries the mark of transcendence: by guiding his behavior by the normative principles of a moral ideal, the individual binds himself to something external, which temporally and numerically surpasses his existence, that is, the group he belongs to.

We return, thus, to the argument formulated in *The Moral Education*, according to which the spirit of group attachment is one of the necessary dispositions for moral action. A moral ideal, therefore, is not just an abstract idea that is logically supported, which an individual is capable of attaining by her own means. A moral ideal, in the strict sense of the term, is always an expression of ideas that circulate collectively and are elevated to a higher category to the extent that they are tied to processes of collective effervescence. When a moral ideal is challenged, whether with actions or arguments, there is a reaction from the group: when something held as sacred is challenged, the group itself is threatened, in its most vital part, its "soul," symbolically crystallized in the form of its values, saints, or heroes.

An exercise of imagination allows us to think about how these concepts apply in the analysis of complex contemporary societies. A society is never a homogeneous collectivity, in which these relations are built in a clear and peaceful way. In a globalized context, not even the borders of a nation-state are enough to circumscribe a closed set of interacting groups, and even the inner core of a nation-state is far from being thought of as a collectivity with more or less homogeneously shared values. Moreover, there is the increasingly common fact that the same individual moves through several groups in the course of his or her life, or even maintains simultaneous belongings. If we take our author's theoretical premises to their ultimate consequences, we can affirm that each group that constitutes the social fabric has the ability to produce its own moral ideals, sacred to the group and, therefore, belonging to a specific category that protects them from criticism, whose threat is taken as a threat to the very essence of the group.

There are numerous possibilities to draw empirical consequences from these considerations, but we return to the illustration given by the COVID-19 pandemic, in order to grasp how all these theoretical elements can fit together in analyzing an empirical phenomenon. Once again, it helps us to understand how disciplines within human sciences can shed some light into some issues. It is important to insist on this example because it helps us realize to what extent the human sciences in general, and the sociology of morality in particular, offer conceptual tools to understand, including, issues that transcend their borders.

If it is true that the biological sciences play a central role in mapping the ways in which the virus is disseminated and in elaborating ways to prevent or mitigate its events, the human sciences have revealed themselves to be fundamental in understanding how human behavior unfolds when faced with the guidelines formulated in the scope of those sciences. The sociology of morality elaborated by Durkheim allows us to identify the processes behind the adherence or refusal of individuals to these guidelines, revealing how even when human life is at stake, moral beliefs are decisive in directing our actions, revealing that the truth of science is not universally shared.

The health crisis caused by the coronavirus had global impacts, spreading rapidly to every corner of the planet, and from one moment to the next, very simple actions, such as washing hands, were elevated to the status of a moral rule. The same goes for the use of a new prop, previously restricted to specific environments: the use of the mask.

When the pandemic broke out, the media circulated arguments anticipating a great wave of solidarity in face of the need to confront the same threat. Few things have the capacity to bring people together as much as the need to confront a common enemy, as in war situations. Nevertheless, after some months, the press was reporting very different ways of coping with the virus in each country. In some cases, Brazil and the United States perhaps being the most conspicuous, polarization regarding the extent of the threat at hand and the ways to fight it contributed to deepen ruptures already present. In these two countries, the diagnoses drawn up by the scientific community and international agencies – such as the World Health Organization (WHO) – were discredited and challenged by a significant portion of the population, including by their then presidents, Jair Bolsonaro and Donald Trump.

Wearing or not wearing the mask, beyond a moral action in itself, became a symbol of divergent values. Understanding this moral phenomenon in its full extent would require in-depth research to go through its multiple nuances, but this exercise in applying Durkheimian moral theory allows us to understand that the refusal to wear the mask cannot be taken simply as a deviation from a socially sanctioned rule. While it is not possible to generalize about all cases, to a large extent the refusal to wear the mask was linked to a broader set of values, cultivated in physical or virtual moral communities. Even arguments such as that of "individual freedom" – much mobilized by critics of the use of the mask – are fostered within groups where a complex moral cosmology is in operation. To sustain a position contrary to the actions recommended by the WHO is not merely an individual decision, but a commitment to everything that the moral community represents: contact networks, affective intensities, symbolic circuits, and a whole set of practices and beliefs, which circumscribe the realm of what gives meaning to existence. This is stronger than the potential fear of death itself.

Taking a Stand: Normative Boundaries

Understanding the social and subjective processes that articulate what we call "moral life" is, therefore, the main task of the sociology of morality formulated by Durkheim. To a certain extent, he recreates something of the pre-Socratic Greek universe, restoring the importance of the group, the ancestors, and myth as the foundation of morality. There is, however, a crucial difference: sociology unveils the *relative* character of this foundation. The authority of the ancestors has no value as truth; it is only a contingent necessity, or rather, a necessary contingency. From his perspective, the process of creation and legitimation of morality can take different forms, from the mention of heroes from a mythical past, to the attribution to prophets or incarnate divinities, or even the democratic premise that bestows authority on an assembly. What matters, in all cases, is that the moral *ought*, the *should*, is always instituted by a collectivity and is sacralized by means of its inscription in processes of collective effervescence and symbolic transfiguration and crystallization.

This does not mean, however, that sociology should be limited to just describing morals. Following Renouvier's precepts, Durkheim believed that sociology would only be worth our efforts if it could contribute to the transformation of reality (Durkheim 2013b, p. 4). It is not a matter of "discovering" what the true morality is, but of critically evaluating existing morality and pointing out the pertinence or problematic character of the various moral forces in movement in a society. It is a complex and ambitious project, of a sociology which has a say not just on the "is," but is also concerned with the "ought."

When a Moral Ideal Is Not Really "Normal"

One of Durkheim's most important texts for thinking about the relation between "is" and "ought to be" is the third chapter of his book The Rules of Sociological Method, which he opens with the following statement:

> Observation conducted according to the preceding rules mixes up two roders of facts, very dissimilar in certain respects: those that are just as they ought to be, and those that ought to be

different than what they are – normal phenomena and pathological phenomena. (Durkheim 2013a, p. 50)

One of the difficulties in Durkheim's discussion in this chapter is precisely the author's ambiguous use of the concept of "normal." In the sentence above, this concept is used as a biological metaphor, as a synonym for "health," that is, absence of disease. Normal facts are, from this point of view, those that do not harm society. However, in the same text, the concept of normality is also used in its statistical sense, opposing not the idea of pathology, but of "deviation." This distinction is important so that Durkheim can state that there are some facts that, although "normal" in a statistical sense, that is, within the normal curve, being general or shared by the "average" society, are not "normal" in the sense of being adequate to society. On the other hand, there are certain values or actions that, although "deviant," that is, outside the normal curve, are not pathological.

As observed by Dominick LaCapra (1972, p. 65), this discussion on the normal and the pathological should have been the maximum expression of intellectual capacity, but it ends up representing a hole, since it would be, in the opinion of this interpreter, the most problematic writing in all of Durkheim's work. However, it is possible to reconstruct the general sense of this proposal by Durkheim, following the interpretation suggested by William Watts Miller (Miller 1993, 1996) and Susan Stedman Jones (Jones 2001), who allow us to understand in another way the concept of normal as an expression of "what is what should be." According to Miller, the idea of "normal" as opposition to that which is pathological must be understood as an adequacy to the logic of reality.

To understand this meaning, it would be necessary, according to Miller, to read The Rules of Method in conjunction with another of Durkheim's texts, written in Latin, *Montesquieu – Quid Secundarus Politicae Scientae Instituendae Contulerit* (Durkheim and Watts Miller 1997). In this thesis, Durkheim formulates a notion of causality as a necessary connection, so that the idea of normality is presented as an adequacy to the logic of reality. This is an internalist evaluation of socially constructed morality, yet it allows us to situate his sociology within a critical tradition, according to which the social sciences should not resign themselves to the explanation of facts, but can state value judgments about them and thus contribute to social justice.

According to Susan Stedman Jones, the idea of normal evoked by Durkheim should be read against the background of Renouvier's theory, for whom the idea of normal should be understood in relation to the general conditions of existence and as a necessary means for a being to achieve its end. From this consideration, the author interprets Durkheim's argument as follows:

> I suggest that the normal, so understood, is concerned with questions of satisfaction of action in the milieu as the only sphere in which significant human life is possible. (...) The normal is part of the project of developing solidarity and underwriting change (...). As such, it is a critical concept, concerned with the viability of social action and its reflexive foundation (...). (Jones 2001, pp. 145–6)

Based on this framework, we can appreciate that the idea underlying the normal-pathological pair opens the way for sociology to take upon itself the task of critically evaluating the multiple moral ideals prevailing in a given social context. It is not a matter of building a hierarchy with the purpose of pointing out which is the best moral ideal, but, above all, of indicating those acts or values that are potentially incompatible with social life. It is also important to keep in mind that it is not about avoiding confrontation or preventing social transformation, as numerous interpreters have suggested (Nisbet 1952), associating Durkheim with a conservative position.

What is at stake is the production of critical evaluations about values that make the formation of social bonds impossible and that go against the logic underlying the existence of a specific society. This becomes clearer when we look at the writings in which Durkheim formulates his critical judgment and moral propositions for the society in which he lived.

Secular Morality and Moral Individualism

In the same lessons on moral education in which the author established duty and the good, and the individual dispositions corresponding to them, as constitutive elements of morality, there was an underlying query that was to be the ultimate point of his discourse. It was about the general meaning and feasibility of a secular morality. As pointed out above, morality acquired prominence in the context of the Third Republic, because the consolidation of this political regime depended on a social anchoring that demanded individuals educated by a republican ethos.

At the time there were many conceptions of secularism in dispute and, as a consequence, many ideas on how a secular morality teaching should be instituted (Stock-Morton 1988; Isambert 1990). It is worth remembering that, historically, moral teaching served to maintain the power of the Catholic Church, since it continued to be taught by churchmen, and its content was oriented around the dissemination of Catholic values. Durkheim took part in these discussions, formulating a laicism we may characterize as "moderate," insofar as he did not attack religion or refuse its right to exist, as in more radical versions, such as Guyau's (1887).

From a sociological point of view, Durkheim asserted, all religions are true. The point is not to dispute or demonstrate the professed faith of religions, but to take them as social phenomena and, as such, entitled to exist. However, considering the new ideals that emerged during the French Revolution, the structural changes in the patterns of sociability, the demand for the institution of the Republic, and, above all, the plurality of cultures coexisting on French soil, the author considered the predominance of Catholicism in moral discourse to be "pathological." It was necessary to find a way of grounding morality that would do without reference to any revealed religion, precisely so that all religions could coexist, as well as properly secular beliefs.

A secular morality, for Durkheim, should maintain the elements of duty and the good found in religious morality, to which a third element, rationality, should be added. This was, according to the author, a modern claim that, in moral terms, can be translated into what William Watts Miller calls the "ideal of transparency," that is, the urge to explain the causes and justifications for all things (Miller 1996). Put simply, this means that regardless of the specific values shared by the French citizen, it would be necessary to render evident the reasons behind them: how they were produced, what justifications are mobilized to defend them, and what purposes they serve. The third disposition of the moral subject, in a secular context, would therefore be the "spirit of autonomy," understood not in the Kantian sense of the subject's self-determination by its rational faculty, but as an enlightened adhesion to existing morality. To this end, Durkheim supported the importance of teaching sociology from elementary school on, as a subject central to the formulation of the lay teaching of a secular morality (Durkheim 1992). After all, sociology – alongside history – would have the prerogative of elucidating the origin of the values then prevalent in the Third Republic, by showing the long path taken in order to reach the present and, above all, by teaching children the meaning of the beliefs sustaining values.

The second pillar of Durkheim's engagement with the moral question in propositional terms is the advocacy of what he called "moral individualism" an expression coined in his article "Individualism and the Intellectuals" (Durkheim 1898), written in the context of the Dreyfus Affair (Lukes 1969). Alfred Dreyfus was wrongly convicted of high treason to the army. After his first trial, a stenographic test revealed that the handwriting on an intercepted note, which contained state secrets and had been intended for the German army, was not from Dreyfus' hand.

That moment in French history was marked by a polarization between Dreyfusards and anti-Dreyfusards. Those opposed to reviewing the case claimed it would weaken the army's authority and therefore bring a threat to France itself. Those in favor of a new trial for Dreyfus, given the new evidence proving his innocence, argued for respect for the rights of the individual. The most emblematic figure of this movement, the writer Émile Zola, became famous for an article published in the newspaper *L'Aurore*, in which he nominally accused several generals and other figures involved in the trial of having committed serious crimes in the name of preserving the army. Known as "J'accuse," this text gave rise to the first manifesto of the intellectuals, in which major names in academia and literature called for a review of the trial of the then army captain.

The text "Individualism and Intellectuals" was written as a response to Ferdinand Brunetière (Brunetière 1898), a literature professor who ridiculed the recently published manifesto, of which Durkheim was one of the signatories. In his article, Durkheim establishes a fundamental distinction between "egoistic individualism" and "moral individualism." The first type is the expression of utilitarianism, which legitimizes individual interest and elevates selfishness to the category of a moral value. It is a doctrine to be combated because, according to the author, no moral life can be based on selfishness. The second type is a derivation of the ideal of the human person defended by Rousseau and Kant, and consecrated during the French Revolution. According to Durkheim, respect for the human person should be recognized as a fundamental value of French society, and required to be deepened, not fought against. Granting Dreyfus the right to a new trial would therefore preserve the very soul of France, without which the army would have no reason to exist.

Written at the turn of the century, this text is considered a complement to the main argument of *On the Division of Social Labor*, in which Durkheim had stated that the solidarity engendered by cooperation in the world of labor would be the centerpiece of the moral order of modern societies, which could no longer amalgamate around the earlier system of shared values. In the text of 1898, the division of labor appears to be insufficient as a support to the Republic. A minimum common ground was needed, an ideal capable of articulating the multiplicity of beliefs and ways of life. The "moral individualism," the respect for human dignity, seemed to fulfill this task very well, to the point that Durkheim affirmed, with his beautiful rhetoric, that "The human person, whose definition is the touchstone by which good is distinguished from evil, is considered sacred, so to speak, in the ritual sense of the term" (Durkheim 1898: 8). Further on, the author characterizes individualism as follows:

> In short, individualism thus understood is the glorification, not of the self, but of the individual in general. What moves it is not selfishness, but a sympathy for all that is human, a greater compassion for all human pains and miseries, as well as a passionate need to overcome and alleviate them, and finally, a greater thirst for justice. Isn't there enough there to make all good consciences partake of it? (Durkheim 1898, p. 9)

At that point in his work, this ideal of moral individualism seemed to him a guarantee for the future, an enthusiastic defense of a society built on the foundations of a progressive sense of justice. Together with his writings on secular morality, rationality appears as an achievement on the verge of inexorable consolidation, with the potential to solve the main human dramas, constituting an adequate substitute for any divinity as a moral foundation.

Conclusion

Having crossed the threshold between the two centuries, as we have seen, Durkheim reconfigured his conception of religion and introduced in his theory elements that relativized the role of reason in weaving the destiny of social life. The practical and affective dimension gained space, changing the way he approached his theoretical challenges. Moreover, the experience of the First World War then stripped away the basis of his certainties in an ever-better future oriented toward deepening the moral ideal of the dignity of the human person. In *The Elementary Forms*, we are faced with a pessimistic diagnosis of the present:

> If today we have some difficulty imagining what the feasts and ceremonies of the future will be, it is because we are going through a period of transition and moral mediocrity. The great things of the past that excited our fathers no longer arouse the same zeal among us, either because they have passed so completely into common custom that we lose awareness of them or because they no longer suit our aspirations. Meanwhile, no replacement for them has yet been created. (Durkheim, 1995 [1912]: 429; our emphasis)

A little further on, this statement is accompanied by a pledge for a somewhat hopeful future, but whose horizon is left open:

> In short, the former gods are growing old or dying, and others have not been born. [...] It is life itself, and not a dead past, that can produce a living cult. [...] A day will come when our societies once again will know hours of creative effervescence during which new ideals will again spring forth and new formulas emerge to guide humanity for a time. [...] As to knowing what the symbols will be in which the new faith will come to express itself, whether they will resemble those of the past, whether they will resemble those of the past, whether they will better suit the reality to be expressed – that is a question that exceeds human faculties of prediction and that, moreover, is beside the point. (Durkheim, 1995 [1912]: 429; our emphasis)

Durkheim did not live long enough to witness the great transformations that the twentieth century was to undergo, let alone the new challenges emerging in the first decades of the twenty-first century. In any case, many of his concerns still resonate in the contemporary world. In particular, his commitment to a secular morality offers a very important perspective for thinking about modernity. While seeking to establish a departure from the normative claims of moral philosophies, his discussion of secular morality indicates in what sense the human sciences can – and should – make a commitment to the ethical principles of our time. Indeed, in the model of secular morality idealized by Durkheim, the various human sciences, in particular sociology and history, play a crucial role for this kind of morality, insofar as these disciplines reveal the fact that all morality is always socially constructed and therefore contingent. While this points to an eventual fragility of the moral foundation, it is also a way of summoning our responsibility for building and maintaining just values that make sense in a plural world.

The increasingly globalized world, due to networked communication processes and economic and political interdependence between countries, imposes issues concerning the whole of humanity, such as the challenge of climate crises, food production, or the recent pandemic. Even with the emergence of international organizations, such as the United Nations, we are far from thinking in terms of moral ideals shared by all humanity. Durkheim was right, however, that beliefs, rites, and symbols have never ceased to exist. Old traditions renew their discourses and practices, so that the past continues to leave its mark on the present, even though new moral ideals multiply in profusion, not only among different societies, but within the same national territory.

Globalization has not produced the same universally shared moral ideal, but it has allowed some specific moral ideals to constitute themselves transnationally. An American who supports Donald Trump has much more in common with a Brazilian follower of Jair Bolsonaro than his neighbor, a queer activist. The latter, in turn, may have a lifestyle much closer to that of a Turkish immigrant living in Berlin than to her cousin converted to a Neo-Pentecostal Church. The complexity of the present makes the task of a sociology of morality even more relevant. Mapping the different moral rules and ideals, identifying the narratives condensed into symbols and slogans, and, above all, understanding the social processes behind the production of multiple expressions of the sacred is of paramount importance. The dynamics of interaction between different moral ontologies is one of the most important challenges to the human sciences of the present. The sociology of morality inspired by the tradition inaugurated by Durkheim has much to contribute.

References

Brunetière F (1898) Après le procès. Revue des deux mondes 146:428–446
Connell RW (1997) Why is classical theory classical? Am J Sociol
Durkheim É (1887) Ethics and Sociology of Morals. (Translated by Robert T. Hall). Buffalo, New York, 1993. 24:33–58, 113–42, 275–84
Durkheim É (1898) Individualism and the Intellectuals. In: Lukes, S., Scull, Andrew. Durkheim and the Law. Palgrave Macmillan, London, 2013 p. 152–163
Durkheim É (1902) Sur le Totémisme. L'Année sociologique 5:82–121
Durkheim É (1907) A propos du conflit de la morale et de la sociologie. Lettres de M. Durkheim et réponses de S. Deploige. Revue néo-scolastique 14:606–621
Durkheim É (1925) L'éducation morale. F. Alcan, Paris
Durkheim É (1975) Introduction à la Morale. Les Éditions de Minuit, Paris, pp 313–331
Durkheim É (1992) L'Enseignement de la Morale à L'École Primaire. Revue Française de Sociologie XXXIII:609–623
Durkheim É (2002) Moral education. Dover Publications/David & Charles, Mineola/Newton Abbot
Durkheim E (2013a) Durkheim: the rules of sociological method: and selected texts on sociology and its method. Macmillan International Higher Education
Durkheim E (2013b) The division of labour in society. Palgrave Macmillan, Houndmills
Durkheim É, (1993) Ethics and the sociology of morals. Prometheus Books, Buffalo
Durkheim É, Watts Miller W (1997) Montesquieu: quid secundatus politicae scientiae instituendae contulerit/Montesquieu's contribution to the establishment of political science [bilingual edition]. Durkheim Press, Oxford
Durkheim É (2010) Sociology and Philosophy. Routledge, New York. Translated by D. F. Pocock.
Glowczewski B (2014) Rejouer les savoirs anthropologiques: de durkheim aux aborigènes. Horizontes Antropológicos 20:381–403
Guyau J-M (1887) L'irréligion de l'avenir: étude sociologique. F. Alcan, Paris
Isambert F-A (1990) Durkheim: une science de la morale pour une morale laïque [A science of ethics for a secular ethics]. Archives des sciences sociales des religions 69:129–146
Jones RA (1994) The positive science of ethics in France: German influences on "De la division du travail social". Social Forum 9:37–57
Jones SS (2001) Durkheim reconsidered. Polity, Cambridge
LaCapra D (1972) Émile Durkheim: sociologist and philosopher. Cornell University Press, Ithaca
Logue W (1993) Charles Renouvier, philosopher of liberty. Lousiana State University Press, Louisian
Lukes S (1969) Durkheim's "individualism and the intellectuals". Polit Stud 18:14–19
Massella AB (2006) O Naturalismo Metodológico de Émile Durkheim. Humanas/Editora UFG, São Paulo/Goiânia
Mbembe A (2017) Critique of black reason. Duke University Press
Miller WW (1993) Durkheim's Montesquieu. Br J Sociol 44:693–712
Miller WW (1996) Durkheim, morals and modernity. University College Press, London
Miller WW (2012) Durkheimian quest, A: solidarity and the sacred. Berghahn Books
Nisbet RA (1952) Conservatism and sociology. Am J Sociol 58:167–175
Pickering WSF (2009) Durkheim's sociology of religion – themes and theories. James Clarke & Co., Cambridge

Rawls AW (2004) Epistemology and practice – Dukheim's the elementary forms of religious life. Cambridge University Press, Cambridge/New York

Renouvier C (1848) Manuel Républicain de L'Homme et du Citoyen. Pagnerre, Paris

Rosati M (2009) Ritual and the sacred: a neo-Durkheimian analysis of politics, religion and the self. Ashgate, Farnham

Rosati M, Weiss R (2015) Tradição e autenticidade em um mundo pós-convencional: uma leitura durkheimiana. Sociologias 17:110–159 [Tradition and Authencity in a post-conventional world: a Durkheimian reading]

Schmaus W (1994) Durkheim's philosophy of science and the sociology of knowledge: creating an intellectual niche. University of Chicago Press, Chicago/London

Schmaus W (2004) Rethinking Durkheim and his tradition. Cambridge University Press, Cambridge

Stock-Morton P (1988) Moral education for a secular society. State University of New York Press, New York

Weiss R (2012) From ideas to ideals: effervescence as the key to understanding morality. Durkheim Stud 18

Problematizing Societal Practice: Histories of the Present and Their Genealogies

7

Ulrich Koch

Contents

Introduction	160
Genealogy as Critique	162
History and Critique	163
Problematizing as Critique	165
Defining the Present	169
Genealogy and the History of Scientific Ideas	170
Praxis and Power	172
Practicing Critique Through "A Historical Ontology of Ourselves"	174
Conclusion	176
References	178

Abstract

The chapter discusses a mode of critical historical work often referred to as the writing of the "history of the present." In the broadest terms, scholars who contribute to this genre place the study of the past in the service of a critical engagement with the present while committed to a radical historicism. To examine the epistemological and ethical-political commitments of this approach, the chapter, first, distinguishes different forms of critique. Genealogical critique in the tradition of Nietzsche and Foucault, on which this critical historical practice heavily draws, represents a variant of critique that does not set out to reject what it critiques but to render it problematic and contestable. Second, the chapter highlights two strands of recent scholarship written in this mode. On the one hand, authors have employed genealogy to study the fractured, nonlinear histories of scientific ideas and thought traditions in the human sciences, whereas another group of scholars has been more committed to Foucault's praxeological

U. Koch (✉)
George Washington University, Washington, DC, USA
e-mail: koch@email.gwu.edu

perspective and the analysis of power relations. Building on these examples and a discussion of the late Foucault's understanding of critique, the chapter, finally, aims to explicate both the broader theoretical and ethical-political stakes that animate this critical use of history.

Keywords

Critique · Genealogy · "History of the present" · "Genealogy of the present" · Foucault · Critical Theory

Introduction

Academic historians are generally reluctant to designate their scholarly pursuits as "critical," whereas members of other disciplines have been far less hesitant to adopt the label. Most prominently, critical theory – a brand of theorizing no longer only associated with intellectuals whose lineage can be traced back to the *Frankfurter Institut für Sozialforschung* – has become a well-established current within social and political philosophy. And the designation has similarly caught on to demarcate fields and subfields, such as *critical* psychology, *critical* legal studies, or, more recently, *critical* medical humanities, from their conventional, "mainstream" counterparts. The reluctance by historians is even more striking given that many of the scholars involved in these and other fields have long embraced historical methods; history and the work of historians remain crucial intellectual resources for contemporary critical engagements across the humanities and social sciences. At the same time, historians themselves have on many occasions critically intervened in several academic and public debates, sometimes in ways that have had rather tangible societal consequences (Smith, introduction to this section).

Yet it seems as though, because history can be so readily harnessed to serve normative projects, – including explicitly political ones – academic historians remain worried about the potential for abuse. Leaving the more obvious methodological challenges that the instrumentalization of history poses aside for the moment, what likely continues to haunt historians is the fear that social critics would attempt to *derive from history* (or its anticipated course) their vision of how things ought to be. The various forms of critical social science that emerged in the middle decades of the twentieth century were indeed either deeply enmeshed with or directly grew out of Marxist intellectual traditions (see, e.g., Birnbaum 1973; Schraube and Osterkamp 2013). Although many of these critical undertakings challenged or at least significantly modified the historical determinism associated with Hegel-Marxism (e.g., Adorno and Horkheimer 1947/1986; Kaye 1984). Moreover, the practice of drawing on stories of the past for moral or political ends certainly has a longer history and is by no means confined to the political left. Whether it is the commemoration of past events deemed significant for a social collective, the idea that presumed historical constants reflect the necessity of a particular social order, or the general notion that historical change reveals the purported endpoint of history's long arc, historical narratives have long been and continue to be deployed to both vindicate and critique societal arrangements in the present.

Clearly, the trouble with such "presentist" uses of history is that the normative claims buttressed, and many times only illustrated, by (meta-)narratives can never be distilled from history alone. Instrumentalizations of history are prone to tendentious readings of the past, which, along with their backward projections of present concerns, their anachronisms, make them an easy target for critics. How, then, can there be a critical history that itself holds up to critical scrutiny? And, conversely, what can serve as the normative footing of a societal critique that not only incorporates but generates historical accounts?

This chapter discusses one mode of critical historical work which is frequently, albeit rather loosely, referred to as the writing of the "history of the present." In the broadest terms, scholars who contribute to this genre place the study of the past in the service of a critical engagement with the present while still committed to a radical historicism. Historians writing in this critical mode, that is, reject the notion that history can be understood as a developmental process – an idea which so often underwrites the various instrumentalizations of history – and that there are certain subjects that stand outside history, as it were, removed from the reach of historical critique. This nonlinear, anti-teleological view of history is closely linked with an approach Michel Foucault (1971/1977), following Friedrich Nietzsche, called "genealogy." Hence, some authors also prefer to describe their critical work in terms of a "genealogy of the present" (see, e.g., Leys and Goldman 2010).

As a method, "genealogy" is still applied within philosophy but also has been taken up by political theorists, anthropologists, and, of course, sociologists and historians, who each have adapted it to meet specific methodological challenges they face in their respective disciplines. Again, only very broadly speaking, it can be said that "histories of the present" are distinct from some other uses of "genealogy" in that they explicitly target an aspect of contemporary forms of life that is seen as particularly disconcerting and at the same time indicative of what is ethically and politically at stake in the present moment, although the extent to which a history of the present draws on the genealogical method, more narrowly defined, varies.

Because of this and the heterogeneity of the genealogical approach as it is applied within various fields, it remains difficult to point to a set of specific conceptual and methodical features as defining for such histories. The work of Foucault, especially from the mid-1970s onward, clearly remains a central reference not just for anyone who sets out to write a "history of the present" but also for any discussion of the ideas motivating such an endeavor – not least because it was Foucault who first used the phrase in the opening chapter of *Discipline and Punish* (Foucault 1975/1977). Inspired by his late work, in the mid-1980s, an academic newsletter carrying the phrase in its title was compiled by a group of academics affiliated with UC Berkeley. And shortly thereafter the notion gained exposure through the work of a group of mostly sociologists, the so-called English Foucauldians, affiliated with the *History of the Present Research Network* founded at Goldsmiths College, London, in 1989 and coordinated by Nikolas Rose (Barry et al. 1995/2005; Rose 1985, 1998). On the other hand, upon its launch in 2011, the editors of *History of the Present: A Journal of Critical History* ensured readers that theirs was not a post-structuralist journal. As a publishing venue, it remains first and foremost devoted to *critical* contributions, which may be informed by a number of theoretical frames, including Marxist and psychoanalytic theory, representing various types of history – social, cultural, political, intellectual, or economic, among others. Further, and more specifically, the journal aims to be a forum for critical reflections upon the discipline of history itself, its often taken-for-granted assumptions about the relation between past and present, and the historicity of its analytic categories and their entanglements with the politics of the present (Scott and Connolly 2011). In contrast, the more recently founded online magazine *Geschichte der Gegenwart* (*History of the Present*), edited by a group of historians and literary scholars based in Switzerland and Germany, is public-facing and concerned with topics that either have already entered public debate or are deemed of potential interest to a broader, nonacademic readership.

In part, these varied usages reflect and extend Foucault's pragmatic approach to the concepts and "analytics" he developed, which were always tailored to meet the demands of a specific project and never exhaustively, authoritatively defined (Garland 2014). They also show that histories of the present address a broad range of audiences: they can be vehicles for critical interventions within disciplinary debates, though more often they transcend such institutionalized boundaries as they seek to engage scholars from various fields and even nonacademic publics. Despite this diversity in usage and conceptual breadth, it is still possible – and indeed more fitting, given its pragmatic approach – to ask what makes this type of engagement with the past *effective* as a form of critique.

To answer this question, it is useful to explicate what differentiates genealogy, as a form of historical critique, from other forms of critique and criticism. The first part, "Genealogy as Critique," thus discusses different notions of critique in philosophy and the sciences as well as, in greater detail, Foucault's and Nietzsche's paradigmatic uses of genealogy and their radical historicist presuppositions. The subsequent part, "Defining the Present," draws on more recent examples of scholarship in this critical vein to explore how, in practice, histories of the present define and theoretically engage with their objects of study to make their historical inquiries relevant for a reflection on the present. In this context, the chapter also attempts to explicate the rarely acknowledged normative impetus that animates this critical use of history. The conclusion provides a summary and briefly points out the inherent limitations and dangers of this type of critical history.

Genealogy as Critique

In the everyday sense of the word, to criticize something amounts to rejecting it by articulating reasons for doing so. Providing reasons, arguing against whatever one is opposing, the philosopher Raymond Geuss has noted, is integral not only to our everyday understanding but ultimately also to the philosophical notion of critique (Geuss 2002). This is so, even though ideas about what may count as a valid reason for someone to reject something will vary depending upon context. To criticize something further implies that we can, in fact, reject it, that we, as critics, have a say in the matter. Critique usually represents an evaluation of humans, their actions, the outcomes, and, in the broadest sense, the "products" of those actions.

Despite the variety of what may count as a sound basis of a critique, beginning with the Enlightenment, philosophers have nevertheless found reasons to place their faith in a form of critique that promised not only to discern truth from what merely appeared to be true but also to reveal the bounds of reason itself. This kind of epistemological critique, of which the philosophy of Immanuel Kant remains the prime example, elevated the practice to a "transcendental reflection" that allegedly exposed the universal validity of a rationality expressed through science and morality. By design, as it were, critique could not be turned against reason as such, calling it into question. In the context of pre-twentieth-century philosophy, exercising critique, then, included not only discerning right from wrong or good from bad but also *analyzing* what could serve as proper grounds for such distinctions (Jaeggi and Wesche 2009).

History and Critique

Scholarly critiques need not always take on the form of an analysis of an entire system of thought, of course, nor is epistemological critique necessarily reflexive as in Kant's case, though some authors have given the term "critique," as opposed to "criticism," a narrower meaning, namely, as referring to the practice of analyzing the presuppositions built into a theoretical system (e.g., Johnson 1994). Generally, an array of critical methods has evolved in the sciences and humanities from their philosophical roots, and it would be wrong to understand genealogical critique as a critique in the narrow sense just mentioned. If a genealogy does indeed function as a critique of a system of thought, it refuses to engage with its premises directly and, rather, delivers its criticism by *revealing* something about this system (Hoy 2008). But, more conspicuously, a genealogy unfolds through historical narrative rather than, say, a series of theoretical claims strung together to form an overarching argument.

Non-historians often perceive the writing of history to be inherently critical, presumably because, in hindsight, one is tempted to evaluate the actions of past generations – though this is not how genealogy functions as a form of critique. Before explicating *how* a genealogy exposes something about its subject, it is worth highlighting that often the mere choice of its topic represents a critical transgression. Employed critically, genealogies historicize aspects of our present that appear to have no history and, hence, seem to lie beyond the reach of criticism. A genealogy, of course, is not the only kind of critical practice that transgresses implicitly or explicitly drawn boundaries around what may become the object of critique. Since critique, in the conventional sense, can only be concerned with what directly springs from human activity, its role appears to be limited in the natural sciences and those branches of social science committed to similar ideals of objectivity; although the history of science amply shows that the line separating what can become the object of critique and what must necessarily elude it is never consistently and lastingly drawn in the sciences (see also Hampe 2009; Smith 2005). Clearly, historians and sociologists of science, using their own respective critical methods, have repeatedly crossed the line between what scientists, at a given moment, would and wouldn't critique.

In the context of the past debates surrounding realism and anti-realism in science, for instance, the social study of science was often made to appear as if it rejected the possibility of scientific knowledge. Most memorably, in the public controversy later dubbed the "science wars," self-styled defenders of science asserted that their opponents, mostly sociologists and literary theorists, had set out to "deconstruct" legitimate scientific claims, presenting them as mere "social constructs." On the other side of this divide, the exposition of the social construction of certain subjects of scientific inquiry, particularly of the social sciences, indeed often went hand in hand with their devaluation and rejection, as Ian Hacking observed (Hacking 1999). In hindsight, however, the rhetoric on both sides seems overdrawn, even harmful, as it obscured rather than elucidated the different points of contention between realist and anti-realist accounts of knowledge (see also Oreskes 2019). David Bloor, proponent of the so-called strong program in the sociology of scientific knowledge, later stressed that relativism should not be placed in opposition to truth or objectivity – which are the opposites of falsehood and subjectivity, respectively – but absolutism (Bloor 2011). On this view, scientific relativism is not necessarily ontologically anti-realist, nor anti-science in its orientation; rather, it merely offers a methodological vantage point that allows social scientists to raise questions about the social embeddedness of scientific practices and ideas. The larger point here is that the practices of scientists and the evolution of scientific knowledge can be studied without *grounding* one's scholarly critique in the *question* of whether the former were successful (or not) and the latter true (or false). With respect to bracketing pertinent scientific knowledge claims, this critical approach, which has become rather common in the more recent historiography of science, coincides with the one genealogists take.

The example also illustrates that, given the variety of its forms, it can become difficult to discern what particular claim, presupposition, belief, or practice a critique rejects, tests, modifies, or affirms. The varieties of critique cultivated within various fields follow conventions and serve purposes specific to the disciplines in which they are practiced. The disciplines clearly "discipline" critique, by channeling its critical impulses toward what one must call constructive criticism, perhaps at times neutralizing them completely. As soon as critique is deployed as an analytic device, moreover, the "technical" aspects of a critic's argument gain in importance, making certain forms of critique even less penetrable to outsiders.

Self-declared *critical* intellectual endeavors, on the other hand, often redirect the aim and broaden the scope of critique. Max Horkheimer, explicating the terms of his Critical Theory, alleged that the "system of the disciplines" ensured that the knowledge produced under its conditions would be seamlessly applied to further pregiven, undisputed ends. The traditional disciplinary matrix was ultimately organized around theoretical and practical demands arising from the "reproduction of life within contemporary society," he charged. Critical Theorists, in contrast, not only set out to examine these ends as they presented themselves within contemporary society but also rejected the idea of them being given and immutable. As Horkheimer put it, Critical Theory has as its subject matter (*Gegenstand*) "humans as the producers of their historical forms of life in their entirety" (Horkheimer 1937/2005, 261); and as such its efforts were also aligned with ongoing "struggles over certain forms of life" in the present (262). The target of this type of critique, then, was no longer defined as well as restrained by disciplinary conventions; rather, Critical Theory was directed at and entailed the rejection of a form of life associated with late capitalist society.

This critique unfolded against the backdrop of a philosophy of history, which Horkheimer together with Theodor W. Adorno exposed in the *Dialectic of Enlightenment*. The book was a sweeping indictment of instrumental reason, which, they argued, made the Enlightenment movement and the forms of rationality it engendered complicit in bringing about the social pathologies which, with two World Wars and the rise of totalitarian forms of government, had come clearly into view in the first third of the twentieth century. The enlightenment, including the rise of modern science, was situated within a broader historical arc that pointed toward increasingly oppressive societal arrangements (Adorno and Horkheimer 1947/1986). Horkheimer and Adorno, thus, still entertained a version of what one could call developmental historicism, albeit one that assumed a negative outcome (Bevir 2008). Adorno also suggested that a decidedly "critical history" had to reveal the unexamined presuppositions that served to legitimate social inequality in the present (Adorno 1998/1963; see also Scott 2007).

Consistent with their far-reaching condemnation, Horkheimer and, more emphatically, Adorno refrained from spelling out what would constitute a better state of human affairs, a non-oppressive, emancipated form of life (Adorno 1966/1973). Their critique was not based on a set of theoretically elaborated normative standards; the first generation of Critical Theorists still aimed to expose, *through* their critique, society's inherent contradictions and the impossibility of rational societal arrangements under the impositions of instrumental reason. (Jürgen Habermas would famously revise this approach and undergird his brand of critical theory with the normative framework developed in *The Theory of Communicative Action* (Habermas 1984).) Critical Theory thus retained "critique" as an analytic procedure, while becoming more emphatically critical by refuting as false what appeared as necessary and self-evidently true to most social actors.

Problematizing as Critique

The type of critique Nietzsche called "genealogy," however, does not follow this pattern of argumentatively grounded rejection, although it also sets out to question self-evident assumptions. Instead, as Geuss has shown, genealogy operates by way of *problematizing* what it critiques (Geuss 2002; see also Geuss 1994).

To take the paradigmatic example, in his *Genealogy of Morals*, Nietzsche did not directly challenge moral norms but instead forced his readers to question their value by presenting a series of speculative historical accounts of multiple developmental lineages, rather than one definitive narrative. Christian morality, the target of his critique, is described as the product of heterogeneous developments – such as the lingering effects of the "resentment" felt by the enslaved toward their masters, the social pressures arising from urbanization letting individuals turn their aggressions inward, or the striving for power of a priestly caste (Nietzsche 1887/1989).

Genealogy, then, does not offer historical explanations that point to a decisive moment, a specific cause, or a continuous development revealing its essential feature. In the beginning pages of his book, Nietzsche contrasted his approach with origin stories that tie the emergence of moral norms to the social utility of corresponding behaviors. This type of explanation – to which evolutionary psychologists still gravitate today – must presuppose a questionable continuity between past and present. So-deemed moral actions can only prove their utility over time if the historical conditions under which they are maintained are relatively uniform, if certain anthropological features and sufficiently stable environmental demands keep them useful. Nietzsche's genealogy of a multitude of possible descents called into question this linear, teleological account of a stable, defining feature of morality that can be traced throughout history, depicting morality's emergence not as matter of utility, and hence as seemingly necessary, but of contingency. Disparate historical events, human struggles whose outcomes are never certain, are among the scattered sources that make up the layers of Nietzsche's genealogy of morality.

Raising the question of the value of morality served a broader critical purpose, of course. Both Nietzsche's approach to history and its subject matter let his genealogy act as a destabilizing force to readers' self-understanding (see Saar 2007). To present – one should add, in a rather dramatic, highly stylized manner – the origins of what commonly is understood as a cultural achievement as contingent, the chance outcome of power struggles, gnaws at fundamental convictions. As it excavates different forms of subjectivity while disregarding transhistorical features in which they can be rooted, the work of the genealogist suggests that the self is variable, perhaps constitutively historical, since a significant aspect of one's own subjectivity can now be read as the effect of past struggles over influence and power.

It is important to note, however, that not all who draw on Nietzsche's method of the genealogy do so with critical intent. Among the genealogies proposed by philosophers, there are also "vindicative genealogies," as David Hoy has called them (Hoy 2008). Perhaps the most prominent example is Bernard Williams's genealogy of *Truth and Truthfulness*. By embarking on an historical account that is openly speculative, Williams did not aim to discredit the value we place in truth but to resolve a paradoxical attitude toward the truth, which he diagnosed as distinctly modern: a tension between not wanting to be deceived and a deep-rooted skepticism toward ever attaining the truth (Williams 2010). Hoy suggested that a vindicatory genealogy such as this aims at helping us better understand and reconcile aspects of ourselves that we may have overlooked.

On the other hand, in his influential essay "Nietzsche, Genealogy, History," Foucault foregrounded genealogy as a method of critique while also abandoning Nietzsche's speculative mode of writing, reconceiving it as a serious empirical undertaking (Foucault 1971/1977). At the same time, he also engaged with Nietzsche's skeptical remarks "On the uses and disadvantages of history for life," tracing a fractured line between his earlier, damning conclusions concerning the value of the study of history and how he would later apply historical methods in the *Genealogy of Morals*. In 1874, Nietzsche had famously declared that the works of academic historians are of no practical use to those trying to go about their lives. Despite his disdain for history as science (*Wissenschaft*), he did mark out "critical history" (*kritische Historie*) as among three possible uses of history with some practical value. Here, "critical history" amounted to a deliberate rejection, a liberation from the past by condemning past injustices from the vantage point of the present. "In order to live," he contended at the time, "a person must sometimes have the courage to dissolve, "break up the past" (Nietzsche 1874/1983, 21). Foucault argued that this initial version of "critical history" "metamorphosized" in Nietzsche's later work from a "critique of the injustices of the past by a truth held by men in the present" to a condemnation of the subject that maintains such truths (Foucault 1971/1977, 164).

In the much-cited opening paragraphs of the essay, Foucault strikingly conveyed the historico-philosophical underpinnings of the genealogical approach, including its dedication to historicizing aspects of subjectivity that are generally presumed to have no history:

> Genealogy is gray, meticulous, and patiently documentary. It operates on a field of entangled and confused parchments, on documents that have been scratched over and recopied many times. [...] [I]t must record the singularity of events outside of any monotonous finality, it must seek them in the most unpromising places, in what we tend to feel is without history—in sentiments, love, conscience, instincts; it must be sensitive to their recurrence, not in order to trace the gradual curve of their evolution, but to isolate the different scenes where they engaged in different roles. Finally, genealogy must define even those instances where they are absent, the moment when they remained unrealized. (Foucault 1971/1977, 139–140)

What this passage does not adequately capture, however, are features, to some degree already present in Nietzsche, that would become more salient in Foucault's own genealogical work: the focus on practices, and the ways through which the forming of subjectivities is tied to the exercise of power. It was indeed this "praxeological" perspective that freed Foucault and Nietzsche from metaphysical commitments to a presupposed "nature" of human subjectivity, without falling into a radical critique that denies the significance of subjective experience altogether or reduces it to other, heterogeneous phenomena (Saar 2009).

Foucault's earlier engagement with the history of the human sciences, which he described as a kind of "archaeology," is often contrasted with his later "genealogical" studies, beginning with *Discipline and Punish*, published in 1975. The archaeological approach still built on more established methods in the history of science in that it focused on the "excavation" and subsequent reconstruction of what one might call the implicit systems of thought characteristic of distinct historical periods (e.g., Foucault 1966/1970, 1969/1972). The "archaeologist's" task is not to detail what led to the transition from one order of knowledge, or "episteme," to the next, but to distinguish them, investigate their inner logics, the ways in which they preconfigure what becomes known and how.

As already mentioned, his first major genealogical study, a history of penal practices and institutions in France since the eighteenth century, was explicitly characterized as a "history of the present." In 1971, Foucault had become involved in a political movement calling for the abolition of prisons, and his work on the book coincided with a period of intense political activism outside the academy (Sarasin 2005). Foucault, the "genealogist," no longer excavated, examined, and then contrasted circumscribed layers of the past; he turned his attention to the emergence and descents of practices and institutions that he deemed problematic in the present. Beyond a history of the prison that challenges conventional rationales of incarceration, *Discipline and Punish*, consistent with the genealogical approach, also offered a history of the self or, rather, of the disciplinary practices and knowledges that, shifting from the body's surface to "inner" predispositions, help constitute modern subjectivity. In conjunction with and organized around novel forms of punishment, "normalizing technologies," disciplinary practices, and new fields of knowledge emerge. Power and knowledge are imbricated, Foucault famously asserted, and "the soul" of "man," he wrote in a rather dark passage in the book's introduction, is but the product or effect of the disciplining of bodies – "an instrument of political anatomy" (Foucault 1975/1977, 32).

This indictment of the prison and "the disciplines," along with his activism, raises the question whether in this paradigmatic case Foucault's genealogy, in fact, amounted to a critique of the type that rejects what it sets out to criticize. One could argue, though, that *Discipline and Punish* does not provide an account of reasons as to why the disciplining of bodies or the prison as an institution should be rejected; it does not directly engage with the arguments for or against reforming or abolishing prisons, for instance, by constructing counterarguments. His indictment of the prisons, instead, relies on an historical account that details the relations between certain practices, bodies of knowledge, and institutions which stand in stark contrast to and raise doubts *about* the reasons usually given to justify incarceration.

During an interview, conducted later in his life, Foucault declared that he never intended to make claims of the kind that "x is wrong" or "false"; rather he was concerned with showing that "x is dangerous" (cited in Geuss 2002). Genealogy as understood by Foucault, Geuss has suggested, is "a way of concentrating attention on a given situation in the context of an imminent danger" (Geuss 2002, 213). Turning away from an "imminent danger" – an approach Nietzsche advocated at times – is often not a good strategy, Geuss further noted in this context. The strategy pursued by Foucault, instead, demands that the threat be identified, discriminated from other, less pressing dangers, and closely examined. The sociologist David Garland also emphasized that a "history of the present" must be quite deliberate about its subject, which is to say, about *what* it problematizes, and therefore always builds on a preliminary analysis (Garland 2014). Histories of the present do not entail a comprehensive historical diagnosis of "the present"; they begin with an identification and critical assessment of a troublesome aspect of something *in* the present.

With regard to its scope, then, genealogical critique differs from the kind of critique the early Critical Theorists of the Frankfurt School engaged in. It does not object to societal arrangements in their "totality," nor does it help identify an all-encompassing ideology that sustains such arrangements by making them appear rational and inevitable. A history of the present delivers a critique that is more targeted and circumscribed, though it still derives its impetus from challenging the inevitability of a given form of life.

This, however, considerably raises the stakes of identifying a topic for historical investigation so that it may serve as a medium for a critical reflection on the present. How is "the present" understood by those making use of genealogy as a form of critique? In practice, what principles guide the preliminary analyses that give critical genealogies "a clear object and direction" without lapsing into a sweeping diagnosis of the times (Garland 2014, 378)? To answer these questions, the following section turns to more recent examples of how genealogy has been employed in the historiography of the human sciences.

Defining the Present

Although histories of the present focus on a specific feature of the present, their respective scopes are often not confined to the phenomena one would conventionally associate with their topics. In the case of Foucault's genealogies, Garland has further pointed out, a particular issue virulent in the present, such as the penal system or sexuality, is studied in relation to the emergence of a network of coordinated practices, power relations, norms, and bodies of knowledge – a "power-knowledge regime" or, as Foucault termed it, a *dispositif*. His later project, a history of sexuality, for example, is not so much concerned with the history of how sexual acts or longings were represented in the past; rather, it effectively functioned as a history of the present because it presented and studied them as part of such a *dispositif* (Foucault 1976/1978). At the same time, it radically reframed contemporary discourses and practices aimed at sexual liberation as only the most recent and visible effects of this sprawling regulatory apparatus that has been in place since the nineteenth century (Garland 2014). In the case of Foucault's genealogies, the practice of problematizing is not only what the genealogist engages in. "Problematization" gains the status of a technical term in his late work, referring to the discursive and nondiscursive practices through which a certain aspect of human conduct or experience becomes presented and treated as a specific kind of problem, one that required scientific investigation, moral consideration, or political analysis (see Foucault 1994). As Foucault remarked at the end of his career, he was interested in the emergence of "modes of problematization," as opposed to the formal conditions of truth or the only relative stability of representations, to expose the contingency, fundamental instability of what "defines objects, rules of action, modes of relation to oneself" (Foucault 1984, 49). Pointing out the reflexivity of his approach, Graham Burchell has aptly noted that a history of the present is in effect a deliberate *re-problematization* of its subject matter, or, as Thomas Lemke has put it, "problematization" becomes both the object and the *objective* of Foucault's critical investigations (Burchell 1993; Lemke 2011).

Genealogy and the History of Scientific Ideas

Not all histories of the present involve an initial analysis of a *dispositif*, however, nor do genealogies always rest on a thorough historical investigation of practices, techniques, and institutions and their relations to power. Some conceptual and intellectual histories, for instance, are conceived as genealogies, insofar as their authors share Foucault's radical historicist commitments. Such genealogies trace the intellectual lineages of an idea, concept, or novel way of approaching a problem not as a progression toward the full realization of an intended meaning or an approximation to the truth but as a fractured, discontinuous movement, marked by contradictions, reversals, or false starts. Even though they do not closely follow the methodological model set by Foucault's genealogies, critical histories of this sort still share three defining features, identified by Hoy: a commitment to *nominalism*, rejecting the idea of universals, and a view of historical change that emphasizes *contingency* and understands it as always *contestable*, and therefore indetermined (Hoy 2008). Even though in the context of intellectual histories the everyday workings of power recede from view, this variant of genealogy still aims to expose the contestations and unsettled questions that point to the multiple, often conflicting descents of a given theory, concept, or claim. Also in this sense, genealogy serves to denaturalize what over time has hardened into a given, a static reflection of a supposedly natural state of affairs. However, to render visible discontinuities, contestation, but also neglected affinities, works in this vein more often employ interpretative methods that draw out implicit models and common presuppositions; their historical accounts often build on or incorporate "structural" analyses, in other words.

Ruth Leys, for instance – whose studies on the histories of trauma, guilt, and "affect" are all written in a "mode" she calls the "genealogy of the present" – comments that she was initially drawn to the genealogical approach, as it allowed her to convey the many "structural repetitions" she had observed in the successive attempts to theorize psychic trauma (Leys and Goldman 2010, 657). *Trauma: A Genealogy* presents a nonlinear narrative, which in Ley's telling is marked by recurring, albeit futile attempts to reconcile two competing ways of conceptualizing traumatic experience, what she calls the "mimetic" and "anti-mimetic" models (Leys 2000). In her subsequent book, she identifies a similar structural opposition, though in *From Guilt to Shame: Auschwitz and After* she also conveys a significant shift, a displacement of the notion of survivor guilt, which draws on "hypnotic-immersive" ideas concerning the relation between perpetrator and victim, by a theory of shame that hinges on the victim's awareness of being shamefully exposed to others (Leys 2009). Although Leys cautions against associating her methodological approach with a particular sociopolitical stance, these books clearly function as critical interventions in several ongoing yet closely related debates in the humanities and social sciences (Leys and Goldman 2010). *Trauma: A Genealogy*, for one, represents a critical response to the emergence of "trauma studies" in the humanities, which had culminated in the 1990s; and the follow-up and her latest work on theories of affect in the sciences and humanities both extend and broaden lines of criticism already present in the previous work. In *The Ascent of Affect: Genealogy and Critique*, Leys, after uncovering the unstable empirical foundations on which the postwar science of emotion rests, mounts a devastating critique of its one-sided appropriation by some social and cultural theorists in the humanities and what she diagnoses as their questionable commitment to "nonintentionalism" – a theoretical tendency to downplay and sometimes deny outright the role of intentions, beliefs, ideology, and rationality in human affairs by endorsing a "nonintentional" theory of affect as something rooted in the body and withdrawn from mental processes (Leys 2017). In this latest work, Leys is outspoken about the broader sociopolitical stakes of theorizing "affect" in ways that foreground personal differences and identity and, she fears, thus limit the possibilities for debate and the role of argumentative persuasion in public discourse. Even though her critique is articulated through an engagement with intellectual history, then, it still raises concerns about a contemporary practice, namely, the devaluation and potential marginalization of thoughtful reflection and deliberation. In other words, Leys's critique also functions as a defense of the conditions of its own possibility – of the *practice* of critique and the assumptions that underwrite its epistemic significance.

Katja Guenther's *Localization and Its Discontents: A Genealogy of Psychoanalysis and the Neuro Disciplines*, to take up another example, operates in a similar methodological vein. Guenther also expounds upon a structural difference, which she finds in the opposition between connectionist and localizationist approaches to studying – and treating – the brain, to invite her readers to rethink the relationship between the neurosciences and the disciplines that study the mind with, in the most inclusive sense, social science methods. She does so genealogically, by showing that psychoanalysis and the modern brain sciences both descend from the late nineteenth-century neuropsychiatry, and that what led to the eventual branching off of psychoanalysis should not be thought of as an abandonment of neurophysiological principles but, instead, as their radicalization (Guenter 2015). Much like Leys's work, Guenther's genealogy can be read as an intervention in ongoing debates concerning the neurosciences and their relation to other fields in the human sciences, particularly concerning their supposedly far-reaching epistemological and ethical implications. Detailing the shifting fortunes of the brain sciences, their long-standing theoretical and practical challenges as well as their imbrication with seemingly incommensurable thought traditions and ways of knowing, reveals how much of their current scientific authority relies on rhetorical maneuvers that affirm the tenuous opposition between "brain" and "mind," or "soma" and "psyche." In this case also, genealogy functions as a critical intervention within the fields whose history it addresses with the potential to inform ongoing efforts to establish a dialogue between them.

Praxis and Power

Apart from such "genealogies of the present" that engage with the history of scientific ideas – and only to a limited extent also with scientific practice – there is another important strand of recent scholarship whose analyses, instead, more consistently extend both Foucault's praxeological perspective and his efforts to analyze relations of power.

A prominent example of this – also because he and his collaborators were quick to embrace the label "history of the present" – is the historical sociology of Nikolas Rose. In his programmatic essay, "How should one do the history of the self?", Rose unambiguously declared that such a history cannot be written by reconstructing the changing philosophical, anthropological, or literary notions of what it means to be a person; it cannot be "a history of ideas: its domain of investigation is that of practices and techniques, of thought as it seeks to make itself *technical*" (Rose 1998, 23). What he called a "genealogy of subjectification" aims to uncover not just different *forms* of subjectivity but the specific practices and processes through which an "individualized, interiorized, totalized, and psychological understanding of what it is to be human" takes form. And he further stressed that his history "of *the relations* that human beings have established with themselves" does not seek to contextualize the emergence of the self within a given "culture"; rather it studies this assembling, the process of "subjectification" in its relation to problems of "government" (Rose 1998, 24). With an incisive focus on sociopolitical changes that have taken place in many so-called advanced democracies in recent decades, some of Rose's studies of this period aimed to show, for instance, how the constitution of the "autonomous subject" and the self-directed practices it engages in are resources for, rather than hindrances or obstacles to, governing individuals and populations.

Drawing on and expanding the late work of Foucault on what he termed "governmentality," Rose and other English Foucauldians thus give their histories of the present a distinct orientation by deploying them within the context of investigations concerned with "the conduct of conduct" within liberal and neoliberal forms of government (Barry et al. 1995). Perhaps also due to discipline-specific conventions and demands, their historical accounts are less often centered on a specific practice, tradition, or idea that usually presents as the primary target of genealogical critique, such as Christian morality, the penal system, sexuality, psychic trauma, or the neuro disciplines. While sharing with Nietzsche's and Foucault's genealogies the commitment to the study of everyday practices in relation to the exercise of power – now more narrowly conceived of as problems of governing – these studies diverge from the earlier examples insofar as they tend to de-emphasize the role of contingency and specific historical struggles, and again broaden the scope of their critique by elevating the role of the theoretical framing of the social practices under study. As the political theorist Thomas Biebricher has argued, this tendency, which deviates from the original genealogical approach, was already present in Foucault's lectures on "governmentality" and then found its continuation in the field of study that evolved in the wake of this work (Biebricher 2008, 2015). "Governmentality studies," he and other critics have therefore charged, often owes "more to modernist sociology with its ideal types than to genealogy" (Bevir 2008, 274; see also Bevir 2010).

The impression, though, that the work of the "Anglo-Foucauldians" could be indicative of a fallback into structuralist forms of analysis may also arise, as has already been insinuated, from disciplinary conventions, particularly regarding the role of theoretical claims in sociological explanations. (Given the examples mentioned above, moreover, it appears that some "structural" elements are often crucial to constructing genealogical narratives.) The more recent work by the historian Jürgen Martschukat on *The Age of Fitness*, for instance, also conceived as a history of the present, similarly employs the notion of "governmentality" to frame the rise of practices and technologies targeted toward augmenting one's physical and mental "fitness" since the 1970s (Martschukat 2019/2021). Here, theoretical claims feature less prominently, and the explanatory role of such claims seems much more bounded. Targeting also a nonacademic audience, the book effectively functions as a history of the present because of its subject matter, because it historicizes societal practices, collective anxieties, and personal convictions that are widely shared in contemporary liberal democracies (the book's focus is on Germany and the United States). Yet it is also the theoretical framing, the understanding it helps to articulate that the ideal of fitness tracks the societal imperative of competition, shaped by the exigencies of government and economics, even as it is practiced as a personal obligation toward oneself, guided by the seemingly independent ideals of health and beauty, that give this history its critical significance.

Also, in their introduction to the collected volume *Foucault and Political Reason*, titled "Writing the history of the present," Andrew Barry, Thomas Osborne, and Nikolas Rose did not fail to emphasize that, generally, the chapters making up their seminal collection "allocate theorizing a more modest role." Consistent with Foucault's own approach, the theoretical concepts they employed (e.g., "strategies, technologies, programmes, techniques") are thought of as "more local" and of limited use when attempting to "sum up the present historical 'conjuncture'" (Barry et al. 1995, 2). With this statement the authors signaled a commitment to a crucial point regarding Foucault's stance vis-à-vis "the present." Foucault's radical historicism also entailed a rejection of the inclination to locate the present moment at a crucial point on a determinable historical path, say, as a break with the past, a moment of historical climax, holding the promise of history's fulfilment, or, instead, a return to more noble origins (see, e.g., Foucault 1984, 33). Attempts at defining the present, Foucault suggested, and Barry and his co-authors concurred, cannot rely on historical periodization, nor does it involve the quasi-heroic act of deciphering what the present moment demands of "us." A history of the present, as already mentioned, is not motivated by a specific idea concerning the significance of the present. A "reflection on 'today'" begins with perceiving "a difference in history," and asks, as Foucault pointed out on various occasions late in his career, "What are we in our actuality?" [...] 'What are we today?'" (Foucault 1988, 145). The "present," Barry and his colleagues explained, is not an epoch but an array of questions that is bound to fragment any coherent notion of what it might entail. Because of this, they argued, an effective history of the present is not just a destabilizing force but may also be a liberating one, creating an opening, "a space for the work of freedom" (Barry et al. 1995, 5).

It is worth presenting this claim more carefully within the context of Foucault's late work, since it raises important questions about the political and ethical stakes of his historical projects, questions which are rarely acknowledged and explicated in the context of a discussion of the history of the present as a critical mode of writing history.

Practicing Critique Through "A Historical Ontology of Ourselves"

Foucault was famously evasive when it came to explicating or even only committing to what appeared to be the normative principles guiding his work, frequently unnerving his interlocutors. Habermas famously accused Foucault of "cryptonormativism" – of failing to publicly acknowledge the norms that tacitly underwrite his critiques (Habermas 1987). Although he repeatedly turned to the theme of "ethics" in his lectures and seminars beginning in the late 1970s, many commentators remained skeptical of the seriousness of his commitment to such topics. Confounding some of his listeners and readers, after having launched scathing critiques of various humanist projects and some of the institutions they engendered, Foucault explicitly aligned his work with the tradition of the Enlightenment, particularly with that of Kant, on several occasions between 1979 and his death (Kolodny 1996; see also Sarasin 2019). The link between his own work and that of several luminaries, "from Hegel through Nietzsche or Max Weber to Horkheimer to Habermas," he proposed, was indeed that they took it upon themselves to critically examine the times in which they lived (Foucault 1984, 32).

This self-diagnosed proximity is most clearly articulated in his posthumously published paper "What is Enlightenment?" (Foucault 1984). Here, Foucault's point of departure is Kant's essay of the same name, "Was ist Aufklärung?", which appeared in a periodical in 1784 (Kant 1784/2000). In the opening sentences of this essay, Foucault declared, Kant proposed a new way of defining the present, namely, in negative terms, as the leaving-behind of a state of *Unmündigkeit* or "immaturity" – what Kant understood to be the inability to make use of one's reason without the guidance of another. (For consistency, this usage adopts the English translation of the term already used in Foucault's published talk. It is, though, a problematic rendering of the meaning Foucault seems to build upon, since "immaturity" carries the connotation of an organic growth process, which is absent from the German word *Unmündigkeit*.) This way of thinking about the present as introducing something new that poses a problem or task, Foucault further suggested, is indicative of a critical stance toward the present, which he assimilated to what he called "the attitude of modernity." Rather than thinking of modernity as a historical period, it should be conceived of as a specific attitude toward the present, a philosophical *ethos*, Foucault argued. And it is this ethos, invoked in Kant's essay, that places later authors engaged in this type of critique, including himself, in the Enlightenment tradition, and not "faithfulness to doctrinal elements," say, the unwavering commitment to a set of theoretical claims or humanist convictions usually associated with the Enlightenment (Foucault 1984, 41).

Whereas in his major philosophical critiques Kant sought to place reason and morals on a sound theoretical footing, he now raised the question of what *Aufklärung* means *in practice*. Or as the philosopher Niko Kolodny emphasized: Foucault can read Kant's essay as a "history of the present," because of an emergent, "new investment of rationality and technology in political and private life that made the present an anxious question. [...] While Kant's critical philosophy laid the foundations of a rational and autonomous science and morality, his history of the present established a form of critique capable of monitoring their *empirical consequences*" (Kolodny 1996, 76, my emphasis). Foucault, in short, situated his own critical projects within a tradition of the enlightenment that calls its *practical effects* into question (see also Geuss 2008). Unlike some of the other philosophers included in this lineage, however, Foucault remained deeply skeptical about the intellectual work of establishing and defending normative orders that might correct enlightenment's practical failings in the future. Rather, true to his radical historicism and the "modernist attitude," he appears to have maintained that the enactments of normative orders themselves must be understood and studied as part of the "empirical consequences" of enlightened rationality and morality. Because of his skepticism toward any form of normative *theory*, which must claim to know what, say, "liberty" or "reason" mean in *practice*, and which, he feared, can only be had for the price of theoretical and practical complacency, Foucault eschewed any form of "transparent normativism" (Kolodny 1996, 79).

This does not mean, however, that Foucault never made the ethical *stakes* of the genealogical work he saw himself engaged in transparent. In a sense, they have not changed fundamentally since the time Kant's essay was written, as they concern the tensions between the expansion of humans,' collective and individual, "capabilities" and the accompanying growth of "autonomy," which became so salient in the eighteenth century. In "What is enlightenment?", Foucault declared that the broader stakes of his critical work came down to this question: "How can the growth of capabilities be disconnected from the intensification of power relations?" (Foucault 1984, 45) With this statement, Foucault signaled an affinity to the critique of instrumental reason in the tradition of the Frankfurt School. At the same time, though, he upheld the view that what is ultimately at stake is the transformation of a societal practice and not a critique of the principles on which it supposedly rests, nor the affirmation of a competing set of principles in which a more rational practice could be grounded. The task of partially disconnecting the "growth of capabilities" from the exercising of power is always limited to "specific transformations," since there are no universal principles that could provide guidance (Foucault 1984, 44). Given the doubts he had concerning the tenuous relationship between theory and practice, it would have been inconsistent to further theoretically explicate or justify the general normative thrust of his critique.

This also sheds light on an important aspect of Foucault's understanding of the philosophical ethos he associates with "modernity" and the enlightenment. As others have pointed out, it demands "character" and entails a certain virtuous practice that involves not only the audacity to dare to know, as Kant emphasized in his essay, but also, as Foucault's critical practice indicates, of taking ethical risks in the face of great uncertainty (e.g., Butler 2004). The philosophical ethos, Foucault remarked, consists in a critique of "what we are saying, thinking and doing through a historical ontology of ourselves." It is a means not only of directly critiquing societal practice but also, by way of shattering preconceptions about who "we" are, empirically testing the limits of "our" historical being, making possible new, still undefined ways of being. "Criticism," Foucault continued:

> indeed consists of analyzing and reflecting upon limits. But if the Kantian question was that of knowing what limits knowledge has to renounce transgressing, it seems to me that the critical question today has to be turned back into a positive one: in what is given to us as universal, necessary, obligatory, what place is occupied by whatever is singular, contingent and the product of constraints? The point in brief is to transform the critique conducted in the form of necessary limitation into a practical critique that takes the form of a possible transgression. (Foucault 1984, 45)

Historical investigation can become a medium for a kind of critique that does not seek to reject but to transform because it can expose the contingencies and past power struggles involved in something becoming part of "our," his readers' and listeners,' being. Sorting through, as it were, the contingencies involved in its problematization, we might stumble upon, Foucault suggested, what once appeared as a necessity, as inevitable, or universally true, but now no longer does, pointing to the possibility of ceasing to be in this particular way.

Conclusion

A history of the present thus involves a genealogical critique of an idea or practice in the present that is generally regarded as necessary, valuable, and sometimes also – particularly if it has emerged only recently – as indicative of a collective achievement or significant advancement. Genealogy calls this idea or practice into question by destabilizing its identity, complicating its meaning, and tracing the struggles that have led to its emergence and continue to shape its enactments in the present. A genealogical critique does not, and indeed often is not in the position to, reject whatever it critiques, since it does not place its trust in the practice of justifying or grounding normative claims, e.g., by developing comprehensive theories. As Martin Saar has pointed out, Nietzsche's as well as Foucault's genealogies instead often employed unique rhetorical means to achieve a desired destabilizing, disorienting effect in their readers (Saar 2007).

As the preceding sections have also shown, however, genealogical critique is not merely based on a specific historical account that is presented for dramatic effect. First, it involves a consideration of the significant "difference" experienced in the present with regard to the past. This is not merely an empirical matter. In practice, historians of the present do draw on preliminary analyses when identifying the targets of their genealogical critique, especially if their histories function as a critical intervention within ongoing scholarly or scientific debates. This should not be at odds, moreover, with Burchell's observation that histories of the present may often be motivated by an initial estrangement, "the experience of not being a citizen of the community or republic of thought and action in which one is, nevertheless, unavoidably implicated or involved" (Burchell 1993, 276). Second, genealogical critique does not dissolve a presumed identity merely by tracing its fractured lineage. Genealogies aren't inherently critical, as we have seen. Conceptual tools, such as Foucault's *dispositif* or the identification of conflicting models, offer ways of bringing into contact historical observations that may otherwise appear unrelated. The effectiveness of a critical genealogy also relies on the plausibility and effective uses of theoretical frames to expose inherent tensions and contradictions that undermine presumed identities. Third, genealogies start out with a theoretically informed notion of subjectivity as historically constituted, distinguishing them from other ways of writing the history of the self (see, e.g., Taylor 1992; Hacking 1995). Fourth, because of their radical historicist commitments, histories of the present can serve as mediums for critical reflection for historians, offering them the possibility of historicizing their own analytic categories and methodological presuppositions. In short, histories of the present are characterized by certain epistemological commitments as well as their thorough theoretical engagements, even though they never aim to establish and defend theories themselves. What Geuss writes of Foucault's genealogy may also apply to the history of the present more broadly: "Genealogy does not lay down the law, nor is it a policing discipline. Rather it is a summons to develop an empirically informed kind of theoretical imagination under the conditions of perceived danger" (Geuss 2002, 213).

Constituting a theoretical as much as an empirical practice, this form of critique entails unique difficulties and risks. The reluctance of historians, especially those committed to a notion of objectivity that implies impartiality, to engage with (post-structuralist) "theory" has been an obstacle for the writing of critical history, Joan Scott has observed (2007). On the other hand, Foucault has vexed philosophers not only because of his refusal to commit to "theory building" but also because of his empirical approach to philosophical problems, his "happy positivism," as he called it on occasion. As a practice, writing the history of the present continues to sit uncomfortably between the disciplines and is thus vulnerable to criticisms from various sides. The common criticism forwarded by philosophers regarding Foucault's refusal to explicate the normative foundation of his critique has been discussed. Historians, also, raised ample criticisms regarding the accuracy of his historical accounts, periodization, or choice of topics (e.g., see Scott 2007; Garland 2014). Foucault's defenders may at times have dismissed such empirical criticism too easily, as besides the more general point he was trying to make. A history of the present must remain reliable in its accounts of historical events, since its effectiveness also depends upon its trustworthiness, even though, given the varieties discussed here, the same standards may not apply for all its forms. Despite their interdisciplinary character and ambitions, then, critical histories of this sort, if they indeed abandon the speculative mode associated with some philosophical uses of genealogy, still adhere to the disciplinary standards of history.

Aside from the depth of their historical engagement, a further distinguishing feature of these variants of genealogical critique is the respective audiences they address. Foucault, a public intellectual, could still hope to speak to an audience that extended beyond the confined worlds of academic specialists. Contemporary histories of the present are often, but not in every case, more targeted interventions within ongoing debates in various fields, although they consistently involve members of several disciplines, even if they do not extend beyond the academy. A practical condition for the possibility of this intellectual practice, then, are spaces of intellectual exchange that transcend disciplinary boundaries. With the diminished role of public intellectuals, the lack of institutional support for interdisciplinary engagements in the humanities, and the simultaneous multiplication of subjectivities in contemporary societies, the preconditions and reach of this practice as well as its transformative potential may be reduced. In light of recent shifts in public discourse, Rose has lately raised doubts about the critical potential of "the now familiar tropes of genealogy and 'histories of the present'," for instance (Rose 2007, 4). When commentators on the present cultural moment are quick to invoke the idea of a world constantly in flux, a critical stance no longer demands destabilizing presumed identities, he suggested, but a destabilization of the future by exposing the many historical paths intersecting in the present, pointing to multiple possible trajectories and outcomes. Such a "cartography of the present," as he called it, would emphasize continuity rather than discontinuity, yet it would also remain true to the spirit of a "history of the present" in that it aims to demonstrate the contestability and hence indeterminacy of societal change.

References

Adorno T (1966/1973) Negative dialectics. Continuum International Publishing Group, New York
Adorno T (1998/1963) Critical models: interventions and catchwords. Columbia University Press, New York
Adorno T, Horkheimer M (1947/1986) Dialectic of enlightenment (trans: Cumming J). Verso, London
Barry A, Osborne T, Rose N (1995) Writing the history of the present. In: Barry A, Osborne T, Rose N (eds) Foucault and political reason: liberalism, neoliberalism and the rationalities of government. UCL Press, London, pp 1–18
Bevir M (2008) What is genealogy? J Philos Hist 3:263–275
Bevir M (2010) Rethinking governmentality: Towards genealogies of governance. Eur J Soc Theory 13(4):423–441. https://doi.org/10.1177/1368431010382758
Biebricher T (2008) Genealogy and governmentality. J Philos Hist 2:363–396
Biebricher T (2015) Governmentality. In: Routledge handbook of interpretive political science. Routledge, pp 153–166
Birnbaum N (1973) Toward a critical sociology. Oxford University Press
Bloor D (2011) The enigma of the aerofoil. University of Chicago Press, Chicago
Burchell G (1993) Liberal government and techniques of the self. Econ Soc 22:267–282
Butler J (2004) What is critique? An essay on Foucault's virtue. In: Salih S (ed) The Judith Butler reader. Blackwell, Malden/Oxford

Foucault M (1966/1970) The Order of things: an archaeology of the human sciences (trans: Sheridan A). Tavistock, London
Foucault M (1969/1972) The archaeology of knowledge (trans: Sheridan Smith AM). Tavistock, London
Foucault M (1971/1977) Nietzsche, genealogy, history. In: Bouchard D (ed) Language, counter-memory, practice. Selected essays and interviews. Cornell University Press, Ithaca, pp 139–164
Foucault M (1975/1977) Discipline and punish: the birth of the prison (trans: Sheridan A). Allen Lane, London
Foucault M (1976/1978) The history of sexuality, volume 1. Random House, New York
Foucault M (1984) What is enlightenment? The Foucault reader. Pantheon Books, New York, pp 32–50
Foucault M (1988) The political technology of individuals. In: Martin L, Gutman H, Hutton P (eds) Technologies of the self. A seminar with Michel Foucault. University of Massachusetts Press, Amherst, pp 145–162
Foucault M (1994) Dits et écrits: 1954–1988, 4 volumes. Editions Gallimard, Paris
Garland D (2014) What is a "history of the present"? On Foucault's genealogies and their critical preconditions. Punishment Soc 16(4):365–384. https://doi.org/10.1177/1462474514541711
Geuss R (1994) Nietzsche and genealogy. Eur J Philos 2:274–292
Geuss R (2002) Genealogy as critique. Eur J Philos 10:209–215
Geuss R (2008) Philosophy and real politics. Princeton University Press, Princeton/Oxford
Guenter K (2015) Localization and its discontents: a genealogy of psychoanalysis and the neuro disciplines. University of Chicago Press, Chicago
Habermas J (1984) The theory of communicative action, 2 vol. Beacon Press, Boston
Habermas J (1987) Some questions concerning the theory of power: Foucault again. In: The philosophical discourse of modernity: Twelve Lectures (trans: Lawrence F). MIT Press, Cambridge, MA
Hacking I (1995) Rewriting the soul: multiple personality and the sciences of memory. Princeton University Press, Princeton
Hacking I (1999) The social construction of what? Harvard University Press, Cambridge, MA
Hampe M (2009) Wissenschaft und Kritik. In: Jaeggi R, Wesche T (eds) Was ist Kritik? Suhrkamp, Frankfurt a. M., pp 353–371
Horkheimer M (1937/2005) Traditionelle und kritische Theorie. Fischer, Frankfurt a. M.
Hoy D (2008) Genealogy, phenomenology, critical theory. J Philos Hist 2(3):276–294
Jaeggi R, Wesche T (2009) Einführung: Was ist Kritik? In: Jaeggi R, Wesche T (eds) Was ist Kritik? Suhrkamp, Frankfurt a. M., pp 7–22
Johnson B (1994) The wake of deconstruction. Blackwell, Cambridge
Kant I (1784/2000) Beantwortung der Frage: Was ist Aufklärung? In: Brandt H (ed) Was ist Aufklärung? Ausgewählte kleine Schriften. Felix Meiner Verlag, Hamburg, pp 20–28
Kaye HJ (1984) The British Marxist historians: an introductory analysis. Polity Press, Cambridge
Kolodny N (1996) The ethics of cryptonormativism: a defense of Foucault's evasions. Philosophy & Social Criticism 22:63–84
Lemke T (2011) Critique and experience in Foucault. Theory Cult Soc 28(4):26–48
Leys R (2000) Trauma. University of Chicago Press, Chicago
Leys R (2009) From guilt to shame: Auschwitz and after. Princeton University Press
Leys R (2017) The ascent of affect. University of Chicago Press, Chicago
Leys R, Goldman M (2010) Navigating the genealogies of trauma, guilt, and affect: an interview with Ruth Leys. Univ Tor Q 79:656–679
Martschukat J (2019/2021) The age of fitness: how the body came to symbolize success and achievement. Polity Press, Cambridge
Nietzsche F (1874/1983) On the uses and disadvantages of history for life. In: Stern JP (ed) Untimely meditations (trans: Hollingdale RJ). Cambridge University Press, Cambridge, pp 57–123
Nietzsche F (1887/1989) On the genealogy of morals. Vintage, New York

Oreskes N (2019) Why trust science? Princeton University Press, Princeton/Oxford
Rose N (1985) The psychological complex: psychology, politics and society 1869–1939. Routledge and Kegan-Paul, London
Rose N (1996/1998) Inventing our selves: psychology, power, and personhood. Cambridge University Press, Cambridge
Rose N (2007) The politics of life itself. Biomedicine, power, and subjectivity in the twenty-first century. Princeton University Press, Princeton/Oxford
Saar M (2007) Genealogie als Kritik: Geschichte und Theorie des Subjekts nach Nietzsche und Foucault. Campus Verlag, Frankfurt a. M.
Saar M (2009) Genealogische Kritik. In: Jaeggi R, Wesche T (eds) Was ist Kritik? Suhrkamp, Frankfurt a. M., pp 247–265
Sarasin P (2005) Michel Foucault zur Einführung. Junius, Hamburg
Sarasin P (2019) Foucaults Wende. In: Marchart O, Martinsen R (eds) Foucault und das Politische. Politologische Aufklärung – konstruktivistische Perspektiven. Springer VS, Wiesbaden. https://doi.org/10.1007/978-3-658-22789-0_2
Schraube E, Osterkamp (eds) (2013) Critical theory and practice in psychology and the human sciences. Psychology from the standpoint of the subject: selected writings of Klaus Holzkamp (trans: Borehain A). Palgrave Macmillan, Basingstoke
Scott JW (2007) History-writing as critique. In: Manifestos for history. Routledge, London, pp 31–50
Scott JW, Connolly B (2011) Introducing the history of the present. Hist Present 1(1):1–4
Smith R (2005) Does reflexivity separate the human sciences from the natural sciences? In: Smith R (ed) Special issue on reflexivity, history of the human sciences 18(4):1–25
Taylor C (1992) Sources of the self: the making of the modern identity. Harvard University Press, Cambridge, MA
Williams B (2010) Truth and truthfulness. Princeton University Press, Princeton

Human Sciences and Theories of Religion

Robert A. Segal

Contents

Introduction	182
Origin and Function	183
Truth	183
Theories from Religious Studies	184
Social Scientific Theories	186
Contemporary Social Scientific Theories	186
The Religionist Argument	187
Anthropologists of Religion	190
Tylor and Frazer	190
Lévy-Bruhl	194
Douglas, Geertz, and Turner	195
Lévi-Strauss	200
Conclusion	203
References	204

Abstract

There are two ways of studying religion: religion by religion and comparatively. To study religions one by one is invariably to stress the individuality of each religion. To study religions comparatively is invariably to emphasize the similarities among religions. It is theologians who seek the uniqueness of their own religion. It is social scientists who seek the commonality of religions. To be sure, there is the hermeneutical, or interpretive, movement, which, starting with Wilhelm Dilthey and culminating so far in Clifford Geertz, focuses on the particularities of the cases of any category studied, such as marriage or revolution or religion. But this chapter concentrates on the mainstream quest for similarities, or generalizations.

R. A. Segal (✉)
School of Divinity, History and Philosophy, King's College, University of Aberdeen, Aberdeen, UK
e-mail: r.segal@abdn.ac.uk

This chapter gives an overview of the main modern theories of religion. Those theories come mostly from the social sciences, a term used interchangeably with the "human sciences." The focus in this chapter is on theories from anthropology, which has provided the largest number of theories. What makes theories theories is the issue. What questions about religion do theories raise, and what questions do they ignore? Twentieth-century theories are distinguished from nineteenth-century ones. In the nineteenth century, religion was deemed incompatible with natural science – not, to be sure, by natural theologians but by social scientists. In the twentieth century, religion came to be deemed compatible with natural science.

Keywords

Douglas, Mary · Frazer, J. G. · Freud, Sigmund · Function · Geertz, Clifford · Jung, C. G. · Levi-Strauss, Claude · Müller, Friedrich Max · Origin · Religion · Science · Theory · Truth · Turner, Victor · Tylor, E. B.

Introduction

This chapter is an overview of explanations of religion in the past two centuries. The explanations presented are theories. They are generalizations about religion per se, not about specific religions or specific periods of religion. Theories are taken as differing answers to the same basic questions: What is the origin, what is the function, and what is the subject matter of religion? For some theories, religion is universal. For others, it is merely "primitive" or, more precisely, premodern. For some theories, nothing can replace religion. For others, religion has been replaced by science or ideology or philosophy. (Secondary sources on theories of religion include Barfield 1997; Barnard 2000; Barth et al. 2005; Bowie 2006; Erickson and Murphy 2008; Freedman 1979; Ingold 1994; Moore and Sanders 2006.)

Theories of religion go all the way back to the Presocratics. Modern theories come almost entirely from the modern disciplines of the social sciences: anthropology, sociology, psychology, and economics. The focus here is on anthropology, from which the largest number of theories has come. Pre-social scientific theories came largely from philosophy and were speculative rather than empirical. What John Beattie writes of modern anthropological theories of culture as a whole holds for theories of religion, and for theories from the other social sciences as well:

> Thus it was the reports of eighteenth- and nineteenth-century missionaries and travellers in Africa, North America, the Pacific and elsewhere that provided the raw material upon which the first anthropological works, written in the second half of the last century, were based. Before then, of course, there had been plenty of conjecturing about human institutions and their origins; to say nothing of earlier times, in the eighteenth century Hume, Adam Smith and Ferguson in Britain, and Montesquieu, Condorcet and others on the Continent, had written about primitive institutions. But although their speculations were often brilliant, these thinkers were not empirical scientists; their conclusions were not based on any kind of

evidence which could be tested; rather, they were deductively argued from principles which were for the most part implicit in their own cultures. They were really philosophers and historians of Europe, not anthropologists. (Beattie 1964, pp. 5–6)

Origin and Function

A theory of religion is an answer to at least two questions: What is the origin and what is the function of religion? The term "origin" is confusing because it can refer to either the historical or the recurrent beginning of religion. It can refer either to when and where religion first arose or to why religion arises whenever and wherever it does. According to the conventional view, nineteenth-century theories focused on the origin of religion, where twentieth-century theories have focused on the function of religion. But "origin" here means historical origin. Nineteenth-century theories sought the recurrent origin of religion at least as much as any historical origin, yet no more so than twentieth-century theories have done. Moreover, nineteenth-century theories were concerned as much with the function of religion as with the origin, and no less so than twentieth-century theories have been. Furthermore, the historical origin proposed by nineteenth-century theories was not that of a single time and place, such as the Garden of Eden, but that of the earliest *stage* of religion – any time and anywhere. Therefore the discussion of even the "historical" origin of religion was as much about the recurrent origin as about any first one.

The questions of recurrent origin and of function are two sides of the same proverbial coin, and few theories theorize about only the recurrent origin or only the function. Ordinarily, the answer to both questions is a need, which religion originates to fulfill and functions to fulfill. Theories differ over what that need is. Any theories that do concentrate on only the recurrent origin or on only the function typically either attribute the origin of religion to an accident or make the function a byproduct. And "accident" and "byproduct" really refer to the means, not the ends. Unless religion, however accidental its origin or coincidental its function, serves a need, it surely will not last and surely will not continually re-arise. Still, origin and function are distinct issues, and to argue on the basis of the sheer fulfillment of a need that religion arises *in order to* fulfill the need is to commit the fallacy of affirming the consequent.

The issues of origin and function can each be divided into two parts: not only *why* but also *how* religion arises or functions. In explaining the ends of religion, theories do not thereby automatically explain the means. Some theories explain how religion arises, others how religion functions, others both, and still others neither.

Truth

Most twentieth-century theorists forswear the issue of the truth of religion as beyond the ken of the social sciences (see Segal 1989, Chap. 7). One exception is the sociologist Peter Berger, who ever since *A Rumor of Angels* (1969) was prepared to use his theory to validate the truth of religion (see Segal 1992, pp. 6–7, 16, 117–118). Most nineteenth-century theorists were not at all reluctant to take a stand on the issue of truth. But they based their assessment on philosophical grounds, not on social scientific ones. Instead of enlisting the origin and function of religion to assess the truth of religion, they assessed the truth on an independent basis and, if anything, let their conclusion about it guide their theorizing about origin and function (see Segal 1992, pp. 15–17). They thereby circumvented the possibility of committing either the genetic fallacy or what I call the functionalist fallacy: arguing that knowledge of either the origin or the function of religion refutes – necessarily refutes – the truth of it. One grand exception is Sigmund Freud in his *The Future of an Illusion* (1964). He argues against the likely truth of religion on the grounds that it originates in wish fulfillment (see Segal 1989, Chap. 4).

Writing at the turn of the twentieth century (1902), William James is open-minded. Focusing on the claims by mystics that their experiences reveal an immaterial reality beyond that of the senses, James considers the arguments pro and con. He concludes not that mystical experience is true or false but that it may be true:

> Mystical states indeed wield no authority due simply to their being mystical states. But the higher ones among them point in directions to which the religious sentiments even of non-mystical men incline.... They offer us *hypotheses,* hypotheses which we may voluntary ignore, but which as thinkers we cannot possibly upset. The supernaturalism and optimism to which they would persuade us may, interpreted in one way or another, be after the truest of insights into the meaning of this life. (James 1902, pp. 419–420)

A common objection by scientists to sociologists of science is that the issue of truth is separate from the issues of origin and function. The best known of the sociologists are the members of the Edinburgh Strong Programme. Whether the claims of the program venture beyond origin and function to truth is a topic for another paper.

Theories from Religious Studies

The key divide in theories of religion is between those theories that hail from the social sciences and those that hail from religious studies itself. Social scientific theories deem the origin and function of religion nonreligious. The need that religion arises to fulfill can be for almost anything. It can be either physical – for example, for food, health, or prosperity – or intangible – for example, for explanation, as for E. B. Tylor (1871), or for meaningfulness, as for Max Weber (1963). The need can be on the part of individuals or on the part of society. In fulfilling the need, religion provides the means to a secular end.

By contrast, theories from religious studies deem the origin and function of religion distinctively religious: The need that religion arises to fulfill is for god. There really is but one theory of religion from religious studies. Adherents to it include Friedrich Max Müller, C. P. Tiele, Gerardus van der Leeuw, Raffaele

Pettazzoni, Joachim Wach, and Mircea Eliade. (This list excludes Rudolf Otto because he does not *account for* religion but instead simply *defines* religion as an encounter with god.) For all of these "religionists," religion arises to provide contact with god. Like many social scientists, many religionists confine themselves to the issues of origin and function and shy away from the issue of truth. Just as social scientists entrust the issue of truth to philosophers, so religionists leave it to theologians.

For religionists, human beings need contact with god as an end in itself: They need contact with god because, put circularly, they need contact with god. An encounter with god may yield peace of mind and other benefits, but the need is still for the encounter itself. The need is considered as fundamental as the need for food or water. Without that contact, humans may not die, but they will languish. Because the need is for god, nothing secular can substitute for religion. There may be secular, or seemingly secular, *expressions* of religion, but there are no secular *substitutes* for religion. Religionists consider the need for god not only distinctive but also universal. To demonstrate its universality, they point to the presence of religion even among professedly atheistic moderns.

Strictly speaking, there are two versions of the single religionist theory. One is the form just described: Religion originates within human beings, who seek contact with god. The exemplar of this version of the theory is Eliade, who stresses the yearning for god or, so he prefers, the sacred: "But since religious man cannot live except in an atmosphere impregnated with the sacred, we must expect to find a large number of techniques for consecrating space" (Eliade 1968, p. 28). Sacred places, or spaces, are one venue for encountering god. Religious sites, such as churches and mosques, are built on spots where god is believed to have appeared – the assumption being that wherever god has once appeared, that god, even if formally omnipresent, will more likely reappear. Sacred times or time is the other venue for encountering god. Myths, which describe the creation by god of physical and social phenomena, carry one back to the time of creation, when, it is believed, god was closer at hand than god has been ever since: "Now, what took place 'in the beginning' was this: the divine or semidivine beings were active on earth.... Man desires to recover the active presence of the gods .... [T]he mythical time whose reactualization is periodically attempted is a time sanctified by the divine presence" (Eliade 1968, p. 92).

This version of the religionist theory seemingly bypasses the issue of the existence of god. The theory is committed to the existence of only the *need for* god. The catch is that if religionists claim that religion actually fulfills the need – and why else would they advocate religion? – then god must exist. Religionists thus prove to be theologians. Still, the emphasis is on the need itself.

The other version of the religionist theory, epitomized by Müller (1867), roots religion not in the need for god but in the experience of god. However indispensable the experience of god may be for human fulfillment, religion originates not in the quest for god but in an encounter with god. Müller himself singles out the sun and other celestial phenomena as the occasion where god or, for Müller, the Infinite is encountered: "Thus sunrise was the revelation of nature, awakening in the human mind that feeling of dependence, of helplessness, of hope, of joy and faith in higher powers, which is the source of all wisdom, the spring of all religion" (Müller 1867, p. 96). James might also be noted as rooting religion in experience, but it was Müller who founded the field of religious studies.

The two versions of the religionist theory are compatible. The quest for an encounter with god may be fulfilled by an uninitiated encounter, and an uninitiated encounter can lead to a quest for further encounters. Still, the approaches differ. One starts with a need. The other starts with an experience. Deriving religion from a need for god makes the religionist theory more easily comparable with social scientific theories, nearly all of which do the same even if the need is a means to a nonreligious end.

Social Scientific Theories

Religionists commonly assert that social scientists, in making religion a means to a nonreligious end, are less interested in religion than they. This assertion is false. Social scientists are interested in religion for exactly its capacity to produce anthropological, sociological, psychological, and economic effects. Many social scientists consider religion a most important means of fulfilling whatever they consider its nonreligious function. Some even make it the key means of doing so.

Moreover, for religion to function nonreligiously, it must be operating *as* religion. The nonreligious effect comes from a religious cause. The power that religion has, let us say, to goad adherents into accepting social inequality stems from the belief that god sanctions the inequality, that god will one day remedy the inequality, or that the inequality is a merely worldly matter. Without the belief in god's part in the inequality, religion would have no social effect. Undeniably, the social sciences approve or disapprove of religion for its anthropological, sociological, psychological, or economic consequences. Undeniably, religion is admired only when it inculcates culture, unites society, develops the mind, or spurs the economy, not when it makes contact with god. But the nonreligious benefit of religion presupposes the efficacy of religion as religion.

Put another way, religion for social scientists functions as an independent variable, or as the cause of something else. In *origin*, religion is indisputably a dependent variable, or the effect of something else, as it, like anything else, must be, unless it creates itself. But in *function*, religion is an independent variable. Even if it is the product of nonreligious causes, it is in turn the cause of nonreligious effects. If religion could not be an independent variable in its effect because it was a dependent variable in its origin, there would be few independent variables around.

Contemporary Social Scientific Theories

Religionists often assert that contemporary social scientists, in contrast to earlier ones, have at last come round to seeing religion the way the religionists do. Contemporary social scientists are consequently embraced by religionists as belated converts. The figures embraced most effusively are Mary Douglas (1966, 1973), Victor Turner (1967, 1968), Clifford Geertz (1973, 1983), Robert Bellah (1970), Peter Berger (1967, 1969), and Erik Erikson (1958, 1969). These social scientists are pitted against classical ones like Tylor (1871), Frazer (1890, 1922), Emile Durkheim (1995), Bronislaw Malinowski (1925), Freud (1950), C. G. Jung (1938), Karl Marx and Friedrich Engels (1957), and Weber (1963).

What is the difference between classical and contemporary social scientists? The difference cannot be over the importance of religion. Classical social scientists considered religion at least as important a phenomenon as any of their contemporary counterparts do. The power of religion is what impelled them to theorize about it. Similarly, the difference cannot be over the utility of religion. While for Frazer, Freud, and Marx religion is incontestably harmful, for Tylor, Jung, and Durkheim it is most helpful. For these three, religion is one of the best and, for Durkheim, the best means of serving its beneficial functions. Contemporary social scientists grant religion no greater due.

The difference between contemporary and classical social scientists must be over the nature of the function that religion serves. In contrast to classical theorists, for whom the functions of religion are nonreligious, contemporary theorists purportedly take the function of religion to be religious. But do they? Where religionists attribute religion to a yearning for god, contemporary social scientists attribute it to a yearning for, most often, a meaningful life. Contact with god may be one of the best means of providing meaningfulness, but even if it were the sole means, it would still be a means to a nonreligious end. For Douglas, humans need cognitive meaningfulness: They need to organize their experiences. For Turner, Geertz, Bellah, Berger, and Erikson, humans need existential meaningfulness: They need to explain, endure, or justify their experiences. Existential meaningfulness as the function of religion is not even new and goes back to at least Weber, who, to be sure, limits the need for meaningfulness to the "higher" religions (see Weber 1963, Chaps. 8–13). But even if this function were new, the need would remain secular. In short, the divide between social scientific theories of religion and the religionist one remains.

The Religionist Argument

What is the case for the religionist theory? The case tends to be presented negatively. It appeals to the inadequacy of social scientific theories. All social scientific theories, whether historical or recurrent, are supposedly inadequate because in deeming the origin and function of religion nonreligious, they necessarily miss the religious nature of religion. Only the religionist theory captures the religious nature of religion.

In actuality, social scientific theories do not miss the religious nature of religion. On the contrary, it is what they *mean* by religion. The religious nature of religion is the starting point of their theorizing. It is the datum to be theorized about. Far from somehow failing to perceive that adherents pray to god, sacrifice to god, and help others in the name of god, social scientists take for granted that adherents do so. The question for social scientists is why they do so. The religious nature of religion may be the starting point of theorizing, but it is not the end point. If social scientists somehow missed, let alone denied, that Christians go to church, sing hymns, take sacraments, read the Bible, and devote their lives to God, and do all of these things because they believe in God, they would be left with nothing to explain. Religiosity, far from being overlooked, is the preoccupation of social scientists of religion.

Against social scientists and others, religionist Eliade, in a famous passage, declares that "a religious phenomenon will only be recognized as such if it is grasped as its own level, that is to say, if it is studied *as* something religious. To try to grasp the essence of such a phenomenon by means of physiology, psychology, sociology, economics, linguistics, art or any other is false" (Eliade 1963, p. xiii). But Eliade conflates description with explanation, not to mention description with metaphysics (essence). No social scientist fails to recognize religion as religion. That is why there exists the anthropology of *religion*, the sociology of *religion*, the psychology of *religion*, the politics of *religion*, and the economics of *religion*. There would be no social scientific theories of *religion* if the distinctiveness of religion went unrecognized. But the *recognition* of religion as religion does not mean the *explanation* of religion as religion.

It is as believers in God that Christians go to church, but it is also, for example, as members of a group that they do so. While acknowledging the difference between a religious group and a team, a family, or a gang, sociologists explain religion in the same way that they explain a team, a family, and a gang: as a group.

At the same time no sociological account of religion can be exhaustive. There is a point at which any sociological account must cease – the point at which a religious group differs from any other kind of group. But to acknowledge a stopping point for sociology is not to concede a starting point. Sociology can account for religion to whatever extent religion does constitute a group. How fully religion constitutes a group, it is up to sociologists to establish. The more group-like they show religion to be, the more successful their account. Sociologists are to be commended, not condemned, for attempting to account as fully as possible for religion sociologically. Their inevitable inability to account for it entirely sociologically marks the limit, not the failure, of the sociology of religion.

Religionists would reply that the attempt to "sociologize" is inherently futile, for the origin and function of religion can only be religious. Otherwise religion ceases to be religion and becomes society. But this conventional rejoinder, offered like a litany, misses the point. Nobody denies that religion consists of beliefs and practices directed toward god rather than toward the group. But Durkheim, for example, is not thereby barred from matching a believer's experience of possession by god with an individual's experience of participation in a group. Durkheim is not barred from asserting that the euphoria and power which individuals feel when they gather and brush up against one another account for religious experience. Participants, thinking their state of mind superhuman, attribute it to possession by god, but Durkheim attributes it to "possession" by the group.

Still, Durkheim is not maintaining that religion originates exclusively through group experience. After all, the group is not itself god, just god-like. The concept of god and attendant practices must still be created. Furthermore, the group comes together in the first place for religious reasons – one of the circularities in Durkheim's argument. Thus Australian aboriginal clans, Durkheim's test case, amass to "celebrate a religious ceremony" (Durkheim 1965, p. 246). Durkheim offers his account of religion as a necessary but not quite sufficient one. What must yet be accounted for is precisely the step from group to god. But to concede that there is more to an account of religion than the group is not to grant that religiosity is all there is to an account. For Durkheim, religion is to be accounted for sociologically *and* "religionistically" – with the sociological element predominant. For all other social scientists, the same is true: A sociological, anthropological, psychological, political, or economic account of religion must be supplemented by a religionist one.

The final religionist rejoinder is the appeal to symmetry. If the effect is religion, the cause must be religionist. There must be a match between cause and effect. A sociological cause can produce only a sociological effect. Explained sociologically, the product is the group, not religion.

This rejoinder, like other ones, misses the point. Of course, there must be symmetry between cause and effect. Causes must be enough akin to their effects to be capable of producing them. But a sociological account of religion does not purport to account for the *nonsociological* aspects of religion, only for the sociological ones. To reply that the sociological aspects are aspects of the group and not of religion is to commit a double fallacy: excluding the middle and begging the question. A sociological account of religion is not an account of something *other than* religion. It is an account of aspects *of* religion. To limit religion to its religionist aspects is to beg the central question at hand: What is the nature of religion?

To be sure, the claim that sociology can explain anything of *religious* beliefs and practices might seem to be asserting that sociology can explain something *nonsociological*. But this concern is misplaced. Sociology takes *seemingly* nonsociological aspects of religion and transforms them into sociological ones, which it only then accounts for. Durkheim matches attributes of god – god's power, god's overwhelming presence, and god's status as the source of values and institutions – with attributes of the group whose god it is. The symmetry between cause and effect is preserved by sociologizing the effect. A gap remains between the sociological cause and the religious effect: The group is still just a group, not a god. But symmetry is not intended to mean identity.

To take an example from another field, Freud contends that a believer's relationship to the believer's father matches the believer's relationship to god. He contends that believers' feelings toward their fathers precede their feelings toward god, parallel those feelings, and therefore cause the feelings. But he proposes only a necessary, if also largely sufficient, cause of religion. No more than Durkheim does, he proposes, to use a redundancy, an altogether sufficient one. What must still be supplied is the step from father to god. The father for Freud, like the group for Durkheim, is godlike but not god. God may be humanlike, but no human being is omnipotent, omniscient, or immortal. God is "father" of the whole world, not just of a family. The adult conception of god may derive from a child's "idolization" of the child's own father, but the conception transforms a godlike figure into a god. Even when Freud brashly declares that "at bottom God is nothing other than an exalted father" (Freud 1950, p. 147), he is still distinguishing a father from a god. The closer the link that Freud draws between father and god, the more convincing his account. But he, like Durkheim and all other social scientists, takes for granted a limit to the link and does so even while ever trying to tighten that link. Again, symmetry does not mean identity.

Anthropologists of Religion

The easiest way to characterize anthropologists of religion is to distinguish them from sociologists and from psychologists. For anthropologists, the key subject is culture. For sociologists, it is society, or the group. For psychologists, it is the mind. The line between anthropologists and sociologists is not always clear, and some early writers now seem to have been as much sociologists as anthropologists. Religion is as much a group enterprise for many anthropologists as for sociologists, but the need being served is that of the individual and not, or not just, that of the group. For sociologists, especially Weber, the need being served is sometimes that of the individual, but the group is indispensable to the fulfillment of that need. For psychologists, the need that religion serves is that of the individual, and the group plays a secondary, if still important, role.

Tylor and Frazer

E. B. Tylor, one of the founding figures of anthropology, epitomizes the purportedly nineteenth-century focus on origin. He roots religion in observations by "primitives" of, especially, the immobility of the dead and the appearance in dreams and visions of persons residing far away. The why of origin is an innate need to explain these observations, which trigger the need rather than implant it. The *how* of origin is the processes of observation, analogy, and generalization. Independently of one another, primitive peoples the world over create religion by these means and for this end. Later stages of humanity do not reinvent religion but instead inherit it from their primitive forebears. They perpetuate religion because it continues to satisfy for them, too, the need to explain observations. Similarly, religion changes not because the need changes but because believers revise their conceptions of god. Religion eventually gives way to science not because the need changes but because science provides a better means of satisfying it. The why of function is the same as the why of origin: a need to explain observations. The how of function is the one issue that Tylor ignores.

One confusion: "primitive" refers both to our long dead ancestors and to present-day societies living in what anthropologists deem a comparably early state. The fieldwork that anthropologists originally did was among living "primitives," though today anthropologists study peoples everywhere and in all states of development.

In the nineteenth century religion was assumed to arise to serve the same need as science, which was taken to be a largely, even an exclusively, modern enterprise. In fact, "modern" and "scientific" were synonyms. The nineteenth-century view is epitomized by the theories of Tylor, the first edition of whose *Primitive Culture* appeared in 1871, and of J. G. Frazer, the first edition of whose *The Golden Bough* appeared in 1890. Both assume that religion arises and functions either to explain (Tylor) or to control (Frazer) the physical world. Religion is the "primitive" counterpart to either scientific theory (Tylor) or applied science (Frazer). Where Tylor focuses on religious beliefs, Frazer focuses on religious practices, or rituals. For both, religion ascribes all physical events in the world, such as the birth and death of vegetation, either to a decision by a god (Tylor) or to the effect of the physical state of the god (Frazer). Either the vegetation god, for whatever reason, decides to cause the crops to grow, or the crops grow as long as the god is healthy. Science, by contrast, attributes the growth of the crops to impersonal, biological, and chemical processes. There are no gods.

Science, it was assumed by Tylor and Frazer alike, renders religion not merely redundant but outright impossible. For the religious as well as the scientific explanation is direct, so that one cannot stack the religious explanation atop the scientific one, with science providing the direct explanation and religion the indirect one. According to religion, the vegetation god acts not *through* vegetative processes but *in place of* them. One cannot, then, reconcile the explanations and must instead choose between them. But the choice has already been made, for moderns by definition accept science. Our forebears had only religion, so that for them, too, the choice had already been made. Religion in modernity is fated to die out, according to Tylor and Frazer, once moderns recognize its incompatibility with science.

Tylor is even more concerned than Frazer with the relationship between religion and science. He subsumes both religion and science under philosophy. He divides philosophy into "primitive" and "modern." Primitive philosophy is identical with primitive religion. There is no primitive science. Modern philosophy, by contrast, has two divisions: religion and science. Of the two, science is by far the more important and is the modern counterpart to primitive religion. Modern religion is composed of two elements, metaphysics and ethics, neither of which is present in primitive religion. Tylor uses the term "animism" for religion per se, modern and primitive alike, because he derives the belief in gods from the belief in souls (*anima* in Latin means soul). Gods are the souls in all physical entities *except* humans, who themselves are not gods. Though there is the in-between category of heroes, humans are assumed to be less than gods. Souls are the source of action in all living things.

Primitive religion is the primitive counterpart to science because both are explanations of the physical world. Tylor thus characterizes primitive religion as "savage biology" and maintains that "mechanical astronomy gradually superseded the animistic astronomy of the lower races" and that today "biological pathology gradually supersedes animistic pathology." The religious explanation is personalistic: The decisions of gods explain events. The scientific explanation is impersonal: Mechanical laws explain events. The sciences as a whole have replaced religion as the explanation of the physical world, so that "animistic astronomy" and "animistic pathology" refer only to primitive, not modern, animism. Modern religion has surrendered the physical world to science and has retreated to the immaterial world, especially to the realm of life after death, that is, of the life of the soul after the death of the body. Where in primitive religion souls are deemed material, in modern religion they are deemed immaterial and are limited to human beings.

Similarly, where in primitive religion gods are deemed material, in modern religion they are deemed immaterial. Gods thereby cease to be agents in the physical world – Tylor assumes that physical effects must have physical causes – and religion ceases to be an explanation of the physical world. Gods are relocated from the physical world to the social world. They become models for humans, just as they were expected to be for Plato. One now turns to the Bible to learn ethics, not physics. One reads the Bible not for the story of creation but for the Ten Commandments, just as for Plato a bowdlerized Homer would enable one to do. Jesus is to be emulated as the ideal human, not as a miracle worker. This view was epitomized by the Victorian cultural critic Matthew Arnold.

This irenic position was also like that of the evolutionary biologist Stephen Jay Gould, for whom science, above all evolution, is compatible with religion because the two never intersect. Science explains the physical world; religion prescribes ethics and gives meaning to life. But where for Gould religion has *always* served a function different from that of science, for Tylor religion has been forced to retrain, having been made compulsorily redundant by science. And its present function is a demotion. Tylor is closer to biologist Richard Dawkins, though Dawkins, unlike Tylor, is not prepared to grant religion even a lesser function in the wake of science.

The twentieth century rejected the nineteenth-century view. Now religion was assumed to be compatible with science, on the grounds that it serves a different need from that of science. The need can be anything – as long as it is other than that of explaining or controlling the physical world. Religion can cease to be about the world altogether and instead be about humans, either individually or collectively. Or religion can retain a link to the world, but be about the *experience* of the world rather than about the world itself. Religion can serve, for example, to try to fend off guilt over human sexuality (Freud), to defuse or redirect aggression (René Girard, Walter Burkert), to help encounter the unconscious (Jung), to instill commitment to the group (Durkheim), or to make life meaningful (Weber). Whatever the need postulated, religion and science can coexist, for they go their separate ways.

While Frazer, the pioneering Scottish historian of religion, is best known for his tripartite division of all culture into the stages of magic, religion, and science, the bulk of *The Golden Bough* is devoted to an intermediate stage between religion and science – a stage of magic and religion combined. In combining magic with religion, this in-between stage also combines ritual with myth. In the stage of sheer magic, there are rituals – the routines involved in carrying out the directions – but no myths, for there are no gods. In the stage of sheer religion, there are both myths and rituals, but they are barely connected. Myths describe the character and behavior of gods. Rituals, such as sacrifices, seek to curry divine favor. Rituals may presuppose myths, which would suggest what activities would most please the gods, but are otherwise independent of myths.

By contrast, in the following stage of magic and religion combined, rituals and myths work together in what is called "myth-ritualism." Frazer actually presents two versions of myth-ritualism, though he never disentangles them. In the first version the myth provides the biography of the god of vegetation, and the ritual enacts it. More precisely, the ritual enacts that portion of the myth which describes the death and rebirth of the god. The myth constitutes the script of the ritual. The ritual operates on the basis of the magical Law of Similarity, according to which the imitation of an action causes it to happen. The ritual does not manipulate vegetation directly. Rather, it manipulates the god of vegetation. But as the god goes, so goes vegetation. The assumption that vegetation is under the control of a god is the legacy of religion. The assumption that vegetation can be controlled, even if only through the god, is the legacy of magic. The combination of myth with ritual is the combination of religion with magic. In the ritual a human being plays the role of the god and acts out what he magically causes the god to do. Even when the actor is the king, the king is only an actor.

In Frazer's second version of myth-ritualism the king is central. In the first version the king may not even participate in the ritual. In the second he must. In the first version the king, even when the actor in the ritual, is merely human: He imitates, not becomes, the god. In the second version the king is divine, by which Frazer means that the god resides in him. Just as the health of vegetation depends on the health of its god, so now the health of the god depends on the health of the king: As the king goes, so goes the god of vegetation, and so in turn goes vegetation itself. Above all, where in the first version the king does not even die, in the second version he is killed. For to ensure a steady supply of food, the community kills its king while he is still in his prime and thereby safely transfers the soul of the god to his successor.

The king is killed either at the end of a fixed term or at the first sign of infirmity. Doubtless the sudden death of a king can never be precluded, but the killing of the king before his likely passing most nearly guarantees the continuous health of the god and so of vegetation. The aim is to fend off winter. The withering of vegetation during the winter of even a year-long reign is attributed to the weakening of the king.

This second version of myth-ritualism has proved the more influential by far, but ironically it in fact provides only a tenuous link between myth and ritual and in turn between religion and magic. Instead of enacting the myth of the god of vegetation, the ritual simply changes the residence of the god. The king dies not in imitation of the death of the god but as a sacrifice to preserve the health of the god. Myth, hence religion, plays a scant part here. Nor does magic play any part. Instead of reviving the god by magical imitation, the ritual revives the god by substitution. By contrast, in Frazer's first myth-ritualist scenario one would not ritualistically enact the rebirth of the god of vegetation without the myth of the death and rebirth of that god.

While Frazer presents two distinct and incompatible versions of myth-ritualism, he combines, or tries to combine, them in his description of the myths and rituals of his key gods. For example, he writes of Adonis:

> There is some reason to think that in early times Adonis was sometimes personated by a living man who died a violent death in the character of the god. Further, there is evidence

which goes to show that among the agricultural peoples of the Eastern Mediterranean, the corn-spirit, by whatever name he was known, was often represented, year by year, by human victims slain on the harvest-field. (Frazer 1922, p. 394)

Here Frazer tries to combine the imitation of the death and rebirth of the god – "Adonis was sometimes personated by a living man ... in the character of the god" – with the killing of a human being – "a living man who died a violent death." But unless the human victim harbors the god within him, the victim's death duplicates, not imitates, Adonis'. There is no magical efficacy, only emulation. Adonis himself is unaffected. Kingship is not even mentioned.

For Frazer, Christianity is just another vegetation cult, simply the most successful of the lot. Christianity takes over from Judaism the primitive practice of killing a human being not for the ethereal purpose of atoning for sin but for the mundane purpose of putting food on the table. Jesus was a human king in whom, it was believed, resided the soul of the god of vegetation. The annual spring celebration of Jesus' death and resurrection was a ritual intended to revive the crops, with Jesus the King of the Jews. At work here is Frazer's second myth-ritualist scenario exclusively, not any combination with the first one. Jesus was not an actor imitating the death and rebirth of a god but a divine king who was himself killed and replaced, though Frazer never discloses who his successor was.

Frazer demotes Jesus not even to a vegetation god or, worse, to a king but to the temporary substitute for the king. Because the substitute is killed, the king himself is thereby spared and so can be "reborn." By no coincidence the case of Jesus, in the third edition of *The Golden Bough*, falls in the volume (number 9) entitled "The Scapegoat." To be sure, Frazer does praise Jesus as a teacher of ethics, but for Frazer it is Jesus' death, not his life, that explains the appeal of Christianity. For Frazer, as for Tylor, ethics is what religion becomes, but it was not the heart of religion before the advent of Judaism and Christianity, and not even till much later in their history. Ethics existed, but in paganism it was not part of religion.

Lévy-Bruhl

Lucien Lévy-Bruhl (1985) was a French philosopher who became an armchair anthropologist. Lévy-Bruhl never asserts, as is commonly charged, that "primitive" peoples are inferior to moderns. On the contrary, he means to defend primitive peoples *against* this charge, made above all by Tylor and Frazer. For Tylor and Frazer, "primitives" think the way moderns do. They just think less rigorously. The difference between primitive and modern thinking is only of degree. For Lévy-Bruhl, the difference is of kind. He maintains that primitive thought is both mystical and "pre-logical," which means not illogical but rather divergent from logic.

Lévy-Bruhl attributes primitive thinking to culture, not to biology. Like other twentieth-century anthropologists, he separates culture from race. What distinguishes "primitives" from us is their "collective representations." By "collective representations" (*représentations collectives*), a term taken from Durkheim, Lévy-

Bruhl means group beliefs, which for him are the same across all primitive societies. Primitive representations, or *conceptions*, shape *perceptions*, or experiences. According to Lévy-Bruhl, primitive peoples believe that all phenomena, including humans and their artifacts, are part of an impersonal sacred, or "mystic," realm that pervades the natural one. To take Lévy-Bruhl's most famous example, when the Bororo of Brazil declare themselves red parakeets, they mean that they are in all respects outright identical with red parakeets.

Mysticism is only the first of the two key characteristics of primitive mentality. The other characteristic, "pre-logicality," builds on the first one but is more radical: It is the belief that all things are not only mystically one but also somehow distinct. The Bororo believe that a human is a parakeet yet still a human. They do not believe that a human and a parakeet are, say, identical invisibly while distinct visibly. That belief would merely be a version of mysticism, itself hardly limited to "primitives." Rather, primitive peoples believe that humans and parakeets are simultaneously both identical and separate in the same respects. Visibly as well as invisibly, humans and parakeets are at once the same and different. According to Lévy-Bruhl, that belief violates the law of noncontradiction – the law that something cannot be both X and non-X at the same time – and is uniquely primitive.

Lévy-Bruhl does not conclude, as is conventionally said of him, that primitive peoples cannot think logically, as if they are mentally deficient. Instead, he concludes that "primitives," ruled as they are by their collective representations, regularly suspend the practice of logic. Primitive thinking is prelogical, but prelogical does not mean illogical. Still, many readers mistook prelogical for illogical, so that Lévy-Bruhl seemed to be making primitive peoples even more hopelessly inferior to moderns than Tylor and Frazer had made them – the opposite of his intent. However, in arguing that primitive thinking differs in nature from modern thinking, Lévy-Bruhl is not arguing that it is equally true. Primitive thinking does make sense in light of its premises, but its premises are still illogical: Something cannot simultaneously be both itself and something else in the same respects at the same time. Where for Tylor and to a lesser extent Frazer primitive thinking is false but still rational, for Lévy-Bruhl primitive thinking is irrational and consequently false.

Douglas, Geertz, and Turner

Rightly or wrongly, the social sciences were long considered hostile to religion. But in the last half of the twentieth century there emerged movements in the social sciences that proved sympathetic to religion and to which, not coincidentally, scholars of religion proved sympathetic in turn. These movements span the social sciences most concerned with religion: anthropology, sociology, and psychology. Among the most notable movements among them was symbolic anthropology.

The difference between these movements and their predecessors was not over the truth of religion. As noted, the social sciences have traditionally considered only the origin and function of religion, not the truth of it. The contemporary movements sympathetic to religion have also shied away from the issue of truth.

The difference between contemporary theorists and their predecessors is neither over the importance of religion. As noted, classical anthropologists, sociologists, and psychologists all consider religion at least as important as any of their contemporary counterparts do.

For Tylor, Frazer, Durkheim, A. R. Radcliffe-Brown (1922), and Malinowski, religion serves not just an important but an outright indispensable function. In explaining the world, religion for Tylor fulfills an indispensable individual need: Humans may not die without an explanation, but *qua* humans they seek one. In purporting to provide food and other necessities, religion for Frazer clearly fulfills an indispensable individual need. In preserving society, religion for Durkheim and Malinowski serves an indispensable need on the part of society itself.

For Marx, the need that religion serves is also indispensable – not, however, to any society but to only an economically oppressed one: the need either to maintain that oppression or to compensate for it. As soon as society ceases to be economically oppressed, the need for religion will dissolve. For the Freud of *Totem and Taboo,* the need that religion serves is similarly indispensable to only a sexually repressed society: the need to maintain that repression. As society becomes less sexually repressed, the need for religion will disappear. Likewise for Jung the need that religion serves is indispensable to only a psychologically unrealized humanity, which realizes the collective unconscious unconsciously rather than consciously. Once humanity becomes conscious of the collective unconscious, the need for religion will end.

Furthermore, there is a difference between the indispensability of the function served by religion and the indispensability of religion itself. Few who consider the function served by religion indispensable consider religion indispensable to the serving of it. For Tylor and Frazer, modern science serves the same function as religion. For Tylor, religion may serve its function nearly as well as science, but for Frazer science alone serves the function that religion claims to serve: putting food on the table. For Radcliffe-Brown and Malinowski, secular ideology likely serves its same function far *less* potently than religion, but neither Radcliffe-Brown nor Malinowski thereby advocates the retention of religion. Only Durkheim, among classical social scientists, does. For him alone is religion itself indispensable: To survive, society needs religion itself, not merely some functional equivalent. No contemporary social scientist goes as far as he. At most, "contemporaries" are like Tylor, Frazer, Radcliffe-Brown, and Malinowski: However effective religion is, only the function it serves, not it itself, is indispensable.

The difference between contemporary social scientists and classical ones is not over the "positiveness" of the function religion serves. Marx and the Freud of *Totem and Taboo* may deem religion harmful to either individuals or society, but all of the other classical social scientists noted deem it helpful. Indeed, they deem it at least as beneficial as contemporary social scientists do.

The true difference between contemporary social scientists and their predecessors is over the eternality of religion. With the conspicuous exception of Durkheim, all of the classical social scientists mentioned deem religion either primitive or premodern. For none of them does religion or will religion serve its function for moderns. Even

Durkheim must find a distinctively modern, if seemingly secular, variety of religion. Contemporary social scientists, by contrast to their predecessors, deem religion compatible with modernity. They do not assume that everyone is religious, but they do assume that everyone *can* be religious.

Mary Douglas, Clifford Geertz, and Victor Turner were the key figures in the movement called symbolic anthropology. The key figure in a separate branch of symbolic anthropology was Claude Lévi-Strauss (1963–1976, 1966, 1967, 1969–1981). Ironically, in the conventional sense of the term "symbol," in which a symbol stands for something else, neither branch of symbolic anthropology deals with symbols. More accurately, for Lévi-Strauss symbols *collectively* stand for something else, but *individually* they do not. The *relationships* among individual symbols give them their symbolic meaning. Individually, symbols mean nothing. For Douglas, Geertz, and Turner, who in this respect are more radical, even collectively symbols do not necessarily stand for anything. Rather than *representations* of the symbolized, symbols are *expressions,* or *manifestations*, of it.

For both groups of symbolic anthropologists, the symbolized is beliefs, beliefs about the place of humans in the world. All of culture expresses those beliefs. For Douglas, the expressions are to be found not only in conventional domains of belief like religion and ideology but above all in seemingly meaningless, rote, random, or trivial ones like the kinds of food one does and does not eat; the kinds of clothes one does and does not wear; the way one categorizes animals, time, and space; and one's attitude toward one's body. For Douglas, these expressions of belief take the form of either classifications – for example, into permitted and forbidden food – or attitude – for example, one's attitude toward one's body.

For Douglas, all of culture not merely unintentionally reflects beliefs but actually propounds them. To eat only certain food is not just coincidentally to evince one's convictions about one's place in the world but, consciously or not, to proclaim those convictions. It is to make a statement.

Yet even as a declaration of beliefs culture does not symbolize the beliefs. It articulates them. Put another way, culture spreads rather than symbolizes beliefs. The way one classifies food parallels, not represents, the way one classifies clothes. The way one treats one's body parallels, not represents, the way one treats others or others treat one. The same set of beliefs, formulated as either classifications or attitudes, pervades all of culture.

Though Geertz as well as Douglas examines unexpected domains of belief – notably, cockfights, funerals, and sheep raids – he, in contrast to her, is equally interested in conventional domains like religion, ideology, common sense, and art. Still, he, like her, argues unconventionally that the beliefs expressed in even the most conventional domains concern neither the world itself, as religion is often taken to do, nor the domain itself, as art taken as irreducibly aesthetic is assumed to do. Rather, the beliefs concern the place of humans in the world. At the same time Geertz is far less concerned than Douglas with classification as a form of expression. He is also far less concerned with parallels between one cultural domain and another, though at times he draws them. Perhaps because he usually considers much broader domains, each of which can seemingly constitute the whole of one's culture, he is less concerned with showing the linkage of beliefs across domains.

Yet Geertz is concerned, indeed preoccupied, with paralleling beliefs about one's place in the cosmos – beliefs that in the case of at least religion he calls the "world view" – with beliefs about one's place in society – beliefs that in the case of at least religion he calls the "ethos." He stresses the need for a fit between the world view and the ethos. Where there is a fit, the two reinforce each other: The world view makes the ethos natural rather than arbitrary by rooting it in the nature of things, and the ethos makes the world view germane rather than irrelevant by constituting a concrete instance of it.

Turner is closer to Geertz than to Douglas. He, too, is less concerned than she with unexpected domains of beliefs yet at the same time explores some, especially colors and trees. Again like Geertz, Turner usually considers not only conventional domains but also broad ones. Unlike Geertz, he concentrates overwhelmingly on the domain of religion. Although Turner sees colors and trees as systems of classification, he, like Geertz, does not accord classification the same prominence as a form of expression that Douglas does.

Turner and Geertz alike stress the difficulty of establishing and maintaining a secure sense of place. Dealing with changing societies rather than, like Douglas, with stable ones, both are far more attentive than she to threats to a secure place. Douglas herself stresses the need to forge one's place, but she is much less worried than they about threats to the place forged.

If for Geertz the prime threat to a secure place comes from the disparity between one's place in the cosmos and one's place in society, for Turner it comes from the precariousness of one's place in society itself. Turner hardly denies the need for a fit between one's cosmic and one's social place. Indeed, Douglas hardly denies the need either: She simply takes its fulfillment for granted. Turner refuses to take for granted one of the parts: one's social place. Where the disparity on which Geertz focuses is that between the way the cosmos purportedly is – the world view – and the way *society* purportedly is – the ethos – the disparity on which Turner focuses is that between the way society *purportedly* is and the way society *actually* is.

For Douglas, one's purported place in the cosmos mirrors one's purported place in society, which mirrors in turn one's actual place. Indeed, Douglas does not really distinguish between one's ideal place and one's actual place. For Turner as well, one's purported place in the cosmos mirrors one's purported place in society, but one's purported place in society by no means mirrors in turn one's actual place. Turner, like Geertz, allows for far more creativity. In forms that he calls "antistructure," "liminality," and "communitas," culture offers a sharp alternative to one's actual place. That alternative can even serve as a model for creating a new "actual" place. For Geertz, one's purported place in the cosmos can likewise serve as a model for creating an ideal and in turn actual place in society.

For Douglas, too, culture is creative: It *creates* one's sense of place. But that place is created out of one's actual social place rather than, as for Turner and Geertz, out of the imagination. Without that place humans for all three would be lost – cognitively for Douglas, existentially for Turner and Geertz.

The place that culture gives one is for all three the symbolized. It is the subject matter of culture. But once again, culture expresses, not symbolizes, that "symbolized," in which case none of the three, despite the collective label given them, is, strictly, concerned with symbols.

Within the social sciences there have been two main views of religious ritual. One view has considered it a matter *of feelings,* which ritual either implants or releases. This view, by far the more common one, is found above all in Durkheim, Radcliffe-Brown, Malinowski, Marx, and Freud. For Durkheim and Radcliffe-Brown, ritual creates feelings: for Durkheim, feelings of dependence on society and of possession by society; for Radcliffe-Brown, feelings of dependence on society and also of love and hatred toward phenomena which, respectively, help and hurt society. For Malinowski, Marx, and Freud, ritual discharges feelings: for Malinowski, feelings of helplessness before nature; for Marx, pent-up economic desires; and for Freud, pent-up instincts.

The other main view of ritual has deemed it fundamentally a matter of *belief,* which ritual applies. This view is found above all in Tylor and Frazer. For both, ritual controls the world by applying prescientific beliefs about it.

Like Tylor and Frazer, Douglas, Geertz, and Turner regard ritual as belief. Unlike them, the three regard ritual as the expression, not the application, of belief. Even more unlike them, they regard the belief expressed as other than the primitive counterpart to science. For Tylor and Frazer, ritual is the primitive equivalent of applied science: For the purpose of controlling the world, ritual puts into practice the primitive belief that personal gods rather than impersonal laws of nature regulate the world. For Tylor and especially Frazer, not just ritual but religion as a whole gives way to modern technology.

For Douglas, Geertz, and Turner, by contrast, ritual is a modern as well as primitive phenomenon. It can be modern exactly because even as part of religion it does not rival science and therefore does not get superseded by science. Rather than either explaining or controlling the world, ritual for all three describes the place of human beings in the world.

Ritual describes the place of humans in not only the cosmos but also society. It describes the place of humans vis-à-vis not only the physical world and god but also other humans. Ideally, the cosmic and social places are in harmony. Ideally, they parallel, if not mirror, each other – a point stressed by Geertz above all.

For Geertz and Turner, the need for a place is existential: a fixed, certain place makes life secure, fair, and tolerable. For Douglas, the need is more intellectual: a fixed, certain place makes life intelligible. Perhaps because Geertz and Turner deal with changing societies, they are more attentive to "existential" anxiety than Douglas, who, dealing with stable societies, is freer to concentrate on purely intellectual issues.

As concerned as Douglas, Geertz, and Turner are with the function of ritual for the individual, they also are concerned with its function for society. As resolutely as they reject Durkheim and Radcliffe-Brown for their "emotivist" view of ritual, they accept the view of Durkheim and Radcliffe-Brown that the function served by ritual is social, albeit individual as well. Turner, Geertz, and Douglas assert that ritual serves at once to uphold society and to give humans places in both it and the cosmos.

Lévi-Strauss

Lévi-Strauss is the key creator of structuralism, which he himself applies overwhelmingly to myths and only secondarily to rituals. For him, the meaning of symbols lies not in any individual symbol but in its relationship to other symbols. The basic relationship – the structure – is that of opposition. A myth at once presents an opposition and, in varying ways, seeks to resolve or, more precisely, diminish it. Individual symbols represent either side of the opposition. The meaning of a symbol is exactly its opposition to its counterpart symbol. For example, the sky might be pitted against the earth. Life might be pitted against death. Human might be pitted against divine. Myths temper oppositions in varying ways – for example, by introducing a third symbol or group of symbols that fall between each side and thereby partake of both. A hero might temper the opposition between human and divine. A myth is to be read not chronologically, from beginning to end, but structurally, as a succession of instances of mediated oppositions. The plot does not count. Only the structure does. Purportedly, all symbols express the fundamental opposition between "nature" and "culture": between the view of humans as animal-like and the view of them as distinctly human.

For Lévi-Strauss, in contrast to Douglas, Geertz, and Turner, symbols do not serve to establish the place of humans in either society or the cosmos. Rather, they presuppose the establishment of the places of humans in both and serve entirely to reconcile those places, which stand in opposition. Like Geertz, Lévi-Strauss is preoccupied with the tension between one's place in society and one's place in the cosmos. That tension is, as for Douglas, cognitive rather than existential, but it is still, as for Geertz, a tension between one's social and one's cosmic place.

What Lévi-Strauss would mean by both "society" and "cosmos" is, however, different from what Geertz means. For Geertz, the "cosmos" means the entire external world, everything beyond humans. "Society" means the human world. For Lévi-Strauss, the "cosmos," for which he uses the term "nature," means the animal world and, even more specifically, the animal side of humans. "Society," for which he uses the term "culture," means the human world – more precisely, the civilized side of humans. Even more confusing is the distinction rigidly drawn by Geertz and others between "social structure" and "culture." Vis-à-vis "cosmos," "society" does not mean social structure. It means ethos, which, like world view, falls under "culture."

For Geertz, there is no inherent tension between one's cosmic place and one's social place. Whatever tension exists stems from the disparity between what one happens to believe about the world as a whole – for example, that it is just – and what one actually experiences in society. For Lévi-Strauss, by contrast, there is an inherently irresolvable tension between nature and culture. The conflict is not, as for Geertz, between what one is taught and what one experiences. It is between the experience of oneself as part of nature and the experience of oneself as part of civilization. Still, the conflict may well coincide with the difference between what one is taught about human nature and what one experiences.

In any case the conflict cannot be overcome. For Lévi-Strauss, in contrast to Geertz, there can never be a fit between culture and nature. But the conflict can be alleviated. Symbols alleviate it in various ways: by juxtaposing to a case of the clash a kindred case which is somehow more easily tolerated, by citing phenomena which manage to encompass aspects of both culture and nature, and by demonstrating that capitulation to either side would prove worse than endurance of the tension between them.

Like Douglas, Lévi-Strauss is intent on showing the pervasiveness of beliefs. For him, as for her, beliefs are everywhere. Cooking, music, art, literature, dress, etiquette, marriage, and economics are among the areas he has analyzed. He considers these areas both in their own right and as topics in myths – for example, the cooking done by mythic characters.

Unlike Douglas, Geertz, and Turner, Lévi-Strauss deems these expressions of belief not only expressions but also actual symbols. At times, he implies that the symbolized is always the opposition between culture and nature, but some of the oppositions surely fall either entirely within culture – for example, matrilineal versus patrilocal kinship – or entirely within nature – for example, life versus death. Still, the prime opposition is that between culture and nature – for Geertz, between society and cosmos. For Turner, the prime, if not sole, opposition is within culture, or society. Douglas, again, does not deal with opposition.

Like Turner, Lévi-Strauss stresses the way symbols not only explicate tensions but also overcome or at least allay them. For Turner, "symbols" – here used loosely for sheer expressions of belief – allay tensions by either the sheer explication of the tensions or the presentation of a better alternative. For Lévi-Strauss, symbols allay tensions only secondarily by the explication of them. Primarily, symbols either present a worse alternative, which thereby makes the existing situation preferable by default, or else partially resolve existing tensions by showing them to be less severe and therefore more manageable than they seem. Lévi-Strauss is here like both Geertz and Turner: Symbols do not merely reflect the existing situation but create a new one.

As concerned as symbolic anthropologists of both camps are with what symbols do for individuals – namely, order their world – they also are concerned with what symbols do for society. Yet the interest of most of them in the social function of symbols declines. Geertz is far more concerned with the effect of religion on society in his earlier writings than in his later ones, when he becomes much more concerned with the effect of religion on individuals. Turner's focus likewise shifts from society to the individual. Still, both remain concerned with the effect of religion on society. Lévi-Strauss, by contrast, becomes almost indifferent to that effect. Conversely, Douglas' interest in the social effect of religion never wanes.

Because for Douglas the ideal cosmic and social places, which symbols provide, mirror one's actual social place, symbols automatically serve to reinforce that place. Because for Geertz the cosmic place ideally parallels the social one, the cosmic place ideally makes the social one appear natural and thereby reinforces it. In addition, symbols help create the social place itself.

Sometimes symbols for Turner merely depict existing society. Because social conflict is otherwise either missed or ignored, the depiction itself is efficacious: It compels members of society to recognize the conflict. Where for Douglas symbols articulate existing society in order to *perpetuate* it, for Turner they do so in order to *remedy* it. Other times symbols for Turner present an alternative society – one not necessarily worse than the present one, as for Lévi-Strauss, but simply contrary to it. The alternative society usually gets depicted merely to make the present one clearer by the contrast, not to spur its rejection.

The approach of symbolic anthropology to religion differs, first, from an irreducibly religious one – the approach of Eliade above all. For Eliade, religion originates and functions to establish and maintain the relationship of humans to the sacred. The relationships of humans to society or the cosmos either are independent of their relationship to the sacred or else result from that relationship.

For symbolic anthropologists, by contrast, the origin and function of religion are secular: Religion originates and functions either to maintain or outright to establish the relationships of humans to society and the cosmos themselves. The establishment and maintenance of a relationship to the sacred are only one means to that end. Even if religion proved the sole means, the end achieved would remain secular. Moreover, that secular end is social as well as individual: Religion serves society itself as well as the individual.

The approach of symbolic anthropology differs, second, from an exclusively intellectualist one – the approach of, notably, Tylor. Approached wholly intellectually, the origin and function of religion are protoscientific: Religion originates and functions to explain the cosmos. Even though the explanation provided is supernatural, the need fulfilled is still secular. Even if religion provided the only explanation, the need it fulfilled would remain secular. Moreover, religion here deals entirely with the cosmos, not with society, and deals with the cosmos itself, not with the place of humans in it.

For symbolic anthropologists, by contrast, the function, not to say origin, of religion is social as well as individual: Religion abets society as well as the individual. In dealing with the cosmos, religion deals not with it in itself but with the relationship of humans to it. For Douglas, Geertz, and Turner, religion serves, furthermore, not only to explain that relationship but, even more, to establish it in the first place.

The approach of symbolic anthropologists differs, third, from an exclusively social functionalist one – the approach, to varying degrees, of Durkheim, Radcliffe-Brown, Malinowski, and, taken negatively, Marx and Freud. Approached functionally, the origin and function of religion are entirely social: Religion originates and functions to preserve society. More precisely, religion either originates to preserve society or, whatever its origin, simply effects the preservation of society. If "social function" is taken to mean intended social effect, none of the figures cited save perhaps the Freud of Totem *and Taboo* qualifies as a social functionalist. If, more typically, "social function" is taken to mean sheer effect, Durkheim, Radcliffe-Brown, Malinowski, and perhaps Marx also qualify as functionalists. Indeed, Durkheim, Radcliffe-Brown, and Malinowski are the founders of the "functionalist" school.

Approving of existing society, Durkheim, Radcliffe-Brown, and Malinowski approve of the role that religion plays in maintaining it. Disapproving of existing society, Marx and Freud disapprove of that same role. For both groups, religion may well explain and justify the relationships of humans to society and the cosmos, but it serves only to keep humans in their social place. Rather than, moreover, first creating that place, religion presupposes it and simply reinforces it. The place of humans in the cosmos serves simply to legitimate their place in society.

For symbolic anthropologists, by contrast, the origin and function of religion are, once again, individual as well as social: Religion serves to abet the individual at least as much as society itself. Religion upholds the place of humans in the cosmos as well as in society and upholds both places as an end in itself, not merely as a means to a social end. For Douglas, Geertz, and Turner, religion creates those places in the first place.

Symbolic anthropologists find symbols in far more domains than religion. Most of human life is, for them, symbolic. They focus on religion because it provides so strong a case for their fundamental tenet: that human behavior is motivated by beliefs. Where other areas of life like kinship and economics presuppose beliefs, religion appears to be fundamentally a set of beliefs themselves. Religious belief, moreover, appears to be the most deeply held of all beliefs. Because it matters so much to humans, it demonstrates the power of belief for them.

Symbolic anthropologists find religious symbols in far more domains than creeds and other conventional expressions of belief. Lévi-Strauss finds them in myths. More strikingly, Douglas, Turner, and Geertz find them in rituals, which, as physical activities, might seem to be the least likely expressions of belief. The three nevertheless interpret rituals as mental activities, as statements of belief.

Conclusion

Religionists see religion as distinct from the rest of life. Social scientists see religion as part of the rest of life. For religionists, whatever explains religion explains religion alone. Religion serves a uniquely religious function. For social scientists, whatever explains religion explains other things as well. Whatever the function of religion, other activities, such as science or ideology, serve it as well.

Yet what unites social scientists and religionists may be more significant than what divides them. First, for both groups, religion is not assumed to be natural. Whether deemed universal or less than universal, it must be explained. It is not taken for granted. And second, for both groups, whatever explains one religion explains another. The goal for both groups is a uniform explanation of religion. The goal is to figure out what makes all religions religions. The quest is for a common account of all religions. The distinctiveness of any religion is never denied. It is simply downplayed. Put another, summary way, the quest is for theories, which means for the explanation of a category.

Modern theories of religion, those from both the nineteenth and the twentieth centuries, come largely from the social sciences. Scholars of religion accept what social scientists maintain. Scholars do not create theories of their own, though they may well favor one theory or one social science over another. Whatever social scientists maintain about culture, society, the economy, or the mind is accepted by scholars of religion. How those views apply to religion is also accepted.

Social scientists are interested in religion as an example of what their theories espouse. For some social scientists, religion may be a minor, even dying, enterprise. For many others, it remains a central one. Religion matters not for what it illuminates about itself, let alone about God, but for what it illuminates about culture, society, the economy, or the mind. The study of religion changes as the social sciences change.

References

Barfield T (ed) (1997) The dictionary of anthropology. Blackwell, Malden/Oxford
Barnard A (2000) History and theory in anthropology. Cambridge University Press, Cambridge
Barth F et al (2005) One discipline, four ways. University of Chicago Press, Chicago
Beattie J (1966 [1964]) Other cultures. Routledge & Kegan Paul/Free Press, London/Chicago
Bellah RN (1970) Beyond belief. Harper & Row, New York
Berger PL (1967) The sacred canopy. Doubleday, Garden City. (Also published as The social reality of religion. Faber and Faber, London, 1969)
Berger PL (1969) A rumor of angels. Doubleday, Garden City
Bowie F (2006) The anthropology of religion, 2nd edn. (1st edn., 2000). Blackwell, Malden/Oxford
Douglas M (1966) Purity and danger. Routledge, New York/London
Douglas M (1973) Natural symbols, 2nd edn. (1st edn., 1970). Barrie and Jenkins/Vintage Books, New York/London
Durkheim E (1965 [1912]) The elementary forms of the religious life (trans.: Ward Swain J). Free Press, New York
Eliade M (1963 [1958]) Patterns in comparative religion (trans.: Sheed R). Meridian Books, Cleveland
Eliade M (1968 [1959]) The sacred and the profane (trans.: Trask WR). Harvest Books, New York
Erickson PA, Murphy LD (2008) A history of anthropological theory, 3rd edn. (1st edn., 1998). University of Toronto Press, Toronto
Erikson EH (1958) Young man Luther. Norton, New York
Erikson EH (1969) Gandhi's truth. Norton, New York
Frazer JG (1890) The golden bough, 1st edn., 2 vols. Macmillan, London
Frazer JG (1922) The golden bough, abridged edn. Macmillan, London
Freedman M (1979) Main trends in social and cultural anthropology. Holmes and Meier, New York/London
Freud S (1950) Totem and taboo (trans.: Strachey J). Routledge & Kegan Paul, London
Freud S (1964) The future of an illusion (trans.: Robson-Scott WD, rev. Strachey J). Doubleday Anchor Books, Garden City
Geertz C (1973) The interpretation of cultures. Basic Books, New York
Geertz C (1983) Local knowledge. Basic Books, New York
Ingold T (ed) (1994) Companion encyclopedia of anthropology. Routledge, London/New York
James W (1902) The varieties of religious experience. Longmans, Green, London/New York
Jung CG (1938) Psychology and religion. Yale University Press, New Haven
Lévi-Strauss C (1963–1976) Structural anthropology, 2 vols (trans.: Jacobson C, Schoepf BG [vol I] and Layton M [vol II]). Basic Books, New York
Lévi-Strauss C (1966) The savage mind (trans.: not given). University of Chicago Press, Chicago
Lévi-Strauss C (1967) The story of Asdiwal. In: Edmund Leach E (ed) The structural study of myth and Totemism. A. S. A. Monographs, No. 5. Tavistock, London, pp 1–47

Lévi-Strauss CE (1969–1981) Introduction to a science of mythology, 4 vols (trans.: Weightman J and D). Harper, New York
Levy-Bruhl L (1985) How natives think (trans.: Clare LA). Princeton University Press, Princeton. Original publ. of tr.: Allen &Unwin, London, 1926
Malinowski B (1925) Magic, science and religion. In: Needham J (ed) Science, religion and reality. Macmillan, New York/London, pp 20–84
Marx K, Engels F (1957) On religion. Foreign Languages Publishing, Moscow
Moore HL, Sanders T (eds) (2006) Anthropology in theory. Blackwell, Malden/Oxford
Müller FM (1867) "Comparative mythology" (1856), in his Chips from a German workshop, vol 2. Longmans, Green, London, pp 1–141
Radcliffe-Brown AR (1922) The Andaman Islanders. Cambridge University Press, Cambridge, UK
Segal RA (1989) Religion and the social sciences. Scholars Press, Atlanta
Segal RA (1992) Explaining and interpreting religion. Peter Lang, New York
Turner VH (1967) The forest of symbols. Cornell University Press, Ithaca
Turner VH (1968) The drums of affliction. Clarendon Press, Oxford
Tylor EB (1871) Primitive culture, 1st edn., 2 vols. Murray, New York/London
Weber M (1963) The sociology of religion (trans.: Fischoff E). Beacon Press, Boston

Historical Studies in Nineteenth-Century Germany: The Case of Hartwig Floto

Herman Paul

Contents

Introduction	208
Social and Educational Background	209
Ranke's Historical Exercises	210
The Priority of Teaching	211
Teachers of the Nation	213
Professionalization	214
Confessional and Political Fault Lines	216
Virtues and Vices	217
A Broken Career	219
Source Editing Projects	220
Conclusion	221
References	223

Abstract

History was a key discipline in what the German philosopher Wilhelm Dilthey called the "human sciences" (*Geisteswissenschaften*). Focusing on the German lands, this chapter surveys what the study of history looked like in the decades prior to the publication of Dilthey's *Einleitung in die Geisteswissenschaften* (Introduction to the Human Sciences, 1883). It does so, somewhat unconventionally, by zooming in on Hartwig Floto (1825–1881), a largely forgotten pupil of the famous Leopold von Ranke. Apart from the fact that this biographical angle adds color and flavor to an otherwise too abstract story, Floto's life and work lend themselves well for discussion of both familiar and not-yet-familiar themes in the history of the humanities: Ranke's historical exercises, historians' middle-class backgrounds, research institutions like the *Monumenta Germaniae Historica*, but also historians' personae as typically described in terms of virtues and vices. This chapter therefore aims to do two things at once: it wants to offer an accessible

H. Paul (✉)
Institute for History, Leiden University, Leiden, The Netherlands
e-mail: h.j.paul@hum.leidenuniv.nl

© The Author(s), under exclusive licence to Springer Nature Singapore Pte Ltd. 2022
D. McCallum (ed.), *The Palgrave Handbook of the History of Human Sciences*,
https://doi.org/10.1007/978-981-16-7255-2_41

introduction to nineteenth-century German historical studies, and also seeks to showcase both older and newer lines of research in the history of the humanities.

Keywords

Historical studies · Historiography · German historical scholarship · Hartwig Floto · Leopold von Ranke · Georg Waitz · Heinrich von Sybel · Jacob Burckhardt

Introduction

In December 1855, the Berlin historian Leopold von Ranke sent a letter of recommendation for one of his former students, Hartwig Floto, to the University of Basel. Another former student of his, Jacob Burckhardt, had taught there for a couple of years, but moved on to a chair in Zurich. Was the 30-year-old Floto, the author of a historical monograph on Emperor Henry IV, a good match for the job? According to Ranke, Floto was a man with "lively zeal for historical studies" and "excellent capacities." Pairing broad historical knowledge with solid methodical techniques, Floto was likely to be a stimulating role model for students. Although Ranke confessed that "I cannot judge his teaching talent," he added that "everything I have heard" testified positively to Floto's teaching qualities (Ranke 1949: 369).

Although Ranke's letter was instrumental in getting Floto appointed to the chair in Basel, its prose was not as glowing as on other occasions. Compared to the enthusiastic letter that Ranke wrote for Wilhelm Wattenbach, just a year before, it described Floto's suitability in rather generic terms. Clearly, this tells us something about the job applicant. As Ranke had confided to a Basel university administrator, just a few weeks earlier, Floto had a "lively spirit" and could boast a "broad education," but was not as thorough a researcher as, for instance, Ernst Dümmler (a man who grew to become a leading figure in the German historical discipline). What Ranke's letter tacitly conveys, therefore, is that Floto, though diligent and talented, belonged to a different league than Burckhardt, Wattenbach, or Dümmler. He was, indeed, a more average talent than some other students of Ranke's – which is one reason, though not the only one, why Floto has become an almost forgotten historian.

Precisely his lack of remarkability, however, makes Floto an appropriate figure for the purposes of this chapter. The goal in the pages that follow is to offer a broad survey of historical studies in nineteenth-century Germany, explored through the prism of Floto's biography. On the one hand, this allows consideration of some familiar themes: Ranke's historical exercises in Berlin, which Floto attended in 1846–1847, the middle-class background of most "professional" historians, and the growing importance of research institutions like the *Monumenta Germaniae Historica*. On the other hand, the case of Floto makes it possible to highlight some newer insights, related to historiographical virtues and vices as well as to the relation between "professional" audiences and non-academic readers. Also, whereas biographies of famous scholars like Ranke sometimes suggest a course of life in which hard work at early age is rewarded with honor and fame at later stages, the case of Floto shows how differently a historian's career could develop under less fortunate circumstances.

On a broader canvas, the case of Floto therefore makes it possible to showcase some of the research being done in the history of the humanities – a small but flourishing field of inquiry that can be regarded as part of the history of the human sciences, even if it is not strongly represented in this handbook (cf. Paul 2022). One might argue that the discipline of history, especially in its nineteenth-century German incarnation, deserves coverage in a volume like this because so many of the human sciences originate in what Wilhelm Dilthey famously called the *Geisteswissenschaften*: a cluster of disciplines to which the field of history was central (Smith 2007: 128). Along these lines, this chapter will describe in some detail the look of German historical studies in the decades prior to the publication of Dilthey's *Einleitung in die Geisteswissenschaften* (Introduction to the Human Sciences, 1883). (Rudolf Makkreel, in his Chap. 5, "Diltheyan Understanding and Contextual Orientation in the Human Sciences," to this volume, discusses Dilthey's own contributions.) Most importantly, however, the case of Floto will be used to illustrate what kind of questions historians of the humanities are currently addressing, what kinds of concepts they are employing, and how a biographical perspective enables us to see how abstract issues of methodology and professionalization played out in the life of an ordinary, not very successful historian.

Social and Educational Background

Friedrich Wilhelm Theodor Hartwig Floto (sometimes also spelled as Flotho) was born in 1825 in Arendsee, a municipality in the German Altmark region, where his father, a high-ranking government official, was responsible for administering the royal lands. When the boy was 7 years old, the family moved to Oschersleben, a town just north of Halberstadt. Floto attended the Stephaneum gymnasium in Halberstadt before being admitted to Schulpforta, the famous boarding school near Naumberg, where Karl Rudolf Fickert and Karl Steinhart were among his teachers. Judging by the fact that Floto recited a self-written poem on the occasion of the school's third centenary, his performance as a student must not have been bad (Kirchner 1843: vii).

With this social and educational background, Floto was fairly typical for historians of his generation. Most of them – 69% of the cohort that entered the professoriate in the 1850s – came from upper-middle-class families, with fathers being employed as professor, gymnasium teacher, pastor, or middle to high-ranking civil servant. Confessionally, the Protestant Floto also belonged to the mainstream: no less than 75% of his cohort had a Protestant background (Weber 1984: 72–73, 84–85). Only his education at Schulpforta, the elite school also attended by Johann Gottlieb Fichte, Ranke, Ulrich von Wilamowitz-Moellendorf, and Friedrich Nietzsche, gave Floto a small advantage over his peers. With its traditional emphasis on classical studies and character development, this neo-humanist school prepared him well for academic study in Berlin, where Floto enrolled in October 1844.

As customary at a time when specialized study did not yet exist, Floto attended courses by professors as diverse as the geographer Carl Ritter, the philosophers Georg Andreas Gabler and Leopold George, the theologian Johann Karl Wilhelm Vatke, and the *Staatswissenschaftler* (scholar of the sciences of state) Wilhelm von Dönniges. However, as Floto would state in the Latin *vita* attached to his dissertation, the teacher who influenced him most was Ranke, "who in public and private teaching opened me the way to a correct understanding of history" (Floto 1847: 66). Given Ranke's central role in the nineteenth-century German historical profession, it is worth looking in some detail at how Floto experienced his study with him.

Ranke's Historical Exercises

Beginning with Ranke, who by the end of the century came to be known as the "founding father" of modern historical scholarship, is not entirely without risk. Such a start is in danger of reproducing some of the historiographical myths that historians around 1900 spun around the Berlin historian – his allegedly "scientific" approach to history, for example – while ignoring Ranke's indebtedness to earlier generations of scholars. It is especially noteworthy that Ranke's critical distance from predecessors like Johann Christopher Gatterer, August Ludwig von Schlözer, and Arnold Heeren has often been overstated. In addition, beginning with Ranke's historical exercises might obscure the fact that his research and teaching habits were not unique but part of a broader transformation of early nineteenth-century German intellectual life (Toews 2004). However, this being said, it cannot be doubted that one of key factors responsible for Ranke becoming the best-known German historian of his time was his successful imitation of the philological seminar, such as the seminar offered in Berlin by the classical scholar August Böckh. In weekly gatherings, Ranke had some of his most talented students familiarize themselves with primary sources with the aim of developing their critical reading skills (Berg 1968: 51–56). Writing in 1856, Floto still remembered how different this teaching format had been from what he had been used to:

> [T]he method of the famous historian surprised me. What we were offered was not a well-delineated course, no well-argued lectures. Ranke *had us read*. Soon we were interpreting the *Germania* [by Tacitus]; soon we were reading two chroniclers who covered one and the same subject – or he shared with us the three *relazioni* [reports] issued by Charles V on the day of his conquest of Tunis, alerting us to the contradictions between them. . . In one word: he showed us from what documents alone an authentic history could be created and taught us how to read these documents. This was all he did (Floto 1856a: 12).

The first thing to notice about this passage is that it depicts Ranke as a man fascinated by primary sources – *relazioni* and other unpublished material in particular. Although Ranke was, of course, not the first historian to recognize the importance of unpublished sources, he provoked both admiration and criticism for emphasizing the importance of unpublished source material to a greater degree than had been customary (Eskildsen 2008, 2019). Secondly, Floto emphasizes how critically the Berlin historian treated this material: attentive to inconsistencies and eager to distinguish between reliable and unreliable testimonies. Although this was not exactly representative of how Ranke was perceived around mid-century – at the time, he was better known for his aversion to moralizing history than for his advocacy of critical methods – Floto's portrayal of Ranke resembled that of Georg Waitz, who turned Ranke into an epitome of "criticism," "precision," and "penetration" (Paul 2019a).

Thirdly, Floto's memories of Ranke's exercises reveal a distinctive feature of this Rankean school. Judging by such different figures as Waitz and Heinrich von Sybel, Ranke's students were not united in their understanding of the historian's task. They shared, however, the experience of having sat at Ranke's feet in the historiographical equivalent to the philological seminar that would soon become a site of high symbolic value. In the 1840s and 1850s, many of Ranke's former students started similar exercises at other German universities, convinced that such *Übungen* (exercises), lovingly compared to scientific laboratories, were an effective means for socializing students into a critical historical ethos (Eskildsen 2015). Floto would do the same: soon after his appointment in Basel in 1856, he would start offering "historical exercises."

Partly because of the exclusivity of these *Übungen* – typically held in the professor's home, in the sanctuary of his private study – the exercises quickly acquired a prestigious aura. Former students expressed their gratitude for having participated in them by organizing festivities on anniversary occasions, with speeches, music, and presents for the man who had initiated them into the historian's craft (Schnicke 2015a). Emphasizing the exclusivity of the bonds created through shared study, many students conceived of themselves as a "family" headed by a "father" (with Ranke, the "father of modern historical scholarship," serving as patriarch *par excellence*). Although Floto did not literally call himself a son of father Ranke, it is significant that he shared his memories of Ranke's exercises in his inaugural address, thereby presenting himself to his Swiss colleagues and students as a representative of Rankean historiography. Tellingly, Floto also dedicated his dissertation and his first book to Ranke, as a sign of what he called his *pietas* (loyalty) to the master.

The Priority of Teaching

Ranke's historical exercises have often been interpreted as a decisive step towards more specialized, research-oriented education than was provided in lecture series of the kind that Floto also attended during his student years in Berlin. It is important to keep in mind, however, that *Übungen* never replaced *Vorlesungen* (lectures) (Lingelbach 2006). Ranke himself, for one, was convinced that teaching broad survey courses was as important a task for historians as initiating talented students into the secrets of source criticism. Also, not all the young men who attended exercises eventually became historians. In 1866, the ancient historian Alfred von Gutschmid reported that eight of the nine students in his exercises were philologists (Liepmann 1916: 368). More importantly, by mid-century, few historians thought of themselves primarily as researchers. Emerging out of an academic tradition in which history classes mostly served propaedeutic purposes, history professors resembled *Gymnasium* teachers in that they were first and foremost educators of the youth. As Hans-Jürgen Pandel (1993: 348) puts it: "Their self-understanding was shaped after the model of the 'teacher' more than after the model of the 'researcher'." As we shall see below, it was only near the end of the century, when large-scale research projects gave an impetus to specialized archival research, that historians would come to grant research a more prominent place in their understanding of the professor's vocation.

Floto's career reflects this priority given to teaching over research in at least two ways. First, like most other historians of his generation, Floto spent some years teaching in non-academic settings before landing an academic teaching position. After finishing his studies with Ranke in 1847, he taught for a while at the knight academy in Liegnitz (nowadays Legnica), a school for sons of the Silesian aristocracy and landed gentry, where Floto proved himself a "promising young teacher" (von Bethusy 1849: 9). Soon, however, political tensions between Prussia and Austria interfered with teaching. Drafted for military service in November 1850, Floto was added to the Emperor Alexander Guard Grenadiers in Berlin. Interestingly, the young historian managed to make a virtue out of necessity: he produced a textbook for aspiring infantry officers, published in 1853 with a laudatory preface by Colonel Gustav von Griesheim (Floto 1853). After this military interlude, Floto returned to teaching, this time as a private family tutor in Berlin (the kind of job that had been common among eighteenth-century historians, but, as Blanke [1989: 357] points out, had become more exceptional by the 1850s). So when Ranke, in the letter with which this chapter began, stated that Floto was reported to have strong didactic skills, he apparently assumed that teaching experience at the *Gymnasium* level was relevant to a university career.

One reason why the University of Basel agreed with this assumption was that its (only) professor of history was also expected to teach the three upper classes of the city's humanistic *Gymnasium*. Accordingly, "a teacher who has a stimulating effect on the youth" matched their job profile better than a historian who devoted his time to specialized research (Roth 1935: 58). Arguably, then, the reason that not Dümmler but Floto got the chair was the expectation that the latter would be best able to win "the fondness of the Swiss youth." Interestingly, even assessments of Floto's research were focused on the author's didactic potential. Judging by his articles, wrote Jacob Burckhardt to the mayor of Basel, Floto would not be "a bad teacher" for the local youth. Likewise, the Basel university officials highlighted his "inspiring teaching talent, vivid spirit, youthful enthusiasm, and broad education." Speaking about Floto's research on Emperor Henry IV, they showed themselves especially pleased by the historian's "fresh manner of expression" (Roth 1935: 60). The *Basler Zeitung* maintained this tone even in reporting about Floto's inaugural address in May 1856. Devoting not a single word to his research, the newspaper observed that the "newly appointed teacher of history" had displayed a heartening talent for teaching: "Any observer competent to judge . . . has been able to see that Floto is the right man to motivate the youth for historical education and study" (quoted in N. N. 1856a).

Teachers of the Nation

Students, however, were not the only audience that German historians at the time sought to reach. Neither was their teaching limited to *Vorlesungen* and *Übungen*. Since the eighteenth century, historians had taken pride in educating the nation. In writing for educated fellow countrymen, they had tried to be "teachers of Germany" (*praeceptores Germaniae*) or public intellectuals who helped their readers understand themselves in the mirror of history. Because of the political dimension of this popular history writing, scholars have often highlighted the extent to which historians were "builders of the nation" (Lenhard-Schramm 2014), even if they limited themselves to writing history books, without participating in, for instance, the Frankfurt Parliament of 1848–1849 (as did Johann Gustav Droysen, Sybel, and Waitz). It is important to keep in mind, however, that not all historians who tried to be teachers of the nation were as fervently patriotic as Droysen, Sybel, Heinrich von Treitschke, and other members of the so-called Prussian Historical School, or as narrowly focused on the history of Germany's political fate as Treitschke in his *Deutsche Geschichte im neunzehnten Jahrhundert* (German History in the Nineteenth Century, 5 vols., 1879–94). Friedrich Christoph Schlosser, for example, wrote his *Weltgeschichte für das deutsche Volk* (World History for the German People, 19 vols., 1844–57) primarily from a moral point of view, even though his anti-elitist identification with the German "people" also had an unmistakable political subtext.

The extent to which Floto aimed to be a teacher of the nation is apparent from his 1850s work on Henry IV, the eleventh-century German ruler whom Pope Gregory VII famously forced to seek absolution for his excommunication in Canossa. In order to reach a wide readership, Floto pre-published lengthy excerpts of his two-volume monograph, *Kaiser Heinrich der Vierte und sein Zeitalter* (Emperor Henry IV and His Age, 1855–6), in cultural magazines such as the *Deutsches Museum*. The book itself was reviewed in more than a dozen periodicals, most of which were cultural monthlies and review journals targeted at a broad spectrum of readers. Although some reviewers offered quasi-professional commentary by pointing out factual mistakes or argumentative flaws, most reviews focused on the readability of Floto's book, arguing that it was eminently suited for a general public because of its attractive writing style. Indeed, if reviewers agreed on one thing, it was that Floto knew how to write. According to the *Schwäbischer Merkur*, even readers "who are not used to attending seriously to an old German emperor" would enjoy *Kaiser Heinrich der Vierte* (N. N. 1855). Another reviewer even found the book so "plain, fresh, and clear" that it could stand comparison with "the best German, English, and French historical works from recent times" (N. N. 1856b – a judgment that Floto's publisher did not fail to use as a blurb in advertisements for the book).

Echoing these judgments, other readers came to similar conclusions. The diplomat and biographer Karl August Varnhagen von Ense noted in his diary that he found Floto's style reminiscent of Ranke's. The Austrian novelist, playwright, and poet Ferdinand von Saar found himself so impressed with the book that he planned to dedicate the second part of his dramatic poem, *Kaiser Heinrich IV.* (Emperor Henry IV), to Floto. Even the Bavarian King Maximilian II, who was an avid history reader, responded favorably to Floto's debut, judging by his attempt to get the author appointed to a history chair in Erlangen (where things worked out differently, though: the chair went to Karl Hegel, another student of Ranke's).

Floto's aim of reaching readers beyond the circle of his colleagues is also apparent from the content of his book, especially from his perhaps surprising habit of emphasizing, time and again, the superiority of "modern" thought over superstitious religious ideas of the kind held by Gregory VII and other eleventh-century clergy. Drawing on stereotypical contrasts between science and religion that circulated widely in nineteenth-century Europe (Ungureanu 2019), Floto presented pre-Copernican geo-centrism as a vivid illustration of "the ignorance and barbarism of the Middle Ages." In even less flattering terms, he described the doctrine of Eucharistic transubstantiation as the "biggest and most ridiculous aberration of the human spirit" that has ever occurred, to which he added that the continuous prevalence of this idea sadly shows that "we clever Europeans . . . in many respects do not stand much higher than the fetish worshippers at the southern border of the Sahara" (Floto 1855: 117, 163). Clearly, Floto did not hesitate to adopt a stance and tell his readers, not merely "how things actually had been" (*wie es eigentlich gewesen*, in Ranke's famous expression), but also, in the words of another great historian, Benedetto Croce, "what is living and what is dead" in Europe's medieval past.

This, of course, gave reviewers ample grounds for accusing Floto of "partiality." Even commentators who shared the author's "historical-political views" doubted whether it was appropriate to articulate these views in a historical monograph. As the *Heidelberger Jahrbücher der Literatur* put it: "[The historian should] keep the pages of history writing as clean as possible and not spatter them with his potential exuberance of patriotic gall and disgruntlement" (N. N. 1856c: 121). Likewise, despite Floto's style being widely praised, several reviewers found his "unpolished" prose, his "striving for popularity," or his penchant for rhetorical effect incompatible with serious historical scholarship. One reviewer grumbled that the book seemed to be written for Berlin salon audiences (a verdict that was actually not wide of the mark, given that Floto had attended such salons when teaching in Berlin). Or as a British reviewer concluded: "Indeed his style altogether savours too much of the newspaper, and is disfigured by frequent instances of vulgarity" (N. N. 1858a).

Professionalization

It is possible to interpret these criticisms as evidence of the "professionalization" of German historical studies? The answer depends on what the term is understood to mean. There is a rich tradition of equating professionalization with the development of a discipline-specific scholarly infrastructure, complete with university chairs, specialized journals, and professional conferences (see, e.g., Porciani and Raphael 2010). This is clearly not what Floto's critics worried about, nor something to which Floto himself actively contributed. Tellingly, he never published an article in Sybel's *Historische Zeitschrift*. Professionalization becomes a more relevant concept, however, if it denotes the rise of specialized training such as offered in historical exercises and the codification of historical methods in textbooks like Ernst Bernheim's *Lehrbuch der historischen Methode* (Manual of Historical Method, 1889) (Torstendahl 2014). Both the growing demand for methodological reflection, known in German as *Historik*, and the spread of Ranke-style exercises showed that historians began to place increasingly higher demands on research. By mid-century, it was no longer plausible to say, as Arnold Heeren had done, that source criticism "is a beautiful and necessary thing" as long as it remains an auxiliary science, subordinate to historical writing. Also, at a time when archival repositories made rapidly expanding amounts of source material accessible to researchers, the Rankean dictum that historians write on the basis of primary sources made it increasingly difficult to cover large topics within the covers of a single book. Historical research was therefore professionalizing in the sense that scholars increasingly expected each other to write in a degree of detail that effectively excluded general readers, thereby creating a demand for specialized journals like the *Historische Zeitschrift* (Jørgensen 2012).

For historians who still primarily saw themselves as teachers, the rise of specialized *Forschung* (research) did not imply that middle-class ideals of Bildung (self-cultivation) became obsolete. Throughout the nineteenth century, German historians continued to write for nonprofessional audiences in the form of newspaper articles, essays for cultural monthlies, and popular books. Just as the emergence of journals as platforms for scholarly communication did not lead to the disappearance of books, so the advances of specialized research did not alienate historians from the educated middle classes that had been their primary audience. As Martin Nissen (2009: 317–9) argues, it is more accurate to say that professionalization increasingly required historians to *negotiate* the demands of their profession and the demands of a wider public sphere.

Although this resulted in some scholars turning their back on popular history writing, many others continued at least occasionally to reach out to nonprofessional readers. One reason for doing so was that historical scholarship and *Wissenschaft* more generally enjoyed high prestige among the German educated middle classes. "Knowing your history" was part and parcel of what it meant to be an educated citizen (Mommsen 1998). Also, in an era of cultural and political nationalism, historians could serve as experts on national identity – a public role that turned historians like Sybel into "political professors" (scholars eager to use their broad knowledge of the past to help the nation understand its present situation and guide it firmly towards an imagined future) (Muhlack 2001). Against this double background, it makes sense to say that until at least the 1880s, "professional history remained, by and large, popular in the sense that the works by professionals were widely celebrated and professional historians saw themselves as national pedagogues" (Berger et al. 2012: 8). However, as illustrated by the reviews that Floto's book elicited, historians trying to reconcile the demands of *Forschung* and *Bildung* could not expect to gain approval from all sides. In an age of professionalization, scholars reaching out to general readers could be seen as insufficiently living up to scholarly standards. They ran a risk of being perceived, at least by those most committed to *Wissenschaftlichkeit* (a scientific attitude), as "outsiders" instead of "insiders" – as happened most famously to Treitschke, the Prussian historian who reached more readers than any of his colleagues but found himself fiercely attacked for violating the research standards of an emerging historical profession (Gerhards 2013).

Confessional and Political Fault Lines

German historians disagreed not only over the relation between *Bildung* and *Forschung*; they were also divided along political and confessional lines. Although virtually all historians were, in one way or another, committed to the German national cause, in a country known as "a nation of provincials" (Applegate 1990), this nationalism took on different forms. For many German citizens, regional and confessional identifications were at least as strong as national ones. Among historians, this diversity resulted in two loosely defined schools, geographically located in the northern and southern German lands, respectively. While the first was committed to a "little" Germany led by Protestant Prussia, the second dreamt of a "big" Germany with a hegemonic role for Catholic Austria (Brechenmacher 1996).

Insofar as scholars of German historiography have focused their attention on the Protestant north – on places like Berlin and Göttingen, where Ranke, Waitz, and others pushed the limits of historical criticism – they have repeated a pattern already visible by the mid-nineteenth century. To the annoyance of their colleagues in Bavaria, Prussian historians made few attempts to hide their sense of professional superiority. Waitz, for instance, openly declared that "north German historians are more learned [and] more objective" than their colleagues in the south – a verdict that an angry critic subsequently denounced as "slander." Likewise, when Sybel, Prussian to the bone, was appointed to a chair in Munich, where he launched a periodical from which "ultramontane" contributors were explicitly excluded, Bavarians agitated against what they perceived as Sybel's "historical sect." Although later generations, weary of the confessional polemics of Otto von Bismarck's Kulturkampf, would adopt more reconciling stances, by the 1850s, the political divide between north and south, reinforced by a confessional divide between Protestantism and Catholicism, was still strong enough to challenge the serene idea of a single historical profession in which scholars of different persuasions could participate on equal footing (Paul 2018: 708–9).

As a proud Lutheran Prussian, Floto did not fail to contribute to these tensions. As we saw above, his book on Henry IV, written for an audience of "we Protestants" (Floto 1855: 163), did not eschew anti-Catholic polemics. On the contrary, the author's dislike of clerical celibacy was such that he saw no harm in digressing from historical analysis to praise Martin Luther for allowing priests to marry. Among other things, this provoked a 27-page rebuttal in the *Historisch-politische Blätter für das katholische Deutschland*, a Catholic periodical that accused Floto of abusing the past for present religious-political purposes ("The poor Salian Henry IV is raised from his grave to make hocus pocus for the party goals of Professor Hartwig Floto": N. N. 1858b: 453). Instead of trying to refute such charges, Floto self-confidently affirmed his religious-political views by stating that he was not at all ashamed of ascribing to Goethe's maxim: "We scarcely know what we owe to Luther" (Floto 1856b: vi).

Unsurprisingly, Floto's anti-Catholicism became most virulent during the *Kulturkampf* of the 1870s. This is especially apparent from a speech delivered in 1877 at the unveiling of the so-called Canossa Pillar in the Harz mountains near Goslar. Marking the eighth centenary of Henry IV's trek to Canossa, this fifteen-meter-high monument was decorated with a portrait of Bismarck, the staunchly anti-Catholic chancellor, who just a few months earlier had uttered the famous words, "Don't worry; we are not going to Canossa, neither spiritually nor physically." Unlike Waitz, who showed himself increasingly critical of such political appropriations of the medieval past, Floto used the occasion to pull out all the stops. According to a newspaper report, he charged the Jesuits with plotting a war against "Protestant Prussia," while aiming for a re-Catholicization of the world that would effectively undo the Protestant Reformation (Dormeier 1990: 238). Although Floto was hardly politically active, at moments like this, he resembled the figure of the "political professor" mentioned above.

Virtues and Vices

All of this – Floto's upper-middle-class background, his Rankean training, his popular history writing, and his political anti-Catholicism – translated into the persona of the historian as portrayed in Floto's inaugural address, *Ueber historische Kritik* (On Historical Criticism). Older scholarly literature has treated this inaugural as a methodological contribution to "historicism," with the term *Historismus* serving as shorthand for historians who tried to study the past as objectively as possible through consistent application of critical methods (Rüsen 1993). Arguing along these lines, Jörn Rüsen has cited Floto as stating that criticism lies at the heart of historical studies, and students of Rüsen have treated Floto's inaugural as evidence of critical methods becoming a means for realizing scholarly objectivity (e.g., Blanke 1991: 259). Although the term "historicism" continues to be used here and there, the habit of interpreting German historical studies in these terms is not as strong anymore as it was 30 or 40 years ago. Whereas a previous generation emphasized the methodological assumptions shared among nineteenth-century historians – among other things with the aim of showing that there were other historical methodologies than those advocated by social scientists in the 1970s and 1980s – recent scholarship has come to recognize the diversity of the field that included Floto among its members. This diversity was not limited to political convictions or religious affiliations, but also manifested itself in disagreement over what "professionalization" meant or what *Wissenschaftlichkeit* required (Middell 2010: 159).

This is neatly illustrated by the virtues (*Tugenden*) that Floto associated in his inaugural with the persona of the historian. On the one hand, he reproduced a set of well-known, almost stereotypical ideas about the virtues that a good historian should possess. Few colleagues would have disputed Floto's claim that historians ought to be driven by a "sincere desire only to investigate the truth." Few would have challenged Floto's argument that scholars should be sufficiently "reasonable" and "honest" to check the accuracy of their interpretations. Likewise, Floto's comparison of the historian to a judge was as conventional as the implication of this image: historians should never be "indolent" or "partial" and guard themselves especially against "ecclesial or political party considerations." Even when portraying the historian as a scholar excelling in criticism, Floto felt that he was stating the obvious: "There has never been a historian who has not, in one way or another, exercised criticism" (Floto 1856a: 14, 16, 15, 21, 17, 7).

Yet, on the other hand, the virtues that Floto advocated were not without implications: they corresponded to a distinct scholarly persona. For him, virtues of criticism distinguished the "real historian" from the "popular" one. Moreover, despite the fact that "'criticism" was the watchword of historians such as Dümmler, who focused their attention on minutiae of source criticism, Floto dissociated himself from them by arguing that criticism encompasses more than determining the reliability of primary source material. *Kritik* includes both "dissecting" and "creating," that is, both analysis and synthetic vision (seeing the past appear "before one's eyes," as he put it in typically Rankean terms) (Floto 1856a: 17, 9). For this reason, Floto highlighted the need for historians to familiarize themselves with "human nature" by participating actively in societal life. Over against the proverbial armchair scholar, he held up the example, not of Ranke, but of Edward Gibbon, whom Floto, the former infantry officer, assumed to have learned more from his year in the South Hampshire militia than he could ever have learned from reading another 20 folios (Floto 1856a: 18). Floto, in other words, depicted the critical historian as a figure in between the popular history writer, on the one hand, and the philological critic, on the other.

Moreover, for Floto, the virtues of the critical historian were colored by confessional allegiance. Drawing on the liberal Protestant view that Luther's Reformation had been a fight for freedom of conscience, Floto argued that Protestants were in a privileged position to exercise historical criticism, because they had "least reason to conceal the truth" (Floto 1856a: 19). This was a thinly veiled way of saying that Catholics, to the extent that they were obliged to obey the church, could not be impartial or objective – especially not in studying an age like Henry IV's, in which the church had been so powerful. Similar views were articulated by Waitz, who argued that Catholics could enter the historical profession only by exchanging the Catholic vice of prejudice for the Protestant virtue of objectivity, and by later scholars such as Max Lenz, who as late as 1902, in response to the appointment of a Catholic historian in Freiburg, repeated that only "the spirit of the Reformation" allowed "the will to objectivity" to flourish (Lenz 1902: 30). Clearly, then, Floto's historiographical virtues were charged with religious and political meaning.

Arguably, this is one of the reasons that virtues and vices are as of late increasingly receiving scholarly attention (Creyghton et al. 2016; Paul 2016). Among other things, the prism of virtues and vices allows historians of historiography to situate Floto and his colleagues in what Manfred Hettling and Stefan-Ludwig Hoffmann (2000: 9) call Germany's "middle class universe of values." It draws attention to overlap and interplay between qualities associated with a good historian, on the one hand, and those cultivated by middle class citizens, on the other. Historians, after all, were not the only ones who valued virtues like "industriousness" (*Fleiß*) and "loyalty" (*Treue*). In addition, Floto's portrait of the historian as a "man of the world" resonated with masculine identities cherished among politicians and more broadly in the public sphere (Schnicke 2015b). All this suggests that the qualities regarded as characteristic of good historians were more than merely "epistemic virtues." In a society where virtues were central to moral, political, and religious discourse alike, no historian could speak about virtues without invoking connotations that reached beyond the realm of knowledge production (Paul 2019b).

A Broken Career

When Floto delivered his inaugural, in May 1856, it looked like his career was developing successfully. He had been appointed to a chair at age 30, 6 years below the average age at which historians entered the professoriate (Blanke 1989: 359). In addition to his regular teaching, which mostly consisted of survey courses on early modern European history, he engaged with broader local audiences in public lectures on Dante's *Divina Commedia*. Published in 1858, the lectures were received warmly by Robert Prutz, the poet and professor of literature in Halle, who spoke highly of Floto's "thoroughness" and "critical sharpness" (P[rutz] 1860: 663). His fame even spread abroad, judging by Floto being elected as corresponding member of a Dutch literary society.

But then misfortunate struck. In May 1857, the rising young scholar was hit by a stroke that left him half paralyzed and unable to speak, due to what a local physiologist diagnosed as cerebral softening. Initially, Floto's prospects did not look bright. Writing to a friend, the classical scholar Johann Jakob Bachofen reported that while bodily recovery was conceivable, "though not likely," "mentally he will never recover." Against all odds, however, Floto sufficiently recovered, at least in his own perception, to resume his work. Confidently, he announced new lectures on the Dutch Revolt, the French Revolution, the Reformation, and even "The Beginnings of Roman History, Critically Treated." Local authorities, however, noted more reservedly that Floto's hope of recovery was "not supported by judgments of doctors." This reservation turned out to be justified. After some difficult years, Floto had to retire (Roth 1935: 69, 79–80).

For the unfortunate historian, this marked the beginning of a wandering existence, marked by personal and professional difficulties that this chapter will not attempt to trace in any detail. A future biographer may want to examine how Floto tried to resume his teaching career at a girl's school in Berlin, while also making a vain attempt to get back into research by working on a prize contest on the history of the Hanseatic League. Likewise, this future biographer may want to investigate what happened to Floto in Göttingen, where the partly recovered historian was fortunate enough to get an honorary teaching position, yet managed to ruin his prospects by running up debts that became the talk of the town (Floto 1872). (What didn't help either was that Floto, not known for his expertise in other areas than eleventh-century history, offered a lecture course on ancient Egypt at a university where Heinrich Brugsch, the famous Egyptologist, already taught an intensive, source-based course on Egyptian monuments. Known as a stronghold of historical criticism, the Georg August University was used to more research-oriented teaching than Floto seemed to realize.)

Source Editing Projects

Despite all of these failures, Floto eventually landed in a project that is relevant to this chapter, as it allows discussion of the emergence of research institutions – "big humanities" projects, as they are sometimes called – that played no small role in the transformation of nineteenth-century historical studies (Saxer 2010). Such projects included the *Monumenta Germaniae Historica*, founded in 1819 with the aim of publishing sources pertaining to Germany's medieval past. Initially a small, private initiative, the Monumenta developed into a publishing enterprise that employed lots of recently graduated historians as research assistants, especially after Waitz took over leadership from Georg Heinrich Pertz in 1875. In the meantime, other source editing projects had been launched, such as the *Deutsche Reichstagsakten* (German Reichstag Records), which had started under auspices of the Bavarian Academy of Science in 1857, and the *Hansisches Urkundenbuch* (Hanseatic Book of Records), for which the Hanseatic Historical Association had been sending out researchers to archives across Europe (Paul 2017). These projects were important not only because they made a wealth of source material available in print, but also because they offered employment to young historians, while socializing them into an ethos of philologically oriented research (Saxer 2014: 146–55). Notably, between the 1870s and the 1890s, the percentage of historians employed in a source editing project prior to completion of their *Habilitationsschrift* (the second dissertation required in German academia) doubled from 16 to 32. In that same period, the percentage of historians who wrote their second dissertation while teaching at a *Gymnasium* dropped from 20 to 3 (Weber 1984: 122). This implies that young historians were increasingly trained in research, at the cost of gaining teaching experience.

Although it was argued above that Floto embodied the priority given to teaching over research that was common in his generation, this priority had begun to change, even for Floto himself, after his retirement from Basel. Arguably, the job market was one contributing factor: job opportunities for researchers were less scarce than those for academic teachers. Indeed, the demand for scholars able to devote themselves for years to research in far-away archives was such that Sybel, the director of the Prussian state archive in Berlin, was able to offer Floto a way out of his professional cul-de-sac. In 1878, he announced that the retired Basel professor would start working on a research project that was expected to result in a three-volume *Geschichte des Deutschen Ordens in Preussen bis 1525: Nach den Acten vornehmlich des Königsberger Archivs* (History of the Teutonic Order in Prussia until 1525: Based on Records Mainly from the Königsberg Archive) (von Sybel 1878: ix). Although funding was not immediately available, 2 years later, Sybel managed to get Floto awarded a 6-year stipend for research in Königsberg. Delivered from his financial struggles, the 55-year-old historian relocated to the East Prussian city and started working on the rich collections of the Teutonic Order. Yet he did not live to complete the project: Floto unexpectedly died in 1881.

Despite this tragic course of events, Floto's move to Königsberg testified to the growing importance of source editing projects. Armed with government funding, such research institutions allowed figures like Sybel to hire staff, edit book series, and thereby shape their field to a greater extent than they could ever have done in university positions. Projects like the *Acta Borussica*, on the history of Prussia, even developed into little academic kingdoms, ruled by men powerful enough to make or break careers (Neugebauer 2000). Work in such hierarchical settings did not fail to leave its marks on young researchers. As critics pointed out, historians used to spending years on medieval charters ran a risk of becoming *Urkundionen* – Burckhardt's term for scholars who are so absorbed in minutiae of source criticism that they "consider themselves superior to everyone if they have found out that Emperor Conrad II went to the toilet at Goslar on May 7, 1030." Other critics feared that the quasi-industrial organization of source editing projects would turn historians into "factory workers," used to obeying orders instead of thinking for themselves (Paul 2013). Thus, whereas editing projects, on the one hand, expanded historians' job opportunities, while strengthening the profession's research orientation, there was also, on the other hand, a chorus of voices that wondered to what extent the industry, perseverance, and sense of duty cultivated in such contexts could rank as virtues, especially if they were practiced to the point of turning historians into narrow specialists unable to teach broad survey courses or to write a book like Floto's *Kaiser Heinrich der Vierte*.

Conclusion

So what does the case of Floto tell us about nineteenth-century historical studies? His life and work illustrate at least six important features of German historical scholarship around the mid- century. (1) His example shows, first of all, to what extent historians were teachers, in the sense that transmission of historical knowledge, to students as well as to audiences outside of the classroom, was what they regarded as their primary task. (2) As demonstrated by the reception of Floto's book on Henry IV, this priority of teaching shaped readers' expectations of historical monographs. Argumentative clarity and an attractive style of writing were valued more highly than critical study of source material, even though it would not take long for these criteria to change. If these findings already suggest that mid-nineteenth-century historians should be situated firmly in their social contexts – they were middle class citizens first, academics only in the second place – the polemics in which Floto engaged also reveal (3) how frequently historians drew on broadly shared stereotypes of North and South, Protestant and Catholic, or middle class and working class. Political and religious fault lines of the kind that would become central to Bismarck's *Kulturkampf* left their mark on the historical profession, to the point of Protestants denying their Catholic colleagues the very ability to be "objective." (4) Even if this virtue of objectivity was a relative late-comer (Daston 2014), it is significant that virtues (*Tugenden*) was historians' preferred idiom for talking about the qualities needed for historical inquiry, even if alternative vocabularies, such as language of "methods," gradually gained in importance. More than anyone else, Ranke was perceived as embodying this virtue of objectivity, although he had initially, around mid-century, been better known for his aversion to moralizing history writing. (5) Floto's memories of his study with Ranke point to one of the most important factors that brought Rankean historiography to prominence: historical exercises that familiarized students with primary source material, while sharpening their critical gaze. Finally, (6) the source editing project to which Floto devoted the last years of his life illustrates the growing significance of "big humanities" projects in historical scholarship. If, in the course of the nineteenth century, the priority of teaching over research was gradually reversed, this was not in the last place due to projects that socialized ever-larger numbers of young historians into an ethos of meticulous source criticism.

There were other developments in German historical studies that we see less clearly mirrored in Floto's life and work. Although he gave some virulently anti-Catholic speeches in the 1870s, Floto was not a "political professor" of the kind that Droysen or Sybel was. Still, even if Floto was not active in the political arena, his commitment to the nineteenth-century project of anchoring national identity in a rose-colored past (Berger and Conrad 2015) was obvious, especially in his monograph on Henry IV. Secondly, Floto's career reveals only little about local and regional historical associations of the kind that emerged almost everywhere in the nineteenth century (organizations that offer yet another example of scholars and "friends of history" occupying one and the same social space [Clemens 2004]). Although Floto joined at least two associations – the Historical Society in Basel and the Society for German Cultural History – he does not seem to have been an active participant. Something similar applies to the archival institutions that became increasingly important players in German historical studies (Müller 2019). Likewise, Floto did not live to see the methodological battles prompted by Karl Lamprecht's forays into social psychology – resulting in a type of history that was much more responsive to the emerging social sciences than traditional political history – or to witness the growing international reputation of German historiography (e.g., Lingelbach 2002).

Finally, the case of Floto hardly touches on a theme that has recently developed into a subject of research: historians' memory cultures, including their habit of honoring deceased scholars with often lengthy obituaries (Tollebeek 2015). In nineteenth-century Germany, social conventions required historians to commemorate especially their former teachers in public, as a sign of gratitude for the education they had received. Floto, however, did not live long enough to write an obituary for Ranke (who died 5 years after his student, at the advanced age of 90). Also, because his teaching in Basel had been too short to yield any doctoral dissertations, Floto did not have any Swiss students to erect a "literary monument" to him. Indeed, it seems as if his passing went largely unnoticed by the profession to which he had belonged. Fourteen years after Floto's death, his name was, painfully, still listed among the corresponding members of the Society of Dutch Literature in Leiden. If few historians of his generation had been more fortunate at the beginning of their careers than Floto, few historians' lives eventually ended as lonely as his.

Acknowledgments Funding was generously provided by the Dutch Research Council (NWO).

References

Applegate C (1990) A nation of provincials: the German idea of Heimat. University of California Press, Berkeley

Berg G (1968) Leopold von Ranke als akademischer Lehrer: Studien zu seinen Vorlesungen und seinem Geschichtsdenken [Ranke as academic teacher: studies into his lectures and his historical thinking]. Vandenhoeck & Ruprecht, Göttingen

Berger S, Conrad C (2015) The past as history: national identities and historical consciousness in modern Europe. Palgrave Macmillan, Basingstoke

Berger S, Melman B, Lorenz C (2012) Introduction. In: Berger S, Melman B, Lorenz C (eds) Popularizing national pasts: 1800 to the present. Routledge, New York, pp 1–33

von Bethusy [Huc] [E] (1849) Bericht über die Königliche Ritter-Akademie zu Liegnitz von Ostern 1848 bis dahin 1849 [Report about the Royal Knight Academy at Liegnitz from Easter 1848 to Easter 1849]. W. Pfingsten, Liegnitz

Blanke HW (1989) Historiker als Beruf: Die Herausbildung des Karrieremusters 'Geschichtswissenschaftler' an den deutschen Universitäten von der Aufklärung bis zum klassischen Historismus [Historian by profession: the creation of a career pattern for historians at German universities from the Enlightenment to classical historicism]. In: Jeismann KE (ed) Bildung, Staat, Gesellschaft im 19. Jahrhundert: Mobilisierung und Disziplinierung [Education, state, society in the nineteenth century: mobilizing and disciplining]. Franz Steiner, Stuttgart, pp 343–360

Blanke HW (1991) Historiographiegeschichte als Historik [The history of historiography as philosophy of history]. Frommann-Holzboog, Stuttgart-Bad Cannstatt

Brechenmacher T (1996) Großdeutsche Geschichtsschreibung im neunzehnten Jahrhundert: Die erste Generation (1830–48) [Greater German historiography in the nineteenth century: the first generation (1830–48)]. Duncker & Humblot, Berlin

Clemens GB (2004) Sanctus amor patriae: Eine vergleichende Studie zu deutschen und italienischen Geschichtsvereinen im 19. Jahrhundert [Holy love for the fatherland: a comparative study of German and Italian historical associations in the nineteenth century]. Niemeyer, Tübingen

Creyghton C et al (2016) Virtue language in historical scholarship: the cases of Georg Waitz, Gabriel Monod and Henri Pirenne. Hist Eur Ideas 42:924–936

Daston L (2014) Objectivity and impartiality: epistemic virtues in the humanities. In: Bod R, Maat J, Weststeijn T (eds) The making of the humanities, vol 3. Amsterdam University Press, Amsterdam, pp 27–41

Dormeier H (1990) 'Nach Canossa gehen wir nicht!' Das Harzburger Bismarck-Denkmal im Kulturkampf ['We are not going to Canossa!' The Bismarck memorial in Bad Harzburg in the Culture Struggle]. Niedersächsisches Jahrbuch für Landesgeschichte [Lower Saxonian Yearbook for Regional History] 62:223–264

Eskildsen KR (2008) Leopold von Ranke's archival turn: location and evidence in modern historiography. Mod Intellect Hist 5:425–453

Eskildsen KR (2015) Private Übungen und verkörpertes Wissen: Zur Unterrichtspraxis der Geschichtswissenschaft im neunzehnten Jahrhundert [Private exercises and embodied knowledge: on educational practices in nineteenth-century historical studies]. In: Kintzinger M, Steckel S (eds) Akademische Wissenskulturen: Praktiken des Lehrens und Forschens vom Mittelalter bis zur Moderne [Academic cultures of knowledge: practices of teaching and research from the Middle Ages to the modern period]. Schwabe, Bern, pp 143–161

Eskildsen KR (2019) Leopold von Ranke (1795–1886): criticizing an early modern historian. Hist Humanit 4:257–262

Floto H (1847) De S. Annone [On Saint Anno]. Gustav Schade, Berlin

Floto H (1853) Handbuch für Subalternoffiziere bei der preußischen Landwehr-Infanterie und für einjährige Freiwillige bei der Linien-Infanterie [Manual for subalterns in the Prussian Landwehr infantry and for one-year volunteers in the line infantry]. Louis Nitze, Berlin

Floto H (1855) Kaiser Heinrich der Vierte und sein Zeitalter [Emperor Henry IV and his age], vol 1. Rudolf Besser, Stuttgart

Floto H (1856a) Ueber historische Kritik: Akademische Antrittsrede gehalten am 2. Mai c. in der Aula zu Basel [On historical criticism: academic inaugural address held on 2 May [1856] in the Auditory in Basel]. Bahnmaier, Basel

Floto H (1856b) Kaiser Heinrich der Vierte und sein Zeitalter [Emperor Henry IV and his age], vol 2. Rudolf Besser, Stuttgart

Floto H (1872) Erklärung [Statement]. Göttinger Zeitung [Göttingen Newspaper] (8 February)

Gerhards T (2013) Heinrich von Treitschke: Wirkung und Wahrnehmung eines Historikers im 19. und 20. Jahrhundert [Heinrich von Treitschke: influences and perceptions of a historian in the nineteenth and twentieth centuries]. Ferdinand Schönigh, Paderborn

Hettling M, Hoffmann SL (2000) Zur Historisierung bürgerliche Werte: Einleitung [Historicizing bourgeois values: introduction]. In: Hettling M, Hoffmann SL (eds) Der bürgerliche Wertehimmel: Innenansichten des 19. Jahrhunderts [The bourgeois universe of values: inside views of the nineteenth century]. Vandenhoeck & Ruprecht, Göttingen, pp 7–21

Jørgensen CM (2012) Scholarly communication with a political impetus: national historical journals. In: Porciani I, Tollebeek J (eds) Setting the standards: institutions, networks and communities of national historiography. Palgrave Macmillan, Basingstoke, pp 70–89

Kirchner C (1843) Die Landesschule Pforta in ihrer geschichtlichen Entwicklung seit dem Anfange des XIX. Jahrhunderts bis auf die Gegenwart: Einladungsschrift zur dritten Säcularfeier ihrer Stiftung den 21. Mai 1843 [The Pforta state school in its historical development from the early nineteenth century to the present: invitation to the third centenary celebration of its foundation on 21 May 1843]. Karl Aug. Klassenbach, Naumburg

Lenhard-Schramm N (2014) Konstrukteure der Nation: Geschichtsprofessoren als politische Akteure in Vormärz und revolution 1848/49 [Builders of the nation: professors of history as political actors before and during the 1848/49 revolution]. Waxmann, Münster

Lenz M (1902) Römischer Glaube und freie Wissenschaft [Roman faith and free science]. Herm. Walther, Berlin

Liepmann M (ed) (1916) Von Kieler Professoren: Briefe aus drei Jahrhunderten zur Geschichte der Universität Kiel [On professors in Kiel: letters from three centuries about the history of Kiel University]. Deutsche Verlags-Anstalt, Stuttgart

Lingelbach G (2002) The historical discipline in the United States: following the German model? In: Fuchs E, Stuchtey B (eds) Across cultural borders: historiography in global perspective. Rowman & Littlefield, Lanham, pp 183–204

Lingelbach G (ed) (2006) Vorlesung, Seminar, Repetitorium: Universitäre geschichtswissenschaftliche Lehre im historischen Vergleich [Lecture, seminar, examination review course: a historical comparison of academic history teaching]. Meidenbauer, Munich

Middell M (2010) Germany. In: Porciani I, Raphael L (eds) Atlas of European historiography: the making of a profession. Palgrave Macmillan, Basingstoke, pp 159–166

Mommsen WJ (1998) Der Historismus als Weltanschauung des aufsteigenden Bürgertums [Historicism as the worldview of the emerging bourgeoisie]. In: Blanke HW, Jaeger F, Sandkühler T (eds) Dimensionen der Historik: Geschichtstheorie, Wissenschaftsgeschichte und Geschichtskultur heute: Jörn Rüsen zum 60. Geburtstag [Dimensions of philosophy of history: historical theory, history of science and historical culture today: for Jörn Rüsen at his sixtieth birthday]. Böhlau, Cologne, pp 383–394

Muhlack U (2001) Der 'politische professor' im Deutschland des 19. Jahrhunderts [The 'political professor' in nineteenth-century Germany]. In: Burkholz R, Gärtner C, Zehentreiter F (eds) Materialität des Geistes: Zur Sache Kultur: Im Diskurs mit Ulrich Oevermann [The materiality of the spirit: on the issue of culture: in conversation with Ulrich Oevermann]. Velbrück, Weilerswist, pp 185–204

Müller P (2019) Geschichte machen: Historisches Forschen und die Politik der Archive [Making history: historical research and the politics of the archives]. Wallstein, Göttingen

N. N. (1855) Review of Floto 1855. Schwäbischer Merkur [Schwabian Mercury] (28 October)

N. N. (1856a) Schweiz. Schwäbischer Merkur [Schwabian Mercury] (7 May)

N. N (1856b) Review of Floto 1855. Leipziger Repertorium der deutschen und ausländischen Literatur [Leipzig Directory of German and Foreign Literature] 14:282–283

N. N (1856c) Review of Floto 1855. Heidelberger Jahrbücher der Literatur [Heidelberg Yearsbooks of Literature] 49:118–121

N. N (1858a) Floto's history of the emperor Henry IV. The Saturday Review:398–399

N. N. (1858b) Kaiser Heinrich IV. und sein Zeitalter, von Hartwig Floto [Emperor Henry IV and his age, by Hartwig Floto]. Historisch-politische Blätter für das katholische Deutschland [Historical-Political Papers for Catholic Germany] 41:445–471

Neugebauer W (2000) Die 'Schmoller-Connection': Acta Borussica, wissenschaftlicher Großbetrieb im Kaiserreich und das Beziehungsgeflecht Gustav Schmollers [The 'Schmoller connection': Acta Borussica, big science in the Wilhelmine Empire, and Gustav Schmoller's network of relations]. In: Kloosterhuis J (ed) Archivarbeit für Preußen: Symposion der Preußischen Historischen Kommission und des Geheimen Staatsarchivs Preußischer Kulturbesitz aus Anlass der 400. Wiederkehr der Begründung seiner archivalischen Tradition [Archival work for Prussia: symposium of the Prussian Historical Commission and the Prussian Privy State Archive on the occasion of the four hundredth anniversary of the establishment of its archival tradition]. Geheimes Staatsarchiv Preußischer Kulturbesitz, Berlin, pp 261–301

Nissen M (2009) Populäre Geschichtsschreibung: Historiker, Verleger und die deutsche Öffentlichkeit (1848–1900) [Popular historical writing: historians, publishers, and the German public sphere (1848–1900)]. Böhlau, Cologne

P[rutz] R (1860) Zur Erklärung Dante's [On explaining Dante]. Deutsches Museum [German Museum] 10:663–665

Pandel HJ (1993) Wer ist ein Historiker? Forschung und Lehre als Bestimmungsfaktoren in der Geschichtswissenschaft des 19. Jahrhunderts [Who is a historian? Research and teaching as determining factors in nineteenth-century historical scholarship]. In: Küttler W, Rüsen J, Schulin E (eds) Geschichtsdiskurs [Historical discourse], vol 1. Fischer, Frankfurt am Main, pp 346–354

Paul H (2013) The heroic study of records: the contested persona of the archival historian. Hist Hum Sci 26(4):67–83

Paul H (2016) The virtues and vices of Albert Naudé: toward a history of scholarly personae. Hist Humanit 1:327–338

Paul H (2017) Hunting for sources: dreams and realities of nineteenth-century archival travel. In: Martin AE, Missinne L, van Dam B (eds) Travel writing in Dutch and German, 1790–1930: modernity, regionality, mobility. Routledge, New York, pp 95–113

Paul H (2018) The virtues of a good historian in early imperial Germany: Georg Waitz's contested example. Mod Intellect Hist 15:681–709

Paul H (2019a) Ranke vs Schlosser: pairs of personae in nineteenth-century German historiography. In: Paul H (ed) How to be a historian: scholarly personae in historical studies, 1800–2000. Manchester University Press, Manchester, pp 36–52

Paul H (2019b) Germanic loyalty in nineteenth-century historical studies: a multi-layered virtue. História da Historiografia [History of Historiography] 12:16–43

Paul H (ed) (2022) Writing the history of the humanities: questions, themes and approaches. Bloomsbury, London

Porciani I, Raphael L (eds) (2010) Atlas of European historiography: the making of a profession. Palgrave Macmillan, Basingstoke

Ranke L (1949) Neue Briefe [New letters], ed. Hoeft B, Herzfeld H. Hoffmann und Campe, Hamburg

Roth P (1935) Aktenstücken zur Laufbahn Jacob Burckhardts [Archival records regarding Jacob Burckhardt's career]. Basler Zeitschrift für Geschichte und Altertumskunde [Basel Journal for History and Archaeology] 34:5–106

Rüsen J (1993) Konfigurationen des Historismus: Studien zur deutschen Wissenschaftskultur [Configurations of historicism: studies on German scientific culture]. Suhrkamp, Frankfurt am Main

Saxer D (2010) Monumental undertakings: source publications for the nation. In: Porciani I, Tollebeek J (eds) Setting the standards: institutions, networks and communities of national historiography. Palgrave Macmillan, Basingstoke, pp 47–69

Saxer D (2014) Die Schärfung des Quellenblicks: Forschungspraktiken in der Geschichtswissenschaft 1840–1914 [Training the eye for sources: research practices in historical scholarship 1840–1914]. Oldenbourg, Munich

Schnicke F (2015a) Rituale der Verkörperung: Seminarfeste und Jubiläen der Geschichtswissenschaft des 19. Jahrhunderts [Rituals of embodiment: seminar anniversaries and jubilees of nineteenth-century historical scholarship]. Zeitschrift für Geschichtswissenschaft [Journal for Historical Scholarship] 63:337–358

Schnicke F (2015b) Die männliche Disziplin: Zur Vergeschlechtlichung der deutschen Geschichtswissenschaft 1780–1900 [The male discipline: on gendering German historical scholarship 1780–1900]. Wallstein, Göttingen

Smith R (2007) Being human: historical knowledge and the creation of human nature. Columbia University Press, New York

von Sybel H (1878) Prospect [Prospectus]. In: Lehmann, M, Preussen und die katholische Kirche seit 1640: Nach dem Acten des Geheimen Staatsarchives [Prussia and the Catholic Church since 1640: based on records from the Privy State Archive], vol 1. S. Hirzel, Leipzig, pp v–x

Toews JE (2004) Becoming historical: cultural reformation and public memory in early nineteenth-century Berlin. Cambridge University Press, Cambridge

Tollebeek J (2015) Commemorative practices in the humanities around 1900. Adv Hist Stud 4:216–231

Torstendahl R (2014) The rise and propagation of historical professionalism. Routledge, New York

Ungureanu JC (2019) Science, religion, and the protestant tradition: retracing the origins of conflict. University of Pittsburgh Press, Pittsburgh

Weber W (1984) Priester der Klio: Historisch-sozialwissenschaftliche Studien zur Herkunft und Karriere deutscher Historiker und zur Geschichte der Geschichtswissenschaft 1800–1970 [Priests of Clio: historical social scientific studies on backgrounds and careers of German historians and the history of historical scholarship 1800–1970]. Peter Lang, Frankfurt am Main

Part II

Visualizations

Anatomy: Representations of the Body in Two and Three Dimensions

10

Anna Maerker

Contents

Introduction	230
A Brief History of Anatomical Representations in Europe	231
Early Historiography	235
The Politics of Representation, Identity, and Boundary Work	236
Gender	237
Race	239
Humans and Animals	243
Normality and Productivity	245
Artists and Anatomists: Iconography and Practices of Making	246
Epistemology and Education	248
Reception and Use	250
The Politics of Class and Professional Identity	252
Art and Anatomy	253
Conclusion	254
References	254

Abstract

This chapter traces the history of anatomical representations in Western medicine and culture since the Renaissance, and assesses the changing analytical perspectives brought to images of the human body by historians of science, medicine, and art. In particular, the chapter focuses on the ways in which representations of the body have been used to define and demarcate humanity, and the role such images have played for creating and challenging boundaries of gender, race, and species. The chapter also investigates the role of anatomical images for articulations of

---

A. Maerker (✉)
History, King's College London, London, UK
e-mail: anna.maerker@kcl.ac.uk

© The Author(s), under exclusive licence to Springer Nature Singapore Pte Ltd. 2022
D. McCallum (ed.), *The Palgrave Handbook of the History of Human Sciences*,
https://doi.org/10.1007/978-981-16-7255-2_21

expertise, class, and professional status, as well as key scholarly works on the epistemology of images. It finishes with a brief account of contemporary interactions of art and anatomy.

Keywords

Anatomy · Art · Images · Models · Representations · Gender · Race · Class · Professionalization · Historiography

Introduction

Modern Western medicine rests on the assumption that knowledge of the interior of the human body is the foundation for understanding the body and its functions, and thus the basis of successful treatments. Anatomy, in particular, has become a foundational element of professional medical education, while images and models of the human interior play a key role in health education for children and lay audiences, and in public health messaging. At the same time, anatomy has played an important part as a path to understanding humanity. Both assumptions are frequently mutually reinforcing, and they raise crucial questions for historians of anatomy and anatomical representations.

1. The assumption that successful healing requires knowledge of the interior of the body is geographically and temporally specific. Many non-Western healing traditions do not accord such a central role to anatomy, and in European history this assumption emerged gradually, locally, and contingently.
2. A subset of this assumption is the claim that images and representations are indispensable (or at least suitable) tools for learning about the structure of the body. This claim is similarly historically emergent. The important textbook *Anathomia corporis humani* by Italian anatomist Mondino de Luzzi of 1316, for instance, did not contain images, but today no anatomical education can do without them. Their use has long engendered debates about the epistemology and pedagogy of representations: What do images and models show? What can they do and what are their limitations?
3. Anatomy played a key role in diverse modern attempts to define "being human." The motto "know thyself" (nosce te impsum) was adopted by European anatomists, and suggested that self-knowledge should include, or even center, an understanding of one's body. Beyond individual introspection, anatomy was a key instrument for defining humanity by articulating categories such as gender and race, the concept of normality, and the boundary between humans and animals. Images, in particular, played a central role for identity formation and boundary work. The present chapter will sketch the European tradition of anatomical representations before considering in more detail how scholars have approached this history and its role for defining the boundaries of humanity. The chapter highlights key concepts such as gender, race, and normality, and

investigates the role of artists and anatomist as well as epistemological and pedagogical issues raised in the production and use of anatomical representations. It also considers the role of representations for medical professionalization and the articulation of boundaries of class and expertise, before closing with observations on recent artistic responses to historical anatomical images.

A Brief History of Anatomical Representations in Europe

Representations of the interior of the body have been made for a wide range of purposes which continue to shape medical practice in different ways, as well as articulating our experience and understanding of being human and being in the world. In Ancient Babylon, for instance, inscribed clay models of the liver were used for the purpose of divination. Ancient Greek and Roman temples held votive offerings in the shape of human organs and body parts, from ears and hearts to wombs and feet. Dissection was largely restricted to animal bodies, but there is some evidence of the use of anatomical images for medical education. While no copy survives of Aristotle's book on anatomy, for instance, later references indicate that the work was illustrated. Mondino de Luzzi's text-based approach to anatomy teaching notwithstanding, numerous images of the body's interior were produced in the Middle Ages, both by Christian and Islamic scholars. These focused mainly on more or less schematic representations of key structures and processes such as the skeleton, organs, blood vessels, and pregnancy. Medieval artists also created early depictions of dissections and autopsies which celebrated scholarly achievements and the human body as a wonder of divine creation. In the Renaissance, scholars and artists revived a range of ancient practices of reading and understanding the body as well as developing new representational conventions and forms of realism. Philosophers, alchemists, and physicians embraced the ancient idea of the human body as a microcosm of creation which should be understood in relation to the body of the earth and the structure of the cosmos. Based on this idea, the practice of physiognomy promised to reveal a person's inner character and temperament through visible signs which could be read in the face of the individual. The early modern era also saw the development of new forms of anatomical representation for medical teaching and scholarship. The sixteenth-century anatomist Andreas Vesalius (1514–1564) purposefully styled himself a revolutionary with his use of detailed images of the body, expressed in the programmatic frontispiece of his magnum opus *De Fabrica* (1543) which placed anatomist and corpse at the very center of a bustling scene of anatomical instruction (Fig. 1). Images were central to Vesalius' scholarly agenda. His carefully designed representations of the human body in various stages of dissection, but alive and expressive, represented access to the body's interior in ways which went on to shape conventions of anatomical imaging for centuries, for instance, by removing evidence of the anatomist's labor from the final image. At the same time, his illustrations were shaped by artistic conventions of the time, and retained elements of the memento mori as moralizing reminders of human mortality (Sawday 1995). Like many of his contemporaries, Vesalius professed the motto

Fig. 1 Frontispiece, from Andreas Vesalius, *De humani corporis fabrica* (1543). Public Domain Mark 1.0

"nosce te ipsum" (know thyself), a maxim which continued to articulate anatomy's claims to moral instruction and self-improvement as well as practical medical knowledge well into the modern age.

This dual function continued to characterize anatomy not just in its stated motto, but also in the practices connected to it. In the centuries following Vesalius's programmatic work European cities became sites of public dissections where physicians turned the bodies of executed criminals into a moralizing spectacle of human mortality and the ingenuity of God's creation, performed before local dignitaries and students while simultaneously demonstrating the medical profession's privileges in accessing the body's interior. By the eighteenth century, anatomy became a central element of medical education, and work with real corpses was accompanied by a diverse array of teaching aids such as images, models, and preparations. The use of images and the act of drawing became important methods for students to familiarize themselves with anatomical structures (Berkowitz 2011). In an age before invasive surgery, however, the question remained what practical purpose anatomical

Fig. 2 Three anatomical dissections taking place in an attic. Colored lithograph by T. C. Wilson after a pen and wash drawing by Thomas Rowlandson, n.d. Public Domain Mark

knowledge served. The argument that the study of the body's interior provided superior understanding of healing was used by physicians to support their claims to professional preeminence over rivalling healers such as barber-surgeons and midwives whose access to dissections was usually limited. Anatomical study was also increasingly regarded as an indispensable element of socializing students into the profession: Dissection, in particular, was considered to provide future doctors and surgeons with the "necessary inhumanity" in their practice, and a sense of group identity among students (Fig. 2). Around 1800, the "Paris School" was widely considered the leading place for medical education, especially for its focus on anatomy and access to bodies for dissection. Other European countries were under pressure to create similar opportunities, for instance with the 1832 Anatomy Act in Great Britain which authorized the use of unclaimed bodies for medical education. Throughout the nineteenth century, technological developments made new details of the body visible, and allowed for their representation in new ways. Microscopic observation, for instance, had long been problematic as the quality of glass lenses led to visual artifacts such as chromatic aberrations, the appearance of colorful, blurred fringes due to the different refractions of light of different wavelengths. The development of new types of glass minimized this phenomenon, and the resulting achromatic microscopy reduced visual artifacts and thus facilitated the observation of minute anatomical structures. Another nineteenth-century technology, the microtome, enabled the creation of extremely thin slices of tissue which could be mounted on microscopy slides for close study. Stereoscopes created the appearance of three-dimensionality, while photography (discussed in more detail later in this chapter) promised a new method of apparently unmediated representation by replacing the eye and hand of the artist. However, educators frequently returned to traditional imaging techniques which allowed for clearer visual differentiation and identification, and illustrations based on drawings remained central to influential educational publications such as *Gray's Anatomy*, published continuously in many revised editions since 1858. New industrial processes of printing and papermaking such as steel engraving, lithography, and the invention of the rotary press which used continuous rolls of paper made illustrated anatomical textbooks and atlases more affordable than ever. Beyond the confines of the medical profession itself, anatomical images found diverse audiences through various popular displays.

Since the eighteenth century, earlier public performances of dissection gave way to displays of collections of delicate wax models which were put on display by enterprising showmen and -women in commercial venues such as Rackstrow's Museum, but also in emerging scientific and medical museums such as the Museo "La Specola" in late eighteenth-century Florence and the contemporaneous collection of the "Josephinum" surgical academy in Vienna (Craske 2011; Maerker 2011). Modern public health initiatives continued to circulate images of the body in a range of innovative formats. Plaster casts, waxes, or papier-mache models of syphilitic sores and the ravages of veneral diseases in popular displays advertised "for men only" drew attention to sexual health among men, and especially soldiers, while offering discreet medical services to the afflicted (Stephens 2011). More palatable were turn-of-the-century schematic images of the human body as a "palace of industry" which enforced individual responsibility for maintaining one's body in health and productivity (Sappol 2017).

The images generated by anatomy and related disciplines played a key role in attempts to define boundaries and hierarchies of humanity, from comparative images of human and animal embryos which foregrounded the similarities between species in the early stages of development to hierarchical arrangements of human difference designed to justify Europeans' claims to racial superiority in the age of empire. The twentieth century saw further development of technologies for visualizing bodily interiors, from the discovery and rapid adoption of x-rays to the introduction of ultrasound and MRI. At the same time, anatomists never abandoned the work with actual corpses and body parts. This continues to raise ethical questions concerning matters such as consent and the origin of specimens. Some university collections still contain the remains of patients who did not consent to donate their bodies, including victims of Nazi eugenics and experimentation, while more recently the "Bodyworlds" exhibition of plastinated corpses caused controversy with its use of bodies of unclear provenance. Such controversies have led to tighter legislation around the uses and display of human bodies across Europe and beyond (for instance, with the European Tissues and Cells Directive, adopted in 2004). Anatomical representations and simulations continue to be at the core of debates about medical education and the epistemology of images: Can imaging techniques replace work with the corpse altogether? Whose bodies should be represented to serve as models for instruction? At the same time, anatomical motifs are now firmly entrenched in popular culture and continue to inform images of humanity, whether it is the concept of the mind as a disembodied "brain in a vat," or the heart as a metaphor for human emotions and affection.

Early Historiography

Historical studies have developed a range of questions and approaches in their exploration of anatomy and anatomical representation through the ages. Medical professionals themselves created the first historical accounts of the discipline, its images, and their development. Early works frequently served as celebrations of medical achievements, or presented historical curiosities to fellow medics, such as Joseph Hyrtl's *Antiquitates anatomicae rariores* of 1835, which charted the earliest attempts to visualize the interior of the body from prehistory to Ancient Rome, described two Chinese anatomical figures, and included a chapter on "Anatome culinaris," the practice of culinary and sacrificial anatomy which used the preparation of animal bodies for cooking and religious ceremonies as an opportunity for anatomical observation.

A more systematic and ambitious early study which provided important impulses for research was the 1852 *History and Bibliography of Anatomic Illustration in Its Relation to Anatomic Science and the Graphic Arts* by physician and medical historian Ludwig Choulant (1791–1861). Choulant's *History* provided a brief account of anatomical illustrations both medical and artistic from antiquity to the early nineteenth century, followed by detailed analyses of individual anatomists and their key publications, selected for their "lasting influence and ... historic significance" (Choulant 1852: x). Choulant's work shaped the agenda of historical scholarship on anatomical images in various ways. His thorough bibliographical information was intended to facilitate future research as well as supporting collectors and connoisseurs. Choulant drew attention to the important role of artists and artisans not only for the historic development of anatomical images, but also for his own publication, which was richly illustrated with copies of historic images produced by a team of wood engravers, painters, and chromolithographers (Choulant 1852: xiv). However, he rejected a "mere copying of facts" (Choulant 1852: xi), and developed a broader analytical framework as well as value judgments concerning the scientific accuracy and artistic merit of different works. In particular, he proposed the classification of anatomical illustrations into three categories: "schematic" representations, exact depictions of a "particular subject," and images of the "ideal human type" based on the synthesis of multiple observations (Choulant 1852: 22). Only the latter, he suggested, was suitable for teaching purposes, and it required numerous observations based both on dissection and on "ardent artistic vision" (Choulant 1852: 23) which could amalgamate specific observations into a composite image of the ideal type. Choulant further proposed a chronology which divided the development of anatomical images into six distinct periods, distinguished by a combination of criteria from images' aesthetic characteristics, methods, and purposes to the relationship between artists and anatomists. Thus he began with the first period, characterized by schematic drawings in the Middle Ages before Berengario da Carpi, and ended with "modern times" since Samuel Thomas von Soemmering, a period he defined by a combination of anatomical truth and artistic beauty, by the adoption of new technologies such as lithography and steel engraving, and the creation of new histologic and microscopic representations which took anatomy beyond the limitations of the naked eye (Choulant 1852: 25–26). Choulant acknowledged that his selection was open to challenge, as his criteria of importance and "artistic merits" were subjective (Choulant 1852: xiii). Overall, his study highlighted a wide range of issues which would become important for future scholarship on anatomical images: the contribution of artists and artisans and their collaborative relationship with anatomists, the impact of new techniques and technologies, the role of pricing and marketing for the distribution and readership of illustrated publications, and the interaction between European imaging traditions and those of the Islamic world and China.

In the wake of Choulant's pioneering study, subsequent historical scholarship of the late nineteenth and early twentieth centuries provided increasingly detailed documentation of existing images, and traced their transmission across space and time. Among the most prominent were the works of Karl Sudhoff, including his study *Tradition und Naturbeobachtung in den Illustrationen medizinischer Handschriften und Frühdrucke vornehmlich des 15. Jahrhunderts* (Tradition and the Observation of Nature in Illustrations of Medical Manuscripts and Incunabula Mostly of the Fifteenth Century, 1907) and his *Beitrag zur Geschichte der Anatomie im Mittelalter speziell der anatomischen Graphik nach Handschriften des 9. bis 15. Jahrhunderts* (Contribution to the History of Anatomy in the Middle Ages, Especially of Anatomical Images After Manuscripts of the Ninth to Fifteenth Centuries, 1908). More recent research has developed a wider range of thematic foci, from the politics of anatomy to the historical epistemology of representation, the relationship between different visual traditions across the globe, and images' multiple uses as instruments of identity formation and boundary work relating to race, gender, and professionalization.

The Politics of Representation, Identity, and Boundary Work

The postwar period saw a rise in historical studies of anatomy and its images by academic historians rather than medical professionals. These scholars generally turned away from evaluating representational accuracy by modern standards. The social and cultural history of medicine emerging since the 1970s, in particular, rejected older traditions which focused on the lives and ideas of "great men," and developed new directions of research on topics such as the role of the patient in the medical encounter (Porter 1985), the development of medicine as a profession and discipline (Gelfand 1980), and the historicity of somatic experience (Duden 1998).

Medicine, from this perspective, was more than a progressive accumulation of ever more accurate knowledge of the human body and of ever more powerful techniques and technologies of imaging and healing. It was embroiled in and inseparable from its political, cultural, and social context. Historians highlighted how medical ideas and practices were used to develop, justify, and implement political agendas, from welfare policies to the support of imperial rule and eugenics. Anatomy, in particular, was reevaluated by social and cultural historians as a discipline whose practices, images, and concepts were deeply emmeshed in political action and ideology. For instance, political power, class, and poverty were key to debates about access to bodies for dissection and the dreaded practice of illegal body-snatching; access to dead bodies continued to be an expression of privilege and power well into the nineteenth century (Richardson 1987). Since the 1980s, historians of medicine increasingly engaged with new approaches to cultural history which broadened their attention beyond elite cultural forms, and with the work of Michel Foucault which drew attention to the politics of images by framing modern medicine as a quintessentially visual discipline and tying vision to new forms of knowledge and new forms of discipline (for the anglophone reception of Foucault see e.g. Rabinow 1991). Scholars interested in the visual culture of medicine challenged simplistic concepts of representational realism, showing that concepts and conventions of "realism," "objectivity," and "mimetic success" are always historically contingent (Sekula 1986; Gilman 1986; Jordanova 1989; Daston and Galison 2007; Maerker forthcoming). Work shaped by feminist and postcolonial approaches continues to highlight how images of the body and their production were inseparable from their political and social context.

As a result, interpretations of anatomical images adopted a more explicitly political focus, one which investigated the role of images for articulating concepts of gender, sexuality, race, and the productive body of industrial capitalism. In particular, this scholarship investigates the use of representations to define humanity through the articulation of boundaries and hierarchies, creating or challenging distinctions between men and women, the "civilised" and the "primitive," the normal and the abnormal, humans and animals, as well as men and machines.

Gender

Carolyn Merchant's influential and programmatic study *The Death of Nature: Women, Ecology, and the Scientific Revolution* (1980) argued that the subjugation of women and the exploitation of the environment were at the core of the birth of modern science. Gender inequality, in particular, was supported by and expressed through anatomical images. Important case studies developed this argument further since the 1980s with detailed analyses of specific anatomical representations.

Londa Schiebinger's 1986 study of Enlightenment anatomy argued that the decline of the traditional humoral concept of the body which had located fundamental differences between men and women in their humoral constitution prompted scientists to search for physical signs of difference between the sexes in their anatomy, and especially in the skeleton. Such efforts, Schiebinger claimed, had to be understood within the context of contemporary attempts to redefine women's place in society more generally. If the new political philosophy of Locke and his contemporaries aimed to develop an optimal social order on the basis of an appeal to nature and natural laws, then the continuing inequality of the sexes, and the exclusion of women from civil rights, had to be justified by natural inequalities. According to Schiebinger, Vesalius had made no explicit distinction between male and female skeletons in his foundational anatomical works in the sixteenth century. By the eighteenth century, however, anatomists paid close attention to signs of sexual difference. Samuel Thomas Soemmering, for instance, claimed to have achieved the representation of the "typical" female skeleton, while Marie Thiroux d'Arconville suggested that women had smaller skulls and a larger pelvis than men, confirming their intellectual inferiority and their natural destination for childbirth and motherhood. Subsequent studies have challenged Schiebinger's chronology (see, e.g., Stolberg 2003 for a study which locates the emergence of the two-body model in the seventeenth century), but her work has given crucial impulses to the historiography of anatomy and its images.

Similarly influential was Ludmilla Jordanova's *Sexual Visions* (1989) which analyzed a range of medical images with a focus on anatomical representations. She documented how eighteenth-century wax models presented images of men and women in strikingly different ways, showing male bodies in active poses, muscular and upright (Fig. 3), while female bodies were frequently depicted in repose, passive, and with a focus on pregnancy and the organs of reproduction. These images, Jordanova argued, both expressed and reinforced contemporary gender roles. Another one of her case studies, William Hunter's *The anatomy of the human gravid uterus exhibited in figures* (1774), further supported this claim. Often cited as a benchmark publication due to the high quality of its large-scale, life-sized images, Hunter's *Anatomy* made aesthetic choices which separated the female patient from the development of the fetus by reducing the female body to the gravid torso, devoid of head and limbs, truncated by raw cuts that enacted the violence of dissection (Fig. 4).

Subsequent studies of anatomical images in this analytical tradition have highlighted that the influence of gender stereotypes continued up to the present, and that it was not restricted to the medium of drawings, prints, and models. Medical photography, a technology which was often presented as objective and untainted by observers' subjective perceptions, was never detached from aesthetic and cultural conventions such as those of contemporary portraiture and the tradition of the freak show (Amirault 1993–4). Applied to the anatomy of the female body, photography was used to make the mother invisible once more in twentieth-century images of the embryo, free floating and detached from the mother's body which nourished and protected it (Jülich 2018). And indeed, attempts to identify visual markers of gender difference have also been traced down to the level of microscopic anatomy with early twentieth-century attempts to provide incontrovertible visual proof of individuals' "true" sexual identities in the face of complex and temporally fluid sexual characteristics at the macroscopic level (Löwy 2013).

Fig. 3 Wax male anatomical figure, Italy, 1776–1780. Attributed to Clemente Susini. Science Museum, London. Attribution 4.0 International (CC BY 4.0)

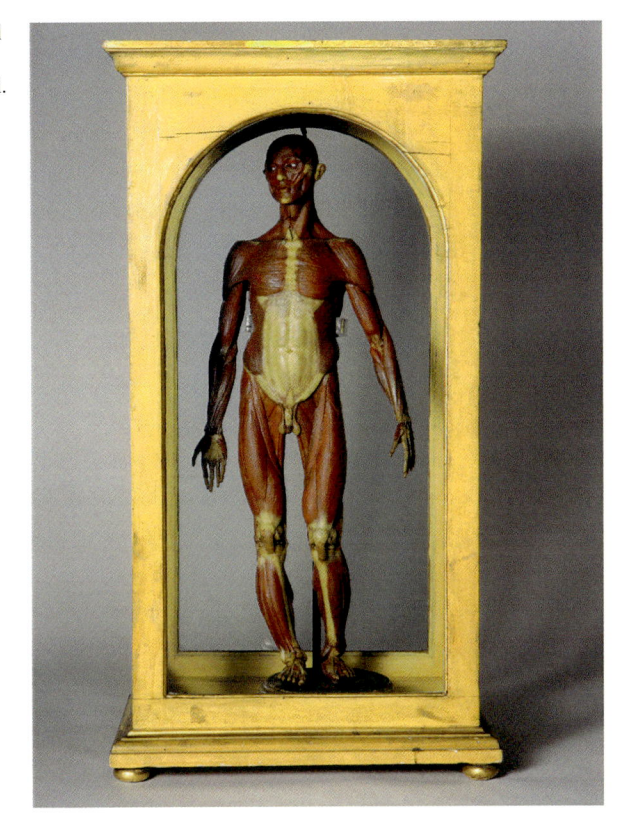

As many scholars have pointed out (see, e.g., Haraway 1989; Schiebinger 1990; Morgan 1997; Curran 2011), the oppression of women was frequently justified by comparing their supposedly inferior anatomy to that of children and "primitives." In the nineteenth century, such assertions were framed within the context of evolutionary theories, and women were considered to represent a less than fully developed human body, a more primitive stage in human evolution than the (White) male. Race as a marker of difference became another key category for the historiography of anatomical representation.

Race

With historians' turn to the political uses and implications of medicine, scholars have documented how Western medicine justified the violence of slavery and colonialism by postulating the existence of racial difference, and a racial hierarchy which placed Whites at the pinnacle of humanity. An early marker of racialized identity was skin color; for instance, in the depiction of South American "castas," degrees of mixed

Fig. 4 William Hunter, *The anatomy of the human gravid uterus exhibited in figures* (1774), plate VI: Dissection of the pregnant uterus, showing the foetus at 9 months. Copperplate engraving by R. Strange after I.V. Rymsdyk. Public Domain Mark

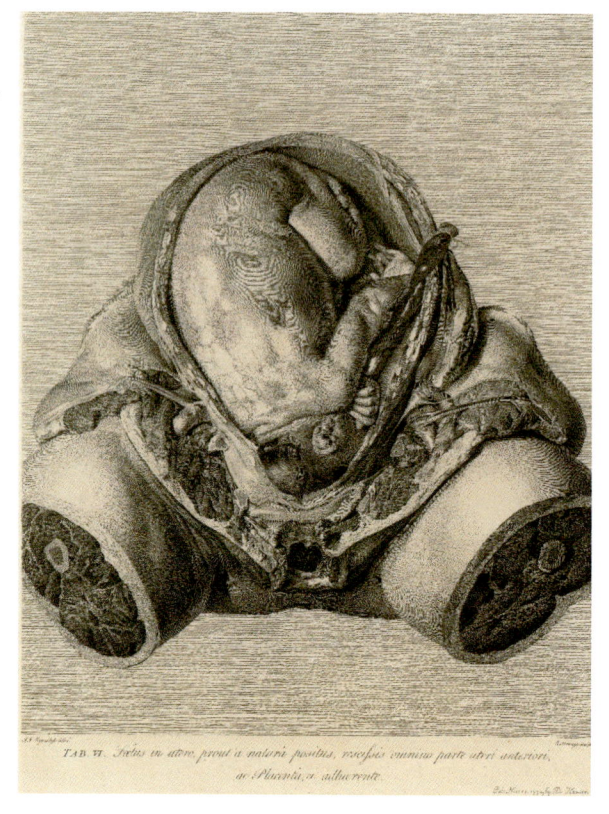

races (Katzew 2005; Martínez 2008). Increasingly, Europeans turned from exterior to interior difference. Anatomy, in particular, became central to arguing that racial difference was deeply embedded in the fabric of the body. As in the case of gender difference, the skeleton became a central focus for locating racial difference in the eighteenth century. The age of Enlightenment saw the adoption of cranial angles as markers to create racial hierarchy (Mosse 1978). The Dutch physician Petrus Camper (1722–1789), an early pioneer of the new discipline of "anthropology," used measurements of skulls to define races, and he used illustrations to reify not just racial difference, but also hierarchy, by suggesting that European cranial structure was closest to the ideal type of classical sculpture, while African and Asian skulls were closer to the cranial angles of apes. Camper presented this image mainly as an aesthetic hierarchy, and suggested that Europeans' perception of other races as less beautiful was due to "customs" (Meijer 1999: 164). However, in its reception the image was widely used to support ideas of innate superiority and inferiority more generally, implying that "inferior" races were more closely related to apes than the "superior" European type (Fig. 5). "Race" became a pervasive but unstable category, and anatomists developed a range of different, potentially conflicting taxonomies. In

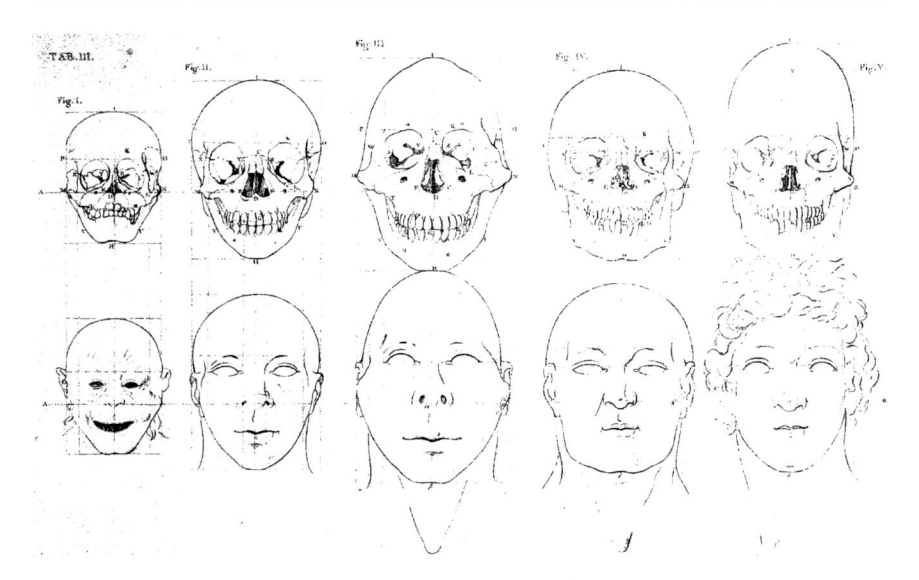

Fig. 5 Petrus Camper, Illustration from *The works of the late Professor Camper, on the connexion between the science of anatomy and the arts of drawing, painting, statuary,* etc. (1821). Public Domain Mark

the mid-nineteenth century, for instance, the anatomist Robert Knox (1791–1862) developed a taxonomy of human races which assigned to each racial type a specific set of characteristics, both physical and temperamental (Biddiss 1976). He illustrated his categories with a haphazard collection of images crudely copied and compiled from other publications, including ancient sculptures, skulls, and silhouettes, and used this classification to support anti-Semitism, and to advocate the extinction of the "Celtic" race whom he considered dangerous. At the same time, unlike many other "race scientists" of the period, Knox argued against slavery and colonialism. During the heyday of European imperialism in the nineteenth century, public displays visualized ideas of racial difference in the mould of earlier displays of monsters and curiosities. A famous example was the Khoikhoi woman Sarah Baartman, whose dehumanizing display as the "Hottentot Venus" to European audiences was used to support claims of African (and female) inferiority by drawing attention to visible differences in Baartman's body, especially those located in the buttocks and genitalia of the subject (Qureshi 2004; see also ▶ Chap. 12, "Visualizations in the Sciences of Human Origins and Evolution," by Sommer in this volume).

Anatomical studies attempted to locate racial difference in various parts of the human body. The structure of the skull, in particular, became central to the nexus of anatomy and humanity in the nineteenth century in different ways. Where measurements of cranial volumes and ratios were used to create racial categories, the brain became metonymic with character. Anatomists documented the key role of the brain as the seat of human knowledge and behavior. This focus was developed in different ways and for different audiences. Around 1800, anatomists Franz Joseph Gall and

Johann Gaspar Spurzheim promoted phrenology, based on the idea of brain anatomy as determining character and abilities, which could be "read" by measuring or feeling the shape of the skull (much as earlier physiognomists had sought those characteristics in the features of the face). Phrenology was quickly denounced by medical professionals and largely abandoned as a scientific theory by the 1840s as experimental evidence contradicted phrenology's central tenets (Staum 2003), but the practice remained popular throughout the nineteenth century as a pastime and form of self-help which generated its own visual conventions and material culture including the iconic phrenological head which mapped character traits onto the surface of the skull (Fig. 6) (Bittel 2019; Poskett 2019). Popular images mapping character features onto the shape of the skull, in particular, were circulated widely to enable laypeople to engage in self-diagnosis. This could be used for practical purposes, such as determining whether prospective spouses were a good match.

Fig. 6 Fowler's phrenological head, Staffordshire, England, late nineteenth century. Science Museum, London. Attribution 4.0 International (CC BY 4.0)

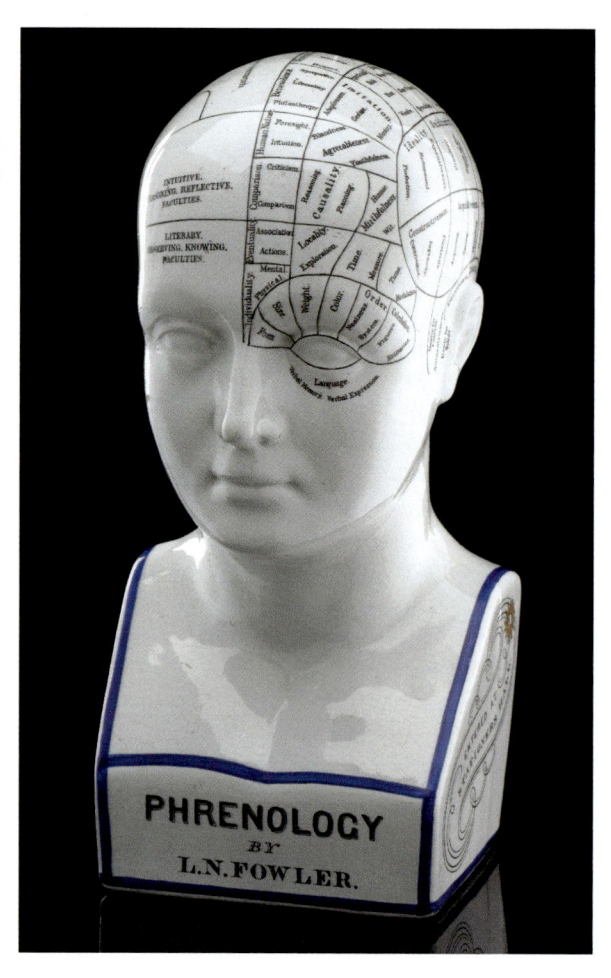

While medical researchers rejected the claims of the phrenologists, the central concept of the relationship between brain, skull, character, and cognition remained at the core of anatomical research, especially in the development of the technique of craniometry, or skull measurement. The French anatomist and anthropologist Paul Broca, whose central interest lay in studying the human capacity for language and its relation to the anatomy of the brain, contended that it was the shape of the brain which determined intellectual capability rather than its weight or volume. To collect data Broca developed craniometry and its instruments. This practice and the creation of large and diverse collections of skulls became central to the formation of the discipline of physical anthropology. Beyond skulls themselves, images and wax models were crucial for nineteenth-century constructions and contentions of the racialized and gendered brain as they were negotiated in scientific publications and articulated in popular displays such as anatomical waxworks and sideshows (Sowerwinea 2003; Blanckaert 2009; Pogliano 2019).

Craniometry was not the only new technique of the nineteenth century which was enrolled in the project of creating categories of human difference. Photography, for instance, was similarly used to generate racialized images of humanity. Much like earlier debates about the creation of anatomical images, photography did not escape the question whose bodies should be depicted, how, and what such photographs could actually represent. In the case of the new medium, it resulted in a conflict between advocates of composite photographs which amalgamated several subjects into attempts to represent general "criminal types," such as the English polymath Francis Galton (Fig. 7), and those who followed the pioneering police officer and biometrician Alphonse Bertillon in using photography as a means to identify individual criminals (Sekula 1986: 17–19).

Humans and Animals

Images of the body were central for the creation and delineation of anatomical categories of gender and race. Beyond such attempts to define difference between humans, anatomical images also played their part in defining the boundaries of humanity as a whole. Comparisons between human and animal bodies had long been central to attempts to capture humanity. The physiognomy of Renaissance scholars such as Giambattista Della Porta had used parallels between human and animal facial features to ascribe character: The fierceness of the lion and the stubborn character of the ox were inscribed in the human face. In this framework, the microcosm of the human body revealed analogous relationships with the macrocosm of creation, visible signs of character which could be read and interpreted to human benefit (Muratori 2017). The emergence of comparative anatomy in the eighteenth century problematized the boundary between humans and animals. Descriptions of monkeys and apes, in particular, offered a fruitful avenue for articulating humanity, variously emphasizing and minimizing anatomical differences between humans and animals in attempts to naturalize different ideas of human nature and social order. In the wake of Darwin's theory of evolution, in particular, similarities between humans

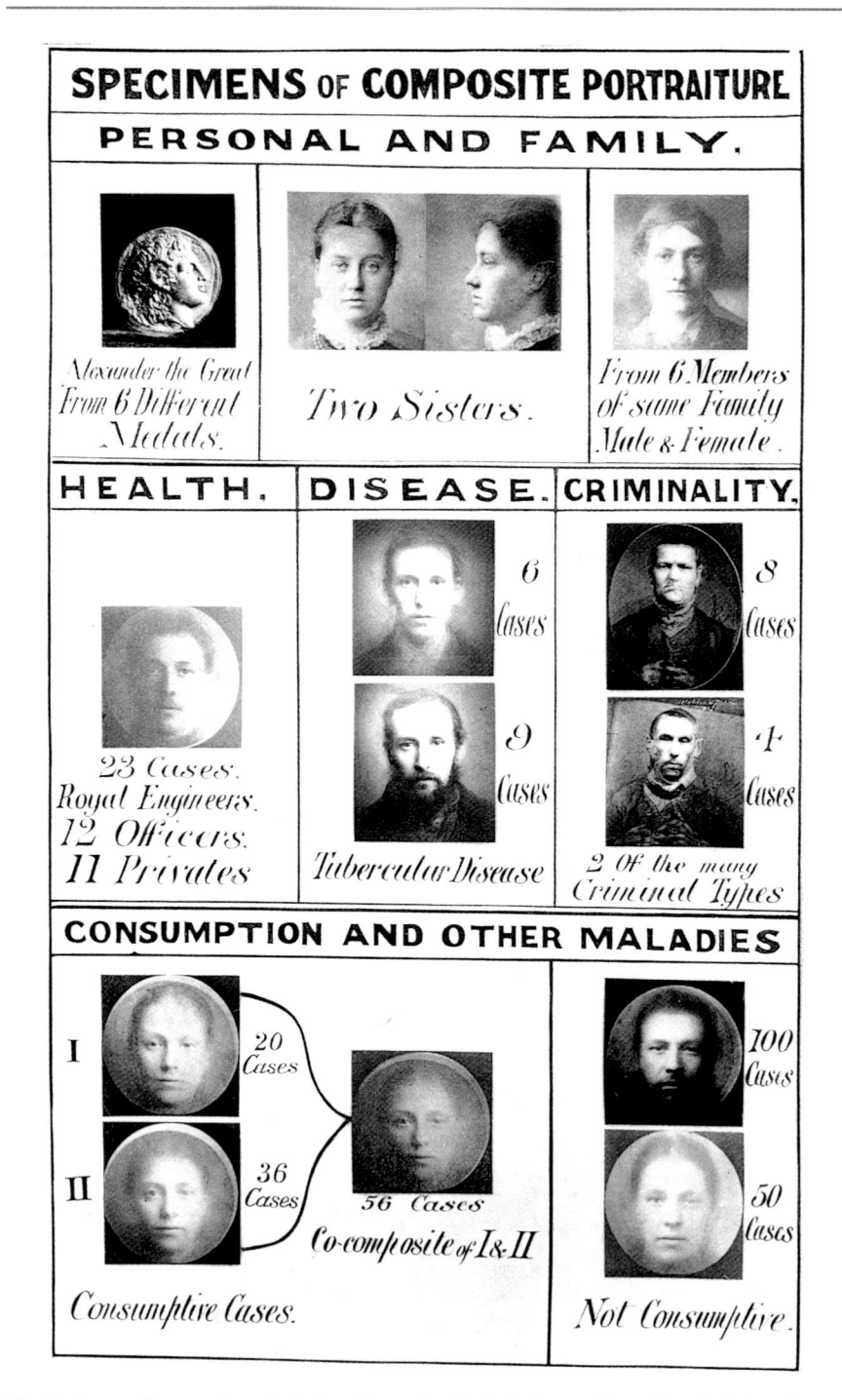

Fig. 7 Composite portraiture, including diseased and criminal types. From Francis Galton, *Inquiries into human faculty and its development* (1883). Public Domain Mark

and primates prompted heated debates about the savage and violent nature of humanity. These debates frequently centered around depictions of the newly discovered gorilla, variously enhancing and minimizing the giant ape's similarities with human anatomy (Voss 2010). As Donna Haraway has shown in her influential study *Primate Visions: Gender, Race, and Nature in the World of Modern Science* (1989), early twentieth-century displays of human and animal bodies staged in photographs and museum dioramas created a fertile nexus between race, gender, humanity, and animality which replicated and reinforced stereotypical assumptions about racial superiority, masculinity, motherhood, and the nuclear family.

Normality and Productivity

Anatomical images could also be used to draw boundaries which did not directly invoke categories of gender and race. In recent decades, historians have analyzed the use of images to chart the development of two central concepts applied to human bodies and human identities: the "monstrous" and the "normal." Studies of medieval and early modern anatomy, in particular, have traced the evolution of European conceptions of "monsters" from evidence of divine playfulness and divine signs (Park and Daston 1998) to an increasingly developmental understanding which ultimately interpreted the monstrous body as a disrupted and defective deviation from normal human development (Hagner 1999).

Critical studies of anatomy and modernity have identified the concept of the "normal" as central to the definition of the modern body (Cryle and Stephens 2017). Emerging in the nineteenth century, the *homme normal* was initially a statistically defined entity (Hacking 1990). However, historians have documented an increasing slippage between the original statistical definition and attempts at visualization which were applied to individual bodies and used in a normative fashion. Images, as studies such as Cryle and Stephens' suggest, were key to the process of rearticulating the concept of the normal body from a statistical and descriptive concept to one that was individualized and normative – and frequently visual. Such images could be used to support particular public health measures and to promote specific beauty ideals, but also collected in archives to generate an image of "society" and to be used as the basis for social policy. In interwar Vienna, for instance, the philosopher and sociologist Otto Neurath and his circle propagated the "Viennese method of pictorial statistics" ("Wiener Methode der Bildstatistik"), a "new proletarian style" of schematic representations of human anatomy and physiology designed to convey new, anti-bourgeois visions of communal urban living (Nemec 2020: 122–123).

In the age of industrialization, the early modern exhortation to "know thyself" was replaced by an ideology of individual responsibility for the maintenance of a healthy (and therefore productive) body. This development was accompanied and supported by the recurring use of the image of the body as a machine. This way of making the workings of the body intelligible had a long tradition – Leonardo da

Vinci, for instance, had used simplified mechanical drawings in his anatomical studies to understand the mechanics of motion. Early modern philosophers and physicians from Descartes to De La Mettrie had used the machine metaphor to assess the relationship between mind and body, while ingenious artisans like Jacques Vaucanson materialized the idea in the form of animated automata of humans and animals (Riskin 2016). In the nineteenth and twentieth centuries, this vision of the body-as-machine was used to create the concept of a specifically "industrial body" (Sappol 2017). This industrial anatomy found its expression in images such as the physician Fritz Kahn's "Man as industrial palace" (Der Mensch as Industriepalast) of 1926 which depicted the body as a factory, and its physiology as a series of carefully directed and supervised mechanical and chemical processes.

Artists and Anatomists: Iconography and Practices of Making

Beyond the politics of anatomical representation scholars have turned their attention to questions of style and iconography. Early analyses in art history compared the depiction of bodies in fine art and anatomy, highlighting stylistic and technical parallels such as the resonances of Bellini's peacefully ecstatic Saint Theresa in female wax anatomies of the eighteenth century (Bucci 1969). A well-established line of scholarship continues to explore the boundaries of art and anatomy with a focus on representational styles, conventions, and materials, as well as the role of anatomy in fine art (Schlosser 1911, Callen 1995, Petherbridge and Jordanova 1997, Didi-Huberman 1999, Kemp and Wells 2000). Other scholars have turned their attention from the style and content of images to their production. Following Choulant's lead, historians have investigated the collaboration between anatomists and artists, and have highlighted the potential tensions when it came to claims to expertise. Already in 1945, the influential medical historian Henry Sigerist acknowledged that "The many illustrators known and unknown have in all probability contributed infinitely more to medical science than we commonly assume" (Sigerist 1945: 207). In 1992, Roberts and Tomlinson continued the long tradition of anatomists exploring the history of their discipline, highlighting the difficulty of defining and achieving accuracy in any medium as well as the complex process of creation. More recently, historians have connected studies of anatomical representations to questions about the nature of expertise and cognitive authority: What was more important for the creation of useful and accurate anatomical images, anatomical knowledge, or artistic skills? Whose expertise was considered crucial for the creation of accurate anatomical representations? (Maerker 2011) Such studies frequently challenged and complicated established narratives about the role of anatomy for demarcations of gender and race. Lucia Dacome's investigation of the practices of female anatomical wax modeler Anna Morandi, for instance, showed how Morandi's claims to expertise were inextricably tied to her femininity. Indeed, the modeler used her identity as a woman to strengthen her anatomical authority, as motherhood had made her experience the malleability and formation of the human body in the most intimate way possible. Sewing and other domestic skills further contributed to the intricate labor of modeling (Dacome 2017).

The production and circulation of anatomical images and representations continues to be a fruitful avenue for historical research. Recent scholarship has drawn attention to aspects such as the role of three-dimensional models as "publications" which shaped the boundaries of the new discipline of embryology and the creation of its disciplinary status (Hopwood 2004;). Hopwood also reconstructed the circulation of Ernst Haeckel's famous comparative images of human and animal embryos to show how their reception and interpretation was shaped by complex trajectories across academic and popular media, and by numerous processes of copying (Hopwood 2015). For the early modern period, a recent census of Vesalius's iconic *De fabrica* by Dániel Margócsy, Mark Somos, and Stephen N. Joffe traced the book's long history as a commercial object, a teaching tool, and collectors' item (Margócsy, Somos, and Joffe 2018). Rather than creating permanent, static representations, images of the human emerge from such analyses as dynamic and historically contingent. Scholars have also highlighted the important role of the preservation and use of anatomical representations. Recent studies have shown the role of anatomical collections as "dynamic and flexible entities" (Huistra 2018: 2), open to modification and reinterpretation which ultimately enabled collections to remain useful through changes in research and teaching agendas and methods.

The turn to practice has also unearthed the role of contributors to the production and maintenance of anatomical representations whose labor was frequently erased from the public record. While individual anatomists, or more rarely artists, claimed authorship of anatomical images, most production processes required a high level of collaboration, and the contributions of those who procured the bodies and body parts, those who preserved and prepared them, artists and artisans, patients, and live models among others. The relationship between these contributors was often complex and shaped the images' epistemological claims to objectivity (Daston and Galison 2007). Artists and anatomists engaged in extensive debates about the perceived tension between the anatomist's quest for "accuracy of representation" and the artist's desire to achieve "elegance of form," as the anatomist and educator John Bell put it (Berkowitz 2015: 68). In public some experts, such as the late eighteenth-century natural philosopher Felice Fontana who directed the production of anatomical wax models for the Museo "La Specola" in Florence, played down artists' roles in the process of image-making by presenting them as mere "tools" in the hands of the learned anatomist (Maerker 2011). Similarly, the nineteenth-century anatomist Henry Gray deliberately minimized the contribution of artist Henry V. Carter, despite the latter's status as a professor of anatomy in his own right (Richardson 2008). In other cases, however, the participation of an accomplished artist was used to support the authority of images. In his *Anatomie pathologique* (1829–42), the French anatomist Jean Cruveilhier praised the brilliant talent of his artist Antoine Chazal with whom he collaborated frequently (Bertoloni Meli 2018).

Epistemology and Education

The search for authoritative anatomical images revolved around two core questions: What is an appropriate subject for depiction and how should it be depicted? Renaissance anatomists debated whether average (here conceived as male and middle-aged) bodies should be chosen for dissection, or whether anatomists should instead strive to observe as wide a range of body types, ages, genders, and ethnicities as possible (Siraisi 1994). The question of human anatomical diversity was raised in one of the most influential works of anatomical description and illustration, Vesalius' *De Fabrica* which combined anatomical illustrations with textual descriptions of the human body. Critics of Vesalius highlighted the inherent tension in his approach: If dissection demonstrated the individuality of human bodies, how could idealized images stand in for them? Indeed, it was not obvious what images represented, and what functions they performed. As Sachiko Kusukawa has shown, the representational strategies used by Vesalius required a wider rethinking of the role of anatomical images for medical teaching and inquiry. What is their status as evidence or teaching tools? What can and cannot images do? What should be the relationship between bodies, images, and words? For Vesalius, images were central to the construction of specifically "visual arguments," which treated anatomical images like geometrical diagrams which would allow readers to draw general conclusions concerning questions such as the best location of cuts for bloodletting (Kusukawa 2012).

Historians of medicine and science have frequently used anatomical images and the debates surrounding them as key examples for claims about the historical epistemology of images more generally. Daston and Galison's influential study *Objectivity* of 2007, for instance, used the genre of the anatomical atlas as a case study for their argument that techniques of production and the subjectivity of image makers were inseparable from claims about the accuracy and utility of images. They also showed how concepts of "objectivity" changed over time, and developed a broad classification of different "epistemic virtues": "truth-to-nature" based on the selection of ideal types, "mechanical objectivity" which resorted to the use of automatic devices, and "trained judgment," which emphasized the importance of education for the creation of expertise in pattern recognition. Historians have qualified Daston and Galison's chronology (see, e.g., Kusukawa 2012); their definition of three distinctive conceptions of objectivity, however, has been widely adopted.

Beyond analyses of the epistemic status of anatomical images, historians have investigated the role of anatomy itself as a model for scientific enquiry more generally, and suggested that anatomy and its practices of investigation and representation became central to European concepts of knowledge production (Stafford 1991). Enlightenment anatomy created rich collections of images and models which afforded a glimpse into the body's interior, and with it an understanding of the functional relationship between different parts. The success of this approach led to a shift in European science from a text-based to a visual epistemology which took "anatomy" as its central metaphor to describe processes of analysis and synthesis which allowed researchers to visualize the interior structures of things which would lead to a truer understanding of nature.

The epistemology of images has been shown to be inseparable from questions about how anatomical knowledge was made in practice, and about the use of images in research and education. Recent scholarship has highlighted the importance of users' hands-on interactions with anatomical representations, and the role of multimediality. Examples include the peculiar genre of "flap anatomies," paper bodies constructed from layers of movable parts which afforded a sense of depth and three-dimensionality to the user (Fig. 8). Such flap anatomies were first created by early

Fig. 8 Anatomical fugitive sheet (female) with movable flaps which can be raised to show cutouts of the viscera attached beneath. Public Domain Mark

modern printers, both for scholarly books and as more widely affordable fugitive sheets (Carlino 1995). The technique continued to be used in medical and lay education with popular examples such as the colorful and popular *Anatomie iconoclastique* published by the French physician and popularizer Gustave Witkowski since 1876. Representations such as these further strengthened a long-standing conception that anatomical knowledge was based on a "reading" of the body, both through the process of dissection and interactions with paper tools (Hallam 2016). Crucially, the body was made legible through combinations of different media. Leading anatomy schools of the late eighteenth and early nineteenth century had already proudly advertised their large anatomical collections, which were supplemented with the parallel use of printed images, drawings, and models (Berkowitz 2011).

New technologies continued to shape anatomical images beyond the advent of the printing press. The introduction of photography and improvements in microscopy such as the development of achromatic lenses once more challenged anatomists to reflect on the epistemology and utility of images. How reliable were representations produced with the help of such new technologies? Technological innovation often heightened rather than solved conceptual and practical problems. Nineteenth-century practitioners came to the conclusion that photography had little value for morbid anatomy as it offered a chaotic wealth of detail rather than a clear and instructive representation of salient features (Curtis 2012; Fox and Lawrence 1988). Medical innovators such as Charcot preferred to use a combination of old and new representational techniques to represent living bodies (Ruiz-Gomez 2019). Pioneers of forensic science who tried to locate criminality and degeneration in the individual and social body grappled with the apparent contradiction that photography "define[d] both ... the typology and the contingent instance of deviance and social pathology" (Sekula 1986: 7). Often presented by its nineteenth-century advocates as a prime example of "mechanical objectivity," photography did not resolve long-standing questions about the role of subjectivity, connoisseurship and cognitive authority in visual representation.

Reception and Use

Anatomical images and displays have served many different purposes beyond medical research and teaching. Such representations of the body were frequently used to simultaneously instill in their audiences a veneration of divine creation, to act as tools for moral instruction, or as opportunities for reflection and the creation of deeper knowledge of the self.

The Greek dictum "know thyself" was first used with reference to anatomy when it was adopted as a motto on sixteenth-century anatomical fugitive sheets and textbooks, and widely cited by anatomists. These fugitive sheets were a key means of bringing anatomical knowledge to a wider public beyond medical practitioners (Carlino 1995). "Flap anatomies" provided a quasi-three-dimensional image of the inner organs which the user could leaf through, as well as a brief textual description of the names of organs and their functions. Early modern publications aimed at a broad readership combined such anatomical representations with the aesthetic conventions of the *danse macabre*, creating images which invited the observer not only to develop a better understanding of the structure of the human body, but also to reflect on the fragility of human life, and the contrast between the mortal body and the immortal soul. Such images were intended to serve as memory aids, but also to encourage others to study anatomy, and to admire the handiwork of the creator. Protestant humanist writers such as Melanchthon stressed the utility of engagement with anatomy as a means for introspection, moral instruction, and illustration of divine power (Nutton 1993). Public dissections performed in early modern Europe were similarly multifunctional (Klestinec 2011), as were anatomical miniatures produced in wood and ivory (Buckley 2020). This interpretation of anatomical displays beyond strictly medical uses continued to shape audiences' responses well into the modern era. In the eighteenth century, enlightened reformers seeking to emancipate medical laypeople from superstition and quackery frequently created visually appealing displays of images and models which, lacking the stench of the corpse itself, were deemed more suitable to encourage engagement and learning. And yet, audiences responded to exhibitions of anatomical waxes in ways reminiscent of earlier traditions, as opportunities for pious reflection and moral improvement (Maerker 2011).

The close relationship between anatomy and morality continued in the nineteenth and twentieth centuries. Gruesomely accurate depictions of syphilitic lesions warned young men of the dangers of promiscuity. Public health campaigners highlighted the agency and responsibility of the individual to maintain their bodies in good health and (re)productivity. To the present day, anatomical displays aimed at lay audiences play with the interpretive flexibility of such representations. Bodyworlds, a globally successful show of plastinated bodies and body parts, advertises itself as educational in ways not dissimilar to the rhetoric of eighteenth- and nineteenth-century reformers. However, the bodies' appeal to audiences goes beyond the educational. Bodyworld's invocation of the "fascination of the real" ostensibly serves to convey anatomical knowledge to lay audiences. And yet, the side-by-side display of the smoker's blackened lung next to the light pink of the healthy organ simultaneously plays with titillation and disgust.

How effective were anatomical images at conveying their intended messages? Historians have highlighted the role of such images for popularizing specific ideas about health (Nemec 2020). In the modern period, in particular, such images often implied a large degree of control over the body, as well as the individual's personal responsibility for maintaining it in a healthy state. Scholars frequently see the role of anatomical images as constitutive and supportive of hegemonic ideological positions on matters of race, class, and gender. Other interpretations have foregrounded the role of anatomical images as agents of change, suggesting that modern imaging techniques contributed to changing sexual attitudes, and inspired new ways to represent the human body in art and literature as well as promoting progressive social movements (Kevles 1997; Maerker 2019).

The reception of anatomical representations frequently poses methodological challenges for the historian, as evidence of responses and uses tends to be less well documented than the programmatic/prescriptive texts which accompanied images. The "life" of anatomical images and their interpretive flexibility has recently been traced in innovative studies which focus on the publishing history of influential images such as the widely circulated but controversial images of human and animal development by Ernst Haeckel (Hopwood 2015). Overall, it was often the "proper" use of anatomical images, or the use of proper images, which was used to demarcate experts from laypeople.

The Politics of Class and Professional Identity

Beyond the articulation of conceptions of gender and race, historians have a long-standing interest in the politics of class and its intersection with the formation of professional medical identities. Portraits of the anatomist posing with the open corpse or with his collection of specimens and images have long been used to articulate claims to anatomical expertise, whether in Vesalius's portrait at the center of his programmatic frontispiece to *De fabrica* (1543), Rembrandt's *Anatomy Lesson of Dr. Nicolaes Tulp* (1632), Joshua Reynold's portrait of John Hunter (1786), or the bust of Wilhelm His (1900) in pensive pose over one of his embryo models (Jordanova 2000, 2003; Hopwood 2012).

However, the long-standing associations of dissection with corporeal and capital punishment in Europe meant that, not surprisingly, there was widespread reluctance to offer one's own body, or those of loved ones, for dissection. Anatomy became ever more central to medical education, and around 1800 the Paris Medical Faculty was celebrated for its ability to provide students with an exceptionally large number of bodies for hands-on learning. At the same time, the popular perception of dissection as dehumanizing and shameful often resulted in shortages of "human material" in the eighteenth and nineteenth centuries, which in turn gave rise to the insalubrious practice of "body-snatching" corpses from morgues and graveyards. Protest and resistance against body-snatching, and against legal measures to make the bodies of the unclaimed poor available to medical schools, produced rich visual artifacts – caricatures of body-snatchers and anatomists wielding the butcher's knife, but also the rather unique object known as Jeremy Bentham's "auto-icon," as the utilitarian philosopher requested the preservation of his body after death to encourage wider acceptance of dissection.

Such anatomical practices and representations shaped the identities of those whose bodies were instrumentalized, but also those who were involved in image production and use. Practices of making and using images of the body became central to the professional and disciplinary formation of medical students, practitioners, and researchers. Medical teachers such as the celebrated William Hunter operated on the assumption that dissection would imbue students with the "necessary inhumanity" required of medical practitioners. The medical schools of the late eighteenth and nineteenth centuries employed a multimedial range of corpses, specimens, models, prints, and drawings to facilitate an emotionally detached interaction with the body which was considered central to professional identity and discipline formation (Cunningham 2010; Berkowitz 2011). Simultaneously, images and preparations were anonymized and described as "cases." The names of patients were rarely recorded, except in the case of celebrity patients whose patronage reflected well on the medical practitioner (Chaplin 2012).

The introduction of new imaging technologies further supported this distancing development. The late nineteenth century saw the development of dissection photographs which staged groups of students and teachers around a corpse on the dissection table. These images were often accompanied by bleakly humorous or straightforwardly brutal dicta – "She lived for others but died for us." They also represented the vast power imbalance and social gulf between medics and their "human material." In the USA, in particular, the dissected were not only poor, but often black, surrounded in the photograph by groups of White men in middle-class clothing. Such staged images strengthened group identity among medical students, while making other actors invisible, especially the socially inferior technicians and janitors who were responsible for the dirty work of corpse procurement and disposal (Warner and Edmonson 2009). Inhumanity thus became central to anatomists' attempts to define the human.

Art and Anatomy

Artists and scholars continue to draw on anatomy for creative explorations of the human condition. Modern imaging techniques have contributed to new ways of representing the human body in art and literature (Kevles 1997). In recent decades, growing interest has also led to a reevaluation of medical representations by art historians (e.g., Fend 2017; Ruiz-Gómez 2019), as well as to new forms of engagement of fine artists and curators with historical and contemporary medical images and models of the body. Growing engagement with the conventions, materials, and techniques of anatomical representations by artists such as Damian Hirst and Marc Quinn have contributed to the rehabilitation of "vulgar" materials and genres including wax modeling and wet preparations (Mike Kelley, *The Uncanny*, Tate Liverpool 2004). Influential exhibitions exploring the intersection between art and anatomy both expressed and reinforced artists' and audiences' interest in representations of the body and in the history of such representations (*The quick and the dead: artists and anatomy*, London, Hayward Gallery,1997–98; *Spectacular Bodies: The Art and Science of the Human Body from Leonardo da Vinci to Now*, London, Hayward Gallery, 2000; *Art et Anatomie: dessins croisés Musée Atger/Musée Fabre*, Montpellier, 2020). Other exhibition projects have focused on the intersections of anatomical and religious art, and the use of hyperrealism in art and science education (*The Sacred Made Real*, National Gallery London, 2009–10; *Die grosse Illusion: Veristische Skulpturen und ihre Techniken*, Liebighaus Frankfurt, 2014; *Surreal Science*, Loudon Collection with Salvatore Arancio, Whitechapel Gallery, 2019). Beyond fine art and established museums and galleries there has been a resurgence of anatomy as spectacle, presented in a variety of formats: a reinvention of the cabinet of curiosities and the freak show for audiences both live and online. Once again in these displays, anatomical representations such as the plastinated smoker's lung displayed in Bodyworlds are presented and perceived in different ways, whether as reminders of mortality, as appeals to personal agency and responsibility, as opportunities for voyeurism, or as occasions to empathize with past sufferers.

Conclusion

The history of anatomy has engaged with the role of anatomical images since its inception. Along the way, it has responded to ideas and approaches from a wide range of disciplines, from art history, social, and cultural history, the history of education, and the history of the book to feminist and postcolonial studies, cultural anthropology, and communication studies. These analyses have engaged with longstanding debates among medical practitioners about what anatomical images actually show, how they should be produced, and what they can and cannot be used for. Attention to the historical specificity of artistic styles and conventions highlights that anatomical images are never unmediated representations of an objective, universal truth, but fundamentally shaped by the social, political, and cultural contexts of their production and use. Images in turn have shaped human activity and the very definition of humanity itself: They have been used to justify the subjugation of women and non-Whites, to support public health measures and eugenics, and to locate agency and responsibility for health. Anatomical representations have also played a central role as expressions of professional status and symbols for the authority of medicine. They could be used to display the medical profession's privileged access to the body, and contribute to the process of medical socialization through dissection as an important rite of passage for medical students. However, scholars have also begun to read anatomical images against the grain, and with a focus on conflict, subversion, and resistance. Attention to the process of image-making and to the role of artisans and artists reveal struggles for authority and authorship, as well as challenges to traditional gendered claims to superior understanding. The recent historiographical focus on practices, audiences, and use thus shows that the function of anatomical images is never completely determined by their design, and that users demonstrate interpretive flexibility which could ignore, challenge, or subvert intended messages and uses of anatomical representations. Humanity, it seems, is never fully captured by its images.

References

Amirault C (1993–94) Posing the subject of early medical photography. Discourse 16(2):51–76
Berkowitz C (2011) The beauty of anatomy: visual displays and surgical education in early nineteenth-century London. Bull Hist Med 85(2):248–271

Berkowitz C (2015) Charles Bell and the anatomy of reform. University of. Chicago Press, Chicago
Biddiss MD (1976) The politics of anatomy: Dr. Robert Knox and Victorian racism. Proc R Soc Med 69:245–250
Bittel C (2019) Unpacking the phrenological toolkit: knowledge and identity in antebellum America. In: Bittel C, Leong E, von Oertzen C (eds) Working with paper: gendered practices in the history of knowledge. University of Pittsburgh Press, Pittsburgh, pp 91–107
Blanckaert C (2009) De la race à l'évolution. Paul Broca et l'anthropologie française, 1850–1900. Harmattan, Paris
Bucci M (1969) Anatomia come arte. Edizioni d'Arte Il Fiorino, Florence
Buckley C (2020) Pathos, eros, and curiosity: the history and reception of ivory anatomical models from the seventeenth century to today. Nuncius Ann Storia della Scienza 35(1):64–89
Callen A (1995) The spectacular body: science, method, and meaning in the work of Degas. Yale University Press, New Haven
Carlino A (1995) Knowe thyself: anatomical figures in early modern Europe. RES Anthropol Aesthet 27:52–69
Chaplin S (2012) The divine touch, or touching divines: John Hunter, David Hume, and the Bishop of Durham's rectum. In: Deutsch H, Terrall M (eds) Vital matters: eighteenth century views of conception, life, death. University of Toronto Press, Toronto
Choulant L (1852) Geschichte und Bibliographie der anatomischen Abbildung nach ihrer Beziehung auf anatomische Wissenschaft und bildende Kunst: Nebst Auswahl von Illustrationen nach berühmten Künstlern. Rudolph Weigel, Leipzig
Craske M (2011) "Unwholesome" and "pornographic": a reassessment of the place of Rackstrow's Museum in the story of eighteenth-century anatomical collection and exhibition. J Hist Collect 23(1):75–99
Cryle P, Stephens E (2017) Normality: a critical genealogy. University of Chicago Press, Chicago
Cunningham A (2010) The anatomist anatomis'd: an experimental discipline in enlightenment Europe. Ashgate, Aldershot
Curran AS (2011) The anatomy of blackness: science and slavery in an age of Enlightenment. Johns Hopkins University Press, Baltimore
Curtis S (2012) Photography and medical observation. In: Anderson N, Dietrich MR (eds) The educated eye. Visual culture and pedagogy in the life sciences. Dartmouth College Press, Hanover, pp 68–93
Dacome L (2017) Malleable anatomies: models, makers and material culture in eighteenth-century Italy. University of Chicago Press, Chicago
Daston L, Galison P (2007) Objectivity. Zone Books, New York
Didi-Huberman G (1999) Ouvrir Vénus. Nudité, rêve, cruauté. Gallimard, Paris
Duden B (1998) The woman beneath the skin: a doctor's patients in eighteenth-century Germany (trans: Dunlap T). Harvard University Press, Cambridge, MA
Fend M (2017) Fleshing out surfaces: skin in French art and medicine, 1650–1850. Manchester University Press, Manchester
Fox DM, Lawrence C (1988) Photographing medicine: images and power in Britain and America since 1840. Greenwood, New York
Gelfand T (1980) Professionalizing modern medicine: Paris surgeons and medical science and institutions in the 18th century. Greenwood, London
Gilman S (1986) Difference and pathology: stereotypes of sexuality, race, and madness. Cornell University Press, Ithaca
Hacking I (1990) The taming of chance. Cambridge University Press, Cambridge
Hagner M (1999) Enlightened monsters. In: Clark W, Golinski J, Schaffer S (eds) The sciences in enlightened Europe. University of Chicago Press, Chicago/London
Hallam E (2016) Anatomy museum: death and the body displayed. Reaktion Books, London
Haraway D (1989) Primate visions: gender, race, and nature in the world of modern science. Routledge, London
Hopwood N (2004) Plastic publishing in embryology. In: de Chadarevian S, Hopwood N (eds) Models: the third dimension of science. Stanford University Press, Stanford, pp 170–206

Hopwood N (2012) A marble embryo: meanings of a portrait from 1900. Hist Work J 73(1):5–36
Hopwood N (2015) Haeckel's embryos: images, evolution, and fraud. University of Chicago Press, Chicago
Huistra H (2018) The afterlife of the Leiden anatomical collections: hands on, hands off. Routledge, London
Jordanova L (1989) Sexual visions: images of gender in science and medicine between the eighteenth and twentieth centuries. University of Wisconsin Press, Madison
Jordanova L (2000) Defining features: Scientific and medical portraits, 1660–2000. Reaktion Books, London
Jordanova L (2003) Portraits, people and things: Richard Mead and medical identity. Hist Sci 41(3): 293–313
Jülich S (2018) Picturing abortion opposition in Sweden: Lennart Nilsson's early photographs of embryos and fetuses. Soc Hist Med 31(2):278–307
Katzew I (2005) Casta painting: images of race in eighteenth-century Mexico. Yale University Press, New Haven
Kemp M, Wells M (eds) (2000) Spectacular bodies: the art and science of the human body from Leonardo to now. Hayward Gallery, London
Kevles B (1997) Naked to the bone: medical imaging in the twentieth century. Rutgers University Press
Klestinec C (2011) Theaters of anatomy: students, teachers, and traditions of dissection in renaissance Venice. Johns Hopkins University Press, Baltimore
Kusukawa S (2012) Picturing the book of nature: image, text, and argument in sixteenth-century human anatomy and medical botany. University of Chicago Press, Chicago
Löwy I (2013) Sex on a slide: Antoine Lacassagne and the search for a microscopic definition of masculinity and femininity. Hist Philos Life Sci 35(3):363–378
Maerker A (2011) Model experts: wax anatomies and enlightenment in Florence and Vienna, 1775–1815. Manchester University Press, Manchester
Maerker A (2019) Papier-mâché anatomical models: the making of reform and empire in nineteenth-century France and beyond. In: Bittel C, Leong E, von Oertzen C (eds) Working with paper: gendered practices in the history of knowledge. University of Pittsburgh Press, Pittsburgh, pp 177–192
Maerker A (forthcoming) Anatomical preparations and mimetic expertise. In: Bol M, Spary E (eds) The matter of mimesis. Studies on mimesis and materials in nature, art and science. Brill, Leiden
Margócsy D, Somos M, Joffe SN (2018) The Fabrica of Andreas Vesalius: a worldwide descriptive census, ownership, and annotations of the 1543 and 1555 editions. Brill, Leiden
Martínez ME (2008) Genealogical fictions: limpieza de sangre, religion, and gender in colonial Mexico. Stanford University Press, Stanford
Meijer MC (1999) Race and aesthetics in the anthropology of Petrus Camper (1722–1789). Rodopi, Amsterdam/Atlanta
Meli DB (2018) Visualizing disease: the art and history of pathological illustrations. University of Chicago Press, Chicago
Morgan JL (1997) 'Some could suckle over their shoulder': male travelers, female bodies, and the gendering of racial ideology, 1500–1770. The William and Mary Quarterly 54(1):167–192
Mosse GL (1978) Toward the final solution: a history of European racism. University of Wisconsin Press, Madison
Muratori C (2017) From animal bodies to human souls: (pseudo-)Aristotelian animals in Della Porta's physiognomics. Early Sci Med 22(1):1–23
Nemec B (2020) Norm und Reform. Anatomische Körperbilder in Wien um 1925, Wallstein, Göttingen
Nutton V (1993) Wittenberg anatomy. In: Grell OP, Cunningham A (eds) Medicine and the reformation. Cambridge University Press, Cambridge, pp 11–30
Park K, Daston LJ (1998) Wonders and the order of nature, 1150–1750. Zone Books, New York

Pogliano C (2019) Unconventional views of racial brains in the 19th century. Nuncius 34(3): 602–634
Porter R (1985) The patient's view: doing medical history from below. Theory Soc 14(2):175–198
Poskett J (2019) Materials of the mind: phrenology, race, and the global history of science, 1815–1920. University of Chicago Press, Chicago
Petherbridge D, Jordanova L (1997) The quick and the dead: artists and anatomy. South Bank Centre, London
Qureshi S (2004) Displaying Sara Baartman, the "Hottentot Venus". Hist Sci 42(2):233–257
Rabinow P (1991) The Foucault reader. Penguin, London
Richardson R (1987) Death, dissection, and the destitute. University of Chicago Press, Chicago
Richardson R (2008) The making of Mr. Gray's anatomy. Oxford University Press, Oxford; New York
Riskin J (2016) The restless clock: a history of the centuries-long argument over what makes living things tick. University of Chicago Press, Chicago
Roberts KB, Tomlinson JDW (1992) The fabric of the body: European traditions of anatomical illustration. Clarendon Press, Oxford
Ruiz-Gómez N (2019) The model patient: observation and illustration at the Musée Charcot. In: Graciano A (ed) Visualizing the body in art, anatomy, and medicine since 1800: models and modeling. Routledge, London, pp 203–232
Sappol M (2017) Body modern: Fritz Kahn, scientific illustration, and the homuncular subject. University of Minnesota Press, Minneapolis
Sawday J (1995) The body emblazoned: dissection and the human body in Renaissance culture. Routledge, London
Schiebinger L (1986) Skeletons in the closet: the first illustrations of the female skeleton in eighteenth-century anatomy. Representations 14:42–82
Schiebinger L (1990) The anatomy of difference: race and sex in eighteenth-century science. Eighteenth-Century Stud 23(4):387–405
Sekula A (1986) The body and the archive. October 39:3–64
Sigerist H (1945) Civilization and disease. Cornell University Press, Ithaca
Siraisi N (1994) Vesalius and human diversity in De humani corporis fabrica. J Warburg Courtauld Institutes 57:60–88
Sowerwinea C (2003) Woman's brain, man's brain: feminism and anthropology in late nineteenth-century France. Women's Hist Rev 12(2):289–308
Stafford BM (1991) Body criticism: imaging the unseen in enlightenment art and medicine. Cambridge, MA, M.I.T Press
Staum MS (2003) Labeling people: French scholars on society, race and empire, 1815–1848. McGill-Queen's University Press, Montreal
Stephens E (2011) Anatomy as spectacle: public exhibitions of the body from 1700 to the present. Liverpool University Press, Liverpool
Stolberg M (2003) A woman down to her bones: the anatomy of sexual difference in the sixteenth and early seventeenth centuries. Isis 94:274–299
von Schlosser J (1911) Geschichte der Porträtbildnerei in Wachs. Translated in Panzanelli R (ed) (2008) Ephemeral bodies: wax sculpture and the human figure. Getty Research Institute, Los Angeles, pp 171–314
Voss J (2010) Darwin's pictures: views of evolutionary theory, 1837–1874. Yale University Press, New Haven
Warner JH, Edmonson JM (2009) Dissection: photographs of a rite of passage in American medicine, 1880–1930. Blast Books, New York

History of Embryology: Visualizations Through Series and Animation

11

Janina Wellmann

Contents

Introduction	260
Seeing and Dating	261
Development in Images	265
Epigenetic Iconography	267
The Rhythm of Becoming	272
Series of Sections	274
Living Organisms and Dead Material	277
Time and Experiment	281
Optical Slicing	282
Embryos, in Silico	283
Algorithmic Animation	286
Conclusion	287
References	288

Abstract

Embryological questions about the origins of life and the nature of becoming are among the most profound questions that human beings have asked, in all cultures and across the millennia. The history of embryological thinking encompasses the long history of religions, philosophies, and cultures, yet the history of the science of embryology first emerged in eighteenth-century Europe. This scientific discipline was founded on the observation that living organisms do not simply exist once and for all but come into being and continue to change throughout their lives.

Becoming is a process whose fleeting essence permanently eludes observation and representation. It only becomes an object of scientific inquiry to the extent that it is analyzable, measurable, and depictable. Understanding development and

J. Wellmann (✉)
Institute for Advanced Study on Media Cultures of Computer Simulation (MECS), Leuphana Universität Lüneburg, Lüneburg, Germany
e-mail: janina.wellmann@leuphana.de

© The Author(s), under exclusive licence to Springer Nature Singapore Pte Ltd. 2022
D. McCallum (ed.), *The Palgrave Handbook of the History of Human Sciences*,
https://doi.org/10.1007/978-981-16-7255-2_22

fixing it in the image is a cognitive and perceptual struggle for the ephemeral "in between," the "not yet" or "no longer" of a process that is regarded as continuous – so as to make that which has just ended, that which proceeds apace or begins anew comprehensible, graphic, and intelligible both in and with the image.

This chapter relates the history of embryology – by way of its central milestones and from its beginnings in the eighteenth century through to modern developmental embryology – as a story of seeing, of media and techniques of visualization and forms of representation, but also as a story of preparing and experimenting with model organisms.

Keywords

Development · Series · Animation · Modeling

Introduction

The human embryo marks the beginnings of each individual life, and since the twentieth century, it has enjoyed a prominent place in the scientific and cultural discourse of the West. Its history is closely related to the medicalization of pregnancy, the politics of motherhood and gender, ethical deliberations and jurisdiction over medical intervention, from assisted reproduction to genetic screening or prenatal surgery (Maienschein 2014; Morgan 2009).

It was not until the late nineteenth and early twentieth centuries that the human embryo emerged as an object of scientific research in attempts to standardize human development, an object of psychologists interested in uterine life, of doctors and scientists putting embryos on display in ambitious collections all over the world (Arni 2018; Hopwood 2000).

Yet embryological inquiries into the origins of life and the nature of development are no recent phenomenon – such questions have been posed in all cultures at all times. The history of embryological thinking thus comprises the long intellectual and global history of religions and philosophy, societies and cultures, habitats and climate zones (Wallingford 2021). On the other hand, the science of embryology, more narrowly defined, is the study of the coming into being and development of living forms. This science took shape in eighteenth-century Europe and was based on the observation that living creatures do not emerge as finished products but become ever anew. Living beings are not fashioned immutably, as if set down perfect and complete in the world, in the moment of God's creation; rather, they are subject to a continuous transformation in which their becoming transpires in and with a world that generates them and which they in turn generate.

Becoming is a process whose fleeting essence permanently eludes observation and representation. But at the same time, it only becomes comprehensible, analyzable, measurable – in short, the object of scientific inquiry – in concrete organisms, with material that is malleable by dint of human intervention and with forms of representation geared to create and facilitate understanding.

From the very beginning, therefore, the scientific enterprise of embryology has been a history of seeing and making visible, observing and manipulating, of media and techniques of visualization and forms of representation. Just like its subject, the science itself has been transformed through its practices and devices, the organisms it has chosen to investigate and the experimental techniques it puts to use. Starting in the seventeenth century with chicken eggs, model organisms over the centuries have come to include mouse and drosophila, more recently zebrafish, and in the twenty-first century, human embryos have emerged from newly digitized preparations made of embryos stored in classic collections, from ultrasound in vivo examination or CRISPR-Cas intervention and stem-cell cultures (Rossant and Tam 2018). Taking the whole of nature into account, a science of becoming also encompasses the diversity of biological development, such as the sexual reproduction of mammals, the metamorphosis of insects, or the parthenogenesis of lower animals.

Furthermore, the question of how to get hold of ever changing forms scientifically is intimately linked to tinkering with the life of the object under investigation. Throughout the history of embryology, living organisms as well as dead specimens have been studied; whole organisms have been captured by cinematography or bits and pieces of embryos exposed to experimental stress; embryos have been put to death and physically sectioned or been kept alive and optically sliced. Embryology is dedicated to living dynamics, but its methods have long depended on dead material. To overcome this paradox at the heart of the investigation of development, from the very beginning science would switch back and forth between experimental methods and innovative pictorial approaches and techniques.

How life develops is a fundamental part of human beings' inquiry into their own existence and the world around them, and embryology is the science of the constructs, artifacts, and inventions used to explore and pin down the permanent struggle between the fleeting "in between," the "not yet" or "no longer" of a life regarded as continuous. Although it is a natural process in a living being, development only becomes scientifically intelligible when demarcations between the visible and invisible, the living and dead are constantly shifted. The twenty-first century sets out to add another protean aspect to this endeavor by blurring the boundary between what is found in nature and what can literally be (man)made.

Seeing and Dating

For centuries, the emergence of new life was an act of divine creation. With onset of the seventeenth century, however, notions of development multiplied. In particular, sundry manifestations of a theory of preformation arose at century's close. What these theories had in common, at their core, was that they postulated a complete miniature embryo present in the germ. This embryo was at first infinitely small and therefore invisible and became larger only in the course of its development. Its evolvement was therefore due to its growth, and development here was synonymous with growth in size, unraveling, and unfolding. Different varieties of preformation allowed for this process to be located in the female ovum (ovism) or the male sperm (animalculism). Panspermism even construed the germs as circulating freely in the air. Regardless of the differences in detail, the variants were congruent with a Deistic view of life in which God created the world but afterward withdrew from it (Needham 1975; Roger 1963).

The empirical analysis of development began toward the end of the sixteenth century when Ulissee Aldrovandi (1522–1605) and his student Volcher Coiter (1534–1576) observed the development of the chick embryo on a day-to-day basis. The first pictorial representations of development appeared in 1625 in *De formatione ovi et pulli* by Hieronymus Fabricius ab Aquapendente (1537–1619) (Fig. 1). Aquapendente's plates depict the daily progress in the incubated chicken egg and are the result of some remarkably exact observation. They establish a tradition of chronological visualization in embryology where the observation and pertaining image refer to the observer's time. Accordingly each of the individual figures in the plate is denoted by a number that indicates the day on which the examined egg had the appearance documented in the plate; reference point of the image is a fixed time of observation and the image's task is to capture the stage of development at a particular moment.

The drawings of the Italian anatomist and physician Marcello Malpighi (1628–1694) follow this chronological tradition. They were produced for his investigations in *De ovo incubato* (1672) and *De formatione pulli in ovo* (1673) (Adelmann 1966, vol. 2) but remained authoritative far into the eighteenth century (Fig. 2). Whereas Aquapendente's images make no reference to one another, Malpighi's detailed descriptions together with his excellent drawings provide a report of the developmental process as he depicts it in individual structures – primarily the embryo and the vascular system along with the heart – not just for entire days but at various hours. For his observations, Malpighi used a microscope. One of the most notable microanatomists of his era, he exposed the evolving organism to an entirely new form of observation and invasive procedures by anatomically dissecting it under the microscope (Adelmann 1966; Meli 2011).

Alongside the animal body and plants, it was particularly insects whose existence was lent new weight by enlargement of the microscope lens – and in its wake so too their genesis. In his mid-seventeenth century work *Exercitationes de generatione animalium* (1651), the English physician and discoverer of the circulation of blood William Harvey (1578–1657) defined the concept of metamorphosis in insects as a change of shape where a form is impressed on the malleable material in a one-time act like a seal.

As alternative to this kind of direct molding of insects, Harvey also brought another type of formation into play – the successive reshaping of structures – for which he coined the term "epigenesis," a word which is taken from the Greek and literally means "new formation" or "growing upon" (Harvey 1965, p. 334). Indeed the core elements of this notion can be traced back to the Aristotelian doctrine of generation in the fourth century BC: structures arise successively one after the other in the course of development, and their differentiation ensues teleologically from the homogeneous initial structuring of the embryo.

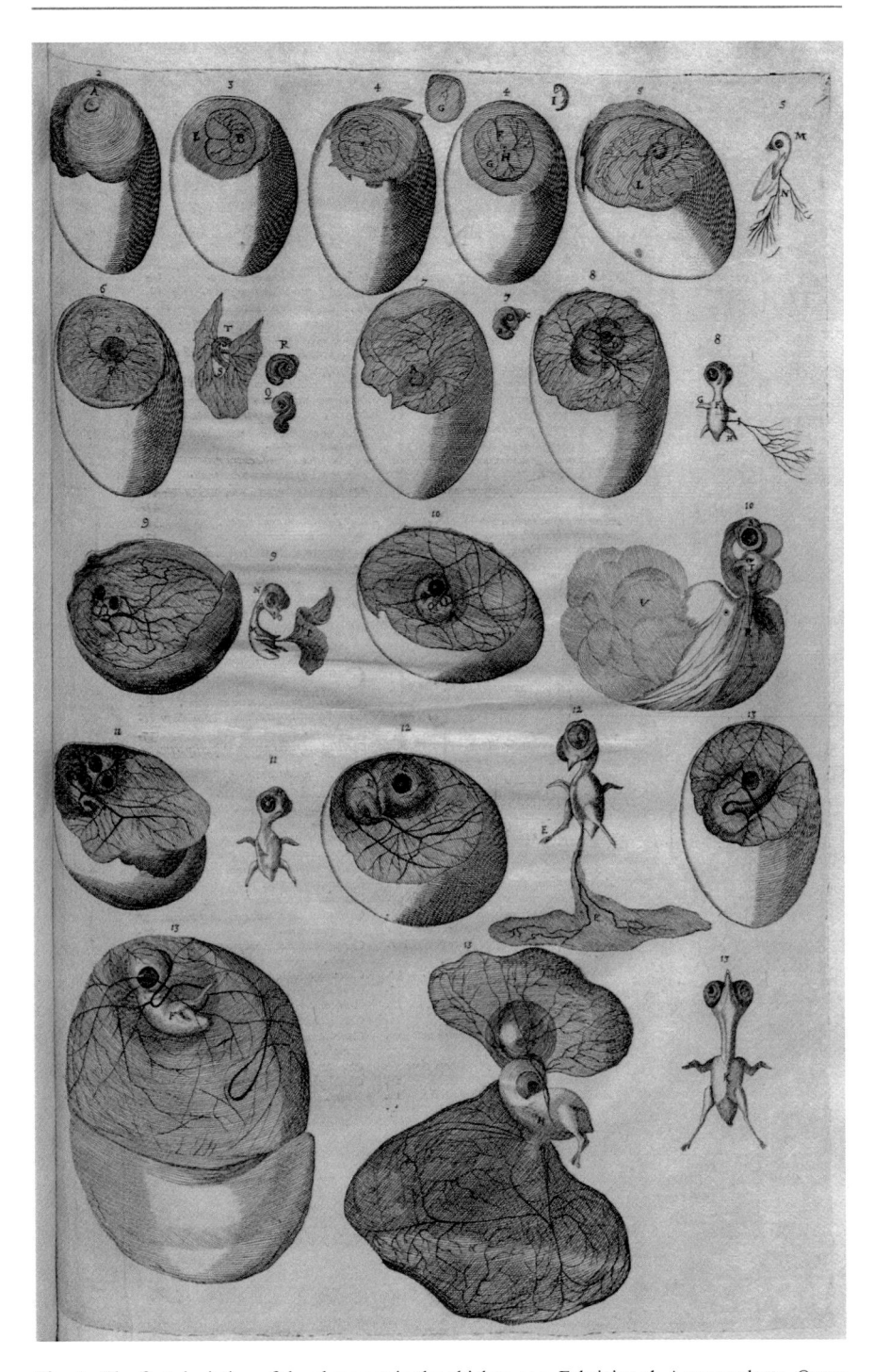

Fig. 1 The first depiction of development in the chicken egg. Fabricius ab Aquapendente, Opera physica anatomica, Padua 1625, plate 3

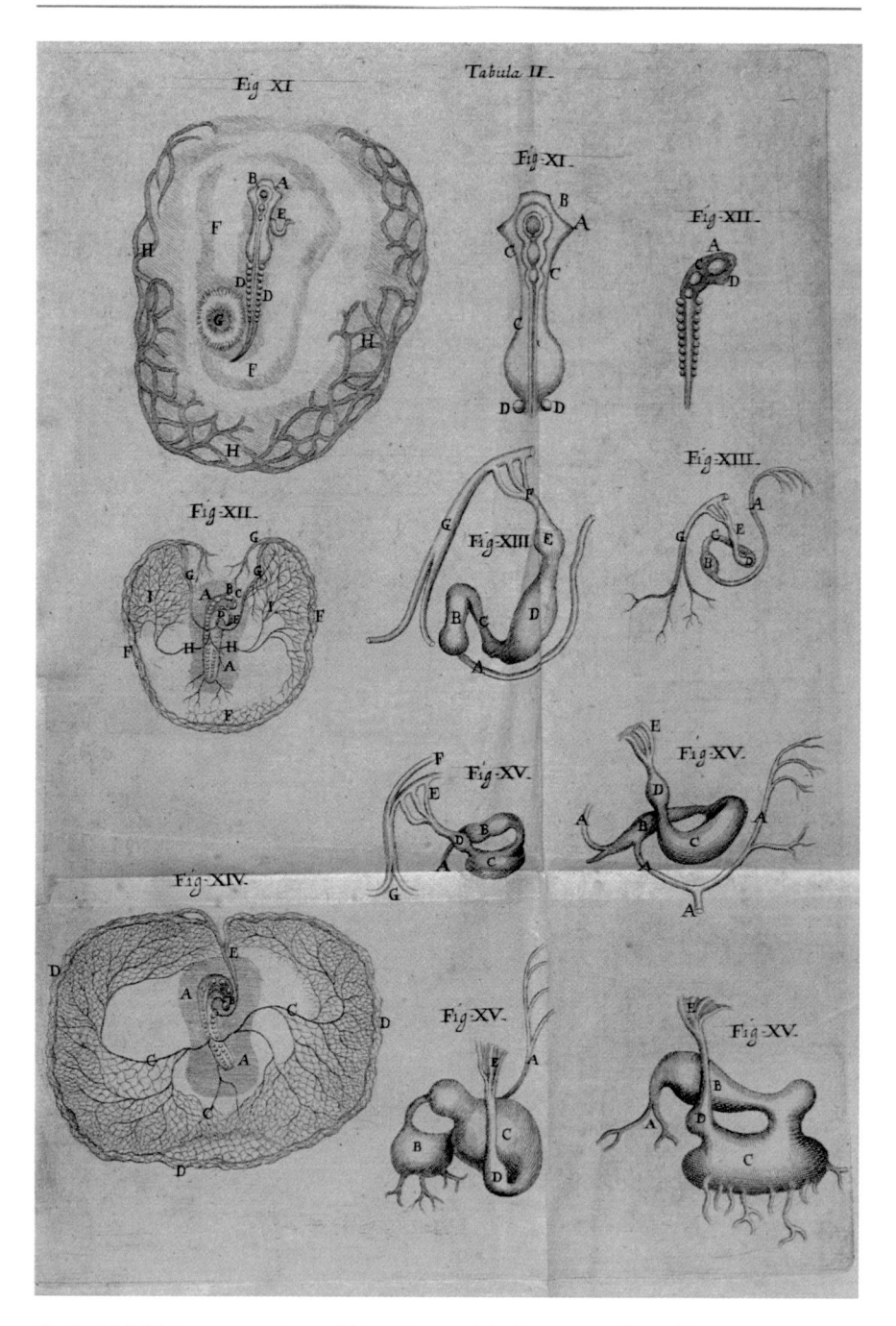

Fig. 2 Malpighi's representations of the embryo and the heart. Marcello Malpighi, De formatione pulli in ovo, London 1673, plate II

Development in Images

The modern theory of epigenesis was established only in the second half of the eighteenth century by Caspar Friedrich Wolff (1734–1794) and his seminal treatise *Theoria generationis* of 1759.

As of the mid-eighteenth century, there was a broadening of the spectrum of developmental concepts that were distinct from preformation. For example, Georges Buffon (1707–1788) took a decidedly different approach. He devised the corpuscular theory of generation which based development on an entirely new concept of organic matter. Buffon envisaged an interior mold to cluster organic molecules and organize them into the whole diversity of organic life (Buffon and Daubenton 1804, vol. 2).

But a new concept of development required not only a new approach epistemically; to conceive development epigenetically, as a gradual coming into being, required a new image regime: Wolff was the first to make pictorial representation an integral component in apprehending development as the gradual emergence of embryological forms.

Understanding formation now had a fundamental reliance on observing processes and their temporal design and how they led to the gradual establishment of structures in the developing embryo. Finding visual strategies to depict them, therefore, proved crucial to a new concept of development that was based on the gradual coming into being of and continuously changing form.

This becomes clear in the dispute that Wolff pursued with the eminent anatomist and physiologist Albrecht von Haller (1708–1777). Haller adhered to the theory of preformation and held that the embryo was preformed in the female ovum, whereas Wolff argued for the gradual shaping of form out of homogenous matter. Both men relied on careful empirical observation of the development of chick embryos. In his 1758 work *Sur la formation du coeur* (*On the Formation of the Heart*), Haller listed no fewer than 300 observations which he had meticulously recorded. In his *Theoria generationis*, Wolff combined observations of plants and animals, and in 1768 added to these with a study on the development of an individual organ in his *De formatione intestinorum* (Haller 1758; Monti 2000; Wolff and Meckel 1812).

While Haller employed images as visual aids – using them in the chronological tradition to precisely determine the shape of an embryo at a particular point during the period of observation – Wolff's depictions were the first in the history of embryology to displace the development he had observed into the image itself. Wolff's revolutionary handling of the image meant that henceforth any representation of the observed form would also reveal how it became such.

To draw up his epigenetic theory of development in and with the image, Wolff took it upon himself to invent a new pictorial form. The means he employed were a recurring pictorial element (Wolff spoke of vesicles and globules), a series of several drawings, and an optical zoom so as to reproduce progressive formation in a sequence of images (Fig. 3). With this step-by-step comparison, from one image to the other, it was now possible to produce visual relations between the forms, to

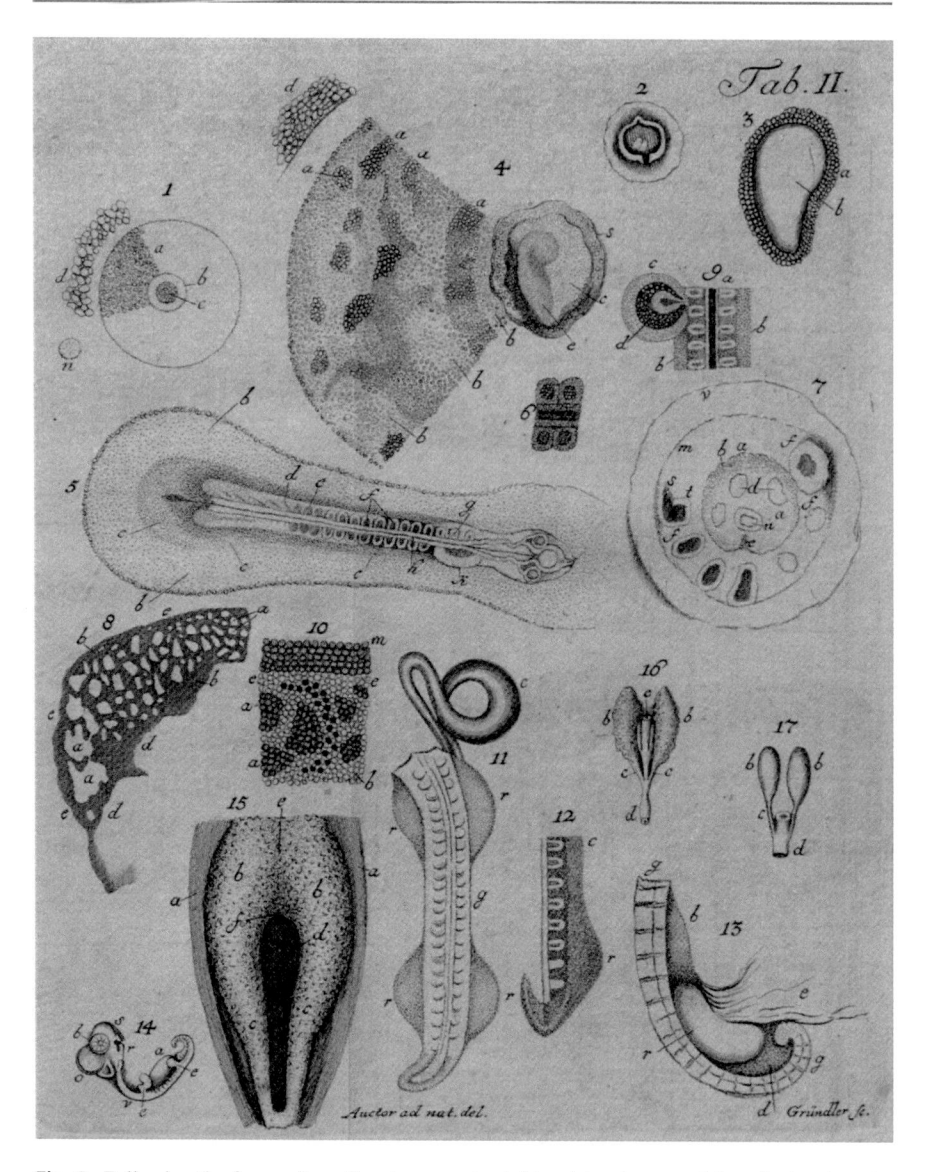

Fig. 3 Following the figures from four to seven to eight and ten the gaze delves further into the tissue with every picture. Caspar Friedrich Wolff, Theoria generationis, Halle 1759, plate II

analogize changes in the structures and thus establish regularities. For Wolff the images were not only an aid to the imagination – not just serving as proof while also functioning as a document (Wolff 1966, p. 81) – but they *were* his theory of epigenesis. By no longer mimetically reproducing what was being viewed at a precisely determined point in time, but by placing observations made at various stages in relation to one another, the images represented both gradual change and revealed the law of their formation.

A theory of epigenetic development required the simultaneous design of the individual and still unknown states of becoming as well as their internal coherence. Wolff's epigenesis is thus a theory on the laws of interaction and organization, according to which organic form emerges from the combined effect of the production of substance and its structural formation. Concretely speaking, Wolff conceived all organic processes as a permanent interplay between flowing and solidifying nutritive fluids. This mutually dependent interplay of flowing and stagnant nutritive fluids was sufficient reason for all structural formation and change in an organism. The process was guided by an unspecified force that Wolff labeled "vis essentialis" (Detlefsen 2006; Roe 1981; Witt 2008).

Generation for Wolff thus encompassed not only ontogenesis but all fundamental organic processes including nutrition and growth. In any event, formation ensued by first producing organic matter, which in a second step was then organized. This was not only how an organism came into being but how the organism sustained itself through growth and nutrition in endlessly repeated cycles throughout its life (Wellmann 2017b).

Epigenetic Iconography

The word *Entwicklungsgeschichte* (developmental history) appeared in the title of three contemporaneous and pathbreaking treatises that launched modern embryology at the start of the nineteenth century: Johann Moritz David Herold's *Entwickelungsgeschichte der Schmetterlinge* (*Developmental History of Butterflies* [1815]), Christian Heinrich Pander's *Beiträge zur Entwickelungsgeschichte des Hühnchens im Eye* (*Contributions to the Developmental History of the Chick in the Egg* [1817]), and Karl Ernst von Baer's two-volume *Über Entwickelungsgeschichte der Thiere* (*On the Developmental History of Animals* [1828 and 1837; reprint 1967]). For modern developmental thinking to emerge around 1800, scientists needed a new conceptual framework along with new observational techniques and experimental practices but above all a new form of representation.

Christian Heinrich Pander (1794–1865) and the more famous Karl Ernst von Baer (1792–1876) were fathers of the "theory of germ layers." This theory served as a foundation for embryology, and its essentials are still valid today. According to the theory, all of embryonic development takes place as a gradual differentiation of primary membranes or germ layers (today called entoderm, ectoderm, and mesoderm). Pander and Baer arrived at their trailblazing theory by conceiving the development of forms – which they, like Wolff, had observed in the chicken egg – as a series of folds.

Among the first observations of organization in the egg was that the homogeneous organic matter differentiates into two membranes after some 12 h, and after a further 10–14 h, a third membrane emerges between the two existing ones. Pander called the outer membranes the serous layer and the mucous layer, while dubbing the middle one the vascular layer. Baer later assigned the two primary membranes the names animal and vegetative layer, which would later differentiate into two further membranes each (Pander 1817a, p. 6) (Baer 1828 and 1837 ([reprint 1967]) vol. 1, p. 9; vol. 2, p. 46).

The subsequent development of all further structures was understood by Pander as the formation and transformation – he spoke of metamorphosis – of these three membranes. Of crucial importance here was that Pander conceived this development or metamorphosis as one of folding: all changes in the egg are the result of a repeated series of folds.

Pander distinguished a total of three different folding processes in the chick. After an initial folding of the germ layer (the mucous layer and the serous layer) where the primitive folds emerge and unite at their upper end, and a second folding where the primitive folds turn inward along their length, there then ensues a third folding at the lower end of the membranes. Emerging from the first fold above is the embryo's head; originating from the second fold on the side are the tubular beginnings of the intestine and heart; the third folding ensues on the dorsal instead of the ventral side of the embryo and turns outward as opposed to inward. Because the folding starts in a different place and moves in a different direction, this creates the extraembryonic structures, the amnion and the chorion. With formation of the primitive folds, the intestinal tract and the heart as well as the membranes encasing the embryo, the elementary organ systems of the early embryo are thus formed (Pander 1817a, pp. 7–25).

Folding is a complex interaction of membranes in which they first approach one another, join up at certain points, sometimes fusing together, and then undergo a complex spatial movement – twisting, bending, folding, turning inward, turning outward – whereby new forms are continuously created. Fundamental to this process is the spatiotemporal coordination of folding. It is the spatial bending and turning at different locations on the membranes, in different directions as well as their temporally differentiated onset and completion which gives order to physiological movement in the process of folding.

With Baer, too, the differentiation of the germ layers – he spoke of separation and distinguished between primary, histological, and morphological separation – is the fundamental principle of all organic development (Baer 1828 and 1837 [reprint 1967] vol. 2, pp. 46, 74–91). His nuanced view of membranes as flexible surfaces brought an additional complexity to the concept of folding. The membrane's movements can also ensue in a sporadic manner, for they are the result of organic growth at certain membrane sites that can be localized in detail. If for example there is a thickening of the membrane at a specific point, then this isolated change results in a cascade of folds – in general the movements and reciprocal displacements of all layers – and every bending, swelling, bulging, or grooving directly changes the spatiotemporal coordinates of the entire embryo.

The folds which shape the membranes into structures are thus temporally and spatially coordinated with one another. They occur in the various dimensions and directions of space, simultaneously and subsequently, in the one or the other membrane, at one or the other end or at concrete locations. Every form that closes off to the inside simultaneously opens a new space to the outside – hence every outside is also an inside. In the course of development the distinction between formed and unformed matter is constantly reversed, every form is solely a temporary one and that which has been formed does not exist in order to exist but rather to disintegrate once more, to change its place, switch sides, transforming the outside into an inside and the surface into a body. Always anew, separated from each other and yet intertwined, differentiating and varying with each repetition, the folds successively peel out the embryo's form.

Pander first described membrane folding in his 1817 Latin-language *Dissertatio inauguralis sistens historiam metamorphoseos* (Pander 1817b). Appearing in that same year was also his *Beiträge zur Entwickelungsgeschichte* (*Contributions to Developmental History*) which in contrast to the dissertation was now a treatise of ten copper plates with the text merely placed alongside: having arrived at the limits of verbal description, Pander now delegated elucidation of the folding process to images. It is one of the outstanding achievements of Pander and Baer to have made the highly dynamic and complex process of folding comprehensible by means of and in the image.

In order to understand the folding, a single image is by no means sufficient. Instead, Pander portrays the folding in a series of images. In serial representation, moving the observer's gaze is crucial, on the one hand directing it to the figure and on the other having it glide from one figure to the next and thus evoking the impression of development – in which the two primitive folds repeatedly shift into and with each other – in the transition from one picture to the other (Fig. 4).

The sequence of images combines a number of pictorial means. Firstly, it constructs a form in the making by multiplying the number of images and bringing an entire series of individual images into play for the depiction; secondly, this sequence of images constructs development as a pictorial relation, i.e., as the ordered and lawful relation of the images to one another; thirdly, in this pictorial relationship, the gap is constitutive for the series. In this third aspect, what is shown and not shown in the series thus goes to making a well-ordered interval – image and empty space constitute each other. The sequence of pictures constructs the change from one form to another as well as each individual form, which can only be isolated as a single form in relation to the sequence. Fourthly, the series is distinguished by the principles of repetition and variation. While the individual images in the series are essentially a repetition of the previous image, the individual depictions nonetheless differ in one or a few central, divergent features. It is therefore not a repetition of an identical but a similar image. Hence the series produces no repetitive pattern but a variable sequence. Fifth and last, development arises solely as the impression that the entire series evokes as a whole.

As Baer writes, he began his own embryological investigations as a commentary to Pander's research (Baer 1828 and 1837 [reprint 1967] vol. 1, pp. XX, VI). The two plates that he subsequently published in the first volume of his developmental history are today icons of embryology.

Baer's schematic representations – arranged in rows, colored and sequentially numbered, showing the magnified embryonic structures and juxtaposing

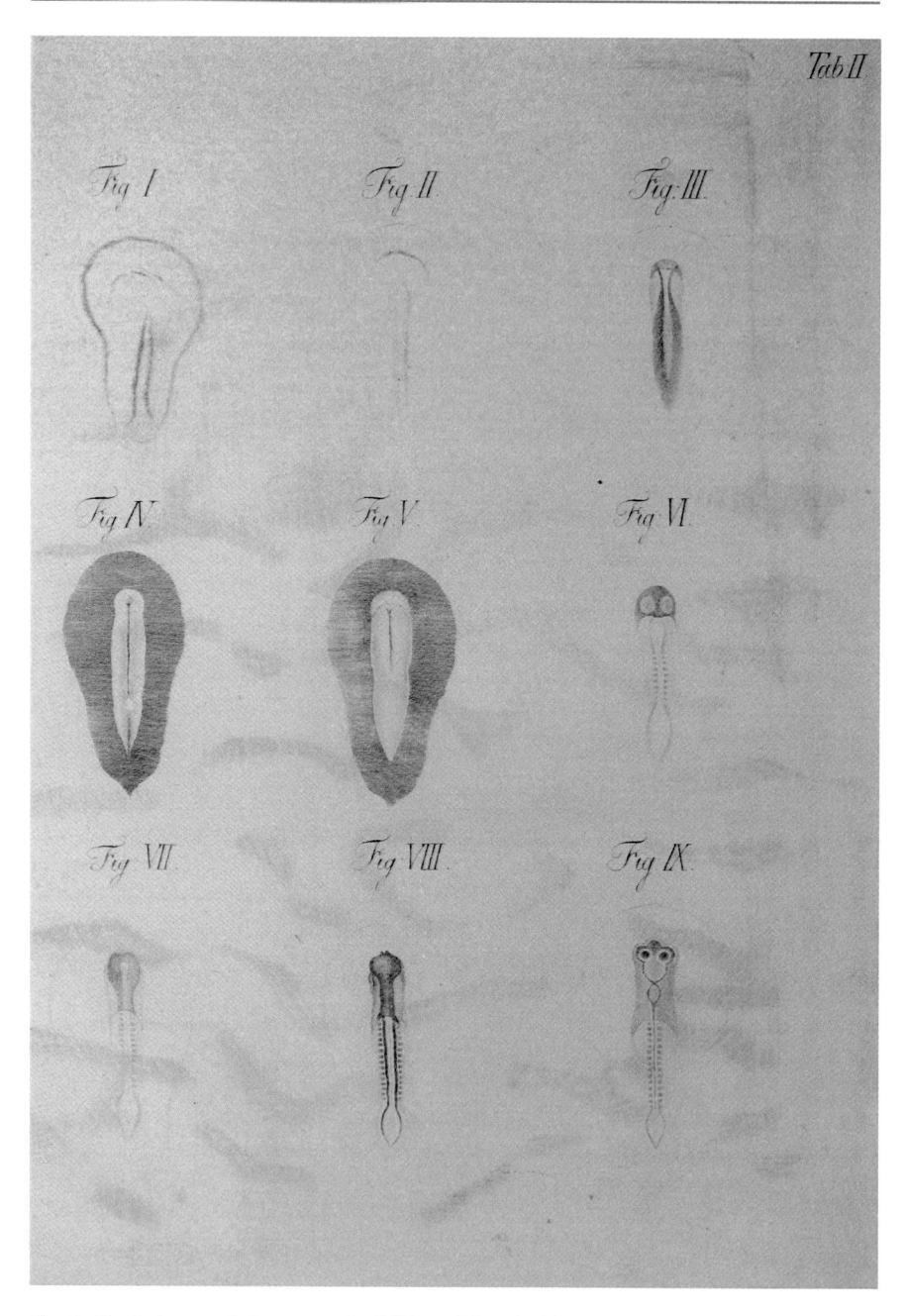

Fig. 4 Pander's second plate traces the folding of the primitive streak in nine consecutive figures. Christian Heinrich Pander, Beiträge zur Entwickelungsgeschichte des Hühnchens im Eye. Würzburg 1917, plate II, courtesy of the Library, Max Planck Institute for the History of Science, Berlin

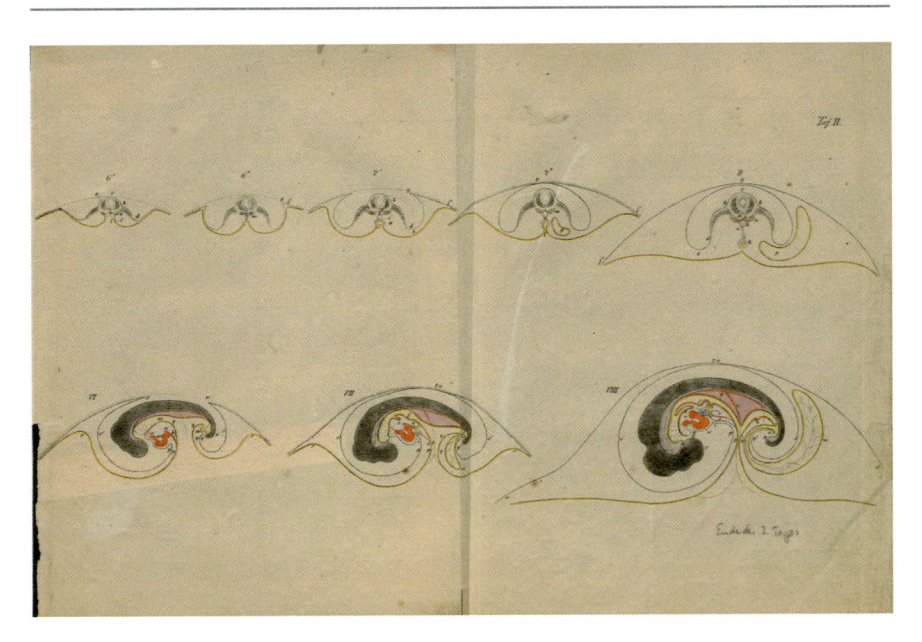

Fig. 5 On Baer's plate, transverse and longitudinal sections are aligned vertically in pairs. Karl Ernst von Baer, Über Entwickelungsggeschichte der Tiere: Beobachtung und Reflexion, vol. 1, plate II

longitudinal and transverse sections – succeed in visually ordering the embryo in its spatial structure (Fig. 5). If the verbal description of folding in his *Developmental History* borders on the cryptic, bringing the reader to the brink of despair, the pictures have an immediate clarity. By placing cross and longitudinal sections on a single plate, combining a horizontal and vertical axis of vision, Baer has our gaze glide in a zigzag pattern through the changes that the embryo undergoes in space and time. Alternating between the longitudinal and transverse views, the observer winches in a manner of speaking the surface of the picture back into space so as to reproduce the foldings and bendings of the membranes. Concomitantly, the forward-pressing gaze extends this torsion into time and advances with the changes that the structures gradually undergo.

At the same time, Johann Moritz David Herold (1790–1862) delivered an almost perfect serial representation. A professor of natural history in Marburg, Herold devoted his research to the metamorphosis of insects. In the early days of biology around 1800, development through metamorphosis played a significant part in the discussion around theories of generation because metamorphosis was still one of the big unsolved riddles in natural history. Despite the wealth of research and anatomical dissections of insects undertaken by Jan Swammerdam, Marcello Malpighi, Maria Sibylla Merian, and Pierre Lyonet, metamorphosis presented the fascinating conundrum as to whether scientists were dealing with entirely different animals or – notwithstanding their different stages – one and the same animal but merely in a new guise. While this preformationist view was often argued for, the opposite position – that such *diverse* guises as larva, caterpillar, or imago were interrelated even if they were entirely different animals – was much more difficult to defend not to mention substantiate empirically.

Yet this is what Herold did when he showed that metamorphosis was no one-time act of imprinting, as Harvey defined it, but rather a form of successive epigenetic transformation. He put forward the twofold argument that reproductive organs are the sole organs available to the adult insect but not to the caterpillar. He observed that the reproductive organs were not only formed from initial rudiments but also migrated in the caterpillar's body. By moving downward in the course of metamorphosis, they supersede other structures which they then replace, so in no case could they have been preformed.

One can trace the caterpillar's "mode of development" through the male and female sexual organs (Herold 1815, p. 12) in over 30 depictions that Herold himself drew and which were subsequently engraved in copper by the Nuremberg miniaturist Jakob Samuel Walwert (1750–1815). The viewer's attention is focused on the red (male) and yellow (female) markings of the sexual organs placed in the picture's center. Events in the center are similarly highlighted through a careful separation into foreground and background, in which all organs uninvolved in the metamorphosis are merely drawn schematically in outline. Isolated in this way and distributed across a series of pictures, the viewer traces the organs as they move from top to bottom through the caterpillar's body (Fig. 6).

In Herold's account, the caterpillar and the butterfly are two entirely different animals yet still connected to each other. Although in insects the starting and end point of development look radically different, from now on they would represent a developmental context because their connection no longer consisted in an external identity of structures but their inner becoming.

The Rhythm of Becoming

The theory of germ layers and their foldings posits development as the choreography of simultaneous spatiotemporal change. By constructing development as a relation of images, the series frees it from the chronology of observation. The individual pictures no longer correspond to a continuous observation at fixed points in time. Rather, the series follows the inner logic of the developmental process. So instead of a chronology of stages, we have relationships of forms in the image – or, to put it in a slightly different way, forms *as* relationships – which emerge visually in the series (Wellmann 2017b).

At this point, a new experimental technique played a central role. Employing a specially designed incubator, Pander and his teacher and mentor, the physiologist Ignaz Döllinger (1770–1841), were able to incubate and graphically depict more than 2000 eggs at once.

Because forms during development are in continual change, the observer is at first presented with only an undifferentiated abundance of material. That is why

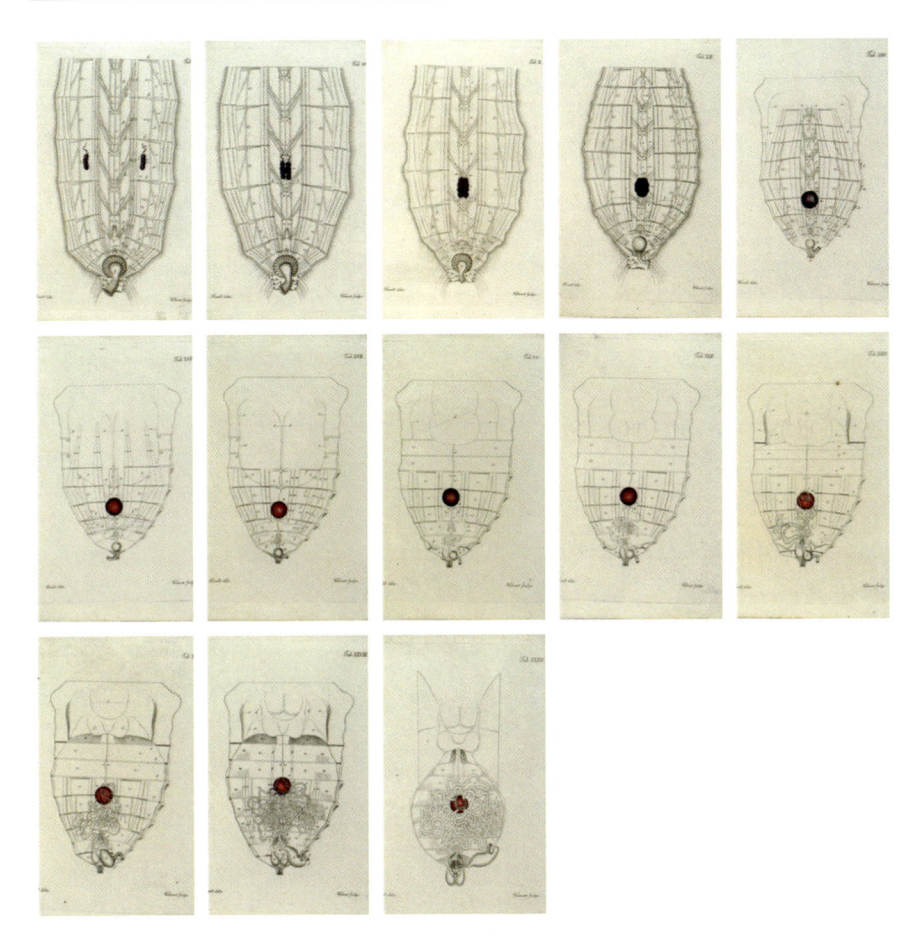

Fig. 6 The development of the male sexual organs of the cabbage butterfly larva. Johann Moritz David Herold, Entwickelungsgeschichte der Schmetterlinge, Marburg 1815

development cannot be observed, neither continuously nor in a single organism. Instead, it is constructed through the conscious selection of all those states and forms in the embryo that can be related to earlier and later forms by viewing and comparing them. It was only due to the wealth of observational material that Pander and his illustrator, the naturalist and copperplate engraver Joseph Wilhelm Eduard d'Alton (1772–1840), were able to select those stages of development that could be placed in relation to one another and together form a sequence. Pander was interested not in the chronology of observation, the timeline of incubation, but the sequence of changes that distinguished a specific structure of the embryo at various randomly chosen moments in development. Only the randomly selected and constructed sequence of images made it possible to make not time but the embryo itself and its sequence of changes the reference point of any investigation.

For this reason, the new epigenetic time of becoming was far more than a consequence of the generally diagnosed temporalization of the world circa 1800, which was reflected in new categories such as progress, revolution, acceleration, and especially the dynamization of nature (Koselleck 1967; Lepenies 1976); rather, epigenetic thinking established a specific temporality for organic matter. This temporality does not detach organic life from the general course of time in the world, for organic processes also take place in time. It is not time that structures organic development, however, but organic matter which structures time. The revolutionary aspect of epigenetic thinking and its pictorial form – the developmental series – did not lie in the fact that it reproduced a linear, chronological progression of time in the linear disposition of images. To the contrary, the developmental series constructed an autonomous, inherent order, a temporality embedded in the chronological flow of time but not identical to it. Organic temporality is a rhythmic one, a regularity that subjects the time of the organism to an oscillation between repetition and variation. Development is thus traced back to a law that binds the emergence of form via repetition to the already existing form, while at the same time designing it for a prospective future through variation. Rhythm establishes a complete autonomy of successive becoming that is based neither on the causal connection of its individual elements nor is it teleologically oriented toward a goal that is fixed from the outset. Instead, at each point in its interrupted sequence, development allows for a variation that issues in something new without abandoning the lawful order inscribed in the rhythm. Moreover, rhythm demonstrates its autonomy by simultaneously positing the individual links and the movement as a whole. The individual stages are underscored as independent units, yet they are no absolute entities but obtain their meaning exclusively within the nexus of relationships, the stages preceding and subsequent to them.

The folding of the germ layers is an equally rhythmic figure. Like a wave, the movement of the folding runs through the developing body in regular repetitions, differentiating and varying with each repetition. The folds therefore gradually make manifest the shape of the embryo in the choreography of their complex spatiotemporal movements.

Series of Sections

Replacing the preformationist idea of development via growth by the epigenetic notion of gradual formation led to the emergence of embryology as a scientific discipline. At the technical level, this was accompanied by a new experimental method that made it possible to re-mark the transition from the invisible to the visible. The microscope not only refuted the preformationist argument of invisibility through mere optical resolution. With the help of a new experimental technique, Pander and Döllinger also succeeded in detaching the embryo first from the shell and then, together with the germ layer, from the vitelline membrane so that it could be observed in the first days of development without injuring it (Baer 1972, p. 199). Hence, like all of their forerunners, Pander and Baer could investigate live chick eggs and embryos that were mostly undamaged and physically whole. This approach led to the establishment of epigenetic thinking and the sequence of images as a standard for the depiction of development in organisms and of continuous change in general.

Yet an image series was a pictorial surface, and in the further course of the nineteenth century, this flatness was to become radicalized, as it were, by directly encroaching on the organism and subjecting its body to the new scientific norm of two-dimensionality. The rise of embryological research in the remainder of the nineteenth century was to be significantly determined by a new technique that forced a completely different approach to development: cutting the organism into slices.

Observation using a microscope works best when the organism is converted as much as possible into a flat object. Various knives and cutting devices were already being deployed in the eighteenth century – for example, the cutting machine that botanist John Hill used for plant observations – to produce thin, even cuts. At the start of the nineteenth century, the word "microtome" began to be applied to just such refined cutting devices. Microtomes could not only produce individual sections but entire series of sections. Although histological sections were already known, the 1830s and 1840s saw major developments in tissue sectioning, driven by intensive cell research, which was epoch-making especially in the cell theory of the zoologist Theodor Schwann (1810–1882) and the botanist Matthias Schleiden (1804–1881) along with new microscopic developments (Bracegirdle 1978; Lawrence 2009).

Sectioning turned the organism into a specimen, and numerous techniques for conserving it emerged for both plants and animals. For instance, tissue sections were hardened and dyed with acids, colors, and resin through chemical processes, cast in Canada balsam, fixed on slides and in paraffin or cut in wax plates. They were viewed not just once but had the advantage of being drawn, photographed, printed in books and sent round the world, placed in collections, and being maltreated by students. Cutting techniques adapted the organism – split into the finest layers, secured now with a slide and cover glass – for observation, the device, the experiment. On the one hand each section served as interface for the next and on the other between organism and technology, between that which was still biological in nature and that which was already an artifact (Rheinberger 2006). Cutting the embryo, by the end of the nineteenth century, had established itself as a standard procedure for morphological investigations. For embryology, this meant locating development in the time between the not-yet and the no-longer and in the space between the many razor-thin slices through a three-dimensional body.

But from the very beginning, this flattening was accompanied by the converse attempt to keep the spatial object in mind. Baer's earlier juxtaposition of transverse and longitudinal sections was already the pictorial attempt to bring those developmental changes – which had been drawn in abstract and schematic form on the flat surface of paper – back into space to have the viewer's gaze rotate since he could not hold the specimen in his hand.

Tissue sections, produced in series with the microtome, could be used for the same purpose. The many individual surfaces could be combined into a body, this time as a three-dimensional specimen that could now in fact be held in the hands.

The Leipzig anatomist and embryologist Wilhelm His (1831–1904) made this procedure the linchpin of his embryological work. It was through His that human embryos found their way into embryology in the late nineteenth century. He made use of embryos which were mostly spontaneous abortions and miscarriages, sent to him by a network of doctors and collectors, in an attempt to standardize development in depictions which he called normal plates (Hopwood 2000, 2007). His began his work with sections and plastic reconstructions through use of that classic organism of embryological study, the chick embryo. But later he also subjected human embryos to a microtome and an adjustable object table to produce tissue sections for what he called "reconstructions of plastic views" (His 1870, p. 231) of the entire embryo or individual organs whose development he studied. He fashioned material models by using the sections, with the help of orienting drawings, to freely model the body. Owing to his tactile skill and intellectual grasp, he was able to reassemble the individual sections into a body (Hopwood 1999, 2004).

But the reconstruction of three-dimensional bodies from series of sections could also be employed in a precise procedure by which body volume and the exact surface of an object could be determined. One such method was Gustav Born's plate modeling, which became the standard in vertebrate embryology around 1900. The goal of wax plate modeling is to achieve not only a spatially correct but enlarged model of the organism as divided in sections. Stated very simply, the process involves drawing a segment from the series of sections on a wax plate (Fig. 7). Thickness of the plate is determined by enlargement of the segment – and the plate's greater thickness in relation to the section is equal in proportion to the greater surface area of the chosen segment after its enlargement. Correctly glued together, one after the other and with the help of lines and planes for orienting and positioning, a spatial model arises from the individual plates which can in turn be used at one's discretion.

Significant in this procedure of spatial modeling is the fact that here, as in the developmental series drawn by hand, time can be compressed, condensed, and

Fig. 7 Wax plate modeling. Karl Peter: Die Methoden der plastischen Rekonstruktion. Jena 1906, Fig. 34

manipulated. For instance, if the structure whose development is of interest changes only slowly, not every sectional plane must be transferred to a plate. Instead, a selection of the morphologically significant moments is made. In the model, this is then reflected in the strength of the plate, which must be all that many times thicker than the omitted plate steps (Peter 1906, p. 97).

Living Organisms and Dead Material

Wilhelm His's predecessors in the seventeenth, eighteenth, and nineteenth centuries had looked at living embryos, whereas His studied histological sections. By combining the slices in the model, the shortcomings in the organism's cut-up corporeality seemed to be compensated for and even given the scientific advantage of additional manipulability. But behind reconstruction of a body from the surface area, there lay a central decision of the embryologist that could hardly be concealed – a decision which impacted research far into the twentieth century: the choice between observation of the intact living organism or the dead specimen.

Embryology after 1900 took an experimental turn and led proponents such as Wilhelm Roux (1850–1924) to favor "developmental mechanics" (*Entwicklungsmechanik*) and invasive methods that operated with live embryos (Mocek 1998). The forms of representation also kept up with the times: instead of time-consuming series of drawings, photography allowed for rapid snapshots and chronophotographic developmental sequences. Added to this was the advent of cinematography at the start of the twentieth century. Embryologists were among the first to adopt cinematic technology. Not only did they have long training in classic histological approaches, staining techniques and dyes, but they were accustomed to crafting and adjusting their devices to suit their delicate objects of study and specific research needs. Consequently they were quick to welcome photographic and cinematographic practices as well as their devices.

The Swiss biologist Julius Ries (1879–1949) is primarily known in the history of biology for his pioneering work with cinematography. His 1909 study *Kinematographie der Befruchtung und Zellteilung* (*Cinematography of Fertilization and Cell Division*) is today regarded as the first known filmic depiction of development (Ries 1909). In the preceding year, he had published the slim and virtually unknown volume *Beiträge zur Histologie und Physiologie der Befruchtung und Furchung* (*Contributions to the Histology and Physiology of Fertilization and Cleavage*) (Ries 1908).

The problem of an organism's corporeality and vitality did indeed prove difficult, and in very different ways, for many questions of developmental biology. Research required variable interfaces between the organism, equipment, and representation. Ries investigated cell division and fertilization with experimental and pictorial means that were almost mutually exclusive. His work allows today's researchers a glimpse into the laboratory of the embryologist, who not only filmed but in whose work observation and experimentation, modeling, drawing, and photography, specimen and live observation never failed to complement one another.

Ries had traveled to the Stazione Zoologica of Naples in 1907 to study development of the sea urchin. To observe fertilization, he first collected and drew the eggs and sperm of frogs, starfish, and salamanders "in the freshest possible state" (Ries 1909, p. 5). He observed the eggs in seawater, to which he added a dye solution of gentiana, rose aniline, or neutral red solution. His goal was not to color the whole egg but merely the region of the egg membrane that was central to fertilization. In this way, Ries was able to observe how the membrane was gradually colored from the edge to the inside and how the spermatozoa moved independently through the dyed mass with assistance of their tail movements until finally a sperm actively – through rotation of its flagellum – bored into the membrane at whatever spot (Ries 1908, pp. 34, 36).

The coloring used to trace the fertilization made the egg's periphery visible, but the procedure was only suitable at low magnification. In order to observe the coming together of ovum and sperm, however, Ries needed a higher resolution so he changed the registers: he no longer investigated living organisms but fixed specimens, and he made photographs instead of drawings.

Ries first dried the dyed specimens and encased them in Canada balsam. He then took a series of microphotographs of these fixed specimens and the spermatozoa movements near the embryonic membrane contained in them (Ries 1908).

But even in deploying these means, Ries was unable to prove his scientific hypothesis. At the start of the twentieth century, the moment of fertilization was still one of the most significant and mysterious phenomena in the developmental process. It marked the beginning of a new living creature and was the trigger for a rapid series of events inside the egg that led to the first cell cleavage. Ries asserted that the sperm did not, as generally presumed, lose its flagellum upon penetration of the egg membrane. To the contrary, according to Ries, it continued to move even subsequent to penetration and was responsible for further development within the egg – more specifically for the radial structure that appeared shortly after fertilization. Yet movement of the flagellum could not be seen in the photographs. Ries needed plasticine for this and he built a model.

Ries was interested in a sperm's movement, more specifically the rotation of its flagellum, which he assumed moved solely of its own accord. It was this rotation that led to the radial organization of the interior. The model that Ries built for his hypothesis consisted of a sphere of plasticine doing service as the egg and a needle with a head representing the sperm. To simulate fertilization, Ries guided the tip of the needle into the sphere as far as its middle while accompanying this forward movement with the needle's conical rotation (Ries 1908, preface). The sperm's rotation thus showed itself as movement tracks in the plasticine. Then Ries concocted a second model for this rotational movement by chronophotographically recording – repeatedly in short intervals – the movements of a stem mounted on a spinning top to chart the formation of rays (Ries 1908, p. 50). But in the end, Ries published neither these chronophotographs nor photos of his model but rather a schematically drawn picture series of the simultaneous forward and twisting movement of the rotation (Ries 1908, plate VI, Figs. 5a–c) (Fig. 8).

Fig. 8 Photography and model of the egg membrane and sperm rotation. Julius Ries, Beiträge zur Histologie und Physiologie der Befruchtung und Furchung. Bern 1908, plate VI

Ries thus developed his scientific hypothesis of the sperm's flagellum movements in fertilization by alternating back and forth between the living organism and material pressed under cover glasses; by dyeing, drying, and steaming living material; by photographing specimens or building models which he transported back in time by means of chronophotographs, or which he abstracted in the schema.

In contrast to fertilization – the main thrust of his research – Ries was not interested in a new scientific hypothesis regarding cell division but rather in bringing already established knowledge to life. Ries was "completely overwhelmed" by the study of cell division in living animals, and the film camera allowed him to reproduce these effects for others (Ries 1909, p. 1). But the effect of aliveness, with which Ries was able to demonstrate division of the cell in the film, could not be transferred so easily to his actual research interest: whether and how the sperm moves onward as soon as it penetrates the ovum's membrane. The fertilization could not be made visible with a living egg and captured on a strip of film. At the same time, back then there was no medium more suitable to depicting movement than film.

The solution that Ries seized upon was highly original: he drew the fertilization in a series of images in the form of a film strip, i.e., he staged the radial organization inside the egg as a film (Fig. 9). In this way, he coupled the drawing with the iconography of film, combining the scientific hypothesis obtained through

16   Julius Ries:

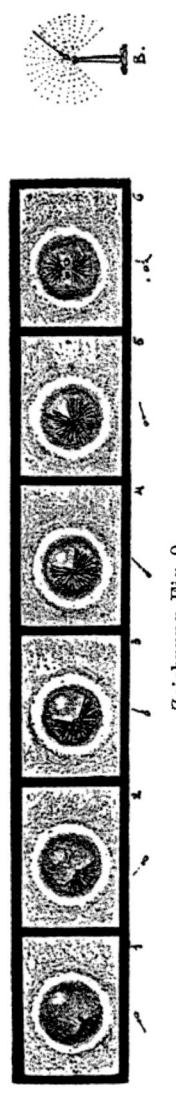

Peripherie zum Zentrum rückt (Textfig. 9). In O. Hertwigs Handbuch der Entwicklungslehre Fig. 165 III, S. 507, sieht man ein Stadium, in welchem die Strahlen schon fast eine Sphäre bilden; doch fehlt noch ein zum Eizentrum gerichteter Kegel. Erst nachdem auch in diesem Kegel sich Strahlen gebildet, sieht man das Sonnenstadium. Auch in anderen Abbildungen vieler Autoren konnte ich solche Kegelbildungen finden, ohne dass dieselben irgendwie beschrieben werden, für meine Anschauungen aber sind dieselben von grosser Wichtigkeit.

Wer sich die Möglichkeit der Entstehung der Strahlung durch die Rotation des Spermienschwanzes und die damit verbundene Bildung solcher Kegelfiguren leicht verständlich machen will, kann dies mit Hilfe eines leicht herstellbaren Modelles erreichen. Eine Modellierwachskugel stellt das Ei vor, eine lange Nadel mit Doppelknöpfchen vertritt das Spermium, wobei also die beiden Knöpfchen Kopf und Zentrosom, die Nadel den Spermienfaden darstellt. Wenn die Nadel mit dem Knöpfchen voran von einem Pole (Nord) bis zum Mittelpunkte der Kugel vorgestossen wird, so dass die Nadelspitze an der Einstichöffnung noch herausragt, so kann man die bei der Befruchtung beschriebene Drehung des Kopfes und Zentrosoms ausführen. Mit der Spitze der Nadel beschreiben wir vom Nordpole ausgehend eine Spirale an der Oberfläche der Kugel über den Äquator zum anderen Pole. Die an der Nadel befestigten Knöpfchen (die wir uns im Zentrum wie in einem Kugelgelenke bewegt denken können) haben der Nadelbewegung folgen müssen, haben eine Drehung vollführt und das Zentrosom ist jetzt dem Südpole zugekehrt.

Wenn wir durch die von dieser Schraubenbewegung im Wachs hinterlassene Spur einen Querschnitt machen, erhalten wir eine Strahlenfigur.

Fig. 9 Film strip drawn by hand. Julius Ries, "Kinematographie der Befruchtung und Zellteilung," *Archiv f mikrosk Anat* 74 (1909), Fig. 9

specimens and models with the immediate visual perception and vitality of film footage. As film had done with other developmental processes such as cell division, a real authenticity and animation were thereby lent to a thesis that Ries would have otherwise been unable to demonstrate through technical and systematic means. The drawn film strip was the pictorial solution for Ries's research, for his scientific hypothesis, and was last but not least an amalgam of the most diverse techniques, materials, and procedures to be placed in service of his investigations.

Time and Experiment

Although film as an investigative tool stayed relegated to the margins of embryological research in the twentieth century – notably on the continent and especially in Germany where it was effectively ended by the Second World War – research using film nevertheless enjoyed a lively presence elsewhere. The Anglophone world in particular established a film-research culture that went beyond the caesura occasioned by the war and was borne by developments such as cellular culture and time-lapse microcinematography (Landecker 2006). In the 1960s and 1970s, filmic methods were still an important research tool, for instance in the work of Michael Abercrombie (1912–1979) on embryology, fibroblasts and cell behavior in general as well as in the work of Lewis Wolpert (b. 1929) and Tryggve Gustafson (b. 1911) on morphogenesis and in the work of Marcel Bessis (1917–1994) on sudden cell death (Landecker 2011; Wellmann 2018).

While those cinematographers among biologists did all they could to keep the embryo alive, embryology unleashed its greatest scientific forces in the twentieth century when it intervened structurally in the organism and tested the limits of what put an end to its life or at least gave its course of development a serious jolt. In famous experiments, Wilhelm Roux (1850–1924), Hans Driesch (1867–1941), and Hans Spemann (1896–1941) treated the embryo at its two-cell stage with shaking and hot needles, laced it in two with a hair, or transplanted tissue at different developmental stages between embryos of various species. Hilde Mangold (1898–1924), Spemann's doctoral student at the Zoological Institute of the University of Freiburg, conducted crucial transplantation experiments in 1922 – experiments that would prove pathbreaking for embryology and for which Spemann won the 1935 Nobel Prize in Physiology or Medicine (Allen 2005).

In their transplantation experiments with newts, Spemann and Mangold discovered a region in the early-stage embryo (today known as the Spemann-Mangold Organizer) which lays down the embryo's axis. They transplanted the dorsal lip in the early stage of the gastrula to the ventral side of an embryo of another darker-colored species and observed that the transplanted tissue there induced the development of a second body axis. These experiments led to the insight that cells transplanted at a very early stage developed according to the new place to which they were transferred and not according to their origin, and that the organizing region could steer the development of other cell groups in the host embryo.

Such experimental interventions in embryology aimed to decipher the mechanisms and causal nexuses of development and, in the second half of the twentieth century, rushed to ever new successes with the new tools of molecular biology, biochemistry, and genetics (Allen 1979).

In the mechanical-causal investigation of development, time is unimportant. The organism is reduced by the time factor because it plays no role in the functioning of the mechanism, which is the main concern. By contrast, film takes the organism back in time. Time is palpable in film, which makes time the *conditio sine qua non* of development by focusing on its perception. But cinematography also exceeds the limits of human perception and turns the body inside-out and simultaneously into time. Because of film's manifold relations to time, historians have distinguished between the time of experiment, the time of recording, and the time of demonstration (Landecker 2006). Most importantly, therefore, time can be manipulated – it can be condensed, stretched out, sped up and the film played forward or backward or stopped.

With their film cameras, the embryologists observed not only how the cells moved and behaved, where they came from and what their purpose in the course of development was, but these moving images broadened the view for a new kind of research and for thinking further about the mechanisms of development under the condition of their being played out in time and, importantly, being played out in different ways.

Optical Slicing

With the new millennium, new optical procedures have revolutionized biology. While at the beginning of embryological research Pander and Baer had to first find a new pictorial convention in order to give the mechanism of development the form of a fold, in the twenty-first century, images have served the converse function, namely finding the mechanism by which they are produced.

The abundance of various optical, physical, and biochemical techniques and their further development has been continuously and expeditiously growing in the first decades of this century. Among the most significant microscopic techniques for embryology are not those which *de facto* cut the organism into razor-thin layers but instead use light. In essence, the microscope focuses the transmitted light in a focal plane. Optimal sharpness in the focal plane cannot be achieved in a normal light microscope because light enters the lens from other parts of the organism under inspection. Fluorescence microscopy, on the other hand, works with marking by way of fluorescent substances and laser light, which means that only selectively illuminated fluorescence can be detected and the organism successively scanned in individual focal planes.

In confocal microscopy, the manufacture of such single focal planes (optical sections) is optimized through elimination of out-of-focus light by means of a pinhole to better control the focus and achieve a sharper focal plane of the fluorescent material. In light-sheet fluorescence microscopy (LSFM), an entire micrometer-thin volume of the specimen is illuminated. The laser comes from the side, scanning happens orthogonally to the fluorescence detection. This produces advances in imaging speed while reducing the light exposure of the specimen. The laser beam moves vertically while the fluorescence emitted by the organism is detected by a camera, which creates an image slice. Once a single section is generated, a series of such sections, made at different focal planes, can be stacked and thus allow for the reconstruction of three-dimensional images. In LSFM, the specimen can be moved slightly albeit continuously along a horizontal line or, using a rotary stage, can be recorded sequentially from multiple views. After having been completely scanned from one viewpoint, the organism is rotated 180 degrees and scanned again. In a more advanced optical setting, called SiMView, even simultaneous multiview imaging is possible by illuminating the embryo from two opposite directions. Finally, scanning the specimen in every position at certain intervals (of some 30–90 s) over a given period will succeed in adding the dimension of time to spatial image acquisition.

By circumventing the diffraction limit for the maximum possible optical resolution in the microscope – calculated by Ernst Abbe in the 1870s – fluorescence microscopy opens up microscopy for investigation into the smallest subcellular and molecular structures. For their work in the field of fluorescence microscopy, which is of enormous potential for research into biological mechanisms, Eric Betzig, Stefan W. Hell, and William E. Moerner were awarded the 2014 Nobel Prize for Chemistry.

Of no less value than the expanded investigative sphere at the now molecular level is vitalization of the microscopic object as made possible by fluorescence microscopy. In the 1960s, the naturally occurring Green Fluorescent Protein (GFP) was discovered in a species of jellyfish; in the mid-1990s, GFP was successfully fused with other proteins and thus deployed as a marker for these proteins or the cell structures tied to them. Now it was possible to obtain a glimpse into the living cell at the subcellular and molecular level, revealing a cosmos of exuberant activity in its interior. For the very first time, one could now directly observe, in real-time under the microscope, processes such as mitosis and meiosis, the work of mitochondria or the transport of molecules (Liu et al. 2018).

Embryos, in Silico

In embryology, the new techniques found prominent application in the project of a "digital embryo," which made a strong impression on the scientific community (Vogel 2008). The aim of the project was twofold. For one, it sought the "in toto representation," the "in vivo" or "live" imaging of embryos. It attempted to represent early embryogenesis in model organisms such as drosophila and zebrafish in animated visualizations of the movement patterns of cells, showing the organism's development by tracking every single cell from every angle and moment by moment. For another, it sought to construct a model of embryogenesis – or, more precisely, a predictive computer model of development – which is why the experimental data obtained from scans and visualizations had to be quantitatively assessed (Keller 2013; Megason and Fraser 2003).

In the methods and goals of such system biology approaches, the experimentation on as well as the observation and manipulation of living embryos is inextricably linked with computation. Observation and experimental regimes, modeling and visualization, materiality and virtuality are telescoped and thus perpetually intertwined (Wellmann 2017a).

The microscope is first and foremost an optical device. It enhances seeing so that the observer can view what was previously undetectable with the unaided eye. But seeing here no longer means the qualitative act of an observer, it does not even presuppose it. The in vivo imaging serves merely as a vehicle to digitize the information (Phillips 2007) and seeing becomes a computational analysis, for the microscope generates data that is registered and followed up by a computer program. Central to this procedure is firstly the process of segmentation. What this means is the process of automatically identifying and tracking a cell – or more precisely a biological "object of interest" (e.g., fluorescently labeled cell nuclei or cell membranes) in an image – over time with the help of software and algorithms. In a 24-hour time-lapse film of zebrafish development, for example, no less than 20 million segmented objects are identified. Segmented objects are referred to as "traces" and are subdivided into different classes: figures (2D), meshes (3D), tracks (4D, i.e., including the time dimension), and lineages (branched trees representing cell lineages). It is by means of segmentation that pixel-based images such as those resulting from the optical slicing of the embryo are then "converted" into collections of traces. As traces – that is, numerical datasets – the segmented objects can be annotated with additional data such as cell width, cell type, cell location, and cell velocity (Megason 2009, p. 329, Megason and Fraser 2003, pp. 1412–1413) (Fig. 10).

With this shift from seeing to collecting data, the transition is one that goes from observing the single cell to observing the behavior of all cells on the systems level, in other words, from the biological experiment to scientific modeling (Xiong and Megason 2015). Powerful software is needed for the millions of cells that have to be identified and subsequently tracked over time during embryogenesis and so as to gather all the information extractable from the images for quantitative analysis. In addition, this huge amount of data relies on a heavy computational infrastructure, especially databases to store and manage the data.

In fluorescence microscopy, there is thus no simple distinction between a biological entity (the cell in the embryo), a visual object (the pixel or pixel-based image), and the computational trace (the cell segmented by an algorithm) but instead a constant back and forth, a flowing conversion and reconversion of one into the other.

In these cycles of conversions, there is yet another one. Not only in the segmentation is the pixel transferred into the trace but also the trace back into the image, i.e., once data is gathered, it can be put back onto the screen. This process is called rendering. The traces can be annotated with color, for example, in order "to mimic the original fluorescent colors of the cells or to visualize an annotation such as cell type or cell velocity" (Megason and Fraser 2003, p. 1415). Rendering happens with the use of commercial software, often adjusted, but tailored to the specific needs and

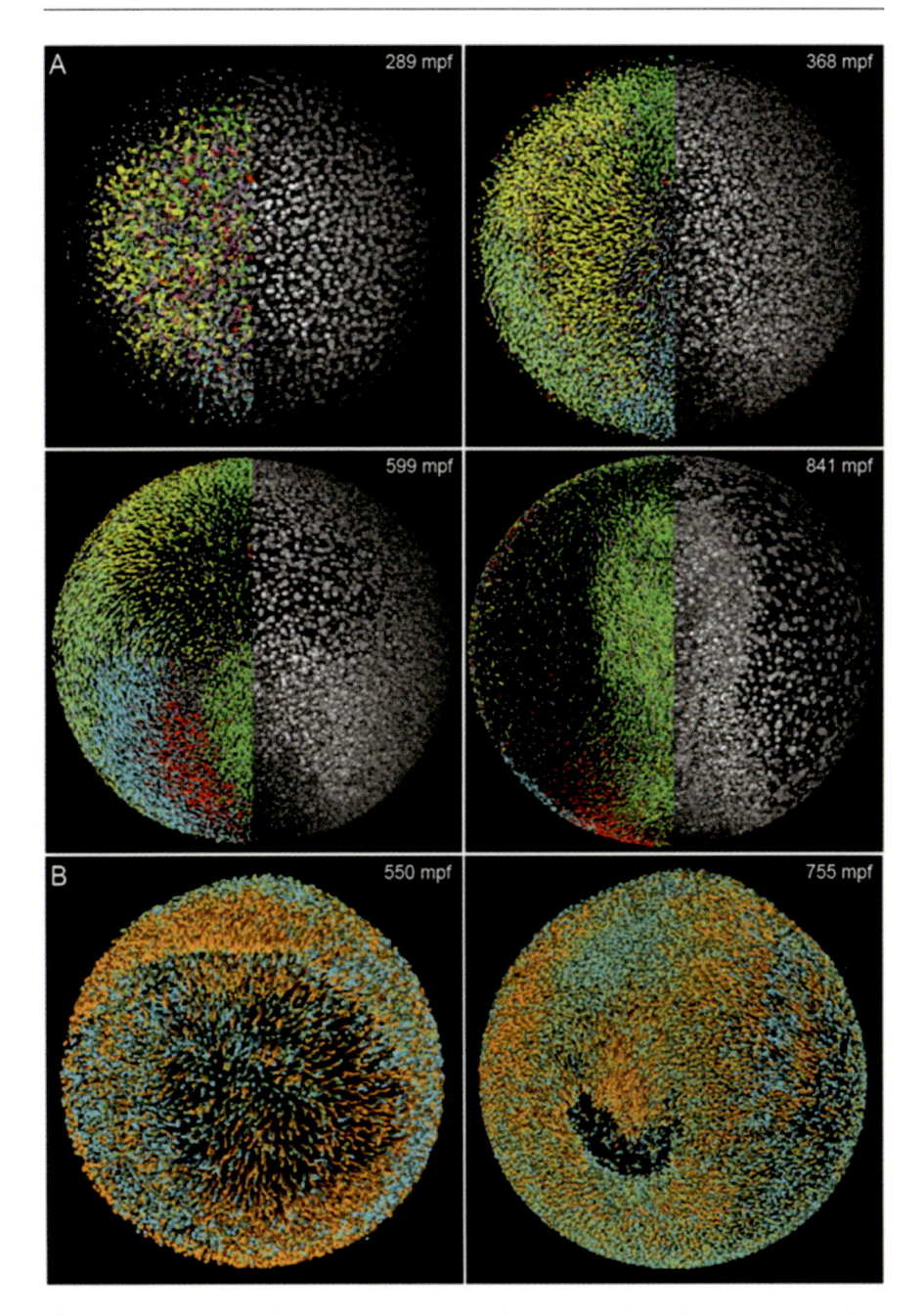

Fig. 10 In toto imaging of the embryo, microscopical and computational views are juxtaposed and annotated with colors. Philipp J. Keller, "In vivo imaging of zebrafish embryogenesis," *Methods* 63/3 (2013), p. 276, with permission from Elsevier

purposes of the research in question. Finally, reconstructing 3D images from stacks of optical sections and rendering them over time as animated representations that permit rotation or zoom is another step that must be taken, and this again demands high-powered computing (Khairy and Keller 2011).

In developmental biology today, the scientific animations of development – i.e., following the trajectory of every single cell over the entire course of embryogenesis – constitute a distinct imagery and visual quality that is indispensable to the field. But also far beyond the disciplinary boundaries of embryology, intensive coloration, complementary colors, strong contrasts, and exuberant activity are a sign of the ubiquitous imagery of contemporary biotechnology.

Algorithmic Animation

The mere unwinding of film has captured development as a process. As a rule, live imaging and in vivo visualizations are also called movies or films, yet computer animations are fundamentally different from twentieth-century film. As the film scholar Tom Gunning puts it, film presents "movement automatically captured through continuous-motion picture photography." Animations, by contrast, are "moving images that have been artificially made to move" (Gunning 2014, p. 40).

In the case of the digital embryo, the continuous movement of the cells is first discretized, otherwise it is not analytically tractable for the computer (Hinterwaldner 2017). Then it is resynthesized into an animate visualization whose parameters are set to the human sensory apparatus. Hence the animations are related to modeling. But unlike computation, which tackles the cognitive problem of how to model organic processes, these animations are visual solutions to the problem of representing constantly changing forms. What distinguishes scientific animations in the computer, then, is a feature that Gunning more generally ascribes to animations: they are not only "displaying but also *playing* with the production of motion" (Gunning 2014, p. 40).

Animations, in other words, draw attention to the conditions of their own making. Other than their names suggest, "in vivo" or "live" imaging fails to capture cell movement or, for that matter, development in a specimen while being observed alive but recreates movement computationally. These images are designed to not only move and capture the quality of continuous change and dynamics that characterize development but also to make the aliveness of such processes palpable. Twenty-first century computer animations have thus traveled a long road from the early cinematographies of development that fascinated researchers such as Julius Ries because they convinced the observer of the "truth of life" in a very intuitive fashion.

The animations of development, like those in the digital embryo, are technical images or, more precisely, high-tech images (Mersch 2006). They long ago abandoned the simple notion of illustrability. Instead the animated pictorial worlds of the digital embryo are operational analyses of development, they are a "digital recreation" (Sean Megason), a program of an embryo that can consequently also be assembled, disassembled, changed, and processed like a computer program and whose development likewise takes place with the same predictability and manipulability as the software running on a computer.

Thus the animations provide vividness without disclosing the complex conditions and algorithmic operations that underlie their production. Furthermore, they constitute scientific evidence – simply *qua* visual perception and pictoriality, whose technological constitution they simultaneously conceal. For development as process, i.e., as a continuous transition from not-yet to no-longer, corresponds to no actual sensory event as suggested by the image of an object and no matter how technical it might be. By definition, development is elusive. Thus the power of images lies precisely in their depiction of processes, more so than in other cases, "as being looked at as pictures without functioning as such" (Mersch 2006, p. 410). Paradoxically, this reverses the relationship of image and nature. Whereas in the past the limits of the image were known and nature seemed limitless, it is now nature which seems to be moving within the limits of the image.

Conclusion

Development takes place ceaselessly before our eyes in all conceivable variety and abundance of forms. As easy as it is to name, describe, and survey as a whole, development still eludes analysis, detailed comprehension, and a unified conception of individual entities.

The first embryologists began their work by subjecting development to their ticking watches. Development was that which took place before their eyes and what the observer saw as a change in relation to time. One can hardly overestimate the epistemic shift that occurred when in the late eighteenth century the change of forms among themselves became a reference system of development. Now development became the relation of morphological forms, and it was these forms and exclusively their reciprocal relationality to one another which carried development through time. No longer was the point in time of the observation selected; rather, the thing to be observed was carefully chosen and the course of time no longer defined the individual forms but bound them all together in the background to create a whole.

From today's perspective, it would seem that the developmental series played an almost trivial role in this shift. Before the advent of Herold, Pander, and Baer, the series was an unknown representational form in the field of biology – but in their wake, development was never depicted in any other way. To this day, the series is the standard iconography for representing any kind of change, transformation, and process in biology and elsewhere.

Yet the picture series was only the most visible expression of a form of investigation of development where over the centuries the boundaries between the objects and their images, between surfaces and bodies, between living and dead ceaselessly shifted, enhanced, and excluded each other.

In the nineteenth century, the picture surface became the physical section – and along with it the intact living embryo became a dead specimen in pieces. Chronophotography and film took development back in time, now under the auspices of its mechanical manipulability. The freely chosen tempo of the recording and playing back of development made visible those processes that would not have existed without the device, while models and three-dimensional reconstructions in turn brought a new tactility and manipulability to the investigation. Both of these together made the living embryo's time and space into the scientific parameters of the experimenter.

In the animated optical slices of contemporary fluorescence microscopy, it is no longer the embryo's time and space that are negotiable but its very aliveness. In developmental systems biology, the embryo is both an intact body and a dissected specimen; a surface, a unit, a section; an experiment and a program; it is injured and unimpaired; accelerated and stopped; physical and virtual at one and the same time.

It is the advent of novel digital-imaging technologies that has also allowed for a "rebirth" of human embryology in the new millennium. In 2009, for the first time, a list of exclusively human embryonic terms was published by the Federal International Committee for Anatomical Terminology. Moreover, in the past two decades, huge amounts of funding and collaborative initiatives have enabled the digitization of embryo sections stored for decades in collections which were started in the early twentieth century all over Europe and Japan – among them the prominent Carnegie Collection, now in Washington, and the Kyoto Collection, now the largest collection of human embryos in the world (Gasser et al. 2014; Hill 2018). These efforts resulted in the creation of massive databases of virtual human embryos (such as http://www.ehd.org/virtual-human-embryo). Imaging technology now makes the human embryo accessible and searchable via "animations, fly-throughs and 3-D reconstructions" in ways that were previously unimaginable (Gasser et al. 2014; Hill 2018; Yamada et al. 2018). Yet the human embryo has not only become visually exposed but has been subjected to invasive scrutiny, not least since CRISPR-Cas technology and jurisdiction now allow for the experimental editing of genes in early human embryos (Rossant and Tam 2018).

For a long time, the scientific image was characterized as an illustration, as a mere explanatory adjunct to a superordinate text and the knowledge it articulated. In the twenty-first century, it seems, by contrast, that it is animations, fly-throughs, or optical dissections of virtual embryos that constitute new knowledge. Or to put it another way: it is these novel imaging options that now send science in search of knowledge that illuminates the image.

References

Adelmann HB (1966) Marcello Malpighi and the evolution of embryology, 5 vols. Cornell University Press, Ithaca
Allen GE (1979) Life science in the twentieth century. Cambridge University Press, Cambridge
Allen GE (2005) Mechanism, vitalism and organicism in late nineteenth and twentieth-century biology: the importance of historical context. Stud Hist Philos Biol Biomed Sci 36(2):261–283. https://doi.org/10.1016/j.shpsc.2005.03.003
Arni C (2018) Pränatale Zeiten. Das Ungeborene in den Humanwissenschaften (1800–1950). Schwabe, Berlin
Baer KE (1828 and 1837 (1967)) Über Entwickelungsgeschichte der Thiere: Beobachtung und Reflexion, 2 vols. Bornträger, Königsberg. Repr Brussels

Baer KE (1972) Nachrichten über Leben und Schriften des Herrn Geheimraths Dr. Karl Ernst Baer, mitgeteilt von ihm selbst, 2nd edn. reprint of the 1886 edition. Hirschheydt, Hanover-Dören

Bracegirdle B (1978) A history of microtechnique: the evolution of the microtome and the development of tissue preparation. Heinemann, London

Buffon GLL, Daubenton JM (1804) Volume 2: Histoire naturelle, générale et particulière: avec la description du Cabinet du roi. Imprimerie royale, Paris

Detlefsen K (2006) Explanation and demonstration in the Haller–Wolff debate. In: Smith JEH (ed) The problem of animal generation in early modern philosophy. Cambridge University Press, Cambridge, pp 235–261

Fabricius ab Aquapendente, H (1625) Opera physica anatomica: De formato foetu, venarum ostiolis, formatione ovi, et pulli, locutione et eius instrumentis et ovi. Roberto Meglietti, Padua

Gasser RF, Cork RJ et al (2014) Rebirth of human embryology. Dev Dyn 234(5):621–628

Gunning T (2014) Animating the instant: the secret symmetry between animation and photography. In: Beckman K (ed) Animating film theory. Duke University Press, Durham, pp 37–53

Haller A (1758) Sur la formation du cœur dans le poulet sur l'œil, sur la structure du jaune &c. 2. Bousquet & Comp, Lausanne

Harvey W (1965) The works of William Harvey (trans: Willis R). Johnson Reprint Corp, New York

Herold JMD (1815) Entwickelungsgeschichte der Schmetterlinge, anatomisch und physiologisch bearbeitet. Kriegersche Buchhandlung, Cassel/Marburg

Hill MA (2018) Two web resources linking major human embryo collections worldwide. Cells Tissues Organs 205(5-6):293–302

Hinterwaldner I (2017) The systemic image: a new theory of interactive real-time simulations. MIT Press, Cambridge, MA

His W (1870) Beschreibung eines Mikrotoms. Archiv f mikrosk Anat 6(1):229–232. https://doi.org/10.1007/BF02955980

Hopwood N (1999) "Giving body" to embryos: modeling, mechanism, and the microtome in late nineteenth-century anatomy. Isis 90:462–496

Hopwood N (2000) Producing development: the anatomy of human embryos and the norms of Wilhelm His. Bull Hist Med 74:29–79

Hopwood N (2004) Plastic publishing in embryology. In: Hopwood N, Chadarevian S (eds) Models: the third dimension of science. Stanford IP, Stamford, pp 170–206

Hopwood N (2007) A history of normal plates, tables and stages in vertebrate embryology. Int J Dev Biol 51(1):1–26. https://doi.org/10.1387/ijdb.062189nh

Keller PJ (2013) Imaging morphogenesis: technological advances and biological insights. Science 340:1234681–12341610. https://doi.org/10.1126/science.1234168

Khairy K, Keller PJ (2011) Reconstructing embryonic development. Genesis 49:488–513

Koselleck R (1967) Vergangene Zukunft der frühen Neuzeit. In: Barion H (ed) Epirrhosis: Festgabe für Carl Schmitt. Duncker & Humblot, Berlin, pp 549–566

Landecker H (2006) Microcinematography and the history of science and film. Isis 97:121–132

Landecker H (2011) Creeping, drinking, dying: the cinematic portal and the microscopic world of the twentieth-century cell. Sci Context 24(3):381–416

Lawrence S (2009) Anatomy, histology, and cytology. In: Pickstone J, Bowler PJ (eds) The Cambridge history of science. Volume 6: The modern biological and earth sciences. Cambridge University Press, Cambridge, pp 265–284

Lepenies W (1976) Das Ende der Naturgeschichte. Wandel kultureller Selbstverständlichkeiten in den Wissenschaften des 18. und 19. Jahrhunderts. Hanser, Munich

Liu T, Upadhyayula S, Milkie E et al (2018) Observing the cell in its native state: imaging subcellular dynamics in multicellular organisms. Science 360(6386). https://doi.org/10.1126/science.aaq1392

Maienschein J (2014) Embryos under the microscope. The diverging meanings of life. Harvard University Press, Cambridge, MA

Megason SG (2009) In toto imaging of embryogenesis with confocal time-lapse microscopy. In: Lieschke GJ, Oates AC, Kawakami K (eds) Zebrafish: methods and protocols. Humana Press, New York, pp 317–332

Megason SG, Fraser SE (2003) Digitizing life at the level of the cell: high-performance laser-scanning microscopy and image analysis for in toto imaging of development. Mech Dev 120 (11):1407–1420. https://doi.org/10.1016/j.mod.2003.07.005

Meli DB (2011) Mechanism, experiment, disease: Marcell Malpighi and seventeenth-century anatomy. Johns Hopkins University Press, Baltimore

Mersch D (2006) Naturwissenschaftliches Wissen und bildliche Logik. In: Hessler M (ed) Konstruierte Sichtbarkeiten: Wissenschafts- und Technikbilder seit der Frühen Neuzeit. Fink, Munich, pp 405–420

Mocek R (1998) Die werdende Form. Eine Geschichte der kausalen Morphologie. Basilisken Presse, Marburg

Monti MT (ed) (2000) Albrecht von Haller. Commentarius de formatione cordis in ovo incubato. Editione critica a cura di Maria Theresa Monti. Schwabe, Basel

Morgan LM (2009) Icons of life: a cultural history of human embryos. University of California Press, Berkeley

Needham J (1975) A history of embryology. Arno Press, New York

Pander CH (1817a) Beiträge zur Entwickelungsgeschichte des Hühnchens im Eye. Würzburg

Pander CH (1817b) Dissertatio inauguralis sistens historiam metamorphoseos, quam ovum inubatum prioribus quinque diebus subit. Nitribitt, Wirceburgi

Peter K (1906) Die Methoden der Rekonstruktion. Jena

Phillips M (2007) Deciphering development: quantifying gene expresion through imaging. Bioscience 57(8):648–652

Rheinberger HJ (2006) Epistemologie des Konkreten. Suhrkamp, Frankfurt/Main

Ries J (1908) Beiträge zur Histologie und Physiologie der Befruchtung und Furchung. Max Drechsel, Bern

Ries J (1909) Kinematographie der Befruchtung und Zellteilung. Archiv f mikrosk Anat 74:1–31

Roe SA (1981) Matter, life and generation: eighteenth-century embryology and the Haller-Wolff debate. Cambridge University Press, Cambridge

Roger J (1963) Les sciences de la vie dans la pensée française du XVIIIe siècle. La génération des animaux de Descartes à l'Encyclopédie. Armand Colin, Paris

Rossant J, Tam PLT (2018) Exploring early human embryo development. Science 260(6393):1075. https://doi.org/10.1126/science.aas9302

Vogel G (2008) Lights! Camera! Action! Zebrafish embryos caught on film. Science 322(5899): 176–176. https://doi.org/10.1126/science.322.5899.176

Wallingford JB (2021) Aristotle, Buddhist scripture and embryology in ancient Mexico: building inclusion by re-thinking what counts as the history of developmental biology. Development 148 (3):dev192062. https://doi.org/10.1242/dev.192062

Wellmann J (2017a) Animating embryos: the in toto representation of life. Br J Hist Sci 50(3):521–535. https://doi.org/10.1017/S0007087417000656

Wellmann J (2017b) The form of becoming: embryology and the epistemology of rhythm, 1760–1830. Zone Books, New York

Wellmann J (2018) Model and movement: studying cell movement in early morphogenesis, 1900 to the present. Hist Philos Life Sci 40(3):1–25. https://doi.org/10.1007/s40656-018-0223-0

Witt E (2008) Form–a matter of generation: the relation of generation, form, and function in the epigenetic theory of Caspar F. Wolff. Sci Context 21(4):649–664

Wolff CF (1759) Theoria generationis. Litteris Hendelianis, Halae ad Salam

Wolff CF (1966) Theorie von der Generation: in zwei Abhandlungen erklärt und bewiesen. Theoria generationis. Reprogr Nachdruck der Ausgabe Berlin und Halle 1764 und 1759. Olms, Hildesheim

Wolff CF, Meckel JF (1812) Über die Bildung des Darmkanals im bebrüteten Hühnchen. Uebersetzt und mit einer einleitenden Abhandlung und Anmerkungen versehen von Johann Friedrich Meckel. Rengersche Buchhandlung, Halle

Xiong F, Megason SG (2015) Abstracting the principles of development using imaging and modeling. Integr Biol (Camb) 7(6):633–642. https://doi.org/10.1039/c5ib00025d

Yamada S, Nakano S et al (2018) Novel imaging modalities for human embryology and applications in education. Anat Rec 301(6):1004–1011

Visualizations in the Sciences of Human Origins and Evolution

12

Marianne Sommer

Contents

Introduction	292
Drawings, Photographs, and Graphs	293
Trees and Maps	299
Life-Scene and Full-Body Reconstructions	305
Expositions and Museums	311
Cartoons and Films	314
Conclusion	316
Cross-References	317
References	317

Abstract

This chapter discusses literature about visualizations in the human origins sciences that were produced and published roughly during the last two centuries in fields like physical anthropology, paleoanthropology, prehistoric archeology, and human population genetics. It engages with scholarship on diagrammatic representations of skulls and other human remains for the purpose of comparative measurement as well as with graphic expressions of results. Following the discussion of such expert visualizations, the chapter moves on to treatments of kinds of images that had and continue to have more interdisciplinary and public impact: phylogenetic trees and migration maps. The origin and history of tree iconography, as well as the controversies surrounding it, are outlined up to the tree-building on the basis of molecular data, statistics, and information technology in human population genetics and its public projects such as the Genographic Project. In the second part of the chapter, popular illustrations and sites, from prehistoric life-scene paintings to museum exhibitions, take center stage. Possibly due to the fact that issues of gender and "racial" stereotypes as well as of

M. Sommer (✉)
Department for Cultural and Sciences Studies, University of Lucerne, Luzern, Switzerland
e-mail: marianne.sommer@unilu.ch

teleological understandings of evolution are most evident in these media, they have gained more attention by scholars. Historians of the human origins sciences have also approached the reception and production of such visualizations, inquiring after the interactions between scientists and artists and/or after the strategies of authentication and authorization. While the collaboration between science and art seems to have presented challenges in reconstruction painting, sculpting, and in exhibition, other genres of prehistoric representation such as cartoons, popular films, video games, etc., are completely beyond the control of the experts.

Keywords

Visualization · Human evolution · Prehistory · Gender · "Race" · Progress

Introduction

This chapter is about visualizations in the sciences of human origins and evolution. It does not include other fields such as classical archeology or ethnography, except where knowledge from these fields inspired views about the human deep past. With regard to the topic of human prehistory, this chapter is shaped by the author's expertise as well as by the limited text length and number of citations. The human origins sciences are and always have been visual endeavors. In fact, understandings of the epistemic role of diagrams have been pioneered by scholars working on the history of geology, a field that established the very prerequisite for a truly evolutionary paradigm: deep time (Rudwick 1976). Visualization is part of the process of knowledge production from the moment geologists, paleontologists, or archeologists start to work on a site. Then there is the kind of visualization usually called *reconstruction*. The role of prehistoric reconstruction as a research tool often goes far beyond the expression of, and engagement with, changing interpretations of an excavation site over time. Reconstructions may be an integral part of the excavation practice, with new questions and possible answers being created in feedback between the two (James 1997: 27–33). However, the term *reconstruction* can be misleading, because it is impossible to rebuild a complete skeleton from fragments, a cave setting as it once appeared, a full-bodied hominin, or even an entire live-group as it once was – much must be left to informed guesses. Even though such reconstructions of the past can take place visually or verbally and are most often a combination of the two, in this chapter, visualization will take center stage.

Visualizations related to human origins are by no means purely scientific endeavors and never have been. From the second half of the nineteenth century onward, popular authors and scientists alike have made use of visualizations to purport their views of human prehistory. And there is a wide range of genres of "making visible," including full-body or scenic reconstructions: anatomical reconstruction, replication, restoration, modeling, drawing, painting, photography, and even reenactment, and increasingly photorealistic graphics, multimedia, and virtual reality computer visualization (Redknap 2002). Similarly, the spaces in which we visualize or interact with visualizations of the human deep past are manifold and encompass laboratories, studies, or lecture rooms. In the context of being taught about human evolution, we may be given plasticine in school to build a cave and its inhabitants, or pupils are handed a sheet of paper with skulls of fossil humans in order to cut them out and arrange them in a phylogenetic tree. People encounter the fascinating worlds of prehistory in museums, in glossy magazines, and in animated movies, and – from time to time – in the form of a cartoon, possibly in a newspaper.

Thus, the chapter starts with more expert images that nonetheless were always intended for more or less extended audiences and moves on to visualizations that have been made in order to communicate knowledge about human prehistory to the public. A second structuring principle is chronology. In the first section especially, it treats the emergence of physical anthropology and moves toward the present. However, since this is a handbook on the *history* of the human sciences, the emphasis lies on past ways of producing and engaging with different kinds of visualizations.

Drawings, Photographs, and Graphs

The onset of physical anthropology was closely interlinked with craniology, and methods and results of measurements on skulls were communicated visually. The provision of skulls from as many human societies as possible involved ethically highly problematic practices, the moral depravity of which seem however to have eluded the collectors who often did not shy away from describing the hunts for skulls in detail. Skull collectors made their "treasures" not only available to other researchers at visits, but they also distributed them on paper. In the wake of Johann Friedrich Blumenbach's six instalments of *Decas (Altera/Tertia/Quarta/Quinta/ Sexta) Collectionis suae carniorum diversarum gentium illustrata* (1790, 1793, 1795, 1800, 1808, 1820), in which a series of ten skulls were represented on copper plates and described, skull atlases grew into a genre of their own. In general, diagrammatic images of skulls that showed the points and lines relevant for comparative analyses formed an integral part of the establishment of the supposed hierarchy from the "white" adult male via the female and "primitive races" down to apes. The American Samuel George Morton's skull atlases – *Crania Americana; or, A comparative view of the skulls of various aboriginal nations of North and South America* (1839) and *Crania Aegyptiaca; or, Observations on Egyptian Ethnography, Derived from Anatomy, History and the Monuments* (1844) – were programmatic for a quantitative approach to great collections of crania that was, however, still also guided by esthetic concerns (e.g., Richards 2018; Sommer forthcoming). Anders Adolf Retzius in Sweden, Paul Broca in France, James Hunt in England, and Rudolf Virchow in Germany, too, were among those central for the development of methods and instruments to the purpose of establishing anthropology as a metric science. The contributing scientists were not necessarily evolutionists, but they used volumes (particularly cranial capacities), ratios like brachy- or dolichocephaly, and angles such as Peter Camper's facial angle to arrive at "racial" classifications that could support the understanding transmitted from the seventeenth century that the different "human races" were separate creations and had the status of species (Sommer 2015b: 45–58). By around 1900, physical anthropology had developed into a metric science that established "races" with a visual repertoire including tables, curves, photographs, and drawings of skeletal parts that could contain a number of points, lines, and angles (Hanke 2007).

Daguerreotypes and photographs were integral to physical anthropology from the beginning of these new media. They constituted another way of collecting different "human types" for comparative analysis – sometimes really in the natural history tradition of front, back, and side view – as in the case of the polygenist Louis Agassiz. The Swiss-born naturalist was not in favor of evolution, and his move to the United States brought him in contact with anthropologists of Morton's thinking. For Agassiz, there were a set number of human groups, the individuals of which varied around a given type. These "racial types" were collected in photo albums as clothed, half-naked, and naked people without any ethnographic information. In this "racial typing," the artistic tradition of portraiture met with the scientific of the specimen. Like plant or animal specimen, the persons photographed by scientists, on plantations at home or during expeditions, often had little say in a process that culminated in their objectification (Banta et al. 2017: 57–61). Darwin was in possession of such a collection of ethnographic photographs, but even before the publication of *The Descent of Man* (1871), such stereotyped photographs of "primitive human types" had great public appeal in the context of the "ape theory" and the mono- versus polygenists debates. They circulated, for example, in the form of *cartes de visite* and were entangled in visual formations with the human "life specimens" exhibited for science and as entertainment in ethnographic shows, the "missing links" of freak shows, and the wooden models of "exotic peoples" that were exhibited at the New York Crystal Palace between 1853 and 1858 (e.g., Edwards 2009; Qureshi 2011).

These "political, economic and social matrices" (Edwards 2009: 168) of representation inextricably connected the science of anthropology with diverse publics. However, anthropologists increasingly "scientized" their photography through the integration of systems of measurement. In the late 1860s, the renowned comparative anatomist and evolutionary biologist Thomas Henry Huxley suggested a collecting project to the Colonial Office to render photography useful for a measuring anthropology in the service of a "racial" classification of the British Empire. In the attempt to achieve standardization, he gave precise instructions on how to proceed because, for physical anthropology, photography came with a caveat. It did promise objectivity, but it posed several challenges to a science that wanted to be metric and comparative. Photographs bring in perspective, they are not to scale or proportion, and they may also lack a clear outline. Especially craniologists therefore preferred drawing, more specifically geometric drawing. In the end, a statistical endeavor such as physical anthropology could not be satisfied with photography (Spencer 1997). While anthropologists tried to reclaim photography for science, the physical anthropological measurements themselves gained public appeal. At the World's Columbian Exposition in Chicago in 1893, the influential American anthropologist Franz Boas invited visitors into a setup laboratory in which their measurements were taken, and they could compare themselves to the "racial types" pictured on the wall. Scientists also increased their own popularity and that of their work through coverage in newspapers and magazines, as in August 1939, when *Life* famously showed the powerful American physical anthropologist Earnest Alfred Hooton surrounded by his collection of human skulls. Up until today, front cover photographs of scientists and even more so of important fossil finds promote the sciences of human origins (Banta et al. 2017: 61–63).

As indicated by Huxley and Hooton's work, by this time, the measurement, calculation, and visualization techniques introduced to compare "modern human races" had been integrated into an evolutionary and phylogenetic framework, and paleoanthropology had become a science of its own. Two hallmarks in this process were the collaborative volumes *Reliquiae Aquitanicae* (1865–1875) and *Crania ethnica* (1882). They followed in the wake of such lavishly illustrated books as John Lubbock's *Pre-historic Times as Represented by Ancient Remains and the Manners and Customs of Modern Savages* (1865) that synthesized knowledge from archeology, ethnology, geology, anthropology, and to a lesser extent history and philology, into a new "Pre-historic Archaeology." *Reliquiae Aquitanicae* was published by the then deceased French paleontologist Édouard Lartet, who had introduced a chronological system for prehistoric cultures, and the gentleman scientist Henry Christy. It described and integrated the archeological industries and fossil bones from the south of France. It may be seen as a precursor to contemporary interdisciplinary publications on the excavation of and data from particular Paleolithic sites, but it was not yet presented in a unitary evolutionary framework. *Crania ethnica*, as its subtitle *Les Crânes des races humaines* suggests, followed in the tradition of Morton's *Crania Americana* and *Crania Aegyptiaca*, but it was a compendium and classification of not only the living but also the fossil "human races." Its appendix contained 100 plates with lithographs from drawings by Henri Formant, and close to 500 illustrations accompanied the texts. The material was compiled at the Muséum d'Histoire Naturelle where Jean Louis Armand de Quatrefages held the chair of anthropology and Jules Ernest Théodore Hamy was his assistant. De Quatrefages was not a proponent of human evolution from simian origins, but he defended human antiquity and monogenism. This was his motivation for carrying out the comprehensive study of the collections of his museum, of the anthropological society in Paris, and other major collections at home and abroad. De Quatrefages' and Hamy's volume achieved the establishment of Neanderthals and Cro-Magnons as "fossil races" (Sommer 2007: 123–130).

With the general turn to evolution, images, particularly diagrams, gained new impact. One such diagram was the stratigraphic series that was used in geology to document the layers of the earth and came to represent the progress of life through the ages, including hominin development. To illustrate the convincing power of the series, let us look at Fig. 1. The English amateur archeologist James Reid Moir published visualizations of serial steps in archeological tool production and of what he thought to be the analogous process of serial tool evolution over time to lend support to the workmanship of the so-called eoliths. These were stones from the

Geological Periods.	Climate.	Deposits.	Cultures.
Recent.	Temperate. ...	Surface soil and latest alluvium of valleys	Neolithic and later.
P	? Cold	Deeper levels of alluvium in valleys ...	Magdalenian and ? Solutrean.
L E	Glacial	Brown Boulder Clay, Hill Washes, and ? Flood Plain Gravel	Derived Implements.
I	Warm in earlier part.	Floors in lateral valleys, Ipswich	Aurignacian and Mousterian.
S	Glacial	Upper Chalky Boulder Clay	Derived Implements.
T	Inter-Glacial ...	Brickearths and Gravel, Hoxne, High Lodge and Ipswich	Clacton III and late Acheulean.
O	Glacial	Kimmeridgic Chalky Boulder Clay and ? Contorted Drift	Derived Implements.
C	Inter-Glacial ...	Corton and ? Mundesley Sands ("Middle Glacial")	?Lower Acheulean.
E	Glacial	Cromer Tills and ? Norwich Brickearth ...	Derived Implements.
N	Inter-Glacial ...	Cromer Forest Bed, ? Norwich Crag in part, and ? Foxhall horizon	Chellean and Pre-Chellean.
E	Increasing Cold ...	Red Crag	Some derived Implements.
Pliocene	Warm	Suffolk Bone Bed	Eolithic.
		Coralline Crag. ...	Some derived Implements.
		Suffolk Bone Bed	Eolithic.
Eocene		London Clay.	

The Red Crag and the Suffolk Bone Bed.

Fig. 1 "Stratigraphical Table of the Implementiferous Deposists of East Anglia." (Republished with permission of John Wiley & Sons – Books: The age of the pre-crag flint implements, Reid Moir J, Journal of the Royal Anthropological Institute of Great Britain and Ireland 65:343–374, 344, 1935; permission conveyed through Copyright Clearance Center, Inc. Discussed in Sommer 2015a: 31–32)

Pliocene in East Anglia that were not universally accepted as human-made. In the stratigraphic table of Fig. 1, Reid Moir integrated his Eolithic cultures into the series of geological epochs with the parallel series of climates, deposits, and archeological cultures. The series of tool cultures condensed his visual documentation of the experimental transformation of eoliths into accepted paleoliths. Integrated into the existing visual language of the serial, eoliths should partake in its evidential power (Sommer 2015a: 23–34).

It is in particular the line or the serial that was used to render the nonsimultaneous simultaneously to convey evolution over time. This was and is not only true for the stratigraphic series, but also for serial arrangements of (representations of) artifacts, parts of skeletons such as skulls, entire skeletons, etc., up to the series of paintings representing the epochs of the earth or successive exhibits in a museum. Through the serial, evolutionary development is conveyed, mostly in the form of progress (Gould 1995: 38–60; Sommer 2010, 2015a). At the beginning of this kind of iconography that has since gone viral might stand the famous frontispiece of Huxley's *Man's Place in Nature* (1863) that featured a series of skeletons from gibbon to man. A similar effect of evolutionary progress taking place in front of the viewer's eyes could be achieved by the superimposition of the outlines of skulls to compare different prehistoric, recent "primitive," recent "civilized races," and sometimes apes – a technique that had been applied to fossils as early as the studies of the Fuhlrott Neanderthal by both Huxley in the same book and Charles Lyell in *The Antiquity of Man* (1863). Comparative alignment of bones from two or more specimens, in particular also jaws, was in general a widely employed visual technique (e.g., Hermann Klaatsch *Entstehung und Entwicklung des Menschengeschlechts* 1902; Arthur Keith *Ancient Types of Man* 1911, *The Antiquity of Man* 1915/1925; William Sollas *Ancient Hunters and Their Modern Representatives* 1911/1915/1924; Marcellin Boule *Les hommes fossiles* 1921). Arthur Keith, the eminent British anthropologist, developed a special skill of arranging skulls, jaws, and teeth in grids opposite each other to highlight their points of divergence (*Antiquity of Man* 1925). The information these images should convey is usually easily grasped, particularly if the viewer is accustomed to their style. What cannot be seen are the controversies that went on about the right reconstruction of the skeletal remains, indeed the correct alignment for comparison, and the most reliable (points of reference for) measurements (Lipphardt and Sommer 2015: 4–8; on the use of diagrams in anthropology, see the contributions to Sommer and Lipphardt 2015 in general).

Like in early physical anthropology, brain size was ascribed particular relevance in evolutionary studies, and tables were published that ordered specimens or taxa according to increasing skull capacity. Sometimes, as in the case of the British geologist William Sollas, even the skulls of supposed genii were included in this practice. Working during the last decade of the nineteenth and the first decades of the twentieth century, Sollas developed a new instrument and method for drawing profiles of skulls mechanically. He also introduced novel ways of diagrammatic representation. He, for example, produced a diagram with curves that supposedly represented the skull size range of "the Tyrolese," of "the Australian," of *Pithecanthropus* (today *Homo erectus*), and of the gorilla. The mutual overlap of the curves of the hominins and the lack of overlap with that of the ape should prove Sollas's argument that *Pithecanthropus* was closer to modern humans than apes. However, this intrahominin overlap was only achieved by differentiating for gender. Only once he included "Australian females" as their own category, did he manage to close the gap between *Pithecanthropus* and humans. So this is an instance of the then common practice of turning women (of particular "races") into missing links. To give another example of a diagrammatic technique, in Fig. 2, Sollas entered skull sizes as dots in a coordinate system and linked them by a line to show the linear increase in brain size in the course of human evolution from a fossil ape, via

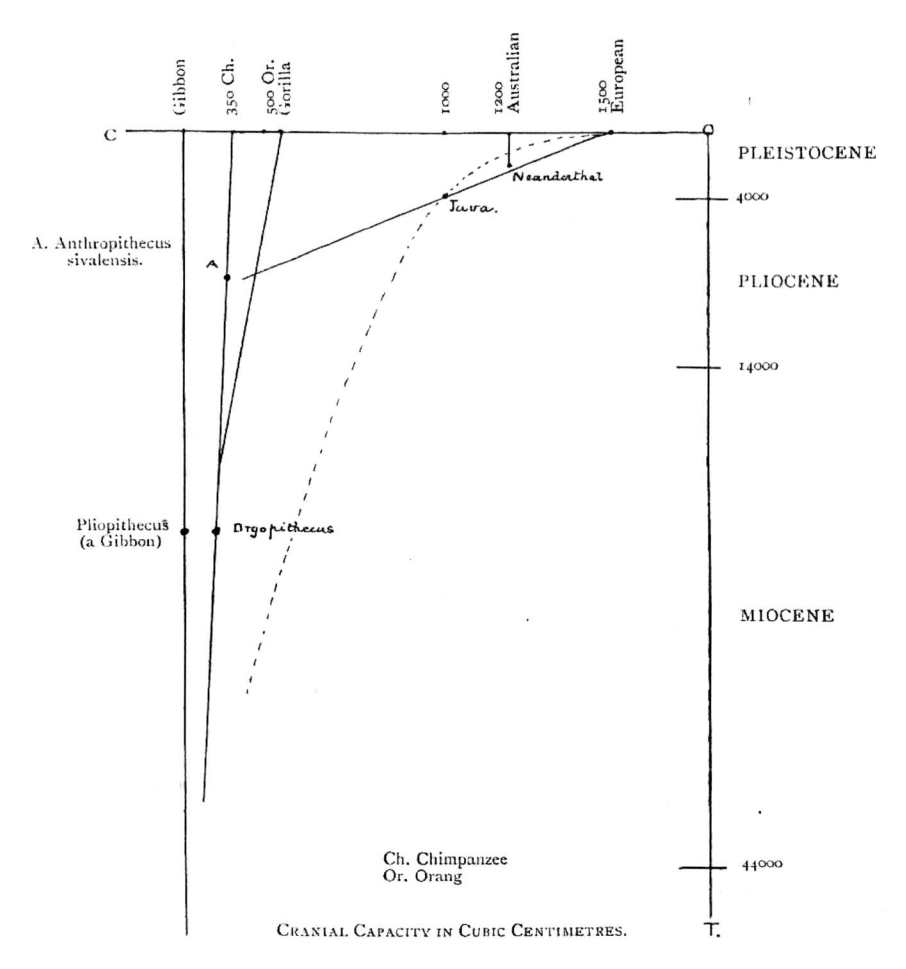

Fig. 2 Diagrammatic visualization of two different scenarios of human evolution on the basis of cranial capacity. (Reprinted by permission from Springer Nature: Nature 53(1364):150–151, 151, "Pithecanthropus erectus" and the evolution of the human race, Sollas W, 1895, discussed in Sommer 2005: 7–10; 2007: 155–159)

Pithecanthropus ("Java"), Neanderthal, to modern European. With the dotted and curved line, the image also hinted at a second possibility, which stood for the view of other experts that there had been an acceleration in brain size increase. This second possibility seemed less beautiful to Sollas because it was less parsimonious – it demanded the assumption of more unknown ancestors. However, both straight line and curve expressed evolution as a function of cranial capacity and time, and both possibilities assumed a linear (not yet a branching) view of human phylogeny. The horizontal line at the top of the diagram simultaneously established a contemporary hierarchy from gibbons to Europeans. However, the Aboriginal Australians were also projected into the past by the line connecting them to the Neanderthals, this time indicating the practice of imagining contemporary "primitive races" as stuck in the past of more "advanced races" (Sommer 2005: 333–340).

The understanding of hominin evolution as unilinear descent was undermined by the renowned French paleoanthropologist Marcellin Boule's expulsion of the Neanderthals from modern human ancestry. The argument for the new paradigm involved a decidedly visual language, including yet another pictorial juxtaposition for anatomical comparison at a glance: that between a Neanderthal and an Aboriginal Australian. Even though the skeleton was well preserved, for the reconstruction of the La-Chapelle-aux-Saints Neanderthal specimen in 1913, Boule had to infer certain parts on the basis of the La Ferrassie specimen that he shaded in the illustration. The resulting visual reconstruction of the Neanderthal, as well as the verbal reconstruction accompanying it, was juxtaposed to a modern Aboriginal Australian skeleton, apparently with the expectation that the viewer would immediately notice the obvious difference. According to this logic, even "the primitives at the peripheries of the earth" were considerably more advanced than the Neanderthal brute, which was closer to the apes than any "human race." The Neanderthal contrasted not only with "modern human races," however, but also with "our fossil ancestors," the Cro-Magnons. Even though this strategy clearly drew on the visual tradition of conveying linear progress, implicit in Boule's comparison of the Neanderthal and Aboriginal Australian skeleton was indeed a redrawing of genealogical relationships: While there had been the understanding that human evolution had proceeded from *Pithecanthropus* to Neanderthals to modern humans, Boule turned this line of descent into a tree structure (Sommer 2006: 213–214).

Trees and Maps

Indeed, when speaking of visualizations in the human origin sciences, the phylogenetic tree must take a central place. However, while the history and debates around "the tree of life" have received considerable attention by scientists and historians of science, there is little literature on specifically human and hominin phylogenies. An exception is provided by Peter Bowler in one of his classics on (human) evolution (1986: 61–146; for another early example, see Brace 1981). In *Theories of Human Evolution: A Century of Debate, 1844–1944*, Bowler gives some overview over the developments in phylogenetic thinking for that period. The focus here is not on the visualization of these theories, but some diagrammatic renderings are reproduced and discussed in the context of the long-standing debates about the relationship between humans and mammals, humans and apes, or modern humans and Neanderthals.

Such trees of life or of the animal kingdom that include "Man," and even the "human races," predate the publication of *On the Origin of Species* (1859) and certainly *The Descent of Man* (1871), even if they were not necessarily meant to represent evolutionary kinship. Besides, for example, the primate trees of the British comparative anatomist St. George Mivart that included *Homo*, Darwin himself drew a primate phylogeny with ink on paper in 1868. Darwin experimented with tree-like structures on many sheets of paper, but he published none of these phylogenies. The image that he did publish in *Origin* is not a phylogeny. It is a diagram in essence in that it represents the contingent processes of natural selection, divergence, and extinction that underlie the descent of species. This omission has been attributed among other things to a certain skepticism toward fleshed out genealogies as too speculative, a caution that might well have been partly due to the work of "the great phylogenist," Ernst Haeckel (Sommer 2021). When *Descent* appeared, Haeckel had already published eight phylogenies in the form of trees in *Generelle Morphologie der Organismen* (1866), among them a "family tree of the mammals" that contained *Homo* at the upper right hand of the image, and tree-like genealogies in *Natürliche Schöpfungsgeschichte* (1868). In the aftermath, he would circulate the famous "family tree of man" in the shape of an oak with humans as the crown in his *Anthropogenie* (1874). Haeckel's trees convey a hierarchical view of the organismic world in accordance with an understanding of evolution as progressive. Furthermore, among Haeckel's phylogenies were racist ones from early on. In fact, Haeckel produced trees that separated living humans into different species and "races" (Sommer 2021; also Livingstone 2010; Archibald 2014: 80–112, 119–129; Sommer 2015b: 40–45).

Haeckel's phylogenetics drew strongly on his analogies between biological evolution and language evolution as well as between phylogeny and ontogeny. Within this scheme, progress even acquired a quasi-religious touch. Indeed, his "family tree of man" has been compared with visualizations of sacred genealogies such as the trees of Jesse and linguistic trees of comparative historical philology, as well as with secular genealogies of important families (Bouquet 1996). Wherever its (multiple) origins may lie, in evolutionary biology, the tree structure lent itself for communicating the differentiation of taxa, be that of hominid genera and species or "human races," and so we find the image multiplying in the early decades of the twentieth century, when paleoanthropology came of age and turned from a linear view of hominin evolution to a branching one – not least due to the influence of Boule's Neanderthal interpretation treated above. This was a time when every new fossil was classified as a new genus, adding more and more branches to the tree. This was also a time when paleoanthropology was decidedly about nation and "race." While an anthropologist might provide his "race" or nation with a long and "noble" pedigree, fossil hominins and "extent races" considered unworthy could be distanced by placing them on separate branches that suggested long independent lines of descent. These theories, histories, and images of human genealogy and evolution were clearly situated within the context of imperialism and war. It was common to associate evolutionary progress with the marginalization if not extermination of "lower races" through geographically migrating and expanding "higher races." The phylogenetic trees could visualize such violent histories and hierarchies by having the "Caucasian," "European," or "Aryan race" at the apex or by aligning the "living races" from left to right or bottom to top according to a presumed ascent in morphology and culture (Sommer 2007: 187–212).

Sometimes, in an attempt to reconcile evolution and religion, evolutionary parallelism was also a means to distance humans from apes, and nothing lent itself better to this purpose than the 'family tree.' As we have seen in the case of Haeckel who conceived of "the human races" as different species or even genera, tree images could also still express a kind of polygenism – an evolutionary kind. Such "polygenist trees" represented high degrees of parallel evolution of similar traits in apes and humans as well as within the human populations by distancing these taxa from each other with long independent branches possibly leading back to different anthropoid origins. Such trees were drawn up to the 1940s, and just as the human family tree was political, so was its critique that could consist in renouncing visualizations of human phylogeny or in counter images that looked more like networks of constant genetic exchange between hominins throughout evolution (Sommer 2015b: 111–134, 2015c: 2–7). Ironically, the trees that expressed parallel evolution also came under attack from new approaches that were very much about tree-building themselves. Novel immunological techniques that suggested a close kinship between humans and apes, and particularly with the African great apes, had already emerged at the beginning of the twentieth century, and in the course of time, the phylogenetic trees that were built on the basis of serological reactions, DNA hybridization, electrophoretic techniques, and a calibrating system called *immunological clock* became more refined (see Fig. 3). In 1962, the term *molecular anthropology* had been introduced to designate the research on primate phylogeny and human evolution on the basis of the genetic information in proteins and polynucleotides. The comparison of amino-acid sequences and the application of the molecular clock, too, suggested that the human and ape lines had separated more recently than paleoanthropologists estimated on the basis of the fossil record (Sommer 2008a).

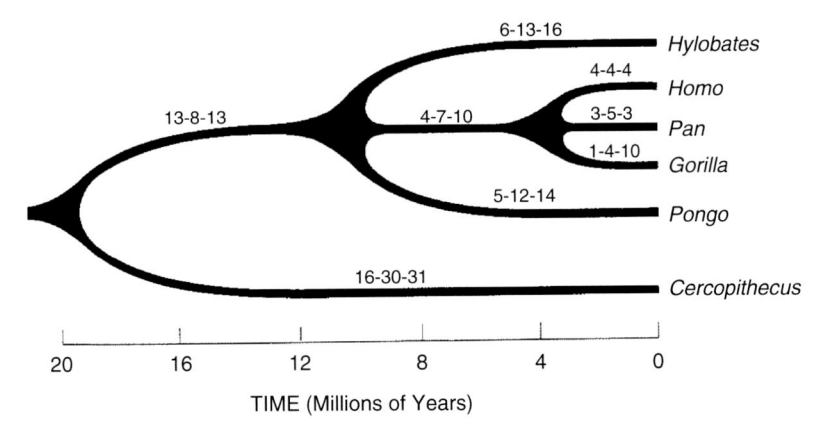

Fig. 3 Albumin (units of change), transferrin (units of change) and DNA (units in dissociation time of hybrids) phylogeny of the *Hominoidea* based on the molecular clock. (Reprinted by permission from Springer Nature: Sarich VM, Cronin JE (1976) Molecular systematics of the primates. In: Goodman M, Tashian RE, Tashian JH (eds) Molecular anthropology. Genes and proteins in the evolutionary ascent of the primates. Plenum, New York, p 141–170, 151, Fig. 7, discussed in Sommer 2008a: 501–506)

Already during World War I, there was furthermore the application of the new knowledge regarding blood types to the question of their distribution among human groups that came to be recognized as a fresh approach to the systematics of "races." Eventually, great amounts of data about blood-type frequencies and other polymorphisms in human populations were collected, tabulated, and visualized on maps in books like *The Distribution of the Human Blood Groups and Other Polymorphisms* (1954 und 1976), *The ABO Blood Groups* (1958), und *Blood Relations: Blood Groups and Anthropology* (1983) (co)authored by Arthur Mourant. While some used the new blood group studies as a novel technology for approaching conventional questions of racial anthropology, Mourant was among those who considered the method to be more objective than the study of bones and thus a way out of the racial science of old. In this vein, as Jenny Bangham (2015) has shown, the new science was also presented in the antiracist campaign of UNESCO. In a picture book for school children of 1952, genetics was popularized in image and text as a neutral way to engage with "human racial difference." Within science, progress in information technology as well as statistical and genetic techniques heralded a new era of tree building. The first population tree computed from blood group gene frequencies was published in 1965 by Luigi Luca Cavalli-Sforza and A. W. F. Edwards (see Fig. 4). By the early 1980s, such trees began to be based on DNA sequences, foremost on mitochondrial DNA. Although it was not the first of its kind, it was the 1987 mtDNA-tree that became famous and also popularly associated with the "out-of-Africa" model of modern human evolution. A team at the University of Berkley compared the mtDNA of about 150 people that were taken to represent the African, Asian, Australian, Caucasian, and New Guinean populations. The tree of these DNA sequences was thus racialized in intricate ways. At the same time, the tree made clear that there were multiple lineages per "race." It did not suggest clear-cut "racial groups," because it visualized individual mtDNA types; it was in the population trees like the early one shown in Fig. 4 that the individual scatter was hidden behind population labels (Sommer 2015c, 2016: Ch. 11).

The 1987-tree also came with a narrative. It suggested that the most recent common modern human ancestor had lived in Africa about 200,000 years ago. No more than 140,000 years ago, modern humans had begun to spread and conquer the globe. The migrants in this process replaced the other hominins like the Neanderthals in Asia and Europe, without interbreeding. This exemplifies that such population-genetic trees may not only entail narratives but are also a kind of map – a map that tells the story of humankind's migration across the earth and its differentiation into subgroups in the process. In fact, the very first such tree of 1965 shown as Figure 4 had initially been published drawn on a map. And both kinds of images – the tree and the tree or paths of migration and differentiation on a map – remained current in the science of human population genetics and in the public and popular realm. This kind of human-population-genetic imagery also accompanied the large projects, the Human Genome Diversity Project and the Genographic Project, that gained much public attention. The call for the Human Genome Diversity Project was issued in 1991 and was linked to the canonical *The History and Geography of Human Genes* (Cavalli-Sforza, Menozzi and Piazza 1994) that was rendered more accessible to the

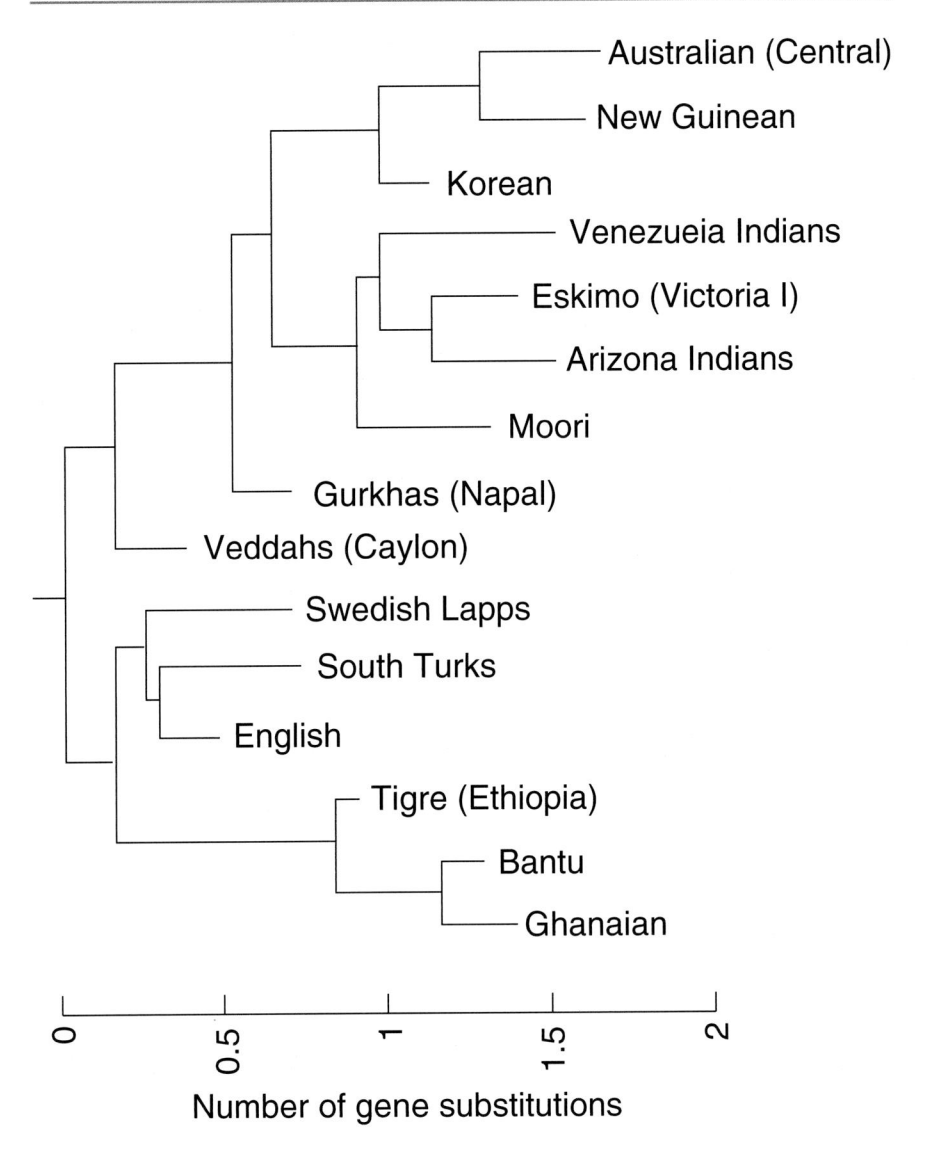

Fig. 4 The evolutionary tree of human populations from blood group polymorphism frequencies produced by parsimony. (Reprinted from Genetics today. Proceedings of the XI. International Congress of Genetics, The Hague, The Netherlands, September 1963, Vol. 3, Cavalli-Sforza LL, Edwards AWF, Analysis of human evolution, p 923–933, 929, 2021, with permission of A.W.F. Edwards, discussed in Sommer 2015c: 117–121)

non-expert in *Genes, Peoples, and Languages* (Cavalli-Sforza 1996). Luigi Luca Cavalli-Sforza and his Italian colleagues Alberto Piazza und Paolo Menozzi reconstructed the history of migration and differentiation of humankind on the basis of the distribution of mostly classical genetic markers in indigenous or isolated populations worldwide and knowledge from linguistics, archeology, (paleo-) anthropology as well as climate, ecological, and human history. Their book contained a range of tables, diagrams, and graphs to present data, models, and results from statistical analyses, most conspicuous of which are (combinations of) trees and maps (Sommer 2015c, 2016: Ch. 11).

The Human Genome Diversity Project should not only include classical markers and the new autosomal and mtDNA systems, but also advance the Y-chromosomal system with its own tree, map, and narrative. With the Genographic Project, the Y-chromosome became a star, and it was especially the Genographic Project and genetic ancestry tracing companies that have made the human-population-genetic imagery known to a wide range of people. The Genographic Project was an endeavor of the National Geographic Society, IBM, and the Waitt Family Foundation and had three pillars: the reconstruction of human population history and kinship; the commercialization of the information-genetic technologies in genetic ancestry tracing; and projects in support of indigenous communities that were also financed with the money from the public testing. It was associated with citizen science projects, and the people who partook in the genetic ancestry testing were referred to as participants – when the project terminated, there were over one million such participants in more than 140 countries. The project was further popularized through a strong media presence and web appearance as well as through popular books and film documentaries, all of which relied on typical imagery. Despite its big-science allure, the Genographic Project came with the (political) legacy of the history of (molecular) anthropology. The Human Genome Diversity Project and Genographic Project shared goals, approaches, and forms of organization, and they showed some personal continuity as in the case of Cavalli-Sforza. Both projects were thus criticized from diverse sides, including from the Indigenous Peoples Council on Biocolonialism. The projects were seen as exploitative in the use of indigenous peoples as "objects of scientific study." The impression was that researchers would collect bodily material from them for the realization of their own goals, such as generating histories that might even contradict tribal knowledge (Sommer 2016: Chs. 13–14; on the legacy and ethical issues see in particular Reardon 2005).

Even though the tree to render the genealogy and history of human populations had a particular revival in human population genetics, alternative visualizations were and are produced especially by scientists who emphasize local evolution in conjunction with genetic exchange over geographical regions and between geographical varieties, and within population genetics and ancient DNA studies, a certain shift toward analyses of processes and states of admixture can be observed (Sommer 2015c). Furthermore, though seemingly an uncharted territory in science studies, novel methods in paleoanthropology, too, have led to a distinctly virtual kind of imagery such as computed reconstructions of fossils for three-dimensional geometric and morphometric analyses and experimentation. Through virtual anthropology, fossils can thus be reconstructed from fragments and complemented, they can be changed; they can be used for simulation to study ontogeny, diet, and locomotion; and they can be statistically analyzed as part of population studies.

Life-Scene and Full-Body Reconstructions

Archeological practices and representations have entered the digital stage, but images of archeological sites, including the human bones and artifacts found there, have been made since the discovery of the first fossil human in 1823 – even if the so-called Red Lady of Paviland was not taken as such at the time (Fig. 5; for a thick description of the skeleton's history of interpretation, see Sommer 2007). Beyond the documentation of sites, there were visual imaginings of the scenes as they might once have been, images that circulated only among the savants of the new science of geology and paleontology (Rudwick 1989; Sommer 2003: 193–197). These early visions of the deep past did not include humans, since human antiquity was not generally accepted in science until the end of the 1850s. Nonetheless, the reconstructions of landscapes populated by extinct animals, also by artists, such as John Martin's "iguanodons," "sea-dragons," and other prehistoric "reptiles" (Rudwick 1992), were a close precedent of the paintings and dioramas of prehistoric human life scenes. Furthermore, as Stephanie Moser shows in her by now classic *Ancestral Images: The Iconography of Human Origins* (1998), artists, such as Emile Bayard, who produced the progressive series of prehistoric scenes for Louis Figuier's *L'Homme primitif* (1870), relied on schemata from the artistic renderings of Greco-Roman origin myths, the topoi of the medieval wild man, of Adam and Eve in Paradise, or of ancient tribes. Yet another influence was travel literature about foreign peoples and human-like animals. The first picture that to the author's

Fig. 5 The sketch of Paviland Cave in Wales with excavators, including the geologist William Buckland and his dog. We see the skeleton of the so-called Red Lady of Paviland close to a mammoth skull. However, Buckland estimated the remains to be postdiluvial. (Courtesy of The National Museum of Wales: Buckland W (1823) Reliquiae diluvianae; or observations on the organic remains contained in caves, fissures, and diluvial gravel, and on other geological phenomena, attesting the action of a universal deluge. Murray, London, pl 21, ©Amgueddfa Genedlaethol Cymru ©National Museum of Wales, discussed in Sommer 2007: 66–68)

knowledge was loosely based on paleoanthropological evidence and expressed the idea of a human prehistory was Pierre Boitard's "L'Homme fossile" in the *Magazin universel* of 1838. Boitard was a believer in human antiquity prior to general acceptance, with radical political intent. The visual reconstruction of human ancestors was linked to politics from its birth, in Boitard's case to the aims of radical socialism. At the same time, Boitard's fossil man had been inspired by imagery of apes and African people as well as by Dumont d'Urville's descriptions of "primitive tribes" (Sommer 2007: 320–321). Finally, as becomes obvious in Boitard's posthumously published *Études antédiluviennes* (1861), the prehistoric fantasy was also already gendered: Reminiscent of medieval illustrations of the "wild" men and women, the female is sitting on the ground holding her child, while the male is standing upright, protecting his helpless family by means of a weapon (see Fig. 6).

Little changed regarding these codes of gender, "primitiveness," and "race" with the general acceptance of human antiquity and the scientific classification of Cro-Magnon, Neanderthal, and so-called *Pithecanthropus*. The reconstruction of a *Pithecanthropus-alalus* pair carried out for Haeckel reproduced the theme of sitting female with child and standing male, even if his gait is still unsure. Similarly, the supposedly first Neanderthal life scene, which appeared in *Harper's Weekly* in 1873, adopted the cave scene with its clear gender division in image space and social role.

Fig. 6 Boitard P (1861) Études antédiluviennes. Paris avant les hommes. (Passard, Paris, discussed in Moser 1998: 135–136)

However, scientists differed in their opinion of which fossil forms deserved to be stigmatized with the markers of primitiveness and which deserved to be imagined as noble ancestors, as becomes evident from the contrast between the La Chapelle-aux-Saints Neanderthal images based on Boule's and Keith's contradictory interpretations. In 1909 and 1911 respectively, the images were widely distributed in *The Illustrated London News*, and while the first othered the Neanderthal as ape-like, the second portrayed a rather sophisticated male hunter and craftsman (Moser 1992, also Moser 1996 on the visual fight for and against australopithecines as "ape-men;" Sommer 2006: 225–231).

Since then, the number of such scenes from human deep time seems to have risen exponentially. Correspondingly, efforts have been made to inquire into strategies of persuasion, to establish genres and artistic as well as scientific traditions, and to disentangle the manifold sociocultural preconceptions that are worked especially into life scenes of prehistoric humans. Due to their irresistible appeal and their often blatant racism and sexism, anthropologists and archeologists as well as historians and sociologists of science have dealt with visual reconstructions of prehistoric life, highlighting the fact that they betray anthropocentric, ethnocentric, and androcentric perspectives. In visualizations of the lives of our prehistoric ancestors, women traditionally constitute a minority. They tend to be shown as close to the home base, often a cave, and mostly seated, possibly by the fire, or on all fours, for example preparing hides. They are passive, even childcare tends to look passive. They may also be ornamental and eroticized. In contrast, the males are the tool users and makers, hunters, protectors, artists, and ritualists, standing, mobile, and in action, who take center stage in the images (Gifford-Gonzalez 1993; also Moser 1993, 1998; Conkey 1997; Wiber 1998; Berman 1999; Solometo and Moss 2013).

Other aspects, such as the production of authenticity, have gained attention, too. In general, a strong claim to truth may be established through details in landscape, flora, and fauna, combined with the familiarity of the prehistoric human bodies, their expressions and gestures (Gifford-Gonzalez 1993: 28–29). But as we have seen, a claim to truth may also be combined with strategies of othering, as in the reconstruction painting based on Boule's interpretation of the La Chapelle-aux-Saints specimen by the Czech painter, engraver, and illustrator Franz Kupka. This rendering of the specimen was obviously situated in Boule's expulsion of the Neanderthals from human ancestry that we have treated above. So the viewer of Kupka's illustration was not intended to be engaged by a familiar scene and being, but was instead to be distanced from the apish creature. The realism of the image draws on a tradition that goes back to the Renaissance, but in the reconstruction painting that first appeared in *L'Illustration* (1909), the familiar realistic style was employed to establish a stark contrast to the values of rationalism, materialism, or technological and social progress a bourgeois viewers of the French Third Republic might hold (Sommer 2006: 225–231).

Further important insights into life-scene reconstructions can be gained by studying the image in-the-making, including the interaction between scientists and artists in the production of scenes of the Paleolithic and Neolithic. The artist's cooperation with the scientists, like the artist's own expertise, can be used to provide an image with authority. Kupka, to continue with this example, explained how he had arrived at the Neanderthal's physical appearance by applying the missing muscles to the fossil bones in order to enhance his reconstruction's authenticity. While this seems unlikely in view of our current knowledge, Kupka did certainly not represent the specimen's true material culture. Instead of making him handle Mousterian tools, he armed the creature with a club – a caveman prop that would certainly stick. It is a far cry to imagine Kupka's brute being ritually buried by his companions. In fact, Boule complained to Kupka about the caption in *The Illustrated London News*, which claimed that the picture had been carried out according the scientist's instruction. Kupka on his part insisted that, although he had told the magazine that Boule had designed a draft to indicate the composition of the scenery, he had not told them of any other drafts. To the contrary, he had tried to ensure that the responsibility for the image would not be placed on Boule. If Boule did provide Kupka with an outline of the scene, he obviously contributed to the production of the reconstruction. He must have welcomed the support such a primitive visual interpretation would lend to his scientific interpretation. At the same time, he did not wish to be associated with the popular picture (Sommer 2006: 225–231).

Studies of the very influential painter of the human deep past, Charles Knight, too, brought to light difficulties in cooperation. Knight, who was strongly influenced by his training in the beaux arts (Cain 2010), worked with the powerful paleontologist and president of the American Museum of Natural History in New York, Henry Fairfield Osborn. From the 1910s, they collaborated on the murals for the series of exhibitions bringing to life the progressive eras of the Age of Reptiles, of Mammals, and of Man. For Osborn, the visual experience of reconstructions of evolutionary history could teach through demonstration the most important law of nature: Progress depended on a strenuous life in close contact with nature. Knight staged this struggle for survival with unprecedented dramaturgy, dynamics, and authenticity. However, a study of the traces the cooperation between the artist and the museum scientists has left behind brings to light that the codification of the production process in contracts could not prevent quarrels and that the strong hand of science could not fully control the messages as they were finally communicated through the serial exhibit of the murals. Besides technical problems, there were constant contests for the prerogative of science respectively art, controversies regarding who had the greater expertise, to whom belonged the power of interpretation, and how much the work was worth in money. And there were once again also "racial" and gender issues involved, as when the paleontologist William King Gregory advised Knight against rendering the Neanderthals too similar to "white men." Osborn instructed that the prehistoric humans should be of a "wild" and aboriginal character. As a model for the Neanderthals' skin color, he referred Knight to Italian working men. In contrast, he considered that Knight had rendered the central male figure in the painting of Neolithic stag hunters too dark. The advising scientists further found Knight's Neanderthal woman too European in looking. Osborn proposed Aboriginal Australian and Tierra del Fuegian women as references for the portrayal of the Neanderthal female, which would make obvious that she would have had no waistline (Sommer 2010: quotes on 474; Sommer 2016: 82–92).

Despite the difficulties in collaboration that ultimately led to its termination, Knight appropriated Osborn's conviction of the importance of struggle for progress, and shared Osborn's view of evolution as overall a progressive development toward humankind. This is evident not only in the many reconstructions he did for scientific books but also in the popular books Knight wrote and illustrated after his work for the American Museum of Natural History had ended, such as *Life through the Ages* (1946). The impact of Knight's images on scientific as well as lay conceptions of prehistoric animals and humans was great indeed, and they continue to shape the imagination of paleontologists and paleo-artists to this day (Sommer 2010). The production of reconstruction paintings or murals also remains a very labor intensive and often difficult process due to the different actors involved. Commissioning archeologists and paleoanthropologists are intent on ensuring that reconstructions take into account the available evidence and certainly do not contradict it, but this still leaves room for imagination – so much so that Simon James, in his contribution to *The Cultural Life of Images: Visual Representation in Archaeology,* has considered replacing the term *reconstruction* with *simulation* (James 1997: 22). *Simulation* can also express the fact that reconstructions of prehistoric life scenes condense time in that an image might stand for an entire period such as the Early Upper Paleolithic. It can similarly capture the fact that reconstructions condense action, too, because everyone is shown as engaged in a certain activity considered typical of the time and place represented.

While illustrators are careful to fill the evidential gaps through informed guessing (or what Wiktor Stoczkowski [1997: 250] refers to as *deductive imagination*), they may be less aware of the way in which they are influenced by visual traditions that can be traced back to antiquity of depicting early humans (Moser and Gamble 1997) – a process which Stoczkowski calls *conditioned imagination*. Reconstructions thus run the risk of becoming idées fixes, of being reproduced again and again, even after novel data is available or the existing evidence has been newly interpreted. At the same time, even today there might exist widely different interpretations and therefore rivaling visualizations of the same moment and scene in prehistory, a fact that can work against the seeming objectivity of an image. The topic of objectivity leads back to the issue of style: Artists can decide against realism and, for instance, opt for the use of watercolor in order to express uncertainty or vision. Other strategies for engaging the viewer to reflect on the fact that a reconstruction tells one story to the exclusion of other legitimate stories on the basis of a restricted set of evidence are possible, such as the application of impressionism rather than naturalism, the blurring of the image, the inclusion of blank spots, or the introduction of alienation by showing an archeologist at work among the Paleolithic peoples portrayed. One also has to consider context: Whether a reconstruction is studied in a museum gallery, exhibition guide, textbook, or scientific report will most likely affect the way it is read. Depending on the social meaning of the space in which it appears, and on the immediate context of caption, related text, and adjacent illustrations, the image will be understood differently (Sommer 2007: 247–262).

Closely related to reconstruction paintings are 3D-reconstructions in the form of busts or entire bodies. In fact, the famous busts of J. Howard McGregor, one of

Osborn's former students who had been brought to the American Museum of Natural History, were used by Knight for his Cro-Magnon and Neanderthal murals. Also in the production of the busts, the method was considered strictly scientific. Osborn had McGregor detail the laborious and careful making of a statue from bones in the museum's bulletin, *Natural History*. McGregor explained and reproduced as photographs the stages the Neanderthal bust went through: the cast of the skull; the skull with the missing parts restored; the skull with muscles modeled out of plasticine, restored cartilages, and plaster eyeballs; and the hairless skull with the soft-tissue applied. Another image showed the restored cast superimposed on the bust to visually prove the match. A photograph of the final figure with facial expression and hair was published for the first time in Osborn's *Men of the Old Stone Age* (1915). By then, McGregor had produced such "racial portraits" of *Pithecanthropus*, Piltdown, Neanderthal, and Cro-Magnon. They were widely reproduced and a great attraction of the Hall of the Age of Man that opened at the American Museum of Natural History in 1924 (Sommer 2016: 45).

Thus, also in the case of "selling" sculptures to the public, the artist is presented as possessing a great amount of scientific expertise. This is true for cases ranging from Benjamin Waterhouse Hawkins, who reconstructed the dinosaurs for the Crystal Palace (1851), to the Parisian sculptor Elizabeth Daynes, one of the most sought-after interpreters of Neanderthals today (Moser 1993: 76). Like Kupka before her, Daynes does not aim at representing types – her images and sculptures are instead intended as resurrections of individual human beings on the basis of fossil finds. She tries to even capture their personality (Sommer 2006: 227–228). Today, craniofacial identification and reconstruction rely on computerized tomography scans, computer modeling programs, 3D-printers, and computer animation. Scholars have argued, however, that in the process of "envisioning" a face from osseous remains, "race" – is it "Caucasoid," "Mongoloid," or "Negroid"? – continues to play a defining role from the first step of studying the skull, to the choice of indexes of soft tissue depth, up to the decision about skin, eye, and hair color (Nieves Delgado 2020).

Finally, beyond bust reconstructions, there is the tradition of full-body models. At the world exhibition in Paris in 1900, a *Pithecanthropus* reconstruction was exhibited, and an early exhibition that focused on that technique was made at the Field Museum of Natural History in Chicago; it opened up a series of eight dioramic scenes from human evolution to the eye of the public in 1933 (Moser 2003: 4–7). More recently, the Smithsonian Institution had the famous paleo-artist John Gurche build 15 hominin sculptures for the David H. Koch Hall of Human Origins that opened in 2010. In *Shaping Humanity* (Gurche 2013), the artist provides his own account of the interplay between science, art, and imagination. Like his predecessors, he not only cooperated with scientists, but also had to engage in scientific research, spending time in dissection rooms and fossil laboratories around the world, and devise a methodology. As shown in Fig. 7, today's 3D models of fossil humans may, too, be integrated into dioramas that carry the same cultural baggage as reconstruction paintings but might be interpreted as scientific truth or, by young museum visitors, even as containing stuffed humans (Scott and Giusti 2006: 54–56; on these kinds of visual reconstruction, see also Hochadel 2013: Ch. 7)! At the same

Fig. 7 Diorama of a Cro-Magnon Family in today's Ukraine in the Hall of Human Biology and Evolution at the American Museum of Natural History (Photo by Dennis Finnin, Scott and Giusti 2006:55, Fig. 3. Image # 2A21120, American Museum of Natural History Library)

time, in all of these genres from reconstruction painting to diorama, there can also be found more or less conscious attempts at transcending stereotypes and conveying different – if not reverse – messages.

Expositions and Museums

The above makes clear that all of the discussed kinds of reconstructions may come together in exhibitions, which historically tend to convey evolutionary advance by immersing visitors in movements through visual arrangements in space. Progress in general was especially the theme of universal exhibitions, and the prehistorian Gabriel de Mortillet of the École d'Anthropologie in Paris illustrated what he believed to be the universal law of progress at the prehistoric section of the Universal Exhibition in Paris of 1867. The series of world fairs organized in the Western metropolises of the nineteenth century themselves epitomized stages in the progress of industry and empire (Stoffel and Wessely 2020). In the case of the oval exhibition building at the Parisian Universal Exhibition, the exhibits on the history of industry (*l'histoire du travail*) occupied the first concentric ring around the central garden. Each nation filled a section of the ring, with the nations deemed most important appearing first and being allotted more space (see image at: http://commons.wikimedia.org/wiki/File:Exposition_universelle_de_1867.png?uselang=de). The halls were ordered chronologically within the territory of each nation, moving from prehistory into the present. When entering through the main gates, the visitors could either start with the section of Great Britain or France, but de Mortillet advised in the guide he wrote for the prehistoric exhibits that the visitors proceed clockwise, thus turning left and beginning the tour with France. In this way, France's technological progress unfolded before the visitors' eyes from the hall "La Gaulle avant l'emploi du métaux" to those celebrating recent innovations. The prehistoric part of this progressive narrative was represented by the Paleolithic, Neolithic, dolmen, lake-dwelling, Celtic, Gaul, and Gallo-Roman halls. Within the Stone Age period, the artifacts were again arranged chronologically on the basis of archeological sites.

In this movement through the inner exhibition circle, with its repetition of the prehistoric epochs in national sections, de Mortillet wanted the visitors to notice the parallel developments in different geographical regions. To illustrate continuity in form or use beyond prehistory, he also directed the guide readers toward more recent technologies and customs shown in the exhibits of the civilized nations. Finally, he instructed on the fact that the universal technological development takes place at different times for different peoples by drawing attention to the galleries on the colonies of France and of other European nations, where visitors were referred to the similarities with objects from Western prehistory. In sum, de Mortillet made use of his guide to make the exhibition architecture and the serial arrangement of the exhibits enact for the visitors what he held to be the great laws of human evolution. They should witness the law of universal human progress, the law of similar developments in all "human races," and the great antiquity of humankind. And de Mortillet emphasized that *l'histoire du travail* staged by the progressive series in material cultures stood for corresponding mental and anatomical progress (Sommer 2015a: 13–14).

Expositions remained important places to communicate progressive human evolution in body and culture, for example, at the Panama-California Exposition in San Diego in 1915 that now included not only skeletal casts but also illustrations and busts of prehistoric humans (Redman 2016: Ch. 4, on exhibitions also Ch. 6). However, visitors might have taken home the message of progress even if this was not the intention of museum curators. This was the case for the American Museum of Natural History exhibits discussed above, which were produced over a long period of time and in the end sent a confusing message through the skeletal mounts, the cases with stones and bones, the busts, the little reconstruction images, and large murals (Clark 2008: Chs. 2, 6; Sommer 2016: Ch. 4; see Fig. 8). Part of the reason for this "misunderstanding" might be the importance of seriality in the visual communication of evolutionary history that has been under concern throughout this chapter, be it in images of the production stages of a tool that supposedly mirrored the development through successive archeological cultures, in the stratigraphic tables showing the series of epochs, cultures, and fauna, or in the arrangement of exhibits, reconstruction paintings, and museum rooms (Sommer 2010, 2015a).

Significant research has also been carried out on the question of how contemporary visitors understand human origins exhibitions, with Monique Scott's *Rethinking*

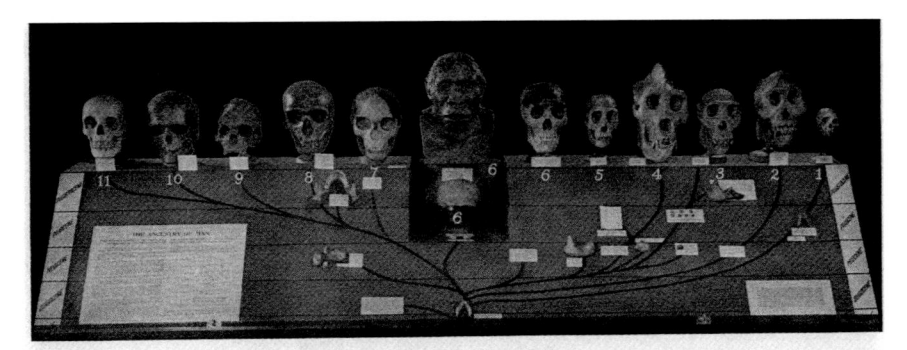

MAN'S PLACE AMONG THE PRIMATES

1, skull of gibbon; 2, orang; 3, chimpanzee; 4, old male gorilla; 5, young gorilla; 6, *Pithecanthropus*, skull reconstructed by Dr. J. H. McGregor; 6, model of bust of *Pithecanthropus* by Dr. McGregor; 6, cast of original skull top of same; 7, Piltdown skull (*Eoanthropus*) reconstructed by Dr. McGregor; 8, Neanderthal (Chapelle-aux-Saints), cast of original; 9, Talgai (Australia) cast of original; 10, old man of Crô-Magnon, cast of original; 11, modern white skull. The Heidelberg jaw (cast) is between 7 and 8. The other specimens are casts of various teeth and jaw fragments of fossil apes: *Propliopithecus* at the bottom, *Sivapithecus* at the left, and several species of *Dryopithecus* and allied genera at the right. The geologic ages (Oligocene, Miocene, etc.) are indicated by the horizontal zones. The black lines indicate the relationships as inferred by Dr. W. K. Gregory.

Fig. 8 This phylogenetic tree was one of the pieces of exhibition at the American Museum of Natural History. From this showcase, the visitor may have learned the following concerning the then existing "knowledge about man's place among the primates": The genealogy of the primates is a branching structure, yet there are no fossils on the line leading from *Propliopithecus* (represented by a jaw), the last common ancestor of all lines from Oligocene times, to living humans. However, the message is ambiguous, because the horizontal series of skulls at the top might be read as a descending ladder from "modern white man" (No. 11 in the legend), via Cro-Magnon, Neanderthal, Piltdown, *Pithecanthropus* (J. Howard McGregor's bust in the middle), gorilla, chimpanzee, and orangutan, down to the gibbon. (From Osborn HF, McGregor JH, Gregory WK, Nelson NC (1923) The Hall of the Age of Man, American Museum of Natural History, 2nd edn. Guide Leaflet Series 52, p 5. Image # 38097, American Museum of Natural History Library, discussed in Sommer 2015a: 16–21)

Evolution in the Museum: Envisioning African Origins (2007) standing out in breadth and insight – it is based on interviews with nearly 500 museum visitors. Now as then, what people take home is a product of negotiation between prior understandings, mostly from the popular media, and the interpretations presented in the museum. And this continues to mean that – also against the museum's intent or scientific insight – visitors may leave with their story of a path of progress out of (a primordial) Africa all the way to the (civilized) West intact, with some people of color even identifying themselves as still closer to the origins than their "white" fellow humans. Another important source from where people receive their knowledge of human evolution is the school. Quite disconcertingly, a study published in 2008 of the visualizations of human evolution in 30 textbooks from 12 countries brought to light little exception from the representation of *Homo* as "white" adult male and the linear series of progress leading to this "apex of evolution" (Quessada et al. 2008). However, suggestions have been formulated of how museum exhibitions on human origins might be improved (Moser 2003), and serious as well as creative attempts have been made to unsettle "racial" stereotypes, as in a traveling exhibition on the newer genetic approach to human evolution that opened in 2011. It contained a 3D-figure of pregnant Mitochondrial or African Eve and a thread of DNA emanating from her connecting all exhibits/visitors as well as an interactive installation of a tree of life that conveyed an experience of pan-organismic kinship and the fact that degrees of kinship are always relative. Although these messages were taken up by many visitors, the exhibits and the meanings visitors made of them remained contradictory (Sommer 2016: Ch. 15).

Cartoons and Films

There were and remain of course yet other sources through which human evolution is conveyed to a larger public by visual means. Newer ways of commercializing prehistory are computer games and reenactments, while among the more traditional spaces of encounter are zoos (Sommer 2016: Ch. 7), and among the more common techniques of (satirical) engagement are cartoons. In the course of the first half of the nineteenth century, advances in printing technologies and the diversification of audiences created the stage for popular experimentation with the new ideas of (human) evolution. In the context of Darwin's publications on the subject, cartoons satirized the possibility of "man's descent from apes" (Browne 2001). Ape ancestry continued to be scandalous, and Constance Areson Clark (2008: Chs. 1, 9; 2009) has studied cartoons about evolution surrounding the Scopes Trial of 1925 that often played with the image of the missing link – the monkey, ape, ape-human-hybrid, or the caveman and cavewoman (see Fig. 9). From the late nineteenth century onward, cavemen have been used for more or less good-humored comments on sociocultural phenomena, and as the figure of the missing link in general, these cartoons

Fig. 9 Cartoon mocking antievolutionism after the Scopes Trial. (Adapted from Evolution, a Journal of Nature (Feb. 1928)3:16, discussed in Clark 2009: 579–580)

transported teleological, racialist, and gendered meanings. Cartoons were often far away from expert opinion, and scientists at times tried to intervene in public caveman perceptions by presenting their knowledge in newspapers. Newspapers thus became a site of encounter between different views on the human deep past and of negotiation about its meanings for the present (Sommer 2006).

Films, too, could be a thorn in the side of experts. Beginning with D. W. Griffith's comic prehistoric love story in the silent *Man's Genesis* of 1912 and the imperialistic theme in the cinema version of Arthur Conan Doyle's *The Lost World* (1925), caveman imagery has saturated films (e.g., Privateer 2005). A typical aspect of animated cartoons and other films, in close reference to prehistoric science fiction novels, is the coexistence of humans and dinosaurs, which was also the case for *The Lost World*. In more recent everyday life, cavemen have been more present than ever, for example, in the figures of *The Flintstones*, one of the most successful animated television series that started in the 1960s, and their multimedia avatars (e.g., Berman 1999: 289). Also Gary Larson's *The Far Side* comics that from the 1980s onward were syndicated worldwide, published in many collections, and turned into animated films have become part of our cultural heritage. The 1980s further brought forth the film *The Caveman* (1981) that features the Beatle Ringo Starr as Neanderthal antihero and is a sarcastic comment on our constant policing of the boundary between "them" and "us" – animal and human, ancient and modern, and Neanderthals and Cro-Magnons – because it turns out that a simple stretch of the back suffices to turn the slouching caveman into the upright modern.

From the known hominins, it is the Neanderthals who are the proverbial cavemen of the popular realm, probably because even in supposedly scientifically based imagery, they used to be represented as stooping and cave-living, as clad in furs and equipped with clubs. Today, one encounters their appropriation in children's learning tool kits, computer games, as plastic figures, on stamps, t-shirts, stickers, and in the visual arts. Neanderthals also continue to be the objects of inquiry and speculation in every genre from the scientific journal article to the science fiction novel. Interesting hybrid forms are the "documentaries," such as the BBC series *Walking with Cavemen* (2003), which succeeded the award-winning *Walking with Dinosaurs* (1999) and *Walking with Beasts* (2001) series. It combined a certain scientific take on, among others, the Neanderthal question (*The Survivors* episode) with a fictional reanimation of the dead bones and stones by means of the newest cinematic technologies. The animals were computer generated or animatronic, while the hominins were played by actors wearing makeup and prosthetics. The reality-fantasy mix is in the form as well as in the content, and the viewer is uncertain about the genre: Is it a natural history documentary in good Attenboroughian fashion or prehistoric science fiction involving time travel? But such series like *Walking with Cavemen* and its Channel 4 predecessor *Neanderthal* (2000) want to be taken seriously, and scientific authority may not only be conveyed through the consultation of many scientists, the docu-style, and technology and makeup, but possibly also through accompanying popular books that explain "the informed guesses" made. Thanks to these strategies of collateral authentication, fictive elements, such as Neanderthal social structure and behavior, including gender relations, and their fatal competition with the Cro-Magnons, appear as scientifically more robust than they actually are (Sommer 2008b: 162–163).

As paintings and dioramas, movies and TV-documentaries – a strange word for films on a largely inaccessible past – at the same time mirror and naturalize social realities of their time, especially with regard to family values and gender roles. Particularly disturbing, in the staging of the encounter between modern humans and Neanderthals (a topic that has caught the imagination of scientists and the public from the start), is how the motif of kidnapping and even rape of women have been endlessly repeated since J.-H. Rosney Aîné's *La Guerre du feu* (1909, made into a film in 1981). This prehistoric topos can be connected to at least nineteenth-century artistic traditions (Moser 1998: 148–153), but there also remain racialist overtones. For the simulation of prehistoric times, films rely on a mix of techniques like the staging of excavation sites, (computer) animation, talking heads, and reenactments – that is contemporary humans playing "our ancestors." In the last case, indigenous communities such as the San are often performed as living fossils. Their seemingly original ways of life are thus presented as a remnant from the common human deep past, and in a genetic history perspective, the people appear as a direct genetic link to "our earliest human ancestors." The stories might therefore also be racially connoted in that the history that is being reconstructed is that of ("white") Europeans and Americans. Furthermore, a Western expert – a paleontologist, archeologist, or geneticist – may lead through the movie, a technique which was used for critical engagement with the above stereotypes especially in the 1980s and 1990s, but that can just as well strengthen ethnocentrism and the justification of social inequality (Koch 2017; for genetic history films, see Sommer 2016: 345–354).

Finally, another way of dramatically animating the human deep past is the play, and already in the pre-Darwinian period, themes of evolution such as the missing link between humans and apes were taken up on the popular stage. Paleolithic themes – obviously including cavemen – remain popular today, but there are also more sophisticated engagements such as by the Prehistoric Body Theater company that, in exchange with paleontologists and beginning in Indonesia, generates a unique experience of "our deep history" when the human bodies of dancers impersonate vertebrates through four stages of evolution; it "offers a dynamic forum for dialogue on humanity's diverse understandings of identity, body, and origins" (Rudenko 2018: 1293).

Conclusion

Overall, we have seen that the human origins sciences are inherently visual fields. From their beginnings, genres of visualization have ranged from more expert practices and circulation to purely popular imaginations, while a certain level of mutual influence between science and art as well as between science and everyday perceptions has always been present. We have also seen how issues of "racial" hierarchies and sexual inequality have been at the core of physical anthropology and have to a certain degree continued to be implicated in more recent representations from phylogenetic trees to 3D-reconstructions, dioramic exhibits, or films. At the same time, the developments in visualization techniques are strongly marked by technological innovation and by an increasing methodological sophistication in paleoanthropology and prehistoric archeology as well as in human population genetics. While quantification is an old endeavor that originates in racial anthropology, the gap between the expert and lay communities has thus widened, including the gaps between disciplines – a tendency that is met with great efforts of public communication and popularization. However, it seems that the widening of the gap between science and the public also renders the explanation of how a visualization is arrived at, what assumptions went into it, and what has been left out even more vital. The same holds true for the dialogue between the practitioners of human origins sciences and the scholars engaging with the interactions between science and society. After all, the subject of the human origins sciences implicates everyone. The knowledge that comes out of these sciences impacts the understanding of self and that of individuals and groups. The sociopolitical should therefore be an aspect of reflection in its production and communication.

Cross-References

▶ Before Fieldwork: Textual and Visual Stereotypes of Indigenous Peoples and the Emergence of World Ethnography in Hungary in the Seventeenth to Nineteenth Centuries
▶ Colonialism and Its Knowledges

References

Archibald JD (2014) Aristotle's ladder, Darwin's tree: the evolution of visual metaphors for biological order. Columbia University Press, New York
Bangham J (2015) What is race? UNESCO, mass communication and human genetics in the early 1950s. Hist Hum Sci 28(5):80–107
Banta M, Hinsley CM, O'Donnell JK, Jacknis I (2017) From site to sight: anthropology, photography, and the power of imagery. Thirtieth anniversary edition. Peabody Museum Press, Harvard University, Cambridge
Berman JC (1999) Bad hair days in the Paleolithic. Modern (re)constructions of the cave man. Am Anthropol 101(2):288–304
Bouquet M (1996) Family trees and their affinities: the visual imperative of the genealogical diagram. J R Anthropol Inst 2(1):43–66
Bowler P (1986) Theories of human evolution: a century of debate 1844–1944. Johns Hopkins University Press, Baltimore
Brace CL (1981) Tales of the phylogenetic woods: the evolution and significance of evolutionary trees. Am J Phys Anthropol 56(4):411–429
Browne J (2001) Darwin in caricature: a study in the popularisation and dissemination of evolution. Proc Am Philos Soc 145(4):496–509
Cain VEM (2010) 'The direct medium of the vision': visual education, virtual witnessing and the prehistoric past at the American Museum of Natural History, 1890–1923. J Vis Cult 9(3): 284–303

Clark CA (2008) God – or gorilla. Images of evolution in the jazz age. Johns Hopkins University Press, Baltimore

Clark CA (2009) 'You are here': missing links, chains of being, and the language of cartoons. Isis 100(3):571–589

Conkey M (1997) Mobilizing ideologies: Paleolithic 'art', gender trouble, and thinking about alternatives. In: Hager LD (ed) Women in human evolution. Routledge, London, pp 172–207

Edwards E (2009) Evolving images: photography, race and popular Darwinism. In: Donald D, Munro J (eds) Charles Darwin, natural science and the visual arts. Yale University Press, New Haven, pp 167–193

Gifford-Gonzalez D (1993) You can hide, but you can't run: representation of women's work in illustrations of Palaeolithic art. Vis Anthropol Rev 9(1):23–41

Gould SJ (1995) Ladders and cones: constraining evolution by canonical icons. In: Silvers RB (ed) Hidden histories of science. New York Review Book, New York, pp 38–67

Gurche J (2013) Shaping humanity: how science, art, and imagination help us understand our origins. Yale University Press, New Haven

Hanke C (2007) Zwischen Auflösung und Fixierung: Zur Konstruktion von 'Rasse' und 'Geschlecht' in der physischen Anthropologie um 1900. transcript, Bielefeld

Hochadel O (2013) El Mito de Atapuerca. Orígenes, Ciencia, Divulgación. Edicions UAB, Barcelona

James S (1997) Drawing inferences: visual reconstructions in theory and practice. In: Molyneaux BL (ed) The cultural life of images: visual representation in archaeology. Routledge, New York, pp 22–48

Koch G (2017) It has always been like that...': how televised prehistory explains what is natural. In: Fries JE, Gutsmiedl-Schümann D, Matias JZ, Rambuscheck U (eds) Images of the past: gender and its representations. Waxmann, Münster, pp 65–84

Lipphardt V, Sommer M (2015) Visibility matters: diagrammatic renderings of human evolution and diversity in physical, serological and molecular anthropology. Hist Hum Sci 28(5):3–16

Livingstone DN (2010) Cultural politics and the racial cartographics of human origins. Trans Inst Br Geogr 35:204–221

Moser S (1992) The visual language of archaeology. A case study of the Neanderthals. Antiquity 66(253):831–844

Moser S (1993) Gender stereotyping in pictorial reconstructions of human origins. In: du Cros H, Smith L (eds) Women in archaeology: a feminist critique. Research School of Pacific Studies, Australian National University, Canberra, pp 75–92

Moser S (1996) Visual representation in archaeology: depicting the missing-link in human origins. In: Baigrie BS (ed) Picturing knowledge: historical and philosophical problems concerning the use of art in science. University of Toronto Press, Toronto/Buffalo, pp 184–215

Moser S (1998) Ancestral images: the iconography of human origins. Cornell University Press, Ithaca

Moser S (2003) Representing archaeological knowledge in museums: exhibiting human origins and strategies for change. Public Archaeol 3:3–20

Moser S, Gamble C (1997) Revolutionary images: the iconic vocabulary for representing human antiquity. In: Molyneaux BL (ed) The cultural life of images: visual representation in archaeology. Routledge, London, pp 184–212

Nieves Delgado A (2020) The problematic use of race in facial reconstruction. Sci Cult. https://doi.org/10.1080/09505431.2020.1740670

Privateer P (2005) Romancing the human: the ideology of envisioned human origins. In: Smiles S, Moser S (eds) Envisioning the past: archaeology and the image. Blackwell, Oxford, pp 13–28

Quessada M, Clément P, Oerke B, Valente A (2008) Human evolution in science textbooks from twelve different countries. Sci Educ Int 19(2):147–162

Qureshi S (2011) Peoples on parade: exhibitions, empire, and anthropology in nineteenth-century Britain. Chicago University Press, Chicago

Reardon J (2005) Race to the finish: identity and governance in an age of genomics. Princeton University Press, Princeton
Redknap M (2002) Re-creations: visualizing our past. National Museums and Galleries of Wales and Welsh Historic Monuments, Cardiff
Redman SJ (2016) Bones rooms: from scientific racism to human prehistory in the museum. Harvard University Press, Cambridge
Richards RJ (2018) The beautiful skulls of schiller and the Georgian girl: quantitative and aesthetic scaling of the races, 1770–1850. In: Rupke N, Lauer G (eds) Johann Friedrich Blumenbach: race and natural history, 1750–1850. Routledge, New York, pp 142–146
Rudenko A (2018) Prehistoric body theater: bringing paleontology narratives to global contemporary performance audiences. Integr Comp Biol 58(6):1283–1293
Rudwick MJS (1976) The emergence of a visual language for geological science 1760–1840. Hist Sci 14(3):149–195
Rudwick MJS (1989) Encounters with Adam, or at least the hyaenas: nineteenth-century visual representations of the deep past. In: Moore JR (ed) History, humanity and evolution: essays for John C. Greene. Cambridge University Press, Cambridge, pp 231–252
Rudwick MJS (1992) Scenes from deep time: early pictorial representations of the prehistoric world. The University of Chicago Press, Chicago
Scott M (2007) Rethinking evolution in the museum: envisioning African origins. Routledge, New York
Scott M, Giusti E (2006) Designing human evolution exhibitions: insights from exhibitions and audiences. Museums Soc Issues 1(1):49–68
Solometo J, Moss J (2013) Picturing the past: gender in *National Geographic* reconstructions of prehistoric life. Am Antiq 78(1):123–146
Sommer M (2003) The romantic cave? The scientific and poetic quests for subterranean spaces in Britain. Earth Sci Hist 22(2):172–208
Sommer M (2005) Ancient hunters and their modern representatives: William Sollas's (1849–1936) anthropology from disappointed bridge to trunkless tree and the instrumentalisation of racial conflict. J Hist Biol 38(2):327–365
Sommer M (2006) Mirror, mirror on the wall: Neanderthal as image and 'distortion' in early 20th-century French science and press. Soc Stud Sci 36(2):207–240
Sommer M (2007) Bones and ochre: the curious afterlife of the Red Lady of Paviland. Harvard University Press, Cambridge, MA
Sommer M (2008a) History in the gene: negotiations between molecular and organismal anthropology. J Hist Biol 41(3):473–528
Sommer M (2008b) Neanderthals. In: Regal B (ed) Icons of evolution. Greenwood, Westport, pp 139–166
Sommer M (2010) Seriality in the making: the Osborn-Knight restorations of evolutionary history. Hist Sci 48(3–4):461–482
Sommer M (2015a) (Net)working a stone into a tool: how technologies of serial visualization, arrangement, and narration stabilized eoliths as archeological objects. In: Eberhardt G, Link F (eds) Historiographical approaches to past archaeological research. Edition Topoi, Berlin, pp 15–45
Sommer M (2015b) Evolutionäre Anthropologie zur Einführung. Junius, Hamburg
Sommer M (2015c) Population-genetic trees, maps, and narratives of the great human diasporas. Hist Hum Sci 28(5):108–145
Sommer M (2016) History within: the science, culture, and politics of bones, organisms, and molecules. University of Chicago Press, Chicago
Sommer M (2021) The meaning of absence: the primate tree that did not make it into Darwin's *The Descent of Man*. BJHS Themes, pp 1–17. https://doi.org/10.1017/bjt.2020.14
Sommer M (forthcoming) Making anthropology diagrammatic: Samuel George Morton's 'American Golgotha' and the contest for the definition of a young field. Hist Hum Sci

Sommer M, Lipphardt V (eds) (2015) Visibility matters: diagrammatic renderings of human evolution and diversity in physical, serological, and molecular anthropology. Hist Hum Sci 28(5): 3–16

Spencer F (1997) Some notes on the attempt to apply photography to anthropometry during the second half of the nineteenth century. In: Edwards E (ed) Anthropology and Photography 1860–1920. Yale University Press, New Haven, pp 99–107

Stoczkowski W (1997) The painter and the prehistoric people: a 'hypothesis on canvas'. In: Molyneaux BL (ed) The cultural life of images: visual representation in archaeology. Routledge, London, pp 249–262

Stoffel P, Wessely C (2020) Exhibiting earth history: the politics of visualization in the second half of the nineteenth century. In: Genge G, Schwarte L, Stercken A (eds) Aesthetic termporalities today: present, presentness, re-presentation. transcript, Bielefeld, pp 57–66

Wiber M (1998) Erect men – undulating women: the visual imagery of gender, 'race' and progress in reconstructive illustrations of human evolution. Wilfrid Laurier University Press, Waterloo

Part III
Self and Personhood

Made-Up People: Conceptualizing Histories of the Self and the Human Sciences

13

Elwin Hofman

Contents

Introduction	324
Theatres of the Self	325
Technologies of the Self	328
Performing the Self	331
Looping Effects	333
Practices of the Self	336
Neurohistory	339
Problems and Questions	342
References	345

Abstract

This chapter discusses the different concepts and theories historians have used to discuss the reflexive relationship between the human sciences and the self, that is to say, how the human sciences have altered human selves and vice versa. It highlights both applications and critiques of these theories. The chapter begins with the Erving Goffman's and Mary McIntosh's sociological theories on the presentation of self, labelling theory and role theory, and discusses their influence on the historiography of homosexuality. The chapter proceeds with a discussion of Michel Foucault's work and historians' uses of the concept of "technologies of the self." Next, the chapter examines Judith Butler's theory of gender performativity and the relational nature of the self. Ian Hacking's influential concepts of "making up people" and "looping effects" are then discussed. Since the new millennium, historians have adopted new approaches inspired by practice theory and by the neurosciences. These theories, their applications for the history of emotions and deep history, and their critiques are analyzed. Finally, the chapter highlights some recurring critiques and remaining questions.

E. Hofman (✉)
Cultural History Research Group, KU Leuven, Leuven, Belgium
e-mail: elwin.hofman@kuleuven.be

© The Author(s), under exclusive licence to Springer Nature Singapore Pte Ltd. 2022
D. McCallum (ed.), *The Palgrave Handbook of the History of Human Sciences*,
https://doi.org/10.1007/978-981-16-7255-2_71

Keywords

Reflexivity · Technologies of the self · Looping effects · Performativity · Practice theory · Neurohistory

Introduction

One of the central tenets of many historians of the human sciences is that their research subject is not only relevant as an intellectual history. The human sciences, they often argue, have had – and still have – an impact in the world. Not only have they had political, economical, and disciplinary effects, they have even reshaped the essence of being human; they have reshaped the self. A classic example is Ian Hacking's (Hacking 1995) analysis of multiple personality: as the diagnosis of multiple personality developed in the 1970s, more and more people started to identify with the diagnosis and displayed more different personalities. Developments in psychiatry therefore seemed to have affected their selves, new kinds of people were "made up." In turn, these people provided new material for psychiatry.

How could the human sciences have such an impact, especially as their stated goal was often to describe the workings of the human mind, rather than alter them (or at least not in the ways historians have suggested)? This chapter explores the concepts and theories historians have used to describe and explain the influence of the human sciences on the self. While most historians agree that the human sciences did have an impact on their subject of study, they disagree on how and to what extent.

At the heart of this chapter lies an examination of the conceptualizations of the relationship between the human sciences and humans. The human sciences develop in interdependence with their subject matter. Human subjects provide these sciences with their source material, but human subjects may also learn the knowledge generated by the human sciences and adapt to this knowledge. This interdependence has been seen as a process of "reflexivity," as a "reflexive circle": humans and the human sciences continuously reflect on one another (Smith 2013, p. 16). It may be conceived of in different ways and in different degrees. The theories and concepts discussed in this chapters provide ways of doing that, with particular attention to the self.

The "self" is conceived in a broad manner in this chapter. Various human sciences have proposed various definitions of the self. Finding a definition that encompasses and surpasses them all is an impossible task. This chapter therefore proceeds with a loose circumscription of the self: the self is about who people are, about how they feel, think, and act as individuals, to some extent apart from others, and to some extent apart from their bodies (Hofman 2016). Above all, this chapter will follow scholars and human scientists in their own conceptions of what the self is. Likewise, this chapter uses a loose definition of the human sciences, including all disciplines that study, classify, and analyze human behavior. As much research on the relationship between the human sciences and the human self has focused on medical, social, and behavioral sciences, this chapter will likewise give less attention to the humanities than to psychiatry.

This chapter does not provide an exhaustive historical overview of theories of how the human sciences affect the self. Its scope is rather more limited: it examines the theories and concepts historians have used to describe the influence of the human sciences on its object, the human self (and vice versa), and analyses how they have used them. Most theories were not developed by historians themselves, but adopted from sociologists or philosophers (or in one case, neuroscientists) and adapted for a historical context. One of the difficulties inherent to this field is that historians *of* the human sciences have to rely on theories and concepts *from* the human sciences – or develop their own *as* human scientists – to analyze the relationship between the human sciences and the self. There is therefore a further reflexive part in this enterprise; a difficulty which will be further discussed in the concluding section.

Not all of the theories that will be discussed are as commonly used among historians of the human sciences. By far the most popular have been Foucault's "technologies of the self" and Ian Hacking's "looping effects." The other concepts and theories that will be discussed are more commonly used by historians of the self who do not see themselves primarily as historians of the human sciences. Nevertheless, these concepts and theories, and their historical applications, also provide room for the human sciences and are therefore relevant to historians of the human sciences interested in its effects on the human self.

The chapter will proceed roughly chronologically, starting with the early constructionist theories that historians referred to, and then discussing commonly used concepts and theories borrowed from the work of Michel Foucault, Judith Butler, and Ian Hacking. Subsequently, the use of practice theory and the rise of neurohistory are addressed. In the final section, the chapter discusses common criticisms and remaining problems and questions regarding the relationship between historians, the human sciences, and the self.

Theatres of the Self

Up to the 1970s, historians of the different disciplines that make up the human sciences did not usually give much attention to the effects of these sciences on selfhood or subjectivity. When they discussed the self, it was firmly as the object of study of the human sciences, not as an effect of the human sciences. At the same time, many early historians of the self focused on the period before 1800 and while they did attend to work that relates to the human sciences, such as Descartes', Locke's, and Kant's writings, they did not usually problematize the reflexivity between the intellectual history of the self and the history of the self or subjectivity itself (e.g., Mauss 1938; Macpherson 1962).

Yet in the 1950s and 1960s, some sociologists were proposing new ways of conceiving the self that would be picked up by historians from the 1970s onwards. These sociologists proposed a "dramaturgical" approach to the self. They used theatrical metaphors, such as "roles," "performance," and "backstage." Among them, American sociologist Erving Goffman (1922–1982) is one of the most influential. Famous for his attention to everyday social interactions, his work ranges from the presentation of self, over asylums and social stigma, to the organization of experience.

For historians of the self and everyday life, Goffman's work on self-presentation of has been particularly inspiring. In *The Presentation of Self in Everyday Life*, he discussed how people perform their self in everyday interactions with others. This performed self is often seen as an organic character, housed within the body of its possessor, but, Goffman argued, it is actually "a dramatic effect arising diffusely from a scene that is presented" (Goffman 1959: 252–3). The self – or at least a part of what was seen as the self – was not given, not a biological essence, but a product of interactions. Nevertheless, Goffman also suggests that behind this performed self, there is an individual as a performer, who was not merely an effect of particular performances. This self (or this part of the self), which has dreams and desires and is capable of learning, is "psychobiological in nature," but still interacts with its own performances (Goffman 1959: 253–4). Goffman has strongly influenced "social constructionist" views of the self, but he did not himself view the self as entirely socially constructed.

Goffman's dramaturgical approach to the self – and similar approaches by other sociologists and anthropologists – have inspired some scholars of the self to study how people "fashioned" their self in various instances (e.g., Greenblatt 1984). The insistence on social interactions was not, however, particularly suitable to analyze the impact of the human sciences on the self: Goffman's analysis was firmly situated on the micro-level, and power, institutions, or larger structures were remarkably absent. Historians of the human sciences have therefore engaged more with two of his subsequent works, *Asylums* (1961) and *Stigma* (1963). Especially the latter work is interesting for its consideration of the reflexive effects of social labelling.

In *Stigma: Notes on the Management of Spoiled Identity* (Goffman 1963), Goffman considered the situation of people who were denied complete acceptance by society, such as people with disabilities, homosexuals, or people without a job. Much of his study dealt with how people coped with and tried to avoid stigmatization in interactions. But Goffman also suggested that stigmatization could reinforce deviant behavior: one way of coping with stigmatization was for people to become exactly that what others expected them to be. Around the same time, Howard Becker, John Lofland, and David Matza put forward similar arguments. When someone is classified in a certain category, they argued, they can in some instances take on this identity, start to see themselves as such and behave accordingly. The category, Lofland argued, could become an "essential nature" (Lofland 1969, pp. 127–8). The self was thus shaped (at least in part) by others' classifications.

While these proponents of so-called "labelling theory" were not particularly focused on the human sciences or on historical developments, the idea that social classifications impacted the self opened up possibilities for historical research into the impact of classifications designed by physicians, psychiatrists, or sexologists. Among historians of sexuality, for instance, British sociologist Mary MacIntosh's study of homosexuality proved to be influential. McIntosh argued that "the homosexual should be seen as playing a social role rather than as having a condition." The human sciences played an important part in creating and sustaining such roles, as "psychologists and psychiatrists on the whole have not retained their objectivity but become involved as diagnostic agents in the process of social labeling" (McIntosh 1968, p. 184). Sexual identity, often considered as a core aspect of one's self, did not come "from the inside" but was created through classification and roleplaying.

Unlike many other labelling theorists, McIntosh also explicitly historicized her claim. She situated the origins of the modern concept of homosexuality as a "condition which characterizes certain individuals and not others" (188–9) in England at the end of the seventeenth century. This early and concise historicization of homosexuality would inspire many historians and historically minded sociologists. Jeffrey Weeks, for instance, built on McIntosh's work to explore the importance of legal and medical discourse for the making of homosexual identities in the second half of the nineteenth century. According to Weeks, psychiatrists and sexologists did not singlehandedly dictate people's identities, but, as the labelling and role sociologists had proposed, people developed a consciousness of homosexuality and a sense of self in relation to the concepts that were applied to them (Weeks 1977).

The dramaturgical, interactionist, and labeling perspectives have had an enormous impact on thinking about human behavior, the self, and identities. They have allowed scholars to denaturalize the self, to question its stable psychobiological nature, and paved the path towards a social constructionist view of the self. But these sociological theories of the 1960s were not without their problems. Critics have pointed out many theorists' lack of attention for complex dynamics of power and resistance, for politics, economics, and institutions, in short, for the historical context in which performances, labelling, and roleplaying took place (Brune et al. 2014). Moreover, some historians rejected the language of casting, labelling, and roles, because it was too rigid: identities were more messy and more fluid than neatly delineated roles and, unlike roles in a play, people cannot simply decide to take up or quit a performance at will. Finally, much of the empirical evidence of Goffman and his colleagues was qualitative and remained unconvincing to critical observers. That the self was "socially constructed" – a consequence rather than a cause of behavior and social identities – remained an inconclusive interpretation, they claimed. It would result, for instance, in tedious debates on whether there existed a biological or psychological "homosexual essence" throughout history or not. While empirical evidence showed that there were definitely variations in homosexual behavior, how these variations related to the self remained a matter for interpretation (Boswell 1980; Greenberg 1988).

Historians continue to be inspired by social interactionist, dramaturgical, and labelling approaches to human behavior and the self, not only in the history of sexuality but also in fields such as disability history (Brune and Wilson 2013), gender history (Griffin 2018), and the history of emotions (Nash and Kilday 2010). Nevertheless, the great breakthrough of historical research on the reflexivity between the human sciences and the self would come with the work of Michel Foucault.

Technologies of the Self

The French philosopher Michel Foucault (1926–1984) has had an enormous impact on historians of the human sciences and on historians of the self. Many have built on or adapted his concepts, claims, and interpretations, and others have ferociously contested them. Although interactionists and labeling theorists had already laid its foundations, Foucault is often heralded as the founding father of social constructionist thought. Because Foucault often wrote about the past to argue and illustrate his ideas about power, knowledge, and subjectivity, he has managed to attract the attention of historians more than the sociologists discussed in the previous section.

In *The Order of Things* (Foucault 1966), Foucault studied the making of the human sciences. He provocatively argued that "man" as a knowing and independent subject was co-constituted with the human sciences around 1800. Only as "man" became a being distinct from the rest of nature, he could (and necessarily would) also construct himself as an object. So not only was the individual self – "man" – a social construct, a "recent invention" in Foucault's words, this invention was also closely tied to the human sciences. Both were the effect of a change, around 1800, in the "fundamental arrangements of knowledge." This denaturalization of the human subject – and the announcement of its impending "death" – was not yet much elaborated. Foucault did not show how "man" was created, nor did he elucidate how the modern subject was constituted. Few historians, moreover, were convinced by his claim of a radical break around 1800. However, his work did inspire historians of the human sciences to reflect on their subject matter, and particularly on how different fields of the human sciences related to each other, to larger systems of knowledge and to the self.

Foucault returned to the question of subjectivity and its relationship with knowledge and power in his works of the 1970s. In *Discipline and Punish* (Foucault 1975), Foucault argued that the making of the individual as an object was a result of *external* discipline. Around 1800, as brutal physical punishments were increasingly contested, "the soul" became the object of penal intervention and knowledge gathering (rather than simply "crimes"). This "specific mode of subjection" gave birth to the human sciences, which took "man" in general as an object of scientific knowledge. The year after, in the first volume of his *History of Sexuality* (Foucault 1976), Foucault discussed the making of the individual as a subject as a result of *internal* discipline. Through the demand of confession – in Catholicism, but also in court, to the doctor, to loved ones – people were stimulated to examine themselves and in this way create an interior self (Foucault 1993: 203; Goldstein 1995: 108). Foucault thus made his position on the self and the human sciences clearer. Both the knowledge generated by the human sciences and the self were constituted as allies through disciplinary and confessional practices.

Foucault's most commonly employed concept for operationalizing the links between the human sciences and the self has perhaps been that of "technologies of the self" (also: "techniques of the self"), which he proposed in several lectures and writings in the early 1980s. His previous works had mainly considered how the self was historically constituted through knowledge and power relations. The "Subject" was a result of "subjection" (*assujettissement*). But in his work of the 1980s, Foucault became more interested in how the self actively constituted itself, a process Foucault called "subjectivation" (*subjectivation*). This subjectivation took place through specific techniques and practices, which he called "technologies of the self." They "permit individuals to effect by their own means or with the help of others a certain number of operations on their own bodies and souls, thought, conduct, and way of being, so as to transform themselves in order to attain a certain state of happiness, purity, wisdom, perfection, or immortality" (Martin et al. 1988, p. 18).

In the second and third volumes of *The History of Sexuality*, Foucault studied technologies of self-care and self-knowledge in antiquity and early Christianity, when they were not yet associated with a modern forms of the self as constituted by discipline and the human sciences. He did not connect these technologies with the structures of knowledge and power in his earlier work. However, in lectures and interviews, Foucault did make this link, arguing that technologies of the self are "not something invented by the individual himself. They are models that he finds in his culture and that are proposed, suggested, imposed upon him by his culture, his society and his social group" (Foucault 1989, pp. 440–1). Technologies are therefore ambiguous: they can serve both to dominate us and to liberate us.

Although Foucault's ideas about how "technologies of the self" would look in the modern age remained vague, the concept was quickly adopted by historians. It was much easier to operationalize than many of Foucault's earlier concepts. As Jan Goldstein has argued, it offers several advantages. Through its links with regimes of knowledge and power, the concept of technologies of the self can bridge the distinction between theory and practice. Technologies combine both intellectual systems and everyday behavior. Moreover, the concept also held rhetorical power, suggesting a sort of "how-to" format that neatly illustrates the historical mutability of the self (Goldstein 1999).

Two influential scholars in the history of the self and the human sciences can illustrate how Foucault's work – on technologies of the self and more generally – was used in historical studies. The first is Nikolas Rose, a sociologist who wrote extensively about the role of psychology in shaping society and the self in late nineteenth- and twentieth-century Britain, most notably in *The Psychological Complex* (Rose 1985) and *Governing the Soul* (Rose 1999). Rose's work was strongly shaped by Foucault's thought. He used Foucault's critical insights in the historicity of knowledge to argue that psychological knowledge in nineteenth- and twentieth-century Britain was generated as a means to manage specific societal problems, particularly in settings such as factories, hospitals, schools, and social administrations. Psychology, for him, was a technology, or a set of technologies, to order, frame, and produce certain modes of existence. And these technologies have shaped the self. He has used these Foucauldian concepts and insights to show how the "psy-disciplines" (psychology, psychotherapy, psychiatry,...) have played a major role in making the current regimes of subjectification and their unification under the label "self."

A second scholar to make extensive use of Foucault's work in a historical context is the aforementioned historian of France Jan Goldstein. In *The Post-Revolutionary Self* (Goldstein 2005), she used of the concept of "technologies of the self" to study the influence of eighteenth- and early-nineteenth-century psychology, particularly the work of Victor Cousin, on ordinary educated people. Inspired by Foucault's historicization of subjectivity and its links with power, she argued that state-organized education was an important means to shape the self in France around 1800. Cousin promoted introspection in the education system, which Goldstein saw as a technology to reshape the self. More than Rose, Goldstein applied Foucault's methods and concepts to a more traditional and specific historical setting to show how intellectual historical trends could have a wider impact in society.

Foucault was a major influence for many historians of the self and the human sciences. Yet he also provoked severe criticism, leading to heated debates between Foucauldians and critics. One line of critique was that the perspective of Foucault and his followers on the human sciences and the self was too much top-down and did not leave enough room for individual agency and resistance. People seemed to be passive victims of technologies, docile bodies rather active individuals. Foucault did not give much attention to the tension between aims and effects: in practice, attempts to impose discipline often failed and people made use of certain technologies and forms of knowledge for their own purposes. Particularly in his earlier work, before the 1980s, critics argued, there was insufficient room for these complexities (Thomson 2006, pp. 5–7).

His later work on technologies of the self, however, has sometimes received the opposite critique. If technologies of the self permit individuals to operate on and transform themselves, how did they obtain this possibility? Lynn Hunt has criticized the formulation of the concept and the unreflective use of the term individuals as a "distinctly modern or post-eighteenth-century formulation" that was insufficiently historicized (Hunt 1992, pp. 80–5). Moreover, she argued, Foucault's individual was a distinctly *male* individual. He all but ignored questions of gender. The self that he studied was an implicitly and sometimes explicitly male self. At the same time, power relations seemed sexless. Foucault hence missed the inherently gendered nature of subjectivity. Hunt and others have therefore critiqued Foucault's masculinist vision on the history of the self.

Finally, as with the interactionists before him, several historians remained unconvinced that the self that was formed through discipline, scientific discourses, and technologies was the only self, or the only self worth talking about. The focus on discourses, bodies, and technologies, some scholars claimed, ignored people's actual psychic make-up. Lynn Hunt has more recently claimed that with Foucauldian methods, we cannot "address what is inside the black box of the self," such as emotions and experiences (Hunt 2014, p. 1578). Other, particularly "essentialist" historians of sexuality, have criticized Foucault's social constructionist claims more generally. Still, since ca. 2000, many historians of self and sexuality alike, wary of the "Foucault wars," have implicitly accepted some of Foucault's central tenets, but moved on to use different concepts and theories.

Performing the Self

In 1990, the American philosopher and feminist thinker Judith Butler (1956–) offered an attempt to move the conversation forward. In *Gender Trouble* (Butler 1990), she used the concept of performativity to throw new light on gender identities. She argued that gender identity does not exist naturally in the world, that the gendered self is not a given, but has to be created, to be performed. Butler thus combats the notion that men and women have distinct, fixed selves. Male and female bodily characteristics are meaningless in themselves: the production of a self and the attachment of gender to bodily characteristics is a "discursive effect" of performances, a result of involvement with cultural and linguistic codes. Only by performing gender identities over and over again, gender identities start to appear natural, biologically determined, and self-evident. Deviations from this naturalized behavior – Butler gives the example of the drag queen, performing extreme femininity – seem like aberrations. They lead to "gender trouble" and reveal the constructed nature of gender.

With this performative approach, Butler reconciled theatrical and Foucauldian perspectives on the gendered self. She built on the English philosopher of language John Austin's speech act theory, in which he investigated how words make things happen, such as marriages ("I pronounce you husband and wife") or criminal verdicts ("You are found guilty of murder"). Similarly, Butler suggested, performative utterances create – rather than simply describe – gender and sexual identities. They bring into being the thing they seemingly describe. Unlike Goffman and his colleagues, Butler did not assume that there was a "real" self behind the performances; there was no preexisting performing subject. The "I" is entirely made up of performances. Not only gender and other identities but also subjectivity itself, she argued, are entirely constituted by their expression and a result of power relations. This view was clearly inspired by Foucault's work on subjectivity. For Butler, the self is inherently unstable, never just "being" but always "becoming" through action.

In *The Psychic Life of Power* (Butler 1997) and *Giving an Account of Oneself* (Butler 2005), Butler elaborated her views on selfhood, power, and relationality. Following Foucault, she argued that "no subject comes into being without power," while that at the same time, this coming into being involves the subject being heralded as the source of power. The subject, she claimed, "is neither fully determined by power nor fully determining of power (but significantly and partially both)" (Butler 1997: 15–17). Moreover, she argued that the self, the "I," is always relational: it has no story that is not the story of a set of relations, not only to others but also to social norms. It does not exist by itself but is always *called* into existence, a response to a demand to give an account of oneself and to take responsibility.

Like Foucault and Goffman before her, Butler destabilizes the self, perhaps even more radically than her predecessors. The self originates in – and can only be understood through – *performances*, in *relations,* and in response to *power*. This approach allows researchers to attend more to difference and nonnormative forms of selfhood and identity. The self is not necessarily unified or consistent, but can take many forms. Seemingly aberrant performances are not lesser but become relevant and useful precisely for their destabilizing potential.

Butler's influence has been pervasive in historical writing, particularly in cultural history. By approaching historical sources as "performative," rather than merely descriptive, historians were able to sidestep discussions about whether a given source's claims were "accurate," but instead show how they could be productive. This allowed them to interpret contradictions and show the wider relevance of seemingly unrepresentative sources.

The human sciences play no privileged role in Butler's work. Unlike Foucault, she does not offer a sustained analysis of the role of medicine, psychology, or other sciences in performing gender and subjectivity. However, her concepts and theories invite us to see the human sciences at work when they legitimize and stimulate certain performances and delegitimize others. Scientific performances can be particularly powerful and can be used when calling on people to give an account of themselves. When the sciences seem to describe a situation, they are – in the right circumstances and given enough repetition – in fact bringing them into being. They influence possibilities for the self and subjectivity. But because Butler relatively rarely explicitly discussed the human sciences, or indeed historical evolutions, historians of the human sciences have usually engaged with Butler in a more limited way than with Foucault. Some historians have adopted a performative approach and cite Butler in passing, but have not engaged with her work more extensively.

Unsurprisingly, Butler's greatest impact has been among gender historians. Her work offered a more inclusive way to address gender and subjectivity and to move away from assumptions about a masculinist stable self (Barclay and Richardson 2013). Using a performative approach inspired by Butler, for instance, historian Josephine Hoegaerts (2014) has studied the performances – from singing to military maneuvers and the construction of homosocial spaces – that led to the making of a common language of masculinity in nineteenth-century Belgium. Scientists such as architects, engineers, and pedagogues played an important role in her story.

Historians of sexuality have also fruitfully used Butler's insights, along with those of other scholars associated with "queer theory" (most notably Sedgwick 1990), to venture to new grounds. Unlike Foucault, Butler and Sedgwick were not interested in searching for great ruptures in the histories of gender and sexual identities. Instead, they emphasized that identities were always inherently unstable, paradoxical, and fluid. In response, some historians of sexuality have rethought their goals and methods (e.g., Halperin 2002; Doan 2013). Whereas earlier historians had often focused on the making of a modern homosexual (or heterosexual) identity, they seek to lay bare the contradictions in sexual discourses through the ages. They accept that nobody can wholly own discourses about sexual identities – not scientists proposing new categorizations and conditions, not their patients and study objects, and not gay and lesbian activists – but study how words and concepts describing sexualities are used, circulate and generate meaning, in all their contradictory ways.

Finally, Butler has also inspired historians focusing on the history of the self and subjectivity, particularly scholars working on marginalized subjects who were visibly called to give an account of themselves in criminal courts (e.g., Kounine 2018;

Hofman 2021). Laura Kounine, for instance, used Butler's work to study how the identity of the witch was created in the process of being on trial. Judges asked people who were tried for witchcraft to give an account of themselves and as such take on – or reject – a particular identity. This performance, if sustained, created this identity. In line with Butler's stress on contradictions and fluidity, Kounine refused to go along with the traditional narrative of the "making of the modern self," but instead stressed that selves were made in different ways according to specific situations, without a clear chronological path (Kounine 2018, pp. 6, 131–2).

While Butler's work has been influential, it has not been without critiques. One major strand of critique has been that Butler, much like Foucault, denies the possibility of individual agency. Her theory of performativity only allows for a limited number of available gender roles and she argues that there is no subjectivity before discourse. Many scholars see this as problematic, even within Butler's own work: she does suggest that there are possibilities of resistance, but individual intention and motivation are mostly absent from her work. Relatedly, scholars have criticized the absence of desires, emotions, and experiences – matters that seem of vital importance to discuss selfhood.

A different group of scholars has taken issue with Butler's perceived denial of biological reality. Butler has argued that biological sex is as much a construction as gender, seemingly denying the physical reality of the body. However, Butler herself and her followers have contested this. They seek to denaturalize discussions of the body, to debunk biological "statements" as, in fact, performative, but they do not deny that the body is real (Butler 1995; Ruberg 2020: 68–9). Finally, scholars have criticized Butler's stress on abstract norms rather than specific practices. Although her theories of the performative and of relational selfhood invite us to think about how norms are put into practice, her focus remains mostly abstract. She does not give much attention to the specific practices that put the self in the world beyond discourses (Mak 2012).

Looping Effects

The Canadian philosopher of science Ian Hacking (1935–) has provided a set of concepts that have been fruitfully adopted by historians to discuss the relationship between the human sciences and the self. More than many of the other theorists described in this chapter, Hacking worked with historical material and applied his concepts to the history of the human sciences. As a consequence, his work has been discussed much more explicitly within the historiography of the human sciences.

Hacking started out as a philosopher of mathematics, logic, and probability. By the 1980s, however, he became interested in the human sciences as well. In 1986, he published an influential essay: "Making Up People" (Hacking 1986). Partly inspired by Foucault, partly by Goffman and the labeling theorists, Hacking argued that "numerous kinds of human beings and human acts come into being hand in hand with our invention of the categories labeling them" (Hacking 1986: 236). New sorts of categories and labels created new sorts of people, behavior, and selves, which in turn influenced and stabilized these categories. In this, he claimed people (he would later speak of "human kinds") are different from things ("natural kinds"). What things or animals do, does not depend on how we describe them. They act or work in the same way regardless. Some of the things that people do, however, are closely connected to our descriptions of them. Adapting a concept from Elizabeth Anscombe, Hacking suggests that when people act intentionally, they must act "under a description." People cannot intend what is indescribable. But they can sometimes adopt, adapt, and transform the descriptions, leading to the need to revise them. As he would later write: people are a moving target (Hacking 2007). Hacking called this view "dynamic nominalism."

In subsequent work, Hacking elaborated on the notion of making up people and proposed additional concepts. In *Rewriting the Soul* (Hacking 1995), Hacking studied the history of multiple personality disorder as a case study for how kinds of people come into being, as a case study for "the dynamics of the relation between people who are known about, the knowledge about them, and the knowers" (Hacking 1995, p. 6). And multiple personality, as discussed in the introduction to this chapter, neatly illustrated the "feedback effect" or "looping effect" between knowledge, knowers, and people known: "People classified in a certain way tend to conform to or grow into the ways that they are described; but they also evolve in their own ways, so that the classifications and descriptions have to be constantly revised" (p. 21).

From the outset, Hacking wrote that there could be no general theory of making up people; each history had its own peculiarities: homosexuals, split personalities, child abusers, and *garçons de café* were all "made up" but in very different ways. Yet in later work, he did endeavor some generalizations about how social, medical, and biological sciences created new classifications, knowledge, and hence people (Hacking 2007). Hacking argued that the making up of people is driven by seven "engines of discovery" (counting, quantifying, creating norms, correlating, medicalizing, biologizing, and relating to genetics) and three additional engines, normalization, bureaucratization, and resistance by the people being made up, who "claim back" their identities.

In *The Social Construction of What?* (Hacking 1999), Hacking addressed one of the critiques we have encountered for the theories discussed above: that "social constructionists" deny the physical and psychical reality of identities, subjectivity, and selfhood. Hacking argued that the metaphor of "social construction" had worn out. While it could and can sometimes still be liberating to realize that something (homosexuality, gender, the self) that seems inevitable is constructed, and therefore not part of the unavoidable nature of things or people, the claim is too vague. For Hacking, we must distinguish ideas from objects. Scholars tend to say that something is constructed (e.g., child abuse) but that it is a real evil. But doing so confuses two things, for it is the *idea* of child abuse that is constructed, but the *object* that is evil. Ideas and objects may interact and influence each other (through looping effects, for instance), but they are different.

In later work, Hacking therefore carefully stipulated what he is talking about. For instance, he specified that when arguing that multiple personality was "made-up," he meant that "in 1955 this was not a way to be a person, people did not experience themselves in this way, they did not interact with their friends, their families, their employers, their counsellors, in this way; but in 1985 this was a way to be a person, to experience oneself, to live in society" (Hacking 2007, p. 299). It is at that level, Hacking claimed, that looping effects occur. It is a much less contentious statement than claiming that there were no multiple personalities in 1955, and one that could resolve heated debates between constructionists and essentialists – a debate that is more a matter of attitude than of historical practice.

Hacking's analyses are mostly about categorizations and identities, but they have important implications for selfhood and subjectivity. In the process of making up people, many classifications become related to the inner self: homosexuality is a prime example. Hacking's concepts and theories have hence been used by many historians who discuss the human sciences and their impact. Historians of sexuality have found the concepts of "feedback" and "looping effects" useful to describe the cross-fertilization between the scientists of sexuality and the people they studied in the late nineteenth and early twentieth century (which Hacking himself had also referred to as an example). Wannes Dupont, for instance, has suggested that it was not only the case that homosexuals adopted the labels designed by sexual scientists, but also that "although few were prepared to admit it in public or even to themselves, many doctors, magistrates and policemen allowed their thinking to be influenced by what those led before them had to say about their inclinations" (see also Brickell 2012). Moreover, he expanded Hacking's notion of feedback beyond the strict confines of the scientific world. Courts of justice, Dupont suggested, were important arenas where the experience of and ideas about the nature of homosexuality were constantly renegotiated (Dupont 2015: 299).

Similarly, Hacking's insights and terminology were taken up by historians of psychology, psychiatry, and medicine (Rose 1999; Smith 2013), fields which Hacking's own research directly related to. For instance, Saskia Bultman has used Hacking's concepts in her study of a Dutch reform school in the 1930s and 1940s to examine how a new conception of the "delinquent girl," revolving around an inner self, was made up, among others through the use of the Rorschach inkblot test (Bultman 2020). Similarly, Nancy Campbell and Laura Stark have used Hacking's concepts to study how the idea of "vulnerable" human subjects for medical experiments was made up in the twentieth century and why some people adopt the self-understandings generally associated with people belonging to a certain kind, while others rejected them (Campbell and Stark 2015). Still others have observed looping effects in domains as diverse as eighteenth-century stomach complaints (Kennaway and Andrews 2019), shrinking penises (Crozier 2012), and psychologists' psychological explanations of psychologists' thoughts (Morawski 2020). Some of these studies stress the relationship to the self of this making up of people, while others limit their scope to the disease or category and its symptoms.

More than many other theories, Hacking's concepts offer an opportunity to not just study the effects of top-down labeling, which is then appropriated, but also the bottom-up influence of the people who take up or reject the labels on the scientists who label them. Moreover, Hacking makes clear, at least in his later work, that he is not making ontological claims about categories, but limits his analysis to the *effects* of the making up of people. Nevertheless, his framework has also been subject to criticism. While Hacking conceptualizes the links between the human sciences and selfhood and identities, his focus remains mostly on the side of the labelers, the scientists, their language, and their changing classifications. There may be room for feedback, but how this feedback works and how the people who are classified experience it remains invisible. The causal trajectory by which subjects change their behavior and self-conceptions in response to classifications is unclear. While Hacking suggests that the human sciences can change the self, his own account of the self is rather vague. Hacking claims that it has agency and a restricted freedom to make choices, but he does not make clear how this agency is related to the effects of classifications (Tekin 2014).

Similarly, scholars have also criticized the lack of conceptual tools to analyze why some classifications affect people's behavior and are taken up as lived identities, while others do not. Hacking seems to suggest that there are *always* looping effects, even if they work in different ways. Yet Saskia Bultman has shown that some classifications did not seem to result in new lived experiences or identities (Bultman 2016). Indeed, Chris Millard has suggested, Hacking's theories themselves should be historicized: they are a product of twentieth-century thought and their applicability outside the twentieth century remains tenuous (Millard 2017).

Finally, Jonathan Tsou (Tsou 2007) and others have argued that Hacking's distinction between kinds where looping effects occur ("human kinds") and those where they do not ("natural kinds," such as water or animals) is misleading and untenable. Tsou argues that Hacking is wrong to assert that there are no stable objects in the human sciences. Certain aspects of, for instance, psychiatric classifications can be linked to specific biological pathologies and hence remain stable, regardless of the development of certain knowledge. While Tsou agrees that there are weak looping effects, to the extent that people's behavior changes in response to classification, he contests that the defining criteria of classification necessarily change. He argues that people are therefore made-up in a more limited way than Hacking suggests: as in the critiques of other social constructionist theories, Tsou implies that we need to do more to acknowledge physiobiological and psychopathological realities.

Practices of the Self

To make the causal links between classifications and selfhood more specific, some historians have resorted to practice theory. Practice theory has had an enormous impact in sociology, anthropology, and history since ca. 2000. The term covers a wide variety of approaches, but three theorists have been particularly influential among historians. The most important one is perhaps the French sociologist Pierre Bourdieu (1930–2002).

Bourdieu used the term habitus, borrowed from the anthropologist Marcel Mauss, to denote the "system of cognitive and motivating structures," which consists of "schemes of perception, thought, and action" that generate and organize what we do and say, "without presupposing a conscious aiming at ends or an express mastery of the operations necessary in order to attain them" (Bourdieu 1990: 53–4). Habitus thus consists of behavior that is learned, not only explicitly but also implicitly, for instance, by mimesis or by a developing a certain "feeling." It is shaped by the communities in which an individual moves: their social class, their subcultures, and their "milieu." Bourdieu shows that pre-reflective, bodily, and automatic practices, such as reflexes, gestures, and dispositions, are influenced by social environments. The habitus becomes a "second nature" which provides a structure within which people can improvise. Bourdieu suggests that what we do and say, over and over again, shapes what we will do and say in the future. Our very bodies and our very selves are altered through practice and habituation, and thus need to be historicized.

The Dutch philosopher and medical anthropologist Annemarie Mol (1958–) shared this interest in practices and embodiment, but took practice theory in a new direction under the heading of "praxiography" – a nod to "ethnography," but with a focus on practice rather than culture. In *The Body Multiple* (Mol 2002), Mol studied medical practices surrounding atherosclerosis and argued that different medical techniques "enacted" multiple bodies and multiple diseases. Bodies or diseases can never simply "be," but must always be "enacted" through different practices and routines. There is no underlying entity: the body is inherently multiple. Practices of coordination, such as grouping techniques in the patient file, are needed to navigate the multiplicity.

Neither Bourdieu nor Mol were particularly occupied with the self. Yet their approach could have important repercussions for the study of the self, as something that is *enacted in practice* and *multiple*. The German sociologist Andreas Reckwitz has theorized how practices affect the mind and the self. Practices are "sets of routinized bodily performances," Reckwitz argued, but also require know-how, interpretation, intentions, and feelings. They require mental patterns. These mental patterns are not inherent characteristics "deep inside" an individual, but "part of the social practice" (Reckwitz 2002). The self is therefore not prior to practice, but the product of practice. It is created in the sequence of habits that require attention to "inner" processes of thought, feeling, and perception. In *The Hybrid Subject* (Reckwitz 2006), Reckwitz applied this theory to the history of the self (though on a macroscopic and theoretical level). He argued that the self is a cultural form that is produced and reproduced through practices such as reading and writing, working, and being intimate. He distinguished different forms of selfhood that have characterized different cultures, but also stressed their hybridity: they all incorporated elements of earlier cultures of selfhood and were always contradictory. These contradictions gave people room, and even required them, to make choices.

Practice theory – or, with Mol, praxiography, or, with Reckwitz, praxeology – differs from the approaches discussed above by its insistence on concrete doings and saying, rather than language, labels, and classifications. Although it highlights different things, practice theory can help to operationalize Foucault's technologies of the self (Burkitt 2002), Butler's performances and Hacking's looping effects: the concept of habitus provides us with a way to understand how particular ideas, classifications, or aspirations are adopted. Moreover, practice theory helps to bridge the gap between body and mind, between nature and culture. Practices cross and shape both.

Historians have used practice theory – or at least adopted the language of practice theory, often without much explicit theorizing – in many domains. In the history of emotions, it has become common to refer to "emotional practices." Emotions, many historians assume, do not naturally emanate from within (we do not just "ex-press" them) but have to be done: they are formed and transformed by writing love letters, shouting at political opponents, or kneeling before a priest. Building on Bourdieu and Reckwitz, Monique Scheer has most explicitly discussed the theoretical underpinnings of this assumption; her work provides the direct methodological foundation for many historians (Scheer 2012). The role of the human sciences is often rather limited in this type of scholarship, but it does not have to be. The human sciences have regulated certain emotional practices by claiming that they were healthy or unhealthy, coined new words to name feelings, and provided therapeutic techniques that allowed people to "do" emotions in new or different ways. For instance, in *History of Science and the Emotions* (Dror et al. 2016), several contributors used the concept of emotional practice to study how techniques such as anatomical dissections, making scientific photographs, and showing health education films in classrooms led to new ways of feeling.

Elwin Hofman has adapted Monique Scheer's framework to look more specifically at practices of the self and to disentangle how the self is shaped in different ways. He has distinguished "technologies of the self" by which people consciously try to change their self (e.g., meditating or following therapy), self-talk (discussions about what the self is), interpretations of the self (e.g., explaining behavior by referring to personality), and regulating practices (norms and expectations concerning selfhood, or attempts to change other people's selves, e.g., through education) (Hofman 2016). Hofman has applied this approach in a study of eighteenth- and nineteenth-century criminal justice, to show how changing practices in and around the criminal courts brought about changing selves (Hofman 2021).

Historians have often found that the focus on practices is a way to make other theories, such as those of Judith Butler and Ian Hacking, more specific and more historical. In her study of hermaphroditism around 1900, Geertje Mak, for instance, was inspired by Butler's ideas on the performance of gender, but found that Butler's theories did not sufficiently show how gender was enacted beyond discourse. By adopting a praxeographic approach inspired by Mol, Mak was able to stay clear of essentialist transhistorical definitions of hermaphroditism and instead to study how doubts about someone's sex were enacted in the nineteenth and early twentieth centuries. New medical techniques and routines led to new and multiple ways of thinking about bodily sex. These techniques and routines produced – or sometimes did not produce – the self of the hermaphrodite (Mak 2012). Other scholars have used praxeographic approaches to study how the Rorschach test brought into being a new understanding of "the delinquent girl" (Bultman 2020), or how medical and legal procedures that identified particular children as "idiots" in the nineteenth century affected these children's ways of being (Van Drenth 2016).

Despite its popularity in a wide variety of fields, practice theory has been subject to critique as well as praise. One of its problems – and one of the problems when discussing these critiques – is the great diversity of theories and methodologies that claim to promote a practice-based approach. Bourdieu's practices are of a very different nature, and operate on different ontological assumptions, than Mol's. One critique that has been levelled against some forms of practice theory is the lack of agency it affords to human beings. In some cases, such as Mol's praxeography, this relates to a decentering of the human being in the analysis, by giving more attention to objects and techniques. In other cases, such as in Bourdieu's theory, this relates to the seemingly irresistible reproduction of habitus through social environment. While most scholars using practice theory leave room for agency, and indeed Bourdieu's form of practice theory does not preclude individual agency, its concepts are less capable of making sense of this agency. As a result, some scholars have argued that practice theory is also less capable of explaining how individuals or groups have been able to bring about historical change (Hunt 2014, p. 1584).

Neurohistory

One persistent critique of many of the theories and concepts discussed above is that they disregard biological aspects of selfhood. Many of these theories were inspired by psychoanalysis on the one hand (even if they often contested some of its central tenets) and by the sociological metaphors of theatre and acting on the other. In response, some historians have turned to the neurosciences, genetics, and evolutionary psychology to find new foundations and concepts for historical research on subjectivity and emotions. They have argued that contemporary neuroscientific insights can help us to understand brains in the past. While their approach remains controversial, some scholars have started to speak of a "neuro-turn": "neurohistory" is but one aspect of a "new materialism" in which subjectivity and selfhood are firmly located in the brain. If "we are our brains," if the self is above all *cerebral*, as many people have come to assume (Vidal 2009), how does the history of selfhood change?

The term "neurohistory" was coined by the medievalist Daniel Lord Smail (2008). Smail argued that certain ways of behaving are so "deep" that they seem transhistorical. Certain emotions, for instance, are physiological, a consequence of natural selection and relatively automated. They have "a universal biological substrate that simply cannot be ignored." Hence, we must study the body and the brain – as complex assemblages of electrical impulses and chemicals – themselves as actors in history. As the body interacts with new cultural forms and social practices, it changes and generates new, unpredictable patterns and trajectories. Historical transformations, Smail suggested, are often unintended consequences or large-scale shifts in cultural practices. He related the decline of ritual and religion in eighteenth-century Europe, for instance, with the rising consumption of stimulants such as coffee, tobacco, and opium and the massive recourse to reading novels. These practices were, he proposed, sources of dopamine and other chemical messengers that replaced those provided by religion and ritual. The increased consumption of certain chemicals, or the increased recourse to certain practices, leads to transformations in the neurochemistry of the brain, which in turn creates an appearance of being hardwired.

In Smail's neurohistorical approach, the role for the history of the human sciences seems much smaller than in the approaches discussed above. When Smail discussed the increasing occurrence of hoarding in the late twentieth century, he did not, as we would expect someone using Hacking's theories, relate this to the development hoarding as a psychiatric condition. Rather, he related it to developments within consumer capitalism (Smail 2014). If there is a place for the history of the human sciences in Smail's form of neurohistory, it is not to study how people responded to the ideas and diagnoses of the human sciences, but to analyze how pharmaceutical drugs and therapeutic practices developed or promoted by the human sciences have changed the neurochemistry of the brain.

Smail himself did not elaborate much on what a neurohistorical approach would mean for the history of the self. This question was taken up by Lynn Hunt, a leading historian of the French Revolution who also turned to the neurosciences at the beginning of the new millennium. For Hunt, neuroscience could shed new light on the causes of the French Revolution and the development of human rights. Rather than looking at economic causes or cultural events, perhaps, Hunt argued, we should look at the brain. Like Smail, Hunt emphasized the importance of reading novels. This new and ubiquitous practice "had physical effects that translated into brain changes and came back out as new concepts about the organization of social and political life." Reading novels activated a "hardwired" capacity of human beings to identify and empathize with others (Hunt 2007: 33–34, 2009). For Hunt, historians of the self should pay particular attention to emotions, because they show up more readily in historical sources than any other "expression of selves." Historical change, new practices, and new selves came about "through the agency of writers and readers whose ever-changing embodied selves interacted with their ever-changing worlds" (Hunt 2014, p. 1585).

Smail and Hunt were both, like many other historians and humanities scholars, attracted to the work of the neuroscientist Antonio Damasio. Damasio adapted the Basic Emotion Theory model that had earlier been proposed by Silvan Tomkins and Paul Ekman. He supposed that certain brain functions, such as emotions, are independent of intentions, innate and automatically triggered behaviors. For Damasio, and for Hunt in his trail, the self is a process, always present when we are conscious, organizing, categorizing, and managing experiences. Its operation is in part hardwired, a result of evolution, and facilitates certain ways of interacting with others and with the world (Damasio 2010).

While Damasio is one of the most well-known neuroscientists, his theories are not uncontested; not among neuroscientists and not among historians turning to neuroscience. The historian Rob Boddice, for instance, has turned to the work of neuroscientist Lisa Feldman Barrett to develop neurohistory in a different direction (Boddice and Smail 2019). Feldman Barrett has contested the assumptions of Basic Emotions Theory and promoted what has been called a "psychological constructionist" model instead. Emotions or the self are not hardwired, she argues, but constructed by the mind out of a range of parts of the brain, which store prior experience and knowledge. Like anger or fear, the self is a goal-based concept, tied to immediate circumstances. There is no neural essence for "the self." Hence, by "tweaking your conceptual system and changing your predictions, you not only change your future experiences; you can actually change your 'Self'" (Feldman Barrett 2017, p. 192).

This "plastic" account of the self, emotions, and experiences is attractive to many historians, who find in her work a neuroscientific legitimation for doing historical research into subjective experiences. For Rob Boddice, Feldman Barrett's and her colleagues' insights highlight the relationship between social and cultural contexts and the brain. Cultural relations "*make* the brain." As such, historians should find out how people in the past *experienced* and *perceived* reality. They should, as Larry McGrath has argued, "explore how the embodied self was not the same in the past as it is now," through the shifting and problematic relations between feeling, intention, and meaning (McGrath 2017, p. 142). They should not use present-day categories to explain people's afflictions but take them at their own words (Boddice and Smail 2019).

The turn to neuroscience enables historians to look at the past in new ways. Neuroscientific findings provide means to combat the idea of a universal, unchanging biological human "nature" that is opposed to a historically specific "culture." No longer need historians classify behavior and identities – as in the debates on the history of homosexuality in the 1980s – as either biological or cultural: both biology and culture are constantly changing. People can be "born this way" and still historically constructed (Boddice and Smail 2019, p. 315). Yet what these insights can mean – practically – for the history of the self and the human sciences still requires much elaboration. Apart from a few examples (Berco 2016; Bourke 2016), most neurohistorical studies remain program-setting rather practical applications (Burman 2012).

One of the reasons for this lack of applications – besides its relative youth compared to the theories discussed above – is that many historians remain wary of neurohistory. Historians such as Roger Cooter and Ruth Leys have written trenchant critiques of the "neuro-turn" in historiography and the humanities in general. For starters, they argue, historians are often ill-informed about the state of the art in the quickly evolving field of neuroscience. They mainly rely on popularizing accounts, such as those written by Damasio and Feldman Barrett, while these views are often strongly contested. Historians do not usually have the capacities to critically evaluate neuroscientific insights: they have to take them on faith (Reddy 2010; Leys 2017).

Even if they are up to date on their neuroscience, historians often lack evidence for brain developments – they cannot scan past people's brains. Their applications of neuroscientific theories are therefore speculative and contradictory neuroscientific explanations of the kind that Smail proposed can be generated for almost any phenomenon (Reddy 2010). Rob Boddice resolved this problem by relying more on traditional historical evidence and takes from neuroscience that we should take historical actors at their word. But if so, Ruth Leys has argued, the "neuro" in "neurohistory" does not add much anymore: "So we seem to be left with a proposal to the effect that historians need to pay attention to what people have said and done in the past, which is what they've always tried to do." (Leys et al. 2020).

Finally, critics see political problems with the application of neuroscientific insights in historiography. The relationship between history and neuroscience is unequal, as neuroscientists do not need historians to make their claims (Millard 2020). Moreover, historians risk being unable to critique neuroscientific methods and insights. Smail, for instance, has been accused of acting as if the neurosciences are objective and value-free, which the historiography of science shows us they cannot be. For critics, the neuro-turn undoes the possibilities for critical histories that frameworks such as Foucault's and Butler's opened (Cooter 2014; Smith 2019). Finally, critics decry that a neuroscientific approach extinguishes human agency: what is important is not beliefs and intentions, but the bodily processes that produce them. There is no human agency, only "neural agency" (Cooter 2014; Leys 2017).

Problems and Questions

The theories and concepts discussed above challenge mainstream assumptions about the self. They highlight that the self is not universal or self-evident. The self is not necessarily (but can be) an essence inside every human; it is not necessarily (but can be) a natural, innate given. The way we conceive of and experience the self is contingent upon historical developments. Social and economic circumstances, cultural and technological developments, and indeed changing knowledge about what it means to be human – the history of the human sciences – can alter the way we think about and feel the self. In different ways, the concepts and theories discussed above allow historians to operationalize this reflexivity between the human sciences and the self; they give historians a language to discuss this relationship.

Early historians of the self often did not explicitly theorize how their histories, often of intellectual developments, related to the lived experiences of the self. In the 1960s, 1970s, and 1980s, the adoption of sociological, particularly social interactionist perspectives, among others by historians of homosexuality, brought about more explicit reflections on this relationship. Role and labelling theories provided historians with metaphors to discuss the impact of developments in the human sciences on identities and human lives. It became common to argue that the self was "socially constructed." By the 1980s and 1990s, as poststructuralist approaches to history became popular, historians turned to new concepts and theories, particularly those formulated by Michel Foucault, Ian Hacking, and Judith Butler. They stressed the role of power in the history of the self and provided critical tools to interrogate the present by studying selfhood in the past. Since around 2000, however, this approach has been criticized for being too focused on language rather than on the body and the material world. Some historians have therefore turned to practice theory or to neuroscience for new ways of conceiving the history of the self that try to bridge that gap.

Throughout all these theories, two critiques frequently recur. The first concerns the oppositions of discourse and experience and of mind and body. Historians of the human sciences are very good at analyzing the methods, categorizations, cases, and conclusions scientists write about. They sometimes suggest that these discourses "make up" the self, relinquishing attention for physical experiences. Some "neurohistorians" turn this approach on its head, privileging the effects of particular practices and drugs on bodily chemistry and neural pathways – and declining to attend to the impact discourses may have had. Even though many historians now claim to transcend these divides, in practice, they often still privilege either discourses or experiences, either mind or body, inciting critiques from their colleagues.

A second recurring critique concerns the lack of agency that many of the theories and concepts of the self imply. Is it possible to retain agency while fully historicizing the self, where agency is usually located? To claim that human agency is always possible is to universalize a part of the self, to place it outside history. Yet not leaving the possibility for agency, or only within specific circumstances, brings forward a determinist view of history. This is not only difficult to carry through in all domains, it is also politically and ethically undesirable.

Apart from these recurring critiques of individual theories and concepts, three more general critiques affect the theories discussed in this chapter. The first concerns the role of historians in developing and applying these theories. Historians studying how the human sciences related to human selves often do not develop their own theories for this relationship: they adopt and adapt theories from philosophers, sociologists, or neuroscientists. While they sometimes alter such theories in important ways, making them more suitable for specific historical applications, their contribution to these theories often remains implicit and they rarely talk back to the theorists they build on. Historians have often been more innovative in developing theories and concepts that show the relationship between the scientist's self (as a subject) and their theories (e.g. Paul 2016), rather than the relationship between the selves of the objects of the human sciences and these theories. Reconciling approaches in the history of the human sciences that focus, for instance, on "scientific persona" and the approaches discussed in this chapter may give opportunities to move the field forward.

The second challenge involves the relation between conceptual tenets and historical practice. Historians are mostly interested in changes through time. It therefore makes sense for historians to adopt a constructionist perspective in their work: they assume that the present state of things is not inevitable but a consequence of earlier historical events. This "historical constructionism" is, as Ian Hacking has remarked (Hacking 1999), not very different from "history" in general – a matter of attitude. While historians may diverge in their claims about how "deep" certain changes are, about whether the self changes or merely its representations or expressions, this does not greatly affect their historical analyses. Ruth Leys has, as we have seen, suggested that Rob Boddice's approach to neurohistory is not very different from regular history; she has similarly argued that Monique Scheer's plea for the use of practice theory in the history of emotions does not change much for historians' methods – it only changes their language (Leys et al. 2020). While some historians have certainly been able to come to new insights or causal explanations by using particular concepts and theories, they also run the risk of providing old wine in new bottles.

A final problem concerns the very principle underlying this whole chapter: that the human sciences affect the self; that reflexivity takes place between the human sciences and humans. To make this claim, historians draw upon insights *from* the human sciences – from anthropology, sociology, psychoanalysis, critical theory, philosophy, or neuroscience; either explicitly or more implicitly, in their use of "folk psychologies." This leads to a tension for historians of the human sciences: on the one hand, they are critically studying scientific theories and placing them in a historical perspective; on the other, they often relinquish this critical attitude when they adopt concepts and theories from the human sciences to address the impact of their findings. While they often stress that scientific theories must be understood within their historical context, they abandon this attitude when they apply present-day scientific concepts to the past they are studying.

This tension has become especially visible with respect to neurohistory. On the one hand, many historians of the human sciences critically study how the neurosciences have become so omnipresent in the late twentieth and twenty-first centuries, how their methods have developed, and how the field remains very diverse. At the same time, neurohistorians are – so their critics claim – uncritically applying provisional theories from neuroscience to do historical research. Neuroscientific theories about "brainhood" might be or become true, more critically oriented historians of the human sciences argue, through the reflexive effects of their increasing purchasing power in the present, but that does not necessarily make them suitable for historical application (Smith 2019).

This tension also applies to other theories. Chris Millard has argued that the whole idea of the malleability of human nature and the self – and therefore the possibility of reflexivity – is a product of twentieth-century anthropology that found its way to philosophy, sociology, and historiography. If, as many historians argue, we cannot assume that aspects of human nature are universal, why assume that human nature is universally malleable? As such, it makes no sense to project concepts developed in the late twentieth century, such as Ian Hacking's looping effects or Michel Foucault's technologies of the self, on earlier historical periods, just as it makes no sense to retroactively project medical diagnoses developed in the twentieth century on the past (Millard 2017, 2020).

Despite the extensive work that has been done, historians interested in the relations between the human sciences and the self continue to face a difficult task. They need not only be aware of the different theories available to conceptualize this relationship, but they also need to critically assess to what extent they can use such theories and their underlying assumptions in specific historical contexts. The theories and concepts discussed in this chapter are therefore often better at helping historians to ask new questions rather than to provide them with solid answers.

References

Barclay K, Richardson S (2013) Introduction: performing the self: women's lives in historical perspective. Womens Hist Rev 22:177–181. https://doi.org/10.1080/09612025.2012.726108

Berco C (2016) Perception and the mulatto body in inquisitorial Spain: a neurohistory. Past Present 231:33–60. https://doi.org/10.1093/pastj/gtw001

Boddice R, Smail DL (2019) Neurohistory. In: Tamm M, Burke P (eds) Debating new approaches to history. Bloomsbury, London, pp 301–325

Boswell J (1980) Christianity, social tolerance, and homosexuality: gay people in Western Europe from the beginning of the Christian era to the fourteenth century. University of Chicago Press, Chicago

Bourdieu P (1990) The logic of practice. Polity, Cambridge

Bourke J (2016) An experiment in "neurohistory": Reading emotions in Aelred's De Institutione Inclusarum (rule for a recluse). J Mediev Relig Cult 42:124–142. https://doi.org/10.5325/jmedirelicult.42.1.0124

Brickell C (2012) "Waiting for uncle Ben": age-structured homosexuality in New Zealand, 1920–1950. J Hist Sex 21:467–495. https://doi.org/10.1353/sex.2012.0050

Brune J, Garland-Thomson R, Schweik S et al (2014) Forum: on the 50th anniversary of Goffman's stigma. Disabil Stud Q. https://doi.org/10.18061/dsq.v34i1.4014

Brune JA, Wilson DJ (eds) (2013) Disability and passing: blurring the lines of identity. Temple University Press, Philadelphia

Bultman S (2016) Constructing a female delinquent self: assessing pupils in the Dutch state reform school for girls, 1905–1975. Dissertation, Radboud University Nijmegen

Bultman S (2020) Seeing inside the child: the Rorschach inkblot test as assessment technique in a girls' reform school, 1938–1948. Hist Psychol. https://doi.org/10.1037/hop0000167

Burkitt I (2002) Technologies of the self: habitus and capacities. J Theory Soc Behav 32:219–237. https://doi.org/10.1111/1468-5914.00184

Burman JT (2012) History from within? Contextualizing the new neurohistory and seeking its methods. Hist Psychol 15:84–99. https://doi.org/10.1037/a0023500

Butler J (1990) Gender trouble: feminism and the subversion of identity. Routledge, New York

Butler J (1995) Bodies that matter: on the discursive limits of sex. Routledge, New York

Butler J (1997) The psychic life of power: theories in subjection. Stanford University Press, Stanford

Butler J (2005) Giving an account of oneself. Fordham University Press, New York

Campbell ND, Stark L (2015) Making up 'vulnerable' people: human subjects and the subjective experience of medical experiment. Soc Hist Med 28:825–848. https://doi.org/10.1093/shm/hkv031

Cooter R (2014) Neural veils and the will to historical critique: why historians of science need to take the neuro-turn seriously. Isis 105:145–154. https://doi.org/10.1086/675556

Crozier I (2012) Making up koro: multiplicity, psychiatry, culture, and penis-shrinking anxieties. J Hist Med Allied Sci 67:36–70. https://doi.org/10.1093/jhmas/jrr008

Damasio A (2010) Self comes to mind: constructing the conscious brain. Pantheon Books, New York

Doan LL (2013) Disturbing practices: history, sexuality, and women's experience of modern war. The University of Chicago Press, Chicago

Dror OE, Hitzer B, Laukötter A, León-Sanz P (eds) (2016) History of science and the emotions. Osiris 31. https://doi.org/10.1086/687590

Dupont W (2015) Free-floating evils. A genealogy of homosexuality in Belgium. Dissertation, University of Antwerp

Feldman Barrett L (2017) How emotions are made: the secret life of the brain. Houghton Mifflin Harcourt, Boston

Foucault M (1966) Les mots et les choses. Une archéologie du sciences humaines. Gallimard, Paris

Foucault M (1975) Surveiller et punir. Naissance de la prison. Gallimard, Paris

Foucault M (1976) Histoire de la sexualité. La volonté de savoir. Gallimard, Paris

Foucault M (1989) The ethics of the concern for self as a practice of freedom. In: Lotringer S (ed) Foucault live. Interviews, 1961–1984. Semiotext(e), New York, pp 432–449

Foucault M (1993) About the beginning of the hermeneutics of the self: two lectures at Dartmouth. Polit Theory 21:198–227

Goffman E (1959) The presentation of self in everyday life. Doubleday, Garden City

Goffman E (1963) Stigma: notes on the management of spoiled identity. Prentice-Hall, Englewood Cliffs

Goldstein J (1995) Foucault and the post-revolutionary self: the uses of Cousinian pedagogy in nineteenth-century France. In: Goldstein J (ed) Foucault and the writing of history. Blackwell, Oxford, pp 99–115

Goldstein J (1999) Foucault's technologies of the self and the cultural history of identity. In: Neubauer J (ed) Cultural history after Foucault. De Gruyter, New York, pp 37–54

Goldstein J (2005) The post-revolutionary self: politics and psyche in France, 1750–1850. Harvard University Press, Cambridge, MA

Greenberg DF (1988) The construction of homosexuality. University of Chicago Press, Chicago

Greenblatt S (1984) Renaissance self-fashioning: from more to Shakespeare. University of Chicago Press, Chicago

Griffin B (2018) Hegemonic masculinity as a historical problem. Gend Hist 30:377–400. https://doi.org/10.1111/1468-0424.12363

Hacking I (1986) Making up people. In: Heller TC, Sosna M (eds) Reconstructing individualism: autonomy, individuality and the self in Western thought. Stanford University Press, Stanford, pp 222–236

Hacking I (1995) Rewriting the soul: multiple personality and the sciences of memory. Princeton University Press, Princeton

Hacking I (1999) The social construction of what. Harvard University Press, Cambridge, MA

Hacking I (2007) Kinds of people: moving targets. Proc Br Acad 151:285–318

Halperin DM (2002) How to do the history of homosexuality. University of Chicago Press, Chicago

Hoegaerts J (2014) Masculinity and nationhood, 1830–1910: constructions of identity and citizenship in Belgium. Palgrave Macmillan, Basingstoke

Hofman E (2016) How to do the history of the self. Hist Hum Sci 29:8–24. https://doi.org/10.1177/0952695116653305

Hofman E (2021) Trials of the self: murder, mayhem and the remaking of the mind, 1750–1830. Manchester University Press, Manchester

Hunt L (1992) Foucault's subject in the history of sexuality. In: Stanton DC (ed) Discourses of sexuality: from Aristotle to AIDS. University of Michigan Press, Ann Arbor, pp 78–93

Hunt L (2007) Inventing human rights: a history. W. W. Norton, New York

Hunt L (2009) The experience of revolution. Fr Hist Stud 32:671–678

Hunt L (2014) The self and its history. Am Hist Rev 119:1576–1586. https://doi.org/10.1093/ahr/119.5.1576

Kennaway J, Andrews J (2019) 'The grand organ of sympathy': 'fashionable' stomach complaints and the mind in Britain, 1700–1850. Soc Hist Med 32:57–79. https://doi.org/10.1093/shm/hkx055

Kounine L (2018) Imagining the witch: emotions, gender, and selfhood in early modern Germany. Oxford University Press, Oxford

Leys R (2017) The ascent of affect: genealogy and critique. University of Chicago Press, Chicago

Leys R, Knatz J, Coamhánach N (2020) The ascent of affect: emotions research and the history of emotions – interview with Ruth Leys (part II). J Hist Ideas blog. https://jhiblog.org/2020/04/20/the-ascent-of-affect-emotions-research-and-the-history-of-emotions-interview-with-ruth-leys-part-ii/. Accessed 15 Dec 2020

Lofland J (1969) Deviance and identity. Prentice-Hall, Englewood Cliffs

Macpherson CB (1962) The political theory of possessive individualism: Hobbes to Locke. Clarendon, Oxford

Mak G (2012) Doubting sex: inscriptions, bodies and selves in nineteenth-century hermaphrodite case histories. Manchester University Press, Manchester

Martin LH, Gutman H, Hutton PH (eds) (1988) Technologies of the self: a seminar with Michel Foucault. University of Massachusetts, Amherst

Mauss M (1938) Une catégorie de l'esprit humain: la notion de personne, celle de "moi". J R Anthropol Inst G B Irel 68:263–281. https://doi.org/10.2307/2844128

McGrath LS (2017) Historiography, affect, and the neurosciences. Hist Psychol 20:129–147. https://doi.org/10.1037/hop0000047

McIntosh M (1968) The homosexual role. Soc Probl 16:182–192. https://doi.org/10.2307/800003

Millard C (2017) Concepts, diagnosis and the history of medicine: historicising Ian Hacking and Munchausen syndrome. Soc Hist Med 30:567–589. https://doi.org/10.1093/shm/hkw083

Millard C (2020) Balance, malleability and anthropology: historical contexts. In: Jackson M, Moore MD (eds) Balancing the self: medicine, politics and the regulation of health in the twentieth century. Manchester University Press, Manchester, pp 314–339

Mol A (2002) The body multiple: ontology in medical practice. Duke University Press, Durham

Morawski J (2020) Psychologists' psychologies of psychologists in a time of crisis. Hist Psychol 23:176–198. https://doi.org/10.1037/hop0000140

Nash D, Kilday A-M (2010) Cultures of shame: exploring crime and morality in Britain 1600–1900. Palgrave Macmillan, Basingstoke

Paul H (2016) Sources of the self: scholarly personae as repertoires of scholarly selfhood. BMGN – Low Ctries Hist Rev 131:135–154. https://doi.org/10.18352/bmgn-lchr.10268

Reckwitz A (2002) Toward a theory of social practices: a development in culturalist theorizing. Eur J Soc Theory 5:243–263. https://doi.org/10.1177/13684310222225432

Reckwitz A (2006) Das hybride Subjekt: eine Theorie der Subjektkulturen von der bürgerlichen Moderne zur Postmoderne. Velbrück, Weilerswist

Reddy WM (2010) Neuroscience and the fallacies of functionalism. Review of Daniel Lord Smail, On deep history and the brain. Hist Theory 49:412–425

Rose N (1985) The psychological complex: psychology, politics and society in England, 1869–1939. Routledge & Kegan Paul, London

Rose N (1999) Governing the soul: the shaping of the private self, 2nd edn. Free Association Books, London

Ruberg W (2020) History of the body. Red Globe Press, London

Scheer M (2012) Are emotions a kind of practice (and is that what makes them have a history)? A Bourdieuian approach to understanding emotion. Hist Theory 51:193–220. https://doi.org/10.1111/j.1468-2303.2012.00621.x

Sedgwick EK (1990) Epistemology of the closet. University of California Press, Berkeley

Smail DL (2008) On deep history and the brain. University of California Press, Berkeley

Smail DL (2014) Neurohistory in action: hoarding and the human past. Isis 105:110–122. https://doi.org/10.1086/675553

Smith R (2013) Between mind and nature: a history of psychology. Reaktion Books, London

Smith R (2019) Resisting neurosciences and sustaining history. Hist Hum Sci 32:9–22. https://doi.org/10.1177/0952695118810286

Tekin Ş (2014) The missing self in Hacking's looping effects. In: Kincaid H, Sullivan J (eds) Classifying psychopathology: mental kinds and natural kinds. MIT Press, Cambridge, MA, pp 227–256

Thomson M (2006) Psychological subjects: identity, culture, and health in twentieth-century Britain. Oxford University Press, Oxford

Tsou JY (2007) Hacking on the looping effects of psychiatric classifications: what is an interactive and indifferent kind? Int Stud Philos Sci 21:329–344. https://doi.org/10.1080/02698590701589601

Van Drenth A (2016) The 'truth' about idiocy: revisiting files of children in the Dutch 'School for Idiots' in the nineteenth century. Hist Educ 45:477–491. https://doi.org/10.1080/0046760X.2016.1177123

Vidal F (2009) Brainhood, anthropological figure of modernity. Hist Hum Sci. https://doi.org/10.1177/0952695108099133

Weeks J (1977) Coming out: homosexual politics in Britain, from the nineteenth century to the present. Quartet Books, London

Inner Lives and the Human Sciences from the Eighteenth Century to the Present

14

Kirsi Tuohela

Contents

Introduction	350
The Roots of Modern Moral Culture	352
Interiority and Confession	353
Interiority as a Process of Reason	356
Sensibility and Human Interiority	357
Childhood as Interiority	362
Interiority and Reflexivity in Late Modernity	364
The Modern Mind, the Autobiographical, and the Inner	366
Conclusion	370
References	371

Abstract

This chapter addresses the question of "inner lives" and human interiority as it has been approached and discussed in the human sciences, mainly not only by historians, but also including writers oriented toward philosophical and literary analysis. The main focus is on Charles Taylor's ground-breaking *Sources of the Self: The Making of the Modern Identity* (1989) and Carolyn Steedman's *Strange Dislocation: Childhood and the Idea of Human Interiority 1780–1920* (1995). Progressing both thematically and chronologically, the chapter also discusses histories of sensibility, the rise of novel, the culture of melancholy, (late) modern reflective selfhood, and autobiographical practices in relation to human interiority. The chapter thus aims to map the key authors discussing "inner lives" in western culture from the perspective of cultural history.

K. Tuohela (✉)
University of Turku, Turku, Finland
e-mail: kirsi.tuohela@utu.fi

Keywords

Interiority · Inner life · Self · Confession · Autobiography · History of the self · Cultural history · History of modernity

Introduction

Humans are self-conscious beings—a notion that researchers and thinkers from several scientific fields and diverse academic traditions have aimed to clarify in terms of the concepts of mind, consciousness, the self, and subjectivity. Philosophers, psychologists, and neuroscientists, as well as scholars within the more traditional fields of the humanities, have pondered what role these aspects play in people's lives, and how these hidden entities inside us might relate to culture and society (see Seigel 2005; Goldstein 2005; Eghigian et al. 2007; Summerfield 2019; Hofman 2021).

The idea of human "interiority," this inner life of ours, is ancient, although it is suggested (see interiority, Merriam-Webster 2021) that the word was printed in English for the first time in 1701 in the Webster's dictionary. It originates from the medieval Latin *interioritas*, and via the Latin *interior*, the adjective of "inner" thus has a clear link to religious life and faith (interiority, Oxford Dictionary of English). "Inner" is a commonly used word in most European languages and, for example, in Swedish, it refers to the "mind" (*sinne*), "heart" (*hjerta*), or "fantasy" (*fantasi*). In religious contexts from the Bible onward, the "inner" (*innersta* in Swedish) may not only refer to the deepest level of mental and emotional life but also to the revelation of some godly truth, when talking about, for instance, "inner light" or "inner words" (*Inre/Svenska Akademiens ordbok*, SAOB /). Historian Monique Scheer (2012) maintains that the concept of "interiority" (*Innerlichkeit*) has played a key role in the emotional religious experience within German culture. She argues that the moral subject of Protestant Christianity was essentially emotional in nature, manifesting itself in teaching people to learn how to feel in the right Protestant way. A relationship with God was achieved through putting religious emotions—such as "love, trust and hope"—actually into practice. The emphasis here was on the need to focus on interiority (or the "inner life," as advocated by the Reformation), rather than on the "outer" religious rituals of the Catholics (180).

If we now turn to the contemporary knowledge-intensive societies of the twenty-first century, which prioritize mental capacity above all else, it would seem this inner focus has only intensified. However, today's interiority is perhaps couched in different terms as "inner space," "inner images," and "inner emotions" such as psychological pain. It can relate to the ability of the mind to focus on itself either in the past or the future, and this usually occurs in a scientific, rational, or analytical context. Sociologist Philip Rieff (2006) argues that the reason why [twentieth-century] "modern man" is no longer religious or spiritual is because he is psychological instead. As such, the culture of Rieff's "psychological man," and consequently the importance it attaches to the "inner life" is perhaps no less than it was previously, even if the modern psychological self differs in making these connections more complex, multilayered, deep, and secretive. What really sets it apart from Protestant interpretations of interiority, however, Rieff argues, is that in the modern secular world, where there is no need for conversion or even a faith, the inner life is no longer the space to search for salvation, but rather an arena for the praxis of therapy.

Rieff was writing about his own secular period more than 50 years ago in the 1960s, and yet some of his observations—such as the idea that the western world has started to follow more "secular visions of comfort"—still hold strong in many respects. After the collapse of the "latest failed god," as Rieff called it referring to the "proletariat" westeners were now creating "techniques [...] to be called 'therapeutic', with nothing at stake beyond a manipulatable sense of well-being. This is the unreligion of the age, and its master science" (10). Rieff's provocative theory of modern psychological man and a therapeutic culture are an attempt to synthesize the major cultural transformations of the twentieth century. His argument also contributes to the debate on how modernity has affected the psychic or inner lives of individuals. In this respect, it joins with other grand narratives that discuss what it means to be modern.

Even though we are living in what for many of us is a secular twenty-first century, we may still use concepts of the soul or spirit to refer to our immaterial sense of consciousness. "Interiority" or "inner lives" are often used loosely to refer to our minds, but more precisely they can also refer to private and viscerally felt experiences. In modern dictionaries such as Merriam-Webster, for instance, they can mean relating "to the mind or spirit" or "existing as an often repressed part of one's psychological makeup; *inner* child; *inner* artist" (inner, Merriam-Webster 2021). Elsewhere—in the literary context, for example—"interior monologue" is described as "the written representation of a character's inner thoughts, impressions, and memories" (Baldick 2008). "Inner life" may thus refer to imaginative or contemplative activities of the mind, and in the literary context this inner world is a lexical one—made of words. This being said, these very personal worlds may nowadays be also understood as also being visceral and physical experiences. These embodied emotions are an essential part of the unique psychological existence of each individual and seem to have become an important part of late modern conceptions of selfhood. So, what "interiority," "inner life," or indeed "inner man" means remains fluid, and certainly for a historian, the concept cannot be defined in its various forms without putting them in their particular cultural and historical context. With contextual reading, the historian can try to identify and distinguish various continuities, transformations, and disruptions. This chapter examines and discusses what has been written by historians of the mind, conceptual historians, cultural historians, and historians of the self and subjectivity on the topic of human interiority and our inner lives. The purpose is to look at how these concepts evolved into their modern form, and to look at their origins.

Certain key scholars in this field, such as the Canadian philosopher Charles Taylor (1931–), would suggest that in western culture the single most important factor in this evolution has been the rise of "unbelief," or abandonment of faith. In other words, the Enlightenment of the late eighteenth century cast doubt on many aspects of the Christian faith, to the extent where the latter began to lose ground against many more secular explanations of both the natural and social world, with the result that Christian theology also lost much of its influence as a source of ethical guidance and good behavior. This meant that a centuries-old system of thought about social order and human conduct was no longer as readily available for the psychic, inner life of individuals either. It is this major upheaval that this chapter also aims to examine.

Several choices were made when evaluating the historiography of human interiority for this chapter. A major role has been given to some key authors—mainly Charles Taylor and the British historian, Carolyn Steedman. They have both set the standards for the field with book-length studies about the history of human interiority. Other important contributions have been made by Dror Wahrman (2004) and Jerrold Seigel (2005). There are now many historians of the self, of personhood, and of subjectivity, and there are still more academic writers who have taken these concepts into the history of philosophy, psychology, literature, and religion (for example, Baggerman 2011; Summerfield 2019; Hofman 2021).

One choice made in this chapter has been to follow Taylor's argument concerning the rise of a modern identity from the Enlightenment onward, which has had major consequences on western culture. Another was to give space to Carolyn Steedman's analysis of human interiority, which charts the historicity and psychologization of culture with the rise of the "inner child" from the late eighteenth century onward. Among the main themes discussed by Taylor and other historians of the self are the cultural practice of confession, the rise of reason, the emergence of sensibility, and the complex character of modern reflexivity which are all covered as themes in this chapter as well. The important subthemes on the focus are the rise of novel, the idea of historicity and modern autobiography, and within the theme of sensibility the increasing importance of melancholy. These topics will be treated as fields where western individuals could fully express—and learn to express—their inner lives.

The Roots of Modern Moral Culture

The concept of "modern identity"—as a widely shared secular orientation to the western world and oneself—has been explored by many scholars. But perhaps the most comprehensive examination of the subject in a single volume was Charles Taylor's landmark study *Sources of the Self: The Making of the Modern Identity* (1989). Taylor started his academic studies with a BA in history but soon moved to philosophy and social sciences, writing on moral and political philosophy and philosophy of mind and action. During the latter part of his career, he devoted his thinking to the complex interpretative question of modernity and the idea of oneself as "self" in the formation of it. Although an industrious writer on a large array of topics, there is also a unifying concern in his thinking. Taylor seems to argue repeatedly for "telos-driven" behavior, seeing humans as beings who are evaluating the ends and purposes of their actions. This idea seems to have motivated his study

Sources of the Modern Self, which, in the words of historian Jan Goldstein, "covers the more than two millennia from Plato to Derrida" (Goldstein 2005, 16; Levy et al. 2020).

Like many who have since followed in his footsteps, Taylor sees the Enlightenment in the late eighteenth century as the period which heralded a cultural shift in the moral lives of not just the academic elite but, slowly and surely, people from every class. In Taylor's eyes, this was a "radical Enlightenment" which abandoned the model of Providence to explain how the world was ordered—the Christian faith, a belief in God, and theism no longer offered a valid explanation for it. In their place came world views and philosophies to prominence such as materialism, atheism, and utilitarianism—based on what Taylor called "modern unbelief" (1989, 317–318).

With the rise of modernity, Taylor argued, the natural world and human society could now be explained without recourse to any accounts of transcendental power, or belief in any other forces than natural causes. Ethical questions regarding the nature and origins of good, for example, were not so easily swept aside however. With the abandonment of God, morality now needed to look elsewhere for guidance, and Taylor notes this quest was conducted on three fronts. The first explored faith and the theistic tradition, the second explored with the dignity of "disengaged reason," and the third looked to the goodness of nature (Taylor 1989, 317)—in other words, the answers lay in either God, reason, or nature. Because a theistic belief in a personal God had run aground for many in the Enlightenment, many now turned to the other two for answers: the dignity of human reasoning and the wholesomeness of nature. As Taylor makes clear from the start, the Enlightenment did not come out of nowhere. It had a prehistory, and it was not as radically opposed to certain religious traditions within Christian spirituality as some would have had it to be. Taylor even goes so far as to suggest that the rational and natural bases for morality—in the context of modern unbelief—are actually "mutations in forms of Christian spirituality" (Taylor 1989, 318).

Without wanting to overly simplify Taylor's complex magnum opus, his main argument for these "modern unbelief" systems nevertheless having a basis in Christianity underlines the importance of the Augustinian idea of interiority, the Cartesian notion of reason, and Locke's notion of the mind. In the following subsections, these three aspects will be explored, with priority being first given to the practice of confession—one of the key elements in the Augustinian idea of interiority.

Interiority and Confession

Human interiority, and the self in this context, certainly goes back further in time than just to the rise of modernity. Charles Taylor's prehistory takes the Church Father Augustine (354–430) as his starting point (Taylor 1989, 127–142). Indeed, scholars of autobiography cannot rate the impact of Augustine's *Confessiones* (written in 397–401) on concepts of human interiority highly enough (Gusdorf 1980; Freeman

1993, Chap. 2; Kosonen 2007, 132–133). Even though Augustine was himself following a precedent of authors who wrote about the personal ramifications of their own conversion—the search for truth and their inner turmoil—Augustine holds the prime spot when it comes to the history of writing about human "insideness." He verbalized the inner struggle of the soul in front of both God and other people—confessing in the way the Christian tradition had certainly suggested. Nevertheless, as philosopher-historian Michel Foucault points out in *The History of Sexuality, Confessions of the Flesh* (Vol.4, 2021), the early Church's theology regarding sin and baptism was already quite complex as early as the second century. Before Augustine, the apostolic fathers had created a laborious procedure, known as "conversion-penance," which made it no easy task to become a true Christian (Foucault 2021, Chap. 2). The unique value of Augustine's *Confessions* lies not so much in the fact that he was carrying out this task, but in the very humble way he did it—as an ordinary, poor, and sinful human being. He reveals the struggle of his soul, but at the same time he is not trying to prove he is any kind of hero. He is simply an example of the Christian mind contemplating himself and his God—the God that also lies within himself.

While his conversion plays a crucial role in the story, Augustine also recounts many aspects of the life he experienced before turning to God. Although he is clearly arguing from a Christian didactic perspective, as someone who now prays to God, the *Confessions* also take readers to the "edge of an abyss" that the modern unbeliever may well recognize themselves; "where a person is left teetering, after slipping through the safety net of their old world, and forced to find a foothold within themselves and their own interiority" (Kosonen 2007, 133–141, cit. 134–135, own translation).

For Taylor (1989), Augustine stands somewhere between Plato and Descartes. Augustine's universe is platonic insofar as it is the "external realization of a rational order"—in this case, God's order (128). Like other platonic thinkers, Augustine saw that the material manifestation of truth in ideas was not always observable, and he also inherited the dualistic aspect of Platonism in a crucial way, highlighting contrasts between spirit and matter, the higher and lower, the eternal and temporal, and the immutable and changing. Taylor (1989, 128–129) rightly draws attention to what all these dualistic views of the world have in common: Augustine was portraying the world "essentially in terms of inner/outer." Taylor elaborates: "Augustine distinguishes between inner and outer man. The outer is the bodily, what we have in common with the beasts, including even our senses, and the memory storage of our images of outer things. The inner is the soul [...] the road from the lower to the higher [...] passes through our attending to our selves as *inner*" (ibid, 129). For Taylor, Augustine seems to claim that again, and again, we need to go inward, because that is where we will meet the truth—and find the way to God.

What Augustine added to the ancient "care of the self" is, according to Taylor, the idea of "radical reflexivity," to look inside oneself. Whereas the moral philosophers of Ancient Greece may well have been content to follow reason in the quest to become a better person (Foucault 1986, 43–54, 2021, 101), in radical reflexivity, the emphasis was now on a "turn to the self in the first person." For Augustine, this turn to the self was absolutely necessary before finding a way back to God; not only was God manifest in the physical world but also within each person, "in the intimacy of self-presence" (Taylor 1989, 132). This is how "we come to God within" (ibid, 136).

Confession is a term that Foucault discusses in some depth in his late lectures from the 1980s (Foucault and Blasius 1993), and in *The History of Sexuality* (1998/ 1976–1984), especially in the abovementioned fourth volume published posthumously in 2021 (subtitled *Confessions of the Flesh*). "As everybody knows," he claims, "Christianity is a confession" (Foucault and Blasius 1993, 211), arguing that western societies have used it as a means to not only uncover truths, but also to actually "produce truth" in the form of knowledge about sexuality—even if sexuality is far from the only theme. From the thirteenth century onward, confession was being used not just in religious contexts, Foucault argues, but increasingly in secular ones too. Indeed, by the latter half of the twentieth century, he claims confession was playing "a part in justice, medicine, education, family relationships, and love relations, in the most ordinary affairs of everyday life, and in the most solemn rites; one confesses one's crimes, one's sins, one's thoughts and desires, one's illnesses and troubles; one goes about telling, with the greatest precision, whatever is most difficult to tell" (Foucault 1980, 57–67, cit. 59). By Foucault's account, the practice of confession went on for centuries and has become so widespread in the western world that "the obligation to confess" is no longer "perceived as the effect of a power that constrains us." Rather, it is expected that "confession frees" and that it is power that "reduces one to silence" (Foucault and Blasius 1993).

Foucault argues that the origins of the idea that "truth is not by nature free" link to sexuality, something which was seen by the Ancient Greeks as a force to be managed in the cultivation of the self, and only later within Christianity became seen as something evil or as a sin to be confessed (Foucault 1980, 60, 1986, 67). In *Confessions of the Flesh* (2021), Foucault offers a detailed interpretation of how the Christian confessional tradition began and its connection to producing self-knowledge and indeed truth. According to him, this tradition had some traits in common with previous philosophical reflection, but one clearly new aspect it introduced was to focus "not on past *acts* but on the thoughts that *occur* [...]" (ibid, 101). In other words, it was the mind and the thoughts in it, not merely deeds, that needed to be brought out and examined. In Christian spirituality, the shifting thoughts of the mind and "the seducer inside us" were deemed to create "the battle within" between good and evil. It was only through examining this mind and the conflicts within it that redemption could be found. Sins lost their force when they were brought out into the light and verbally expressed—confession freed the soul. It is this power to free that Foucault also emphasizes: "confession has a performative force that is peculiar to it: it tells, it shows, it expels, it frees" (ibid, 108). Although confession was a practice first introduced within a monastic context, it eventually became far more widespread, and essential to Christian spirituality and the inner lives of Christians.

Interiority as a Process of Reason

As well as "insideness," another important concept in Taylor's account of the development of modern identity and moral culture is the idea of "disengaged reason" (Taylor 1989), as expounded by another epoch-making figure in the history of the self, Réné Descartes (1596–1650). Taylor points out that whereas the platonic ideal—particularly when it comes to looking for the source of morality—was to love reason and be in awe of the rational order of ideas outside the self, Descartes began this quest from the inside, leading to a significant change in the way we understand ourselves, the world, and morals (ibid, 143; also Goldstein 2005, 14). The scientific, mechanistic, and objectifying approach of Cartesianism no longer took it for granted that the world of ideas would materialize or manifest itself in "the disenchanted world of matter," as Taylor puts it (1989, 152). As a consequence, finding the basis for moral orientation in the eternal order of things was thought to be impossible—this order simply did not represent reason. One had to disengage from the old tradition of thought and rely on the power of human reason alone. Reason had become instrumental, as the tool with which to master the world and one's morals, and most importantly, this sense of reason could only be found within the self. "[T]he sense of the superiority of the good life, and the inspiration to attain it," Taylor notes, "must come from the agent's sense of his own dignity as a rational being" (ibid).

Enlightened ideas of reason, dignity, and self-esteem meant "maintaining our sense of worth in our own eyes," so that all these aspects started to play an important role in moral reflection and thoughts about human interiority during the eighteenth century (ibid, 143–153). Furthermore, the route to "modern unbelief," a key concept in Taylor's argument about modern identity, would henceforth always remain open. This was followed by a distancing from romanticism as science and technology gathered pace, especially in the twentieth century (Taylor 1989, Chap. 24, Epiphanies of Modernism).

Modern unbelief is one of the consequences of the Enlightenment that has been discussed also by historians such as Roy Porter and Christopher Lane. In the British context, Porter describes this as a process of "rationalizing religion" or secularization—meaning that a previously all-pervasive religiosity ceded ground to the more mundane concerns of commercial society. It also meant that "God's will" was no longer sufficient as an answer to all life's questions (Porter 2000, 205); it was now a matter of both rational analysis and individual choice (Porter 2000, 99). As the nineteenth century wore on, rapid industrialization and the progress of science in countries such as Britain, for example, threw many aspects of religious certainty into doubt for an increasing number of people. Indeed, this "Age of Doubt," as Lane (2011) has dubbed it, was not a straightforward process of secularization but a "complex and unsettling" period with the loss of belief accompanied by a certain degree of anxiety. Lane cites the poems of Anne Brontë (1820–1849) as one example of someone who had grown up in an evangelical home and expressed the fear of losing her faith, or the "agony of failing to feel the presence of God" (ibid, 7, 76), and certainly she was not the only Victorian to fear this.

From the Enlightenment onward, reason became increasingly important as scientific progress picked up speed. Taylor (1989) understandably chooses to discuss John Locke (1632–1704) at some length with regard to modern theories of the self, as does Jerrold Seigel (2005, 88) although he has quite a different reading in this context, which we will discuss later. For Taylor, Lockean empiricism aimed to get at "the way things really are" by giving rationality a procedural meaning—it was "above all a property of the process of thinking, not of the substantive content of thought" (168). The antiteleological Locke, author of *An Essay Concerning Human Understanding* (1689), and seen by many as the father of psychology, was certainly a key figure in this transformation; a reformer who, Taylor argues, demolished many philosophical ideas and rebuilt them from a perspective of history of the self – perhaps the most significant of which was his "objectifying theory of the mind" (Taylor 1989, 166). With Locke, the idea of the scientific principle to "think it yourself" started to mean freeing reason from customs or authorities. It was expressed strongly, and this idea of independence in reflection has proven long-lived. Also, with Locke, the concept of personhood, the self, and consciousness got some of their modern flavor that have remained important ever since—about being a "rational, self-responsible subject" capable of grasping the laws of the material world and capable of reflecting on one's own activity and "the processes which form us" (Taylor 1989, 165–176; see also Seigel 2005, Chap. 3).

Sensibility and Human Interiority

For a philosopher-historian like Taylor, the shift from the early modern to modern is massive—even if it happened only gradually and over a period of centuries. In his opinion, reason certainly played an important role in that shift, but this was not simply a philosophical shift—it manifested itself on the broader cultural level: in a "new valuation of commerce, the rise of the novel, the changing understanding of marriage and the family, and the new importance of sentiment" (Taylor 1989, 285).

One cultural shift that was of particular importance for Taylor was the new relationship with nature (see also2000, Chap. 13); another was the rise of the autonomous individual who valued self-expression and personal commitment. Inner feelings that could be examined, explored, and reflected in one's self became "morally crucial" and played an important social role too. Family or family-like relationships started to be seen as communities of love, and finding that love soon became an important part of one's search for fulfillment in life, as did benevolence, which soon became one of the most important social virtues (Taylor 1989, 305; Porter 2000, 207–209).

Like Taylor, the historian Dror Wahrman (2004) discusses the "birth of modernity" and the modern self. He too portrays a new emerging moral culture with new ways of sensing, thinking, and being in the world. He discusses cultural phenomena, such as the rise of the novel in the eighteenth century, as a means through which we can intellectually approach the transformation of self from old to new or, as he puts it, from "the ancien régime of identity to the modern regime of identity." In this respect, he argues that novels provided the "birth ground of inner selfhood," insofar as they "supposedly contributed to the development of interiority not only in their fictional characterizations but also in their readers" (Wahrman 2004, 185; Taylor 1989, 285).

But rather than stressing the role of the novel in transforming literary culture, like Taylor, Wahrman draws attention to its role in the context of changing the culture of feelings. For Wahrman, it is the emotion of sympathy especially that was at the core of this new, broader language of eighteenth-century sensibility. Taylor and Wahrman also discuss at length the emerging role of commerce (and political economics) in this cultural transformation as a self-regulating system that became essential to this new, modern subjectivity (Taylor 1989, 286; Wahrman 2004, 207–212). However, from the perspective of "inner lives," sensibility is the aspect of subjectivity that appears essential (Barker-Benfield 1992; Porter 2000, Chap. 12).

The history of sensibility is at the point where the history of emotions intersects with the history of the self. It looks beyond intellectual history, however, into other aspects of culture such as gender, for example, in the study by G.J. Barker-Benfield. He writes that sensibility in the eighteenth century "was a form of religion," but that "the evidence suggest it was overwhelmingly a religion of women" (Barker-Benfield 1992, 262–266). For Goldstein, sentimental fiction was an important part of the culture at that time, and it aimed to transform the morals, feelings, and indeed consciousness of readers by giving practical examples to learn from. Although women like Mary Wollstonecraft (1759–1797) criticized sentimentalism if overly exaggerated, it was often female characters with moral power and the right degree of sentimentalism that were its real heroines. In many respects, this reformation of manners and cultivation of emotions unfolded in a way that would resonate strongly in later centuries (ibid., 285–286, Porter 2000, 281–286).

According to Charles Taylor (1989), the modern novel embodied a "new consciousness," and like other scholars of the Enlightenment, he claims that essential aspects of this culture of feelings were the new narratives of ordinary life, their particularity, and a new time-consciousness. In other words, novels told stories of particular people at a particular moment in their everyday lives (see also Watt 2000, Chap. 1; McKeon 1991, 52–55; Steedman 1996, 77–79; Meretoja and Mäkikalli 2013, 26–29). This applied to the readers of novels as much as their authors, if they could relate to the time and place of the novel's setting. The self that people interpreted as theirs was born from experience and made up of events that people had most probably genuinely lived through. At the same time, however, there was clearly some sort of "inner seed" in readers and writers that the novel catalyzed—something latent that was waiting to develop, to be told, and to be identified with. Taylor sees this as a demand for narration, or an urge for construction of the subject through storying, or a "disengaged, particular self, whose identity is constituted in memory" (Taylor 1989, 286–289).

By relating the rise of the modern novel to broader transformations in the culture, Taylor follows Ian Watt and Eric Auerbach's ideas. He sees the *Bildungsroman* as the form of novel that offers a story of the self *par excellence*, while Ian Watt, for his part, claims that "the novel in general has interested itself much more than any other literary form in the development of its characters in the course of time" (2000/1957, 22). Meanwhile, Roy Porter writes that "novels span "humanitarian narratives" exploring moral predicaments and social dilemmas" (2000, 288). Lives and experiences are thus particular, which means that physical, emotional, and social milieux are portrayed in a detailed setting of time, place, and environment (also Burke 2011, 17).

Historicity (see below), or at least a consciousness of the time period one is in, is of key importance to the way people think and tell who they are (Baggerman 2011, 467–469). And this time-consciousness is what historian Peter Fritzsche (2004) touches on in relation to how such time perspectives can transform the inner self. For instance, consciousness of time for the individual in the nineteenth century, Fritzsche claims, meant "the inclination to think about the individual self in historical and developmental terms (the impulse to reflect on and write about life time)" (160–200, cit. 161). Nations now had individual histories like people, and these individual histories could then become part of larger historical process to which can be referred to with the term "historicity" (Freeman and Brockmeier 2001, 77–79; also Steedman 1996, 11; Burke 2011, Fritzsche 2011). For Fritzsche, it is not only the processuality, but also the emotional landscape itself, dominated by the romantic idea of loss and nostalgia for a past that is gone forever. What might also have been particular to the time-consciousness of many people in the nineteenth century was a sense of hardship and feelings of uncertainty, as instead of feeling they were mastering their path or heroically dealing with the flow of events, they most likely felt lost or caught up in the brutality of historical events. According to Fritzsche (2004), "it was precisely this awareness of injuries that laid the foundation for the new discovery of the self" (179). Dutch historian Arianne Baggerman (2011) analyzes the new time consciousness in the nineteenth century Dutch "ego documents" such as diaries and autobiographies and argues that the widening gap between past and present, the "rupture of continuity" as Reinhardt Koselleck called it, was bridged by writing about oneself. In the Dutch material she has studied, however, nostalgia and intense emotional language are initially lacking. Diaries and autobiographies "documented [an] unique era" rather than explored feelings. Only in the later part of the nineteenth century, feeling as inner life started to play a crucial role in the Dutch autobiographies, Baggerman notices (484, 492–494, 503).

Vulnerability, in the form of melancholy, was historically seen as another route to inner truths and deeper self-knowledge. Anne Brontë's aforementioned fears are said to have stemmed from a "religious melancholy" or "religious doubt" (– Lane 2011, 78). Similarly, her brother Branwell Brontë (1817–1848) expressed pessimism and doubt in his poems, and Lane argues that he turned "the doubt to opportunity, a psychology, and even a creative endeavour" (ibid, 91). Before melancholy was fully medicalized in the nineteenth century, it referred to any number of symptoms: "physical illness, moral deviance, intellectual distinction, gastric dysfunction, spiritual sensitivity, godly piety, or sinful corruption" (Sullivan 2016, 1). Cultural historian Erin Sullivan elaborates on what she calls "emotive selfhood," showing the many ways in which the emotional and existential turmoil of Renaissance melancholy connects to the material, spiritual, and social self (ibid, 3). From a conceptual point of view, this last aspect of Sullivan's culturally produced self, as "a kind of social craft," follows in the footsteps of Stephen Greenblatt and Michel Foucault: One becomes the self through "operations on bodies and souls, thoughts, conduct, and a way of being." For Sullivan, it is important that the self is an ontological term in the Renaissance context, too, and it has three aspects: the mind (or natural soul), the body, and the (immortal) soul (18–19).

Although controlling one's passions was encouraged (becoming overwrought, for example, with grief was seen as very bad for the health), certain emotions, such as "godly sorrow," were actively encouraged in the early modern period. Feeling sorrow for the sins one had committed was seen as an important step in a spiritual life that aimed for renewal and salvation, and after the Reformation, godly sorrow became more crucial in Protestantism for galvanizing people to atone for their sins (which now also included original sin); and this also meant not falling into the one unforgivable cardinal sin of despair (Sullivan 2016, 30–38). Sorrow thus became an essential part of the spiritual self—of the sinner contemplating not only their sins, but also the hope of salvation and the mercy of God.

It is not just about sin then; melancholy and the sorrowful mind have been linked to the contemplative life since Aristotle and the Renaissance revival of the Ancient Greek humors onward (for example, Lawlor 2012, 32–46, Chap. 2). In the Renaissance, perhaps the golden age for melancholy, it was seen to be the temperament of inward-looking types and was seen to be inextricably linked to their personality. Erin Sullivan cites John Donne as being a good example of one such personality. Donne asked whether he was destined to be melancholic—whether it was part of his self and identity, his calling, his way of life, or simply the will of God. Through Donne, among other examples, Sullivan portrays melancholy in the Renaissance as a means of achieving "bodily, even earthly sort of emotive selfhood," and equally how some literature of the period show that there was a Renaissance culture of "melancholic selfhood" (Sullivan 2016, 87–88).

This is perhaps most clear in the "godly sorrow" that appears in life writing and autobiographies of the time. Sullivan (2016) discusses, for instance, not only John Donne's *Devotions*, but also Elizabeth Isham's *My Booke of Rememberance* (1639), among others. Describing herself in the opening of this autobiography in terms of "ugly sins, staining filth, spotted soul," Isham accepts her sorrowful self, Sullivan argues, and actually prays for more sadness to come her way. Among other examples, Isham's confession shows that unlike extreme expressions of "worldly sorrow"—which were seen as a passion to rein in—being "spiritually sad was far from a shameful thing." For Protestant believers, Sullivan maintains, godly sorrow offered one way to construct a "holy selfhood," an inner identity that was spiritual in nature and often one and the same thing as a "melancholy selfhood" (Sullivan 2016, Chap. 4, 127).

This spiritual or reflective self reinforced by grief, melancholy, or sorrow did not disappear with scientific progress and the rise of secularism. In fact, as the nineteenth century wore on, an increasing number of autobiographies, memoirs, and diaries were published, including a range of personal and family histories (Fritzsche 2004). One of the most widely read *journals intimes* of the late nineteenth century was that of the Swiss diarist, Henri-Frederic Amiel (1821–1881). It is often described as a masterpiece of self-analysis and a marvelous record of the contemplative life. Amiel wrote it from the age of 25 until the year of his death at the age of 60 and, according to the literary scholar Päivi Kosonen (2020), comes across as "an introvert who, in his weakest moments, is a stranger to himself, seeing himself as futile dreamer slowly committing his own murder with "depressed stoicism"" (45, own translation). Amiel compared himself to the well-known melancholic figures of the time such as the main protagonist in Chateteaubriand's short novella *René* (1802). Even though he blamed himself for dwelling too much on these troubled thoughts, he could find no other emotional escape than that of the reflective space of the *journal intime* (Kosonen 2020, 44–47).

Eighteenth- and nineteenth-century literature has countless examples describing a life of solitude and contemplation, and as a genre, the novel is particularly suited to this kind of self-oriented subjectivity. These novels seem to delight in going beyond the social and physical environment of their characters, to delve into their inner worlds of dream, desire, thought, and feeling. Ian Watt (2000) highlights the private nature of these experiences in the novel by referring to the domestic context of Richardson's *Clarissa* (1748): "the domestic life and the private experience of the characters who belong to it: the two go together—we get inside their minds as well as inside their houses" (175). In another example, this time from the remote fringes of Europe and about a century later, we find ourselves in the Grand Duchy of Finland in the far west of the Russian empire with the first novel of the female author Fredrika Wilhelmina Carstens (1808–1888). Published in 1840, *Murgrönan* is an epistolary novel, which recounts the experiences of a recently orphaned young woman sent to Sweden to be educated in a noble family—in a series of letters back home to her female friend in Finland. Novels were becoming a fashionable literary genre at this time, and the book's publication also happened at a time when the National Romantic movement was gathering pace in Finnish culture, so the fact that it was the first novel of "Finnish origin" clearly helped its popularity (Grönstrand 2007, 6). The novel is a portrait of domestic life and education, contrasting the finer qualities of middle-class morality with the soulless life of the nobility. The narrator clearly suffers, but she shows self-control and most importantly develops her personality through a process of reflection and self-analysis. She opens her correspondence by warning the reader that there will be "no adventures" in her account, only "feelings and thoughts, plus some events" (Carstens 2007, 19). In other words, this will be a personal account that will relate the troubles of her heart more than anything else. As a humble orphaned female, she feels she could not be further from being a typical romantic or historical hero. Many other examples of nineteenth-century life writing—usually in the form of early autobiographies and diaries—similarly do more than simply recount the life experiences of particular persons in an everyday setting and also probe this "inner turmoil" (see Tuohela 2017).

Childhood as Interiority

Historian Carolyn Steedman has followed a quite different approach to the history of interiority. In *Strange Dislocations* (1996), she traced "the development of an idea or concept of the self from the 18th century onwards" and combined this with an analysis of how childhood was understood during this period (ibid, 1). Her main conclusion is that from the late eighteenth century onward, a cultural transformation took place which changed the way people saw both themselves and the figure of the child. By the nineteenth century, she argues, Victorians started to see childhood quite differently: Many began to make it the core of their identity and centered their psychic life around the idea of a "lost past" (ibid, 4).

Steedman starts her analysis with the literary figure of the androgynous child acrobat "Mignon," from Goethe's novel *Wilhelm Meister's Apprenticeship* (1795–1796). For some peculiar reason, this figure from literature found its way into a British article on social reform in 1911 about improving the situation for the poorer children in society. Mignon provides the cultural lens through which Steedman is able to trace 150 years of transformations in the meaning of childhood in the period 1780–1930. However, Steedman focuses not so much on the idea of childhood itself, but on the new "form of subjectivity" that arose with it (Steedman 1996, 1–4). The 1920s seems a logical place for her to stop her study, because by then "a certain understanding of selfhood had been formalised, most typically in the "discovery" of the unconscious and its connection as a formulation to the idea of the lost child within all of us" (ibid, 4). In this respect, Steedman's story is also an analysis or genealogy of the cultural meaning of the Freudian unconscious.

To start with, Steedman discusses the concept of interiority as her aim is to historize it rather than take it as a universal concept. She accepts its meaning as the "inner space," "private emotional terrain within," but finds it important to also connect interiority to the concept of the past. She wants to see how different writers in different times have used "the historical past" when ascribing it to interiority. For Steedman, it seems that the concept of interiority is in many ways a product of modernity (from the late eighteenth century onward); "history" helped create the concept insofar as "the individual and personal history that a child embodied came to be used to represent human "insideness" in the period under discussion." In other words, "the belief itself, that each individual self has a history," was, Steedman argues, crucial to the development of the "interiorised self" (ibid, 4–5).

What was new here was not the idea of childhood as such, Steedman argues, but the way it was being used to frame the self and one's personal history—making it clear she is *not* writing the social history of childhood. However, one problem here is that these ways of understanding and representing the child cannot be simply separated from the real lives of children (possibly the most temporary subjects in our society)—even for a historical enquiry. Steedman acknowledges that symbols and social reality do clearly interact but maintains that she is trying to "understand how adults used the idea of childhood or the figure of the child" during this period. It is thus essentially a history of thought (ibid, 5).

"Sensibility" is a term which Steedman refers to when discussing how her study also scrutinizes feelings. She notices that romantic writing in the eighteenth century portrayed the child in emotional terms via a language of loss—especially in texts portraying personal journeys in time such as novels and autobiographies—and this framework of sadness also affected the scientific approaches that were to follow.

Steedman's story starts with exploring the various possible origins of the Mignon figure who took on numerous forms of littleness, as "almost always a sympathetic figure with a terrible or mysterious past who is meant to evoke pity" (ibid, 29). The feeling this evoked in adults—a plea for sensibility—was crucial to the figure's ongoing cultural significance when Mignon traveled beyond Germany. Because Steedman continues into the twentieth century, however, she also takes the reader there via scientific breakthroughs that happened in cell theory and physiology. In this sense, a child's growth poses the physiological question of what happens to interiority as the child develops (ibid, 76). In the nineteenth century also, the thoughts and theories of material processes also had an important effect, via the emerging discipline of psychoanalysis developed by Sigmund Freud (1895–1920). According to Steedman, Freud was inspired by cell theory and the idea of childhood as a personal past. His idea was that childhood, though lost and gone, "has left behind memories and traces" (ibid, 77) at the physical level of cells, which would therefore represent the smallest units of personality.

Steedman also points to psychological inspirations for Freud's theory of the unconscious. Perhaps the most important of these is *Mental Evolution in Man* (1888) by George Romanes—Darwin's collaborator and pupil. In his quest to discover what it was that made the consciousness of humans and animals different, Romanes wanted to examine "the psycho-genesis of a child" (Steedman 1996, 84). Steedman argues that Freud, along with many others in the psychological sciences, interpreted the emerging child sciences he was reading about in evolutionary terms:

> The older, pre- or non-Darwinian biology that is now understood to have shaped much late nineteenth-century evolutionary thinking was also used in the construction of childhood in the new child-study movement, from which Freud learned so much. Non-Darwinian evolutionary theory used by psychologist of the child-study movement expressed an inherent teleology, with the idea of progress being embedded in the idea of development. In this way, the child's developing body and mind could be understood as an epitome of a more general historical process. When W.B. Drummond published his popular and summative Introduction to *Child Study* in 1907, he suggested in his epigraph that "child-study marks the introduction of evolutionary thought into the human soul." (Steedman 1996, 85)

Steedman is highlighting here how Freud was inspired by several sources at once—contemporary ideas of historical consciousness and emotional culture: the latest progress in medicine, physiology, and evolutionary theory, and most importantly the new discipline of child psychology. The human soul could now be explained in a scientific way that also took into account the theory of evolution. This way was, according to Steedman, a fundamentally historical phenomenon, because growth takes place over a period of time. However, the traces of childhood experiences—mooted by Freud's theory of the unconscious—were hidden from the conscious mind so, at the same time, could offer a place outside time. Freud's significance for Steedman thus lies in this conceptualization of childhood as a form of interiority that began to represent a "timeless interiority of the unconscious" (Steedman 1996, Chap. 5, cit. 93).

Interiority and Reflexivity in Late Modernity

Interiority in the Late Modernity is a complex issue discussed by cultural theorist and sociologists as a concept linked to "self" and "identity." Anthony Giddens (1991), for instance, who sees modernity as more of a contemporary than historical term, approaches the "modern self" is from the basis of experimental and developmental psychology of the twentieth century. Even though his widely known book *Modernity and Self-Identity: Self and Society in the Late Modern Age* is an analytical study rather than exactly historical, it has interesting trajectories in time bringing it to the dialogue with, for example, Taylor.

One of the key terms used by Giddens when discussing the modern self is "ontological security"—referring to that which gives a fundamental sense of safety, or at least intelligibility, to being in the world. It also conveys a "sense of "invulnerability" [which] blocks off negative possibilities in favour of a generalised attitude of hope." This sense of security or "trust," Giddens (1991) writes, originates from the interaction of infant and carer in the early years (40). Establishing this basic sense of trust is fundamental to creativity and self-identity. According to Giddens, only once a human trusts in the reality and routines of daily life, they can emotionally accept the external world and learn about the other that is "not-me" and so achieve a sense of self. Otherwise the "feelings of unreality which may haunt the lives of individuals in whose early childhood basic trust was poorly developed may take many forms" Giddens warns. "They may feel that the object-world, or other people, have only a shadowy existence, or be unable to maintain a clear sense of continuity of self-identity" (Giddens 1991, 43). So although Giddens discusses trajectories in time, there are also some transhistorical ideas of self-formation here.

Apart from developmental psychology, Giddens also draws on existentialist writers such as Søren Kierkegaard (1813–1855) to posit "existential anxiety" as the antithesis of "ontological security." This angst represents the struggle of the human individual against the threat of nonbeing, the search to find one's "mode of being-in-the-world." For Giddens, this anxiety of being wiped out is a necessary part of being conscious of our mortality. It may lead individuals to moral pondering in their search for an "authentic life," before their time runs out, and so is clearly an important part of the inner life of the modern individual (Giddens 1991, 48–50).

Also for Charles Taylor, Kierkegaard is one of the key thinkers through which we can understand the birth of complex modern identity, human interiority in which "we choose ourselves," become unique selves "worthy to be loved." For Kierkegaard, this higher being is not linked to nature or reason but to the radical choice of "becoming what I really am, my true self." By choosing oneself, one accepts the infinite dimension beyond the finite, beyond "despair and dread," Taylor sums up. In addition, for Taylor, Kierkegaard's "I," in its fullest form, is an ethical "I," and to become "the true, the ethical I" means a process of change, transforming of ones "stance towards oneself" (Taylor 1989, 449–451). This seems to be a very "inner" process, happening "inwardly."

In contrast, Giddens' reflections on the dynamics of selfhood include thoughts about the other, and not only in terms of the outer world, but also in terms of language. Self-identity follows from the idea that it is a form of self-awareness, which gives us the ability to reflect, and be conscious of our own actions. This leads to an idea that the self is "reflexively understood by the person in terms of her or his biography" (Giddens 1991, 53). Furthermore, this self is universal, he argues, whereas "what a "person" is understood to be certainly varies across cultures" (Giddens 1991, 53).

Despite some universal features, Giddens acknowledges that self-identity is in an important way dependent on temporality and biography. One's biographical self may be fragile, and this selfhood requires the continual job of narrativization, or as Giddens (1991) puts it, the individual "must continually integrate events which occur in the external world, and sort them into the ongoing "story" about the self" (54). Giddens also acknowledges the idea of a narrated self, but for Giddens the self is temporal in another, "real," way; this is "choronic work" that involves being able to situate oneself in the time and flow of events (54). This biographical work, he notes, varies very much according to the social and cultural situation—especially in the modern context.

This modern context is important, as Giddens sees himself less as a historian and more as a social theorist of modernity and late modernity. However, "who to be?" is for him among the most important questions an individual of late modernity will ask themselves, and indeed it is a question which all of us also answer on some level (Giddens 1991, 70). To be an individual or a self in the modern world (rather than earlier times) is quite simply different, and Giddens uses the terms "therapy" and "self-observation" to make that distinction clear. According to him, having a modern reflexive identity of oneself is essential, and to be able to always adapt in order to optimize one's life and live it to the full means undergoing the continuous and pervasive practice of self-observation; the logic of development thus seems unavoidable. In this way, Giddens sees the practice of autobiography as typical of modernity and its discourse: We think autobiographically. Giddens summarizes this idea by saying that "autobiography—particularly in the broad sense of an interpretative self-history produced by the individual concerned, whether written down or not—is actually at the core of self-identity in modern social life" (Giddens 1991, 76).

Even though Giddens does not use the term "interiority," he does discuss the reflective self at some length and offers "autobiography" as an answer to the question of "individuation," the question of "who am I." It may be that he has abandoned interiority as a term because it seems too spiritual or religious for the secular age. Giddens contemplates the landscape of modernity and especially late (or high) modernity as a world of complex social systems demanding reflexive selves, but when it comes to the question of ethics, he deems it one which has failed to address the existential and moral questions. For Giddens, the answer is not to be found in the "inwards gaze" or an attempt to be true to one's authentic self, but in the politics of emancipation and life, and through engaging in processes that have moral meaning for the larger human community.

Where Giddens has chosen to focus on social systems with reflective individual minds as part of it, Charles Taylor who throughout his *Sources of the Self* writes about the "radical reflexivity" continues to claim that our sense of inwardness does not vanish in modernity but on the contrary even intensifies when the modern identity transforms toward more secular and plural forms. He argues that despite the focus on structures outside the self such as language, we are linked to the legacy of the Romantic in our search for deep or hidden. This can happen only through the personal. "In this sense," Taylor writes, "the depths remain inner for us as much as for our Romantic forebears." Similarly, we may think of consciousness as multi-leveled or decentered, transpersonal or relational; we, however, access the impressions and visions through the personal. For Giddens, this might be the "autobiographical," but for Taylor this means "the inner" (Taylor 1989, 428, 480–481).

The Modern Mind, the Autobiographical, and the Inner

Historian Jan Goldstein (2005) has discussed the concept of "interiory," religious and secular, in connection to the modern psyche in the French context. She wants to enrich the interpretation of the self in the postrevolutionary era we get, for example, through Charles Taylor's magnum opus. According to Goldstein, Taylor pays all too little attention to the several "subnarratives" in the philosophical culture of the modernity one of them being "sensationalism." Goldstein sees the modern "psychological interiory" emerging after Locke and shows how discussions on the psycho, the soul, and the "moi," as well as the question of the unity versus fragmented nature of the "self," became central in the intellectual life of the nineteenth century—in France at least. The reflective subjectivity in the modernity, already from the early nineteenth century onward, was more complex than Taylor's metanarrative allows us to discern, she argues (Goldstein 2005, 16–17).

For Goldstein, a valorizing example of the nineteenth century person who expressed the importance of "inner life" (vie intérieure), and the complexities it meant for intellectuals of the nineteenth century, is historian-philosopher Ernst Renan (1823–1892). Renan saw the introspection as "the central symptom of modernity" and wrote in his notebook in 1845 that "What characterizes the modern era is radical reflexivity, the folding back on oneself. In the realm of philosophy, psychology is everything. In literature, we no longer tell stories in the manner of Homer and the Bible; instead we paint the impressions and sentiments suppressed by the ancients, [...] When I love, when I suffer, when I think, I have a model in mind or, at least, if my sensibility runs ahead of me, I reflect [...]" (Goldstein 2005, Chap. 6, cit. in Goldstein 2005, 235–6). To be oneself was to think, to reflect, and to fold back to oneself. And instead of turning to religious truths, people searched for answers in sciences such as psychology. Like for many, also for Renan, the creation of the self had its roots in the religious practice that in the course of time was, however, challenged by the psychological and more secular theories of the mind/ psyche. People of the nineteenth century felt ambivalence and anxiety about the pure materialism and ended up with many mixtures of scientific and religious. They witnessed both the psychological theories that fragmented the soul to several faculties of the (material) mind and the revival of the idea of the "unitary moi," but "the free, unified, and morally accountable self" was hard to resist—and this form of "moi" maintained well in the twentieth century (Goldstein 2005, Epilogue).

Feminist historians have also explored the concept and history of the self both in methodological terms and via case histories. Penny Summerfield (2019), for example, has focused on personal histories, and she is one of many who acknowledge the cultural turn of the 1980s as an important turning point in the historiography of the self. With poststructuralist notions about the centrality of language and embeddedness of thoughts and ideas in sociocultural discourse came a change in historical practice too, she argues. Concepts such as the self became "the subject of, and subject to, power and regulation through discursive and institutional practices," raising "the possibility that subjectivity, emotion, and memory, all central to "the personal", have a history" (7). Historians could therefore—like some literary theorists in the 1990s—feasibly treat the lives narrated in autobiographies and diaries as examples of the "technology of self" rather than fiction (Freeman 1993, 7–8; Steedman 2000). Nevertheless, historians usually went for something in the middle, with the aim to keep the "real" alive, and Summerfield (2019) too, suggests that we should read personal memoirs and autobiographies both objectively, for facts and data about an external social life (in the workplace, for instance), and subjectively, for the inner life of thoughts and feelings (Chap. 4).

The self that Summerfield (2019) introduces as a theme for historical analysis is certainly situated in time and place, and it is also textual and discursive. "Inner life" as a concept does not get special attention in her writing, but she does cover it in the context of certain material—such as working-class autobiographies. Importantly, she reminds us that the history of the self must also take into account the aspects of class and gender. For her, and other historians like her, gender is a social and cultural construct that changes with time, and only one aspect in our personal or subjective lives. Another attempt to address the history of the self from a fresh angle is that of Anna Clark (2017), who has chosen to work with case studies from a "deviant self-fashioning perspective" and in this way dig deep into the identities that are constructed in personal documents. In her study, the focus is on the "unique self," "a distinctive self unlike all others," and on individuals between the late eighteenth and early twentieth century. In the context of Clark's study, the inner/outer division plays a role and the "inner" refers to private with religious being only one aspect of it. More than anything, "the inner" is emotional, personal life often including elaboration on the secrets of sexuality and forces of passions. The division between public and private self is clear and can be seen as one alteration of the division of outer/inner life.

In the analysis of the modern culture, literature and modernism especially are often in focus. In general, modernism can be outlined as something interested in the mind and the self, the question of consciousness and subjectivity. The troubles of the individual, what happens in the subjective consciousness, can be seen as key features in the modernist fiction. Writers such as Virginia Woolf wrote how the aim of the modernist writers was "to put into words what life 'feels like': the way the fragmented impressions of the world are shaped into a continuous stream of consciousness" as literary scholar Anna Ovaska has described Woolf's key role. The novelist should "share this experience of being alive, of being conscious, as it emerges, and to free the portrayal of the mind—'the unknown and uncircumscribed spirit'—from the old literary conventions" (Ovaska 2020,21–22). The fictive worlds are "inner" as well as personal supporting perhaps Taylor's idea that modernity does not stop but continues to turn inward.

Along fiction, another important literary field of self-exploration that continues to attract readers and writers in the twenty-first century is autobiography. If we break the word down into its constituent parts—*auto* (self), *bios* (life), and *graphein* (writing)—it clearly refers to questions of identifying the self, placing this in a life story, and setting the story down, and this will inevitably lead to a rich range of stories about human experience (Marcus 2018, 1–2). Autobiography overlaps with other kinds of writing but usually tackles issues such as confession, conversion, and testimony and concepts of truth, person, and time, writes Laura Marcus (2018) in her brilliant *Autobiography: A Very Short Introduction*. In this way, autobiographies offer an excellent insight into the history of inner lives.

According to Georges Gusdorf (1980), one of the founders of autobiography as a field of research, autobiography is not a universal genre nor is it very old. "Autobiography is not possible in a cultural landscape where consciousness of self does not, properly speaking, exist," he maintains, before then launching into a definition of autobiography in western culture (30). He argues that the "metaphysical preconditions" are that the culture must have a historical consciousness, and stories of "historic personage," before autobiography is possible. Besides historicity, Gusdorf underlines how the autobiographical self is created as part of the narrative process—it is not just a case of recollection. Autobiography is clearly linked to "inner truth" for Gusdorf when he describes it as "a true creation of self by the self" or "an attempt and the drama of a man struggling to reassemble himself in his own likeness at a certain moment in history" (Gusdorf 1980, 43, 44).

Marcus (2018) also gives examples of how "romantic confessions" started to play "a central part in the formation of modern understandings of selfhood" (14). Jean-Jacques Rousseau's *Confessions* was, for instance, a very influential work that claimed to divulge the painful truth of himself. Marcus also points out that he was one of the first to emphasize childhood experiences and to portray childhood as a time of innocence and happiness (14–15). Later, autobiographies have not necessarily taken such a rosy view of their early years, however. "My true self, my character and my name were in the hands of adults," confessed Jean-Paul Sartre (1904–1980); "I had learnt to see myself through their eyes; I was a child, this monster they were forming out of their regrets" (*Les Mots*, cit. in Marcus 2018, 66). Rosy or not though, these accounts are held to be "true."

For Philippe Lejeune (1989), autobiographies are published texts that recount someone's life, and like Gusdorf, he believes the first examples of the genre appeared in the late eighteenth century. In Lejeune's opinion, to write such a text requires an "autobiographical pact" to be made (4). Although he does not directly discuss "inner lives," Lejeune touches on the "truth of "human nature"" while acknowledging the complex nature of "truth" (27).

Lejeune's pact includes a contract between the reader and autobiographer about the nature of the text as "a retrospective prose narrative produced by a real person concerning his own existence [...]" (Lejeune 1989, 4; Marcus 2018, 3; on "autobiographical space" as a conceptual means for studying stories of the self, see also Angelini 2001; on Lejeune also Burke 2011, 16; Baggerman 2011, 457). Historical consciousness as a precondition seems to be important for Lejeune as well. The aim of recollecting or looking back, of telling a personal truthful story, thus remains, even if no final definition for autobiography is forthcoming.

Narrative scholars Freeman and Brockmeier (2001) suggest that the autobiographical identity, in the modern west, is linked not that much to truth solely but to the ideal of a "good life." This means binding an "ethical rationale to one's life." They argue that "constructions of autobiographical identity always aim at some form of narrative integrity," so identity is never free from ethics (75–77). Interestingly, Freeman and Brockmeier discuss Augustine's *Confessions* not only as a classical model for "representations of the self in a retrospective literary structure" but also as a model for giving an account of oneself in the ethical sense. For early Christians, giving an account of oneself meant that "each person must answer, in the face of God, for his own life," (80) but in the secular age "postulating the meaning of the past is a psychological and moral function of the here and now" (83). Moral here thus seems to refer to shared sociocultural norms and not to any "godly order." Thus, according to Freeman and Brockmeier, the normative orientation, or what it means to live "the good life" guides the way the story of the self is told. In other words, the norms and values of the shared social reality provide criteria for the good life and also to the narrative that is to be told.

It seems that autobiography as a genre and practice has expanded into myriad forms of life writing on the self, and in other forms of media than just writing (Smith and Watson 2010, 1–5; Chap. 6; Poletti 2020, 12–14). Autobiography is certainly close to the practices of testimony or confession, but many scholars remind us that it is also performative—an act that is addressed "to an audience/reader," to a "you," or to some other (Chiper 2020; Smith and Watson 2010, Chap. 3).

Autobiography or the autobiographical subject is also bodily material and has agency. Recently, for example, Anna Poletti has suggested that more attention should be paid to the material nature of the narrative act in autobiography, to the way it is mediated in the material. This would offer an alternative perspective and challenge to the "sovereign human subject," which has long been at the center of autobiographical thinking. According to Poletti, one can give an account of oneself "within the constant generative flow of matter," in materials and media different from the traditional text and words. This can take place, for example, through a cardboard box that becomes "a narrative act" (Poletti 2020, 23, Chap. 1). In this kind of thinking, lives are no longer "inner" or personal in the traditional way, but fragmentary, momentary, and material. They can still nonetheless be identified as asking—both in time and in relationship to others—the same question of "who am I?"

Conclusion

This chapter has charted how the topic of "inner lives" has been encountered by historians—as a question about the human condition and about human existence in different cultural and temporal situations. The focus is on those who have aimed to see the meaning of "human interiority" embedded in the thought and practice of a particular time and place, and how this idea has changed through time. The long history of ethics, of how to do the right thing, and of how to live a good life forms the thread of essential studies in this field.

The theme of inner lives in history includes both continuities and ruptures, slow transformations and quicker changes. One major cultural shift in the west was the gradual loss of faith in divine providence and the emergence of a scientific worldview to replace it. A very different kind of subjectivity from that of a "true believer" started to take its place, and the interior of the sinner now made room for introspective reflection on a rich variety of feelings and thoughts.

The romantic notion of trust in the goodness of nature vanished with the onset of modernity, taking with it the existing grounds for moral thinking too. In modernity (and postmodernity), scattered, fluid, flux of identities started to offer solutions that were also an escape from what had become a restrictive "unitary self." In the search for some solid ground upon which to build identity, it was often language that "was there before us" which then provided the answers. Identities became lingual, and narrative, and inner lives were portrayed in a burgeoning literary culture of memoirs and novels from the late eighteenth century onward.

The replacement of a coherent self with the flow of experience in the modern context meant that the concepts of time and memory played an increasing role. New time-consciousness emerged hand in hand with history both as an academic practice and as a way of pondering the question of "who am I?" History got inside; it became part of our personal stories and identities. We became creatures in a flow of time that was now uncertain, with unexpected events and sometimes troubles that made our journey often a struggle.

Paradoxically perhaps, it was in the nineteenth century, which experienced this upheaval to the flow of history most brutally, that ideas coalesced to form the theory of evolution. This, in turn, gave rise to the idea of evolution for the better in the name of progress. The child was given new potential, in the very terms of growth which also became crucial for the narratives of the personal "I." Autobiographies and personal histories in novels depicted both the process of growth in a way the *Bildungsroman* had foreseen, portraying lives as flowing through obstacles, in episodes—a life that cannot be brought together as a story of achievements but remains fragmented.

Today, inner lives as a concept may be more at home in historical studies than in the fields of cultural studies, literary theory, or philosophy. The terms of self, subjectivity, identity, and agency appear in these fields more often than "interiority." The concept of "experience" is however vital, and even though it might be seen as the bodily or material effect of a discourse or language, it is also something happening in the space we refer to with the concept of "mind´. In its historiography, 'interiority' might have reached its saturation point or lost its 'unity" as an analytical concept. However, just as efforts to find a meaning to our temporary existence through contemplation continue to this day, so too must its history continue to be written.

References

Angelini EM (2001) Strategies of "writing the self" in the French modern novel: C'est moi, je crois. Queenston, Lampeter, The Edwin Mellen Press, Leviston

Baggerman A. (2011) Lost time: temporal discipline and historical awareness in nineteenth-century Dutch Egodocuments. In Baggerman, Arianne., et al. Controlling time and shaping the self: developments in autobiographical writing since the sixteenth century. Leiden, Brill. 455

Baldick C (2008) interior monologue. In The Oxford dictionary of literary terms: Oxford University Press. https://www.oxfordreference.com/view/10.1093/acref/9780199208272.001.0001/acref-9780199208272-e-601. Accessed 13 August 2021

Barker-Benfield GJ (1992) The culture of sensibility. Sex and Society in Eighteenth -century Britain. The University of Chicago Press, Chicago and London

Burke, Peter (2011) Historizing the self, 1770-1830, In Baggerman, Arianne., et al. Controlling time and shaping the self: developments in autobiographical writing since the sixteenth century. Leiden, Brill. 13–32

Carstens FW (1840/2007) Muratti [Ivy (Hedera)]. Turku, Faros. [In Finnish]

Chiper S (2020) Performative selves, performative poses: Gertrude stein. Norman Mailer and Philip Roth as Autobiographers, Iași, Institutul European din România

Clark A (2017) Alternative histories of the self: a cultural history of sexuality and secrets, 1762–1917. Bloomsbury, London

Eghigian G, Killen A, Leuenberger C (2007) Introduction: the self as project: politics and the human sciences in the twentieth century. In: Eghigian G, Killen A, Leuenberger C (eds) The self as project: politics and human sciences, pp 1–25. Osiris 22, A Research Journal Devoted to the History of Science and Its Cultural Influences, Georgetown University

Freeman M (1993) Rewriting the self: history, memory. Narrative, London and New York, Routledge

Freeman M, Brockmeier J (2001) Narrative integrity: autobiographical identity and the meaning of the "good life". In: Brockmeier J, Carbaugh D (eds) Narrative and identity: studies in autobiography, self and culture. Amsterdam/Philadelphia, John Benjamins Publishing House, pp 75–99

Fritzsche P (2004) Stranded in the present: modern time and the melancholy of history. Harvard University Press, Cambridge, Mss and London

Fritzsche P (2011) Drastic history and the production of autobiography. In: Baggerman A et al (eds) Controlling time and shaping the self: developments in autobiographical writing since the sixteenth century. Brill, Leiden, pp 77–94

Foucault M (1980) The history of sexuality, volume I: an introduction. Vintage Books, New York

Foucault M (1986) The care of the self: volume 3 the history of sexuality. The Penguin Press, London etc.

Foucault M (2021) Confession of the flesh: the history of sexuality, volume 4. [orig. 2018, transl. Robert Hurley]. Pantheon Books, New York

Foucault M, Blasius M (1993) About the beginning of the hermeneutics of the self: two lectures at Dartmouth. Political Theory 21(2):198–227. https://doi.org/10.1177/0090591793021002004

Giddens A (1991) Modernity and self-identity: self and Society in the Late Modern age, 1st edn. Polity Press, Chichester

Goldstein J (2005) The post-revolutionary self: politics and psyche in France, 1750–1850. Harvard University Press, Cambridge MA, London

Grönstrand H (2007) Esipuhe [preface]. In: Fredrika Wilhelmina Carstens, Muratti. Turku, Faros, pp 5–13. [in Finnish]

Gusdorf G (1980) Conditions and limits of autobiography. In: Olney J (ed) Autobiography: essays theoretical and critical. Princeton University Press, Princeton, New Jersey, pp 28–48

Hofman E (2021) Trials of the self murder, mayhem and the remaking of the mind, 1750–1830. S.l, Manchester University Press. http://search.ebscohost.com.ezproxy.utu.fi/login.aspx?direct=true&db=nlebk&AN=2917824&site=ehost-live&ebv=EB&ppid=pp_1

Inner (2021) Merriam-Webster dictionary. https://www.merriam-webster.com/dictionary/inner. Accessed 13 August 2021

Inre [Inner] (2021) Svenska Akademiens ordbok [The dictionary of Swedish Academy], SAOB, https://www.saob.se/. [in Swedish]. Accessed 13 August 2021

Interiority (2021) In Oxford dictionary of English. Ed. Stevenson A. Oxford: Oxford University Press. https://www.oxfordreference.com/view/10.1093/acref/9780199571123.001.0001/m_en_gb0416430. Accessed 13 August 2021

Lejeune P (1989) On autobiography. University of Minnesota Press, Minneapolis

Kosonen P (2007) Isokrateesta Augustinukseen: Johdatus antiikin omaelämäkerralliseen kirjallisuuteen [From Isocrates to Augustine: an introduction to the ancient autobiographical literature]. Jyväskylä, Atena. [in Finnish]

Kosonen P (2020) Päiväkirja itsestä ja maailmasta huolehtimisen välineenä [Diary as a tool for self care and the concern of the world]. In: Leskelä-Kärki M, Sjö K, Lalu L (eds) Päiväkirjojen jäljillä: Historiantutkimus ja omasta elämästä kirjoittaminen [Tracing the Diary: Historical research and life writing]. Tampere, Vastapaino, pp 39–54. [in Finnish]

Levy, Jacob, Jocelyn Maclure and Daniel M. Weinstock (2020) Interpreting modernity: essays on the work of Charles Taylor. Montreal: McGill-Queen's University Press. https://search-ebscohost-com.ezproxy.utu.fi/login.aspx?direct=true&db=nlebk&AN=2619604&site=ehost-live. Accessed 1 Sep 2021

Marcus L (2018) Autobiography: a very short introduction. Oxford University Press, Oxford

McKeon M (1991) The origins of the English novel 1600–1740, 2nd edn. The Johns Hopkins University Press, Baltimore

Meretoja H, Mäkikalli A (2013) Romaanin historian ja teorian monimutkainen vuorovaikutussuhde antiikista nykypäivään [The complex interplay of the history and theory in the novel from the Antiquity to the present]. In Meretoja H, Mäkikalli A (eds) Romaanin historian ja teorian kytköksiä [Connections of the history and theory in the novel]. Suomalaisen kirjallisuuden seura, Helsinki, pp 15–48 [in Finnish]

Ovaska A (2020) Fictions of madness: shattering minds and worlds in modernist Finnish literature. University of Helsinki, Helsinki

Poletti A (2020) Stories of the self: life writing after the book. New York University Press, New York

Porter R (2000) The creation of the modern world: the untold story of the British enlightenment. Norton, New York, W.W

Rieff P (2006) The triumph of the therapeutic. Uses of faith after Freud. 40th Anniversary Edition. ISI Books, Wilmington, Delaware

Scheer M (2012). Protestantisch fühlen lernen: Überlegungen zur emotionalen Praxis der Innerlichkeit. Z Erzieh, vol. 15, no. S1, VS-Verlag, pp. 179–93. https://doi.org/10.1007/s11618-012-0300-1

Smith S, Watson J (2010) Reading autobiography: a guide for interpreting life narratives, 2nd edn. University of Minnesota Press, Minneapolis and London

Steedman C (1996) Strange dislocations. Childhood and the idea of human interiority 1780–1930. Harvard University Press, Cambridge, Massachusetts

Steedman C (2000) Enforced narratives: stories of another self. In: Cosslett T, Lury C, Summerfield P (eds) Feminism and autobiography: texts, theories, methods. London, Routledge, pp 25–39

Seigel J (2005) The idea of the self: thought and experience in Western Europe since the seventeenth century. Cambridge University Press, Cambridge

Sullivan E (2016) Beyond melancholy: sadness and selfhood in Renaissance England. Oxford University Press, Oxford

Summerfield P (2019) Histories of the self: personal narratives and historical practice. Routledge, London and New York

Taylor C (1989) Sources of the self: the making of the modern identity. Cambridge University Press, Cambridge

Tuohela K (2017) The rise of inner subjectivity: childhood in early nineteenth-century Finnish autobiographies. In: Ahlbeck J et al (eds) Childhood, literature and science: fragile subjects, 1st edn. Routledge, London

Wahrman D (2004) The making of the modern self: identity and culture in eighteenth-century England. Yale University Press, New Haven and London

Watt I (2000/1957) The rise of novel. Studies in defoe, richardson and fielding. Pimlico, London

Michel Foucault and the Practices of "Spirituality": Self-Transformation in the History of the Human Sciences

15

Nima Bassiri

Contents

Introduction	376
Histories of the Malleable Self	380
Spirituality among the Human Sciences	384
Labor and Revolution	386
Pathology and Health	388
Psychoanalytic Spirituality	390
Violence and Decolonization	394
Conclusion	396
References	398

Abstract

Prompted by scholarship which has proposed that Western selfhood beginning in the nineteenth century was largely defined as a stable and static form of self-identity, this chapter turns to the late writings and lectures of Michel Foucault and his account of "spirituality," or the ethical practices of conversion and transformation. While Foucault posits that spirituality was a hallmark of ancient ethical traditions, he proposes nonetheless that vestiges of a self-transformative ethics continue to be evident in two particular post-eighteenth-century doctrines of thought – namely, Marxism and psychoanalysis, which tacitly espoused a political and medical form of spirituality, respectively. This chapter considers the modern perseverance of the notion of spirituality in the context of radical political and medical doctrines.

Keywords

Selfhood · Transformation · Spirituality · Marxism · Psychoanalysis · Decolonization

N. Bassiri (✉)
Duke University, Durham, NC, USA
e-mail: nima.bassiri@duke.edu

Introduction

In the second essay of *On the Genealogy of Morality*, the German philosopher Friedrich Nietzsche presented his abridged and speculative gloss on what he dubbed the "long history of the origins of responsibility" [*Verantwortlichkeit*]. Responsibility, Nietzsche contended, was a moral value ultimately secured through the violence of conscience and the internalization of the debtor-creditor relationship – a form of self-punishment that ultimately conditioned one's capacity to remember, to make promises, and to freely oblige oneself to another. The subject who occupied the endpoint of this long history was "a human being ... whose *prerogative is to promise*," idealized in the form of the "sovereign individual" [*souveraine Individuum*] who was morally compelled to remain, above all else, self-accountable, reliable, regular, and enduring (Nietzsche 2017, 37–38; Drochon, 2018; Sedgwick, 2009; Tanner, 2001). While a steadfast sense of autonomy, responsibility, and abiding self-identity have become central hallmarks of the modern Western self, what Nietzsche proposed was that these moral attributes were little more than the effects of a profound brutality underlaying the formation of the modern subject. Nietzsche not only impugned the unquestioned value uncritically attributed to these moral traits, but he also underscored the contingency of the entire history of the moral self – that this particular unfolding of subjectivity was neither essential nor fated. After all, the *Genealogy* was not a comprehensive philosophical history of moral subjectivity, nor was it intended to be. Despite its characteristic depiction of the unbridgeable chasm dividing ancient from modern moral cosmologies, a more targeted periodization is latent in the text, one which seems to lay the brunt of the historical critique on the nineteenth century itself, which, perhaps more than any other modern period, witnessed the rise and enthronement of sovereign individualism. The autonomous self – unified, stable, and unchanging – was a product of Nietzsche's own historical moment; and so, in that sense, the *Genealogy* could be called the earliest iteration of what would eventually acquire the moniker of a "history of the present" (Foucault 1995, 31; Foucault 1977).

Over the past few decades, cultural and intellectual historians have lent some credibility to elements of Nietzsche's tacit hypothesis, that a consistent and unchanging sense of self did not predominate European conceptions of personhood prior to the nineteenth century. The literary critic Stephen Greenblatt famously proposed that the sixteenth century, for example, was a period during which one noticed "an increased self-consciousness about the fashioning of human identity as a manipulable, artful process" (Greenblatt 1980, 2). The cultural historian Dror Wahrman has argued that an "*ancien régime of identity*" pervaded late-seventeenth to mid-eighteenth-century England, a period "before the self," defined by way of the fluidity and malleability of identity formations and performances. After the end of the eighteenth century, however, in what Wahrman calls the "new identity regime," it "suddenly became much harder for people to imagine identities as mutable, assumable, divisible, or actively malleable." It was in this subsequent period, primarily over the course of the nineteenth century, where a stable, interiorized, and essentialized sense of self and immutable self-identity emerged and solidified (Wahrman 2004, 168, 274–75). The historian Jan Goldstein (2005) has provided an analogous narrative of post-Revolutionary France through her account of the installation and dissemination of Victor Cousin's psychology and philosophical eclecticism and his attendant conception of *le moi* – a distinctly bourgeois sense of selfhood defined by its coherence, uniformity, and moral autonomy.

It would, in some measure, make sense that dominant conceptions of selfhood after the eighteenth century would prioritize the value of a unified and stable self-identity, given the novel legal and institutional demands associated with the administration of emergent industrial populations as well as the growing sense of economic contractualism that increasingly pervaded the commercialized North Atlantic; the normative ideal of a stable and essential self-identity would have particularly behooved a select population who most stood to benefit from the political and economic landscapes of nineteenth-century Europe and the United States. Nevertheless, such a conception of post-eighteenth century personhood raises a crucial question for the historian of the human sciences: how rigidly should we accept this history and periodization of the self? Is it indeed not possible, in general terms, to consider, say, transformation or malleability to be core attributes of subjectivity after the start of the nineteenth century?

This chapter will attempt to answer these questions by appealing to the work of the philosopher and historian Michel Foucault. Foucault's writings have functioned as core conceptual pillars for the history and theory of the human sciences (Gutting 2019; Elden 2017; Eribon 1991; Dreyfus and Rabinow 1983). Indeed, Foucault's *The Order of Things,* translated into English in 1970 from the 1966 *Le Mots et les chose,* was among the first and most popular systematic investigations into the history of the human sciences, which helped to prompt after the 1970s the establishment of the field as a significant and critical topic of study in its own right (Smith 1997, 2007). Foucault was elected to the faculty of the prestigious Collège de France in 1970, where he lectured for the remainder of his life, and during which time he published some of his most significant books, including *Discipline and Punish* (1995) and *History of Sexuality* (1990). In 1978, however, and prompted largely by the events of the Iranian Revolution, Foucault turned his attention to the study of what he dubbed political "spirituality" (Bremner 2020; Ghamari-Tabrizi 2016; Afary and Anderson 2005).

For Foucault, "spirituality" referred to the dominant practices, ethical tenets, and general belief systems "by which the individual is displaced, transformed, disrupted, to the point of renouncing their own individuality, their own subject position" (Foucault 2020, 124). This turn toward the topic of spirituality coincided with a more wide-ranging shift in Foucault's own scholarly interests, from modern nexuses of power and knowledge to ancient ethical traditions, spanning from Greek antiquity to early Christianity. Such ethical traditions, Foucault contended, displayed a commitment to practices of spiritual self-transformation. In his 1981 Collège de France lecture course, *The Hermeneutics of the Subject,* Foucault presented perhaps his most extensive treatment of ancient spirituality by focusing on what he claimed to have been its earliest historical tradition, namely, the ancient Greek practice of *epimeleia heautou,* the demand to care for the self – a notion which, Foucault argued, was forever associated with Platonic thought through the figure of Socrates (Foucault 2005, 8). Variations of the tradition of *epimeleia heautou* survived, Foucault argued, in subsequent Hellenistic, Roman, early Christian, and Medieval pastoral ethics, until eventually diminishing by the eighteenth and especially nineteenth centuries. Foucault's long history of spirituality, in other words, presented a familiar narrative of the post-eighteenth-century self, insofar as it was characterized by the historical decline of the possibilities for spiritual self-transformation. Bearing the rough outlines of Max Weber's thesis of modern disenchantment, the loss of spirituality as an attribute of modern subjectivity corresponded to the attenuation of transformative possibilities for the self, and thus its solidification into a structure of fixed identity.

And yet, just as scholars have troubled and even rebuked the Weberian thesis of a disenchanted modernity (Josephson-Storm 2018; Coleman 2021), so too did Foucault complicate the proposition that spirituality had indeed fully disappeared from the scene of an increasingly secularized Europe beginning in the nineteenth century. It was not uncommon, after all, for Foucault to frustrate the historical periodizations that he himself popularized; the so-called break, for example, that Foucault had once posited as separating the classical age from the modern era was to some degree unsettled when he claimed that the antecedents to both disciplinary and biopower could be found in the governing techniques of the pastorate (Foucault 2007). Furthermore, at the end of *Confessions of the Flesh,* the final volume of *History of Sexuality,* Foucault explicitly argued that the emergence of an ostensibly modern juridical subject was evident as early as the fifth century, in the writings of Augustine. Augustine's conception of free will, Foucault claimed, was somewhat paradoxical, in that it remained deeply entwined with the involuntary forces of desire and the libido. Augustinian freedom was not freedom *from* the libido but, rather, a qualified freedom, *in light of* and always *in relation to* the involuntary forces of desire (Foucault 2021, 265–285). In that sense, Augustine's account of freedom and the libido resembled Foucault's descriptions of the mired relationship between freedom and "power." In the same year that Foucault completed *Confessions,* he asserted in his famous essay, "The Subject and Power," that freedom was in principle the very "condition for the exercise of power" (Foucault 1994a, 342); in other words, that the notion of freedom – the preeminent political, legal, and economic virtue of the liberal age – has remained ineluctably bound to imperceptible vectors of regulation and control from as early as the fifth century. And so, by the same measure, while Foucault would claim that the nineteenth century in particular witnessed the emergence of an increased hostility toward structures of spirituality, it was nevertheless the case that the *possibility* of spirituality persisted in the North Atlantic well after the eighteenth century, in doctrines and forms of thought that held that the very being of the self was capable of undergoing transformation, at least to some degree and by way of some kind of reflexive practice or work; and, furthermore, that only by way of a transformative process could the truth of the self be finally manifest.

In *The Hermeneutics of the Subject* Foucault explicitly named Marxism and psychoanalysis as the two dominant modern doctrines that still made use, or at least presented the vestigial possibilities, of a spirituality of the self (Foucault 2005,

25–30). They were certainly not the *only* doctrines harboring a remnant spiritualism, Foucault admitted; they were simply the most prominent and recognizable. In the case of Marxism, a possible spirituality endured in the "revolutionary subjectivity" that Marxist forms of thought broadly enabled and encouraged, a form of subjectivity which demanded a process of radicalization, a transformative turn, a "conversion to revolution" (Foucault 2005, 208). The transformative potential of psychoanalysis was much more evident in the context of the therapeutic practice on which the theory was in the first instance grounded; already in 1905, Sigmund Freud maintained that among the various therapeutic protocols available for the treatment of psychopathology, psychoanalytic therapy was the one that realized "the most extensive transformations" [*ausgiebigste Veränderung*] in patients (Freud 1904, 260).

The transformative potentials of both Marxism and psychoanalysis lay not, therefore, in their theoretical precepts but, rather, in the practices that they either indirectly or quite expressly espoused – practices that were ultimately directed toward a remaking of the self. In his lecture course, Foucault went on to suggest that one particular attribute of the possible spiritualities respectively latent in both forms of thought is that they were and continue to remain the sources of profound resistance. The intense vitriol, for example, often accompanying the political and scientific rebukes of both Marxism and psychoanalysis could very well be a reaction to the threats these forms of thought were perceived to pose, in their transformative potential, to the de facto value of the permanence, continuity, and regularity of the self – that is, to a structure of subjectivity particularly crucial to both the epistemic and economic commitments of political liberalism. On the other hand, both Marxism and psychoanalysis have a history of rejecting their perceived metaphysical trappings in favor of elevating a scientific self-image; in this sense, they seem to demonstrate a kind of *internal* resistance to their own possible spiritual and transformative potential.

This chapter begins by situating the idea of spirituality in a broader historiography of selfhood, before considering what relationship spirituality and other "technologies of the self," as Foucault called them, might have to the history of the human sciences. This is followed by close examinations of some of the writings of Karl Marx and Sigmund Freud, in an effort to substantiate Foucault's proposition that practices of spirituality were indeed embedded in both Marxist and psychoanalytic doctrines – and thus, to some of the most formative doctrines of thought of the late-nineteenth and twentieth centuries. Foucault's suggestion that Marxism and psychoanalysis could exhibit spiritual possibilities remains, unfortunately, underexamined. Foucault himself did not go into much detail into precisely how these doctrines harbored any sort of spiritual possibilities, as his attention was squarely focused on ancient spiritual practices. The later sections of this chapter, therefore, function primarily as speculative efforts to flesh out and build upon a proposal that Foucault only tersely advanced. The chapter concludes with a further reflection on how the decolonial writings of Frantz Fanon, whose ideas were heavily informed by Marxism and psychoanalysis, might be reexamined by way of the notion of spirituality, as a way of considering the utility of the concept in a global anticolonial context.

Histories of the Malleable Self

Foucault's discussion of the waxing, waning, and latent modern resurgence of spirituality as a form of ethical self-transformation is itself part of a prevalent historiographic motif that has accompanied scholarship on the history of modern selfhood for decades. This motif takes the form of a tension between two dichotomous views of the self – namely, that selfhood is either characterized through the language of fixity, stability, integration, and coherence or, on the other hand, through the language of malleability, variation, transition, and disintegration. Although the terms themselves are not always synonymous – after all, to be a malleable self is not necessarily to be a fragmented self – still, what remains common among recent scholarship on the emergence of modern Western selfhood is the dramatization of a conflict between a very general sense of self-sameness (e.g., unity, stability, balance) and one of self-difference (e.g., fragmentation, self-transformation, pliability); many historians have exhibited a shared tendency, in other words, to chronicle the self's historical unfolding by virtue of this polarity.

For some scholars, such as Greenblatt, Wahrman, and Goldstein, the nineteenth century largely ushered in more concrete conceptions of stable self-identity; self-sameness, in other words, became a much more prevalent attribute of modern North Atlantic subjectivities after the eighteenth century. Other scholars, however, such as the intellectual historian Gerald Izenberg, offer somewhat more dialectically interwoven histories of the relationship between self-sameness and self-difference. Izenberg (2016) has argued that turn-of-the-century European modernism provoked a crisis of identity by virtue of an attraction to the aesthetics of self-fragmentation and a repudiation of the philosophical ideal of absolute self-identity. He proposes that it was precisely this desacralizing crisis of individual identity that incited after the Second World War efforts to resurrect *collective* forms of identity, around race, sexuality, ethnicity, and class. Stefanos Geroulanos and Todd Meyers (2018) have provided more substance to the thesis of self-fragmentation by arguing that the early twentieth-century disintegrations of the self were not simply philosophical or aesthetic preoccupations but medical realities as well. The authors describe that as a consequence of the pervasive and harrowing violence of the First World War, physiologists, neurologists, and psychologists across Europe and America increasingly began to view the human body as an individuated whole that was at once self-integrated and yet fragile, always vulnerable to its own internally prompted disintegration. Historian Jerrold Seigel has laid the blame for the fragmentation of the self – as both a philosophical critique and historiographic propensity – squarely at the door of "poststructuralist" thought. Seigel, in particular, has sought to counter the narrative of self-fragmentation by advocating for a historical theory of selfhood in which the modern Western self is at once autonomous, materially concrete, and socially embedded and reliant (Seigel 2005; Izenberg 2005).

This is all to say that while the binary of self-sameness and self-difference has been a durable feature of many histories of modern selfhood, its particular permutations have been diversely narrated. Indeed, one such prevailing version of the binary has taken the form of histories of the developing recognition of the self's underlyingly malleable nature – that subjectivity is inherently plastic, but only after the nineteenth century did the scientific and political tools arise to properly acknowledge the mutability of the self as both an object and a "project," capable of being acted upon, managed, and governed (Eghigian, Killen, and Leuenberger 2007). Such a view was most appreciably fueled by Foucault himself, who, in *Discipline and Punish*, described the modern institutional processes by which individuals, bodies, and interiorities were modeled, trained, engineered, and effectively "normalized" (Foucault 1995, 293–295).

The tendency to historicize selfhood according to its allegedly inherent malleability has been more recently stoked by the popularization of the concept of plasticity, which despite its current appeal among contemporary scholars is a relatively longstanding concept whose origins can be traced to nineteenth-century psychological and neurophysiological doctrines of habit. In 1890, the philosopher William James claimed that the modifiability of human habits were ultimately grounded on the "plasticity" of organic nervous tissue and neural matter (James 1890, 105), and that the "plastic age of childhood" (James 1890a, 407) denoted the period during which most habits coalesced. Following James, John Dewey also defined "plasticity" as the "ability to form habits of independent judgment and of inventive initiation" (Dewey 1922, 97), while pointing out that the theory of the "alterability of human nature" (106) had its antecedents in the sensuous empiricist philosophies of John Locke and Claude Adrien Helvetius who advocated the value of education in the molding of the human form. The notion of neural plasticity, which remained a topic of neuroscientific interest throughout the twentieth century (Stahnisch 2003), became something of a scientific sensation in the 1990s with the discovery of neuro-cellular embryogenesis, or continued neuronal development and growth, in adulthood (Rees 2016), and has since become an alluring lens through which to reimagine theories of subjectivity (Malabou 2012; Malabou 2009). On the other hand, plasticity has not always been taken up as an object of adulation or positive valuation. The gender and sexuality scholars Kyla Schuller and Jules Gill-Peterson have argued that plasticity denotes a political, economic, and racializing regime which "seizes the malleable body as a means to engineer the individual and the population" (2020, 2). While Schuller and Gill-Peterson concede that bodies are in principle malleable, they distinguish the body's malleability from its plasticity, which they understand to signify the political values that underwrite processes of racialization.

Such diverse appraisals of the status of the malleable self nevertheless operate according to the familiar motif of an underlying dichotomy between self-sameness (fixity) and self-difference (malleability, plasticity). The historian Chris Millard has suggested that the historiographic tendency to view the self as malleable emerged between the 1960s and 1980s, as a consequence of the scholarly appropriation of a normative conception of human nature – originally promulgated by early-to-mid twentieth century anthropologists and psychologists —that human nature was susceptible to being molded and shaped by cultural contexts and forces (Millard 2016, 2020). Post-structural thinkers were most responsible, Millard argues, in tacitly accepting this vision of human nature, which has since dispersed widely into humanistic and social theory, and consequently influenced historical analyses in such a way that malleability has come to be uncritically accepted as a human trait capable of being projected backwards into all historical contexts. Millard's primary concern is that historicizing selfhood according to its presumed malleability risks attributing transhistorical value to a quality of human nature that was itself only a product of twentieth-century thought.

While it would be worth acknowledging that numerous anthropological traits – such as the inherent rationality, sociality, and symbolic/linguistic capacity of human beings – are almost inevitably treated as transhistorical realities, often simply for the sake of being able to write viable histories of human actors, Millard's observation concerning the status of the malleability of the self is crucial, though should be qualified somewhat. For if the dichotomy between malleability and fixity is itself simply part of the historical motif of the dramatization of a crisis between self-sameness and self-difference, then this motif can be traced to an even earlier historical moment, one which corresponds to the very emergence of the human sciences themselves. For example, between 1822 and 1826, the German neuroanatomist Franz Joseph Gall, founder of "organology" or what later became known as phrenology, published his multivolume *On the Functions of the Brain*, in which he described how adverse modifications to brain matter can result in peculiar instances of the pathological transformation of the self. "There are cases," Gall writes, "where, by an alteration of the [cerebral] organ, the ME is transformed into another ME; for instance when a man believes himself transformed into a woman, a wolf, etc.; there are cases where the old ME is entirely forgotten or replaced by a new one; not an uncommon accident after severe disease, especially in cerebral affections" (Gall 1835, 3:76; Finger and Eling 2019).

For Gall, the thesis concerning the changeability in subjective states and senses of self was an explicit rebuttal against an eighteenth-century tendency among European nervous physiologists to be overly philosophical in their view of the nervous system by correlating the apparent homogeneity of the brain's white matter to the presumed unity of the mind and self (Bassiri 2013). It was not uncommon for prominent medical thinkers and clinicians during the early nineteenth century to counterpose their observations and theories about human nature against what they perceived to be unempirical errors of their philosophical forebears. When the French psychiatrist Philippe Pinel first presented in 1801 his diagnosis of "reasoning madness" [*folie raisonnante*] or manic insanity without accompanying delusional states [*manie sans délire*], a genre of mental illness in which patients could be both insane and yet also quite lucid, he was clear to point out not only the novelty of his diagnosis but also its incompatibility with prior philosophical appraisals of madness. Reasoning madness, Pinel wrote, "would have appeared completely enigmatic if we were to follow the ideas that [John] Locke and [Étienne Bonnot de] Condillac have given about the insane" (Pinel 1801, p. 81; Goldstein 1987). Even Freud would claim that Immanuel Kant's views on the transcendental principles governing subjective experience should be rehabilitated in light of his discovery of the unconscious (Freud 1915, 187; 1920, 28).

In their earliest assertions, then, the human sciences and psychological medicine in particular attempted, among other things, to negotiate away from philosophical discourse the authority to make truth claims about human nature. Part of that negotiation took the discursive form of a specific dichotomy concerning the nature of the self, one that explicitly distinguished the philosophical devotion (even among the empiricists) to a universal, timeless, transcendental, and principled theory of mind from medical and scientific observations that subjectivity and one's sense of self was actually quite dynamic, adaptable, and inconstant. This discursive trend was as much part of an effort to seize epistemic authority and professional recognition as it was the tacit recognition that the nineteenth century represented a moment when the proximity between science and politics began to develop in earnest, and that human nature and subjectivity were not only objects of scientific interest but simultaneously a focus of political governance, administrative influence, and institutional molding. Eventually Gall's own thesis concerning the pathological modifiability of the self, which he advanced in explicit contradistinction to the philosophically inflected medical theories of the eighteenth century, became something of a medical presupposition in its own right. At the end of the nineteenth century, the British neurologist John Hughlings Jackson would describe neuropathological conditions as having a twofold effect on subjective states. In cases of brain disorder, Jackson explained, the neurological dissolution and disintegration of the self was actually met with the successive pathological manifestation of an entirely "new person," or novel personality state. Neuropathic patients were no longer themselves; they were, in point of fact, someone else (Jackson 1958; Bassiri 2016). By the time Jackson was reaching these medical conclusions, mind and brain medicine and the human sciences in general had acquired not only a substantial degree of epistemic authority but also a considerable amount of institutional and administrative power in managing individuals and populations.

The dichotomy between self-sameness and self-difference is, in other words, not a timeless or transhistorical contrast, to be sure. It would be better to think of that dichotomy as a discursive invention, one that was framed precisely *as a conflict* in need of resolution by the human sciences and psychological medicine in particular. In order to wrest control of the self away from philosophical discourse, medical thinkers had to turn selfhood into a *problem,* an irresolution concerning the very nature of subjectivity. For it was not enough simply to declare that certain philosophical conclusions were erroneous. Only by making the self into a problem could scientific and medical thinkers hope to demonstrate that metaphysical approaches to the mind and to subjectivity simply lacked the epistemic tools necessary to provide any sort of adequate resolutions concerning the conundrums that clinical observations were yielding with respect to human nature. Psychological medicine in particular deployed a unique strategy by portraying philosophy as a discourse that presupposed what human nature was from the outset, while for Gall, Pinel, and other medical thinkers, the human subject was, in the first instance, nothing more than a question – or a set of questions, one of which happened to take the form of whether humans were fundamentally fixed and stable or mutable and inconstant. The conflict between self-sameness and self-difference, therefore, was simply one of the ways in which selfhood was problematized by the human sciences, from the earliest moments of its historical formation. It is a problematization whose legacy continues to bear out today, as the recent interest into whether selfhood is malleable and plastic or stable and fixed is in fact animated by a genealogy that precedes the 1960s and the rise of post-structuralism by over a century.

Spirituality among the Human Sciences

If, therefore, the question of self-sameness/difference was a problematization that the human sciences relied upon in order to secure epistemic and administrative power, then for Foucault the notion of spirituality was an attempt both to think *within* that problematization and also to think *beyond* it. When historians and philosophers of the human sciences consider the extent to which people are fashioned, or capable of being fashioned (Hacking 2002), the work of Foucault remains a common scholarly referent – particularly Foucault's idea of the "technologies of the self." Foucault's notion of the technologies of the self derives from his conviction that the human sciences, fields such as psychiatry, medicine, economics, and penology, as they emerged at the end of the eighteenth century were not simply benign knowledge fields but were instead epistemic techniques, or "truth games," that human beings diversely deployed in order to understand themselves. Some of these techniques, or technologies, functioned coercively in a dominating or disciplinary fashion, in an effort to regulate bodies, orchestrate conduct, and shape interiorities. Yet in addition to disciplinary techniques, other techniques, what Foucault called "technologies of the self," functioned as mechanisms by which people were able to act upon themselves, in order to orient their own behaviors and styles of being. A core postulate that proliferated Foucault's work in the 1970s was that technologies of domination and technologies of the self increasingly operated in tandem after the end of the eighteenth century and became deeply intertwined, such that disciplinary techniques were mapped onto strategies of self-making. The phenomenon that Foucault sought to analyze, which he dubbed "governmentality" (Foucault 1994, 224–225), was precisely this seemingly paradoxical entanglement, when individuals freely and willingly fashioned themselves, yet did so according to normative and disciplinary constraints, incitements, and demands.

Spirituality, on the other hand, was a notion that Foucault first made use of to consider whether self-fashioning could ever take place beyond the boundary of modern forms of discipline and domination. Were there, Foucault wondered in his discussion of spirituality, technologies of the self that could exceed the remit of governmentality? In that sense, spirituality was not simply any technology of the self; rather, it was a technology that possessed a distinctly liberatory potential, a central concept in Foucault's otherwise shrouded emancipatory politics. In his earliest discussion of it, Foucault portrayed spirituality as a practice that could best be analogized to forms of uprising. "Rising up," he explained, "must be practiced, by which I mean one must practice rejecting the subject status in which one finds oneself, the rejection of one's identity, the rejection of one's own permanence, the rejection of what one is" (Foucault 2020, 133). Uprising and resistance were not only struggles against sites of political authority; they were also practices of insurrection *against oneself*, against one's normative formation as a specific subject of modernity. "[T]he political, ethical, social, philosophical problem of our day," Foucault wrote just two years before his death, "is not to try to liberate the individual from the state, and from the state's institutions, but to liberate us both from the state and from the type of individualization linked to the state. We have to promote new forms of subjectivity through the refusal of this kind of subjectivity that has been imposed on us for several centuries" (Foucault, 1994a, p. 336).

For Foucault, subjectivity at its core was not an essence but an artifact, the consequence of processes or techniques of self-making and subject-formation (what Foucault would call "subjectivation" [*assujettissement*]). Selfhood, in other words, is not something an individual *is*, but rather something an individual *does* or, conversely, a process that is enacted upon an individual. One of the effects of governmentality, that is, when techniques of self-making are aligned with techniques of power, is that the processual quality of subjectivity is hidden or obscured. The self is subsequently naturalized as a timeless essence rather than as a product of historically and geopolitically situated possibilities. On the other hand, the distinctive value of spirituality for Foucault was that it was a technique of self-making that explicitly recognized selfhood to be a *practice* through and through. Spirituality, in other words, highlighted the artifactuality of the self because, as a practice of self-transformation, it posited that the truth of the self lay precisely in the ability to perform oneself differently, otherwise and anew, and therefore not necessarily according to dominant norms and prevailing rules of conduct.

As Foucault's turn to antiquity demonstrates, spirituality was bound to a distinctly premodern legacy, particularly to the practices of self-adjustment and self-attunement that defined ancient virtue ethics. The idea, for example, of "habituation" [*ethismos*] in Aristotle's ethical writings presumes that people have some part to play in modifying the intensity and degree of their feelings and the nature of their actions, and that one's character, like an instrument, can be regulated for the purpose of developing virtue (Aristotle 2000). At the same time, spirituality was likewise bound to the modern legacy of the human sciences, to the extent that Foucault observed a resurgence of spiritual possibilities within Marxism and psychoanalysis. While the human sciences in many ways foregrounded the adjustability and modifiability of the self, what much of Foucault's work on sexuality, medicine, and penology demonstrates is that this presumed modifiability was deployed for disciplinary ends. Even Marxism and psychoanalysis were themselves guilty of harboring and deploying techniques of regulation and domination – techniques that supposed from the outset that self-difference was an inherent and governable trait of human nature. "[T]he relation between manipulating things and domination appears clearly in Karl Marx's *Capital*," Foucault writes, "where every technique of production requires modification of individual conduct – not only skills but also attitudes" (Foucault 1994, 225). And as Foucault's 1973 course at the Collège de France devoted entirely to the topic of "psychiatric power" (Foucault 2006a) evinces, psychoanalysis and psychiatry more broadly were among the most disciplinary expressions of modern medicine.

And yet, to imagine that practices of spirituality were conceivable within Marxist and psychoanalytic traditions was to suggest that these doctrines could enable techniques of the self that were somehow capable of exceeding or eluding the technologies of power that those very same doctrines otherwise facilitated. Foucault in effect suggests that Marxism and psychoanalysis were ambivalent modern practices of selfhood, to the extent that they could be deployed for disciplinary ends but also, depending on how they were utilized, for emancipatory purposes as well. Practices of spirituality effectively made use of the very same suppositions of malleably and docility exploited by techniques of power, though for drastically different aims. Foucault in a sense amends the question posed by earlier figures like Gall; his question was not, are we stable or tractable? It was, instead, to what ends or aims should the malleability and transformability of the self be directed? Within the context of the regimes of power that permeated the epistemic techniques instantiated and formalized by knowledge fields such as psychiatry, medicine, and political economy, spirituality in its specifically modern resurgence introduced an anomalous and disruptive possibility, namely, that individuals could potentially take advantage of their formability in order to transform themselves beyond the normative demands of modern regulatory governance.

Labor and Revolution

When the human sciences emerged and formalized over the course of the nineteenth century, they did so at a moment when a univocal conception of the natural world, and thus human nature, was coming undone. According to the turn-of-the-century German philosopher Wilhelm Dilthey (1988, 83-88), who famously distinguished the human sciences [*Geisteswissenschaften*] from the natural sciences in 1883, no matter how free human activity might have been from the axiomatic strictures of the natural world, the human was still an undeniably natural entity (Makkreel 1992, 35–73). And yet, the *naturalness* of human nature denoted something very different than a simple and reductive physical determinism. A crucial discovery of nineteenth-century biology was that the vital order did not unconditionally reproduce the rules and processes of the physical universe. The postwar French historian and philosopher of the life sciences Georges Canguilhem famously argued that while the physical world could answer to mathematical principles, the world of living beings was instead organized according to a more indeterminate "order of properties," perennially subject to anomaly and variation (Canguilhem 2008, 125; Elden 2019; Lecourt 1975).

And so, over the course of the nineteenth century, as human nature was increasingly rooted within a vital order of physiological functions and processes, the explanatory force of metaphysical doctrines and principles steadily diminished as hitherto dominant ontological precepts concerning human nature were revised. Human essence was no longer defined solely by recourse to a substance or principle, whether it be the soul or an immutable doctrine of matter. To know what the human *was* instead became the task of knowing what the human *did*, and it was precisely the human sciences that undertook such a charge. Foucault proposed just this point at the end of *The Order of Things* – that it was the *function*, not the substance, of human nature which was at stake for the human sciences. To live, to speak, to work, to act – whichever functions were prioritized, the fact remained that for the human sciences, the meaning of human nature was to be found in "conduct, behaviour, attitudes, gestures already made, sentences already pronounced or written" (Foucault 1994b, 354). The human sciences opened up the analytic possibility that the forms and expressions of modern selfhood were not actually the effects of what human beings *were* in any a priori sense but, instead, that such expressions were simply the effects of particular actions, comportments, and dispositions. After all, it was Nietzsche's own observation that humans achieved a stable sense of sovereign individualism not because they finally actualized their ontological essence; instead, autonomy, self-identity, and the overvaluation of responsibility were simply consequences of a violence that humans *performed,* first upon each another and eventually upon themselves (Nietzsche 2017, 58–59).

The historical development of the human sciences serves as a useful lens through which to explore what Foucault might have meant by a Marxist political spirituality and, furthermore, to recognize why Foucault would say of Marxism that it "exists in nineteenth-century thought like a fish in water ... unable to breathe anywhere else" (Foucault 1994b 262). The strongest signs of a Marxian spirituality are especially evident in Karl Marx's earliest writings and fragments, which feature a pronounced metaphysical humanism and which are often divided, by way of an alleged epistemological break in 1845, from a more mature Marxian political-economic science (Musto 2015). In the *Economic and Philosophical Manuscripts of 1844*, for example, and in keeping with the style of thought most emblematic of the human sciences, Marx equates the "species-life" [*Gattungslebe*] of human beings with their "productive life" [*produktive Leben*], that is to say, with the human capacity to work and produce (Marx 1994, 64). Through labor, Marx writes, the human "produces himself not only intellectually, as in consciousness, but also actively in a real sense and sees himself in a world he made." Labor functions, above all else, to fulfill the quintessential vital, social, and even ontological function of human nature, namely, to create and produce; for it is through the laborious creation [*Erzeugung*] of the world and its history that a human being could manifest her own self-formation [*seiner Geburt durch sich selbst*] (Marx 1994, 78). In the first instance, then, what Marx's earliest conception of labor most fundamentally engendered was human nature itself, and not a productivist "craft ideal" (Sayers 2007). The Frankfurt School political philosopher Herbert Marcuse wrote of the *Manuscripts* that its conception of labor, as an act of self-creation, was "the activity through and in which man really first becomes what he is by his nature as man." Labor, in other words, was an affirmational form of self-actualization, "in which human existence is realized and confirmed" (Marcuse 1973, 13).

Indeed, according to Marcuse, the *Manuscripts* were chiefly concerned with the project of revolution, in terms of providing a foundation for revolutionary theory and, ultimately, with "revolutionary *praxis*" (Marcuse 1973, 26). The revolutionary spirit of the *Manuscripts* lay in its recognition that labor, in its presently alienated form, had essentially become incapable of "reconciling man with the world" (Axelos 1976, 56). By yielding only a splintered and disjointed self-existence, alienated labor introduced a schism separating humans from their authentic species-life. It was therefore the task of revolution, indeed the *work* of revolutionary action, to return and restore human labor to its properly creative and self-engendering condition. Revolution was, in other words, itself a kind of labor, a kind of productive form of conduct, a radical praxis whose primary function was to effectuate the "complete and conscious restoration of man to himself" [*Rückkehr des Menschen für sich*] (Marx 1994, 71). Revolution was the practice of reinstating the productive and authentically self-creating vitality of the human being. To *turn* toward revolution, therefore, was itself a kind of transformative praxis, a modification of the self which initiated a prolonged process, the final endpoint of which was to reestablish the self as a thoroughly self-fashioning being. The "conversion to revolution" – to use Foucault's own shorthand for politically spirituality – was simply the first self-altering and creative act of an otherwise alienated subject.

Pathology and Health

In the 1960s, the Frankfurt School psychologist and psychoanalyst Erich Fromm, another sympathetic reader of Marx's early writings, characterized the depiction of alienation in the *Manuscripts* as "the most fundamental expression of psychopathology" (Fromm 2009, 35; Thomson 2009). Indeed, Fromm went so far as to propose that Marx had produced a tacitly medicalized form of political critique. "Alienation then, is," Fromm maintained, "for Marx, *the* sickness of man" (Fromm 2009, 38). From Fromm's point of view, it was not the milieu of postwar structuralism, as it is often believed, that brought Marx into the orbit of Freud and psychoanalytic theory (Robcis 2019). With the undeniable partiality of a psychiatric clinician, Fromm instead insisted that Marx and Freud were the two preeminent diagnosticians of the pathologies of modern human life. "Surveying the discussion of Freud's and Marx's respective views on mental illness," Fromm wrote, "it is obvious that Freud is primarily concerned with individual pathology, and Marx is concerned with the pathology common to a society and resulting from the particular system of that society" (Fromm 2009, 45). According to both authors, Fromm maintained, these social and individual pathologies were not chance occurrences but intrinsically inevitable developments of human history. For Marx, the inevitability of alienation was set into motion by the earliest divisions of labor. For Freud, on the other hand, the unavoidability of pathology arose as a consequence of an intrinsic supposition within the psychoanalytic theory of disease – namely, that the boundary between normality and pathology was "fluid," and, as such, "we are all a little neurotic" [*wenig nervös*] (Freud 1901, 278).

Fromm's overly medicalized view of the pathologies of modern society and politics, while perhaps misplaced with respect to Marx, nevertheless signaled a crucial observation in what might be called the history of the concept of health. As Canguilhem has argued, after the eighteenth century "the truth of the body in an ontological sense" was no longer simply defined according to a doctrine of materialism or mechanism but, rather, by virtue of the status of the body's *health* – that is, the capacity of an organic body to endure, to thrive, and to continue to constitute itself in the face of external forces and environmental fluctuations (Canguilhem 2012, 48, 52). The health of a body, however, did not imply an imperviousness to injury or disease; quite the contrary, Canguilhem explained, for "[d]isease is the risk of the living as such" Canguilhem 2012, 35). To be healthy meant not simply to live but to live in a way that was open to harm and infirmity, for health denoted not the avoidance of pathological states but, rather, the capacity to heal from them and to reestablish a flourishing vitality. If health, therefore, was inaugurated over the course of the nineteenth century as the most fundamental truth of the body, then pathology was no mere natural error or aberration but, instead, constitutive of the underlying logic of health itself, the condition against which health was continuously defined and opposed. Pathology became the perpetual shadow of health, the condition upon which the body's vital truth was not only possible, but possible in an adaptive and dynamic way. "Adaptation to the personal milieu," Canguilhem writes, "is one of the fundamental presuppositions of health" (Canguilhem 2008, 129). Health, in other words, was not an attribute a body passively and statically possessed but a condition that needed to be perennially re-achieved, reestablished, and renewed.

Throughout the nineteenth century, then, the status of pathology shifted from an irregular and unnatural residuum of human life to an inescapable reality built into the very essence of human nature. And as Freud above all evinced, there was perhaps no better example of the inescapability of pathology than the psychiatric recognition of the ubiquity and unavoidability of psychopathological states. It was not, however, solely within the domain of psychological medicine that the profound inescapability of mental disease was recognized. In his *Encyclopedia of the Philosophical Sciences,* published and revised between 1817 and 1830, the German philosopher G.W.F. Hegel effectively countered a long tradition in the history of philosophy which framed mental illness as something either to be disavowed or only supplementarily discussed, as Immanuel Kant had done in his 1798 *Anthropology from a Pragmatic Point of View.* Hegel not only upended a protracted philosophical silence on the topic of madness, but he in fact posited madness [*Verrücktheit*] as a pivotal and necessary moment in the dialectical development of an objectively conscious form of subjectivity (Hegel 1971, 122–139). Hegel sought to reframe the relationship between pathology and health that typically underwrote the empirical observation that a rational subject was capable of falling mentally ill, that madness was something that happened to a healthy individual. Inverting the causality between health and illness, Hegel posited instead that madness was a metaphysical possibility that actually had to *precede* the emergence of a healthy and rational mind. We must be able to say, Hegel claimed, that a rational subject is capable in principle of becoming insane; the capacity to become insane, the "*privilege* of folly and madness" [*das* Vorrecht *der Narrheit und des Wahnsinns*], must be posited in advance of the healthy and rational subject, as a precondition and logical prerequisite. Empirically speaking, health is commonly understood to be the organic state present prior to the onset of disease. But according to Hegel's dialectical reasoning, the logical order of health and disease had to be inverted. Disease – in this case, mental pathology – needed to function as a *metaphysical antecedent* to health, even as health remained the *empirical precursor* to illness. Hegel, perhaps more than any other nineteenth-century philosopher, accentuated what Canguilhem would characterize as the theoretical primacy of pathology, the idea that the actuality of health was built upon the a priori prospect of illness and injury. For Hegel, the purely philosophical possibility of madness was a necessary precondition upon which rational subjectivity was grounded, even if very few of us actually became mad. Anything, therefore, which was incapable in principle of genuine mental pathology – a computer, a nonhuman organism, etc. – was denied the capacity for becoming a viable self. A healthy and rational subjectivity was only that which in theory could become insane.

Over the course of the nineteenth century, then, pathology became more than just an organic contingency to which human beings might occasionally succumb. To the extent that health acquired a novel ontological status with respect to the corporeal specificity of human nature, so too did pathology – serving in this case as the covert force, the surreptitious logic, that tirelessly catalyzed the novel vital truth of human nature as a dynamic state of health perpetually in need of adaptation and renewal. If health, therefore, were not a static or inert condition but a dynamic process of continued revitalization, then to simply designate health as an *attribute* of a living being would be rather insufficient. Prompted by the ubiquity of pathological possibilities, health instead denoted a perpetual movement away from morbidity, an exercise in self-renovation, an enduring act of self-transformation.

Psychoanalytic Spirituality

There was perhaps no better formalization of the furtive pervasiveness of pathology, and its incitement for self-transformation, than the psychoanalytic recognition that no one, no matter how healthy, was free from some degree of psychopathology. For Freud, neuroses and other disorders were the solicitations that prompted the turn toward a transformative medical rehabilitation. And yet, at the same time, health never meant the total eradication of every vestigial trace of disease in the nervous system. Indeed, Freud remained somewhat ambivalent as to what precisely psychoanalytic health actually denoted. "The distinction between nervous health and neurosis," he explained, "is thus reduced to a practical question and is decided by the outcome – by whether the subject is left with a sufficient amount of capacity for enjoyment and of efficiency [*Genuß- und Leistungsfähigkeit*]" (Freud 1917, 457).

What we can say for certain, however, is that therapeutic health did not refer to a simple restoration to the state of mind and body that preceded the onset of mental illness. As Freud would discuss most explicitly in *Beyond the Pleasure Principle,* restoration as a model for medical treatment was actually quite fraught, since the most complete and accomplished form of restoration represented a return to an inorganic state. Restoration, in other words, was itself the very function of the death drive, and so a therapeutic treatment organized around the impulse to restore was itself a certain form of morbidity. The curative power of psychoanalysis, the turn toward psychoanalytic health, had to be framed differently, and it is here that we can consider what Foucault might have meant by a vestigial medical spirituality in psychoanalytic thought – a practice which was as much curative as it was transformative. When the ancient Greek tradition of *epimeleia heautou* implicitly resurfaced in the context of psychoanalytic thought, it did so, Foucault somewhat laconically claimed, through the writings and seminars of Jacques Lacan (Foucault 2005, 30). What Foucault does not consider, however, is that it is equally possible to observe traces of the tradition in an even earlier psychoanalytic moment, namely in the writings of Freud himself. In fact, despite the vast historical disparity that separated ancient Greek, specifically Platonic, ethics from early twentieth-century psychiatry, Freud's writings are actually quite conducive to a loosely Platonic construal and analysis, if only as a way of highlighting the vestigial spirituality and tradition of *epimeleia heautou* latent therein (Ferrari 2007, 176–178). Freud's writings, in other words, were not simply reducible, as Foucault otherwise maintained, to the historical culmination of an increasingly diffuse "psychiatric power" beyond the boundaries of the asylum (Foucault 2006, 510–511; Foucault 2006).

For example, both Plato's ethics and Freud's metapsychology were essentially organized around the health of the soul/psyche. In the *Republic,* Socrates's opening encounter with the figure of Cephalus, who laments his "state of apprehension" of having perpetrated many possible improprieties throughout his long life, is met later with a definition of justice as the internal harmony of the soul, of the harmonious balance, agreement, and proper performance of its parts. When asked why the soul should on its own be balanced – what benefit justice in this sense serves – Socrates answers that the harmony of the soul lends itself to a kind of internal health and vigor, an absence of internal strife or turmoil, the feeling that life is indeed worth living (Plato 2014, 127–138, 142–143). A similar motif of reconciliatory balance is taken up by Freud, for whom the ego's primary function is to balance the demands of the id and the superego (Freud 1938, 146). Of course, the ego often fails in this regard, and those failures correspond to the onset of neurosis. Psychopathology corresponds essentially to the interruptive imbalance of unconscious impulses spilling into the domain of everyday life, and it is, therefore, the practical task of psychoanalytic therapy to facilitate the ego's reconciliatory aims.

For both Plato and Freud, however, the disruptions and imbalances of the soul/psyche did not merely indicate a deviation from optimal health. They instead functioned as a kind of solicitation, not simply to heal, but to heal in such a way that demanded and depended upon a fundamental modification to one's own subjectivity. At the center of both doctrines, at the heart of what it meant to care for oneself either ethically or medically, was the mandate to undergo a process of conversion. This mandate is perhaps most evident in the way both doctrines formalized a relationship between the figure of a leader (the philosopher or the analyst) and the figure of the governed (the political masses, the student, or the patient). It is most famously in the *Republic* where Socrates explains that the true function of the philosopher was not only to apprehend the forms but also to establish a particular relationship with the polity, with governance over the polity, as a philosopher-prince. (Indeed, both Plato and Freud (1930, 123-124, 136-137) describe the superimposed relationship between the psyche and the polity.) Unfortunately, Plato concedes, such a relationship between the philosopher and the polis was impeded by the fact that the philosopher, while perhaps the one most fit to rule, was perceived by the political masses to be the most useless to society. The polity writ large was effectively incapable of apprehending the political value of a philosophical disposition (Plato 2014, 201).

The only way that the philosopher could hope to lead the polity was as a consequence of some kind of divine occurrence, a profoundly interruptive event within the very fabric of political life. Plato never explicitly identifies what such an event might be, but one particular phenomenon is intimated to fulfill this function: a philosophical education. A philosophical education was, after all, no simple pedagogical undertaking. While organized around the eventual recognition of "the good," the *site* of that education was the soul itself, its divine immortality. It was the imperative of the student to learn how to radically recollect what the soul had always known, namely the eternal forms (Plato 1981). A philosophical education involved, therefore, learning, presumably for the first time, how to turn oneself inward, toward one's own soul, in order finally to see the truth of oneself; such an act comprised the interruptive event that might very well set the stage for a philosophical form of political governance. The *Republic's* allegory of the cave dramatizes precisely the challenges involved in this educative "turning." For the when the philosopher attempts to free the enchained prisoners trapped within the cave, their response is one of immediate violence, a reflection of the everyday attitude of the polity toward the philosopher.

In order to provide a philosophical education, Socrates explains, the teacher cannot therefore "put knowledge into souls where none was before," since such an overwhelming effort would be met only with great hostility. Instead, something else had to be done in order to free the prisoners from the cave, in order to instruct them in a proper philosophical fashion, and thus to free them from the world of becoming. "The entire soul has to turn," Socrates insists; education is the art of directing the soul, "of finding the easiest and most effective way of turning it around," converting it, pointing it toward the light of the good, and drawing it toward "the world of what is" (Plato 2014, 223–228). The student is not turned, however, in the sense of being passively manipulated; the task of the philosophical, indeed *spiritual,* education is to incite the prisoners to turn on their own accord. The pseudo-literality of turning the prisoner toward the light of the good figuratively signifies the task of turning the student inward toward herself, toward the underlying ontology and divine immortality of her soul, which was itself a practice of self-conversion – a turning away from the falsehood of what the student believed herself to be in order finally to experience the truth of what she truly was. It was, in fact, the philosopher's responsibility to educate the masses accordingly, to the extent that such a feat was possible, in order to assume the mantle of the ruling pedagogue of the soul.

The entire practice of psychoanalytic psychotherapy could be read through this lens of a Platonic spiritual conversion toward the ontological truth of the self. As Freud would insist, for example, psychoanalysis did not replicate the standard medical model common to most psychiatric settings wherein a doctor merely diagnoses a disease and subsequently attempts either to manage or to cure it. Psychiatric psychotherapy introduced a very different sort of approach and relationality with the patient. "We avoid telling [the patient] at once things that we have often discovered quite early," Freud explains, "or we avoid telling him the whole of what we think we have discovered." The practical objective of psychoanalytic therapy was to incite patients to turn toward themselves, to turn to their own psyches and unconscious psychical processes, often by way of rather unconventional forms of medical provocation including, dream analysis, free association, and the comprehensive attribution of meaning to ostensibly trivial symptoms.

In many respects, the analyst-patient dyad replicated the relationship between Plato's philosopher and prisoner (who was at once a member of the masses and as well as a potential student of the spiritual conversion toward philosophical truth). Not only did patients play a role in their own treatment, they played the central role. For unlike other forms of medical treatment, such as surgical intervention or the prescription of medication which rely upon a certain degree of passivity on the part of patients, psychoanalytic therapy could not but function without the patient's agency and active participation. Freud tells us that patients, not unlike Plato's prisoners, cannot simply be informed of the nature of their condition. "If we ... overwhelmed [the patient] with our interpretations before he was prepared for them," Freud warns, "our information would either produce no effect or it would arouse a violent outbreak of *resistance*" [*einen heftigen Ausbruch von* Widerstand] (Freud 1938, 178). The therapeutic patient was directed, not in the sense of being forcefully steered, but in the sense of being guided and led toward her unconscious self, as it were. The Freudian unconscious, again not unlike the Platonic soul, represented the site of a person's ontological authenticity – the foundations of subjectivity, the repository of ageless remembrances, impervious to time. At the same time, the unconscious, much like Plato's soul as the site of access to the good, was equally opaque in that it was a truth that could never be apprehended as such.

What mattered, therefore, for psychoanalytic practice, what organized its curative potential, was not the eradication of neurosis per se but the incitement to turn toward the ontological core of oneself. The turn toward the unconscious was, in the first instance, the initiating act of what can be called a psychoanalytic spirituality. For at stake in the therapeutic practice was neither the management of symptoms nor the restoration to a pre-neurotic state of health but, instead, the imperative to turn – to turn away from what the patient thought she was in order to bear witness to a more underlying truth of herself. The very turn toward the unconscious self was already an act of conversion, one which preceded and instigated any of the subsequent modifications to the psyche that an ongoing psychoanalytic therapy would doubtlessly engender. The turn was the very conversion that made possible an ensuing practice of self-transformation.

Violence and Decolonization

The forms of subjectivity facilitated by both revolutionary Marxism and psychoanalysis located the truth of the self in some act of conversion and self-transformation. The true self lay, these doctrines held, precisely in the practice of turning – toward radical insurgency, for example, or toward an embrace of the unconscious life of the psyche. If, following Fromm, both doctrines of subjectivity were responses to morbid political and psychical forces, then we might conclude that pathology did not solely denote the deprivation *of* the self, thereby demanding a curative restoration to a pre-pathological norm. Instead, pathology functioned as a kind of solicitation *for* the self, opening up the possibility of a post-pathological state of transformed being. Through the analytic of spirituality, in other words, Foucault was able to attribute to both Marxism and psychoanalysis a transformative and liberatory potential, one that was concurrent with their otherwise disciplinary forms of subjectivation. Foucault is often understood to have offered little more than criticism and reproach against both Marxist politics and psychoanalytic medicine. It is the suggestion of this chapter that this is not entirely true, and that while Foucault maintained that neither doctrine was *inherently* liberatory, the practices of spirituality they each enabled ultimately furnished the conditions for some forms of emancipatory conduct.

Although the respective political and medical spiritualities of revolutionary Marxism and psychoanalysis represented separate and distinct vectors of the modern resurgence of an ancient tradition of the ethics of self-transformation, there were instances in which these two vectors found themselves united into a single doctrine of medical-political spirituality. Perhaps no better example exists than in the decolonial writings of the psychiatrist and political philosopher Frantz Fanon. Born in French colony of Martinique, Fanon began his study medicine, with a specialization in psychiatry, in the French city of Lyon. He quickly joined the psychiatric hospital of Blida-Joinville in Algeria in 1953 – the epicenter of French colonial psychiatry (Keller 2007) – just a year prior to the onset of Algeria's war for independence from France, a political struggle with which Fanon's thought has become inextricably linked (Macy 2000).

Fanon understood European colonialism to be in all respects the enactment of a profoundly pathogenic form of political and psychical violence perpetrated upon the minds, bodies, and social lives of indigenous populations. Fanon's writings embodied the amalgamated attitude of both a physician and political theorist. As such, from the standpoint of colonialism and its attendant forms of violence, the political realm and the psychiatric clinic were deeply interwoven spaces. To truly appreciate the annihilating devastation wrought by the colonizer, it was necessary to view political life through the lens of psychiatric health and unhealth and, conversely, to understand the psyche as an eminently political site (Robcis 2021, 48–73). The indigenous subject of colonial occupation was, as the social theorist Achille Mbembe has argued, conceptually synonymous with "the figure of the patient" (Mbembe 2012, 25). This was an assumption that Fanon held in the strictest medical regard. "The human brain," he wrote, "has enormous potentialities, but these potentialities must be able to develop in a coherent milieu." If any milieu, however, frustrates such a coherence, if it "does not authorize me to reply, I will atrophy, I will be halted" (Fanon 2018, 520–521). The political violence of the colonizer was, in other words, always a form of the psychical destruction of the colonized subject; likewise, the therapeutic revitalization of the colonized psyche could only take place through the process of a salubriously decolonized form of political renewal.

It is in the specific account of the practice of decolonial violence, however – a form of political insurgency that Fanon believed to be both medically remedial and politically self-actualizing – that we observe a combined medical-political spirituality at work. In *The Wretched of the Earth,* his examination of the political and medical turmoil of the Algerian war for independence, Fanon would describe the practice of engaging in violent insurgency against colonial rule as a form of "absolute praxis" [*la praxis absolue*]. Violence was, for Fanon, a genre of self-creative labor in the early Marxian sense. "To work [*travailler*] means to work towards the death of the colonist," Fanon explained. "The colonized man liberates himself in and through violence" (Fanon 1963, 44). Marcuse attributed a degree of revolutionary and liberating value to Marx's early conception of labor in the *Manuscripts.* "Man becomes free in his labour," Marcuse explained. "He freely realizes himself in the object of his labour" (Marcuse 1973, 25). For Fanon, it was imperative to conceive of decolonial violence not simply as a form of defense or rebellion, but as a self-actualizing form of creative labor. "Labour," Fanon elsewhere insists, "must be recovered as a humanization of man." He continues: "Man, when he throws himself into work, fecundates nature, but he fecundates himself also." The recovery of a fundamental and authentic form of labor, in true early Marxian fashion, was the means by which human beings could attend to, and thereby shape, their own human nature (Fanon 2018, 530). Decolonial violence was the first instance of the recovery of creative labor and the reconstitution of oneself in light of the ravages of colonial eradication.

At the same time, Fanon described decolonial violence as a "cleansing force" [*désintoxique*], a rehabilitative practice of "reintroducing man into the world" no longer as a fractured self but as a "total man" (Fanon 1963, 51, 62). Much like the ethical imperative that underwrote psychoanalytic therapy, decolonial violence also possessed an ethical dimension, even as it involved the destruction of the colonizer. Decolonial violence was precisely the practice by which colonized subjects experienced a conversion of sorts – through which, in other words, they were able to turn toward themselves, not necessarily in an effort to bear witness to their unconscious impulses but, rather to bear witness to their own legitimate humanity, to turn away from the colonizer's racist and inferiorizing fictions in order to see the truth of themselves as properly self-determining and self-enacting beings. It was precisely by way of the conversional nature of violence that decolonization was "truly the creation of a new man" [*création d'hommes nouveaux*] (Fanon 1963, 2).

Decolonial violence was, therefore, at once a therapeutic practice and a revolutionary labor, an effort to reestablish an authentically self-creating human vitality. A single act of insurgent violence was itself the very first instance of a conversion, the first (self-)creative act of an otherwise colonized and subjugated being, the first step in a rehabilitative chain of events that would eventually lead to a new and transformed human condition. The pathologies instituted by colonialism were the very provocations by which colonized subjects could, through a "creative violence," step beyond the brutal racialization of colonial rule and seek to become "the origin of the future" (Mbembe 2019, 118, 129, 141). And precisely because of how it functioned as a transformative and self-creative form of work, how it facilitated a profound conversion and turn toward the self, decolonial violence was also a spiritual practice, in both a political and medical sense. After all, the violence of decolonization was not simply an action directed at the forces of colonial rule, a means for "launching a new society." Decolonial violence was also something that colonized subjects performed upon themselves, a practice through which "a radical transformation [*bouleversement*] takes place within [the colonized] which makes any attempt to maintain the colonial system impossible and shocking." Violence was not simply a tool for the sake of freedom; violence was instead *freedom in practice*, a liberatory work in and of itself, and the activity which made possible "the spiritual [*spirituelles*] and material conditions for the reconversion [*reconversion*] of man" (Fanon 1965, 179). Fanon had in effect refashioned violence into an insurrectionary ethics and practice of the self. By adopting both the conversional and revitalizing force of a therapeutic logic and the creative spirit of a revolutionary labor, decolonial violence was more than militant uprising; it was a veritable form of medical-political spirituality, a transformative practice of actualizing the truth of oneself.

Conclusion

This chapter has explored and elaborated on Foucault's pithy suggestion that spirituality or practices of self-transformation were not solely restricted to ancient forms of ethical thought but were latently present in the modern doctrines of Marxism and psychoanalysis. Through extended analyses of the writings of Marx, Freud, Fanon, and others, the preceding sections have sought to fill out and develop what remained for Foucault only an outline of an idea. Furthermore, in situating Foucault's discussions of spirituality within the larger historiographic motif of the conflict between self-sameness and self-difference – a motif genealogically tied to modes of inquiry largely inaugurated by the human sciences during the nineteenth century – this chapter argues that while spirituality was for Foucault a largely premodern form of ethical practice, it was also an attempt on Foucault's part to reimagine the emancipatory potentials within modern technologies of the self. Spirituality, in other words, remains among the few overtly discussed liberatory concepts in Foucault's political thought. And while Foucault never detailed precisely how a Marxist or psychoanalytic spirituality could in fact function in an openly emancipatory way, this chapter has proposed that Fanon's decolonial writings on insurrectionary violence can function as an exemplary instance of what Foucault might have meant.

As mentioned above, Foucault first began exploring the notion of spirituality in his writings on the Iranian Revolution and Islamic subjectivity beginning in 1978. Hence spirituality was, in the first instance, a concept that described counter-colonial and anti-imperial forms of political resistance and uprising – in distinctly non-Western contexts – before it was recast as part of a long tradition of radical ethical thought in the West. Unfortunately, some recent scholars have attempted to bind Foucault's turn to spirituality and the ancient practices of the self to an alleged sympathy for neoliberalism; Foucault's emancipatory politics have been questionably interpreted as an oblique form of advocacy for free-market capitalism (Zamora and Behrent 2016; Sawyer and Steinmetz-Jenkins 2019). This is, in itself, a peculiar construal of, among other things, Foucault's Collège de France lecture course, *The Birth of Biopolitics* (2008), and the descriptions of German and American neoliberalism therein. Scholars have mistaken for advocacy what is instead a critical exploration of the ways in which the market has become the preeminent site of governmentality, or the entanglement of power and freedom. To read *Birth of Biopolitics* as a defense or valorization of the market is tantamount to reading *The History of Sexuality* (1990) as a defense or valorization of sexuality, that is, to confuse a critique and genealogical analysis for a defense. It is, therefore, an odd and unfortunate elision that the emancipatory potential apparent in the discussion of spirituality has been effectively maligned instead of being viewed as potentially linked to global practices of radical resistance and political self-transformation.

Indeed, the very fact that Foucault selected 1978 Iran rather than, say, 1968 France to epitomize what it meant to rise up against one's own subject status suggests that the most salient expressions of spiritualty might very well have been found in distinctly decolonial and anti-imperial struggles, as Fanon's writings on decolonization – as well as Fanon's own, albeit limited, proximity to Iran and Islamic Marxism by way of his acquaintance with the Iranian political theorist Ali Shariati (Fanon 2018a; Davari 2014) – perhaps best exemplify. Certainly, the global dissemination of Marxism and psychoanalysis may have helped shape to some degree the spiritual features of anti-imperial and decolonial resistance. After all, some of the most powerfully deployed and violently enforced imperial and colonial norms of self-identity emerged from sites of both political and medical authority. And so, while the dominant expression of post-eighteenth century North Atlantic selfhood was defined by a stable and static form of self-identity, such a model of the self was nevertheless cut through with an undercurrent of transformative possibilities, much of which played out in forms of resistance against Western global political and epistemic hegemony.

While spirituality to some readers might smack of latent religiosity and thus signify a practice incompatible with forms of secular politics, this chapter has suggested that the notion need not entail any religious conception at all (though it certainly can) and that, anyway, the real value of spirituality rests in its emancipatory implications. Spirituality indeed embodies what we might call a revolutionary practice of the self – an affirmation, in other words, that selfhood is *nothing but* a style of conduct, a mode of comportment, and so susceptible at all times to conversion and alteration; and that there is nothing more radical and unsettling than when the very craft of selfhood is directed toward the task of self-modification, as a way of undercutting the nodal anchors the prop up the most vicious and exclusionary global norms of the self.

References

Afary J, Anderson KB (2005) Foucault and the Iranian revolution: gender and the seductions of Islamism. University of Chicago Press, Chicago, IL

Aristotle (2000) Nichomachean ethics. Cambridge University Press, Cambridge

Axelos K (1976) Alienation, praxis, and Techne in the thought of Karl Marx. University of Texas Press, Austin, TX

Bassiri N (2013) The brain and the unconscious soul in eighteenth-century nervous physiology: Robert Whytt's sensorium commune. J Hist Ideas 74:425–448

Bassiri N (2016) Epileptic insanity and personal identity: John Hughlings Jackson and the formations of the neuropathic self. In: Bates D, Bassiri N (eds) Plasticity and pathology: on the formation of the neural subject. Fordham University Press, New York

Bremner SV (2020) Introduction to Michel Foucault's 'political spirituality as the will for alterity'. Crit Inq 47:115–120

Canguilhem G (2008) Knowledge of life. Fordham University Press, New York

Canguilhem G (2012) Writings on medicine. Fordham University Press, New York

Coleman C (2021) The Spirit of French capitalism: economic theology in the age of enlightenment. Stanford University Press, Stanford, CA

Davari A (2014) A return to which self?: Ali Shari'ati and Frantz Fanon on the political ethics of insurrectionary violence. Comp Stud South Asia Afr Middle East 34:86–105

Dewey J (1922) Human nature and conduct: an introduction to social psychology. Modern Library, New York

Dilthey W (1988) Introduction to the human sciences: an attempt to lay a Foundation for the Study of society and history. Wayne State University Press, Detroit, MI

Dreyfus HL, Rabinow P (1983) Michel Foucault, beyond structuralism and hermeneutics. University of Chicago Press, Chicago, IL

Drochon H (2018) Nietzsche's great politics. Princeton University Press, Princeton, NJ

Eghigian G, Killen A, Leuenberger C (2007) The self as project: politics and the human sciences in the twentieth century. Osiris 22:1–25

Elden S (2017) Foucault: the birth of power. Polity Press, Cambridge

Elden S (2019) Canguilhem. Polity Press, Cambridge

Eribon D (1991) Michel Foucault. Harvard University Press, Cambridge, MA. trans., Betsy Wing

Fanon F (1963) The wretched of the earth. Grove, New York

Fanon F (1965) A dying colonialism. Grove, New York

Fanon F (2018) The meeting between society and psychiatry: Frantz Fanon's course on social psychopathology at the Institut des Hautes Études at Tunis. Notes taken by Lilia ben Salem, Tunis. In: Khalfa J, Young R (eds) Alienation and freedom. Bloomsbury, London, pp 1959–1960

Fanon F (2018a) Letter to Ali Shariati. In: Khalfa J, Young R (eds) Alienation and freedom. Bloomsbury, London

Ferrari GRF (2007) The three-part soul. In: Ferrari GRF (ed) The Cambridge companion to Plato's republic. Cambridge University Press, Cambridge

Finger S, Eling P (2019) Franz Joseph Gall: naturalist of the mind, visionary of the brain. Oxford University Press, New York

Foucault M (1977) Nietzsche, genealogy, history. In: Language, counter-memory, practice: selected essays and interviews. Cornell University Press, Ithaca

Foucault M (1990) The history of sexuality, volume 1: an introduction. Vintage Books, New York

Foucault M (1994) Technologies of the Self. In: Rabinow P (ed) Essential works of Foucault, 1954–1984, volume one: ethics. New Press, New York

Foucault M (1994a) The subject and power. In: Faubion J (ed) Essential works of Foucault, 1954–1984, volume three: power. New Press, New York

Foucault M (1994b) The order of things. Vintage, New York

Foucault M (1995) Discipline and punish: the birth of the prison. Vintage, New York

Foucault M (2005) The hermeneutics of the subject: lectures at the Collège de France, 1981–1982. Picador, New York

Foucault M (2006) History of madness. Routledge, London

Foucault M (2006a) Psychiatric power: lectures at the Collège de France, 1973–1974. Palgrave, New York

Foucault M (2007) Security, territory, population: lectures at the Collège de France, 1977–1978. Palgrave, New York

Foucault M (2008) The birth of biopolitics: lectures at the Collège de France, 1978–79. Palgrave, New York

Foucault M (2020) Political spirituality as the will for alterity: an interview with the Nouvel Observateur. Crit Inq 47:121–134

Foucault M (2021) Confessions of the flesh: the history of sexuality, vol 4. Pantheon, New York

Freud S (1901) The psychopathology of everyday life. In: Strachey J (ed) The standard edition of the complete psychological works of Sigmund Freud, vol VI. Hogarth, London

Freud S (1904) On psychotherapy. In: Strachey J (ed) The standard edition of the complete psychological works of Sigmund Freud, vol VII. Hogarth, London

Freud S (1915) The unconscious. In: Strachey J (ed) The standard edition of the complete psychological works of Sigmund Freud, vol XIV. Hogarth, London

Freud S (1920) Beyond the pleasure principle. In: Strachey J (ed) The standard edition of the complete psychological works of Sigmund Freud, vol XVIII. Hogarth, London

Freud S (1917) Introductory lectures on psycho-analysis. In: Strachey J (ed) The standard edition of the complete psychological works of Sigmund Freud, vol XVI. Hogarth, London

Freud S (1930) Civilization and its discontents. In: Strachey J (ed) The standard edition of the complete psychological works of Sigmund Freud, vol XXI. Hogarth, London

Freud S (1938) An outline of psycho-analysis. In: Strachey J (ed) The standard edition of the complete psychological works of Sigmund Freud, vol XXIII. Hogarth, London

Fromm E (2009) Beyond the chains of illusion: my encounter with Marx and Freud. Continuum, New York

Gall F (1835) On the functions of the Brian and each of its parts., 6 volumes. Marsh, Capen & Lyon, Boston, MA

Geroulanos S, Meyers T (2018) The human body in the age of catastrophe: brittleness, integration, science, and the great war. University of Chicago Press, Chicago, IL

Ghamari-Tabrizi B (2016) Foucault in Iran: Islamic revolution after the enlightenment. University of Minnesota Press, Minneapolis, MI

Goldstein J (1987) Console and classify: the French psychiatric profession in the nineteenth century. Cambridge University Press, New York

Goldstein J (2005) The post-revolutionary self: politics and the psyche in France, 1750–1850. Harvard University Press, Cambridge, MA

Greenblatt S (1980) Renaissance self-fashioning: for more to Shakespeare. University of Chicago Press, Chicago, IL

Gutting G (2019) Foucault: a very short introduction, 2nd edn. Oxford University Press, New York

Hacking I (2002) Making Up People. In: Hacking I (ed) Historical ontology. Harvard University Press, Cambridge MA

Hegel GWF (1971) Hegel's philosophy of mind: part three of the encyclopedia of the philosophical sciences (1830). Oxford University Press, Oxford

Izenberg G (2005) The self in question: on Jerrold Seigel's the idea of the self. Mod Intellect Hist 2:387–408

Izenberg G (2016) Identity: the necessity of a modern idea. University of Pennsylvania Press, Philadelphia, PA

Jackson JH (1958) The factors of insanities. In: Taylor J (ed) Selected writings of John Hughlings Jackson, vol 2. Basic Books, New York

James W (1890) The principles of psychology, vol 1. Henry Holt and Co., New York

James W (1890a) The principles of psychology, vol 2. Henry Holt and Co., New York

Josephson-Storm JA (2018) The myth of disenchantment: magic, modernity, and the birth of the human sciences. University of Chicago Press, Chicago, IL

Keller RC (2007) Colonial madness: psychiatry in French North Africa. University of Chicago Press, Chicago, IL

Lecourt D (1975) Marxism and epistemology: Bachelard. Canguilhem and Foucault, London

Macy D (2000) Frantz Fanon: a biography. Verso, New York

Makkreel RA (1992) Dilthey: philosopher of the human sciences, 2nd edn. Princeton University Press, Princeton, NJ

Malabou C (2009) Plasticity at the dusk of writing: dialectic, destruction, deconstruction. Columbia University Press, New York

Malabou C (2012) The new wounded: from neurosis to brain damage. Fordham University Press, New York

Marcuse H (1973) Studies in critical philosophy. Beacon Press, New York

Marx K (1994) Economic and philosophical manuscripts. In: Simon L (ed) Selected writings. Hackett, Indianapolis, IN

Mbembe A (2012) Metamorphic thought: the works of Frantz Fanon. Afr Stud 71:19–28

Mbembe A (2019) Necropolitics. Duke University Press, Durham

Millard C (2016) Concepts, diagnosis and the history of medicine: historicizing Ian Hacking and Munchausen syndrome. Soc Hist Med 30:567–589

Millard C (2020) Conclusion: balance, malleability and anthropology: historical contexts. In: Jackson M, Moore MD (eds) Balancing the self: medicine, politics and the regulation of health in the twentieth century. Manchester University Press, Manchester

Musto M (2015) The 'Young Marx' myth in interpretations of the economic-philosophical manuscripts of 1844. Critique 43:233–260

Nietzsche F (2017) On the genealogy of morality and other writings. Cambridge University Press, Cambridge

Pinel P (1801) Traité médico-philosophique sur l'aliénation mentale ou la manie. L'Harmattan, Paris

Plato (1981) Meno. In Grube, G.M.A. (trans) Five dialogues: Euthyphro, apology, Crito, Meno, Phaedo. Hackett, Indianapolis, IN

Plato (2014) The republic. Cambridge University Press, Cambridge

Rees T (2016) Plastic reason: an anthropology of brain science in embryogenetic terms. University of California Press, Berkeley, CA

Robcis C (2019) Structuralism and the return of the symbolic. In: Gordon P, Breckman W (eds) The Cambridge history of modern European thought, vol 2. Cambridge University Press, Cambridge

Robcis C (2021) Disalienation: politics, philosophy, and radical psychiatry in postwar France. University of Chicago Press, Chicago, IL

Sawyer SW, Steinmetz-Jenkins D (eds) (2019) Foucault, neoliberalism, and beyond. Rowman & Littlefield, New York

Sayers S (2007) The concept of labor: Marx and his critics. Science & Society 71:431–454

Schuller K, Gill-Peterson J (eds) (2020) Biopolitics of plasticity. Social Text 38(2):143

Seigel J (2005) The idea of the self: thought and experience in Western Europe since the seventeenth century. Cambridge University Press, Cambridge

Sedgwick PR (2009) Nietzsche: the key concepts. Routledge, New York

Smith R (1997) The Norton history of the human sciences. W.W. Norton, New York

Smith R (2007) Being human: historical knowledge and the creation of human nature. Columbia University Press, New York

Stahnisch FW (2003) Making the brain plastic: early neuroanatomical staining techniques and the pursuit of structural plasticity, 1910-1970. J Hist Neurosci 12:413–435

Tanner M (2001) Nietzsche: a very short introduction. Oxford University Press, New York

Thomson A (2009) Erich Fromm: explorer of the human condition. Palgrave Macmillan, Basingstoke

Wahrman D (2004) The making of modern self: identity and culture in eighteenth-century England. Yale University Press, New Haven, CT

Zamora D, Behrent MC (eds) (2016) Foucault and neoliberalism. Polity, London

Human Sciences and Technologies of the Self Since the Nineteenth Century

16

Fenneke Sysling

Contents

Introduction: Technologies of the Self	402
Physiognomy and Phrenology	404
IQ Tests	408
Projective Tests	410
Ultrasound Scans	411
Brain Scans	412
Genetic Ancestry Testing and Its Reception	415
Conclusion	416
References	418

Abstract

This chapter outlines a history of self-making through human-science technologies. These technologies of the self introduced the human sciences into everyday life and made it possible for individuals to incorporate facts and ideas from the human sciences into a process of subject formation. The chapter follows technologies from the silhouettes in physiognomy from around 1800 to genetic ancestry testing in the twenty-first century. Each of these practices propagated the existence of an autonomous self and in most cases encouraged a search for hitherto unknown aspects of the self. The technologies of this chapter also point to varying outcomes with respect to how proactively people used these technologies, how the relation between body and self was mediated, whether the self was conceptualized as one or as compartmentalized, how important comparisons were with a wider community, and with respect to the location of the self. Finally, the chapter highlights human flexibility vis-à-vis technologies of the self. Whether living in

F. Sysling (✉)
Leiden University, Leiden, The Netherlands
e-mail: f.h.sijsling@hum.leidenuniv.nl

© The Author(s), under exclusive licence to Springer Nature Singapore Pte Ltd. 2022
D. McCallum (ed.), *The Palgrave Handbook of the History of Human Sciences*,
https://doi.org/10.1007/978-981-16-7255-2_74

the early nineteenth century or in the 2000s, individuals have combined human-science inspired self-making with everyday ideas about their character, intelligence, or brain functions.

Keywords

Technologies of the self · History of science · Phrenology · Intelligence testing

Introduction: Technologies of the Self

As historians and sociologists of the self have shown, individual practices of self-making range from confessions, reading self-help books, discussing one's feeling with friends, to using drugs. In the terms of Michel Foucault (1926–1984), these practices are all "technologies of the self." For Foucault, technologies of the self were activities or practices or enactments that constitute selfhood. Technologies of the self, he says in his lectures at the University of Vermont in 1982, "permit individuals to effect by their own means or with the help of others a certain number of operations on their own bodies and souls, thought, conduct, and way of being, so as to transform themselves in order to attain a certain state of happiness, purity, wisdom, perfection, or immortality" (Foucault 1988: 18).

Some of these technologies of the self have clear links to the human sciences: going to a psychologist, reading books about the brain, or having yourself tested when entering a school or an institution. It is these practices that are central to this chapter. It looks, for example, at how people's selves were impacted if they were confronted with a psychological diagnosis, but also at individuals with an interest in self-knowledge and self-development, who used the tools of the human sciences to assess themselves. While practices of selfhood in the modern era were varied, and Romanticism or religion were or continued to be important influences, the human sciences have played a key role in expanding the toolbox of available strategies of self-making in his period.

Foucault first wrote about "technologies of the self" in 1980–1981 (Goldstein 1999: 42), and it indicated a new avenue in his work. Foucault's earlier work was preoccupied with how power should be defined and with how modern selfhood developed as an effect of these power structures. *Discipline and Punish* (1977), for example, showed how, in the modern world, individuals were produced by disciplinary procedures such as those in the school classroom or military training. According to Foucault, the human sciences played an important role in the construction of the modern subject, with their power to define what and who was normal and who was not, and with the growing reliance of the modern state on scientific expertise. In this model, there was little agency for the individual to fight the structures of education, policing, prison, and science around him.

Foucault's definition of technologies of the self gives the self some more agency in its own constitution. As he says in 1982, "Perhaps I've insisted too much in the technology of domination and power. I am more and more interested [now] in the interaction between oneself and others and in the technologies of individual domination, the history of how an individual acts upon himself, in the technology of self" (Foucault 1988: 19). According to Jan Goldstein, there are also hints in *The Use of Pleasure* that Foucault started to use "subjectivation" as a term to indicate subject-making in a more reflexive mode, with more openings for human agency, but he left this term unpacked (Goldstein 1999: 43). Foucault's examples from the history of the ancient world in 1982 suggest that the individual in that period is relatively free to pursue self-knowledge, but the question to what extent this term is applicable to the modern world was not taken up by Foucault.

This chapter looks at what historians and sociologists have written about individual encounters with the tools and knowledge from the human sciences in the nineteenth and twentieth centuries. "Technologies of the self" is a useful term for bringing together these works: it is a term that can be used to unearth or emphasize individual agency, even though it is never studied without reference to the larger structures of science and power. As we will see later on, authors vary in the degree of agency they give individuals. As Elwin Hofman writes, "The concept of 'techniques of the self' makes possible the analysis of both disciplining and liberating techniques of forming and transforming the self" (2016: 17). In general, early followers of Foucault have emphasized the coercive aspect. Nikolas Rose, for example, defines "techniques of the self" as "practices by which individuals seek to improve themselves and their lives and the aspirations and norms that guide them" (Rose 1996: 95). In Rose's work, however, it is the norms that are the focus of his research, rather than individual practices. Works published more recently (whether they use the term "technologies of the self" or not) are keen to escape from the Foucauldian power/knowledge framework to emphasize how individuals discard medical, genetic, or phrenological knowledge, or use it to their own advantage in their own unique ways.

There are two more advantages of using the term "technologies of the self" to survey this particular body of scholarship. The first is that the term, already in Foucault's use, highlights activities, performances, or practices. It is therefore a useful concept to describe work that has aimed to steer away from intellectual histories of the self. Of course, the history of the self has been done very productively by intellectual historians who traced ideas of the mind, brain, or selfhood through the history of ideas in medicine, philosophy, and adjacent disciplines. They have been able to point to shifts in our thinking about ourselves that came with scientific thinking about the self. Philosophers such as Rene Descartes (1594–1650) and Jean-Jacques Rousseau (1712–1778), for example, introduced examples and concepts for individuality, human interiority, and the possibility of self-knowledge. Sigmund Freud (1856–1939) opened up vast unknown depths of the self with his division of the human mind into a conscious and an unconscious part, and into an id, ego, and superego, while brain research in the second half of the twentieth century promoted the message that "we are our brains" (Taylor 1989; Seigel 2005; Raymond and Barresi 2006; Porter 1997).

On the other hand, studies that focus more on micro-history and local practices, as we will see in the following, have steered away from the suggestion that there is a linear development towards a modern self. Few people disagree with the fact that the modern era saw an expanding sense of interiority and individuality but histories of local practices show the pace and intensity of this development varied (among different social classes, for example), and that there was a wide variety in conceptions of the self: whole or divided; knowable or unknowable, changeable or unchangeable (Hofman 2016; Ruberg 2020).

A second advantage is that a focus on technologies of the self makes it possible to connect to the history of science. The word "technology" already paves the way for an interpretation of the term in scientific and material terms. This is also how Peter Galison, in a publication about Rorschach-tests, interprets the term. He defines technologies of the self as "highly refined apparatuses for defining, in fine and in large, the nature of 'interior life'" (Galison 2004: 258). It is the word "apparatus" here that suggests a material, technical interpretation of Foucault's technologies. In the history of science, the practice turn of the last decades has generated important work on material practices in the human sciences, from the use of patient files in the history of medicine to the travels of skulls in the history of phrenology (Hess and Mendelsohn 2010; Poskett 2019). These studies show how the material practices of science influence the outcome of scientific work but also indicate the connections between scientific work and quotidian knowledge practices. This chapter also centralizes technologies of the self that are, in the words of Galison, "more local, more material. It means following the micro-establishment of the self, not in the abstract, but in the routinized procedure followed in thousands of ordinary [Rorschach] tests" (Galison 2004: 274). In Foucault's wider sense of the term "technologies of the self," the verbal exchanges between psychologists or psychoanalysts and clients of course have an important role too, but since the chapter focuses foremost on material technologies, these are outside of its boundaries.

In the following sections, this chapter gives a chronological overview of several technologies of the self, each tied to specific human sciences. The chapter takes off with the technologies in physiognomy and phrenology around 1800. It then takes a big step into the twentieth century with two sections on psychological testing: one on intelligence tests and one on Rorschach tests. Then the chapter continues with sections on prenatal scans in the second half of the twentieth century, brain scans in the late twentieth century, and genetic ancestry testing in the twenty-first. By way of a conclusion, this chapter ends with a closer look at the kinds of selves produced by applying the technologies from this chapter.

Physiognomy and Phrenology

One early practice that has been seen as crucial in the making of a modern self is physiognomy. Physiognomy was based on the assertion that facial features were interpretable manifestations of internal aspects of character (Woods 2017). Physiognomy has a history that goes back to the ancient world and it had a continuing popularity into early modern Europe. If a young man wrote in his diary in 1670 about the moles on his face, as Martin Porter (2005) describes, this was because of his knowledge, from books about physiognomy, that these moles signified something else, just as red hair was a sign of anger and deep-set hollow eyes were the sign of a liar (Porter 2005: 12).

Physiognomy had a revival in the eighteenth century when the Swiss Protestant pastor, poet, and physiognomist Johann Kaspar Lavater (1741–1801) gave an impulse to the practice. He achieved fame with the publication of his lavishly illustrated works *Physiognomische Fragmente*, which promised its readers that if they learned to read people's faces (and to a lesser extent bodies), they would be able to judge other people's character at first sight. Physiognomy is now usually considered as a social practice and art rather than a science, but Lavater's physiognomy thrived on the suggestion of scientificity and on the success of the medical sciences, and should therefore be seen as part of the history of the human sciences too. Its popularizer Lavater appealed to the authority of science when writing about physiognomy and so did many of its adherents.

While the older tradition of physiognomy used archetypal faces to convey certain universal human characteristics, such as pride or gluttony, Lavater's approach made physiognomy an art applicable to the individual. Lavater insisted that all faces differed from each other, and had "as little ability to become another self as to become an angel" (Lavater, quoted by Wahrman 2004: 297). According to Dror Wahrman, this stands in stark contrast with the way in which people thought about themselves before around 1780. In what he calls the *ancien régime* of identity, people saw themselves as members of groups and identity was relatively mutable. If a person would put on a costume in a masquerade, for example, this could signify a real change of person. In the late eighteenth century, however, an individual modern selfhood developed, characterized by psychological depth and interiority (Wahrman 2004: xi). This kind of self was innate and immutable, and in this period a mask would be seen as concealing the real individual's true countenance and self. Because of Lavater's individualizing approach, Wahrman sees Lavater's physiognomy as a turning point in the history of the self (Wahrman 2004. See also Rose 1996: 108–109).

Wahrman's history is not an intellectual history, but one based on all sorts of popular cultural practices such as the masquerade. A focus on technologies of the self in physiognomy would highlight practices such as the making of silhouettes. The physiognomic study of people's character, according to Lavater, could be carried out on living people but also by means of cut-out silhouettes of their faces (Fig. 1). This technique, which became fashionable around 1800, promised to retain the most telling features of the face, conveying the character of the individual (Wahrman 2004: 299). A skilled artist could cut out a portrait of an individual's head, in profile, from black paper, in a couple of minutes, to present the sitter with a portrait of themselves (Naeem 2018).

These silhouettes are perhaps physiognomy's most important technology of the self: they were seen as objective and replicable, even if they still needed to be interpreted (Lyon 2007). They were easy to make and easy to communicate. They also made physiognomy a practical thing to do for individuals and translated abstract physiognomical knowledge into self-knowledge. And as Dror Wahrman notes, these

Fig. 1 Lavater's silhouette. (W. H. Mewes, *Portrait of Johann Caspar Lavater.* Wikimedia commons)

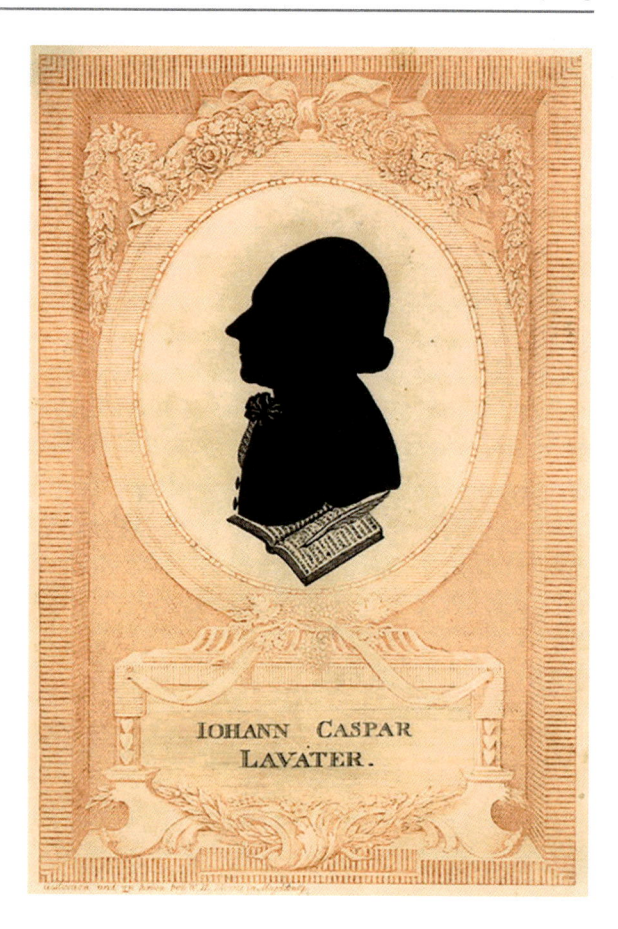

silhouettes were objects that taught people that their self was unique, stable, and individual, and interpretable from the outside. Lavater received hundreds of letters from all over Europe with silhouettes that he then interpreted, including silhouettes of the letter writers themselves or silhouettes of others that they provided (Schooneboom 2003). They also found their way into booklets with physiognomic advice for people ranging from young women looking for a husband to men starting a new business.

The science of phrenology had a similar impact. Phrenology originated with the work of the Viennese physiologist Franz Joseph Gall (1758–1820). Phrenologists claimed that the human mind consisted of different mental faculties, such as the faculty of language or the faculty of conscientiousness, with each particular faculty located in a different area of the brain. The brain, phrenologists argued, pushed the skull outwards and the strength of these faculties could therefore be read by analyzing the bumps on the head. As an academic pursuit, the science of phrenology was soon pushed to the margins, and many contemporaries called it pseudoscience already from the early nineteenth century. More so than physiognomy, however, phrenology was based on an actual theory on the relation between brain, bumps, and mind, and its rhetoric insisted that it had withstood scientific testing. The impact of phrenology increased when it was translated and popularized for the English-speaking world by Johann Gaspar Spurzheim in the early 1800s and it reached people in western Europe but also in the United States, (British) India, and China (Poskett 2019).

Historians have recognized the role of phrenology in the making of the self, and its importance in making the brain central to the self. According to Jan Goldstein, for example, phrenology furnished "the conceptual basis and the language for a distinctive experience of interiority" (Goldstein 2005: 310). It was the attraction of phrenology, wrote Sherrie Lynn Lyons, that people could now be investigated according to scientific standards and would acquire knowledge about themselves that was truer than what they already knew (Lyons 2009: 83). Learning the concepts of phrenology, people could start to see themselves and others as having an individual self, located in the brain and reflected in the shape of the head, a self that was also knowable and malleable. And even though the shape of the head is rather static once we reach adulthood, phrenology should not be mistaken as deterministic: phrenologists insisted that self-knowledge would lead to self-improvement and that individuals had some agency in changing the shape of their heads.

Recent historians of phrenology have gone beyond the history of phrenological ideas and their implication in wider society and have connected phrenology with the latest trends in the history of science, such as the focus on materials and technologies and their travels, and on individual encounters with phrenology. Phrenology thrived on the assessment of individuals. It was a popular pastime in the nineteenth-century Anglo-American world to visit a phrenologist who then analyzed the client by observation, measurement, and a manual examination of their heads. After these sessions, clients received a written description of their character or, more often, a standardized chart. These charts contained lists of all the mental faculties, with the score of the client going from small to average to large, words referring to the size of the bumps and with it to the prominence of the specific mental traits in people's character. Because most traits (such as benevolence or the talent for calculation) were described as neutral or positive, above-average scores were seen as desirable, and shrewd phrenologists in fact sold every individual a chart with an above-average score. These charts looked both scientific, with numbers and tables as symbols of objectivity, and resembled certificates, with seals and decoration. This made them mediators between the larger theories of phrenology and individual self-making: they were tools to make phrenological knowledge authoritative, while also tailoring it to the individual (Sysling 2018).

We can suggest how consumers of phrenology internalized its principles, saw themselves with new definitions and were encouraged to believe in self-improvement, but of course that tells us little about how individuals received their phrenological assessments. Based on American diaries and notes in the margins of phrenological booklets, Carla Bittel (2013, 2019) concludes that individuals balanced the new phrenological knowledge with what they already knew, and at times also took it with a grain of salt. For example, in December 1838, when Phoebe George Bradford from Delaware went to a phrenological reading, she later noted in her diary that she was "much amused." The phrenologist had told her that her organ (or faculty) of order was unusually developed. Bradford's response was: "Quite a mistake." But when the phrenologist had said "she jumped quickly from one topic to another" she wrote that "he hit the truth exactly." (Bittel 2019: 353).

With a similar focus, Frans Lundgren connects the role of scientific paper tools with the making of selfhood in his study of the records with measurements received by clients of Francis Galton's Anthropometric Laboratory at the International Health Exhibition in London in 1884. Galton, anthropologist, statistician, and eugenicist, had opened the laboratory to obtain data for his studies of human growth and of the comparison of human physique across Britain. He also believed that continuous self-assessment would make better and healthier people. Half a century earlier, the Belgian mathematician Quetelet had argued something similar when he wrote that each patient should keep a record of his personal statistics, to be able to know the difference between his usual bodily characteristics and an anomalous state of health (Cryle and Stephens 2017: 110).

Visitors to Galton's laboratory had 17 different measurements on their bodies taken and were presented with their personal records of which Galton also kept a copy. Though these measurements did not focus on the mind or brain, clients were made aware of the value of self-observation and its close companion, self-improvement, if the results were disappointing. "The civic selfhood of this project," writes Frans Lundgren, "was about getting actively involved, committed to learning the lessons of the new methods and knowledge" (Lundgren 2013: 465–466).

IQ Tests

The history of psychology and its influence on wider culture provides us with many examples of technologies of the self, from psychological tests to self-help books. Of these technologies, IQ tests are some of the best-known and pervasive instruments of the twentieth century self. Intelligence became part of people's lives from the early twentieth century through public discussions about intelligence and the implications of test scores in education, but also through self-testing. Intelligence tests were often part of state institutions: pupils would encounter them at school and there they became a starting point for people to think of themselves as individuals with a certain intelligence.

Much of the literature on intelligence testing focuses on the concepts of intelligence, the biases of these tests, and on the notion of racial and national intelligence (Gould 1981; Chapman 1990; Sokal 1990; Zenderland 2001; Carson 2007; Porter 2017). These were important themes before the First World War, but, as John Carson (2007: 183) writes, at that time interest in intelligence was still restricted to anthropology, comparative psychology and discussions about those at the ends of the mental ability spectrum. After the war, testers turned to those considered "normal," and intelligence and testing became more standardized and accepted in broader culture, although it was not conducted more systematically until the 1950s. Looking at broader culture, Nikolas Rose places intelligence tests in a broader context of the assessment of children and the way individuals become measurable and manageable, rather than focusing on the function of the test for individuals (Rose 1990, 1996: 109–113).

As Jim Wynter Porter has shown in an attempt to emphasize the individual impact of intelligence testing, in the United States in the 1950s, grouping by race gave way to individual categorization in a school context, with intelligence as a guiding principle. Intelligence was measured by a standardized test. This shows, writes Porter, "the potential power intelligence had to differentiate and individualize students and contribute to the internal and external processes of self-making that shaped their identities in school, at home, and in their communities" (Porter 2020: 193). According to those who advocated testing and those who did the counselling at schools, self-knowledge of their intelligence, and with it their future potential, could give the pupils direction, and provide motivation (in case of a reasonably good result of course). Teachers, peer groups, and family could support this process (Porter 2020).

Yet, intelligence tests could also be used for individual self-knowledge more proactively. As Susanne Schregel shows for the United Kingdom, from the 1950s, individuals eager to know their intelligence, and especially those with an inkling that they were highly intelligent, could do tests through the Mensa foundation. Mensa, a society that limited its membership to those who scored highest on the intelligence tests, was founded in 1946. People who were interested in how they scored could go to test locations, organized by Mensa, but could also practice with or do tests from newspapers and magazines. In a period that saw the emergence of a knowledge economy, in which the prestige of science and scientists increased, Mensa members presented themselves as highly intelligent and promoted the use of tests as an objective criterion for intelligence and an important technology for self-knowledge (Schregel 2020).

Even though their membership was never very substantial, Mensa functioned "as a kind of lobbying group for intelligence measurement and IQ valuation, and played a pioneering role in promoting intelligence testing for adults," according to Schregel (2020: 17). Advertisements started to encourage people to take part in organized tests, and newspapers cooperated with Mensa to promote intelligence games, puzzles, and tests. Mensa members also published intelligence self-testing books, adapting the intelligence test and commercializing it. For example, Hans Eysenck, a psychologist and Mensa associate, published the successful *Know Your Own I.Q.* in 1962 and long-term Mensa member Victor Serebriakoff wrote *How Intelligent Are You? – Test Your Own IQ*, first published in 1968. These books contained puzzles and tests, and a key to turn the test scores into an IQ score.

In the nineteenth century, practical phrenologists told every single individual that they were above average. This was a very early example of the idea that an individual could be compared with his peers, that there was an average score, and that is was a good thing to score better than that. Historians and philosophers such as Foucault have demonstrated how the human sciences created social categories, and how new statistical definitions of what was "average" and "normal" gave way to social norms about what was normal and abnormal, that average was best in certain circumstances (human growth, for example) while above average was best in other (such as IQ) (Foucault 1978; Ernst 2006; Cryle and Stephens 2017).

Mensa too was and is focused on above-average individuals and these highly intelligent people were the ones to benefit from this organization. How individuals who failed the Mensa entry test articulated their identity with the new-found information is not found in their archives. It is likely, however, that these individuals, like those who went to a phrenologist and those who had a genetic ancestry test done (see below), were flexible in balancing these results and other input in their processes of self-making.

Projective Tests

Another example of a technology of the self from the history of psychology is the projective test, a personality test in which the taker offers responses to ambiguous images (or words). They became popular from the 1920s onwards as a way to capture the taker's immediate and unfiltered response to the stimulus, and to access to the unconscious part of the mind. In the Word Association Test, for example, respondents are asked to respond to a word or several words with anything that comes to mind. With the Thematic Apperception Test respondents have to tell a story based on a set of ambiguous figures. The most iconic and influential of these tests is the Rorschach test, or inkblot test, developed by the Swiss psychiatrist Hermann Rorschach (1884–1922). Rorschach saw the test as a way to understand the personality of his patients through an analysis, not of what the patient saw in the blots but of *how* they described them, in terms of color or movement, for example. The Rorschach test gained fame both inside and outside professional psychological circles. The test became the subject of heated debate in many countries (Hubbard and Hegarty 2016; Lemov 2011) but also had a popular appeal. One of the central characters of the graphic novel *Watchman* (1987), for example, is Rorschach, a "crime-fighting, ink-blot-mask-wearing vigilante" (Hubbard and Hegarty 2017).

Historian of science Peter Galison described the Rorschach test as a perfect example of a technology of the self. The term "technology" is particularly apt, he writes, since the interpretation of the Rorschach test involves a procedure detailed in a 741-page long handbook with possible responses. Today, it comes as a piece of software (Galison 2004: 284). Analyzing the kind of self that inkblot tests presume and produce, Peter Galison found that Alfred Binet saw the inkblot test as a way to test one mental faculty: imagination. Binet is best known for his work on IQ testing, but he was also interested in the inkblot test. Binet saw the self as an aggregate of faculties, which meant that each different domain (imagination, intelligence, etc.) necessitated a different test. (Galison 2004: 259–260). Rorschach, however, who made the test famous, was interested in perception rather than imagination, and in the moment perception became apperception. According to Galison, the Rorschach test promoted an apperceptive self "picked out by its insistence on relations of depth and surface, inner and outer life, and the inseparability of ideation and affect" (Galison 2004: 277): a self not compartmentalized by different faculties. The inkblot tests gave Rorschach the possibility to access this self, even if the subject said nothing at all. The Rorschach test became a test that measured but also reinforced this specific notion of the self (Galison 2004: 259), for the subject who took the test but also because the Rorschach had such a great cultural impact.

Yet very few people took a Rorschach test out of interest in their own selves. As the literature shows, more often than not the tests were done to try to define those who were considered abnormal such as gay men or defeated Nazi's (Hegarty 2003; Brunner 2001). It was also used in the 1970s in Belgium for heterosexual couples who wanted to use donor insemination (Claes 2021). Saskia Bultman studied how the Rorschach test was applied to delinquent girls at a Reform School in the Netherlands, and concluded that through the test, these girls were conceptualized as having an "inner realm, populated with drives, complexes and neuroses" (Bultman 2020: 312). These interiors of the girls were described in remarkably spatial terms, as if they had "inner depths that the psychologist could see into" (325). As Bultman emphasizes, the Rorschach could create a situation in which the psychologist knew something that the "patient" did not, which in this case strengthened the power inequality between the girls and the psychologists at their institution (Bultman 2020). More than in the case of phrenology, or in the case of IQ tests, projective tests foregrounded the idea of an unconscious part of the self that was only accessible through the mediation of the psychologist. Test takers usually knew very little about what to expect from the test, but they could learn to see themselves in a different light, *if* they were informed about the outcome.

Ultrasound Scans

The postwar era saw the wider application of IQ tests and Rorschach tests but also the development of medical imaging technologies, such as the medical ultrasound, the PET scan, and the MRI scan. These imaging techniques introduced new ways of perceiving the body and conceptualizing selves. For example, ultrasound images, a technique of high-frequency sound waves, have become an important experience during pregnancy, especially in the United States where most of the literature is about, but elsewhere too. Ultrasound technology makes it possible to see the unborn child as a shape on a display screen or a printout. These images also trickled into popular literature. Rosalind Petchesky (1987: 268) saw the earliest one in a 1962 issue of the magazine Look, which featured a story about a book *The First Nine Months of Life*.

These images make the child visible at a stage when in earlier times women would only just know they were pregnant from several missed menstruations and perhaps other physical symptoms. Before the mid-twentieth century, babies only became more real when the "quickening" happened, when the mother first felt the child move, and finally when the baby was born (McLaren 1984; Duden 1991; McClive 2002). Ultrasound images have therefore changed how an unborn child is experienced: with the introduction of ultrasound, the unborn child becomes "human" already at an earlier stage. According to Margaret Sandelowski, this is because the ultrasound shows the baby as an entity separate from the mother's womb: "as if it were floating free in space: as if it were already delivered from or outside its mother's body" (Sandelowski 1994: 240). Sociologists and anthropologists, often feminists, have argued that these technological mediations influence the ontological status of the unborn baby. This phenomenon has been called "foetal personhood": the fetus is increasingly seen as an autonomous being, an individual person, on which hopes and dreams and an identity are projected (Mitchell 2001; Boucher 2004; Zechmeister 2001).

The autonomous self that is projected onto the baby through technology has influenced how people experience pregnancies. For example, "prenatal bonding" with the fetus has been made easier. It has also made possible greater involvement of the fathers, permitting them access to the baby that they cannot otherwise perceive yet. Or, phrased more negatively, in Sandelowski's words, the ultrasound has changed the "epistemological privilege" towards men: now they too, can know the fetus, while women are the passive body on which the ultrasound is being performed (Sandelowski 1994: 239). Another result some authors have emphasized is the development from "foetal personhood" to "foetal patienthood," the idea that the unborn child is made into the primary patient, for whom the mother has to care. This places the burden on the mother to live healthily and to give the baby its best possible start while the medical institutes provide the surveillance apparatus.

And finally, "foetal personhood" has been important in abortion debates, especially in the United States, where imagery of fetuses (pro-life campaign groups would prefer the term "baby") is used to place fetal rights over women's reproductive rights (Petchesky 1987; Boucher 2004). The use of fetal images in these campaigns suggests that the fetus is already as much a person as a newly born baby, and are a medical addition to other, religious, arguments against abortion. According to Julie Palmer, three-dimensional (3D) technology will only strengthen this line of thinking, as 3D images of the fetus are easier to interpret for nonmedical audiences (Palmer 2009: 177).

Brain Scans

Other medical imaging technologies such as brain scans have spurred discussion about their impact on how we see our selves. Sociologists and historians have shown that brain scans and the neurosciences in particular have been influential drivers of self-making since the 1990s but build on a much longer tradition.

For individuals suffering from diseases of the brain or from brain damage, a brain scan showing anomalies can have a direct impact on their self-perception. But also the wider visual appeal of brain images and the rhetoric of the brain sciences, with the often heard slogan that "we are our brains" (Swaab 2014), have influenced our selfhood. The suggestion that our brain does things before we realize that we are going to do these things, has important implications for the question of the extent of our free will: in this rhetoric biological reactions take pride of place at the expense of free will.

As often when new medical technologies are introduced, the first sociological studies are based mostly on this rhetoric, while later studies bring nuance with a closer look on the experiences of individuals. As an example of the former, Nikolas Rose and Joelle Abi-Rached (2013: 22) conclude that with the new developments in the neurosciences, mental processes are understood as the outcome of processes in the brain and individuals are taught that their selves are shaped by their brains, as well as shaping their brains in return. The vocabulary of the neurosciences, they write, gives individuals "a rich register for narratives of self-fashioning," a "neurobiological self," that does not replace the psychological conceptions of the self, but builds on it with neurobiological insights (Rose and Abi-Rached 2013: 223). In earlier work, Rose used "neurochemical selves" (Rose 2007: 186–187) to describe how our desires, emotions, and behavior have been mapped on the brain.

Fernando Vidal takes the success and promissory capital of brain research in the last decades and the discourses that surround it as an argument to contend that we are living in an age of "brainhood." Central to this age of brainhood, according to Vidal, is that the brain is the location of the self, an idea with historical roots in phrenology and early modern brain research (Vidal 2009; Vidal and Ortega 2017). Michael Hagner and Cornelius Borck too emphasize continuities and call the new conceptions of the brain "a cohabitation of new visualization techniques with old psychological parameters" (Hagner and Borck 2001: 508).

Where phrenologists had to work with how the brain shaped the skull, brain scans visualize the interior of the skull and promise information about illness and (mental) health. Brain scans come in great variety, among them MRI scans, CT scans, and PET scans. According to sociologists, the images that these scans produce have a special appeal. They appear as objective, legible, and aesthetic. They sometimes are color-coded with certain colors specifying areas of more or less activity. In the words of Sarah de Rijcke and Anne Beaulieu (2007: 737), they look like "rainbow colours superimposed on a floating brain." Katja Guenther and Volker Hess (2016: 301) would call them "soul catchers," "technologies [that] attempt to capture, render visible for study, and manipulate what otherwise eludes our physical grasp," comparable to earlier attempts to capture the soul.

However, these images are hardly photographs but were made through interpretation of very complex data in which many decisions influence what we get to see (Beaulieu 2002). Yet, for the wider public, the suggestion that we get to see the "brain 'at work' producing mental states" is fascinating, and these images therefore have a wide circulation outside their scientific context. But of course, writes anthropologist of science Joseph Dumit, brain images "confuse the part with the whole – even though brain images show only a slice of the brain, they show the slice as representing the whole brain, which in turn *is* the person" (Dumit 2004: 162–163, see also Beaulieu 2000).

People's individual identification with these scientific images can be called "objective self-fashioning," according to Dumit (2004: 164). Dumit followed the making and discussing of PET scans as an anthropologist among brain scientists. In his book *Picturing Personhood*, he shows how images were interpreted by the media and by patients. Because scientists increase contrast between images, to emphasize differences, "they make the additional argument that each brain kind is easily distinguishable" (Dumit 2004: 8). In particular, the way in which these PET scans circulated and were discussed, created categories through which people started to understand themselves. For people who are mentally ill, this meant, for example, that they started to see their brain as diseased and concluded that their illness was a physical thing and not the result of a lack of will.

Thus, objective self-fashioning is "how we take the facts about ourselves (...) and incorporate them into our lives" (Dumit 2004: 164). Scientific "objective" information, says Dumit, becomes part of how people shape their everyday notions of personhood. Individuals combine notions about bodies and brains that we see as objects of science and medicine and therefore as more objective information about ourselves. These "received-facts function as particularly powerful resources because they bear the objective authority of science" (Dumit 2004: 162).

While several authors insist on the increasingly widespread idea of an overlap between the self and the brain, there is also room for a more ambivalent relation between the two (Brenninkmeijer 2015, 2016; Slaby and Gallagher 2015; see also Casper and Gavrus 2017; Bassiri 2019). When, for example, people differentiate between their brain that is diseased and their self which is not, this suggests that brain and mind are different things. Jonna Brenninkmeijer's work, for example, looks at individuals who try out brain enhancing methods such as neurofeedback, a practice she conceptualizes as a technology of the self in the Foucauldian sense. Neurofeedback is a practice in which individuals are presented with feedback from their brain activity. Aware of the changes, the subject will then aim to improve the desirable brain activities and suppress undesirable ones. People using this therapy share the idea that it is possible to change the brain and that this is something individuals can take care of. Advertisements for neurofeedback also promote that idea. One ad from Brenninkmeijer's research said: Why go to therapy when "you can also send your brain in for treatment?" (Brenninkmeijer 2016: 88).

This kind of thinking creates an interesting ambiguity between the brain and the self: users of neurofeedback believe they are (in) their brains, as they want to change themselves by manipulating their brains, but they also see the brain as something that keeps them away from their real self (Brenninkmeijer 2016: 4). Neurofeedback users, writes Brenninkmeijer, make a clear distinction between the (unconscious) brain and their (conscious) selves, but do not necessarily see this as problematic. Brenninckmeijer conceptualizes this as a new mode of selfhood, a self extended with a brain. This fits well with the way Dumit's patients incorporate scientific information about the brain as one aspect of themselves. It suggests that with brain techniques comes an "extended self," a self, scaffolded with (knowledge about) the brain, but also with other common-sense notions of selfhood.

Genetic Ancestry Testing and Its Reception

While brain science has major consequences for our ideas about cognition, free will, and the mind, it is less obvious that a science such as genetics has similar drastic influences on selfhood. However, as historians and sociologists have realized, the body is an important site for work on the self. All sort of options and decisions need to be taken around health and fitness that influence (ideas about) self-identity. Nikolas Rose writes that the new genomic and molecular sciences come with new vocabularies of ourselves. Diagnoses, for example, do not function as mere labels, but can constitute patients' forms of identification and the ways in which they conduct their lives. As Rose argued with Carlos Novas (2000), in genetics, persons who are diagnosed as "genetically at risk" come to have new "genetic responsibilities" and use techniques of the self, such as chat forums where they share experiences with other genetically challenged individuals. These decisions around health and disease demand constant monitoring of bodily and mental processes, self-scrutiny, and risk management. These activities make up a new kind of "somatic individuality" (Novas and Rose 2000; Gibbon and Novas 2008) or "biological self" (Rose 2007: 4).

The advent of genetic ancestry testing since the 2000s has added an extra option of self-making. Genetic ancestry testing works by comparing an individual's DNA, taken from a sample of saliva, with larger DNA databases of populations of the world. People who send in their DNA receive their results with a colored table or with a map that indicates where those people live with whom they share DNA, and to which percentage.

Alondra Nelson (2008) studied how African American users of ancestry tests dealt with the results. She described an initial "genealogical dislocation," if the information was not as expected, but individuals turned out to be creative with the results. They were selective in those parts they incorporated in their own identity, a process that Nelson calls "affiliative self-fashioning." She writes: "Whereas objective self-fashioning highlights the epistemological authority of received-facts that become resources for self-making (...) affiliative self-fashioning attends as well to the weight of individual desires for relatedness (...) and how this priority shapes evaluations of the reliability and usability of scientific data. In the process, received-facts are also reconciled with a complex of alternative identificatory resources" (Nelson 2008: 771). Shim et al. (2018) add to this more specifically that many women whom they followed after an ancestry test, made a distinction between the new information that they considered meaningful and their sense of self that they claimed had not changed after digesting the information. The new facts "seemed to hold relatively insignificant weight in the face of lifelong biographical, social experiences as members of communities and social groups" (Shim et al. 2018: 59; see also Sommer 2010; Roth and Ivemark 2018; Abel and Schroeder 2020; Panofsky and Donovan 2019).

Conclusion

In the historiography of the human sciences and the self, scholars have often followed the earlier Foucault who was preoccupied with power and knowledge and the close relation between the two. The subject produced in that context had few possibilities to resist the disciplinary forces of society. Following Foucault, scholars of the human sciences have taken on the task of revealing the political and moral meanings of knowledge and its potential to discipline individuals into docile bodies without them even realizing. The focus on technologies of the self, however, builds on the Foucault of the early 1980s and his new interest in what individuals do themselves to become better and happier selves. It also follows recent shifts in the history of science that look at how people *do* science, and at the role of instruments and materials in knowledge production and dissemination.

Foucault said that different technologies of the self produced different kind of selves: "As there are different forms of care, there are different forms of self" (in Martin et al. 1988). Indeed, this chapter shows that a history of technologies of the self lays bare how different tools and different types of knowledge produce different conceptions of self. This suggests that conceptions and experiences of inner selves have developed differently for different groups and in different places: a very uneven "democratization of self-fashioning," as Nicholas Rose terms it, from around 1800 onwards.

In all cases, tools from the human sciences provide knowledge that was hitherto unknown to the individual. This suggests that individuals do not necessarily know themselves fully. There are aspects of the self that are unknown, to the individual and to the scientist, before they are unveiled by techniques from the human sciences. With the help of the human sciences, and with the help of an interpreter (the scientist or specialist), it is possible to gain self-knowledge. Or, in the case of prenatal scans, the scans help define the moment that the embryo is seen as a new self.

There are several other aspects of the self that we have come across in this chapter. The role of an expanding sense of interiority and individuality is of course a starting point for histories of modern selves. The popularity of Lavater's physiognomy and cut-out silhouettes of people's heads which could be read for their character, reveal how people learnt this new knowledge and applied it in their own lives. In physiognomy, one's character is destiny, and it is with phrenology (which incorporated many aspects of physiognomy) that individuals are told that they can influence their character, that they can work on themselves. Whereas Lavater marketed his knowledge by stressing how physiognomy was useful to read *other* people's character, phrenology broadened this argument with the insistence that it was useful to know one*self,* and that self-knowledge was the starting point for self-development.

The various historical practices from this chapter also show how individuals had a varying degree of individual agency in how proactively they used technologies of the self and how interested they themselves were in the outcomes. In some practices, such as phrenology, and genetic ancestry testing, most of its users were consumers who themselves decided to visit a phrenologist or take an ancestry test. In other practices, such as brain scans or Rorschach tests, new information about the self needs to be absorbed, because it is a new given in the life of an individual, whether they like it or not.

Physiognomy also points to the relation between body and self. Enthusiasts in physiognomy insisted that the self was readable on the outside of the body, most notably on the face. In phrenology too, people's interior was believed to be discernible on the outside, but the place where it could be assessed moved to the top of the head. Flash forward two centuries, and DNA testing is now marketed for individuals. Genetic ancestry testing uses hereditary material that is present in nearly every cell of the body to tell people where their ancestors were from. Though it is a tool that has little to say about personality or character, ancestry is an important part of how people identify themselves, and surprises in the outcome of the ancestry test can give takers a difference sense of self.

Phrenology emphasized that people consisted of a variety of mental faculties, each better or less developed in individuals. It thus enacted a self that consisted of many selves, lodging in different areas of the brain, each pushing the skull outwards in different places. This stands in stark opposition to the self that was constructed with Rorschach testing, as Peter Galison argued. From the early practices where inkblot testing was said to test the imagination, it became a tool to study the entire (unconscious) self. Intelligence testing on the other hand tests only one faculty of people's self, but an important one in modern society, especially since the Second World War. Modern brain research and its visualization techniques suggest again a compartmentalization of brain functions, though less clear-cut (and less comprehensible for nonspecialists) than in phrenology.

In phrenology but also, a century later, in intelligence testing, individuals were compared against their peers. In phrenology (pseudoscience as it was), this meant that everyone who went for a phrenological assessment received the reassuring message that they scored above average. This did give people the idea that their selves stood in relation to a wider community, and that comparison with that community could be favorable or unfavorable. This was perhaps even more true for intelligence testing, where scores were and are always in relation to peers, and where people are often tested in the context of school and the psychologist's office. One of the few practices where individuals test themselves out of interest (Mensa entrance tests) is very much focused on the high ends of the test scores. These practices do fit in Foucault's older framework, which emphasizes how people become part of knowledge practices which define who is normal and who is not, between who is mediocre and who is fit to become a leading figure in society.

The history of phrenology also points to the centrality of the brain in our thinking about ourselves. Phrenology has located the self firmly in the brain, and modern brain research tends to do the same in its rhetoric, but as Jonna Brenninckmeijer has shown, individuals who do neurofeedback have a remarkable flexibility when it comes to their self-conceptions and envisage a self that is outside the brain and has to live with an unruly brain.

Scholars of neurofeedback and brain visualization have suggested that perhaps the best way to conceptualize the role knowledge from the human sciences plays in self-making is with the concept of an extended self. The extended self is a flexible self, that uses any extra information (from the human sciences) to broaden their conception of self. So the "objective self," as Dumit terms it, becomes just one of the "scaffolds" of people's selves, besides all sorts of common-sense convictions people have about themselves. This fits with recent historical and sociological work on how individuals deal with this new information. Whether living in the early nineteenth century or in the 2000s, individuals have combined objective self-making with everyday ideas about their character. Both historians and sociologists point to the same human flexibility vis-à-vis technologies of the self.

References

Abel S, Schroeder H (2020) From country marks to DNA markers. The genomic turn in the reconstruction of African identities. Curr Anthropol 61(S22):S198–S209

Bassiri N (2019) What kind of history is the history of the self? New perspectives from the history of mind and brain medicine. Mod Intellect Hist 16(2):653–665

Beaulieu A (2000) The space inside the skull: digital representations, brain mapping, and cognitive neuroscience in the decade of the brain. Ph.D. thesis, University of Amsterdam

Beaulieu A (2002) Images are not the (only) truth: brain mapping, visual knowledge, and iconoclasm. Sci Technol Hum Values 27(1):53–86

Bittel C (2013) Woman, know thyself: producing and using phrenological knowledge in 19th-century America. Centaurus 55:104–113

Bittel C (2019) Testing the truth of phrenology: knowledge experiments in Antebellum American cultures of science and health. Med Hist 63(3):352–374

Boucher J (2004) Ultrasound: a window to the womb? Obstetric ultrasound and the abortion rights debate. J Med Humanit 25:7–19

Brenninkmeijer J (2015) Brainwaves and psyches: a genealogy of an extended self. Hist Hum Sci 28(3):115–133

Brenninkmeijer J (2016) Neurotechnologies of the self: mind, brain and subjectivity. Palgrave Macmillan, London

Brunner J (2001) Oh those crazy cards again: a history of the debate on the Nazi Rorschachs, 1946–2001. Polit Psychol 22:233–261

Bultman S (2020) Seeing inside the child: the Rorschach Inkblot test as assessment technique in a girls' reform school, 1938–1948. Hist Psychol 23(4):312–332

Carson J (2007) The measure of merit: talents, intelligence, and inequality in the French and American republics, 1750–1940. Princeton University Press, Princeton

Casper ST, Gavrus D (2017) Introduction. In: Casper ST, Gavrus D (eds) The history of the brain and mind sciences: technique, technology, therapy. University of Rochester Press, Rochester, pp 1–24

Chapman PD (1990) Schools as sorters: Lewis M. Terman, applied psychology, and the intelligence testing movement, 1890–1930. New York University Press, New York

Claes T (2021) Obedient men and obsessive women: donor insemination, gender and psychology in Belgium. Gend Hist. Advance online publication. https://doi.org/10.1111/1468-0424.12531

Cryle P, Stephens E (2017) Normality. A critical genealogy. University of Chicago Press, Chicago

de Rijcke S, Beaulieu A (2007) Essay review: taking a good look at why scientific images don't speak for themselves. Theory Psychol 17(5):733–742

Duden B (1991) The woman beneath the skin: a Doctor's patients in eighteenth-century Germany. Harvard University Press, Cambridge, MA
Dumit J (2004) Picturing personhood. Brain scans and biomedical identity. Princeton University Press, Princeton
Ernst W (ed) (2006) Histories of the normal and the abnormal. Social and cultural histories of norms and normativity. Routledge, London
Foucault M (1978) The history of sexuality: vol. 1. An Introduction (trans: Hurley R). Pantheon Books, New York
Foucault M (1977) Discipline and punish. The birth of the prison. Pantheon Books, New York
Foucault M (1988) Technologies of the self. In: Martin LH, Gutman H, Hutton PH (eds) Technologies of the self: a seminar with Michel Foucault. University of Massachusetts Press, Amherst, pp 16–49
Galison P (2004) Image of self. In: Daston L (ed) Things that talk. Object lessons from art to science. Zone Books, New York, pp 257–296
Gibbon S, Novas C (eds) (2008) Biosocialities, genetics and the social sciences: making biologies and identities. Routledge, New York
Goldstein J (1999) Foucault's technologies of the self and the cultural history of identity. In: Neubauer J (ed) Cultural history after Foucault. De Gruyter, New York, pp 37–54
Goldstein J (2005) The post-revolutionary self: politics and psyche in France, 1750–1850. Harvard University Press, Cambridge, MA
Gould SJ (1981) The mismeasure of man. Norton, New York
Guenther K, Hess V (2016) Soul catchers – a material history of the mind sciences. Med Hist 60(3): 301–307
Hagner M, Borck C (2001) Mindful practices: on the neurosciences in the twentieth century. Sci Context 14(4):507–510
Hegarty P (2003) Homosexual signs and heterosexual silences: Rorschach research on male homosexuality from 1921 to 1969. J Hist Sex 12:400–423
Hess V, Mendelsohn JA (2010) Case and series: medical knowledge and paper technology, 1600–1900. Hist Sci 48(3–4):287–314
Hofman E (2016) How to do the history of the self. Hist Hum Sci 29(3):8–24
Hubbard K, Hegarty P (2016) Blots and all: a history of the Rorschach Ink Blot test in Britain. J Hist Behav Sci 52:146–166
Hubbard K, Hegarty P (2017) Rorschach tests and Rorschach vigilantes: queering the history of psychology in Watchmen. Hist Hum Sci 30(4):75–99
Lemov R (2011) X-rays of inner worlds: the mid-twentieth-century American projective test movement. J Hist Behav Sci 47(3):251–278
Lundgren F (2013) The politics of participation: Francis Galton's Anthropometric Laboratory and the making of civic selves'. Br J Hist Sci 46(3):445–466
Lyon JB (2007) "The science of sciences": replication and reproduction in Lavater's physiognomics. Eighteenth-Century Stud 40(2):257–277
Lyons SL (2009) Species, serpents, spirits and skulls: science and the margins in the Victorian Age. SUNY Press, Albany
Martin LH, Gutman H, Hutton PH (1988) Introduction. In: Martin LH, Gutman H, Hutton PH (eds) Technologies of the self: a seminar with Michel Foucault. University of Massachusetts Press, Amherst, pp 3–8
McClive C (2002) The hidden truths of the belly: the uncertainties of pregnancy in early modern Europe. Soc Hist Med 15(2):209–227
McLaren A (1984) Reproductive rituals: the perception of fertility in England from the sixteenth century to the nineteenth century. Methuen, London
Mitchell L (2001) Baby's first picture: ultrasound and the politics of fetal subjects. University of Toronto Press, Toronto
Moore A, Gibbons D (1987) Watchmen. DC Comics, New York
Naeem A (2018) Black out: silhouettes then and now. Princeton University Press, Princeton

Nelson A (2008) Bio science: genetic genealogy testing and the pursuit of African ancestry. Soc Stud Sci 38(5):759–783

Novas C, Rose N (2000) Genetic risk and the birth of the somatic individual. Econ Soc 29(4):485–513

Palmer J (2009) Seeing and knowing: ultrasound images in the contemporary abortion debate. Fem Theory 10(2):173–189

Panofsky A, Donovan J (2019) Genetic ancestry testing among white nationalists: from identity repair to citizen science. Soc Stud Sci 49(5):653–681

Petchesky RP (1987) Foetal images: the power of visual culture in the politics of reproduction. Fem Stud 13(2):263–292

Porter R (1997) Introduction. In: Porter R (ed) Rewriting the self: histories from the renaissance to the present. Routledge, London, pp 1–14

Porter M (2005) Windows of the soul: the art of physiognomy in European culture, 1470–1780. Oxford University Press, Oxford, UK

Porter JW (2017) A "precious minority": constructing the "gifted" and "academically talented" student in the era of Brown v. Board of Education and the National Defense Education Act. Isis 108:581–605

Porter JW (2020) Guidance counseling in the mid-twentieth century United States: measurement, grouping, and the making of the intelligent self. Hist Sci 58(2):191–215

Poskett J (2019) Phrenology, race, and the global history of science, 1815–1920. University of Chicago Press, Chicago

Raymond M, Barresi J (2006) The rise and fall of soul and self: an intellectual history of personal identity. Columbia University Press, New York

Rose N (1990) Governing the soul: the shaping of the private self. Routledge, London

Rose N (1996) Inventing our selves: psychology, power, and personhood. Cambridge University Press, Cambridge, UK

Rose N (2007) The politics of life itself: biomedicine, power, and subjectivity in the twenty-first century. Princeton University Press, Princeton

Rose N, Abi-Rached JM (2013) Neuro: the new brain sciences and the management of the mind. Princeton University Press, Princeton

Roth WD, Ivemark B (2018) Genetic options: the impact of genetic ancestry testing on consumers' racial and ethnic identities. Am J Sociol 124(1):150–184

Ruberg WG (2020) Health and illness, the self and the body. In: Arcangeli A, Rogge J, Salmi H (eds) The Routledge companion to cultural history in the Western world: part 3: the Western world and the global challenge, from 1750 to the present. Routledge, London, pp 406–419

Sandelowski M (1994) Separate, but less unequal: fetal ultrasonography and the transformation of expectant mother/fatherhood. Gend Soc 8:230–245

Schooneboom M (2003) "Mag ik u mijn vriend noemen?". Fysiognomische correspondentie van Lavater [Can I call you my friend? Lavater's physiognomic correspondence]. Krisis 3:33–42

Schregel S (2020) "The intelligent and the rest": British Mensa and the contested status of high intelligence. Hist Hum Sci 33(5):12–36

Seigel JE (2005) The idea of the self thought and experience in western Europe since the seventeenth century. Cambridge University Press, New York

Shim JK, Rab Alam S, Aouizerat BE (2018) Knowing something versus feeling different: the effects and non-effects of genetic ancestry on racial identity. New Genet Soc 37(1):44–66

Slaby J, Gallagher S (2015) Critical neuroscience and socially extended minds. Theory Cult Soc 32(1):33–59

Sokal MM (ed) (1990) Psychological testing and American society: 1890–1930. Rutgers University Press, New Brunswick

Sommer M (2010) DNA and cultures of remembrance: anthropological genetics, biohistories and biosocialities. BioSocieties 5:366–390

Swaab DF (2014) We are our brains: from the womb to Alzheimer's. Allan Lane, London

Sysling FH (2018) Science and self-assessment: phrenological charts 1840–1940. Br J Hist Sci 51(2):261–280
Taylor C (1989) Sources of the self: the making of the modern identity. Harvard University Press, Cambridge, MA
Vidal F (2009) Brainhood, anthropological figure of modernity. Hist Hum Sci 22(1):5–36
Vidal F, Ortega F (2017) Being brains: making the cerebral subject. Fordham University Press, New York
Wahrman D (2004) The making of the modern self: identity and culture in eighteenth-century England. Yale University Press, New Haven
Woods K (2017) "Facing" identity in a "faceless" society: physiognomy, facial appearance and identity perception in eighteenth-century London. Cult Soc Hist 14(2):137–153
Zechmeister I (2001) Foetal images: the power of visual technology in antenatal care and the implications for women's reproductive freedom. Health Care Anal 9:387–400
Zenderland L (2001) Measuring minds: Henry Herbert Goddard and the origins of American intelligence testing. Cambridge University Press, Cambridge, UK

The Sex of the Self and Its Ambiguities, 1899–1964

17

Geertje Mak

Contents

Introduction	424
Subject/Object	425
From Gay and Lesbian to Trans History	425
Body/Psyche	433
Intermediacy-Binaryness	440
Conclusion	449
References	450

Abstract

The conceptualization of a psychological gendered self as independent from sexual preferences or physical sex is relatively new. Since Robert Stoller coined the concept "core gender identity" in 1964, the idea of an irreversible deep psychological identification with either sex has become increasingly accepted as the basis for "gender-affirmative treatment" in the West. This chapter traces the roots of medical, psychiatric, and psychological conceptualizations of such a "sex of the self" before this seminal moment, concentrating on the scientific practices and methods by which such a "sex of the self" was established. The deliberately alienating expression "sex of the self" brings together distinct historical concepts under one umbrella, without equaling them too easily with "gender identity" as we currently understand it in the West.

This chapter concentrates on the period between the moment in which one of the founders of sexology, Magnus Hirschfeld, coined the sexological concept of "transvestite" in 1910 until Robert Stoller in 1964 famously conceptualized the notion of "(core) gender identity" in cases of transsexualism. It brings together the

G. Mak (✉)
NL-lab, KNAW Humanities Cluster, Amsterdam, The Netherlands

Amsterdam School of Historical Studies, University of Amsterdam, Amsterdam, The Netherlands
e-mail: geertje.mak@huc.knaw.nl

history of three different disciplines in relation to the "sex of the self": the history of sexology and sexual "deviancy," the history of medicine with regard to sex, and the history of psychology with regard to femininity and masculinity or "gender."

It examines this historiography with a particular interest in the question of *how* medical doctors, psychiatrists, sexologists, and psychologists have attempted to assess psychological sex or a "sex of the self." The work of Magnus Hirschfeld will serve as a point of departure for analyzing three inherent ambiguities that permeated scientific knowledge practices pertaining to the sex of the self: subject and object, body and psyche, and fluidity and binaryness. Each of these sections will be introduced by a short general description of the historiography.

Keywords

Gender identity · M-F personality test · Gender fluidity · Transvestite · Transgender · LGBTI · Intersex · Sex change

Introduction

From the late twentieth century, Western societies have come to consider it self-evident that people have an essential inner gender identity. Psychological gender identity is supposed to be established somewhere between birth and the first 18 months as a stable core element of someone's inner self. It is considered so basic or "true," that when it does not correspond with someone's physical sex or assigned gender, this is considered psychologically harmful. The concept of a "true gender self" undergirds Western transgender policies as well as compassionate help and gender-affirmative surgery in the clinic.

The aim of this chapter is to historicize this notion of a "sex of the self" by analyzing the historiography of human sciences dealing with its coming into being from the turn of the twentieth century to the 1960s. The deliberately alienating expression "sex of the self" is meant to defamiliarize contemporary Western assumptions. It combines several distinct historical concepts under one umbrella, without equaling them too easily with "gender identity" as we currently understand it in the West. This historiography is thus fundamentally influenced by Michel Foucault's genealogy of sex and self: to be cautious about projecting current identity categories onto the past is one of his most fundamental insights (Foucault 1979, 1980).

The history of the sex of the self brings together three different historiographical strands: the history of sexology, the history of medicine with regard to sex, and the history of psychology with regard to femininity and masculinity or "gender." As the reader will soon notice, these histories are often deeply entangled. One of the essential elements of the history of the human sciences pertaining to the sex of the self is therefore the question of how the concept of gender identity has come to be isolated from sexual orientation and physical sex. Interestingly enough, however, the strand of psychological research into femininity and masculinity has in general hardly touched sexological insights into the sex of the self, and vice versa.

This chapter is concentrated on the first half of the twentieth century in Europe and the United States. It thus covers the period between the moment in which one of the founders of sexology, Magnus Hirschfeld (1868–1935), coined the sexological concept of "transvestite" and John Money (1921–2006) and Robert Stoller (1924–1991) conceptualized the notion of "(core) gender identity" in cases of intersex and transsexualism. However, as this chapter discusses, the historiography often takes a "true gender self" for granted rather than problematizing or historicizing it.

To open up the question of the sex of the self, this chapter examines the historiography with a particular interest in the question of *how* medical doctors, psychiatrists, sexologists, and psychologists have attempted to assess psychological sex or a "sex of the self." How did scientists come to know the sex of the self? In what kinds of practices, through which techniques and routines, and with what conceptual or theoretical tools did they manage to create an object of knowledge, the "sex of the self," which could analytically be separated from sexuality and bodily sex? How did they deal with subjective experiences, insights, and impressions as necessary sources of information in the light of scientific standards of objectivity? How did they separate sociocultural structures and discourses of gender from individual identifications? This chapter thus offers a genealogy of an essential inner gender identity within the human sciences, spanning (German-speaking) Europe until 1933 and the United States until the 1960s.

The work of Magnus Hirschfeld will serve as a point of departure for analyzing three inherent ambiguities that permeated scientific knowledge practices pertaining to the sex of the self: subject and object, body and psyche, and fluidity and binaryness. Each of these sections will be introduced by a short general description of the historiography.

Subject/Object

From Gay and Lesbian to Trans History

Only a few historians have started to historicize gender (identity) as a topic in its own right. However, lesbian, gay, bisexual, trans, and intersex (LGBTI) historiography has continuously touched upon the issue. A brief long-term glance demonstrates a profound shift in the LGBT historiography of the nineteenth- and early-twentieth-century sexology. In the 1970s through the 1990s, much effort was made to assess whether examples of cross-dressing or gender-nonconforming behavior could count as truly "homosexual" or "lesbian" (Bonnet 1981; Smith-Rosenberg 1985; Hacker 1987; Vicinus 2004; Hacker 2015). However, from the late 1990s onwards, the reverse started to happen: some of the very same or similar sources were (re)interpreted as historical examples of gender-crossing (Bauer 2006; Feinberg 1996; Mak 1997; Halberstam 1998; Mak 2004; Stryker 2017; Manion 2020). That both interpretations are perfectly possible has to do with the fact that from the very beginning – with the work of Viennese psychiatrist Richard von Krafft-Ebing's (1840–1902) *Psychopathia Sexualis* –medico-psychiatric conceptualizations had grouped these phenomena under the umbrella term of "sexual inversion" (Krafft-Ebing 1886; Chauncey 1982; Hekma 1994; Müller 1990; Oosterhuis 2000).

Thus, as Heike Bauer noted, lesbian and gay historians, feminist scholars, and trans historians alike have insisted on the necessity to (analytically) distinguish between (nonconforming) sex, gender, and sexuality. They have done so, each from their own meaningful but not always congruent perspectives (Bauer 2009, 84). However, departing from the concept of "sexual inversion" within this sexological scientific field, this leaves open the question to be answered here: how was a concept of sex as true, inner self carved out, and how did scientists develop ways of perceiving, knowing, or measuring it?

The generation of scientists after Krafft-Ebing, considered to be the founders of sexology, started to split the diagnostic concept of "sexual inversion" into "sexual orientation" and "psychological sex." Berlin-based physician Magnus Hirschfeld coined the category of "transvestite" as separate from "homosexual," making the important claim that not every homosexual was "effeminate" and not every "effeminate" was homosexual in his monumental study *Die Transvestiten: ein Untersuchung über den erotischen Verkleidungstrieb* (*Transvestites: A research into the erotic drive to cross-dress*) (Hirschfeld 1910; Mak 1998; Herrn 2005). In 1928, the British psychologist Havelock Ellis (1859–1939) also published a separate volume about transvestism – which he called eonism – after Chevalier d'Eon (1728–1810) who dressed and lived as a woman (Ellis 1928). Not only did Hirschfeld's term "transvestite" live on, but his theoretical work continued to be influential as well. Hirschfeld started to systematically disentangle (variance in) bodily sex, psychological sex, and sexual orientation. This, at the time, innovative classification is still very similar to the distinction we currently make, for example, in the categories covered by the LGBTI acronym (Hirschfeld 1910; Hirschauer 1993, 82–98; Mak 1997, 144–54; 316–51; Meyerowitz 2002, 21–34; Herrn 2005; Hill 2005; Sutton 2012; Stryker 2017, 98–103). Between the publication of Ellis' seminal work in the 1920s and the 1960s, a nonconforming sex of self was slowly carved out from the larger concept of "sexual inversion" within the fields of sexology and psychiatry.

Nevertheless, the (often implicit) entanglement of gender and sexuality within psychiatric, psychological, and sexological scientific work as well as in the public imagination has been remarkably persistent since the days of Hirschfeld. Hirschfeld himself was certainly not always consistently disentangling them (as we will further see below). In current psychological M-F testing and the medico-scientific studies of gender in the brain, for example, sexual and gender identities are often still mixed up in underlying conceptual frameworks or measures (Terry 2010, 168–77; Fausto-Sterling 2000; Constantinople 2005, 391–93; Jordan-Young 2011; Fine 2011; Dean and Tate 2017, 648–49). Attempts to disentangle gay- and lesbian-affirmative psychology from gender variance affirmative psychology are still in progress (Hegarty 2018, 89–99).

Subject Versus Object

One of the main issues underlying much of LGBTI historiography is the question of the relation between expert "objective" diagnoses or knowledge and "subjective" experience or identification. On what authority are claims about the sex of the self based, who has the right to decide on the assessment of a person's sex, medical treatment, or surgery? What is more important, the subjective feeling of the "patient" or the objective knowledge of the medico-psychiatric expert? By asking *how* the self appeared in medical cases in the past, this subject/object question becomes less dichotomous.

From the start medico-psychiatric knowledge was based on (collections of) case histories. Communication between professionals in the field happened by publishing clinical cases in journals and collections – such as Krafft-Ebing's famous collections of sexual diversions *Psychopathia Sexualis*. Such case histories were structured as classical medical cases: an anamnesis (the history of the "illness" as told by the patient), a diagnosis (the observations of the doctor), and a prognosis (the estimation of how the "illness" would proceed) (Epstein 1995). In the case of "sexual inversion," homosexuality, or transvestism, the narrative of the "patient" was central to the case history, but restructured in order to frame it as a medical case with a diagnosis on the basis of the reported "symptoms." As Ivan Crozier has argued, one of the major knowledge technologies used within sexology was thus formed by a dynamic interaction between the clinical (psychiatric, medical, or sexological) expert and the gender or sexual nonconforming subject or "patient" (Crozier 2008).

Crozier refers to the work of the philosopher and historical statistician Ian Hacking to describe this as a looping process between subjects and objects of knowing. Labeling people, or classifying them, Hacking argues, may have real consequences in "making them up." In particular, when administrative bodies introduce a new category, or when a new label appears within psychiatry or medicine, this might have concrete consequences: it might entail rights, help to legitimate or explain oneself, allow access to medicine, help, subsidies, or therapy, and the like. Being interpellated by a (new) category might well create concrete possibilities or offer advantages. But people do not just subject themselves to a category. With the concept of "looping effects," Hacking points to the interaction that comes into being between the labeling or categorizing "authority" and the subjects that were interpellated by it. This interaction will change the label so that it fits the subjects' experiences or situation better. Identity categories are thus not simply imposed on subjects, nor is there a simple "authentic" experience suddenly "recognized" by experts or doctors, but a process in which subjects mold themselves according to labels, and labels are adjusted to get subjects to fit better (Hacking 2002, 2006). While this has not been given explicit attention, much of the LGBTI historiography has actually dealt with describing such processes. Putting the various studies together shows how looping processes were responsible for the shift from the concept of sexual inversion to the distinction between gender and sexual variety that was made from the 1960s onwards.

Since the 1980s a rich gay historiography has been built up, showing loops in which sexually deviant subjects told their stories, psychiatrists collected and published ever-increasing volumes with cases while (re)creating taxonomies, which in turn molded the individual life stories of subjects who (partly) recognized their situation in these published accounts. Because every new publication of cases invited other subjects to tell their stories, and because these might in certain respects differ from the existing material, psychiatrists kept finding new "symptoms" and creating new (sub)categories (Müller 1990; Oosterhuis 2000; Crozier 2008; Kennedy 2005; Hill 2005; Paolo Savoia 2010). Others have critically noted, however, that gay self-conceptualization and self-narration did not start with forensic or psychiatric interest. Research into early modern sodomite trials demonstrates the emergence of a self-understanding in terms of an inborn characteristic as well as mutual recognition within specific urban subcultures (van der Meer 2007; Trumbach 1998). One aspect of the emerging subcultural recognition and self-explanation was effeminacy, for example, in London's molly houses. In Germany, such a self-explanation was extensively elaborated in the work of Karl Heinrich Ulrichs (1825–1895), who described (his) inclination towards men as being "a female soul trapped in a male body" (Ulrichs 1864; Kennedy 2005).

Building on his ideas, psychiatrists discussed cases of both gender-crossing and sexual orientation towards one's own sex as "sexual inversion." Many taxonomic subdivisions were presented, but the umbrella term "sexual inversion" remained the basic category until Hirschfeld's proposal for a new taxonomy in which homosexuality and transvestism were distinguished. Lesbian historiography demonstrates a more complex pattern, as both female masculinity and female-to-female intimate relations were often related to general gender issues (women's emancipation, feminism, female sexuality). For example, some famous radical feminists in masculine attire were pathologized as illustrative cases in Hirschfeld's Transvestites (Bonnet 1981; Mak 1997; Vicinus 2004; Hacker 2015; Leng 2018).

An "Objective" Questionnaire

The relation between the subject and object of sexological inquiry entered a new phase in 1899, when Hirschfeld published "The Objective Diagnosis of Homosexuality." This included the first version of a questionnaire consisting of 85–127 questions, pertaining to the respondent's sexual desires and experiences and – more prominently – their (masculine or feminine) attitudes, habits, or preferences. Revised and extended versions appeared separately under the title of *Psychobiologische Fragebogen* (Psychobiological Questionnaire) and other titles over the next decades, and many thousand copies were distributed to all visitors of the Institute for Sexual Science (Hirschfeld 1909, 1915, 1921, 1930; Müller 1990, 303–5; Herrn 2014, 218–24; Mak forthcoming). The word "objective" in the title of Hirschfeld's article indicates the intended break with existing diagnostic methods based on the loops described above. Theoretically, a set of standard questions as presented by Hirschfeld could make the diagnosis of homosexuality *the same* among different experts (doctors, psychiatrists) and patients. "Objective" in Hirschfeld's title referred to as what Lorraine Daston and Peter Galison have labeled "communitarian" objectivity, which would calibrate the medical and psychiatric examinations of "sexual intermediates" and thus erase individual scientists' subjectivity (Daston and Galison 2007, 139–63, 265–72).

The questionnaire's questions were firmly anchored in the existing collection of case histories of sexual inversion and its established symptoms. Interestingly, most of the questions of the questionnaire pertained to what we would now refer to as gender, not sexuality. At the time, Hirschfeld still understood male homosexuality as strongly related to effeminacy – a standpoint he explicitly left behind with the publication of *Transvestites* a decade later. Topics varied from comportment (way of walking, looking, use of one's voice, pitch of one's voice) to dress and adornment, interests at school, work and leisure, sexual fantasies and desires, and character. These were all binarily gendered according to contemporary traditional Victorian bourgeois gender roles.

The first edition of the questionnaire demonstrated an ambiguity as to whom got the authority to answer the question, the respondent or the (observing) physician. Later editions left it more clearly to the respondent (Mak forthcoming). While on the one hand giving the impression of a quantifiable list of questions about (gendered) elements of one's attitude, habits, desires, gestures, and inclinations, in practice it invited respondents to tell their life stories in systematically ordered fragments. Compared to a medical case history, this offered less space for respondents to avoid or add items, and, importantly, did not offer a plot of self-discovery. After all, the interpretation of the combined fragments was now left to the expert. Besides its scientific use, Hirschfeld claimed that the questionnaire also helped his patients and clients to come to terms with themselves. Yet, he also inserted items with the specific intention of making the respondents unaware of its diagnostic meaning. This was also true for his request to add a photo of the subject in sleep in order to study the face more objectively. These attempts to keep respondents unaware of the purpose of the questionnaire can hardly be taken seriously, however: at first glance its deeply traditional bourgeois female/male binary is completely obvious (Mak forthcoming).

While it is clear that Hirschfeld meant the questionnaire to ground his emancipatory practice in science ("justice through science" was his slogan), and while we know that the thousands of questionnaires collected at his Institute must have been used in statistical studies on minor subthemes, the data were never fully statistically analyzed and published (Herrn 2014, 219). The 1933 Nazi destruction of the Institute's library and collections has thwarted a historical analysis of the dynamics between respondents and experts with respect to the questionnaire over the decades since. From published case histories, the influence of the questionnaire's systematic approach can be inferred, however, as well as the way in which respondents used the terminology offered to understand and legitimate themselves – sometimes also before the law. The dynamics between Hirschfeld's theoretical and diagnostic practices and cross-dressing clients of all sorts have been well analyzed in various studies (Mak 1997, 1998; Herrn 2005; Hill 2005; Herrn 2014). These studies show that, besides the (attempt to) systematize inquiries into the "sex of the self" through a questionnaire, the "looping dynamics" were still fully at work.

Looping from Transvestite to Transsexual

In *Schnittmuster des Geschlechts*, Rainer Herrn carefully reconstructs the early German history of transvestism and transsexualism around Hirschfeld's work and Institute. Herrn describes the interactions between Hirschfeld and his colleagues of the Institute for Sexual Science with their transvestite clients and the transvestite communities emerging both around the Institute and elsewhere. He also shows how the emergence of such communities was influenced by media coverage about transvestite people and scientific discoveries in the field of hormones between the appearance of *Transvestites* and the destruction of the Institute (and all its collections) by the Nazis. Hirschfeld's view on "transvestism" profoundly changed under the influence of the input of transvestite clients themselves.

First of all, Herrn describes the influence of the appeal made by quite some female and male transvestites to Hirschfeld and his colleagues to help them acquire formal permission to cross-dress in public, with the so-called *Transvestitenschein* (transvestites' identity paper). For female cross-dressers, there was a long history of "passing" as male. For them, the *Transvestitenschein* offered a solution to the increasing requirement to show identity papers in daily life (Mak 1997, 1998). But male transvestites who expressed their will to fully live and love as a woman were quite far removed from what Hirschfeld had initially presented as an "erotic drive to cross-dress." Almost none of the male transvestites in his study were actually trying to live as a woman, and they were all heterosexual. The other major influence on Hirschfeld's conceptualization of transvestism consisted of published cases and new clients who wanted to eradicate elements of their physical sex at birth (castration, removal of facial hair, amputation of the penis, or removal of breasts and/or ovaries). In a later phase, they also hoped for possibilities to create the physical traits of their preferred sex. In particular the published accounts of Eugen Steinach's experiments with the implantation of ovaries in castrated guinea pigs triggered a hope for similar possibilities in some (male-born) transvestites.

What did "transvestites" gain from their identification with the category of "transvestite" and their exchanges with Hirschfeld? Both offered medical explanations and legitimations helpful in acquiring police authorizations for public cross-dressing or court cases, therapeutic help, and advice in which "self-affirmation" was a central goal, as well as (support for) surgical and other help in changing physical characteristics. Moreover, the creation of a community alleviated their feelings of loneliness and isolation. Emerging subcultural spaces such as Hirschfeld's Institute, emerging transvestite clubs, and transvestite media allowed for the exchange of experiences, knowledge, and strategies (Herrn 2005; Sutton 2012). Scientists, for their part, also profited from these exchanges. From 1910 to 1933, Hirschfeld and his colleagues at the Institute for Sexual Science were thus able to collect an increasingly rich collection of cases around the sexological category of "transvestite." From this category, a subcategory was carved out, the "homosexual transvestiate" and (later) "extreme transvestite" – with demarcation lines both in the transvestite community and in therapeutic treatment (Herrn 2005; Sutton 2012).

Remarkably, however, when surgeons approved the wish to carry out a castration or sterilization, or to amputate a penis or breasts, these decisions were not supported by an accompanying diagnostic proof of the extreme degree of "sexual intermediacy." There is no evidence of a kind of threshold based on, for example, the questionnaire. Surgeons often only refer to the case as being a "(homosexual) transvestite." Instead, the degree of suffering – the prognosis of possible self-mutilation or suicide – was decisive. This is true for all the cases of castration and amputation Herrn has managed to reconstruct (Herrn 2005, 167–218). Hirschfeld had always advised against such surgeries, until he learned that some "extreme transvestites" would otherwise commit suicide (Herrn 2005, 184–85). Suicide was a strong trope in all of Hirschfeld's case histories – also pertaining to intersex and homosexuality (m/f) – and was included as topic in his questionnaire. As Heike Bauer has noted, LGBTI histories need to address these issues of suffering and death in their accounts of emancipation (Bauer 2017, 37–56). In the context of the question of how the "extreme transvestite's" self was known, however, the threat of suicide or dangerous bodily self-mutations mainly functioned to *foreclose* further inquiries: other attempts at therapeutic help stopped, and surgery was felt to be an urgent matter of life and death.

This is not to say that Hirschfeld did not conduct a diagnostic interview based on the questionnaire. Herrn made a careful analysis of the physicians involved in the famous Lili Elbe case, the first full "sex change" operation which was internationally published with anonymized names. Herrn's research reveals that the "inquisitorial" interview Lili Elbe underwent was carried out by Hirschfeld at his Institute. In contrast to what Hirschfeld claimed, she did not experience the questionnaire as an incitement to tell her story. How the biopsychological assessment of her sex related to the agreement to perform the surgeries she sought is not clear. Neither do we know of people rejected for surgery on the basis of his questionnaire.

In conclusion, until 1933 looping effects between Hirschfeld's new classification "transvestite," the people who consulted him, media coverage on legal transvestite permissions and "sex changes" (including cases of intersex and Steinach's experiments) led to a further subclassification of "extreme transvestites" who (via Hirschfeld's consults) were afforded surgery. This did not lead to a more refined, distinguishing medico-psychological technique to establish these people's feminine or masculine selves. Rather, their willingness to undergo the surgery *itself* (under the threat of dangerous self-mutilation or suicide) was the key symptom.

Developments in the United Kingdom and the United States concerning the first "sex change" operations and the coinage of the classification "transsexual" also point to looping effects, but with a difference: they increasingly centered around media coverage of "sex change" surgery – both in cases of intersex and otherwise (Meyerowitz 2002, 51–97). In many ways, cases of intersex surgery paved the way for other "sex change" surgery – as we will also see below (Hausman 1995; Germon 2009; Griffiths 2018a). The publication of the (pseudo)autobiography of Lili Elbe – first in Danish in 1931, then in German in 1932, and the following year in English under the title *Man into woman: An authentic record of a change of sex*

(1933) – was very influential, mainly via media attention in the United States (Elbe en Hoyer 1933). Concerning this, Sandy Stone and Bernice Hausman argued that published stories and autobiographies molded the ways in which cross-identifying people narrated their stories, which in turn influenced the diagnostic system (Stone 1992; Hausman 1995; Meyerowitz 2002, 14–97; Stryker 2017). David Griffiths showed how stories about the "sex change" of intersex persons incited other people – who were not (so clearly) intersexed – to demand sex surgery. Some physicians were adamantly opposed to such operations on the basis of a sharp distinction between physical intersex and psychological pathology, while others were more willing to turn a blind eye – such as in the famous case of the first female-to-male surgery with phalloplasty of Michael Dillon (1915–1962) in the 1940s (Griffiths 2018a).

Besides the media coverage on Lili Elbe, another link to the earlier loop by which "extreme transvestite" subjectivities came into being in Germany was endocrinologist Harry Benjamin (1885–1986), who had met Hirschfeld on several occasions and regularly corresponded with him. Based on Hirschfeld's ideas (as we will see below) but without acknowledging this, he used the same concept of "extreme transvestism" when he defended transsexual surgery "in the last resort" (Herrn 2005, 219–20; Meyerowitz 2002, 103). With the Gender Identity Project at UCLA in 1958, the new concept of "transsexualism" – coined by David O. Cauldwell (1897–1959) and further propagated by Benjamin – emerged. A key member of this project was psychoanalyst Robert Stoller. In 1964, he successfully proposed the concept of "core gender identity," based on John Money's concept of gender identity/role (G-I/R). Stoller's most pertinent influence has been to introduce a strict sex/gender distinction (Germon 2009, 63–83; Meyerowitz 2002, 98–129). Precisely the practical and theoretical difficulties of distinguishing between the two will be discussed in the next section.

Beyond Self-Narration

For decades, Hirschfeld's questionnaire remained the only attempt at "de-narrativizing" knowledge about variance in sex, gender, or sexuality. It was a tool to be shared by professionals, in which a limited set of questions ordered the information and left the diagnosis to the medical or psychological expert (Mak forthcoming). Whether its use extended beyond measuring people visiting his Institute for Sexual Science in Berlin is not clear. In the context of the examination of sex- and gender-variant people, it was not until the 1940s that a knowledge technique was used which was not primarily based on self-narration. In 1936 a test to measure psychological sex was introduced in the United States: the Terman-Miles M-F scale, meant to measure degrees of femininity and masculinity in the general population. As will be discussed below, this test became one of the standard tests in cases of intersex in the 1940s (Terry 2010, 168–75; Redick 2004, 158–223).

The difference between this test and Hirschfeld's questionnaire was not only a much more refined psychometric and statistical technique with which people could be actually scored according to a validated M-F scale. More importantly, respondents were kept entirely unaware about the purpose of the test. The test was referred to as AIST – attitude-interest analysis test – in order to mask its purpose (Terman and

Miles 1936, 4). This is a significant difference with Hirschfeld's questionnaire, of which the purpose was not only very apparent but which was also emphatically meant to enhance people's self-expression and self-understanding. The fundamental assumption behind the AIST test, on which all M-F tests in the following decades were based, was that masculinity and femininity were a real core aspect of personality but not readily observable. Moreover, issues of gender were so charged with value, that subjects had to be deceived about the test's true intention in order to avoid socially desirable answers (Morawski 1985, 206–11). M-F tests are still used today in cases of intersex and trans people. The subject/object issue is still of relevance, as shown in, for example, Naeema Taher's comparison of a self-concept test with a M-F test (Taher 2007).

Body/Psyche

Doubtful Sex, Doubtful Gender

Another strand of the historiography with regard to the sex of the self deals with medical and psychological conceptions of its relation to physical sex. Early modern medical perceptions of the sexual body were not engaged with the question of whether or how biology defined destiny. Bodies were rather interpretated metaphysically, as reflections of a divine or cosmic order (Laqueur 1990; Duden 1991; Park 2006). As Thomas Laqueur argues in his groundbreaking work *Making Sex*, medical treatises stressed the morphological sameness between the sexes, the uterus and vagina being the less well-developed, inwardly turned, versions of the penis and scrotum. This medical conception of *gradual* sex difference (which Laqueur termed the "one-sex system") stressed the divinely ordered hierarchy between the sexes. Physical sex was not causally linked to the sexual order or an inner sense of sex. Yet, legal and social categories of sex were strictly binary. From the second half of the eighteenth century, the *difference* between the sexes was increasingly stressed – in what Laqueur calls the "two-sex system." As the new political order was based on natural equality, medical research into the natural and absolute differences between men and women became urgent in order to legitimate their different positions in society. From that period onwards, the question of how the body caused the different "natures" of men and women became a central issue within the (medical) study of humans (anthropology) (Laqueur 1990; Honegger 1991; Schiebinger 2004).

Central to any discussion of the relations between bodily sex and psychological sex are cases of doubtful physical sex, termed (pseudo)hermaphrodites until the first decades of the twentieth century, and since then increasingly referred to as "intersex." Laqueur shows how hermaphroditism under a "one-sex system" served as medical proof for sexual difference being gradual: hermaphrodites offered proof that intermediate developments between women and men were possible. Under a "two-sex system," the reverse was the case: theoretically, "true" hermaphrodites could not exist, as the difference between men and women was considered absolute. After all, no human was ever reported to have reproduced themselves. From the late eighteenth century, the sexual glands, or gonads, were increasingly considered the scientific essence of sex, defining a person's proper position in society (Foucault 1980; Laqueur 1990; Dreger 1998). However, as we will see below, in clinical practice, sexual glands were far from decisive. What's more, the question of whether and how to deal with a patient's own perception or sense of their sex started to be discussed from 1900 (Mak 2012).

The twentieth-century historiography on (pseudo)hermaphroditism is pertinent to the analysis of the relation between the sex of the body and the sex of the self (Hirschauer 1993; Hausman 1995; Dreger 1998; Mak 2012; Reis 2012; Germon 2009; Griffiths 2018a). This history started to be entangled with the history of transsexualism from 1900 onwards. Hirschfeld's work crucially included cases of pseudohermaphroditism (now called intersex), which meant that he brought LGBT and intersex issues together (Mak 2012, 193–99; 205–19). He was particularly interested in the function of hormones in relation to gender and sexual nonconformity (Sengoopta 1998). Nelly Oudshoorn's seminal work *Beyond the Natural Body* – a history of the medical inventions of (sex) hormones – explains why these new insights complicated rather than clarified the relations between sexual bodies and selves.

David Andrew Griffiths revealed how in the United Kingdom intersex treatments formed the basis on which the first trans "sex change" surgeries were carried out in the decades around World War II (Griffiths 2018a). In the United States, as Bernice Hausman, Alison Redick, and Jennifer Germon have shown, the diagnosis and treatment of intersex between 1920 and 1950 was fundamental to the development of the concept of "gender" and "core gender identity" (Germon 2009, 65–71). The invention of an easy chromosomal test deeply influenced practices of sex determination and assignment in the 1950s (Griffiths 2018b). After decades of idiosyncratic solutions to each case of intersex in the United States and an increasing confusion about what constituted the essence of physical sex, psychiatrist John Money proposed the concept of G-I/R as the basis for the John Hopkins intersex treatment protocols which, in modified form, are still in use. His concept of gender can also be considered a core building block in the then upcoming diagnosis and treatment of transsexualism (Hausman 1995, 72–109; Redick 2004, 158–222; Germon 2009, 23–62; Eder 2010; Rubin 2012).

The subject/object ambiguities in knowledge pertaining to the "sex of the self" discussed above relate to body/psyche ambiguities. The underlying assumption is that the body is the domain of the (objective) medical doctor and the self the domain of (subjective) self-knowledge. Hirschfeld's ambiguity vis-à-vis bodily or biological sex can be posited precisely in the heart of this opposition: what he hoped to achieve was to scientifically objectify the voice of "sexual intermediates" *through* biology. People who did not feel, behave, desire, or identify according to the sociocultural expectations attached to their sexed bodies should be accepted because they constituted natural variations. How this position led to many inconsistencies within his work has been thoroughly analyzed by Chandak Sengoopta (Sengoopta 1998). In the late 1950s, Harry Benjamin almost literally took over Hirschfeld's concepts and theories in his cooperation with Robert Stoller in their Los Angeles-based Gender Identity Project designed to study transsexualism.

In between Hirschfeld's work and Robert Stoller's famous sex/gender distinction lies a complex and dynamic history of the relation between sexed bodies and sexed selves. While for cross-identifying people the problem of how to align their morphological sex with their sense of self was central, for scientists the question of how to analytically distinguish them was pertinent in order to be able to examine their causal or functional relation. Cases of pseudohermaphroditism and intersex played a key role in these scientific discussions (Hausman 1995; Dreger 1998; Meyerowitz 2002; Redick 2004; Germon 2009; Rubin 2012; Mak 2012; Griffiths 2018a).

Hirschfeld and the Body

(...) as paradoxical as it might seem, there exist men with female and women with male genitals. (Hirschfeld, 1899, 4)

Hirschfeld analytically distinguished five levels of sexual differentiation:

- Sexual glands
- Genital apparatus
- Secondary sex characteristics appearing with puberty
- Mental or psychological differences
- Differences in the orientation of the sexual drives

During the human development from fetus to adult, sexual differences appeared chronologically in the order given above. The differences that developed first were thought to result in the least variation from the absolute man or woman; the later the development, the more variation. Accordingly, Hirschfeld postulated that the examiner would most likely find connections between secondary sex characteristics, psychological sex, and sexual orientation because these appeared simultaneously during puberty (Hirschfeld, 1899, 8–9; Sengoopta 1998, 452–256; Mak forthcoming).

Hirschfeld's encounter with several cases of a mistaken sex assignment (*erreur de sexe*) had a clear impact on his understanding of these relations – dismissing the warning of the contemporary expert on hermaphroditism, Franz Ludwig Neugebauer, that these were not that simple (Sengoopta 1998, 453). He literally imagined future discoveries of "internal secretions" (*männliche Keimstoffe*) in a female transvestite even before Eugen Steinach published the outcomes of his experiments with ovarian grafts. By doing this, he could present transvestites as cases of intersex individuals legitimately demanding legal change of sex (Mak 1997, 321–325; Sengoopta 1998, 464–65; Herrn 2005, 87–93). Thus, while analytically separating the "sex of the self" from the anatomical, visible, and functional morphology of sex, he reconnected it to biological sex via the, at the time, imperceivable inner functioning of biological processes or neural structures. Against his own separation of homosexuality and transvestites, Hirschfeld strongly held to the assumption that homosexuality was connected to physical or mental effeminacy. Even if this was not always obvious to the untrained eye, he claimed this to be the case among the fifteen hundred homosexuals he had seen – without further precise empirical or statistical evidence (Sengoopta 1998, 452).

From 1914 onwards Hirschfeld and Eugen Steinach started to collaborate in order to substantiate Hirschfeld's hunches experimentally. Steinach had, famously, successfully grafted ovaries in castrated male guinea pigs, and claimed to have "cured" a homosexual with a similar procedure (a "success" others were not able to repeat). In his attempts to find biological causes for "sexual intermediacy," Hirschfeld aimed to legitimize it as natural. However, Sengoopta has pointed to the fact that these biological explanations pathologized rather than naturalized homosexuality. What is more important here is that, in his assessment of a person's psychological makeup, Hirschfeld did not question the traditional bourgeois qualifications of "masculine" or "feminine" (Sengoopta 1998, 454–56; 469–73; Mak forthcoming).

Body and Self in Cases of Intersex

Some histories of intersex frame the body/psyche relation as an outright suppression by "objective" medical standards of "subjective" sexual feelings or gender identification, pointing to cases in which medical assessments of bodily sex prevailed over gender identity and in which heteronormativity prevailed over a patient's sexual orientation (Epstein 1995; Dreger 1998; Reis 2012). However, such empathy with the fragile subject position of pseudohermaphrodite patients vis-à-vis medical professionals takes the existence of "gender identity" itself for granted. Does a "sex of the self" appear in these intersex case histories, and if so, how?

For the period until about 1900, the answer is negative, not because there was no subjective will of the patient but because physicians could not yet clearly distinguish between the person's body and the person's self. After all, the body was known through declarations of the patient: their statements about menstruation, ejaculation, and sexual experiences were crucial to the assessment of (functioning) sexual glands or "gonads." However, from about 1900, the exponential rise of surgeries in general led to a quickly increasing amount of cases in which the inner "truth" of gonads was established without reference to the patient's subjective accounts about their bodies. All of a sudden, cases of patients whose gonads "contrasted" with the way they lived, looked, and loved appeared in the clinic. Doctors themselves found this extremely confusing. Therefore, they started to conceptualize the notion of a psychosexual self-awareness. With this, the first discussions about this psychosexual self-awareness also started to appear: how to establish this "sex of the self," what weight the voice of the patient should have in sex (re)assignments, what its relations to "the body" or to sexual drives were, and whether to acknowledge it at all (Mak 2005, 2012).

From 1900 onwards the question of the (causal) relation between physical sex and the "psychosexual" development of the person was central. Ludwig von Neugebauer's attempt to statistically understand the relation between psychosexual identification and the late discovery of an *erreur de sexe* on the basis of more than 1250 cases histories in 1908 demonstrates this early interest – but also the sheer impossibility of finding relations in such utterly different cases (Neugebauer 1908, 698–702; Mak 2012, 165–204). With respect to intersex treatments in the United

States, Alison Redick has designated the period after 1900 an "Era of Idiosyncrasy." In this period, the feelings of the person certainly did count, especially with regard to sexual orientation, but there was a deep fear of (legitimating) homosexuality (Dreger 1998; Reis 2012; Redick 2004, 26–91). However, with uncertainty about physical sex, it was impossible to define homo- or heterosexuality (Mak 2015). As long as it was not clear how gonads and hormones (or something biological not yet perceptible) defined a person's psychosexual development, the decision, for instance, to amputate the penis of an 18-year-old woman with undescended testicles who expressed her wish to do so was considered a risk. After all, the girl *might* start to develop sexual desires for women after her testicles had descended (Mak 2005). In other words, the problem was not always simply "objective medical science" against "subjective understanding": subjective feelings themselves were supposed to be able to change in relation to changes in the body.

Meanwhile, bodily sex became evermore "multiple" over time (Mol 2015). Gonads had already started to lose their status as absolute essence around 1900 but continued to be referred to. From Steinach's experiments onwards, hormones promised to be able to account for discrepancies between gonads and outer sexual appearance or sexual orientation (Redick 2004, 26–158). However, as science historian Nelly Oudshoorn has shown, it soon became clear that a strict dual understanding of the presence and function of "sex hormones" was untenable. Yet, the concept of "sex hormones" did something else as well: "Before the introduction of the concept of sex hormones, the categories male/masculine and female/feminine could only be conceptualized as two opposite pairs. Since the 1930s, scientist have considered the concepts of masculinity and femininity as separable from the categories of male and female," an insight which extended to psychological and anthropological research (Oudshoorn 1990, 184–85; 2005; see also Terry 2010, 159–77).

However, in the historical literature, no special diagnostic tools are mentioned to explore an intersex patient's psychological gender, nor any method to distinguish it from sexual orientation until the 1940s. The patient's own and/or their parents' accounts and the impression of the physician were the main basis for this aspect of the assignment (Redick 2004; Reis 2012; Hausman 1995; Germon 2009). Through this, issues of comportment, the impression a person made, the work they were able to do for a living, and their own identification as male or female could also be taken into account. In other words, in the important question of the possible relation between multiple aspects of physical sex and a person's psychosexual makeup, the latter was only referred to in terms of "masculine" or "feminine," and possibly "homosexual" or "heterosexual." Neither in the quickly developing field of endocrinology, nor in case histories of intersex, can any further refinement of these psychosexual classifications be found; a biologically or medically complex establishment of physical facts is thus related to a rather blunt m/f psychological classification.

From the 1940s onwards, psychologists and psychoanalysts were increasingly involved in cases of intersex, as well as different forms of psychological testing. In the United Kingdom in the 1930s and 1940s, the involvement of psychoanalyst Clifford Allen in the work of endocrinologist Lennox Broster only reiterated the idea that "atypical biology could lead to atypical psychology; if adrenal glands were thought to masculinize the biological, then they would be expected to influence sex roles, aims and object choice as well" (Griffiths 2018a, 479). Pediatric urologist David Williams used surgery as a psychological therapy, believing that enabling standing urination for boys before school age was "obviously important" for healthy male development (Griffiths 2018a, 486). What either "atypical psychology" consisted of or what "healthy male development" looked like was not further defined, or measured.

In the United States, during the 1940s and 1950s, "psychosexual role" became leading in questions of sex assignment (Hausman 1995; Redick 2004, 158–223; Germon 2009, 33–35). In 1932 the gonadal standard had already been rejected by researchers of the National Research Council; Emil Novak elaborated a theory – not unlike Hirschfeld's – on biological and psychological sex as a continuum (Redick 2004, 57–58; Novak 1935). The early 1940s showed "a widespread disagreement over a number of factors, including the aetiology of intersex, the relevance of the gonads to diagnosis and rearing, questions of disclosure, the relationship of libido to true sex, and ultimately the social consequences of uncertainty itself" (Redick 2004, 70). This opened the way for a stronger psychological perspective. Germon concludes that, as a result of this situation, clinicians began to rely more and more on psychosocial factors as the primary grounds for (re)assigning a sex, particularly in older subjects (Germon 2009, 34). For the physician involved in most pre-war intersex cases, the urologist Hugh Young, one case of dramatic misassessment leading to suicide probably led to his shift towards psychology (Redick 2004, 70–74; 149–53). From this moment onwards, the Terman-Miles M-F test is mentioned as being used in several cases (Redick 2004, 158–223).

In a key publication on the basis of a large collection of intersex case material, Albert Ellis demonstrated that biological factors were not decisive determinants of masculinity or femininity but the person's sex of rearing (Ellis 1945). In this publication, masculinity and femininity were primarily defined by the person's sexual orientation, and thus in an implicitly heterosexual manner (Redick 2004, 80–84; Germon 2009, 34). Whatever the (many) flaws of this study, it formed the basis for John Money's decisive proposal for a radical shift in intersex management. This shift also related to the decreasing average age of intersex patients and the new possibility for an easy test of chromosomal sex (Griffiths 2018b). Gender identity/role (G-I/R), as he called it, should not be the basis for sex assignment of an adult, but a successful gender identification should become the *aim* of the treatment of intersex children; the so-called "optimum gender of rearing" model. Thereto, both rearing and outer morphology had to be unambiguously steered towards one or the other gender. He further proposed the theory that gender identity was ingrained in a person's identity in the same way native language was – acquired during the first 18 months of a child's life – and something one could never unlearn. This was what he called gender. In fact, the (chromosomally sexed) body now had to be adapted to the gender for which the best options in terms of genital morphology, heterosexual drives, and (reproduction-like) sexual functioning were possible

(Kessler 1998; Fausto-Sterling 2000; Karkazis 2008). A more complex analysis of medical practices and discussions during the coming into being of these protocols for the treatment of intersex or DSD (disorders of sex development) patients, as well as their dissemination in Europe, can be found in the work of Sandra Eder (Eder 2010, 2018).

In the twenty-first century, Money has become infamous as the subject of heated debate after the suicide of one of his patients, John/Joan, or David Reimer (as was his true name) – which he claimed was the successful reassignment of a boy who lost his penis and was raised as a girl following Money's advice. This discussion has often been framed in terms of nature/nurture: whether a person's gender identity is based on their biology or on the way they have been raised (Colapinto 2000; Butler 2001). Others intervened in this discussion by showing that Money precisely tried to resist such a simple opposition (Hausman 2000; Germon 2009; Jordan-Young 2011).

In his theory, Money had not detached gender from biological sex – on the contrary, he was very much opposed to a Cartesian body/psyche binary, proposing a much more complex interaction between them; therefore, in his view, intersex children did not only need one clear gender of rearing but also an unambiguous genital morphology (Redick 2004, 158–222; Germon 2009, 23–62). Yet, both within emerging clinical help for transsexuals and within feminism, the concept of gender became increasingly detached from sexual morphology altogether. With the concept of "core gender identity," Robert Stoller proposed a clear distinction between sex and gender (Germon 2009, 63–83). However, despite this resolute split between gender and morphological sex, "biology" as a legitimation entered again via the backdoor. In his defense of gender-confirmative surgery against attacks from psychologists and psychiatrists who considered transsexualism a psychological pathology to be cured by therapy, Stoller's position is strikingly similar to Hirschfeld's in referring to possible as yet "unknown" internal biological causes to explain transsexualism:

> Stoller's invocation of biology slipped all over (and through) the body: from hormones to chromosomes, to the central nervous system, neurons, and back. While readily admitting that he had no idea of what form such forces might take, Stoller steadfastly refused to relinquish a role for them in gender. (Germon 2009, 76)

No wonder Stoller worked with endocrinologist Harry Benjamin, who on several occasions had met Hirschfeld and corresponded with him (Herrn 2014, 219–20). From Meyerowitz' descriptions of Benjamin's theories it becomes clear that they were almost literally copied from Hirschfeld:

> "It is well known," he wrote, "that sex is never one hundred per cent 'male' or 'female.' It is a blend of a complex variety of male-female components." The "more or less pronounced irregularities in genetic and endocrine development" resulted in "'intersexes' of varying character, degree and intensity," including not only hermaphrodites and pseudo-hermaphrodites but also homosexuals, transvestites, and transsexuals. His ideas were based on the assumption that 'no one is 100% male or female' – or the theory of human bisexuality. Moreover, he considered transsexuals the 'most extreme' cases of transvestism;

the latter could be 'psychogenetic', but in extreme cases it was constitutional 'perhaps due to a chromosomal sex disturbance.' (Meyerowitz 2002, 102)

This was only a slight adaptation of Hirschfeld's hunches to more recent biological discoveries. While Meyerowitz offers a detailed description of the heated discussions between psychoanalysts and Stoller's Gender Identity Project about their different insights into the *cause* of the phenomenon of transvestism, the *content* of what denoted "core gender identity" in such cases remains obscure (Meyerowitz 2002, 98–129).

Like we saw in Hirschfeld's work, the ambiguities between sexed body and self in Stoller's and Benjamin's work are threefold: first an affirmation of the possibility of psychologically not "conforming" to one's morphological sex; second, an assumption of some rather vague, interior "biological cause" for this phenomenon; and, finally, the confirmation that a change of morphological sex could possibly bring the sex of the self and the body into harmonious "accordance." With new endocrinological and surgical technologies available, Stoller and his colleagues paved the way for sex-affirmative treatments for transsexuals. Thus, a *creative* body/subject relation emerged. How the idea that surgery could be of psychological help in cases of intersex transformed into possibilities to form new subjectivities through transformations of the sexed body, and how this renewed ideas about the "sex of the self" is up to further historical research (Plemons and Straayer 2018).

As argued by Hausman, Meyerowitz, and Germon, Money's and Stoller's use of the concept of gender had far-reaching consequences for thinking about gender outside the clinic (Hausman 1995; Meyerowitz 2002; Germon 2009). The split between psychological gender and biological sex stood at the basis of the emerging field of feminist psychology in the 1960s. In her personal look back on these years and later developments, historian of psychology Jill G. Morawski points out in particular the pitfalls of this divide, as it left "the body" too much to the biosciences (Morawski 2005; see also Mol 2015). In later decades, cases of intersex continued to play a pivotal role in discussions on gender-sex relations (Liao 2005).

Intermediacy-Binaryness

Testing Typical Femininity and Masculinity

Most of the developments described above refer to sciences that make use of individual case histories. Hirschfeld was the first sexologist to develop a technique that strove towards more "objectivity" and quantification in this field: he developed a psychobiological questionnaire with 85 (1899) to 127 (1908) questions to be filled in by his patients and other visitors to his Institute for Sexual Science in Berlin (Müller 1990; Herrn 2014; Mak forthcoming). This attempt towards a more systematic and quantitative investigation of people "deviant" from ordinary sex or gender will here be related to psychological research into sex differences among "ordinary" people. The twentieth century saw a tendency in psychology to test, measure, and quantify aspects of "normal" human behavior and personality on a mass scale via psychometry (Schilling 2015). With regard to the sex of the self, this trend started with comparisons between men and women on many different psychological terrains (intelligence, behavior, interests) and developed into testing an individual's degree of masculinity and femininity on M-F scales.

At the time Hirschfeld published his questionnaire, the Dutch pioneer of experimental psychology, Gerard Heymans (1857–1930), published a book on the "psychology of women" – internationally the first study to be based on a large questionnaire (Heymans 1910; van Drenth 2018; Mak forthcoming). This questionnaire was similar to many of the tests measuring differences between men and women that were developed in the United States during the first decades of the twentieth century. The shift towards measuring femininity and masculinity as a core aspect of personality was developed by Lewis M. Terman (1877–1956) and Catherine Cox Miles (1890–1984) in 1936 (Terman 1936; Morawski 1985; Terry 2010, 168–75; Dean and Tate 2017). Terman and Miles's test used male and female homosexuals as "intermediates," who constituted a counterpoint for the M-F scores of "normal" men and women; the M-F test was therefore fundamentally heteronormative (Terry 2010, 168–75; Constantinople 2005, 391–92; Hegarty and Coyle 2005, 381–82). This, again, shows the difficulty of drawing a sharp line between the history of the sex of the self and of sexuality. The Terman-Miles test was used until the 1950s, and set the standard for later tests such as the MMPI and MMPI-2, the latter still being in use in, for example, cases of intersex (Ercan et al. 2013). Feminist scholars started to criticize the fundamental assumptions of these tests, starting with the criticism of Anne Constantinople in 1973, reprinted in a special feature of *Feminism and Psychology* in 2005 (Constantinople 2005; Hegarty and Coyle 2005). Constantinople's criticism was followed by a feminist-inspired new "androgyny" test developed by Sandra Bem in 1974, which, according to a later historical assessment of M-F tests by Jill Morawski, only shifted the normativity from an ideal of gender bipolarity to an androgynous ideal (Morawski 1985). Recently, however, Bem's feminist work has been reappraised as an important historical intervention (Dean and Tate 2017). This history of M-F psychometrics reveals the gender fluidity paradox, which will be discussed in the next section.

The Gender Fluidity Paradox

Hirschfeld famously proposed the "theory of sexual intermediaries." He considered a "complete man" (*Vollmann*) and a "complete woman" (*Vollweib*) to be no more than the fictional ends of a scale. In reality, everyone had both male and female characteristics, differing only gradually. Moreover, in theory all these characteristics existed independently of each other. In an almost hilarious section of *Die Transvestiten*, Hirschfeld calculated, with the help of Prof. K. F. Jordan, how many varieties of sexual intermediates were possible. He first divided the variations into four groups: (A) primary sexual characteristics; (B) secondary sexual characteristics; (C) sexuality, and (D) psychological sex. He then proposed four different elements within of each of these (see Fig. 1), and claimed that all could vary independently: "Any imaginable combination, all possible connections between all possible male and female characteristics do exist." This led him to calculate the amount of possible

Eigenschaftsgruppe A. (Primäre Geschlechtsmerkmale.)		Eigenschaftsgruppe B. (Sekundäre Geschlechtsmerkmale.)	
1. Keimstock:	A ^{I}	1. Haarkleid:	B ^{I}
2. Ei- oder Samenleiter:	A ^{II}	2. Kehlkopf:	B ^{II}
3. Geschlechtshöcker:	A ^{III}	3. Brust:	B ^{III}
4. Geschlechtsrinne:	A ^{IV}	4. Becken:	B ^{IV}
Eigenschaftsgruppe C. (Tertiäre Geschlechtsmerkmale.)		Eigenschaftsgruppe D. (Quartäre Geschlechtsmerkmale)	
1. Richtungsart:	C ^{I}	1. Gefühlsleben:	D ^{I}
2. Annäherungsart:	C ^{II}	2. Denktätigkeit:	D ^{II}
3. Gefühlsart:	C ^{III}	3. Beschäftigung:	D ^{III}
4. Betätigungsart:	C ^{IV}	4. Kleiduug:	D ^{IV}

Fig. 1 Hirschfeld (1910, 289)

variations, which amounted to $3^{16} = 43.046.721$ or "1/33 of the world population." This was, by the way, a low calculation, as four variations per group were a minimum (Hirschfeld 1910, 286–91, cit. 286).

A. Primary sex	B. Secondary sex
1. Gonads	1. Bodily hair
2. Ovarian tube/vas deferens	2. Larynx
3. Clitoris/penis	3. Breast
4. Vagina/scrotal raphe	4. Pelvis
C. Sexuality	D. Psychological sex
1. Orientation	1. Emotional life
2. Approach	2. Intellectual activity
3. Feeling	3. Occupation
4. Activity	4. Dress

At first glance the concept of sexual intermediates seems to offer a surprisingly innovative, fluid, and nonbinary concept of sex and gender. However, in actuality Hirschfeld did not consider these separate elements as independent from each other. For example, in his theories he kept connecting effeminacy and male homosexuality to the influence of "internal secretions" or hormones – long after publishing *Transvestites*. Moreover, theoretically his theory of sexual intermediaries applied to all people in the world and promised to de-pathologize all sexual varieties, but in practice "sexual intermediate" only referred to nonconforming genders or sexualities. To put it in present-day terms, he drew a clear line between queer and transgender versus cisgender. The difference between his theoretical and his clinical approach can also be noted in his diagnostic classifications of these "sexual intermediates," which reduced the number of more than 43 million variations to only 4 categories: true hermaphrodites (characteristics of both gonads), pseudo-hermaphrodites (some physical sexual characteristics incongruent with gonads), homosexuals (m/f), and transvestites (m/f). As we have seen, he also created further subclassification for transvestites. Such classifications had all kinds of real consequences for real people (Hacking 2006; Bowker and Star 2000). Detailed studies of Hirschfeld's case histories clearly show the tensions and troubles between the stories of the people who consulted him and the way he interpreted and classified these (Mak 1998; Hill 2005; Herrn 2005, 2014).

Finally and most importantly, the theory of "sexual intermediates" was based on a firm sexual binary (Hirschauer 1993, 82–85; Mak forthcoming). Every single characteristic for "measuring" someone's intermediary position was based on a firm division into "masculine" or "feminine," which can easily be seen from the way he calculated the amount of intermediates (see Fig. 2). As Sengoopta commented:

> The coexistence of masculinity and femininity in the body and mind of the homosexual did not involve any redefinition of those qualities. Even though sexual intermediacy was

A_m B_m C_m D_w	A_w B_m C_m D_w	A_{m+w} B_m C_m D_w
A_m B_w C_m D_w	A_w B_w C_m D_w	A_{m+w} B_w C_m D_w
A_m B_{m+w} C_m D_w	A_w B_{m+w} C_m D_w	A_{m+w} B_{m+w} C_m D_w
A_m B_m C_w D_w	A_w B_m C_w D_w	A_{m+w} B_m C_w D_w
A_m B_w C_w D_w	A_w B_w C_w $D_w{}^{3)}$	A_{m+w} B_w C_w D_w
A_m B_{m+w} C_w D_w	A_w B_{m+w} C_w D_w	A_{m+w} B_{mw} C_w D_w
A_m B_m C_{m+w} D_w	A_w B_m C_{m+w} D_w	A_{m+w} B_m C_{m+w} D_w
A_m B_w C_{m+w} D_w	A_w B_w C_{m+w} D_w	A_{m+w} B_w C_{m+w} D_w
A_m B_{m+w} C_{m+w} D_w	A_w B_{m+w} C_{m+w} D_w	A_{m+w} B_{m+w} C_{m+w} D_w

Tabelle II.

------

[1]) Die nachfolgende Berechnung der Zahl der Zwischenstufentypen habe ich in Gemeinschaft mit Prof. Dr. K. F. Jordan ausgeführt.
[2]) Vollmann.   [3]) Vollweib.

Fig. 2 Hirschfeld Die Transvestiten, 287. In this calculation table, each separate element is labeled "m" masculine, "w" feminine, or "m+w" masculine and feminine 'universal,' male and female remained distinct, separate, complementary qualities, endowed with all their traditional attributes. It was the distribution of maleness and femaleness in individuals that Hirschfeld tried to redefine: the qualities themselves he left strictly alone and, indeed helped reinforce. (Sengoopta 1998, 470)

At a time during which the women's movement fought against those traditional definitions, this was a sociopolitical position which also had concrete consequences for (feminist) nonconforming women (Mak 1997, 1998; Leng 2018).

This problem of a fluid conceptualization of the sex or gender of a person in terms of a rigid dichotomy of gender or sex characteristics is what is labeled "the paradox of gender fluidity" in this chapter. This problem is not of Hirschfeld's making alone. In *An American Obsession*, Jennifer Terry analyzed how Hirschfeld's (as well as Havelock Ellis' and Freud's) work inspired similar "fluid" conceptions of gender in various disciplines in the human sciences in the United States during the 1930s, then referred to as "theories of human bisexuality": endocrinology, anthropology, and psychometrics (Terry 2010, 161, 175–77). Biologists working on "sex hormones" had begun to describe masculinity and femininity as non-dichotomous; rather than being two distinct poles, they overlapped, cooperated, or were gradual in difference (Oudshoorn 1990, 179–85). Emil Novak, too, described an intermediary area between "normal" and "intersex" in his work on intersex, postulating that "there are few 100 percenters among human beings" (...) "there being a bit of the feminine in all men and a corresponding tinge of the masculine in all women" (Redick 2004, 59–60).

Thus, these scientists were referring to a binary in terms of masculine or feminine characteristics. For instance, Novak referred to "masculinity in women—'the 'virago' type—and effeminacy in men—'the 'pansy' type'" (Redick 2004, 59). Oudshoorn summarizes: "With the new concept of relative sexual specificity, endocrinologists constructed a biological foundation for a definition of sex in which an individual could be classified in many categories varying from 'a virile to an effeminate male' or from 'a masculine to a feminine female,' as gynaecologist Robert Frank described it" (Oudshoorn 1990, 184). As Terry concludes about these sex continuum or "human bisexuality" models:

> With the advent of sex continuum models, differences between the sexes, even as they were reinscribed along fluid spectra, remained mapped in a dichotomous arrangement that suited the general norm of male dominance and female subordination upon which heterosexual marriage and reproductive families were based. (Terry 2010, 176)

In the 1950s and 1960s, the "theory of human bisexuality," which was at the time also popularized through media attention, resurfaced in the United States via Benjamin and Stoller (Meyerowitz 2002, 99–111). Sandra Bem's androgyny scale, developed in the 1970s, can also be seen as an extension of the idea of gender fluidity. The "Bem Sex-Role Inventory" (BSRI) tested the extent to which a person combined both masculine and feminine traits – a balance which was considered healthy at the time. It is interesting to note that it soon was criticized because "the newer models retain, even if unintentionally, certain values associated with masculine and feminine, and thus contribute to their ossification as universals" (Morawski

1985, 213; Dean and Tate 2017). With the emergence of trans studies, new versions of gender fluidity as well as criticisms of cisgenderism are now entering the field of psychology (Hegarty 2018).

Psychometrics

Whereas Hirschfeld's theory of "sexual intermediaries" mainly concerned non-conforming cases, the paradox of gender fluidity also permeated the emerging psychometric technologies. The first major European empirical-statistical inquiry into sex differences in personality was carried out by the Dutch experimental psychologist Gerard Heymans. Heymans was an internationally renowned pioneer of experimental psychology at the University of Groningen in the Netherlands and a member of the women's suffrage movement. At the basis of his 1910 publication, *Die Psychologie der Frauen* (*The Psychology of Women*), which appeared in German, Dutch, and French, was a questionnaire developed to inquire into the heredity of psychological characteristics in general (Heymans 1910; Van Strien 1993; van Drenth 2018). Rejecting all kinds of more subjective or less systematic research, the questionnaire was sent out to general practitioners who filled in 90 questions for all members of one family they knew well. Heymans' over 2500 objects of research were unproblematically divided into a binary of women and men, but in the results the measured characteristics appeared as gradually more or less male or female. Therefore, paradoxically, the design of his questionnaire demonstrated the fluidity of each measured gender characteristic: these were never purely male or female (Mak forthcoming).

The accompanying interpretative text does not systematically analyze the data. Heymans used the outcomes in a descriptive analysis of what characterized "average" femininity. Heymans did not use statistical techniques to discriminate between characteristics which were statistically relevant and those which were not. However, clearly overlapping characteristics were simply left aside in the interpretative text (Heymans 1910; van Drenth 2018). Such operations which exclude sameness between men and women strengthened the binary (Fig. 3).

In her insightful overview of psychological research on men and women in the United States, Jill Morawski describes how psychological research aiming at establishing sex differences in intelligence, skills, attitudes, or mind failed to be conclusive. These studies, based on a design similar to Heymans' questionnaire, failed to ascertain an essence of gender (Morawski 1985, 200). A major shift in psychological research with regard to gender took place in 1936, when Lewis Terman and Catherine Cox Miles published their study Sex and Personality. Instead of measuring characteristics of men and women, they designed a test to establish degrees of "femininity" and "masculinity." In contrast to previous research into the differences between men and women, this test was based on theories of the fluidity of the sex of the self (Morawski 1985, 204; Terry 2010, 168–75). In their opinion, masculinity and femininity should "not be thought of as lending to [personhood] merely a superficial coloring and flavor; rather they are one of a small number of cores around which the structure of personality gradually takes shape" (Terman 1936, 451). This can be seen as a prelude to later definitions of gender as a core of personhood.

> Männern eintreten. Dementsprechend haben dann auch die Berichterstatter unserer Hereditätsenquete Veranlassung gefunden, 59.8% der von ihnen beschriebenen Frauen und bloß 45.9% der von ihnen beschriebenen Männer als „emotionell", dagegen von jenen bloß 26.5% und von diesen 39.3% als „nicht emotionell" (Fr. 9) zu bezeichnen. Und in den von weiblicher Hand herrührenden Berichten hat, indem jene Prozentsätze sich auf 70.9, 48.5, 20.3 und 39.7 erhöhen bzw. erniedrigen, der betreffende Unterschied einen noch bedeutend entschiedeneren Ausdruck gefunden.

Fig. 3 Example of how Heymans presented his data in a narrative. (Heymans 1910, 67)

Terman and Miles undertook an extensive project to set up their M-F scale test, known as AIST, published in 1936. Their measures were based on what they thought to be a representative population in the United States in the 1930s (Terman 1936, v–vi; Lewin 1984). They first selected test items from seven different subtests which they proved to be statistically relevant in discriminating between women and men. Subsequently, the selected 910 items were used to measure masculinity and femininity in new groups or individuals. In its final form, the test contained seven subtests: word association, inkblot association, information, interests, introversion, emotional and ethical attitudes, and opinions.

The AIST test became the model for M-F testing until the 1960s. As discussed above, all tests shared the idea that masculinity and femininity existed on a level that could not be simply observed (Morawski 1985, 206). All M-F tests based on Terman and Miles tried to prevent contamination with cultural bias or a subject's faking. From the 1930s onwards, grounded in depth psychology, femininity and masculinity were increasingly located beyond the awareness of the person. Several tests aimed at identifying "unconscious" masculinity or femininity, such as the Draw-A-Person or word association tests. Yet all fit equally well in the traditional Victorian gender scheme (Lewin 1984, 169–70; Morawski 1985, 208–12). Moreover, as Terry demonstrated, effeminate homosexual males functioned as the abnormal "other" against which the standard of "normal" masculinity and femininity was set (Terry 2010, 171–75; Hegarty 2007).

In her razor sharp analysis of the history of M-F testing, Constantinople showed how bipolarity (or what we now often refer to as the gender binary) was both assumed and reified. She wrote:

> In M–F test construction, the assumption of bipolarity is evident in at least three ways: (a) the dependence on biological sex alone as the appropriate criterion for an item's M–F relevance, since item selection is usually based solely on its ability to discriminate the responses of the two sexes; (b) the implication that the opposite of a masculine response is necessarily indicative of femininity, especially in tests where only two options are provided; and

(c) the use of a single M–F score which is based on the algebraic summation of M and F responses and places the individual somewhere on a single bipolar dimension. (Constantinople 2005, 390)

For the Terman-Miles test, Constantinople described how M-F was based on items discriminating between male and female responses, which was further simplified by not taking the *degree* of M-F discrimination into account, but simple unit scoring (M or F). Furthermore, the M = + and F = − calculations of the scores – indicated the absolute polarity of M and F. As Terman and Miles themselves also acknowledged, the test systematically exaggerated differences between the sexes because all items which showed psychological overlap between females and males were erased (Terman 1936, 6, 18–19; Constantinople 2005, 390–91). In the Terman-Miles test, moreover, association tests (words, inkblots) were adapted so that they could be more easily scored. Instead of free association, a multiple-choice question consisting of four words, two feminine, two masculine, was offered (see Fig. 2) (Terman 1936, 23–24). These operations, mostly legitimated as more convenient test practices, thus sharpened the binary: *even* when the measures were originally gradual, they were counted as absolute. This is the paradox of fluidity in a different form.

EXERCISE 2.—Continued

			Omission					Omission
5.		brush + centipede + comb − teeth −	0	9.		fence + letter "E" 0 ship − tree		+ 0 0
6.		ax + boat + chopper − moon +	0	10.		babies 0 cloud − dancers − lovers +		0
7.		dish − ring 0 target + tire 0	0	11.		Indian 0 man hanged + scarecrow − tassel −		0
8.		bird house 0 flagpole + sword − torch +	+	12.		chimney − coil + smoke − thread −		0

Unidimensionality

Another sharp point of critique formulated by Constantinople was the assumed "unidimensionality" of the measures for F or M. This problem can already be traced in Hirschfeld and Heymans, and returns in the more scientifically sophisticated psychometrical tests from Terman and Miles onwards. In all these tests, an overwhelming amount of very *different* items serve as signifier for a *singular* F or M. Hirschfeld inquired into, for example, a weak or strong will, fearfulness, or courage (Q74); an analytic or rather a receptive mind (Q58); an inclination to cooking, cleaning, or hairdressing or rather to sports, hunting, and fighting (Q62); one's way of walking (Q28); whether one could whistle (Q31); and one's gaze. Heymans asked questions like the following: liking to go out or preferring to stay at home (Q71); an inclination to talk about people, issues, or oneself (Q72); being occupied with philanthropic work, only offering philanthropic gifts, or neither (Q56); practical (in solving a problem) or unpractical (Q29)?

As to Terman and Miles, Constantinople observed that, despite the fact that there was only a very low correlation between the seven different subtests, the scores of these tests were added up. In other words, "(...) we find evidence for the multi-dimensionality of the trait [M or F, gm] while unidimensionality is implicit in the scoring procedure." Or, to put it somewhat differently, the "self" – and the sex of self – was (and is) extraordinary multiple. In F-M questionnaires and testing, there appears to be an extreme elasticity in moving back and forth between a sheer endless multitude of things associated with psychological masculinity and femininity and the singular signifier "masculine" or "feminine" as a one-dimensional degree. It is ironic to bring to mind here, again, why physicians involved in intersex cases from the 1940s, in particular Money cum *suis*, thought that gender identity/role (G-I/R) should become the predominant guideline for intersex sex assignment and treatment: they considered physical sex to be too confusing....

Money persisted in his assumption that (inner) identity and (outer) role were intrinsically connected: "(...) gender identity is the private experience of gender role and gender role is the public expression of gender identity" (Germon 2009, 66; Redick 2004, 164–71). In her analysis of the heated discussion about John Money and the David Reimer case (see above), Georgia Warnke observes that in the initial discussion, the concept of "gender identity" itself was not so much questioned; both parties used the same criteria for what counted as a "boy" or "girl" (Warnke 2007, 15–48). Following up on her analysis, one might add, based on the perspective taken in this chapter: how did Money and his colleagues Joan and John Hampson *know* an individual's (prospective) gender role and identity? After all, they needed a test to measure the result of their decision. According to Germon, they rejected the Terman-Miles test because they considered the psychological or "mental" aspect it measured to be too restricted. Instead, Money and the Hampsons used indicators which rested much more – again – on the impressions of the assessor: "[G]eneral mannerisms, deportment and demeanor, play preferences and recreational interests, spontaneous topics of talk in unprompted conversation and casual comment; content of dreams, daydreams and fantasies; replies to oblique inquiries and projective tests; evidence of erotic practices; and the individual's own replies to direct inquiry, heterosexual drives" (Germon 2009, 52–53). Just as Money refused to disconnect biological sex from gender entirely, he did not conceptualize gender identity as something that could be understood in isolation from its (historically and culturally contingent) expression. To a certain extent, he thus kept the door open to all the complexities and varieties in which gender identity is expressed, lived, and embodied in ordinary life and in the outer world. Yet, at the same time, with G-I/R he unified a set of complex indicators for gender and sexual identity into one F-M classification. This involved an enormous reduction of the complexity of psychological masculinity and femininity.

With "core gender identity," Robert Stoller further detached identity from Money's combined concept, distilling it into a stable and purified essence of the "I," which is completely located inside the subject's interior (Germon 2009, 63–83). The reduction of a masculine or feminine self to a sheer "I," to a gendering of the inner center from which the subject acts and speaks, is echoed in Sahar Sadjadi's article on contemporary practices concerning gender-nonconforming children in a hospital in the US. Sadjadi shows how in this context the nature/culture opposition has been replaced by an internal/external opposition: gender is not so much opposed to biological sex, but conceived as an inner, true, authentic self that is contrasted with an "outside" sociocultural (normative) notion of gender. This "true gender self" is somewhat vaguely situated in a dematerialized concept of "the brain" (Sadjadi 2019). In this way, the concept of a "true gender self" itself escapes historicizing and cannot be analyzed as culturally specific – such as modern, Western, and white. Or, to put it slightly differently, it is isolated from gender as a contested social, cultural, and political domain in which power is at stake.

Conclusion

This overview and analysis of the historiography of scientific inquiries into the sex of the self between the turn of the twentieth century and the 1960s has left three ambiguous issues unresolved. It showed that there was no clear line between subjective self-understanding and expert knowledge but a constant looping between subjective narratives, (new) diagnostic classifications, public discourse, and (new) identifications. The second unsolved ambiguity pertains to the body/psyche distinction. The sexed body appears multiple throughout this period in medical, scientific, and clinical cases of intersex and transvestism or transsexualism – as gonads, as outer or inner sexual morphology, as hormones, as chromosomes, as hair on one's face. The sex of the self could at any one point be decisively seen or treated as detached from one such version of the body, at the same time as it was (re)connected to another version of it. The history of this relation shows how this relation differs from case to case and from period to period, and is never clearly defined or explained. Finally, conceptions of psychological sex have been radically fluid and extraordinary rich in content and variety, at the same time as they were based in rigid binarism and complete reduction. This third ambiguity in particular demonstrates the deep underlying problem of separating the sex of the self from societal gendered structures and discourses. None of these ambiguities with regard to the "sex of the self" have been definitively resolved during the history of the coming into being of a concept of a "true sex of the self" in Western culture and society. Yet, it has gained increasing credibility and legitimacy and functions on many levels in current practices and policies.

References

Bauer H (2006) Women and cross-dressing 1800–1939. Routledge, New York
Bauer H (2009) Theorizing female inversion: sexology, discipline, and gender at the fin de siècle. J Hist Sex 18:84–102
Bauer H (2017) The Hirschfeld archives: violence, death, and modern queer culture. Temple University Press, Philadelphia/Rome/Tokyo
Bonnet M-J (1981) Un choix sans équivoque: recherches historiques sur les relations amoureuses entre les femmes, XVIe-XXe siècle. Denoël, Paris
Bowker GC, Star S (2000) Invisible mediators of action: classification and the ubiquity of standards. Mind Cult Act 7:147–163
Butler J (2001) Doing justice to someone: sex reassignment and allegories of transsexuality. GLQ 7: 621–636. https://doi.org/10.1215/10642684-7-4-621
Smith-Rosenberg C (1985) Disorderly conduct : visions of gender in Victorian America. AA Knopf, New York
Chauncey G (1982) From sexual inversion to homosexuality: medicine and the changing conceptualization of female deviance. Salmagundi:114–146
Colapinto J (2000) As nature made him: the boy who was raised as a girl. Toronto Star, Toronto
Constantinople A (2005) Masculinity-femininity: an exception to a famous dictum? *fem. Psychol 15:385–407. https://doi.org/10.1177/0959-353505057611
Crozier I (2008) Pillow talk: credibility, trust and the sexological case history. Hist Sci 46:375–404. https://doi.org/10.1177/007327530804600401
Daston L, Galison P (2007) Objectivity, Zone Books. MIT Press, New York/Cambridge, MA
Dean ML, Tate CC (2017) Extending the legacy of Sandra Bem: psychological androgyny as a touchstone conceptual advance for the study of gender in psychological science. Sex Roles 76: 643–654. https://doi.org/10.1007/s11199-016-0713-z
Dreger AD (1998) Hermaphrodites and the medical invention of sex. Harvard University Press, Cambridge; London
Duden B (1991) The woman beneath the skin: a doctor's patients in eighteenth-century Germany. Harvard University Press, Cambridge, Mass
Eder S (2010) The volatility of sex: intersexuality, gender and clinical practice in the 1950s. Gend Hist 22:692–707. https://doi.org/10.1111/j.1468-0424.2010.01615
Eder S (2018) Gender and cortisone: clinical practice and transatlantic exchange in the medical management of intersex in the 1950s. Bull Hist Med 92:604–633. https://doi.org/10.1353/bhm.2018.0073
Elbe L, Hoyer N (1933) Man into woman An authentic record fo a change of sex: the true story of the miraculous transformation of the Danisch painter Einar Wegener (Andreas Sparre). Dutton, New York
Ellis H (1928) Studies in the psychology of sex, vol VII. F. A Davis, Philadelphia
Ellis A (1945) Reviews, Abstracts, Notes, and Correspondence: The Sexual Psychology of Human Hermaphrodites. Psychosom Med 7 (2): 108–25
Ercan O, Kutlug S, Uysal O, Alikasifoglu M, Inceoglu D (2013) Gender identity and gender role in dsd patients raised as females: a preliminary outcome study. Front Endocrinol Lausanne 4:86–86. https://doi.org/10.3389/fendo.2013.00086
Fausto-Sterling A (2000) Sexing the body: gender politics and the construction of sexuality. Basic Books, New York, NY
Feinberg L (1996) Transgender warriors: making history from Joan of arc to RuPaul. Beacon Press, Boston
Fine C (2011) Delusions of gender: how our minds, society, and neurosexism create difference. Icon Books, London
Foucault M (1979) The history of sexuality, The will to knowledge, vol 1. Allen Lane, London
Foucault M (1980) Herculine Barbin: being the recently discovered memoirs of a nineteenth-century French hermaphrodite. Vintage Books, New York

Germon J (2009) Gender: a genealogy of an idea. Palgrave Macmillan, New York, NY
Griffiths DA (2018a) Diagnosing sex: intersex surgery and 'sex change' in Britain 1930–1955. Sexualities 21:476–495. https://doi.org/10.1177/1363460717740339
Griffiths DA (2018b) Shifting syndromes: sex chromosome variations and intersex classifications. Soc Stud Sci 48:125–148. https://doi.org/10.1177/0306312718757081
Hacker H (1987) Frauen und Freundinnen Studien zur "weiblichen Homosexualität" am Beispiel Österreich 1870–1938. Weinheim, Beltz
Hacker H (2015) Frauen* und Freund innen: Lesarten weiblicher Homosexualität, Österreich 1870–1938. Zaglossus, Wien
Hacking I (2002) Historical ontology. Harvard University Press, Cambridge, Mass
Hacking I (2006) Making up people. Lond Rev Books 28:23–26
Halberstam J (1998) Female masculinity. Duke University Press, Durham
Hausman BL (1995) Changing sex: transsexualism, technology, and the idea of gender. Duke University Press, Durham
Hausman BL (2000) Do boys have to be boys? Gender, narrativity, and the John/Joan case. NWSA J 12:114–138. https://doi.org/10.2979/NWS.2000.12.3.114
Hegarty P (2007) From genius inverts to gendered intelligence: Lewis Terman and the power of the norm. Hist Psychol 10:132–155. https://doi.org/10.1037/1093-4510.10.2.132
Hegarty P (2018) A recent history of lesbian and gay psychology: from homophobia to LGBT. Routledge, London
Hegarty P, Coyle A (2005) An undervalued part of the psychology of gender canon? Reappraising Anne Constantinople's (1973) "masculinity–femininity: an exception to a famous dictum?". Fem Psychol 15:379–383. https://doi.org/10.1177/0959-353505057610
Hekma G (1994) "A female soul in a male body": sexual inversion as gender inversion in nineteenth century sexology. In: Herdt G (ed) Third sex, third gender: beyond sexual dimorphism in culture and history. Zone Books, New York, pp 213–239
Herrn R (2005) Schnittmuster des Geschlechts: Transvestitismus und Transsexualität in der frühen Sexualwissenschaft. Psychosozial-Verl, Gießen
Herrn R (2014) Die falsche Hofdame vor Gericht: Transvestitismus in Psychiatrie und Sexualwissenschaft oder die Regulierung der öffentlichen Kleiderordnung/the fake lady on trial: transvestitism in psychiatry and the sexual sciences, or the regulation of public dress-code. Med J 49: 199–236
Heymans G (1910) Die Psychologie der Frauen. Winter, Heidelberg
Hill DB (2005) Sexuality and gender in Hirschfeld's die Transvestiten: a case of the "Elusive Evidence of the Ordinary". J Hist Sex 14:316–332. https://doi.org/10.1353/sex.2006.0023
Hirschauer S (1993) Die soziale Konstruktion der Transsexualität: über die Medizin und den Geschlechtswechsel. Suhrkamp, Frankfurt am Main
Hirschfeld M (1899) Die objektive Diagnose der Homoseksualität. Jahrb Für Sex Zwischenstufen Unter Bes Berücksichtigung Homosex 1:4–35
Hirschfeld M (1909) Psychoanalytischer Fragebogen, Berlin
Hirschfeld M (1910) Die Transvestiten: ein Untersuchung über den erotischen Verkleidungstrieb : mit umfangreichem casuistischen und historischen Material. Alfred Pulvermacher & Co., Berlin
Hirschfeld M (1915) Psychobiologischer Fragebogen, Berlin
Hirschfeld M (1921) Psychobiologischer Fragebogen. Institut für Sexualwissenschaften, Berlin
Hirschfeld M (1930) Psychobiologischer Fragebogen. Institut für Sexualwissenschaft, Berlin
Honegger C (1991) Die Ordnung der Geschlechter. Campus-Verl, Frankfurt aM
Jordan-Young RM (2011) Brain storm: the flaws in the science of sex differences. Harvard University Press, Cambridge, Mass
Epstein J (1995) Altered conditions: disease, medicine, and storytelling. Routledge, New York
Karkazis KA (2008) Fixing sex: intersex, medical authority, and lived experience. Duke University Press, Durham
Kennedy HC (2005) Karl Heinrichs Ulrichs: pioneer of the modern gay movement. Peremptory Publications, Concord CA

Kessler SJ (1998) Lessons from the intersexed. Rutgers University Press, New Brunswick, N.J

Laqueur TW (1990) Making sex: body and gender from the Greeks to Freud. Harvard University Press, Cambridge, MA

Leng K (2018) Sexual politics and feminist science: women sexologists in Germany. Cornell University Press, Ithaca, 1900–1933

Lewin M (1984) In the shadow of the past: psychology portrays the sexes: a social and intellectual history. Columbia University Press, New York

Liao LM (2005) III. Reflections on 'masculinity-femininity' based on psychological research and practice in intersex. Fem Psychol 15:424–430. https://doi.org/10.1177/0959-353505057614

Mak G (1997) Mannelijke vrouwen: over grenzen van sekse in de negentiende eeuw. Boom, Amsterdam

Mak G (1998) "Passing women" im Sprechzimmer von Magnus Hirschfeld: warum der Begriff "Transvestit" nicht für Frauen in männerkleidern Eingeführt wurde. ÖZG 9:384–399. https://doi.org/10.25365/oezg-1998-9-3-5

Mak G (2004) Sandor/Sarolta Vay: from passing woman to sexual invert. J Womens Hist 16:54–77. https://doi.org/10.1353/jowh.2004.0030

Mak G (2005) So we must go behind even what the microscope can reveal'. The Hermaphrodite's "self" in medical discourse at the start of the twentieth century. GLQ J Lesbian Gay Stud 11:65–94. https://doi.org/10.1215/10642684-11-1-65

Mak G (2012) Doubting sex: inscriptions, bodies and selves in nineteenth-century hermaphrodite case histories. Manchester University Press, Manchester

Mak G (2015) Conflicting heterosexualities: hermaphroditism and the emergence of surgery around 1900. J Hist Sex 24:402–427. https://doi.org/10.7560/JHS24303

Mak G (forthcoming) Hirschfeld's 1899 psychobiological questionnaire. The paradoxes of de-narrativizing sexual and gender nonconformity. Eur Hist Q

Manion J (2020) Female husbands: a trans history. Cambridge University Press, Cambridge/New York

Meyerowitz JJ (2002) How sex changed a history of transsexuality in the United States. Harvard University Press, Cambridge, Mass

Mol A (2015) Who knows what a woman is... On the differences and the relations between the sciences. MAT 2:57–75. https://doi.org/10.17157/mat.2.1.215

Morawski JG (1985) The measurement of masculinity and femininity: engendering categorical realities. JOPY J Personal 53:196–223

Morawski JG (2005) I. Moving gender, positivism and feminist possibilities. Fem Psychol 15:408–414. https://doi.org/10.1177/0959-353505057612

Müller K (1990) Aber in meinem Herzen Sprach eine Stimme so laut: Homosexuelle Autobiographien und medizinische Pathographien im 19. Jahrhundert, Verlag Rosa Winkel, Berlin

Neugebauer FL v (1908) Hermaphroditismus beim Menschen. Klinkhardt, Leipzig

Novak E (1935) Sex determination, sex differentiation and intersexuality: with report of unusual case. J Am Med Assoc 105:413–420

Oosterhuis H (2000) Stepchildren of nature. Krafft-Ebing, psychiatry, and the making of sexual identity. University of Chicago Press, Chicago

Oudshoorn N (1990) Endocrinologists and the conceptualization of sex 1920–1940. J Hist Biol 23:163–186

Oudshoorn N (2005) Beyond the natural body: an archeology of sex hormones. Routledge, London

Park K (2006) Secrets of women: gender, generation, and the origins of human dissection. Zone Books, MIT Press, New York/Cambridge, MA

Plemons E, Straayer C (2018) Introduction reframing the surgical. TSQ 5:164–173. https://doi.org/10.1215/23289252-4348605

Redick A (2004) American history XY: the medical treatment of intersex 1916–1955. Diss. New York University

Reis E (2012) Bodies in doubt: an American history of intersex. Johns Hopkins University Press, Baltimore

Rubin DA (2012) "An unnamed blank that craved a name": a genealogy of intersex as gender. Signs J Women Cult Soc 37:883–908. https://doi.org/10.1086/664471

Sadjadi S (2019) Deep in the brain: identity and authenticity in pediatric gender transition. CUAN. Cult Anthropol 34:103–129. https://doi.org/10.14506/ca34.1.10

Savoia P (2010) Sexual science and self-narrative: epistemology and narrative technologies of the self between Krafft-Ebing and Freud. Hist Hum Sci 23:17–41. https://doi.org/10.1177/0952695110375040

Schiebinger LL (2004) Nature's body: gender in the making of modern science. Rutgers university press, New Brunswick

Schilling RC (2015) Of psychometric means: Starke R. Hathaway and the popularization of the Minnesota multiphasic personality inventory. Sci Context 28:77–98. https://doi.org/10.1017/S0269889714000337

Sengoopta C (1998) Glandular politics: experimental biology, clinical medicine, and homosexual emancipation in fin-de-siècle Central Europe. Isis 89:445–473. https://doi.org/10.1086/384073

Stone S (1992) "The empire" strikes back: a posttranssexual manifesto. Camera Obscura: Feminism, Culture, and Media Studies 10:150–176

Stryker S (2017) Transgender history: the roots of today's revolution. Seal Press, New York

Sutton K (2012) "We too deserve a place in the sun": the politics of transvestite identity in Weimar Germany. Ger Stud Rev 35:335–354. https://doi.org/10.1353/gsr.2012.0095

Taher NS (2007) Self-concept and masculinity/femininity among normal male individuals and males with gender identity disorder. Soc Behav Personal 35:469–478. https://doi.org/10.2224/sbp.2007.35.4.469

Terman LMCC (1936) Sex and personality studies in masculinity and femininity. McGraw-Hill Book Company, New York

Terry J (2010) An American obsession: science, medicine, and homosexuality in modern society. University of Chicago Press, Chicago. https://doi.org/10.7208/9780226793689

Trumbach R (1998) Sex and the gender revolution, vol 1, 1st edn. University of Chicago Press, Chicago

Ulrichs KH (1864) Forschungen über das Räthsel der mannmännlichen Liebe. Zweite Schrift. Self-published, in commission at Heinrich Matthes, Leipzig

van der Meer T (2007) Sodomy and its discontents: discourse, desire, and the rise of a same-sex proto-something in the early modern Dutch republic. HIST REFLECTIONS 33:41–67

van Drenth A (2018) Van zielenroerselen tot sekseverschillen: de vrouw als katalysator in het denken over de psyche in Nederland 1900–1960. In: Everard M, Jansz U (eds) Sekse: een begripsgeschiedenis. Verloren, Hilversum, pp 223–240

van Strien PJ (1993) The historical practice of theory construction. Ann Theor Psychol 8:149–227

Vicinus M (2004) Intimate friends: women who loved women 1778–1928. University of Chicago Press, Chicago

von Krafft-Ebing R (1886) "Psychopathia sexualis", eine klinisch-forensische Studie. Enke, Stuttgart

Warnke G (2007) After identity. Rethinking race, sex, and gender. Contemporary political theory. Cambridge University Press, Cambridge, UK

Part IV

Anthropology

Economic Anthropology in View of the Global Financial Crisis

18

Timothy Heffernan

Contents

Introduction	458
Crisis, Stability, and Economic Neoliberalism	459
Theoretical Imperatives: The Influence of Social and Moral Philosophy	462
"Noble Savage" and "Economic Man" in European Social Thought	462
The History and Key Approaches of Economic Anthropology	464
Individualist Versus Systemic Approaches	464
Formalist and Substantivist Debates	466
The Cultural Turn in Economic Anthropology	468
Virtualism and Neoliberal Economic Transformation in Iceland	470
The 2008 Crisis: A Postmortem of Neoliberal Economic Policy	473
Making Sense After Crisis: Ethnography, Reflexivity, and World History	476
Concluding Remarks	478
Cross-References	479
References	479

Abstract

Economic anthropology is the study of how individuals and communities understand and engage with economic life, broadly conceived. This chapter provides an overview of central debates and approaches used in the subdiscipline over the past century. These debates – ranging from the form and substance of the economy, the impact of the cultural turn, and the rise of neoliberal economic policy – are explored amid changing relationships with credit and debt following the global financial crisis (GFC). Positioned between anthropology and economics, the field of economic anthropology has long sought to understand notions of exchange, ownership, consumption, value, reciprocity, production, and labor and considers how these relate to the function and maintenance of distinct cultural

---

T. Heffernan (✉)
School of Built Environment, Faculty of Arts, Design and Architecture, University of New South Wales, Sydney, NSW, Australia
e-mail: t.heffernan@unsw.edu.au

© The Author(s), under exclusive licence to Springer Nature Singapore Pte Ltd. 2022
D. McCallum (ed.), *The Palgrave Handbook of the History of Human Sciences*,
https://doi.org/10.1007/978-981-16-7255-2_14

worlds. Analyzing central debates in historical perspective, this chapter asks how practitioners continue to engage with key ideas after the GFC. What is more, it decenters key theoretical approaches by examining the experience of the GFC from outside the global centers of finance. Through a case study of the Icelandic banking collapse as part of the GFC, questions of how credit and debt are understood in light of crisis are pursued, particularly after the collective prosperity of Iceland's "economic miracle" in the early 2000s. It concludes with a discussion of the harms of neoliberalism and economic "virtualism" and charts emerging inquiries in economic anthropology that boast flexibility for examining economy in a changing world.

Keywords

Economic crisis · Europe · Debt · Stability · Neoliberalism, Virtualism, Iceland

Introduction

This chapter examines how individuals and communities relate to the economy in an uncertain world. It takes the 2008 global financial crisis (GFC), an event remembered for the collapse of national and international financial systems, as a moment of collective disruption. By tracing the economic fallout of the GFC in Europe, where crisis was felt acutely, the impact of this event on how people understand and participate in the economy is explored. As a subdiscipline of anthropology, *economic anthropology* uses the ethnographic toolkit to examine the organization, function, and significance of economic phenomena within and between social groupings. It examines how such groups provision and sustain themselves along cultural lines. So too, it enquires into what it means to be a participating community member through a focus on the role and importance of economic activities in daily life and across the life course. Etymologically, it is worth noting, "economy" stems from the Greek *oikos*, which referred to household management. Its ancient roots compared householders' self-sufficiency against the threat of the market in the context of the Athenian state, first explored by Aristotle. The term has transformed over time in anthropology. It has grown to encompass the "primitive economics" of nonmarket societies among whom anthropologists worked in the early twentieth century (Herskovits 1940) as well the growth of new states, regional economies, and the transition of socialist regions in the decades after the world wars (Hann 2002). Moreover, the term has come to explore the experience of national capitalism, globalized finance, and lived effects of neoliberal economic policies in comparative context (Dunn 2004).

Economic anthropologists examine the values and meanings individuals, groups, and organizations attribute to economic life. This includes attitudes toward subsistence and gainful work, livelihood management, trade, and participation in markets. It concerns the relationships between credit and debt, householding, property ownership, and exchange. What is more, it examines the effects of globalization on local and international economies, and the differences, comparisons, and tensions between capitalist and noncapitalist societies. Taking stock of the aftereffects of the GFC, this chapter focuses on the experience of national capitalism and globalized finance, with attention paid to the effects of crisis on property ownership, credit, and debt in the country of Iceland. Following several years of national economic prosperity, fueled by Iceland's rising status as an emerging player in international banking and finance, the country's banking sector completely collapsed in late 2008. With combined assets of US$182 billion, the crash of Iceland's banks is the third largest bankruptcy. When viewed relative to the size of the national economy, it comprises the largest crash in modern times (Sigurjónsson and Mixa 2011). How this event was experienced for Icelanders is explored ethnographically amid communal efforts to negotiate the constraints of underemployment, asset repossession, currency devaluation, and loss of trust in globalized finance.

It is through a case study of crisis in Iceland and the recapitulation of core theoretical debates that this chapter demonstrates how economic anthropology is distinguished from economics through anthropologists' embrace of diverse historical, cultural, temporal, and geographic points of view. It is this wider – and often critical – approach to studying economic life in its full cultural and geographic specificity that contributes to anthropology's mission of understanding the particularistic and universal aspects of what it means to be human. This can involve many things, including insights on the production, consumption, circulation, valuing, and regulation of culturally sanctioned objects, resources, processes, and knowledges by individuals and groups. Certainly, while economy may be understood differently across discrete settings, insights are brought into larger conversation using theory, including those explored below on the formalist–substantivist divide, the cultural turn, and the rise of economic virtualism. With reference to these debates, a focus on crisis is pursued through an examination of how changes to accepted economic processes and structures have tangible and persistent effects on group flourishing. The chapter closes with a discussion of the role economic and world history ought to play in anthropological theory and praxis after the GFC. To this end, it is argued that engagement with world history is particularly important after an international crisis that left permanent marks on how individuals and communities experience and understand the economy, now and in comparative terms (Hart and Ortiz 2014).

Crisis, Stability, and Economic Neoliberalism

Methodologically, this chapter is informed by the author's positionality and training as an Australian anthropologist and ethnographer working in northern Europe. Since 2016, the author has conducted research in Iceland and reflected on economic crisis within Europe's social geography (Heffernan 2020). Principle among this research is the understanding that Iceland, much like Australia, occupies a geographic position that is peripheral to the centers of metropolitan life in Europe and North America. For practitioners living and working outside the geographic metropole, one notices a theoretical orientation toward the interests and ideas vested in life at the center. Yet, a decade ago in the context of economic insecurity and increased geopolitical tension, Comaroff and Comaroff (2012) noted after the work of Said (2014) and Chakrabarty (2008) that if Europe has historically been the progenitor of social theory, this is no longer strictly the case. Indeed, with national and regional economic insecurity across Europe since 2008, the imposition of structural adjustment measures by supranational bodies, such as the International Monetary Fund (IMF), and the recent attacks on democracy mounted by populist politics, there are now vocal calls for greater disciplinary reflexivity in economic anthropology (Rakopoulos 2018; Hart and Ortiz 2014). In turn, a reliance on theoretical work generated *by* metropolitan thinkers *about* those living outside the geographic center (and thus comparing the sociopolitical and economic milieu of the center to its peripheries) has begun to shift alongside fiscal crisis in Europe.

Variously described within economic and academic circles as a "historical" moment through its market-stopping and disruptive force, the GFC wreaked havoc on national and international economies. This event resulted from increased risk appetite among financial service providers and the overleveraging of the US housing market in the mid-2000s, setting-off economic repercussions across the global economy. Mortgage foreclosures for borrowers who were unable to service their loans ultimately weakened the position of banks and financial service firms, culminating in the oft-cited domino-like collapse of firms, beginning with Lehman Brothers in late 2007. The lead up to this event resembled something of a high point in neoliberal economics based on record company profits, increased personal wealth, and bumper gross domestic product indices. When economic anthropologists look at neoliberalism, they do so by engaging with the wider effects of government deregulation on private enterprise (including on the conduct of banks and financial firms), the privatization of public assets, institutions, and services, and the general rolling back of government intervention in the market in favor of free-market principles (Greenhouse 2010). As a result, a belief in the market's "invisible hand" to influence production, consumption, competition, and innovation dominates economic thinking, resulting in the necessity of the flexible, agile, and resilient individual within a globalized economy (Scheppele 2010; Dunn 2004).

In the context of Europe, Daniel Knight and Charles Stewart (2016: 2) posit that the onset of crisis led to pressing questions over how people ought to "cope with losing their assumed standard of living, their trappings of social status, and their day-to-day structure of life." The moral underpinning of this question, of how communities ought to live amid crisis, bookends this chapter's discussion as it explores the subdiscipline of economic anthropology and highlights emerging research trends after the GFC. Historically, the European continent has boasted tremendous economic strength, a position that has been revised in recent years through the full effects of economic crisis. In this vein, sudden crisis has come to challenge the dominant model of national capitalism, a model which has informed mass, capitalist societies after modernity. It is built on "the institutional attempt to manage money, markets and accumulation through central bureaucracy within a community of national citizens that is supposed to share a common culture"

(Hann and Hart 2011: 30). With international crisis, precipitated by the decline of leading US and European financial services firms beginning in late 2007, many communities were devastated by economic disruption and the subsequent introduction by governments of fiscal austerity.

The development of neoliberalism as the dominant economic policy globally can be traced through four key processes (Hann and Hart 2011). This includes the international shift away from the gold standard in the 1970s as a unit of economic valuation, when gold ceased to be the economic unit used to value fiat currencies, following the floating of the US dollar. Further, the introduction of the Washington Consensus, a set of international policy directives in the 1980s promoting tax, legal, and trade reform through economic liberalization, only hastened the spread of neoliberalism. At the same time, the US political and corporate support for Glass-Stegall (or the *1933 Banking Act*) gradually began to weaken in the 1980s, a law introduced after the 1930s Great Depression mandating the separation of commercial and investment banking, which led to the introduction of speculative ventures (explored below). Finally, for many developing and less advanced economies, structural adjustment was introduced through IMF and World Bank intervention as a means of regulating and controlling regional economic performance, a key moment in the neoliberalization of markets internationally. It is through disentangling these processes that neoliberalism is shown to be less a social fact than the combination of politico-legal and socioeconomic factors masked in quantified and theoretical opacity (Davies and McGoey 2012). In *Liquidated: An ethnography of wall street* (2009), Karen Ho analyzes the corporate world wherein key economic decisions are made inside many of the world's largest financial firms, illustrating the ideologies and working practices behind the scenes of money markets. Between 2007 and 2009, it was these very firms that were bailed out by governments, who had come to the aid of big business – oftentimes over that of the people – during the GFC.

The austerity measures introduced in response to the crisis included cuts to spending on critical infrastructure, education, and welfare, increased taxation, and limits to the use and movement of local currencies (often with a view to paying down government debts or to bailout private corporations). While a sense of crisis now appears to be a common aspect of contemporary life, seen through the effects of viral contagion, border politics, geopolitical tension, and environmental decline, the significance of financial crisis stems from the fact it highlighted the limits of neoliberal economics as it was practiced in capitalist markets. Anthropology's interest in neoliberalism as an economic system and policy approach has been spearheaded by examining the intersection of collective prosperity and individualized misfortune, and the interplay between austerity measures and maintaining social security (Greenhouse 2010). Yet anthropology and neoliberalism now intersect somewhat more personally through the effects of the GFC on practitioners' home societies. This has led to a critical reflexive view of how people and systems affect economic phenomena through their daily participation in local, regional, and international markets. In turn, with severe economic crisis shaping aspects of global society over the past decade, scholars have begun exploring how individuals and communities affect markets through their participation in them, but also how people can be adversely affected by markets (Muir 2021; Tsing 2005).

Theoretical Imperatives: The Influence of Social and Moral Philosophy

By highlighting the need for greater reflexivity through, placing the theoretical legacies of dominant thought under an ever-increasing critical gaze, an understanding of the history and practice of theory-making has in turn gained vital currency (Siniscalchi 2012). Theory, it is well understood, allows for thinking about *how* and, importantly, *why* individuals and communities conceive of the economic phenomena in ways they do. This is the critical nature of social science knowledge production, which duly encompasses the practice of operationalizing localized understandings about one's place within the world and their immediate environment, such as family, social, religious, and employment environs, and one's place within national collectives and the broader international community. Oftentimes, this includes thinking about systems of value and ethics as they are linked to economic life, particularly ideas of wealth, exchange, value, and reciprocity. From the scholarship of Marx and Weber to more contemporary theorists, crisis and its bearing on economy has been a feature of social thought. To understand how crisis and economy are understood after the GFC, it is important to highlight the theoretical environment in which conceptual thinking about economy developed, namely within the European Enlightenment.

"Noble Savage" and "Economic Man" in European Social Thought

The ethics and values of economic relations in anthropology began with interpretations of the political philosopher Jean-Jacques Rousseau (1712–1778) on inequality in medieval and early modern European society. Born in Geneva but spending time throughout Europe's cultural hubs, Rousseau's social philosophy was influential during periods of sweeping social and political change. In *A discourse on inequality* (1984 [1754]), Rousseau tried to understand what was behind inequalities in individual wealth. So concerned was Rousseau (2010 [1762]) with what he viewed as the unsustainable social environment of the eighteenth century, he proclaimed society would be plagued by a series of crises and revolutions. This gave birth to critical works by Rousseau and peers on how inequality stemming from the political and economic order of the day might be eventually overcome, ultimately suggesting it could be found on what humanity shared – our human nature (Hann and Hart 2011, 2012). Rousseau invoked a preindustrial, pre-social state of nature in his writings, a state in which humans (or as he referred, "noble savages") enjoyed freedoms: freedom of will, freedom from rules-based society, and freedom from hierarchical oppression. Rousseau's thesis was fleshed out in his influential text *The social contract* (2012 [1762]).

Through this extended thought experiment, Rousseau demonstrated how inequality and alienation have become prevalent as part of industrialized civilization, to the greater detriment of many.

Rousseau's experiment and the social and political philosophy it produced went on to influence leading thinkers in the nineteenth and twentieth centuries, especially on how to conceive of the way discrete societies operate and the inherent *worthiness* of persons. Such writings were produced at the same time as expanding colonial missions, the rise of the nation-state, and market society, and influenced the thinking of Marx and Engels. It was in this context, too, that crude evolutionary models representing different societies were made, of which early ethnology and the ethnographic toolkit very much played a part. Writing at a similar time to Rousseau, the Scottish moral philosopher, Adam Smith (1723–1790) examined the wealth and prosperity of national markets in *The wealth of nations* (2008 [1776]). Smith stressed in much the same way as Rousseau had about wage labor that productivity and therefore profit was key to collective prosperity and reduced inequality. As the father of neoclassical economics, Smith's work had a similar influence on liberal ideologies, explaining that markets do best when they are open and when domestic and international trade is unimpeded. The main actor in this dynamic was the rational actor *homo economicus*. "Economic man," as it was termed, was premised on four key traits, including that it was (1) a self-interested individual (2) who used market choice and calculation to maximize utility amid budget constraint and (3) took an objective view of such choice in the market, (4) thereby expressing the character of being highly rational (Dixon and Wilson 2011).

While stimulating theoretically, there are two main critiques of this model today which concern contemporary understandings of crisis and economy. The first is that this model has become unattainable as it presupposes *homo economicus* to be atomistic and asocial. The second critique is that it overlooks the significance of human morality in favor of advantageous market participation and calculation. Humans are moral entities, a point which many modernist philosophers would agree upon, meaning human action goes beyond mere self-interest. Yet in neoclassical economic circles, it was held that morality has little economic significance because it is about outcomes while rationality is about means (Dixon and Wilson 2011). Indeed, suggestion that human action was premised on "trucking and bartering" is a common association in the neoclassical doctrine, with such actions seen to be facilitated by markets, the very functioning of which serve to increase "the wealth of nations" (Hann and Hart 2011: 26). In the present time, given our confrontation with social and environmental crisis, discussions about climate change and the Anthropocene largely carry both a rational calculation and broad appeal to moral thinking and action. A prime example involves discussion of climate action as part of The Green New Deal (a revision of the US Great Depression policy suit named The New Deal, which aimed at social and economic reform). In many ways, the pairing of morality and rationality highlights how it is in crisis that new knowledge practices and epistemic habits are generated out of necessity or convenience, as communities set about contending with rapid changes to the status quo.

The History and Key Approaches of Economic Anthropology

Crisis, as a constituent part of contemporary life under the auspices of globalized capitalism, is defined as a particular impasse, when certain customs and traditions are adversely affected by internal or external causes. The crisis faced by communities in 2008 uncovered the flaws in neoclassical and neoliberal economic models, which emphasized the role of individuals within self-regulating markets of supply and demand. Indeed, the crisis highlighted the role of elite cultures of corruption within an international financial system awash with cheap credit and, afterward, the key role of government as the lender of last resort to failing (private) financial institutions. When examining crisis, practitioners stress the importance of institutional order and proper functioning, insights that are gained empirically through the ethnographic toolkit, especially immersive fieldwork based on interviews and observations. The economy was included in early anthropological analyses of core systems and processes, giving rise at the turn of the twentieth century to the subdiscipline of economic anthropology. A focus on order and function harkens back to early anthropological endeavors and the development of the ethnographic methodology through attempts to understand economy as part of studying the total lifeworld of a community (Carrier 2019). This includes anthropological endeavors where attention was given to the economy "to place people's economic activities, their thoughts and beliefs about those activities and the social institutions implicated in those activities, all within the context of the social and cultural world of the people being studied" (Carrier 2012: 2).

This area of study has not been without its fair share of subdisciplinary debates, which will now be outlined, to demonstrate how practitioners sought to understand the place and content of economic life in early anthropological research. This includes individualistic and system approaches to examining the economy, formalist–substantivist debates concerning its nature and functioning, and the impact of the cultural turn in anthropology in the latter half of the twentieth century on how economic life was accounted for and understood. The debates explored below map onto three discrete historical periods in Euro-American intellectual life (Hann and Hart 2011), which may prove useful to readers approaching them for the first time. The periods include the prewar years, when a focus on the universality of economic rationality was a key dimension of anthropological study, the postwar years when the economy's form and substance was examined, and the tensions emanating from the cultural turn in the 1980s during the height of neoliberal economic policy. If the economy is a constituent part of a community's lifeworld and a discrete area of study within anthropology, how practitioners engage with economic life varies considerably. As James G. Carrier (2012) notes, two approaches are common: a focus on the individual or the system.

Individualist Versus Systemic Approaches

A focus on the individual prioritizes thoughts, values, and practices as they relate to economic participation. The systemic approach, which is elaborated below, places social systems and institutions at the center of research, noting how they affect economic participation and understanding. This debate began with research by the key disciplinary figures Bronisław Malinowski (1884–1942) and A.R. Radcliffe-Browne (1881–1955). It is from these early practitioners that anthropology transitioned from a speculative science into a fieldwork-oriented discipline. In early fieldwork in the Trobriand Islands (in present-day Papua New Guinea), Malinowski traced the exchange of culturally significant objects (for instance, handmade necklaces and armbands) across an archipelagic area. In the text *Argonauts of the western pacific* (1922), this became known as the *Kula* ring, a defined trading sequence of interisland exchange. An early pioneer of sustained ethnographic practice, Malinowski's observations, conducted over several years while working among the Trobriand people, were central to understandings of trading systems among non-European groups, while at the same time being used to "test" the universality of rational systems of trade beyond the metropole. Between 1870 and 1940s "most anthropologists were interested in whether the economic behaviour of 'savages' was underpinned by the same notions of efficiency and 'rationality' that were taken to motivate economic action in the West" (Hann and Hart 2011: 2). The trained anthropologists of this era "initially devoted themselves to assembling compendious accounts of world history conceived of an evolutionary process" (Hann and Hart 2011: 2).

Through his work, Malinowski questioned why the exchange of what appeared as mere trinkets was ardently pursued through interisland trade, leading him to develop an individualist theory of exchange. Tracing the specifics of this system, Malinowski focused on the actions of individuals as they participated in exchange, ultimately suggesting that exchange satisfies an individual's needs and drives, yet also highlights the political and cultural significance of interisland trade. This showed how trade can occur as a function of individual and community life, rather than simply being a component of market society (that is, trade powered by supply and demand forces), which was prominent in European theorizations. What is more, it complemented observations from North American societies on the potlatch. In contrast, an individualist approach differed to that of Malinowski's contemporary, Radcliffe-Browne, who sought to locate the function of exchange in the structures of society. Radcliffe-Brown was influenced by the work of French sociologist Emile Durkheim in *The division of labour in society* (2014 [1893]), which noted division lay within collectives or society at large, rather than an individual's actions. Durkheim found that the nature and function of society could be studied by looking at the "degree of their division of labour," which is linked to other societal trends and attributes than to individuals (Carrier 2012: 5).

In this key text, Durkheim sought to outline the total social phenomenon through the interpenetration of the individual, the collective, and systems within that which we call society. Going further, he stressed that the kind of society that is produced will depend on the root of social solidarity, seen either as mechanical or organic in nature. Mechanical solidarity referred to a sense of social homogeneity in *collectivist societies* that did not rely on an intermediary but was found in a common belief and sentiment in community membership. Organic solidarity, on the other hand, refers to settings where difference exists and is plainly visible, especially in *industrialized societies*, with social homogeneity found is a reliance on one another to perform specific tasks. Durkheim's influential ideas, the significance of which is explored below, were developed through the analysis in Europe of collected material about community trade and sociality in the Pacific, New Zealand, and Australia, through which theory about the generalizability of human exchange was created. The development of early anthropological theories in societies peripheral to Europe, such as the Trobriand Islands and greater pacific area, occurred amid significant social, political, and economic change in Europe. To this end, the effects of industrialization and urbanization at the turn of the twentieth century were radically changing life in cities at the same time as seeing the emergence of mass society and the modern formulation of the family unit.

In a rapidly shifting environment in Europe, Durkheim's ideas of society and systemic relations became popular in anthropology, particularly so for Marcel Mauss, the nephew of Durkheim. In his development of exchange theories in *The gift* (1966), Mauss drew on the *Kula* ring to argue that exchange practices develop social relationships, challenging previous assumptions. It is through a focus on exchange that Mauss highlights the necessary components inherent within the gift: to give, to receive, and, most importantly, to reciprocate. This demonstrates the total phenomena of the gift, critiquing dominant ideas at the time about the utility of exchange relationships. Rather, he suggested that across a great many societies, reciprocity builds and maintains social solidarity as well as contributing to wealth creation and important sources of alliance, both within and between communities. Mauss had a lasting impression on influential mid-century French anthropologist, Claude Levi-Strauss. It is through ideas of reciprocity that social relations (and social indebtedness) is most pronounced, an idea Levi-Strauss (1969) explored through food and the practice of eating, noting how the received company of others is returned through a meal that is given.

Formalist and Substantivist Debates

Following the world wars, the theoretical focus turned toward the form and substance of the economy, principally explored by Karl Polanyi. Born in Hungary to an intellectual family and later trained in law, Polanyi (1886–1964) sat between the fields of economics and anthropology, having contributed to understandings of the market economy as well as conducting ethnological research into the economy of archaic societies. In his view, market society was not, nor should it necessarily be, a universal phenomenon. Rather, the context of market relations bear on exchange and distribution. In his influential text *The great transformation* (1944), Polanyi analyzed sociopolitical tensions during the rise of the free market in nineteenth-century England, arguing that the nation-state and market economy were not separate domains, but constitutive of the kind of market society many are now used to seeing today. Whereas kinship and other social relations might guide economic behavior in some settings, including distribution and exchange, in market societies, decisions about investment, production, and distribution are informed by the price of a good or service ("price signal"). With the rise of market society in Europe and the solidification of capitalist trading routes through colonial expansion, market society became the dominant model, as goods and services reached new markets and were traded in larger circulation. In turn, ideas of equality and justice once again took on resonance in economic anthropology.

In intellectual debates, formalists held that economic life could be traced through the rational behaviors and thought processes exhibited by individual players in a given context. Substantivists, on the other hand, posited that context – more so than rationality and behavior – was important for how people understood and participated in economic life. For Polanyi, a substantivist, the context of the medieval world and modernist social, political, and economic processes were capable of "throwing light on matters of the present" (Polanyi 1944: 4). In this earlier context, there was a focus on managing household wealth, not necessarily scheming in capitalist terms to *exponentially* increase wealth as is the case currently. Like scholars before him, Polanyi was influenced by notions of equality and reciprocity, and looked to classical Greek and modernist philosophy to understand their place in economic life. In *The nicomachean ethics*, Aristotle examined virtue ethics within society: of building people of good character and, therefore, generating good communities through appeal to the virtuousness of human reason. Examination of archaic societies, such as ancient Greece, boasted opportunities to make distinctions between the role of economics in society and vice versa. It is important to consider the inherent social aspects of economy that have contributed to how we understand economic life as well as social life. Indeed, much like Mauss, Polanyi (1944) stressed that archaic societies enable us to see that economic theory is wholly Eurocentric and disciplines like anthropology are capable of cross-cultural analysis of market and nonmarket practices, the kind that might position the principles of dominant economic theory as a kind of construct.

Writing in the mid-twentieth century following the Great Depression, Polanyi's (1957) reflections on household management and the social and political aspects of exchange in ancient Greek society, led to furtive debate about the quality of the term economy, one which emphasized the difference between form and substance. Polanyi's engagement with economic behavior in market and nonmarket settings highlights the fact that market societies have only enjoyed prominence for a short period of time in history, namely from the nineteenth-century onward. The formalist approach, dominant in such societies, stresses the universality of ideas and their use and operation; the primacy of neoclassical economics, based on a supply and demand logic, is one such example. Such a position posits that the "exotic" trading practices that early economic anthropologists observed were "underdeveloped versions" of European systems (see Sahlins 1969). Substantivist debates, on the other hand, emphasized the content of economy, including the exchange practices of non-European societies. Through this division, Polanyi was interested in demonstrating how the economy is *embedded* in society rather than the other way around.

It is with industrialization, according to Polanyi, that the economy becomes *disembedded* as the focus moves from reciprocity to redistribution through too greater focus on the market as the answer for circulation. In a recent engagement with the discipline of economic anthropology, Hann and Hart (2011: 57) state: "reciprocity was a symmetrical form of exchange between persons or groups of equal standing, as in the Trobriand *kula* ring. Redistribution reflected a principle of centricity, whereby resources were pooled and handed out through a hierarchy." Continuing, they suggest that "reciprocity was dominant in 'primitive' egalitarian societies with simple technologies, whereas redistribution usually presupposed the possibility of storing a surplus and some degree of social stratification" (Hann and Hart 2011: 57). Based on significant debate on redistribution and reciprocity, scholars have gone on to apply social theory to study a variety of "primitive," capitalist, socialist, and communist societies, especially with reference to modern economic and development theories, industrialization, globalization, and contemporary capitalist accumulation (Maurer 2006). In many instances, this has involved appreciating the interplay between formal and informal governance and economic systems, cash and subsistence economies, the role of civil society, religion, politics, and cultural life in economic activities, and the importance of the public and private spheres in economic life.

Chris Hann and Keith Hart (2011) cogently suggest that, in retrospect, the formalist–substantivist debate that waged throughout the mid-twentieth century was a high point in economic anthropology, because it brought economists and anthropologists into meaningful and productive dialogue, which is today no longer the case. In a recent essay, Jane I. Guyer (2017) stresses the importance of *thinking with* the social aspects of economy through an appreciation for the spatial and social engagements of the economy in real terms. Recollecting the market in ancient Greece, "people met in designated public spaces where varieties of goods and their specialists came together regularly, commented on the conditions of economic life, invented novel terms and commodities, and even novel modes of association and political expression" (Guyer 2017: 43). Indeed, in the context of ancient Greece, the physical market space (*agora*) "brought public life, debate, and the market into one place of assembly" (Guyer 2017: 51). Exchange was based on the idea of upholding key traditions, thus highlighting, as Malinowski (1922) did in more recent times, the productivity in focusing on the imponderabilia of everyday life. It is the disruption of these traditions – the physical market and the embeddedness of the market in society – that highlights the alarming quality of crisis in society today: How the imponderabilia of life is disrupted through the influence of neoliberal economic policy that, in turn, is wholly dissembedded from society.

The Cultural Turn in Economic Anthropology

A key aspect of economic anthropology since the mid-twentieth century moved away from debates specifically on economy's embeddedness in society and toward understanding the implications of the cultural turn in anthropology, and its effect on studying economic life. While twentieth-century scholarship across the discipline of anthropology has been influenced by cultural relativism, the cultural turn of the 1980s sought to further embedded cultural critique into anthropological praxis and theory. This shift emphasized cultural aspects for understanding the lifeworld of a community. For Stephen Gudeman (1986), a contributor to early cultural interpretations of economic life, "economies" and "economics" are social constructions used to shape economic reasoning and participation. Indeed, "the material modes of operation of economic systems are driven by cultural worldviews, all encompassing perspectives that are rarely examined but are...in great need of close scrutiny" (Trawick and Hornborg 2015: 2). Discussion after the GFC within affected countries about robust economies and social welfare programs, including in the UK, focused on "getting a job," "staying employed," and "keeping off welfare," demonstrating the degree to which formal participation in the economy is built on *ideas* of hard work connected to agreed-to ideas about labor and economic participation (Powell 2017).

In later works, Gudeman (2008) made plain the connections between people's material lives and the practices employed by economists, highlighting the tension between community and market. The meeting of this tension is observable through what Gudeman (2012: 94) surmises as economy's base: "the social and material space that a community or association of people make in the world. Comprising shared material interests, it connects members of a group to one another, and is part of all economies." A key example might include, for instance, guilds "in which apprentices gain knowledge in mechanical circumstances, build on a base, but the skills are limited to the group as private property" (Gudeman 2012: 98). A key issue here is the limited good, and the fact that wealth creation always has a cost to be borne. In peasant societies, it was held that the generation of new wealth came at the expense of the commonwealth among society. This concern was successfully denied by Europeans during expansionist colonialism, whereby, for example, "Britain was proving itself able to transcend those limits by substituting labor and capital for land while outsourcing its land requirements to other continents" (Trawick and Hornborg 2015: 1), such as the settler colonial states of Australia, New Zealand, and Canada.

In recent decades, a connection between economy's base and the idea of a moral economy has gained in appeal, particularly through work of Edward P. Thompson (1968) and James C. Scott (1979). Scott's scholarship, in particular, propelled the argument that cultural worldviews are indeed bound up within understandings of economic transactions and one's participation in these. Rather than production and consumption being viewed through a material or systems lens, a moral economy is one that focuses on the moral considerations of economic participation. Focusing historically on colonial rebellion in Vietnam and Burma, Scott (1979) considered how uprisings actually violated the conditions of the peasant class engaged in subsistence agriculture and the fulfillment of their cultural roles and obligations. He noted, therefore, that mobilization against larger structural forces entails a moral calculation of the content of a subsistence ethic, especially to what is deemed tolerable in terms of economic justice given that creating a living "is attained often at the cost of a loss of status and autonomy" (Scott 1979: 5). Taking inspiration from Polanyi's work on premarket societies, Scott deftly illustrates how peasants make use of moral logic (e.g., the subsistence ethic) and market logic (e.g., capitalism) to negotiate livelihoods and social processes. In turn, petty and formal participation follow along a continuum, with both "necessary for the survival of small-scale producers" (McCormack 2007: 63). Scott shows how peasants employed safety-first principles as part of economic engagement.

In our own century, the cultural aspects of economy have been further complicated by the rise of "the virtual." Indeed, since the end of the Cold War and the global spread of neoliberal economic policy, new engagements with economy have emerged in connection with the defeat of socialism by capitalism. Economic policy changes led to the expansion of the free market and its insistence on unfettered transfer into all corners of global society. Of the virtual, Daniel Miller (2001: 299) asserts "it is not that the economic model of the market represents capitalism, but that capitalism is being instructed to transform itself into a better representation of that model." In this instance, economics as a course of study and domain of life have become supreme, often informing how we understand contemporary life. A kind of abstraction has occurred in the process and the rise of virtualism has ensued. James G. Carrier (1998), a primary contributor to this theory, posits that virtualism accords with the disembedding of economy and the abstraction of a worldview that conforms to market principles in daily life. Economic models no longer seek to measure and describe the world they operate within, but rather this world becomes measured against such models, with life made to conform to this reality (Carrier 1998). "The abstraction," Carrier (1998: 2) asserts, "occurs in practical activity, what Polanyi (1957) referred to as the realm of substantive economy, which is the ways that people, firms and other agents organise and carry out their activities of the production and circulation of objects and services." Continuing, he suggests, "it occurs as well at a conceptual level, what Polanyi refers to as the realm of the formal economy, which is the ways that people think about and understand their economic lives" (Carrier 1998: 2). The neoliberalization of daily life, seen in expectations of the economic subject (*homo economicus*) to be flexible, agile workers capable of responding to the market and to profit-driven agendas, is a case in point.

Virtualism and Neoliberal Economic Transformation in Iceland

The antecedents of Iceland's banking collapse can itself be traced back to the virtualism thesis and the implementation of neoliberal economic policy by conservative governments and their supporters in the early 1990s. This came particularly after the country began modernizing and industrializing its economy after gaining independence from Denmark in 1944. Indeed, the introduction of a neoliberal economic policy distinguished by an adherence to the "free market mantra" was a radically new cultural and economic experience. Following the footsteps of larger international actors, a neoliberal policy suite was introduced by government to further strengthen the country's postindependence economy through increased economic efficiency. This agenda was pursued through the exploitation and commodification of natural resources, including the development of renewable energy and privatizing access to the nation's fisheries to identify new territories for economic growth and to stimulate new sources of profit for economic maturation

(Heffernan and Pawlak 2020). Critical to economic development was the commodification of fishing resources, including the introduction of a fishing quota system, whereby common use rights over natural resources were privatized. This began through a quota system that later transitioned to become a more rigid system of individual transferable quotas (ITQs).

Agnar Helgason and Gísli Pálsson (1998: 117) contend that "the implementation of ITQ management typically requires extensive changes to the institutional and conceptual framework of resource use, the most significant of which concerns the quantification and commodification of resource rights and their subsequent allocation to individuals." What began by assigning quotas to individuals led to a transferable system, whereby quotas ended up in the hands of a few large operators with superior social and economic capital (Helgason and Pálsson 1997). There is significant anthropological literature on this transition (see McCormack 2017; Einarsson 2015). What interests us here is the implication of the privatization of natural resources in Iceland with the virtualization of the economy, leading to a *moral conundrum* over shared resources and the neoliberalization of Iceland's economy and its ultimate decline with the 2008 crisis. There is a history of boom-and-bust fishing success within Icelandic coastal communities, with ITQs giving supposed economic stability and environmental sustainability by placing limits on fishing. Following the introduction of this mechanism, however, there has been a move from "following fleshy fish swimming in nearby seas," to "following virtual fish; ones that have been taken out of the sea and brought to life in exchanges and electronic marketplaces and upon which extraordinary sums of money can be and have been made" (Maguire 2015: 122). In this way, economic virtualism and neoliberal economic policy initiatives have effects on social connections and the commodification of once common natural resources. Reflecting on fieldwork in northern Iceland after 2008, James Maguire (2015: 126–127) relays in *Virtual fish stink, too*:

> Since the virtual fish can be rented out on a yearly basis, sold for life, or mortgaged to raise finance, lower stock levels equate to higher quota demand and hence increased rental, sale, and mortgage values. The effects of virtualization on the fishing industry in general and on smaller communities and small-scale fishermen in particular have been extensive. The escalating price of virtual fish becomes both a barrier to entry into the fishing industry for new fishermen who cannot afford the sums required to buy into the system as well as a seductive point of exit for current small-scale holders who can sell out for vast sums of money. This is still a highly contentious issue within small communities, as those who sell the quota are often considered to be 'betraying the town,' given the reduction in baiting, fishing, fish processing, and hence employment that ensues from any such sale. The risk to these small locations, where families or tightly networked groups tend to hold the quota, can and has been devastating. In 2007 the primary quota holder in Flateyri, a small town in the...West Fjords, sold his entire holdings at the highest recorded price ever of 4000 krónur per kilo. Since then, the town has been in a state of terminal decline.

The accumulation, transfer, and renting out or mortgaging of quotas to preserve social and economic capital led to a culture within the business and banking community of market activity being driven by speculative finance, through the virtualization and financialization of future fishing stocks. In turn, the fate of a nation rested on this practice, which became untenable when financial firms in America and Europe began to falter, and investor confidence was completely lost across national and regional stock markets. Prior to this, Maguire highlights how "the virtual" came to inform "the real" in Icelandic society:

> The overall trend in Iceland has therefore been for quotas to accumulate in the hands of an ever-decreasing number of large fishing companies whose power to control the industry has been greatly enhanced. In 2010, 73 percent of all quotas were held by fewer than twenty large companies, and it was these companies that tended to rent out a high proportion of their quotas to others, creating social unease and in many cases outright resentment. The general macro-level effects of this combined ability to sell and collateralize virtual fish were well illustrated in a paper by a prominent fisheries economist written just several months before the crisis...[Which] argues that the ability to use virtual fish...to raise financial – or what many would call speculative – capital created up to $5 billion in wealth 'where none existed before.' Moreover, [the report] contends that there is a direct correlation between the creation of this wealth, or 'living capital'...and the growth of the Icelandic economy during the economic boom years (Maguire 2015: 127).

A key area of growth was the banking sector, which was privatized and sold off to political allies in the early 2000s (Pálsson and Durrenberger 2015: xvi). Capitalizing on the millennial boom of international banking and finance, the likes of which had resulted from neoliberal processes, the banks' owners quickly expanded their operations abroad, becoming leaders in this sector and setting off a "financial miracle" in Iceland through neoliberalization.

While neoliberalism is explored in connection with the social and economic policies that contributed to the global financial crisis, neoliberalism is often remarked upon by scholars as difficult to define in real terms (see Collier (2012) for a review). It is helpful, however, in light of the example of crisis in Iceland, to position the term as one relating to a structural force and ideology as much as a form of discourse (Ganti 2014). As a force, neoliberalism shapes individual and collective subjectivity (Ganti 2014), while as ideology, neoliberalism encompasses ideas, beliefs, and values about the world, with "ideological effects of naturalness and inevitability...produced through the affective investments in particular significations that grants them the claim to represent the world" (Anderson 2016: 737). Neoliberalism as discourse, on the other hand, stands in for and highlights the ways that "affective investments and attachments are organized" through everyday interactions with structures and ideologies (Anderson 2016: 737). In this way, while neoliberalism "seems to mean many different things depending on one's vantage" (Ong 2006: 1), the term is useful for showing how we come to understand and become shaped by neoliberal policy forces, often concerning establishing a "better" path from the hardships of the past.

Through increased participation in global markets in the 2000s, Iceland enjoyed tremendous economic stability and visibility globally. In turn, the fervor that characterized this period produced an atmosphere charged with optimism, national pride, and conspicuous consumption, described as the *manic millennium* by cultural economist Már Mixa (2009). When viewed in relation to Iceland's history as one of Western Europe's poorest countries at the turn of the twentieth century, this period of economic success led to a feeling that Iceland had finally "made it" on the international stage (Loftsdóttir 2010). It followed that increased material wealth and improved access to capital came to affect how Icelanders thought about the present and invested in the future. Indeed, despite the hast with which the banking and business community made their fortunes, Iceland's newfound success was legitimized by politicians via recourse to the nation's ancestral Viking history and the ability to capitalize on emerging opportunities (Loftsdóttir 2010). In the early 2000s, this class of *nouveau riche* entrepreneurs acquired overseas hotel chains, fashion houses, and sports teams, flouting their wealth at home and abroad. This was made palatable to the Icelandic public through the availability of cheap credit and encouragement to consume at home. Signs of growing affluence also became more pronounced between 2006 and 2008. A key example included the presence of cranes positioned across the Reykjavik cityscape amid unprecedented levels of urban construction in a city with less than 120,000 inhabitants.

The 2008 Crisis: A Postmortem of Neoliberal Economic Policy

In late 2008, several years of economic growth abruptly ended when the country plunged into recession with the reach of the GFC, bringing an end to the nation's newfound prosperity. With loan sheets more than ten times Iceland's GDP, the government was forced to intervene and save the domestic branches of the banks. As the government faced the difficult task of stabilizing the economy, half-finished housing and building developments became the literal and figurative ruins of a future that never came to pass (Loftsdóttir 2019). As Irene Sabaté (2016: 107) notes in Spain, a similar overdevelopment and investment in the housing market in the lead up to the GFC was "founded on common perceptions of secure income, endless economic growth, rising housing prices, and investing for the future." The presence of economic crisis means moral calculations of how one reconciles their understanding of economic life amid market forces and economic virtualism is perhaps now ever more pressing. In the years since the GFC, many national and international economies and systems of finance suffered greatly, and the shoehorning of economic participation into a virtual model has been questioned. Indeed, government-imposed responses to crisis have illustrated how "austerity throws the issue of human dignity into high relief as people set about deciding on the new minimum requirements for an acceptable life" (Knight and Stewart 2016: 2). For this reason, debates about the moral economy and virtualism have taken on new resonance in economic anthropology since 2008 (Spyridakis 2018).

Sitting in a café in Reykjavik with Dóra, who worked as a translator and project manager during the boom years in Iceland, the reality of the collapse amid the hope for a bright future was laid bare during a series of interviews conducted by the author. Reflecting on her own aspirations for a brighter future prior to the collapse,

Dóra stated her decision to purchase a larger apartment was motivated by this sense of unending growth:

> I had a 2-bedroom place and wanted an extra room. I looked at the prices and thought I might buy a house [building] with two apartments so I could rent one out and then, later, my daughter could have the other apartment. It was a bit of a crazy idea, but people were thinking big and that was my big thinking. I had 30% equity when I bought it, so I thought I was being responsible. I thought I would only earn more assets in this block of concrete over time. But just after the purchase, in Spring 2008, when the mortgage repayments started coming, the loan began to grow by five thousand krónas [US$374] more than I thought each month, which is a lot. I called the bank and said, 'I don't understand, the mortgage is increasing...and that's hurting more and more each month.' She said, 'Eh, that's how it is.'

Icelanders came to believe in the market and its benevolent capacity to provide for their future. Money rallies us together during economic prosperity, and yet debt sets us apart. In turn, while the privatization and expansion of the banks promised to grow the national economy and, therefore, increase the private fortunes for all Icelanders, the collective financial losses that accumulated after the crash were interpreted as intense, personal misfortune. This sense of collective prosperity and individual misfortune demonstrates how crisis is never simply economic but also social and political (Tsing 2005), with the tension between the personal and collective figuring in how Icelanders view themselves historically as a small, postcolonial country (Loftsdóttir 2019). In response to economic collapse, Icelanders responded through mass anti-government protest. Organizers sought to highlight the government's blind faith in neoliberal economic policy and highlight how this had brought concern and disadvantage to ordinary Icelanders. Such protests mirrored other protests around the world between 2008 and 2011, including the global Occupy movement, anti-austerity protests, and the Arab Spring. While in these contexts the conduct of the 1% was censured by society's 99% in the context of globalization and neoliberal finance, in Iceland protests were geared toward questioning the morality of the events which lead to the ultimate demise of the country's economy and banking system, along with questions over what society and politico-economic relations among citizens and elected representative ought to look like, thereby bringing into focus an open question of equity.

Dóra's reference to thinking "big" in a moment marked by future orientation and the growth of the national economy is an example of a style of thought which encourages people to live beyond their means, one which puts into tension their ability to carve out a living and uphold accepted morals in society. Indeed, there is a "cruel optimism" (Berlant 2011) inherent in buying into a future that places people in more and more debt, which only works to individualize a collective economic crisis when decline sets in. In this way, the "fictitious" commodities of mortgaged fishing quotas within a disembedded economy are not tangible assets as they "are not produced but bought and sold as if they were real" (Pálsson and Durrenberger 2015: xx). There is a staying power to the repercussions of when these kinds of commodities turn sour. During ethnographic fieldwork between 2016 and 2018, conducted by the author several years after the economic crisis, it was common for suburban Icelanders living in Reykjavík to speak about *still* serving their obligations on housing and personal loans.

Noting Scott's (1979) work on the moral economy, Catherine Alexander et al. (2018) focus on the tension between participating in economic boom periods, and the expectation of secure and affordable housing, which in turn effects the household economy through practices of scrimping and saving to make ends meet. Indeed, this tension manifests in the desire "for adequate and safe housing" and "aspirations to build and maintain their homes" which "are often out of sync with, or undermined by, political rhetoric, state officials, loan terms and the law" (Alexander et al. 2018: 112). They suggest that this can result in a conundrum through "the complex and often overlooked ways in which people claim allegiances to particular moral communities and how they (re)constitute themselves as deserving of secure tenure and of what they consider to be proper homes, often in the face of stigma, harsh laws or policies that construct them as the very reverse" (Alexander et al. 2018: 112). In the context of national economic crisis in Spain, Sabaté (2016: 118) illustrates the collapse of the housing market led to a tension between economic systems and those who engage with these systems:

> First, it has affected the commercial relationship between banks and/or real-estate agencies and customers, mortgage debtors, and home buyers. Second, the reciprocal relationship between the welfare state and citizens, comprising a set of rights and duties assumed by citizens in exchange for state responsibilities of basic provision, also seems to have been upset. Moreover, the right to subsistence (Scott 1976: 176) has also been undermined in the eyes of growing sectors of the Spanish population. As the chances of securing a livelihood decrease, the right to subsistence becomes endangered and, as a result, prior hierarchies of needs are challenged. The obligation to repay mortgage debts is called into question or, at least, loses its top priority within household budgets.

This vignette and Dóra's experience of the economic and material repercussions of national crisis demonstrate a disembedded economy: of economic activities becoming removed from their social relationships. Indeed, measures and societal norms that previously defined the Icelandic economy, such as a small currency market, the absence of outside capital, and protectionist measures that limited the number of actors importing goods into the country, meant "Icelanders regarded the savings as collective earnings to be either redistributed internally or used to pay for collective goods" (Jónsson and Sæmundsson 2015: 27). With individual debt and misfortune after several years marked by shared prosperity, collective goods and a nominal sense of equality among compatriots has come under intense strain. In this context, classed notions of those with and those without now pervade Icelandic society. For this reason, the free market has now become a victim of the GFC through suggestion that markets no longer operate "free" from political intervention (if that was ever really the case). In the wake of Iceland's economic crisis and across much of the Global North since 2008, citizens have looked to their governments and social and political institutions for fiscal support and for guidance on how to contend with economic insecurity. The lack of direction offered by governments has only entrenched calls for political and economic overhaul.

Making Sense After Crisis: Ethnography, Reflexivity, and World History

Of the Icelandic banking collapse, Kalman Applbaum (2009: 26) notes "the Icelandic government's policies of monetary deregulation and its reckless trust in the expansion of international finance markets provide a radical example of the influence of global financial trends on the domestic life of a people." This continues in Iceland through a burgeoning tourism boom that has brought, on average, over one million tourists to the country from 2015 until international travel was impeded by the coronavirus. As Icelandic anthropologist Kristín Loftsdóttir (2019) stresses, the onset of crisis brings about new relationships between the global and the local, with Iceland's small economy within a global arena that is home to an even smaller population (360,000). Appreciating these new relationships highlights the importance of ethnography for understanding the particularities of local experience. Yet in the vein of Mauss and Polanyi, a focus on social relations and the embeddedness of the economy in society shows how world history also needs to factor into anthropological analysis (Hart and Ortiz 2014). The geographic and social determinants of neoliberalism, largely developed within the financial centers of Euro-America, needs to be seen as one particular history – or perhaps a particular episode of history that invokes the particularities of metropolitan thought and practice. This is important for, as Bruce Kapferer (2004: 152) argues, a focus on individualistic or collective reductionist elements "has displaced a concern with social complexities and the differentiated and differentiating structurating processes that are integral to the conceptions of the social and of society."

In the decade following the GFC, economic anthropology has taken stock of common debates and economic norms and pursued new theoretical and empirical ends. This includes a larger focus on the ways theory and method are deployed to understand economic life. Julia Elyachar and Bill Maurer (2009) argue in favor of analyses that ethnographically follow economic phenomena, including loans. To this end, Fabio Mattioli (2019) takes as his focus the everyday life of credit and debt and encourages a more expansive focus on the different forms debt takes and the roles that industry and political institutions play. Such an approach embraces the ethnographic method as well as the strengths of anthropology, such as cultural theory and inquiry. Indeed, "because so much of the global economy is virtual, it is important for anthropologists and others to develop theoretical understandings of the transformation of capitalism to virtualism and the global and local implications of this transformation for culture and society" (Pálsson and Durrenberger 2015: xx; see also Carrier 2019). For Keith Hart et al. (2010), there is a need to *re-people* the economy both within anthropology and the discipline of economics. This requires moving away from axioms focused on the supremacy of the free market and neoclassical economics (e.g., to go from spreadsheets to city streets) to understand economic life through a "human" economy (Hart et al. 2010).

Keith Hart and Horacio Ortiz (2014), two leading voices in economic anthropology on the role that history could play, acknowledge that if history and ethnography are able to "throw more than superficial light on society, we must transcend the categories that shape media discourses of the 'crisis' and try to understand our shared human predicament as a moment in the history of money" (Hart and Ortiz 2014: 466). There is, therefore, a need to move beyond mid-century debates about the form and substance of the economy, in favor of recalling the "openness to ethnographic discovery with a global vision of economic history" earlier examined through the works of Mauss and Polanyi (Hart and Ortiz 2014: 466). Appreciating, for example, Iceland's geographic position in the social imaginary of Europe and Iceland's postcolonial history, the country's "marginal" position (*mitt á milla*) has become a frame through which the economy is understood in cultural terms. It is a framing, moreover, that can extend understandings of how economy is understood without simply referring to neoliberal policy and its societal implications. The banking collapse has given rise to affective conditions that amount to a collective sense of unabated anticipation of another crisis occurring (or at least leaving open the idea that the stability of the national economy cannot be taken for granted) (Heffernan and Pawlak 2020). This has led to new localized understandings of how economy engenders social relations, leading to new spatial and temporal relations between the local and the global.

Crucially, a human economy must comprise four parts, including: possessing people, first and foremost, at the center, through which the economy is (re)made; it must represent and work for a range of individuals, groups, contexts, and lifeways; expand beyond market principles to envelop a majority of needs and interests; and, finally, work across different settings to accommodate world society in its heterogeneity (Hart et al. 2010).

In a separate, albeit linked, debate, J.K. Gibson-Graham (2014) has argued for a reconceptualization of economic life through thick description – after Clifford Geertz (1973) – and weak theory. This approach combines the depictions of a diversity of economic phenomena in myriad settings with weak theory to "produce a performative rethinking of economy centred on the well-being of people and the planet" (Gibson-Graham 2014: 147). Importantly, given the economic insecurity experienced across Europe, such insights seek to understand and theorize about life as it is lived in local settings. An insistence on diverse economies and the practices that underpin them in given locations call to mind foci whereby "the dynamics of change are an open empirical question, not a structural imperative..." (Gibson-Graham 2014: 149). "A thick description attends to multiple transactions that are bound up in the cash payment" (Gibson-Graham 2014: 148). Indeed, "a weak theory of diverse economies opens [attention] to these and a myriad of other motivating forces that not only confined to so-called nonmainstream practices" (Gibson-Graham 2014: 151). These might include "a much wider range of social relations [that] bear on economic practices, including to name just some, trust, care, sharing, reciprocity, cooperation, divestiture, future orientation, collective agreement, coercion, bondage, thrift, guilt, love, community pressure, equity, self-exploitation, solidarity, distributive justice, stewardship, spiritual connection, and environmental and social justice" (Gibson-Graham 2014: 151).

Across both a human and diverse economic approach, these authors argue in favor of bringing world history into analysis. Anthropological works that can bring together global history, local experience, and cogent analysis of the economy are needed, all without necessarily making recourse to universalist readings of such things as "money," "market," and "social relations," among others. David Graeber's *Debt: The first 5,000 years* (2011) is a fine example of how history, experience, and analysis invoke the human in all its diversity, while still commenting on the commonness of debt, credit, money, and markets in a comparative view. Jane I. Guyer is also credited for embarking on a similar feat in *Marginal gains: Monetary transactions in Atlantic Africa* (2004). Deftly, Guyer "shows that Atlantic Africans developed a plural framework for commerce, where social rank, multiple scales of monetary valuation, and circuits of exchange intersect and diverge" (Hart and Ortiz 2014: 474). Guyer's historical examination takes in local developments, engagements with colonialism, and participation in the global economy. In turn, her work "contrasts this flexible regional culture with the reductive oversimplifications of economic historians and the parochialism of local ethnographers" (Hart and Ortiz 2014: 474).

Concluding Remarks

A key emerging trend, then, following the GFC has been a dual focus on the cultural components of economic life as well as its structural and processual aspects, with the latter two becoming separated as part of the cultural turn in the 1980s. The growing focus on anti-austerity responses across Europe and North America after 2008 demonstrates this through practitioners' focus on local responses to national and international processes and systems. In other areas, the rise of territorial, populist, and reactionary responses in connection with the rise of far-right political parties or citizens' push to reject "globalist" agendas constitutes another. In contrast to the subdiscipline in the early twentieth century and its focus on small-scale societies, economic anthropologists are now studying large societies responding to the dynamics of the free market, or else highlighting the ways people make sense of new pathways forward in light of market failures (Carrier 2019), particularly through repositioning the human in economic study. Throughout this chapter, the key subdisciplinary debates from economic anthropology have been recapitulated amid the shifting economic and sociopolitical context that local and global society now finds itself.

Through exploring these key debates, including the form and content of economy, the cultural components and material consequences of economic life, and the rise of virtual economies, this chapter has shown how practitioners have come to enquire into economy in different parts of the world and at different points in time. Methodologically, this has included embracing individualist and systemic approaches to understanding economic phenomena. The case of Iceland, and other European countries who were adversely affected by the GFC, serves to illustrate the ways that economic phenomena are continuously tweaked and refined, usually by elites, the reality of which oftentimes disproportionately affects ordinary classes of people. This example and the recapitulation of the above explored debates has shown the subdiscipline of economic anthropology to be dynamic, albeit constrained by academic pigeonholing and geopolitical tensions that prioritize understanding economy through the metropolitan center. The human economy and diverse economy show tremendous promise for continuing to understand economic life in its full cultural, temporal, and geographic specificities. This is particularly so as crises outside of economics and finance (e.g. in health and climate) continue to bring individuals and communities into new relationships with commodities and markets, and stoking existing inequities or creating new concerns over social justice, public welfare, resource allocation, and collective equality.

Cross-References

▶ Durkheimian Revolution in Understanding Morality: Socially Created, Scientifically Grasped
▶ History of Thought of Economics as a Guide for the Future
▶ Neoclassical Economics: Origins, Evolution, and Critique

References

Alexander C, Bruun MH, Koch I (2018) Political economy comes home: on the moral economies of housing. Crit Anthropol 38(2):121–139
Anderson B (2016) Neoliberal affects. Prog Hum Geogr 40(6):734–753
Applbaum K (2009) Free markets and the unfettered imagination of value: a response to Hart/Ortiz and Gudeman. Anthropol Today 25(1):26–27
Berlant L (2011) Cruel optimism. Duke University Press, Durham
Carrier JG (2019) Introduction to a research agenda for economic anthropology. In: Carrier JG (ed) A research agenda for economic anthropology. Edward Elgar, Cheltenham, pp 1–9
Carrier JG (2012) Introduction. In: Carrier JG (ed) A handbook of economic anthropology, 2nd edn. Edward Elgar, Cheltenham, pp 1–9
Carrier JG (1998) Introduction. In: Carrier JG, Miller D (eds) Virtualism: a new political economy. Routeldge, London, pp 1–24
Chakrabarty D (2008) Provincializing Europe: postcolonial thought and historical difference. Princeton University Press, Princeton
Collier SJ (2012) Neoliberalism as big leviathan, or...? A response to Wacquant and Hilgers. Soc Anthropol 20(2):186–195
Comaroff J, Comaroff JL (2012) Theory from the south: or, how euro-America is evolving toward Africa. Routledge, Abingdon
Davies W, McGoey L (2012) Rationalities of ignorance: on financial crisis and the ambivalence of neo-liberal epistemology. Econ Soc 41:64–83
Dixon W, Wilson D (2011) A history of homo economicus: the nature of the moral in economic theory. Routledge, London
Dunn E (2004) Privatising Poland: baby food, big business, and the remaking of labor. Cornell University Press, New York
Durkheim E (2014/1893) The division of labour in society. The Free Press, New York
Einarsson N (2015) When fishing rights go up against human rights. In: Durrenberger EP, Palsson G (eds) Gamblilng debt: Iceland's rise and fall in the global economy. University Press of Colorado, Colorado, pp 151–160

Elyachar J, Maurer B (2009) Retooling anthropology: a response to Hart/Ortiz and Gudeman. Anthropol Today 25(1):27
Ganti T (2014) Neoliberalism. Annu Rev Anthropol 43(1):89–104
Geertz C (1973) The interpretation of culture. Basic Books, New York
Gibson-Graham JK (2014) Rethinking the economy with thick description and weak theory. Curr Anthropol 55(s9):s147–s153
Graeber D (2011) Debt: the first 5,000 years. Melville House, New York
Greenhouse CJ (2010) Introduction. In: Greenhouse CJ (ed) Ethnographies of neoliberalism. University of Pennsylvania Press, Philadelphia, pp 1–12
Gudeman S (2012) Community and economy: Economy's base. In: Carrier JG (ed) A handbook of economic anthropology, 2nd edn. Edward Elgar, Cheltenham, pp 95–108
Gudeman S (2008) Economy's tension: the dialectics of community and market. Berghahn, London
Gudeman S (1986) Economics as culture: models and metaphors of livelihood. Routledge, London
Guyer JI (2017) Money is good to think: from "wants of the mind" to conversation, stories, and accounts. In: Hart K (ed) Money in a human economy. Berghahn, Oxford, pp 43–60
Guyer JI (2004) Marginal gains: monetary transactions in Atlantic Africa. Chicago University Press, Chicago
Hann C (ed) (2002) Postsocialism: ideals, ideologies and practices in Eurasia. Routeldge, London
Hann C, Hart K (2011) Economic anthropology: history, ethnography. Polity Press, Cambridge, Critique
Hart K, Laville J-L, Cattani AD (2010) Building the human economy together. In: Hart K, Laville J-L, Cattani AD (eds) The human economy: a citizen's guide. Polity Press, Cambridge, pp 1–18
Hart K, Ortiz H (2014) The anthropology of money and finance: between ethnography and world history. Annu Rev Anthropol 43:465–482
Heffernan T (2020) Crisis and belonging: protest voices and empathic solidarity in post-economic collapse Iceland. Religions 11(1):22
Heffernan T, Pawlak M (2020) Crisis futures: the affects and temporalities of economic collapse in Iceland. Hist Anthropol 31(3):314–330
Helgason A, Pálsson G (1998) Cash for quotas: disputes over the legitimacy of an economic model of fishing in Iceland. In: Carrier JG, Miller D (eds) Virtualism: a new political economy. Routledge, London, pp 117–134
Helgason A, Pálsson G (1997) Contested commodities: the moral landscape of modernist regimes. JRAI 3(3):451–471
Herskovits MJ (1940) The economic life of primitive peoples. Alfred Knopf, New York
Ho K (2009) Liquidated: an ethnography of wall street. Duke University Press, Durham
Jónsson ÖD, Sæmundsson RJ (2015) Free market ideology, crony capitalism, and social resilience. In: Durrenberger EP, Pálsson G (eds) Gambling debt: Iceland's rise and fall in the global economy. University Press of Colorado, Colorado, pp 23–32
Kapferer B (2004) Introduction: the social construction of reductionist thought and practice. Soc Anal 48(3):151–161
Knight DM, Stewart C (2016) Ethnographies of austerity: temporality, crisis and affect in southern Europe. Hist Anthropol 27(1):1–18
Lévi-Strauss C (1969) The elementry structures of kinship. Eyre and Spottiswoode, London
Loftsdóttir K (2019) Crisis and coloniality at Europe's margins: creating exotic Iceland. Routledge, Oxford
Loftsdóttir K (2010) The loss of innocence: the Icelandic financial crisis and colonial past. Anthropol Today 26(6):9–13
Maguire J (2015) Virtual fish stink, too. In: Durrenberger EP, Palsson G (eds) Gambling debt: Iceland's rise and fall in the global economy. University Press of Colorado, Colorado, pp 121–136
Malinowski B (1922) Argonauts of the Western Pacific: an account of native enterprise and adventure in the archipelagoes of Melanesian New Guinea. Waveland Press, Illinois
Mattioli F (2019) Debt, financialisation and politics. In: Carrier JG (ed) A research agenda for economic anthropology. Edward Elgar, Cheltenham, pp 56–73

Maurer B (2006) The anthropology of money. Annu Rev Anthropol 35(1):15–36
Mauss M (1966) The gift: forms and functions of exchange in archaic societies. Cohen and West, London
McCormack F (2017) Private oceans: the enclosure and marketisation of the seas. Pluto Press, London
McCormack F (2007) Moral economy and Maori fisheries. Sites 4(1):45–69
Miller D (2001) A theory of virtualism. In: Miller D (ed) Consumption: critical concepts in the social sciences. Routledge, London, pp 298–308
Mixa MW (2009) Once in khaki suits: Socioeconomical features of the Icelandic collapse. In: Hannibalsson I (ed) Rannsóknir í félagsvísindum X: Hagfræðideild og viðskiptafræðideild. Félagsvísindastofnun HÍ, Reykjavík, pp 435–447
Muir S (2021) Routine crisis: an ethnography of disillusion. University of Chicago Press, Chicago
Ong A (2006) Neoliberalism as exception: mutations in citizenship and sovereignty. Duke University Press, Durham
Pálsson G, Durrenberger EP (2015) Introduction: the banality of financial evil. In: Durrenberger EP, Pálsson G (eds) Gambling debt: Iceland's rise and fall in the global economy. University Press of Colorado, Colorado, pp xiii–xxix
Polanyi K (1957) The economy as instituted process. In: Polanyi K, Arensberg C, Pearson H (eds) Trade and market in the early empires: economies in history and theory. Free Press, Chicago, pp 243–269
Polanyi K (1944) The great transformation: the political and economic origins of our times. Farrar and Rinehart, New York
Powell K (2017) Brexit positions: neoliberalism, austerity and immigration – the (im)possibilities of political revolution. Dialect Anthropol 41:225–240
Rakopoulos T (2018) Introduction: austerity, measured. In: Rakopoulos T (ed) The global life of austerity: comparing beyond Europe. Berghahn, New York, pp 1–16
Rousseau J-J (2012/1762) The social contract. Penguin, London
Rousseau J-J (2010/1762) Emile, or on education. Dartmouth College Press, Hanover
Rousseau J-J (1984/1754) A discourse on inequality. Penguin, Harmondsworth
Sabaté I (2016) The Spanish mortgage crisis and the re-emergence of moral economies in uncertain times. Hist Anthropol 27(1):107–120
Sahlins M (1969) Economic anthropology and anthropological economics. Soc Sci Inf 8(5):13–33
Said EW (2014) Orientalism, 25th Anniversary ed. Knopf Doubleday, New York
Scheppele KL (2010) Liberalism against neoliberalism: resistance to structural adjustment and the fragmentation of the state in Russia and Hungary. In: Greenhouse CJ (ed) Ethnographies of neoliberalism. University of Pennsylvania Press, Philadelphia, pp 44–59
Scott J (1979) The moral economy of the peasant: rebellion and subsistence in Southeast Asia. Yale University Press, New Haven
Sigurjónsson ÞO, Mixa MW (2011) Learning from the "worst behaved": Iceland's financial crisis and the Nordic comparison. Thunderbird Int Bus Rev 53(2):209–223
Siniscalchi V (2012) Towards an economic anthropology of Europe. In: Carrier JG (ed) Handbook of economic anthropology. Edward Elgar, Cheltenham, pp 553–567
Smith A (2008/1776) The wealth of nations. Oxford University Press, Oxford
Spyridakis M (ed) (2018) Market versus society: anthropological insights. Springer, Cham
Thompson EP (1968) The making of the English working class. Penguin, Hardondsworth
Trawick P, Hornborg A (2015) Revisiting the image of limited good: on sustainability, thermodynamics, and the illusion of creating wealth. Curr Anthropol 56(1):1–27
Tsing A (2005) Friction: an ethnography of global connection. Princeton University Press, Princeton

Anthropology, the Environment, and Environmental Crisis

19

María A. Guzmán-Gallegos and Esben Leifsen

Contents

Introduction	484
Human Adaptation	486
Natural Resource Extraction	488
Toxicity	491
Temporalities and Nonhuman Agency	493
The Damaged Planet and Persistent Colonial Orders	496
Conclusion	500
References	501

Abstract

This chapter provides an overview of anthropological research on the environment and socio-environmental and geosocial transformations which often are associated with crises and the Anthropocene. It presents ways anthropologists and researchers from related disciplines study and write about these transformations and considers their proposals to develop analytical tools, frameworks, and innovative methods that scrutinize and redefine the relations between humans and the environment. The chapter consists of five interlinked thematic discussions that span a long period in the study of human-environmental interrelations, from the early perspectives on cultural ecology to the most recent work on more-than-human relations. The thematic discussions consider human adaptation; the political ecology of natural resource extraction; research on toxicity; perspectives on

---

M. A. Guzmán-Gallegos (✉)
VID Specialized University, Oslo, Norway
e-mail: maria.guzman-gallegos@vid.no

E. Leifsen
Department of International Environment and Development Studies, Norwegian University of Life Sciences (NMBU), Aas, Norway
e-mail: esben.leifsen@nmbu.no

temporalities and more-than-human agency; and finally, recent approaches to the study of life on a damaged planet.

Keywords

Environmental crisis · Anthropocene · Human adaptation · Resource extraction · Toxicity · More-than-human anthropology

Introduction

At the time of finalizing the work with this chapter, the broadcaster brings news about an Australian federal court judgment: It declares that the approval of the expansion of a large-scale coal extraction project in New South Wales is contingent on the environmental minister's duty to take reasonable care that young people are not harmed by carbon dioxide emissions. In this case brought by eight school children and an 86-year-old nun, the judge, under the Environment Protection and Biodiversity Conservation Act, establishes a probable relation between increased industrial coal extraction in Australia and global average temperature rise. The duty of care principle refers to environmental consequences of global climate change which is already experienced by the Australian population and which most probably will intensify in the future: extreme heat, extensive and repeated bushfires, the disappearance of The Great Barrier Reef, and the destruction of most of the country's eastern eucalypt forests (cf. The Guardian 2021). Resounding an international discourse backed by the scientific work of the UN Intergovernmental Panel on Climate Change (ICCP), the federal court judgment confirms that continued coal extraction and $CO2$ emissions have *planetary consequences*. Interventions of this kind will most probably accelerate the environmental crises the earth's inhabitants already live by.

Awareness about environmental crises has been with us at least since the 1960s. Often, this awareness is considered to emerge from the scientific work and environmental activism of the biologist Rachel Carson. Her 1962 publication, *Silent Spring* (Carson 1962), on the damaging environmental effects of indiscriminate use of chemicals in US agriculture, marks the beginning of a political movement and an interdisciplinary field of study. The anthropologist, David Bond, identifies in a paper on the history of the concept "environment" a change in the late 1960s and early 1970s. In this period, "the environment shifted from an erudite shorthand for the influence of a context to the premier diagnostic of a troubling new world of induced precarity" (Bond 2018:1). The precariousness of life that came to the fore by this shift of meaning was in subsequent decades inextricably tied to a series of crisis phenomena – the extraction and combustion of fossil hydrocarbons, industrial and urban contamination, radioactive fallout, waste disposal, and chemical industrial food production.

Despite a conceptual redefinition in the 1960s and 1970s, this troubling new world had already been in the making for a much longer time. In a literature with divergent orientations and approaches, environmental crises are notwithstanding closely associated with the epoch of colonialism and the proliferation of capitalism (and other systems of imperial expansionism), and with industrial forms of production and economic organization. These economic forms have a decisive hold on life on earth. Social critique of the commodification of nature (Castree 2003) has taught us that not only solutions to the environmental crisis – such as carbon capture and storage, and forest and biodiversity conservation – can be adapted to the market logic and form part of new "green" deals, but also environmental harm itself can be turned into market products and business opportunities, for example, by making environmental disasters profitable (Adams et al. 2009), and by tying environmental restoration to market mechanisms through the exchange of ecosystem service equivalents (Sullivan 2013).

Around the turn of the millennium, and a result of several decades of climate change research, human activity related to industrial production and dominant forms of economic organization was tied to a new concept – the Anthropocene. First introduced by natural scientists in 2000, the concept addresses a range of serious environmental challenges: climate change, biodiversity loss, the acidification of oceans, the thinning of the ozone layer, changes in the global nitrogen and phosphorus cycles, aerosol pollution in the atmosphere, irreversible changes to fresh water sources and flows, and extensive transformation of landforms (cf. Clark and Szerszynski 2021). The notion of the Anthropocene highlights a new dimension regarding human efforts to generate economic growth and achieve "progress," namely that humans not only leave traces on the Earth's surface, but their activity also has harming consequences on earth systems. As indicated in the case this chapter started out with, the human quest for earth materials, and the ways these materials circulate and are transformed through their extraction and uses, is now considered to affect the earth as a whole and alter the conditions for life on it.

This chapter discusses ways anthropological research has responded to the intellectual and political challenges brought about by the new consciousness of environmental crises and the Anthropocene. This response includes how anthropologists write about socio-environmental and geosocial transformations, how they apply new analytical tools and frameworks, and how they make use of innovative methodological strategies in order to understand these transformations. Additionally, it considers how approaching and theorizing earth problems has fostered interdisciplinary collaboration in anthropological research. Such collaboration is often foregrounded by researchers working with environmental harm and the damaged planet. Some of their contributions will form part of this account.

This overview is a nonexhaustive review of relevant perspectives. Given the substantial and varied literature on topics related to environmental crisis, and the many inter-disciplinary routes these studies take, this overview is necessarily selective. It is informed by the authors' research on topics related to socio-environmental issues and conflicts, extractivism, toxicity, and Indigenous livelihoods and lifeways, and by debates and lines of inquiry they find especially important. An assumption is made about the state of the world in this discussion, namely that one can write about it in terms of the disrupted earth and the damaged planet: Human activity has seriously altered the conditions and possibilities for survival of life on earth as we know it. Moreover, this state of the world makes it analytically difficult to maintain a distinction, persistent in dominant Western thought, between society and nature. In the review of anthropological studies and related research that address environmental problems and crises, particular attention is given to the academic search for alternative concepts and analytical resources to understand the present conditions on Planet Earth.

The chapter is knitted together by a set of loosely connected themes, which are viewed in continuation and contrast to each other. It starts out considering a main theme in environmental anthropology, namely *human adaptation*, and ways climate change is addressed within this approach today. This focus is followed by a discussion of the anthropology of *natural resource extraction*, and especially the extraction of hydrocarbons and minerals, and how this research has contributed to the interdisciplinary study of political ecology. After this, a growing and heterogenous field of research is presented, and which is core to the study of the Anthropocene, namely *toxicity*. A fourth section highlights some of the main challenges regarding *temporalities* in the study of transformations generated by anthropogenic activity. Of special concern here are studies that include time frames in their analysis that exceed human time, and which consider *more-than-human agency*. A final section titled *the damaged planet* considers some interdisciplinary tendencies that offer new perspectives and conceptual tools to understand the conditions of life in disturbed and damaged environments.

Human Adaptation

A long-standing theme in anthropological research has been human *adaptation* to the natural environment. Within materialist approaches in the study of human populations and their relations to the natural environment, social organization (of production in particular) and cultural conceptualizations (and especially environmental knowledge) are main dimensions of inquiry. In the cultural ecology currently associated with Julian Steward and Leslie White, especially the former inspired several new generations of researchers to explore the idea of adaptation and pose questions about social and cultural stasis and change (cf. Orlove 1980). Anthropology studying human-environment interrelations focused in early stages (loosely delimited to the 1930s – 1970s) on the dynamics of human collectives. Nature, as the counterpart, was in this mode of thought considered to have its proper systemic dynamics based on the idea of "balance of nature." An understanding of equilibrium, drawn from ecology and ecosystem theory, assumed "stasis, homeostatic regulation and stable equilibrium points or cycles" (Scoones 1999:481). Consequently, debates, disagreements, and analytical innovation in the social sciences, anthropology included, revolved around the issue of how nature influences human life and not vice versa.

Basic assumptions regarding human adaptation started to be questioned during the 1970s – a time of reorientation in many respects regarding the environment, within academia as well as in the wider field of international policy formulation and global environmentalism. Processual approaches within ecological anthropology opened for more nuanced understandings of human behavior and decision-making, which among a range of other issues facilitated "the examination of the ways in which human action affects ecosystems and environmental constraints influence human decision-making" (Orlove 1980:249). Still incipient at the time of the writing of an Annual Review of Anthropology article in 1980, Orlove indicates the emergence of an approach in which adaptation addresses a dynamic two-ways interaction. This suggested additional problems with separating realms (humans vs. nature), which consequently also raised questions about dichotomous thinking as analytical strategy. Social sciences were informed by systems ecology in this respect, which often included humans in their models.

Another important influence from biological research regards the new ecology emerging in the 1970s. At this time, researchers started to question the equilibrium and homeostasis theses of conventional ecological systems theory. A wide specter of new empirical studies drew on nonequilibrium theory in the research of ecosystems, and emphasized nonlinear, stochastic, and chaotic processes and dynamics. Variability, scale, and temporal dimensions informed the framing of how systems and systems' transformations were understood. Scoones argues that social sciences were slow in integrating these insights from new ecology research. Too often, he writes, "social sciences analysis – whether in anthropology, sociology, geography or economics – has remained attached to a static equilibrial view of ecology" (1999: 483).

The study of environmental stresses and hazards indicated, nevertheless, a new direction in the social sciences that highlights the unstable character of natural environments. This view was strengthened in combination with new understandings of the socio–environmental interactions and their role in generating transformations that condition human livelihoods. A broadening of perspective came together with a widening of the scope of study beyond the local context to also include external contexts and influences. Political economy and later on political ecology, discussed in the next section, provided here the analytical apparatus to handle these additional dimensions. Over time, studies within this current constituted a robust field of research of environmental disasters with a focus on the social origins of environmental changes (Orr et al. 2015). Interestingly, the understanding of nonequilibrium in social sciences tends to be conceptualized as the result of modern interventions. Disruptive transformations are oftentimes viewed as results of market forces, economic growth policies, and the expansion of capitalist and neoliberal "development." An underlying assumption seems to be that environmental stability preceded these man-made influences (Scoones 1999).

Current disaster studies direct their attention to global climate change issues (Orr et al. 2015). Research in this field acknowledges that human activity affects global temperature rise. There is a general understanding that intensification of natural disasters and the many expressions of extreme weather and related phenomena are related to anthropogenic influences. The main concern of disaster studies, however, is not to explore the complexities of these socio-environmental transformations, but rather to understand people's adaptive responses to climate-related change.

Equipped with the conceptual framework of adaptive capacity, primary adaptive strategies, coping, vulnerability, resilience, flexibility, diversity, and action, the American Anthropological Association's Global Climate Change Task Force states in their final report that "adaptation is and has been a core concept of anthropology since the emergence of the field" (Fiske et al. 2014:41). During the time when the Anthropocene was constituted in academia as the current geological epoch, influential scholarship brought adaptation back into the core of anthropological knowledge production – now directing the attention toward human responses to changes in an unstable and troubled natural world.

Natural Resource Extraction

The focus in this section is on another strand of studies where the issue of human relations to environmental transformations is treated differently than in human adaptation perspectives. This concerns the anthropology of resource extraction, and more specifically the anthropology of minerals and hydrocarbons. Minerals and hydrocarbons are vital resources to human societies and are essential to value generation in the increasingly globalized world economy. Extraction of subsoil materials brings important revenue to resource-rich countries and generates astronomical levels of surplus value to transnational companies. Oil and oil derivatives are essential substances for human transport and for industrial production, and a range of minerals are crucial components in the technology we live by and which politicians envision we will depend on in the future. Moreover, modes of extraction and different uses of minerals and hydrocarbons have impacts on the environment, which are considered to contribute to climate change and other transformative planetary processes. Drawing from diverse theoretical and interdisciplinary frameworks, anthropology problematizes a well of themes related to these dimensions in their studies (for articles reviewing literature on mineral and oil extraction, see Gilberthorpe and Rajak 2017; Jacka 2018; Rogers 2014).

However, in anthropological studies that more specifically address environmental issues and problems around resource extraction, political ecology is a main theoretical orientation. Political ecology constitutes an interdisciplinary field of critical social study, in which anthropology is in conversation with other disciplines such as geography, ecological economics, and agrarian and global development studies. It addresses human-environmental relations and conflicts related to issues regarding access and management of resources, and transformations of the resource bases that provide people's livelihoods. Moreover, dynamics around vital resources are viewed in relation to wider institutional contexts and not least in relation to external structures of power.

This branch of research took form in the 1980s and combined ecological anthropology and political economy, the latter with specific contributions from Marxist and world systems theory, and with economic geographer, David Harvey, and anthropologist, Erik Wolf, as influential thinkers. As part of a theoretical exploration in the 1990s, political ecology was influenced by the "poststructuralist" critique inspired by Michelle Foucault's work. From this strain of critique, political ecology added perspectives on the role of knowledge in the exercise of power and authority (Foucault 1980), and on life as an area of governance through the exercise of "biopolitics" (Foucault 2003). In later decades, the understanding of socio-economic conflicts within political ecology has been influenced by ecological economics. From Joan Martinez-Alier's book *The Environmentalism of the poor* (Martinez-Alier 2003), this current has integrated a perspective on ecological distribution conflicts and broadened its understanding of how marginalized people defend their livelihoods through mobilizations around environmental issues. Finally, the influence from environmental justice perspectives should be mentioned – both related to core foci in this literature on distribution, recognition, and participation issues (Fraser 2000; Schlosberg 2013), regarding environmental racism (Voyles 2015) and in relation to violence (Peluso and Watts 2001; Nixon 2011).

The anthropological and wider literature on mineral extraction offers insight into the entanglements of mining with economic development and "expectations of modernity" (Ferguson 1999). It also outlines the many and serious environmental impacts related to the removal of earth material, the storage of mineral waste, and the processing and transportation of mineral ore. This includes deforestation, alteration of landscapes, and ecosystems and biodiversity loss, as well as contamination to water, soil, and air. In large-scale mining, a main problem-complex is related to the extensive volumes of waste produced – tailings, sterile rock, and overburden – and to the long-term effects of acid mine drainage and sedimentation on forests, land, and hydro-landscapes. In middle and artisanal small-scale mining (ASM), special attention has been given to the use of mercury and cyanide in the processing of gold, and to the problems related to the bioaccumulation and biomagnification of mercury up through the food chain. Anthropological studies on ASM direct their attention to the informal – formal / illegal – legal gray zone that this kind of activity tends to operate in (Verbrugge 2015; High 2012; Kolen et al. 2018; Peluso 2018). In the wider literature, there is an emphasis on issues related to the formalization of this sector and to authorities' efforts to ban mercury use (Hilson et al. 2017). Research on large-scale mining approaches environmental governance differently as part of power structures that formalize and legitimize extractive interventions, and which generate socio-environmental conflicts and popular mobilization (Golub 2014; Jacka 2015; Li 2015).

A work that in many ways guides and inspires current research on large-scale mining is Stuart Kirsch's *Mining capitalism* (Kirsch 2014). Informed by two decades of ethnographic research, Kirsch offers in this book a complex analysis of social, political, legal, and corporate science practices constituting and over time maintaining the Ok Tedi open-cut copper and gold mining project in Papua New Guinea. As part of a multidimensional analysis of the relationships between the mining industry and its critics – including people affected by the extractive intervention and activists forming part of environmentalist and human rights transnational action networks – Kirsch explores how mining corporations continuously negate the socio-environmental impacts of their activity. This detachment from responsibility conditions the life of affected Indigenous people, and the ways resistant to mining capitalism are formulated, enacted, and reinvented. Kirsch writes about this socio-environmental conflict in terms of "colliding ecologies," referring with this concept to distinct ways affected peoples and the mining company make use of and exploit natural resources. When they interact, "one system may limit the viability of the other" (Kirsch 2014:16). In this case, the viability of a subsistence economy is severely limited by the effects of the contamination from the Ok Tedi mine. As this study emphasizes, industrial mineral extraction impoverishes and dispossesses people by contamination (for similar perspectives, see Perrault 2013; Leifsen 2017).

While anthropological and related research on socio-environmental mining transformations move within and occasionally beyond the analytical boundaries of political ecology, the main body of corresponding research on hydrocarbons has another analytical orientation. There are important studies of oil extraction that could be placed within the broad political ecology umbrella, such as Susanne Sawyer's early work, *Crude Chronicles* (Sawyer 2004), and also Behrends et al.'s (2011) edited volume on "crude domination" and Watts' (2013) study of "blood oil." However, the environmental impacts of the extraction and combustion of hydrocarbons, and more generally of the activity of petroleum-dependent industries, are analytically explored, particularly within toxicity studies, and are, as Bond (2018) argues, strongly associated with environmental crises.

Before an outline of perspectives on toxicity in the next section, it should be mentioned that the study of resource extraction has a wider scope than oil and minerals. Scholars working with extraction in Latin America deploy the concept of "extractivism." This concept draws on dependency theory (Svampa 2016) and is the product of collaborations between intellectuals and activists. "Extractivism" expands the scope of analysis associated with extraction not only by including hydrocarbons and minerals, but also by highlighting the unevenness of global capitalism, and by implicitly foregrounding a critical stance toward the politics, policies, and ideologies that promote different kinds of extraction (Acosta 2016; Gudynas 2015). Beside the extraction of nonrenewable resources, "extractivism" refers to industrial fishing and extensive and intensive monoculture agriculture. Importantly, "extractivism" points not only to "colliding ecologies," but also to the structural reproduction of cores and peripheries made possible through the continuous expansion of extractive frontiers, on which capitalism always has depended. "Extractivism" likewise refers to the topographical and chemical recompositions of landscapes: the removal of mountains, the rerouting of rivers, the transformation of forests into bare landscapes, and the use of a wide diversity of chemicals, which change aquatic and terrestrial ecologies. Alluding to the powerful technologies that such transformations presuppose, which make evident the human geological forces of extractive activities, "extractivism" is described as an expression of the Anthropocene (de la Cadena and Blaser 2018). With this shift in focus, the concept of "extractivism" frames as well a fundamental critique of hegemonic understandings of development, of forms of ordering the world that constantly sacrifice spaces of living, and that make both human and nonhuman life sacrificial.

Toxicity

The environmental and political concerns of scholars working with extraction and extractivism resonate with the interests of those working with toxicity. A core theme in the latter is the interrelations between human agency, structures of power and dispossession, and the creation of chemically disrupted environments. A starting point for scholars working with toxicity is that we live in a permanently polluted world from which nobody can escape (Murphy 2017; Liboiron et al. 2018; Nading 2020). Industrially produced chemicals together with the pollutants from extractive industries are continuously released in waterways, in soils and into the air. These pollutants, also called toxicants, are ubiquitous and pervasive and affect human and nonhuman bodies alike. Testing indicates that industrial chemicals are present in human bodies anywhere in the world. Scholars underline that the production and release of toxicants are intrinsically part of and reproduce different orders of life that are based on and maintain inequality and destruction. Murphy (2008) calls them, "chemical regimes of living" while Walker (2011:xi) argues that they constitute "a new age of toxicity." Both descriptions allude to the notion of the "Anthropocene" in the sense that they underline toxicity's human origin and planetary scale.

These assertions of the toxic condition of the world acknowledge both a shared situation of chemical alterations, and the analytical and political impossibility of separating the natural environment from social organization and structures of power. This acknowledgment has fostered collaborations within academia between anthropology and other disciplines – Science and Technology Studies (STS), history and health sciences. Similar to scholarship on extraction and extractivism, collaborations have also extended beyond research environments and include affected communities and social movements. How scholars deal with the implications of living in a permanently polluted world reveals nevertheless distinct orientations and tensions within anthropological research on toxicity.

One of the major questions raised in this literature is who the "we" living on this disrupted planet are. Recognizing the uneven distribution of exposure, a shared understanding among anthropologists is, as Liboiron et al. (2018) point out, that planetary is not the same as generality. There is not a general human "we" or a general condition of exposure. Toxicity and the experience of exposure are highly situated, and "premised upon and reproduced by systems of colonialism, racism, capitalism, patriarchy, and other structures that require land and bodies as sacrifice zones" (Liboiron et al. 2018:332).

These anthropological insights are particularly salient in research carried out in the USA on toxic exposure. Influenced by environmental justice concerns, toxicity studies have focused on native, minority, and low-income communities, and on their varied, slow, and long-term exposure to radioactive minerals, such as uranium or plutonium and other industrial chemicals. Taking into account the structures of settler colonialism, as well as industrial and military power, they explore how these structures cocreate racialized bodies and degraded landscapes (Bohme 2014; Kuletz 1998, Lerner 2010, Voyles 2015). Concepts such as "sacrifice zones," coined first by US government officials and redefined by Lerner, "slow violence" (Nixon

2011), and "slow disaster" (Knowles, cited in Liboiron 2018) are central to this analytical approach. In earlier work and focusing on disaster events, such as the 1984 gas leakage in a pesticide plant in Bhopal, or the 1986 nuclear accident in Chernobyl, anthropologists argue that disasters of this kind can engender enunciatory communities. As Fortun (2009) explains, these communities emerge as a response to profound change – they do not exist before the occurrence of a disaster. In a similar vein, Petryna referring to Chernobyl describes how those affected sought to gain state recognition of their suffering, engendering what Petryna (2002) calls "biological citizenship."

Another central question regards the im(perceptible) character of toxicity. While anthropologists broadly speaking agree that a chemical's toxicity is not easily determined, explanations about why vary. Nuances here are consequential. Some scholars assert that "toxicity is elusive" (Nading 2020:11) because toxicity emerges in situated biologies and ecologies, that is, in particular bodies and places. Thus, the toxicity of a chemical depends on dosages, on the duration of exposure, and on the genetic composition of bodies and nutritional factors. Building on STS studies, other scholars underline that toxic effects of chemicals and of radiation are brought into being through chains of associations between molecular processes, legal regulations and procedures, and scientific knowledge practices. The imperceptibility of toxicity is far from natural since it is often actively produced as such through the intertwining between research's methodological design, what the units of analysis are (individuals or populations), political processes, and state and industry's interests (Guzmán-Gallegos 2021; Michaels 2008; Murphy 2006; Sawyer 2015). Bauer (2018) demonstrates, for instance, how the political changes in Soviet-Union that followed Glasnost involved a redefinition of affected populations to affected individuals. It also meant the introduction of evidential criteria, common in Western academia and global institutions, that underline single causality rather than epidemiological correlation.

Scholars underline, more broadly, that science and politics bring about different definitions and manifestations of toxicity (Cram 2016; Liboiron et al. 2018) and contribute to sustain the structures and practices that make and maintain the world toxic. What is defined as toxic depends on regulations that determine threshold limits. While a premise of threshold limits is that bodies and ecologies can assimilate certain amounts of toxicants (chemicals or radiation alike), these limits are often the result of political negotiations and are measured in relation to effects on a standardized human (often white and male) body. Additionally, pollution regulations operate with cost-benefit metrics that justify the use of technologies not too expensive to industry and the state, even if the implication of this is the release of toxicants in the environment (Cram 2016). Focusing closely on the operations of science, Murphy (2017:495–497) argues furthermore that certain epistemic habits permeate all work about chemical exposure. One is to portray chemicals as discrete and isolated entities without context, while another is to document and measure the amount of single chemicals in individuals and the damage these chemicals do. Both, she suggests, characterize North American environmental biomedical research and environmental justice contestation. As it has been widely documented, however, establishing single causal relations between exposed individual bodies and health effects is not easy. State authorities or industry often dismiss the claims of affected populations adducing that their demands do not comply with scientific, statistical evidentiary criteria or regulatory metrics (Bauer 2018; Brown 2013; Lyon 2018; Sawyer 2015). Moreover, Tuck (2009) points out that seeking to provide evidence, scientists and activists can end up rendering Indigenous bodies and lands pathologically tainted and doomed, often resuscitating racist portraits of already dispossessed communities.

Critical to the intertwining between politics, science, and toxicity, scholars propose to reconceptualize both what the political is and what constitutes evidence. Instead of mainly conceiving of politics and political action as based on claims and counterclaims, and building on feminist science studies, Liborion (2016), Tironi (2018), and Lyons (2018), among others, suggest that the generation of spaces of ethics and care can be included in the political. The main focus here is not on oppositional events, or on the struggle to obtain political power, but rather on quotidian practices of coping, resisting, and creating alternatives. Tironi calls these practices "intimate activism." Seeking to go beyond state-based knowledge production and the limits of toxicology, and questioning hegemonic "evidence regimes," Lyons discusses the need of new analytical tools and proposes the concept of "evidentiary ecologies." This concept is based on her work in Colombia on aerial fumigation of glyphosate of illicit crops, forests, and soils – a central component of US-Colombia antidrug policy – and the demands for compensation from rural communities. "Evidentiary ecologies" refers to multiple layers of exposure and violence, which are registered in the tissues of plants, the organisms of animals, the sediments of soils, and in the webs of forest life that also include their human inhabitants. "Evidentiary ecologies," or "ground truths" (Weizman 2017), have the capacity to register present and past damage, revealing at the same time the capacity of regeneration after destruction.

Recognizing the toxic condition of the world and focusing on the possibilities and alternatives that this condition creates characterize anthropological research that explores emerging "chemo-socialities." Shapiro and Kirksey (2017) use chemo-socialites to refer to current chemically altered social ways of living which, they suggest, allow us to ask new questions about personhood, bodies, and sexuality (Chen 2012). Chemical alteration makes evident the permeability of bodies, and this understanding makes it possible to conceptualize disabilities, entities, and mixtures in new ways (Chen 2012). Yet, Hecht (2018) warns that a joyful connotation of toxicity and the acceptance that purity never existed run the risk of disregarding the brutal histories, structures of power, ontologies, and epistemologies that have contributed to current planetary conditions.

Temporalities and Nonhuman Agency

An increasing awareness about the human species as a geophysical and metrological agent challenges anthropologists and other social scientists to include other than human time frames into their analysis. Researchers who explore the conditions and governance of life, in environmentally altered places, see the relevance and need of expanding their analytical time horizons both into the future and the past. The uncertainty that surrounds environmental futures has made the work of anticipating – "modelling, planning, and interpolating the future of resources and environments" – central to environmental politics (Mathews and Barnes 2016:10). As a result, practices of imagining and prospecting the future have both become an important research focus and are integral parts of current research designs. Human-induced environmental change also challenges researchers to rethink their approaches to the study of the past. In environmental history research, there is a growing interest in more-than-human agency. Starting out by linking to the former section on toxicity, the following presentation highlights some of the studies that address the future and the past in extended ways.

Toxic exposures often have long-lasting consequences due, for example, to substances' slow processes of decomposition, sedimentation, and bioaccumulation, and to volatile chemical compounds reconfiguring into new toxic materiality. Acids leaking from sterile rock "graveyards" of an industrial mining project (Leifsen 2020), aromatic hydrocarbons filtering out into waterways from abandoned oil installations (Guzmán-Gallegos 2019a), radioactive uranium waste stored in the landscape of a former nuclear research and testing ground (Kuletz 1998), and methylmercury bioaccumulating and biomagnifying up the food chain (Tschirhart et al. 2012) are examples of environmental exposure that exceeds the temporalities of industrial production systems, the capacity of waste storage infrastructure, and planned restoration schemes. The longevity of chemical relations invites researchers to rethink their analytical time frames. Murphy (2013) precisely introduces the concept of "latency" to describe how past exposure marks present conditions and already alters the future.

Similarly, forecasting environmental risks and envisioning risk management are state and corporate practices now studied through what Mathews and Barnes (2016) call "prognostic politics." One example is the making of Environmental Impact Assessments (EIAs). An EIA is the outcome of a technical-scientific knowledge production controlled by the industry, structured by national and international standards, and approved by the state. It could be viewed as a product of epistemological politics resembling what Kirsch (2014) describes as "corporate science" production. The EIA establishes "the baseline against which potential disruptions could be measured and managed" (Bond 2018:6) and, at the same time, limits the responsibilities a company has in relation to the impacts of its intervention. As a core instrument in environmental management and governance, EIAs turn future impacts and potential crises into manageable risk scenarios – they forecast potential damage within frameworks of manageable risks (Bond 2018; Hébert 2016; Leifsen 2017; Li 2009). Anthropological research on this issue relates the making of a diagnosis and prognosis to the production of externalities not recognized by those causing them. Critical analysis of this kind provides us with insights into state and corporate practices of containment. Through these practices, environmental problems are construed in ways that rule out the possibility of an environmental crisis that humans cannot handle.

Moving now from the future to the past, the environmental impacts of current human activity on the natural world also challenge researchers to rethink the time depth of their studies. Little (1999) indicates this through the illustrating example of frontier expansion in the Ecuadorian Amazon due to oil extraction, colonization, deforestation, and conservation initiatives. Understanding transformation of this kind implies, according to him, to consider "at once the geological time frame of the formation of underground oil deposits, the biological time frame of the establishment of world-record levels of plant and animal diversity, and the cultural time frame of developmentalist frontier expansion" (Little 1999:262). In relation to the study of environmental consequences of the expansion of industrial capitalism, and even more so regarding research on the Anthropocene, one can register analytical ambitions to include other time horizons than those delimited by human activity.

One prominent example is the work of historian Dipesh Chakrabarty. In a much-debated article, he associates what he calls a collapse in the natural and human history divide with the academic and political awareness of global climate change and to the age of the Anthropocene. A main characteristic of this age and global situation is, according to him, that "the human being has become something much larger than the simple biological agent that he or she always has been. Humans now wield a geological force" (Chakrabarty 2009:206). The perspective that humans have become geological or geophysical agents, and as species alter the conditions for life on Earth, challenges the way scholars can think and write environmental history. It induces us to work with other time horizons than those related to human activity. Temporalities (the geological, biological, and cultural times in Little's perspective above) that natural and social sciences have separated intersect through the impacts of anthropogenic activity – or as Chakrabarty underlines, through the unintended consequences of human interventions staged by industrial capitalism.

Part of scholars' ambition, to rethink temporalities, is to extend their understanding of other than human historical actors. More-than-human agency is not a focus introduced to anthropology by research on environmental transformations and crises. In the Amazonian context, for example, anthropologists have for a longer time grappled with Indigenous perspectives on sociality that exceeds the human (e.g., Descola 2014; Guzmán-Gallegos 2019b; Kohn 2013; Viveiros de Castro 2004). In addressing more than human agency in human-disturbed landscapes and connecting this focus to various coexisting temporalities of transformation, newer anthropological studies resonate with the study of ecological interrelations within the history discipline: Both in environmental history (cf. McNeill 2010) and historical ecology (cf. Balée 2006; Crumley 2017), human events are viewed "as part of a larger story in which humans are not the only actors" (McNeill 2010: 347).

In order to give a sense of diversity, a few studies within this growing body of scholarly work should be mentioned: Proposing multispecies history writing "at the intersection between ethnography and natural history," Tsing (2013: 28) takes the Japanese *Satoyama* forest interrelations as one of her case studies. In another study, her focus is on the *Matsutake* mushroom's capacity to interrelate with other species, humans included, in environments disturbed and damaged by industrial capitalism (Tsing 2015). Working with another time frame, but still within the historical epoch of capitalist expansion, Ficek (2019) explores the agency of cattle in her study of colonial transformations in Latin America. Resembling the orientation of environmental history inspired by Crosby's *Colombian exchange* (Crosby 1972), she views cattle as historical transformative actors that entered "into relations of independence with humans and other species" (Ficek 2019:S260).

Extending the understanding of transformations in ways indicated here implies to increase sensitivity toward the agentive capacities of other than human beings. By implication, this motivates the anthropologist to take into account other kinds of knowledge – science knowledge about the natural world, and Indigenous knowledge about other lifeworlds. Moreover, some researchers working within this field include other than human sources, again resembling an orientation in environmental history to make use of data from f.ex. bioarchives (such as pollen deposits) and geoarchives (such as soil profiles) (cf. McNeill 2010). A compelling example is the work of historian and STS researcher, Kate Brown (2019), on environmental transformations in the Ukrainian Pripyat Marshes caused by industrial chemical agriculture, nuclear bomb testing, and the 1986 Chernobyl catastrophe.

Brown argues for the need to train and strengthen the researcher's ability to notice the effects – in her case – of toxic exposure and chronic radioactivity on bodies and ecosystems, by combining different kinds of human and other than human sources and listening to and collaborating with different kinds of knowledge holders and producers. Included in her method repertoire is the study of transformations of plant morphology brought about by radioactive isotopes that change organisms' cellular structure. Including plant morphology, says Brown, enhances her sensitivity as a social researcher toward environmental transformations and improves her *literacy* in reading and understanding more-than-human damaged landscapes. Andrew Mathews similarly emphasizes "the art of noticing" processes of change in more-than-human sources and assemblages in a research method he calls "landscape ethnography." Writing about the Monti Pisani pine and chestnut forests in central Italy, he seeks to trace past relationships of the Anthropocene, between humans and nonhumans and between nonhumans and other nonhumans, through the study of "forms of trees, areas of forest, banks, terrace walls, and drainage systems" (Mathews 2018:386). Through a constant shift between archival document studies, exchanges with scientific conversation partners, talks with people related to the forest, and walks and elaborate observations, he combines research on the biography of specific organisms with the study of changing landforms.

Work with the history of environmental transformations indicated in this section blur in different ways the long-established distinction between natural history and human history. More than crossing the distinction between the human and non-human, these studies focus on relations and trajectories that connect spheres – the human, the bio, and the geo.

The Damaged Planet and Persistent Colonial Orders

With a focus on environmental transformations and temporalities, the previous section reviewed academic work that explores new analytical dimensions and vocabularies in the study of more-than-human life and in the reporting of modes of living on a planet seriously affected and altered by anthropogenic activity. Recently, an approach of this kind was thoroughly reflected upon in the volume *Arts of living on a damaged planet* (Tsing et al. 2017). With contributions from leading researchers within anthropology and related disciplines, the volume is organized in two parts with two beginnings mirroring each other. One is associated with the image of "ghosts," addressing the traces, vestiges, and signs of past histories of more-than-human life present in current inter-species interaction and sociality. Contributors to this part are interested in the remnants of ruination – of historical processes related to industrial capitalism that shape current modes of living. In the other part associated with the image of "monsters," the authors discuss the assemblages and entanglements that currently constitute life in ruinated environments. The focus on assemblages and entanglements is ambitious since it includes interspecies interaction, communication, and relationality across nested scales of micro- and macrolife, from bacteria to water and soil, including humans as well.

Writing in a time of rupture and accelerated change, the contributors to *Arts of living* examine life possibilities in contexts of human-induced environmental change that threatens multispecies existence. The art of living in damaged places is also about modes of dying and the risks that species and interspecies relations are extinguished. Damaged places are within this academic conversation viewed as contexts – or patches – "increasingly dominated by industrial forms" (Tsing et al. 2019:S186). Industrial forms refer in this perspective to the modes of designing, organizing, and practicing industrial production and processing within the global market regime. One of the main characteristics of these forms is that they constitute controlled spaces of industrial production based on processes of ecological simplification, i.e., the reduction of species diversity is used as a device to increase the volumes of some profitable products. One example is the plantation system based on extensive monocrop agro-industrial production. At the same time, these modes of production generate unpredictable "feral effects," the uncontrolled proliferation of pathogens, and other harming (f.ex. nuclear and toxic) agents. The global spread of the coffee rust fungus due to coffee plantation production is one example (cf. Tsing et al. 2019), the spread of the Covid 19 virus another (cf. Brown 2020). These and a wide range of other similar effects cannot be contained within the industrial production systems. Harming agents form part of new interspecies networks of relations that have general effects on life and life-sustaining practices, and at the same time are unevenly distributed; some human and more-than-human assemblages are more heavily burdened than others.

Writing and thinking about the altered conditions of life in the Anthropocene, and about the fragility, endurance, and generative capacity of life in damaged places, motivate some researchers to formulate propositions of hope. Demanding to comprehend, these seem to share the view that optimism is found elsewhere than in the "revived modernist hope for capitalism and humanity to reinvent itself in a "greener" and "better" form in the face of crisis and disruption" (Tsing et al. 2019:S192). Of these alternative propositions, the authors of this chapter wish to emphasize Donna Haraway's employment of a new analytical vocabulary in storytelling about the disturbed and damaged states of the earth. One might read the book *Staying with the trouble* (Haraway 2016) as the envisioning of a new conceptual landscape meant to inspire new intellectual and political behaviors that can foster restorative forms of interspecies communication and coexistence. Trouble, in Haraway's thinking, refers to multiple dimensions; to the political economy of environmental disaster and mass extinction and its threatening effects on multispecies life; and, furthermore, to dominant knowledge practices that inform and sustain destructive designs, organization forms; and modes of intervention. She is not primarily interested in this trouble, but in what she calls critters' (all kinds of creatures, humans included) ability to respond to and live with the trouble. Moreover, she associates this "response-ability" with a kind of collective knowing and doing, an ecology of practices that cultivate conditions for survival.

Haraway's storytelling makes use of an innovative conceptual vocabulary which is too extensive to present in any detail here. Hence, only one important aspect of *Staying with the trouble* will be mentioned here, namely "tentacular thinking." The starting point, or model, for tentacular thinking about current troubled world/earth affairs is a spider – the *Pimoa cthulhu*, living under stumps in the Redwood forest in California. The name of this spider is taken from the language of the Indigenous Goshute people of Utah and refers to denizens living in the depths – the "chthonic." "This spider is in place, has a place, and yet is named for intriguing travels elsewhere" (Haraway 2016:31). With its many body parts, it finds its ways and makes its trails by sensing, trying, attaching to, and detaching from others. Its tentacular life serves as a model for thinking open-ended and changeable interspecies relations and assemblages, which are both situated in specific places and not bounded or contained by these places. Haraway makes use of the concept "becoming-with" to indicate the importance of more-than-human kinship making. Existence in precarious environments, often on the edge of extinction, as well as species' regenerative possibilities, depends on their ability to create what she calls "odd-kin" relationships. With tentacular thinking, Haraway offers us a model to think with, built on other than human life and more than (but including) scientific knowledge. She argues that it matters which concepts we think the world with and emphasizes the need to change the story about life on Terra. Storytelling, she maintains, should be decentered from the anthropocentric modes of conceiving the social and the transformative earthly processes.

Along a similar line, human geographer Nigel Clark and sociologist Bronislaw Szerszynski (2021), in their book *Planetary social thought*, explore the potential of making use of the diverse knowledge generation about life worldwide to think through planetary change in a moment in time when scientists predict that earth systems are rapidly approaching tipping points into new physical, chemical, and biological states. In formulating questions about planetary changes, Clark and Szerszynski question the idea of the Earth as a single unity, and they emphasize multiplicity in two fundamental ways: on the one hand, what they call *earthly multitudes* – "the different ways this planet is engaged with, experienced, known and imagined" (2021:9) by collectives around the world - including Indigenous people, transnational activist, and academic networks and the diverse scientific collectives; and on the other hand, *planetary multiplicity* – the capacity of the

Earth "at every scale, from the microscopic to the entire Earth system – to become other to itself, to self-differentiate" (pp. 8). In their reasoning, the world's multiple knowledge practices (earthly multitudes – including modes of knowing of Australian Aborigines, and of Pacific, Amazonian, and Andean peoples) can inform social science research on the Anthropocene. Clark and Szerszynski acknowledge, however, that there is a tension in their proposal of acknowledging the validity of other knowledge practices and narratives that arises from the historical colonizing role of Western Science in suppressing these other practices. Moreover, through this interlocking, science has contributed to reduce the possibilities of planetary multiplicity.

Taking a step further in this direction, several anthropologists call our attention to the insights from other (and often Indigenous) understandings of life, and other ways of making worlds that neither assume the human-nature divide nor that between life and the inert. Building on her work with Australian Aborigines communities, Elisabeth Povinelli (2016) questions the dominant partition in Western modern thought between what she calls biontology and geontology. Biontology refers to life while geontology refers to nonlife. Povinelli seeks with these terms to challenge a fundamental assumption in Western metaphysics: that all forms of existence can be reduced to and measured by one form of existence, that related to bios and to being. She proposes that Australian Aborigines knowledge practices about the landscapes they inhabit not only have analytical purchase, but also reveal that the partition of the bio from the geo sustains the expansion of capitalism and colonialism, and the extinctions they provoke.

Expanding on Povinelli's critique, Marisol de la Cadena and Mario Blaser, who both have worked in the Americas, look closer at social scientific "understandings" of politics and of environmental transformations. They question what the political might entail if social research acknowledges the existence and validity of, to use Clark and Szerszynski's words, the practices of *earthly multitudes*. Based on conversations with members of a Quechua community in the Peruvian highlands, De la Cadena introduces the idea that abiotic formations can have political agency. She explores the theoretical implications of taking seriously that her interlocutors relate to a mountain, *Ausangate* – threatened by large-scale mining operations – as an "earth being" which influences political events and the course of history (De la Cadena 2015). Building on his work with the Yshiro people in Paraguay and the Innu in Canada, Blaser (2013, 2019) similarly shows that the Yshiro and the Innu conceive of their worlds and their agency as dependent on and inseparable from the world and agencies of more-than-humans. For Blaser and De la Cadena, assuming such intrinsic interrelatedness and more-than human agencies questions not only the nature-social divide, but also requires rethinking scientists' understandings and construction of a common world. They argue that the construction of a common world or the planet, which admits multiplicity as its core, cannot longer build on colonial, singularizing designs that make other ways of knowing and of existence unviable. In their view, approaching our ecological compromised planet requires not only to undo the natural-social divide, but also to counteract colonial patterns of destruction.

Conclusion

There is a huge span between the cultural ecology perspective of Julian Steward and Leslie White mentioned in the beginning of the adaptation section, and the extended more-than-human perspectives presented in the last section on the damaged planet. Comparing these perspectives, a prominent difference regards the ways scholars within these orientations understand and deploy the social-natural divide in their analysis. In anthropological work that focuses on interspecies relations, scholars no longer view "nature" as a separate domain, nor as a stable context for human adaptation and intervention. Sociality is furthermore not considered a defining quality of the human. Besides humans, the "social" includes other species that authors working within cultural ecology and related orientations in environmental anthropology placed under the category of "nature."

Anthropological scholarship that deals with environmental transformations raises questions about the validity of this divide, both as a description of the world and as an analytical tool. Throughout this chapter, the authors emphasize that the making visible of the interweaving and inseparability of the social and the natural is essential. Notions of the Anthropocene, "a new toxic age," or "polluted or damaged planet" exemplify this point. The redefinition of the divide between the social and the natural, which also includes questioning the distinction between bio (life – the biotic) and geo (nonlife – the abiotic), involves questioning the assumed stability of "nature." There is a greater emphasis on the dynamic and changing conditions of ecologies in plural – an emphasis that to some extent resonates with earlier non-equilibrium perspectives.

Admitting that extractive and chemical technologies have altered the conditions for human and nonhuman life leads anthropologists to radically reconsider their methods and tools of analysis. Paying methodological attention to other-than-humans, whether these are plants, insects, animals, geological strata, rock formations, or other kinds of beings, and learning to read how long-standing exposure and other forms of violence are inscribed in their tissues and layers, might equip us better to question current evidentiary regimes and modes of knowing. This attentive reading entails acknowledging, Barad (2010:26) suggests, that the "world 'holds' the memory of all traces." The recent emphasis on "multiplicity" has however wider implications. By attending to the connections between scientific knowledge-production and colonialism, social researchers are challenged to question the epistemic habits that inform research, and to rethink what the categories of analyses used allow them to see and not see.

Scholars suggest that living on a damaged, chemically altered planet, in which the futures of its inhabitants and ecosystems might be compromised, requires us to explore the possibilities of living and surviving that nonetheless exist. Yet, people and nonhumans around the world live and experiment damaging planetary conditions differently, to put it mildly. To account for the structures that produce these differences is undoubtedly needed, and appears at the same time as insufficient. If an academic ambition is to generate knowledge that also could lead to other and less damaging forms of living, it seems necessary to engage in the work that some social thinkers now do: to let the concepts and categories with which we build our analysis be guided by the lives and experiences of humans and more-than-humans that we learn to know through our fieldwork and research.

References

Acosta A (2016) Las dependencias del extractivismo: aporte para un debate incompleto. Aktuel Marx 20:1–22

Adams V, Van Hattum T, English D (2009) Chronic disaster syndrome: displacement, disaster capitalism, and the eviction of the poor from New Orleans. Am Ethnol 36(4):615–636

Balée W (2006) The research program of historical ecology. Columbia University Press, New York

Barad K (2010) Quantum entanglements and hauntological relations of inheritance: dis/continuities, spacetime enfoldings, and justice-to-come. Derrida Today 3(2):240–268

Bauer S (2018) Radiation science after the cold war: the politics of measurement, risk and compensation. In: Zvonareva O, Popna E, Horstman K (eds) Health, technologies and politics in post-soviet settings: navigating uncertainties. Palgrave Macmillan, p 225

Behrends A, Reyna SP, Schlee G (2011) Crude domination: an anthropology of oil. Berghahn, New York

Blaser M (2013) Ontological politics and the people in spite of Europe: towards a conversation on political ontology. Curr Anthropol 54(5):547–568

Blaser M (2019) On the properly political (disposition for the) Anthropocene. Anthropol Theory 19 (1):74–94

Bohme SR (2014) Toxic injustice: a transnational history of exposure and struggle. University of California Press, Berkeley

Bond D (2018) Environment: critical reflections on the concept, SSS occasional papers 64. Institute of Advanced Study, Princeton

Brown K (2013) Plutopia: nuclear families, atomic cities, and the great soviet and American plutonium disasters. Oxford University Press, New York

Brown K (2019) Learning to read the great Chernobyl acceleration: literacy in the more-than-human landscapes. Curr Anthropol 60(20):198–208

Brown K (2020) The pandemic is not a natural disaster: The coronavirus isn't just a public health crisis. It's an ecological one. The New Yorker: Annals of Inquiry, June 2021. Retrieved from: https://www.newyorker.com/culture/annals-of-inquiry/the-pandemic-is-not-a-natural-disaster

Carson R (1962) Silent spring. Houghton Mifflin Company, Boston

Castree N (2003) Commodifying what nature? Prog Hum Geogr 27(3):273–297

Chakrabarty D (2009) The climate of history: four theses. Crit Inq 32(2):197–222

Chen M (2012) Animacies: biopolitics, racial mattering, and queer affect. Duke University Press, Durham

Clark N, Szerszynski B (2021) Planetary social thought: the Anthropocene challenge to the social sciences. Polity Press, Cambridge

Cram S (2016) Living in dose: nuclear work and the politics of permissible exposure. Publ Cult 28 (3):519–539

Crosby AW (1972) Colombian exchange: the biological and cultural consequences of 1492. Greenwood, Westport

Crumley CJ (2017) Historical ecology and the study of landscape. Landsc Res 42(S1):565–573

De la Cadena M (2015) Earth beings: ecologies of practice across Andean worlds. Duke University Press, Durham

De la Cadena M, Blaser M (2018) Pluriverse: proposals for a world of many worlds. In: De la Cadena M, Blaser M (eds) A world of many worlds. Duke University Press, Durham, pp 1–22

Descola P (2014) Modes of being and forms of predication. HAU J Ethnograp Theory 4(1):271–280

Ferguson J (1999) Expectations of modernity: myths and meaning of urban life on the Zambian copperbelt. University of California Press
Ficek RE (2019) Cattle, capital, colonization: tracking creatures of the Anthropocene in and out of human projects. Curr Anthropol 60(20):S260–S271
Fiske SJ, Crate SA, Crumley CL, Galvin K, Lazrus H, Lucero L, Oliver-Smith AL, Orlove B, Strauss S, Wilk R (eds) (2014) Changing the atmosphere: Anthropology and climate change. Final report of the AAA Global Climate Change Task Force. AAA, Arlington
Fortun K (2009) Advocacy after Bhopal: environmentalism, disaster, new global orders. University of Chicago Press, Chicago
Foucault M (1980) Power/knowledge: selected interviews and other writings 1972–77. Vintage
Foucault M (2003) Society must be defended: lectures at the college de France 1975–76. Picador, New York
Fraser N (2000) Rethinking recognition. New Left Review 7:107–120
Gilberthorpe E, Rajak D (2017) The anthropology of extraction: critical perspectives on the resource curse. J Dev Stud 53(2):186–204
Golub A (2014) Leviathans at the gold mine: creating indigenous and corporate actors in Papua New Guinea. Duke University Press, Durham
Gudynas E (2015) Extractivismos. Ecología, economía y política de un modo de entender el desarrollo y la naturaleza. CEDIB, Cochabamba
Guzmán-Gallegos M (2019a) Controlling abandoned oil installations: ruination and ownership in northern Peruvian Amazonia. In: Ødegaard CV, Andia JJR (eds) Indigenous life projects and Extractivism. Palgrave Macmillan, Cham, pp 53–75
Guzmán-Gallegos M (2019b) Philippe Descola: Thinking with the Achuar and the Runa in Amazonia, in special issue of Ethnos., Guest editors: Lien M. & Pálsson G., Ethnography Beyond the Human: The 'Other-than Human in Ethnographic Work, special issue of Ethnos, https://doi.org/10.1080/00141844.2019.1580759
Guzmán-Gallegos M (2021) Modes of knowing: perceptibility of toxicity and living in the Ecuadorian Amazonia. Anthropol Today 37(4):23–26
Haraway DJ (2016) Staying with the trouble: making kin in the Chthulucene. Duke University Press, Durham
Hébert K (2016) Chronicle of a disaster foretold: scientific risk assessment, public participation and the politics of imperilment in Bristol Bay. J R Anthropol Inst 22(S1):108–126
Hecht G (2018) Interscalar vehicles for an African Anthropocene: On waste, temporality and violence. Cult Anthropol 33(1):109–141
High MM (2012) The cultural logics of illegality: living outside the law in the Mongolian gold mines. In: Change in democratic Mongolia. Brill, pp 249–270
Hilson G, Hilson A, Maconachie R, McQuilken J, Goumandakoye H (2017) Artisanal and small-scale mining (ASM) in sub-Saharan Africa: re-conceptualizing formalization and 'illegal' activity. Geoforum 83:80–90
Jacka JK (2015) Alchemy in the rainforest: politics, ecology, and resilience in a New Guinea mining area. Duke University Press, Durham
Jacka JK (2018) The anthropology of mining: the social and environmental impacts of resource extraction in the mineral age. Annu Rev Anthropol 47:61–77
Kirsch S (2014) Mining capitalism: the relationship between corporations and their critics. University of California Press, Oakland
Knowles SG (2018) Deferred maintenance: slow disaster and American infrastructures. In: Davis Center seminar (Lecture). Princeton University, Princeton, 23 February
Kohn E (2013) How forests think: toward an anthropology beyond the human. University of California Press, California
Kolen J, de Smet E, de Theije M (2018) "We are all Garimpeiros:" settlement and movement in communities of the Tapajós small-scale gold mining reserve. J Latin Am Caribbean Anthropol 23(1):169–188

Kuletz V (1998) The tainted desert: environmental and social ruin in the American West. Routledge, New York

Leifsen E (2017) Wasteland by design: dispossession by contamination and the struggle for water justice in the Ecuadorian Amazon. Extracting Ind Soc 4:344–335

Leifsen (2020) The socionature that neo-extractivism can see: practicing redistribution and compensation around large-scale mining in the Ecuadorian Amazon. Polit Geogr 82:1–11

Lerner S (2010) Sacrifice zones: the front lines of toxic chemical exposure in the United States. MIT Press, Massachusetts

Li F (2009) Documenting accountability: environmental impact assessment in a Peruvian mining project. PoLAR 32(2):218–236

Li F (2015) Unearthing conflict: corporate mining, activism, and expertise in Peru. Duke University Press, Durham

Liboiron M (2016) Redefining pollution and action: the matter of plastics. J Mater Cult 21(1):87–110

Liboiron M, Tironi M, Calvillo N (2018) Toxic politics: acting in a permanently polluted world. Soc Stud Sci 48(3):331–349

Little PE (1999) Environment and environmentalisms in anthropological research: facing a new millennium. Annu Rev Anthropol 28:253–284

Lyons KM (2018) Chemical warfare in Colombia, evidentiary ecologies and senti-actuando practices of justice. Soc Stud Sci 48(3):414–437

Martinez-Alier J (2003) The environmentalism of the poor: a study of ecological conflicts and valuation. Edward Elgar Publications

Mathews AS (2018) Landscapes and throughscapes in Italian forest worlds: thinking dramatically about the Anthropocene. Cult Anthropol 33(3):386–414

Mathews AS, Barnes J (2016) Prognosis: vision of environmental futures. J R Anthropol Inst 22 (S1):9–26

McNeill J R (2010) The state of the field of environmental history. Annual Review of Environment and Resources 35:345–374

Michaels D (2008) Doubt is their product. How industry's assault on science threatens your health. Oxford University Press, Oxford

Murphy M (2006) Environmental politics, technoscience, and women workers. Duke University Press, Durham

Murphy M (2008) Chemical regimes of living. Environ Hist 13(4):695–703

Murphy M (2013) Distributed reproduction, chemical violence, and latency. Scholar Fem Online 11(3). Retrieved from: http://sfonline.barnard.edu/life-un-ltd-feminism-bioscience-race/distributed-reproduction-chemical-violence-and-latency/

Murphy M (2017) Alterlife and decolonial chemical relations. Cult Anthropol 32(4):494–503

Nading A (2020) Living in a toxic world. Annu Rev Anthropol 49:209–224

Nixon R (2011) Slow violence and the environmentalism of the poor. Harvard University Press, Cambridge

Orlove BS (1980) Ecological anthropology. Annu Rev Anthropol 9:235–273

Orr Y, Lansing JS, Dove MR (2015) Environmental anthropology: systemic perspectives. Annu Rev Anthropol 44:153–168

Peluso N (2018) Entangled territories in small-scale gold mining frontiers: labor practices, property, and secrets in Indonesian gold country. World Dev 101:400–416

Peluso NL, Watts M (eds) (2001) Violent environments. Cornell University Press, Ithaca

Perrault T (2013) Dispossession by accumulation? Mining, water and the nature of enclosure on the Bolivian Altiplano. Antipode 45(5):1050–1069

Petryna A (2002) Life exposed: biological citizens after Chernobyl. Princeton University Press, Princeton

Povinelli EA (2016) Gentologies: A requiem to late liberalism. Duke University Press, Durham

Rogers D (2014) Oil and anthropology. Annu Rev Anthropol 44:365–380

Sawyer S (2004) Crude chronicles: indigenous politics, multinational oil and neoliberalism in Ecuador. Duke University Press, Durham

Sawyer S (2015) Crude contamination: law, science and indeterminacy in Ecuador and beyond. In: Appel H, Mason A, Watts M (eds) Subterranean estates: life worlds of oil and gas. Cornell University Press, pp 126–146

Schlosberg D (2013) Theorizing environmental justice. Theories, movements, and nature. Oxford University Press, Oxford

Scoones I (1999) New ecology and the social sciences: what prospects for a fruitful engagement. Annu Rev Anthropol 28:479–507

Shapiro N, Kirksey E (2017) Chemo-ethnography: an introduction. Cult Anthropol 32(4):481–493

Sullivan S (2013) After the green rush? Biodiversity offsets, uranium power and the 'calculus of causalities' in greening growth. Hum Geogr 6(1):80–101

Svampa M (2016) Debates Latinoamericanos: indianismo, desarrollo, dependencia, populismo. Edhasa, Buenos Aires

The Guardian (2021) Australian government must protect young people from climate crisis harm, court declares. The Guardian, 8th July: Retrieved from: https://www.theguardian.com/environment/2021/jul/08/australian-government-must-protect-young-people-from-climate-crisis-harm-court-declares

Tironi M (2018) Intimate activism: hypo-interventions and the politics of potentiality in toxic environments. Soc Stud Sci 48(3):438–455

Tschirhart C, Handschumacher P, Laffly D, Bénéfice E (2012) Resource management, networks and spatial contrasts in human mercury contamination along the Rio Beni (Bolivian Amazon). Hum Ecol 42:511–523

Tsing AL (2013) More-than-human sociality: a call for critical description. In: Hastrup K (ed) Anthropology and nature. Routledge, New York, pp 27–42

Tsing AL (2015) The mushroom at the end of the world: on the possibility of life in capitalist ruins. Princeton University Press, Princeton

Tsing AL, Swanson HA, Gan E, Bubandt N (eds) (2017) Arts of living on a damaged planet. Ghosts and monsters of the Anthropocene. University of Minnesota Press, Minneapolis

Tsing AL, Mathews AS, Bubandt N (2019) Patchy Anthropocene: landscape structure, multispecies history and the retooling of anthropology. Curr Anthropol 60(20):S186–S196

Tuck E (2009) Suspending damage: a letter to communities. Harv Educ Rev 79(3):409–428

Verbrugge B (2015) The economic logic of persistent informality: artisanal and small-scale mining in the southern Philippines. Dev Chang 46(5):1023–1046

Viveiros de Castro E (2004) Exchanging perspectives: the transformation of objects into subjects in Amerindian ontologies. Common Knowl 10(3):463–484

Voyles TB (2015) Wastelanding: legacies of uranium mining in Navajo country. University of Minnesota Press, Minneapolis

Walker JS (2011) Toxic archipelago: a history of industrial disease in Japan. University of Washington Press, Seattle

Watts M (2013) Blood oil: the anatomy of petro-insurgency in the Niger delta, Nigeria. In: Behrends A, Reyna SP, Schlee G (eds) Crude domination: an anthropology of oil. Berghahn, New York, pp 49–80

Weizman E (2017) Forensic architecture: violence at the threshold of detectability. Zone Books, New York

On the Commonness of Skin: An Anthropology of Being in a More Than Human World

20

Simone Dennis

Contents

Introduction	506
Before (?) Multispecies Ethnography	507
Multispecies Ethnography, Multiple Politics	512
Conviviality	516
Being with (Some) Animals: The Taking of Animals Literally	518
Cuticulae	519
Laboratory	523
Conclusion	527
References	528

Abstract

The point of departure of multispecies ethnography is that animals are good to be with, a proposal that seeks to destabilize human primacy and reveal new orders of human-nonhuman relations and becomings. This chapter explores the possibilities and limitations of "good to live with." Close examination of how the undergirding theoretical principles informing multispecies ethnography have been operationalized reveals a somewhat romanticized research imaginary. This has manifested in the exploration of a limited and localized range of nonhuman life that does not include those animals who have become necroavailable to humans on an industrial scale. While there is pressure for multispecies ethnographers to take up animal rights agendas for the (meat and laboratory) animals that have the most to gain from decentering the human, there are quieter potentials that might be realized by multispecies ethnographers. These potentials might be attained if ethnographers recognized how the most unlikely of environments offer opportunity to trouble the ontological distinctions that they attempt to

---

S. Dennis (✉)
School of Archaeology and Anthropology, College of Arts and Social Sciences, Australian National University, Canberra, ACT, Australia
e-mail: simone.dennis@anu.edu.au

© The Author(s), under exclusive licence to Springer Nature Singapore Pte Ltd. 2022
D. McCallum (ed.), *The Palgrave Handbook of the History of Human Sciences*,
https://doi.org/10.1007/978-981-16-7255-2_19

destabilize. These include those between nature and culture, human and nonhuman, and other binaries that compartmentalize messily entangled human and nonhuman lives. This chapter rehearses the possibilities that come available for realizing the potentials of multispecies ethnography in the laboratory, as well those that dwell on the cuticle that envelopes plant, animal, and human beings. These potentials are explored in this chapter in two case studies that provide readers with the full gamut of possibilities and limitations occasioned and entailed by doing multispecies anthropology in the Anthropocene.

Keywords

Skin · Cuticle · Multispecie · Ethnography · Nonhuman animal

Introduction

A sense of growing urgency attends anthropological critiques of its own profoundly humanist epistemology (see, for example, Brown and Nading 2019; Locke and Muenster 2016). Certainly, bearing the prefix anthro- into the (unofficial unit of) geological time defined by the (negative) effects of human activity on the planet's climate and ecosystems provides powerful compulsion to destabilize the centrality of the human. While the Holocene that began 11,700 years ago after the last major ice age is the official name of the current epoch, the Anthropocene (a deeply political term coined by atmospheric chemist Eugene Stormer and limnologist Paul Crutzen in 2000 – see Steffen 2013; Trischler 2016) recognizes the recent significant impact of humans on the planet's climate and ecosystems (Crutzen and Stoermer 2000). It would be a brave and seriously out of touch discipline that refused to address the implications visited upon it by its own name. It would be an even braver one that continued to stand unapologetically on the grounds of human exceptionalism during continuous coverage of human impact on the earth including, but not limited to, anthropogenic climate change. The intention to trouble and discompose the centrality of the human, which can be pursued in a variety of ways and modes, is often loosely described as "multispecies ethnography" (Kirksey and Helmreich 2010), a term that captures not only anthropologists but many humanities and social sciences scholars who seek to challenge the humanist epistemology upon which many disciplinary systems and practices (including of course conventional ethnography) are predicated.

At the very heart of the destabilization agenda lies the work of removing the ontological distinctions between nature and culture, human and nonhuman, subject and object, and any and all other (related) binaries that seek to render still and bounded the human and nonhuman lives that are in fact restlessly entangled. The use of relational perspectives that allow multispecies theoreticians to recognize the ways in which human and nonhuman figures encounter one another in context, undermines anthropocentrism by multiple means. These include recognizing the effects of classifying animals in western terms and the related obscuration of indigenous relations with animal partners, acknowledging the agency of nonhuman others, and focusing closely in on the intersections between ecological relations, political economy, and representation. All of this allows life to unfurl toward or to recoil from other life as the local context occasions and entails. If one tends properly and diligently and reflexively to the politics of representation of those entanglements, the outcome of multispecies ethnography is a sense of the interconnectedness and inseparability of humans and other life forms, and the extension of ethnography beyond the human center.

The point of departure of multispecies ethnography is that animals are "good to live with"; as opposed to being good to think, or good to eat (Kirksey and Helmreich 2010: 552). The purposefully disruptive insertion of "good to live with" into Levi-Strauss' original (1963) claim that animals are "good to think with" (as opposed to the utilitarian proxy "good to eat") effectively proposes that fleshy, localized proximity with nonhuman beings permits new orders of relations and becomings to emerge. Living with animals and indeed a vast variety of nonhuman others can then, at least theoretically, be ethnographically explored in any context.

This chapter explores the possibilities and limitations of "good to live with." It does so by first examining the way in which membership as a multispecies ethnographer is both excitingly broad, and extremely contentious. The contentious positions that are taken up within the "movement" (a term that reflects the generalized ethical motivation shared by multispecies ethnographers, to decentralize the human) reveal the limited circumstances and nonhuman beings to which multispecies principles and practices have thus far been applied. This in itself is revelatory and allows the most strident critique of multispecies ethnography to be rehearsed alongside the most persuasive of its potentials. These potentials are explored in this chapter in two case studies that provide readers with the full gamut of possibilities and limitations occasioned and entailed by doing multispecies anthropology in the Anthropocene.

The case studies, concerning analyses of the cuticle that animals and plants share in the first instance, and relations between experimental animal (rodent) models and human scientists in immunology, virology and neurology laboratories in the second, permit the strong connections (and issues) between earlier anthropological attempts to engage "others" differently by means of paradigmatic irruption. Of particular interest here is the disciplinary mainstay of kinship which has always involved speculation about the nonhuman animal and now, in the time of the Anthropocene, highly vexing questions about in and exclusion of nonhuman others in disciplinary reckonings of relatedness.

Before (?) Multispecies Ethnography

While multispecies ethnography is distinguished from that which went before it by its intention to decentralize the human, it may not represent a straightforward rupture. The following section relies heavily on White and Candea's very detailed Cambridge Encyclopedia of Anthropology entry on the topic of multispecies ethnography (2018). Serving as an authoritative resource on the current state of play, the entry represents canonical knowledge on multispecies scholarship and its precedent roots. As such, it is deferred to here; significant departures from position of multispecies scholarship are extensively made in the latter parts of this chapter.

White and Candea (2018) submit that there are continuities between ostensibly different schools of thought concerning the animal leading up to the emergence of multispecies ethnography. Although it is clear that multispecies ethnography does signal a new intention to regard animals beyond their utility and beyond the conceptual, the endurance of some key ideas over time calls into question the occurrence of hard breaks between named bodies of theory. As discussed later on in this chapter, some of those who have been described as multispecies ethnographers, such as Ingold (2013) explicitly draw out those endurances by demonstrating that animals have been recognized as active agents by anthropologists well before the invention of the term "multispecies ethnography"(see also Ingold and Pálsson 2013).

From the nineteenth century, Anthropology's interest in nonhuman animals:

> ran along one or both of two very broad thematic axes. One asked about the role of animals as material, economic, and political resources for humans in society. Another investigated the role of animals in human cultural, symbolic, and conceptual schemes. The opposition between the two was famously cast by Claude Lévi-Strauss, a prominent proponent of the second line of enquiry, as one between seeing animals as 'good to eat' and seeing them as 'good to think' with (Lévi-Strauss 1963: 89). The oscillation between the two approaches and their recombination can be seen by briefly scanning the role of animals in successive anthropological paradigms. (White and Candea 2018: np)

In the nineteenth century, totemism and animism were understood to prefigure modern religious organization. Frazer interpreted the practice of venerating particular animals and plants as gods as religious precursor (see, for example, Frazer 1887). Certain plant or animal species were associated with particular human groups (clans, for instance). Different groups could be emblematically distinguished from the generalized society, and this compartmentalizing structure could serve to regulate relations between the various totemic groups (as well as within them; see Tylor 1871). Configured as a much more generalized tendency to see the nonhuman world as animated by spiritual forces, the concurrently described "animism" referred to how non-white people interpreted nonhuman objects, plants, and animals and endowed them with souls (see, for example, cf. McLennan 1869–70, 1899; Bird-David 1999). The failure of non-white humans to understand human exceptionalism and dominion over animals (described by missionaries and intrepid sojourners) saw them firmly placed in the "primitive" stage of a "singular progressive historical path leading to the 'advanced' or 'modern' societies of Europe" (see White and Candea 2018: np; see also Kuper 2005).

While described as "leaving behind the evolutionist question of the historical origins of religion" that had nourished debate during the preceding century, White and Candea (2018) also note that Durkheim (1915) relied heavily on totemism. Where Tylor had declared animism a window onto the primitive and childlike sensibility and capacity of the non-western human, Durkheim rejected the notion that animal totems or gods derived their power from some mystical capacity. He instead proposed that the social groups for which those animals stood as totems held the real power (see Bird-David 1999). The trick was to see beyond and behind the totemic animal itself to appreciate that beneath and beyond it lay the power of social solidarity. A White and Candea observe,

> members of a clan would unite around their totemic emblem through collective rituals which created a powerful sense of togetherness. This grounded Durkheim's broader functionalist theory of religion. In worshiping god(s), Durkheim suggested, people unknowingly worshipped and maintained the structure of society itself: totems reflected and maintained the sub-divided structure of clan-based societies, just as monotheistic gods became a single focus for a broader, undivided church (2018: np).

It is important to note that in attending as Durkheim did to the social functions of totemism, and as Malinowski and Radcliffe-Brown did to the pragmatic basis of non-white peoples' concern with animals (and other nonhuman entities) the functionalists simultaneously continued a founding concern with religious and social organization, but equally radicalized the terms upon which that investigation was conducted. Malinowski was plain about it, remarking that:

> In totemism we see not the result of early man's speculations about mysterious phenomena, but a blend of a utilitarian anxiety about the most necessary objects of his surroundings, with some preoccupation in those which strike his imagination and attract his attention, such as beautiful birds, reptiles and dangerous animals. (Malinowski 1925: 4)

Harris's Cultural Materialism, which proposed superstructural safety nets for protecting infrastructural assets, otherwise known as a variant of ecological determinism (published as *The Rise of Anthropological Theory: A History of Theories of Culture*, in Harris 1968), is often considered to share a utilitarian sensibility with functionalist analyses; Levi-Strauss certainly thought so. Indeed, Levi-Strauss' now cliched line, animals are good to think with (1963) was issued as a direct assault on functionalist thinking, proffered in eminently quotable shorthand. His view was that animals were part of a structure of human thinking that was common across all societies. It was levelled at functionalists on the basis that their insights were reducible to simple utilitarianism. While there is no doubt that structural-functionalists such as Radcliffe-Brown were profoundly interested in apportioning explanatory weight to the material and economic bases of social arrangements, none of the functionalists was interested in explaining cultural phenomena exclusively or even principally through singular utilitarian analysis (as some might accuse Harris' profoundly and unapologetically materialist theories of doing; see Harris 1968). While Evans-Pritchard's classic ethnography *The Nuer* (1940) notes at the outset that cows are doubtless the economic foundation of Nuer life, they are also and equally the foundation of every other part of life, too. Space, time, politics, murder, marriage, aesthetics – cows orient everything. Nothing sums up their centrality better than Evans-Pritchard's own 1940 observation of the Nuer: "[t]heir social idiom is a bovine idiom" (1940: 19).

Lévi-Strauss's own structuralist interpretation of totemism probably owed a great deal to Radcliffe-Brown's; one could argue that the difference really comes down to Radcliffe-Brown's interest in and attendance to multiple kinds of functional interrelations versus Levi-Strauss' tighter focus on the conceptual structure of totemism. White and Candea put this case succinctly:

> The key thing about animals in totemic systems, Lévi-Strauss noted, was that they formed a series of natural entities which corresponded to a series of social entities, in the same way that flags, for instance, form a series of colours and patterns which stands for a series of nations. To bring this picture into view, one had to stop looking for relationships between particular clans and particular totems – be it in terms of use value or aesthetic proclivity. Totemism had nothing to do, for Lévi-Strauss, with a privileged relationship between particular people and a specific animal – such as that maintained, for example, between the Nuer and their cows. He noted that in many totemic systems, the list of entities invoked as totems included animals, plants, and objects of no discernible value or even aesthetic significance, while many such 'pragmatically' significant animals were omitted. Once the particular relationship between individual totems and individual groups of people was seen as entirely arbitrary, one could suddenly bring a broader logic into view: it was the entire system of totems and the entire system of clans which related to one another (2018: np).

Of course, Mary Douglas shared Levi-Strauss's anxious intention to right wrongheaded utilitarian tendencies, nowhere more famously than in her analysis of food prohibitions in Leviticus (Douglas 1966). Disputing proffered medical, economic and ecological explication of food prohibitions that put her work in diametric opposition to Harris's materialist position and formed the basis of the Sacred Cow Controversy (see also Harris 1978) Douglas closely considered those to whom Leviticus would be addressed. White and Candea explain that,

> behind the seemingly random list of prohibited animals – the pig, but also the camel, shellfish, and so on – Douglas detected (or devised) a logical structure according to which excluded animals were each in their own way being singled out as anomalies from a broader type or rule... The master logic of the entire system was one of categorization and perfection – a setting apart of the perfect from imperfect, which echoed the setting apart of the chosen people to whom Leviticus was addressed, from the other people surrounding them (2018: np).

Geertz, too, rejected functionalist terms outright in favor of an unapologetically messy account of the cockfight. Its purpose was not to function to assuage social passions, or bring them into existence, but in fact to *show* them (see Geertz 1973: 443–444). Geertz's attendance to the animal brought sociological matters of meaning and emotion into the interpretive frame and was as sharply critical of utilitarian understandings of the animal as Levi-Strauss had been. Despite his evident difference from the functionalists, however, Geertz was deeply interested in picking out those public interpretations that his own interlocutors would pick out, unlike structuralists, who were interested in revealing the deeper hidden patterns that informants could not see. The position of those to whom the animals mattered and meant were, then, of high importance – as one might argue they were to Evans Pritchard as he penned *The Nuer*.

The endurance of functionalist analyses and their appearance in ostensibly very different streams of anthropological theory notwithstanding, all the proponents of "Good to Think With" shared in common a concern not with actual flesh-and-blood animals, but with mythical, symbolic, or ritual animals, rather than participants in social interactions with humans. Cassidy (2007) concludes that this

> anthropological posture mirrored zoological thinking, which emphasized human control over plants, animals and "primitive" peoples and incorporated them to the "civilized" world for purposes of exploitation. (Segata and Lewgoy 2016: 27)

However, as Segata and Lewgoy (2016: 27) note, these criticisms are not new, existing in seed forms in *Anthropology and the Colonial Encounter*, edited by Talal Asad (1973) and *Women, Culture and Society*, edited by Michele Rosaldo and Louise Lamphere (1974). These were crucially important in destabilizing the ethnocentric and androcentric premises upon which anthropology was founded, and appear coincident with Singer's *Animal Liberation*, in 1975. The reformulation of the ethical and political horizon of the problem of animals in the West in the late twentieth century, the rising post-colonial feminist (continuing) struggles, and the (continuing) erasure of white privilege have leached into the ways in which anthropologists have approached nonhuman animals over time, making it hard to identify a hard break between multispecies ethnography and that which preceded it.

Despite the pressure building in related anthropological fault lines that indubitably produced critical disciplinary reworking of the relations between human and nonhumans, an intention to think beyond the conceptual engagement of nonhuman others was indeed taken at the turn of the twenty-first century (Kirksey and Helmreich 2010: 552; White and Candea 2018). Arising from the intersection of environmental studies, science and technology studies and animal studies, multispecies ethnography sought to simultaneously take inspiration and play truant from all three, taking its initial cues from the possibilities arising from "living organisms, artifacts from the biological sciences, and surprising biopolitical interventions" (Kirksey and Helmreich 2010: 575), all of which were present at the founding art exhibition, the Multispecies Salon.

The burgeoning multispecies ethnography must be contrasted against the prospect of braiding materialist and conceptual analytic strands together, as was canvassed in Mullins' Mirrors and Windows (1999) in the *Annual Review of Anthropology* (see also Shanklin 1985). In it, Mullins recognized that the veracity of the arguments between "materialists" and "idealists," perhaps most vigorously set out in Roger Keesing's introductory textbook *Cultural Anthropology: A Contemporary Perspective* (Keesing 1981), had effectively disguised their complementary potential. A case in point is Palsson's exploration of Icelandic fish, which explores the water dwellers in structuralist as well as economic and practice terms (see Palsson and Helgason 1995). Another is Michael Stewart's complex account of the equally important economic and symbolic roles of horses in the lives of Hungarian Roma (Stewart 1997). There are, besides, many others (see for example Descola 1992; Brightman 1993) – certainly sufficient to make the claim that the last days of the twentieth century were dedicated to fusing what had once appeared untraversable theoretical division. Productive as this work was, animals remained in the positions of hierarchical difference they had always occupied relative to humans in the discipline. And, even though as White and Candea (2018) note, antecedent for entwining human animal lives is to be found in Radcliffe-Brown's suggestion that, through totemism, "a system of social solidarities is established between man and nature" (Radcliffe-Brown 1952: 131), a lot of work had to be done to abandon the nature-society (or nature-culture) dualism which had oriented a good deal of this and earlier work:

> where anthropologists once sought to explain away the 'erroneous' ideas of social solidarity between humans and nonhumans, recent work often regards such thinking as a timely corrective to the 'dualism' characteristic of western thought, which allegedly lies behind contemporary environmental catastrophe, as well as the brutalities of the 'animal industrial complex' (White and Candea 2018: np).

Such timely possibilities are part of Ingold's call to regard beings as if they pre-existed (ontology), a project that:

> must be based on their production in a continuous event in their environment and in their entanglements (ontogenesis). Therefore, and taking another route, more than a definition of borders between humans and animals, anthropology has been interested in the joint forms of conviviality (Segata and Lewgoy 2016: 28).

Multispecies Ethnography, Multiple Politics

It would not require any particular astuteness or even extended reading to work out that the project of exploring how nonhuman beings are "good to live with" attracts both great enthusiasm and fierce criticism. Indeed, a very cursory look at the works that attempt to capture the state of the "movement" indicates that not everyone who is interested in undoing dualisms and theoretically integrating the entwinements of all kinds of being into accounts of existence wants to be called a multispecies ethnographer.

White and Candea (2018) suggest for instance that it is best to regard the descriptor "multispecies ethnography" not only as an approach to ethnography that seeks to epistemologically recontextualize the human in relation to other life forms and not only as a sensibility or orientation rather than an agenda or program, but also – and even principally – as an elastic point of reference. They note in particular that some scholars included in its expansive envelope are in fact actively critical of it. White and Candea yet find it productive

> to use 'multispecies ethnography' in a capacious way to capture a loosely shared orientation among these anthropologists, and an explicit sense of rupture from earlier approaches to animals (2018: np).

The grounds of the objections to inclusion are various. Despite being described as one of its ancestral influencers (see Segata and Lewgoy 2016:27; see also Locke and Muenster 2016; White and Candea 2018). Ingold objects to his own inclusion in a new, named "turn" toward the nonhuman (see Ingold 2013: 19). For Ingold, there really isn't anything new to speak of about multi-species ethnography; it is nothing more than a glossy brand name whose main effect has been to obscure the fact that animals *have* been taken seriously as actors in human social systems for decades, (and at least since the 1970s, when his own research on reindeer and their herders in Finland was published; see Ingold 1974).

Kohn's membership as a multispecies ethnographer has been subject to question by others. In 2013, Kohn argued that semiosis should remain central to undermining human exceptionalism, on the grounds that the nonhuman animal world continuously makes meaning in and on non-linguistic terms (see also Bateson 1972). His alternative understanding of semiotics (which owed much to von Uexküll's precedent ([Von Uexküll 1934]1992) biosemiotics which draws an essentialist view of the human body as a boundary into the work) produced a highly contentious ethnography that, for its many critics, tended much more to the human informants in the mix (see, for example, Descola 2014). According to Schroer (2019), it failed to think through the ways in which the world is cocreated and enacted in concert, in social, temporal, and material ways and forms. While these critiques recognize and reject the importation of biosemiotics into Kohn's work, another accusation lies in wait, in the form of political quietism.

Best et al. (2007), for example, criticize multispecies ethnography on the grounds that it effectively extends the political quietism already firmly established in animal studies – for Best et al. that silence is all too manifest in practitioners' evident willingness to remain wedded to speciesist values. These values might appear in a various and subtle ways. For example, a 2021 special issue of *The Australian Journal of Anthropology* edited by Fijn and Kavesh "strives to extend beyond human embodiment to engage with the senses beyond the human" (Fijn and Kavesh 2021: 10) and seeks to examine human-animal relations "beyond language" (Fijn and Kavesh 2021: 13). Yet, some of the claims made therein might be open to accusations of retaining a commitment to the human as the central figure in the enterprise. One of the editors of the volume, Fijnn, notes that her own contribution to including animals as agents rather than actors in the Anthropocene, has employed multispecies storytelling through observational filmmaking (see, for example, Fijn 2015, 2019):

> In a recent exhibition, More Than Human: The Animal in the Age of the Anthropocene (2020), Fijn included two different forested landscapes in Mongolia and Australia through film segments played within a gallery context…By using a GoPro camera that was strapped to the helmet of horse riders, Fijn's footage compares the extreme climates of the Khangai Mountains of Mongolia with the drought-stricken landscape of New South Wales in Australia, while riding through both landscapes on horseback. The viewer in adjacent film segments experiences how both horses and humans encounter a whitened, snowstorm landscape in Mongolia… in comparison with a blackened, fire-ravaged landscape in Australia. … Fijn's two video pieces sitting adjacent to one another in an exhibition space

demonstrate in a corporeal sense how the engagement with the land is changing for both horses and humans with the advent of extreme weather events caused by climate change. (Fijn and Kavesh 2021: 17)

A critique based in political quietism might protest that the horse remains an object of utility, put to the use of riders who need to be conveyed through the landscapes filmed with the intention of including the animal as coparticipant in experiences of climate change. The unchanged position of the horse relative to the human may well mean that multispecies-al ethnography does very little to fulfill one of its key aims, as articulated by Anna Tsing, precisely to work together toward a "collaborative survival" (Tsing 2015: 20). That is, the horse may not realize that it is understood differently –for the horse, a human still sits astride it, just as one did back when horses were good to use. Horses might be good to be with for the human, but humans might not be as good to be with, for the horse. Indeed, one might say that the relationship between the horse and the rider is not so much a collaboration as a continuation of a power relationship in a new politico-environmental atmosphere. This sense is undergirded by the fact that the GoPro is strapped to the head of the rider, not the horse and thus cannot capture the perspective of the horse, rather only that of the horse rider. Indeed, one might argue that even if it were to be strapped to the horse, a GoPro would still be unable to feel, smell, see, taste, hear that which came to greet the horse's sensory receptors. Notwithstanding the notion that horse and rider might become a single entity – something that is phenomenologically required for competent horse-riding — the horse encounters the ground, the rider does not; the horse encounters the climate under investigation in the films through hair-covered skin, and through eyes, ears, mouth, and nostrils oriented to the world in different ways from the human rider. Recognition of these things might lead to the further critique, that the visual method employed has not been examined for its political import in these and potentially other ways. Thus, the accusation of political quietism may well be put through the channel of the Western primacy of the visual.

Kopnina (2017) is perhaps the loudest critic of political quietism, attacking multispecies ethnography in far less subtle terms than might be brought against horse-rider collaborations on film. Kopnina's question, to which she seeks urgent heretical response, is: why doesn't multispecies ethnographers *do* anything to address the actual conditions under which (particular kinds of) animals suffer the most? Kopnina's question immediately indicates that only certain kinds of nonhuman being have thus far been explored under multispecies hand; mushrooms, dogs, bird, horses, and insects abound; pigs, sheep and cows, rats, and mice are far less present. Partly for this reason, mutispecies ethnographies collectively appear to Kopnina to leave the relations of power largely untouched, and instead offer, at best, a more sympathetic approach to the animal (Kopnina 2017: 351).

Kopnina opens her 2017 provocation against multispecies ethnography with a vignette. In it, she recalls listening to the presentation being given by Nancy Scheper Hughes at the Plenary session of the German Anthropology Association Annual Meeting in 2015. Scheper-Hughes used the occasion to call for a radicalized applied anthropology that actually makes a difference to those desperately entailed as donees in the organ trade. Kopnina's shock at the bloody images Sheper Hughes shows on an oversize screen during her talk turns to outrage as her fellow conference goers later speculate about the obvious solution of making more animal bodies available to satisfy the need for organs and diminish the illegal and tragic human side of the trade. Kopnina wonders why Scheper Hughes' devastatingly unsettling paper is not regarded to be as urgently applicable to animals as it is to the humans of whom she speaks; do animals not constitute just another disempowered minority whose own organs belong even further down the scale of the most vulnerable humans in the story? The proffered inclusion of animal (and predominantly pig) organs as a solution to minority human suffering leaves Kopnina to conclude that all we have managed in respect of the animal is a deeper sympathy:

> Much of the recent multispecies scholarship has remained rooted in comfortable intellectual and ethical spaces, addressing political issues in a local context and leaving the immensity of the global nonhuman abuse outside the scope of engaged anthropology (Kopnina 2017: 350).

The remark draws into the frame the stunning absence of millions of sheep, chicken, pigs, cows, and rodent research animals whose unwilling entailment in the industrial coordinates of egg, meat, and milk and pelt production and consumption do not come up in pine forests replete with mushrooms, or in accounts of the negotiations traditional hunters make with individual prey to render the latter edible.

The self-restriction of multispecies ethnographers to "comfortable intellectual spaces," their addressing of local political issues, and their consequent failure *to be with* most animals suggests that multispecies ethnographers have remained firmly in conceptual terrain. It is in this sense that the accusation Kopnina makes is very serious.

The pursuit of certain kinds of beings in multispecies ethnography has been purposeful; those nonhuman animals, plants fungi, and microbes that have been entirely subsumed in anthropological accounts as, variously, food, or symbol, and consigned to the categories of life that are killable (Agamben 1998) are reinterpreted and brought into political life, alongside humans. Those animals, plants and fungi that are to be found "amid [the] apocalyptic tales about environmental destruction" have attracted especial attention; they are the "modest examples of biocultural hope" upon which multispecies ethnography is founded and sustained (see Kirksey and Helmreich 2010: 545; see also (Harding 2010). Key examples have become paradigmatic of the political, ethical, and intellectual founding of multispecies ethnography, including Paxson's 2008 work that enlivens and values microbial cultures, Tsing's 2015 work on the Matsutake mushroom, and Raffles' 2010 exploration of insect love.

The combination of impulses and intentions set out above has led to the consistent elucidation of the animals who might benefit most from ethnographic attention. Moreover, as Kopnina's accusation, above, suggests, that elucidation indicates that nonhuman beings have become conceptually useful, as metonymic representatives of an intellectual turn, rather than fleshy beings. To right the wrongs that flow from excluding the fleshy plights of some animals, Kopnina calls for both practical and intellectual action, namely, for multispecies ethnographers to, "take a stance on AR [Animal Rights]...While the commitment to the advocacy of ecological justice, deep ecology, and AR seems radical and heretical, so was the commitment to minority [human] rights in the past" (Kopnina 2017: 350). One might say of Kopnina's position that there is enormous luxury involved in advising an end to any and all laboratory research involving animals (see Kopnina 2017); after all, it is not white Western academics who would likely suffer the worst consequences of, for instance, the vaccines that would not come available as a result of ceasing animal experimentation. It is much harder, though, to dismiss her accusation, that multispecies ethnography has not done much to address the human-animal hierarchies in which animals suffer in the greatest numbers, the least visibly. That is especially frustrating when one of the main compulsions of multispecies ethnography is to challenge the consequences of western logics and patterns *for animals*; the abattoir floor thick with the blood of the animals who live the lives entirely within human-designed coordinates is not similarly thick with ethnographers trying to make sense of them in the convivial terms favored by multispecies ethnographers. It is even more confounding when another aim is to, with nonhuman beings, work collectively to mutually assured survival against dire environmental damage. Given that the cessation of industrial forms of farming for meat that terminates in animal death in the modern slaughterhouse would have a far greater and more tangible impact on the environment than considering what landscapes look like for horses, it is difficult to dismiss Kopnina's position as merely intemperate political interjection.

Conviviality

Kopnina's remarks in particular suggest that multispecies ethnography is limited by its own imaginary of a waiting utopia. Ogden, Hall and Tanita (2013) for instance, describe multispecies ethnography as bringing about "speculative wonder." Wonderment might be pulled up sharp in the thick of laboratory relations or in the abattoir, however. Several theoreticians, such as Given (2018) have attempted to address this overwhelming sense of positivity by offering up critiques of conviviality, an important support pillar of multispecies ethnography. As Tsing notes, conviviality is frequently pressed into the service of transgressing "the boundaries that cordon nature from culture" (Tsing 2012: 141). When Ivan Illich (1973) introduced conviviality, he intended it to be an irruption of the coordinates of industrial productivity (see also Illich 1978). Modern capitalist systems governed by radical monopoly, reduced individuals to mere consumers who could not choose what, or how, things were produced, and divided them into experts who could use the tools and nonexperts, who could not. Situated as the opposite of industrial productivity, conviviality signaled a society in which individuals were distinguished not by their relations to the means of production but instead by their autonomy and creativity. Purposefully eutrapeliaic in both its (Aristotelean) emphasis on the mean, and its focus on (joyful) simple living and (apparently equally joyful) localization of production systems, the term has, as Given writes, acquired an overtly jovial tone that makes it "easy to fall into an idealist or romantic sense of conviviality as some sort of vaguely defined 'harmony' between people and the environment" (Given 2018: 131; and see for Bassey 2012). As Given notes,

> Competition, tension and conflict are as much part of convivial relations as symbiosis and collaboration. And we cannot assign human value judgements to these relationships. When a goat eats a cyclamen flower, it is irrelevant that this is 'good' for the goat and 'bad' for the flower: what matters is the continuance of the cycles of matter, nutrients and life. A goat eating a flower and returning its nutrients to the soil by defecation and decay maintains the conviviality; it works within the limits of the symbiosis (Given 2018: 131).

As Given notes, tarmacking over the place where goats and cyclamens once encountered each other tends to put paid to conviviality – and not just for goats and flowers, but for soils, worms, grubs, weeds and the rest. It is vitally important to include the events and contexts that stomp atop convivial relations – and, as will be suggested later in the chapter, there is much to be learned about human-nonhuman relations by situating oneself squarely in their midst. So doing can reveal not only those beings that persist at the edges of the tarmac, to whom we might attend with hope as Kirksey and Helmreich say we ought, but equally to those who have been covered over but yet emerge to say something.

Conviviality has also garnered an equally overt antidotal quality that has resulted in its enthusiastic adoption by a range of academic and social movements, including sociology (see for example Neal, Bennett, Cochrane and Mohan 2019); social geography (see, for example, Elshater 2020; Anoop 2017), as a pillar of degrowth theory and practice (see, for example, D 'Alisa et al. 2015; Kerschner et al. 2018; Muraca and Neuber 2018; Pansera et al. 2019; Vetter 2018; Kallis 2011) and a host of others alongside multispecies ethnography.

In its antidotal manifestations (see, for example, Costa 2019) conviviality tries to draw the world back into academia. For instance, Given agues we need to use it to get the world back from the abstractions we have made of it, noting that,

> We need an approach that integrates a stimulating theoretical framework with a specific 'ecology of practices' that allows us to immerse ourselves in the world we have externalized through research and analysis. (Given 2018: 130; see also Pétursdóttir: 578)

A crucial undergirding of Actor Network Theory, itself a major influence of multispecies ethnography, conviviality

> acknowledges the agency and centrality of assemblages of a wide range of players (or actants) such as soils, animals (including humans), plants, materials, artefacts, environmental processes and human technologies and memories...Heterogeneity is an inevitable but central consequence of this all-encompassing diversity, as is the participation of local histories of action, collaboration and tension... Similarly, conviviality enthusiastically supports the Latourian philippic against the human/non-human dualism .... What conviviality brings ... is a commitment to the central role of non-human and non-human-made players (2018: 131; see also McMaster and Wastell 2005, 176–177).

All of this indicates that conviviality allows us to politically, rhetorically, practically, theoretically, analytically, and materially tend to the "living together" we do as entailed beings, things, atmospheres. But its emphasis on inextricable entailment – no matter how inevitable that most assuredly is – might effectively obscure our ability to examine some of the potentials consequences of "living together in difference."

Being with (Some) Animals: The Taking of Animals Literally

The importance of being ethnographically present in the thick of the coordinates within which multiple evident and hidden species meet is of course indicated in the use of "ethnography" in "multispecies ethnography." It is equally evident in ensuring that the ethnographer's position is not the principal one. This extends to trouble the classifications the ethnographer might otherwise make of the nonhuman others she encounters that render their moving (and therefore multiply meaningful) being taxonomically still (and stable) in definitive Western terms. That might involve taking literally, and not metaphorically, the relationships that locally resident humans have with animals. Paul Nadasdy (2007) notes the historical tendency of Western anthropologists to treat indigenous accounts of fleshy animal reciprocity as metaphorical, rather than accepting that they might be literally true. Having carried out extensive ethnographic fieldwork with hunters from the Kluane First Nation in the Yukon, Nadasdy came to understand that the animals he would hitherto have ascribed the status "prey" engaged in reciprocal agreement with human hunters (see also Stepanoff 2012, 2017). The possibility (which is actually quite a longstanding one, being registered by Hallowell, for instance, in Hallowell 1960) has produced excitement, particularly around the idea lauded by some scholars (e.g., Avelar 2013; Danowski and Viveiros de Castro 2016) that, "it is the continued elevation of indigenous thought beyond the analytical level to that of the political which offers the best hope of salvation in a time of environmental crisis and multispecies extinction" (White and Candea 2018: np).

While it is certainly high time that Indigenous perspectives were taken seriously, and literally, the bounds of what multispecies ethnographers are prepared to take seriously, as Kopnina (2017) notes, seems highly romanticized. Utilizing Kopnina's logic, if we are to revisit what Indigenous people know about animals, we must acknowledge that those perspectives were once considered, in disciplinary explorations of totemism and animism, as childlike and naïve. If we are to revisit animals, which we have hitherto understood to be conceptually and materially in the service of humans, the same kind of seriousness is required. While it is unthinkable to regard Indigenous perspectives as childlike now, it remains entirely thinkable for animals to remain in the service of humans. As Kopnina notes, the recognition of minority rights emerges only after a great deal of suffering has already been endured; suffering that appears to be subsumed under enthusiastically pursued studies of horses (Fijn 2015), mushrooms (Tsing 2015), and rivers and their multiple partners in being (e.g., Doron 2021). For Kopnina, such elucidation of industrially begotten (and ethnographically forgotten) animal bodies begs the question: what else could and should be taken literally? She suggests that their invisible suffering could and should, and that if it were, an evident impact on the environmental crisis that multispecies ethnographers claim to worry over might actually be achieved.

The work of this chapter is not, however, to advance Kopnina's call, to immediately protect animal rights in the lab and in the abattoir, instead of waiting to do so when those rights to become valid as they once did for non-white, (nonmale) humans. It is instead to examine the potential of explorations guided by the foundational notion of multispecies ethnography, "good to live with." As the next (case-focused) sections demonstrate, It is in the direst of contexts, such as the laboratory, that we might understand the consequences for all kinds of parties of removing, erecting, or maintaining boundaries. We might learn in the laboratory, too, that the destruction of boundaries does not always bode well for human or nonhuman parties, just as the erection of boundaries between them does not necessarily bring about disastrous consequence. By examining the cuticle, we might include a much larger range of beings in multispecies explorations, and we might also realize some of the consequences of having become distanced from nonhuman beings – including human survival.

Cuticulae

Application of multispecies ethnographic principles and techniques can be applied to the cuticle and, when they are, the value of thinking beyond the human – and the possibility of being collectively entailed in survival – is revealed. The cuticle is any of a variety of tough but flexible, non-mineral outer coverings of an organism, be it dog, human, insect or mushroom.

The alert reader will have perhaps felt a flash of recognition reading through this only apparently random list of entities. Dogs, humans, insects and mushrooms should indeed feel familiar, because each of them has been enthusiastically explored in multispecies-al frame; the dog and the human are of course references to feminist philosopher of science and technology Donna Haraway's influence on multispecies ethnography (see Kirksey and Helmreich 2010) – Haraway's effect on thinking about the agency of nonhumans is matched only by that of Latour (e.g., Latour 2005). Haraway (2008) used the relationship between herself and her dog to think about the ways in which humans and animals are mutually shaped by interactions across species boundaries. The insect on the list references Raffles' 2010 *Insectopedia*, which implores readers to make the explicit connections of insects to human culture. Insects are not categories, Raffles asserts, and cannot be approached as though they were "merely the opportunity for culture"; they are instead, "its co-authors" (Raffles 2010: 100). The mushroom, is, of course, in reference to Anna Tsing's 2015 *The Mushroom at the End of the World*. Tsing uses the mushroom and its dependence on human disturbed environments to describe the multi-species "assemblages" that permit the matsutake mushroom to thrive under pines:

one could say that pines, matsutake, and humans all cultivate each other unintentionally. They make each other's world-making projects possible. This idiom has allowed me to consider how landscapes more generally are products of unintentional design, that is, the overlapping world-making activities of many agents, human and not human. The design is clear in the landscape's ecosystem. But none of the agents have planned this effect. Humans join others in making landscapes of unintentional design. As sites for more-than-human dramas, landscapes are radical tools for decentering human hubris. Landscapes are not backdrops for historical action: they are themselves active. Watching landscapes in formation shows humans joining other living beings in shaping worlds. (Tsing 2015: 152)

There is no doubt that each of these cases permits reconceptualization of the place of the nonhuman in and through the various pokings and proddings delivered to the line that would otherwise sit between dog, cricket or mushroom, and human, even if that line permitted each to border and salute the other, as Rilke poetically observed of lovers; he presciently forecast that,

Some day there will be girls and women whose name will no longer signify merely an opposite of the masculine, but something in itself, something that makes one think, not of any complement and limit, but only of life and existence: the feminine human being. This advance will (at first much against the will of the out-stripped men) change the love-experience, which is now full of error, will alter it from the ground up, reshape it into a relation that is meant to be of one human being to another, no longer of man to woman. (Rilke 1904 [2001]: 26)

The poetic insight might just as easily be pressed into the service of describing multispecies agenda. Although the aforementioned multispecies-works indubitably attend to some key principles of multispecies ethnography, such as the constitutive significance of nonhuman species; the involvement of nonhuman species in multiple aspects of human life; the decentering of the human; attendance and the asymmetrical becomings of human-nonhuman assemblages, they ironically reaffirm the reviled western influence over nonhuman worlds *in their narrow restriction to some few nonhuman beings*.

The focus on (among some few others) dogs, fungi, insects is remarkable because, as vulnerable as they are to human decisions and their consequences, they are not the beings who suffer the worst of human passions. But more than a simple refocusing appears to be required. Where Kopnina (2017) argues for a rights-oriented multispecies ethnography that makes its applications wherever they are most urgently needed, precisely as interventions, it is far more intellectually robust and antidotally effective to make an expansive embrace of life – wherever it is to be found. So doing ensures a broad diversity of experiences are included, and equally guards against the descent of multispecies ethnography into the political and conceptual abstraction against which it initially railed, and of which it presently stands accused (e.g., Kopnina 2017; see also Smart 2014).

One possibility for such broad engagement persists in an analysis of the cuticle. The cuticle stretches to encompass life in teeming expansiveness, because the cuticle appears everywhere, skinning up pigs, sheep, chickens, gum trees, peaches, cats, sparrows – as well as the much favored horses, fungi, dogs and insects. The cuticle offers up the potential of flattening not the distinctive differences between species that are so crucial to conviviality and multispecies ethnography, but rather, human biases and preferences (see Smart 2014). Here though, the cuticle is engaged to demonstrate the consequences of maintaining borders that separate humans from nonhuman actants. These consequences are especially dire for the humans in the equation.

The cuticle is a lively membrane. Used here in place of the more prosaic "skin" because of its capacity to encompass a broader diversity of nonhuman and human bodies, the cuticle covers everything from arthropod exoskeletons to the human epidermis, from hydrophobic leaves to a mushroom's basidiocarp. The cuticles of mushrooms, spiders, sheep, humans, horses, cows and plants confound such stark divisions of the world into "inside" and "outside," operating instead in a way akin to breathing. Just as breathing is cyclic rather than divisible into discrete acts of inhalation and exhalation, the cuticle simultaneously protects against and extracts from the worlds in which spiders, mushrooms, humans, pigs, chickens and sheep dwell. Indeed, the cuticle is defiantly non-homologous, manifesting in different structures, chemical compositions and originating of different geneses to perform different functions in the lives of a dazzling array of life.

In humans, for instance, the cuticle appears as several different structures, including the superficial layer of overlapping and dead cells covering the hair shaft (cuticula pili) that locks the hair into its follicle (see James, Berger, Elston and Odom, 2006). Most would be familiar with the term as it is applied it to the thickened layer of skin surrounding fingernails and toenails (called the eponychium), but it is much more generalized than that; the cuticle is that outer layer of an organism that comes in contact with the environment. In most invertebrates the dead, noncellular cuticle is secreted by the epidermis (see for example Page and Johnstone 2007). In simple biological terms, in humans the cuticle is the epidermis, the very outer layer of the human skin that at first encounters the world (for an example of this basic biological explication, see Encyclopedia Britannica 2019).

A more rigid cuticle, composed chiefly of chitin, encircles arthropods, and epicuticular wax covers hydrophobic leaves in plants that benefit from the way water, which rolls off the leaf, carries away harmful pollutants and dust (see for example Onda et al. 1996). More generally, plant cuticles prevent water loss and reduce the opportunities for pathogens to penetrate the plant. The hotter, drier, and saltier the environment in which the plan dwells, the thicker and waxier the plant's cuticle, a feature that slows the inevitable inbound movement of heat and toxic salts and preserves the water inside the plant's cellular structures (see, for example, Mulroy 1979; Jordaan and Kruger 1998). And, in the mycological world, the pileipellis, Latin for "skin" of a "cap," protects the trama, the inner fleshy tissue of a mushroom from the brutality of the world, and equally from its own spore-bearing tissue layer, the hymenium (see Jaeger 1959).

The fact that the cuticle is common to us all provides a very fruitful foundation upon which to base analysis of different life engaged in world building in common. Such as prospect is not entirely new and brings to mind Merleau-Ponty's observation, that the main differences between animals could be best described as differences in fleshy style as dictated by how the world interacts with their various forms, and penetrates their various skins (see Merleau-Ponty 1964).

The cuticle, in the form of the mammalian skin, has long been understood by biologists to be held in common. Humans and other species share some of the same genetic information, but humans are now unique among mammals when it comes to the types and diversity of microorganisms present on the skin. Professor of Biology Joshua Nuefeld, leader of a 2020 study examining the mammalian microbiome – the collection of microorganisms such as bacteria, fungi, and viruses that naturally occur on our skin – registered his surprise at its results. They revealed "just how distinct we humans are from almost all other mammals, at least in terms of the skin microbes" (see Ross, Müller, Weese and Neufeld 2018). The study, the most comprehensive survey of mammals to date found that the human microbiome was somewhat impoverished compared with other mammalian examples. It reveals as well that the human skin tracks and traces the recoiling humans have made from the broadest manifestations of the world and our more constrained domestic and other revolutions around the suburbs. A withdrawal from the worlds our mammalian kin continue to inhabit comes available in comparative analysis of mammalian microbiomes which, in the context of habitat reduction and loss, and temperature changes might also be expected to become impoverished. Far from being simply metaphorical statements of loss and impoverishment, microbiomic penury serves as an index of our distance from a range of nonhuman and nonhuman made companions. And, it is highly consequential.

Microbiomes are shaped by a combination of biological and genetic factors. They are immediately postnatally transferred from mother to baby to encircle the infant cuticle with a share of protective microbial defenders. They are in this sense a kind of enduring cuticular habitus. Beyond being post natally given, they are impacted by the encounters the individual cuticle has with food, drink, air, water, hand sanitizer, clothes, shampoo, dogs, cats, dirt, pollution, the inside of the house or the inside of the local supermarket, forests, seas, mountains, cigarettes, black coffee and pure water, plastic bags, the dining table, another human skin – *everything*. The microbiome is a ceaseless registrar, recording each encounter in the lives and deaths of individual cuticle dwellers: the funguses, bacteria, and viruses collectively termed the "human microbiota." Our traversals and our meetings are all highly impactory on the microbiota, which have en masse registered that western dwelling humans generally occupy homes made of a narrow range of manufactured materials, wear a similarly narrow range of manufactured clothes, make tight circulations between a limited number of manufactured places, and bathe more than is strictly necessary. These narrow participations in the world have caused compositional shifts ('dysbiosis') in microbiomes that are consequentially linked to changes in human disease states; the cuticular and gut microbiome exhibit distinct composition and diversity patterns in disease states ranging from diabetes and gastrointestinal disorders, to autism and psoriasis, and similarly exhibit distinctive patterns in healthy human subjects (see Ross et al. 2018).

The upshot of all of this is that restricting ourselves to the human made worlds of the home, the office. and the gym and so on result in poor health outcomes for humans (as well as their silent and invisible phylosymbiotic bacterial, viral, and fungal partners).

Antidotal advisements have been issued by biologists that include broadening our circulations to include a greater diversity of nonhuman made environments and nonhuman companions, including those that might be found in soil, on the family dog, who really might serve to improve health, as has long been anecdotally suggested. Not suggested are the tree that grows a great distance away from the home, the fauna that live with it, or the lamb, cow or pig prior to its appearance in the home on a plate. Those beings are very likely not suggested because they are inconveniently geographically and otherwise distanced from humans, which rather confirms the problem of the impoverished human cuticle. That is, health-conferring gut and skin microbial communities can be transferred between people and environments, and other beings when we interact with them. The inverse is also true; a paucity of commingling opportunities and encounters makes for poor health.

It is possible, then, to suggest that 'health' cannot be conceived of exclusively or even principally in the sociological compartment called the human world. Indeed, the consequences of not interacting with other beings and environments would be nothing short of fatal. It is in this sense that it is crucial we recognize the sharedness of the permeable, absorptive and transmitting cuticle: we all depend on picking up the living material that looses itself to travel outbound from the one being to the impoverished skin of the other.

The consequences of bearing an impoverished skin are of course readily apparent in the time of COVID 19, which forces us to even more tightly contain ourselves to the home and the zoom screen, to refuse relations with others bearing microbial gifts, and to kill benign and maleficent cuticle dwellers alike. The registrations of even tighter constraints on movements and encounters with nonhuman others will be made on the cuticle, that stretchy membrane that unites us all in our difference. But only one cuticle, the human skin, registers a level of alienation from other life that might be fatal. Multispecies ethnographic engagement with the cuticle permits full rehearsal of the benefits of being with nonhuman others.

Laboratory

Laboratories that carry out experiments on mouse and rat bodies (and, in the case of the latter animal, minds) provide a very interesting context for exploring the possibilities of destroying and erecting divisions between different modes of being. Ostensibly, laboratories separate nature from culture, specie from specie, human from animal; on the face of it, they are the ultimate multispecies ethnographer's nightmare. Indeed, the laboratory can easily be considered in the terms of the classic Baconian configuration of the strictest of human–animal divisions. Bacon's God scientists appear on one side of the divide in a Judeo-Christian heritage of human supremacy over nature. From their side of the boundary, humanist God scientists regard the animal as both biological and genetic mirror for self-reflection, and the raw material for self-reproduction:

God scientists inhabit the ontotheological domain that the union of science and technology has produced; as Heidegger (1962) insisted, under the banner of modernity, science itself is arrogated to the place of Plato's Good and the Christian God. God scientists also represent the fulfillment of Bacon's call to the mastery of nature through its ontological transformation. This is particularly evident in the production of transgenic animals, as God scientists here claim not only omniscience but ultimate creative power; as Bacon wrote, "On a given body to generate and superinduce a new nature, is the work and aim of human power." (Bacon 1999, p. 148; Dennis 2013: 507)

Laboratory animals are, in other words, very good for setting to utilitarian purpose.

They are also very, very good to think with. Concealed under a thick layer of multiple meanings, it is difficult to access rats and mice as flesh and blood animals beyond their utilitarian and conceptual identities. Both creatures occupy opposing reaches of human imaginations in Western society; as similar to humans as they are in the terms of their mammalian identities, their sociality, their habits of domestic occupation and food consumption, their habits of hand, indeed their use of their 'hands' and in that they serve as homologies for our minds and our genes in laboratory settings, they are also and equally situated as our symbolic opposites. Where humans are figuratively and literally creators of conceptual and material worlds, rats and mice destroy them: they are underminers of purity and underminers of the dry wall, appearing simultaneously in western contexts as the filthy bearers of diseases harmful to people and as pure white weapons in the fight against human diseases when they occur in laboratory contexts:

> Their capacity to occur simultaneously in more than one category of meaning, suggests that mice and rats exist not only at the polar reaches of our imaginations, but are also loci of slippage and indistinction between polar categories. Polarity, ambiguity, and ambivalence set up a number of possibilities for rat/mouse–human relations generally. They also mark the relationships between scientists and rodent research animals that occur within the confines of the research laboratory. Here they embody directly opposing positions and meanings, and present simultaneous and conflicting meanings, which are sometimes reconciled. (Dennis 2013: 506)

They are, in other words, very, *very* good to think with.

Could animals be *good to be with* in the laboratory? A multispecies ethnographic orientation to the laboratory could reveal new insights into the ways in which humans and animals relate to one another, beyond the material and conceptual. It might be able to do so by leveraging the possibilities of thinking with that anthropological mainstay, kinship, and the careful revisions that have been made to it to render it fit for encountering non-white (and nonhuman) others.

As Haraway notes, to flourish in the Anthropocene, we ought to make kin with nonhuman others: "Who and whatever we are, we need to make-with—become-with, compose-with" (Haraway 2016: 102). Such an advisement holds true for laboratory context, in several ways. First, nonhuman bodies must be sufficiently similar to human bodies for them to be in the laboratory in the first place, and for the outcomes of experimentation to have application to human bodies. The required sameness of nonhuman animal bodies and human bodies is accomplished in and through the subsumption of the specie differences of rodent research animals and humans to a shared mammalian membership. The effect is that these animals appear in the laboratory as human kin, in genetic and biological terms:

> Mice, for instance, on the basis of their close biological and genetic relatedness to humans, are critical models for experimental investigation of human immune diseases without putting human individuals at risk. Immunological research, for instance, leverages mouse models for studies on diabetes, bone development, autoimmunity, infectious diseases, and transplantation tolerance. Using rodents as bio and gene kin, scientists try to predict the answers they would obtain directly if they could perform the experiments on/in humans. Rodent models are, then, powerful research tools to in-vivo decipher [human] physiology. (Dennis 2013: 508)

Those versions of kinship are not very satisfying for multispecies ethnographers. All they really do is express that the utility of rodents to humans comes available because of our basically similar body plans.

Conceptual versions of kinship are not very helpful to those who want to get at fleshy interactions, either. Indeed, thinking conceptually about kinship in the laboratory elides any fleshy relationship between human and animal parties in favor of setting out how and when mammalian bodies become different for the purposes of experimentation and, ultimately, disposal. This is best demonstrated by recourse to the importance of "sacrifice" in the laboratory. Mammalian membership not only situates rat and mouse animals as human kin in biological and genetic terms, it also qualifies them for entry into the laboratory's sacrificial economy. To qualify for sacrifice, a substantive equivalency with human being must be met (see, for example, Govindrajan 2015; Willerslev et al. 2014). Mammalian membership locates rats and mice as human stand-ins in the laboratory. This is vividly demonstrated in Haraway's description of the mammalian homology between the transgenic-breast-cancer-animal-model Oncomouse and humans, which is based on "Oncomouse's essence." In common with humans, her "essence" is, "to be mammal, a bearer by definition of mammary glands, and a site for the operation of a transplanted, human, tumor producing gene" (1997: 89). This mammalian homology is the basis upon which the mouse must be sacrificed, in order "that I and my sisters [the human mammals in the equation] might live" (Haraway 2008: 76). In a very detailed analysis, Lynch (1988) sets out precisely how the anthropologically well-rehearsed conditions for sacrifice (as opposed to killing) are met in the laboratory by rat and mouse animals, and Dennis (2011) further demonstrates how specie-al differences between humans and rodents emerge out of mammalian homology at critical moments in experimental temporalities to make specie bodies necroavailable.

There *is*, though, one way of thinking about kinship that does get at "being with" animals, however, and is ripe for the application of multispecies thinking. This entrée into the laboratory could be made by recourse to anthropological reconsiderations of kinship, to get us beyond accessing only the conceptual and material utility of having rodents as kin.

Anthropologists have closely reconsidered the scope for the study of kinship in post-Schneiderian terms:

> According to Schneider (1968, 1984), kinship studies were, overwhelmingly, the expression of anthropologists' ethnocentric biases and disciplinary preoccupations. Biological reproduction, for instance, was taken to be foundational to kinship; a view not shared by all cultures, and one that might be missed by a predefined interpretive scheme. Post-Schneiderian kinship is characterized by an increased understanding of a researcher's own cultural assumptions and practices. The inclusion of the study of Western kinship systems in and on their own terms has also required investigation of the varying relationships between culture and nature, and changing conceptions of nature. This has been applied to Western systems of kinship, too, in postmodernity (see Carsten (2000) (Dennis 2013: 508).

As a principal tool of deconstruction, critical versions of kinship analysis are useful for encountering animals in the laboratory. This is especially because such an approach could get beyond presumptions of the biological terms in which rodents come to be human kin and trouble the presumptions that ethnographers might hold about how scientists and animals encounter one another in the laboratory context. As Dillard-Wright (2009) notes, such a possibility might only seem unusual or impossible, if we are prepared to accept the exceptionalist argument that human communications differ in kind and in practice from those of other animals, even in laboratories (see also Davis and Balfour 1992).

The kinds of communications that happen in laboratories are often presumed to be unilineal and detached (see Merleau-Ponty 1964). Merleau-Ponty critically questioned such presumptions, and particularly the notion that the scientist be detached from her animal subject, who sits firmly on the other side of the human-animal divide. Merleau-Ponty argued instead that the practicing of the "sciences of man" requires the scientist to interpret, "as best he can the acts of [animal] others, reactivating from ambiguous signs an experience which is not his own." So doing occasions a reliance on the simultaneous giveness of animality and humanity that Merleau-Ponty called "strange kinship." This term is meant to capture how the world is fleshily shared among and generally available to beings, despite their evident differences (see also Godway (1998: 50). Flesh makes communication possible because it is reversible, in that we are all sensing and sensible — this enables intercorporeal being and founds transivity between bodies, including between animal and human bodies. This kinship,

> neither erases difference nor similarity, makes us neither identical nor separate; for Merleau-Ponty, animals and people are at once strangers and kin (see also Oliver 2007; Dennis 2013: 514).

Haraway describes this kinship as the ways in which "species of all kinds are consequent on a subject and object shaping dance of encounters" (2008: 4).

Such dances occur in the laboratory; Dennis for example has documented how rats and neurologist researchers negotiate participation in experiments, and the ways in which laboratory veterinarians communicate with immunology mice about their physical well-being. In both cases, researchers and vets must become fluent in the fleshy communications they each referred to as being able to 'speak rat' or 'speak mouse' (Dennis 2011). Unlike Kohn's (2013) bio semiotically informed work that privileges human interpretations of communication, fluent interspecies communication entails negotiation; for instance, if rats involved in neurological studies did not want to participate in certain experimental phases, they could effectively refuse to do so by making themselves unavailable to the hands of researchers. Dennis explains how rats positioned their bodies in ways that prevented researchers from taking them from their enclosures; researchers were not permitted to extract the animals in any way but by mandated handling techniques. The rats knew this and had to be convinced with treats and other cajoleries to participate, something that ethnographically necessitated spending time with rats as much as with human scientists. As every laboratory veterinarian can attest, the communications that domestic pets and private practice veterinarians make with equal fluency do not often result in cats and dogs refusing cancer treatment or the extension of their lives under any circumstances; their communications are invariably obscured under what could sometimes be the unbearable weight of love.

Some analysts have argued that such communicative exchanges might enable the decentering of the human in and through the at least partial reworking of human-animal relations in the laboratory. The possibility of ratty negotiations, and mouse refusals to participate in immunology trials on the basis of ill health are powerful - not least because mice and rats who are ill or distressed skew experiments. Alongside them sit other discoveries – such as those documented by Shanor and Kanwal (2009), that mice can, and do, giggle. But beyond the enticing immediacy of such discoveries and what they mean in situ and beyond, they also indicate a much bigger set of ideas that recognize, call out to and engage animals in the search for an understanding of Being – not simply human being. As Wolf-Meyer (2006) suggests, this is already manifest in the realm of posthuman biopolitics, in which the human genome project is located; here, an understanding of being, not simply species-specific being, is sought. And, that becomes a scientific mandate that encourages contingency of species and counters the over-determination of humanity.

Conclusion

It is initially difficult to recognize the laboratory as a place in which strange kinship emerges to decenter human exceptionalism. But, as Haraway (2008) has reminded us, it was science that put Homo sapiens firmly in the world of animals – Freud's second of the three great wounds dealt to the narcissism of the self-centered human subject specifically picks out the devastatingly injurious Darwinian blade. If ethnographers were thicker on the ground of the laboratory, they may well find that such strange kinships do nought to alleviate the kinds of concerns registered by Kopnina; they may well feel compelled to activate exactly the rights agenda for which she calls. Or, they may find that the multispecies relations in the laboratory provide more potential for decentering the human than first blush indicates. Whatever unfolds, it is clear that ethnographic tending to the relationships in that unfurl in the most challenging of environments is required. As Nigel Rapport argued in his I Am Dynamite, it is critical to look for the results of ostensibly unilineal power in the very thick of its operations: investigating these always turns up the most productive material for thinking up how relations are lived in situ – the place of every ethnographer, especially when ethnography is conceived of as doable in every place, even the laboratory.

A close look at the cuticle reveals the potential reach of multispecies ethnography. The impact of living beyond the lively thrall of what we might once have called "nature" is registered in the paucity of the human skin; the antidotes for our failure to circulate among our plant and animal fellows are readily indicated in the prescription issued by biologists: return to the world. Such advice is very difficult to follow under the conditions of COVID 19 that constrict the already tight circulations of humans even further, but the advisement does other kinds of work. Principally, it indicates the interconnection of all beings and proposes a single entry point into the exploration of that interconnectedness. The narrow range of beings included in multispecies ethnographic explorations are, in the end, profoundly human constrictions of that entry point; the paucity of range, as has been indicated in this chapter, translates to a paucity of analytic and practical potential. And the antidote? Return to the world – including the world in its grubby, unpleasant, unromantic manifestations, for even here, we all have skin in the game.

Acknowledgments This chapter contains excerpts from Ambiguous Mice, Speaking Rats: Crossing and Affirming the Great Divides in Scientific Practice, Simone Dennis, Anthrozoös, copyright © International Society for Anthrozoology, reprinted by permission of Taylor & Francis Ltd, http://www.tandfonline.com on behalf of International Society for Anthrozoology.

References

Agamben G (1998) Homo Sacer: sovereign power and bare life. Daniel Heller-Roazen, trans. Stanford University Press, Stanford
Anoop N (2017) Purging the nation: race, conviviality and embodied encounters in the lives of British Bangladeshi Muslim young women. Trans Inst Br Geogr 42(2):289–302
ASAD T (1973) Anthropology & the colonial encounter. London, Ithaca Press. (7th ed.)
Avelar I (2013) Amerindian perspectivism and non-human rights. Alter/nativas: Latin Am Cult Stud 1:1–21
Bacon F (1999) The new Oregon. In: Sargent R (ed) Francis Bacon: selected philosophical works. Hackett Publishing Co., Indianapolis
Bassey M (2012) Convivial policies for the inevitable: global warming, peak oil, economic chaos. Book Guild, Brighton
Bateson G (1972) Steps to an ecology of mind: collected essays in anthropology, psychiatry, evolution, and epistemology. University Press, Chicago
Best S, Nocella A, Kahn R, Gigliotti C, Kemmerer L (2007) Introducing critical animal studies. Animal Liberat Philos Policy J 5(1):4–5
Bird-David N (1999) 'Animism' revisited: personhood, environment, and relational epistemology. Curr Anthropol 40:567–591
Brightman RA (1993) Grateful prey: rock Cree human-animal relationship. University of California Press, Berkeley

Brown H, Nading A (2019) Introduction: human animal health in medical anthropology. Med Anthropol Q 33(1):5–23
Carsten J (2000) Cultures of relatedness. Cambridge University Press, New York
Cassidy R (2007) Introduction: domestication reconsidered. In: Cassidy R, Mullin M (eds) Where the wild things are now: domestication reconsidered. Berg, Oxford, pp 1–26
Costa S (2019) The neglected nexus between conviviality and inequality. Novos Estudos CEBRAP 38(1):15–32
Crutzen PJ, Stoermer EF (2000) The Anthropocene. Glob Change Newslett 41:17–18
D'Alisa G, Demaria F, Kallis G (eds) (2015) Degrowth: a vocabulary for a new era. Routledge, London
Danowski D, Viveiros de Castro E (2016) The ends of the world. Polity Press, Cambridge
Davis H, Balfour D (1992) The inevitable bond: examining scientist–animal interactions. Cambridge University Press, Cambridge
Dennis S (2011) For the love of lab rats. Cambria Press, New York
Dennis S (2013) Ambiguous mice, speaking rats: crossing and affirming the great divides in scientific practice. Anthrozoös 26(4):505–517
Descola P (1992) Societies of nature and the nature of society. In: Kuper A (ed) Conceptualizing society. Routledge, London, pp 107–126
Descola P (2014) All too human (still): a comment on Eduardo Kohn's how forests think. HAU J Ethnograph Theory 4(2):267–273
Dillard-Wright D (2009) Thinking across species boundaries: general sociality and embodied meaning. Soc Anim 17:53–71
Doron A (2021) Stench and sensibilities: on living with waste, animals and microbes in India. Aust J Anthropol 32(1):23–41
Douglas M (1966) The abominations of Leviticus. In: Douglas M (ed) Purity and danger: an analysis of the concepts of pollution and taboo. Routledge & Kegan Paul, London, pp 41–57
Durkheim É (1915) The elementary forms of the religious life (trans. JW Swain). Allen & Unwin, London
Elshater A (2020) Food consumption in the everyday life of liveable cities: design implications for conviviality. J Urban Int Res Placemak Urban Sustain 13(1):68–96
Encyclopedia Britannica (2019) Cuticle. Encyclopedia Britannica, https://www.britannica.com/science/cuticle. Accessed 18 July 2021
Evans-Pritchard EE (1940) The Nuer: a description of the modes of livelihood and political institutions of a Nilotic people. University Press, Oxford
Fijn N (2015) Yolngu homeland. Ronin Films, Canberra
Fijn N (2019) The multiple being: multispecies ethnographic filmmaking in Arnhem Land, Australia. Vis Anthropol 32(5):383–403
Fijn N, Kavesh M (2021) A sensory approach for multispecies anthropology. Aust J Anthropol 32(1):6–22
Frazer JG (1887) Totemism. A. & C. Black, Edinburgh
Geertz C (1973) Deep play: notes on the Balinese cockfight. In: Geertz C (ed) The interpretation of culture: selected essays. Basic Books, New York, pp 412–454
Given M (2018) Conviviality and the life of soil. Camb Archaeol J 28(1):127–143
Godway E (1998) The being which is behind us: Merleau-Ponty and the question of nature. Int Stud Philos 1:47–56
Govindrajan R (2015) The goat that died for family: animal sacrifice and interspecies kinship in India's Central Himalayas. Am Ethnol 42(3):504–519
Hallowell AI (1960) Ojibwa ontology, behavior, and world view. In: Diamond S (ed) Culture in history: essays in honor of Paul Radin. Columbia University Press, New York, pp 19–52
Haraway D (1997) Modest_Witness@Second_Millennium.FemaleMan©_Meets OncoMouse™. Routledge, New York
Haraway D (2008) When species meet. University of Minnesota Press, Minneapolis
Haraway (2016) Donna Haraway, Staying with the Trouble: Making Kin in the Chthulucene. Durham, NC: Duke University Press

Harding S (2010) Get religion. In: Gusterson H, Besteman C (eds) The insecure American: how we got here and what we should do about it. University of California Press, Berkeley, pp 345–361
Harris M (1968) The rise of anthropological theory: a history of theories of culture. AltaMira Press, Walnut Creek
Harris M (1978) India's sacred cow. Hum Nat 1(2):28–36
Heidegger M (1962) Being and time. Trans. John Macquarie and Edward Robinson. Harper and Row, New York
Illich I (1973) Tools for conviviality. Calder and Boyars, London
Illich I (1978) Toward a history of needs, 1st edn. Pantheon Books, New York
Ingold T (1974) On Reindeer and Men. Man 9(4):523–538
Ingold T (2013) Anthropology beyond humanity. Suomen Anthropologi 38(3):5–23
Ingold T, Pálsson G (eds) (2013) Biosocial becomings: integrating social and biological anthropology. Cambridge University Press, Cambridge, UK
Jaeger EC (1959) A source-book of biological names and terms. Thomas, Springfield. ISBN 0-398-06179-3
James W, Berger T, Elston D, Odom R (2006) Andrews' diseases of the skin: clinical dermatology, 10th edn. Saunders Elseiver, Philadelphia
Jordaan A, Kruger H (1998) Notes on the cuticular ultrastructure of six xerophytes from southern Africa. S Afr J Bot 64(1):82–85
Kallis G (2011) In defence of degrowth. Ecol Econ 70(5):873–880
Keesing R (1981) Cultural anthropology: a contemporary perspective. CBS Publishing, New York
Kerschner C, Wächter P, Nierling L, Ehlers M (2018) Degrowth and technology: towards feasible, viable, appropriate and convivial imaginaries. J Clean Prod 197:1619–1636
Kirksey E, Helmreich S (2010) The emergence of multispecies ethnography. Cult Anthropol 25(4): 545–576
Kohn E (2013) How forests think: toward an anthropology beyond the human. University of California Press, Berkeley
Kopnina H (2017) Beyond multispecies ethnography: engaging with violence and animal rights in anthropology. Crit Anthropol 37(3):333–357
Kuper A (2005) The reinvention of primitive society: transformations of a myth. Routledge, London
Latour B (2005) Reassembling the social: an introduction to actor-network theory. University Press, Oxford
Lévi-Strauss C (1963) Totemism. Beacon Press, Boston
Locke P, Münster U (2016) Multispecies ethnography. Entry for Oxford University Bibliographies Online. Retrieved from: Bibliographieshttp://www.oxfordbibliographies.com/view/document/obo9780199766567/obo97801997665670130.xml?rskey=fapCAD&result=1&q=Multispecies&print#firstMatch. Accessed 7 July 2021
Lynch M (1988) Sacrifice and the transformation of the animal body into a scientific object: laboratory culture and ritual practice in the neurosciences. Soc Stud Sci 18(2):265–289
Malinowski B (1925) Magic, science and religion. In: Science, religion and reality. Sheldon Press, London, pp 19–84
McLennan JF (1869–70) The worship of animals and plants. Fortnightly Rev 6, 407–427, 562–582; 7, 194–216
McMaster T, Wastell D (2005) The agency of hybrids: overcoming the symmetrophobic block. Scand J Inf Syst 17:175–182
Merleau-Ponty M (1964) Sense and Non-Sense. Northwestern University Press, Evanston
Michele Rosaldo Z, Louise L, Joan B (1974) Woman, culture, and society. Stanford, CA: Stanford University Press
Mullin MH (1999) Mirrors and windows: sociocultural studies of human-animal relationships. Annu Rev Anthropol 28(1):201–224
Mulroy T (1979) Spectral properties of heavily glaucous and non-glaucous leaves of a succulent rosette-plant. Oecologia 38(3):349–357

Muraca B, Neuber F (2018) Viable and convivial technologies: considerations on climate engineering from a degrowth perspective. J Clean Prod 197:1810–1822
Nadasdy P (2007) The gift in the animal: the ontology of hunting and human-animal sociality. Am Ethnol 34(1):25–31
Neal S, Bennett K, Cochrane A, Mohan G (2019) Community and conviviality? Informal social life in multicultural places. Sociology 53(1):69–86
Ogden L, Hall B, Tanita K (2013) Animals, plants, people and things: a review of multispecies ethnography. Environ Soc 4(1):5–24
Oliver K (2007) Stopping the anthropological machine: Agamben with Heidegger and Merleau-Ponty. PhaenEx 2:1–23
Onda T, Shibuichi S, Satoh N, Tsujii K (1996) Super-water-repellent fractal surfaces. Langmuir 12(9):2125–2127
Page AP, Johnstone IL (March 19, 2007) The cuticle. WormBook, ed. The C. elegans Research Community, WormBook. https://doi.org/10.1895/wormbook.1.138.1. http://www.wormbook.org. Retrieved from http://www.wormbook.org/chapters/www_cuticle/cuticle.html. Accessed 06 July 2021
Pálsson G, Helgason A (1995) Figuring Fish and Measuring Men: The Quota System in the Icelandic Cod Fishery. Ocean and Coastal Management 2(8):1–3: 117–46
Pansera M, Ehlers M, Kerschner C (2019) Unlocking wise digital techno-futures: contributions from the degrowth community. Futures 114:102474
Paxson H (2008) Post-pasteurian cultures: the microbiopolitics of raw-milk cheese in the United States. Cult Anthropol 23(1):15–47
Radcliffe-Brown A (1952) Structure and function in primitive society. Cohen & West, London
Raffles H (2010) Insectopedia. Pantheon, New York
Rilke R (2001, 1904) Letter VII: Rome, May 14 1904. In Soulard R Jr (ed) Letters to a young poet mass: burning man books. Scriptor Press
Ross A, Müller K, Weese J, Neufeld J (2018) Comprehensive skin microbiome analysis reveals the uniqueness of human skin and evidence for phylosymbiosis within the class Mammalia. Proc Natl Acad Sci. Retrieved from https://www.pnas.org/content/115/25/E5786. Accessed May 25 2021
Schneider D (1968) American kinship: a cultural account. Prentice-Hall, Upper Saddle River
Schneider D (1984) A critique of the study of kinship. University of Michigan Press, Ann Arbor
Schroer S (2019) Jacob von Uexküll: the concept of umwelt and its potentials for an anthropology beyond the human. Ethnos. Retriueved from: https://www.tandfonline.com/doi/abs/10.1080/00141844.2019.1606841?journalCode=retn20. Accessed May 05 2021
Segata J, Lewgoy B (2016) Presentation: Vibrant. Vibrant Virtual Brazil Anthropol 13(2):27–37
Shanklin E (1985) Sustenance and symbol: anthropological studies of domesticated animals. Annu Rev Anthropol 14(1):375–403
Shanor K, Kanwal J (2009) Bats sing, mice giggle: revealing the secret lives of animals. Icon Books, London
Smart A (2014) Critical perspectives on multispecies ethnography. Crit Anthropol 34(1):3–7
Steffen W (2013) Commentary on Paul J. Crutzen and Eugene F. Stoermer "The Anthropocene" (2000). In: Robin L, Sörlin S, Warde P (eds) The future of nature: documents of global change. Yale University Press, New Haven, pp 486–490
Stépanoff C (2012) Human-animal 'joint commitment' in a reindeer herding system. HAU J Ethnograph Theory 2(2):287–312
Stepanoff C (2017) The rise of reindeer pastoralism in Northern Eurasia: human and animal motivations entangled. J R Anthropol Inst 23:376–397
Stewart M (1997) The time of the gypsies. Westview Press, Oxford
Trischler H (2016) The Anthropocene. NTM 24:309–335
Tsing A (2012) Unruly edges: mushrooms as companion species. Environmental Humanities 1:141–154
Tsing AL (2015) The mushroom at the end of the world: on the possibility of life in capitalist ruins. University Press, Princeton

Tylor EB (1871) Primitive culture: researches into the development of mythology, philosophy, religion, art, and custom, vol 2. Murray, London

Vetter A (2018) The matrix of convivial technology – assessing technologies for degrowth. J Clean Prod 197:1778–1786

Von Uexküll J (1934, 1992) A stroll through the worlds of animals and men: a picture book of invisible worlds. Originally published in Claire H. Schiller (ed.) as Instinctive behavior. International Universities Press, Madison, 1957:5–80. Semiotica 89(4)(1992):319–391

White T, Candea M (2018) Animals. Cambride Encycolpedia of Anthropology accessed 19 August. https://www.anthroencyclopedia.com/entry/animals

Willerslev R, Vitebsky P, Alekseyev A (2014) Sacrifice as the ideal hunt: a cosmological explanation for the origin of reindeer domestication. J R Anthropol Instit (NS) 21:1–23

Wolf-Meyer M (2006) Review essay: Agamben's the open. Reconst Stud Contemp Cult 6-(1) Retrieved from: http://reconstruction.eserver.org/BReviews/revTheOpen.htm. Accessed June 30 2021

Indigeneity: An Historical Reflection on a Very European Idea

21

Judith Friedlander

Contents

Introduction	534
Indigeneity and the Baptism of Pagan Customs	540
The "Noble Savage"	546
Nation-States and Minority Nationalism	549
The League of Nations, The United Nations and The Rights of Minorities	552
Conclusion	554
References	556

Abstract

The word indigeneity may still not exist in mainstream English-language dictionaries, but the concept itself has deep historical roots in European religious, philosophical, and political thought — in foundational ideas about culture, land ownership, the nation-state, and ethnic minorities. After introducing the concept and discussing its significance in the current academic and political climate, the chapter traces those roots back to the first century AD, to the days of the early Christians and their proselytizing mission, which, over the next 2,000 years, would transform the world. The chapter then describes key moments in the history of Europe, its colonies, and former colonies, highlighting some of the major theological and philosophical debates that took place during the second half of the second millennium, debates that went on to shape indigenous policies for global institutions like the Catholic Church and the United Nations. These debates also shaped our understanding of indigeneity today, perhaps distracting us as well, the chapter concludes, from the social and economic challenges indigenous communities have faced for more than 500 years.

J. Friedlander (✉)
Hunter College of the City University of New York, New York, NY, USA
e-mail: jfriedla@hunter.cuny.edu

© The Author(s), under exclusive licence to Springer Nature Singapore Pte Ltd. 2022
D. McCallum (ed.), *The Palgrave Handbook of the History of Human Sciences*,
https://doi.org/10.1007/978-981-16-7255-2_110

Keywords

Christianity · Culture · Indigenous rights · League of Nations · Minorities rights · Nation-state · United Nations

Introduction

The Oxford English and *Merriam-Webster* dictionaries do not recognize indigeneity as a word in the English language. No doubt they will in the not too distant future. Either way, the term is very much in use and has been circulating widely in recent years.

Definitions of indigeneity abound: on the internet; in documents produced by human rights activists and jurists; in scholarly works by social scientists and philosophers; in declarations published by the United Nations (UN) and other supranational organizations; and in the laws adopted by member states. While opinions vary among those who write on the subject, everyone agrees that indigeneity serves as a noun for the adjective indigenous. Derived from the Late Latin *indigenus,* indigenous is a synonym of native and modifies the place of origin of a living organism (animal or vegetable). As in, kangaroos are indigenous to Australia. In its noun form, however, indigeneity refers exclusively to specific groups of human beings, who, by virtue of their origin, lay claim to certain rights.

According to the philosopher Jeremy Waldron, indigeneity identifies descendants of the first human inhabitants of a particular land, or the descendants of people who were living there at the time of European contact. Sometimes they are one and the same, but not always (Waldron, 2003:55). The Aztecs, for example, were relative newcomers to Central Mexico in 1521, at the time of the Spanish conquest.

Sweeping into power 93 years earlier, the Aztecs built a vast empire over 20 million people, the majority of whom belonged to ethnic groups that despised the invaders. Known for their cruelty, the Aztecs' subjects needed little incentive to abandon the tyrants and their blood-thirsty god, Huitzilopchtli. And so they did, when the opportunity presented itself, rising up by the tens of thousands to join Hernán Cortés and his ragtag army of 508 soldiers, 16 horses, 32 crossbows, and a few pieces of artillery (Wolf 1959: 152). Historians agree that Cortés would never have succeeded in conquering the Aztecs had it not been for the help his troops received from indigenous allies (*Proceso* 2021). Not that this stopped the Spaniards from betraying their friends, forcing them to serve yet another foreign ruler and deity, together with their Aztec enemies.

According to the anthropologist Francesca Merlan, definitions of indigeneity fall essentially into two categories: "criterial" and "relational." The first offers a universal set of criteria, shared by all peoples identified as indigenous; the second describes the structural relationships forced upon marginalized aboriginal peoples by the European colonists who laid claim to what had once been their land (Merlan 2009: 305). In her introduction to Oxford's online bibliography of works concerned with indigeneity, Paulette Steeves reminds her readers that the term itself is relatively new, gaining traction in the early 1990s, when the indigenous rights movement received international attention. Although activists had been organizing globally since the 1970s, they only began making headlines in October 1992, as they mounted a vigorous campaign against celebrating the 500th anniversary of the "discovery" of the New World on the holiday known politely today as Indigenous Peoples' Day, but is still called Columbus Day in the United States and, more defiantly, el Día de la Raza in Latin America.

Expressing solidarity with the indigenous rights movement, the Nobel Prize Committee announced on October 16 that it had awarded the 1992 Peace Prize to Rigoberta Menchú, the internationally acclaimed indigenous activist from Guatemala. On December 10, during the annual celebration of the UN's International Human Rights Day, Secretary General Boutros Boutros-Ghali proclaimed that he was designating 1993 as the International Year of the World's Indigenous Peoples. Soon after, the General Assembly declared, in Resolution 49/124, that the UN would henceforth recognize August 19 as International Day of the World's Indigenous People. A year later, the UN launched its First Decade of the World's Indigenous Peoples. A Second Decade followed in 2014. Then, most recently, the United Nations designated 2022–2032 as the Decade for Indigenous Languages.

Founded in 1945, the United Nations has played a central role in supporting indigenous rights as part of its wider campaign to combat racism. With that goal in mind, UNESCO asked the French anthropologist Claude Lévi-Strauss, in the early 1950s, to write what became his path-breaking essay, *Race and History* (1952). By 1957, the International Labor Organization (ILO) had drafted the Indigenous and Tribal Populations Convention, which it updated and strengthened in 1989 (No. 169). This, in turn, inspired ceremonial gestures during the 1990s, on the part of Secretary General Boutros-Ghali and the General Assembly, as they waited for member states to settle on the wording for the UN Declaration on the Rights of Indigenous Peoples (UN Indigenous People's Website).

Decades in the making, this landmark document came to a vote in 2007 and passed easily, with 144 nations in favor, 11 abstentions and 4 against. In a shocking act of defiance, the four negative votes came from Australia, Canada, New Zealand, and the United States, all of which had large indigenous communities and strong indigenous rights movements. Over the next few years, the four dissenting voices endorsed the declaration as well, while continuing, occasionally, to obstruct the efforts of their indigenous citizens to claim their cultural and economic rights. For example, in one highly publicized case, the provincial government of Alberta, Canada, supported the Transcanada Keystone XL Pipe Line Project, which proposed laying 1,700 miles of pipelines, partially through treaty-protected lands in the US, to ship a noxious, climate-polluting product from Hardisty, Alberta, to Steele City, Nebraska. Pressured by indigenous rights activists and environmentalists more broadly, President Barack Obama rejected the proposal; then President Donald Trump asked Congress to reconsider the idea, but nothing transpired before he left office, at which point, President Joe Biden rejected it again in 2021, prompting Alberta's government to back away as well (Native American Rights Fund). In 2020, the Canadian government endorsed a similar project proposed by the fossil fuel company, Coastal Gaslink (*The Economist*, February 20, 2020).

While they continued to waver on their promises occasionally, Canada and the United States made a second commitment in 2016 to honor the rights of indigenous peoples by signing the American Declaration on the Rights of Indigenous Peoples, drafted by the Organization of American States (OAS). The last article in this second declaration states: "The rights recognized in this Declaration and the United Nations Declaration on the Rights of Indigenous Peoples constitute the minimum standards for the survival, dignity and wellbeing of the indigenous peoples of the Americas (XLI), who have faithfully served as stewards of the Earth for millennia." Then in April 2021, a few months before this chapter went to press, the UN's Permanent Forum on Indigenous Issues (UNPFII) called on the UN, to recognize officially the "role of indigenous peoples in implementing Sustainable Development Goal, #16."

The United Nations placed the UNPFII under the auspices of the Department of Economic and Social Affairs, in recognition of the fact that indigenous peoples rank among the most impoverished inhabitants of every country where they live. On the Forum's website, however, UNPFII's description of these economically endangered peoples speaks more eloquently and expansively about the challenges they face in preserving their cultures than it does about their dire economic circumstances:

> Indigenous peoples are inheritors and practitioners of unique cultures and ways of relating to people and the environment. They have retained social, cultural, economic and political characteristics that are distinct from those of the dominant societies in which they live. Despite the cultural differences, indigenous peoples from around the world share common problems related to the protection of their rights as distinct peoples.
> Indigenous peoples have sought recognition of their identities, way of life and their right to traditional lands, territories and natural resources for years, yet through history, their rights have always been violated. Indigenous peoples today, are arguably among the most disadvantaged and vulnerable groups of people in the world. The international community now recognizes that special measures are required to protect their rights and maintain their distinct cultures and way of life (UN Indigenous People Website).

The United Nations is not alone in paying more attention to the threat indigenous peoples face from losing their cultures than to the social and economic challenges they struggle with every day. The same is true of other supranational organizations, like the OAS, as well as human rights activists, and anthropologists. Almost everyone focuses more on the need to protect traditional languages and cultures than on improving the material conditions of people living on what is designated as indigenous land, much of it uninhabitable, given the ravages of climate change. In the few paragraphs dedicated in policy papers, declarations, and scholarly reports to the impoverished conditions of indigenous peoples, even those appear in the context of a broader campaign to save endangered cultures (e.g., Articles 26, 28, and 29 in the United Nations' Declaration). Understandably so, cynics might add. A commitment to fund projects that help indigenous communities preserve their languages and cultures is considerably less costly than one to create the kinds of opportunities desperately needed to confront the social and economic problems plaguing these impoverished and marginalized people.

Is it therefore trivial to support campaigns that defend indigenous ways of life? Of course not, but such campaigns should not be the top priority. Programs to preserve languages and cultures should be incorporated into far more ambitious initiatives to address the material needs of aboriginal peoples, not the other way around. And to do that effectively, we need to reach a more coherent understanding of what we mean by indigeneity in the first place, grounded in history and embedded in serious conversations with those struggling to live full and productive lives on the shamefully few acres still set aside for indigenous peoples in the countries where they live, a tiny fraction of what they once had before European colonists dispossessed them of virtually all of their land many centuries ago.

As scholars and activists quibble over definitions of indigeneity and cultural rights, the stakes keep rising for the indigenous peoples themselves. Addressing the urgency of the ongoing crisis, political theorist Claire Timperly noted, "Defining indigeneity has at least two important consequences: it affects who has access to resources for indigenous peoples; and it shapes the kinds of privileges available to indigenous peoples" (Timperly 2020: 38). And while scholars, political activists, and governmental/supra-governmental agencies debate the issue, we prolong the agony of people who, depending on the year, fall in and out of definitions of who is or is not indigenous. Not to mention the confusion many of these same people face when official and unofficial agencies assign them to an identity they neither recognize as their own nor wish to accept – a problem Carmen Martínez Novo has poignantly described in *Who Defines Indigenous: Identities, Development, Intellectuals, and the State in Northern Mexico* (2006).

In sum, as we turn now to look at the history of the concept of indigeneity and ongoing debates about the term, let us not lose sight of the fact that those identified as indigenous know perfectly well what others mean when they talk about indigeneity, even if the word itself does not yet exist in English dictionaries. Whether or not the people so designated recognize themselves as indigenous, they still understand, and have understood for centuries, what the non-indigenous have in mind when they call an individual or community aboriginal, autochthonous, First People, indigenous, or by a host of other, less flattering names meaning essentially the same thing. Leaving ethnic slurs aside, condescending and confusing ideas still prevail, even among those who are passionate advocates of indigenous rights – and that often includes leaders of communities that self-identify as indigenous with pride.

Others have made similar arguments. Perhaps most provocatively, Adam Kuper, in his *Current Anthropology* article, "The Return of the Native" (2003), the sardonic title of which sets the tone for his take on the subject. Although Kuper's views have understandably offended advocates of indigenous rights, they deserve serious attention all the same:

"The rhetoric of the indigenous-peoples movement," Kuper begins, "rests on widely accepted premises that are nevertheless open to serious challenge, not least from anthropologists." For example, the "initial assumption" that the descendants of a country's original inhabitants "should have privileged rights, perhaps even exclusive rights, to its resources," over "immigrants," who are "simply guests" and should behave accordingly. This problematic assumption, notes Kuper, with more than a touch of irony, is also "popular with extreme right-wing parties in Europe, although the argument in that case is seldom pushed to its logical conclusion given that the history of all European countries is a history of successive migrations." By which Kuper means, the extreme right in Europe worries little about giving priority to the descendants of the Celts and Anglo Saxons over the Romans and Visigoths. Their goal, more simply, in the Nazi tradition, is to rid the continent of Jews, Muslims, and other undesirables, grouped together as "people of color." Similarly, Kuper continues, indigenous rights advocates worry little about giving priority to the descendants of paleolithic peoples, except, perhaps, in the few countries where hunters and gatherers still exist – not that this makes any sense to him either.

When talking about foragers and nomadic herders, Kuper acknowledges that they "represent not merely the first inhabitants of a country, but the original human populations of the world." But does it therefore follow, as indigenous rights advocates have proclaimed, that these peoples live in "the natural state of humanity" and that "their rights must take precedence"? The answer, says Kuper, is no. "While Upper Paleolithic hunters and gatherers operated in a world of hunters, every contemporary community of foragers or herders lives in intimate association with settled farmers." In the case of the Kalahari Bushmen and Congo Pygmies, for example, they have had interactions with farmers and traders for centuries and these interactions have been "crucial for their economy" and other aspects of their culture: "All of this suggests," Kuper continues, "that the way of life of modern hunters or herders may be only remotely related to that of hunters and herders who lived thousands of years ago."

The arrival of European settlers only complicated the picture further: "Local ways of life and group identities have been subjected to a variety of pressures and have seldom, if ever, remained stable over the long term." So, how is it possible, asks Kuper impatiently, for advocates of indigenous rights to insist that "each local native group is the carrier of an ancient culture"? Or, "in familiar romantic fashion, [that] this culture is associated with spiritual rather than with material values" and that it therefore expresses the unique "genius of a native people"? (Kuper 2003:390).

After debunking this line of argument, which he attributes to European intellectual and political fashions in the first place, Kuper rejects efforts on the part of advocates of indigenous rights to "rely on obsolete anthropological notions and on a romantic and false ethnographic vision," when defending land claims cases in national and international courts of law (395). To which, Adam Kuper's critics have responded in kind, accusing him, in the strongest language possible, of blaming the victim. As the Brazilian anthropologist Alcida Rita Ramos puts it in the comments section following his article, how dare Kuper associate the indigenous rights movement with the racism and cultural essentialism of European parties on the extreme right! And then, go on to suggest, she continues, that their land claims are based on bogus anthropological notions (397)!

But even for those sharing Ramos' commitment to indigenous rights, a weak argument is a weak argument. Indigenous peoples may have legitimate claims to specific tracts of land, but not because they have preserved their peoples' ancient cultures, but because the treaties their forefathers had signed have been broken. Adding insult to injury, these breaches of contract occurred after their ancestors had already been tricked out of nearly all of their land by European settlers hundreds of years ago.

In sum, even if Kuper has caricatured the arguments made by the defenders of indigenous land claims, he has not misrepresented their views about traditional cultures, which indeed resonate with those espoused by members of the extreme right about their own cultures. Not only are claims such as these politically dangerous, no matter who makes them – be they white supremacists or indigenous activists –they are also wrong. Little remains of the ancient traditions of peoples identified as indigenous and the little that does, continues to disappear. Far more resistant to the forces of change is the persistence of poverty in indigenous communities, a plague that dates back to early colonial times. And this is hardly a secret. The UN's indigenous peoples' website warns us that "Indigenous peoples today are arguably among the most disadvantaged and vulnerable groups of people in the world." So, why have the United Nations and indigenous rights activists focused more on cultural preservation than on improving the social and economic lives of the world's most "disadvantaged and vulnerable"?

When Adam Kuper writes about indigenous peoples, he limits his discussion to the few remaining groups of foragers and nomadic herders who are eking out a living on virtually uninhabitable lands. Others are more inclusive. In documents produced by the United Nations and its member states, they recognize as indigenous members of agricultural communities as well, and, in some cases, extend the definition to the children of indigenous peasants who live in cities or have left their countries of origin to work as undocumented laborers overseas. Advocates of indigenous rights are equally inclusive, often going out of their way to welcome anyone who claims that he or she is indigenous, though not always, as we have seen in the United States, where false claims by wannabe Native Americans have recently been exposed (Viren 2021). Inclusive or exclusive, either way, the concept itself has more to do with European ideas about identity and culture, imposed on native communities hundreds of years ago, than it does with ideas emanating from the ancient traditions of the first inhabitants of colonized lands, whose cultures, we are asked to believe, have miraculously survived centuries of persecution and indoctrination.

This chapter traces the idea of indigeneity back to the first century AD, when the disciples of Jesus set out to convert pagans to Christianity. Breaking with Jewish tradition, they began actively recruiting strangers into the faith, a strategy they expanded exponentially during the Age of Discovery, as theologians, philosophers, and writers debated passionately among themselves about whether those strange creatures, inhabiting the lands that Europeans had just conquered, were human at all. Assuming that they were, Christian missionaries would have the obligation to save their souls and convert them to the one and only true religion.

Indigeneity and the Baptism of Pagan Customs

When the early Christians set out during the first century AD to convert the pagans of Europe, they may have had different strategies for spreading the Gospel, some more gentle than others, but they all shared the same mission: to create a universal religion for a single, all-inclusive humanity. Christianity, at least in theory, embraced all the peoples of the world: the uncircumcised and the circumcised; those who ate pork and those who forbade it; the repentant sinner and the eternally innocent; the rich and the poor. After they converted pagan kings and emperors to Christianity, religious missionaries began accompanying their imperial armies to save the souls of the vanquished, marching west from Rome in the early Middle Ages, to the island capitals of the Celts and the Saxons; then south into the Middle East and Africa and east into Asia. Finally, by the late fifteenth century, they were sailing off across the Atlantic, taking part in the conquest of the New World.

As missionary priests faced the daunting task of saving pagan souls, they tried to ease the conversion process by incorporating some of the pagans' customs into Christian rituals, encouraging their initiates, for example, to continue eating their traditional foods, dancing their favorite dances and speaking their native languages. This strategy came to be known as "baptizing" customs.

In *La Conquête spirituelle du Mexique,* Robert Ricard explained how the priests justified the practice of using elements of the pagans' culture as a way to help them convert indigenous Mexicans to Christianity: "To become a true Christian," the historian wrote, "the Mexican did not have to stop speaking his native language because to become a true Christian, as everyone knows, it is not at all necessary to become Spanish. It is permitted, even recommended, that the Indian remain Mexican. The Church...does not ask her sons to betray their country, nor turn against their race" (Ricard 1933: 338).

But, and this "but" is critical, the Church insisted that her newly converted sons and daughters abandon their gods and pray to God the Father and His Son Jesus Christ, to the Virgin Mary and the Church's canonized saints, who revealed themselves on occasion to newly converted Christians at the sites of their former pagan deities.

The idea of welcoming people from different cultures into the same universal human family was a radically new and progressive idea, no matter how problematic we might find that missionary project today. As Lévi-Strauss noted in *Structural Anthropology,* such glimmers of tolerance rarely appeared in other societies, in the past or present. Most cultures, the anthropologist continued, not only rejected people who spoke other languages and practiced customs different from their own, they did not even consider them worthy of the name human – and this was particularly true of members of hunting and gathering and horticultural societies, the people anthropologists traditionally studied, whose societies have been rapidly changing, if not disappearing entirely, over the last 100 years. As Lévi-Strauss explained it, throughout human history and for most cultures around the world,

The idea that humanity includes every human being on the face of the earth does not exist at all. The designation stops at the border of each tribe, or linguistic group, sometimes even at the edge of a village. So common is the practice that many of the peoples we call primitive call themselves by a name which means 'men' (or sometimes—shall we speak with more discretion?—the 'good ones,' the 'excellent ones,' the 'fully complete ones'), thus implying that the other tribes, groups, and villages do not partake in human virtues or even human nature, but are, for the most part, 'bad people,' 'nasty people,' 'land monkeys,' or 'lice eggs.' They often go so far as to deprive the stranger of any connection to the real world at all by making him a 'ghost' or an 'apparition.' Thus, curious situations arrive in which each interlocutor rejects the other as cruelly as he himself is rejected (Lévi-Strauss 1983, vol.2: 329).

Presumably the same prejudices existed in Paleolithic and Neolithic times. Not that things improved significantly, with the "rise of civilization," as we learned from the ancient Greeks, who called non-Greeks barbarians, and the ancient Hebrews, whose God may have loved all His children, no matter whom they worshipped – as He reminded a disgruntled Jonah in Nineveh – but Jehovah ominously warned His "chosen people" of the dangers of mixing in intimate ways with those of other faiths, who did not embrace the one and only God and obey the Laws of the Torah. The Christians, in contrast, not only welcomed pagans into the fold as long as they converted, they actively recruited them, accepting many of their customs along the way. A project, admittedly, that did not protect indigenous peoples from the cruelty of Europe's zealous missionaries, rapacious soldiers, and land-hungry settlers.

The murder and rampage that followed were apocalyptic. In the name of Jesus and tolerance for others, European Christians devastated the world's peoples and their cultures. Yet they did so while also introducing a major paradigm shift, one that turned early Christians, ironically, into champions of cultural diversity.

Saint Paul led the way, with defiant proclamations that good Christians did not have to observe the hundreds of commandments (*mitzvot*) imposed on observant Jews by Halachic Law – the number of which, for men, reached 613 by the third century AD. On the contrary, as the disciples of Jesus recruited pagans into the faith, they encouraged their new converts to continue practicing customs and rites familiar to them – that is, <u>after</u> the priests had carefully "baptized" them, to eliminate any hint of sacrilege. Today, as a result, we find widely divergent rituals among observant Christians in different parts of the world, in particular among isolated and marginalized descendants of the first inhabitants of lands colonialized by Europeans hundreds of years ago.

By the time missionaries from Rome were converting northern Europeans to Christianity, the Church had become quite adept at "baptizing" pagan customs. In a famous example, they introduced a new holiday into the Christian calendar in the third century AD to compete with the pagans' celebrations of Winter Solstice and the New Year. When they arrived in northern Europe, a century later, they borrowed pagan customs to mark the newly invented holiday of Christmas, among them the tradition of hanging fragrant boughs of evergreens over the doorways of individual homes, during the darkest days of the year. Recognizing its intrinsic appeal, the priests turned this custom into a celebratory symbol of the Virgin birth of Jesus.

Since the Gospels mentioned no date for the Virgin birth, the way they had for the Crucifixion of Christ, the Roman Church chose the 25th of December to mark the miracle, eight days before Rome's celebration of the pagan New Year on January 1. The precise day was not incidental: Since Jesus was born into the Jewish faith, he had to be circumcised eight days later (The Venerable Bede, 731).

By the late eleventh century, Europe's monarchs were collaborating closely with the Church's hierarchy and welcoming the help of missionary priests in advancing their colonial ambitions. In 1095, on the eve of the first Crusade, Pope Urban II issued a decree (papal bull) that authorized Christians to occupy and exploit the resources on lands inhabited by infidels (Jews and Muslims) and pagans. Other bulls followed in the mid-fifteenth century in support of Portugal's rapidly expanding empire in Africa and the far East. Then, in 1493, the year after Columbus "discovered" the Americas, Pope Alexander VI pronounced yet another decree with the aim of defining clear spheres of influence among competing Christian empires, as the rulers of Portugal, Spain, and other Christian nations staked their claims to different parts of the globe. Known together as the Doctrine of Discovery, these papal bulls proclaimed that only Christians could own land. But with this right, came the obligation to save the souls of infidels and pagans who were living on these newly conquered lands.

In 1517, after Luther defied the authority of the pope, fierce theological arguments followed, but not over the Doctrine of Discovery. On the contrary, Protestants continued to respect these papal bulls for centuries, on both sides of the Atlantic – even after the United States declared its independence from England and abolished the Church's authority over the State. Ignoring the First Amendment of the US Constitution, Chief Justice John Marshall used the Doctrine of Discovery to justify the Supreme Court's decision in 1823 (Johnson vs. McIntosh) to expand the borders of the new republic, from "sea to shining sea," basing his argument on the Christian principle that pagans could not own lands. Not that the court took any interest in the pagan souls of Native Americans, but it did not stand in the way of the nation's spiritual leaders. Following in the footsteps of the early Christians, Protestant missionaries marched across the continent, converting Indians to Christianity and baptizing indigenous customs along the way. Perhaps, most impressively, after 1823, American missionaries joined a worldwide effort to translate the Bible into indigenous tongues (The Christian History website).

As successful as Protestants became in converting indigenous peoples across the globe, they could never compete with the work of the Catholics during the early years of colonization, in particular of the Spanish, who had the good fortune of conquering densely populated agricultural societies. When they took possession of indigenous land, instead of expelling the native populations, the way the English would do a century later, they turned the Indians into serfs, incorporating them into the cultural, social, and economic fabric of colonial life. And they did this at a time when the King of Spain had a privileged relationship with the pope, which, in turn, further facilitated the conversion process (Pope Alexander VI was Spanish). Important as well, at least for later historians, Spanish priests kept detailed records of their impressive missionary achievements, including careful descriptions of the way they baptized pagan customs.

Although some members of the clergy were fearful that the Aztecs, for example, would remain pagans if they practiced baptized versions of their old traditions, many others embraced the strategy with enthusiasm, going so far as to authorize converting sites associated with ancient deities into shrines for the Virgin Mary and Catholic saints. Idolatrous customs certainly persisted in Mexico, as a result, but by and large the strategy worked. As the anthropologist Pedro Carrasco noted in the late 1960s, while describing Tarascan religious practices in the State of Michoacan, these Indians (now known as the Púrepecha), "will pass their Christianity test with honors if we compare their religion to the folk religion of Southern Europe" (Carrasco 1970: 6). By baptizing indigenous customs, the priests also made it easier to establish a clearly identifiable caste system in New Spain, where converted Indians, with exotic languages and customs, remained clearly visible at the bottom of New Spain's socioeconomic ladder – while their indigenous belief system, social organization, and economic way of life had disappeared almost entirely.

One of the most dramatic examples of this baptism process occurred in Mexico in 1531, ten years after the Spanish conquest, at the desecrated site of Cihuacoatl (Serpent Woman), which was located on the outskirts of Mexico City, on top of the sacred hill, Tepeyac. Although the Spaniards had destroyed the deity's shrine soon after the conquest, the Aztecs' fertility goddess returned, only this time as the Virgin Mary, in the first apparition of the mother of Jesus in the New World.

According to legend, the Virgin appeared several times to the converted Indian, Cuauhtlatoatzin, better known by his baptized name, Juan Diego. The miracle took place in early December 1531, as Juan Diego was on his way to Mexico City to celebrate the day of the Immaculate Conception (December 8). But instead of going directly to the cathedral, Juan Diego made a detour to pay his respects to Cihuacoatl, where, much to his surprise, the Virgin Mary appeared before him.

When the converted Indian arrived at the cathedral, he asked to see the bishop, a Franciscan by the name of Juan de Zumárraga, to inform the head of the Mexican Church of what he had just seen. A skeptical Zumárraga urged Juan Diego to return to Tepeyac and ask the Virgin for a sign. When the Virgin reappeared, she instructed the Indian to gather all the flowers on top of the hill, wrap them up in his *agave*-fiber cloak and carry them back to the bishop. Juan Diego did as he was told, then returned to the cathedral, where upon opening his cloak, the flowers had disappeared. In their place, the assembled saw a painted image of the Virgin. Acknowledging the miracle, Zumárraga kept the cloak in the cathedral until a small chapel was built at Tepeyac for Our Lady of Guadalupe. Two years later, the bishop moved the cloak to the chapel, displaying it as indisputable evidence that the Virgin had appeared to a newly converted Indian at the former shrine of Cihuacoatl.

Nearly 100 years later, the Franciscan missionary and chronicler Bernardino de Sahagún, proclaimed that Cihuacoatl had demonstrated the universality of Christianity. Reinterpreting the meaning of the Aztec goddess, Sahagún claimed that her presence had "proven," as historian Jacques Lafaye later put it, that even before the Spanish conquest, the Indians knew about "our mother Eve who was abused by the serpent." It took only another small step to turn the Aztec serpent woman into the Mother of Christ, for the Virgin Mary was also known as the New Eve. Helpful as well was the fact that the Aztecs called their goddess Cihuacoatl "Our Mother" in the Náhuatl language (*Tonantzin*). Having made the connection between Cihuacoatl and the Virgin Mary, Sahagún provided the theological justification necessary for designating a hill associated with an Aztec goddess as the sacred site for the chapel of Our Lady of Guadalupe, the Virgin of indigenous Mexico (Lafaye 1974: 287, Friedlander 2006: 99–100).

Bartolomé de las Casas was one of the most beloved priests in colonial Mexico. Soon after arriving in the New World to convert the Indians to Christianity, he quickly became their champion, boldly speaking out against the cruelty of the Spanish colonists and some of his fellow missionaries. Known by his admirers as "the great gatherer of Indian tears," in 1542, he described the brutality of the conquest in lurid detail, in his highly controversial *Devastation of the Indies* (Casas, 1542). Eight years later, in a famous debate with the Aristotelian philosopher Ginés de Sepúlveda, las Casas made the case against enslaving the Indians, speaking eloquently about the humanity of the pagans' traditional cultures and daily customs, aspects of which were being incorporated into Mexican Christianity. Nevertheless, and again this is critical, the bishop firmly maintained that the Indians had to abandon their idolatrous ways and embrace Christianity. He took his mission of saving their souls very seriously.

Although las Casas lost the debate, he had many sympathizers in Europe. Influenced by the bishop of Chiapas, others began to speak out in defense of pagans and infidels, even in defense of the most "primitive" among them, who dressed scantily and wore feathers. Yes, Christians had an obligation to save their souls, but they should still treat them as human beings. Horrified by the reports trickling back to Europe from the colonies, people of conscience raised their voices in protest, among them philosophers and playwrights we still read today. Deeply troubled by what he was hearing, the sixteenth century French philosopher, Michel de Montaigne, observed in *Of Cannibals*, "From what I have heard, there is nothing barbarian or savage in these people [the peoples of the New World], except that everyone calls barbarian that which he does not do himself; it seems that we have no other measure of truth and reason than the beliefs and customs of the country from which we come." (Montaigne 1897:37, cited in Ellingson).

Shakespeare, too, raised concerns at the end of the sixteenth and beginning of the seventeenth century, putting words of wisdom into the mouths of reviled infidels and pagans. In *The Merchant of Venice*, Shylock the Jew famously asks: "Hath not a Jew eyes? Hath not a Jew hands, organs, dimensions, senses, affections, passions? Fed with the same food, hurt with the same weapons, subject to the same diseases, healed by the same means, ... as a Christian is? If you prick us, do we no bleed? If you tickle us, do we not laugh? If you poison us, do we not die?" (Shakespeare, *The Merchant of Venice*, Act 3, Scene 1). Then again in *The Tempest*, Shakespeare's monstrous Caliban (whose name is a play on the word cannibal), raises poignant questions about slavery, resistance, and culture: The depraved slave, without any semblance of culture, by the standards of early seventeenth century Europe, has learned the language of his master ("tyrant") and uses it exquisitely against him (Shakespeare, *The Tempest,* Act 1, Scene 2).

The more tolerant some Europeans became about respecting cultural differences, the more their critics objected, dismissing the possibility of creating a unified humanity. How could the world's peoples, in all their diversity, share a single faith and live together as one? Was the world not populated by three irreconcilable types? Civilized peoples of the Christian faith; infidels who have recognizable but dangerous religious beliefs (Jews and Muslims); and pagan "savages," who have no culture at all? And, when it comes to this last group, are they even fully human? Should they be recognized as equal, before God and other men, even after the priests have converted them to Christianity? These were the questions facing "barbaric" Europe, between the sixteenth and eighteenth centuries, while their soldiers were savagely killing one another in religious wars and social revolutions, both at home and in their nations' colonies overseas.

In the midst of all these upheavals, Europeans and their descendants continued conquering the lands of infidels and pagans. In the newly constituted republic of the United States, politicians made uplifting declarations about equality and freedom for all, while they continued enslaving Africans and fulfilling the nation's Manifest Destiny. By the middle of the nineteenth century, heads of state no longer justified acts of plunder in the name of religious salvation, but they still set off with missionary zeal to fulfill their economic and political ambitions, now in the name of democracy – yet another "universal" calling, that continues to inspire Europeans and their descendants well into the twenty-first century.

To repeat, over the last 500 years, critics have raised their voices to challenge the way Europeans and their descendants have treated indigenous peoples. These included, most famously, in the sixteenth and early seventeenth centuries, las Casas, Montaigne, and Shakespeare. But others quickly followed, exposing readers to savage descriptions of European acts of brutality against the original inhabitants of colonized lands and introducing them to sympathetic pictures of the cultures of so-called primitive peoples. By the late eighteenth and early nineteenth centuries, these more sympathetic narratives became myths of their own. Rejecting the idea that primitives were ferocious barbarians, dissidents described the savages as innocent, peaceful, and loving, living in complete harmony with nature. This idyllic picture of indigenous life is still with us today, admittedly in more sophisticated forms, and they continue to influence contemporary debates about indigeneity.

While some may associate harmonious descriptions like these with the words handed down by indigenous ancestors, they evoke more persuasively the writings of the eighteenth century philosopher Jean-Jacques Rousseau and those of novelists identified with nineteenth century romantic literature, like James Fenimore Cooper and Chateaubriand. Perhaps, in the end, these sentiments come down to us from both cultural traditions, but their European roots are undeniable.

The "Noble Savage"

During the seventeenth and eighteenth centuries, as philosophers and writers read reports about exotic peoples overseas, some of them tried to incorporate new insights about human behavior into their reflections about politics and society. And as they learned more about these so-called savages, who lived in what some called a state of nature, they began to wonder about what life might have been like for their own ancestors as well, before they came together and signed a "social contract." The most influential descriptions of this state of nature – in addition to the Book of Genesis – were *Leviathan* (1651), by the English philosopher Thomas Hobbes, and *The Social Contract* (1762), by the Swiss philosopher Jean-Jacques Rousseau.

In *Leviathan,* Hobbes described man's existence in the state of nature as a condition of perpetual war: what he called a "war against all": "Life," he wrote, in a frequently cited phrase, was "solitary, poor, nasty, brutish and short." Given how dangerous it was to survive in the wild by one's own devices, people gave up their freedom in exchange for living under the protection of a higher earthly authority, preferably a king.

A century later, Rousseau described life in this same state of nature as peaceful and harmonious. Corruption and evil were the products of human society. It did not exist in the natural order of things: "Man is born free and he is everywhere in chains."

Many identify the idea of the "Noble Savage" with Rousseau. As it turns out, Rousseau neither invented the term nor used it. In the early 1600s, about 150 years before Rousseau wrote *The Social Contract,* a French lawyer and writer by the name of Marc Lescarbot coined the term, after spending a few years living in what we call Nova Scotia today. During that trip, he met the Mi'kmaq Indians, an indigenous group that had befriended the French. As he observed their customs, Lescarbot was fascinated to see that male members of the tribe had the right to hunt, a privilege enjoyed only by the nobility in Europe. Hence, with a touch of irony, he called the Mi'kmaq "noble savages" (Ellingson 2001: 35ff).

Although Lescarbot introduced the term in the seventeenth century, as a playful joke on European nobility, the term took on a life of its own. Before long, it became short hand for referring to Rousseau's idyllic description of life in the state nature, where people lived in blissful harmony with one another, uncorrupted by the evils of society. Then, with the rise of anthropology during the second half of the nineteenth century, scholars provided empirical evidence of life in primitive societies that supposedly confirmed Rousseau's evocations. The works of the anthropologist Lewis Henry Morgan were particularly influential with some of the major theorists of the day, including Karl Marx and Friedrich Engels.

Like many of their contemporaries, Marx and Engels, had already been reading Darwin, whose work had inspired them to think about the implications of evolutionary theory on human societies. This, in turn, led them to read descriptions of prehistoric and present-day primitive cultures being published at the time by archeologists and ethnologists, whose scientific reports were replacing less reliable accounts by religious missionaries and explorers.

In 1877, Morgan published *Ancient Society*, an ambitious attempt on the part of the anthropologist to develop a grand theory for explaining the rise of civilization. Marx died before he and Engels had the chance to apply Morgan's scheme to their political theory, leaving the task to Engels alone, whose *Origin of the Family, Private Property and the State* was published in 1884. Relying on Morgan's scholarship, Engels argued that human societies ranged from a kind of primitive communism where everyone lived happily together, sharing everything with one another, in a harmonious state of nature, to a highly competitive form of social organization that valued the accumulation of private property.

By the last quarter of the nineteenth century, anthropologists and philosophers, who followed this line of thinking, understood that the Noble Savage lived in society, not outside of it, as Rousseau had originally suggested. But they still embraced the idea that primitive peoples lived peacefully with one another and in harmony with nature. Born in innocence, and with a clear moral compass, they knew the difference between right and wrong and had the courage to fight for what was right. And like Rousseau, they believed, "What crimes, wars, murders, what miseries and horrors would the human race have been spared, had someone pulled up the stakes or filled in the ditch and cried out to his fellow man: 'Do not listen to this imposter. You are lost if you forget that the fruits of the earth belong to all and the earth to no one!'" (Rousseau 1755, cited in Winchester 2001).

During the nineteenth century, novelists and poets belonging to the romantic literary movement made similar arguments. They used the idea of the Noble Savage to challenge the still prevailing opinion of the day that Native Americans and indigenous peoples in other parts of the world were lawless and violent, even those, like the Mohicans, who had sided with the English during the French and Indian War and had willingly sold some of their land to the colonists! James Fenimore Cooper and François-Auguste-René, Vicomte de Chateaubriand described the fate of these rapidly vanishing indigenous peoples on the eastern seaboard of the United States, with sadness and sympathy.

Descriptions in this same romantic vein are still with us today, in less patronizing language than we found in nineteenth century novels, but the message remains the same. While champions of indigeneity reject terms like the Noble Savage, they proudly defend the idea that, despite centuries of persecution and destruction, indigenous peoples have preserved their cultures and their special relationship to the land. Living, as they still do, in harmony with nature, they have a great deal to teach the rest of us about implementing sustainable development. Frequently referred to as stewards of the land, indigenous people, according to this narrative, know how to renew the natural environment from the ravages of global capitalism, which have led us to where we are today – fleeing floods and forest fires of apocalyptic proportions.

An eloquent spokesman for this position in 2021, is the writer Simon Winchester, who published a powerful condemnation of the impact of European colonialism on indigenous peoples worldwide. Limiting himself, for the most part, to the damage caused by the British Empire, Winchester describes the devastation in breathtaking detail, across the centuries and around the globe. Near the end of his book, he acknowledges that he has chosen to conclude on a naïve, idealistic note.

Perhaps, Winchester hopes, the dire effects of climate change might finally give pause to the rich and powerful descendants of colonialism. Perhaps they will finally understand the dangers of privatizing huge tracts of land on fragile terrain, near the ocean, for example, where the rising seas threaten to wash away their costly mansions. Given the impact of harsh weather events, perhaps the very rich will come to accept that sharing land is better than hoarding it – a conclusion that indigenous peoples had reached long ago:

> The aboriginal Australians, the Maori, the Canadian First Nations populations, the Inuit who inhabit the high latitudes from Siberia to Alaska and back again, the Aztec, the Incas, the North American Indians — to all and each of these, land was a commodity so precious and life-giving that it was indeed to be shared by all, and owned by none. (Winchester 2021: 402)

Winchester then adds to the list of "such a kindly, philosophical approach to the world's surface," the words and beliefs of the Ashanti of West Africa, the Goldi of far eastern Russia, and others, before ending with the speech that Chief Seathl gave in 1854 to members of the Duwamish and Suquamish tribes of the Pacific Northwest, informing them that the American Government had asked (forced) him to sell their lands. Seathl began his speech as follows: "The President in Washington [Franklin Pierce] sends word that he wants to buy our land. But how can you buy or sell the sky? Buy or sell the land? If we do not own the freshness of the air and the sparkle of the water, how can you buy them?" (cited in Winchester 2021: 404–405).

Activists in the indigenous rights movement join Winchester in believing, to repeat Rousseau's words, "that the fruits of the earth belong to all and the earth to no one!" But since they are forced to live in the modern world, shaped by people who hunger to own land, they have had little choice but to fight for the right to hold on to what they can still legally call their own. And as they make their claims in national and international courts of law, they rely on arguments based persuasively on a democratic principle, known as the rule of law. But also, less persuasively, perhaps, on their cultural rights –rights for which they have had very little protection in nation-states, until very recently.

The indigenous rights movement has joined a campaign, first initiated by other ethnic and racial minority movements, to fight for the right to practice their own cultural traditions as citizens of modern democratic nation-states. And in doing so, they have been challenging a foundational philosophical idea, which dates back, once again, to eighteenth century Europe. When Rousseau and other Enlightenment thinkers proposed abolishing hereditary monarchies and replacing them with bourgeois democratic states, their revolutionary idea would only succeed, they argued, if citizens rallied around a single national culture. This, in turn, required inventing what Benedict Anderson has called an imagined community (2006). All members of such a community had to speak the same language, practice the same customs and identify with a common heritage, a process that forced ethnic minorities to abandon their own cultures and blend in – that is, if they could – an option denied members of different races.

As minorities today defend their cultural rights in modern-day nation-states, they do so, curiously, with the support of supranational organizations, founded by European and American victors of the last two world wars, the League of Nations and the United Nations. That support deserves closer scrutiny. When the League of Nations recognized the rights of minorities after World War I, it did so only for those living in the fourteen new states, carved out of the Austro-Hungarian, Ottoman, and Russian Empires; not for those living in England, France, and the United States, where the Allies firmly maintained the eighteenth century idea of one nation within one state. Since the late twentieth century, however, the UN has expanded its support of for the rights of minorities to include those living in the West as well. The question is, why?

Nation-States and Minority Nationalism

Historians often trace the idea of the nation back at least as far as medieval times. Modern nationalism, however, dates to a much later period. It took shape in Western Europe during the seventeenth and eighteenth centuries, as political philosophers imagined building new democratic states. Recognizing the power of ethnic attachments, progressive theorists argued that people would willingly embrace this new form of government, if they identified with the culture and heritage of the state.

Rousseau was perhaps the first to develop a program for educating people to become loyal citizens of nation-states, capable of participating actively in a democratic society. He did so in the early 1770s, at the invitation, ironically, of a political faction of aristocrats (Bar Confederation) in Poland, a country where the nobility elected their king. Following Plato's description of the ideal city state, Rousseau began with the education of children, recognizing that he had a more complicated problem on his hands than did Plato in Ancient Greece, involving a state whose population spoke different languages, practiced different customs, and lived scattered over a large territory. To turn people like these into dependable citizens with a common set of goals, they had to learn to identify with the same history and take part in the same rituals and traditions: "National education belongs to the free; they alone live a national life, they alone are truly bound by law. . . . At twenty a Pole should be nothing else; he should be a Pole. When he learns to read, he should read of his country; at ten he should know all its products, at twelve he should know all its provinces, roads and towns, at fifteen all its history, at sixteen all its laws; there should not be in all Poland a noble deed or a famous man that he does not know and love, or that he could not describe on the spot" (Rousseau 1964 [1771–1772], 65).

Soon, other voices joined in. Democracy required cultural uniformity: "one nation within one state." Before legislating that all men were created equal, everyone had to become the same. With this principle in mind, the revolutionary government in Girondist France came up with a plan for emancipating the Jews – the quintessential other in Europe at the time – and making them good French citizens. In 1791, Count Stanislas de Clermont-Tonnerre made his famous declaration in defense of the Jews: "We must refuse the Jews everything as a nation and give them everything as individuals; they must constitute neither a political group nor an order within the state; they must become citizens as individuals" (Clermont-Tonnerre, cited in Poliakav, vol. 3: 224).

During the late eighteenth and early nineteenth centuries, enlightened Jews understood the limitations of this emancipation, but they eventually accepted the terms. They agreed to compromise, to give up their collective cultural autonomy in order to gain rights as individuals. Although Jewish communities across the country varied considerably – culturally, socioeconomically, and politically – some favoring emancipation and others vigorously opposing it, the French Revolution did not offer Jews much of a choice. Emancipation was thrust upon them. They became "Frenchmen of the faith of Moses" and agreed, for the most part, to do away with many of their own traditions, so that they might conform to the national model, which allowed them to practice their religion, but only discretely and in private.

The same assimilationist model was adopted in England, not only for Jews but also for indigenous minorities and immigrants who were living in what had become a constitutional monarchy by the end of the seventeenth century. In the colonies, however, things were different. While the British treated immigrants from other European countries the way they treated them back home, encouraging them to assimilate to the dominant national culture, they and their descendants pushed the aboriginal peoples off their lands, into segregated enclaves. Then, when the American colonists declared their independence in 1776, they denied the Indians citizenship. Finally, in 1851, with the Indian Appropriations Act, Congress segregated Indians on reservations, creating isolated nations within the state. Not that this policy stopped ongoing efforts to convert Indians to Christianity and to educate their children in private and state-run boarding schools, located far from home, cutting them off from their cultures while depriving their parents from gaining citizenship and the economic means for living productive lives (Demos 2014). In 1924, Congress finally granted citizenship to all Native Americans born in the United States, but did little to help raise the standard of living of those living on reservation lands, where levels of poverty have remained the highest in the country.

Even after the United States granted citizenship to Native Americans, it continued to treat residents living on Indian reservations as members of different nations. And as such, the federal government did little to help them assimilate into mainstream America. Immigrants from Europe, on the other hand, were expected to shed their old ways and blend in as quickly as possible.

During the early years of the twentieth century, many of these same immigrants came from East and Central Europe, where their families, like those on Indian reservations, lived cloistered together in segregated communities in the Austro-Hungarian and Russian empires. They too were identified as separate nations within the state. Although there were many differences between the challenges facing ethnic minorities in East and Central Europe, and those facing indigenous nations on reservation lands, there were some interesting similarities as well. Both groups, for example, mobilized politically to demand their cultural rights as the "empires" in which they were living began crumbling – East and Central European minorities in the late nineteenth and early twentieth centuries and the indigenous movements in the late twentieth and early twenty-first centuries. Both groups, what is more, gained the support of supranational organizations that intervened on their behalf against the wishes of member states – the League of Nations in the years following World War I and the United Nations since World War II, most aggressively, since 2007.

As East and Central European minorities demanded their cultural rights, they did not always agree with one another about identity and culture, raising interesting questions about what constituted their authentic culture in the first place. And that question persists to the present day, among indigenous groups as well. But the biggest conflicts that East and Central European minorities and Native Americans have had, both then and now, are with members of their country's majority cultures.

During the early years of the nineteenth century, Western European ideas about nationalism and democracy began moving east, holding out promises for a more enlightened future based on the nation-state model. These aspirations inspired members of minorities living in the Russian and Austro-Hungarian Empires to begin learning Russian, German, sometimes Polish as well and educating their children in the traditions of the wider society. Then, by the middle of the century, members of these very same minorities began forming nationalist movements of their own, with the hope of breaking away from Russia and Austria-Hungary and forming their own nation-states or, if they were socialists, of creating multinational democratic states. These nationalist movements also provided opportunities for minorities to free themselves from the constraints of their own cultures by inventing new national identities.

In the case of the East European Jewish minority, progressive members of this ethnic group split into several different factions. Some remained assimilationists, embracing the French nation-state model, and tried to become Russians or Austro-Hungarians of the faith of Moses. Others became Hebrew nationalists, a decision that stimulated the development of a secular Jewish culture, posing a serious threat to Orthodox Jews, who vigorously objected to the idea that one could remain ethnically Jewish without practicing the religion. Many Hebrew nationalists, it is true, remained committed to the Jewish faith, but they still insisted on the importance of defining Jewish traditions outside the synagogue, giving rise to the idea that one could be a nonpracticing Jew. By the end of the nineteenth century, Hebrew nationalists had become Zionists, who dreamed of emigrating to Palestine, and Yiddish movements had begun emerging as well, devoted to creating a secular Jewish culture in the Diaspora in the vernacular of East European Jews.

It is easy to see how Jews transformed their cultures as they accommodated French, German, or Russian national traditions, but the process is less clear when we turn to Jewish national movements. If we look closely, however, we see that they too embraced Western European ideals about what constituted a national culture in general and what their own "authentic," if newly imagined, culture should look like. Even social democrats, like the Jewish Bund, who opposed territorial nationalism, shared similar beliefs about the need to develop their new cultural traditions in ways that engaged the cultures of Western European nation-states.

Reviewing the history of Yiddish letters, the literary critic Benjamin Harshav summarized the contributions of members of this new literary canon in the following way:

> In a short period, dozens of important writers created a literature [in Yiddish] with European standards, moving swiftly from Rationalist Enlightenment through carnivalesque parody to Realism, to Naturalism and psychological Impressionism, and then breaking out of these conventional European modes into the general literal trend of Expressionism and Modernism....This became possible [Harshav continued] because of the secularization of the Jewish masses and the trend to join the general world of modern culture and politics in the language they knew. (Harshav 1990:84).

In other words, the proliferation of writing in the Yiddish language was culturally cosmopolitan and politically social-democratic. It had very little to do with the ancient traditions of the Jewish people. And the few works that did, had a singularly modernist feel about them, like *The Dybbuk,* a play written by the folklorist S. Ansky that challenged the repressive traditions of Hasidic Judaism.

By the beginning of the twentieth century, similar activities were also taking place in the lives of other ethnic minorities throughout East and Central Europe. Then, with the fall of Austria-Hungary, Russia, and the Ottoman Empire after World War I, the League of Nations established fourteen new states, providing much welcomed support to the national aspirations of many. But even these were not enough to satisfy the dreams of every ethnic group. A compromise had to be reached, which they accomplished by abandoning their commitment to the nation-state model in the newly constituted republics, while they rigidly held on to it for themselves.

The League of Nations, The United Nations and The Rights of Minorities

The details of the compromise appeared in what came to be known as the Minorities Treaties. Appended to the Versailles Treaty, the Allies imposed a series of laws to protect the cultural rights of minorities living in the fourteen newly constituted states in ways that similar minorities were not protected in Britain, France, and the United States. The new states had to provide primary school education in languages of their ethnic minorities and staff hospitals and other public facilities with personnel fluent in those idioms as well. The treaties also included a series of minority-specific regulations that prevented passing laws that might interfere, for example, with the Jewish Sabbath and dietary practices. Elections could not take place on Saturdays.

Although the United States, in 1918, had a reputation for welcoming immigrants from around the world, the American government did nothing to support the languages and cultures of ethnic minorities. On the contrary, children of immigrants quickly learned to abandon their families' traditions and blend into the great melting pot. Yet, when Woodrow Wilson returned from Paris to campaign for the Minorities

Treaties, he spoke as if he were president of a country that encouraged cultural diversity:

> We of all peoples in the world, my fellow-citizens, ought to be able to understand the questions of this treaty and without anybody explaining them to us; for we are made up out of all the peoples of the world. I dare say that in this audience there are representatives of practically all the peoples dealt with in this treaty.
> You don't have to have me explain national ambitions to you, national aspirations. You have been brought up to them; you learned of them since you were children, and it is those national aspirations which we sought to realize, to give an outlet to in this great treaty. (Woodrow Wilson 1920:76)

Offering a more honest assessment of the intentions of the Minority Treaties, the British diplomatic historian H.W. V. Temperly explained why the League of Nations would only protect the rights of ethnic and racial minorities in the fourteen newly constituted states:

> The objection was raised at the time in Poland and in other quarters that it was difficult to justify procedures by which the Polish State, a friendly and allied Power, was subjected to an invidious control of its internal affairs, from which Germany herself was exempt....
> The ultimate truth is this. If the principle of these Treaties had been applied to Germany, it would have been very difficult eventually to refuse a demand that it should be applied universally to all established States, but to do this would have been, as we have seen, a quite unprecedented innovation. No one with any knowledge of the condition of opinion on this matter can believe that such a proposal would have had any chance of acceptance or that it would have been wise to press it. This principle, if once adopted, could have been interpreted in such a way as to bring the Negroes in the Southern States of America under the protection of the League; it could have been applied to the Basques of Spain, to the Welsh and the Irish. (Temperly 1921: 141–142)

Undeterred by the the League of Nations' decision to limit its support of minorities to those living in the fourteen new states, two indigenous leaders, from Canada and New Zealand, appealed to the League in 1923 and 1924, allegedly on behalf of their peoples: Levi General Deskaheh, Cayuga Chief and Speaker of the Six (Iroquois) Nation Hereditary Council, and T.W. Ratana, a Maori religious leader, who played a prominent role in national politics. When they arrived in Geneva, the League refused to grant either one of them a hearing.

Without defending the League's treatment of Deskaheh and Ratana, the two leaders were controversial figures back home. It is not at all clear that they had the authority to speak for their communities. Even in the eyes of the most enthusiastic advocates for indigenous rights, their motives for going to Geneva looked suspicious (Deskaheh, Ratana). Nevertheless, Deskaheh and Ratana were pioneers in their day, paving the way for future indigenous leaders to ask supranational organizations to intervene on their people's behalf in disputes they were having with members of their own communities and with their national governments. Although indigenous minorities would have to wait until the end of a Second World War, they now receive a warm welcome in Geneva and New York, where the UN sponsors declarations, celebrations, and decade-long programs to support indigenous rights. But still, many problems persist.

Conclusion

Given the nature of those persisting problems, why has the United Nations dedicated more time and resources to protecting endangered languages and cultures than to defending the political, social, and economic rights of indigenous communities, as those rights are defined in the UN's Universal Declaration of Human Rights? And why, after making the choice to focus more on cultural rights than on social and economic rights, has the UN paid so little attention to the cultures themselves? Specifically, to the ethnographic and historical details of what activists call indigenous cultures today? As this chapter has argued, what many identify as indigenous customs are "baptized" traces of the past, transformed and incorporated centuries ago into cultural, social, and economic systems established to serve the needs of European colonial empires. As a result, when indigenous communities pay tribute to their cultural heritage, they often find themselves celebrating traditions that are not only European in origin, but were introduced to their communities centuries ago in order to destroy the indigenous customs of their ancestors.

As the British-Ghanaian philosopher Kwame Anthony Appiah has put it: "the cause of cultural nationalism in Africa has been to make real the imaginary identities to which Europe has subjected us" (Appiah 1992: 62). In the same vein, as Clifford Geertz has observed, in his characteristically ironic and iconoclastic fashion, "Like nostalgia, diversity is not what it used to be; and the sealing of lives into separate railway carriages to produce cultural renewal or the spacing of them out with contrast effects to free up moral energies, are romantical dreams, not undangerous" (Geertz 2000: 78).

All cultures, Lévi-Strauss, has reminded us, are "a mishmash, borrowings, mixtures." People have been exchanging traditions with one another for centuries, while finding ways to carve out unique national identities for themselves all the same: "There is no country more the product of mixture," Lévi-Strauss continues, "than the United States, and nonetheless there exists an 'American way of life' that all inhabitants of the country are attached to, no matter what their ethnic origin" (Lévi-Strauss and Eribon 1988: 152–153). And those inhabitants include the country's Native Americans. But this "American way of life" has indeed been fragmenting in recent years into many different identities, as people look for new ways to divide themselves up into ever smaller intersectional communities. We see the same thing occurring in other Western democracies as well.

But why is this happening today? Until very recently, Western capitalist democracies agreed with what Rousseau had written, in his *Treatise on the Government of Poland*, that nations flourish best in states where everyone identifies with a single culture. After separating Church from State, the authors of the American and French constitutions respected religious diversity as well, provided that citizens observed their faith in the privacy of their homes and places of worship. But that was about as far as they would go. People like the Jews had to make serious compromises, and most of them accepted the terms, agreeing with the French philosopher Jean-Paul Sartre that in spite of it all, the democrat was the minority person's "best friend." The democrat, wrote Sartre in 1946, in his famous condemnation of anti-Semitism, "affirms that Jews, Chinese, Negroes ought to have the same rights as other members of society, but he demands these rights for them as men, not as concrete individual products of history" (Jean-Paul Sartre 1948 [1946]: 55,117).

And what Sartre proclaimed in 1946 remains true to the present day. The protections granted minorities in democratic nation-states are immeasurably better than those granted to them in any other political system. Not in theory, perhaps, but in practice. Whatever its limitations, the US Constitution, for example, "has a built-in self-correcting mechanism: It can be amended" (Stengel 2021: 11), thanks to which it eventually outlawed slavery in the nineteenth century and gave women the vote in the twentieth. But this political system has also exacted a heavy price on its ethnic and racial minorities, one that many are no longer willing to pay. Minorities rights activists want democracies to recognize them as members of communities with specific cultural characteristics, not as abstract individuals, a demand that the UN seems willing to support as well. But again, the question is why, or rather, why now?

Cynics might suggest that, with the rise of global capitalism, strong nation-states interfere with the flow of money and products by imposing stiff tariffs, sanctions, and other punitive taxes on the wealthy. Better to let them splinter into ethnic factions as the Allies did after World War I in East and Central Europe – a fact Temperly acknowledged openly and, apparently, without embarrassment. When the Allies gathered in Paris in 1918, they refused to give ethnic majorities residing in the fourteen new states the same authority that they kept for themselves to build strong national cultures, an authority that they deeply believed was critical to the health of democratic states.

Yes, there are a few exceptions to the rule of one nation within one state, for example, Belgium and Switzerland in Western Europe and Canada in the Americas, but they are exceptions, and problematic ones at that. It is also true that left-leaning scholars have made eloquent cases in favor of multi-cultural democracies, beginning with the works of Central European social democrats in the late 19th/early 20th centuries, most impressively Otto Bauer. Then again, in the 1970's, with the rise of identity politics in the West, a new generation of scholars began promoting multiculturalism. And their numbers grew after the fall of Communism in 1989. Among the most important theorists working on the question in recent years are two Canadians, Will Kymlicka and Charles Taylor, both of whom have championed indigenous rights as they call for reimagining democracies in ways that recognize diversity in the public arena.

The fundamental problem, however, remains the same: How do we defend the collective rights of ethnic and racial minorities with laws written to protect the rights of abstract individuals? Complicating matters further, the rights of these abstract individuals, as summarized in the UN's Universal Declaration of Human Rights, often clash with so-called traditional values, espoused by indigenous peoples, some of which, ironically, are not "indigenous" at all, but date to the colonial period, for example, practices known in Latin America as customary laws (usos y costumbres).

Looking ahead, as the United Nations launches its Decade for Indigenous Languages, why not give priority to defending the basic human rights of indigenous peoples? Their languages may be endangered but that is the least of their problems. Why not allocate the funds dedicated to UNESCO for the upcoming decade to provide indigenous children with the kind of education they need to rise out of poverty?

And yes, provide that education in their own indigenous languages, but also in the country's national language. And, if that national language is not Arabic, Chinese, English, Russian, or Spanish, then in one of the UN's five languages as well.

References

Books and Plays

Aguirre Beltrán G (1944) La Población Negra de México, 1510–1810, Estudio Etnohistórico. Ediciones Fuente Cultural, México

Anderson B (2006) Imagined communities. Verso, London [1983]

Ansky S [SZ Rappoport] (2002) The Dybbuk and other writings, trans. Gerda Werman, ed. Roskies D. Yale University Press, New Haven [1916]

Appiah KA (1992) In My Father's house: Africa in the philosophy of culture. Oxford University Press, Oxford

Bauer O (2000) The question of nationalities and social democracy, trans. O'Donnell J, ed. Nimni E. University of Minnesota Press, Minneapolis [1907]

Bede The Venerable (1955) Ecclesiastical history of the english people, trans. Sherley-Price L. Penguin Books, London [731]

Blackhawk N (2006) Violence over the Land: Indians and Empires in the Early American West. Harvard University Press, Cambridge

Casas B de la (1992) The Devastation of the Indies, trans. Briffault H. Johns Hopkins University Press, Baltimore [1542]

Deloria V (1969) Custer died for your sins. Avon Books, New York

Deloria PJ (1994) Playing Indian. Yale University Press, New Haven

Demos J (2014) The Heathen school: a story of hope and betrayal in the age of the Early Republic. Random House, New York

Ellingson T (2001) The Myth of the Noble Savage. University of California Press, Berkeley

Engels F (2010) The origin of the family, private property and the state: In Light of the Researches of Lewis Henry Morgan. Penguin Classics, London [1884]. See also Engels F (1972) The Origin of the Family...., ed Leacock EB. International Publishers, New York [1884]

Ewen A (ed) (1994) Voices of indigenous peoples. Clear Light Publishers, Santa Fé

Friedlander J (2006) Being Indian in Hueyapan. Palgrave, New York [1975]

Geertz C (2000) Available light: anthropological reflections on philosophical topics. Princeton University Press, Princeton

Harshav B (1990) The meaning of Yiddish. University of California Press, Berkeley

Hobsbawm E Ranger T (eds) (1984) The invention of tradition. Cambridge University Press, Cambridge

Kymlicka W (ed) (1995) The rights of minority cultures. Oxford University Press, Oxford

Lafaye J (1974) Queztalcoatl et Guadalupe. La formation de la conscience nationale au Mexique (1531–1813). Éditions Gallimard, Paris
Lévi-Strauss C (1952) Race et histoire. UNESCO, Paris. Available online in English: https://www.scribd.com/document/51310018/Levi-Strauss-Race-and-History
Lévi-Strauss C (1983) Structural anthropology, vol. 2, trans. Layton M. University of Chicago Press, Chicago
Lévi-Strauss C, Eribon D (1988) Conversations with Claude Lévi-Strauss, trans. Wissing P. University of Chicago Press, Chicago
Martínez Novo C (2006) Who defines indigenous? Identities, development, intellectuals and the state in Northern Mexico. Rutgers University Press, New Brunswick
Montaigne M de (1897) Of the Canniballes ["Of Cannibals,"] The Essays of Michael, Lord of Montaigne, trans. Florio J, vol 2, pp 32–54. J.M. Dent, London [1580]
Morgan LH (1977) Ancient society. World Publishing, New York
Niezen R (2003) The origins of indigenism: Human Rights issues and the politics of identity. University of California Press, Berkeley
Poliakav L (1968) Histoire de l'antisémitisme, vol 3. Calmann-Lévy, Paris
PROCESO (2021) Mentiras y verdades de la Conquista de México. Special issue of Mexican magazine PROCESO, 8/30
Ricard R (1933) La Conquête spirituelle du Mexique. Travaux et mémoires de l'Institut de l'Éthnologie, l'Université de Paris, vol.10
Rousseau J-J (1964) "Treatise on the Government of Poland," chap 4. In Excerpts from Emile, Julie and Other Writings. Baron's Educational Series, Woodbury, NY [1771–1772]
Sartre J-P (1948) Anti-Semite and Jew, trans. Parshley HM. Schocken Books, New York [1946]
Shakespeare W (1953a) The Merchant of Venice. In: Sisson CJ (ed) William Shakespeare: the complete works. Harper &Brothers, New York. [1600]
Shakespeare W (1953b) The Tempest. In: Sisson CJ (ed) William Shakespeare: the complete works. Harper &Brothers, New York. [1623]
Taylor C (1994) In: Gutmann A (ed) Multiculturalism and the politics of recognition. Princeton University Press, Princeton
Temperly HVW (1921) A History of the Peace Conference in Paris, vol 5. Henry Frowde, Hodder & Stoughton, London
Wilson W (1920) The hope of the world: messages and addresses by the President, July10,1919–December 9, 1919. Harpers and Brothers, New York
Winchester S (2021) Land: How the Hunger for Ownership Shaped the Modern World. HarperCollins, New York
Wolf E (1959) Sons of the Shaking Earth. The University of Chicago Press, Chicago

Articles

Beteille A (1998) The idea of indigenous people. Curr Anthropol 39:187–191
Cadena M de la (1995) Women are more Indian: ethnicity and gender in a community in Cuzco. In: Larson B, Harris O (eds) Migration and markets in the Andes. Duke University Press, Durham
Carrasco P (1970) Tarascan Folk Religion, Christian or Pagan? In: Goldschmidt W, Hoijer H (eds) The social anthropology of latin america: essays in honor of ralph beals, vol 14. UCLA Latin American Studies, Los Angeles, pp 3–16
Clifford J (2007) Varieties of indigenous experience: diasporas, homelands, sovereignties. In Cadena M de la, Starnberg O (eds) Indigenous experiences today. Oxford University Press, Oxford
Dombrowski K (2002) The Praxis of indigenism. Am Anthropol 104(4):1062–1073
Kenrick J, Lewis J (2004) Indigenous peoples' rights and the politics of the term 'Indigenous'. Curr Protocols 20(2):4–9
Kuper A (2003) The return of the native. Curr Anthropol 44:389–402

McIntosh I, Colchester M, Bowen J, Rosengreen D (2002) Defining onself and being defined as indigenous. Anthropol Today 18(3):23–35

Mcrlan F (2009) Indigeneity: global and local. Curr Anthropol 50(3):303–333

Nowell C (2021) The Red Nation Wants Its Land Back: Indigenous-led leftist collective is committed to freeing the world from capitalism and colonialism. The Nation Magazine, 8/10. https://www.thenation.com/article/activism/red-nation-new-mexico/

Stengel R (2021) Two of America's Leading Historians Look at the Nation's Founding Once Again – to Understand Its Complexity. New York Times Book Review, 9/21. https://www.nytimes.com/2021/09/21/books/review/the-cause-joseph-j-ellis-power-liberty-gordon-s-wood.html

Timperly C (2020) Constellations of indigeneity: the power of definition. Contemp Polit Theory 19(1):38–60

Uh GH (2021) Truth and justification: on the cruelties against indigenous people. The Nation Magazine, 9/23. https://www.thenation.com/article/society/indignous-residential-boarding-schools-canada/

Viren S (2021) The Geneaology of a Lie. The New York Times Magazine, 5/28. https://www.nytimes.com/2021/05/25/magazine/cherokee-native-american-andrea-smith.html

Declarations (listed chronologically)

The United Nations Declaration on the Rights of Indigenous Peoples (2007). https://www.un.org/development/desa/indigenouspeoples/declaration-on-the-rights-of-indigenous-peoples.html

The American Declaration on the Rights of Indigenous Peoples of the Organization of American States (2016). https://www.oas.org/en/sare/documents/DecAmIND.pdf

Universal Declaration of Human Rights (1948). https://www.un.org/en/about-us/universal-declaration-of-human-rights

UN Decades for indigenous peoples (2019). https://www.un.org/development/desa/indigenouspeoples/news/2019/11/ga-third/. https://www.un.org/en/about-us/universal-declaration-of-human-rights

Websites

Christian History Institute. https://christianhistoryinstitute.org/magazine/article/bible-translation-since-john-wycliffe

Coastal Gaslink Pipeline. https://www.economist.com/the-americas/2020/02/20/a-pipeline-through-historically-native-land-has-sparked-protests-in-canada

Deskaheh Dictionary of Canadian Biography. http://www.biographi.ca/en/bio/deskaheh_15E.html

Indigenous Peoples, United Nations, Department of Economic and Social Affairs: https://www.un.org/development/desa/indigenouspeoples/

Indigenous Rights. https://www.un.org/en/chronicle/article/conference-diplomacy-united-nations-and-advancement-indigenous-rights

Keystone Pipeline. https://www.narf.org/cases/keystone/

Oxford Bibliographies on Indigeneity Updated 2018. https://www.oxfordbibliographies.com/view/document/obo-9780199766567/obo-9780199766567-0199.xml

Ratana: "Story: Ratana, Tahupotiki Wiremu." https://teara.govt.nz/en/biographies/3r4/ratana-tahupotiki-wiremu

The Doctrine of Discovery: GilderLehrman Historical Resources: https://www.gilderlehrman.org/history-resources/spotlight-primary-source/doctrine-discovery-1493

UNESCO Mission and Mandate. https://en.unesco.org/about-us/introducing-unesco

*Note to readers: In addition to works and websites alluded to in this essay, I have included a number of others that have influenced the argument I made and that I highly recommend for further consideration on the issues raised here. Judith Friedlander

Part V

Historical Sociology

The Past and the Future of Historical Sociology: An Introduction

22

Marta Bucholc and Stephen Mennell

Contents

Introduction	562
Telling a Story of Historical Sociology	562
A Brief History of Historical Sociology	565
The Golden Age: The Beginnings of Historical Sociology	566
The (not so) Dark Ages: Insulation and Marginalization	567
The Renewal: Revival and Revision	570
State of the Discipline and Its Future Outlooks	573
The Themes of Historical Sociology	574
Conclusion	578
References	579

Abstract

The chapter provides an overview of the development of historical sociology as a sociological subdiscipline. The authors argue that over the decades of institutionalization of sociology as an academic discipline, historical imagination was gradually forsaken in many if not all sociological traditions, and the revival of historical sociology in the recent decades is but a return to the origins of sociology as a historically informed science of society. The chapter is organized as a chronological narrative of the history of historical sociology following the three-waves model, starting with the early classics, through the breakthrough of the 1960s, up to the contemporary state of the discipline. An overview of the main themes and research problems of historical sociology is followed by a brief review of the contents of the chapters included in the *Historical sociology* section,

---

M. Bucholc (✉)
University of Warsaw, Warsaw, Poland
e-mail: bucholcm@is.uw.edu.pl

S. Mennell
School of Sociology, University College Dublin, Dublin, Ireland
e-mail: stephen.mennell@ucd.ie

which are devoted to historical sociological insights in the fields of research of state and power, war and violence, emotions, sport and leisure, gender relations, collective identities, law and legal cultures, and memory studies.

Keywords

Historical sociology · Three waves of historical sociology · History of sociology

Introduction

It has become a time-honored tradition to start any introduction to historical sociology with a statement of a difficulty implied by its very name. Historical sociology is located somewhere between history and sociology, in a highly contested area claimed by two academic disciplines, each of them with its distinct way of studying the social world. One of them deals with the past, and historians who draw on their knowledge of the past to declaim on problems of the present are often viewed with disapproval by their fellow historians. In contrast, the discipline of sociology tends to be mainly preoccupied with present-day social problems, to which the past is seen as no more than "background." Moreover, relations between history and sociology are often complicated by academic hierarchies, with historians – whose academic discipline became established within universities from the first half of the nineteenth century – feeling superior to sociologists whose subject was not widely found in universities until perhaps the second half of the twentieth century (Elias 2009 [1982]).

However, there is another puzzle that we believe should be solved before addressing the issue of interdisciplinary cooperation – or disciplinary trespassing – between sociology and history. It concerns the reason why the need to delimitate the subdiscipline of historical sociology ever even arose. Instead of asking why there is historical sociology, we must inquire how sociology that is not historical could ever become possible or, indeed, dominant. Once this puzzle has been solved, the affinity between sociology and history is much more readily understood, and the development of historical sociology can be put into perspective by a reminder of the discipline of sociology having almost forsaken its historical imagination for a few ponderous decades.

Telling a Story of Historical Sociology

There are many narratives of how historical sociology developed. In the beginning, there are always the classics. But who are the classics? From about the 1970s to the 2000s, the sociological pantheon seemed to be almost deserted, with only three major occupants, the Holy Trinity of Karl Marx, Emile Durkheim, and Max Weber. Before that, many other deities had been worshipped. For example, in British universities in the 1960s it was common for undergraduate courses in the history of sociological thought to begin with Montesquieu and work forward through Rousseau, St Simon, Comte, Tocqueville, and Spencer before arriving at Marx, Durkheim, and Weber. For the first great American Department of Sociology in the University of Chicago, and the tradition it established, Simmel was a primary influence. From Harvard, Talcott Parsons's *The Structure of Social Action* (1937) played a large role in introducing Durkheim and Weber into Anglophone sociology; yet the prominence he also gave to Pareto failed eventually to secure the Italian's lasting place among the greats, and he said little about Marx.

At the beginning of the twenty-first century, the repertoire of the classics broadened out again to include other names, less widely influential and at first more grudgingly acknowledged, but no less significant in their contribution to the further development of the discipline. A recent textbook of sociological theory (Loyal and Malešević, 2021) includes such names as Hintze, Gumplowicz, Ratzenhofer, Small, Ward, Du Bois, G. H. Mead, Gramsci, Lukács, and Schutz, as well as many figures from the second half of the twentieth century; some of these are regaining their former place in the pantheon, others breaking into it for the first time. And, notably, among the traditional list mainly of dead white males, female sociologists can now be found: Harriet Martineau, Arlie Russell Hochschild, Patricia Hill Collins, Judith Butler, Simone de Beauvoir, Ann Oakley, and Dorothy Smith.

What the classics have in common – no matter how they are defined – is, however, that their mindset was inherently historical, even if history played many different roles in their theorizing. But the story of what happened to this initial entente of sociology and history afterward can be told in various ways. One of the main factors shaping the tale is the teller's idea as to what sociology is.

Sociology was born in the nineteenth century, because other disciplines of older pedigree were unable to provide answers to questions posed by the new social form emerging at the time: the modern industrial mass society (Elias 2009 [1962] pp. 43–69). In the nineteenth century, "society" as an actor entered the great drama of humankind, and a new science arose to address the new problems raised by its sudden and revolutionary appearance. Even though this science claimed from the very beginning that its goal was to study human society as such – any society, all societies, in any time or place – both its conceptual apparatus, its main theories, and its research methods were embedded in the reality of the "Modern Age." As a result, sociology turned out to be incapable of studying non-Western and non-modern societies efficiently, and had to cede this research field to other disciplines, notably to anthropology. The division of labor between sociology and anthropology was soon institutionalized. Hitherto sociology has largely remained a science of modern industrial-commercial society, with other societies often falling outside of its scope.

If we understand sociology in that way, then historical sociology will be the branch of sociology studying the historical process of "modernization" of industrial societies. Indeed, among the most prominent proponents of historical sociology, this problematic seems to prevail. How the West has become modern is the question that unites the classics and contemporary writers, such Michael Mann, Theda Skocpol, Charles Tilly, Immanuel Wallerstein, and many, many others. However, the sociological quest for historical forces of modernization has had its ups and downs. Julia

Adams, Elisabeth Clemens, and Ann Shola Orloff (2005) distinguish no fewer than three waves of historical sociology, starting with the beginnings in late nineteenth century, the second wave in the 1970s and the third which is still in process. Some other authors prefer only to speak of two waves (Smith 1991), and yet others distinguish a number of approaches developing in parallel, each with its own internal dynamics, but focusing invariably on the genesis of modernity (Delanty and Isin 2003).

However, the framework of historical sociology as a study of modernization was challenged early on. It could be argued that in the works belonging to what Adams, Clemens, and Orloff see as the "second wave" of historical sociology, a less West-centered and idiosyncratic research program was already forthcoming by way of international comparison. While some of the classics of sociology, like Max Weber, were eminent comparativists, most of them never launched any large-scale empirical comparative studies. But in the historical sociology of the 1970s and 1980s the comparative framework was omnipresent. Even though most of the studies in this period focused on the developments taking place in the West, the rest of the globe was no longer a big blank. Sociology did not remain untouched by the rise of postcolonial theory in the discipline of history and in literary studies. The dominance of the Western perspective was contested, and the position of non-Western societies in historical sociology slowly changed. Initially, it was but a negative reference for understanding Western modernity, as in the writings of Max Weber, who described himself as "a child of modern European civilization" (1930 [1904–5]: 13). Later on, it became an object of comparison for the phenomena identified in the West within a globally framed reflection on modernization, as in the work of Immanuel Wallerstein or Shmuel Eisenstadt: an ambivalent position of being a part of modernization theory, but not as an equal of the West (Wallerstein 2000; Eisenstadt 1974; Bhambra 2016). In the late twentieth century non-Western societies were emancipated and the polycentric view of human history was declared to prevail, even if it did not yet do so in research practice and in writing. Consequently, the calls for a "global historical sociology" have been heard, overcoming the trap of Eurocentrism both as far as the research material and the theory are concerned (see Go and Lawson 2017).

The revival of historical sociology and its subsequent expansion into new areas corresponds to a vision of sociology that is not limited to studying the historical background of modernization. Sociology as a science of human societies is not limited to any particular, historically contingent form of the social, such as modern Western society. Moreover, since the critique of and reflection on the foundation of the social sciences reveals their deep involvement with the history of their own home society, the only way to move beyond this limitation is to accept that a study of human societies must account not only for their diversity, but also for the differences in their paths of becoming what they are, and for the unfinished and infinite nature of this process. A return to historical sociology freed from its original sins of West-centeredness and preoccupation with modernity means a turn toward dynamic narratives combining theoretical claims with rich research material coming from various sources of knowledge about social life.

An inspiration for this kind of historical sociology can be found particularly in one project of historical sociology, advanced by Norbert Elias (1897–1990). While Elias himself was not entirely innocent of the original sins of historical sociology, his work is an example of how comparative and interdisciplinary research approach can be combined with theoretical consequence and commitment to a well-defined disciplinary identity of sociology. Elias's way of doing historical sociology was unique, but it was at the same time a model of its main facets as exemplified in the works of its best representatives: a long time-perspective, stepping beyond the predefined historical epochs in search of continuities and discontinuities; using various sources and data to reach a potentially rich description of the past; and taking the people of the past seriously, with their ways of acting, feeling, and thinking that are so very different from our own. In his theory of the process of civilization, Elias connected a study of the institutionalization of the state with its monopolies of violence and taxes with habits observable in everyday, private lives, and standards of behavior propagated by education, art, and literature. Elias's work, initiated in 1939 with the monumental study *On the Process of Civilisation* (2012) and concluded – or not really concluded – with his death in 1990, lies at an intersection of many streams and waves of historical sociology. The structure of this volume owes much to Eliasian inspiration, valued critically as stretching throughout all the phases of the continental tradition of historical sociology and representative of both its accomplishments and its shortcomings.

A Brief History of Historical Sociology

The story of historical sociology in this introduction is told according to a simple chronological scheme, without engaging with the question of how the subsequent periods could be characterized in a more sophisticated theoretical manner. The story starts with a glance at the classics in the nineteenth and early twentieth centuries, in the (first?) Golden Age of historical sociology. Later, it moves toward the vicissitudes of historical sociology in the decades after World War II, and the reasons why the flourishing historical imagination of the first generation of sociologists withered in the 1950s. Once the historical narrative reaches the period of renewal of historical sociology in the 1970s, it proceeds toward mapping the main research problems of what was only then being established as a sociological subdiscipline, and the divergences between the meaning and practice of historical sociology in continental Europe and in the Anglo-Saxon world. The final point is a discussion of the future of historical sociology. Do its past achievements hold enough potential to preserve historical imagination into the third century of sociology? The contents of the chapters which make up this section give reasons to expect that historical sociology not only has a brilliant though complicated past, but also bright perspectives for the future. In each of them evidence can be found of the pertinence and relevance of historical sociology, and of the fruitfulness of combining historical and sociological imagination.

The Golden Age: The Beginnings of Historical Sociology

To most if not all of the efforts of the classics the formula applies, which was once used by Stephen Mennell: "in attempting to put their own society and its recent transformations in the perspective of the history of humanity as a whole, [they] actually succeeded in putting the whole history of humanity in the perspective of their own society" (Mennell 1990: 55). Ever since the beginning of the Enlightenment, history had been a central problem of continental philosophy in the late eighteenth and nineteenth century, and it coincided with articulations of the self-consciousness of the West as the driving force of modernity. The monumental *Esquisse d'un tableau historique des progrès de l'esprit humain* (1795) by Jean Antoine Nicolas Caritat de Condorcet is exemplary of the way history was conceived of in clear-cut stages of progress and development leading to equality, fraternity, and liberty of all nations based on the use of emancipated reason and following the Western model.

The name of Georg Wilhelm Friedrich Hegel (1770–1831) stands for the nineteenth-century's philosophical effort to conceptualize history of humankind as a meaningful, intelligible, and comprehensible process (1991 [1822]). Hegel focused very strongly on the political sphere and on the state as the highest achievement of social development. The impact of Hegelian inspiration on the philosophy of history was tremendous. Much of it was mediated through the later reception of Hegel, notably by Karl Marx and his historical materialism. But Hegel's mode of thinking also reverberated in the writings of Friedrich Nietzsche, whose fervent philippic on the "uses and disadvantages of history for life" (Nietzsche 1983) well expressed the *Zeitgeist* around the year 1900, the portentous date of Nietzsche's death: the growing consciousness of and concern for the existential and social meaning of history.

The end of nineteenth century was marked by gloomy premonitions, and in a retrospective these would seem to have been prophetic indeed, with two world wars shortly coming to change the face of the world once and for all. Immensely popular thinkers such as Oswald Spengler (1918–22), who at the turn of nineteenth century were primarily occupied with the fate of Western civilization, preached its imminent breakdown and fall. However, sociology, born in this transitional climate, bore very few marks of its origin. Despite the frequently critical view of modernization, industrialization, and capitalism expressed by Émile Durkheim, Max Weber, Ferdinand Tönnies, Georg Simmel, Werner Sombart, and many others who would fill the sociological hall of fame, there was a general consensus among them that modernity was, if not an historical necessity, then at least a historical fact. However, one excellent win for the historical imagination of the emergent discipline of sociology was that the uniqueness of European modernity could only be brought to light by an extensive use of comparative method, both in comparisons between Western and non-Western societies and between various stages in the history of Western societies themselves.

An unsurpassed monumental comparative synthesis exemplary of this period is Max Weber's project on "The economic ethics of the world religions" (see Ertman 2017). This unfinished multi-volume work was intended as a comparison of the historical development of the economic principles and practices produced by the great world religions. The most renowned part of the overall project is Weber's *The Protestant Ethic and the Spirit of Capitalism* (1930 [1904–5]), an endlessly cited epic story of the emergence of capitalism out of the spirit of Protestantism, and probably the most influential self-narrative of the capitalist West. But Weberian analyses of the Chinese literati (Weber 1951) or of the so-called kadi-justice (Weber 1946: 221) also found their way into comparative historical scholarship and they remain a subject of reception and critique. However, not only Weber's sociology of religion, but also his theory of modernization – including such highlights as the concept of bureaucracy or the idea of charismatic leadership – were a result of systematic, and West-centered, comparative endeavors. The turn of mind exemplified by Weber was also characteristic, though to varying extents, of other sociological pioneers, who tended to belong to German and French continental philosophical traditions. Modernity, the way they saw it, could only be plausibly explained through historical comparison.

The (not so) Dark Ages: Insulation and Marginalization

The historical imagination of the classics did not last long. One contributory cause of its decline was the Americanization of social science (Steinmetz 2007: 1). It is a paradox that Talcott Parsons, who had been the principal agent of Weber's reception in the USA, was himself a proponent of the systems theory that virtually did away with the historical questions by theorizing social change away (Wright Mills 1959). Even before he went to study in Weimar Germany, Parsons had spent a year at the London School of Economics, encountering the "structural functionalism" of Bronisław Malinowski; there may have been some excuse for anthropologists adopting a static, ahistorical theory when trying to understand pre-literate societies where evidence about their past was (or seemed) hard to come by, but it is less easy to comprehend why it should have appealed as a way of looking at twentieth-century advanced societies. Parsons came to focus exclusively on modern Western society, of which by this time middle-class America was emblematic. Even in his late "evolutionary" period, when he appeared to be returning to an interest in the long-term development of human society rather in the manner of Herbert Spencer, Parsons (1966, 1971) explicitly saw the USA as the "lead society." As Orloff, Clemens, and Adams note, "the mid-twentieth century was the apex of presentism in US sociology as well as the moment of highest confidence in modernity" (2005: 4). It was also the period when American sociology was at its hegemonic peak over the discipline across the world.

Parsonian systems theory for a time almost submerged the influence of the older Chicago School, but that had largely become just as ahistorical. The world's first department specifically of sociology had been established at the University of Chicago in 1895, under Albion W. Small who had studied in Germany and was especially influenced by Georg Simmel – not the most thoroughly *historical* of the classic German sociologists. The mature Chicago School was most concerned with the social ecology of the city and study of subcultures. On to it was grafted the influence of the pragmatic philosophy and social psychology developed by William James, and more locally through the philosopher George Herbert Mead. Nearly all Mead's work was published after his death in 1931. His book *Movements of Thought in the Nineteenth Century* (1936) shows a profound interest in the history of ideas, but sociologists were more in thrall to *Mind, Self and Society* (1934). Out of this came what is generally known as symbolic interactionism, of which perhaps the most celebrated exponent was Erving Goffman. And whatever his undoubted virtues, Goffman was not an exponent of historical sociology. His central interest in questions of "the presentation of self in everyday life" was one that he shared with Norbert Elias for instance, and the difference between an historical and an ahistorical treatment could scarcely be clearer. Later still, in the 1960s and 1970s, a further admixture of phenomenological philosophy led on to even more timeless approaches such as ethnomethodology and discourse analysis. What all these "movements of thought" in twentieth-century sociology have in common, it might be argued, are their basically Kantian epistemological foundations. They were central to Simmel's thought and via that to Chicago, and Parsons (1970) acknowledged his reading of Kant as a student as his own point of departure. Norbert Elias explicitly rejected Kantian epistemology, as probably do many other historical sociologists implicitly.

It is not easy to point to any single reason why American sociology was not interested in history. As George Steinmetz observed:

> There was some interest in historical topics and epistemologies before 1914, but this largely disappeared after World War One. Many of the founders of US sociology came out of fields like the natural sciences and economics (the latter was not dominated by historicism in the US, in contrast to Germany). Most of these founders took economics or one of the natural sciences as their model. The intellectual resources that that were required in order to imagine sociology as a *Geisteswissenschaft* ... were missing in early American sociology. (Steinmetz 2007: 16)

The resources then arrived with the wave of refugees from the Nazi Germany and wartime Europe, but it was not enough to attract many American colleagues to what was then perceived as an essentially European, and soon also somewhat antiquated, mode of thinking about society.

And not just antiquated but also in the view of some philosophers and social theorists actually dangerous. This warning was expressed in the most extreme way by the philosopher of science Karl Popper. In *The Open Society and its Enemies* (1945), Popper launched an onslaught on Plato, Hegel, Marx, and Karl Mannheim's sociology of knowledge; most scholars specializing in each of those thinkers are unconvinced. Then came *The Poverty of Historicism*, dedicated "In memory of the countless men and women of all creeds or nations or races who fell victim to the fascist and communist belief in Inexorable Laws of Historical Destiny" (1957: iii). Popper's argument is roughly that the course of human history is strongly influenced by the growth of human knowledge, which we cannot predict and thus are also unable to predict the future course of human history (1957: v–vi). That, claims the former physicist Popper, proves that a "theoretical history" comparable to theoretical physics is impossible. Of course, a "theoretical history" is not the goal of historical sociologists today, but it is easy to understand why such an argument had considerable appeal in the post-war, post-Holocaust, and almost-post-Gulag world. Popper had a marked influence on British sociology in those decades, and there were not dissimilar voices like Leo Strauss in the USA.

In the USA and in the UK, the ahistorical trend coincided with the prevalence in sociology of quantitative research methods and of social surveys in particular. While the classics of historical sociology did from time to time refer to such quantitative data as were available, their use of statistics did not correspond to the sophisticated methodological standards first worked out in the 1950s and 1960s. Moreover, primary historical data do not usually come in forms satisfying high demands for representativeness, accuracy, and validity. Historical data force the researcher to apply an interpretative approach, because the past is very seldom if ever available to us in an unmediated, unprocessed form. This was the basis of a trenchant attack on historical sociology by John Goldthorpe (2000 [1991]) – himself originally a history graduate who became a champion of quantitative research – who argued that history was the study of "relics," the representativeness of which could not be firmly ascertained, so sociologists should go out and generate their own "relics" through quantitative survey research. This attracted equally trenchant responses from historical sociologists, including Michael Mann (1994).

Interestingly enough, in European continental scholarship the period after World War II was marked by a rising tide of social history, marked first and foremost by the presence of the *Annales* school. This group of historians whose interest in long-term social processes resulted in a new approach to history was founded in France in the 1920s by Marc Bloch and Lucien Febvre. It famously focused on the *longue durée*, the perspective of many centuries, in its description of the social change. While its research subjects initially included the traditional problems of class, modernization, capitalism, war, and the state, it subsequently developed toward the history of mentalities, everyday life and everyday practices, and the history of culture. The school reached the peak of its influence after World War II, and its sway in the social sciences and humanities alike was one of the reasons why continental sociology never parted ways with history so fundamentally as the American one. However, the quantitative turn and prevalence of survey research in continental European sociology was one of the reasons why social history (based in history departments and exercised by academic historians) took over the torch of historical-social research, especially in Germany, with the emergence of Bielefeld school, to which belonged Reinhart Koselleck (1923–2006), the proponent of conceptual history as a study of historical semantics (see Brunner et al. 1972–1992). And so, in continental Europe, in some universities, the historical imagination in social studies survived or even thrived, but beyond the disciplinary boundaries of sociology.

The perseverance of historical imagination was even more conspicuous in Latin American sociology, where the presentism and the satisfaction with modernity were far less prevalent than in the North. Paradoxically enough, this was due to the Latin American sociology's public profile. It was characterized by a more direct engagement with the political questions of the moment, including those related to decolonization and social and economic development agendas of the countries all over the continent. The public engagement of sociology was reflected in its involvement with the international programs, including those sponsored by the United Nations, such as the Comisión Económica para América Latina y el Caribe (CEPAL). Interestingly enough, it is in this context that the turn toward a historically informed and interdisciplinary outlook was conspicuous. As Manuel Antonio Garretón and Naim Bro-Khomasi wrote:

> A first critical stance towards modernization theory came in the late fifties from CEPAL's 'historical-structural' perspective and its related 'integrated' approach. In opposition to the predominantly ahistorical modernization perspective, historical-structuralism emphasized the need to attend to historical context in the search of explanations, while the integrated approach stressed the need to consider economic, social, political and cultural factors beyond the limits of particular disciplines. An early work within this perspective was 'El desarrollo social de America Latina en la postguerra' (CEPAL 1963), which raised the challenges of urbanization and industrialization in the region and the role of the middle classes in championing this process. While some of its main themes were still part of the modernization theory agenda, this work went further by urging social scientists to look into social history in order to analyse social processes (...) (2013: 16)

In the 1950s and 1960s, the sociological imagination also managed to keep some of its historical hue in other regions whose contribution to sociology was far from negligible, including socialist Europe, notably Poland and Czechoslovakia (see Bucholc et al. 2022). This was despite the conflicting pressures of Sovietization, which some countries and academic milieus resisted more efficiently than the others, and Americanization in both theory and methodology, which gained pace in some of the Eastern Bloc sociologies, notably in Poland, following the – restricted and discontinuous – restoration of academic mobility and international connectivity in the second half of the twentieth century (see Bucholc 2016). While in some post-socialist countries, notably in Russia, historical sociology played and still plays a minor role, generally it could be stated that in the Eastern Bloc the historical imagination preserved over the post-war period served as a resource for the revival of historical-sociological research agendas after the fall of the Iron Curtain (Skovajsa and Balon 2017).

The Renewal: Revival and Revision

For both the global North and the global South, or – at the time – both for the socialist and the capitalist world, the end of the 1960s was the time of great upheaval, from which a renewal of interest in history was reborn. As Andrew Abbot wrote in connection with the developments in American academia:

> Theoretically, historical sociology was for them a way to attack the Parsonian framework on its weakest front – its approach to social change – and a way to bring Marx into sociology. (...) As for the Chicago interactionists, ... historical sociology simply bypassed them. Central to historical sociology was its invocation of the ponderous respectability of

'History'. . . . Above all, History's respectability redeemed the radical politics of the historical sociologists. Even as an evolutionist, Marx was more historical than Parsons. (Abbott 2001, pp. 94–5)

What happened in America in the 1960s cannot be taken as a simple model for a global intellectual change, but it is a marker of the *Zeitgeist* bridging the ideological divides and cultural hostilities of the world partitioned by the Iron Curtain. A perception of society – each society – as a dynamic historical being subject to constant change was equally dangerous to the emergent neoliberal capitalism and to socialism and communism (see, in the context of legal history and Marxism, Bucholc 2020). Ideologies, even those built on evolutionary theoretical premises, are inherently anti-historical, hence the unlikely alliance of historical sociology with the youthful revolutionary forces both in the West and in the East.

As a result, historical sociology flourished. Most of the books that are habitually listed in encyclopedic accounts of historical sociology were published in the 1970s and 1980s. They include Wallerstein's *The Modern World System* (the first volume of which appeared in 1974), three seminal works by Charles Tilly (1975, 1978, 1984), Theda Skocpol's *States and Social Revolutions* (1979), and the first volume of Michael Mann's *The Sources of Social Power* (1986). Historical sociology's development in this period was marked, on the one hand, by a return to the classical research agenda. On the other hand, some distinct innovations appeared in its research focus, methods, and manner of theorizing.

As Steinmetz remarked:

> To understand the ebb and flow of historical interest among sociologists it is necessary to pay attention to *extra-scientific* changes such as macrosocial crisis and stabilization, as well as *intra-scientific* processes such as the varying relations between history and sociology in different periods and countries and the internal hierarchies within each disciplinary field (Steinmetz 2007: 1).

In the period after 1968 both the intra-scientific and extrascientific factors worked toward a revival of interest in the attempt to make a socio-theoretical sense of the past. The old macro-social problems reemerged with wars, revolutions, totalitarian, and authoritarian power and the rampant global capitalism. Within academia, in many countries sociology was at the peak of its institutional influence in the late 1960s. As Wallerstein put it:

> The golden era of sociology as a discipline was probably 1945–1965, when its scientific tasks seemed clear, its future guaranteed, and its intellectual leaders sure of themselves. Since 1965, sociologists have scattered along many, quite divergent paths. This has created much dismay within sociology about the presumed future of the field and has led to much external social critique (2000, p. 25).

One of these divergent paths along which sociology scattered was that of historical sociology. However, in a paradoxical manner, Wallerstein's diagnosis also explains how sociologists in the early 1970s could afford to coalesce with history: they had no fear of losing their disciplinary identity and they might even have perceived that identity as a burden, longing for more openness and interdisciplinary collaboration.

The revival of the classical agenda included an intense interest in the institutionalization of power with the state as the protagonist. But it was a state whose nature and functions were no longer self-evident: it was changing dynamically and underwent a series of transformations affecting politics, economy, and culture alike. The focus on social change – though not always going under that very name – directed attention not only toward the ongoing and usually relatively slow consolidation and transformation of state power, but also toward the abrupt social change such as a revolution, be it a political or an economic one. While the impact of some classics, such as Karl Marx or Max Weber, was reinforced by this macro-sociological tendency to study change, some others never made a great return, including Pitirim Sorokin with his studies of Russian revolution (Sorokin 1960, 1967). The case of Alexis de Tocqueville is particularly interesting: his masterly account of historical interplay of politics and society remains regarded as a classic of political science much more than of historical sociology. One reason for this is that most attention is paid to the first volume of *Democracy in America*, published in 1835, to the neglect of the second volume published in 1840. The earlier part of Tocqueville's total work not only concentrates particularly on political arrangements in America, but also portrays them mainly (though not uncritically) in a favorable light. This part is thus itself viewed in a favorable light in America. The later volume focuses rather on the whole range of cultural life in America, and is rather more critical in tone. Stone and Mennell (1980) attempted to redress this balance, but with limited success.

Some core themes of classical historical sociology, while still important, underwent a profound revision in recent scholarship. We have witnessed a turn toward the study of groups and societies heretofore marginalized in the largely Eurocentric or, rather, West-centered sociological narratives (Bhambra 2016, Goody 1996). Women, children, working classes, poor people and people living on social margins, the elderly, the non-heteronormative, as well as national, racial, and ethnic minorities, and peripheral societies marginalized in the world system, have joined the ranks of those whose experience is being subjected to historical enquiry and for whom agency and voice has been claimed in the historical narrative. This coincided with a wave of sociological self-reflection, aimed at revisiting the genealogy of the discipline and the limitations imposed on basic concepts and paradigms by their embedding in a singular historical experience. Vindication of everybody's right to history, advocated most of all by postcolonial studies, has enriched historical sociology and endowed it with a new critical edge. At the same time, it contributed to an increase in the internal diversity of historical sociology.

Stephen Mennell (2015: 548) provided a representative list of books in the so-called "figurational paradigm" inspired by Elias, which had been written after 1970, to demonstrate the polyvalence of this particular brand of historical sociology. What Mennell also notes is the interdisciplinary character of the listed works: almost none of them can be unequivocally qualified as belonging to the sociological mainstream. However, it is not only the figurational variant that "appeals more to people working in the interstices of the social sciences" (Mennell 2015: 548). The transdisciplinary flair is characteristic of contemporary historical sociology in general. And it is not only the intersection of history and sociology: many other disciplines contribute to the landscape of historical sociology nowadays, including anthropology, social psychology, archaeology, literary studies, cultural studies, art history, linguistics, economics, gender studies, law, and jurisprudence. Much can also be expected from cognitive science and evolutionary psychology, as demonstrated in Steven Pinker's epic narrative of the "better angels" of human nature (2011).

While openness to interdisciplinary collaboration is not, in itself, strange to sociology and never was – apart from the relatively short though momentous time of disciplinary mainstreaming in the 1950s and 1960s – today's interdisciplinarity is somewhat different from that of the 1970s. Instead of interdisciplinarity, it is usually the label of multi-disciplinarity or trans-disciplinarity that is used to designate the endeavors that are not really located within any single disciplinary model. Methodological opportunism (see Feyerabend 1975) seems to be a mark of many contemporary historical sociologists: they seek out problems in the world and they design their research tools correspondingly, drawing on various resources when and how they become available. Skocpol's distinction between "theory-driven" and "problem oriented" scholarship (1984: 16–17), while useful, is far from a binary dichotomy: the two orientations mix all the time.

State of the Discipline and Its Future Outlooks

Damon Mayrl and Nicholas H. Wilson (2020) have identified no less than four "analytic architectures" used by historical sociologists, showing how they negotiate the boundary between history and sociology, how they differ in their critical and constructive engagement with theory and evidence, how they value primary and secondary sources, etc. Even though their illuminating analysis has a heavy American bias owing to their sampling criterion being the prizes awarded by the American Sociological Association, their conclusion seems valid enough: as a result of the diversity of the analytic architectures, historical sociology is becoming more and more "substantively fragmented" and "methodologically pluralist" (Mayrl and Wilson 2020: 1346). Historical sociology is now more heterogeneous than ever – as much as sociology at large. It is also becoming more diversified, as the objects of study are no longer limited to large-scale Western social processes. The convergence of interdisciplinary research agendas was followed by a new methodological opening in the digital age. Digital humanities, digital archaeology and other similar areas of study make it possible for historical sociology to be more precise, more accurate, and more rigorous than ever, though how much of a win that is exactly remains debatable. Nevertheless, easier access to new primary sources is certainly a gain, and the substance of more than 50 years of human history has now been largely digitalized. This reduces the costs, the time and the energetic expenditure of archival research, at least in the most recent history. The impact of technological change on historical sociology is still, however, a big unknown.

An overview of the themes featuring in the historical-sociological scientific journals nowadays, including (no mention just a few who proclaim their disciplinary commitment in their titles) *Journal of Historical Sociology* (published by Wiley), *Historical Social Research/Historische Sozialforschung* (published by GESIS – Leibniz Institute for the Social Sciences), *Historická sociologie* (Charles University in Prague), *Sociohistórica* (Universidad Nacional de La Plata), or *Sociología Histórica* (University of Murcia) confirms the thesis about fragmentation and pluralism. However, the influence of the classics seems to underpin a form of sub-disciplinary identity: debates on Marx and Weber do not cease, even though they are conspicuously less vehement than in the 1970s. However, in many regions, notably in Latin America, the revival of socio-historical agendas has created a new angle from which to perceive the regional sociological tradition, including the disciplinary boundaries between sociology and history. For example, the work of Waldo Ansaldi and his collaborators at the Universidad de Buenos Aires has contributed to the reinforcement of the program of comparative studies in historical sociology and to a debate on the goals and methods of socio-historical research (see, e.g., Ansaldi 2007; Astarita 2006).

The evergreens such as states, empires, nations, power, war, capitalism, and revolution do not seem to lose attraction. The interest in the themes, which emerged in a big way in the 1970s, including emotion, sexuality, and the body, still remain among the top themes, as are the fates of the social underdogs of various societies, including especially women, children, racial, and ethnic minorities. The new topics include, among other things, big data and the new financial markets. If anything is underrepresented, these are historical studies of cultural phenomena. As far as the time span of research is concerned, most of the studies published in these journals are located in the twentieth and nineteenth centuries, and there are precious few articles dealing with pre-modern period, whether in Europe or elsewhere. These tendencies are unlikely to change in the future, for a number of reasons ranging from the availability of sources and data usable to sociologists through the necessary investment in the acquisition of technical and linguistic skills necessary to use them. Indeed, the expansion of historical sociology nowadays seems to be largely inhibited by methodological considerations. However, the proliferation of mixed methods research designs may give some hope in this respect (see Tierbach et al. 2020), and an argument has been made by Nina Baur and Stefanie Ernst for the use of recently developed sociological research methods in studies of historical processes (Baur and Ernst 2011).

The Themes of Historical Sociology

This section combines an overview of classical motives of historical sociology with new insights into emergent research areas. While handbook editors always have to choose and find a compromise between their own interests and a fair representation of the field in question, it seemed reasonable to prioritize research fields and subjects over chronological ordering, theoretical traditions, methods, or any other alternative ordering criterion. The authors in this section come from various academic cultures and traditions, and they each of them represent a distinct socio-scientific research field, which sometimes stands right in the center of historical sociology, sometimes overlaps with it significantly, and sometimes seems to lie parallel to it. As a result, each chapter presents a microcosm of historical sociological research in its own right, focusing on the uses of the past in sociological research perceived through different lenses. Some of these microcosms are as old as sociology itself, and some others are relatively new additions. All of them, however, participate in the recent "historical turn" (Morawska 1989) of sociology.

Two chapters deal with the classical themes of the second-wave historical sociology. Helmut Kuzmics takes a broad view of *Power and Politics: State Formation in Historical Sociology*, tracing the origins of the main categories in historical-sociological theorizing of state and power. He leads the reader through the three main stages of state formation, stressing the dynamics of power-relations. From the early states and civilizations, through the early empires, the early modern and modern state systems, up to the industrial nation-state and supra-national associations of states in late modernity, Kuzmics demonstrates the variety of the reflection on state, power, and politics in historical sociology in a very broad meaning of the term. He reviews the main paradigms in historical sociology of the state and state formation, stressing their interdisciplinarity and the multiple theoretical inspirations from which they draw, to arrive at the final point that seem to summarize the goal of the whole section:

> It is only in combination with psychology, cultural anthropology, economics, literary studies and several other branches of the Human Sciences that it [historical sociology] can show its fertility; but without the acknowledgment of the processual nature of human societies, these disciplines would miss their very essence.

Chapter 24, "Organized Violence and Historical Sociology," by Christian Olsson and Siniša Malešević discusses socio-historical studies of organized violence. The authors begin with an explanation of the relative reluctance of sociology to address violence as such, despite the focus on the intrinsic connection between violence and state formation, social change, and especially revolutions and warfare. They argue that the interest in organized violence as a multifaceted historical phenomenon is only a few decades old, and they propose to split their overview of the field into sections dedicated to war and warfare, clandestine political violence, revolutions, and, finally, genocides. Their reflections connect at many points to those of Kuzmics, as they stress the link between state formation and state agency, and the new forms of organized violence. Olsson and Malešević critically assess the biases in the violence scholarship, which – for example – have led to revolutions being a central topic of historical sociology, whereas other forms of organized violence have been largely unrecognized as such until the late twentieth century. They conclude with three suggestions for future research to overcome state-centricity by broadening the picture of organized violence, to expand on "the situational logic of particular violent events," and to meet the challenge of applying "the tools of historical sociology beyond the West without necessarily taking the latter's historical trajectory as point of reference."

Chapter 25, "Historical Sociology of Law," by Marta Bucholc tackles the problem of developing historical sociology of law as a full-fledged subfield of historical sociology. The author stresses the difficulty of entering an inherently historical field that is law, and she ponders on the mutual relation of legal history and historical sociology of law, indicating some contemporary efforts to discharge the apparent competition between the two. She further discusses the deceptive historicity of the classical sociology of law, pointing out the weaknesses of socio-historical understanding of law and legal change in Marx, Weber, and Durkheim. From there, she moves on to an overview of the three lines of social theorizing, which offer nondeceptive models for studying law as a historical phenomenon displaying mechanisms of development of its own, embedded in the modes of legal thinking, communication, and institutional setup. Following the theories of Michel Foucault, Pierre Bourdieu, and Niklas Luhmann, Bucholc depicts a broad landscape of historical-sociological theory and empirical studies today. She provided illustrations from the studies of criminal law, procedural law, constitutional law, transitional justice, and international law to point out some of the main directions in the development of historical sociology of law as a multiparadigm, interdisciplinary research field.

Dominic Malcolm's chapter presents a review of research in historical sociology of sport and leisure. While not a classical theme in historical sociology, the field has been the object of attention of classics such as Thorstein Veblen and Johann Huizinga, who shaped the early intuitions in the study of leisure, with sport as one particular kind of leisure activity. Malcolm's presentation of the subdiscipline of sociology of sport distances itself from these early endeavors and focuses on the sociology of sport having emerged as a self-standing sociological subdiscipline in the 1960s, and gaining momentum in the late twentieth century. He analyses three stages in the development of historical sociological research on sport: pre-1990s, 1995–2005, and early twenty-first century. He focuses on the forces within the field of sociology of sport, and on the difficulties of defining the boundaries of sociological, historical sociological, and historical research. In his conclusion, Malcolm identifies the main factors that have shaped sociology of sport as a historically informed research field. He ends with a glance into the future of its interdisciplinary relations and theoretical orientations.

Robert van Krieken's chapter is devoted to a subject whose historicity has caused much philosophical argument in the past and much methodological debate in the present: emotions. The author observes that the historical sociology of emotions has not as yet developed fully as a distinct field of human sciences. He points out the affinity of historical sociology of emotions to the rather better developed fields of sociology of emotions, the history of emotions, and historical sociology in general, and he discusses the dynamics of cooperation and competition between these fields and the historical sociology of emotions. He draws on the classical contributions to history and sociology of emotions, and he analyses the causes of historical sociology's relative immunity to the "affective turn" in social sciences. In his discussion of the theoretical setup of historical sociology of emotions, van Krieken focuses on the influence of Freudian psychology on sociology of emotions and in particular on the approach to emotions advocated by Norbert Elias. The chapter follows the theoretical adaptations of the psychological concept of emotion in social sciences, stressing the critical perspective on the historical development of emotions.

Chapter 28, "Identity, Identification, Habitus: A Process Sociology Approach," by Florence Delmotte discusses the problems of identity and identification in historical sociology. The author insists on the necessity of approaching identities and identifications as processes, and she stresses the particular affinity of identity studies with socio-historical program of tracing group belonging to collective imaginaries. The first part of the chapter reviews the highlights of the process-sociological approach to identities. Delmotte stresses the problematic nature of the notion of identity and the controversies linked to its use by the social sciences. In the second part, she offers a detailed review of Norbert Elias's theory as a paramount example of socio-historical study of the problem of identity, combining a sophisticated theoretical apparatus with extensive historical evidence. Finally, in the third part, she offers a case study of the problem of European identity and the European Union as an illustration of the historical process of identity formation and its socio-scientific analysis.

Stefanie Ernst completes the thematic scope of the section with a chapter discussing the juncture of gender studies and historical sociology. She tackles the hidden gender orders to show the socio-historical dynamics of power and inequality between men and women and the main lines along which it has been conceptualized. She points out that research on gender relations reaching out to include the studies of family and business life is characterized by fragmentation as well as a certain tendency toward ahistoricity. She argues for a need to synthesize the knowledge in this field in order to understand the embeddedness and the interdependencies of gender, work, and private life. She begins with a summary of the integrative potential of socio-historical and empirical perspective for Gender Studies. Next, she reiterates central terms, including power, informalization and social change, and she outlines an integrative, socio-historic approach to gender. In the next sections, she offers an example of an application of this approach in the studies of changing gender relations. Her case studies come from research on the German society, and they draw on the material from private life (including marriage and family) as well as working life.

Joanna Wawrzyniak takes on the task of connecting memory studies with historical sociology. She stresses the fact that the encounters of the two disciplines were relatively rare despite their many overlaps in themes, methods, and research questions. She argues that the dynamically growing field of memory studies has much to offer historical sociology, including its contribution to reinterpretations of the concept of collective memory. She starts the chapter with a review of the current state of sociological reflection on collective memory, and she stresses the impact of the recent rise in memory studies for some key discussions in historical sociology. Wawrzyniak points out the ubiquity of memory research in history and its inherent interdisciplinarity, offering particular promise to the synergies of socio-historical approaches. She discusses the main research traditions, and she focuses specifically on continuity and change as two notions, which best illustrate the intrinsic connection of memory and historical sociology particularly. She concludes with mapping new research directions, which can be expected to bring memory studies and historical sociology even closer in the future. Memory studies and historical sociology face similar challenges, they share many of their intellectual roots, and their future paths will in all probability continue to run parallel, with collective memory as a research subject playing an ever more important role in culture-oriented historical sociology.

Conclusion

The chapters of this section of the *Handbook* provide a useful and accurate overview of some of the main themes in the field of historical sociology at the end of the first quarter of the twenty-first century, and it is to be hoped that they will remain of value to scholars for many years. The thematic array of this section cannot pretend to be exhaustive. It is not only owing to the richness of historical sociology, but also to the mental habits of its readers and writers. Other topics not normally considered typical of the field, such as studies in the sociology of food and eating (like Mennell 1985), could equally well be considered as examples of historically informed sociology (or sociologically informed history), yet they tend to be pigeon-holed elsewhere in sociology's astonishingly diverse range of subdisciplines. Similarly, Jason Hughes's *Learning to Smoke: Tobacco Use in the West* (2003) was probably sorted into a box labeled "social problems" or even "sociology of health and illness," but its long-term perspective over the four centuries since tobacco found its way to Europe added greatly to its explanatory power. While some of the topics chosen for this section, such as state formation or at least some forms of organized violence, indisputably belong to the realm of historical sociology, some others, like emotions, have only been a recognized subject in some socio-historical paradigms; and some others, like collective memory, are infrequently if ever associated with historical sociology. However, the combination of the reiterating and new themes offers a broad view of the robust and dynamically growing discipline, which may struggle with defining its own identity at the intersection of history and sociology (and, frequently, other research areas), but which is never lacking new inspirations for expanding its empirical scope and theoretical framework.

Indeed, arguably *all* sociology is or ought, in a certain sense, to be historical sociology. That is because *time* is an indispensable element in sociological theories and thus of sociological explanations. True, the span of time to which it may be necessary or reasonable to refer in many sociological studies is often quite short and so would scarcely be recognized as a justification for such studies to claim the place within historical sociology. Yet we would argue, and hope that this section helps to demonstrate that a historical perspective enriches a whole range of sociological research, beyond the business of academic labeling. In order to ensure a fruitful and mutually beneficial collaboration between sociology and history, three points can usefully be borne in mind. First: it must always be remembered that the stock of relevant information contained in existing libraries, archives, and databases always dwarves any new evidence that can be gathered by current surveys or interviews. Second, history does not yield mere "background" to sociological findings: it most often provides key ingredients in *sociological* understanding. Third, by extension, historical sociologists search libraries, archives, and databases for two rather distinct purposes: yes, to discover what earlier social scientists and historians have previously had to say on one's topic, but also to seek out empirical evidence about the recent or more distant past that can be used in constructing better *sociological* theories and explanations about the world we live in.

While this section can never aspire to be an exhaustive review of the whole field of historical sociology, it is just conceivable that it may do something more important. It may help sociologists at large to recognize that historical sociology is not just yet another specialist subfield among sociology's myriads of such subfields, but that it offers a perspective that is useful – maybe essential – across all the specialties. We believe that historical imagination needs to be restored to the central place in sociology.

Acknowledgments Marta Bucholc acknowledges the support of Polish National Science Centre (2019/34/E/HS6/00295).

References

Abbott A (2001) Chaos of disciplines. University of Chicago Press, Chicago
Ansaldi W (2007) A mucho viento poca vela. Las condiciones sociohistóricas de la democracia en América Latina. Una introducción. In: Ansaldi W (ed) La democracia en América Latina, un barco a la deriva. Fondo de Cultura Económica, Buenos Aires
Astarita CAT (2006) En las tradiciones de Weber y de Marx. Reflexiones sobre un artículo de Waldo Ansaldi. Sociohistórica 19/20:159–187
Baur N, Ernst S (2011) Towards a process-oriented methodology: modern social science research methods and Norbert Elias's figurational sociology. Sociol Rev 59:117–139
Bhambra GK (2016) Postcolonial reflections on sociology. Sociology 50(5):960–966
Brunner O, Conze W, Koselleck R (eds) (1972–1992) Geschichtliche Grundbegriffe: Historisches Lexikon zur politisch-sozialen Sprache in Deutschland. Ernst Klett Verlag, Stuttgart
Bucholc M (2016) Sociology in Poland: to be continued? Palgrave Macmillan, London
Bucholc M (2020) Juliusz Bardach and the agenda of socialist history of law in Poland. In: Erkkila V, Haferkamp H-P (eds) Socialist interpretations of legal history. Routledge, pp 115–135
Bucholc M, Kolasa-Nowak A (2022) Historical sociology in Poland: transformations of the uses of the past. East Eur Polit Soc Cult (fortcoming)
CEPAL [United Nations Economic Commission for Latin America and the Caribbean] (1963) El desarrollo social de América Latina en la postguerra [The social development of Latin America in the post-war period]. https://repositorio.cepal.org/handle/11362/14734. Retrieved 13 October 2021
Condorcet, Marie Jean Antoine Nicolas de Caritat, Marquis de (1795) Esquisse d'un tableau historique des progrès de l'esprit humain. Paris. [Sketch for a Historical Picture of the Progress of the Human Mind. Weidenfed & Nicolson, London, 1955]
Delanty G, Isin EF (eds) (2003) Handbook of historical sociology. Sage, London
Eisenstadt SN (1974) Studies of modernization and sociological theory. Hist Theory 13(3):225–552

Elias N (2009 [1962]) On the sociogenesis of sociology. In: Essays III: on sociology and the humanities [Collected works, vol. 16]. UCD Press, Dublin, pp 43–69

Elias N (2009 [1982]) Scientific establishments. In: Essays I: on the sociology of knowledge and the sciences [Collected works, vol 14]. UCD Press, Dublin, pp 107–160

Elias N (2012) On the process of civilization: sociogenetic and psychogenetic investigations [Collected works, vol 3]. UCD Press, Dublin

Ertman TC (2017) Max Weber's economic ethics of the World's religions: an analysis. Cambridge University Press, Cambridge

Feyerabend P (1975) Against method. New Left Books, London

Garretón MA, Naim B-KN (2013) Sociology in Latin America: does historical sociology exist? Global Trans Sociol (G&TS) Newslett 3:16–19

Go J, Lawson G (eds) (2017) Global historical sociology. Cambridge University Press, Cambridge

Goldthorpe JH (2000 [1991]) The uses of history in sociology: reflections on some recent trends. In: On sociology: numbers, narratives and the integration of research and theory. Oxford University Press, Oxford, pp 28–44

Goody J (1996) The east in the west. Cambridge University Press, Cambridge

Hegel GWF (1991 [1822]) The philosophy of history. Prometheus, Buffalo

Hughes J (2003) Learning to smoke: tobacco use in the west. University of Chicago Press, Chicago

Loyal S, Malešević S (2021) Classical sociological theory and contemporary sociological theory. Sage, New York

Mann M (1986) A history of power from the beginning to AD 1760. Cambridge University Press, Cambridge

Mann M (1994) In praise of macrosociology: a reply to Goldthorpe. Br J Sociol 45(1):39–52

Mayrl D, Wilson NH (2020) What do historical sociologist do all day? Analytic architectures in historical sociology. Am J Sociol 125(5):1345–1394

Mead GH (1934) Mind, self and society. University of Chicago Press, Chicago

Mead GH (1936) Movements of thought in the nineteenth century. University of Chicago Press, Chicago

Mennell S (1985) All manners of food: eating and taste in England and France from the middle ages to the present. Blackwell, Oxford

Mennell S (1990) The sociological study of history: institutions and social development. In: Bryant CGA, Becker HA (eds) What has sociology achieved? Palgrave Macmillan, London. https://doi.org/10.1007/978-1-349-20518-9_4

Mennell S (2015) Sociogenesis and psychogenesis: Norbert Elias's historical social psychology as a research tradition in comparative sociology. Comp Sociol 14(4):548–561. https://doi.org/10.1163/15691330-12341357

Morawska E (1989) Sociology and 'historical matters'. J Soc Hist 23:440–444

Nietzsche F (1983) On the uses and disadvantages of history for life: untimely meditations. Retrieved from http://leudar.com/library/On%20the%20Use%20and%20Abuse%20of%20History.pdf

Orloff AS, Clemens ES, Adams J (2005) Introduction: social theory, modernity and the three waves of historical sociology. In: Adams C, Orloff (eds) Remaking modernity: politics and processes in historical sociology. Duke University Press, Durham, pp 1–73

Parsons T (1937) The structure of social action. McGraw Hill, New York

Parsons T (1966) Societies: evolutionary and comparative perspectives. Prentice-Hall, Englewood Cliffs

Parsons T (1970) On building social system theory: a personal history. Daedalus 99(4):826–881

Parsons T (1971) The system of modern societies. Prentice-Hall, Englewood Cliffs

Pinker S (2011) The better angels of our nature. A history of violence and humanity. Penguin Books, New York

Skocpol T (1979) States and social revolutions: a comparative analysis of France, Russia, and China. Cambridge University Press, Cambridge

Skocpol T (ed) (1984) Vision and method in historical sociology. Cambridge University Press, Cambridge

Skovajsa M, Balon J (2017) Sociology in the Czech Republic: between East and West. Palgrave Macmillan, London

Smith D (1991) The rise of historical sociology. Temple University Press, Philadelphia

Sorokin P (1960) Mutual convergence of the United States and the U.S.S.R. to the mixed sociocultural type. Int J Comp Sociol 1(2):143–176

Sorokin P (1967) The Sociology of Revolution, New York: H. Fertig

Spengler O (1918–22) Der Untergang des Abendlandes. 2 vols. Beck, München [The decline of the west. Oxford UP, New York, 1991]

Steinmetz G (2007) The historical sociology of historical sociology: Germany and the United States in the twentieth century. Sociologia 1:3

Stone J, Mennell S (eds) (1980) Alexis de Tocqueville on democracy, revolution and society. University of Chicago Press, Chicago

Thierbach C, Hergesell J, Baur N (2020) Mixed methods research. In: Atkinson PA, Delamont S, Williams RA, Cernat A, Sakshaug JW (eds) SAGE research methods foundations. Sage, London/Thousand Oaks/New Delhi

Tilly C (1975) The formation of national states in Western Europe. Princeton University Press, Princeton

Tilly C (1978) From mobilization to revolution. Addison-Wesley, Indianapolis

Tilly C (1984) Big structures, large processes, huge comparisons. Sage, New York

Wallerstein I (2000) From sociology to historical social science: prospects and obstacles. Br J Sociol 51(1):25–35

Weber M (1930 [1904–5]) The protestant ethic and the spirit of capitalism, trans T Parsons. Scribner, New York

Weber M (1946) Bureaucracy. In: Gerth HH, Wright Mills C (eds) From Max Weber. Oxford University Press, New York, pp 196–244

Weber M (1951) The religion of China: Confucianism and Taoism. Free Press, New York

Wright Mills C (1959) The sociological imagination. Oxford University Press, New York

Power and Politics: State Formation in Historical Sociology

23

Helmut Kuzmics

Contents

Introduction	584
The State in European Social Thought: A History of Riddles and Perspectives	585
The Formation of Early States and Civilizations	591
Developmental Paths Toward the Early Modern European State	603
The Industrial Nation-State and Beyond	612
Conclusion	619
References	621

Abstract

This chapter first outlines the history of questions and research problems predating and constitutive of the historical sociology of state formation as it exists today. It deals with three main stages of state formation and the development of both internal and external power relations, starting with the joint process of the formation of early states and civilizations taking place in Neolithic societies, including the rise of empires. It then shifts focus to the diverse developmental paths toward the early modern European states, from more liberal to more absolutist states. Finally, theories are discussed that attempt to explain the rise of the industrial nation-state entangled in a worldwide web of interstate relations with the tendency to transcend nation-states by supranational associations of states. Since historical sociology assembles various approaches – Marxist, Weberian, structural functionalist, Eliasian, and others – this chapter characterizes them according to five criteria:

What is the theories' dominant problem or question related to the power aspect of the state? How far back do they follow the processes that are responsible for today's institutions, patterns, problems, or solutions? Do they predominantly treat the relevant processes in a comparative way, or do they rather study them as

H. Kuzmics (✉)
University of Graz, Graz, Österreich
e-mail: helmut.kuzmics@uni-graz.at

historical individuals on their own? What is the place they give to firsthand historical sources and their "emic" interpretations by their contemporaries, or do they put more emphasis on secondary, theoretically modeled "etic" interpretations? Is their historical, processual understanding based on the uniqueness of a specific development with universal relevance, as "universal history"?

Keywords

States · State formation processes · Marxist theories · Weberian theories · Structural functionalism · Neo-evolutionism · Empires

Introduction

Statehood can be seen as one of the many stages in the development of "survival units" (Elias 2010b [1987]: 137–208; Mennell 1990) of all humankind. The survival units on which people rely for their most basic needs (including protection against violence) and to which they feel the strongest bond have tended to become bigger throughout human history. Before states emerged, hunter-gatherer societies, horticulturalists, and early agrarian societies had evolved from small bands without institutionalized and hereditary leadership into big man societies, chiefdoms, and more complex, stratified, and interrelated tribal structures of dominance, embedded into broader networks of units that have been described by cultural anthropologists since the foundation of their discipline. Eventually, their work was complemented by archaeologists. Thus, "state formation" can refer to the development of the first states in human history, taking place even before the invention of writing, which can rightly be called "prehistoric." However and more conventionally in the specific context of Europe, historical sociology has been credited with the analysis of a quite particular process, namely, the formation of the modern territorial state from its origins in "feudalism," with the advantage of the availability of written sources and documents. Since it was the early modern state that laid the foundations for the later transition to the nation-state (often conceptualized as the industrial nation-state), it would also make sense to start here, since history can be defined by its reliance on written sources. We can, therefore, distinguish between at least three different meanings of "historical sociology": (a) as a sociological interpretation of history in this narrow sense, (b) as a broader, long-term processual analysis of human societies including prehistory, and (c), in the more specific understanding of the contemporary subdiscipline of "historical sociology" that was established from the 1960s, as a counter-reaction to the present-centered, "hodiecentric" (Goudsblom 1977) sociological paradigms, from structural functionalism and phenomenological approaches to symbolic interactionism and quantitative empirical studies. The latter is the definition used here. Finally, the nation-state of today – which must always be understood as embedded in a "system of states" – seems slowly to give way to even larger political structures with a supranational character. The concept of "state formation" must, therefore, be extended to deal with these new processes, as well.

This chapter will first outline the history of questions and research problems predating and constitutive of the historical sociology of today. It will then deal with three main stages of state formation and the development of both internal and external power relations, describing in section "The Formation of Early States and Civilizations" the joint process of the formation of early states and civilizations taking place in Neolithic societies, including the rise of empires. In a subsequent step (section "Developmental Paths Toward the Early Modern European State"), it shifts focus to the diverse developmental paths toward the early modern European states. They do not come alone, but are embedded in larger systems of states with a range from more liberal to more absolutist states. Finally, the discussion arrives (section "The Industrial Nation-State and Beyond") at the industrial nation-state entangled in a worldwide web of interstate relations with the tendency to transcend nation-states by what we may refer to as supranational associations of states.

The State in European Social Thought: A History of Riddles and Perspectives

We can understand the development of European thinking on state and state-related power by linking it both to the history of state formation and to the problems it posed to the various groups of people dealing with it. Their frequently normative theories depended on their social position – as an advisor of princes, a passionate citizen, or an intellectual figure siding with potential victims of state power who have become rebellious – that influenced their perspective as well. It is no mere coincidence that thinking about the state was often carried out by jurists since they represented the group most closely linked to the business of the state. But this was not always the case. Sometimes, philosophers with a formal training in medicine could also be involved, as in the cases of Michel Foucault and Norbert Elias. Their metaphors and insights would differ from those developed by specialists in legal thinking, such as Max Weber. In a comprehensive account of changing state ideals during the course of world history, Berber (1973) distinguished the following stages in this process: ancient civilizations, from their origins in the Neolithic period, in Egypt, Mesopotamia, Israel, China, and India; Greek state ideals from the city-state to Hellenism; the Roman idea of the state; early Christianity and the state; ancient Islamic notions of the state; the beginnings of Germanic concepts of the state; the struggle for the ideal state in the Middle Ages, referring to the conflict between Emperor and Pope; Reformation and its consequences; and the breakthrough to the modern ideas of the role of the state, circling around absolutism (Machiavelli, Bodin, Hobbes). Then came the English, American, and French revolutions and their impact on the respective theories of the state, from Locke, Hume, Bentham, Burke, Carlyle, Mill, and Spencer to intellectual figures of the twentieth century, while in France a development took place, from Montesquieu, Voltaire, Rousseau, and Robespierre to French philosophy of the nineteenth and twentieth centuries. In Germany, the path led from the state philosophers in service of princely states like Althusius (born as Johannes Althaus, 1557–1638; he was Calvinist and advisor of the German town of

Emden and favored an early contractualism, binding the prince to the people's will; Berber 1973: 366–9), Pufendorf (1632–1994, legal counselor in Sweden and Brandenburg and advocate of a moderate absolutism), and Leibniz (1646–1716, who saw the necessity of a theocratical, moral basis for his conception of the ideal absolutist state) to the German enlightenment tradition of Kant, turning to the radical subjectivism of Fichte and to the idea of reason as embodied in the state in Hegel. Finally, the state ideals of socialism follow, from the early Marx to the Russian revolution and beyond.

It can easily be seen that Plato as a citizen of a Greek city-state belonging to its upper strata was more interested in good governance by "the best" than in participation of the masses. Near permanent warfare between these states was accepted by him as natural as was the role played by a warrior elite. Medieval theories of the distribution of power between the church and the secular ruler focused on whose right it was to rule – the Emperor's or the Pope's. Legitimacy was central. Both parties in this struggle had their advocates. When we are approaching the formation of "the" state in its narrower, modern meaning (as a territorial unit under the monopoly of violence), different questions arise that have remained relevant until today: Why a central government at all? (In order to protect life and property.) How can states survive in the competition with other states? What are the best strategies a ruler has to follow in order to secure his authority? These questions were asked during the process that gave birth to the first really modern states and state apparatuses in European history: for instance, in England of Thomas More, Thomas Cromwell, and Thomas Hobbes when the late feudal feuds of the Wars of the Roses had ended in the new strength of central state power – throughout its history, England was perhaps never closer to absolutism than under Henry VIII – and in the Italian city republics of the fifteenth to sixteenth centuries when the concept of the state of the republic or princedom (*lo stato* meaning the changing state of affairs of a *civitas*, as Hobbes used the latter term in the seventeenth century; see Elias 2009a [1985], on More's *Utopia*) became an expression for the whole. The issue centered on how to best organize the resources of a territory in order to strengthen its influence in competition with other princely states (in French mercantilism and/or German cameralism). During the long and winding path of Enlightenment, the problem arose of how to protect the interests of the working bourgeois against unrestrained and unchecked power of the state embodied in the monarch and the church. The experience of the Industrial Revolution led the impoverished working classes to be aware that the state could be seen as an instrument of power directed against them and so forth.

The development of European thinking about the state and state power was also very much an affair of various national lines of thinking, in Italy, France, England, and Germany (similar to the national differences in the relationship between the established discipline of "history" and the emergent field of "sociology"). For views of the state, Jonas (1976) delineated a French path of Enlightenment thinking where the subject of "emancipation" was central from Montesquieu, the Physiocrats, and the Encyclopedists to Rousseau. English and Scottish liberalism focused on the ideas of the contract and property from Hobbes to the acknowledgment of a "Civil

Society" and the market as non-intentionally evolved institutions in Locke, Smith, and Ferguson and from here to the more conservative turn of Burke and Malthus. The German contribution was, according to Jonas, "German idealism," influenced by French and British ideas, replacing older notions and ideals of the state (Althusius' early commitment to democratic control of the monarch, Pufendorf's vision of an absolutism tempered by natural law) by referring, in Kant and later Fichte, to individual reason as opposed in contradiction to all human structures, in particular that of the state which can only be solved by linking the state to the law. Jonas sees the completion of this line of thinking in Hegel's idea of the state as the embodiment of revolutionary, emancipatory reason by way of a Civil Society embedded in the (Prussian) state. Turning to the theories of socialism and industrial society of the nineteenth century, Jonas manages to show how they built on the various national roots in their thinking: Marx combined French socialism, German idealism, and British liberal economics. There is a German line of thinking leading from romanticism to the historicism of Droysen (in the older *Historische Schule der deutschen Nationalökonomie* with Roscher and Knies, Jonas 1976: 288) and to the works of Schmoller and Sombart, until Weber tried to bridge the gap between the historical school and the economics of Menger, whereas in England liberalism took the form of utilitarianism. The more radical French line of emancipatory thinking leads, in the age of industrialization, to the socialism of Fourier and Babeuf. But sociology originated in France from the positivism of Comte to the empiricism of Durkheim and his school.

The British line of utilitarian thinking advances, according to Jonas, from liberalism to the social Darwinism of Spencer and his followers. But there was also a socialist strand that had started with the levelers and diggers (one of their proponents, Gerrard Winstanley in his *The Law of Freedom in a Platform*, published in 1652, formulated the right to cooperative work as something naturally given; the diggers claimed free access to land) in the seventeenth century and which proceeded to Paine, Thompson, and Owen (Jonas 1976: 190–9). According to Lepenies (1985), the development of sociology as a scientific discipline can be located between the paradigms of natural science, literature, and history, with different emphasis and weights placed on the respective pole in France, England, and Germany. As Delanty and Isin (2003) also note, in Germany, the creation of sociology as a discipline was always linked to history, in contrast to France and Britain. And it was also Germany and Austria-Hungary where the focus of historical and institutionalist *Staatswissenschaft* had for a long time been on the state. In a major volume on sociology and its neighboring disciplines in the Habsburg Monarchy before 1918, Acham (2019: 221–400) distinguished six different roots for the formation of sociology: social reform and social policy; statistics, social geography, and ethnography; physics and biology; economics, law, and *Staatswissenschaften*; history, historical anthropology, and history of art; and psychology and philosophy. A short list of central figures contains the names of Lorenz von Stein, Albert Schäffle, Ludwig Gumplowicz, and Gustav Ratzenhofer as founding fathers of sociology in Austria-Hungary. They were, with the exception of Wilhelm Jerusalem, all linked to the *Staatswissenschaften* (political economy with a focus on *öffentliches Recht*, that is, public law, as opposed to private law or *Privatrecht*). This was also Max Weber's background, so it is, therefore, no accident that Max Weber's approach to exploring the causes of the rise of occidental rationalism (as an approximation to "modernity") concentrates heavily on "bureaucratic domination" and the rational state, replacing patriarchal and patrimonial domination as one of its main achievements. (Von Stein saw the state as mainly responsible for social policies that were able to correct inequality and risks generated by the market in a kind of state socialism, while Gumplowicz and Ratzenhofer argued for the state as the only means to control anarchic struggle, although they saw the origin of the state as precisely the product of such ethnic struggles.) The list of historical sociologists of the state (in all but name) in the Central European tradition is long (Max and Alfred Weber, Werner Sombart, Emil Lederer, Edmund Bernatzik, Norbert Elias, Franz Oppenheimer, Otto Hintze, Carl Schmitt, Karl Polanyi, and many others). It is not possible here to give even a rough sketch of their contributions, with the exception of those famous figures who have found recognition in the circle of "historical sociologists" established in an attempt to "de-Parsonise" Weberian sociology (Skocpol 1984: 4) in the late 1970s: Max Weber, Marx, Polanyi, and Elias.

It was precisely the American situation of an equilibrium-centered timeless structural functional systems theory (Parsons 1951) that had dominated (now also as a modernization theory with the focus on differentiation and integration) sociological thinking not only in the United States but also in other regions of the Western world until the 1970s. Their manifold political crises (Cold War, Vietnam, student unrest) were discussed and felt by an intellectual, sociologically educated elite. The topics of class and inequality reentered the sociological discourse and, with them, a reorientation toward Marxist historical materialism. However, at the same time, as Skocpol rightfully maintains, the experiences of "real socialism" discredited a naive understanding of Marxism and led to a focus on a broader and deeper historical-sociological interest in the formative processes of "modernity" including the perspectives of a whole range of classical authors, in order to better understand the trajectories and different paths to a global, quite heterogeneous present. What were the questions – or riddles – that were thought to be answered by a renewed historical sociology? The number one problem was certainly capitalist commercialization and industrialization: How to explain their roots? Why was it the European west that generated both? And why was it linked to so much inequality? What about the political conflicts related to this development? What about the fate of moral values? Would modern industrial society morally collapse or, in contrast, even be able to create new forms of solidarity, in the face of the history of social development having taken place so far? Given its previous history, how would European expansion influence the further path of global development? The agenda set by Philip Abrams in his *Historical Sociology* (1982) was even broader: he treated "sociology as history" by identifying the diachronic perspective in classical sociological thinking (Durkheim, Marx, Weber), in the areas of industrialization and anomie, class and state formation, rationalization, and development of individual identities. Dennis Smith (1991b) distinguished three phases of postwar historical sociology, (1) starting in the 1950s with a focus on social change from structural functionalist sociology

(Parsons and Smelser; Eisenstadt and Lipset), T. H. Marshall's (1950) *Citizenship and Social Class*, and the Weberian view of Bendix, proceeding to (2) in the 1970s when Bloch and Elias' writings on feudal society and the civilizing process were rediscovered and Marxism had turned to the questions of injustice and domination (Moore 1966 on the *Social Origins of Dictatorship and Democracy* and Thompson 1963 with *The Making of the English Working Class*), also with a focus on states and social revolutions (Skocpol) and with Tilly's *Coercion, Capital, and European States, AD 990–1992*, including two Marxian perspectives – Perry Anderson from ancient Greece to absolutist monarchy and Wallerstein on the capitalist world economy – Smith's inclusion of Fernand Braudel's (1979a,b) comprehensive studies of the longue durée of Mediterranean civilization and its rise to capitalism, and Michael Mann's (1986) *The Sources of Social Power* which covers a period of 7,000 years. Runciman and Giddens are discussed as theoreticians. The last stage (3) is historical sociology in the 1990s and its sophisticated discourse on exploration, generalization, and strategies of explanation around the development of democracy and capitalism.

In their introduction to a handbook, Delanty and Isin (2003) also stressed the riddles of the present (the crisis of modernity and of capitalism, the role of the state and of revolutionary movements) that gave rise to the enterprise of historical sociology, and like Skocpol, they locate its origins in the United States: although some of its important roots lie in German thinking, a German "historical sociology" was never coined as a term since sociology as a whole understood itself to be historical. (Delanty and Isin draw a similar conclusion with French historiography.) They argue, therefore, that it was an Anglo-American development of the postwar era that led to its explicit constitution as reaction to a foregoing retreat from history – using words similar to those of Norbert Elias in his article (Elias 2009b [1987]) on "The retreat of sociologists into the present," postulating a return to the processual perspective. Delanty and Isin also see the godfathers of this renewed historical sociology in Marx, Weber, and the *Annales* school, but they list many additional disciplines, paradigms, and names they regard as relevant – from social history to modernization theory, from a highly theoretical "Grand Historical Sociology" that only uses historian's knowledge for sociological purposes to blends of a (new) history of knowledge, death and dying, childhood, health, everyday life, and pop culture. The list contains the names of many famous historians of the Middle Ages as well as those of representatives of "postcolonial studies" (against Eurocentrism), "cultural turn" (Foucault, Hayden White, Habermas, and Gellner), cultural memory studies, and the history of ideas.

Even more pronounced is the viewpoint of Hobden and Hobson (2002) who edited a book with the intention of bringing historical sociology into the field of international relations, a branch of political science. In his introduction, Hobson (2002) argues most vehemently against what he calls "chronofetishism" and "tempocentrism" as an underlying, implicit condition of a whole range of approaches dealing with the power relationship between political units, in particular states that always emerge in systems of such units, be they city-states, tribes, empires, or nation-states. He means it is necessary to overcome three illusions, namely, that of (1) the reification of a present artificially separated from the past, using static categories, treating the current state of affairs as self-constitutive, autonomous, and lacking historical context; (2) the illusion of naturalization when the present is regarded as a spontaneous emanation of ongoing forces without recurring to the process of power formation, of the making of identities and norms; and (3) the illusion of unchangeability, when the present is seen as eternally the same and resistant to any change. The major culprit is the school of "defensive realists," with its most pronounced spokesman Kenneth Waltz (but one could also easily include the "offensive realism" of John Mearsheimer). In particular, it is the Westphalian peace settlement of 1648 which is taken as a kind of eternal norm for the regulation of hegemonic struggles through and after war; or the idea that great power rivalry is something "natural" and essentially the same from the conflict between Athens and Sparta to the Cold War between the USSR and the United States, with the implication that attempts to overcome these conflicts by peaceful means are futile.

Keeping all this in mind, several questions can be posed that can enable us to classify or at least discuss the historical sociology of power and state formation along five lines:

a) What is the dominant problem or question related to the power aspect of the state demanding a historical-sociological answer? It may, of course, vary over time as well.
b) How far back should we follow the processes that are responsible for today's institutions, patterns, problems, or solutions?
c) Do we have to analyze the relevant processes in a comparative way, or is it sufficient to study them as historical individuals on their own?
d) Must we turn to historical sources and their "emic" interpretations by their contemporaries, or can we rely on secondary, theoretically modeled "etic" interpretations of interesting structures, experiences, and events?
e) Is our historical, processual understanding of the development of state and state-related power based on historical parallels between different periods, or is it rather based on the uniqueness of a specific development, albeit with more general or even universal relevance?

Since in this chapter the focus is on the understanding of the contemporary subdiscipline of "historical sociology" established since the 1970s as an Anglo-American counter-reaction to then prevalent "hodiecentric" sociological paradigms (while classical sociology had indeed always been historical), the following sections will deal with the research agendas of their main directions plus an open, residual category of relevant works shedding light on the issues at hand. The selection of approaches comprises the structural functionalist school turned neo-evolutionist (Parsons, Eisenstadt), variants of Marxist and post-Marxist thinking (Wallerstein, Perry Anderson, Barrington Moore), Weberian historical-comparative analyses (Weber, Bendix, Mann), the Annales school of social historians (from Bloch to Braudel), and finally the process-sociological approach founded by Elias and his theory of civilizing processes. In addition, there are contributions of authors who cannot easily be classified as members of any of these schools – for example, Mumford, Pinker, and others.

The Formation of Early States and Civilizations

The most common, current definition of a state is Weber's, as "an organization which successfully upholds a claim to binding rule-making over a territory, by virtue of commanding a monopoly of the legitimate use of violence" (Weber 1978 [1920]: I, 54). This power is exercised both within this space and also against enemies from without. Modern states may be highly bureaucratized, industrialized, and ethnically/ linguistically homogeneous nation-states. However, these qualities are the product of later developments and not characteristic of the "first" states in history and prehistory, but even the criterion of the "monopoly of (legitimate) violence" does not always fit. Early states and civilizations are the subject of several scientific disciplines; archaeology, history, cultural anthropology, and linguistics, and even, as in the case of Steven Pinker's historical psychology of violence (Pinker 2011), of a research tradition that includes a lot of neuroscience. The explanatory goals of a historical sociology dealing with such remote times differ from most of these approaches since they lie in the present: Marxists wanted to know about the path leading from slave labor societies to exploitative feudal structures of serfdom and capitalist domination over the working classes in industrial societies; or with the apparently separate path leading to the oriental mode of production (Wittfogel 1977 [1957]). Weberians have always been interested in the preconditions for the rise of rationality and the West, including rational, legal, bureaucratic domination. In theories of superimposition and military conquest, early states were held to be imposed by mounted warriors on Neolithic farmers (Rüstow 1950; similar views could be found in Gumplowicz, Ratzenhofer, Small, Ward, etc.). However, cultural anthropologists argue (Harris 1983) that even in relatively egalitarian hunter-gatherer societies, the functions of violence control and internal pacification can be served without formal structures of authority being imposed from outside: institutions of leadership with coercive powers do not exist, but their function is performed by informal sanctions and shaming, largely the effect of public opinion. Weakly institutionalized leadership, chiefs, and shamans, none of them possesses the coercive powers that are – since the invention of states – thought to be necessary in order to enforce conformist behavior and control deviance. Malinowski's studies on the Trobriand Islands (Malinowski 1966 [1935]) referred to a horticulturalist society and reported a low degree of role differentiation between leaders with magical and pragmatic knowledge and their ability to guide and control the people largely by nonviolent means. Although most hunter-gatherer and horticultural societies are not observed to be engaged in overt "war," most of them are now believed to be involved in regular inter-group violence with small numbers of victims in absolute terms, but quite substantial proportions of people killed in combat if calculated as percentages (Pinker 2011: 47–55, who refers both to archaeologists and anthropologists for statistical estimates, drawing, among others, on Keely 1996). Nevertheless, the rise of first states in Mesopotamia, Egypt, the Indus Valley, and Mesoamerica and South America seems to have radically transformed structures not only that of the "mode of production" but also that of domination within and between these societies. Not every historical-sociological author or school has dealt with this stage in the developmental process of mankind. Attention here is drawn to the following: the neo-evolutionist thinking of Parsons and Eisenstadt, Weber's programmatic remarks and the approach of Michael Mann; Marxist approaches; and the grand designs of Lewis Mumford and Steven Pinker.

Let us start with Talcott Parsons. His systems theory (Parsons 1951) is the very epitome of a static and ahistorical approach and, together with the mainstream of American empirical sociology, can be held accountable for the retreat of sociology to the present. It is quite surprising that the third and last phase of his work saw a return to exactly that kind of evolutionism he had renounced with the words, quoting Crane Brinton, "Who now reads Spencer?," when he gave priority to his voluntaristic theory of action in his famous book (Parsons 1937: 3). Although he had become familiar with Max Weber, whose large shadow still loomed over Heidelberg when Parsons was there during the 1920s, Parsons had criticized his methodology (Weber's polemic against Roscher and Knies) as still too close to German *Kulturwissenschaft* and historicism and too distant from the lawlike model of economics in the style of the natural sciences. He had favored an "analytical" understanding of what he saw as Weber's one-sided interpretation of the function of laws in historical sciences stressing the "particular" at the expense of the general both in natural sciences, like geology, and in social sciences, like history and anthropology. According to Parsons, even if we want to explain the historically given specific formation of a landscape, the role of physics or biology for an adequate explanation is indispensable, and their role as "analytical sciences" can also serve as a model for theoretical sociology. In *The Structure of Social Action* (the first phase of Parsons' thinking), he developed a vocabulary to express these quite a priori, "lawlike" categories for his voluntaristic theory of action. It is exactly here where Friedrich Tenbruck (Tenbruck and Homann 2002) saw Weber to be misunderstood: for Weber, it was the general *Kulturbedeutung* of a phenomenon like the money economy or occidental rationalism which would lead to questions referring to their *uniqueness* in their character (the *So-Geworden-Sein*), but with consequences for most of all humankind and with lawlike generalizations as part of the causal explanation as mere auxiliar constructions (necessary but *trivial* as information). This is the meaning of historical sociology as *Wirklichkeitswissenschaft* (literally, reality science), a term borrowed from the older German historical school, but with a different shift of orientation. In the second phase of Parsons' work, the notion of a social system with the inbuilt tendency to equilibria and a central explanatory role for "values" complemented his theory of action, much criticized by C. Wright Mills, Alvin W. Gouldner, Melvin Tumin, Ralph Dahrendorf (listed in Holton/Turner 1986), and Elias (2009b [1987]), partly also because of its neglect of processes. So Parsons' return (third phase, Parsons 1966, 1971) to (neo)evolutionism in the 1960s might have come as a surprise. It is remarkable for at least two reasons: First, it is a return to Weber's historically comparative analysis of the path to occidental rationalism by focusing on ancient Egypt and Mesopotamia, China, India, Islam, Rome, Israel, and Greece (Vol. 1 includes "primitive" societies as a starting point, like Australian aborigines, and widens the scope by referring to African kingdoms). Second, Parsons adds to the abstract, though dynamized categories of the AGIL classification (processes of social development with four dimensions: inclusion, value generalization, differentiation, and adaptive upgrading), the evolutionist concepts of "seedbed societies" (innovation analogous to mutation takes place in various cultures), and the idea of selection at the level of cultural codes (variations can be chosen and stabilized according to the evolutionary gains on the highest cybernetic level of information). Using power can, therefore, mean to communicate a decision to the parties involved. The implications of these decisions will bind their collectives and their members. Parsons thought that we can distinguish between "primitive" or "advanced primitive" societies and later "intermediate" or "modern" societies according to their level of complexity (functional differentiation) and integration by pointing to the relationship between the social and the cultural system. The latter's internal dynamics in the direction of cumulative growth was enabled through the invention of writing. Written tradition is characteristic for the progression from "primitive" to "intermediate" societies, in particular if it is followed by the distribution of the ability to read and write among the male members of an upper class. A greater distance between "gods' and the human situation in a formal religion, as is the case in the "historical religions," would be typical for an "intermediate" society. The separation of an economic sphere of money and markets from technology and kinship (Greece was more advanced than Mesopotamia) would mean a more advanced level of it. What does this mean for the development of the state and the related structures of domination? In the Orient, early states embedded in a system of states were characterized through two-class divisions, and only the upper classes (often as priests and literates) were part of the positively valued societal community; whereas, by contrast, in the Occident, citizenship would later be universally extendable, because a world-oriented societal community could develop without this restriction. According to Parsons, this happened in a most significant way in Rome, but progression to this (premodern) stage was already made in Israel and classical Greece. Israel developed a societal community which did not necessarily depend on any concrete political-territorial unit, and Parsons explains this by the special covenant of Yahweh with his people. In Greek city-states, laws favored the rise of a societal community of equal citizens and also formed the basis for the right to citizenship in the Roman Empire. Therefore, crucial developments in such "seedbed" societies came to determine the future of "modern" forms of social systems. The violence aspect does not figure very prominently in these considerations, and since the area of politics was always treated by Parsons as serving the functions of goal attainment (and protection), this was not surprising. Parsons was a long way from doing historical research himself and relied on secondary or even tertiary interpretations according to his theoretical aims. Whereas historians of ancient Egypt, like Jan Assmann (2000), dealt with the religious legitimation of pharaonic and priestly rule and its role of preserving a very conservative cultural memory by studying hieroglyphic texts to gain an "emic" understanding, Parsons was certainly heavily lopsided toward the "etic" perspective of his neo-Kantian approach to moral development. The distinction between "emic" and "etic" was coined by linguist Kenneth Pike (1967 [1954–1960]) who saw "phonetic," related to sound, as different from "phonemic," related to meaning. From there, the terms entered the field of cultural anthropology: the emic perspective enabled the understanding of culturally specific beliefs and practices, whereas the etic perspective was thought to capture comprehensive information from outside, often of a theoretical kind (Harris 1976). The notion of an emic perspective is also similar to the everyday life view of actors in their life world, interpreted in the dialectics of "Selbst" – and "Fremdverstehen" according to the phenomenological sociology of Schütz (1932).

It was Shmuel N. Eisenstadt who took the sociological, theoretical research on early civilizations to a new level with his *The Political Systems of Empires* (1963). Although he remained within the confines of the structural functional systems approach, his analysis was both more comprehensive and concrete. He distanced himself from the lawlike qualities of the Parsonian AGIL scheme and the causal relevance of an evolutionism based upon it (Hamilton 1984: 90) and saw himself explicitly in the tradition of Max Weber, particularly of his historical-comparative work, but not in line with Bendix who, he thought, erred too much to the side of historicism. For an answer to the two criteria mentioned above (What is the dominant problem or question related to the power aspect of the state that demands a historical-sociological answer? And how far back do we have to trace the process that is responsible for today's institutions, patterns, problems, or solutions?), Eisenstadt goes far back indeed, in a comparative way, by turning to bureaucratic empires like Egypt, China, Rome, and Byzantium, but also to more recent absolutist states in Europe, including Russia and the Ottoman Empire. The overall goal of this research is to find out which conditions lead either to democratic or to dictatorial political regimes with implications for the present. The method is twofold: Eisenstadt seeks to describe specific developmental paths, but he also sees parallels between societies in quite different periods of time. Empires are characterized as centralized political structures with bureaucratic administration and institutionalized channels regulating conflicts between elites and other social groups, in particular landowners and burghers, tradesmen, and religious specialists, but also between the latter themselves. Empires rise because rulers pursue their own goals, often in open resistance to tradition and relying on their charismatic qualities. Social differentiation allows the exploitation of new resources which contribute to the strengthening of central power (this mechanism strongly resembles the notion of a "monopoly mechanism" which was used by Norbert Elias to understand the transition from feudalism to absolutism in France and much of Europe; Elias 2012a [1939]). Empire formation takes place in situations of transition; Eisenstadt develops a rich classificatory typology in a quite formal way, focusing on endogenous driving forces and less on war. This is also the case in "patterns of exceptional change" (Eisenstadt 1978), for example, in the processes taking place in Greek and Roman city-states or around the tribal societies and their transition toward Judaism and Islam. Here, revolutionary political conflicts within these entities shape new political orders, but they still differ from modern revolutions (toward the ideals of equality, freedom, and participation) which create a totally different, new type of society.

The second major line of historical sociology has its roots in Marx and various Marxisms. The main obstacle to every Marxist interpretation of state-related power and their premodern history has always been that Marx himself saw in his most famous work, *Das Kapital*, the state exclusively as an agent of the ruling class to facilitate the control of the exploited working classes. The economic base demands a specific type of legal and political superstructure. Of course it is necessary to correct this impression by pointing to the complex layers of Marx's own intellectual development in the tradition of the German understanding of *Rechts- und Staatswissenschaften* that preceded the "anatomy of the bourgeois society" as a critique of industrial capitalism and that which followed it in his later political writings on the French restoration period of Napoleon III (Kelly 2003). However, as Shaw (2002) has argued, it is precisely in Marx's framework that the importance of technology and economy is overrated at the expense of military and political factors of globalization if one wants to use the classics for the purpose of diagnosing some state-related problems of contemporary societies. Another influential perspective on the sociogenesis of the state can be found in Engels' (2014 [1884]) treatise on *The Origin of the Family, Private Property and the State*. Here, the state (early Athens) was the late product of a long-term development of kinship, toward gentile constitutions and patriarchal monogamy, that came to be thoroughly shattered by a rising money economy and was consequently replaced by a territorially organized authority needed to regulate individual property. Engels, too, explained the state in terms of endogenous, economic factors. The great exception is the topic of "oriental despotism" and its relationship to the "Asiatic mode of production" as the fourth type that does not fit in with the path from ancient slaveholders to feudal serfdom and capitalist exploitation. Marx had come to acknowledge this different path, influenced, according to Wittfogel (1977 [1957]: 28), by J. S. Mill and Richard Jones, as the dominant form of economic development in India, China, Ancient Egypt, and Russia, where the state had become the ultimate owner of land as agrarian property. Wittfogel not only showed in great detail how the "hydraulic" economy prevented class struggle between the two main classes according to the criterion of property, replaced by the nearly total power of the state and its bureaucracy (a "monopoly bureaucracy"), but he was also more open to the military codeterminants and their consequences for the social stratification of "conquest societies" (with some remarks referring to Rüstow, Gumplowicz, and Oppenheimer). Since Lenin also shared the view of Tsarist Russia as a half-Asiatic, despotic system, but did not hesitate to try to develop socialism there, it was, in Wittfogel's interpretation, the key to an understanding of Russia's dictatorial, totalitarian turn to point to its despotic past.

The consequences for a Marxist (or post-Marxist) historical sociology of early state formation are manifold. First, the ultimate political goal to further the emancipation of the oppressed lower (working) classes (P. Anderson, Wallerstein, B. Moore, and E. P. Thompson were also political activists) led some – but not all – to develop models for very long-term processes and to refer to Greece and Rome

(Anderson) or to oriental empires (Wittfogel). Michael Mann with his more Weberian self-understanding also turned to the first states in history (not so Wallerstein, Moore, and Thompson, who chose more recent starting points). Second, these attempts had to come to terms with the limitations of Marx's most popular ideas: their endogenous and economistic character and their neglect of interstate rivalry and military, exogenous forces. The 24th chapter of the first volume of *Das Kapital* treats the lords of the manor who are responsible for the enclosures of the Middle Ages not as warriors but as early businessmen (or robber barons) and the central institutions of the monarchy as their instrument. Third, the subject of emancipation of the oppressed changed subtly toward the agenda of democracy. Fourth, a comparative perspective was needed. Fifth and final, the analyses had to come to terms with the transition of capitalist exploitation from the absolute immiseration of the nineteenth-century European working class to its new role as a kind of much better-off imperialist, colonialist worker aristocracy facing the paupers of a more global kind.

Perry Anderson's *Passages from Antiquity to Feudalism* (1974) does not begin with the oriental empires, though it follows some of their traces in the analysis of Hellenism and in the chapter on the Ottoman Empire of the companion volume on the *Lineages of the Absolutist State* (1979). The whole project aims to understand the very different state structures on the eve of industrial society in Europe: Roman antiquity plus Germanic-Celtic institutions create state and democracy in the West, while Eastern Roman antiquity plus Slavic elements lead to Eastern types of absolutism. The enterprise is, therefore, a comparative one, but Anderson gives sufficient emphasis to both pathways to see them as historical individuals on their own. It is rich in institutional detail (relying on the work of, among others, Finley, A. H. M. Jones, Weber, Andrewes, and Vernant; Anderson 1974: 18–28). It uses mostly secondary, theoretically modeled "etic" interpretations of other writers and relies less on the "emic" interpretations of contemporaries; these can be found, for example, in the work of Vernant (1955) who reconstructs the semantics of Greek expressions for types of work in a very detailed manner.

Anderson's basic argument is that the Greek city-states and the Roman Empire were in a unique way tied to slavery which was the most central cause of their rise and fall, much more so than in the ancient empires of the Middle East that were based on alluvial agriculture. Paradoxically and approaching the solution already found by Weber, it was the greater reliance on slave labor in the Western part of the Roman Empire than in the Hellenized East (with its state property of land and a reduced role for slave labor) which resulted in its earlier decay and final downfall. Political democracy in the Greek city-states was always limited to the community of free warriors, and thus the state was a matter of a minority. The lack of a larger territory (coast-bound as they were) provided no stimulus for the creation of a bureaucratic state apparatus, although this was to change with the transformation of the Roman city republic into an Empire, until its Western part collapsed under the weight of a superstructure that could no longer be carried by an agrarian economy with productivity at levels that were permanently too low. Again, in similar fashion to Weber, Anderson sees the establishment of the late Roman coercive state as an obstacle not only to capitalistic economic progress but also to the defensive power of the army. Administration became too expensive, and the loyalty of the senatorial nobility sank, while the burden of the clerical apparatus of the Christian church, when it had become the state religion, grew too heavy to shoulder. The main culprit in Anderson's account is the spread of the system of latifundia, particularly in the West (Gaul), which became unproductive since Rome had stopped its slave wars of conquest. As Weber (1988) had already mentioned in his essay of 1896, the institution of lease-holding *coloni* who were poor but free to start a family and to work for themselves replaced the former slave barracks. However, the class of immensely rich landowners would soon have to share power with the Germanic invaders, and this already marked the transition to feudalism.

Anderson's account of the development of the state and state-related power is here not so much based on historical parallels between different periods than it is on the uniqueness of a specific development. Its general or universal relevance lies in the future of European feudalism as both a political and an economic precondition for the Western absolutist state, which is not oriental despotism but instead a system mixed with aristocratic and town-based democratic participation. War does not figure very prominently, although it is not omitted completely, but Anderson's approach deals with the interrelational aspect of state development and does not reduce it to the result of mere endogenous forces. Religion, in particular its emotional aspect (Greeks sacrificed hecatombs of bulls before commencing a battle, for instance), the control of which can be seen as a central power source, is not given a central role. The same is true for the acknowledgment of emotions in general.

The third branch of the renewed interest in historical sociology is explicitly linked to the work of Max Weber (as we have seen, he also tacitly influenced Parsons and Anderson). If we want to know what he saw as the dominant historical-sociological question related to the power aspect of the state, we should note the fact that it certainly varied over time. From the beginning of his career, one main object of his research was to reflect on the origins of the Roman legal notion of individual property, a path that would lead him to the study of capitalism. Another point of departure was the comparative study of ancient and medieval towns and the shift from the military and political autonomy of the Mediterranean *polis* as city-state to the militarily impotent, but economically highly competitive and politically partly autonomous urban settlement of the European Middle Ages. Grappling with the question of the causes of the fall of the Western Roman Empire and the focus on feudalism, war and its influence on structures of political domination also became important. However, it was by broadening his access to the role of religion in economic development (1904/1905) in the direction of a comparative study of world religions that Weber came to a comparative understanding of the ancient empires (from 1915 to 1918). In his late *Vorbemerkungen zu den Gesammelten Aufsätzen zur Religionssoziologie* (Weber 1992 [1920], translated as *Author's Introduction*, the last essay written in his own hand), the overall explanatory goal is to understand the genesis of "occidental rationalism," and this is also the aspect under which he analyzes the modern state – as a "rational, bureaucratic" form of domination against earlier, less rational structures (patriarchal, patrimonial). In Chapter X of

Economy and Society: An Outline of Interpretive Sociology, his "sociology of domination" (Weber 1978 [1920]) deals with the different pathways to the state in the broadest possible way. Thus, Weber's process sociological understanding develops different points of historical departure: sometimes, he goes back to the fifteenth and sixteenth centuries (in the case of the Protestant ethic), and sometimes, he follows the traces back to the Roman Republic and in other cases to the palace economies of Ancient Egypt or China. Corresponding to the different types of rationalization in the religious sphere (with consequences for the development of a rationally controlled *Erwerbsstreben*) – *rational world mastery* against rational *adaptation* to (*Weltanpassung*) or rational *denial* of the world (*Weltablehnung*; literally: *religious rejections of the world*) – there are also different tracks toward legitimate domination via the state. There is always a strain between various sources of power (charisma of the chieftain, the warrior king, or the priest, the force of tradition, or the bureaucratic legal state apparatus), with manifold types of patriarchalism and patrimonialism and transitional stages like the patrimonial bureaucracies which could already be found in the ancient empires of the Middle East.

How comparative is Weber's method, or is it sufficient to study these subjects as historical individuals on their own? Weber's own words have become famous and point to the direction of the historically singular:

> A child of modern European civilization (*Kulturwelt*) who studies problems of universal history shall inevitably and justifiably raise the question (*Fragestellung*): what combination of circumstances have led to the fact that in the West, and here only, cultural phenomena have appeared which – at least as *we* like to think – came to have *universal* significance and validity. [Weber 1920/1992, 13: translation amended]

However, as he noticed in his methodological works (*Objektivität, Verstehen, Sinn der Wertfreiheit*), this does not imply ignoring laws: only by identifying them is individual causal attribution possible. And the more complex Weber's thinking became, the more comparative his view, until the similarities between different patterns led to nontrivial classifications of several cases of a potentially open character, which is typical for laws.

In terms of historical sources, Weber's preference was largely "secondary" (he relied on the expert knowledge of historians and anthropologists of his time; some of it may today be seen as Eurocentric). He certainly tried to be as close as possible to the "emic" interpretations of the historical actors involved and to consider not only their rationality (whose understanding privileges the cultural scientist above the natural researcher whose objects neither behave rationally nor can tell about their motives) but also their emotions and all social structures with inbuilt, frozen emotions. However, the selection of meaningful data from an ocean of available, senseless information had always to follow theoretical modeling (*ideal types*) under the premise of an explanatory agenda and the criteria of relevance according to the general *Kulturbedeutung* of the phenomena in question. The complete logic of sociological explanation combines the postulate that it is necessary both to *understand* individual action (as long as it is possible; unconscious forms of social habitus and spontaneous affects cannot be understood as well as reflected ones) and to *causally explain* it by also taking into account the enabling and limiting nature of material and symbolic forces. Max Weber is additionally a theoretician of historically developing types of social habitus; in the case of the Protestant ethic, the behavioral aspects can be explained by deep-seated emotional controls.

Weber's historical and processual understanding of the development of early states and state-related power was based on historical parallels between different periods as much as it was on parallels between different places. Therefore, for instance, he saw patrimonial bureaucracies as entrepreneurs of a state socialist orientation in Egypt of the pharaohs (Weber 1978: 1044–7), the Chinese, the late Roman Empire, and in the Byzantine Empire, but also in patrimonial European states like Austria, Prussia, and Russia (Weber 1978: 1098). Charisma can be found in warriors, prophets, and democratic leaders and can also be "plutocratized" in a money economy (ibid.: 1121–2). Therefore, on the one hand, it is more or less an ahistorical category, but, on the other, unique elements characterize China's bureaucracy because it lacks any feudal power (ibid.: 1145) and is highly devoted to magical thinking. Apart from this, the Chinese mandarin is still quite far from being genuinely rationally, professionally educated and trained. Unlike in the works of Karl Marx, Weber's interest in state-related forms of domination is not politically motivated by the struggle for emancipation and liberation of the working class, but he was committed to the idea of liberty and political participation of the working bourgeois type of citizen. Neither "caesaropapism" nor an all-suffocating state bureaucracy matched his own political ideals.

A historical sociologist who sees himself in the line of thought that started with Max Weber is Michael Mann. The first of his four volumes *The Sources of Social Power* deals explicitly with the beginnings of civilization during the extended "Neolithic Revolution." Distinguishing four main sources of social power (economic, military, ideological, and political) and distancing himself from teleological and harmonizing, functionalist neo-evolutionism, he tries to sketch the important milestones of early state development by describing how prehistoric peoples who managed to "evade power" turned into "caged" systems of successful agriculturalists and pastoralists whose population had grown so irresistibly that it was no longer possible to escape to empty spaces if they came under external pressure. The relatively egalitarian structure of hunter-gatherer or horticultural societies gave way to that of territorially bound and stratified early states – a process that accelerated in irrigation cultures like Mesopotamia, Egypt, China, and the Indus Valley (less so in America). Mann discusses both theories of superimposition by conquest and Wittfogel's hydraulic approach, but he does not find them satisfying enough to explain the two related phenomena of rising "quasi-private property and the state" (Mann 1986: 82). In the materialist tradition, he points to the economy of such societies, but he nevertheless directs attention also toward military, religious, and political networks, thereby avoiding the restriction of one-sided, economically, or politically endogenous explanations for the rise of state-related power of kings and priests. Their enhanced chances to dominate city-states or whole empires do not, in Mann's view, follow from a pre-given functional necessity but are the result of unintended consequences of sometimes contingent developments, creating, nevertheless, a path dependency: leading to correlations between militarism and despotism in a process of slow transformation of patron-client relations of a temporary kind to more permanent coercive power. Organization and coordination become necessary in the face of attack from outside. It was periods of conflict and instability between "survival units" that created structures of legitimate authority because of the urgency to take defensive measures (building fortifications, for instance). Empires could be formed, but the monopolistic tendency could be held in check by more or less stable equilibria of power between various states – resulting in "multipower actor civilizations." Aristocracies as "decentralized elites" of landowners could also form a kind of group monopoly within these empires or states (in the Near East), often legitimized by genealogical descent and ideology. From here, Mann traces the further path of development from empires of domination to the Greek polis (a triple power network based on armed, city-bound peasant proprietors, embedded in a commercial, trading network of agricultural production at land and sea, the latter by generating strong naval powers). Like Marx, Weber, and Anderson, he places slavery at the center of attention. The Roman Territorial Empire inherited both traditions – that of the Greek city-state and that of the oriental empires like Persia or Egypt – when it had to learn that the days of the Republic were over. Mann focuses on the legionary economy of Rome in particular, an enduring and a successful system, but also one with some inherent and eventually fatal weaknesses: spiraling out of control by an interaction of several resource-limiting forces. Mann's analysis of the one power network (Christianity, both ideological and material) that filled and bridged the void between the fall of Rome and the arrival of feudalism is remarkable: according to Mann, as a church, it was "the leading agent of translocal extensive social organization" (Mann 1986: 337), and by its ideological merits, it prevented the regression to a state of complete barbarity. The West was Christendom.

To date, Mann's agenda has led him to four volumes (Vols. III and IV appeared in 2012 and 2013, respectively) and to the global state system of the present. His approach combines endogenous and exogenous aspects of state power and the power relations between states from the slow transformations of their early origins until today. What is its central research question, and what is the value of possible answers related to the application of sociological knowledge? To provide better orientation or to shape policy? If we start with Vol. I, the interest in power seems to be purely academic. In the subsequent volumes, we see that one practical interest lies in the comparative analysis of different paths to democracy and/or "dictatorship" within states and in the decipherment of hegemonic structures in the international relations within the system of states. It seems that Mann sides rather with the weaker parties than with the stronger rulers of states and the world.

It is not clear beyond any doubt that such an agenda really has to go back as far as Mann does in order to explain the institutions, patterns, problems, or solutions of the present. It is justified if we can detect a clear developmental path that indicates at least the necessary, if not sufficient, conditions for the later stage as dependent on the earlier, former steps. If not, the interest in the far distant past might be legitimate in terms of intellectual, scientific curiosity – an interest which, in any case, guides archaeologists and historians. Feminist historian Gerda Lerner (1986) has dealt with the origins of institutionalized male power by going back to the early patrimonial-bureaucratic states of the Near East and their laws regulating behavior of women by excluding them from the public sphere if they were not in the company of husbands, fathers, or brothers. One symbol for this was the veil. Long before the arrival of monotheism, especially Islam, a regime was put in place that is today regarded as a religious norm and custom typical of Islam. By showing the true origin of the veil, categorical mistakes are avoided, and something which is conventionally seen as trivial and self-evident can now be treated as human construction open to deliberate change. In Mann's work, similar findings can be detected relating to territoriality, the concept of individual property and property rights and the reified naturalness of inequality, but also in the seemingly unvaried and eternal nature of war.

Do we have to analyze the relevant processes in a comparative way, or is it sufficient to study them as historical individuals on their own? Mann is indeed against lawlike explanations as they are used in neo-evolutionism but favors a comparative perspective of commonalities in different societies separated by time or region. In this sense, his approach resembles that of Weber in his later years.

Regarding the closeness to historical sources and their "emic' interpretations by their contemporaries, in particular their emotional experience, Mann relies exclusively on secondary, theoretically modeled "etic" interpretations of structures and events; emotions do not figure prominently in these sources, and so they are also lacking in Mann's account. This is quite visible in the notion of "ideology" (here, it is mostly religion; a part of it can be seen as management of fear and anxiety), but also for the emotional experience of power and powerlessness, in triumph, resentment, or humiliated fury. Magical thinking is never just thinking; the offer to overcome fear and uncertainty is, as Elias has shown, a power source of its own and can be exploited by shamans, priests, and the state.

The psychohistorical approach of Steven Pinker (2011) to the history of violence constitutes a fourth paradigm that deals with the transition of prestate societies to statehood predominantly under the aspect of the emotional and embodied experience of violence. He not only reconstructs Elias' theory of civilizing processes (2012a [1939]) in a dedicated chapter but also shares Elias' main insight of pacification as a result of state development. While Elias (as we shall see further in section "Developmental Paths Toward the Early Modern European State") focused his attention on the rise of absolutist court societies since the Middle Ages, Pinker extends this analysis back to hunter-gatherer societies and their transition to early civilizations and states. Referring to statistical information generated by archaeologists (who worked out the percentage of people who met violent death from skeletons in prehistorical graves) and cultural anthropologists (who calculated death rates within and in conflicts between prestate societies), the number of violent deaths falls in state societies to a small fraction of the average death rates in prestate societies. However, this comes at a price: with the agricultural revolution came urbanization and therefore anemia, more infectious diseases, caries, and a reduction in body size. Pinker reports that in centralized societies, more women are killed in combat, more people are enslaved, and human sacrifices are more common. Also, nonviolent deviance (blasphemy, sex, lacking loyalty, and witchcraft) is criminalized more often, and there are more torture and mutilation, enslavement, and capital punishment. The price of pacification seems, therefore, to be paid with more authoritarianism or "despotism," including a worsening of conditions for women in extended harems. However, the reduction of violence within early state societies exceeds the growth of interstate violence in wars – although the list of the most dramatic human catastrophes contains (Chap. 5 in Pinker 2011: 228–67) excessive absolute numbers, like 55 million in the Second World War, these are dwarfed by the An Lushan Rebellion in eighth-century China which would now, corrected by a factor that brings these numbers into proportion with the size of the population in the mid-twentieth century, have amounted to 429 million victims. But the comparative data (Pinker 2011: 98) on war-related deaths of prestate and state societies (524/100.000/year vs. the 60/100.000/year for the whole world of the twentieth century) show a lesser reduction than that of internal violence from even the most peaceful tribal societies (Inuit, !Kung, Semai in Malaysia), compared with the Western European standard of the late twentieth century (Pinker 2011: 101).

For Pinker, the result is, as it was for Norbert Elias, the rise of a more pacified social habitus in the direction of greater peacefulness under the conditions of statehood. Violence is seen by Pinker to have a biologically founded function in the struggle for survival – between different species according to the position in the chain of nutrition, within the same species as a means to secure control over territory, food, and access to sex with females for men – for rank and status in the hierarchy of dominance, and for revenge with the aim of retribution. But he acknowledges the huge difference between biologically grounded violence in human face-to-face interaction and that of an anonymous, abstract type of violence in more bureaucratic or technical machine-like contexts, set in motion by ideologies and executed sometimes by killing on a huge distance. Neither the practice of human sacrifices in early states nor genocidal violence in war or state-ordered torture can be simply explained by referring to biologically founded affects. Lewis Mumford had given this machine-like savagery, often enabled or demanded by magical thinking in the face of unnameable fear and terror, the most vivid form of expression (Mumford 1966). Both types of violent behavior can be "civilized," as Elias saw it (but referring to the period extending from the High Middle Ages to the early modern period), although they demand different types of checks: the civilizing of war between states was more difficult to achieve than the civilizing of affects within state societies.

All in all, Pinker's main question is how to explain global pacification as a progress made by mankind and to defend it as human progress against all skeptics who would argue otherwise, namely, the idea of unchanged or even enhanced aggression throughout the times as part of the *conditio humana* (or because of capitalism, fascism, and communism). The answer lies in the state: its merits outweigh its faults. He goes far back in time because he wants to address quite basic, usually taken-for-granted qualities of life in the majority of more secure state societies of today. Although he cites ample evidence for the unbelievable power excesses of tyrants in state and prestate societies alike (from the pastoralist Kurgan chiefs whose funerals required the human sacrifice of dozens of women and hundreds more of their entourage, repeated in even greater scale by China's First Emperor with his terracotta army and the slaughtered servants of his personal household assembled in his tomb), the focus is less on the historically specific but more on broader tendencies of a potentially nomothetic character. Referring to historical sources, Pinker relies more on secondary interpretations (of archaeological records, field reports) than on "emic" interpretations by contemporaries, but he also draws extensively on psychological, and even neurophysiological, knowledge based on experiments in order to develop theoretically modeled "etic" interpretations of interesting structures, experiences, and events (e.g., what regions of the brain – such as the gratification center – are involved in sadistic violence of a torturer?). Finally, his historical and processual understanding of the development of state and state-related power is based both on historical parallels between different periods and on the uniqueness of a specific development – the humanitarian revolution of enlightenment and the revolution of rights. In order to compare trends, great masses of data are presented to confirm developments of very general or even universal relevance.

Developmental Paths Toward the Early Modern European State

Since the revival of historical sociology in the 1960s and 1970s can be seen, as shown, as a reaction to the then dominant structural functionalist theoretical school of Parsons and its neo-evolutionism, it is interesting to look at its version of the early modern state in order to mark the differences between it and the Marxist, Weberian, and other approaches. Parsons (1966, 1971) subsumed the *power* aspect of the state under the *media of interchange*, differing, for instance, from money. It served the main function of *goal attainment*. The corresponding *value principle* was *effectiveness*. The *mode of communication* was *commandment*, differing, for instance, from persuasion. The basic institutions were those of *political leadership or authority*. The *coordination standard* in the case of states is their *sovereignty*, and their *security base* is guaranteed by *means of coercion* (physical force). For Parsons, it was not unchecked, dictatorial violence that led the path to successful modern statehood; the early parliamentarism of the English Magna Carta was to turn into the model for greater political participation of the masses (inclusion). For the more absolutist European powers, Parsons explained their character partly by the role they played as protector for the whole European state system against the threat from outside, in particular from the East. Referring to the way physical force can be exerted, Parsons was certainly aware that it could also be limitless and unjust as well, but he did not seriously treat excessive state violence, in particular against the lower classes of peasants, as systematic and expectable element of statehood. Therefore, for the critics of Parsons, conflict and dysfunction were underrated and replaced by assumptions of harmony and teleology.

Marx's basic model for the explanation of the "original accumulation of capital" (Chap. 24 of *Das Kapital*; Marx 1999 [1867]) places it in the context of the late feudal society of the fifteenth and sixteenth centuries, not of the industrial society of the eighteenth and nineteenth centuries. It is the expropriation of the peasants from their land and soil in the process of violently enforced "enclosures" that creates the conditions for a capitalist mode of production to develop – by means of robbery and not as a result of a Protestant ethic. Old common property is privatized; capitalist leaseholders and early agricultural workers enable a revolution of the agrarian economy but accompanied by a punitive legislation of harshest and bloodiest measures against the dispossessed (Rusche/Kirchheimer 1939). Wages are kept low, and the early modern state is instrumental in the suppression of agricultural and early industrial workers. In this picture, there is no harmony, but feudal relationships are, first of all, economic relationships. Neither war nor knowledge plays a vital role. Marxist historical sociology wanted to exploit the strengths of Marx's simple formula, but they were also forced to correct its weaknesses. Perry Anderson's *Lineages of the Absolutist State* (1979 [1974]) did it by softening the exploitative character of feudal and absolutist rule on the one hand and by turning to the multiplicity of paths to early modern statehood on the other hand, showing that the ruling classes were themselves often prisoners of a geopolitical situation also characterized by war. The various European states were also to different degrees open to further democratization. This was also Barrington Moore's message in *Social Origins of Dictatorship and Democracy* (1966): The outcome of the massive conflicts between peasantry and nobility in the fifteenth and sixteenth centuries also decided the further path to democracy, fascism, or authoritarian socialism four centuries later. With Wallerstein (1974), the difference between the plight of the peasants and workers in areas far from the center of capitalist development and the comfortable life organized workers could enjoy in core capitalist countries stood in the focus of attention – not only economies but also states contributed to that. But Wallerstein, in the tradition of Rosa Luxemburg and Lenin's models of the way a worker aristocracy in England or Germany could escape absolute immiseration by profiting from the imperialist expansion by exploiting cheap labor in poorer regions of the world, still largely sticks to the economy and less to war and violence as power sources.

If we compare this picture with that of the great sociology-oriented social historians of feudal society, known as the *Annales* school, it is striking how differently they perceive the nature of feudalism as a system of rule. For Marc Bloch, kings are charismatic figures (*Les rois thaumaturges*; Bloch 1973, refers to the healing power, ascribed to kings in the Middle Ages), and the European nobility (with its core in Carolingian France) gained its specific character in a long process in which relationships of personal dependence between vassals and overlords in a warrior society became dominant (Bloch 2014 [1940]). This personal element of bondage is distinct from the anonymity of market and bureaucracy. It consists in loyalty or fealty, confirmed by rituals, Germanic and Christian, and is lifelong, and submission to an overlord is compatible with honor (an aspect always accentuated by Max Weber; the vassals were of upper-class origin). The transformation of this feudal bond into a hereditary form means also transformation of the fief into permanent possession and control of land, and by the fusion of fiefdom with office, military leadership, and judicial rights, economic, military, political, and law enforcement powers are united in a single hand. Hereditary transmission also meant decentralization of power to the point of the complete breakdown of central authority. Seen from this perspective, the fate of the peasants looks different; as another *Annales* historian, Georges Duby wrote (1984), these feudal knights were, first of all, warriors who terrorized their peasants by the force of their sword and according to their passions. These descriptions, close to "emic" witnesses of their age, lack the calculated economic rationality of a feudal "mode of production."

By referring to the work of the *Annales* historians on feudalism, Anderson tried to widen the narrowly Marxist economic perspective by including noneconomic factors and to acknowledge the reality of war and interstate relationships also for his explanation of the absolutist state. His basic theoretical conviction was still Marxist: absolutism is a system of renewed feudal domination over their rural subjects and thus prepares the frame for the future class struggles of industrial societies. Writing against Friedrich Engels' interpretation of absolutist rule as consequence of an equilibrium between land-based feudal nobility and town-based bourgeoisie (giving the state or monarch as mediator freedom for independent action against both groups; Norbert Elias' term "royal mechanism" expresses a similar idea), by stressing the notion of "renewed feudalism," he nevertheless saw political rule still predominantly as an epiphenomenon of economic power. Anderson's comparative analysis of Western and Eastern absolutism stresses a kind of path dependence from their divergent passages from antiquity. Western absolutism, rooted in feudalism, depends on the early liberation of the peasants from serfdom in the fourteenth century and from the existence of strong cities as a precondition of capitalism, whereas in the East without Western-type feudalism, the rural masses experience a second serfdom, dominated by the state and the landholding magnates. Towns are weak and the nobility is in service of the monarch. The result is neither capitalism nor democracy. But this conclusion is both too general and too abstract: By telling the stories of state formation for the most powerful European states, including Russia and "the house of Islam" (meaning the Ottoman Empire), the picture is much more nuanced, and the peculiarities count more than the commonalities between the diverse variations of "Western" and "Eastern" absolutism. For instance, the Polish magnates with their high power ratio in relation to the peasants – an Eastern element – developed an aristocratic republican style of rule (a Western element) which failed miserably in the great power struggle against Prussia, Russia, and Austria because of the weakness of their army. Here, the military factor is of central importance, for Anderson too. Untypical is also Austria: the Habsburg Monarchy combined as an empire both Western and Eastern structures since her beginnings in the sixteenth century – the Tyrolean Diet had since the fifteenth century included representatives of the peasants, unlike almost anywhere else in Europe; the German lands were clearly shaped by a feudalism of the Western type. Bohemia after the Thirty Years' War, Hungary, and Galicia were different cases; and while Austria's great power status was soon endangered by her military weakness (Anderson gives clear reasons for this), her economic success (Good 1984) is undisputable. England's path to a crisis-ridden capitalism (Anderson 1987) begins with an early capitalist rural economy, a strongly commercialized aristocracy, is not impeded by too strong a state (indeed it was rather weak), develops a financial sector outclassing manufacture, but profits from empire and international trade and from the fact that no second political revolution takes place after 1689 (in contrast to France). (In comparison with Germany or Japan, the British state would also be weak in the nineteenth and twentieth centuries.)

Wallerstein's (1974, 1980) paradigm of a spiral of cumulative advantages for some countries and cumulative disadvantages for others who depend on the former, taking place in and creating an economic world system of uneven development, has another solution for the relationship between state and economic power. In his work, there are states, linked to each other in a system of states; nations in the making, linked somehow to these states; classes with and without awareness of themselves; and households, labor, and income somehow tied to these classes. External relations influence the internal power distribution of the classes and vice versa. But for Wallerstein, the state is scarcely defined by its monopoly of violence, in particular as it is exerted in war. For him, feudalism presents no bureaucratic obstacle to the movement of capital and to commercial activity in general; that feudal princes saw themselves as warriors is not important: states are defined by their position in the emerging global markets. Anderson, on the other hand, transcends economistic reasoning by accepting that interstate conflicts – including wars – influence the internal power structure of states; feudalism (and absolutism) is linked to state competition, to war, and to a precondition for the rise of nation-states. But he still clings to the idea that it is the mode of production and the way the state is interested in its functioning that predominantly define the position of the state. It is the absolutist state that protects the property rights of the possessing classes through a system of renewed feudalism, in France more than in Britain, in Russia more than in Poland. What other authors (Norbert Elias, for instance) saw as a process of (functional) democratization[1] and power-sharing between aristocracy and town-based and administrative bourgeoisie is – theoretically, not in empirical institutional detail – reduced by Anderson to a functionalist-instrumentalist cunning of the ruling classes.

Max Weber's contribution to an understanding of the genesis of the modern state is multidimensional, complex in institutional detail, but follows two main ideas that set it apart from Marxist reductionism. One is rationalization of domination, in the direction of professionalization and compatibility with economic rationality; the other is the violence-related aspect of state domination, from feudal to patrimonial and patrimonial-bureaucratic rule, and its expression in war. The second aspect is particularly visible in Weber's treatment of charismatic authority with the special dimension of the king as a warlord (*Kriegsfürst*). But Weber tries to combine both rationalization and state monopoly of violence related to legitimate domination

---

[1] Elias (2012b [1970]: 63) uses this term in order to grasp a development that "is not identical with the trend towards the development towards 'institutional democracy'. It refers to a shift in the social distribution of power, and this can manifest itself in various institutional forms, for example in one-party systems as well as in multi-party systems."

(nonlegitimate domination characterizes the cities and towns) in his concept of "discipline" which is most closely related to military discipline. Seen against this general background, the early modern European state arising from feudalism can appear in many different shapes. The English state governed by "gentlemen" (*Honoratiorenherrschaft, rule of notables*) differs from the French, Habsburg, Prussian, and Russian state by its low degree of bureaucratic rationality; whereas German patrimonial bureaucracy generates in the subjects a social habitus of fixation to authority (*Untertan*; Weber 1978: 1108).

The one scholar who regarded himself most explicitly as Weberian was Reinhard Bendix. In his work *Nation-Building and Citizenship* (1964), he was dedicated to the study of authority relations in different nation-states (in Western Europe, Russia, Japan, and India) as a result of foregoing historically particular paths with consequences for the power distribution between ruling and lower classes. He followed Weber's approach most closely in order to understand the process leading from royal patrimonialism to political participation of the masses in the nation-state of the twentieth century. In the Middle Ages, this patrimonialism was linked to localized feudal jurisdiction. In early modernity, the road to absolutism (when royal power overcame the resistance of the aristocratic estates) led to the near-complete exclusion of the broader masses from the political sphere. Absolutism (and patrimonialism) had to give way, during the nineteenth century, to a highly developed, formally rational state bureaucracy. Traditional feudal obligations were replaced by commercialization, and state authority lost legitimacy. But civil societies were formed which interacted with states (now nation-states) to form new types of political communities. How they evolved depended on their roots long before industrialism: different in Germany, Western Europe, Russia, India, or Japan.

For Michael Mann who sees himself as Weberian, European state formation is very much a product of changes in the conduct of war in the late Middle Ages. The path leading to the new central state which replaces the looser feudal princedoms with standing armies (*miles perpetuus*), firearms, courtly splendor, and despotism is as much a result of economic and technological changes as it is caused by geopolitical constraints. But although most European states later follow the path set by the Western European dynamics of geopolitical rivalry, Christian norms, capitalist/mercantilist mode of production, industrialism, and nation-states, differences between them can be traced back to their respective constitution in the early modern period: the comparatively liberal English, later British, state can be distinguished from the authoritarian monarchies of France, Italy, and Sweden who later become liberal reformist; the equally authoritarian monarchies of Austria, Germany, or Japan will later turn into fascist states; and Russia's particular despotism prepares the ground for authoritarian socialism. Geopolitically, power centers move from the European southeast to the Atlantic northwest, with marcher lords sitting at the margins becoming great powers because they protect the open flanks of Europe (Russia, Habsburg monarchy, Prussia). The tendency to a power monopoly of a dominant empire is stopped by the mechanics of multipower actor civilizations. The internal power distribution (classes) in these states is related also to their geopolitical, exogenous fate.

Elias' magnum opus (Elias 2012a [1939]) has a general subject: How was it possible for our contemporary woes of civilization to have come into existence – from neurotic inhibition to boredom and lack of pleasure – while life has become more peaceful every day, at least within the confines of "states"? This question was asked by a man who as a soldier in the First World War had been deeply traumatized by the lasting experience of brutal, mechanic violence. He gave his answer in two volumes. The first dealt with the changes in the sensation of members of the secular European upper classes in a period stretching from the peak of the Middle Ages to the twilight of the royal courts at the end of the eighteenth century. The second volume focused on the process of the rise of European states as political and military units shaping and transforming emotions with war and violence at the center of this development – from feudal to princely territorial states. Both volumes made use of theoretical models. In order to grasp the psychodynamics of the emotional experience, Elias modified Freud's theory of drives. The development toward "states" with their monopoly of violence within society and against outer enemies was caught with the help of macro-sociological models – referring to centrifugal and centripetal processes, to competition, to monopolization and constitutionalization of the means of physical violence – of the means to torture and kill.

In order to make his theoretical understanding of the psychic processes more suitable for observation, Elias introduced the notion of a "psychic habitus" that was closely related to the concept of "habits," referring to the common philosophical notion of "habitus," deriving from Greek thinking (Aristoteles) and revived by Thomas Aquinas. Body posture, mimics, and gesture correspond to relatively stable inner forces – "self-constraints" in Elias' language. A threshold of shame and embarrassment secures the internalization of constraints exercised from above or at least from outside. They form a kind of "second nature" (E. Dunning and Mennell 1996; Wouters and M. Dunning 2019). If "civilization" can have growing inhibition, refinement of needs, and pacification as its central three dimensions, then the latter can take a twofold shape: (1) as a particular "civilizing thrust" which turns warriors into courtiers and (2) as a more general process in which all kinds of people and not only courtiers feel "inner" constraints that make it difficult for them to act in a physically violent way.

The distinctive quality of Elias' approach to the explanation of the (early) modern, dynastic state consists in its subordination to an even greater riddle, namely, a specific civilizing process, although with broader consequences for the European branch of mankind. It combines theoretical modeling with a historical narrative of French political and military history. The central driving force is the competition between warrior lords which leads under the condition of a barter economy without money, but based on land, to a weakening of central power, since the gains from former conquest have to be distributed among the vassals in order to reward them for their military services and can, thus, be accumulated to form an independent power base versus the prince or king. The result is near anarchy or, as it can also be called, a very fragile equilibrium between competing forces. Under the auspices of a town-based, monetary economy, this process goes in another direction, namely, the strengthening of central power. Here, step-by-step the accumulation of land becomes irreversible; the competing territorial rulers lose their own power base because the central overlord can mobilize resources exponentially. This is the meaning of the so-called monopoly mechanism (Elias 2012a [1939]: 301–11). The power balance between feudal vassals and princes changes in favor of the latter. Princely households become bigger and bigger (what Weber called patrimonialism), and courtly functions are slowly transformed into administrative functions for a larger area. The development of a money economy is the symbolic equivalent for the lengthening of action chains in an ever denser network of reciprocal dependencies, functional differentiation (*Funktionsteilung*). The central overlord is gliding into a position which can be seen as "functional" for the satisfaction of the needs of the people in his territory – among these, the offer of protection from violence is the most important: The monopoly of violence means pacification and needs the monopoly of taxation for which a growing bureaucratic apparatus is necessary. Finally, the position of the prince or king and his household becomes gradually that of the "state" – now in its narrower, present definition. What was for a certain time the seemingly unlimited nature of the king's sovereignty (wrongly explained by Hobbes as constituted by a deliberate contract) is the unintended side effect of a shift in the power balance between the aristocracy and the rising bourgeoisie with administrative functions (*noblesse de robe* against the older *noblesse d'épée*) for the larger whole. These cooperating and at the same time antagonistic groups neutralize each other's ambitions and give room to the particular space of power which becomes, for about 150 years, the privileged position of the absolutist monarch (the equilibrium of forces between these antagonistic groups enables the functioning of what Elias called "royal mechanism" by multiplying his differential weight; Elias 2012a [1939]: 347–79) before it, finally, gives way under the pressure of the functional dependence on broad strata to the democratically controlled collective monopoly of the modern nation-state. One further important term was coined by Elias referring to the power sources of the parties involved: *Gesellschaftliche Stärke* (social strength) – among these, physical power is the first; it shows in battles. Economic power is second, but its nature changes from land to money. The church and its clerics enjoy the advantage to exploit the need for orientation and salvation. A high power ratio is the companion of an uneven condition of dependence: the more dependent (also functionally), the more powerless is a person or a group. Elias manages, thus, to reconcile two conflicting views of functional power: that of domination achieved by previous force and that of power as instrumentally functional for the pursuit of common goals.

The difference from Weber's model of patrimonial domination can be grasped in the character of the theoretical model. Elias' model is very exact and could be rendered in terms of mathematical game theory, from one equilibrium (anarchy) to another (monopoly). The language of action chains, webs of interdependence, and functional differentiation is very general. On the other hand, the models are contextualized in the situations of the concrete French monarchs, from Hugh Capet to Louis XVI. Weber's explanations are more sketchy and fragmentary, lack the coherence of Elias' models, but draw on a wealth of institutional, detailed information, in particular referring to the areas of law, law enforcement, political representation of a comparative kind, ways and methods of armament, and economical techniques. The respective merits and weaknesses of both ways of thinking can be deduced from the differing backgrounds of both authors: Weber was trained in law and in *historische Nationalökonomie* while Elias in medicine and philosophy with a strong leaning to history. Therefore, the term "function" as it was used by Elias (2012b [1970]: 72–3, 121–3, 147) has a certain proximity to "function" in medicine; whereas Weber was highly aware of the differences between the English common law and the Roman law tradition of the continent. In this respect, Michel Foucault (1977) is closer to Elias than to Weber, since he was also trained in medicine and philosophy.

Although Elias' strength is not comparison, at least on the institutional level, he has nevertheless developed some general lines of argument referring to the different paths of development taken by France, Germany, and England. Elias' main works dealt with the development of France (Elias 2006 [1969]; 2012a [1939]). After the breakdown of the Carolingian Empire, a highly centralized state emerges gradually, in which the court is the most important institution shaping upper-class models of behavior. Not only does it help to generate a refined, diplomatically cautious, and softly civilized nobility, but it also influences, directly or indirectly, even bourgeois formations. The courtly heritage has become, according to Elias, a strong and vivacious component of French national character, forming it to this day. Elias' own analysis of England does not reach that of France in terms of scope and empirical detail. But there are some remarks that concentrate on the differences between English, French, and German feudalism.

> In the conduct of workers in England, for example, one can still see traces of the manners of the landed noblemen and gentry and of merchants within a great trade network, and in France the airs of courtiers and a bourgeoisie brought to power by revolution. In the workers too, we find a stricter regulation of conduct, a type of courtesy more informed by tradition in colonial powers which have for a long period had the function of an upper class within a large network of interdependences. (Elias 2012a [1939]: 426)

There is an English history of parliamentarianism opposed to France's way into absolutism: The "royal mechanism" is not fully effective in the English case. Why? Elias offers a number of reasons. England did not experience the knockout competition of territorial rulers as in France or the Holy Roman Empire, at least after the Norman conquest, but was, instead, centralized early, largely because of the small size of the territory (Elias 2012a [1939]: 294–301). The early centralization meant also a monopoly of taxation at a time long before it could be enforced in France. The ruling nobility was relatively homogeneous in terms of their interest and was soon able to turn against the central lord, thus laying the foundations for the later parliamentarization process. For a long time, England appeared as not more than a semicolonial area that belonged to the West Frankish crown, in competition with other territorial rulers there. Only after her elimination in the contest for the French crown did England become "insular," but it was a country already thoroughly centralized and pacified. Finally, both the "monopoly mechanism" and the "royal mechanism" developed differently in England: While in France the rise of the town, of the monetary sector, and of the town-based merchants and craftsmen led to an unstable balance of these groups with the land-possessing warrior caste and the king, in England, those classes formed a coalition with the aristocracy and were, thus, able to limit royal power. However, in Elias' opinion, Tudor rule comes quite close to the absolutist model. Under Henry VIII, the merchants of the city of London were still far inferior to the wealthy landowning classes (Elias 2009a [1985]).

In the early modern period there exists an English counterpart to the French civilizing process. According to Elias and Dunning (2008 [1986]), there is an elective affinity between Parliament and sport. In one "civilizing thrust," the English gentleman learned to be "fair" in peaceful sporting events and to bow to self-imposed rules of political conduct in Parliament. Here "civilizing" means in a rather pure form the control of destructive violence and of feelings of hate while the threshold of shame and embarrassment advanced. That this de-escalation after Civil War and the Glorious Revolution of 1688 became effective had partly to do with the common class basis of "Whigs" and "Tories," of which they represented only factions, not opposing classes. In this respect, it was important that there was a "gentry," a lower aristocracy which emerged and found its place between urban craftsmen and merchants, on the one hand, and the landed nobility, on the other. The English aristocracy was also the first to lose its military function. But with a large navy and a very small standing army, the king could still not impose his will on the aristocracy. This is part of the secret of English "liberty." The gentry way of life led to an exchange of rural patterns of living with those of the capital through the London "season" of the higher aristocracy; a house in the country and a house in the city formed part of it. The eighteenth century, therefore, saw a massive rise of Parliament and the status of wealthy landowners, who turned England into a fiscal-military state (Brewer 1989) without generating a repressive warrior mentality.

Different again, according to Elias, was the path to statehood in Germany. As the Eastern part of the Frankish Carolingian Empire, the Holy Roman Empire became a much larger territorial unit than France or England, originally more successfully centralized – but because of its size and the tribal character of its parts, a slow yet irresistible process of disintegration took place which ended after the indecisive result of the Thirty Years' War, in nearly complete fragmentation of power. In the High Middle Ages, the emperor had also to rely on the church for administrative purposes, and the conflict between the pope and the emperor over recruitment to these posts also weakened the emperor's standing fatally. As Elias says, the Habsburgs, as German emperors over a huge area, were always in the situation of lacking financial resources. Their obligations which resulted from the competition on a larger, European scale led normally to financial ruin and blackmail by the various Diets. More effective units of domination were formed at the lower levels of *Landesbildung*, the formation of larger territories that split from the Empire in a centrifugal process. Among these, the absolutist monarchies of Austria and Prussia, ruled by Habsburgs and Hohenzollern, were to fight the struggle for hegemony in Germany until 1866.

According to Tilly (1992), after the year 1500, the number of states in Europe dropped from some hundreds to approximately 30. Nearly all modern states are nation-states dominated by a major ethnic group (even within the huge Russian

Federation, the dominant Russians represent 82 percent of the population). Tilly assumes that the nation-state was the most appropriate solution in the struggle of survival between states in Europe. All other types of states, like city-states or empires, were eliminated. Tilly isolated three ways how European states were able to increase their power. One path led to the permanent increase of coercion. This mostly happened in the huge empires of Central and Eastern Europe. The other path led to the accumulation of enormous sums of capital in the trade-based city-states. Most successful, however, was the Western European model of combining coercion and capital. This path led in the end to the foundation of the modern nation-state.

The Industrial Nation-State and Beyond

The closer we come to the present, the more acute is the pressure on scholars of historical sociology to take sides when they deal with the conditions and consequences of the rise of a system of industrialized nation-states for the experience of power and powerlessness both within and between these states and also when they have to judge the hegemonic relationships of the ascending European-American West with the global "Rest." The twentieth century and the first two decades of the twenty-first century saw not only interstate rivalry with wars on a global scale but also the emergence of new global powers challenging European and now, American dominance.

Since historical sociology as an academic subdiscipline was founded as and motivated by critique of an affirmative "hodiecentrism," research focused on the repressive or corrective effects of state power over the rising classes of the bourgeoisie or the exploited classes of the workers internally and on the simultaneity of nationalism and imperialist colonialism externally – that is, in an ever-expanding global system.

What we have to be aware of is the huge difference between the European society largely consisting of agrarian, dynastic, "princely" states on the eve of the manifold revolutions (scientific, technological, industrial, informational, and political: accompanied or preceded by the avant-garde project of philosophical Enlightenment) and the global, unevenly developed high-tech mass consumer societies of today. The availability of energy, productivity in agriculture and factory, destructive capacity of weapons of war and political suppression, chances of surveillance, geographical mobility, and degree of the division of labor within and between economies and states of this world – all of these have risen beyond anything imaginable in the times of Hobbes or Rousseau. Just a few teams of machine gunners would have swept away any Napoleonic army; drones might have killed his generals. That more than 99 percent of all newborn infants would survive now to become adults (albeit under close surveillance) and not die from infectious diseases would have been regarded as a miracle. And so also would the peaceful, self-policing character of the people living in the happier parts of the modern global village. But our forebears might also have had difficulty in understanding some of the social problems of today: the struggle for social recognition of LGBT people, the wars conducted by suicide bombers, the wish of women for commanding positions in NATO armies, and the whole fragility of modern chains of production endangered by an invisible biological threat originating in a distant part of the globe. It is with these differences in mind that we can now check the riddles and answers provided by historical sociology.

The one great subject evolving from the accelerated development or revolution in industry was the birth of the "social question" and the role played by the state in economical, political, or military terms. Another big topic was the choice between liberal democracy and authoritarianism in its socialist or "fascist" variation. The third big bundle of puzzling state- and power-related problems circles around nationalism and the militarized nation-state, both as internal affairs of the state and also in terms of the hegemonial conflicts between these states. A fourth focus of growing attention was directed on the imperialist, colonialist dimension of these hegemonial struggles (Wallerstein 2011) and the formation of new, supranational power blocs. Again, the range of possible answers is huge – between empathy for the "Rest" against the "West" and the praise for the American Empire, dominant for now. State formation gives way to empire formation (Mann 2003; Ferguson 2004).

We start again with the structural functionalist, neo-evolutionist view of Western modernization. According to Parsons, modern society was the product of three revolutions (Parsons 1971) that took place in the northwest corner of Europe – in Britain, Holland, and France – namely, the Industrial Revolution, the Democratic Revolution, and the Educational Revolution. The latter could build on the earlier achievements of Reformation and Renaissance; against the dangers that might stem from the towering influence of an almighty state bureaucracy (Weber's fear), Parsons saw a remedy in increased "associationism," reviving features of *Gemeinschaft*, and in "participatory democracy" (Parsons 1971: 116–7). Much of this was realized in the United States, spearheading human progress, with the exception of the still existing problem of Afro-American inclusion, or better the lack of it. T. H. Marshall (1950), who was not committed to the structural functionalist "grand theory," saw the struggle for civil rights in the seventeenth century, for political rights in the eighteenth and nineteenth centuries incomplete, if they were not complemented by the social rights fought for participation in welfare and reluctantly guaranteed by the state during the nineteenth and twentieth centuries.

Marx and Marxist authors saw these developments more critically. The riddle that greatly enhanced productivity did not automatically lead to welfare of the masses but often to renewed misery was famously addressed by Marx. What was the role of the state? Did it support suppression or emancipation? For the nineteenth century, three different answers exist: E. P. Thompson (1963) saw the complicity of religion, church, factory owners, and state in creating a diligent, punctual, and obedient working population that was disciplined and self-policing. A second answer was given by Theda Skocpol (1979) who scrutinized revolutions (France, Russia, China) in which state apparatuses, weakened by defeat in wars against external enemies, were found out to lack the military means necessary for successful suppression of worker (or rather peasant) rebellions. Charles Tilly, who was no Marxist, dealt with the history of social movements in his *Popular Contention in Great Britain, 1758–1834* (1995), stressing the importance of collective campaigns challenging public authorities and promoting various claims, but he noticed that these campaigns would not automatically aim at democracy. A third solution was presented, in the same year as Friedrich A. Hayek's dystopian view of the socialist *The Road to Serfdom* (1944) was published, by Karl Polanyi, whose *The Great Transformation* (1944) saw the state as the only savior from the civilizational catastrophe created by an imperialist market for "fictitious goods" like land or labor that would have destroyed society by creating anarchic insecurity for all. Here, the power of the state is benevolent, and Polanyi's paradigm has become particularly influential in the theoretical and political debate with neoliberalism (Pixley/Flam 2018), in particular since Lehmann Brothers and the financial meltdown of the 2008 crisis.

Immanuel Wallerstein's *World-Systems Analysis: An Introduction* approach links the analysis of domination and exploitation from the core industrial countries to economically linked underclasses in the whole world. In his fourth volume (Wallerstein 2011), he treated the emergence of "centrist liberalism," developed in the aftermath of the French Revolution, as an ideology to cope with the reality of class conflict on a national level, and highly consequential for the further, also social scientific dealing with uneven global development. The nation-state is, in Wallerstein's whole account, only an agent of the larger forces of capital, even when it enforces by military means unequal rates of exchange between the core and the periphery. War, culture, and "habitus" are no prime causal factors for its explanation; Wallerstein does not consider these factors as decisive in the case of those countries that manage to leave periphery status and become new parts of the core. But the rise of East Asian countries like South Korea or the new super power China (for Fernand Braudel (1979b), India and China in the early modern period had been caught in a super world economy of its own) might be explained by their long-lasting habitus of hard work and industriousness. The two topics – the dynamics of European and later global state competition and changes in the power distribution within these states from largely aristocratic and princely domination to a participation of broader strata – are certainly interrelated.

Perry Anderson, too, has extended his research to a new level – that of European integration (Anderson 2009). In order to come to terms with effects of the European Union on the development of capitalism and democracy, he discussed a "neo-functionalist" theory (Haas 1958) concentrating on the economy and a "neorealist" interpretation (Milward 1984) which also stresses predominantly the economic advantages open markets offer for the nation-states. They appear strengthened rather than weakened, especially also in terms of their ability to deliver wealth and economic security for the working class. His own theory of 1995 transgresses the narrow economic boundaries and deals with the hegemonial relationships between states, also including US domination after the collapse of the Eastern bloc. Summarizing the subsequent developments in the year 2008, Anderson saw geopolitically a shift in the power balance between Europe and the United States in favor of the latter and a shrinking of the chances of participation both of the lower classes and the Eastern and Southern European states.

Max Weber's multidimensional contribution to an understanding of politics and the modern state consists in his classical distinction between "Class, Status and

Party" (Weber 1978 [1920]: 926–38), his analysis of the shortcomings of state bureaucratic power as an "iron cage," his skepticism over the possibilities of rational democratic leadership, and his conviction (expressed also in his political writings during the First World War; Weber 1994 [1920: 75–80]) that nation-states were condemned to merciless geopolitical rivalry. Max Weber's (1978 [1920]: 921–6) strangely forgotten outline on the causes for and types of nationalism had already discussed all possible variants and had found none of it sufficient to explain strong national feelings: language, culture and religion, ethnic origin, and the community of fate experienced by people even if they were thrown together quite arbitrarily. But he was convinced that these we-feelings exist. Conflicts or wars between such entities would rather lead to national sentiments than the other way round. Weber also saw the potential to explain the role of "national character" and "national habitus" (without using this term) by the influence of distant historical experiences.

Since the publication of Michael Mann's second volume on power, social classes, and the nation-state (1993), a lot of things have happened; in particular, the acceleration of the process of European unification and the integration of countries of former Eastern bloc countries have changed the discourse also for historical sociologists. Since then, Mann has written three more volumes dealing with globalization and the rise of the American Empire (Mann 2003, 2012 and 2013). In his third volume, he took up the work he ended in Vol. II with 1914 and described developments until 1945, the end of the Second World War which ushered in a breakthrough for two new superpowers in a system of *capitalism*, *empires*, and *nation-states*. The result was, according to Mann, a fractured, politically segmented globalization of rivaling great powers. Volume IV dealt with globalizations in the plural, gaining momentum since 1945. Here, too, Mann linked class conflict with international state rivalry, the (provisional) triumph of neoliberalism and the fall of the Soviet system, but he denied the United States the full right to call itself a true "empire" (Mann 2003), since it proved to be unable to really pacify and permanently integrate the militarily defeated countries as members of an "axis of evil." Mann's work is rich in institutional detail, gives each dimension of social power (also ideology) sufficient attention, but does not care much about the emotional side of it and even less about the way social (and state) development shapes the affective experience as its companion.

It was Norbert Elias who treated the "organization of society in the form of states" (Elias 2012a [1939]: 405) as a stage of a largely unplanned process of civilization in which civilized forms of conduct and feelings are generated. In his essay "Changes in the We-I Balance" (Elias 2010b [1987]: 137–208), Elias developed the concept of a "survival unit" in accordance with his process sociological understanding of the development of human societies along three lines: control of violence, control of nature by economic means, and production and control of the means of orientation, meaning knowledge. States can serve the corresponding needs for physical safety; they can act in order to persist in military, economic, and scientific competition; and they secure continuity and tradition in the memory of the coming generations. The disappearance of a state can, thus, be experienced as a kind of "collective dying," generating deep feelings of mourning. These different layers of the "survival" function correspond to equally differing forces that shape we-feelings, we-images, we-I balances, forms of national habitus, and we-identities.

The dominance of the nation-state was, according to Elias, a quite recent European phenomenon anyway. It had replaced older dynastic states with loyalties toward prince, king, or emperor and with bonds between aristocrats of several European states being stronger than toward the lower classes of their own state. Since feudal warfare had been replaced by what Van Creveld (1991) called – attributing the notion to Clausewitz – the "Trinitarian war" of dynastic state competition, with the affairs of the population, the army, and the government being kept separate, loyalty was extended to the ruler, the regiment, and comrades. For the common foot soldiers, things had been very different. Mechanical drill and massive coercion guaranteed their "discipline." Armies still relied more on foreign mercenaries than on their "own" country's drafted personnel. The open Machiavellism of absolutist states was mirrored in the cynicism of many of their soldiers. Things began to change with the Napoleonic wars, when French armies started to rely on drafted conscripts only and on their loyalty to the cause of revolution. Paradoxically, this was also the starting point for the nationalization of war: Only since then, conscript armies have turned nations into "nations in arms." The "nationalization" of these states went hand in hand with the growing power of the bourgeoisie, and their language would be transformed into the national languages of France, Italy, or Germany. New and higher-level survival units will rise, according to Elias, if growing interdependence between units of lower level makes them useful and necessary, replacing the more and more fictitious autarchy or self-sufficiency of nation-states. But Elias treats national we-feelings as real and not entirely "constructed" or "imagined," as several authors, famous for their contribution to the study of nations and nationalism, did (like Marxist historian Eric Hobsbawm (1990) or Ernest Gellner (1983) and Benedict Anderson (1983); deviating from this perspective and acknowledging ethnic *nuclei* was Anthony D. Smith (1986, 1991a)). National "we-feelings" must be distinguished from national "habitus." They can overlap (national pride can be transferred from generation to generation and become, thus, "habitualized"), but not always: Polished manners do not necessarily refer to feelings of national belonging, and these emotions can also be aroused spontaneously. French "courtization" and English "parliamentarization" and "sportization" (see above, p.29) have shaped national habitus until today; the German "dueling fraternities" (Elias 2013 [1989]: 49–134) had an impact on German national habitus which can be seen as an important link in the long causal chain to explain Germany's readiness to go to war 1914. There might be continuity of habitus formation, but also discontinuity: Elias knew that modern industrial society was not simply an extrapolation of the face-to-face society of the princely court and aristocracy; the anonymous forces of money, professions, and goods could also generate new standards of behavior. In the second half of the twentieth century, social constraints underwent a process toward less direct social control from above and outside, allowing and necessitating more individual space for negotiation – "informalization" (Cas Wouters and Michael Dunning 2019). Continuing the line of Eliasian thinking in another direction, Helmut Kuzmics and Roland Axtmann (2007) compared English and Austrian formations of "authority" in the period between 1700 and 1900, showing how processes involved in the development of markets and Parliament (Britain) produced long-lasting forms and relations of authority different from church, court, and state offices (Austria-Hungary). Helmut Kuzmics and Sabine A. Haring (2013) extended this analysis to the rise and failure of a Habsburg military habitus of hesitation and fatalism, originating in multinational Austria in the eighteenth century and contributing to military defeat in World War I. Stephen Mennell's work, a study on the making of an empire and the American "civilizing process" (2007), can also be understood as research on American national habitus; the nation-building process that formed the United States could be understood as outcome of a knockout competition quite similar to the monopoly mechanism of European states and not as the "manifest destiny" of American exceptionalism. Mennell has also recently demonstrated (Mennell 2020) how an American habitualized self-perception of imagined superiority has developed, in particular since the Second World War, as product of a series of victorious wars and hammered into broader masses with the help of films and other mass media.

The dynamics of state competition and power relations between states are certainly also linked to the internal power dynamics within survival units at the level of nation-states or composite superstates, too. The hegemonial struggle between the United States and the USSR was also the conflict between socialism and capitalism (Elias 2010a [1985]) and, thus, a fight between social classes that had become global, with consequences for the buildup of state power within these superstates. These processes influenced also the chances of a strong Civil Society (Keane 1988) arising that is able to limit the aspirations of central power within states. With the excesses of Stalinism, national socialist rule, and other examples from the twentieth century, the instrumental use of state power for mass murder has also become a special topic of scrutiny (Arendt 1951; De Swaan 2015). Bendix (1964) had already dealt with the long road to citizen rights. The scope of analyses could change between strictly state-centered (in the singular) or more broadly European or even globally oriented.

Pinker (2011) elaborated his thesis of the "decline of violence" as a result of such a global process that does not end with the state of the early modern period, but can be extrapolated to the age of the nation-state and beyond. States are involved in the pacification of violent behavior; with the humanitarian revolution as a consequence of the Enlightenment, it is guaranteed that their pacifying strategies become less violent themselves. But people succumbing to the ideologies of nationalism and revolutionary totalitarianism (communist) can instrumentalize the state monopoly of violence at enormous human costs (quoting Rummel's 1994 *Death by Government* with the number of nearly 170 million of victims of genocidal violence in the twentieth century). The overall balance of the trend for and against state-centered violence is, according to Pinker, still positive: Even the extremely costly wars of the last century cannot outweigh the civilizatory gains, in particular, since wars between modern nation-states have become very rare. Capitalism and democracy complement the average rise of empathy and a revolution of rights (citizen rights, women's rights, children's, LGBT, and even animal rights). This is a clear counter-position to Wallerstein, Perry Anderson, and also Mann's approaches: these authors have certainly a much less friendly picture of global capitalism and see the dominance of the American West in a very critical light. Pinker's strong emphasis on the causal, negative relevance of ideology may also underrate the foregoing processes of exploitation and dominance of the ruling classes (in Tsarist Russia) or the experience of defeat (Germany; Elias (2013) [1989] also treated the national socialist ideology as a highly emotional, pseudo-religious belief system), and it is only by gaining control of the state apparatus that ideological phantasies become reality. On the level of personality, though, Pinker's biopsychologically supported views of the decrease in male, testosterone-based dominance as a consequence of all these counterforces effecting more even power balances are compatible with Elias' insights on "functional democratization." How strong these processes are – in the face of ethnic exclusion and rising inequalities of wealth within state societies and hegemonial struggles between them – is still open to discussion. There may be a certain paradox in the fact that war in general and throughout the ages (Scheidel 2017) and in particular, after 1945 (Klausen 1998), has generated catastrophic effects but has bound survival units closer together (by we-feelings) and reduced inequality by taxing the rich and providing welfare for the poor.

If we want to understand the dynamics of state competition and empire formation, at least two historically informed analyses can stand for a larger genre. Paul Kennedy's (1988) book *The Rise and Fall of the Great Powers* and John Mearsheimer's *The Tragedy of Great Power Politics* (2001) as a plea for his doctrine of "offensive realism" offer many insights in the constraints states find themselves caught in when they are involved in a bitter struggle for the bid of mastery (the United States vs. China, at its most recent stage). The actors (whole countries respectively their elites) were portrayed as rational; in Mearsheimer's view, to fear (the only emotion mentioned) the possible opponent's potential is realistic, and the move to higher expenses for armament in anticipation is justified. The geopolitical losers were simply too weak. What is lacking in these narratives is any mentioning of emotions beyond "rational" fear and of a state- or nation-induced habitus that might be related to success or failure in avoiding or managing wars. The same applies to the narratives of class conflict and domination structure characteristic for power balance within states. Here, the approaches of Elias and Pinker are certainly helpful.

Another problem in the contemporary discussion is how to perceive "the" nation-state with respect to the rise of new survival units – as guarantor of welfare and inclusive economic co-development or as obstacle to supranational integration. Contrasting American military adventurism in the Middle East and advocating the model of European welfare states against American neoliberalism, Europe was seen as a peaceful moral hegemon in the making (Rifkin 2004; Leonard 2005), transgressing European nation-states that appeared as atavisms. But many sociological subdisciplines and sub-discourses had for a long time tended increasingly to treat nations and nation-states as something approaching an empirical "taboo" – one that would be better avoided. The end of the nation-state (Guéhenno 2000) was announced – not only for the European context, but worldwide – because of the development of modern information technologies and their influence on the global division of labor. This judgment might be premature, since the new global powers of

China and India can certainly rely on a sufficient ethno-cultural base and also on quite nationalistic sentiments. And for Europe, it can be shown that the dilemma between democratic participation and geopolitical influence on a nation-state transcending, European level is still there and might continue to remain so in the foreseeable future (Kuzmics 2019).

Conclusion

If we want to assess the contribution the historical sociology of state and political power has been able to make, it is useful to reflect on the pragmatic functions of sociological reasoning as a whole. There is, first of all, the idea that the social sciences deliver instruments to improve politics or advise better planning in present societies – what may be called the goal of social engineering. If the agenda is broadened from mere technocracy to the myth-destroying and myth-enlightening role they can play in social/political discourse, it is still the pressure from the manifold issues of the present (environment, injustice, and so on) that decides sociology's value. The peculiar status of historical sociology with its long-term processual orientation means that it enjoys the advantages of greater detachment, but at the cost of seeming irrelevance. To bridge this dilemma was the intention of many highly engaged historical sociologists, not least of the Marxist tradition. In order to overcome their and other normative biases, a sociology-of-knowledge perspective is needed that permits us to relate their views to their place of ideological location. If this can be done successfully, the merit of the historical sociology of state and power should also be weighed against the background of other disciplines and subdisciplines of the Human Sciences. In particular, this means the avoidance of those categorical mistakes that are linked to the "hodiecentrism" underlying many of their branches.

Historical sociologists have chosen quite different points of departure for their analyses: from 5000 B.C. (Michael Mann) or, respectively, starting with the enclosures of Tudor England (Karl Polanyi). That the formation of early states can be seen as result of "caging" is certainly quite surprising if compared with the unreflected normality of statehood of today. The view of the state as the only savior from the high risk created by market forces is also not trivial. The assumed starting point depends on what is to be believed the time for a critical juncture, for the "switch" that determines the path for the future: Fernand Braudel (1979a), for instance, saw the rules guiding dressing behavior in public places very differently developed in medieval Europe (quick fashion change, laxity in concealing nudity) from Asia and the Middle East (little fashion change, rigidity in concealing nudity). Gerda Lerner (1986) has shown that such a regime was, as a result of patrimonial state formation, already in place in Hammurabi's time, 1800 years B.C. And scholars dealing with the American Empire do not shrink from comparisons with Rome – although there is certainly no direct path leading from one to the other. "Path dependence" has to be proven empirically in every case, according to the main puzzle to be solved.

Do we have to analyze the relevant processes in a comparative way, or is it sufficient to study them as historical individuals on their own? Is our historical, processual understanding of the development of state and state-related power based on historical parallels between different periods, or is it rather based on the uniqueness of a specific development, albeit with more general or even universal relevance? There are famous dilemmas in historical-sociological research that deals with the so-called *Universalgeschichte*: Should we treat the human propensity to trade by the exchange of goods as the natural, sufficient determinant of market capitalism, or is the latter contingent on different particular preconditions inviting comparative study? The answer tells us something about the choices people may have also when faced with the options of the present. Is occidental rationalism the historically individual result of contingent conditions, or is it a Eurocentric version of white supremacy, to be corrected by the comparative study of Arab, Chinese, or Indian achievements? On the other hand, does the comparative, lawlike perspective do sufficient justice to the complexity of each individual process under scrutiny? That England, for instance, could become the "workshop of the world" as the first industrial state might have depended on the availability of coal and iron (Wrigley 1988), providing the power resources necessary for the acquiring of an empire.

Must we turn to historical sources and their "emic" interpretations by their contemporaries, or can we rely on secondary, theoretically modeled "etic" interpretations of interesting structures, experiences, and events? There is a big difference between a methodological position which demands firsthand knowledge of empirical sources in order to draw empirical generalizations and a position which relies on secondary sources for historical evidence collected by archaeologists or historians and on narratives based on them. Randall Collins (2008) on violence, for instance, often relied on videotapes and voice-recorders as empirical sources. Such sources do not exist before the twentieth centrury; if only these would provide credible information, a histoy of emotions and social habitus would indeed become very difficult. On the other hand, even the theoretical input of sociology for the explanation of contemporary social problems (for instance, the sociology of migration lists push and pull factors between areas of different wealth) can be helpful to analyze events and structures of the past (the movements of people and tribes during the wave of barbarian migrations in late antiquity; Heather 2009). With the help of an "emic" perspective, cognitively structured emotions and plain affects can be found that guided human behavior; Elias relied on novels which helped him to explain the drift toward the legitimation of violence of right-wing groups in Weimar Germany. But it was only with a theoretical synthesis based on his knowledge of the German state-formation process since the Thirty Years' War that an explanation was possible. So this is the last insight to be gained from the long-term perspective on state and political power in historical sociology: It is only in combination with psychology, cultural anthropology, economics, literary studies, and several other branches of the Human Sciences that it can show its fertility; but without the acknowledgment of the processual nature of human societies, these disciplines would miss their very essence.

References

Abrams P (1982) Historical sociology. Cornell University Press, Ithaca/New York
Acham K (ed) (2019) Die Soziologie und ihre Nachbardisziplinen im Habsburgerreich. Böhlau, Vienna/Cologne/Weimar
Anderson P (1974) Passages from antiquity to feudalism. Verso, London
Anderson P (1979 [1974]) Lineages of the absolutist state. Verso, London
Anderson B (1983) Imagined communities: reflections on the origin and spread of nationalism. Verso, London
Anderson P (1987) The figures of descent. New Left Rev 161/1
Anderson P (2009) The new old world. Verso, London
Arendt H (1951) The origins of totalitarianism. Schocken, New York
Assmann J (2000) Religion und kulturelles Gedächtnis. C. H. Beck, Munich
Bendix R (1964) Nation-building and citizenship. John Wiley and Sons, New York
Berber F (1973) Das Staatsideal im Wandel der Weltgeschichte. C. H. Beck, Munich
Bloch M (1973) Royal touch: sacred monarchy and Scrofula in England and France. Transl. J E Anderson. McGill-Queen's University Press, Montreal
Bloch M (2014 [1940]) Feudal Society. Transl. L A Manyon. Routledge, London/New York
Braudel F (1979a) Civilisation matérielle, économie et capitalisme, XV-XVIII siècle. Les structures du quotidien: Le possible et l'impossible. Librairie Armand Colin, Paris
Braudel F (1979b) Civilisation matérielle, économie et capitalisme, XV-XVIII siècle. Le temps du monde. Librairie Armand Colin, Paris
Brewer J (1989) The Sinews of power. War, money, and the English state, 1688–1783. Unwin Hyman, London
Collins R (2008) Violence. A micro-sociological theory. Princeton University Press, Princeton/Oxford
De Swaan A (2015) The killing compartments. The mentality of mass murder. Yale University Press, New Haven/London
Delanty G, Isin EF (2003) Introduction:Reorienting Historical Sociology. In: Delanty G, Isin EF (eds) Handbook of historical sociology. SAGE, London et al., pp 1–8
Duby G (1984) L'Europe au Moyen Age. Librairie Ernest Flammarion, Paris
Dunning E, Mennell S (2013 [1996]) Preface. In: Elias, N (2013 [1989]) Studies on the Germans: power struggles and the development of habitus in the nineteenth and twentieth centuries. Transl. and ed. by S Mennell and E Dunning, Collected Works, Vol. 11: UCD Press, Dublin
Eisenstadt SN (1963) The political systems of empire. Free Press of Glencoe, New York
Eisenstadt SN (1978) Revolution and the transformation of societies. Free Press of Glencoe, New York
Elias N (2006 [1969]) The court society. Transl. by E Jephcott, ed. by S Mennell. Collected Works, Vol. 2, UCD Press, Dublin
Elias N (2009a [1985]) Thomas More's critique of the state – with some thoughts on a definition of the concept of utopia. In: Elias N Essays I: on the sociology of knowledge and the sciences. Ed. by Kilminster R and S Mennell, Collected Works Vol. 14, UCD Press, Dublin
Elias, N (2009b [1987]) 'The retreat of sociologists into the present', in: Elias, N Essays III: on sociology and the humanities. Collected Works, Vol. 16, UCD Press, Dublin
Elias, N (2010a [1985]) The loneliness of the dying and Humana Conditio, Collected Works, Vol. 6, UCD Press, Dublin
Elias N (2010b [1987]) The society of individuals. Transl. by E Jephcott; ed. by R van Krieken, Collected Works, Vol. 10, UCD Press, Dublin:
Elias N (2012a [1939]) On the process of civilisation: sociogenetic and psychogenetic investigations. Transl. by E Jephcott; ed. by S Mennell, E Dunning, J Goudsblom and R Kilminster, Collected Works, Vol. 3, UCD Press, Dublin
Elias N (2012b) [1970] What is sociology? Transl. by G Morrissey, S Mennell and E Jephcott. With a foreword by R Bendix. Ed. by A Bogner, K Liston and S Mennell. Collected Works, Vol. 5, Dublin: UCD Press

Elias N (2013 [1989]) Studies on the Germans: power struggles and the development of habitus in the nineteenth and twentieth centuries, Transl. and ed. by S Mennell and E Dunning, Collected works, Vol. 11, UCD Press, Dublin

Elias N and Dunning, E (2008 [1986]) Quest for excitement: sport and leisure in the civilising process. Ed. by E Dunning, Collected Works, Vol. 7, UCD Press, Dublin

Engels F (2014 [1884] The origin of the family, private property, and the state. Transl. Ernest Untermann. Penguin Classics, London et al.

Ferguson N (2004) Colossus: the rise and fall of the American Empire. Penguin Press, New York

Foucault M (1977) Discipline and punish: the birth of the prison. Penguin, Harmondsworth

Gellner E (1983) Nations and nationalism. Basil Blackwell, Oxford

Good DF (1984) The economic rise of the Habsburg Empire, 1750–1914. University of California Press, Berkeley-Los Angeles

Goudsblom J (1977) Sociology in the Balance. Blackwell, Oxford

Guéhenno J-M (2000) The end of the nation-state. Transl. V Elliott. University of Minnesota Press, Minneapolis

Haas E (1958) The uniting of Europe. Stanford University Press, Stanford

Hamilton GG (1984) Configurations in history: The historical sociology of S. N. Eisenstadt. In: Skocpol T (ed) Vision and method in historical sociology. Cambridge University Press, Cambridge et al., pp 85–128

Harris M (1976) History and significance of the emic/etic distinction. Annu Rev Anthropol 5(1): 329–350

Harris M (1983) Cultural anthropology. Harper and Row, New York

Hayek F (1944) The road to serfdom. University of Chicago Press, Chicago

Heather P (2009) Empires and barbarians. Macmillan, London

Hobsbawm E (1990) Nations and nationalism since 1780: programme, myth, reality. Cambridge University Press, Cambridge

Hobson JM (2002) What's at stake in 'bringing historical sociology back into international relations': Transcending 'chronofetishism' and 'tempocentrism' in international relations. In: Hobden S, Hobson JM (eds) Historical sociology of international relations. Cambridge University Press, Cambridge, pp 3–41

Holton RJ, Turner B (1986) Talcott Parsons on economy and society. Routledge and Kegan Paul, New York

Jonas F (1976) Geschichte der Soziologie, vol. 1. Rowohlt, Reinbek bei Hamburg

Keane J (ed) (1988) Civil society and the state. New european perspectives. Verso, London/New York

Keely L (1996) War before civilization: the myth of the peaceful savage. Oxford University Press, New York

Kelly D (2003) Karl Marx and Historical sociology. In: Delanty G, Isin EF (eds) Handbook of Historical Sociology. SAGE, London et al., pp 1–26

Kennedy P (1988) The rise and fall of the great powers. Economic change and military conflict from 1500-2000. Random House, New York

Klausen J (1998) War and welfare. Europe and the United States, 1945 to the present. Palgrave, New York/Houndmills

Kuzmics H (2019) The fall of the Habsburg monarchy and the crisis of contemporary Europe. In: Arnason JP (ed) European integration. Historical trajectories, geopolitical contexts. Edinburgh University Press, Edinburgh, pp 64–88

Kuzmics H, Axtmann R (2007) Authority, state and national character. The civilizing process in Austria and England, 1700—1900. Ashgate, Aldershot/Burlington

Kuzmics H, Haring SA (2013) Emotion, Habitus und Erster Weltkrieg. Soziologische Studien zum militärischen Untergang der Habsburger Monarchie. Vandenhoeck & Ruprecht unipress, Göttingen

Leonard M (2005) Why Europe will run the 21st century. Atlantic, London

Lepenies W (1985) Die drei Kulturen. Soziologie zwischen Literatur und Wissenschaft. Hanser, Munich/Vienna
Lerner G (1986) The creation of patriarchy. Oxford University Press, New York
Malinowski B (1966 [1935]) Coral gardens and their magic, Vol. 1: Soil-tilling and agricultural rites in the Trobriand Islands. Allan and Unwin, London
Mann M (1986) The sources of social power, vol. I. Cambridge University Press, Cambridge
Mann M (1988) States, war and capitalism: studies in political sociology. Basil Blackwell, Oxford
Mann M (1993) The sources of social power, vol. II: the rise of classes and nation-states, 1760–1914. Cambridge University Press, Cambridge
Mann M (2003) The incoherent empire. Verso, London
Mann M (2012) The sources of social power, vol. III: Global Empires and Revolution. 1890–1945. Cambridge University Press, Cambridge
Mann M (2013) The sources of social power, vol. IV: globalizations, 1945–2011. Cambridge University Press, Cambridge
Marshall TH (1950) Citizenship and social class and other essays. Cambridge University Press, Cambridge
Marx K (1999 [1867]) Capital. A critique of political economy. vol I. Marx/Engels Internet Archive (https://www.marxists.org/archive/marx/works/1867-c1/)
Mearsheimer J (2001) The tragedy of great power politics. Norton, New York/London
Mennell S (1990) The globalisation of human society as a very long-term social process: Elias's theory'. Theory Cult Soc 7(3): 359–371
Mennell S (2007) The American civilizing process. Polity Press, Cambridge
Mennell S (2020) Power, individualism, and collective self perception in the USA. In: Historical Social Research 45 (1) (Special Issue: Emotion, Authority, and National Character), 309–329
Milward AS (1984) The reconstruction of western Europe, 1945–1951. Methuen, London
Moore B (1966) Social origins of dictatorship and democracy. Lord and peasant in the making of the modern world. Beacon Press, Boston
Parsons T (1937) The structure of social action. McGraw-Hill, New York
Parsons T (1951) The social system. Free Press of Glencoe, New York
Parsons T (1966) Societies: evolutionary and comparative perspectives. Prentice Hall, Englewood Cliffs
Parsons T (1971) The system of modern societies. Prentice Hall, Englewood Cliffs
Pike KL (1967 [1954-60]) (ed) Language in relation to a unified theory of structure of human behavior (2nd ed.). Mouton, The Hague/Netherlands
Pinker S (2011) The Better Angels of our Nature. Why Violence Has Declined. Viking, New York
Pixley J, Flam H (eds) (2018) Critical junctures in mobile capital. Cambridge University Press, Cambridge
Polanyi K (1944) The great transformation. Farrar and Rhinehart, New York
Rifkin J (2004) The European dream: how Europe's vision of the future is quietly eclipsing the American Dream. Jeremy P Tarcher/Penguin, New York
Rummel RJ (1994) Death by government. Transaction Publishers, New Brunswick
Rusche G, Kirchheimer O (1939) Punishment and social structure. Columbia University Press, New York
Rüstow A (1950) Ortsbestimmung der Gegenwart. Eine universalgeschichtliche Kulturkritik in 3 Bänden, vol. 1: Ursprung der Herrschaft. Eugen Rentsch Verlag, Erlenbach-Zürich/Stuttgart
Scheidel W (2017) The great leveler: Violence and the history of inequality from the stone age to the twenty-first century (The Princeton Economic History of the Western World). Princeton University Press, Princeton
Schütz A (1932) Der sinnhafte Aufbau der sozialen Welt. Eine Einleitung in die verstehende Soziologie. Springer, Vienna
Shaw M (2002) Globality and historical sociology: state, revolution, and war revisited. In: Hobden S, Hobson JM (eds) Historical sociology of international relations. Cambridge University Press, Cambridge, pp 82–98

Skocpol T (1979) States and social revolutions: a comparative study of France, Russia and China. Cambridge University Press, Cambridge

Skocpol T (1984) Sociology's historical imagination. In: Skocpol T (ed) Vision and method in historical sociology. Cambridge University Press, Cambridge et al., pp 1–21

Smith AD (1986) The ethnic origins of nations. Basil Blackwell, Oxford

Smith AD (1991a) National identity. Penguin Books London, New York

Smith D (1991b) The rise of historical sociology. Temple University Press, Philadelphia

Tenbruck F, Homann H (eds) (2002) Das Werk Max Webers: Gesammelte Aufsätze zu Max Weber. Mohr Siebeck, Tübingen

Thompson EP (1963) The making of the English working class. Victor Gollancz, London

Tilly C (1992) Coercion, capital, and European States, AD 1990–1992. Blackwell, Oxford

Van Creveld M (1991) The transformation of war: the most radical reinterpretation of armed conflict since Clausewitz. Free Press, New York

Vernant JP (1955) Travail et Nature dans la Grèce Ancienne. J Psychol 52:18–38

Wallerstein I (1974) The modern world-system: capitalist agriculture and the origins of the European economy in the sixteenth century. Academic Press, New York

Wallerstein I (1980) The modern world-system, vol. II: mercantilism and the consolidation of the European World-Economy. Academic Press, New York

Wallerstein I (2011) The modern world-system, vol IV: centrist liberalism triumphant 1789–1914. University of California Press, Berkeley

Weber M (1978 [1920]) Economy and society. An outline of interpretive theory. Roth G, Wittich C (eds). University of California Press, Berkeley/Los Angeles/London

Weber M (1988 [1896]) Die sozialen Gründe des Untergangs der antiken Kultur. In: Weber M, Gesammelte Aufsätze zur Sozial- und Wirtschaftsgeschichte, J. C. B. Mohr (Paul Siebeck), Tübingen, 289-311

Weber M (1992 [1920] Author's introduction (Vorbemerkung to GARS). In: The protestant ethic and the spirit of capitalism. 1904–05/1992. Parsons T (trans.), A. Giddens (intro), Routledge, London.

Weber M (1994 [1920]) Weber: political writings (Cambridge marx) (Lassman P, ed.; Speirs R, Trans.). Cambridge: Cambridge University Press. https://doi.org/10.1017/CBO9780511841095

Wittfogel KA (1977 [1957]) Die Orientalische Despotie: Eine vergleichende Untersuchung totaler Macht. Ullstein, Frankfurt/Berlin/Vienna

Wouters C, Dunning M (eds) (2019) Civilisation and informalisation. Connecting long-term social and psychic processes. Palgrave Macmillan, Cham/Switzerland

Wrigley EA (1988) Continuity, chance and change: the character of the industrial revolution in England. Cambridge University Press, Cambridge

Organized Violence and Historical Sociology

24

Christian Olsson and Siniša Malešević

Contents

Introduction	626
War	627
Clandestine Political Violence	631
Revolutions	636
Genocides	641
Conclusion	646
References	647

Abstract

This chapter briefly reviews and analyzes the key contributions on organized violence within historical sociology. It explores both the macro- and micro-level studies that have influenced recent debates within the field. The first section looks at war and warfare, the second section analyzes the clandestine political violence, the third section explores the revolutions, and the final section engages with the scholarship on genocides.

Keywords

Organized violence · War · Revolution · Genocide · Clandestine political violence

---

C. Olsson
Université libre de Bruxelles, Bruxelles, Belgium
e-mail: Christian.Olsson@ulb.be

S. Malešević (✉)
School of Sociology, University College, Dublin, Ireland
e-mail: sinisa.malesevic@ucd.ie

Introduction

There is no doubt that organized violence has shaped much of human history. Wars, revolutions, genocides, uprisings, rebellions, riots, clandestine political violence, and many other forms of organized violent action have been the defining drivers of social change and have ultimately created the contemporary world. Wars have played a decisive role in transforming empires and patrimonial kingdoms into nation-states (Wimmer 2013), revolutions have inaugurated variety of modern social orders and have contributed to the development of democracy (Lawson 2019, Moore 1966), and genocides have impacted on the global institutionalization and standardization of human right regimes (David 2020), while the proliferation of the clandestine political violence has transformed security systems and has increased coercive capacities of modern states (Mann 2013).

However, much of the comparative historical sociology has focused on other themes, and organized violence has only recently become a prominent research topic within this field. The conventional and the dominant understanding was that these topics stand outside of sociology's scope and that they should be regarded as the legitimate research domain of other academic disciplines such as political science, security studies, or military history. The pervasive methodological nationalism combined with "the retreat of sociologists into the present" (Elias 1987) tended to obscure the centrality of organized violence in history.

Some of this reluctance to focus on organized violence was also a direct legacy of the post-WWII era as many sociologists working in the 1950s, 1960s, and 1970s felt uncomfortable to revisit the unprecedented bloodshed and devastation caused by the revolutions, rebellions, genocides, and two total wars of the early twentieth century (Malešević 2010a:17–18). Hence, it was only from the 1980s that historical sociologists have started producing influential theoretical analyses and comprehensive empirical studies on organized violence.

This is not to say that the classics of sociology completely ignored violence. On the contrary, both Marx and Weber have discussed nineteenth- and early twentieth-century revolutions, while early historical sociologists such as Otto Hintze, Gustav Ratzenhofer, Franz Oppenheimer, and Lester Ward produced valuable sociological analyses of war (Malešević 2010b). Norbert Elias has also explored the historical dynamics of organized violence in Europe in *On the Process of Civilisation* (1939), but the two volumes of this book were not widely circulated and became available in English only in 1969 and 1982, respectively (Mennell 1998).

Nevertheless, these classical contributions had little impact until well into the 1980s and 1990s when a number of historical sociologists have started revisiting their ideas and have developed new approaches to the study of organized violence. The initial focus was on the state formation and the role revolutions and warfare have played in the transformation of social orders, while the more recent scholarship has also zoomed in on the genocides and the clandestine political violence. In the last few decades, organized violence has become a significant research topic within the comparative historical sociology, and historical sociologists have produced a plethora of comprehensive comparative and theoretical studies. In many respects, the contemporary historical sociology is a vibrant and expanding research field that is characterized by prolific and innovative output.

This chapter briefly reviews and analyzes the key contributions on organized violence within historical sociology. It explores both the macro- and micro-level studies that have influenced recent debates within the field. The first section looks at war and warfare, the second section analyzes the clandestine political violence, the third section explores the revolutions, and the final section engages with the scholarship on genocides.

War

War, defined as sustained organized violence between political units, has long been neglected as an object of study in sociology (Joas and Knöbl 2012). Neo-Weberian historical sociologists have however done much in the 1980s to rehabilitate it as a legitimate object of study, often in a constructive dialogue with Marxism.

In the 1970s, Perry Anderson contributed to a loosening of the Marxist postulate of the economic determination of political superstructure (1974). He argued that the European absolutist state was both the expression of the aristocracy's class interests and the consequence of increased geopolitical struggles. It hence had a "relative autonomy." Pursuing on this idea, historical sociologists inspired by Weber such as Tilly (1985), Mann (1986), or Giddens (1985) analyzed war as a trigger, catalyst, and symptom of large-scale social and political transformations. Without underplaying the role of economic factors, they insisted on the autonomous role of political structures and power struggles. Most of this literature considers war in the context of the advent of national states in Europe. War is accordingly seen to have shaped, and as having been shaped, by the rise of the modern state.

The idea of a close relation between war-making and state-making has been a recurring leitmotiv since the seminal work of Weber (1978), Hintze (1975), and Elias (2012). It is true that the defining elements of modern wars, the pitched battles between well-drilled battle formations, military discipline in the face of mass onslaught, have been pushed to their paroxysm by the advent of bureaucratic states extracting increasing resources from their national (and sometimes colonial) economies.

How have national states crystallized out of the context of war-making? By giving a structural advantage to polities with the most financial and coercive resources, territorial wars have from the end of the renaissance onward pushed toward the fiscal and military monopolies defining the territorial state. In the process, alternative forms of political organization were eliminated: the decentralized authorities of feudalism, city-states, composite states, etc. Many entities that had significantly contributed to the modernization of warfare in the gunpowder age, such as the Duchy of Burgundy and the Kingdom of Piedmont, were also swept away in the process.

The relation between war-making and state-making is however not a linear or even a purely circular one. It is here more useful to think in terms of contingency.

Indeed, some of the features one associates with modern interstate wars, such as well-drilled infantry units, existed long before, including outside of Europe. It must however be noted that the periods seeing significant developments in infantry drills were all characterized by high levels of political organization. The well-drilled Roman legions disappeared with the end of the Roman Empire. Inversely, the return of complex infantry drills in the wars of religion of the sixteenth and seventeenth centuries is linked to the advent of bureaucratic states. In the same way, as highlighted by T. Andrade (2016), the very early appearance of military drill in China during the "Warring States Period" (475–221 BC), and its development under the Tang, Song, Ming, and Qin dynasties, is inseparable from the bureaucratized nature of these Chinese empires.

Why then has the modern state emerged in Europe and not elsewhere? The answer cannot be straightforward. The dispersal of military power characteristic of feudalism has played a role by creating a structural competition between multiple power centers. As highlighted by D. Nexon (2009), religious Reformation and Counter-Reformation also played its part by undermining the composite monarchies of the fifteenth and sixteenth centuries.

Historian G. Parker (1996) has shown that a set of autonomous technological and military innovations also played a crucial role: the invention and improvement of canons able to break down the forts of feudal lords, the emergence of powerful broadside ships, the bastion fortifications of the late renaissance, etc. The increasing costs of these military technologies tilted the advantage toward the polities most able to afford them: the highly extractive national states. The invention of the industrial revolutions further accentuated this tendency, ultimately leading into the century of total wars and totalitarian regimes.

The history of early modern war-making is however not one of the state centralizations alone. As highlighted by Elias (2012), it is also one of the "domestic pacifications" and of the "socializations of the monopoly": the first refers to the fact that as states monopolize violence, interpersonal violence is increasingly repressed domestically; the second describes the situation in which monarchs needed to negotiate with ever wider societal sectors to access the resources required for war without triggering uprisings. As highlighted by S. Tarrow (2015), the advent of parliamentarism and the political inclusion of the bourgeoisie are hence tightly linked to the history of warfare. Sometimes, the requirements of these compromises led to the overturn rather than the reform of states. This can also be observed outside of Europe. The Meiji revolution in Japan in 1868 largely accounts for the military efficiency of the Japanese state in comparison to the Qin dynasty's China toward the end of the nineteenth century: while the Japanese emperor freed himself from his dependency on the traditional warring class to modernize his military, China remained bogged down in its dependency on "warlords" resisting military modernization (Andrade 2016).

What about the role of war in the shaping of the international order? In sociology, it is mostly the Marxian and Braudelian inspiration that, through Wallerstein's world systems theory (2004), further developed by Chase-Dunn (1999), has focused on the global implications of hegemonic wars. The latter are struggles between states at the core of the "world system." The outcome of these wars purportedly redefined the asymmetric relations between core and periphery in the world economy. Wallerstein hence sees three world systemic hegemonies: the Dutch (1625–1672), British (1815–1873), and American (1945–1965). Each one has been marked by 30-year-long hegemonic wars: the Thirty Years' War leading to the Treaties of Westphalia, the French revolutionary and Napoleonic Wars leading to the Concert of Europe, and WWI and WWII leading to the creation of the UN.

Where does this leave the historical sociology of war today? Does its state-centricity not question its relevance in a world in which non-state actors, "failed states," and transnational networks are often said to have fundamentally transformed the nature of war?

State-centricity does not necessarily imply an exclusive focus on interstate wars. Civil wars are not necessarily less state-centric. Indeed, on the one hand, they often can be analyzed as elimination struggles, multiple actors trying to reconstitute a monopoly over legitimate violence. On the other hand, the armed organizations emerging from the collapse of state authorities often build and expand on the organizational infrastructure of former state bureaucracies (Malešević 2017). Just as the dispersed struggles of feudalism set the stage for the modern state in Europe, such civil wars might set the stage for an increasingly bureaucratized post-conflict state as highlighted by post-genocide Rwanda. In other cases, such as in Chad, the successive civil wars revolve around the distribution of official positions inside the state (Debos 2016).

It is however true that the focus on the link between war-making and state-making has prevented historical sociologists from exploring alternative relations between war and social organizations. Anthropologists such as P. Clastres have, for example, has shown that amongst the Guayaki, Guarani and Yanomami native populations of South America, warfare participates actively in the reproduction of stateless societies adverse to differentiated power structures (1989). E.E. Evans-Pritchard shows a similar function of warfare in traditional Nuer society in Sudan (1987): in such segmentary societies, war prevents one segment from becoming more powerful than all others combined, thus precluding the advent of a proto-state.

State-centrism has however not prevented historical sociologists from doing a good job in highlighting the role of non-state actors in the early modern European wars. This is highlighted by the work of J. Thompson (1994; see also Olsson, 2016) on mercenaries and pirates and of D. Nexon (2009) on transnational religious dynamics.

Finally, by analyzing war in colonial, multinational, or transnational settings, historical sociologists are increasingly moving away from the state-centricity and Eurocentricity of early works (Barkawi 2017). In fact, the transnational circulation of military power is what has led authors like M. Mann to distinguish military and political sources of power. Mann sees political power as essentially territorial and centralized, while military power often defies the territorial borders of (pristine) states (1986). As highlighted by M. Shaw, military power asymmetries lead to the creation of "international 'conglomerates' of state power" such as NATO (Shaw 2002:89). T. Barkawi has shown that wars are often moments of accelerated globalization (2006). He further shows that the multinationality of imperial armies hardly was an obstacle to military cohesion (2017).

Although historical sociology tends to privilege the macro-sociological level, there is a prolific literature in sociology on the micro-dynamics of war. As interstate wars become increasingly lethal, what motivates soldiers to fight? Traditionally, in mass charges, collective arousal is believed to play an important role. With the advent of automatic weapons, such mass charges however constitute easy targets. To respond to this challenge, dispersed and technically more demanding infantry tactics have been developed in the beginning of the twentieth century (King 2013). The latter however highlight the difficulty of maintaining combat motivation in the absence of peer pressure: soldiers tend to cower in fear, panic, or flee in the face of the mass slaughter of industrial war (Collins 2008). The use of group-operated weapons and long-distance killing does not allow overcoming this problem. The question of why soldiers fight has hence gained in saliency throughout the twentieth century.

Some authors have in this regard highlighted the role of coercion and military discipline: soldiers fight because they otherwise face disciplinary or penal measures. These explanations are however insufficient at best: soldiers often show great creativity in collectively circumventing disciplinary rules; if the alternative is a gruesome death, why would they fear punishment? More sociological approaches have insisted on the role of micro-solidarity within small units (Malesevic 2017). According to this strand of research, large-scale military organizations are abstract and impersonal and therefore unable to directly motivate soldiers to fight and die. They only do so by mobilizing primary group associations, networks of solidarity based on face-to-face interactions between soldiers (Shils and Janowitz 1948). Another strand of research highlights the role of ideologies imbuing soldiers with a sense of social superiority. This sense compels them to fight to avoid the shame of desertion or defeat. The importance of norms of masculinity and racist ideologies is here of particular interest (Goldstein 2003; Bartov 2001).

Situational emotional dynamics also play an important role. R. Collins has highlighted the extent to which the activity of killing runs counter to the emotional flows of normal social interactions (2008). In situations of armed conflict, the human aversion to killing becomes a grandiose obstacle. In the confrontation between military units, the tension and fear of confrontation build up on both sides. Competent action becomes extremely difficult as a result. If however one of the sides crumbles, a sudden release of the accumulated tension on the opposing side is likely to lead to "forward panics," sudden mass charges, or unilateral killing sprees. Even in these instances, it is typically only a minority of individuals who use force competently (Collins 2008).

A. King has highlighted that in professional armies, the value of professionalism and compliance to professional norms explains much of the competent action of soldiers on the battlefield (2013). King also highlights the extent to which military drills allow conditioning military behavior. Over time, it builds confidence in the collectively performed moves, thus maintaining competent action in spite of "tension/fear."

With the development of modern means of transport, military marches have largely become a ceremonial practice. Why are close-order marching drills then still performed in most armies? T. Barkawi highlights the role of parade ground marching and other military rituals in creating an embodied sense of belonging that favors cohesion in combat (2017). Drawing on Durkheim's work on religion, he highlights how collectively performed rituals create a sense of transcendence, community, and collective meaning in spite of social and cultural differences. Barkawi highlights that although military casualties often are analyzed as a demotivating factor in military units, they at the same time function like a sacrificial rite galvanizing soldiers to continue fighting until the bitter end.

Clandestine Political Violence

What in the news has become an obvious word, terrorism, is a contested concept in historical sociology. As highlighted by Charles Tilly among others, the category conflates distinct phenomena, and its pejorative connotation makes it part of the very conflict from which it springs. For Tilly, the use of disruptive forms of asymmetrical violence is common to different types of actors, and most of these combine different repertoires of action. To characterize a whole organization as terrorist would accordingly not make much sense (Tilly 2004; Bigo 2005; Porta 2013).

Even among those endorsing the concept, there is no consensus on definitions. Among the disagreements, one can mention whether its use shall be restricted to non-state armed groups or can include states, if it can be considered independently from armed conflict or revolutionary processes, whether it by definition targets civilians or if attacks against military targets also can qualify, if it necessarily is indiscriminate or can also target specific civilians.

The term originates in the rule of "terror" (*terreur*) of the French revolutionary government in 1793–1794. It was first systematically used in the nineteenth century to refer to revolutionary anarchists. Some authors like David C. Rapoport have used it to study a much more distant past, seeing the Ismaili Hashashin (Assassin) order of the medieval Levant, the Jewish Zealot-Sicarii under the Roman Empire, or the Hindu "Thugs" under the British Raj as the forebears of "religious terrorism" (Rapoport 1984). Such analyses are however often a-sociological. In the case of the "Thugs," they also take the colonial discourse at face value, as most historical sources serve to justify the British Empire. The parallels drawn (long before the advent of "Jihadi terrorism") between contemporary "terrorists" and past millenarist groups are however interesting from a narrative point of view. They highlight that the rhetoric of many modern groups, irrespective of ideological persuasions, draws on a common religious trope: the destruction of the current corrupt world announces the birth of a new and better one. Narrative parallels alone are however insufficient to identify a historically coherent set of practices.

The elements that are the most systematically used to define terrorism are the intentional killing of civilians to instigate fear and the indirect and disruptive use of violence against "soft targets" to reach political goals. These are however relatively frequent features of warfare as highlighted by commerce raiding, strategic bombing, or the use of nuclear weapons. For this reason, the concepts of clandestine political violence and clandestine armed group are to be preferred to the ones of terrorism and terrorists. These descriptions are indeed more precise and less normatively charged (Porta 2013).

This precision however implies that when referring to clandestine armed groups, major parts of the history of Hezbollah, Hamas, or the Taliban must be disregarded. These groups are indeed not clandestine in their heartlands. They rather take on parastatal characteristics linked to the modern state capacities that are brought to bear against them. Although they all three have used clandestine political violence outside of this heartland, they also have engaged in warfare against conventional armies.

The type of violence instigated in the 1970s–1980s by (nearly) fully clandestine organizations such as ETA in Spain, the Red Brigades in Italy, the RAF in Germany, or the Provisional IRA in the UK also has distinctly modern features. This point is generally recognized by terrorism experts. They have been eager to highlight how contemporary technological developments (miniaturization, etc.), and the global reach by the international media, make disruptive forms of violence less costly and more rewarding. The transnationalization and networking of groups such as Al-Qaeda are also often highlighted. The modernity of such organizations however also must be analyzed against the backdrop of more structural factors affecting political organizations. Three interrelated elements are of interest from a historical sociological perspective.

The first concerns clandestinity. Systematic clandestinity becomes a necessity once national territories are controlled homogenously by Weberian states (Sinno 2011). Secret groups have obviously existed well before the modern state, but generally these were only very contextually secret (in certain cities or areas) or were secret in relation to a dominant religion. Armed groups that operate clandestinely over large territories are quite specific to the modern state. It is not for nothing that "terrorism" is often defined as an urban form of guerrilla. In states that are infrastructurally weak and poorly urbanized, armed groups are generally able to come out of clandestinity in sparsely populated rural areas: they develop into guerrilla movements. Reversely, fully clandestine organizations develop in urbanized and/or infrastructurally strong states. In the "AfPak" region, the Taliban guerrilla's heartland is in rural Afghanistan, while the fully clandestine Al-Qaeda central is rather based in big Pakistani cities.

The second concerns "terrorism" seen as a "weapon of the weak" targeting civilians to challenge official political authorities. The impact of this indirect strategy toward political authorities is linked to the claim on the part the latter to protect their population. This is a specific modern claim that Foucault locates toward the end of the eighteenth century when governments take over the "pastoral power" of the church and shoulder responsibility for the well-being of their population (Foucault 2003). It is only when the "security" of the population becomes an explicit mission of the state that the random killing of civilians becomes an efficient means of challenging political authorities. It is not for nothing that the Zealot-Sicarii or the

Assassin Order targeted influential political and religious leaders rather than random civilians (Rapoport 1984).

The third element concerns spectacular violence. A. Tocqueville has highlighted a peculiar phenomenon: the more the formal egalitarianism of political modernity becomes pervasive, the more the subjective feeling of the ever presence of inequality grows (Tocqueville 2002). The same can be said of violence. The lesser the physical violence is part of the everyday experience of ordinary people, the more the sudden irruption of violence in everyday situations inspires "shock and awe." Consequently, asymmetric violence against civilians is all the more rewarding as societies are domestically pacified. The political impact of the rarely lethal attacks of the PIRA in the UK has hence to be put in the context of the long-term decline of interpersonal violence highlighted by historians of violence (Muchembled 2012).

In order to highlight some of the debates on clandestine political violence in sociology, this chapter will focus on four issues.

The first deals with the question of the instrumental or expressive nature of this violence. Two main perspectives can be distinguished. According to the first, often inspired by Weber or rational choice theory, political movements or organizations deploy political violence instrumentally, as part of their wider effort to mobilize their support base, challenge the incumbents, and maintain control compliance (Tilly 2004; Kalyvas 1999; Oberschall 2004; Malešević 2019). According to the second, more neo-Durkheimian perspective, this form of violence arises in response to social anomy. Violence here serves to express a sense of belonging to an imaginary community. Rather than being instrumental, clandestine political violence is largely used for its immediate "consummatory rewards" linked to its ability to make an identity claim and to reverse the stigma of political exclusion (Wieviorka 2009; Alexander 2011).

The notion of expressive violence is however confusing. Violence, used as a mode of expression and political communication, does not contradict the notion of instrumentality. The two dimensions come together in the notion of "propaganda of the deed" used by revolutionary anarchist groups in the nineteenth century. Violence is a propaganda tool used to mobilize and galvanize the masses. Indeed, when targeting incumbent authorities, it highlights their weaknesses and exposes their repressive nature should they respond indiscriminately. When targeting civilians, it might be part of a conflict strategy of polarization. It seeks to activate boundaries between social groups, for example, by inviting retaliation on the part of the targeted group. While war involves the use of organized violence between political units, clandestine organizations rather use violence to become political units able to engage in war.

The possibilities of using violence to convey a message are in fact diverse. The political communication might be destined to out-group (distant enemies or allies) or in-group (intimate enemies or the support base) audiences. The violence itself might primarily target this audience or the presumed enemy of this audience. When combining these two dimensions, four types of clandestine political violence can be distinguished. They are called here, respectively, rule by intimidation, strategy of polarization, proactive guerrilla, and courting foreign assistance.

	Targeting the audience	Targeting the "enemy" of the audience
In-group audience	1. *Rule by intimidation*: GIA in Algeria	1. *Strategy of polarization*: AQI attacks on Shiite civilians, revolutionary uprisings
Out-group audience	2. *Proactive guerrilla*: attack on Marine Barracks in Beirut 1983	2. *Courting foreign assistance*: Armenian nationalist revolutionary attacks on ottoman civilians to spur state repression and Western intervention in the beginning of the twentieth century

A second debate concerns the causes that lead individuals to join clandestine armed groups. According to a first approach, poverty plays a central role in triggering political violence (Falk et al. 2011). No statistics have however corroborated this claim, and existing studies rather show that people from educated middle classes are overrepresented. A second approach rather points at the role of education in fuelling social aspirations contradicted by limited perspectives of social promotion (Sageman 2004). A third approach focuses on the role of networks of micro-solidarity in the recruitment of activists. Members of clandestine armed groups indeed tend to be recruited by people whom they know closely from other social activities or through family relations (Porta 1988, 2013; Bosi and Porta 2012).

A third question concerns the nature of clandestine groups. Scholars using the framework of contentious politics and social mobilization highlight that clandestine groups emerge (and break way) from wider political movements that use a variety of mostly nonviolent repertoires of action. In this context, "radicalization" occurs as a result of multiple interactions involving policing, competition between divergent groups of the political movement, and the progressive insulation of a network of increasingly radicalized individuals (Porta 1988, 2013). On the contrary, scholars from political science and Weberian sociologists rather tend to focus on the role of hierarchical and bureaucratized clandestine armed organizations in maintaining internal cohesion in the face of state repression and recruiting preexisting networks in order to commit acts of violence (Hassan 2014). These organizations might benefit from (outside) state support or from defectors from states bureaucracies (Sinno 2011; Weinstein 2006; Staniland 2014; Malešević 2017, 2019).

Fourthly, a frequent question concerns the role of religion. While it is common to refer to "religious terrorism," the question is what exact role religious ideologies play. According to some authors, rigoristic religious ideologies serve to shore up combat motivation in asymmetrical conflicts (Kalyvas and Balcells 2010). Others show that it serves a strategy of strategic signaling by which armed groups convey their commitment in situations characterized by high degrees of uncertainty as to the combat motivation of diverse groups (Walter 2017). A third approach highlights how "religious extremism" emerges from strategic outbidding between rival oppositional groups in countries in which ideological offer is severely restrained by political authorities (Toft 2007).

The study of the Afghan civil war in the 1980s and 1990s, from which groups as Al-Qaeda and the Taliban emerged, however reveals an alternative explanation. In Afghanistan, political networks in the countryside (often around influential khans or other bigger notables) tend to be strong but at the same time parochial. On the contrary, the logistical networks of religious leaders tend to be diffuse yet span huge distances: they are nationwide or even transnational. This is linked to the role of madrassas (irrespective of their ideological orientation) in the structuration of long-distance solidarities between students from widely diverse localities. This was in the 1980s even more the case of the madrassas that, because they were located in Pakistani border regions, were able to recruit war refugees from all over Afghanistan who had been forcibly displaced by Russian aerial bombardment.

As a result, any broad armed mobilization initiated by local political leaders tended to fragment geographically and lead to infighting between localities. Armed organizations structured by networks of "students in religion" were infrastructurally more robust for reasons that have not much to do with religious beliefs but everything to do with the logic of recruitment of religious schooling (Dorronsoro 2005). They were able to mobilize on a national, rather than only local, basis without fragmenting. While the Afghan uprising in 1979 hence began with widely inefficient local groups, the more efficient organizations were the ones structured around translocal or transnational religious networks able to command allegiance over wider geographical distances (Rubin 2002).

One of the peculiarities of the literature clandestine political violence is that it more often deals with its micro-foundations than with macrostructures. This linked to a moral or legal focus on criminal intent and a practical concern for individual rehabilitation. Part of this literature deals with the psychological profile (and mental health) of identity-seeking political activists. Most authors however agree that individuals joining clandestine armed groups generally are not any different from other members of their societies. More sociological-oriented authors such as Porta and Bosi however focus on the role of closed networks; some would call them primary group associations, in the recruitment into clandestine groups (Porta 1988; Bosi and Porta 2012). This is less the result of choice than of a permanent fear of infiltration by government services. By giving priority to people, such as school friends, former colleagues, or family members, the recruiter reduces the risk of compromising the organization's security. This is even more the case as the recruits generally come from broader political movements and only progressively are selected into the clandestine organization based on their proven commitment. Although the overexposed phenomenon of "self-radicalization" on the Internet exists, it rarely gives direct access to the organization without prior proof of allegiance.

Closed recruitment and a systematic suspicion encouraging militants to cut preexisting social ties ensure organizational closure and strong emotional bonds between militants. While this process of encapsulation favors ideological orthodoxy and interpersonal solidarity, it also favors groupthink and can lead the organization to lose ground with the wider movements of potential sympathizers. Violence sometimes seeks to restore this missing organic link to wider audiences, often without significant success as highlighted by the RAF. More successful organizations however combine clandestine armed branches with more open front organizations (Berti 2013). The latter allow the organization to remain open to sympathizers or occasional supporters. The former guarantee operational security as well as military efficiency. Organizations like Hezbollah are known, in their recruitment policy, to staff their military units following geographical location. This means that preexisting networks of micro-solidarity based on place of residence, clan, and family are directly recruited into its military organization, a fact that contributes to its units' efficiency in combat.

Revolutions

While wars, genocides, and clandestine political violence largely remained on the margins of historical sociological research, the same thing cannot be said about revolutions. In fact, both classical and contemporary historical sociologists have devoted a great deal of scholarly attention to the study of revolutions. For example, Marx and Weber have written extensively about the revolutionary upheavals of their time and have also explored revolutions in a comparative historical sense. Marx is today better known for his programmatic theory that envisages and advocates the proletarian revolution with the working class successfully overthrowing the bourgeoisie, as depicted in the Communist Manifesto: "Let the ruling classes tremble at a Communistic revolution. The proletarians have nothing to lose but their chains" (Marx and Engels 2005[1848]: 89). Nevertheless, in his more sociological work, he analyzed the social dynamics of 1848 revolutions in Germany, France, Italy, and Austria as well as the legacy of the 1871 Paris Commune. In the "Civil War in France" (1871) and other writings on the Paris Commune, Marx reassessed his earlier views of revolutionary action arguing that the tragic defeat of the Paris uprising indicated that violence is central to revolution and that proletariat cannot use the same coercive methods of the bourgeoisie: "the working class cannot simply lay hold of ready-made state machinery, and wield it for its own purposes."

Weber also studied revolutions as they unfolded. His focus was primarily on the Russian revolutions of 1905 and 1917. He learned Russian to study the revolutionary situation and has written several essays that link the Bolshevik seizure of power to his theories of bureaucracy and rationalization. For Weber, revolution was an illegitimate usurpation of authority that transpires in the context of state breakdown. He explained the revolutionary outcome through the prism of geopolitical forces (i.e., defeat in war), disagreements between political and military elites, and the mobilization of interest groups that have been excluded from power. He emphasized the role of military power which in 1905 was severely paralyzed: "if even a tenth of the officer corps and the troops remain at the disposal of the government...then any number of rebels would be powerless against them" (Weber in Collins 2001:186). By 1917, the war was lost by Russia after heavy casualties which severely delegitimized the old order and contributed to elite polarization, an unwillingness on the part of the military to intervene in support of the government and society-wide dissatisfaction. Weber successfully predicted that Bolshevik victory would bring about bureaucratic dictatorship that would establish a monopoly of party officials instead of the promised "dictatorship of the proletariat."

While the classics of historical sociology have made initial steps toward understanding the revolutionary processes, the first full-fledged theories of revolution emerged in the late 1960s, 1970s, and 1980s. Barrington Moore and Theda Skocpol developed a novel structuralist model of historical change where revolution was identified as the key catalyst of social transformation. For Moore (1966, 1978), revolutionary violence played a decisive role in the birth of modernity. Nevertheless, unlike the conventional interpretations which associated modernity solely with the liberalism of the French and American revolutions, Moore (1966) demonstrated convincingly that both state socialism and fascism were also modern projects built on top of distinct revolutionary experiences. He contrasted the "three routes to the modern world" arguing that the presence of a strong bourgeoisie was a precondition for the development of liberal democracy. Thus, whereas the bourgeoisie prevented the dominance of the aristocracy in England, France, and the USA, the revolutionary experiences of Russia, China, Japan, and Germany were very different as the weakness of the bourgeoisie forced coalitions with the aristocracy or peasants, respectively. These coalitions ultimately brought about communism or fascism.

Skocpol continued this line of argument but focused more on the power of the state. In her comparative historical analysis of the French, Russian, and Chinese revolutions, Skocpol (1979) identified two primary causes of revolutionary upheavals: the weakened state apparatus and the continuous presence of social discontent. However, in her interpretation, revolutionary agents play only a secondary role, while the revolutions are more likely to happen when the profound political crisis, war defeats, natural disasters, or poor economy generate elite polarization and the state authorities are unable to raise taxes and control the social order. In Skocpol's view, political revolutions do not necessarily transform into social revolutions – while the former relates only to the takeover of the state institutions, the latter is also associated with a profound change in class structure. The French, Russian, and Chinese revolutions were social revolutions par excellence as all three involved the social and political ascent of the new, class-based, ruling groups.

Structuralist theories of revolution reached their pinnacle with the contributions of Tilly (1978, 1995), Goldstone (1991), and Goodwin (2001). These approaches followed in the footsteps of Moore and Skocpol as they too emphasized the centrality of state power in revolutions. However, these authors shifted the focus from class to geopolitics and social movements. Tilly and Goodwin argue that revolutions are not only caused by internal factors such as elite disunity, economic collapse, or fiscal incapacity on the part of the state but also by external factors such as changing geopolitical contexts including protracted warfare, political pressures from powerful states, and global socioeconomic downturns. Goodwin (2001) also identifies political oppression and state violence as factors that can foster political mobilization of dissent. For Tilly (1995) geopolitical context is often linked with internal politics as domestic social movements that challenge the state are often influenced and supported by external groups. The weakening of the state apparatus combined with the strengthening organizational capacity of the social movements can generate revolutionary situations, but only a small number of such situations result in revolutionary outcomes. In addition to these internal and external processes,

Goldstone (1991, 2014) also singles out the role of demography and in particular intense population growth which can strain a state's ability to provide resources and, in this way, delegitimize the ruling strata. For example, the revolutions that transpired during the Arab Spring were in part fuelled by the demographic explosion and the dissatisfaction of young educated groups who experienced substantial downward mobility.

More recent research on revolutions have been critical of the structuralist perspectives. Hence, Selbin (2010), Lane (2009), and Foran (2005) among others insist that structuralism overemphasizes the role of the state and geopolitics while neglecting the role of agency and civil society in fermenting revolutionary situations. Moreover, these new approaches argue that revolutions are not generated by economic and political processes alone but that revolutionary events are predominantly shaped by cultural and ideological factors that mobilize various groups within civil society. Foran (2005) and Lane (2009) understand revolutions through the prism of collective action mobilized through specific interpretative frames. Drawing on their research on the 1989–1991 revolutions that brought down state socialism and more recent work on the color revolutions of the early 2000s, Foran and Lane point out that these revolutions took place in settled geopolitical conditions and stable state structures where there was no pronounced elite polarization. Furthermore, these revolutions were largely peaceful and were spearheaded by ad hoc civil society groups rather than by established revolutionaries and social movements. Lane and Foran argue that material and political factors were of secondary importance in the rhetoric of the protesters and that the emphasis was on shared cultural idioms and collective memories and symbols. Selbin (2010:78) develops this approach further by arguing that revolutions cannot happen without believable narratives of change: "stories are the reason why revolutions are made…without them, there is no resistance, no rebellion, no revolution." In his influential book, Selbin differentiates between four types of revolutionary narratives: (1) the stories of civilization and democratization that underpin the foundational myths of liberal democracies; (2) the narratives centered on questions of social justice, poverty, and inequality; (3) the narratives of national liberation and anti-colonial struggles; and (4) the tragic stories of failed uprisings. In his view, many revolutions draw upon several of these narratives to justify and mobilize public support.

This cultural turn in the study of revolutions has been balanced by other new perspectives which aim to go beyond structuralism and culturalism. Hence, Lawson (2016, 2019) develops a relational approach which shows that revolutions are not standardized phenomena with stable and immutable features but are highly contingent, erratic, and historically framed "entities in motion." This approach moves away from the conventional, mostly essentialist and substantialist, views and advocates a process-oriented analysis that views revolutions as intersocietal phenomena shaped by external geopolitical forces, internal politics, social inequalities, and status disparities. Nevertheless, the objective inequities are unlikely to transform into revolutionary action until they become couched in the language of political injustice. In other words, for Lawson, economic inequality is rarely a direct cause of revolutionary upheavals: instead, asymmetry in the individual access to resources has to be articulated as a political exclusion to spur social mobilization. In a series of empirically meticulous studies, Wimmer (2013, 2018) also shows that political rather than economic exclusion, often combined with geopolitical pressure, has proven crucial in mobilizing large-scale unrests and uprisings. In this context, revolutions often transpire when authoritarian and patrimonial regimes attempt to liberalize or democratize and, in the process, open the space for the voicing of political grievances. Mann (2013) also offers a comprehensive theory that goes beyond the structuralist and culturalist accounts. For Mann (1986:1–2), social order cannot be reduced to singular factors such as economy or culture, it is constituted and operates through the intersecting and overlapping of socio-spatial networks of power. In this context, he analyzes the interdependency of four principal sources of social power (economic, political, ideological, and military) and insists that "all four sources of social power provided necessary preconditions for revolution" (Mann 2013:247). For Mann (2013:246), revolution is a "popular insurgent movement that overthrows a ruling regime and then transforms substantially at least three of the four sources of social power." Analyzing the history of the twentieth-century revolutions, Mann (2013:247) contends that authoritarian regimes that were defeated in wars and were divided by class struggle were more likely to experience revolutions (i.e., Russia, China, Korea, Vietnam, Cuba, Nicaragua, Laos, and Nepal) and that war "continued to determine form of the revolution itself." Coercive power remains important for revolutionaries as they often have to contain the counterrevolutionary forces and in this context use violence to control populations. Since democracies allow for compromise and reform, they are less likely to experience revolutions (Mann 2013:266–7).

These new perspectives have pushed the debate forward recognizing the complexity and historical contingency of revolutionary processes. Nevertheless, what is still missing in these accounts is the analysis of the micro-dynamics of revolution. There is no doubt that revolutions are shaped by and dependent on large-scale macro-historical processes. Yet every revolution involves a distinct microworld: the actions of individuals and small groups that spearhead the uprisings, the people who resist the revolutions and those who join enthusiastically, as well as the numerous ordinary individuals who remain bystanders. The revolutionary experiences also create their own emotional dynamics as individuals forge new forms of group solidarity and new sense of identification. Revolutionary uprisings are often characterized by unpredictable events through which old rules, regulations, and systems of order are transgressed and suddenly replaced by new social realities. In this process, some individuals and groups lose their place in the social hierarchy, while others rapidly climb the social ladder. Thus, revolutions also involve a degree of strategic action as large-scale social change provides new opportunities for some individuals to improve their economic and political position.

If one focuses on revolutionary leaderships, it is clear that shared micro-realities have contributed significantly to revolutionary situations. Many nineteenth- and early twentieth-century revolutionary movements were fronted by clandestine revolutionary cells that were proscribed and persecuted by various governments. In this environment, they had to develop highly disciplined and hierarchical yet flexible and decentralized organizational structures to evade the police. Operating in secret and in hostile conditions meant that such clandestine cells had to rely on trust and loyalty of their members. Hence, such revolutionary units tended to develop a strong sense of micro-group attachment with heightened emotional bonds between members of the cell. Living separate from the mainstream society has also contributed to the idea that the revolutionaries are exceptional and ethically superior individuals who have a moral responsibility to enact a revolution or to eliminate those who are considered to be counterrevolutionaries (Malešević 2017:205–10). For example, the nineteenth-century European revolutionary groups included variety of secret nationalist, communist, anarchist, and other clandestine groups. Some of these organizations were successful in bringing about revolutions or in contributing to revolutionary upheavals – the Young Italy, the Greek Society of Friends, or Turkish Committee of Union and Progress. The close emotional ties forged during the years of struggle between the members of revolutionary cells were later instrumental in political alliances that were formed in the postrevolutionary periods. Although the more recent revolutions rely less on clandestine movements, they also forge emotional and social ties between revolutionaries that are later used as a springboard for building strong political alliances and even new political parties. This has been the experience of the color revolutions on the early 2000s, the Tunisian Jasmine Spring of 2011, and the Ukrainian Euromaidan of 2014 (Lawson 2019; Foran 2005).

Nevertheless, revolutions cannot happen without the ordinary individuals, and it is crucial to explore the emotional dynamics of everyday reality of revolution. Many revolutionary events contain a degree of carnivalesque where ordinary individuals can suddenly transgress the established order. In this context, revolutionary situations give birth to temporary moments of collective effervescence where, as Durkheim was already well aware, individuals attain heightened emotional responses ranging from fear and angst to excitement and pride. With the onset of violence, revolutions forge new forms of solidarity as individuals experience heightened emotional valence – killing and dying become interwoven into the revolutionary narrative and as such create new emotionally infused ideological frames. In Collins's (2004) interpretation, these situations of concentrated emotional interaction are building blocks of interaction ritual chains. These chains entail not only shared emotions but also physicality of interaction – it is no accident that the heightened emotional states involve rhythmic coordination of human bodies. For example, many revolutions have been defined by scenes of people gathering on the public squares and holding hands, hugging each other, or singing and dancing in unison.

By focusing on this microworld of revolutions, one can acquire a better understanding of the collective action and social meanings that transpire before, during, and after revolutionary events. Since revolutions often involve extraordinary acts of individuals who willingly sacrifice their lives for others or vigorously destroy other human beings, it is crucial to analytically penetrate this microworld. It is only by digging deep into this microcosm and by connecting it with the larger macro-historical structures that researchers will be able to explain the social and historical dynamics of revolutions.

Genocides

The concept of genocide has traditionally been used as a legal category and has only recently become an object of sociological research. The term was coined by a Polish-Jewish lawyer, Raphael Lemkin, during the WWII, and he successfully campaigned for this concept and the crime of genocide to become enshrined in the UN as a General Assembly Resolution 260. Hence, the convention on the Prevention and Punishment of the Crime of Genocide was adopted by the UN in 1948 and has been ratified by 152 states. However, the definition of genocide that was agreed at the UN did not reflect fully Lemkin's original proposal. As the Soviet delegation objected to class and political ideology being incorporated into the list of categories associated with genocidal acts, the final definition was a political compromise (Üngör, 2012, 2016). Hence, Article 2 of the Convention states that genocide is an act "committed with intent to destroy, in whole or in part, a national, ethnical, racial or religious group." The following five types of activity are specified in the Convention: "Killing members of the group; Causing serious bodily or mental harm to members of the group; Deliberately inflicting on the group conditions of life calculated to bring about its physical destruction in whole or in part; Imposing measures intended to prevent births within the group; and Forcibly transferring children of the group to another group" (https://www.un.org/en/genocideprevention/genocide.shtml). This rather vague definition simultaneously excluded some social categories (i.e., class, gender, political affiliation, etc.) while also leaving the scale of destruction ambiguous. It is not clear what the intent to destroy in part means – a few individuals, a small village, a city, all male population, etc. Furthermore, the compromise definition reduced genocide to the physical and biological destruction of groups. Nevertheless, as Shaw (2007) rightly points out, Lemkin's original understanding conceptualized genocide in wider terms as set of practices that go beyond the destruction of human bodies. In fact, the systematic killings associated with genocides are often only the last phase of the much broader ideological project: "The Nazis did not aim simply to kill subject peoples, even the Jews; they aimed to destroy their ways of life and social institutions…when physical destruction came…this was an extreme development of pre-existing Nazi policies of social destruction" (Shaw 2007:22).

These political influences have shaped the origins of the UN Genocide Convention, and the political motives continue to frame public debates on genocide. Since the Convention came into force in 1951, it has been used by numerous governments and political groups to make a claim that their groups have been victims of genocide. The vague wording of the definition has allowed proliferation of claims and counterclaims about one's own victimhood and the unprecedented criminal behavior of one's enemies. To be recognized as a group that has experienced genocide has become a parameter of greatest suffering. Moreover, attaining this label has also automatically linked the perpetrator nation with the ultimate crime.

This dominance of legal and political uses of the concept has prevented many social scientists from engaging with this type of organized violence. However, in the last few decades, historical sociologists have attempted to move away from these normative debates in order to develop more universalist and historically grounded theories of genocide. Instead of focusing on identifying the individual culprits of genocide, sociologists have shifted the emphasis on tracing the historical dynamics and social conditions that make genocides possible.

One of the key questions raised by historical sociological research is as follows: Are genocides modern or ancient phenomena? While much of legal and historiographic scholarship does not differentiate between the pre-modern and modern forms of organized violence, many historical sociologists argue that genocide is not only a new concept but also a novel historical phenomenon that only arises under modern conditions. Hence, unlike Smith (1987:21) or Kuper (1981:9) who argue that "the word is new, the concept is ancient" or that "genocide has existed in all periods of history," respectively, many sociologists insists genocide is distinctly a modern phenomenon.

Thus, Zygmunt Bauman in his pioneering *Modernity and the Holocaust* (1989) argues that unlike the traditional pogroms and persecutions of Jews that have been present throughout history, the Holocaust was distinctly modern in its organization, technology, and ideology. While pre-modern forms of violence against Jews were sporadic, situational, and largely disorganized events, the Holocaust was systematic, ideologically articulated, well-planned, and a highly organized project. Unlike the previous instances of religious bigotry that periodically targeted the unprotected and visible religious minorities, the Holocaust was grounded in clearly articulated ideological blueprints. The Holocaust was underpinned by the engineering ambitions that aimed to create a new man and new world where there was no room for what the Nazis identified as the racially inferior others. In this sense, the notion of racial purity was not a throwback to a more primitive past but was in fact an idea developed within the heritage of Enlightenment. The classification of human beings into firmly demarcated racial groups associated with different biological, physical, and mental qualities, as constituted by the race science and eugenics, was not some relic from the barbarian past – it was a profoundly novel and modern way of understanding the social world. This thinking about difference was farmed and legitimized by and with the most advanced scientific theories of its time. Hence, the genocide of six million Jews and many other groups was not perpetuated in an ad hoc manner, and out of religious animosity or simple bigotry, this was a state-led and state-organized project that was pursued and prioritized until the very end of the Nazi state. For Bauman, the Holocaust was not just modern in terms of the ideological narratives that underpinned its realization. It was also distinctly modern in its organization and technology. This unparalleled genocide was implemented using the most advanced bureaucratic system, infrastructure, transport, and communications as well as the most sophisticated industrial and technological knowhow. The science, technology, and administration developed to operate modern manufacturing were deployed to run extermination camps and gas chambers. As Bauman (1989:8) emphasizes, the same factory model that produced commodities for everyday use was adopted for mass murder: "Rather than producing goods, the raw material was human beings and the end product was death, so many units per day marked carefully on the manager's production charts. The chimneys, the very symbol of the modern factory system, poured forth acrid smoke produced by burning human flesh. The brilliantly organised railroad grid of modern Europe carried a new kind of raw material to the factories."

Mark Levene (2005, 2013) develops a similar argument but widens his geographical and historical focus. In his view, genocides transpire in the specific historical context where several structural processes coalesce together: intensive state formation and expansion of its coercive power, ideological radicalization, and the onset of protracted wars. In this approach, genocide is a product of modernity and the widespread aspirations to establish sovereign and culturally homogenous nation-states. Nevertheless, Levene also links the origins of genocide to the imperial expansions and argues that many colonial genocides including the Herero and Namaqua in South West Africa, Belgian Congo Free State, French conquest of Algeria, and genocide of native populations in Americas and Australia were inspired by similar ideological goals. He draws parallels between the mass killings of civilians by the French revolutionary army in Vendée and the genocides throughout the colonial possessions of European empires.

The most comprehensive recent modernist historical sociological theory of genocide was formulated by Michael Mann in his books *The Dark Side of Democracy: Explaining Ethnic Cleansing* (2005) and *Fascists* (2004). For Mann, murderous ethnic cleansing – his term for genocide – is a product of changing geopolitical conditions, often combined with intensified ideological rivalries. Genocides are typically associated with wars and as such are shaped by the changing dynamics of armed conflicts. In his view, genocide is rarely a premeditated and well-planned event but something that gradually develops in the environment of increased radicalization. In most cases, genocidal policy emerges in the context "where powerful groups within two ethnic groups aim at legitimate and achievable rival states, "in the name of the people" over the same territory, and the weaker is aided from outside." In Mann's view, the stereotypical perceptions of genocide as something that only authoritarian or pre-modern states do are completely unfounded. On the contrary, he argues that neither authoritarian nor traditional polities (i.e., ancient empires, patrimonial kingdoms, or city-states) had interest or capacity to implement genocidal projects. The pre-modern polities were culturally heterogeneous but deeply hierarchical entities where difference was class-based, and as such they had no benefit from eliminating cultural minorities. The modern authoritarian states prefer stability over cultural uniformity and as such are unlikely to engage in genocidal politics which would inevitably bring about instability. The partial exception here might be the onset of war which can destabilize the state. In the *Dark Side of Democracy*, Mann advances the argument that genocide is more likely to happen in situations where authoritarian states embark on the process of democratization and liberalization. In other words, murderous ethnic cleansing is often a by-product of incomplete or unsuccessful democratization: the shift in political legitimacy toward the idea of popular sovereignty can bring about genocidal outcomes when the notion of people's power is understood in ethnic rather than demotic sense. In Mann's (2005:3) own words, "cleansing is a hazard of the age of democracy since amid multiethnicity the ideal of rule by the people began to intertwine the demos with the dominant ethnos, generating organic conceptions of the nation and the state that encouraged the cleansing of minorities." Hence, Mann is not arguing that democratic states are more prone to genocide as his argument has sometimes been misinterpreted (i.e., Laitin, 2006). Instead he is adamant that "stabilized institutionalized democracies" are least likely to conduct genocides. Nevertheless, he also argues that a number of contemporary democratic states have a long history of colonial murderous cleansing and that the contemporary democratic stability has often been built on top of previous instances of mass murder. In particular, Mann (2005:4) identifies settler democracies as being particularly prone to genocidal actions: "The more settlers controlled colonial institutions, the more murderous the cleansing. . . . It is the most direct relationship I have found between democratic regimes and mass murder." However, what is common to all instances of murderous ethnic cleansing is that they are modern phenomena that are framed by modern forms of political power. The genocides transpire in environments of competing ideological and state-building projects where the rulers embark on modernizing and nationalizing their polities. For example, one of the first twentieth-century genocides, that of Armenian minority, did not happen in a multiethnic, authoritarian, and pre-modern Ottoman Empire but in modernizing, secularizing, and liberalizing new polity that will eventually become the Turkish nation-state. The ideologues of the genocide that involved the murder and expulsion of over 1.5 million people were not old imperial and authoritarian sultans but secular and reformist liberal intellectuals who established and ran the organization that masterminded the genocide in 1915 – the Committee of Union and Progress.

Historical sociologists have also analyzed genocide as a form of warfare. Both Shaw (2003, 2007) and Ugur Ümit Üngör (2012, 2016) explore the transformation of genocide through the prism of interstate conflicts. For Shaw, genocide is an illegitimate type of war that targets the enemy civilians rather than military organizations. While civilians have historically experienced violence in wars, Shaw demonstrates how in the twentieth and twenty-first centuries civilian populations have gradually become principal foci of war projects. He argues that over the course of human history, warfare has become more destructive and centered on civilians. He terms these new types of conflicts as "degenerate wars" where civilians are rarely distinguished from soldiers and where the question of the legitimate use of violence is often blurred or not raised at all. For Shaw (2007), genocides are integral to wars as they are organized by the state, utilize the coercive apparatus of the state (i.e., military, police, intelligence services, etc.), and are executed almost exclusively during war. Üngör (2016) also explores genocides through the context of warfare. He argues that genocides are often a by-product of wars where new nationalizing states focus on ethnic homogenization. In this context, genocide becomes a form of "nationalist population policy" centered on transforming culturally heterogeneous polities into ethnically homogeneous nation-states. In his analysis of Eastern Turkey (1913–1950), he shows how the Young Turk ruler ethnically cleansed non-Turkish populations in order to create an ethnically homogenous Turkish territory (Üngör 2012).

Perhaps even more than other forms of organized violence, genocides are macro-level processes initiated and implemented by the state and other large-scale social organizations. Nevertheless, the implementation of genocide is ultimately dependent on the social dynamics of the microworld: the motivations of direct perpetuators, the resistance of victims, the passivity of bystanders, and the changing interrelationships between these groups and individuals. Although leading historical sociological theories recognize the significance of agency in genocide, they do not explore extensively the micro-dynamics of mass atrocities. For example, both Bauman and Mann discuss the motivations of genocide organizers and perpetrators. Thus, Mann (2000) analyzes the "core Nazi constituencies" and the biographies of 1,581 people involved in the genocide. Bauman too dwells on the motivations of perpetrators and invokes the Milgram experiments on conformity to suggest that violence is linked to the obedience to authority. However, these analyses do not go beyond the identification of static categories attributed to the specific individuals or groups. Instead of tracing the social dynamics of the microworld and the changing social relations and individual actions, these analyses just point to categories of individuals that were more represented among the perpetrators – authoritarian personalities that were obedient to authority and prone to imitation (Bauman) or the young educated *Volksdeutsche* men with the longer career in the Nazi movement (Mann). Nevertheless, as has been argued previously to understand the changing dynamics of participation in genocide, it is crucial to explore the emotional and moral micro-order of *genocidaires* (Malešević 2017). Instead of assuming that genocide perpetrators are natural born killers or suffer from mental illness, it is germane to trace the gradual transformation and radicalization of ordinary individuals. As Bartov (2018), Fulbrook (2012), and Browing (1992) among others show, most individuals who might initially be reluctant or even hostile to killing civilians can gradually become perpetrators, as the practices of killing become normalized and legitimized not only by the state authorities but also by their peers, friends, comrades, and family members. While genocidaires are perceived by outsiders as ruthless and merciless killers, their own view of themselves is usually very different as it is firmly attached to the shared perceptions and actions of their own micro-groups. In other words, the genocide perpetrators inhabit a different moral and emotional micro-universe where their actions are self-justified in reference to their alleged protection of their own groups. For many genocidaires, killing women and children was emotionally difficult but necessary as the enemy group was conceived to be a continuous threat to one's group existence. This was clearly stated in Himmler infamous 1943 Poznan speech when he explains to his soldiers why they have to kill the Jewish children: "I did not consider myself justified to exterminate the men [only]. . .and allow the avengers of our sons and grandsons in the form of their children to grow up. The difficult decision had to be made to have this people disappear from the earth" (Smith and Peterson 1974:169). To fully understand how genocides happen, it is important to dissect these micro-sociological realities and link them to the broader organizational and ideological structures. Historical sociologists with their understanding of the long-term social and historical trends and awareness of the micro-dynamics of social action are best placed to provide comprehensive theories of genocide.

Conclusion

This chapter has scanned through the most important studies on organized violence in historical sociology. The stock has also been taken of the literature in the wider social sciences that is relevant from the point of view of a historical sociological approach to violence.

Although violence has traditionally not been a central preoccupation for sociologists, historical sociology has come a long way since the 1980s in filling the gap. It has allowed identifying long-term trends, such as the role of the growth of infrastructural power in increasing the potential for mass violence (Malešević 2017). Contemporary forms of violence are generally the outcome of such long-term trends. Caution is consequently warranted when contemporary forms of violence are claimed to flow from radically new conditions. Historical sociology also sheds light on the distinctly modern features of contemporary forms of genocide, clandestine, political violence, and war. Notions of a timeless human proclivity toward brutality, so often used to explain violent events, are hence equally questionable (Malešević 2010a).

A lot however remains to be done. Three avenues of future research have here been identified. Firstly, the state-centricity of many seminal works in historical sociology sometimes seems at odds with the ever presence of "non-state violence" today. The issue is however here less to "throw out the state" (after having "brought it back in"), as it is to broaden the picture: historical processes like bureaucratization affect organizations other than the state, be it non-state armed groups or private military companies; and the latter develop in close interaction with states. Most importantly, due consideration for state organizations does not warrant methodological nationalism: as highlighted by historical sociologists, states are the contingent products of long-distance interactions and relations (Mann 1986). In this sense, the transnationalization of organized violence, rather than challenging historical sociology, seems to vindicate some of its initial intuitions.

Secondly, the focus of historical sociology has, in line with its structuralist inspirations, been on material conditions of possibility of organized violence. There has been relatively less interest in explaining the situational logic of particular violent events. What processes accompany the actualization and effectuation of organized violence? This dimension, the one of the microworld of violence, is of increasing interest in sociology (Collins 2008). It supposes to understand the unfolding of violence as lived experience, including its bodily practices, emotional dynamics, and social meaning. Integrating this dimension into historical sociology allows for a more complex analysis of social organizations, one that leaves more place for bottom-up logics and agency.

Thirdly, Eurocentrism has always been a problem in the social sciences. In the case of the historical sociology of organized violence, it can however find a partial justification in the violence projected from the "West" throughout the world from colonization onward. The prevalence of armed conflicts in the Global South cannot be analyzed independently from the latter's position in a world order largely shaped by the "West" (Ayoob 1995). At the same time, it is important not to underestimate geographical variations or to overstate the exceptionalism or influence of the "West." The challenge is hence how to apply the tools of historical sociology beyond the West without necessarily taking the latter's historical trajectory as point of reference. Many recent works have successfully met this challenge, thus taking historical sociology in new and challenging directions (Barkawi 2017).

References

Alexander J (2011) Performance and power. Polity Press, Cambridge
Anderson P (1974) Lineages of the absolutists state. Verso, London
Andrade T (2016) The gunpowder age. In: China, military innovation, and the rise of the west in world history. Princeton University Press, Princeton
Ayoob M (1995) The third world security predicament: state making, regional conflict, and the international system. Lynne Rienner Publishers
Barkawi T (2006) Globalization and war. Rowman & Littlefield, Lanham
Barkawi T (2017) Soldiers of empire. Cambridge University Press, Cambridge
Bartov O (2001) The Eastern Front 1941–1945. German troops and the barbarization of warfare. Palgrave Macmillan, Oxford
Bartov O (2018) Anatomy of a genocide: the life and death of a town called Buczacz. Simon & Schuster, New York
Bauman Z (1989) Modernity and the holocaust. Polity Press, Cambridge
Berti B (2013) Armed political organizations: from conflict to integration. JHU Press
Bigo D (2005) L'*impossible cartographie* du terrorisme. Cultures & conflits. L'Harmattan 2005:2–6
Bosi L, Porta DD (2012) Micro-mobilization into Armed Groups: ideological, instrumental and solidaristic paths. Qual Sociol 35:361–383
Browning CR (1992) Ordinary men: reserve police battalion 101 and the final solution in Poland. Harper Collins, New York
Chase-Dunn CK (1999) Global formation: structures of the world economy. Basil Blackwell, London
Clastres P (1989) Society against the state. MIT Press, Cambridge, MA
Collins R (2001) Weber and the sociology of revolution. J Class Sociol 1(2):171–194
Collins R (2004) Interactional ritual chains. Princeton University Press, Princeton
Collins R (2008) Violence: a micro-sociological theory. Princeton University Press, Princeton
David L (2020) The past Can't heal us: the dangers of mandating memory in the name of human rights. Cambridge University Press, Cambridge
de Tocqueville A (2002) Democracy in America. Folio Society
Debos M (2016) Living by the gun in Chad: combatants, impunity and state formation. Zed Books, London
Dorronsoro G (2005) Revolution unending. In: Afghanistan, 1979 to the present. Hurst & Company, London
Elias N (1987) The retreat of sociologists into the present. Theory Cult Soc 4(2–3):223–247
Elias N (2012/1939) On the process of civilisation. UCD Press, Dublin
Evans-Pritchard EE (1987) The Nuer: a description of the modes of livelihood and political institutions of a Nilotic people. Oxford University Press, New York

Falk A, Kuhn A, Zweimüller J (2011) Unemployment and right-wing extremist crime. Scand J Econ 113(2):260–285

Foran J (2005) Taking power: on the origins of third world revolutions. Cambridge University Press, Cambridge

Foucault M (2003) Society must be defended: lectures at the Collège de France, 1975–1976. Allen Lane, London

Fulbrooke M (2012) A small town near Auschwitz: ordinary Nazis and the holocaust. Oxford University Press, Oxford

Giddens A (1985) The nation-state and violence. University of California Press, Berkeley

Goldstein JS (2003) War and gender: how gender shapes the war system and vice versa. Cambridge University Press, Cambridge

Goldstone J (1991) Revolution and rebellion in the early modern world. University of California Press, Berkley

Goldstone J (2014) Revolutions: a very short introduction. Oxford University Press, Oxford

Goodwin J (2001) No other way out: states and revolutionary movements, 1945–1991. Cambridge University Press, Cambridge

Hassan R (2014) Understanding suicide terrorism: psychosocial dynamics. SAGE, Thousand Oaks

Hintze O (1975) Military organization and the organization of the state. In: The historical essays of otto hintze. OUP, Oxford, pp 178–215

Joas H, Knöbl W (2012) War in social thought: Hobbes to the present. Princeton University Press, Princeton

Kalyvas S (1999) Wanton and senseless? The logic of massacres in Algeria. Ration Soc 11(3):243–285

Kalyvas S, Balcells L (2010) International system and technologies of rebellion: how the end of the cold war shaped internal conflict. Am Polit Sci Rev 104(3):415–429

King A (2013) The combat soldier: infantry tactics and cohesion in the twentieth and twenty-first centuries. Oxford University Press, Oxford

Kuper L (1981) Genocide: its political use in the twentieth century. Penguin, Harmondsworth

Laitin D (2006) Mann's dark side: linking democracy and genocide. In: Hall JA, Schroeder R (eds) An anatomy of power: the social theory of Michael Mann. Cambridge University Press, Cambridge

Lane D (2009) 'Coloured revolution' as a political phenomenon. J Communis Stud Transition Polit 25(2):113–135

Lawson G (2016) Within and beyond the 'fourth generation' of revolutionary theory. Sociol Theory 34(2):106–127

Lawson G (2019) Anatomies of revolution. Cambridge University Press, Cambridge

Levene M (2005) Genocide in the age of the nation state, vol. 2: the rise of the west and the coming of genocide. IB Taurus, London

Levene M (2013) The crisis of genocide, Vol 2: Annihilation: the European Rimlands 1939–1953. Oxford University Press, Oxford

Malešević S (2010a) The sociology of war and violence. Cambridge University Press, Cambridge

Malešević S (2010b) How pacifist were the founding fathers? War and violence in classical sociology. Eur J Soc Theory 13(2):193–212

Malešević S (2017) The rise of organised brutality. Cambridge University Press

Malešević S (2019) Cultural and anthropological approaches to the study of terrorism. In: Chenoweth E et al (eds) The Oxford handbook of terrorism. Oxford University Press, Oxford, pp 177–193

Mann M (1986) The sources of social power I: a history of power from the beginning to A.D. 1760. Cambridge University Press, Cambridge

Mann M (2005) The dark side of democracy: explaining ethnic cleansing. Cambridge University Press, Cambridge

Mann M (2013) The sources of social power IV. Cambridge University Press, Cambridge

Marx K, Engles F (2005) The communist manifesto. Haymarket Books, Chicago

Mennell S (1998) Norbert Elias: an introduction. UCD Press, Dublin
Moore B (1966) Social origins of democracy and dictatorship. Beacon, Boston
Moore B (1978) Injustice: the social bases of obedience and revolt. Macmillan, London
Muchembled R (2012) A history of violence: from the end of the middle ages to the present. Polity, Oxford
Nexon DH (2009) The struggle for power in early modern Europe: religious conflict, Dynastic Empires and International Change. Princeton University Press, Princeton
Oberschall A (2004) Explaining terrorism: the contribution of collective action theory. Sociol Theory 22(1):26–37
Olsson C (2016) Coercion and Capital in Afghanistan: the rise, transformation & fall of the afghan commercial security sector. In: Berndtsson J, Kinsey C (eds) The Routledge research companion to outsourcing security. Routledge, London, pp 41–51
Parker G (1996) The military revolution: military innovation and the rise of the west, 1500–1800. Cambridge University Press, Cambridge
Porta DD (1988) Recruitment processes in clandestine political organizations: Italian left-wing terrorism. In: Klandermans et al (eds) International social movement research, vol Vol. 1. JAI Press, Greenwich, pp 155–172
Porta DD (2013) Clandestine political violence. Cambridge University Press, Cambridge
Rapoport DC (1984) Fear and trembling: terrorism in three religious traditions. Am Polit Sci Rev 78(3):658–677
Rubin BR (2002) The fragmentation of Afghanistan. Yale University Press
Sageman M (2004) Understanding terror networks. University of Pennsylvania Press, Philadelphia
Selbin E (2010) Revolution, rebellion, and resistance: the power of story. Zed Books, London
Shaw M (2002) Globality and Historical Sociology: State, Revolution and War Revisited. In S. Hobden, J.M. Hobson (eds), Historical Sociology of International Relations, Cambridge: Cambridge University Press, pp. 92–98
Shaw M (2003) War and genocide: organized killing in modern society. Polity, Cambridge
Shaw M (2007) What is genocide? Polity, Cambridge
Shils EA, Janowitz M (1948) *Cohesion* and *disintegration* in the *Wehrmacht* in *World War II*. Public Opin Q 12:280–315
Sinno AH (2011) Organizations at war in Afghanistan and beyond. Cornell University Press, Ithaca
Skocpol T (1979) States and social revolutions: a comparative analysis of France, Russia, and China. Cambridge University Press, Cambridge
Smith R (1987) Human destructiveness and politics: the twentieth century as an age of genocide. In: Dobkowski M, Walliman I (eds) Genocide in the modern age. Greenwood Press, New York
Staniland P (2014) Networks of rebellion: explaining insurgent cohesion and collapse. Cornell University Press, Ithaca
Tarrow S (2015) War, states, and contention: a comparative historical study. Cornell University Press, Ithaca
Thompson JE (1994) Mercenaries, pirates and sovereigns, state-building and extra-territorial violence in early modern Europe. Princeton University Press, Princeton
Tilly C (1978) From mobilization to revolution. Addison-Wesley, Reading
Tilly C (1985) War making and state making as organized crime. In: Evans P, Rueschemeyer D, Skocpol T (eds) Bringing the state Back. Cambridge University Press, Cambridge
Tilly C (1995) Popular contention in Great Britain: 1758–1834. Harvard University Press, Cambridge, MA
Tilly C (2004) Terrorism, terrorists. Sociol Theory 22(1):5–13
Toft MD (2007) Getting religion?: the puzzling case of Islam and civil war. Int Secur 31(4):97–131
Üngör U (2012) Studying mass violence: pitfalls, problems, and promises. Genocide Stud Prev 7(1):68–80
Üngör U (2016) Genocide: new perspectives on its causes, courses and consequences. Amsterdam University Press, Amsterdam
Wallerstein I (2004) World-systems analysis: an introduction. Duke University Press, Durham

Walter BF (2017) The Extremist's advantage in civil wars. Int Secur 42(2):7–39
Weber M (1978) Economy and society. University of California Press, Berkeley
Weinstein JM (2006) Inside rebellion: the politics of insurgent violence. CUP, Cambridge
Wieviorka M (2009) Violence: a new approach. SAGE, Thousand Oaks
Wimmer A (2013) The waves of war: nationalism, state formation and ethnic exclusion in the modern world. Cambridge University Press, Cambridge
Wimmer A (2018) Nation building: why some countries come together while others fall apart. Princeton University Press, Princeton

Historical Sociology of Law

25

Marta Bucholc

Contents

Introduction	652
Focus: Official Law	654
A Glimpse at the Prehistory of Sociohistorical Studies of Law	656
The Deceptive Historicity of Early Sociology of Law	658
Law and the Social Change: Theoretical Trends from the Twentieth Century	661
Michel Foucault's Approach to Socio-Legal Change: The Epistemic Dimension of Law	662
Pierre Bourdieu's Sociology of the Juridical Field	664
Niklas Luhmann's System Theory of Law	666
Future Perspectives in Historical Sociology of Law	669
Conclusion	671
References	672

Abstract

The chapter offers and overview of historical sociology of law, focusing on the sociological studies of the official law. It unfolds the main themes in historical sociology of law ordered by research traditions and demonstrates the plurality of paths along which historical-sociological studies of law developed. The introduction briefly outlines the philosophical prehistory of sociohistorical approaches to law centers on the connection between the law and the state, on the one hand, and the law and morality, on the other, as a durable motif of ancient philosophy reiterating in contemporary social sciences. It is followed by a presentation of three paradigms in sociology of law, inspired by Michel Foucault, Pierre Bourdieu, and Niklas Luhmann. The role of historical material and the scope of sociohistorical insights in each of these distinct theoretical approaches is exemplified by a brief discussion of the main contributions of each of the three theorists, indicating the further research directions they inspired and the main

M. Bucholc (✉)
University of Warsaw, Warsaw, Poland
e-mail: bucholcm@is.uw.edu.pl

branches of law which they influenced. The final section includes a synoptic summary of the path of historical sociology of law and concludes with a tentative glance into the future, arguing for a cultural turn in the historical sociology of law and the necessity of its further interdisciplinary connectivity.

Keywords

Sociology of law · Legal history · Michel Foucault · Pierre Bourdieu · Niklas Luhmann

Introduction

When entering the field of historical sociology of law, we encounter a well-established tradition of historical thinking in its own right, practiced by legal historians long before the official birth of sociology around the middle of the nineteenth century. On the one hand, this can be an important advantage to sociological analysis. On the other hand, there are some hindrances to sociology's joining efforts with legal history. It is, on the whole, a complex relationship, which is now being revised, as has been recently shown by Chloë Kennedy, who argued that a rapprochement between the two domains of study is now taking place (2020). In this chapter, I offer an overview of historical sociology of law as a self-standing research field, but this cannot be done without addressing a few preliminary problems.

The first and foremost problem with historical sociology of law is the very notion of law. On the one hand, law is frequently defined as a system of norms supported by an institutionalized power which can use violence to ensure that the norms are obeyed, according to Max Weber's famous formula: "An order will be called law if it is externally guaranteed by the probability that coercion (physical or psychological), to bring about conformity or avenge violation, will be applied by a staff of people holding themselves specially ready for that purpose" (Weber 1954: 5). It is an understanding of law that focuses on coercion as the sanction applied by an institutionalized power as a definitional characteristic of law, which makes it different from other normative social orders such as religion, morality, or good manners.

On the other hand, there is the following influential definition of law by an influential American legal scholar Karl Llewellyn:

> Doing of something about disputes, this doing of it reasonably, is the business of law. And the people who have the doing in charge, whether they be judges or sheriffs or clerks or jailers or lawyers, are officials of the law. What these officials do about disputes is, to my mind, the law itself. (Llewellyn 1930: 3)

Llewellyn stresses the practical dimension of law. He puts in the center of his definition of law its relation to dispute and conflict, and indeed, it has been very frequent in the history of legal theory and sociology of law to define law primarily as the instrument of conflict resolution and settlement of disputes. However, he also insists on the pivotal role of legal professions in making the law – not as an abstract set of rules, but as a regulatory practice. By the same token, even though Llewellyn stresses the "reasonability" of the legal practice, it is a particular, juristic reason that characterizes the activities of the legal profession. The law, in this sense, belongs to lawyers, who know more about it than anyone else does.

But the view according to which law is "what the lawyers do" seems somehow incomplete. After all, we are each and every one of us constantly affected by law, and law transpires in all domains of our lives, even though we are not all legal professionals. Law is literally omnipresent and directly woven into the very fabric of the social world. In acknowledgment of this fact, another understanding of law emerged, advocated by an Austrian legal scholar commonly acknowledged to be the founding father of sociology of law, Eugen Ehrlich. His work was based on an extensive reconstruction of the historical development of law in the West. But, moving beyond pure history of law, Ehrlich insisted that apart from the law made by the state (or other institutionalized power) and pursued by lawyers, there is also a "living law": "the law that dominates life itself, even though it has not been printed in legal propositions" (Ehrlich 1936: 493). There is a law in the lives of people, which seldom if ever corresponds to either legal texts or the ideas of legal professionals, and it is that living law that is binding from the point of view of the social actors themselves even though it may not be known to the legal professionals or recognized by the official power ready to back it with the sanction of coercion.

The plurality of meanings of law results in a vast diversity of historical studies and in a certain confusion as to what these studies may have in common. It does not make the matter any easier that most of them do not go under the institutional heading of "historical sociology of law." In this sense, the field is far less established than historical sociology of state, or that of war and revolutions. Mikael Rask Madsen and Chris Thornhill are essentially right in saying that "[w]ith few exceptions, none of the more recent traditions of socio-legal analysis (...) have tried to resume the quest for a genuine historical sociology of law" (Madsen and Thornhill 2014: 4). Some form of historical reflection is bound to occur in almost any sociological study of law, and it is sometimes difficult to say which of them are "genuinely historical."

While all social phenomena are inherently historical, since they happen as a part of chronologically sequenced social processes, law has a particular temporality of its own, one that is related to its regulatory goal. Laws govern reality; hence the question arises of the temporal relationship between the processual reality and the law. How do the chronologies of law and of the social life relate to each other? It might be an additional reason why a genuinely historical sociology of law is so difficult to conceptualize: Law as a social practice, at least in some of its cultural forms, includes reflection on its own historicity. As I once wrote:

> [It is] the temporality of law, which makes it different from other subjects of historical inquiries. Current mathematical knowledge results from a long and winding path of historical development, but no awareness of this path is required to learn and apply state-of-the-art mathematics. Historical consciousness may sometimes be an enhancement, but it is always a luxury. However, historicity is inherent to law, at least in its form dominant in the modern West, because this form of law states the conditions of its own validity in temporal terms.

> Legal order is a succession of norms which are either in force, no longer in force, replacing each other, or repealing and amending each other. Finding one's way round in this normative nexus requires some form of historical consciousness. (Bucholc 2020: 115)

To sum up: Historical sociology of law constantly trespasses on the institutionalized grounds of legal history, on the one hand, and unqualified sociology of law, on the other, and it deals with a sphere of social practice characterized by a high level of time-related reflection resulting in a particular form of historical consciousness of its own. Historical sociology of law must position itself in respect of all these difficulties.

In order to cope with this task, this chapter begins with a delimitation of the concept of law which is applied here. The next section contains a discussion of the main themes in historical sociology of law ordered by research traditions. While the list is far from exhaustive, it fairly represents the plurality of paths along which historical-sociological studies of law developed. The final section includes a synoptic summary of the development of historical sociology of law and concludes with a tentative glance into the future. The final question of the chapter deals with the expansion of new phenomena in law in the recent decades: how do they translate into the new research agendas in sociology of law, what harbingers of new theoretical and empirical developments can be identified in them, but also, and more importantly, how historical sociology of law can help us find our way around in the world governed by laws which are growing less intelligible by the hour.

Focus: Official Law

This chapter will focus on one particular kind of law: the official law. According to the definition coined by Polish-Russian legal scholar Leon Petrażycki, the "official law" is "applied and sustained by representatives of the state authority in accordance with their duty to serve society" (Petrażycki 1955: 139) (Even though representatives of the state authority do not always serve the society, on certain normative premises it could be argued that it is always their duty to do so.). Roger Cotterrell remarks that: "Petrażycki seems to be identifying official law as an objective social reality (the state representatives and their use of law can be generally observed) and this makes it a phenomenon that jurists can recognise as their normal province of law" (Cotterrell 2015: 6). In the case of the official law, the laywoman's intuition about the law, the lawyer's professional turn of mind, and the definitions employed by legal scholars, historians, and sociologists all largely coincide. It is, so to say, the smallest common denominator of most of the domains and disciplines dealing with the law. While probably all sociologists of law would argue that the official law is not the only game in town, it would be difficult to find anyone who would maintain that under normal conditions it should not be considered law at all. However, to opt for the smallest common denominator is a considerable restriction.

The first limitation is that it leaves out the whole range of nonstate societies, including in particular the traditional non-Western societies, both past and contemporary. This field is mostly being served by legal anthropology. A number of seminal studies pertained to normative orders of traditional nonstate societies, including in particular Bronisław Malinowski's *Crime and Custom in Savage Society* (1926). Today's research in legal anthropology has rejected the use of words like "savage" along with the ethnocentric and evolutionary mindset of many of its founders. Moreover, it is no longer limited to traditional societies on which social anthropology focused in the past. Legal anthropology today is a domain of cross-cultural study of various forms of normative orders which would not fit into Petrażycki's definition of official law. In this sense, for a comprehensive view of law in the global age we should depend on legal anthropology as much as on historical sociology of law, whose main preoccupation was the official law of the modern West. Much could be said for the need to have global historical sociology of law as a constitutive part of the project of global historical sociology such as advocated by Julian Go and George Lawson (see Go and Lawson 2017). However, law seems to be somewhat late in becoming a part of this agenda.

The second limitation resulting from the decision to focus on the official law is related to the concept of legal pluralism. Legal pluralism is defined as "the idea that in any one geographical space defined by the conventional boundaries of a nation state, there is more than one law or legal system" (Davies 2010). The concept is far from uncontested, and its critics point out that the idea of having more than one legal system in the same state is actually based on a conflation of law (as a state-made or at least state-recognized normative order) with other normative orders, which can only be defined as "legal" at the price of depriving the concept of law of all specificity. Nonetheless, the body of studies of legal pluralism in contemporary socio-legal studies is growing in size, covering the relations between state laws and diverse other legal orders in multicultural societies. The nonstate legal orders may include not only religious laws, especially in the case of those religions operating their own courts, but also customary laws of Indigenous peoples and ethnic or cultural minorities. Legal pluralism is also a framework for analyzing the recognition struggles of the collectivities abiding by the nonstate laws. Legal pluralism is also a common framework for analyzing the interferences between religious and secular state law in modern secular nation-states (Sandberg 2017). No matter how we define legal pluralism, it is a concept stressing the nonexclusivity of state-made law in regulating social life, which is one of the reasons why Eugen Ehrlich's notion of the living law is frequently evoked by the scholars of legal pluralism. Hence, there is a definite tension between the focus on the official law and the readiness to acknowledge the existence of many legal orders within the same society living within the institutional framework of the same state: A consequent endorsement of the concept of legal pluralism would require a shift of attention toward nonstate legal orders and their relation to the state law. To avoid a mistake of leaving out the whole area of sociological studies of pluralist regimes, the problem of legal pluralism will be addressed by way of an example in the section on the system and world society theories later in this chapter.

The third major limitation induced by focusing on official law only is that official law has a structure of its own, dictated by legal practice, education, jurisprudence, custom, and established intellectual habits. The divisions into public and private law, into criminal, civil, and administrative law, or into substantive and procedural law, are meaningful to those who look at the law from the inside, but often perfectly incomprehensible to the outsiders. Most of these classifications are adapted by legal historians. However, a number of large-scale, intersectional issues potentially of interest for historical sociology cannot be grasped using such categorizations. In fact, while most legal historians stick to them, some legal sociologists either do not or they only take these categorizations as a starting point in order to reduce or deconstruct them. This is especially true of socio-legal scholars representing the Critical Legal Studies movement that views the law as a pillar of the power dynamic causing social inequalities and consequently advocates its deconstruction is a major step toward social betterment (see Kelman 1991). Legal categorizations and classifications may thus be deconstructed as an element of the professional ideology of lawyers, their "collective hypocrisy," as Pierre Bourdieu put it (1994).

This chapter takes the third way between following the legal classifications uncritically and entirely disregarding the own self-descriptions worked out inside law. After a brief outline of the philosophical prehistory of the discipline, the presentation of the matter of historical sociology of law is organized into three sections, each of them dedicated to a single historically oriented paradigm of sociology of law, inspired, respectively, by Michel Foucault, Pierre Bourdieu, and Niklas Luhmann. In these sections, the historical sociological matter is displayed according to its juridical internal ordering, with the distinction between material and procedural law, as well as civil, criminal, and administrative law, observed in order to relate the gist of the theoretical argument in the authors' own categories.

A Glimpse at the Prehistory of Sociohistorical Studies of Law

The debate on the origins of law has been going on in European philosophy since its very beginnings. The Ancient Greek and Roman philosophy of law was characterized by a number of assumptions: a virtually inseparable connection between the law and the state, on the one hand, and the law and morality, on the other. Its state-centered understanding of social life was completed by an insistence on the limits of its conventional regulation set forth by human nature and by the laws of nature. While some key themes of ancient philosophy succumbed over the ages, many bear on social sciences until today.

One particularly durable theme is the tension between two intuitions regarding the origins of law. On the one hand, lawmaking is clearly a human political activity producing norms which are volatile, terminable, and imperfect. On the other hand, there is a sense of order about human laws, imperfect as they may be, which for many legal philosophers over the ages indicated the existence of a higher and more perfect law, one tainted neither by mundane political agendas nor by the incurable human fallibility. Since late Antiquity, this basic tension has found innumerable solutions. Some, like Augustin of Hippo, proposed an image of two cities. One of them, the earthly one, was governed by the imperfect human laws. Those could nevertheless serve the good cause of peace, unity, stability, and welfare provided that they are modeled after the eternal, perfect *City of God*. The ideal of an almost-perfect earthly state governed by laws informed by moral knowledge proved to be extremely time-resistant, and it contributed to establishing the view of the law as a factor of social cohesion, coordination, and well-being.

The tension between the contingent, diverse, changeable, and imperfect human-made laws and the postulates of a reasonable setup and governance of the society were characteristic of the state of nature theorists of the early modern period, when the concept of the transcendent divine model for good laws faced the competition of more secular approaches. The concept of good law was subjected to the test of reason guided by a more psychological, social, and historical understanding of human nature. Good laws would be those corresponding to the setup of the people as they really were. The novelty was that the latter could be found out by observation, which would be gradually subjected to the methodological rigors of which our contemporary scientific usage largely stems. Montesquieu's concept of *The Spirit of the Laws* (1949[1748]) could be seen as a part of this process. So was the argument for the humanistic reform of criminal law famously made by Cesare Beccaria in his treatise *On Crimes and Punishments* in 1764. While philosophical speculation long prevailed in the business of establishing what people really were, notably in the utilitarian approach to law founded by Jeremy Bentham, the idea of an empirical science of human societies was bound to foster an idea of an empirical study of law.

However, the empirical study need not in itself be historically oriented. An important impulse for studying law as a sociohistorical phenomenon was the emergence of the nation-centered vision of law which either replaced or completed the state-centered one in the nineteenth century. An interest in the past of nations as a source of knowledge about legal developments would be especially pronounced in the German historical school of jurisprudence, whose most renowned representative was Friedrich Carl von Savigny. The school connected *Volksgeist* (the spirit of the people, but with a distinct national hue) to the essential characteristics of the legal regulation of social life. Thus, it could be argued, law first connected to a recognizably though unacknowledgedly sociological mode of thinking.

Madsen and Thornhill thus characterized the connection between law and sociology at this earliest stage of their liaison:

> The focus on law and history was clearly key to the first invention of sociology as a theoretical discipline; in fact, the rise of sociology was prefigured in part by theorists such as Montesquieu, Savigny, Hegel and Bentham, in whose works – however divergent in other respects – inquiry into law, history and society was inextricably interwoven. As modern sociology emerged as an academic discipline in the nineteenth century, then, the question of the role of law and legal practitioners in the construction of society – not surprisingly – assumed decisive importance for the pioneers of sociological research. Marx, Tönnies, Durkheim and Weber all shared an interest in explaining the precise interface between law, legal technology and the formation of modern society. (Madsen and Thornhill 2014: 4)

To the list of presociological and early sociological students of law as a sociohistorical phenomenon, further names could be added those of Otto von Gierke,

Alexis de Tocqueville, Henry Maine, or Georg Simmel. It is noteworthy that the solid foundations for historical sociology of law were laid by scholars born and bred in Germany (Marx, Tönnies, and Weber). Their ideas were shaped not only by the legal education which many of them absorbed, but also by the sense of a close relationship between law and national cultural identity. Sociology arrived in Europe in which nationalistic feeling was on the rise, and it had a deep impact on the development of sociology (see Lepenies 1998). This created an additional tension in sociohistorical study of law. It would be spread between the ambition to explain the rise of modernity as a universal phenomenon and what was later called "methodological nationalism," envisaging the nation-state as a container of social processes (Wimmer and Schiller 2002). The difficulty of overcoming the focus on national law would explain the bias evident not only in the early sociology of law, but also in most of the contemporary approaches. The situatedness of scholarly thinking in this field, though evident to an outsider observer, often seems perplexingly opaque to the interested parties.

The Deceptive Historicity of Early Sociology of Law

History was the basic research material of early sociology of law for all the reasons discussed in the previous section, but also because, very simply, the methodical data-gathering on contemporary societies was only slowly gaining pace when Durkheim and Weber created their sociological systems. However, the ubiquity of historical analysis in early sociological thought was frequently deceptive: The view of the past which it presented was subordinated to the task of providing evidence for the preconceived vision of the historical process. The past was used to explain the present, but the range of explanations which it was allowed to provide was limited by how the past must have happened to fit into the theoretical blueprint.

Karl Marx's contribution to sociohistorical studies is an excellent example of this property of the early theorizing. Even though he was primarily interested in philosophy and economy, Marx also made an important contribution to philosophy of law and, by extension, to historical sociology of law as well. His historical materialism assumed that the material conditions of social life – the economic processes – were the motor of historical and social change. The resulting vision of history was reductionist (its goal was to explain the social in economic terms) and deterministic (the general goal of the historical change was preset by the theoretical assumption). In Marx's theory, the developments in law corresponded to power relations in the society, which in turn depended on the state of the economic relations. Therefore, Marx never identified law as a "major theoretical problem" (Spitzer 1983: 104): Law was an instrument, not a motor of sociohistorical processes.

Marx's own writings contain valuable analyses of legal phenomena, for example, on the development of property law in the West or on the legal frames of land ownership (Cahan 1994). However, Marxist paradigm in sociology of law was only developed later on, and it was marked by an assumption of an analogy between legal categories and economic formations. In thinkers such as Evgeny Pashukanis, legal development – both in private and in public law – followed consistently the line of economically driven revolutionary changes, the expected final outcome being that "the withering of the categories of bourgeois law will ... mean the withering away of law altogether" (Pashukanis quoted in Spitzer 1983: 105). Marx's approach to law, while secondary to his other theoretical concerns, did not seem to offer much in terms of stimulating theoretical discussion on the past, which was mainly a reservoir of arguments in support of the theory. However, it gave a powerful push to critical thinking about law in the second half of the twentieth century, and in this new intellectual context, it reinforced the interest in historical explanation of law as a social phenomenon. Marxism could be argued to have been deceptively historical, because it did not allow history to be what it was, pushing it into allegedly infallible theoretical schemes. Nonetheless, the sociology of law inspired by the Marxian concept proceeded to develop a genuine interest in historical evidence of the connection between law, power, inequality, and exploitation (see Deflem 2008).

Another classic of sociology of law to whom the notion of deceptive historicity applies is Émile Durkheim. His main contribution to the field is his celebrated study of social solidarity, of which he distinguished two varieties, the mechanical (based on resemblance of individuals) and the organic (based on the difference resulting from a division of labor) (Durkheim 2014). Each of them was connected to one kind of law defined by the type of sanction characteristic of it, being either a repression (for instance, a fine in the criminal law) or a restitution (like damages in the civil law). This part of the theory in itself does not imply any historicity at all: What we have here is a classification of legal norms claimed to reflect the classification of the types of social bond. However, Durkheim also postulated that, over time, a shift is taking place from the mechanical toward the organic solidarity, corresponding to the growing internal complexity and differentiation of the society. In each society, the law is an indicator of the nature of social solidarity. By the same token, the differentiation of various branches of law is an indicator of the direction of a postulated evolutionary change toward more restitution and less repression. Therefore, Durkheim's theory potentially opened a number of ways for studying the historical development of law, even though in itself it was not based on historical evidence, which was merely used as an illustration of the theoretical thesis. Furthermore, by combining Durkheim's insights into law with his sociology of religion, sociology of knowledge, and anomie (see Durkheim 2005, 2008), a broad theoretical perspective on law could be construed, far beyond the prevailing focus on the link between law and morality (Lukes and Prabhat 2012). Nonetheless, as Cotterrell put it "Durkheim's writings (...) remain the last neglected continent of classic theory in the sociological study of law" (1991: 923). Since the 1990s, many new insights into Durkheimian sociology of law have been produced (see Deflem 2008: 56–74; Gephart and Witte 2017). Nonetheless, Durkheim's contribution to historical analysis was largely neglected, and sociology of law was no exception (Emirbayer 1996).

It would be simplistic to say that the marginality of law in Marx's thought and the weak reception of Durkheim's thesis created more favorable conditions for the expansion of Max Weber's historical sociology of law. To project the competition which we today construe between the classics as textbook figures onto the past would be misleading. Nonetheless, Weberian grand comparative project of the study of economy and society was both expressly historically oriented and expressly focused on law, which gave it a definite advantage, especially since the American and then global reception of Weber's theory gained momentum after the Second World War. However, Weber's project was also, at least to an extent, deceptively historical. Its goal was to explain the process of rationalization and its connection to the emergence of modernity. To substantiate his claims regarding the connection between societal values, in particular the religious ones, the structure of social action, in particular in the economic sphere, and the institutional setup of societies, Weber gathered evidence from all various cultures and historical epochs. His analyses spread from ancient Judaism to twentieth-century Russia, from the Spring and Autumn period in China to European medieval monasticism, and from the Indian caste system to agrarian capitalism of the modern German empire (see Ertman 2017).

This imposing range resulted in studies which are, admittedly, uneven in quality, and most of them unfinished. However, in all of them Weber spoke – as he phrased in the opening sentence of his most famous work, *The Protestant Ethic and the Spirit of Capitalism* – as "a product of modern European civilisation, studying any problem of universal history" (Weber 2003) with a very specific agenda on his mind: "to what combination of circumstances the fact should be attributed that in Western civilization, and in Western civilisation only, cultural phenomena have appeared which (as we like to think) lie in a line of development having universal significance and value" (ibid.). Weber's drive to know other cultures, or indeed even the history of the Western culture, is a consequence of his commitment to the understanding of the universal significance of the Western line of societal and cultural development. This entails some obvious limitations, the most important of which is that the questions which he asks about historical reality are informed by modern Western concerns, and the implicit reference for understanding is modern European reality. This is especially evident in his observations on non-Western legal orders such as the Muslim law, which he dismisses as irrational, arbitrary, and unpredictable (Rabb 2015), or to his understanding of the traditional Chinese law (Marsh 2000).

Whatever the failures of Weber's grandiose enterprise, its great merit is the full integration of law into the historical explanation of social change. Law is more than an indicator of the state of society (though it may undoubtedly be used as such): It is a normative order connected to politics, religion, economy, and morality, but not reducible to any of them, having its own historical dynamics. This is evidenced especially by Weber's sociology of law, where he engages in theoretical-legal debates of his time from a unique position in which the historical, the legal, and the sociological viewpoint coalesce (Weber 1954). Incidentally, despite his interdisciplinarity, Weber found the mixing of juristic and sociological research agendas problematic, which was the core of his critique of Eugen Ehrlich's project of sociology of law (see Bucholc and Komornik 2019). Weber situated his own endeavors in sociological study of law quite unequivocally on the part of sociology. Nonetheless, several comparative interdisciplinary endeavors certify to its force to inspire new explorations in other disciplines and in interdisciplinary studies (Gephart 2015).

Law and the Social Change: Theoretical Trends from the Twentieth Century

The deceptive historicity of early sociology was reproduced in the subsequent development of historical sociology of law: Historical analysis was conducted specifically in order to provide evidence supporting the theoretical models. However, in the course of the twentieth century, sociohistorical approaches to law gradually became more nuanced and heterogenous, reflecting the growth of the field of sociology of law and the differentiation of major sociological and sociolegal paradigms after the Second World War.

The historical sociology of law remained a field dominated by theory, with a relatively weaker empirical profile. This was partly due to the competition the historical sociology of law was facing from legal history: In the course of the academic institutionalization of sociology, the borderline between history of law and sociology of law was becoming ever more pronounced. An exception to this process of separation of disciplinary fields was the United States legal academia, whose strong affinity with social scientific imaginaries dates before the First World War. An example is Harvard's Roscoe Pound's sociological jurisprudence whose programmatic expression was a 1911–1912 essay on "The scope and purpose of sociological jurisprudence." Pound's sociological jurisprudence combined the inspirations coming from European sociology of law (notably from Ehrlich) with the ambition to increase the sensitivity to social conditions of the operations of law (Stone 1965). The legacy of the entente between legal science and sociology was preserved in the law and society movement born in the USA after the second world war and established in the 1960s. The goal of the movement – as one of its chief representatives, Lawrence Friedman, put it – was to "study how the legal system actually operates" (1986: 764). "Actually" in this case meant: as established by research using the methods of social science applied to the data beyond the usual array of statutes, caselaw and legal treatises. Law and society movement's ambition was to contribute to social science and not to the juristic debates in the first place.

Apart from the developments in American legal science, the separation of sociology, history, and law had a decisive impact on the research in the historical sociology of law in the postwar period.

The historical element contained in the legacy of the classics was often disregarded in the reception of some influential theoretical paradigms in the sociology of law. Such was the case of the system theory of Talcott Parsons, which was a major step in the reception of Weber's theory in the USA. Contrary to some of its critics, Parsons' theoretical framework was not entirely devoid of potential for historically informed understanding of social change, including the role of law in it. Nonetheless, system theory developed in a distinctively ahistorical direction, weighing heavily on the whole path of American sociology, which was then slowly rising to the rank of a global hegemon.

However, in the so-called "second wave" of historical sociology starting in the late 1960s, which at least in the USA directly contested the ahistoricity of postwar American sociology, the legal problematic often featured very prominently, albeit only as a part of the larger analysis and not as a self-standing problem (Adams et al. 2005: 1–72). The historical sociology of the second half of the twentieth century, which was characterized by the domination of elaborate and robust systematic comparisons focusing on big structures and long-term processes, was especially concerned with social change in its violent forms: revolutions, wars, and rapid mobilizations of social movements. The new modes of using and claiming rights, the introductions, amendments, and abandonment of constitutions and statutes, the recognition of new legal statuses, emancipation, and empowerment by legal means, and the institutional upheavals accompanied by an emergence of new regulatory frameworks, these and similar phenomena were part and parcel of the study of social processes, but law was usually just an appendix to structural social change. It is true of most of the scholarship that since the 1990s has been classified as belonging to historical institutionalism, beginning with Karl Polanyi's *The Great Transformation* (first published 1994, on historical institutionalism). They focus on the relation between institutions and historical paths of societies, studying long processes in order to explain specific occurrences and events. Even though the occurrence in question may include a profound legal change, such as the integration of the European Union, it is not the centerpiece of research (see Christiansen and Verdun 2020).

Despite this general tendency, a recognition of the specific historical dynamics of law emerged in the second half of the twentieth century in several influential theories. Three of them will be discussed here: Michel Foucault's concept of the changing regimes of truth and punishment, Pierre Bourdieu's field theory, and Niklas Luhmann's system theory of law. Between them, the three theories offer a good overview of the main topics which emerged from the classical sociologists of law of the nineteenth and early twentieth centuries.

Michel Foucault's Approach to Socio-Legal Change: The Epistemic Dimension of Law

Michel Foucault was an inquisitive and highly critical student of power. His novel analytical strategy involved examining the margins of the society in order to identify the forces which hold it together, including the image of the world and the kind of knowledge on which they operate, and the practices which they produce, including those which they are producing inadvertently. Therefore, although his work is not sociological, the convergence of his research goals with the more power-oriented sociological approaches, especially in sociology of knowledge is striking (see Power 2011: 38). Foucault's specific object was the process of normalization: of producing the model of behavior imposed as a muster by way of social control punishing the deviation from it. Consequently, law as an instrument of normalization has been in the center of many of Foucault's studies, covering, apart from theoretical considerations, a wide range of historical topics, including sexuality, mental health and mental illness, history of medicine, state and politics, and many others (Foucault 1967, 1978, 1979, 1980).

Foucault was particularly interested in criminal law, and his most renowned contribution is *Discipline and Punish*, which has also been described as his "most accessible monograph" (see Power 2011: 39). In this book, he describes the changes in the function and form of criminal punishment since Early Modernity, stressing the shift from external, public, and often exceedingly cruel corporeal punishment to the action oriented toward internally reforming the delinquents. The reform was achieved by subjecting them to strict discipline, characteristic of what Foucault referred to as "disciplinary power," whose goal is to control the population more efficiently than could be achieved by external control (Kallman and Dini 2017).

Foucault's reconstruction of the historical change in the approach to punishment has for some time been criticized both from the perspective of legal science and history. The early sociological reception mostly focused on the question of punishment, control, and disciplining applied to state power. For example, in a 1985 book *Punishment and Welfare: A History of Penal Strategies,* now a classic of law and society studies, David Garland used the Foucauldian discourse analysis to tackle the connection between welfare state and punishment in the twentieth century, arguing that the welfare concept contributed to controlling marginal subpopulations by way of normalization, correction, and segregation (Garland 2018).

Garland's book focuses on British sources from the nineteenth century onward. But Foucauldian inspiration has also been extended to reach beyond Europe, and to include the context of Western colonial domination as an instance of disciplinary power. Owing to Foucault's importance for postcolonial theory, there have been many approaches to use his work in the analyses of postcolonial societies (Hiddleston 2009). Indian social history seems to be the most important field of critical application of Foucauldian interpretation (Samaddar 2013). Reading the Indian social and socio-legal realities of the colonial and postcolonial era through the lens of *Discipline and Punish* (Sampath 2015; Kaplan 1995) has confirmed the force of his conceptual apparatus, but it has also shown limitations to moving it beyond its initial conceptual limitations and geographical framework (Legg 2007). Foucault has also been used extensively in studies on other colonial and postcolonial areas, including Sub-Saharan Africa (Mbembe 2003), Latin America (Rigo 2002), British Columbia (Clayton 2000), and Australia (Dean and Hindess 1998). Last but not least, if we choose to perceive the situation of the Soviet satellite countries before 1989 as a variant of colonization, Joachim Savelsberg conducted comparative historical studies of domination and knowledge production involved in the criminal punishment in totalitarian regimes of Eastern Europe (Savelsberg 1999).

In the wide-spread reception of Foucault's concept of discipline and punishment, law has usually been a complementary rather than the main subject. Relatively little attention has been granted to his probably most original contribution to the understanding of the historical development of law as a social phenomenon, which is his study of law as a regime of truth. In "Truth and juridical forms" (1996), Foucault depicts the developments in the procedural law, focusing on the changing concept of truth and the corresponding shift in the ways of getting as the truth in the court proceedings, beginning with the ancient Greek institutions, through Roman and Germanic laws, the feudal era, and into the modern age. Foucault's analysis, at times rather sweeping in its historical documentation, has been read as "putting the question of capitalism front and centre" (Toscano 2015: 30): an analysis designed to unveil the modern mechanisms of control and governance. However, it is also exemplary in combining an insight into the legal framework with a study of the judicial and administrative practice underpinning it, though this dimension has been appreciated more in the legal scholarship (Tassi 2011). Foucault sets various historical legal systems together to examine what kind of a test of truth and truthfulness was admitted by each of them, and he analyses in detail the judicial practices emerging as a collateral of the legal procedures increasingly imposed by the state. These changes in turn are put in the broader framework of the economic and political transformations of selected European societies. As the latter "violently imposed its dominion on the entire surface of the Earth" (Foucalt 1996: 40), the products of Western legal history come to bear on the developments virtually in every area of the globe: The flow of ideas in Foucault's reconstruction is clearly unilateral, his optic that of one-sided imposition with little space for adaptation or innovation by the colonized and dominated societies. With the truth of modern legal systems, the model of control emerges which becomes globalized.

What also becomes globalized that way is a modern Western mode of subjectivation – of becoming a subject. Foucault's most renown empirical historical study of the interplay between the construction of a subject in its historical and biographical concreteness is a book produced by a seminar which he conducted on a nineteenth-century case of multiple homicide committed by a young peasant by the name of Pierre Riviere, based on archive material including the perpetrator's confession (Foucalut 1975). This book is another example of the variety of uses to which Foucault's work can be put in historical sociology of law: An insight into the documented details of a court case is accompanied by a critical reflection on modern social order and an overview of the main characteristics of the legal system which provides its epistemic scaffolding.

Alan Hunt and Gary Wickham argued in 1994 that Foucault's work could provide an impulse for the emergence of a new sociology of law as governance (Hunt and Wickham 1994). Whereas this does not seem to have taken place quite as expected, Foucault remains a powerful influence in socio-legal studies, facilitating a rapprochement between a study of historical data, discourse analysis, and insight into the internal dynamics of legal systems.

Pierre Bourdieu's Sociology of the Juridical Field

It could be said that while Foucault studied legal epistemologies, Pierre Bourdieu put the structures produced by these epistemologies in the center of his sociology of law. It is, in many ways, a unique area of study in Bourdieusian legacy, as it is empirically underdeveloped compared to Bourdieu's sociology of education (Bourdieu and Passeron 1990), of the academia (Bourdieu 1988), of literature (Bourdieu 1996), or of art (Bourdieu 1987b). Consequently, it was never developed into a full sociological theory of law – it remained a theoretical contribution offering little illustration of how it can be translated into an empirical application. Why this should be so is not entirely clear, but Pierre Guibentif (2019) mentions two principal factors which had worked adversely to the interest in law. First, Bourdieu did not accord the juridical field (the way he reconstructed it) much autonomy, which discouraged investing research forces, because few new insights could be gained by studying it. Second, as in most of his research work with the notable exception of the early studies of Kabylia, Bourdieu's sources and inspirations came almost exclusively from metropolitan France, one particular nation-state-based legal culture entailing a particularly strong and manifold connection between legal training, the social roles of legal professionals, and the state power. However, since the 1970s, the global and regional international connections of the French state made it more and more difficult to study the developments in France without accounting for the international interdependencies, and this would mean going beyond the national juridical field toward an international one, whatever its size. This could not have been done without a thorough revision of the initial theoretical setup configured to reflect the French reality.

The core of Bourdieu's theoretical contribution to sociology of law are two sizeable essays: "The force of law: toward a sociology of the juridical field" (1987a) and "Les juristes, gardiens de l'hypocrisie collective" [Lawyers, guardians of collective hypocrisy] (1991). The main focus of these works are the legal professions and their relation to law. This is not to say that other writings of Bourdieu do not contain important socio-legal insights, for example, into the intrinsic connection between state, power, and law, but these seem to be supplementary contributions (see Dezalay and Madsen 2012). Bourdieu's approach as it emerges from the two central essays is a practice-centered one: He opposes the essentializing tendencies of jurisprudence and sociology of law alike, and he opts for approaching law as a practice pursued by a particular group of professionals. Their actions are subject to the forces which pattern their behavior, and by subjecting themselves to those forces, they in turn make the forces ever more powerful. In a sense, this subordination is, however, in the lawyers' best interest, as they thus create a niche in the social world in which those who are able to conform with the reproduced patterns have the most power to dominate and exclude others. Thus, a juridical field emerges: a field controlled by jurists, in which their specific knowledge forms the basis of the field-specific capital. Bourdieu identifies the skill to interpret the legal text as the key skill determining the position in the juridical field: Lawyers are the people who know their way around the legal texts (it is, admittedly, an exceptionally logocentric view of the legal practice). The extent to which those in possession of the special juridical skills are left to practice it, and position themselves in the field with no intervention from the outside (from politics, religion, literature, etc.), is a measure of the field's autonomy.

In order to keep power to themselves, those in the juridical field need to control the law: Bourdieu's goal would be to reveal the power structures behind the meaning of law, and the interests which legal professionals have in imposing their understanding of law onto their social surroundings. His study focuses on the strategies of the actors in the juridical field, in particular on the way they are using their particular language and conceptual apparatus to legitimize their interpretations of law, to expand the autonomy of their field versus other fields in the society, and to prevent nonlawyers from exercising power in the field (Terdiman 1987: 808).

Probably as a result of insufficient saturation with empirical findings, the historical dimension of Bourdieu's sociology of law has long been underestimated, and that is despite the fact that both his conceptual apparatus – especially the concepts of habitus, field, and capital – are "inherently historical" (see Steinmetz 2011: 45). However, while the point made by Yves Dazalay and Mikael Rask Madsen in 2012 about the weak reception of Bourdieu's sociology of law largely remains true today, the interest in its historical dimension is on the rise among socio-legal scholars, especially in continental Europe (Böning 2017; Dębska and Warczok 2016; Kretschmann 2019; Voutat 2014; Wrase 2017). In the studies of the legal professions, the Bourdieusian insistence on studying the genesis of law as a practice of a specific social group has been successfully applied (Dezalay and Garth 2011). The long globalization processes which transformed nationally based lawyers into global entrepreneurs were analyzed in the context of the emergence of the international law as a separate field (Dezalay and Madsen 2009, 2017), and the international legal institutions have been studied as the loci of professional group and habitus formation (e.g., Avril 2018). The historical approach has been expanded into biographical research on members of the legal professions (Hammerslev 2005) and the studies of historical forms of legal habitus, as in Ruben Hackler's study of the judges in the Weimar Republic (2019). Apart from habitus and group formation, the language of law and the communication within the legal field has also increasingly been studied from the Bourdieusian historical perspective, for example, in the studies of the reproductive function of legal education (Jewel 2008).

A dynamically developing field of application of Bourdieu's historical sociology of law are the studies of postcolonial and decolonizing societies. For example, Robert van Krieken has combined Bourdieu's insights with those of Niklas Luhmann (see next section) to examine the production of knowledge in Australian jurisprudence since the 1970s (Van Krieken 2004). George Steinmetz has proposed analyzing the pre-1914 German colonial state as a social field generating a particular form of capital, which he called "ethnographic capital" (Steinmetz 2008). This is accompanied by a critical revisiting of Bourdieu's experience with French colonialism in Algeria and his early thoughts on colonialism (Go 2013).

Niklas Luhmann's System Theory of Law

Niklas Luhmann's sociology of law is in many ways a unique enterprise. It belongs to the paradigm of system theories, which have been accused of being intrinsically ahistorical. However, Luhmann's sociological project, of which sociology of law (Luhmann 2013) is a major part next to general social theory (1995), sociology of religion (1977), politics (1990), or morality (1996), is in fact based on a view of social change in a deeply historical, though rather abstract, perspective. Luhmann combines system theoretical insights with a theory of social development and social differentiation, to show history as an evolutionary process by which modern societies have arrived at the point in which they are now (Luhmann 1978). He envisages the modern society as a complex system displaying a variety of internal subsystems, functionally distinct but connected by what he calls coupling effects, allowing for them to react to their environment, to process information using their characteristic communication codes, and to change adaptatively. While his work has been variously criticized by historians for missing the standards of the historical reasoning, it has also been rightly argued that there is significant potential to the convergence of system theory in its Luhmannian version with the empirical studies of the past (Schlögl 2001).

Luhmann's sociology of law raises a number of questions which were also pivotal for Bourdieu's theory: the particularities of the communication of legal professionals, the genesis and properties of juridical concepts, and the relation of law to politics and power in general. Luhmann's key question, just like Bourdieu and Foucault's, is about the emergence of modern Western law: what processes of functional differentiation provided the basis for law as what Luhmann calls an autonomous, autopoietic system, self-producing and self-reproducing in the processes of communication using the binary code of "legal-illegal" to organize the information it processes.

Apart from the general outline of the system theory of law in *A Sociological Theory of Law* (2013), Luhmann also specifically considered some of the particular properties of the modern Western law. His main focus was the evolution of the basic characteristics of the legal system, constitutional law, and the interplay between law and politics, as well as the legitimacy of law. He developed a major historical narrative showing how modern (continental European style) constitutionalism emerged as a final step of the process of law becoming more and more positivized, which means the perception of law as a human-made normative system constituted in and legitimized by a particular procedure. Luhmann's definition of law famously states that *"Recht ist was das Recht als Recht bestimmt,"* law is whatever the law defines as law (Luhmann 1993, 143f): This formula captures the autopoietic nature of law as a system. However, this form of law is a historical phenomenon, a product of hundreds of years of social change.

The influence of the Luhmannian variant of system theory differs greatly between various sociological traditions, and so does the weight attached to the historical-evolutionary perspective inherent to it. One influential line of the critical reception of Luhmann was related to the rise of world society theory, in which global institutions and processes are in the center of the understanding of social change (McNeely 2012). In world society theory, Luhmann's ideas are brought into fruitful dialogue with the alternative, particularly Marxist and neoinstitutional historically informed approaches to systemic reading of globalization (see Stichweh 2000). Luhmann's theory inspired historical studies of a number of global legal phenomena, among which human rights and trans–/supranational constitutional orders are the most important (Luhmann 1997).

In the sociology of human rights, Luhmann's writings have provided impulse for studies of the evolution of the global system for the protection of human rights

(Verschraegen 2002). In current German sociology of human rights, historical-qualitative approaches to human rights discourses have combined the perspective of world society with the study of the changing form of the nation-state and state power, and various forms of supranational laws (Koenig 2016). For example, Fatima Kastner connects the human rights system with transitional justice as one particular product of the world society. The concept of transitional justice refers primarily to the processes taking place in post-transitional societies, which consist in seeking legal (and other) remedies to the wrongs done by authoritarian and totalitarian regimes all over the world (see Teitel 2000). The societies in which transitional justice measures are applied include not only the postcolonial and postauthoritarian societies of Africa, South-East Asia, or the Americas, the postwar societies such as the former Yugoslavia, but also postsocialist countries of Eastern Europe. Kastner argues that over the decades from the end of the second world war a distinct global legal order emerged, which she describes as *lex transitus*, a transitional law (Kastner 2015).

Human rights, which are in the center of the transitional justice procedures, belong to a range of phenomena that point out toward the limitations of state-centered approach to law and legal systems. This entails a changing approach to the constitutions which have been heretofore conceived as central to statehood and nationhood alike. A British socio-legal scholar, Chris Thornhill, arguing for a new sociology of constitutions based on the Luhmannian approach, pointed out its potential to become central in the studies of constitutions in the future, owing to its consolidating an historical research tradition with accommodating new global phenomena (Thornhill 2010). Thornhill's comparative study of the changing legitimacy of state constitutions over a wide range of European countries from the Middle Ages up to 1990 provides evidence for the applicability of Luhmannian optic to long-range processes of socio-legal change (Thornhill 2011). Going beyond the comparative historical approach, a German system theorist, Gunter Teubner, has advocated a shift from state-centered constitutionalism to a societal one, arguing that the constitution of the global society emerges "in the constitutionalisation of a multiplicity of autonomous subsystems of world society" (Teubner 2004: 5). Teubner's vision of a new global societal constitutionalism is in line with his thesis regarding the coexistence of multiple legal orders in the world society leading to a development of one single pluralist and complex legal reality unprecedented in human history. His theorizing is inspired by the seminal studies of legal pluralism in the Austro-Hungarian rural region of Bukowina, conducted by Eugen Ehrlich in the early twentieth century. Teubner (1996) even used the expression "global Bukowina" as a metaphor for a new reality of the emergent global legal pluralism.

One way to summarize the reception of Luhmann's theory in historical socio-legal studies would be to point out that this theory brings with it a chance to overcome the limitations of old paradigms focusing on the role of the state as lawmaker, and on law as a top-down mechanism of social control. It is not the only sociohistorical approach striving to explain the particularities of the legal orders beyond the framework of a nation-state: To name but one example, Norbert Elias' theory of social processes has long been represented by a British scholar Andrew

Linklater in the field of international relations, thematizing, and among other things, the role of international law in the making of the global legal order (Linklater 2020). However, despite the growing competition in the field of system-oriented sociohistorical theorizing, Luhmann's theory, although very abstract, seems perfectly suited to research agendas operating in large-scale international comparisons of long-term processes of legal evolution reaching backward beyond the modern age.

Future Perspectives in Historical Sociology of Law

While the three theoretical paradigms discussed in the previous section do not even begin to exhaust the output of the historical sociology of law today, together they offer a good overview of the prevailing research interests. The foundational Weberian problem of the emergence of Western modern legal order does not cease to occupy the imaginaries of historical sociologists of law, but the explanatory models worked out by the nineteenth and early twentieth centuries have been widely questioned since. This is largely due to the new problems arising in late twentieth and early twenty-first century, owing both to the unexpected turn and pace of the social change and to the emergence of new ways to explain it, which seemed to have increased the salience of law in the general explanatory models with which historical sociologists have come up in this period.

The general point about law being an integral part of social development has long been made; to cite Lawrence M. Friedman (1969: 29):

> Legal systems are clearly a part of political, social, and economic development, just as are educational systems and other areas of the culture. No major social change occurs or is put into effect in a society which it not reflected in some kind of change in its laws. ... Many basic questions of the relationship of law to social change and to cultural development are completely neglected. Does the type of legal system and legal institutions that a society uses help or hinder that society in its march towards modernization? How does law influence the rate of economic growth? How does law brighten or darken the road to political wisdom or stability? How can a society improve its system of justice? What happens when laws are borrowed form more advanced countries?

The phrasing of the questions indicates their firm connection to the modernization paradigm and progressivist view of history, and since the words were first spoken in 1968, it does not really come as a surprise. However, while the phrasing might have become obsolete in some ways and inadequate in some others, the questions themselves have largely remained pertinent. In many respects, the work on long-term processes of legal change subscribes to a larger field of transformation studies as comprehensive studies of social change, including its institutional, political, economic, and cultural dimensions.

In the late twentieth and early twenty-first century, some new socio-legal developments have developed into whole research fields covered by extensive historically informed scholarship: the high wave of supranational, transnational, and literally global legal orders after second world war, the role of international judiciary, especially in human rights matters, arising awareness of the impact of European colonialism on legal cultures all over the globe, the increasing privatization of law making the role of the state more marginal versus big business corporations, or the interplay between global capitalism and its legal framework. But there are also other problems, such as the new forms of surveillance and social control in the digital society and their impact on the legal subjectivity, individual rights and their protection and enforcement, the expansion of the capitalist market and its concept of law into the sphere of genetics, or the socio-legal dimension of the environmental disaster toward which the humanity seems to be steering today. In her book *The Code of Capital* (2019), Columbia's Katharina Pistor successfully undertakes to combine a sociohistorical perspective with addressing these problems at the intersection of the present and the future of law, combining early and midmodern developments in property law with contemporary discussion regarding the regulations and legal practice of digital currency and genome engineering. Pistor's work is also emblematic of a revival of the critical theory in the Marxist vein, stressing the pivotal role of law in the revolutionary development of capitalism as we know it today from its preindustrial histories.

New impulses generated by social change have also led to a revival of the areas of study which have been relatively underdeveloped for a variety of reasons. A revival of interest in the long historical trends of law and violence could in all probability be related to the sense of rising instability and insecurity in today's world, hit by violent war and war-like crises (see Rousseaux and Verreycken 2021). On the other hand, the historical-sociological theories that focus on the long-term processes can often serve the cause of disabusing us of the notion that our times are unique in their concentration of violence, as shown by Steven Pinker's impressive and controversial account of human history since prehistorical times (Pinker 2011). Sociohistorical studies of law have also contributed to our historically informed understanding of other pressing problems of our times related to the perceived and ofttimes only too real spread of violence and violent conflicts, social polarization, and culture wars. Responding to the demand for a historical explanation, Robert van Krieken has systematically reviewed the applications of Norbert Elias' theory of the civilizing process in the field of the sociology of law, and particularly in the studies of the rule of law, populism, and the rise of illiberal constitutionalism (Van Krieken 2019; Bucholc 2015, 2021a).

Apart from the application of sociohistorical theories which had not been present in sociology of law before, there is also a change in the approach to the role of law in social change observable in the more directly empirical turn in the existing domains, such as the transformation studies. For example, in the research of the transformations of socialist regimes in the region of East Central Europe, the focus on legal systems, legal consciousness, and legal culture has been a constant in both theory and the empirical research. The installment of free-market democracies in postsocialist societies and the related problems of more or less successful legal transplants emerged in the early phase of democratization and the introduction of capitalism but persisted until the 2000s and the European accession of many of the former Eastern Bloc countries. Grażyna Skąpska's (2011) analysis of the impact of the Stalinist constitutionalism on postsocialist constitutional development is an example of combining legal history with sociohistorical perspective in the understanding of systemic transformation. Many studies have come to connect the mid- and long-term socio-legal experience of the socialist period with the explanation of the progress of democratization, developments in the rule of law, and protection of rights in the postsocialist countries (see Kurczewski 1993; Czarnota et al. 2005; Bucholc 2021b).

The prevalence of neoinstitutional approaches in the transformation studies has limited the interest in law as a normative order as distinct from the institutional setup created as a result of social change. It would seem that law was inevitably limited to the position of a sign or an instrument of change more than a factor endowed with its own efficacy. However, in accordance with Adams, Clemens, and Orloff's thesis about the cultural turn of the most recent wave of historical sociology, law seems to be in the course of emancipating itself from its ancillary role by reclaiming its cultural relevance. Comparative research of legal cultures has been steadily gaining importance as a self-standing field with a strong historical profile, highly relevant for the transformation studies of the contexts in which law has become alienated, culturally disowned, and delegitimized. Legal cultures have come to be understood as complexes of values, attitudes, imaginaries, and behavioral patterns related to law which underpin the operations of the legal systems and their social embedding.

The case of socialist societies, in which a certain form of law was implemented as a result of the process of political subordination and disenfranchisement, can thus be reframed as but one of the many examples in which the official law clashes with or diverges from the cultural experience of social groups and of the whole societies. The clash not only may result in distrust, lack of legitimacy, but it also inevitable causes changes in the perception of law and in the long-term behavioral patterns. The understanding of the origins of the legal cultures and their distinct ways to shape human perceptions and behavior have caused a shift in the historical studies of the cultural dimension of the official state laws. Debunking the chasm between legal culture as a lived reality and the legal imposition of the official law has been vital for the move toward decolonizing law: dismantling the colonial, imperially enforced structures of domination of foreign legal cultures with their corresponding conceptual apparatus and philosophical frameworks (see Xavier et al. 2021). Decolonizing law requires painstaking historical work, identifying the cultural courses and pathways of norms, institutions, and regulatory mindsets. This work is currently being done both on the law of the former colonies and in the former colonial empires. What has been referred to as the emergent "global legal history" (Duve 2020) has a lot to gain by the convergence of sociological and legal perspectives.

Conclusion

The events and processes of the twentieth century which gave rise to the second and third wave of historical sociology are now further and further in the past: Much of our knowledge about the role of law in the social transformations of the twentieth century has imperceptibly become a part of history, thus enriching the domain of historical sociology of law. Many of the statements formed in the classical and early modern Western philosophy have been transposed into our thinking about how law relates to other normative orders, including morality and religion, and how it relates to power, both the symbolic and the more tangible one, in particular institutionalized in the form of a nation-state. On the other hand, the cultural dimension of law has been reflected upon more in the recent decades, and the understanding of its historical dynamics has contributed to the appreciation of law as a distinct sphere of human creativity and self-expression.

However, new problems of our times call for new approaches. In sociology, they would usually be building upon the existing theoretical paradigms, such as one of the three discussed in this chapter: the Foucauldian, the Bourdesian, and the Luhmannian. Nonetheless, it can be anticipated that the trend will continue for the historical sociology of law studies to venture toward the interdisciplinary intersections with other disciplines. These will in all probability include not only legal philosophy and history, but also other fields which have been more distant until now, such as not only literary studies, social geography, or linguistics, but also other sciences, to name just medicine, ecology, and epidemiology as the most obvious runners-up. Sociohistorical studies of law are increasingly conducted as a part of area studies, especially of the countries beyond the narrow club of Western societies whose histories have dominated the scholarly imagination for a long time. Western legal history has provided the philosophical assumptions which have formed the foundation of the theoretical paradigms in historical sociology of law, which are now being heavily questioned all over the world, including those very central ones pertaining to the very definition of the law.

Multiple laws exist in our world today, their forms changing and their interdependencies varying in proportion to the complexity of our realities. Moreover, we are only slowly realizing that multiple laws might have just always been there and that our narratives about the growth of complexities need to be revisited. We are getting reconciled with the legal pluralism, and we combine our disciplinary powers to study the multiple laws adequately. In many ways, it is getting more difficult now than ever to draw a precise line between historical sociology of law and its neighboring disciplines. However, similarly to a "legal turn" in history (Sugarman 2017), an "historical turn" in socio-legal studies seems to be upcoming, and it can be expected to bring an enhancement in our understanding of law as a social phenomenon, in all its variety and versatility beyond its apparent solid façade.

Acknowledgments The author acknowledges the support of the Polish National Science Centre (2019/34/E/HS6/00295).

References

Adams J, Clemens E, Orloff S (2005) Introduction: social theory, modernity, and the three waves of historical sociology. In: Adams J, Clemens E, Orloff S (eds) Remaking modernity: politics and processes in historical sociology. Duke University Press, Durham

Avril L (2018) Lobbying and advocacy: Brussels's competition lawyers as brokers in European public policies. Sociologický časopis/Czech Sociol Rev 54(6):859–880

Böning A (2017) Jura studieren: eine explorative Untersuchung im Anschluss an Pierre Bourdieu. Beltz Juventa, Munich

Bourdieu P (1987a) The force of law: toward a sociology of the juridical field. Hastings Law J 38: 209–248

Bourdieu P (1987b) Distinction: a social critique of the judgment of taste. Harvard University Press, Cambridge Massachusetts

Bourdieu P (1988) Homo Academicus. Polity Press, Cambridge

Bourdieu P (1991) Les juristes, gardiens de l'hypocrisie collective. In: Chazel F, Comaille J (eds) Normes juridiques et régulation sociales. LGDJ, coll. Droit et société, Paris, pp 95–99

Bourdieu P (1996) The rules of art: genesis and structure of the literary field. Stanford University Press, Stanford

Bourdieu P, Passeron J-C (1990) Reproduction in education, society and culture. Sage, New York

Bucholc M (2015) A global community of self-defense. Norbert Elias on normativity, culture, and involvement. Klostermann, Frankfurt am Main

Bucholc M (2020) Juliusz Bardach and the agenda of socialist history of law in Poland. In: Erkkila V, Haferkamp H-P (eds) Socialist interpretations of legal history. Routledge, pp 115–135

Bucholc M (2021a) Figurational sociology of the rule of law: a case of central and Eastern Europe. In: Delmotte F, Górnicka B (eds) Norbert Elias in troubled times: Figurational approaches to the problems of the twenty-first century. Palgrave Macmillan, London, pp 63–81

Bucholc M (2021b) Die Politisierung des Rechts in Polen: Über den Prozess der Entzivilisierung. In: Schmidt C, Zabel B (eds) Politik in Rechtsstaat. Nomos, Baden, pp 195–214

Bucholc M, Komornik M (2019) Eugen Ehrlich's failed emancipation and the emergence of empirical sociology of law. HISTORYKA Studia Metodologiczne 49:15–39

Cahan JA (1994) The concept of property in Marx's theory of history: a defense of the autonomy of the Socioeconomic Base. Sci Soc 58(4):392–414

Christiansen T, Verdun A (2020) Historical institutionalism in the study of European integration. Oxford Research Encyclopedia of Politics 19 Nov. 2020; Accessed 29 Nov 2021. https://doi.org/10.1093/acrefore/9780190228637.001.0001/acrefore-9780190228637-e-178

Clayton D (2000) Islands of truth: the Imperial fashioning of Vancouver Island. University of British Columbia Press, Vancouver

Cotterrell R (1991) Review: the Durkheimian tradition in the sociology of law. Law Soc Rev 25(4): 923–946

Cotterrell R (2015) Leon Petrazycki and contemporary socio-legal studies. Int J Law Context 11(1): 1–16

Czarnota A, Krygier M, Sadurski W (2005) Rethinking the rule of law after communism. Central European University Press, Budapest

Davies M (2010) Legal Pluralism. In: Crane P, Kritzer HM (eds) The Oxford handbook of empirical legal research. Oxford University Press, Oxford

Dean M, Hindess B (1998) Governing Australia: studies in contemporary rationalities of government. Cambridge University Press, Cambridge

Dębska H, Warczok T (2016) The social construction of femininity in the discourse of the polish constitutional court. In: Mańko R, Cercel C, Sulikowski A (eds) Law and critique in central Europe questioning the past, resisting the present. Counterpress, Oxford, pp 106–130

Deflem M (2008) Sociology of law: visions of a scholarly tradition. Cambridge University Press, Cambridge

Dezalay Y, Garth BG (2011) Introduction: lawyers, law, and society. In: Dezalay Y, Garth BG (eds) Lawyers and the rule of law in an era of globalization. Routledge, New York

Dezalay Y, Madsen MR (2009). Espaces de pouvoir nationaux, espaces de pouvoir internationaux. Nouveau manuel de science politique. A. Cohen, B. Lacroix and P. Riutort. Paris, La Découverte, pp 681–693

Dezalay Y, Madsen MR (2012) The force of law and lawyers: Pierre Bourdieu and the reflexive sociology of law. Annu Rev Law Soc Sci 8(1):433–452

Dezalay Y, Madsen MR (2017) In the 'field' of transnational professionals: a post-Bourdieusian approach to transnational legal entrepreneurs. Professional networks in transnational governance, pp 25–38

Durkheim E (2005) Suicide. A Study in sociology. Routledge, New York
Durkheim E (2008) The elementary forms of religious life. Oxford University Press, Oxford
Durkheim E (2014) The division of labour in society. Free Press, New York
Duve T (2020) What is global legal history? Comp Leg Hist 8(2):73–115. https://doi.org/10.1080/2049677X.2020.1830488
Ehrlich E (1936) Fundamental principles of the sociology of law. Harvard University Press, Cambridge, MA
Emirbayer M (1996) Durkheim's contribution to the sociological analysis of history. Sociol Forum 11(2):263–284
Ertman TC (ed) (2017) Max Weber's economic ethic of world religions. An analysis. Cambridge University Press, Cambridge
Foucalut M (ed) (1975) I, Pierre Riviére, having slaughtered my mother, my sister and my brother... A case of parricide in the 19th century. University of Nebraska Press, Lincoln
Foucault M (1967) Madness and civilization: a history of insanity in the age of reason. Tavistock, London/Sydney
Foucault M (1978) Security, territory, population: lectures at the Collège de France 1978. Palgrave Macmillan, New York
Foucault M (1979) The history of sexuality volume 1: an introduction. Allen Lane, London
Foucault M (1980) Power/knowledge: selected interviews and other writings, 1972–1977. Harvester Press, Brighton
Foucault M (1996) Truth and juridical forms. Social Identities 2(3):327–342
Friedman LM (1969) Legal culture and social development. Law Soc Rev 4(1):29–44
Friedman LM (1986) The law and society movement. Stanford Law Rev 38(3):763–780
Garland D (2018) Punishment and welfare: a history of penal strategies. Quid Pro Books, New Orleans
Gephart W (2015) Law, culture, and society. Max Weber's comparative cultural sociology of law. Klostermann, Frankfurt am Mein
Gephart W, Witte D (2017) The sacred and the law. The Durkheimian legacy. Klostermann, Frankfurt am Mein
Go J (2013) Decolonizing Bourdieu. Sociol Theory 31:49–74
Go J, Lawson G (eds) (2017) Global historical sociology. Cambridge University Press, Cambridge
Guibentif P (2019) Pierre Bourdieu und das Feld des Rechts. Lehren einer unbequemen Beziehung. In: Kretschmann A (ed) Das Rechtsdenken Pierre Bourdieus. Velbrück Wissenschaft, pp 96–111
Hackler R (2019) Soziologie der Klassenjustiz revisited. Der Richterhabitus in der Weimarer Republik. In: Kretschmann A (ed) Das Rechtsdenken Pierre Bourdieus. Velbrück Wissenschaft, pp 203–221
Hammerslev O (2005) How to study Danish judges. Hart Publishing, Theory and Method in Socio-Legal Research, pp 203–214
Hiddleston J (2009) Foucault and said: colonial discourse and orientalism. In: Understanding Postcolonialism. Routledge, New York, pp 76–97
Hunt A, Wickham G (1994) Foucalt and law: towards a sociology of law as governance. Pluto Press, London
Jewel LA (2008) Bourdieu and American legal education: how law schools reproduce social stratification and class hierarchy. Buffalo Law Rev 56(4):1157–1224
Kallman M, Dini R (2017) An analysis of Michel Foucalt's discipline and punish. Routledge, New York
Kaplan M (1995) Panopticon in Poona: an essay on Foucault and colonialism. Cult Anthropol 10(1):85–98
Kastner A (2015) Lex Transitus: Zur Emergenz eines globalen Rechtsregimes von Transitional Justice in der Weltgesellschaft. Zeitschrift für Rechtssoziologie 36(1):29–47
Kelman M (1991) A guide to critical legal studies. Harvard University Press, Cambridge Massachusetts
Kennedy C (2020) Sociology of law and legal history. In: Přibáň J (ed) Research handbook on the sociology of law. Research handbooks in law and society. Edward Elgar Publishing, pp 31–42

Koenig M (2016) Weltgesellschaft, Menschenrechte und der Formwandel des Nationalstaats/world society, human rights and the transformation of the nation-state. In: Heintz B, Munch R, Tyrell H (eds) Weltgesellschaft: Theoretische Zugänge und empirische Problemlagen. De Gruyter Oldenbourg, Berlin, pp 374–393

Kretschmann A (ed) (2019) Das Rechtsdenken Pierre Bourdieus. Velbrück Wissenschaft, Weilerswist

Kurczewski J (1993) The resurrection of rights in Poland. Clarendon Press, Oxford

Legg S (2007) Beyond the European Province: Foucalt and Postcolonialism. In: Crampton J, Elden S (eds) Space, knowledge and power: foucault and geography. Ashgate, London, pp 265–288

Lepenies W (1998) Between literature and science: the rue of sociology. Cambridge University Press, Cambridge

Linklater A (2020) The idea of civilization and the making of global order. Bristol University Press, Bristol

Llewellyn KN (1930) The bramble bush: some lectures on law and its study. Columbia University School of Law, New York

Luhmann N (1977) Funktion der Religion. Suhrkamp, Frankfurt am Mein

Luhmann N (1978) Geschichte als Prozeß und die Theorie sozio-kultureller Evolution. In: Faber K-G, Meier C (eds) Historische Prozesse. dtv, Munich, pp 413–440

Luhmann N (1990) Political theory in the welfare state. De Gruyter

Luhmann N (1993) Das Recht der Gesellschaft

Luhmann N (1995) Social systems. Stanford University Press

Luhmann N (1996) The sociology of the moral and ethics. Int Sociol 11(1):27–36

Luhmann N (1997) Globalization or world society: how to conceive of modern society? Int Rev Sociol 7(1):67–79

Luhmann N (2013) A sociological theory of law. Routledge

Lukes S, Prabhat D (2012) Durkheim on law and morality: the disintegration thesis. J Class Sociol 12(3–4):363–383

Madsen MR, Thornhill C (2014) Law and the formation of modern Europe: perspectives from the historical sociology of law. Cambridge University Press, Cambridge

Marsh RM (2000) Weber's misunderstanding of traditional Chinese law. Am J Sociol 106(2):281–302

Mbembe A (2003) Necropolitics. Publ Cult 15(1):11–40

McNeely CL (2012) World society theory. In: Ritzer G (ed) The Wiley-Blackwell encyclopedia of. https://doi.org/10.1002/9780470670590.wbeog836

Montesquieu CS (1949 [1748]) The spirit of the laws. Hafner, New York

Petrażycki L (1955) Law and morality. (trans: Babb HW). Harvard University Press, Cambridge, MA

Pinker S (2011) The Better Angels of Our Nature. A History of Violence and Humanity, New York: Penguin Books.

Pistor K (2019) The code of capital. How the law creates wealth and inequality. Princeton University Press, Princeton

Pound R (1911–1912) The scope and purpose of sociological jurisprudence. Harv Law Rev XXIV (8):591–619; XXV(2):140–68; XXV(6):490–516

Power M (2011) Foucault and sociology. Annu Rev Sociol 37(1)

Rabb IA (2015) Against Kadijustiz: on the negative citation of foreign law. Suffolk University Law Review 48:343–374

Rigo B (ed) (2002) Foucault and Latin America: appropriations and deployments of discourse analysis. Routledge, New York/London

Rousseaux X, Verreycken Q (2021) The Civilising process, decline of homicide and mass murder societies: the history of violence and Norbert Elias. In: Delmotte F, Górnicka B (eds) Norbert Elias in troubled times: Figurational approaches to the problems of the twenty-first century. Palgrave Macmillan, London

Samaddar R (2013) Michel Foucault and our postcolonial time. In: Mezzadra S, Reid J, Samaddar R (eds) The biopolitics of development. Springer, New Delhi. https://doi.org/10.1007/978-81-322-1596-7_3

Sampath R (2015) Using Foucault's discipline and punish to reinterpret the gap between the Indian constitution and Indian society: homage to Ambedkar. J Soc Incl Stud 2(1):118–133

Sandberg R (ed) (2017) Religion and legal pluralism. Routledge, New York

Savelsberg JJ (1999) Knowledge, domination and criminal punishment revisited: incorporating state socialism. Punishment Soc 1(1):45–70

Schlögl R (2001) Historiker, Max Weber and Niklas Luhmann. Zum schwierigen (aber möglicherweise produktiven) Verhältnis von Geschichtswissenschaft und Systemtheorie. Soziale Systeme Zeitschrift für Soziologische Theorie 1(7):23–45

Skąpska G (2011) From "Civil society" to "Europe": a sociological study on constitutionalism after communism. Brill, Boston-Leiden

Spitzer S (1983) Marxist perspectives in the sociology of law. Annu Rev Sociol 9:103–124

Steinmetz G (2008) The colonial state as a social field: ethnographic capital and native policy in the German overseas empire before 1914. Am Sociol Rev 73(4):589–614

Steinmetz G (2011) Bourdieu, historicity, and historical sociology. Cult Sociol 5(1):45–66

Stichweh R (2000) On the genesis of world society: innovations and mechanisms. Distinktion: J Soc Theory 1(1):27–38

Stone J (1965) Roscoe pound and sociological jurisprudence. Harv Law Rev 78(8):1578–1584

Sugarman D (2017) Promoting dialogue between history and socio-legal studies: the contribution of Christopher W. Brooks and the 'legal turn' in early modern English history. J Law Soc Spec Issue 44(5):37–60

Tassi S (2011) Gesetzgebung als Herrschaftstechnik: Eine Studie auf der Grundlage von Michel Foucaults Werk "Die Wahrheit und die juristischen Formen". Tectum Wissenschaftsverlag, Marburg

Teitel RG (2000) Transitional justice. Oxford University Press, Oxford

Terdiman R (1987) Translator's introduction: the force of law: toward a sociology of the juridical field. Hastings Law J 38:805–814

Teubner G (1996) Global Bukowina: legal pluralism in the world-Society. In: Teubner G (ed) Global law without state. Dartsmouth, pp 3–28

Teubner G (2004) Societal constitutionalism: alternatives to state-centered constitutional theory? In: Joerges C, Sand I-J, Teubner G (eds) Constitutionalism and transnational governance. Oxford University Press, Oxford, pp 3–28

Thornhill C (2010) Niklas Luhmann and the sociology of the constitution. J Class Sociol 10(4): 315–337

Thornhill C (2011) A sociology of constitutions. Constitutions and state legitimacy in historical-sociological perspective. Cambridge University Press, Cambridge

Toscano A (2015) What is capitalist Power? Reflections on 'truth and juridical forms'. In: Fuggle S, Lanci Y, Tazzioli M (eds) Foucault and the history of our present. Palgrave Macmillan, London

Van Krieken R (2004) Legal reasoning as a field of knowledge production: Luhmann, Bourdieu and Law's Autonomy

Van Krieken R (2019) Law and civilization: Norbert's Elias as a regulation theorist. Ann Rev Law Soc Sci 15:267–288

Verschraegen G (2002) Human rights and modern society: a sociological analysis from the perspective of systems theory. J Law Soc 29:258–281

Voutat B (2014) Penser le droit avec Pierre Bourdieu. Swiss Polit Sci Rev 20(1):31–36

Weber M (1954) On law in economy and society. Simon and Schuster, New York

Weber M (2003) The Protestant Ethic and the Spirit of Capitalism, Courier Corporation, 2003.

Wimmer A, Schiller NG (2002) Methodological nationalism and beyond: nation-state building, migration and the social sciences. Global Networks 2(4):301–334

Wrase M (2017) Rechtsinterpretation als soziale Praxis – eine rechtssoziologische Perspektive auf juristische Methodik. In: Frick V, Lembcke OW, Lhotta R (eds) Politik und Recht: Umrisse eines Politikwissenschaftlichen Forschungsfeldes, pp 63–84

Xavier S, Jacobs B, Waboose V, Hewitt JG, Bhatia A (eds) (2021) Decolonizing law: indigenous, Third World and Settler Perspectives. Routledge

Sport and Leisure: A Historical Sociological Study

26

Dominic Malcolm

Contents

Introduction	678
Mapping the Field	679
Pre-1990s: Harmonious Inequality	682
1995–2005: Diminishing Contrasts and Increasing Varieties	686
The Twenty-First Century: The "Habits of Good Sociology"	691
Conclusion	694
References	695

Abstract

This chapter provides a developmental account of the historical sociological study of sport and leisure. Adopting a framework informed by the broader approach of Norbert Elias, it begins by presenting an overview of the development of the sociology of sport subdiscipline, arguing that the broader characteristics of this field – its location within the multi- and interdisciplinary sports sciences, its manifest status insecurities stemming from the cultural perceptions of sport as "low brow," and the relative influence of Elias himself on the study of sport – have shaped the tension balances between historically oriented sociologists of sport and sports historians. The chapter charts three distinct developmental phases of relations, starting with a period of relatively harmonious separation in which the pioneers of these respective fields were largely supportive and encouraging of work which fundamentally advanced the study of sport. This was followed by a period of heightened tensions as the research in the respective fields increasingly converged due in part to the maturation of historical sociological studies of sport, and the challenge to history as a discipline posed by postmodernism. In the third and final phase, cross-disciplinary tensions have declined and historical sociological research on sport has been reinvigorated,

D. Malcolm (✉)
School of Sport, Exercise and Health Sciences, Loughborough University, Loughborough, UK
e-mail: d.e.malcolm@lboro.ac.uk

© The Author(s), under exclusive licence to Springer Nature Singapore Pte Ltd. 2022
D. McCallum (ed.), *The Palgrave Handbook of the History of Human Sciences*,
https://doi.org/10.1007/978-981-16-7255-2_59

expanding in both quantity and quality. The chapter concludes with some reflections on the degree to which the historical sociological study of sport and leisure, and the Eliasian study of this field, has distinct if not unique characteristics.

Keywords

Figurational sociology · History · Marxism · Physical culture · Sport · Sport sciences

Introduction

Sport can be defined as organized, playful, nonutilitarian, contests that have a relatively significant physical (as opposed to intellectual) component. The formally organized character of sport makes it a somewhat distinct social activity, in that it requires participants to undergo a relatively explicit engagement with rules prior to and during participation. These rules determine such things as who is and is not included in the "game," over what terrain or area the contest will take place and what kinds of activities and behaviors are legitimate. Sport thus provides both the focus of this chapter and a convenient analogy for its introduction, for before moving to a review of the development of the historical sociological study of sport and leisure, it is necessary to clarify three "rules" which structure this review.

First, this definition of sport – derived from the work of a foundational historian of sport, Allen Guttmann (1978) – is accompanied by two qualifications. It is recognized that the multidimensional character of this definition means that what "counts" as sport is always subject to debate. While sport in the more general sense of nonwork physical contests appears to be a cultural universal, the definition of sport used here applies to phenomena distinct in time and place. For instance, while the most socially significant contemporary sporting event – the Olympic Games – is nominally derived from the sport-like activities of the Ancient Greeks, the two are qualitatively distinct in terms of their structural forms and organizing ideologies. Equally, in recent times many scholars have become uneasy with the confines this definition of sport places on intellectual exploration as, during the late twentieth century, these activities increasingly merged (in an analytical sense) with a range of more loosely organized physical activities pursued for highly utilitarian ends, such as body image or health, and even as an antidote to the rule-bound character of modern sport (i.e., lifestyle sports such as surfing or skateboarding). Additionally, the notion of sport defined here is in some respects the antithesis of leisure. Leisure evokes a sense of freedom, or slackening of rules and the field of "leisure studies" incorporates the analysis of recreation including, for example, tourism and the arts. This chapter does not examine sport *and* leisure in this broad sense, nor the premodern forms of sport pursued, e.g., in Ancient Rome, but rather, a group of activities that emerged during the eighteenth century, initially in the British Isles and its imperial offshoots (Elias 1971). "Leisure" is included in the chapter title in recognition of the processual character of how humans understand sport and in order to reflect the field's shifting focus toward leisure time physical activity in the round.

A second "rule" is the consideration of what counts as a historical sociology of sport and leisure. As illustrated in the next section, sport and leisure studies have a number of distinct interdisciplinary characteristics. Debates between those more closely aligned to history and sociology, respectively, have not only influenced how the field has developed but are also both important in revealing the self-identification of scholars and so provide an underpinning rationale for the distinctions necessarily drawn here between sociologists of sport, historical sociologists, and historians of sport. As will be shown, a peculiar feature of the broader field in which these three groups operate is the degree of structural or organizational interdisciplinarity which, in turn, has contributed to the conflict and harmony, collaboration and critique, between differently oriented scholars. Inevitably, these debates do not fall into distinct dates so the periodization used here is for heuristic purposes only.

Third, the terrain covered in this review is in part determined by the factors enabling and constraining its author. I write as both someone who is now a relatively senior figure within the sociology of sport, has relatively consistently drawn on an Eliasian or figurational sociological perspective, and become embroiled in some of the wider politics of interdisciplinary debates. While this experience partly shapes the works that are given most emphasis, this interpretation of the field is perhaps even more constrained by language limitations. It is likely that in this representation of the field some important non-English texts in particular will have been neglected. While a clear Anglophone dominance has been identified within both sociology (of sport) and history, the reliance on English language sources, and the neglect of the historical development of a broader range of related body cultural practices (perhaps most notably not only German Turnen and Ling's gymnastics in Sweden, but also the development of practices like yoga and Thai Chi), encapsulates the structural biases within both sport scholarship and sport as a cultural form. Indeed, the very definition of sport employed represents a Euro-American, protestant, and British bias.

The pages that follow present a historical sociological account of (English language) historical sociological studies of sport (broadly defined). The chapter begins with a brief discussion of the formation and structural characteristics of the field in question. This provides the backdrop for the analysis of three distinct phases in the production of historical sociological research on sport: pre-1990s, circa 1995–2005, and early twenty-first century. Throughout, the analysis reports on tension balances – the centripetal and centrifugal forces – in the field, as a way of charting and defining the boundaries of sociological, historical sociological, and historical research. The chapter concludes by identifying distinct contextual factors, which have significantly shaped this area of study, and shows how these interdisciplinary relations and theoretical orientations are likely to continue in the future.

Mapping the Field

Despite some significant earlier texts (e.g., Veblen's *Theory of the Leisure Class*; Huizinga's *Homo Ludens*) and some which explicitly used the title (e.g., Risse's *Soziologie des Sports*), the sociology of sport emerged as a distinct field in the mid-1960s (Malcolm 2012) through the collaboration of physical educators and sociologists. While Franklin Henry's seminal paper (1964) had fueled the fragmentation and specialization of physical education around disciplinary homes (biomechanics, psychology, physiology, etc.), sociologists were inspired by leading theorists such as Theodor Adorno, Norbert Elias, and Charles Page who had begun to conduct research into aspects of sport and leisure, and for whom the home discipline of sociology was similarly fracturing into a range of subdisciplines. Subsequently, backed by UNESCO, the International Council of Sport and Physical Education (ICSPE) and the International Sociological Association (ISA) collaborated to form the International Committee for the Sociology of Sport (ICSS). Indicatively the two British representatives in these early meetings were Eric Dunning (University of Leicester, a former student of Norbert Elias, and about whom much more will be said) and Peter McIntosh, a pioneering historian of sport and physical education, responsible for establishing the UK's first physical education degree at Birmingham University.

The balance between sociology and physical education varied across national contexts. As Loy and Kenyon (1969) noted, "nearly all" of the "sport in society" courses in the USA in the late 1960s were located in physical education departments, while in Europe, "no simple pattern exist(ed)," with sociology of sport courses evenly divided between physical education and sociology departments (Loy and Kenyon 1969: 7, 8). Over time, however, those with stronger sociological orientation became predominant, the sociological influences became more pronounced, and physical educators became increasingly marginalized within the ICSS (Malcolm 2012). This can be seen in the types of literature identified as "sociologies" of sport, the organizational politics, and debates over appropriate modes of analysis which led to the rejection of physical education scholarship traditions and alignment more centrally with the sociological orthodoxy of the day. Highly exclusionary boundaries were formed in establishing this field. For instance, CLR James' (1963) *Beyond a Boundary* – a seminal study of sport, postcolonialism, and Caribbean society – was never cited by those working in the developing field, probably because James neither identified as a sociologist, held an academic post, nor conducted research via a method that was (then) defined as sociological (the text being largely autobiographical).

An organizational divide between the historical and sociological study of sport was also established at this time. The International Committee for the History of Physical Education and Sport and the International Association for the History of Physical Education were founded in 1967 and 1973, respectively (merging to form the International Society for the History of Physical Education and Sport or ISHPES in 1989). National and regional organizations followed, with the foundation of the North American Society for Sport History (NASSH) in 1972, and the British Society for Sports History (BSSH) in 1982 as compared, respectively, with the North American Society for the Sociology of Sport (NASSS) (1978) and British Sociological Association's Sport Study Group (1995) (the internationally oriented Leisure Studies Association was founded in the UK in 1975). Concurrently, however, the sociology of sport and sports history were drawn closer together by a shared underlying sense of status insecurity. Barry McPherson was the first sociologist to explicitly raise this, describing the founding fathers of the subdiscipline as "marginal men, neither sociologists nor physical educators," people whose early careers were "characterized by role conflict as they sought to legitimate both the introduction of new courses and the type of research they were undertaking" (McPherson 1975: 57). Pierre Bourdieu (1987) would subsequently comment that the sociology of sport was "doubly dominated," marginal to the sociological "mainstream" and resisted by sportspeople who assumed that these academic outsiders could not fully understand "their" world. Equally, while sports historians have sought to distance their work from the "amateur enthusiasts," there remains "[little] evidence that sport history is highly valued by the wider historical discipline" (Johnes 2004: 148). Echoing Bourdieu, Johnes (2004: 158) notes that, "Sport History ... needs to secure its location within two disciplines [history and sports science], to ensure it engages with and is taken notice of by both."

While self-identifying as distinct in their perspectives and organizationally quite separate, what historians and sociologists of sport shared in terms of status insecurity they also shared in occupational and intellectual opportunities. Occupationally, most sociologists of sport now work in multidisciplinary sports science/human movement/kinesiology departments where they may teach both sociological and historical content, but are simultaneously required to also engage with management topics to meet the student market demand. Similarly, while there are a small number of "lone historians nurturing their individual interest, working their passage as historians of sport by fulfilling other teaching roles in the mainstream syllabus" (Hill 2003: 356), younger colleagues are "required to teach socio-cultural aspects of sport ... [in order to] find broader employment opportunities" (Johnes 2004: 149). Consequently sociologically oriented student textbooks routinely contain chapters or whole sections devoted to the history of sport. Historians and sociologists of sport also publish their research in similar outlets. The *Journal of Sport and Social Issues*, *Sport in Society*, *Soccer and Society*, and *Leisure Studies* are four explicitly cross-disciplinary journals, while the *International Review for the Sociology of Sport* invites historically oriented studies and the *International Journal of the History of Sport* publishes contemporary analyses.

This overview of the field provides the backdrop and a logic to the forthcoming analysis. While advocates of historical sociology largely seek to contest the meaningfulness of the academic distinction between history and sociology, in the study of sport these boundaries are often explicitly blurred in the "core business" of teaching and publication. Concurrently the relations around and across those boundaries have distinctively shaped the development of the historical sociological study of sport which here is divided into three distinct chronological periods. The initial period (up to the late 1980s) is characterized by the production of landmark texts and relatively harmonious cross-disciplinary relations. The late 1990s and early 2000s see the continued emergence of new empirical studies but are perhaps most notable for the presence of cross-disciplinary critiques. The final period, from the mid-2000s onward, is characterized by a return to more harmonious relations, elements of convergence (especially over theory), and the generation of a new wave of historical sociological research on sport. Partly reflecting the author's theoretical orientations, but partly also in recognition of the degree of influence Eliasian theories have had over this field, three terms used by and/or in relation to figurational sociology are used to describe these periods: harmonious inequality, diminishing contrasts and increasing varieties, and the habits of good sociology.

Pre-1990s: Harmonious Inequality

The divisions between sociology and physical education, and the differences between North American and European branches of the sociology of sport, meant that historical sociological studies of sport were always evident rather than prevalent within the emerging subdiscipline. Marxism and the figurational sociology of Norbert Elias represented the main theoretical approaches. Across disciplines, sociologists and historians of sport were largely distinct but also relatively appreciative of each other's contributions to the understanding of the otherwise marginalized subject of sport; there was, we might say, a form of harmonious inequality.

The aforementioned Eric Dunning was a key figure both not only in the early organizational aspects of the field but also in its intellectual development. Significantly, he was also an influential champion of historical sociological analyses of sport. Dunning's work began at the University of Leicester in 1959, and his first publications, derived from his MA thesis on the development of football supervised by Norbert Elias, were published in *History Today* and *New Society* (Dunning 1963, 1964). This supervisory experience seemingly instigated Elias' interest in sport as a vehicle for exploring central aspects of his developing figurational sociological theory. Thus Elias (with Dunning) wrote early papers on "The dynamics of sports groups with special reference to football" (Elias and Dunning 1966) and "The quest for excitement in unexciting societies" (Elias and Dunning 1969) as well as producing a more historically oriented empirical work – "The genesis of sport as a sociological problem" – which provided an early conceptualization of "modern" sport and contrasted this with the sport forms evident in Ancient Greece and Rome (Elias 1971). Elias and Dunning (1971) further collaborated on a study of football in medieval England.

Subsequently Elias and Dunning published a collection of joint and individual essays within a volume titled *Quest for Excitement* (Elias and Dunning 1986). The text combined revised versions of some of these earlier publications, studies by Dunning on commercialization, gender, and player and spectator violence in sport, plus a conceptual introduction authored by Elias. Probably the most significant new work, however, was Elias' historical sociological study, "An essay on sport and violence," which charted the concomitant processes of the parliamentarization (of political conflict) and "sportization" (the codification and standardization of sports rules). Elias showed that sportization invariably entails restrictions on the use and experience of relatively violent aspects and, simultaneously, a greater emphasis on the generation of more prolonged forms of pleasurable excitement). His argument was not that one caused the other, but that one particular social strata came to organize their political business and leisure time under similar types of social norms and explicit rules which, due to the high status of those involved and the particular power structures of England at that time, were subsequently widely diffused across the nation. Empirically, Elias demonstrated this through a study of fox hunting. Combined, Elias' works effectively provided answers to three of the most fundamental questions in the field, illustrating what was distinctive about modern sport forms, why modern sport first emerged in England at this particular time, and why sport holds such a socially significant place in contemporary societies. Even in comparison to Bourdieu who described himself as "virtually alone among major sociologists ... to have written seriously on sports" (Bourdieu and Wacquant 1992: 93), Elias was unique among major social theorist in that he developed both his sociological theory *through* sport as well as a distinct sociology *of* sport (Malcolm 2012).

While Elias seemingly withdrew from the study of sport at this point, Dunning's influence (and thus the prominence of figurational sociology) on the historical sociological study of sport was evident through the 1970s and 1980s, and indeed beyond. He supervised Christopher Brookes' MPhil study, "Cricket as a vocation," which was subsequently turned into a book. While aimed at a more popular readership, the book provided a scholarly analysis of cricket and social class relations from the early 1700s onward. Dunning also coauthored two highly significant historical sociological studies of sport. The first was produced in collaboration with Ken Sheard whom he supervised for an MA on the development of rugby. Publishing articles on "sport as a male preserve" and the development of amateurism in British sport, their work culminated in *Barbarians, Gentlemen and Players* (Dunning and Sheard 1979), a study explicitly informed by Elias' theoretical framework, and empirically informed by the authors' respective studies of football and rugby. The book illustrates how Rugby football (named after a small town in the English midlands) emerged from a variety of violent, local, informal folk games played across preindustrial Britain. Sport played a fundamental role in the reform of British public schools and Rugby football developed in contradistinction to what would become association football (or soccer), driven by "status-rivalry" or class conflict manifest during the nineteenth century (notably the bourgeois dominated Rugby School's rivalry with the more aristocratic Eton and Harrow). The development of these sports into unified national games was marked by the foundation of the Football Association in 1963 and the Rugby Football Union in 1871. The latter subsequently split into union and league codes (1895) as class and regional tensions continued.

The second landmark text was *The Roots of Football Hooliganism* (Dunning et al. 1988). Working primarily with not only Patrick Murphy and John Williams, but also PhD student Joseph Maguire, this formed the middle and, in many ways, the most academically significant in a trilogy of coauthored monographs. The group became known as the "Leicester School" and were said to have enjoyed a "hegemony" in football sociology in the late 1980s. A central motif of this approach was the detailed historical sociological research which illustrated the temporal variability of the phenomenon and the reactions to it from sports organizations, the media, and state officials. This body of work not only was academically influential, but also gained considerable influence in political circles as football hooliganism came to be seen as "the English disease" and threatened the very existence of the sport after a series of stadium disasters during the 1980s.

From the mid-1980s through to the early 1990s, a number of other highly significant, Marxist-informed, historical sociological studies of sport were produced. In Canada, Richard Gruneau (1983) provided an account of the emergence and development of sports in Western capitalist societies which emphasized problems of "class inequality and domination" (Gruneau 1983: 16). Gruneau's research illustrated the different capacities of social classes to structure the organization and meanings of sports. While aristocratic ideals and developing bourgeois values of the disciplined enjoyment of colonial games flourished, the "undisciplined" play practices of the working classes were suppressed. Amateurism was a "regulative strategy" serving to reinforce the dominance of the ruling classes (Gruneau 1983: 134), and entrepreneurial capitalism provided a foundation for the emergence of commercial sport. While state programs for sport reflected the interests of middle class volunteers, commercial entrepreneurs, and state bureaucrats, there is evidence of resistance as the number of opportunities for increasingly diverse groups of people to play a greater range of sports expanded. Canadian leisure thus developed in parallel with broader patterns of class difference.

John Hargreaves' *Sport, Power and Culture* (1986: 2) offered a comparable study of sport in the British context. Similarly focusing on class inequality and domination, Hargreaves emphasized the concept of hegemony as a means by which to understand how sport served as a site for "accommodating" the British working class into the dominant social order. Hargreaves argued that, during the nineteenth century, popular sports forms founded upon traditional values of community and kinship, catharsis and entertainment, were replaced by sports based on bourgeois ideals of discipline, sobriety, self-reliance, and hygienic/therapeutic technologies. Sport was used as a tool for educating and disciplining the masses. Through education, religion, and youth and community groups, attempts were made to incorporate the working class into the bourgeois model of social life. Hargreaves concludes that bourgeois hegemony emerged and developed through twin processes – *recomposition* and *accommodation* of the working class – and that sport was an important tool through which this was achieved.

John Sugden and Alan Bairner's *Sport, Sectarianism and Society in a Divided Ireland* (Sugden and Bairner 1993) shared many of the theoretical premises of the above works. While more explicitly a political sociology of sport, the work sought to illustrate how, rather than being divorced from politics, sport is particularly susceptible to political manipulation, not only reflecting but also helping to sustain broader social conflicts. However, providing a greater emphasis on social resistance, Sugden and Bairner's analysis of the continued armed conflict over Ulster/Northern Ireland's relationship with the United Kingdom and Ireland concludes that aspects of civil society (in this case sport) cannot simply or automatically be mobilized by the state for its own ends. Indeed, paradoxically, the state's facilitation of sports participation served to fuel sectarian differences which in turn threatened to undermine the authority of the state.

Finally, partly extending and partly diverging from this figurational-Marxist dominance were two further bodies of work. In the first, Grant Jarvie demonstrated elements of theoretical synthesis. In *Class, Race and Sport in South Africa's Political Economy,* Jarvie (1985) acknowledged the intellectual guidance of the aforementioned Gruneau on his historically framed analysis of the sport-related conflicts under a system of apartheid. Jarvie subsequently completed a PhD under Dunning's supervision from which *Highland Games* (Jarvie 1991) was published. The study identified four stages or phases of development of Highland Games, which parallel broader social structural transformations in Scotland, and highlights Scotland's dependency on England, the uneven development of the British Isles, and how the development of Scottish national identity was predicated upon the English-dominated British state. The second, Helen Lenskyj's (1986) *Out of Bounds,* marked the arrival of feminist historical sociological approaches to the study of sport. Lenskyj, a Canadian scholar based in Toronto, drew primarily on data from North America to explore the limits imposed on women's participation in sport through medical and social narratives about "female frailty," the potential harm to women's reproductive organs, and "compulsory heterosexuality."

It was during this 12-year period (1979–1991) that many of the foundational English language historical sociological monographs examining sport were produced. This was, it should be noted, also the period in which many of the foundational sport history texts were produced (e.g., Guttmann 1978; Holt 1989; Mangan 1981; Vertinsky 1990). The distinct difference between these two bodies of work was consistently recognized by those working in the field. As Holt, for instance, neatly summarized it:

> Sociologists frequently complain that historians lack a conceptual framework for their research, while historians tend to feel social theorists require them to compress the diversity of the past into artificially rigid categories and dispense with empirical verification of their theories. (1989: 357)

Yet despite these differences, historical sociological studies were widely (if not universally) welcomed within the sport history community. Tony Mangan, for instance, described *Barbarians* as "innovative," "irradiant," and "pioneering" (2005: vii), and Martin Polley (2007: 47) argued that both *Barbarians* and *The Roots of Football Hooliganism* were "seminal ... in linking sociology and history," and praised Jarvie for exemplifying how a scholar could "work comfortably in both history and sociology camps" (Polley 2003: 60). Similarly Gruneau's text was described as "the key work" which "inspired a new generation of sport historians to analyze their subject as a medium of class, gender, racial domination, and resistance" (Pope 1998: ii). *Sport, Sectarianism and Society in a Divided Ireland* (Sugden and Bairner 1993) received the BSSH's inaugural Lord Aberdare Literary Prize for Sports History.

Among British sports historians, Richard Holt was perhaps the most receptive to sociological analyses, while Tony Mason was a particularly persistent and critical voice. Holt concluded his seminal *Sport and the British* (1989), with an appendix in which he commended the influence of sociological thinking on shaping concepts of class and urban relations within sports history, and encouraged sports historians to critically engage with these works. While Holt was critical of some aspects of sociology – the tendency to make unsubstantiated "assertions" (1989: 358), the use theory to produce a "blanket "explanation"" (1989: 364), and the production of interpretations which were ""one-dimensional", lacking context"" (1989: 361) – the tone is of healthy respect for disciplinary difference forged by a shared commitment to the value of the study of sport and recognition of the productivity of cross-disciplinary dialogue.

It is for these reasons that this phase can be defined by the manifestation of harmonious inequality. The production of cross-disciplinary edited collections is indicative of the degree of collaboration. Alan Ingham and John Loy's (1993) *Sport in Social Development* provided a collection of work that emphasized the importance of historically orientated studies of social, political, and cultural analysis of sport, Dunning et al.'s (1993) *The Sports Process* is particularly notable for its advocacy of historical/developmental and cross-cultural analysis and its combination of sociologists and historians' work, while Jarvie and Walkers' (1994) *Scottish Sport and the Making of the Nation* combined historical and sociological analyses to produce "one of the most imaginative of recent studies of British sport" (Hill 1996: 16). This was a period of pioneering sociological, historical, and historical sociological research on sport in which disciplinary boundaries were relatively clearly defined, but a shared experience of status insecurity, empirical focus, and sense of the intellectual value of studying sport led to relatively high levels of mutual respect.

1995–2005: Diminishing Contrasts and Increasing Varieties

During this second period, a combination of "internal" and "external" factors led the balance of centripetal and centrifugal forces between sports history and the sociology of sport to shift. While the notable historical sociological works produced in this era demonstrated elements of theoretical continuity, some of the authors of historical sociological works began to explicitly identify the diminishing contrasts in scholarship. Combined with some broader theoretical trends and, specifically, the impact of the "cultural turn" on (sports) history, elements of the sociology and history of sport were simultaneously forced apart as differences (or varieties) of approach were exposed. In this section, we first review the key historical sociological studies produced in this period and then focus on tensions exhibited in disciplinary relations.

First, while the number of Marxist-informed historical sociological analyses of sport grew, most of these authors were drawn toward cultural studies in part, no doubt, due to the legitimacy this paradigm presented for the social scientific study of sport. Sugden published *Boxing and Society* (1996) which combined ethnographic research with a discussion of boxing in ancient and medieval societies and the popularization of bare-knuckle prize fighting in Regency Britain. Sugden explored boxing's role in the development of physical skill for attack and self-defense within the context of broader traditions of dispute resolution via violent physical means. He went on, in collaboration with Alan Tomlinson, to publish *FIFA and the Contest for World Football* (Sugden and Tomlinson 1998) which explained the ways that key personnel in the international governing body of the sport, the Fédération Internationale de Football Association (FIFA), enabled the emergence of a global political economy of football dominated by corporate sponsorship, market forces, and media relations. Bruce Kidd's (1996) *The Struggle for Canadian Sport* similarly invoked notions of political economy to look at the material and ideological battles that determined how Canadian sport was played, by whom, and with what outcomes. Mention should also be made of David Andrews who would subsequently become a leading figure in the sociology of sport, notable initially for his theoretical expositions which brought to the attention and popularized the work of Baudrillard, Derrida, and Foucault among sport scholars. One of Andrews' earliest empirical studies focused on the historical development of rugby and masculinity in Welsh culture.

Feminist scholars also grew in both number and influence. Most notable in this regard was Jennifer Hargreaves (1994) who, in *Sporting Females,* illustrated how from the late nineteenth to the middle of the twentieth century women's sports reflected and reinforced popular views about women as physically inferior to men. An ideology of the natural weakness of women pervaded the principles and practices of physical education, exercise, and sport for females in Great Britain during the period. Despite such narrow definitions of femininity, some (most often middle-class) women experienced new forms of corporeal freedom through active participation in sports such as tennis and croquet and in physical activities such as medical gymnastics. Yet female emancipation through sport existed at the same time as persistent opposition. Hargreaves further explored the gendered history of the Olympic Games and examined contemporary gender relations and sporting masculinities and femininities. Nothwithstanding the enduring legacies of nineteenth-century versions of femininity, since the middle of the twentieth century processes of contestation and change have resulted in increasing opportunities for female participation in a range of sporting activities and a degree of diversity in what counts as "acceptable" in terms of female appearance, behavior, and emotional expression. The advocacy principles of feminism (plus, ironically, norms of gender segregation) led like-minded researchers to be particularly active within physical education, and Sheila Scraton's (1992) *Shaping up to Womanhood* was a landmark text, providing the first historically informed, feminist critique of physical education with the UK. However, it was Jennifer Hargreaves in particular whose work was a catalyst for the "burgeoning" of scholarship on sports history and gender relations, helping to develop a women's sports history which, in particular, "proceeded from women's experiences and which placed women at its center" (Vertinsky 1994: 23).

But perhaps most of all, figurationally informed historical sociological studies of sport also continued to develop. First, historical sociological studies of individual sports that forwarded Elias' work on civilizing processes were published. Sheard

(1997), for instance, demonstrated how prize fighting had undergone a sportization process, through which combat became increasingly regulated and violence declined. In drawing attention to the significance of the appearance of control rather than the decline of physical harm per se (e.g., the use of gloves and headguards reduces cuts and abrasions but may increase damage to the brain) (Sheard 1997), this work added nuance to our understanding of civilizing processes. Moreover, Sheard's (1998) work illustrated the role of the medical profession in both challenging and defending boxing practices which test the limits of public tolerance of violence, while in a further and somewhat neglected study, Sheard (1999) argued that the development of birdwatching, as a form of symbolic hunting, again illustrated aspects of civilizing processes. The trajectory of birdwatching, increasingly, emphasized preserving rather than killing birds, while aspects of routinization and a controlled decontrolling of emotional controls experienced by the birdwatcher or "twitcher" could be located within Elias and Dunning's (1986) ideas about the quest for excitement. Additionally, and in light of a rising number of critiques, Malcolm (2002) re-examined the premises of Elias' account of the emergence of modern sport. In this study, he showed how the rules/laws of cricket – a game widely perceived as relatively genteel and thus in a conventional rather than technical sense "civilized" – had been developed by English parliamentarians and exhibited a trajectory toward greater internal self-regulation and the restriction of more violent aspects of play. In so doing, Malcolm essentially showed the validity of the principle theoretical messages outlined in "An essay on sport and violence." If one includes an unpublished PhD thesis on horseracing, figurational sociologists had now "tested" Elias' theory about the role of parliamentarization and sportization in the emergence of modern sport in the context of all the relevant eighteenth-century sports except golf and found a significant degree of conformation for Elias' overarching thesis for the development of distinctly modern forms of sport.

Second, Maguire and van Bottenburg contributed figurational accounts that provided distinctly developmental perspectives of sport to ongoing debates about globalization which were then prominent in the social sciences. For instance, Maguire (1999) identified five phases of the global development of sport, starting from Elias' notion of sportization and culminating in a post-1960s phase of creolization of sport cultures. Drawing on Elias, he proposed a theory of "diminishing contrasts and increasing varieties" to steer a path between an overemphasis on either the homogenization of global sport or the growth of heterogeneity, and illustrate the "double-bind" character of relationships in a global society whereby different parties (albeit unequally) exerted influence on each other culminating in a commingling or interpenetration of cultures. Van Bottenburg (2001) charted the differential popularization and diffusion of sport from four main centers: Britain, Germany, the USA, and Japan. Charting first national developments then aspects of imperialism, van Bottenburg showed that the primary social structural factors that shaped the adoption of sports over a number of centuries included the relationship between donor and adopting country, and the absolute and relative social status of adopters and diffusers of sports forms.

But in retrospect, this period was perhaps most notable for the publication of an edited collection called *Sport Histories: Figurational Studies of the Development of Modern Sports* (Dunning et al. 2004). The book contained sport-specific case studies extending previous analyses (e.g., football, rugby, cricket, baseball, and boxing) and exploring new sports such as tennis, motor racing, shooting, Japanese martial arts, and gymnastics. These case studies highlighted the following: the significance of the role of sports rules (or laws) as expressions of attitudes, behaviors, and broader social norms (in relation to violence); the dynamics of more subtle, intra- rather than inter-class, distinctions; the role of violence and civilizing processes; international diversity in civilizing processes; and the central rather than peripheral role of the unintended outcomes of human action. These studies showed the importance of rejecting the "great men" approach evident in (some) sports history, in which individuals were widely accredited with "inventing" certain sports (i.e., William Webb Ellis and rugby, Abner Doubleday and baseball, and James Figg and boxing), and instead placed these people within the networks of interdependence that enabled and constrained their actions. The book's introduction also included a distinct theoretical chapter which set out the case for an Eliasian historical sociological approach to the study of sport, and a conclusion which reflected on the Marxist-oriented historical sociological studies of sport that had appeared in the earliest phase of this literary development. Elias, it was argued, had been the most vociferous and persistent advocate of a developmental of historical sociology of sport. His sociological conceptualization of time as a means of human orientation, his focus on the historical development of knowledge, and considerations of evolutionary biological change provided a distinct approach, culminating in the demonstration that modern sport emerged through specific social class relations accompanied by a change in habitus and conscience, and set within a series of changes within English society at the social structural level. The text also claimed to be "representative of sociological "best practice" in the field of historical sociology" (Dunning et al. 2004: 12) and critiqued Marxist historical sociologists of sport such as Hargreaves and Gruneau not only (predictably) for economic determinism, but also for a failure to deal with the context-specific aspects of the development of sport, and for lacking detailed primary empirical research.

These particularly bold statements need to be understood in a context of broader theoretical developments and distinct disciplinary relations. On the one hand, as noted above, the interdependence of history and sociology and the desirability of cross-fertilization were well-established at this time. Leading sports historians called for the more widespread use of social theories "fashioned outside our discipline" (Hill 1996: 19), the need for "a dose of intellectual rigour," and "an injection of relevance" to broaden the appeal of history within the sports sciences (Booth 1997: 192). Others warned colleagues against closing ranks by treating sociological theory as a threat because sports history needed to develop links with other sport-related disciplines in order to survive (Nauright 1999). Disciplinary boundaries appeared to become particularly permeable at this time with Mike Marqusee (a journalist and political activist), and John Bale (a geographer), joining Sugden and Bairner as recipients of the Aberdare Literary Prize for Sports History. Historians such as Hill and Johnes explicitly expressed their debt to various social theorists. Holt (1998: 11) reflected that "sports history partly grew out of a new social history fed by the boom in sociology." Organizationally too, disciplinary orientation did not seem prohibitive. The Australian Society for Sports History appointed a political scientist (Colin Tatz) as its first president, the aforementioned Grant Jarvie became president of the BSSH, and Gertrud Pfister simultaneously held the leading administrative role in both the International Sociology of Sport Association (ICSS's successor body) and ISHPES.

Undoubtedly fueling these changes in sports history was the challenge of postmodernism, and concerns that these ontological shifts which could lead to "an end to sports history as we know it" (Nauright 1999: 5). This, in turn, reopened or simply exposed the divisions between physical education and the historical "mainstream," and the amateur and the academic sides of sports history, the "old school" and "new school" of sports history (Malcolm 2012). In contrast to the greater disciplinary and theoretical openness noted above, Maguire's (1995) arguments that there were areas within the history and sociology of sport where the respective bodies of research were barely distinguishable, and that disciplinary convergence was thus desirable and logically possible (hence diminishing contrasts), led some sports historians to "express[ed] resentment at being "preached at" by those who were seen to argue that "history was illegitimate unless sociologically driven"'" (Rowe and Lawrence 1996: 14).

Resistance became particularly manifest in the development of historians' critiques of figurational sociological research which identified the variety of ways of "doing" history and/or historical sociology. Previously, Reid (1988) had contested Dunning and Sheard's (1979) contention that the aristocracy withdrew from folk football and became more status exclusive in the late eighteenth and early nineteenth centuries, thus precipitating the emergence of "new" sport forms. Lewis (1996) had argued that football crowd disorder in nineteenth century Lancashire was far more significantly related to gambling than previously believed, thus critiquing what he termed Dunning et al.'s "continuity thesis" of spectator violence. Goulstone (2000) and Harvey (2001) had argued that Dunning and Sheard's "public school status-rivalry" thesis was undermined by new historical evidence which showed that: a) "modern forms" of football, played in the Sheffield area in the nineteenth century, developed independently of the public schools; b) modern notions of equality in football stemmed from working class forms of the game, in which gambling was a central component; and c) the bifurcation of football codes could be seen in folk forms of football which predate codification in the public schools.

But in response to the publication of *Sport Histories,* two further, more wide-ranging critiques were published. First, Collins (2005) suggested the following: a) that Dunning and Sheard overstate the originality of their work in relation to the Webb Ellis myth; b) that civic rather than public school rivalries drove the bifurcation of football and rugby; and c) that Dunning and Sheard falsely depict rugby as being viewed as more violent than football at the end of the nineteenth century. Collins went on to argue that use of the theory of civilizing processes in the work of

Dunning et al. (1988), Malcolm (2002), and Sheard (1997) led to teleological explanations, while Vamplew (2007) also critiqued the empirical basis of Malcolm's work on cricket, and the role of theory in (mis-)guiding this research. Implicitly – rather than explicitly – both effectively contested the claims made in *Sport Histories* about the relative merits of figurational approaches to the historical sociological study of sport.

Figurational sociologists produced numerous responses to these critiques, the details of which have been outlined elsewhere and do not need to be discussed in this context (see Malcolm 2012). But of particular interest is the contrast between the relatively constructive engagement with Marxist and feminist sociology, and the degree of hostility expressed toward figurational sociologists. While historians' critiques of the role of primary empirical research, the analytical de-centering of sport in historical sociological accounts, and core ontological principles could equally if not *more legitimately* be raised against other historical sociologists (Malcolm 2012), this was somewhat obscured by a theoretical empathy for cultural studies of "sport "from below"" (Hill 1996: 3) focusing on marginalized female and working class sporting cultures. Never had sports history and the historical sociological study of sport been more deeply interdependent, but never had relations been as openly hostile. It was, ironically, where the contrasts had most diminished that the different varieties were most explicitly highlighted.

The Twenty-First Century: The "Habits of Good Sociology"

In a 2010 review of the relations between sports history and the sociology of sport, Mansfield and Malcolm (2010: 109) argued that, "the historical sociology of sport is, itself, fast becoming history" and noted the paradox whereby "the number of sociologists of sport who now unquestioningly accept the value of a historical contextualization seems inversely proportional to the number actually engaged in historical-oriented research." They further predicted that broader trends toward the ascendency of natural science within academia would push sociologists of sport ever further away from historical studies. As illustrated in this final section, assessment and prophecy have come to be fundamentally flawed. In recent times, it could be argued, the habits of good sociology live on and have, to some extent, experienced revival.

One indicator of the continuing strength of the historical sociology of sport was the production of the first sport-specific issue of the *Journal of Historical Sociology* (*JHS*), edited by Alan Tomlinson (aforementioned) and Cambridge historian Christopher Young in 2011. The journal had been the outlet for a number of previous sport-related studies both prior to and after this third period. Extending the legacy of previous interdisciplinary collaborations, this special issue brought together a founding figure in the sociology of sport and leisure studies (Alan Tomlinson) and a historian of German literature and culture whose *The 1972 Olympics and the Making of Modern Germany* (Schiller and Young 2010) had won both the BSSH's Lord Aberdare Prize and the NASSH book award. The *JHS* special issue – "Sport in

Modern European History" – explored aspects of consumerism, identity/nationalism, language and cultural power, and space/time dynamics across a range of sports including mountaineering, boxing, cricket, and Basque pelota. While recognizing the "immense" contributions of Elias and Dunning to the historical sociology of sport, the collection positioned itself as theoretically eclectic and thus an antidote to "the single theoretical framework and conceptual imperative" (Tomlinson and Young 2011: 411) of figurational studies. However, with the exception of a contribution by Orlando Patterson, who was interviewed by the editors for this special issue, it could reasonably be suggested that all the contributors would (either largely or entirely) self-identify as historians rather than sociologists, or at least more closely align with history as a discipline.

A second indication, however, was the growth in the range of personnel and sport-related topics explored by figurational sociologists. Malcolm (2001, 2006) and Bloyce (2008) used a figurational framework to analyze the cultural diffusion of cricket and baseball, respectively, the former from Great Britain to the Caribbean and the USA, and the latter from the USA to England. Four other studies focused more centrally on violence and civilizing processes. Lake's (2009) study of real tennis described the interdependence of courtization and the development of the sport and charted how this sportization process led social status to increasingly be linked to the exercise of self-restraint and the greater use of foresight in social relations. Connolly and Dolan (2010) similarly charted sportization and civilizing processes in Gaelic football while Sanchez-Garcia and Malcolm (2010) explored the emergence and developmental trajectory of mixed martial arts (a kind of hybrid combat sport which combined elements of western boxing and wrestling, with Asian and Brazilian fighting codes). Rivero Herraiz and Sanchez Garcia (2016) subsequently explored the "civilizing" impact of the English model on sporting practices in Spain, identifying how the regenerationist movement, in the first third of the twentieth century (a dual response to the loss of Empire and a desire for modernization), led to the increasing interest in (British models of) sport and physical education. As social elites increasingly participated and adopted cultural mores of greater self-restraint, and the working classes were incorporated through spectatorship, practices deemed to have contributed toward Spain's "degeneration," poor health and social backwardness became increasingly marginalized and/or reformed (e.g., bullfighting was linked to Spain's backward traditions, and some of its more violent practices were curtailed). Finally, van Gestel (2019) explored violence and violence control in the context of karate in Belgium, arguing that the value of violent contact had been superseded or replaced by no-touch and light-impact techniques.

Many of these works were expanded into book length projects. Malcolm (2013) analyzed the global development of (men's) cricket from its emergence and sportization phase, its subsequent close association with notions of Englishness, and Imperial and postimperial negotiation in which violence and cross-cultural behavioral expectations intertwined. Velija (2015) subsequently examined the social processes that shaped the emergence and development of women's cricket as a global game, focusing particularly on the impact of nationalism, cultural diffusion, colonialism, and masculinity on the development of women's cricket. Curry and

Dunning (2016) published *Association Football: A study in figurational sociology* which was partly designed as a supplement to Barbarians, partly designed to test the public school "status-rivalry" hypothesis first proposed by Dunning in 1961, and partly a vehicle designed to respond to the critics of this body of work discussed earlier (i.e., Goulstone (2000), Harvey (2001), and Collins (2005). See section on "1995–2005: Diminishing Contrasts and Increasing Varieties"). Sanchez-Garcia (2019) grappled with some interesting cross-cultural challenges in explaining how Japanese martial arts became part of the global sport system (notably the (ir)relevance of the inherently western term "sport" in this context, and the less clear-cut distinctions between the religious/spiritual sphere, politics, war, and art and everyday Japanese life). In so doing, he shows how (with some significant de-civilizing spurts along the way) the main concerns shifted from combat effectiveness in war to questions of etiquette, self-perfection, and even entertainment and thus that the term "martial arts" has become tightly bound to notions of civilizing processes. Finally, Connolly and Dolan (2020) examined the development of the Gaelic Athletic Association in Ireland, charting changes in player and spectator violence, changing tensions around the organization and distribution of economic resources within the game (e.g., amateurism-professionalism, and the emphasis given to youth sport development), and changing identities relative to other sports, nationalism, the assumed "charisma" of the Irish, and international developments.

Concomitantly, the number and range of nonfigurationally informed historical sociological studies of sport has also grown. A notable figure in this regard was Victoria Paraschak, whose career has been primarily devoted to the study of the aboriginal peoples and which, while often foregrounding advocacy, has nevertheless been built on an explicitly sociohistorical approach designed to reveal the historic roots of contemporary ethnic inequities in sport in Canada (see Paraschak 2013 for some autobiographical career reflections). Second, during this period NASSH prize winners included anthropologist Susan Brownell's (2009) interdisciplinary anthology on the 1904 St Louis Olympic games and Mary Louise Adam's (2011) analysis of how contemporary conceptions of figure skating have been shaped by the historical development of ideas about masculinity and femininity. Third, Jeffrey Montez de Oca has explored the role of sport in America during the Cold War. His work initially focused on the role of sport in addressing American cultural insecurities about the physical and mental strength of the American (white, male) population (Montez de Oca 2005), and subsequently the role of college football at this time in these processes. Identifying the interdependence of media, education, and the military, he argues that during this period the sport provided a celebration of the American way of life and a focal point for the development of American masculinity, themes that have reemerged and/or become more pronounced post 9/11 and the war on terror (Montez De Oca 2013). Montez de Oca lately collaborated with French sociologist Christophe Brissoneau (Brissoneau and Montez de Oca 2017) to produce an oral history of doping in French sport which reveals the extreme physical demands placed on competitors, the extensive medical networks developed to provide support, and the creation of a distinct cultural practice and ethical codes which foster the use of performance-enhancing drugs. Other recent studies have looked at the early twentieth century trials of two Canadian ice hockey players who were accused of killing opposition players during matches, both of whom were acquitted on the basis that violence was deemed intrinsic to the sport (Lorenz and Osborne 2015), and the development of sex testing and gender policing in sport (Peiper 2016). In an analysis embracing the 1930s to the early 2000s, Pieper argues that women's sporting performance has historically been constrained by factors such as Cold War tensions, gender anxieties, and controversies around doping, and Peiper (2016) thus updates and complements the analysis to Lenskyj's (1986) *Out of Bounds* published 30 years earlier.

In drawing this section to a close, two bodies of historical sociological research in particular should be highlighted. The first is Rob Lake's work on tennis which, as noted above, began with an analysis of (real) tennis and civilizing processes but has expanded into the culturally more central sport of lawn tennis, and largely departed away from a figurational sociological perspective. That Lake should be drawn toward historical sociology is perhaps not surprising, in that he began his career studying under Jennifer Hargreaves and latterly at the University of Leicester with people such as Dunning, Sheard, and Malcolm. Mirroring in some respects the career of Grant Jarvie, Lake has subsequently held administrative positions with both the BSSH and NASSS. His book, *A Social History of Tennis* (Lake 2015), won the Aberdare prize, and he continues to publish across sociology and history journals. Finally the most recent historical sociological work to win this prize is Paul Campbell's (2016) *Football, Ethnicity and Community: the life of an African-Caribbean Football Club* which draws on historical records to chart the development from the arrival of the Windrush generation, the establishment of "Meadebrook Cavaliers" in the 1970s, the forging of disparate people from various West Indian islands into a more singular but contested Caribbean identity, and the role of football in that process. While neither Campbell's work nor career has been particularly shaped by Elias or Dunning, he now works at the University of Leicester which, as can be seen, provides a significant thread through the development of historical sociological studies of sport.

Conclusion

While an analysis of the distinctiveness of the historical sociological study of sport and leisure must be left to the reader, there do seem to be three particularly significant factors that have shaped the development of this body of work. The first is the multi- and inter-disciplinary character of the broader "sports sciences" in which so much of this work has been conducted. Second, the status insecurities which stem from a commitment to the importance of, and concentration upon, the empirical study of this particular sphere of social life have forged a distinct set of relations between sociologists and historians. Exhibiting various degrees of conflict and cooperation, on the whole these have been more mutually supportive than destructive and currently rest at a position of relatively productive cross-border recognition. Indeed, when, in 2015, Vamplew edited a special issue of the *International Journal for the*

History of Sport on Methodology in Sports History, Malcolm (of whom he had previously been highly critical, see Vamplew 2007) contributed an article on history, historical sociology, and sociological method (Malcolm 2015). The third factor is the theoretical orientation of the field which has no doubt been shaped by Elias' recognition of sport as worthy of study. He and Dunning were keenly aware that "knowledge about sport was knowledge about society," and he hoped that through his collaboration with Dunning, "we helped a little" to make sport a respectable subject for academic study (Elias 1986: 19). "Elias' work on sport... constituted his main attempt to contribute to the understanding of English social development" (Dunning 1992: 98). Indeed a series of overviews and state-of-the-art reviews of the sociology of sport published in recent years have repeatedly shown Elias to be one of the two most significant theoretical influences (Dart 2014; Tian and Wise 2020; Gomes et al. 2021). The two fields in which Elias is perhaps most prominent are the study of sport and leisure and historical or developmental sociology.

The publication of two recent texts illustrates this final point perfectly. In 2018, Haut et al. edited *Excitement Processes*, a collection which included four of Elias' previously unpublished works focused on the emotions, sportization, the role and control of violence in boxing and dueling, and the body in relation to studies of sport and leisure. Accompanied by commentaries from various contemporary researchers, it was hoped that publication would not only augment previous analyses but also open up new fields of enquiry. A year later, Malcolm and Velija (2018) edited *Figurational Research in Sport, Leisure and Health*, which not only included chapters explicitly addressing the development of sport, and chapters that developmentally addressed more contemporary sporting issues, but also included a chapter on the methodology of developmental, figurational, sociological research. Yet perhaps, this point about the role of Elias in the development of historical sociological studies of sport can be made even more pithily; for while it has become perfectly possible to be a Marxist, feminist, or even a poststructuralist sports historian, the notion of a figurational historian of sport is simply oxymoronic.

References

Adams ML (2011) Artistic impressions: figure skating, masculinity and the limits of sport. University of Toronto Press
Bloyce D (2008) "Glorious rounders": the American baseball invasion of England in two world wars – unappealing American exceptionalism. Int J Hist Sport 25:387–405
Booth D (1997) Sport history: what can be done. Sport Educ Soc 2(2):191–204
Bourdieu P (1987) Programme for a sociology of sport. In: Bourdieu P (ed) In other words. Polity Press, Cambridge, pp 156–167
Bourdieu P, Wacquant L (1992) An invitation to reflexive sociology. Polity Press, Cambridge
Brissonneau C, De Oca JM (2017) Doping in elite sports: voices of French sportspeople and their doctors, 1950–2010. Routledge
Brownell S (2009) The 1904 anthropology days and Olympic games: sport, race, and American imperialism. University of Nebraska Press, Lincoln
Campbell P (2016) Football, ethnicity and community: the life of an African-Caribbean Football Club. Peter Lang

Collins T (2005) History, theory and the 'civilizing process'. Sport Hist 25(2):289–306
Connolly J, Dolan P (2010) The civilizing and sportization of Gaelic football in Ireland: 1884–2009. J Hist Sociol 23(4):570–598
Connolly J, Dolan P (2020) Civilising processes, players, administrators and spectators. Palgrave Macmillan
Curry G, Dunning E (2016) Association football: a study in figurational sociology. Routledge, Abingdon
Dart J (2014) Sports review: a content analysis of the international review for the sociology of sport, the journal of sport and social issues and the sociology of sport journal across 25 years. Int Rev Sociol Sport 49(6):645–668
Dunning E (1963) Football in its early stages. History Today (December)
Dunning E (1964) The evolution of football. New Soc 30:14–15
Dunning E (1992) A remembrance of Norbert Elias. Sociol Sport J 9:95–99
Dunning E, Sheard K (1979/2005) Barbarians, gentlemen and players: a sociological study of the development of rugby football. Martin Robertson/Routledge, Oxford/Abingdon
Dunning E, Murphy P, Williams J (1988) The roots of football hooliganism: an historical and sociological study. Routledge, Kegan, Paul, London
Dunning E, Maguire J, Pearton R (eds) (1993) The sports process: a comparative and developmental approach. Human Kinetics, Champaign
Dunning E, Malcolm D, Waddington I (2004) Sport histories: figurational studies in the development of modern sports. Routledge, London
Elias N (1971) The genesis of sport as a sociological problem. In: Dunning E (ed) The sociology of sport: a selection of readings. Frank Cass, London, pp 88–115
Elias N (1986) Introduction. In: Elias N, Dunning E (eds) Quest for excitement: sport and leisure in the civilising process. Oxford, Blackwell, pp 19–62. [Revised and enlarged edition, collected works, vol 7. UCD Press, Dublin, 2008]
Elias N, Dunning E (1966) Dynamics of sports groups with special reference to football. Br J Sociol 17(4):388–402
Elias N, Dunning E (1969) The quest for excitement in leisure. Soc Leisure 2:50–85
Elias N, Dunning E (1971) Folk football in medieval and early modern Britain. In: Dunning E (ed) The sociology of sport: selected readings. Frank Cass, London, pp 116–132
Elias N, Dunning E (1986) Quest for excitement: sport and leisure in the civilizing process. Blackwell, Oxford. [Revised and enlarged edition, collected works, vol 7. UCD Press, Dublin, 2008]
Gomes L, Moraes L, Marchi Júnior W, Moraes e Silva M (2021) A mapping of JLASSS: the academic consolidation of the socio-cultural studies of sport in Latin America. Int Rev Sociol Sport 56(2):276–296
Goulstone J (2000) The working class origins of modern football. Int J Hist Sport 17(1):135–143
Gruneau (1983) Class, sports and social development. University of Massachusetts Press/Human Kinetics, Massachusetts/Champaign
Guttmann A (1978) From ritual to record: the nature of modern sport. Columbia University Press, New York
Hargreaves J (1986) Sport, power and culture: a social and historical analysis of popular sports in Britain. Polity, Cambridge
Hargreaves JA (1994) Sport females: critical issues in the history and sociology of women's sport. Routledge, London
Harvey A (2001) 'An epoch in the annals of national sport': football in Sheffield and the creation of modern soccer and rugby. Int J Hist Sport 18(4):53–87
Haut J, Dolan P, Reicher D, Sanchez Garcia R (eds) (2018) Excitement processes. Norbert Elias's unpublished works on sports, leisure, body, culture. Springer VS, Wiesbaden
Henry FM (1964) Physical education: an academic discipline. J Health Phys Educ Recreat 35: 32–33
Hill J (1996) British sports history: a post-modern future. J Sport Hist 23(1):1–19

Hill J (2003) Introduction: sport and politics. J Contemp Hist 38(3):335–361
Holt R (1989) Sport and the British: a modern history. Clarendon Press, Oxford
Holt R (1998) Sport and history: British and European traditions. In: Allison L (ed) Taking sport seriously. Meyer and Meyer, Aachen, pp 7–30
Ingham A, Loy J (eds) (1993) Sport in social development: traditions, transitions and transformations. Human Kinetics, Champaign
James CLR (1963) Beyond a boundary. Hutchison, London
Jarvie G (1985) Class, race and sport in South Africa's political economy. Routledge & Kegan Paul, London
Jarvie G (1991) Highland games: the making of the myth. Edinburgh University Press, Edinburgh
Jarvie G, Walker G (eds) (1994) Scottish sport and the making of the nation. Leicester University Press, Leicester
Johnes M (2004) Putting the history into sport: on sport history and sport studies in the UK. J Sport Hist 31(2):145–160
Kidd B (1996) The struggle for Canadian sport. University of Toronto Press
Lake RJ (2009) Real tennis and the civilising process. Sport Hist 29(4):553–576
Lake RJ (2015) A social history of tennis in Britain. Routledge, Abingdon
Lenskyj H (1986) Out of bounds: women, sport and sexuality. Women's Press, Ontario
Lewis RW (1996) Football hooliganism in England before 1914: a critique of the Dunning thesis. Int J Hist Sport 13(3):310–339
Lorenz SL, Osborne GB (2015) "Nothing more than the usual injury": debating hockey violence during the manslaughter trials of Allan Loney (1905) and Charles Masson (1907). J Hist Sociol 30(4):698–723
Loy J, Kenyon G (1969) The sociology of sport: an emerging field. In: Loy J, Kenyon G (eds) Sport, culture and society: a reader on the sociology of sport. Macmillan, New York, pp 1–8
Maguire J (1995) Common ground? Links between sports history, sports geography and the sociology of sport. Sport Tradit 12(1):3–25
Maguire J (1999) Global sport: identities, societies, civilizations. Polity Press, Cambridge
Malcolm D (2001) 'It's not cricket': colonial legacies and contemporary inequalities. J Hist Sociol 14(3):253–275
Malcolm D (2002) Cricket and civilizing processes: a response to Stokvis. Int Rev Sociol Sport 37(1):37–57
Malcolm D (2006) The diffusion of cricket to America: a figurational sociological examination. J Hist Sociol 19(2):151–173
Malcolm D (2012) Sport and sociology. Routledge, London
Malcolm D (2013) Globalizing cricket: englishness, empire and identity. Bloomsbury, London
Malcolm D (2015) Durkheim and sociological method: historical sociology, sports history and the role of comparison, Int J Hist Sport 32(15):1808–1812
Malcolm D, Velija P (eds) (2018) Figurational research in sport, leisure and health. Routledge, London
Mangan AJ (1981) Athleticism in the Victorian and Edwardian public schools. Cambridge University Press, Cambridge
Mangan JA (2005) Series editor's foreword. In: Dunning E, Sheard K (eds) Barbarians, gentlemen and players: a sociological study of the development of rugby football. Routledge, Abingdon, pp vii–ix
Mansfield L, Malcolm D (2010) Sociology. In: Pope S, Nauright J (eds) Routledge companion to sports history. Routledge, London, pp 99–113
McPherson B (1975) Past, present and future perspectives for research in sport sociology. Int Rev Sport Sociol 10:55–72
Montez De Oca J (2005) "As our muscles get softer, our missile race becomes harder": cultural citizenship and the "muscle gap". J Hist Sociol 18(3):145–172
Montez De Oca J (2013) Discipline and indulgence: college football, media, and the American way of life during the cold war. Rutgers University Press

Nauright J (1999) 'The end of sports history?' From sports history to sports studies. Sport Tradit 16(1):5–13

Paraschak V (2013) The road (not) taken: academic-life relations in the sociocultural study of sport. Sport Hist Rev 44:77–90

Peiper L (2016) Sex testing: gender policing in women's sports. University of Illinois Press

Polley M (2003) History and sport. In: Houlihan B (ed) Sport and society: a student introduction. Sage, London, pp 49–64

Polley M (2007) Sports history: a practical guide. Palgrave Macmillan, Basingstoke

Pope S (1998) Sport history into the 21st century. J Sport Hist 25(2):i–x

Reid D (1988) Folk football, the aristocracy, and cultural change. Int J Hist Sport 5(2):224–238

Rivero Herraiz A, Sánchez García R (2016) Sport versus bullfighting: the new civilizing sensitivity of regenerationism and its effect on the leisure pursuits of the Spanish at the beginning of the twentieth century. Int J Hist Sport 33(10):1065–1078

Rowe D, Lawrence G (1996) Beyond national sport: sociology, history and postmodernity. Sport Tradit 12(2):3–16

Sanchez Garcia R, Malcolm D (2010) De-civilizing, civilizing or informalizing? The international development of mixed martial arts. Int Rev Sociol Sport 45(1):1–20

Sanchez-Garcia R (2019) The historical sociology of Japanese martial arts. Routledge, Abingdon

Schiller K, Young C (2010) The 1972 Olympics and the making of modern Germany. University of California Press

Scraton S (1992) Shaping up to womanhood: gender and girls' physical education. Open University Press

Sheard K (1997) Aspects of boxing in the Western "civilizing process". Int Rev Sociol Sport 32(1):31–57

Sheard K (1998) '"Brutal and degrading": the medical profession and boxing, 1838–1984'. Int J Hist Sport 15(3):74–102

Sheard K (1999) A twitch in time saves nine: bridwatching, sport, and civilizing processes. Sociol Sport J 16(3):181–205

Sugden J (1996) Boxing and society: an international analysis. Manchester University Press, Manchester

Sugden J, Bairner A (1993) Sport, sectarianism and society in a divided Ireland. Leicester University Press, Leicester

Sugden J, Tomlinson A (1998) FIFA and the contest for world football. Who rules the people's game? Polity, Cambridge

Tian E, Wise N (2020) An Atlantic divide? Mapping the knowledge domain of European and North American-based sociology of sport, 2008–2018. Int Rev Sociol Sport 55(8):1029–1055

Tomlinson A, Young C (2011) Sport in modern European history: trajectories, constellations and conjunctures. J Hist Sociol 24(4):409–427

Vamplew W (2007) Empiricist versus sociological history: some comments on the 'civilizing process'. Sport Hist 27(2):161–171

Van Bottenburg M (2001) Global games. University of Illinois Press, Urbana

Van Gestel J (2019) Violence and violence control in karate: has there been a sportization process? Int Rev Sociol Sport 54(5):557–576

Velija P (2015) Women's cricket and global processes: the emergence and development of women's cricket as a global game. Palgrave Macmillan

Vertinsky P (1990) The eternally wounded woman: women, doctors and exercise in the late nineteenth century. Manchester University Press, Manchester

Vertinsky P (1994) Gender relations, women's history and sport history: a decade of changing enquiry, 1983–1993. J Sport Hist 21(1):1–24

Norbert Elias and Psychoanalysis: The Historical Sociology of Emotions

27

Robert van Krieken

Contents

Introduction	700
Sociology of Emotions	700
History of Emotions	702
Historical Sociology	704
Norbert Elias	704
Defining "Emotion"	705
Elias and Historical-Sociological Psychology	706
The Civilizing Process	708
Foundations in Freud	709
Beyond Freud	711
Emotions and Social Relations	712
From Libido to "Valencies"	715
"Homo Clausus" Versus Figurations and Group Analysis	716
The Superego and Its History	717
Conclusion	719
References	721

Abstract

This chapter outlines the ways in which the historical sociology of emotions should be seen as positioned between three allied fields of study: the sociology of emotions, the history of emotions, and historical sociology more broadly. It examines the work of the leading historical sociologist who has been highly influential across all these fields, Norbert Elias, explaining the main elements of his analysis of emotions, highlighting the relationship between his conceptual

R. van Krieken (✉)
University of Sydney, Sydney, NSW, Australia

School of Sociology, University College Dublin, Dublin, Ireland
e-mail: robert.van.krieken@sydney.edu.au

© The Author(s), under exclusive licence to Springer Nature Singapore Pte Ltd. 2022
D. McCallum (ed.), *The Palgrave Handbook of the History of Human Sciences*,
https://doi.org/10.1007/978-981-16-7255-2_58

approach and that of Sigmund Freud and later psychoanalytic theory. Outlining the nuances and complexities of Elias's relationship to Freud and psychoanalysis functions as a useful window onto the core issues and debates characterizing the historical sociology of emotions, being a leading example of how the disciplines of sociology, history, and psychology can be brought to bear on the analysis of emotions in a variety of ways. Three conceptual issues are examined: (1) how the "drivenness" of emotions should be understood with reference to Freud's theory of drives and Elias's revision of that theory; (2) Elias's critique of the "closed personality" image of human beings and his emphasis on the constitution of emotional experience within social relations; (3) the question of how the psychic agency responsible for the management and control of emotions, Freud's "super-ego," can be seen as having developed over time.

Keywords

Historical sociology · Emotions · History · Sociology · Elias · Freud · Psychoanalysis

Introduction

The historical sociology of emotions is, oddly enough, only in its infancy as a distinct field of knowledge in the human sciences. This is surprising because one would think that the three very closely allied fields – the sociology of emotions (Bendelow and Williams 1997; Bericat 2016; Hopkins et al. 2009), the history of emotions (Boddice 2017, 2018), and historical sociology (Delanty and Isin 2003) – which are in very good health and constantly expanding, should constitute a very productive foundation and stimulus for the historical sociology of emotions as the bridge between all three. However, each of these fields has particular characteristics which continue to restrain the development of an equally robust historical sociology of emotions linking the three together.

Sociology of Emotions

In relation to sociology, emotions were a concern running through the work of protosociological philosophers like Adam Smith, Adam Ferguson, Alexis de Tocqueville, Montesquieu, Rousseau, Mandeville, as well as classical sociologists like Marx, Weber, Durkheim, and Simmel, who were often, as Michael Rustin points out, "developing a sociology of emotions by another name" (Rustin 2009: 23; also Hochschild 2009: 29). In the work of Frankfurt School thinkers like Max Horkheimer, the argument for paying attention to emotions was bound up with the broader argument for the importance of Freud's analysis of the unconscious motivations for human action. The ways in which human action can be poorly aligned with social reality required the analysis of unconscious psychological compulsions, drives, and needs among humans "to employ their aggressive powers, to gain recognition and affirmation as persons, to find security in a collectivity," as well as the sexual drive and the drive for self-preservation (Horkheimer 1993 [1932]: 124). "Our understanding of a variety of world-historical phenomena," wrote Horkheimer, "would be much enhanced if psychology could demonstrate that the satisfaction of these needs is a psychical reality no less intense than that of material gratifications" (p. 125).

However, the reception of psychoanalysis in sociology beyond the Frankfurt School was for some time heavily slanted toward an ego-psychological emphasis which tended to approach emotional life as an aspect of psychic structure to be mastered as quickly as possible. A self-conscious sociology of emotions only emerged from the mid-1970s, in the work of Arlie Hochschild (1975, 1979), Thomas Scheff, Randall Collins (1975), and others (Kemper 1990: 3). Hochschild's formulation of what the core concerns of a sociology of emotions could and should be, countering the cognitive bias of sociological thought since the 1930s (Barbalet 2004: 16 – with some exceptions, such as Goffman's work on embarrassment), remains as good an outline of the core conceptual issues as it was in 1975. She argued then for a movement beyond the opposition between the image of the person as a rational, cognitive, performative actor, and that of the unconsciously "driven," entirely "feeling" actor, in order to develop a third position, based on an understanding of the *sentient* actor who is both rationally aware and "driven" by feelings. As she put it, "Human beings, as sentient actors, are aware of their experiences and consciously respond to their feelings and the cultural expectations concerning them" (1975: 283).

This concern with "drivenness" operating somehow "beneath" the level of cognition and rational awareness is what leads to Freud and psychoanalysis, given the emphasis on cognition, awareness, or simply behavior in other schools of psychological thought. Jonathan Turner has remarked on the tendency in the sociology of emotions to rely on Gestalt and cognitive psychology rather than psychoanalysis, and observed that "there is a kind of cognitive bias in much psychology and sociology that sees individuals as trying to maintain consistency and congruity between self-conceptions and identities, on the one side, and behavioural outputs, reactions of others, and interpretations, on the other" (J. Turner 2006: 279). For Turner, Freud is important for his theorization of the ways in which emotional life operates unconsciously, in ways that are not immediate accessible to rational, conscious thought, and he argues that drawing on psychoanalytic theory would add significantly to the robustness of the sociology of emotions (2014: 293). Even where there is a fruitful interaction between psychoanalysis and sociology, it usually displays little historical sensibility. For example, in Elliott Jacques's (1955) important essay on unconscious psychological mechanisms within organizational live, a further important question is how the dynamics he describes have changed over time, and if there's any sort of long-term learning process at play. With some exceptions, then, on the whole the historical dimensions of most work done in the sociology of emotions are relatively shallow (Newton 1997; Stearns 1989).

History of Emotions

Although it could be said that historians have always been concerned with emotions in one way or another, the arguments of Dutch historian Johan Huizinga, in his 1919 book *Herfsttij der middeleeuwen* (*The Autumn of the Middle Ages*, 1996), constituted an important stimulus toward giving the emotional tenor of life, and how it had changed since the fourteenth and fifteenth centuries, a much more central place in historical understanding. He was perhaps the first to spell out the problem characterizing a focus on political and economic factors motivating human action, remarking that any study of the violent conflicts characterizing Western Europe in the fourteenth and fifteenth centuries will be hopelessly inadequate if it concentrates solely on "economic-political causes." He argued that "one might well be justified in asking whether a political-psychological view would not offer greater advantages... for an explanation of late medieval party conflicts" (1996: 17; in the 1921 Dutch edition, Huizinga wrote "sociological" view rather than "political-psychological" – 1997: 26). He emphasized "the fervent pathos of medieval life" and highlighted "the passions that inflame every sphere of life" (p. 9). People in the twentieth century, he felt, have "as a rule, no idea of the unrestrained extravagance and inflammability of the medieval heart" (p. 15). Although Huizinga could see that there is "a passionate element remaining in contemporary politics," he thought it was generally confined to extreme situations such as civil war, and otherwise subjected to a greater range of checks and "led in hundreds of ways into fixed channels by the complicated mechanisms of communal life." In the fifteenth century, in contrast, "the immediate emotional affect is still directly expressed in ways that frequently break through the veneer of utility and calculation" (1996: 15). However, despite his sensitivity to the importance of the emotional dimensions of social and political life, and his interest in examining psychological factors, he had no psychological or sociological theory to speak of, and his history was primarily descriptive.

Building on Huizinga's innovation, the French historian Lucien Febvre called for a collaboration, a "whole network of alliances" (1973 [1938]: 10) between psychologists, sociologists, and historians to develop a "historical psychology" (1973 [1938]: 5) that would grasp the ways in which human emotional life had changed over time, arguing that a psychology based on observations of people in the twentieth could not simply be applied without any modification to interpret human action in the past. He referred to Huizinga's account of the intense emotional volatility of life prior to the sixteenth century, and observed: "Yes, of course, but what we have to do above all is account for these things, and the explanation is not easy. It touches upon a host of facts which up to now historians have never bothered to bring together as a whole and have always grossly neglected" (1973 [1938]: 7). Febvre was very good, observes André Burguière, at identifying the central problems of historical method in highly innovative ways, but "without suggesting what the necessary conceptual tools to solve them might be" (1982: 436). It was this absence of a dialogue with any developed body of psychological theory, but with psychoanalysis in particular, that proved to be a major barrier to the development of the history of mentalities, and emotions, until the demise of Sartrean existentialist thought, notoriously hostile to Freud, and the rise of structuralism in the 1960s stimulated the reception of Freudian ideas – both positively and critically – in French historiography.

However, the tendency was more toward a "primary focus on political history, usually defined in terms of rational actors and the careful plotting of policy and strategy" (Stearns 2015). Febvre's project of a historical psychology failed to take off in French historiography, and in 1956 Febvre noted that the "history of feelings" still remained "almost virgin territory" (p. 247). The gradually increasing work being done in the sociology of emotions appears to have prompted historians likewise to pay more attention to the emotional dimensions of their research, to the point of developing a distinct field of the history of emotions building on the earlier work of Huizinga, Febvre, and Bloch. By 1985, roughly ten years after Hochschild's intervention, historians Peter and Carol Stearns (1985) had coined the term "emotionology" to distinguish between collective emotional standards and individual emotional experience. The term was never widely adopted, but the basis historical interest in how shared emotional standards change and are related to other lines of social and cultural change did. Initially it was concentrated in family history, but more broadly "emotional change needs to be woven into the historical fabric seems unquestionable" (Stearns and Stearns 1985: 820). In 1989 Peter Stearns was observing that it remained the case that "Sociological scholarship on emotion easily dwarfs explicit historical inquiry into the same subject" (1989: 594), but that has since changed significantly (Stearns 2015).

Historians of emotions are often in fact addressing sociological questions: a great of the *Annales* School history of mentalities (Burguière 2009) in fact engages with historical-sociological questions, interweaving transformations of subjective experience with changes in the structure of social relations, without attaching that label to their work (Duby 1961). It is the case, as Peter Stearns argues, that "[s]ociologists and historians of emotion in fact largely share a research agenda, despite some differences in disciplinary perspectives, in dealing with emotions as factors in and products of social institutions and relationships" (1989: 597). However, it is also true that this shared research agenda remains to be pursued systematically, and engaging with it is more the exception that the rule (Stearns 1989). The sociology remains very subservient to the history, and it is not done a systematic way, with only a weak linkage between history and theory. Most historical accounts focus on particular case studies – as is appropriate for the discipline – to provide particular examples which either support or somehow "disprove" any attempt at a theoretical account, without indicating the ways in which particular examples may be the "exceptions which prove the rule." Hardly ever is there a sense of long-term processes – indeed the conceptual inclination among historians is always to deny such an understanding, emphasizing instead the particular and localized exceptions to any overall historical pattern. As Rob Boddice points out, in spite of "the great explosion of work being produced by historians on feelings, passions, emotions and sentiments, few have attempted general coverage, and none have attempted a narrative from antiquity to the present, to unfold a story of the history of emotions across historical time" (2019: 9). As a result there is very little work being done that could be called a "history of the present," in the sense of reflections on the historical roots of contemporary emotional life, or even a genuine historical sociology of emotion.

Historical Sociology

Historical sociology, the field with which this book is concerned, appears to have remained immune to the "affective turn" that has characterized the human sciences more generally (Greco and Stenner 2008), tending to avoid the question of the emotional dynamics of social life, perhaps because such a concern requires greater familiarity with psychological theories, and above all a specifically historical psychology, still not a significant field of knowledge. In the *Journal of Historical Sociology*, for example, over its history from 1988, there are a scattering of articles that engage with the emotional dimensions of their topics, but there is no discussion of the key concerns and conceptual issues of a historical sociology of emotions. A number of the entries in *The Handbook of Historical Sociology* (Delanty and Isin 2003) do address the question of emotions in their discussions of Weber, the *Annales* School, and Norbert Elias, but there is no distinct sustained examination of the historical sociology of emotions.

So there is a robust history of emotions which often deals with sociological issues in valuable ways, and an equally dynamic sociology of emotions that can also draw on important historical material, so that the two fields overlap in many ways. However, there is still only weak interaction between the two, and relatively little work done that could be called the historical sociology of emotions. Peter Stearns (2009: 293) has remarked on the continued absence of sustained dialogue between history, sociology, and psychology, with each discipline adhering to its own internal disciplinary concerns and debates, and only history interested in change.

Norbert Elias

The one historical sociologist who has developed a compelling combination of historical sociology and historical psychology is Norbert Elias, whose work remains a reference point – if not a model – for the study of emotions in history, sociology, and historical sociology (Greco and Stenner 2008: 26). Michael Rustin (2009: 24) points out that Elias's work was an important exception to the neglect of the emotions in the social sciences prior to the "affective turn." As he argued in an essay originally written around 1939, it is important to find:

> a point from which it is possible easily to demolish the artificial fences we erect today in thought, dividing human beings up into various areas of control: the domains of, for example, the psychologists, the historians, the sociologists. The structures of the human psyche, structures of human society and structures of human history are indissolubly complementary, and can only be studied in conjunction with each other. In reality they do not exist and move with the degree of isolation assumed by current research. They form, with other structures, the subject matter of the one human science. (Elias 2010 [1939]: 38)

His writing – or rather, his most influential book, *On the Process of Civilization* – has worked as either a paradigm or a critical reference point for both historical and historical-sociological studies of emotions. Elias's analysis of the history of emotional expressions such as fear, anxiety, shame, embarrassment, disgust, anger, and aggression (*Angriffslust*, literally "attack-pleasure") has been crucial for transcendence of individual/society dichotomy in the history and sociology of emotions, and his formulations remain today as central as they were in the 1930s. As David Lemmings and Ann Brooks have indicated, *On the Process* is generally regarded as "the most substantial example to date of a historical grand narrative that relates changes in emotional styles and rules to changing social and political contexts" (Lemmings and Brooks 2014: 5). His utilization and sociological reworking of Freud and psychoanalytic concepts has been central to a receptivity to the collective, social nature of emotional experience, and to an understanding that "...the constitution and control of affective relations in society is the essence of cultural reproduction and power" (Lemmings and Brooks 2014: 17).

Some historians are keen to reject various aspects of Elias's approach, but even these writers will nonetheless recognize that his conception of the interweaving of social and psychological forces, indeed his whole project of a historical sociology of emotions, "does in fact fit a persuasive model of biocultural emotional change" (Boddice 2018: 28). Most importantly, Elias does what few others have done in this field, which is to connect the historical development of regimes of emotional standards to shifting patterns of relations of *power* (Boddice 2018: 211) and *interdependence* – this is what makes it *historical sociology* rather than either history or sociology (van Krieken 2014). Even Barbara Rosenwein (2010, 2018), perhaps the most enthusiastic of Elias's critics among historians, in attempting to develop an alternative line of explanation, ends up admiring his achievement in addressing the question of emotions over the *longue durée* (in Plamper 2010: 252).

Defining "Emotion"

A useful way to get a sense of the central importance of Elias's particular approach to the historical sociology of emotions is to take a brief look at the possible definitions of "emotion"; since many discussions effectively talk past each other because they are working from differing starting points (Epstein 1984; Fox et al. 2018). Carroll Izard's 2010 review of a broad range of emotion specialists notes that at the very least the difficulty is the sheer variety of meanings (Izard 2010a, b) stubbornly resistant to any conclusive reconciliation with each other, to the extent that there are arguments to dispose of the concept in psychology altogether. Different disciplines generate differing understandings of what emotions are, but one aspect they all seem to share is a kind of "flooding" of one's subjective experience, an experience of "vivid feelings" (Dixon 2012: 340), the source of which is difficult to pinpoint, but tends to have a bodily or somatic dimension, and that operates in some degree of agonistic interaction with cognitive and rational concerns, although there will be considerable disagreement about how that "agonistic interaction" should be understood.

In sociology, Jack Barbalet insists emotions are a crucial aspect of all human action (2002: 2), proposing that "emotion simply indicates what might be called an experience of involvement" (2002: 1). In this sense, the issue is motivation for action, with emotion a necessary corollary to cognitive, rational, conscious aspects of human subjectivity. Historian Thomas Dixon suggests that emotions can be understood as "a set of morally disengaged, bodily, non-cognitive and involuntary feelings" (2003: 3). Even in the middle of an argument for the importance of cognition, Nico Frijda emphasizes that "tracing the affective effect of a stimulus to its cognitive conditions allows one, in the end, to find a stimulus that evokes affect without cognitive mediation" (1994: 199). The key issue appears to be that reactions are driven by some arousal other than cognition or rational interests, so that one cannot explain human action purely in cognitive or rational terms. As Elias put it himself, indicating fear as the archetypal emotion, "In the wider sense the term emotion is applied to a reaction pattern which involves the whole organism in its somatic, its feeling and its behavioural aspects, as exemplified by a fear reaction" (2009b [1987]: 154).

Elias and Historical-Sociological Psychology

The conceptual concern at the heart of *On the Process* was precisely to historicize psychology from a sociological perspective, and Elias termed his approach "historical social psychology" (2012a [1939]: 449). In writing to Raymond Aron as a potential reviewer, he observed that his book constitutes some early "tentative steps" toward a complexly new field of knowledge, that of "historical psychology," which he also emphasizes is inseparable from "a processual way of thinking," that is, an historical-sociological psychology (Joly and Deluermoz 2010: 100). To Walter Benjamin he wrote in 1938 that the endeavor that concerned him was not the writing of some kind of cultural history, but that of "making the rules of the historical change in the psychical accessible to our understanding" (Schöttker 1998: 55), and to "overcome the hitherto dominant static conception of psychical phenomena" (p. 57), a task that "has today been recognized by very few people" (p. 57).

Elias and Febvre were thus both calling for a sociologically informed historical psychology at roughly the same time, but they were unaware of each other's work, and considered very different kinds of psychological thought. For Febvre it was Charles Blondel, Henri Piéron, and Henri Wallon, in addition to his own kind of Huizingian folk-psychological analysis of the contrasts in everyday experience between different periods of history. For Elias it was Freud and Gestalt psychology. Referring to the similarities and differences between Febvre and Norbert Elias, Burguière speaks of "the vast distance between German thought, heavily informed by Freud's work, and French thought, which was loftily unaware of Freud" (1982: 435–6).

It was Freud and psychoanalytic theory, then, that enabled Elias – and enables historical sociologists today – to comprehend the affective dynamics of social processes in a way that made it possible to link the long-term transformations of social structures to transformations of psychic structures. As a theory of unconscious mental life, psychoanalysis offers the primary means of correcting the cognitive bias of much historical sociology, by analyzing the unconscious dynamics of emotional experience. In Elias's words:

> any investigation that considers only people's consciousness, their "reason" or "ideas", while disregarding the structure of drives, the direction and form of human affects and passions, can from the outset be of only limited value. Much that is indispensable for an understanding of human beings escapes this approach. The rationalization of people's intellectual activity itself, and beyond that all the structural changes in the ego and super-ego functions, all these interdependent levels of people's personalities ... are only very imperfectly accessible to thought long as enquiries are confined to changes in the intellectual aspects of people, to changes of ideas, and pay little regard to the changing balance and the changing pattern of the relationships between drives and affects on the one hand and drive and affect-control on the other. (2012a [1939]: 451)

Psychoanalytic theory is fundamentally about the history of subjective experience over both the short and the longer term, about how one generation's sensibilities, dispositions, and orientations are transmitted – and transformed – from one generation to the next. As Goudsblom has pointed out, Elias's thought can usefully be understood as an integration of history, psychology, and sociology, as exemplified in the writings of Huizinga, Freud, and Weber (Goudsblom 1984: 130), and in this sense Elias was precisely an early exponent of the historical sociology of emotions. "The linking of psychoanalysis with historical sociological research," continues Goudsblom, "has both given the latter a broader psychological base and increased the sociological reach of the former" (1984: 141). Peter Burke also describes Elias's work as the most successful effort to link history, sociology, and psychoanalysis, observing that "Elias's study remains exemplary for its synthesis of three disciplines" (Burke 2007: 10).

However, Elias's relationship to Freud and psychoanalysis was complicated. Throughout his writings there is an ongoing tension between, on the one hand, his very thoughtful and considered absorption of key Freudian and psychoanalytic concepts and modes of thinking, and an ambition to go "beyond Freud" that is often very suggestive and productive, but at other times threatens to lose valuable aspects of Freud's approach. In all three arenas – his mobilization of Freudian and post-Freudian concepts, his efforts to develop psychoanalysis in a more historical-sociological direction, and those times when he abandons or simply overlooks some key Freudian ideas in a debatable way – working through nuances and complexities of Elias's relationship with psychoanalysis constitutes a useful window onto the most important conceptual issues in the historical sociology of emotions.

Those issues include addressing the problem of relationship between biological determination of emotions and their cognitive elements, for which one needs at least to address the psychoanalytic concept of "drive," the history of emotions and their management or control, and exactly how the social constitution of subjectivity should be analyzed. Elias focuses particularly on the concept of "drive economy" in order to understand the experience and management of emotional life, but in ways that rework Freud's conception of drives and which crystallize some important issues in the historical sociology of emotions. Elias's particular reading of Freud also laid the foundations for a central issue in the historical sociology of emotions, the question of the emotional basis of social bonds and collective identity beyond the family at various levels, including the community and the nation.

Before these arguments are examined, it will be useful briefly to outline Elias's overall account of the long-term development of human emotional life in terms of a "civilizing process." A detailed discussion of Elias's approach to the history of emotions can be found elsewhere (van Krieken 2014), here the focus is on the main elements of his historical sociological approach.

The Civilizing Process

Writing in the 1930s, Elias placed at the center of his work the observation that a central element of the emotional makeup of people in Western Europe was their self-perception as being not just civilized, but superior in that respect to other peoples in the world, as well as within their own societies to individuals and groups regarded as inferior in any way. The civilized/uncivilized distinction is always, implicitly of not explicitly, the lens through the experience and management of emotions is perceived. If the expression of anger is suppressed or merely subjected to social conventions, breaching those conventions is framed as being "uncivilized." The lack of restriction of sexual relations between adults and children that has characterized much of human history had gradually become understood as abusive and a fundamental breach of the expectations of any civilized community.

This feeling-state of being supremely civilized had to be understood, argued Elias, as closely bound up with a long-term historical process of gradually increasing social complexity, and "lengthening chains of interdependence," changing patterns of competition between lower and higher status groups, the formation of nation-states exercising a monopoly over the means of violence and generated more stable, pacified conditions of life, leading to related differentiated changes in the affect-economy of differing social groups.

For Elias, a long-term process like rationalization, as Max Weber understood it, needs to be seen as "only one side of a transformation affecting the *whole* personality, the level of drives and affects no less than the level of consciousness and reflection" (2012a [1939]: 484). Elias emphasized the linkages between larger processes of state-formation, economic development, and group relations on the one hand, and people's psychological and emotional makeup and experience on the other hand, arguing for a sensitivity to "how deeply the stratification, the pressure and tensions of our own time penetrate the structure of the individual personality" (2012a [1939]: 488).

Elias's portrayal of the process of civilization is not in fact about the increasing repression or even constraint of emotional life, but about increasing "psychologization" (Elias 2006 [1969]) in the sense of self-observation and reflexivity, harnessing emotions for increasingly complex networks of social interdependency. He spoke of "[t]he increased tendency of people to observe themselves and others," and the ways in which one could observe in Western European history that "people moulded themselves and others more deliberately than in the Middle Ages" (Elias 2012a [1939]: 86).

He saw the structure of social life among the European elite, in court society (2006), as the institutional core of the process of civilization, with the dynamics of court society producing a particular kind of psychological and emotional disposition and code of conduct. The basic psychological principle of court society was that one's conduct and emotional expression need to be regulated in the service of maximizing one's competitive position in an increasingly complex and volatile network of social relations. For Elias court society displayed the emergence of a form of mutual and self-observation which he referred to as a specifically "psychological" form of perception, and which we would today refer to as reflexive self-awareness.

Elias stressed the primacy of the state's monopolization of violence, and also the effects of unlimited economic competition, in that the fears and anxieties associated with unpredictable and everyday violence as well as economic uncertainty tend to dominate all other emotional experience (2012a [1939]: 484–90). "Fears," wrote Elias, "form one of the channels – and one of the most important – through which the structure of society is transmitted to individual psychological functions" (2012a [1939]: 485). Tensions at the level of politics, society, and economic relations manifest themselves *within* individuals as anxieties and fears, as well as particular demands placed on emotional expression and constraint.

The volatility of economic precarity produces differing kinds of anxieties in different social strata: fear of loss of work, poverty, and "unpredictable exposure to those in power" among the lower strata, and among the middle and upper strata, "fears of social degradation, of the reduction of possessions or independence, of loss of prestige and status" (2012a [1939]: 487). For the middle class in particular, anxiety about social position and prestige play a central role in the formation of emotional makeup. Partly unconscious and almost automatic, the parents' anxieties "continuously add fuel to the fiery circle of inner anxieties, which holds the behaviour and feelings of the growing child permanently within definite limits, binding him or her to a certain standard of shame and embarrassment, to a specific accent, to particular manners, whether he or she wishes it or not" (2012a [1939]: 487).

Foundations in Freud

In a footnote in *On the Process*, Elias wrote that "it scarcely needs to be said, but is perhaps worth emphasizing explicitly, how much this study owes to the discoveries of Freud and the psychoanalytical school. The connections are obvious to anyone acquainted with psychoanalytical writings." However, he thought it would have been too distracting for the purposes of developing his own perspective to enter into both the similarities and the "not inconsiderable differences between the whole approach of Freud and that adopted in this study," which would have meant "digressing into disputes at every turn" (2012a [1939]: 570).

Despite the critiques of Freud as being biologistic, sometimes also argued by Elias, in fact what makes Freud's approach to psychic life so important for the historical sociology of emotions is his understanding of how interwoven any individual's emotional life is with that of others, beginning with one's parents, his profoundly relational and intersubjective conception of human subjective experience. Elias's integration of psychoanalytic concepts is, as Stefan Beuer notes, "one of the stronger aspects of the theory of civilization" (1991: 407), enabling him to capture precisely the historical sociology of emotional experience in greater depth than accounts relying on other theories of psychic life, spelling out the centrality of the shifting relations between emotions as expressions of noncognitive drives, and the structure of social relations within which they are embedded.

The debt was perhaps more extensive, then, than Elias's acknowledgments have suggested. For example, the core thesis of *On the Process*, and an argument for a historical approach to psychology, was articulated by Freud in 1927, in *The Future of an Illusion,* where he rejected the idea that the human mind has remained the same throughout history, and highlighted one particular way in which it had changed:

> It is in keeping with the course of human development that external coercion gradually becomes internalized; for a special mental agency, man's superego, takes it over and includes it among its commandments ... Such a strengthening of the superego is a most precious cultural asset in the psychological field. Those in whom it has taken place are turned from being opponents of civilization into being its vehicles. The greater their number is in a cultural unit the more secure is its culture and the more it can dispense with external measures of coercion. (Freud 1927: 11)

As Bernard Lahire has noted, in Freud's 1932 letter to Albert Einstein (Freud 1933b), it is striking how many of the themes central to Elias's approach to the process of civilization are evident there. Freud observes the centrality of violence to human existence throughout history, and the long-term process of a transition from brute, unregulated violence to law, understood as communally organized rather than individual violence. The core issue, wrote Freud, was "violence overcome by the transference of power to a larger unity, which is held together by emotional ties between its members" (1933b: 205). Freud argued that the process of the human evolution of culture, or the process of civilization – unlike Elias, he rejected any distinction between "culture" and "civilization" – is accompanied by important psychical changes, consisting of "a progressive displacement of instinctual aims [*Triebziele*] and a restriction of instinctual impulses [*Triebregungen*]. Sensations which were pleasurable to our ancestors have become indifferent or even intolerable to ourselves" (1933b: 214; see also Freud 1933a: 178–9).

Earlier Freud had also emphasized a historical perspective when he wrote that "[i]n the last resort it may be assumed that every internal compulsion which makes itself felt in the development of human beings was originally – that is, in the history of mankind, only an external one," and it is in this sense that "human being is subject not only to the pressure of his immediate cultural environment, but also to the influence of the cultural history of his ancestors" (Freud 1915b: 282–3).

There are also points in Elias's writings where, despite his concern to go "beyond" Freud, he provides a thoroughly Freudian account of human subjective experience over every individual's lifetime, sometimes almost in contradiction to his efforts to distinguish his approach from Freud's. For Freud, the reduction of aggression between people demanded by the development of civilization needed to be seen as associated with an increase in aggression that people directed internally toward themselves, and this was a central "discontent" of civilization. Elias elaborates on exactly this in *On the Process* when he writes that with the establishment of a "pattern of near-automatic habits" and a "specific 'super-ego'," "the battlefield is, in a sense, moved within . . . the drives, the passionate affects that can no longer directly manifest themselves in the relationships *between* people, often struggle no less violently *within* the individual against this supervising part of themselves" (2012a [1939]: 414).

Elias indicates here a very thoughtful sensitivity to the complexities and indeed painfulness of psychic life, using words like "wounds," "scars," "pain," and "suffering." He talks about the civilizing process as "difficult," mentions the possibility of an "unsuccessful" civilizing process, the mingling of positive and negative features, and he uses the word "resistance" in a way that he rejects at other points. "[T[his semi-automatic struggle of the person with him or herself does not always," wrote Elias, "find a happy resolution; the self-transformation required by life in this society does not always lead to a new balance between drive-satisfaction and drive-control" (2012a [1939]: 414). In this wording he is sensitive to the powerful ambivalence of the civilizing process in a way that he tends to slide over in most of his writing as the detached social scientist. It is a part of the book that "channels" Freud and his sense of the difficulties of human existence more than most of his other writing, especially those where he explicitly rejects Freud's "unrelenting pessimism of the later Enlightenment" and the "bitterness of Freud's existential despair" (Elias 2014: 14; 22).

Beyond Freud

In a number of important ways, then, there are aspects of Freud's work which laid the foundations not just for a for an improved understanding of the psychology of emotions, but also the historical sociology of emotion. The influence is important not just when he mentions Freud explicitly. We have seen how an early version of many of Elias's core concerns can be found in parts of Freud's work, and very often Elias's writing constitutes a debate with the ghost of Freud even when he does not mention his name, developing his own particular approach to drives, libido, the superego, and the relationships between difference emotions. Elias's concern was, as Goudsblom observed, that "[h]ow the long-term sociogenesis of self-coercion took place ... remains largely out of consideration as a historical-sociological problem" (Goudsblom 1984: 139), and this was the question to which Elias addressed himself.

Elias was able to connect Freud's initial observations about the long-term trend toward the increasingly complex interaction of external with internal compulsion, the historical development of the superego, with a rich body of sociological, historical, and comparative reflection on the dynamics and structure of modern social life. In other words, he explained in detail exactly how the process briefly and occasionally flagged by Freud actually operated, unpacking its vicissitudes in differing sociohistorical contexts (Berard 2013).

Where Elias sought to go beyond Freud, not all of his efforts were equally successful. Foulkes indicated in his review of *On the Process* that he thought it would have benefited from his involvement that Elias's use of psychoanalytic concepts was rather loose, despite the important contribution the book was making to psychoanalysis (Foulkes 1941). Later a number of critics (Breuer 1991; König 1993; Erdheim 1982, 1996) have pointed out that there are various ways in which very important and valuable aspects of Freud's thinking were abandoned or misunderstood by Elias in ways that weaken rather than strengthen the analysis of processes of civilization and decivilization.

Three of the issues emerging from Elias's utilization of Freudian concepts that are particularly important for the future development of historical sociology of emotions that genuinely integrates the concerns of both historians and sociologists are: first, what underpins the "drivenness" of emotions, or more specifically, how should the Freudian concept of "drive" be approached, revised or developed? Second, is it fair to say, as Elias did, that Freud continued to see individual psychic life as isolated from the social world around, and what does an assessment of this argument of Elias's mean for a movement beyond the individual/society dichotomy in the historical sociology of emotions more broadly? Third, what does it mean to "historicize" Freud's conception of the psychic agency that supervises, regulates, and controls emotional life, the "superego," how exactly does Elias do that, and does his approach retain all that is valuable about Freud's understanding of the dynamics of the superego?

Emotions and Social Relations

An important preface to any discussion of the concept of "drive" – or "instinct" as Freud's German term *Trieb* has often been translated – is a clarification of how Freud understood the term. He saw "drives" as consisting of three components: the *source* in a somatic stimulus, the *aim* of elimination the tension generated by this stimulus, and the *object* with which the drive can achieve that aim (Laplanche and Pontalis 1973: 214). As Laplanche and Pontalis emphasize, translating the term *Trieb* as "instinct" rather than "drive" risks losing a sense of the distinctiveness of Freud's approach, particularly "the thesis of the relatively undetermined nature of the motive force in question, and the notions of contingence of object" (1973: 214–15).

The concept of "instinct," in contrast, tends to imply tight relationship between the drive and its object, quite contrary to Freud's approach, and this has been the source of a great deal of misunderstanding of Freud's theories. As Freud put it, a drive's object is "what is most variable about a drive and is not originally connected with it" (1915a: 122). In psychoanalysis, as Dennis Wrong argued, the drives, "far from being fixed dispositions to behave in a particular way, are utterly subject to social channelling and transformation and could not even reveal themselves in behaviour without social moulding any more than our vocal chords can produce articulate speech if we have not learned a language" (Wrong 1961: 192).

Against this background, there is a tension between those parts of Elias's writing where he seems to suggest, like Freud, that drives operate in opposition to the requirements of social relations, subject to constraint, regulation, and repression in social relations, and others where he insists that drives are "always already socially processed" (2012a [1939]: 452), constructing Freud's sense of the conflictual nature of human psychology as anchored in biologism and overly pessimistic.

On the one hand, Elias's formulations appear to adhere to the more classically Freudian concept of an agonistic relationship between human drives and the requirements of social existence, beginning with the constant need to deny and renunciate pleasure in infancy and childhood. Elias's wording throughout *On the Process* consistently conveys an impression of a conflictual model. He refers to drives as "manifestations of human nature under specific social conditions" (2012a [1939]: 157). All social life depends on "a channelling of individual drives and affects" (2012a [1939]: 486) and the "moderation of spontaneous emotions, the tempering of affects" (2012a [1939]: 408). "Pleasure-promising drives and pleasure-denying taboos and prohibitions, socially generated feelings of shame and repugnance," wrote Elias, "come to battle within the self" (2012a [1939]: 185). Medieval psychic life was characterized by "the greater spontaneity of drives" (2012a [1939]: 409). He stresses the importance of achieving "the optimal balance between his or her imperative drives claiming satisfaction and fulfilment and the constraints imposed upon them (and without which humans would remain brutish animals and a danger as much to themselves as to others)" (2012a [1939]: 490), and at one point appears to argue for a distinction – as Marcuse (1972 [1955]) argued later – between necessary and surplus repression of drives and affects (2012a [1939]: 486).

Elias explained his position at one point as follows, arguing that in "psychoanalytic research":

> No distinction is made between the natural raw material of drives, which indeed perhaps changes little throughout the whole history of humankind, and the increasingly more firmly wrought structures of control, and thus the paths into which the elementary energies are channelled in each person through his or her relations with other people from birth onward. (2012a [1939]: 451–2)

Here Elias is acknowledging, contrary to what he says at other times, that some aspects of human drives can be regarded as biological and resistant to historical change. He is also misrepresenting psychoanalytic thought: the distinction is in fact central to psychoanalysis, that is the whole point of the development of psychic life over any individual's lifetime. One could accuse Freud and psychoanalytic theorists more broadly of paying less attention than they should to the social dynamics surrounding psychological development, and how those dynamics have a history to them, but that is a different point.

On the other hand, Elias also insists that there is nothing about human drives that escapes social formation, citing approvingly Morris Ginsberg's 1934 comment that "inborn tendencies ... have a certain plasticity and their mode of expression, repression or sublimation in, in varying degrees, socially conditioned" (cited in Elias 2012a [1939]: 569). Elias sees the psyche as a *mediating* instance between somatic stimuli and social demands, in a way that transforms both in their interaction with each other "drive dialogue." The "repression" of drives in Freud gets replaced by the "regulation" of drives in Elias, accompanied by the assumption, again contrary to other passages in *On the Process*, that human happiness and freedom can emerge from "*a more durable balance, a better attunement, between the overall demands of people's existence on the one hand, and their personal needs and inclinations on the other*" (2012a [1939]: 490; original italics).

Helmut König (1993: 206) notes the occasional turn in Elias to behaviorist concepts like "learning" (2012a [1939]: 415) and "imprinting" (2012a [1939]: 415, 570) that run counter to the Freudian understanding of drives, and this is aligned with Elias's frequent inclination to overlook the continued active presence of drives and emotions even after they have been socially processed or regulated. For example, Elias writes that self-restraint comes, in the course of the process of civilization, to be "ingrained so deeply from an early age that a kind of relay-station of social standards, an automatic self-supervision of the drives, a more differentiated and more stable 'super-ego' develops within them and a part of the forgotten drive impulses and affect inclinations is no longer directly within reach of the level of consciousness at all" (2012a [1939]: 413). Dennis Wrong has explained what is problematic about such an "oversocialized" conception of human beings:

> when a norm is said to have been "internalized" by an individual, what is frequently meant is that he habitually both affirms it and conforms to it in his conduct. The whole stress on inner conflict, on the tension between powerful impulses and superego controls the behavioural outcome of which cannot be prejudged, drops out of the picture. And it is this that is central to Freud's view. (Wrong 1961: 187)

In Elias's work that sense of conflict, ambivalence, and paradox recedes into the background, leading Helmut König to argue that Elias "designs an all too harmless harmony model of 'civilisation' that is blind to its ambivalences" (1993: 207). However, just because drives are "socially processed" does not mean they are fully socialized. One of the better Freudian explanations of the relationship between drives and the requirements of social life, which might in fact constitute the best resolution of the problems raised by Elias's formulations, is that of Donald Carveth, who emphasizes that drives are *both* socially shaped *and* socially inhibited or repressed: "Not only is nature repressed by culture, but *nature as shaped and channelled by culture* is repressed by culture" (Carveth 1977: 80, emphasis added).

From Libido to "Valencies"

Elias's frequent inclination to see drives as entirely social "by nature" also leads him to suggest an alternative to the Freudian concept of "libido," the energy or stimulus that underpins emotions and drives, turning to the concept of "valencies." This concept originates in chemistry to refer to attraction or repulsion between atoms, and found its way into psychology in the work of Gestalt psychologist Kurt Lewin (1931: 101) as the imperfect translation of the German term *Aufforderungscharakter*, to refer to the forces which either attract humans toward or repel them from other objects or people (Colombetti 2005). As Bernard Lahire explains it, the term captures the sense in which every individual "bears within him, from childhood, a series of points of attractions and rejections, affinities and indifferences, sympathies and antipathies, which condition his future relations with the many other individuals he will encounter and the multiple situations he will have to face" (Lahire 2013: 82).

For Elias "valencies" serve to distinguish his approach from Freud's use of "libido." He argued that the problem was that Freud "was not very interested in the fact that libido, as he described it, was in many of its aspects directed from one human being to another" (Elias 2009a [1969]: 175). Elias thought that Freud placed "the configurations that individuals form with each other" very much in the background, and in the process neglected the essential orientation of drives toward other people. Elias wanted to stress "everyone's fundamental directedness to other people," and "the deeply rooted emotional need of every human being for the society of other members of his species" (2012b [1978]: 131). One could say that Freud regarded relations with other people primarily as a means to the end of satisfying drives, whereas Elias considers that relationality is a central aspect of the drive itself, making it by nature constitutive of and constituted by affective bonds.

Elias is inclined to see relationality as itself a drive: people, he wrote, are "made up by nature as to be ... obliged ... to enter into relationships with other people and things" (2010 [1939]: 37). He also argued for the essential directedness of drives and emotions toward particular objects as effectively constituting the drive itself, rather than distinguishing drives from their objects, as Freud was more inclined to do.

> Even in psychoanalytic literature, one sometimes finds statements to the effect that the "id" or the drives are unchanging *if one disregards changes in their direction*. But how is it possible to disregard this directedness in something as fundamentally *directed* at something else as human drives? What we call "drives" or the "unconscious" is also a particular form of self-regulation in relation to other people and things, a form of self-regulation though one which, given the sharp differentiation of psychical functions, no longer directly controls behaviour but does so by various detours. (2010 [1939]: 36–7)

Some commentators have noted the affinity of this conception with post-Freudian object relations theory, which pursues a similar shift of emphasis toward seeing relations with others as "the fundamental building blocks of mental life" (Greenberg and Mitchell 1983: 3) and drives as "somatic-psychical excitations that are inherently object-relational and object-seeking" (Cavaletto 2007: 224; Roseneil and Ketokivi 2016: 150–1).

It is fair to say, then, that Elias's conception of drives, of libido as the expression of "valencies," is largely consistent with any psychoanalytic thought that is not firmly wedded to Freud's original theory of drives and willing to adopt a more relational understanding. As Greenberg and Mitchell put it, "[t]he common landscape of psychoanalysis today consists of an increasing focus on people's interactions with others, that is, on the problem of object relations" (1983: 2).

"Homo Clausus" Versus Figurations and Group Analysis

A line of argument that is connected with the question of how drives are understood is Elias's concern with the tendency in Western social and political thought to think of human individuals as closed, autonomous entities – he used the term "*homo clausus*" (Elias 2012a [1939]: 512–19) – which he thought needed to be replaced with a conception of humans as "open personalities," "fundamentally oriented toward and dependent on other people" (2012a [1939]: 535) within varying social and interpersonal figurations, or networks of interdependency. As Elias put it, "[t]he concept of open emotional valencies which are directed towards other people helps towards replacing the image of the human being as a *homo clausus* with that of 'open people'" (2012b [1978]: 130). In Elias's view, Freud did indeed grasp "that the group process of a father–mother–child relationship has a determining influence on the patterning of a person's elementary drives and the formation of his self-controlling functions in early childhood" (Elias and Scotson 2008 [1965]: 27). In this respect Elias agreed with Peter Gay's assessment that "Freud's central idea [is] that every human is continuously, inextricably, involved with others and that individual and social psychology are at bottom the same" (Gay 1985: 147–8).

However, once the basis elements of an individual's psychological makeup, or habitus, were established in early childhood, the self-controlling functions "appeared to him to work on their own, quite independently of the further group processes in which every person continues to be involved from childhood to old age" (Elias and Scotson 2008 [1965]: 27). Elias's observation was that Freud placed individual experience very much in the foreground of his concerns, and allowed the social forces acting on self-formation to recede so far into the background that they played hardly any role in his analysis, which is what leads him to suggest that Freud "conceptualised his findings largely in a manner which made it appear that every human being is a self-contained unit – a *homo clausus*" (Elias and Scotson 2008 [1965]: 26–7).

Here, too, Elias's position is closely aligned with that argued by psychoanalytic object relations theorists, who proposed similar modifications of Freud's original conception of the trajectory of psychological development over an individual's lifetime. Joan Riviere, for example, outlined a similar critique of what Elias called the *homo clausus* conception of human beings, describing the concept of the isolated individual as a "convenient fiction."

There is no such thing as a single human being, pure and simple, unmixed with other human beings. ... That self, that life of one's own, which is in fact so precious though so casually taken for granted, is a composite structure which has been and is being formed and built up since the day of our birth out of countless never-ending influences and exchanges between ourselves and others ... These other persons are in fact therefore parts of ourselves, not indeed the whole of them but such parts or aspects of them as we had our relation with, and as have thus become parts of us. And we ourselves similarly have and have had effects and influences, intended or not, on all others who have an emotional relation to us, have loved or hated us. We are members one of another. (Riviere 1952: 166–7)

So while Freud's theory is very much intersubjective, the problem for Elias, and for the object-relations theorists, is that he has a relatively narrow conception of the "field" within which the intersubjective formation of the self operates, and this was what Elias want to add – expanding the field of the familial formation of subjectivity to include a much broader range of networks of social interweaving, extending over any human being's whole lifetime. One can understand why Freud and most psychoanalysts focus on infancy and early childhood – it has a certain ontological and foundational primacy – but greater weight could also be placed on the later periods of any individual's biography. Freud does do this sometimes, in comments on adolescence and mass psychology, but from Elias's perspective Freud's analysis was relatively tentative and in need of more detailed development, which is why Elias felt that Freud thought mostly in terms of the "I" dimension of human experience, whereas it is important also to analyze the "we" dimension of personality structure, social interaction and social structures.

The Superego and Its History

The superego was for Freud that part of psychic structure that was responsible for judging and repressing desires, drives, and impulses that are prohibited in one way or another. It is grounded in the resolution of the Oedipus Complex, the first arena of conflict between a child's desires and societal and cultural demands, although it is also subsequently developed by the impact of social and cultural expectations, in the context of schooling, religion, and broader moral frameworks (Laplanche and Pontalis 1973: 437). As the psychic agency responsible for repressing, controlling, and regulating emotional life, then, how it and its historical development is understood is central to the historical sociology of emotions.

Elias claims that Freud regarded the superego as not having a history, and although that is incorrect – Freud did remark on occasions about the long-term transformation of external into internal constraints – it is fair to say that he did not go into that aspect of the superego in any detail, except very indirectly in later works like *Totem and Taboo* and *Moses and Monotheism*, works that are too complex to discuss in any detail here. As Bernard Lahire observes, Freud was "not sensitive to the historical and social variations of the forms and modalities of the exercise of power and their consequences for the functioning of the psychic economy" (Lahire 2013: 87). Elias also placed more emphasis than Freud did on the importance of the dimension of power, arguing that today "many of the rules of conduct and sentiment implanted in us as an integral part of our conscience, of the individual superego, are remnants of the power and status aspirations established groups, and have no other function than that of reinforcing their power chances and their status superiority" (2012a [1939]: 489).

However, while Elias's elaboration of the structural dynamics of superego formation over the course of an individual's lifetime is an important point of linkage between psychoanalysis on the one hand, and history and sociology on the other, in an odd way he loses an important aspect of Freud's theory, which was his emphasis on the "drive-dynamics" of the constitution of the superego, the emotional foundations precisely of emotion management in social life, in turn an important aspect of the ambivalent nature of human existence. As König put it, Freud's efforts "revolve around the attempt to make a drive-theoretical justification of civilisation plausible, they revolve around the attempt to understand the drive nature of man not only as the opponent of civilisation to be tamed, but at the same time as its basis" (König 1993: 207, my translation).

For Elias, the superego is the outcome of the "imprinting" of social constraints, leading writers such as Breuer to point out fact that Elias approached the superego in terms that approximated those of behaviorist psychology rather than psychoanalysis (1991: 408), and Foulkes to complain that Elias "lets the superego arise almost purely from its internalization" (Foulkes 1941: 318, my translation). For Freud, the superego depends on an *alliance* with the id and drives. It is not merely the product of social constraints, but,

> the heir of the Oedipus complex, and thus it is also the expression of the most powerful impulses and most important libidinal vicissitudes of the id. By setting up this ego ideal, the ego has mastered the Oedipus complex and at the same time placed itself in subjection to the id. Whereas the ego is essentially the representative of the external world, of reality, the super-ego stand in contrast to it as the representative of the internal world, of the id. (Freud 1923: 36)

The Freudian conception, which Elias appears to have abandoned, is that the superego, or the management of emotions, relies precisely on those emotions themselves, the rational response to the requirements of social life needs to become "emotionalised" in order to be effective as the automatic internalization of social constraints. As a result, Freud also emphasized that the relationship between the superego did not simply "transmit" cultural demands. In the first placed because it was the superegos of parents that formed the child's superego, but also because the demands of the superego often exceed those of society and culture, taking on a life of their own. In other words, the constraint and management of emotional expressions comes as much "internally" from human emotional concerns as it does "externally" from culture and society – indeed the whole distinction between "internal" and "external" is unstable and volatile.

Conclusion

The various issues and questions that emerge in reflecting on the interrelationships between Freud, Elias, and later psychoanalytic theory highlight many of the core concerns of a historical sociology of emotions that can bring about the same "affective turn" in historical sociology as has been more clearly evidence in history and sociology separately. These questions include how the "drivenness" of emotional life should be analyzed, how emotional experience can best be understood in terms of social relationships as well as individual psychological makeup, and how the regulation of emotional life, and its internalization within psychic structure, has changed over time, in connection with various other lines of social development.

With that conceptual foundation in place, the historical sociology of emotions will be able to move a variety of directions. For example, the gendered dimensions of the regulation of drives and the management of emotions deserve greater attention using Elias's overall framework than can be found in his own work, in order to identify the ways in which the historical development of being an "open personality" has taken, and continues to take, quite different pathways for males and females (van Krieken 1990: 364–5; Brooks 2014). The development of Arlie Hochschild's (1983) account of "emotional labour" in relation to longer historical time frames, and drawing on a broader range of psychoanalytic concepts, Elias's work with Eric Dunning on sport as the "controlled decontrolling" of various emotions (Elias and Dunning 2008 [1986]), and Cas Wouters's (2007) analysis of the process of "informalization" of emotion management since 1890 are only some of the other important themes in the historical sociology of emotions.

There is, however, only space here to conclude with a very brief reference to two of the further steps that can be taken to develop the historical sociology of emotions. The first would be to reinstate Freud's insistence on the conflictual character of emotional life, and to devote greater analytic attention to the ways in which self-formation functions unevenly for different groups in any society and during any historical period. Although Elias is sensitive to the ways in which relations of power structured human habitus and emotional experience, in his concern to distance himself from Freud's conflictual model, he leans toward assuming that the regulation of drives and emotions functions more or less smoothly, As Bryan Turner has argued, "Elias takes for granted the resignation of the disprivileged social strata to the cultural legitimacy and political power of their superiors, thereby neglecting resentment and rage toward them" (2014: 178) – unlike Freud. Indeed, Freud was not only conscious of the historicity of the superego, but also its variability across differing social groups in any one historical period; he noted that contemporary European civilization remained marked by profound inequalities, and that under such conditions "an internalization of the cultural prohibitions among the suppressed people is not to be expected" (Freud 1927: 12).

Secondly, the roots of Elias's understanding of psychoanalytic concepts in group analysis, and the connections between his work and that of Foulkes (Blomert 1992) – and, by implication, the broader range of group analytic ideas (Dalal 1998; Waldhoff

2007a, b) – as well as object relations theory, could usefully be considered in greater depth in relation to key questions in the historical sociology of emotions. Peter Gay, for example, outlines the importance of the emotional and psychological dimensions of the social institutions structuring human experience "by constructing forbidding fortifications of honour and indignation, brimming moats of shame and self-reproach – so many stratagems serving to contain the invasion of disorderly, possibly destructive passions" (Gay 1985: 165–6). A large proportion of collective behavior can be understood as driven by unconscious dynamics with deep historical roots, so that the emotional dynamics of collective behavior at a variety of levels can only be properly grasped with a conceptual apparatus that brings together psychoanalysis and historical sociology.

Elias mobilizes this approach productively in his analysis of the historical sociology of German authoritarianism, going further back in time than the German defeat in World War I, and taking into consideration a much broader range of factors than most accounts of National Socialism (Berard 2013: 224). He wrote that it was not possible adequately to understand the rise of National Socialism in Germany "if one does not also take into account its place and function in the wider context of Germany's long-term development" (Elias 2013 [1989]: 287), including, for example, "Germany's lingering decline as a European power" (Elias 2013 [1989]: 288), generating a collective emotional yearning for a lost medieval "Great Germany" (Elias 2013 [1989]: 7). A related line of argument was developed in Elias's 1960 talk on British public opinion (2008 [1960]), where he mistakenly predicted that Britain would not join the European Economic Community, but correctly identified the strength of a "we" identity based on collective nostalgia for former national greatness, that was eventually to lead to the United Kingdom leaving the European Union in 2020 (Dunning and Hughes 2020).

The question of "how the fortunes of a nation over the centuries become sedimented into the habitus of its individual members" is, Elias argued, one that is analogous to Freud's identification of "the connection between the outcome of the conflict-ridden channelling of drives in a person's development and his or her resulting habitus." At the level of "we" relations, "there are often complex symptoms of disturbance at work which are scarcely less in strength and in capacity to cause suffering than the neuroses of an individual character" (Elias 2013 [1989]: 24). As Berard (2013) has argued, Elias's historical sociology of authoritarianism builds on and develops Freudian ideas in a very imaginative and productive way, being receptive to the distinctiveness of the psychological and emotional dimensions of social history, including factors like honor, pride, shame, self-esteem, rather than tying social and cultural life closely to material and class interests, as insisted upon by theorists more strongly influenced by Marx or Weber.

The future development of the historical sociology of emotions, by building on and critically extending the foundational work of Elias and others who have endeavored to link sociology, history, and psychoanalytic theory, has enormous potential, then, for an improved understanding of the history of the present, in relation to collective fantasies at the level of the nation or the ethnic group, the narcissism of political leaders, the continuing and changing role of nationalism in collective identity and action, the eruption of expressions of *ressentiment* in populist politics, and many other issues.

References

Barbalet JM (2002) Introduction: why emotions are crucial. In: Barbalet J (ed) Emotions and sociology. Blackwell, Oxford, pp 1–9

Barbalet JM (2004) Emotion, social Theory, and social structure. Cambridge University Press, Cambridge

Bendelow G, Williams SJ (eds) (1997) Emotions in social life: critical themes and contemporary issues. Routledge, London

Berard TJ (2013) Under the shadow of *the authoritarian personality*: Elias, Fromm, and alternative social psychologies of authoritarianism. In: Dépelteau F, Landini TS (eds) Norbert elias and social theory. Palgrave Macmillan, Basingstoke, pp 209–243

Bericat E (2016) The sociology of emotions: Four decades of progress. Current Sociology 64(3):491–513

Blomert R (1992) Foulkes und Elias – Biographische Notizen über ihre Beziehung. Gruppenanalyse: Zeitschrift für gruppenanalytische Psychotherapie, Beratung und. Supervision 2(1):1–26

Boddice R (2017) The history of emotions: past, present, future. Revista de Estudios Sociales 62:10–15

Boddice R (2018) The history of emotions. Manchester University Press, Manchester

Boddice R (2019) A history of feelings. Reaktion, London

Breuer S (1991) The denouements of civilization: Elias and modernity. Int Soc Sci J 128:405–416

Brooks A (2014) 'The affective turn' in the social sciences and the gendered nature of emotions: theorizing emotions in the social sciences from 1800 to the present. In: Lemmings D, Brooks A (eds) Emotions and social change: historical and sociological perspectives. Routledge, London, pp 43–62

Burguière A (1982) The fate of the history of *Mentalités* in the *Annales*. Comp Stud Soc Hist 24(3):424–437

Burguière A (2009) The Annales school: an intellectual history (trans: Todd JM). Cornell University Press, Ithaca

Burke P (2007) Freud and cultural history. Psychoanalysis & History 9(1):5–15

Burke P (2012) Afterword. In: Alexander S, Taylor B (eds) History and psyche: culture, psychoanalysis, and the past. Palgrave Macmillan, New York, pp 325–332

Carveth DL (1977) The disembodied dialectic: a psychoanalytic critique of sociological reductionism. Theory & Society 4(1):73–102

Cavaletto G (2007) Crossing the psycho-social divide: Freud, weber, Adorno and Elias. Routledge, London

Collins R (1975) Conflict sociology: towards an explanatory social science. Academic Press, New York

Colombetti G (2005) Appraising valence. J Conscious Stud 12(8–10):103–126

Dalal F (1998) Taking the group seriously: towards a post-Foulkesian group analytic Theory. J. Kingsley, London

Delanty G, Isin EF (eds) (2003) Handbook of historical sociology. Sage, London

Dixon T (2012) 'Emotion': the history of a keyword in crisis. Emot Rev 4(4):338–344

Duby G (1961) Histoire des mentalités. In: Samaran C (ed) L'Histoire et ses méthodes. Gallimard, Paris, pp 937–66

Dunning M, Hughes J (2020) Power, habitus, and National Character. Hist Soc Res/Historische Sozialforschung 45(1):262–291

Elias N (2006 [1969]) In: Mennell S (ed) The court society. The collected works of Norbert Elias, vol 2. UCD Press, Dublin

Elias N (2008 [1960]) National peculiarities of British public opinion. In: Kilminster R, Mennell S (eds) Essays II: on Civilising processes, state formation and National Identity. The collected works of Norbert Elias, vol 15. University College Dublin Press, Dublin, pp 233–255

Elias N (2009a [1969]) Sociology and psychiatry. In: Kilminster R, Mennell S (eds) Essays III: on sociology and the humanities. The collected works of Norbert Elias, vol 16. University College Dublin Press, Dublin, pp 159–179

Elias N (2009b [1987]) On human beings and their emotions: a process-sociological essay. In: Kilminster R, Mennell S (eds) Essays III: on sociology and the humanities. The collected works of Norbert Elias, vol 16. University College Dublin Press, Dublin, pp 141–158

Elias N (2010 [1939]) The Society of Individuals. In: van Krieken R (ed) The Society of Individuals. The collected works of Norbert Elias, vol 10. University College Dublin Press, Dublin, pp 7–62

Elias N (2012a [1939]) In: Mennell S, Dunning E, Goudsblom J, Kilminster R (eds) On the process of civilisation: sociogenetic and psychogenetic investigations. The collected works of Norbert Elias, vol 3. University College Dublin Press, Dublin

Elias N (2012b [1978]) In: Bogner A, Liston K, Mennell S (eds) What is sociology? The collected works of Norbert Elias, vol 5. University College Dublin Press, Dublin

Elias N (2013 [1989]) In: Mennell S, Dunning E (eds) Studies on the Germans: power struggles and the development of habitus in the nineteenth and twentieth centuries. The collected works of Norbert Elias, vol 11. University College Dublin Press, Dublin

Elias N (2014) Freud's concept of society and beyond it. In: Joly M (ed) Supplements and index. The collected works of Norbert Elias, vol 18. University College Dublin Press, Dublin, pp 13–52

Elias N, Dunning E (2008 [1986]) In: Dunning E (ed) Quest for excitement: sport and leisure in the civilizing process. The collected works of Norbert Elias, vol 7. University College Dublin Press, Dublin

Elias N, Scotson JL (2008 [1965]) The established and the outsiders. The collected works of Norbert Elias, vol 4. University College Dublin Press, Dublin

Epstein S (1984) Controversial issues in emotion Theory. Rev Pers Soc Psychol 5:64–88

Erdheim M (1982) Die gesellschaftliche Produktion von Unbewußtheit. Eine Einführung in den ethnopsychoanlytishcen Prozess. Suhrkamp, Frankfurt a.M.

Erdheim M (1996) Unbewusstheit im Prozeß der Zivilization. In: Rehberg K-S (ed) Norbert Elias und die Menschenwissenschaften. Suhrkamp, Frankfurt aM, pp 158–171

Febvre L (1973 [1938]) History and psychology. In Burke P (ed) A new kind of history: from the writings of Febvre (trans: Folca K). Routledge & Kegan Paul, London, pp 1–11

Foulkes SH (1941) Referat von Über den Prozess der Zivilisation, zweiter Band: Wandlungen der Gesellschaft. Entwurf zu einer Theorie der Zivilisation. Verlag Haus zum Falken, Basel, 1939, 490 S. Internationale Zeitschrift für Psychoanalyse 26(3–4):316–319

Fox AS et al (eds) (2018) The nature of emotion: fundamental questions. Oxford University Press, Oxford

Freud S (1915a) Instincts and their vicissitudes. In: Strachey J (ed) The standard edition of the complete collected works of Sigmund Freud, vol 14. Hogarth Press & the Institute of Psychoanalysis, London, pp 110–140

Freud S (1915b) Thoughts for the times on war and death. In: Strachey J (ed) The standard edition of the complete collected works of Sigmund Freud, vol 14. Hogarth Press & the Institute of Psycho-analysis, London, pp 273–300

Freud S (1923) The ego and the id. In: Strachey J (ed) The standard edition of the complete collected psychological works of Sigmund Freud, Vol XIX (1923–1925): the Ego and the Id and Other Works. Hogarth Press & the Institute of Psycho-analysis, London, pp 1–66

Freud S (1927) The future of an illusion. In: Strachey J (ed) The standard edition of the complete psychological works of Sigmund Freud, Volume XXI (1927–1931): the future of an illusion, civilization and its discontents, and other works. Hogarth Press, London, pp 1–56

Freud S (1933a) New introductory lectures on psycho-analysis. In: Strachey J (ed) The standard edition of the complete psychological works of Sigmund Freud, Volume XXII (1932–1936): new introductory lectures on psycho-analysis and other works. Hogarth Press & the Institute of Psycho-analysis, London, pp 1–182

Freud S (1933b) Why war? In: Strachey J (ed) The standard edition of the complete psychological works of Sigmund Freud, Volume XXII (1932–1936): new introductory lectures on psychoanalysis and other works. Hogarth Press & the Institute of Psycho-analysis, London, pp 203–215

Frijda NH (1994) Emotions require cognitions, even if simple ones. In: Ekman P, Davidson RJ (eds) The nature of emotion: fundamental questions. Oxford University Press, Oxford, pp 197–202

Gay P (1985) Freud for historians. Oxford University Press, Oxford

Goudsblom J (1984) Zum Hintergrund der Zivilisationstheorie von Norbert Elias: Das Verhaeltnis zu Huizinga, Weber und Freud. In: Gleichmann P, Goudsblom J, Korte H (eds) Macht und Zivilisation: Materialien zu Norbert Elias' Zivilisationstheorie 2. Suhrkamp, Frankfurt a.M., pp 129–147

Greco M, Stenner P (2008) Emotions, history and civilization. In: Greco M, Stenner P (eds) Emotions: a social science reader. Routledge, Abingdon, pp 25–28

Greenberg JR, Mitchell SA (1983) Object relations in psychoanalytic Theory. Harvard University Press, Cambridge, MA

Hochschild AR (1975) The sociology of feeling and emotion; Selected possibilities. Sociol Inq 45(2–3):280–307

Hochschild AR (1979) Emotion work, feeling rules, and social structure. Am J Sociol 85(3): 551–575

Hochschild AR (1983) The managed heart: commercialization of human feeling. University of California Press, Berkeley

Hochschild AR (2009) Introduction: an emotions lens on the world. In: Hopkins D, Kleres J, Flam H, Kuzmics H (eds) Theorizing emotions: sociological explorations and applications. Campus, Frankfurt, pp 29–37

Hopkins D et al (eds) (2009) Theorizing emotions: sociological explorations and applications. Campus, Frankfurt

Horkheimer M (1993 [1932]) History and psychology. In: Between philosophy and social science: selected early essays. MIT Press, Cambridge, MA, pp 111–128

Izard CE (2010a) The many meanings/aspects of emotion: definitions, functions, activation, and regulation. Emot Rev 2(4):363–370

Izard CE (2010b) More meanings and more questions for the term 'emotion'. Emot Rev 2(4): 383–385

Jacques E (1955) Social systems as defense against persecutory and depressive anxiety. A contribution to the psycho-analytic study of social processes. In: Klein M, Heimann P, Money-Kyrle RE (eds) New directions in psycho-Anlaysis: the significance of infant conflict in the pattern of adult behaviour. Marsfield Library, London, pp 478–498

Joly M, Deluermoz Q (2010) Un échange de lettres entre Raymond Aron et Norbert Elias (juillet 1939). Vingtième Siècle. Revue d'histoire 106:97–102

Kemper TD (1990) Themes and variations in the sociology of emotions. In: Kemper TD (ed) Research agendas in the sociology of emotions. State University of New York Press, New York, pp 3–23

König H (1993) Norbert Elias und Sigmund Freud: Der Prozeß der Zivilisation. Leviathan 2: 205–221

Lahire B (2013) Elias, Freud, and the human science. In: Dépelteau F, Landini TS (eds) Norbert elias and social theory. Palgrave Macmillan, Basingstoke, pp 75–89

Laplanche J, Pontalis J-B (1973) The language of psychoanalysis. Hogarth, London

Lewin K (1931) Environmental forces in behaviour and development. In: Murchison C (ed) A handbook of child psychology. Clark University Press, Worcester, pp 94–127

Marcuse H (1972 [1955]) Eros and civilization. Abacus, London

Newton T (1997) The sociogenesis of emotion: a historical sociology? In: Bendelow G, Williams SJ (eds) Emotions in social life: critical themes and contemporary issues. Routledge, London, pp 61–80

Plamper J (2010) The history of emotions: an interview with William Reddy, Barbara Rosenwein, and Peter Stearns. History & Theory 49(2):237–265

Riviere J (1952) The unconscious phantasy of an inner world reflected in examples from literature. International Journal of Psycho-Analysis 33:160–72

Roseneil S, Ketokivi K (2016) Relational persons and relational processes: developing the notion of relationality for the sociology of personal life. Sociology 50(1):143–159

Rosenwein BH (2010) Thinking historically about medieval emotions. Hist Compass 8(8):828–842

Rosenwein BH (2018) Controlling paradigms. In: Rosenwein BH (ed) Anger's past: the social uses of an emotion in the middle ages. Cornell University Press, Ithaca, pp 233–247

Rustin M (2009) The missing dimension: emotions in the social sciences. In: Sclater SD, Jones DW, Price H, Yates C (eds) Emotion: new psychosocial perspectives. Palgrave Macmillan, London, pp 19–35

Schöttker D (1998) Norbert Elias and Walter Benjamin: an exchange of letters and its context. Hist Hum Sci 11(2):45–59

Stearns PN (1989) Social history update: sociology of emotion. J Soc Hist 22(3):592–599

Stearns PN (2009) Preface. Emot Rev 1(4):291–293

Stearns PN (2015) Why do emotions history? Rubrica Contemporanea 4(7)

Stearns CZ, Stearns PN (1985) Emotionology: clarifying the history of emotions and emotional standards. American Historical Review 90(4):813–36

Turner JH (2006) Psychoanalytic sociological theories and emotions. In: Stets JA, Turner JH (eds) Handbook of the sociology of emotions. Springer, New York, pp 276–94

Turner BS (2014) Norbert Elias and the sociology of resentment. In: Lemmings D, Brooks A (eds) Emotions and social change: historical and sociological perspectives. Routledge, London, pp 175–195

van Krieken R (1990) The Organization of the Soul: Elias and Foucault on discipline and the self. Archives Europeénnes de Sociologie 31(2):353–371

van Krieken R (2014) Norbert Elias and Emotions in history. In: Lemmings D, Books A (eds) Emotions and social change: historical and sociological perspectives. Routledge, London, pp 19–42

Waldhoff H-P (2007a) Unthinking the closed personality: Norbert Elias, group analysis and unconscious processes in a research group: Part I. Group Analysis 40(3):323–343

Waldhoff H-P (2007b) Unthinking the closed personality: Norbert Elias, group analysis and unconscious processes in a research group: Part II. Group Analysis 40(4):478–506

Wouters C (2007) Informalization: manners and emotions since 1890. Sage, London

Wrong DH (1961) The oversocialized conception of man in modern sociology. Am Sociol Rev 26(2):183–193

Identity, Identification, Habitus: A Process Sociology Approach

28

Florence Delmotte

Contents

Introduction	726
From Identity to Identification	728
Grandeur and Misery of a Concept	728
Beyond Constructivism	730
The Eliasian Approach	733
Involvement and Detachment Facing Dangerous Emotions	734
Habitus and "the Filo Pastry of Identity" (Mennell 1994)	735
Interdependence and the Changing Balance of Power Between Established and Outsiders	736
From Nationalism to Cosmopolitanism? Enlarging Identification and Changes in the "We–I" Balance	738
The Case of European Identity	741
What European Identity?	741
European Identification as a Process	742
Conclusion	744
References	745

Abstract

Identity is a polysemic, politically saturated, and even polemical notion. Sometimes considered a formidable problem when it is associated with fundamentalism and extreme nationalism, identity is seen at other times as a precious asset, a project to be built or a treasure to be regained, for example when it is a question of a European identity. However, the term identity remains difficult to replace. The many phenomena it evokes are nothing less than existential, for individuals and groups, in that they refer to the questions "Who am I?", "Who are we?" and "Who are they?" as Charles Tilly pointed out. By focusing more on processes of identification than on given identities, historical sociology reveals the relational

---

F. Delmotte (✉)
Fund for Scientific Research (FNRS) / Université Saint-Louis – Bruxelles, Bruxelles, Belgium
e-mail: florence.delmotte@usaintlouis.be

© The Author(s), under exclusive licence to Springer Nature Singapore Pte Ltd. 2022
D. McCallum (ed.), *The Palgrave Handbook of the History of Human Sciences*,
https://doi.org/10.1007/978-981-16-7255-2_53

and changing character of these phenomena and avoids the trap of essentialism. But it also avoids the pitfalls of constructivism, of seeing identities everywhere or nowhere. In particular, Norbert Elias' sociology of figurations and processes considers the long-term transformations of political and psychic structures. Reconciling macro- and microsociological perspectives, this approach focuses on interdependent relationships and power differences between social groups as well as on the place of affects, notably when political identities are at stake. Eliasian-inspired historical sociology thus makes it possible to question the feelings of belonging and the process of identification with a postnational Europe. In the end, process sociology allows for a better understanding of the resistance that the tenacity of "national habitus" continues to put up against it on the part of the citizens of the European Union Member States.

Keywords

Individual and collective identity · Identification processes · Norbert Elias · Habitus · European identity · Belonging · Emotions · Nationalism

The tradition of all dead generations weighs like a nightmare on the brains of the living. (Marx 1852, 5)

Part of the self-love of individuals, it seems, can attach itself to one of the groups with which they identify themselves, most of all to nations and other types of survival groups. (Elias 2007, 7–8)

Introduction

In the spring of 2020, at the beginning of the so-called containment period imposed by the "Covid 19" pandemic, in most European cities flags burgeoned on facades, on the balconies of buildings: national flags in the majority, but also regional, local, more rarely European, sometimes flags of different colors, and of different sizes, side by side. All these flags were often handmade. Some were accompanied by messages. Most were messages of unity in the face of adversity, of solidarity with the countries most affected by the epidemic, or of support for care staff in hospitals and retirement homes, and for other frontline jobs such as garbage collectors, delivery drivers, postmen and cashiers. Some of the banners were less politically committed than others, here denouncing capitalism and the destruction of the environment, and there, the budget cuts inflicted on the health care systems over the past decades. But all were evocative of different dimensions of politics.

Is it not strange to display (most often national) flags in the face of global fear and disease? Not so much. These flags, and some of the messages and symbols raised by the crisis, however discreet they may have been – nothing comparable to the outpouring that can be seen during certain sporting competitions, particularly the football World Cup – remind us of the strength of social attachments and affiliations. They have, in other words, an identity dimension. They say something about those who hold them, about the way they define themselves, both individually and collectively, in the home and in the public space of the street, as belonging to one or more political entities, identifying with one or more groups. These flags and banners are a reminder that this dimension of identity is everywhere, even where it is not necessarily expected. It is present, in the background, or on a few facades, making visible for a few months this "banal nationalism," the one that, by definition, we do not notice most of the time (Billig 1995).

In what follows, we will first turn to the reticence that one may have about the use of the term "identity." This notion invites caution in all cases. Its ambiguities justify for some authors renouncing it. For others, the term identity remains unavoidable. It designates, with rare evocative power, a bundle of phenomena that are, so to speak, consubstantial with social experience. The approaches of historical sociology, beyond their plurality (Bucholc and Mennell 2022; Déloye 2003), share the same aversion to reifying nouns. Logically, therefore, they will favor the study of *identification processes*, from a perspective that is interested both in the transformations of social life and in the "dead hand of the past" that weighs on this moving reality. This is the idea, so clearly expressed by Marx, that "[m]en make their own history, but they do not make it as they please; they do not make it under self-selected circumstances, but under circumstances existing already, given and transmitted from the past" (Marx 1852, 5).

The figurational or process sociology of Norbert Elias, which we will focus on in the second part, is particularly good at understanding and explaining the processes of identification of individuals and groups in the long run – about what, and how one comes to say "I" and "we." In a major work, first published in 1939, Elias (2012a) clearly showed that the development of the mental or psychic structures of individuals and the development of the social and political entities that these connected individuals form together are linked. These developmental processes are not necessarily synchronous and harmonious, but rather interconnected and interacting. This is fundamental to understanding how individuals sharing a certain "habitus," a social knowledge that is shared and incorporated to the point of becoming unconscious, more or less consciously perceive themselves as belonging to groups that are more or less important to them, with which they "identify," and which therefore constitute part of their individual and collective "identity."

This historical sociology also makes it possible to understand that the meaning of these identifications and the importance of these groups – not only families, villages, clans, tribes, or nation-states, but also professions, generations, or genders – are not always or everywhere the same. According to Elias, however, we will see that there are major trends that can be observed over the long term, namely a growing trend toward individualization, which goes hand in hand not only with the expansion of the so-called "survival units," but also with the expansion of the circle of "mutual identification" between human beings as human beings. Driven by the powerful movement of interdependence of all kinds (economic, geopolitical, ecological, etc.)

that has led to what is commonly known as "globalization," these trends simultaneously tolerate particularities and generate tensions, blockages, and backtracking.

The question of European identity – and identification with Europe – is therefore, through the question of the sense of belonging, a fine field of investigation for an Eliasian analysis. One could even say that the two enlighten each other. At the end of his life, Elias (2010 [1987]) was indeed interested in the question of obstacles to postnational political integration, which continue to preoccupy researchers in European studies and political theory. In the third and last part of this chapter, we will see that his reflections have lost none of their relevance. They resonate, less unexpectedly than it seems, with a Kantian-inspired philosophical reflection on European identity and citizenship (Ferry 1992; Habermas 2001). Above all, Elias' analyses may prove to be a valuable tool to accompany the sociological turn of European studies (Favell and Guiraudon 2009; Majastre and Mercenier 2016) and guide qualitative approaches toward paths that remain underexplored.

From Identity to Identification

> Stories and identities intersect when people start deploying shared answers to the questions "Who are you?", "Who are we?" and "Who are they?" (Tilly 2003, 608)

In this first part, we return to the problematic and inescapable character of the notion of identity and the controversies linked to its use by the social sciences. First, we look back at its history and the history of its success, and second, at the criticisms that have been addressed to it.

Grandeur and Misery of a Concept

As Brubaker and Cooper (2000) conceded in an important article for our topic, "identity" is first and foremost one of the terms that has always been used "to address the perennial philosophical problems of permanence amidst manifest change, and of unity amidst manifest diversity" (Brubaker and Cooper 2000, 3). Identity is what remains while everything is changing, what unifies and "makes identical" diverse elements, which otherwise tend to disperse. Like the notion of "person" – a founding concept for political theory, which designates both what is unique and the character of what is common, what is deep, authentic, and hidden (the personality), and what is visible, apparent, and shown (the mask, the face) – "identity" refers at the same time to what makes things unique and what amalgamates (Martin 1994). This tension, between the shared and the singular, and of permanence in change, this contradiction almost, is even constitutive of the notion (Déloye 2010). Identity is, in other words, *"ce qui fait que deux ou plusieurs choses ne sont qu'une mesme"* [what makes two or more things one and the same], says the dictionary of the French Academy in 1694. It is therefore understandable that the encyclopedists did not predict any spectacular posterity or popularity for the notion.

In fact, while the problem of permanence and change was raised in antiquity by Parmenides and Heraclitus, the success of the concept and the multiplication of its uses are recent (Martin 2010). Its uses come first, logically, from the scholarly world. After philosophy many centuries earlier, it is toward psychoanalysis and psychology that we must look. In a landmark work, Erikson (1968) drew on Freud's work on the role of identifications, and the idea that the relations between the person and society are constitutive of the self. On this basis, Erikson defines identity "as consciousness and as a process: both the feeling that the individual has of himself and the affiliation of the individual to social groups" (Martin 1994, 15). Identity appears there in a relational situation and concerns first of all a crisis situation: that of adolescence. It cannot be conceived as a permanent situation. Both isolation and immobility are banished. In history (Braudel), in anthropology (Lévi-Strauss, Mauss), and of course in sociology (Simmel, Mead), we find the idea that the formation of the "I" – or the "we" – cannot take place if there is no "other." In the philosopher Paul Ricœur (1990), identity is transformed in the course of narratives, where "Oneself" is considered "as Another."

While Erikson gives the term identity its conceptual dimension, the notion has other channels of dissemination. On the one hand, it is associated with "ethnicity." On the other hand, in his theory of reference groups, Merton (1957 [1949]) had already stated that identification with ensembles is one of the important mechanisms for building the identity of individuals and collectives. Soon after, social constructionism (Berger and Luckmann 1967) and symbolic interactionism (Goffman 1959) appropriated the notion, insisting on the importance of the way individuals present and represent themselves. The substitution of "identity" for "self" in Erving Goffman's (1963) study of stigma clearly marks a turning point in the history of the notion. In the social sciences, its uses have multiplied to the point where identity has become an inescapable part of "pop sociology" (Gleason 1983).

It is true that it had a remarkable resonance in the 1960s. "Identity" seemed able to apprehend the mass society, the generational revolt at the end of the decade, the Black Panther movement, and other groups in their wake in the struggle for black American civil rights. Brubaker and Cooper (2000) point out that its formidable spread has been facilitated by the weakness, in the United States, of a social and political analysis in terms of class, which leaves room for the proliferation of identity claims and the transposition of the problem of individual identity to the group level. As decried and attacked as the notion subsequently became, its success continued unabated, rapidly transcending borders and disciplines. In the 1970s, the word was already "driven out of its wits by over-use" (Mackenzie 1978, 11; Brubaker and Cooper 2000, 3) and in their turn gender studies, writings on sexuality, race, religion, or nationalism, were invaded by identities, to the point of saturation. Fuzzy, versatile, in its social uses, identity is both practical – in every sense of the word, a conceptual Swiss army knife according to Brubaker and Cooper – but soon also, and for this very reason, highly problematic.

In fact, we can point to several types of problems that are difficult to untangle and which, according to many authors, call into question the use of the concept of identity by the social sciences. First of all, its use is said to have been increasingly guided by political perspectives, whether militant or even insurrectionary, whether institutional or even conservative. It is difficult to speak of "identity" without legitimizing – or condemning – what we are talking about, depending on whether we are talking about a valued belonging or a sectarianism to be defeated. Second, the problem with the notion of identity in many analyses, which indiscriminately consider it as explanatory, analytical, or even descriptive (Surdez et al. 2009), is the indecision of its status. As in the case of other ambivalent and problematic concepts, such as nation, ethnicity, or race, the use of identity always runs the risk of promoting a reifying reading of collective membership, of helping to validate the existence of the groups, communities, or categories to be studied, or even of endowing them with an "essence," with "immutable" characteristics. In other words, the invocation of identity loses its character as an object of analysis and runs the risk of replacing the analysis itself, bringing with it its share of commonplaces, judgments, prejudices, or ideological preferences.

Beyond Constructivism

All these difficulties easily explain why approaches claiming to be constructivist and rejecting essentialism have subsequently prevailed in the social sciences, at least in the intention of many writers in the humanities. In such work, collective identity is seen as the product of social activities, which do not exist a priori. However, some of these approaches err on the side of relativism, and there is a risk of "dissolving" identities, of no longer seeing them anywhere, or of seeing them everywhere. Constructivist approaches thus struggle to account for the strength of the essentialist claims of contemporary identity politics. Another position is concerned about the risk of reducing identities to pure illusions (Surdez et al. 2009). Brubaker and Cooper (2000) propose getting rid of the term in order to distance themselves from its political and essentialist uses. They propose, in other words, to replace the word without abandoning the study of what it concerns: belonging, sense of community, self-understanding, and self-identification – in other words, a bundle of phenomena that are both distinct and often intertwined, and that sociological analysis could not grasp from a single term. Other authors would rather advocate retaining its use "conditionally," particularly within cultural studies (Hall 1996), but sometimes acknowledging that the word, inseparable from a certain idea of permanence, seems ill-suited to processual analysis (Melucci 1995).

In a nutshell, identity is a "vast and polysemous concept," "used in extremely varied and contradictory terms, mixed up with essentialist and constructivist understandings" (Duchesne 2008, 7). For Charles Tilly (2003), however, one should not throw the baby out with the bath water. It is clear that individuals are constantly confronted with the need to produce socially relevant responses to the question of their own definition in relation to the groups to which they belong (Surdez et al. 2009). In his response to Brubaker and Cooper, who proposed to "expunge 'identity' from our analytic lexicon," Tilly proposes instead "that we get identity right" (Tilly 2003, 608). He writes on this subject:

> We can escape the search for inner selves about which Brubaker and Cooper rightly complain by recognizing that people regularly negotiate and deploy socially based answers to the questions 'Who are you?', 'Who are we?' and 'Who are they?' Those are identity questions. Their answers are identities – always assertions, always contingent, always negotiable, but also always consequential. Identities are social arrangements.

He adds: "To be more precise, identities have four components:

1. a boundary separating me from you or us from them;
2. a set of relations within the boundary;
3. a set of relations across the boundary;
4. a set of stories about the boundary and the relations."

To support his argument, Tilly takes the eloquent example of the politics of Henry VIII, his imposition of a boundary between Anglicans and Catholics in England in 1536, and its impact, first on the politicization of the religious identities of the parishioners of Morebath (Devon), and then in the longer term (Tilly 2003, 609 ff.).

Several elements are important in this passage, which well illustrates some major characteristics of a historical sociology of politics (Déloye 2003). First of all, the question of the definition of oneself, of us, and of others is unavoidable and reveals a relational point of view on the social world. Individuals and groups are not defined by their intrinsic characteristics but through the relationships (of interdependence and power, first and foremost) that structure the social world. This vision of things makes it possible to escape from the alternative already mentioned between the reification of categories and criteria of belonging, and their reduction to linguistic fictions, to illusions. Tilly demonstrates that we can be interested both in the construction of identities – in their constructed character – and in their impact, their effects. This also makes it possible to take into account the multiple dimensions of identity, or "identification." A historical sociology or social history approach (Noiriel 2007) would indeed tend to use the term "identification," in order to avoid the often ideological or partisan debate on the substantial dimension of identity, and to focus on "the study of the complex processes of identity construction and projection" (Déloye 2010, 404).

If identification is first understood as "a process which accounts for the way individuals develop the feelings of belonging to a group" (Duchesne 2008), for historical sociology this dimension (identifying oneself) is intimately linked to a process of identification and categorization of other groups (identifying). Here lies another "constitutive tension" of identity phenomena or processes, this time corresponding to the interaction between the objective and subjective dimensions of identifications, in a Weberian reading (Déloye 2010). Max Weber (1978 [1921]) actually distinguishes between "socialization," defined by its "objectivity," on the one hand, and "communalization," which is distinguished by its "subjective" dimension, on the other hand. According to him, it is the sense of belonging to the community that defines and produces the community, not the sharing of supposedly objective characteristics such as language, religion, blood, or custom. And it is, again according to Weber, the relationship to other communities that reveals the community to itself and makes it define the elements that will be held common to its members and by its members, and distinctive from other communities. Language, religion, blood, and custom are therefore not in themselves. Simmel, for his part, emphasizes the importance of opposition to the other and conflict in the making of borders, which he says are "always psychological" (Simmel 2009 [1908]).

Thus self-identification (of "me," of "us") is intimately correlated, in the relational perspective of the historical sociology that interests us, with the identification of others (of "you," of "them"). This self-identification may contain a more or less conscious dimension, and it may be articulated in a discursive way, in a narrative, but this is not necessarily the case. We will see this with Elias through the notion of habitus. For his part, Michael Billig (1995) has clearly shown that the strength of "banal nationalism" lies in its diffuse, tacit, involuntary, and very often unconscious character.

Processes of identification – identification of others, of other groups, and identification of oneself by others, of one's own group, or of one's groups by other groups – are the other side of the same identity coin. Historical sociology allows us to consider that these processes are not only relational but also categorical (Calhoun 1997). Identification processes do not only refer to the existence of networks of relationships, more or less concrete or abstract, close or distant, in relation to which individuals and groups define themselves. They also refer to "categories," that is, groups of individuals who share categorical attributes, or considered as such in a given context – race, gender, nationality, language, sexual orientation, ethnicity, and so on. These attributes are defined in a power relationship between those who are in a position to produce these categories and those on whom the former want to impose them (Bourdieu 1984), with varying degrees of success. In this respect, the modern state, as studied in line with the work of Max Weber, Pierre Bourdieu, and Michel Foucault, has clearly been one of the most important agents of identification and categorization. Bourdieu points out that it claimed not only a monopoly on legitimate physical violence, but also a monopoly on symbolic power. It is not, according this time to Foucault (2007), that the state "creates" identities. Indeed, state is incapable of doing so, a point to which we shall return in discussing Elias. Rather, it has ended up, long enough, with incomparable material and symbolic resources to impose its classifications as well as its modes of accounting. Gérard Noiriel (1991) speaks in this respect of a real "revolution in identification."

To sum up, one may have the impression that "identity" refers above all to an emotionally charged form of "self-understanding," both to a vague feeling and to a narrative of belonging. The problem then is that the use of the notion evades the question of what constructs this feeling or narrative – the processes, necessarily relational, and the power relations from which they emerge. This is why a relational and processual approach involves a preference for, in a word, the idea of "identification." This notion, however, does not necessarily temper the "conscious" and "articulated" dimension of the sense and narrative of identity but tends on the contrary to accentuate it. This is why Eliasian-inspired historical sociology definitely prefers the notion of habitus (Mennell 1994, 177), attentive both to the sedimentation and transformation of affiliations and identity markers over time, and to their inscription in bodies as well as in hearts and minds.

The Eliasian Approach

> The remarkable propensity of people for projecting part of their individual self-love into specific social units, to which they are linked by strong feelings of identity and of belonging, is one of the roots of the dangers which human groups constitute for each other. (Elias 2007, 8)

Generally speaking, the so-called "historical" sociology of Elias – who actually thought that there was no other, that all sociology had to be "historical" – is marked by two fundamental convictions, from which its ambitions derive. First, the individual and society are not two separate and distinct entities. Rather, what affects what we call "the individual" also affects the society in which they live, or to which they are said to "belong." Conversely, what affects society also affects the individuals who are said to "compose" it. Second, everything that concerns sociology, the way in which human beings live together, is of a "processual" nature. What is observed now, what was observed yesterday, and what will be observed tomorrow are necessarily connected in one way or another, and the whole thing is necessarily linked to what happened the day before yesterday and what will happen the day after tomorrow. These two characteristics allow Elias' sociology of figurations and processes, on the one hand, to overcome the micro–macro dilemma, by showing that it is totally abstract (Calhoun 1994, 4), and, on the other hand, to reject "the retreat of the sociologists into the present" (Elias 2009a [1987]), without returning to the historical determinism of nineteenth-century social theories.

On this basis, Elias' sociology provides a very compelling approach to identity phenomena. It avoids many of the pitfalls typical of identity studies while proving to be compatible and complementary with their advances, particularly on issues related to nationalism, along the lines of Anderson (1983) and Hobsbawm (1990). In what follows, we present the characteristics of Elias' approach that make it original and give it added value for thinking about identity, the processes of identification, and the formation of identities, and that explain a rich and increasing posterity (Mennell 2007; Kuzmics et al. 2020). Elias' strong and inspiring points include his conception of habitus, the theory of established–outsider relations, and the study of the "changes in the 'we–I' balance" in the long duration of the civilization process. Before we get to that, the point of view still needs to be situated and clarified.

Involvement and Detachment Facing Dangerous Emotions

In Elias' writings, too, the term identity appears in the 1960s and expressions such as "we-identity," "we-image," "we-perspective," and "they perspective" are frequently used together (Elias 2007, 8, editor's note 1; Elias 2012b, 117–123). The general idea is that the we-image is the collective equivalent of the self-image of an individual person. In *The Established and the Outsiders*, Elias and Scotson (2008 [1965]) further demonstrated from their study of the dynamics between two urban communities in a small industrial town in the north of England that the we-image of the members of a group is inseparable from their they-image of the other group.

Twenty years later, Elias gave these issues a central and urgent character in the long English Introduction for the volume *Involvement and Detachment* (Elias 2007 [1987], 3–67). In the context of the Cold War between the Soviet Union and the United States, this opens with the origins of intergroup warfare and violence. Elias writes that there is a "self-perpetuating propensity of long-lasting cycles of violence" (2007, 7), expressed in civil and ethnic wars, revolutions, and wars between states. Yet war, or the threat of war alone, is rooted in the relationships that human groups have with one another, in their attitudes toward one another, more precisely in the feelings they have for one another.

This dimension, the emotional origin of the dangers that human groups represent for each other, deserves even more attention as it is largely underestimated. Why is this? Because it refers to a narcissistic component of attachment to the group, commonly less valued, more likely to be criticized than its altruistic and disinterested component – the latter we have in mind, for example, when we talk about "patriotism." Why is this narcissistic component of attachment to the group poorly regarded, nearly taboo? Because this self-love, this valuing of ourselves, often goes hand in hand with feeling bigger, better, and stronger than "others" (Elias 2007, 8). In other words, it refers to a frequent and difficult to assume confusion between power (a higher power ratio) and human value, the same confusion which explains both the propensity of human groups to dominate, and the potentially devastating effects of decline. If the weakening, fall, or defeat of the group can be nothing less than synonymous with death and a sense of destruction, both individually and collectively, this is so given the fantasy confusion between power and value and the strong emotional dimension of individuals' identification with the group, especially when it is a "survival unit" (clan, tribe, village, or state, depending on the context).

On this subject, Elias (2007, 11) writes further: "The stronger the hold of involved forms of thinking, and thus of the inability to distance oneself from traditional attitudes, the stronger the danger inherent in the situation created by people's attitudes towards each other and towards themselves." This is, moreover, the key problem studied in the various texts brought together in *Involvement and Detachment*. Elias (2007, 11–12) there explains:

> One difference between a more detached and a more involved approach [...] is a difference in the time perspective. [...] No doubt a long-term perspective demands a greater capacity

for distancing oneself for a while from the situation of the moment. But it also opens the way towards greater detachment from the wishes and fears of the moment, and thus from time-bound fantasies. It increases the chance of a more fact-oriented diagnosis.

For Elias' historical sociology, questions of identity, far from being merely fashionable or sporadic, are questions that affect the lives of human groups, literally and figuratively – and fundamental, structural questions. They still largely escape understanding, including by researchers, who are also identified individuals and who identify with groups. It is therefore important to take a historical distance. This is all the more necessary because, if these questions of identity and identification are eternal, at the same time they are not: These processes do not escape history; they *are* history. The structure of their development has to be brought to light. We shall come back to this after having presented the notion of habitus.

Habitus and "the Filo Pastry of Identity" (Mennell 1994)

In 1994, in another important text for our topic, Stephen Mennell recalls the classical division of the theories on identity formation into two bodies. On the one hand, there are those interested in self-identity and how the self is constructed, in line with the work of George Herbert Mead (2015 [1934]), which was extended into psychoanalytic theory. On the other hand, there are studies that look at how certain categories of people (men, women, whites, blacks, gays, this or that social class, etc.) come to share a collective identity. Mennell emphasizes the obvious nature of this partition, but also its strangeness. For as far back as we can go, humans have never been solitary animals: "their self-images and we-images have always – since the acquisition of the uniquely human capacity for self-reflection – been formed over time within groups of interdependent people, groups that have steadily increased in size" (Mennell 1994, 176). Mennell (1994, 177) also recalls the powerful formula (and dated sexist phraseology) of Kluckhorn and Murray (1948), according to which "every man is in certain respect (a) like all other men, (b) like some other men, (c) like no other man." According to Mennell, the term habitus popularized by Pierre Bourdieu (1984) concerns the second level, focusing on "the characteristics which all human beings share in common with *certain* other human beings in the particular groups to which they belong."

In Elias, habitus appears in the preface, written in 1936, to the first edition of *On the Process of Civilisation* (Elias 2012a, 4–7). Its first merit is not to cut off reflection on the individual from reflection on the collective: "the social habitus of individual forms, as it were, the soil from which grow the personal characteristics through which an individual differs from other members of his society" (Elias cited in Mennell 1994, 179). Through habitus, the processes of individualization and socialization are inextricably linked.

Compared to identity, the notion of habitus refers more directly to the idea of a "second nature" (Elias 2013a) that was already found in Max Weber and others. Thus, habitus has the already mentioned advantage over identity of being more encompassing. It does not imply conscious awareness of belonging to a group and of distinguishing oneself from other groups (Mennell 1994, 177). People may share the same habitus without paying much attention to it, without having thought about its characteristics or the feelings it inspires, or having expressed them discursively, shaped into a narrative. The habitus is, in other words, social knowledge that is shared and embodied in the strong sense, including the literal sense. Before Elias, we find a magnificent illustration of this in an anecdote reported by the anthropologist Marcel Mauss (1935, cited in Déloye 2010). Mauss draws it from his experience of the First World War, which he has in common with Elias:

> The Worcester Regiment, having achieved considerable glory alongside French infantry in the Battle of the Aisne, requested royal permission to have French trumpets and drums, a band of French buglers and drummers. The result was not very encouraging. [...] The unfortunate regiment of tall Englishmen could not march. Their gait was completely at odds. When they tried to march in step, the music would be out of step. With the result that the Worcester Regiment was forced to give up its French buglers.

Another important feature of habitus and identification to which it refers is that they are always, especially in the modern world and in complex societies, multilayered: people "belong" to multiple groups, whether they are, so to speak, concentric or overlapping. The multiple layers of identity of people living in the same era may also date from different periods (Mennell 1994, 177). Even when the sense of national identity dominates, in the sense that it is strongest, emotionally, for the majority of people, and regarding its consequences, this level of identification is profoundly historical: It is over the centuries that "the fortunes of a nation become crystallized in institutions which are responsible for ensuring that the most different people of a society acquire the same characteristics, possess the same national habitus" (Elias 2013a, 23). However, even in this specific case of sedimentation over a very long period of time, the habitus does not imply anything fixed or static: "habitus changes over time precisely because the fortunes and experiences of a nation (or of its constituent groupings) continue to change and accumulate" (Dunning and Mennell 1996, ix).

At this stage, two questions are important for an approach to historical sociology. First question: What is it that makes identities and habitus transform, apart from simply the succession of generations of individuals who give them life? Second question: If the processes of identification have a direction, in what direction do we observe them evolving?

Interdependence and the Changing Balance of Power Between Established and Outsiders

As well as the terms "processes" and "figurations," the idea of "interdependence(s)," singular or plural, marks Elias' sociology and sums up its gesture. Interdependence (s) or interdependent relations might be defined by what they are not. They are not "interactions" in the sense that they do not link two (or more) independent individuals first existing by their own and who would "create" interactions in a second time (Lacassagne 2008, 273). For Elias, such individuals and "interactions" are purely fictitious. Human beings have always been linked from before birth to other human beings by various types of dependency – not only affective, but also economic, political, and cultural ones – which evolve over the course of individuals' lives, and over generations and centuries.

Interdependencies between individuals and between groups are not necessarily "egalitarian" relationships. Most of the time, they are not. For Elias, however, the history of any society up to the present day is not (only) that of class struggles as Marx thought; societies are not ultimately divided between *the* group of those who dominate and *the* group of those who are dominated. What we call "power" is indeed "a structural characteristic of human relationships – of all human relationships" (Elias 2012b, 70) although relationships can be more or less unequal.

Here we can see the very important character of established–outsider relations for a better understanding of identity issues. We remember that questions of identity took off in humanities literature at the time of a crucial episode in the struggle of black Americans for recognition of their civil rights in the 1960s. Starting from Elias' historical sociology implies placing this process in the context of the long-term evolution of the power differential between whites and "people of colour." It also invites us to consider more broadly the dynamics that affect the balance of power between different groups: men and women, rich and poor, gay and straight, colonizers and (de)colonized (Mennell 1994, 180). This means those around which the most difficult identity problems crystallize. In short, Elias' sociology shows the fundamentally relational and processual nature of these problems. It is, for example, sociologically correct to say that women's problems, and their transformations, are also those of men (Liston 2018, 363; Elias 2009b). And the problems of blacks are also those of whites, as America and the world, whites, are perhaps beginning to realize that they are. (On 25 May 2020 in Minneapolis, a 46-year-old African-American man, George Floyd, died by suffocation when a white police officer knelt on his neck for eight minutes: another episode in the history of racism and police violence in American society. However, this caused protests on an unprecedented scale in the United States and around the world. Alongside black activists, a large number of white people and demonstrators of all origins marched chanting "Black lives matter," the slogan of a movement born in 2013.)

It has already been pointed out that Elias is concerned with the remarkable propensity of people to link self-love with that of certain groups with which they identify *and* to confuse human value with superiority in the balance of power between groups. In *The Established and the Outsiders*, Elias and Scotson (2008) look at the relations between the inhabitants of two working-class neighborhoods that nothing a priori radically distinguishes, for example, in terms of income. The only notable difference is that one group is composed of inhabitants who have been "established" for two or three generations, the other of newer inhabitants. The authors show that the sociological seniority favorable to the first group, the "established," is the basis of their privileged access to positions of power and their superiority in terms of symbolic prestige. This dominant position is reflected in their positive collective "we-image," in the negative collective "they-image" they have of the newcomers, and in the negative "we-image" these "outsiders" have of themselves. The dominated have integrated and assimilated the unfavorable "we-image" corresponding to the image of "them" that the established are in a position to impose on them.

The greater the power differential on several levels, the more difficult it is for the oppressed to emerge from a dominated position, and to make their self-image evolve positively. However, "unchanging power ratios are very much the exception" (Mennell 1994, 182). When groups occupying a very unequal position in a power relationship become more interdependent, when dependence becomes more reciprocal, the result is a shift, more or less gradual or dramatic, toward a less unequal power differential. This is what happened in an accelerated way in the twentieth century and led to a relative and partial improvement in the fate of workers, women, and migrants. Elias relates this phenomenon to what he calls "functional democratization" (Elias 2012b, 62–64): "the increasing division of social functions and lengthening chains of interdependence lead to greater reciprocal dependency and thus to patterns of more multi-polar control within and among groups" (Mennell 1994, 183).

The process of translating this development into identities may be slow. In "Changes in the 'we–I' Balance" (2010 [1987]), Elias speaks of a "drag effect." For him, this is a structural characteristic of the dynamic relationship between interdependencies and habitus: The latter are resistant to change and can take a long time to adapt. Let us take once again the example of race relations to illustrate how a change in the shift of power ratio is reflected in the construction of we-identities or we-images. As Mennell reminds us, the abolition of slavery in 1865 in the United States did not change much from this point of view. Black Americans long retained an inferior image of themselves, one that was imposed on them by white people during centuries of slavery (and afterward). The affirmation of Black Power and the Black is Beautiful movement are in fact a very belated development. And the process of asserting a "positive" identity is clearly not complete, nor is it fully accepted by nonblacks.

From Nationalism to Cosmopolitanism? Enlarging Identification and Changes in the "We–I" Balance

Another contemporary expression of this resistance of habitus and identities to change concerns the tenacity of national habitus. There is empirical evidence that, over the long term, human beings become more and more interdependent and tend to form ever larger "survival units" (Elias 2012a, 479). Enlarging social entities are also more complex, as the differentiation of social functions increases. Elias argues that the same is generally true of feelings of belonging and collective identities, of we-images: When we look back over the long history, we see that they tend to expand and multiply even if some of them come to predominate. Furthermore, the current difficulty of moving to forms of post-national political integration that would require broader and less affective forms of identification is striking.

Although it adopts an inclusive and historical perspective, the sociology of Elias remains, as is more or less all sociology in the twentieth century (Bourdieu 2012), centered on the state, and in his own way Elias assumes it. It is not that every society is necessarily state-centered, and the state scale is far from being the most relevant for studying social phenomena. We cannot any more separate what happens within states from what happens between them. The point is rather that, for the Western civilizing process and the recent history of human societies, the "sociogenesis" and "psychogenesis" of the state are crucial phenomena (Elias 2012a, vol. 2). The state has consecrated a double movement – of interdependence and competition – that has pushed individuals and groups to live in peace within an enlarged territory under the monopoly of legitimate violence. In the process, the vast majority of individuals and groups have come to identify with, and become attached to, the state. But this took time, and in many cases, democratization played a very specific role in this process. Elias originally argues that in many European countries it is democratization that makes it possible to develop a "national we," and not the other way around:

> The emergence of the European states as we-units happened gradually and in stages. [...] The dictum ascribed to Louis XIV, *l'État, c'est moi*, 'I am the state', shows a specific fusion of 'we' and 'I' in relation to the dynasty and the incumbent of the throne, and only to them. [...] Only in conjunction with the parliamentary representation of all classes did all members of the state begin to perceive it more as a we-unit and less as a they-group. Only in the course of the two great wars of this century did the populations of more developed industrial states take on the character of nation states [...]. (Elias 2010, 185–186)

Regarding political and social democratization and access to rights (Marshall 1950), great things have been achieved within the framework of the state at certain times. As Elias acknowledges, "the self-esteem of nations and other survival groups need not be undeserved [...]. They may have achievements to their credit which are great benefit to humanity" (Elias 2007, 9). It is above all for this reason that subjects, having become citizens, became more attached to the state – and why they become less when they feel abandoned by the state. That said, collective self-praise is rarely synonymous with realism, and a well-orchestrated conflict is often very helpful in reestablishing the identity of the group.

Hence the remarkable resilience of national identity. The nation-state may have long been overtaken by the globalization of risk, the arms race, and rising ecological perils. Interdependencies may have reached the ultimate level and humanity may be the "true" survival unit: It is difficult for most people to attach themselves to entities beyond their state borders, and even more difficult to identify with humankind (Elias 2010, 195; 203). This level of integration seems too distant and too loose and very few people feel that belonging to humanity is a social bond. Even at a lower level, the continental integration of Europe has been facing "the tenacious resistance of emotive ideas which give the integration the character of ruins, a loss that one cannot cease mourning" (Elias 2010, 201). This is because, "in relation to their own group identity and, more widely, their own social habitus, people have no free choice. These cannot be simply changed like clothes" (Elias 2010, 200). However, Elias speaks of a "drag effect" (Elias 2010, 188–196; see also Ernst et al. 2017). If the habitus has changed, it can change again, and it is not "unrealistic to suppose that in the future terms like 'European' or 'Latin American' will take on a far stronger emotive content than they have at present" (Elias 2010, 202).

This view does not imply a belief in progress (Mennell 1994, 189). Retrogressive steps backward – identity-based retreats – are most probable, because the push for integration implies an increased concentration of power and generates tensions and new inequalities. Nor is this view particularly optimistic. Like Stinchcombe (1975, cited in Mennell 1994, 189), Elias might have written: "it is the great tragedy of social life that every extension of solidarity from family to village, village to nation, also presents the opportunity of organizing hatred on a larger scale." And he never ceased to criticize nationalism, all nationalisms, may it appear extreme or banal, as one of the greatest perils that humanity has ever known (Elias 2013a).

As part of a broader civilizational process, the transformations of the "we–I" balance are thus characterized by their ambivalence and incompleteness. On the one hand, the "I-identity" tends to take precedence over the "we-identity": The growing integration of humanity is accompanied by increasing individualization. References made by Elias to the rise of human rights claims, to the rights of individuals against the state, suggest it is not necessarily a bad thing (Elias 2010, 207–208). On the other hand, this movement feeds strongly affective resistances, hindering the development of a mutual identification enlarged to the scale of humanity. For its part, this broader mutual identification would be based on a historically increased sensitivity to the suffering of others more and more distant and different from oneself (De Swaan 1995). In the end, it refers less to the regulatory ideal of a disembodied cosmopolitanism than to a condition of survival in an interdependent world (Elias 2011).

While recalling the importance of long duration and global transformations, historical sociology, in particular that of Elias, reconciles the methodological detachment that must characterize sociological work and a deep commitment to a sociology devoted to its emancipatory mission: "hunting myths" (2012b, 46–65), developing a nonideological vision of society (Elias 2013b) and thereby increasing the orientation capacities of individuals (Elias 2007; Mennell 2021). If, in its own way, this historical sociology studies identities, it is ultimately to ensure that people are less trapped by them, and that identities set them less against each other and against themselves. Concretely, this implies putting related individuals – their stories, their emotions, their representations, and their practices – back at the center of sociological analysis. This gesture will prove to be particularly valuable in changing the way we look at the question of European identity.

The Case of European Identity

> For you folks, identity is something nice; it's all about institutions, deliberations and elites. Where I study identity, people die for it! (A Middle East specialist decrying the way in which Europeanists study identity, quoted in Checkel and Katzenstein 2009, 10)

Much has been written about the issue of European identity, and it is beyond the scope of this chapter to examine it in detail. Rather, we propose in this last part to recall the elements that frame contemporary discussions on European identity and to summarize how historical sociology, and in particular Elias' perspective, allows us to approach the subject, in comparison with classical or more recent debate.

What European Identity?

"United in diversity": The motto of the European Union (EU) first came into use in 2000, and it sums up well the eternal philosophical problem of identity, its constitutive tension, and its self-contradictory character (see Delanty 2003 for a criticism of this motto). In the European case, as in all cases, the debate on identity raises the question "who are we?" and/or "who are they?" However, because of the obvious multiplicity of national cultures and languages in the EU, "Europe" is a fine case because it casts doubt on the very possibility of the existence of a single, unified collective political identity, beyond a pure rhetoric accompanying governance (Shore 2006).

The question of European identity took off in political life and in the academic field at the beginning of the 1990s, around the signing of a treaty considered important for the history of European construction, the "Maastricht Treaty." Signed in 1992 after encountering greater than expected resistance, this treaty introduces the notion of European citizenship and recognises it for any citizen having the nationality of one of the EU member states. European citizenship is accompanied by the right to free movement within the Union and the right to vote and to be elected, at least in European elections, in the state in which one resides.

In the 2000s, the politicization of issues related to European identity became more pronounced (Checkel and Katzenstein 2009). The decade was marked by the entry into the EU of 12 new countries from the former "Eastern bloc" in 2004 and 2007 and by the failure of a project for a constitution, following the negative results of referendums in two old member states, France and the Netherlands. At the same time, the tensions caused by increased internal mobility and the challenges raised by the securing of the EU's external borders are drawing new logics of exclusion and questioning the very meaning of community (Duez 2014). Hitherto implicitly dominant, the cosmopolitan conception of the European identity – elitist and sometimes inconsistent, for example, in the matter of migration – is increasingly contested not only by sovereigntist parties more or less hostile to Europe, by national populist parties, but also on the left. On the side of the European institutions and the pro-European currents, the new problem is fast becoming the solution. "European identity," "European culture," "common European history," "European memory," the "feeling of belonging to Europe," and "pride in being European" are often invoked and confused as remedies for the EU's real or supposed "democratic deficit" and the "lack of legitimacy" of the European project.

Studies on Europe accompany these developments. Since the early 1990s, a whole theoretical literature has been developing on the question of integration and European citizenship which is no longer of the same type as the major normative theories of integration (federalism, intergovernmentalism, etc.) which had dominated until then. In political theory, the "postnational" model, following the "Kantian way" (Ferry 1992, 2005) and extending the debate on the constitutional patriotism (*Verfassungspatriotismus*) born in Germany, defends the necessity and possibility of dissociating democratic citizenship from national belonging in a multicultural society (Habermas 1998, 2001). The postnational model is opposed not only to national-sovereigntists, for whom nationhood and democracy are inseparable, but also to some advocates of a European federal state, who conceive of Europe as a supernation. Philosophically anchored in the rationalist paradigm, this postnationalism, nevertheless, underestimates the weight of emotions and resistance to Europe and overestimates the identifying potential of the deliberative ideal and democratic institutions.

Once again, it is impossible to review here all the literature dealing with the question of European identity at the turn of the millennium – from historical institutionalism to studies on Europeanization, via constructivism and more classical political sociology. What is certain is that in less than a decade, work dedicated to identity is taking precedence over attitudinal approaches in the growing body of research on the relationship between citizens and the European political system (Duchesne 2008; Belot 2010). However, many of these studies can be criticized for reverting to a conception of identity borrowed from social psychology (Tajfel 1974; Bruter 2005), a conception that focuses too much on the notions of in-group and out-group and on the "othering," "without appropriate theoretical adaptation to the political democratic nature of the social group that a European polity would be" (Duchesne 2008).

These two decades are also marked by a double turning point. The first is that of the "normalisation" of European studies – that is, the reappropriation of European objects by the disciplines and methods of the social sciences, the rediscovery of the classics and even of the long term (Bartolini 2005). In studies on citizens, the qualitative turn of the sociology of European integration (see for example White 2011) then de-sacralizes the importance of electoral behavior, opinion polls, and Eurobarometer to focus on what cannot ultimately be measured: representations, relations, and affects. Sociohistorical approaches are part of this double evolution (Déloye 2006).

European Identification as a Process

For its part, Eliasian historical sociology reminds us that it is possible and necessary to go beyond the false micro-/macroopposition in a relational approach. Such perspective is compatible with studies that mix (qualitative and quantitative) methods (Van Ingelgom 2014). Moreover, processual and long-term approaches favor the reciprocal enlightenment of empiricism and theory and show the relative, not absolute, originality of the question of identification with Europe (Delmotte 2012). By focusing in the 1980s on identification to Europe through "obstacles" to European integration, Elias (2010) anticipates by a generation the development of studies on Euroskepticism (see Szczerbiak and Taggart 2008). At the same time, he draws attention in advance to the limits that these will encounter by focusing primarily on political opposition and political parties. Elias is more interested in latent, less explicit, or even unconscious oppositions: those that relate to the resistance, of an affective nature, of national habitus, and feed a deficit of identification to Europe and a form of widely shared indifference. In so doing, Elias also anticipates the renewed interest of political science in general and European studies in particular for the role of emotions in politics (see Belot and Bouillaud 2008).

The interest in affects and feelings does not imply per se any valuation of these as far as political identities are concerned. Following Elias, it is rather important to recognize their weight, explain and understand their role in the conduct of human affairs. One cannot simply consider that further postnational integration, in Europe and beyond, would be rational and beneficial. Nor is it crucial to "love Europe" as a homeland or even "humanity" as one big family. Rather, it must be seen that at a very high level of relational complexity – as is necessarily the case with a federation or confederation of nation-states – it is difficult for people to find their way around, *a fortiori* in a context of uncertainty. It is then tempting – and dangerous – to listen to the most immediately reassuring and comforting speeches, those that speak more to emotions or know how to mobilize beliefs.

The place of affects, feelings, and emotions is therefore very important in identities and identification processes in general. However, it is not a question of opposing passions to reason. If we rely on historical sociology, we are led to reject such dualistic way of thinking, based on antitheses that do not correspond to any reality. "The heart has its reasons," which reason does not necessarily ignore. Democratization and the development of social solidarity have directly contributed to attachment to the nation. Although we have to be attentive to the diversity of national trajectories, there is therefore no opposition or flagrant contrast between the level of belonging to the nation and a higher, more encompassing level, but a form of continuity. It also means that there is no more equality or homogeneity in identification with Europe than with the nation. As Duchesne (2008) points out, for those who have been left behind by European integration, who do not have the opportunity to experience it "as an empowering idea in their lives," it is reduced to a "supranationalist narrative," "where the political community is imagined as limited and sovereign" (Anderson 1983). For these people, identifying with Europe means giving up the nation they have learned to love since childhood (Billig 1995), unlike the more direct beneficiaries of European construction, whose attachments do not compete but can complement each other.

Contrary to what Inglehart (1970) thought, identification with Europe will not necessarily take place in a more abstract way than identification with nations. It is too early to say whether it will simply take several generations to change inherited affiliations. Nevertheless, over the long term, the recomposition of human groups into ever larger units is a strong trend, which does not exclude the existence of centrifugal effects. The identification with an ever more distant and different other is also a strong trend. Not only the evolution of sensitivities contributes to this, but also the evolution of the awareness of the interdependencies that bind us together.

Historical sociology can provide a powerful critique of nationalism. It also gives grist to the mill of the postnational political project (Delmotte 2012). It thus redefines within itself another relationship – a critical and reciprocal one – between theoretical, philosophical and political normativity, on the one hand, and empirical, sociological and historical observation, on the other. Not only are these two strands not opposed, but also they complement each other and can dialogue. Whether the current European project is doomed to "success" or "failure" changes nothing. Investigating contemporary political identities and identification with Europe as an ongoing process whose origins are lost in the mists of time involves taking an interest in it both in the long term and here and now, both from above, from the macrosocial point of view of the political and economic structures that frame this process, and from below, from the point of view of the individuals who give life to it (Delmotte, Mercenier, and Van Ingelgom 2017). The qualitative shift in the sociology of Europe tends to take seriously the voice of the actors, their feelings, and their life experiences. This gesture is perfectly compatible with the ambition to contribute to the development of a sociology that is both "reality congruent" and emancipatory. And Elias is particularly rich in hypotheses to be tested on the "changes of the 'we–I' balance" which concern Europeans, among others.

Conclusion

Benjamin had always assumed that he would grow old and die at home; that he was bound to end his life by returning to the country of his childhood. But he was starting to understand, at last, that this place had only ever existed in his imagination.
Adieu to old England, adieu
And adieu to some hundreds of pounds
If the world had been ended when I had been young
My sorrows I'd never known (Coe 2019, 419)

We began with the image of national flags hanging on the facades of houses in European cities during the first months of a pandemic, which, in the spring of 2020, had paralyzed part of the world economy and banned half of the world's population from going out (https://www.euronews.com/2020/04/02/coronavirus-in-europe-spain-s-death-toll-hits-10-000-after-record-950-new-deaths-in-24-hou). Some did not hesitate to speak of a "war" against an invisible enemy, a virus, on a planetary scale. We pointed out the apparent paradox of displaying the nation's colors as a sign of resistance to a scourge that knows no borders. While situated as such an image may seem, we felt that it nonetheless evoked the strength and relevance of some identity attachments which may seem out of date or totally irrelevant. We chose the episode of the Coronavirus pandemic as a starting point, namely an object not yet identified by the social sciences. We might as well have begun by mentioning the process, completed in 2020, of the United Kingdom's withdrawal from the European Union, which it joined in 1973. The role played in the so-called Brexit by identities, habitus, affiliations, and identificatory affects is undeniable. Historical sociology of Eliasian inspiration makes it possible, in the end, to realize that these are never insignificant and that it is important to explain them and to better understand them. For doing so, this approach offers a number of assets that set it apart from many others that have taken an interest in identities and identification phenomena – a central object for the human sciences since the 1960s, which has become a veritable political obsession. To conclude, we recall one last time what these most characteristic assets consist of.

First of all, it is a matter of restoring the historical dimension of identity phenomena, and often of placing them in the long term. Identities, habitus, almost always have a long history and evolve slowly. Individuals and groups inevitably inherit, in their own defense, the stories lived by the generations that preceded them. The identities specific to each individual and each group are thus composed of multiple layers and based in part on the oblivion of these. The work of a historical sociology may consist, above all, in finding their origins and uncovering the process of their sedimentation. This approach thus avoids "reifying" identities, all the more so as it reminds us that any identity, any identification process, first of all involves human beings in relationships at different levels, from the most intimate to the most global, and in different temporalities. In doing so, it also takes into account the fact that identification, of oneself and of the group, always takes place in relation to others. Finally, it recalls the importance of power ratios, of the affective dimension, and of the evolution of these two aspects, for the study of the processes of identification.

Such a perspective disenchants identities. But it is not limited to denouncing their potentially murderous character or their sectarian tendencies, nor to deploring the blindness of those who remain prisoners of their identity(ies) while waiting for science to free them. Rather, the historical sociology we are interested in takes seriously the reasons for our attachments and helps us to look at them with an empathic and critical eye.

References

Anderson B (1983) Imagined communities: reflections on the origins and spread of nationalism. Verso, London
Bartolini S (2005) Restructuring Europe: centre formation, system building and political structuring between the nation state and the European Union. Oxford University Press, Oxford
Belot C (2010) Le tournant identitaire des études consacrées aux attitudes à l'égard de l'Europe: Genèse, apports, limites. Polit Eur 30(1):17–44
Belot C, Bouillaud C (2008) Vers une communauté européenne de citoyens: pour une approche par les sentiments. Polit Eur 26(3):5–29
Berger PL, Luckmann T (1967) The social construction of reality: a treatise in the sociology of knowledge. Penguin, London
Billig M (1995) Banal nationalism. Sage, London
Bourdieu P (1984) Distinction: a social critique of the judgment of taste. Routledge, London

Bourdieu P (2012) Sur l'État: cours au Collège de France 1989–1992. Raisons d'Agir/Le Seuil, Paris
Brubaker R, Cooper F (2000) Beyond "identity". Theory Soc 29:1–47
Bruter M (2005) Citizens of Europe? The emergence of a mass European identity. Palgrave Macmillan, Basingstoke and New York
Bucholc M, Mennell, S (2022) The past and the future of historical sociology: an introduction. In: McCallum D (ed) The Palgrave handbook of the history of human sciences. Palgrave Macmillan
Calhoun C (1994) Preface. In: Calhoun C (ed) Social theory and the politics of identity. Blackwell, Oxford, pp 1–7
Calhoun C (1997) Nationalism. University of Minnesota Press, Minneapolis
Checkel JT, Katzenstein PJ (2009) The politicization of European identities. In: Checkel JT, Katzenstein PJ (eds) European identity. Cambridge University Press, Cambridge
Coe J (2019) Middle England. Penguin, London
De Swaan A (1995) Widening circles of identification: emotional concerns in sociogenetic perspective. Theory Cult Soc 12:25–39
Delanty G (2003) Europe and the idea of "Unity in diversity". In: Lindahl R (ed) Whiter Europe: borders, boundaries, frontiers in a changing world. Cergu, Gothenburg, pp 25–39
Delmotte F (2012) About post-national integration in Norbert Elias's work: towards a sociohistorical approach. Hum Fig 1(2). http://hdl.handle.net/2027/spo.11217607.0001.209. Accessed 30 June 2020
Delmotte F, Mercenier H, Van Ingelgom V (2017) Belonging and indifference to Europe: a study of young people in Brussels. Hist Soc Res 42(4):227–248
Déloye Y (2003) Sociologie historique du politique. La Découverte, Paris
Déloye Y (2006) Éléments pour une approche socio-historique de la construction européenne: un premier état des lieux. Polit Eur 18(1):5–15
Déloye Y (2010) En-deçà de l'identité ou le miroir brisé de l'identification. In: Martin D-C (ed) L'identité en jeux: pouvoirs, identifications, mobilisations. Karthala, Paris, pp 403–418
Duchesne S (2008) Waiting for a European identity… reflections on the process of identification with Europe. Perspect Eur Polit Soc 9(4):397–410
Duez D (2014) A community of borders, borders of the community: the EU's integrated border management strategy. In: Vallet E (ed) Borders, fences and walls: state of insecurity? Ashgate, Farnham, pp 51–66
Dunning E, Mennell S (1996) Preface. In: Elias N (ed) The Germans (1st English ed). Polity Press, Cambridge, pp vii–xvi
Elias N (2007) Involvement and detachment. UCD Press, Dublin [Collected works, vol 8]
Elias N (2009a) The retreat of the sociologists into the present. In: Essays III: on sociology and the humanities. UCD Press, Dublin, pp 107–126 [Collected works, vol 16]
Elias N (2009b) Foreword to "Women torn two ways". In: Essays III: on sociology and the humanities. UCD Press, Dublin, pp 270–276 [Collected works, vol 16]
Elias N (2010 [1987]) Changes in the "we–I" balance. In: The society of individuals. UCD Press, Dublin, pp 137–208 [Collected works, vol 10]
Elias N (2011) The symbol theory. UCD Press, Dublin [Collected works, vol 13]
Elias N (2012a [1939]) On the process of civilisation: sociogenetic and psychogenetic investigations. UCD Press, Dublin [Collected works, vol 3]
Elias N (2012b) What is sociology? UCD Press, Dublin [Collected works, vol 5]
Elias N (2013a [1989]) Studies on the Germans: power struggles and the development of habitus in the nineteenth and twentieth centuries. UCD Press, Dublin [Collected works, vol 11]
Elias N (2013b) Interviews and autobiographical reflections. UCD Press, Dublin [Collected works, vol 17]
Elias N, Scotson JL (2008 [1965]) The established and the outsiders. UCD Press, Dublin [Collected works, vol 4]

Erikson E (1968) Identity: youth and crisis. W.W. Norton, New York
Ernst S, Weischer C, Alikhani B (eds) (2017) Changing power relations and the drag effects of habitus (special issue). Hist Soc Res 42(4)
Favell A, Guiraudon V (2009) The sociology of the European Union: an agenda. Eur Union Polit 10(4):550–576
Ferry J-M (1992) Pertinence du postnational. In: Dewandre N, Lenoble J (eds) L'Europe au soir du siècle. Esprit, Paris, pp 39–58
Ferry J-M (2005) Europe, la voie kantienne: essai sur l'identité post-nationale. Cerf, Paris
Foucault M (2007) Security, territory, population: lectures at the College of France 1977–1978. Palgrave Macmillan, New York
Gleason P (1983) Identifying identity: a semantic history. J Am Hist 69(4):910–931
Goffman E (1959) The presentation of self in everyday life. Anchor, New York
Goffman E (1963) Stigma. Penguin, London
Habermas J (1998) The inclusion of the other: studies in political theory. MIT Press, Cambridge, MA
Habermas J (2001) The postnational constellation: political essays. MIT Press, Cambridge, MA
Hall S (1996) Introduction: who needs "identity"? In: Hall S, du Gay P (eds) Questions of cultural identity. Sage, London, pp 1–17
Hobsbawm EJ (1990) Nations and nationalism since 1780. Cambridge University Press, Cambridge
Inglehart R (1970) Cognitive mobilization and European identity. Comp Polit 3(1):45–70
Kluckhorn C, Murray HA (eds) (1948) Personality in nature, society and culture. Knopf, New York
Kuzmics H, Reicher D, Hughes J (eds) (2020) Emotion, authority, and national character: historical-processual perspectives (special issue). Hist Soc Res 42(4)
Lacassagne A (2008) Une reconstruction éliasienne de la théorie d'Alexander Wendt: pour une approche relationniste de la politique internationale. PhD Thesis, Institut d'études politiques de Bordeaux
Liston K (2018) Norbert Elias, figurational sociology and feminisms. In: Mansfield L, Caudwell J, Wheaton B, Watson B (eds) The Palgrave handbook of feminism and sport, leisure and physical education. Palgrave Macmillan, New York, pp 357–373
Mackenzie WJM (1978) Political identity. St. Martin's Press, New York
Majastre C, Mercenier H (2016) Construire un espace français de recherches sociologiques sur l'Europe: un tournant, trois versions? Polit Eur 2:8–31
Marshall TH (1950) Citizenship and social class, and other essays. Cambridge University Press, Cambridge
Martin D-C (1994) Introduction. Identités et politique: récit, mythe et idéologie. In: Martin D-C (ed) Cartes d'identités: comment dit-on "nous" en politique? Presses de la Fondation nationale des sciences politiques, Paris, pp 13–38
Martin D-C (2010) Écarts d'identité, comment dire l'Autre en politique? In: Martin D-C (ed) L'identité en jeux: pouvoirs, identifications, mobilisations. Karthala, Paris, pp 13–134
Marx K (1852) The Eighteenth Brumaire of Louis Bonaparte. https://www.marxists.org/archive/marx/works/download/pdf/18th-Brumaire.pdf. Accessed 26 May 2020
Mauss M (1950 [1935]) Les techniques du corps. In: Sociologie et anthropologie. Presses universitaires de France, Paris. Transl: Techniques of the body https://monoskop.org/images/c/c4/Mauss_Marcel_1935_1973_Techniques_of_the_Body.pdf. Accessed 30 June 2020
Mead GH (2015 [1934]) Mind, self and society. University of Chicago Press, Chicago
Melucci A (1995) The process of collective identity. In: Johnston H, Klandermans B (eds) Social movements and culture. University of Minnesota Press, Minneapolis, pp 41–63
Mennell S (1994) The formation of we-images: a process theory. In: Calhoun C (ed) Social theory and the politics of identity. Blackwell, Oxford, pp 175–197
Mennell S (2007) The American civilizing process. Polity, Cambridge
Mennell S (2021) Some political implications of sociology from an Eliasian point of view. In: Delmotte F, Górnicka B (eds) Norbert Elias in troubled times: figurational approaches to the problems of the twenty-first century. Palgrave Macmillan, New York, pp 335–352

Merton RK (1957) Social theory and social structure. Free Press, New York
Noiriel G (1991) La tyrannie du national: le droit d'asile en Europe (1793–1993). Calmann-Lévy, Paris
Noiriel G (ed) (2007) L'identification: genèse d'un travail d'État. Belin, Paris
Ricœur P (1990) Soi-même comme un autre. Le Seuil, Paris
Shore C (2006) "In uno plures" (?) EU cultural policy and the governance of Europe. Cult Anal 5: 7–26
Simmel G (2009 [1908]) Sociology: enquiries into the construction of social forms. Brill, Boston
Stinchcombe AL (1975) Social structures and politics. In: Polsby NW, Greenstein F (eds) Handbook of political science, vol 3. Addison-Wesley, Reading, pp 557–622
Surdez M, Voegtli M, Voutat B (2009) Introduction: à propos des identités politiques. In: Surdez M, Voegtli M, Voutat B (eds) Identifier – s'identifier. Éditions Antipodes, Lausanne, pp 9–45
Szczerbiak A, Taggart P (eds) (2008) Opposing Europe?, 2 vols. Oxford University Press, Oxford
Tajfel H (1974) Social identity and intergroup behaviour. Soc Sci Inf 13:65–93
Tilly C (2003) Political identities in changing polities. Soc Res 70(2):605–620
Van Ingelgom V (2014) Integrating indifference: a comparative, qualitative and quantitative approach to the legitimacy of the European integration. ECPR Press, Colchester
Weber M (1978 [1921]) Economy and society, 2 vols. University of California Press, Berkeley
White J (2011) Political allegiance after European integration. Palgrave Macmillan, Basingstoke

29. Hidden Gender Orders: Socio-historical Dynamics of Power and Inequality Between the Sexes

Stefanie Ernst

Contents

Introduction	750
The Sociogenetic and Psychogenetic Approach Toward Social History	752
Central Terms	754
The Dawning of Social-Historical Gender Research and the Power Balances Between the Sexes	756
Power Relations Between the Sexes in Marriage and Working Life	760
Marriage	761
Work and Professional Life	765
Conclusion: Changing Gender and Relationship Ideals?	768
References	769

Abstract

The chapter aims to reconstruct the socio-historical dynamics of power and inequality between the sexes in order to understand the enduring long-term development of gender inequality in private and professional life. These developments include ambivalences that express power struggles that are mostly hidden and very often romanticized as "natural" difference, such as that women are the "better angels of our nature," that they civilize and pacify society. On the surface the #MeToo campaign against sexism drastically demonstrated these problems of inequal power relations between the sexes. In focusing the deep, ambivalent and dynamic structure behind relevant emancipation gains, such constructions can be explained in detail. On the one hand we find official agreements of solidarity, gender equality, and the acceptance of women's rights; on the other, in practice a fragile hidden gender order of violence, power, and oppression throughout history up to the present is relevant. Pinker especially analyses the intensity of violence throughout history in a fundamental,

S. Ernst (✉)
Institut für Soziologie, University of Münster, Münster, Germany
e-mail: stefanie.ernst@uni-muenster.de

© The Author(s), under exclusive licence to Springer Nature Singapore Pte Ltd. 2022
D. McCallum (ed.), *The Palgrave Handbook of the History of Human Sciences*,
https://doi.org/10.1007/978-981-16-7255-2_52

evolutionary perspective, stating that in Western societies in the long run we can observe a decline of violence. Women seem to play a central and specific part within this process, which is worth reconstructing, and also comes into mind in the face of renewed stereotypes on women's place in history and the present romanticization of the nuclear family in the course of the Covid-19 pandemic. These ambivalent behavioral ideals and standards are expressed mostly in contemporary and historical guidance literature. Here, the genesis of the private space illustrated with marriage and family life on the one hand, and the public world of work on the other, shows that the balance of power and degrees of social formalization and informalization also affect the dynamic power relations between the sexes. Moreover, this perspective delivers an overdue synthesis about these gender arenas and helps in understanding the embeddedness of gender, work, and private life and its interdependencies.

Keywords

Historical sociology · Gender studies · Women's studies · Discrimination · Inequality · Marriage

And this arrangement of rooms ... is thus an expression of the *co-existence of constant spatial proximity and constant social distance, of intimate contact in one stratum and the strictest aloofness in the other.* (Norbert Elias, *The Court Society* (2006 [1983]: 53)

Introduction

Before the current Covid-19-pandemic threw us out of secure figurations of space and time to an unknown physically distanced, more or less digital life, hardly any current debate like the #MeToo-Campaign had made so drastically clear that patterns of discussion and practices of sanctioning violence, power and sex abuse, and crimes have changed due to social transformation processes over the last 30 years. This ongoing latent or open gender struggle relatively starkly affects the gains and antifeminist countermovements in professional and private gender equality. It not only indicates that sudden backlashes (for example, worldwide epidemics) are possible under the surface of civilized behavior at any time if the "veneer of controls fails." Moreover, "the fragile nature of normative codes" still needs to be reasserted and reformulated (Delmotte 2019: 2). The Covid-19 crisis, which of course will be analyzed more precisely by future socio-historical researchers, is an obvious example for hidden gender norms in family and business life, as well as in society in general, dividing social spheres into the so-called "system-relevant" ones and "irrelevant" ones. Home-schooling and home office work were taken for granted as well as (by tradition) housework. Conflicts of blurred boundaries between work and home life increased, the traditional "arrangements between the sexes" came to a limit and the risk of violence in families increased too, whereas others reported advantages in efficiency through co-working spaces. These new private arrangements are omnipresent but hidden and preconditioned. On the one hand, the Covid-19-pandemic makes these ambivalences drastically obvious, while on the other its management impressively showed an increased civilized awareness of humanity and solidarity with weaker social groups. But, between official agreements of solidarity, gender equality, and the acceptance of women's rights in practice, a fragile hidden gender order of violence, power, and oppression throughout history comes to play too.

Furthermore, the intensity of violence throughout history is analyzed in a fundamental, evolutionary perspective stating that in Western societies in the long run we can observe a decline of violence. Women seem to play a central and specific part within this process that is worth reconstructing; and it also comes into mind in the face of renewed stereotypes about women's place in history and the present romanticization of the nuclear family. Thanks to marriage and family as social institutions, women as the "better angels of our nature" (Pinker 2011) seem to have civilized and "domesticated male violence in the preceding decades" (Pinker 2013: 5); one really wonders about this normative as well as naïve construction. This issue is not very simple, because western gender orders show big social, national and historical varieties in structured and ambivalent back and forth movements. Women's and Gender Studies impressively show that for centuries it was mostly women who had to pay the price with extreme self-control and subordination in marriage and public life, for this very pre-conditional view of (male) civilization processes (Tilly 1989; Scott 1986; Ernst 1996). Life for educated and non-educated men and women was very different in many ways, as not everybody had the right to marry. The construction of women as "peaceful angels of civilisation" neglects their story with its ups and downs of increasing and decreasing power. Moreover, gender relations have become flexible throughout the last 40 years when traditional patriarchal gender orders eroded (Ernst 2019c; Walby 1997; McRobbie 2009). Therefore, the shock of the sudden Covid-19 health crisis seems at first sight to have been managed in favor of the traditional model of (male) production and (female) reproduction we believed had been overcome. But how do these reproductive and productive arenas come to exist and why are they primarily interwoven with the way how public and private life is organized?

Research on gender relations, family, and business life are unfortunately often discussed separately, ahistorically, and even organized in special research sections that ignore each other with boundaries well maintained. A synthesis of all of these arenas is therefore overdue and useful as it helps to understand the embeddedness of gender, work, and private life and their interdependencies.

Against this background, one can unfold the social history of the Fordist and post-Fordist working society as well as family or private life as a powerful basis of gender relations. Moreover, in what follows it will be demonstrated that when dealt with separately these arenas imply a specific inner order, a specific socio- and psychogenetic development that can be understood with a social-historical, comparative method that analyses changing patterns of behavioral codes. Over centuries, each arena – family as well as business life – has been dealt with intensively in the so-called etiquette books, which is to say in explicitly normative literature. So, to adequately contextualize the assumptions implied in the quotation at the beginning, it is useful to select a comparative-historical perspective on long-term developments of contemporary gender relations. Furthermore, in seeking a higher scale of integration, the socio- and psychogenetic approach helps to realize the complexity and interconnectedness of the emotional dynamics between the sexes. Especially for the development of a power theory of gender relations, an interdisciplinary, historical-sociological view will be demonstrated below. Its potential lies in the analysis of sociogenetic and psychogenetic dynamic developments and their inner contradictions. Together with the gender issue it helps to understand the "simultaneity of continuity and change" as well as "capturing the gender inequality order beyond simple attributions of positions of power and powerlessness" (Meuser 2010: 158).

As much potential as the figurational concept of power actually offers for the analysis of gender relations, its systematic application remained ambivalent and hesitant for a long time. In some cases, process theory solidified even more stereotypical ideas. In this context, the debates about power and violence, standards of sexual behavior as well as the theory of formalization and informalization processes play a central role. The arenas of marriage and family life, as in hardly any other area, demonstrate the mutual dependence of the sexes in matters of reproduction. But especially in contemporary Western societies the spheres of work, occupation, and education should be considered. First, therefore, there is a brief summary of the integrative potential of a socio-historical and empirical perspective for Gender Studies, despite the reservations of women's and gender research toward a processual and dynamic concept of gender relations (Ernst 2019c). Then, central terms like power, informalization, and social change will be reiterated in order to extrapolate some central elements for an integrative, social-historic gender approach, before this central work is applied to selected research examples of changing gender relations. Especially private life (marriage and family) as well as professional working life will be reconstructed through the centuries concentrating on Germany and the changing power relations between the sexes.

The Sociogenetic and Psychogenetic Approach Toward Social History

While the concept of process theory has become established as a macro-sociological label, the sociological term *figuration* emphasizes in particular group-related, dynamic networks of mutually dependent individuals, with their own logic and constraints. This approach also became popular with the social upheavals at the end of the 1960s, but the sociological-historical perspective succeeded most in the context of the increased attention of sociologists to historical science since the 1980s. In contrast to many "modernisation theories" that rely on ever more complex differentiation processes, this perspective is not only not evolutionary. Rather, with the rise of the recognized classic of sociology, as it was in the third generation of figurationists, a normalization of process-theoretical thinking has taken place in which traditional reception barriers have been overcome (Ernst 2015).

Nevertheless, nowadays, "in sociology, the range of theories suitable for the explanation of contemporary societal transformation processes or problems is relatively limited" (Ernst et al. 2017: 8). This certainly has to do with overspecialization and the "retreat of sociologist to the present" (Elias 2009a [1987]). However, the theoretical approach of process-theory being emancipated from the classical philosophical tradition of sociology helps to connect structural macrostructures and individual scopes of action. The term of social habitus and drag effects creates an understanding for the genesis of contemporary conflicts as well as for the hidden, ambivalent medium- and long-term transformation processes with its forward and backward movements. Then, by contextualizing empirical data in a socioeconomic and socio-psychogenetic perspective, long-term development and backgrounds of contemporary conflicts can be explained. Especially etiquette books (nowadays substituted by modern counseling books) offer insights into changing figuration *ideals*. A wider set of empirical data including quantitative and qualitative date has been collected over years as important process-produced data (Baur and Ernst 2011: 118). Also keeping the sociology of knowledge in mind, it is obvious that historical and social-historical research have to reflect their speaker's position and the social embeddedness of its interest in knowledge (Elias 2007). Therefore, each epoch has its typical data types, for example, etiquette books, medieval stud books, ego-documents, or official documents (Ernst 2019b). Picking popular etiquette books, one might argue that publishers can only afford long print runs of copies if there is a demand for them. A high demand for books in turn is a sign that the topic and the way it is discussed within the book are important for the figuration. Analyzing these widely circulated documents thus enables researchers to reconstruct issues central to the figuration at the time (Baur and Ernst 2011: 133).

A comparative analysis of typical or untypical topics in manners books about marriage, business, or sex informs us about specific contemporary gender and power relations in family or work life, etc. Through this empirical, historical-comparative approach, one recognizes, for example, that phenomena such as individualization were already visible in the Renaissance and relaxation of behavioral standards have not only been localized since the student protests of the 1960s (Burke 1989: 149 f.). Rather, a relational and comparative perspective is opened, so that gradual, unspectacular advancements and forward and backward movements of social change can be described. Referring to Treibel (2008) and my own approaches to organizational and work sociology (Ernst 2010) a complex process- and figurational analysis without anticipatory dogmatism (Elias 2009d [1977]) should identify the figuration as a whole with its (also nationally shaped) norms and rules, and analyze social tensions and power struggles. Furthermore, the formalization and informalization range as well as the change of self-regulation in relation to the means of satisfying elementary physical and social needs as well as the respective central control and communication means of society with its different national orders should be shown. To understand the almost blind game of long-term dominant trends and counter-trends, changing power balances and functional equivalents have to be studied (Elias ibid.; Treibel 2008; Ernst 2010: 77). These social developments are visible in many directions, which were not exactly planned by individuals, but show a recognizable pattern retrospectively. Capturing this complex polyphony by analyzing the balance of power we can realize permanent instead of disruptive negotiation processes between stronger and weaker social groups.

As a reaction to the advance of weaker groups into positions that were previously closed to them, the varieties and social differences increased. Increased efforts to distinguish established groups from groups that are moving upward in turn evoke new social dynamics of change. While historians discovered social history with delay, conversely historians also found understandable "reactions against statistics, against determinism and against functionalism" (Burke 1989: 153) in contemporary sociology, thus further separating the disciplines. These neighboring disciplines need each other more than ever, because both "require the comparative method" (ibid.: 40). The now renowned social history and historical sociology are equally interested in "understanding from the inside and explaining from the outside, for the general and for the special" (ibid.: 35). Illuminating this inner life and the outside view of the socio- and psychogenetic development of western European societies from the ninth to the early nineteenth century, the specific character of the last pre-bourgeois figurations of the occident becomes obvious: in doing so we understand contemporary professional and commercial life and its transformation (Ernst and Becke 2019). This epoch still had an impact on bourgeois industrialism, but also differed significantly from it. Because, for example, family ties still played a central role at court, in contrast to the time after the Enlightenment: "Family ties and rivalries, personal friendships and enmities were normal components of governmental and of all official affairs" (Elias 2006[1983]: 3). Incidentally, in some parts this tendency can currently be observed again within some autocratic western regimes. Current processes of change, in comparison can be signed more informalized. These also include the power relations between the sexes (Elias 2009b), an approach that should be synthesized more systematically.

Central Terms

If one chooses the perspective of interdependencies, one can reconstruct the socio- and psychogenetic dimensions of gender relations from a higher level. Thus, the central question of the sociality of the individual, which is in constant engagement with social norms of behavior and his or her individual desires, that is, having to balance one's own private pleasure and discomfort, can be answered in a more adequate way. However, a further perspective can be worked out emphasizing long-term spurts of informalization and formalization of emotional and behavioral standards. Specific emotional cultures and love semantics can be observed in connection with spatial sociological perspectives, for example, in the courtly or semipublic bourgeois salon, in company or in the private room (Dibie 1994; Wouters and Dunning 2019; Ernst 2019a). In the bourgeois age, a new social class was formed, which successfully struggled for political emancipation and participation, but now formally includes men and women of all classes to different extents.

This shifting power balance also applies to the leading upper classes and the controlled lower classes. The questioning of traditional behavioral standards, such as a strong sense of shame and embarrassment toward physical and emotional expressions of pleasure and displeasure, strict table manners, and an excessive hierarchical approach at work also forms part of these revolutionary upheavals. This transformation process is widely supported, encompasses all layers, and has been particularly evident since the 1960s. But the strict Wilhelmine or Victorian (Gay 1986) rules and regulations for men and women, adults and children, superiors and subordinates also loosened up in the "Roaring Twenties," but without completely shattering the level of civilization once achieved. This "controlled decontrolling of controls" (Wouters 2007) is ambivalent and can only be more reliable at a high level, for example, in social self-regulation that is provided in the form of legislation. Women who were previously strictly regulated and relatively outside the law experienced an increase in power that had forced men to exercise more self-restraint. Hierarchical gender relations became an outdated ideal and were transformed into arenas of permanent negotiation and self-regulation. The modeling of external and self-constraints plays a special role for the dynamic emotional balance of men and women in the different arenas too. Moreover, in today's complex societies we face an ambivalence of informalization processes, "diminishing power differences incite informalisation because a decline in social distance does indeed prompt people towards greater informality, but this does not mean that all informalisation 'is' or 'expresses' democratisation" (Wouters and Mennell 2015: 10).

For the description of informalization processes, Wouters (2014) starts from striking levels of the civilization process in order to establish control over oneself, over others, and over nature, the mutual reduction of social contrasts while increasing variety. The degree of a relaxed or more strict relationship with each other informs us about the relationship between informalization and formalization. This becomes clear if one looks at forms of professional cooperation or competition, sexuality and feelings of pleasure, mutual empathy and distance between established and outsiders, women and men, old and young, parents and children, managers and employees, and so on. Here, the relationship between public and private behavioral ideals is particularly clear but should be clarified for the affected groups. For example, besides all assumptions the 1960s sexual revolution brought relatively more advantages for men than for women as men by tradition had more sexual freedom (Ernst 1996: 160 ff).

As it is hoped to show in this chapter, contemporary conflicts between established groups and social climbers cumulating in recent emotional debates on discrimination, sexisms, racisms, and so on explicate a hidden order: this means that groups are dependent on formalization processes in law (EU-Charta Diversity, Anti-Discrimination Law – see https://www.antidiskriminierungsstelle.de/DE/Home/home_node.html; https://ec.europa.eu/info/policies/justice-and-fundamental-rights/combatting-discrimination/tackling-discrimination/diversity-management/diversity-charters-eu-country_en) as well as on informalization processes like networks to gain more rights, democratic participation, equality, and acceptance. This "structural insecurity" of human relationships goes hand in hand with growing individualization processes of previously less powerful groups. For example, educational and love ideals have shifted from parenting in the course of nation building to a growing focus on child welfare since the pedagogical age (Waterstradt 2015; Fertig 1984), whereas in court society the "mesh of direct relationships was tighter in his case, the social contacts more numerous, the *immediate* ties to society greater than for the professional or business bourgeois for whom contacts *mediated* through profession, money or goods have overwhelming priority" (Elias 2006 [1983]: 65).

Individualization processes as part of these arenas have particularly influenced women's liberation, work, and family life. In this context, civilization means that increasingly male, often physically and economically practiced, hegemony is being questioned or sanctioned. On the other hand, the multilayered loosening of strict standards of behavior (informalization) successfully (re-)negotiates gender relations until they can finally become part of an advanced legal awareness and a new legal practice (for example, in gender equality policy). These open-ended negotiation processes unsettle structural, traditional gender orders in public and private space, which, like in the present case, has gradually made women in particular legal entities since the bourgeois age. This process is dynamic, fluid, and is still far from complete; rather, the balance of power between the sexes is constantly being negotiated, because power relationships fluctuate. It depends on the interdependent relevance people have for each other in different social arenas:

The Dawning of Social-Historical Gender Research and the Power Balances Between the Sexes

By now it should have become clear that these powerful macro-sociological processes do not run linearly, but mostly lead to a reconfiguration of self and external constraints in advance and retraction. However, speaking of social strength relationships instead of rule helps to focus the dynamic element of changing power relationships. Gender relations therefore are more diverse and a specific field worth to be analyzed in their own socio-historical perspective. However, it is a misconception that power is only ever unilaterally perceived as superior or inferior and that people can live like a *homo clausus* isolated from social structures. Though, on the one hand the gender struggle differs from other conflicts as women's struggle for social equality had started from a relatively outsider position being excluded or marginalized from central resources of power. On the other hand, men and women are very dependent on each other as individuals as well as social groups (Elias 2009c [1983]: 272). This comparatively early reflection of gender inequality in the history of disciplines was initially unlikely for the first women researchers in the 1970s, since they accused the "theory of civilization" of male bias by developing an androcentric understanding of the state and society, as a part of general criticism of scientific androcentrism (Gravenhorst 1996; Ernst 2019c). For decades, early women's history studies in many countries had not only been completely ignored by institutionalized history (Bock 1988; Davies 1976) but also were only understood as a part of other histories "adding material on women and gender without analysing its implications" (Tilly 1989: 439). Moreover, Tilly criticizes Women's Studies for having for a long time followed too narrow a perspective on women by only including perspectives and standardized biographies of middle- and upper-class women, which were then generalized into a shared perspective on women (Tilly 1989: 443ff.). By the mid-1980s, the perspective of white working-class women was acknowledged and incorporated into the bigger (US-American) research picture more and more, even implying intersectional approaches. Therefore, the relational idea of fluid power balances between the sexes was not particularly attractive for the *first generation* of women researchers who struggled for autonomy and were committed to feminism and often to Marxism too.

Meanwhile gender research also differentiated itself. But while one direction is focusing on discursive, interactive, or symbolic gender relations in present-day society from a deconstructive, empirical, or micro-sociological way, socio-historical gender studies ask for the perception and configuration of gender at different times and places. Central to this approach, gender norms and behavioral codes throughout time are also analyzed. Whereas, for example, Hausen (1976) famously developed the concept of *gender character* in describing the binary essentialization of men's and women's abilities and social roles in the development of bourgeois German society, Frevert (1995, 2011) focuses on the emotional side of historical and modern gender differences and Gerhard (1981) already from the beginning reconstructs legal relationships and practice in gender history. Thanks to the social emancipation processes since the 1960s and current debates about the social construction of sex and gender, the introduction of gender as a social and historical category then relieved early women's history research of its one-dimensionality and normativity, and broadened its perspective with an analytical and theoretical research base (Scott 1986: 1054 f.).

Connell especially (1995; Connell and Messerschmidt 2005) impressed with a convincing study about hegemonic masculinity in the Western world, delineating the *patriarchal dividend* and the interconnected forms of marginalized and marginalized masculinity. For American gender history, Brown (1993) asks for studies of the role women played in the Western New World since the eighteenth century, whereas Hagemann and Harsch (2018) look toward the youngest gender history of European societies. Following Scott's approach, there are three different theoretical positions of working with gender in the research field of history: 1) Gender is used "to explain the origins of patriarchy" (ibid.: 1057), which is considered a feminist perspective; 2) gender as an analysis category "locates itself within a Marxian tradition and seeks there an accommodation with feminist critiques" (ibid.); 3) gender is utilized "to explain the production and reproduction of the subject's gendered identity" (ibid.: 1058). This briefly summarized variety of socio-historical gender studies has led to many contributions but the need for a special issue that focuses on the social-history of the "civilisation of the female ego" (Klein and Liebsch 1997; *author's translation*) is still given. In this context, the emerged social-historical figurational gender studies focus informalization and formalization processes, changing behavior codes as well as the nuances in the "power relations between the sexes" (Ernst 1996) in several fields such as politics, organizations, education, sexuality, etc.

In a socio-historical perspective, power is not set absolutely, or just a medium that operates in the political system. Rather, it can be seen as a structural peculiarity of all human and thus public as well as private relationships that are changeable (Ernst 1996). Power relations are subject to a dynamic that power is not unilaterally assigned to men as the established, or to social subsystems such as politics, while outsiders or weaker groups such as women have no power potential at all. By paying attention to the mutual, instead of one-sided, fundamental dependency of people, process and figuration theory deals with a topic relevant to women's and gender studies, which has often been criticized by the particularly relational perspective: the structures of power and dependency. Because supposedly superior powers also depend on weaker powers, the gradual transformation of close-meshed external constraints into self-constraints is dynamic and accompanied by "numerous setbacks" (Treibel 2009: 143; *author's translation*). Controversial issues (such as #MeToo and sexism) imply a double-bind effect that is remarkable in terms of knowledge and sociology (Elias 2007). Power games have become more complex and can be observed with new moves and players (Elias 2012a [1978]). The shades and nuances of relatively successful, conflicting emancipation processes of disadvantaged outsiders, seen in the long term, tend to be less focused. In order to illustrate precisely this attractiveness and synergies of the process sociology in the context of gender studies, the following is therefore considered using the example of power and violence, the pushes of informalization and formalization as well as equality in private and public spheres of the sexes. They show shifts and advances between the sexes and social life situations. Therefore, it is necessary to examine the Christian marriage system since the Middle Ages in order to follow the gendered history of violence, power, and state building surrounding the most successful social institution of society in a socio-and psychogenetic perspective. While in feudalism, social superiority was primarily based on physical strength and resistance, in industrial society the pacified economy became of the most important means of production and the state monopoly on violence became increasingly effective. At least on the normative level, physical force lost weight because it was sanctioned by the legislature, so that legal awareness, if not immediate legal practice, gradually changed (Ernst 1996: 18, 32). This development is accompanied by complex social differentiation of different areas of power with longer mediating chains. Therefore, the long-term development of gender relations is characterized not by a gradual degradation or progress, but by a peculiar up and down pointing to "several spurts towards a lessening of the social inequalities between women and men – mostly within single social strata and, maybe, with simultaneous or subsequent counter-spurts." (Elias 2009d: 243) Power as a structural peculiarity of social relationships not only affects gender relationships centrally (Ernst 1996; Treibel 1997, 2012; Hammer 1997; Klein and Liebsch 2001). It can be located in many places and is not attributed to either side. In the long run, however, the history of European civilization has long shown a clear primacy of men. A power balance means that in social reality it is not an absolute relationship between people, but always a relation between interdependent groups.

Through this relational focus, one can well see that an "utter subjection of women" cannot always be assumed (Elias 2009b: 243). For example, the position of the courtly lady under French absolutism was more or less relatively powerful (Opitz 1997). Sports and gender sociology in this context (Pfister 1997; Liston 2005) also identify a gender-stereotypical construct of bundling male aggressiveness and drive channelization in sports, which, in addition to the initial "angels of our nature" (Pinker 2011), indicates a similar risk of essentialization. Not only has the one-sidedness in the concept of (male) aggressiveness evoked criticism, but also the connection with a narrow concept of socialization processes as one-sided conditioning process in the transformation of external into all-encompassing, unconscious self-constraints (Klein and Liebsch 1997).

Previously to Connell's concept of hegemonic masculinity, Dutch social-historical researchers discovered (Van Stolk and Wouters 1987b) a specific double-bind of "harmonious inequality" pointing to traditional patriarchal habitus and female inferiority power differences maintained in partnerships. They asked (Wouters 1986, 2004, 2007; Wouters and Dunning 2019) about the inner, psycho-genetic and sociogenetic order of sexual desire, the difficult balance of feelings of pleasure and discomfort and the contradictory change in behavioral ideals since the 1960s. While the psychosocial and sexual dimension of gender relations as well as the politicization of gender in public space concerning sexual harassment and feelings of shame in the comparatively informal Dutch society were studied (Brinkgreve 1999; Brinkgreve et al. 2014; Dekker 2019), in Germany innovative developments in reaching for new levels of gender as power relationships were done with the "Civilisation history of dance" (Klein 1992; *author's translation*), and "Reflections on the scientific world" (Treibel 1993; *author's translation*). The power balance concept has been applied to marriage and family (Ernst 1996; Kunze 2005), gender equality policy (Treibel 1997), gender, gossip, and leadership (Ernst 1999, 2003). Then, in *Sex and Manners* (Wouters 2004) the international comparative analysis of figurational gender ideals and female emancipation in the West followed. With this and the next generation of process theorists, the fields of social-historic and comparative gender research expanded significantly. Recent research reconstructs the role of emotions and bodies in the gendered transformation of work (Ernst 2019d) in industrial and service jobs as well as the interwoven areas of consumption and the division of labor within the family (Baur et al. 2019) as well as the interdependencies of increasing precariousness, gender, and social milieu (Norkus 2019). And while power relations between the sexes in rugby sports were studied (Liston 2005), research on "women's involvement in sport-related fitness-activities" (Mansfield 2008) was extended. Finally, the question of female "strategies of coping with oppression on the institutional, symbolic and proxemic level" (Bucholc 2011: 425) rounds this now broadly accepted social-historical approach in internationally various fields of gender research. Therefore, we learn that current gender conflicts are rather asymmetrical and ambiguous and can therefore be connected especially through process-theoretical approaches (Treibel 2012: 88).

The more complex game and the double-sidedness of modern gender relationships can thus be reconstructed in many of the described research fields in the medium and long term. Here, the established and outsiders still struggle for equality in certain spheres, in some cases they have successfully advanced in positions of the established or are still excluded or have lost privileges. In private and in public there are often changed egalitarian figurative ideals, but often still traditional figurative networks:

> At present the complex polyphony of the movement of rising and declining groups over time - of established groups which become outsiders or, as groups, disappear altogether, of outsider groups whose representatives move as a new establishment into positions previously denied them, or as the case may be, which become paralysed by oppression – is still largely concealed from view. (Elias and Scotson 2008: 20)

These dynamics of ascent and descent of current conflicts may appear to be a blind interlocking game; they are being researched successively in figurational gender research and are structurally related to typical double-bind situations (Elias 2007). The shadows and nuances of long-term successful and conflictual emancipation processes of relative outsiders (that is, functional democratization and informalization) are less explored, but recent dynamic gender research, for example, shows its fascinating potential. New (female) players emerged and disadvantaged groups more or less successfully fought against discrimination. These conflicts also include gratification crisis leading to an integration paradox (Treibel 2016), that is not only typical of migrant but also for social climbers.

Power Relations Between the Sexes in Marriage and Working Life

The historical genesis of this presupposed formation of differences and dynamics of ascent and descent will now be reconstructed with a reference to gender relations in marriage and at the workplace. It will be argued that the demand that the relevance of a sociogenetic and psychogenetic perspective that was neglected for a long time still has to be extrapolated into research in the sociology of emotions, and that the social-history of passionate love, neglected in family research, has to be contextualized especially in view of its power aspect (Senge 2013; Burkart 2018). Current research in emotion sociology, be it in the world of work (Ernst 2019d) or in partnerships (Illouz 2011; Burkart 2018), mostly or explicitly operate with the background distinction between a private sphere as a place of emotion and a public sphere as a place of rationality. Nowadays, borderline phenomena show that the workplace becomes home and only homework, that is, housework or family work are waiting (Hochschild 1997). This processing of private activities is not only induced by digitization but has already been created on a diverse level with the informalization of social behavior standards latest since the 1960s and constantly challenges the traditionally "bourgeois-patriarchal" (working) cultures and ways of life.

The focus lies on the mutual social external and self-constraints in their ambivalence and socio-and psychogenetic dimension expressed in the counseling literature. Social differences are continuously marked through the strategy of naturalzsation and culturalization, as gender and "class" differences. The repertoire of behavior offered varied the more, the more varied the types of play or the way of distinction and the smaller the social class differences. Social climbers put privileged classes under pressure, which in turn developed new distinctive practices in order to maintain their lead in power. In particular, the question of "moral and moral behaviour" and successful social advancement drove the social mobility dynamics.

Marriage

Nearly every third marriage ends in divorce, and there are fewer and later marriages than 40 years ago. Marriage has become less important and is no longer an imperative to ensure basic needs or material supplies. Marriage, sexuality, love, romance, and reproduction are no longer related. According to analyses of family research, pregnancy, birth, or the desire to have children determine the marriage behavior of heterosexual couples (Burkart 2018). In addition to the spread of "illegitimate cohabitation" (called "wild marriage" in 1980s Germany), monogamy, bi- and homosexuality, polyamory or triangular relationships, traditional marriage and family, despite all the supposed shocks, have successively asserted themselves as the preferred concept of life. Until the recent legalization of same-sex marriage in several European countries, which is still a socio-political, emotionalized attraction, marriage was defined as the "uniting of a man with a woman in a long-term companionship" (*Bundesverfassungsgericht* 2002; *author's translation*). The deep anchoring and meaning of heterosexual marriage in the cultural customs and practices of a society on which the legal definition relies reveals the symbolic excess of meaning of this form of life favored by German tax law. The developments in marriage and sexuality can clarify what lies behind the much touted and ultimately seemingly inevitable talk about the customs and traditions of a society. In addition to studies on homosocial marriage behavior, child-oriented marriage, and the duration of marriage, further perspectives are opened up on the context of a life form that contains deep-seated emotions and habits. If one disregards the not inconsiderable socioeconomic advantage of marriage and moral considerations, one can ask whether marriage has any special emotional and psychological significance. If, according to Pinker, the bondage of male aggressiveness in marriage would also work, what unrivalled social integrative function does it still fulfill today in pacified societies? First of all, it is necessary to understand the success story of the marriage which is characterized by its enormous social diffusion. Class and gender were interwoven throughout the centuries in a specific way and context. Educating sophistication, mastering the "art of pretence," adequately dealing with nudity and shame are social modeling methods that have gradually sunk from the feudal courts into broader social strata. In this context, gender relationships were largely constructed through the intersection of *noble women* (ladies, in German *Dame*) versus ordinary women (in German *Weib*) versus commoners (in German *Frau*). For example, the noble lady had relatively more freedom or power compared to the maid or petty bourgeois housewife. In the bourgeois age, not only were nobles disempowered *per se*, but noble women also gave way to bourgeois women: constructed "natural gender characters" leveled social differences.

> The lady is and should be the nurse, keeper, priest of good manners. Just as she is naturally endowed with the grace of the body, culture has assigned her the beautiful mission to delight with noble customs and to have a refining effect on the world of men. (Adelfels 1895: 96; *author's translation*)

From this perspective, the easy misunderstanding of some modernization theory can be cleared up, as well as the current apodictic debate about commodification and instrumentalization of intimacy and emotion, which are put on public display. These approaches often represent "pre-modernity" in a simplified way, in that linear gains in power and autonomy are only estimated for modern society. The so-called "whole house" was transformed, and only in "late modern" times were women seemingly released from their class and gender fate (Beck 1986; Burke 1989; Ernst 1996; Ernst 2022). Likewise, the thesis of extended access to emotions and intimacy that become commodities for capitalist modernity (Illouz 2018) inevitably leads to the search for dimensions of comparison. In contrast, the socio-historical approach, with the transition from courtly to professional civil society, can open up both the marital-familial (Ernst 1996) and the professional and political separation of spheres of life (Ernst 1999). A relational and historically comparative perspective can also enrich work research. Already in the courtly age, divergent coding occurred, which successively merged into the bourgeois separation of rationality versus emotionality, publicity and privacy, masculinity and femininity. In the relationship between these spheres, the power relations between the sexes have been specifically linked to each other. It is precisely the history of marriage and family in European civilization (Ernst 1996), which is anything but hidden in private, as the social order of intimate relationships that proves a long-lasting formal superiority of men. Older men and fathers in particular, whether as feudal lords, guardians, or middle-class family heads, were economically and legally superior in the long term. "In society at large, however, men as a social group commanded much greater resources of power than women" (Elias 2009a: 241). Therefore, patriarchy was the term that fitted for that era. Walby (1990, 1997, 2009) analyzes changing patriarchal gender orders and their manifestation in working and private life, depending on the way welfare regimes are organized in modern societies. Here, however, the focus is more on the psycho- and sociogenetic and dynamic concept of power within the long-term transformation of gender.

A closer look at medieval marriage stud books, for example, reveals a character of open violence in the marital power struggle, when the stubborn Sibote (1961) or the fighting Brunhild are subjected to the marriage in the "Unruly Taming" or the *Nibelungenlied* (1970). Unilateral awarding of the bride and the matrimonial power with the husband characterized the female subordination. In the bourgeoisie, on the other hand, casual self-submission is evident, first to the father then to the spouse as the ideal of behavior of the higher middle-class daughter. Now new, more subtle constraints were formed, which were expressed in the gender-specific behavioral ideals of shyness, chastity, and modesty. With the ambivalent process of civilization and formalization of marriage (Ernst 1996), as part of the bourgeois culture of emotion, mutual affection was made par excellence as a reason for marriage. This gave women a relatively stronger legal status at a time when marriage was only recognized as an acceptable social relationship between the sexes and illegitimate relationships were punishable as concubinage. Therefore, marriage was transformed into a matter of state, which protected the only natural and legal way of life against individual intentions. Love also compensated for inequalities in the marital financial rights. Complete devotion was demanded of the woman and she had to hand over her complete fortune to her husband. The radical intention of love was therefore replaced by the revival of the medieval husband's command. Not only the patriarchal choice of residence and patriarchal naming clearly expressed the one-sided male privileges (Weber 1907: 416ff.); the punishment for divorce and adultery was cruel as well. It included the ban on marrying again and imprisonment in a fortress the duration of which was twice as long as for women as it was for men (Ernst 1996). Knigge precisely stated that "in bourgeois society the woman is not at all a person in her own right" (Knigge 1788: 363; *author's translation*).

Therefore the social reality of the nineteenth century conflicted with the *idealized* expectations of love and happiness. Marriage still had the function of fathering heirs and proving the man's potency: the offspring was still a good resource for alliances between the upper-class families. The appropriate development of long-term plans was the behavioral ideal of the bourgeois classes *par excellence*. Overwhelming and uncontrolled love and desire therefore had to be controlled. Romantic love was a prerequisite of marriage, and therefore became a power resource. But marriage itself was supposed to control passion and desire in the right way. Thus in the marital quest for an exciting and satisfying balance between sex and romance or love, the tension level increased. On the other hand, the love semantics that have developed since the Enlightenment were due to intimization and informalization processes as well as to formalization spurts that have been in effect since the 1960s (Wouters 1986; Wouters and Dunning 2019). They left behind a fragile, "structurally unsettled," emotionalized structure.

In this specific network, for example, the aristocratic *Salonière* was not without power and influence. Rather, it took on a relatively more influential position in courtly salon culture, while the bourgeois housewife, imitating and reinterpreting noble standards, represented the semipublic bourgeois parlor (Elias 2013; Ernst 2022), valued as art of human observation and rehearsed at court (Ernst 1999: 210). At first, there was the comparatively greater scope for women to act in the upper classes, because private and public spheres were not yet so strictly separated:

> Flanked by a complementary gender model, the spouses not only moved closer to one another in terms of space. Rather, a 'fundamental gender polarity was stated, which again

limited the ideas of sexual self-determination and autonomy of the Enlightenment. (Ernst 1996: 128; *author's translation*)

In the bourgeois etiquette books, they also gave way to ridicule about the noble scholarly woman who strives for public recognition and does not live in accordance with her "gender character" (Frevert 1995, see below). The search for the "better half," of the "elm tree" and the "swaying ivy," should now be based on a "harmonious whole" (von Humboldt 1903: 320; *author's translation*) that provided for the man to be the conjugal head. From then on, the homely and loving mother symbolized the dedicated and resigned care worker in the family's private immanence as a suitable counterpart. The strict father was "increasingly stronger – or again – responsible for the income, protection and maintenance of discipline within the family" which *he* represented (Fertig 1984: 23; *auther's translation*). The ambivalent processes of informalization (Wouters 2004, 2007) as moderate and sometimes radical easing of external constraints or formal behavioral standards within an achieved civilizational level, subsequently successfully challenged these patriarchal relations of rule and shook the patriarchal dividend of hegemonic masculinity (Connell 1995; Treibel 2012: 89). On the other hand, they also required that the boundaries in motion be renegotiated. This process, however, continued almost until the widespread spurts of informalization in the 1960s. Until then, however, the additional idea of marriage was considered the only legitimate and socially recognized form of intimate gender relations (Ernst 1996). In Germany, sex between fiancés was deemed to be unnatural until 1954, and only since 2017 have homosexuals been allowed to marry legally, but with restrictions such as limited rights to adoption or in vitro fertilization. In 1976, women in Germany gained the right to work independently of their husband's or father's permission, and since 1994, they have been allowed to keep their own name after marriage. Behind the transfigured ideal of "harmonious inequality" (Van Stolk and Wouters 1987a), marital violence still exists, and for a long time it was treated as a private matter and not as a crime. The husband's right to command and even beat his wife was tolerated as *ultima ratio* until 1997, and it still is in many countries. Informalization processes as seen in the relaxation of strict female external constraints and standards of conduct opened up questioning of patriarchal power relationships. These gains in emancipation led to the requirement of all-round self-compulsion of the individual (man) in a formalized way to an equality-oriented social legal consciousness; however, it only brought about partial changes in behavior.

In the secular age, marriage itself seems to have taken on almost para-religious characteristics, since it fulfills the growing ritual needs of people in search of orientation. In the meantime, one can even speak of a socio-emotional overstrain of marriage. The hope of lifelong fulfillment, passion, and security is exaggerated and leads to a higher willingness to divorce if they remain unredeemed.

Patriarchal thinking and acting prevail for a long time even with progressive forces. It even takes us into the present to publicly address internalized sexual shame and the taboo against sexual assault. The hopes "that mutual attraction becomes integrated with mutual consent to the extent that both develop into a taken-for-granted condition for all sexual pleasure" (Wouters 2019: 77) probably will need time to be fulfilled. Wouters's comment on the '#MeToo-Campaign' as a restriction of sexual self-determination, however, lacks the focus on the targeted exploitation of professional dependency relationships through sexualized violence at the workplace, which the development of the industrial working society will be dealt with below.

Work and Professional Life

The social history of capitalist industrial societies shows that an extra-occupational professional activity in the public arena of Fordist society is a gain in female emancipation that should not be underestimated, especially for unmarried women and *middle-class mothers* (Ernst 2016). The narrow 1980s debate especially about the supposed end of working society connected with the cultural turn of gender research led to a loss of the socioeconomic perspective of gender orders. Finally, the emotional offensives as well as the social history of work have led to questioning the rational ideal-typical construction of work organizations and rediscovering the forgotten gendered factory as well as female service work. Moreover, looking at the socio- and psychogenetic development of gender, work, and leadership, there are initially gender-segregated approaches to the labor market in which women in top business post are assumed to lead, decide, and act basically different to men (Ernst 2003). Since the 1920s, the female employment rate has been increasing interconnected with the growing welfare states and the world of work instead of the courtly or bourgeois salon appeared as a central place of socialization and an arena for gender conflicts at the same time:

"This holds in particular for the relationship between external and self-constraints in societies at different stages of development" (Elias 2013 [1996]: 37). However, the world of work is fundamentally linked to forms of private life. The working world in the factory, office, and open-plan office not only became the scene of "permanent small wars" (Gorz 2000: 43), but also of gestures of superiority, fear of shame and status. In this seemingly rational world, the logic of private and public conflicts, real rather than ideal, at the same time, feelings of superiority and inferiority, formal and informal exercise of power as well as gossip, shame, and (sexual) harassment toward subordinates or outsiders can also be found here (Ernst 2003). The rationalized professional man, with his emotions, affects, and drives, is plagued by social and thus historically highly changeable fear of shame about job-related violations of standards, failure, and lack of success, if one takes a closer look at the literary subject of the office novel (Ernst 2019c; Frevert et al. 2011). At the turn of the twentieth century, women were faced with the ambivalent legacy in the bourgeois professional world of only being allowed to work until they were married and at the same time pursuing decent, serving professions. In this service age, middle-class female professions emerged and brought together the operator, secretary, nurse, and conductor, while in the circles of the socialist labor movement female emancipation per se was defined by the employment of women (also in male professions).

And also, in the sciences and universities, women only came to public presence in Europe after the nineteenth century (Ernst 2003). First of all, however, it was deemed necessary to test the intellect of women with "expert reports by outstanding university professors, women teachers and writers on the ability of women to study science and professions" (Möbius 1897; *author's translation*). Even earlier, in the eighteenth century the *intellectual woman* was invented as both a hated and idealized figure at the same time. The most popular women are, for example, Ninon de Lenclos and Olympe de Gouges in France, Mary Wollstonecraft in England, and Charlotte von Stein in Germany, who were heavily attacked for daring to question male dominance in society with their fight for equality and right to work. They figured as an ambivalent construction of separation and stigmatization. Intellectual women represented the "minority of the worst" (Elias and Scotson 2008: 159): Knigge, one of the most popular German authors of behavior manuals, blamed the intellectual woman for being a "torture for every man." Although she might try to "perfect her writing and speech by her own studies and chaste literature (...)" a woman should not dare to "make a profession of it" or to "stray in all kinds of erudition" causing "not only disgust but compassion." The reason is that women should not enter the world and business of "great men" who "during all the centuries made troublesome" but incomparable great research (Knigge 1788: 196; *author's translation*). In contrast, the young working *men's* ideal stands with this insofar as "the doctor, the lawyer, the teacher (...) must be in their place." They could not live a life of "leisure and lightness" like women can; men should "never delay their work" (Lafontaine 1804: 110; *author's translation*). From these power and class conflicts of the gender groups, a bourgeois concept of *harmonious difference* has prevailed, which declared a "female cultural sphere" to be a particularly noble task of women. None other than Marianne Weber (the wife of Max Weber), who held a doctorate in law, believed that women are equal but different:

> There exists no women's view in natural- and literature-science. The things men and women are doing here are the result of equal not different intellectual capabilities. It is another case in the historical cultural science. There a woman is by a special psychical capability able to understand the feelings of others and because of this understand the motives of their actions. She is able to make developmental contributions to science in this case as we can recognise now in biographical studies and in the history of Arts and literature. (Weber 1906: 22–3; *author's translation*)

The author of one of the best selling etiquette books, Biedermann, was convinced that women lack the "organ of deepening," logical thinking, in order to "persistently pursue" certain ideas: Arguing from the perspective of *the minority of the best* he thinks that "therefore, women have never created anything outstanding, groundbreaking in philosophy, history or the exact sciences" (Biedermann 1856: 93; *author's translation*). But women can at least work in special services in the growing capitalist society:

> So, for jobs like barbers, hairdressers, pastry chefs, coffeemakers, for the production of clothes and plaster of all kinds for the female gender, for trimmings, cardboard work and even most parts of the bookbindery, and also for all branches of the retail trade women are

right well, yes, better than men, and not just as workers, but also as independent business entrepreneurs. (Biedermann 1856: 83; *author's translation*)

Compared with this initially progressive seeming view, which assigns women a job and may grant them a little autonomy, most working women until the 1980s and 1990s held the position of serving and assisting at the side of the man or below him, and not above him. This phenomenon of stereotypical female and male jobs, known today as the segregation of the labor market, was therefore established from the very beginning of the early capitalist labor society. Professional explanations in another etiquette book prove that care work especially counts as work for women without weakening the "natural" male dominance:

> Women have always been active in nursing and in the training of individual medical disciplines, such as obstetrics, and they have recently been widely used as assistants in operations (...) (Franken 1898: 2; *own translation*).

Jumping into the time after World War II we can realize that a difference-oriented and ambivalent gender order has been handed down equally in both areas of life. Despite her social and professional emancipation as a result of the forced political and legal formalization in the education sector and gender equality policy since the 1960s, representations predominate that negotiate and trivialize the increase in power of women in everyday working life, especially on the private level of sexual attraction (Ernst 1999: 270–300). "Female" abilities are above all emotional, because everyday office life as an arena for the public should be designed comfortably in analogy to the actual, limited range of action for women. Motherly, peaceful women should civilize the competition and appease male competition fears. With the ideal of the *career and business woman* and female heads of government since the 1980s, an antagonistic construct returned, which appeared briefly in the post-war and reconstruction era. Especially successful women, female employees in 1955 were warned that if they "try to imitate their male colleagues, not only in the case of performance but also in the case of behaviour" considering their talk, their "posture and their movement" (Oheim 1955: 386; *author's translation*) will lose femininity. Business women had to be careful in order to create a good working atmosphere especially if they were the bosses of men: they should employ "grace and agility" instead of playing with "feminine coquettishness" and they should consider that men "like real women with motherly instincts more than outmoded piles of files, grouchy old spinsters who are eaten away by ambition" (Andreae 1963: 114; *author's translation*). This means that, regardless of a woman's intention, her actions are interpreted in the narrow framework of infringing or confirming femininity. She should expect that her boyish behavior implies the loss of male courtesy.

In the management debates from the end of the 1980s, the *career woman* was intensively addressed, who was not only "somehow" different, but apparently an exotic woman who "with competence and high heels" (Manager magazine 1986; *author's translation*) "broke into the men's club" (ibid.: 1988). Although she is "highly qualified, smart and combative," on the "success course in politics, economy and culture" (Stern magazine 1997; *author's translation*), "more courage to rise" (Management Wissen 1988; *author's translation*) is recommended. The question of "adapting or alighting" (ibid.: 1990) is particularly morally charged to this day. It is therefore particularly important to go into the power play with a cool head, not to go too emotionally (Management Wissen 1991). Working women are still primarily addressed as a weaker, loss-making group. In line with this, as in the nineteenth century, the gradual admission of women to study raised the first fears of competition. Not only is fear of the loss of privileges for men at the top when the so-called "quota women for business" (Wirtschaftswoche 1989; *author's translation*) are recruited, but fears about the domestic gender order are also fueled. To this day, the so-called compatibility problem is primarily addressed as a problem for women and the promotion of women in business is mostly reduced to family support or social freezing.

Conclusion: Changing Gender and Relationship Ideals?

The "ideal of harmonious inequality," despite or perhaps because of its antiquity, remains in the private and professional arena to this day and makes intimate and professional relationships conflictual at the present time. In many cases, changed egalitarian *behavioral ideals* can be identified, but often clearly traditional behaviors. Up to the current debate on sexism, and apart from the feminist rebellion between 1960s and 1980s, the current young generation avoided open gender conflict and, despite all the gains in equality, still sometimes reproduced stereotypical ideas, especially with regard to the issue of children and upbringing.

If one remembers the formula of the "reduction of social contrasts," which evokes an "enlargement of the varieties" (that is, efforts at distinction as a reflection of the advancement of weaker groups with their own (subcultural) practices), the current ambivalence becomes understandable. Nowadays it describes these, on the one hand, successful informalization processes against the background of a gender-regulating formalization process. The discourse on plurality and diversity, on the other hand, shows new and further lines of conflict and social differentiation. The "controlled decontrolling of controls" (Wouters 2004) in the informalization of emotional, partnership, and professional ideals requires a reflective, situational handling of external and self-constraints. This manifests itself historically in legislation, changed public norms, state equality policies, contemporary texts, and recent debates on discrimination and sexism. The relationship model of "harmonious *difference*" (Ernst 2003: 297), formerly harmonious *inequality*, is only gradually disappearing. 'However, one likely consequence of this is that we will move from harmonious inequality toward inharmonious equality" (Liston 2005: 81) with never ending complex and permanent negotiation relationships. These negotiations in particular are part of social spurts and counter-spurts embedded in informalization or formalization processes.

In particular, the concept of balance of power, as has been shown here, can be successfully used in socio-historical research in a variety of contexts to explain the socio- and psychogenetic order of social processes. It is currently unclear whether and to what extent less freedom of movement, informalization and relaxation of behavioral ideals, or new formalization tendencies will result in further emancipation and civilization or decivilization. It also remains to be seen to what extent actual equality will succeed in the sense of an emphatic ability to identify with weaker groups. The range of informality and formality is specific: the "normalisation of women's emancipation" (Treibel 1997; *author's translation*) is sometimes accompanied by an astonishing persistence toward the demands of women *not* to be harassed. This contradiction keeps the "blind game" of traditional and egalitarian figurative ideals running full of tension until today.

References

Adelfels K (1895) Das Lexikon der feinen Sitten, 11th edn. Schwabacher, Stuttgart
Andreae I (1963) Die Kunst der guten Lebensart. Spielregeln im Umgang mit Menschen beschrieben und zusammengestellt, 3rd edn. Herder, Freiburg
Baur N, Ernst S (2011) Towards a process-oriented methodology. Modern social science research methods and Norbert Elias' concepts on figurational sociology. Sociol Rev Monogr 59(7).: Norbert Elias and Figurational Research: Processual Thinking in Sociology, pp 117–139.
Baur N, Fülling J, Hering L, Vogl S (2019) Verzahnung von Arbeit und Konsum: Wechselwirkungen zwischen der Transformation der Erwerbsarbeit und den Transformationen der milieuspezifischen innerfamiliärer Arbeitsteilung am Beispiel der Ernährung. In: Ernst S, Becke G (eds) Transformationen der Arbeitsgesellschaft. Prozess- und figurationstheoretische Beiträge. Springer VS, Wiesbaden, pp 105–132
Beck U (1986) Risikogesellschaft. Suhrkamp, Frankfurt a.M
Biedermann K (1856) Frauen-Brevier. Kulturgeschichtliche Vorlesungen für Frauen, Leipzig
Bock G (1988) Geschichte, Frauengeschichte, Geschlechtergeschichte. Geschichte und Gesellschaft. Sozialgeschichte in der Erweiterung 14(3):364–391
Brinkgreve C (1999) Old boys, new girls. Over de beperkte toegang van vrouwen tot maatschappelijke elites. Amsterdams Sociologisch Tijdschrift 26(2):164–184
Brinkgreve C, Gomperts W, Meeuwesen L, Westerbeek J (2014) Changing patterns of self-presentation by depressed clients: from shame to self-respect. Human Figurations. Long-term perspectives on the human condition 3(1). http://hdl.handle.net/2027/spo.11217607.0003.102
Brown KM (1993) Brave new worlds: women's and gender history. William Mary Q. Early Am Hist 50(2):311–328
Bucholc M (2011) Gendered figurational strategies in Norbert Elias's Sociology. Polish Sociol Rev 176:25–436
Burkart G (2018) Soziologie der Paarbeziehung. Eine Einführung. Springer, Wiesbaden
Burke P (1989) Soziologie und Geschichte. Junius, Hamburg
Connell RW (1995) Masculinities. Polity Press, Cambridge
Connell RW, Messerschmidt JW (2005) Hegemonic masculinity: rethinking the concept. Gend Soc 19(6):829–859
Davies NZ (1976) 'Women's History' in transition: The European case. Fem Stud 3(3/4):83–103
Dekker M (2019) Encouraging bystander intervention in street harassment situations. A sociological study of the politicization of gender relations in the public space. Politix 125(1):87–108
Delmotte F (2019) Norbert Elias, Catherin Deneuve and gender equality. https://booksandideas.net/Norbert-Elias-Catherine-Deneuve-and-Gender-Equality. 23 May 2020.
Dibie P (1994) Wie man sich bettet. Die Kulturgeschichte des Schlafzimmers. Klett Cotta, Stuttgart
Elias N (2006 [1983]) The Court society. The collected works of Norbert Elias, vol 2. University College Dublin Press, Dublin

Elias N (2007 [1987]) Involvement and detachment. The collected works of Norbert Elias, vol 8. University College Dublin Press, Dublin

Elias N (2009a [1987]) The retreat of sociologist into the present. In: Elias N (ed) Essays III: on sociology and the humanities. The collected works of Norbert Elias, vol 16. University College Dublin Press, Dublin, pp 107–126

Elias N (2009b) The changing balance of power between the sexes – a process-sociological study: the example of the ancient Roman state. In: Elias N (ed) Essays III: on sociology and the humanities. The collected works of Norbert Elias, vol 16. University College Dublin Press, Dublin, pp 240–265

Elias N (2009c [1983]) Foreword to 'Women Torn Two Ways'. In: Elias N (ed) Essays III: on sociology and the humanities. The collected works of Norbert Elias, vol 16. University College Dublin Press, Dublin, pp 270–275

Elias N (2009d [1977]) Towards a theory of social processes. In: Elias N (ed) Essays III: on sociology and the humanities. The collected works of Norbert Elias, vol 16. University College Dublin Press, Dublin, pp 9–39

Elias N (2012a [1978]) What is Sociology? Collected works of Norbert Elias, vol 5. University College Dublin Press, Dublin

Elias N (2013 [1996]) Studies on the Germans. The collected works of Norbert Elias, vol 11. University College Dublin Press, Dublin

Elias N, Scotson JL (2008 [1965]) The established and the outsiders. The collected works of Norbert Elias, vol 4. In: Wouters C (ed). University College Dublin Press, Dublin

Ernst S (1996) Machtbeziehungen zwischen den Geschlechtern. Wandlungen der Ehe im 'Prozess der Zivilisation'. Westdeutscher Verlag, Opladen

Ernst S (1999) Geschlechterverhältnisse und Führungspositionen. Eine figurationssoziologische Analyse der Stereotypenkonstruktion. Westdeutscher Verlag, Opladen

Ernst S (2003) From Blame Gossip to Praise Gossip? Gender, leadership and organizational change. Eur J Women's Stud 10(3):277–299

Ernst S (2010) Prozessorientierte Methoden der Arbeits- und Organisationsforschung. Eine Einführung. VS Verlag für Sozialwissenschaften, Wiesbaden

Ernst S (2015) The 'Formation of the Figurational Family': Generational chains of process-sociological thinking in Europe. Cambio: Rivista sulle trasformazioni sociali 5(9):65–78

Ernst S (2016) Issues and aspects of comparative long-term studies in youth unemployment in Europe: biographical constructions of 'Generation Y'. Cambio: Rivista sulle trasformazioni sociali 6(12):167–183

Ernst S (2019a) Zivilisations- und Prozesstheorie: Elias und die Geschlechterforschung. In: Kortendiek B, Riegraf B, Sabisch K (eds) Handbuch Interdisziplinäre Geschlechterforschung. Springer VS, Wiesbaden, pp 399–408

Ernst S (2019b) Literarische Quellen und persönliche Dokumente. In: Baur N, Blasius J (eds) Handbuch Methoden der empirischen Sozialforschung. VS Springer, Wiesbaden, pp 829–840. https://doi.org/10.1007/978-3-658-21308-4_81

Ernst S (2019c) Zivilisations- und Prozesstheorie: Elias und die Geschlechterforschung. In: Kortendiek B, Riegraf B, Sabisch K (eds) Handbuch Interdisziplinäre Geschlechterforschung. Springer VS, Heidelberg, pp 829–840

Ernst S (2019d) Fit for life – fit for work? Prozessorientierte Zugänge zu Körper und Emotion in und bei der Arbeit. In: Ernst S, Becke G (eds) Transformationen der Arbeitsgesellschaft. Prozess- und figurationstheoretische Beiträge. Springer VS, Wiesbaden, pp 51–77

Ernst S (2022) Salons, Büros und Privatgemächer als Geschlechterarenen im Zivilisationsprozess. Zur Genese privaten und öffentlichen Lebens. In: Kahlert H, Cichecki D, Degele N, Burkart G (eds) Privat/öffentlich: Gesellschaftstheoretische Relevanz einer feministischen Debatte. VS Springer [in press]

Ernst S, Becke G (2019) Transformationen der Arbeitsgesellschaft. Prozess- und figurationstheoretische Beiträge. Einführung. In: Ernst S, Becke G (eds) Transformationen der

Arbeitsgesellschaft. Prozess- und figurationstheoretische Beiträge. Springer VS, Wiesbaden, pp 1–20

Ernst S, Weischer C, Alikhani B (eds) (2017) Changing power relations and the drag effects of Habitus. Theoretical and empirical approaches in the twenty-first century. GESIS, Köln

Fertig L (1984) Zeitgeist und Erziehungskunst. Eine Einführung in die Kulturgeschichte der Erziehung in Deutschland von 1600 bis 1900. Wissenschaftliche Buchgesellschaft, Darmstadt

Frevert U (1995) Mann und Weib, und Weib und Mann. Geschlechterdifferenzen in der Moderne. Beck, München

Frevert U, Scheer M, Schmidt A, Eitler P, Hitzer B, Verheyen N, Gammerl B, Bailey C, Pernau M (eds) (2011) Gefühlswissen. Eine lexikalische Spurensuche in der Moderne. Campus. Frankfurt/New York

Gay P (1986) Erziehung der Sinne. Sexualität im bürgerlichen Zeitalter. Beck, München

Gerhard U (1981) Verhältnisse und Verhinderungen. Frauenarbeit, Familie und Recht der Frauen im 19. *Jahrhundert*. Suhrkamp, Frankfurt a.M.

Gorz A (2000) Arbeit zwischen Misere und Utopie. Suhrkamp, Frankfurt a.M

Gravenhorst L (1996) Entzivilisierung und NS-Deutschland. Einige feministische Beobachtungen zu einer zentralen Idee bei Norbert Elias. In: Modelmog I, Kirsch-Auwärter E (eds) Kultur in Bewegung. Beharrliche Ermächtigungen. Kore Verlag, Freiburg, pp 165–181

Hagemann K, Harsch D (2018) Gendering central European history: changing representations of women and gender in comparison, 1968–2017. Central Eur Hist 51(1):114–127

Hammer H (1997) Figuration, Zivilisation und Geschlecht. Eine Einführung in die Soziologie von Norbert Elias. In: Klein G, Liebsch K (eds) Zivilisierung des weiblichen Ich. Suhrkamp, Frankfurt a.M., pp 39–76

Hausen K (1976) Die Polarisierung der 'Geschlechtscharaktere'. Eine Spiegelung der Dissoziation von Erwerbs- und Familienleben. In: Conze W (ed) Sozialgeschichte der Familie in der Neuzeit Europas. Neue Forschungen, Stuttgart, pp 363–393

Hochschild AR (1997) The Time Bind. When work becomes home and home becomes work. New York: Metropolitan Books

Illouz E (2011) Warum Liebe weh tut. Suhrkamp, Frankfurt a.M.

Illouz E (ed) (2018) Wahre Gefühle. Authentizität im Konsumkapitalismus. Suhrkamp, Frankfurt a.M.

Klein G (1992) FrauenKörperTanz. Eine Zivilisationsgeschichte des Tanzes. Quadriga, Weinheim/Berlin

Klein G, Liebsch K (eds) (1997) Zivilisierung des weiblichen Ich. Suhrkamp, Frankfurt a.M.

Klein G, Liebsch K (2001) Egalisierung und Individualisierung. Zur Dynamik der Geschlechterbalancen bei Norbert Elias. In: Knapp G-A, Wetterer A (eds) Soziale Verortung der Geschlechter. Gesellschaftstheorie und feministische Kritik. Westfälisches Dampfboot, Münster, pp 225–255

Knigge A (1788) Über den Umgang mit Menschen (Reprographischer Nachdruck). Hannover (Darmstadt 1967)

Kunze J-P (2005) Das Geschlechterverhältnis als Machtprozess. Die Machtbalance der Geschlechter in Westdeutschland seit 1945. VS Verlag für Sozialwissenschaften, Wiesbaden

Lafontaine A (1804) Sittenspiegel für das weibliche Geschlecht, Band 3. Prag, Wien

Liston K (2005) Established-outsider relations between males and females in the field of sports in Ireland. Ir J Sociol 14(1):66–85

Management Wissen (1988) Mehr Mut zum Aufstieg (5):79–83

Management Wissen (1991) Mit kühlem Kopf ins Powerplay (7):14–23

Manager Magazin (1986) Mit Kompetenz und Stöckelschuhen, 16(4):176–180

Mansfield L (2008) Reconsidering feminisms and the work of Norbert Elias for understanding gender, sport and sport-related activities. Eur Phys Educ Rev 93–121

McRobbie A (2009) The aftermath of Feminism: gender, culture and social change. SAGE, Los Angeles/London

Meuser M (2010) Geschlechtersoziologie. In: Kneer G, Schroer M (eds) Handbuch Spezielle Soziologien. VS Springer, Wiesbaden, pp 145–162

Möbius PJ (1897) Über den physiologischen Schwachsinn des Weibes. München

Norkus M (2019) Intersektionale Dynamiken der Prekarisierung: Geschlecht und soziale Lage. In: Ernst S, Becke G (eds) Transformationen der Arbeitsgesellschaft. Prozess- und figurationstheoretische Beiträge. Springer VS, Wiesbaden, pp 211–234

Oheim G ([1955] 1964) Einmaleins des guten Tons, 34th edn. Bertelsmann, Gütersloh

Opitz C (1997) Zwischen Macht und Liebe: Frauen und Geschlechterbeziehungen in Norbert Elias "Höfischer Gesellschaft". In: Klein G, Liebsch K (eds) Zivilisierung des weiblichen Ich. Suhrkamp, Frankfurt a.M., pp 77–100

Pfister G (1997) Sport – Befreiung des weiblichen Körpers oder Internalisierung von Zwängen? In: Klein G, Liebsch K (eds) Zivilisierung des weiblichen Ich. Suhrkamp, Frankfurt a.M., pp 206–248

Pinker S (2011) The better angels of our nature: why violence has declined. Penguin Random House, London

Pinker S (2013) From *The better angels of our nature*: why violence has declined. Hum *Figurations* 2(2). http://hdl.handle.net/2027/spo.11217607.0002.206

Scott JW (1986) Gender: a useful category of historical analysis. Am Hist Rev 91(5):1053–1075

Senge K (2013) Die Wiederentdeckung der Gefühle. Zur Einleitung. In: Senge K, Schützeichel R (eds) Hauptwerke der Emotionssoziologie. Springer VS, Wiesbaden, pp 11–32

Sibote (1961) Der vrouwen zuht. Frauenzucht. In: Hagen F vd (ed) Gesamtabenteuer. Hundert Altdeutsche Erzählungen. Ritter- und Pfaffen-Mären. Stadt- und Dorfgeschichten. Schwänke, Wundersagen und Legenden. Bd.1, pp 37–57

Stern Magazine (1997) Topqualifiziert, klug und kämpferisch – Frauen auf Erfolgskurs in Politik. Wirtschaft und Kultur, Nr 23:36–46

Stolk B, Wouters C (1987a) Power changes and self-respect: a comparison of two cases of established – outsider relations. Theory Cult Soc 4:477–487

Stolk B, Wouters C (1987b) Frauen im Zwiespalt. Beziehungsprobleme im Wohlfahrtsstaat. Suhrkamp, Frankfurt a.M.

Tilly LA (1989) Gender, women's history, and social history. Soc Sci Hist 13(4):439–462

Treibel A (1993) Engagierte Frauen, distanzierte Männer? Überlegungen zum Wissenschaftsbetrieb. In: Treibel A, Klein G (eds) Begehren und Entbehren. Bochumer Beiträge zur Geschlechterforschung. Centaurus, Pfaffenweiler, pp 21–38

Treibel A (1997) Das Geschlechterverhältnis als Machtbalance. Figurationssoziologie im Kontext von Gleichstellungspolitik und Gleichheitsforderungen. In: Klein G, Liebsch K (eds) Zivilisierung des weiblichen Ich. Suhrkamp, Frankfurt a.M., pp 306–336

Treibel A (2008) *Die Soziologie von Norbert Elias. Eine Einführung in ihre Geschichte, Systematik und Perspektiven. Lehrbuch: Hagener Studientexte zur Soziologie*. VS Verlag, Wiesbaden

Treibel A (2009) Figurations- und Prozesstheorie. In: Kneer G, Schroer M (eds) Soziologische Theorie. Handbuch soziologischer Theorien. VS Verlag für Sozialwissenschaften, Wiesbaden, pp 133–160

Treibel A (2012) 'Frauen sind nicht von der Venus und Männer nicht vom Mars, sondern beide von der Erde, selbst wenn sie sich manchmal auf den Mond schießen könnten' – Elias und Gender. In: Kahlert H (ed) Zeitgenössische Gesellschaftstheorien und Genderforschung: Einladung zum Dialog. Springer VS, Wiesbaden, pp 83–104

Treibel A (2016) Integriert Euch! Plädoyer für ein selbstbewusstes Einwanderungsland. Campus, Frankfurt a.M.

von Humboldt WF (1903) Werke: 1785–1795. In: Leitzmann A (ed) Gesammelte Schriften, Bd.1. De Gruyter, Berlin

Walby S (1990) Theorizing patriarchy. Blackwell, Oxford, UK/Cambridge

Walby S (1997) Gender transformations. Routledge, London

Walby S (2009) Globalization and inequalities. Complexity and contested modernities. SAGE, London

Waterstradt D (2015) Prozess-Soziologie der Elternschaft: Nationsbildung, Figurationsideale und generative Machtarchitektur in Deutschland. MV-Verlag, Münster

Weber M (1906) Die Beteiligung der Frau an der Wissenschaft. In: Jahrbuch der 'Hilfe'. Wochenschrift für Politik, *Literatur und Kunst*. Berlin Schöneberg, pp 19–26

Weber M (1971 [1907]) Ehefrau und Mutter in der Rechtsentwicklung. Neudr. d. Ausg. Tübingen. Aalen: Scientia

Wirtschaftswoche (1989) Quotenfrauen für die Wirtschaft? (9):46–64

Wouters C (1986) Formalization and informalization: changing tension balances in civilizing processes. Theor Cult Soc 3(2):1–18

Wouters C (2004) Sex and manners, female emancipation in the West 1890–2000. SAGE, London/Thousand Oaks/New Dehli

Wouters C (2007) Informalization: manners and emotions since 1890. SAGE, London

Wouters C (2014) Universally applicable criteria for doing figurational process sociology: seven balances, one triad. Hum Figurations 3(1). http://hdl.handle.net/2027/spo.11217607.0003.106

Wouters C (2019) Informalisation and Emancipation of Lust and Love: Integration of Sexualisation and Eroticisation since the 1880s. In: Wouters C, Dunning M (eds) Civilisation and Informalisation. Connecting Long-Term Social and Psychic Processes. Palgrave Macmillan, Cham, pp 53–80

Wouters C, Dunning M (2019) Civilisation and informalisation. connecting long-term social and psychic processes. Palgrave Macmillan, London

Wouters C, Mennell S (2015) Discussing theories and processes of civilisation and informalisation: criteriology. Hum Figuration 4(3). http://hdl.handle.net/2027/spo.11217607.0004.302

Collective Memory and Historical Sociology 30

Joanna Wawrzyniak

Contents

Introduction	776
Situating Historical Sociology and Memory Studies: A Brief Overview	778
Collective Memory: Foundations	782
Maurice Halbwachs and His Legacy: "Social Frameworks of Memory" and "Collective Memory"	783
Stefan Czarnowski's Theory of Heroes and the Social Functions of the Past	786
From Collective Memory to Collective Forgetting	788
The Second Wave of Memory Studies: Nations, Power Struggles, Media	789
Collective Memory and Nation States	790
Collective Memory and Power	792
Collective Memory, Media and the *Longue Durée*	793
The Third Wave: Transnationalization and Memory Activism	794
Transnational Turn	794
Activist Turn	796
Conclusion	798
References	800

Abstract

This chapter argues that collective memory studies contribute important insights into the most enduring concerns of historical sociology, notably epochal social transformations, modernization processes, class formation and dissolution, and the origins and decay of states structures. However, memory studies, along with their conceptual toolkit created amid contemporary transnational turn and grassroots struggles over uses of the past, are also inspiring new waves of research in historical sociology. Not only memory studies aid critical reflection on Western-centric categories of historical inquiry, but the alliances between scholars of memory and memory activists make the field sensitive to the diverse application

---

J. Wawrzyniak (✉)
University of Warsaw, Warsaw, Poland
e-mail: wawrzyniakj@is.uw.edu.pl

© The Author(s), under exclusive licence to Springer Nature Singapore Pte Ltd. 2022
D. McCallum (ed.), *The Palgrave Handbook of the History of Human Sciences*,
https://doi.org/10.1007/978-981-16-7255-2_56

of uses of the past around the world. Memory studies remain at the forefront of humanities and social science today and deserve the close attention of historical sociologists.

Keywords

Memory studies · Nation state · Decolonization · Transnational turn · Regions of memory · Memory activism

Introduction

In 2018, a report titled *The Restitution of African Cultural Heritage: Toward a New Relational Ethics* by Senegalese economist, Felwine Sarr, and French art historian, Bénédicte Savoy, made European headlines (Sarr and Savoy 2018). Although the report was criticized for being the "brainchild of academia" rather than "the balanced submission of practitioners" in a heritage field struggling with such practical issues as gaps in knowledge about the provenance of objects or the lack of legal methods for returning them (Herman 2019), the significance of Sarr and Savoy's work was huge. In particular, it caused the French President Emanuel Macron to declare, quickly and publicly, the return of some ceremonial items to Africa. In November 2021, France gave back to Benin over 20 important cultural objects that had been looted during the colonial period. The President of Benin, Patrice Talon made it clear that he saw the handover as the first step toward a large-scale restitution process. More generally, however, Sarr and Savoy's report identified issues central to the worldwide debate on the decolonization of museums. This debate has recently been gaining new momentum because many ethnographic, art, and history collections around the world have been challenged by diverse civil society groups making it plain that such collections originated from the structural violence of white Europeans and the harm they caused in different parts of the globe over the centuries. In effect, newly founded institutions outside Europe, like the Museum of Black Civilization in Dakar, have requested the repatriation of objects, and such well-known European institutions as the Tropenmuseum in Amsterdam, the Royal Museum for Central Africa in Tervuren, Belgium, or the Humboldt Forum in Berlin, have started to reinvent themselves – with varying degrees of success – by overhauling their exhibitions and changing their narratives to address the problematic origins of their collections.

In 2021, the German readers of several key opinion-forming newspapers were confronted with the Holocaust comparability debate, the background to which was paved by a growing public recognition of Germany's colonial-era genocide of the Herero and Nama in South West Africa. Although postcolonial historians like Jürgen Zimmerer (in his book *From Windhuk to Auschwitz*, German edition 2011) have been gathering evidence of structural entanglements between German expansion into Africa and into Eastern Europe for over a decade, more precise terms for the debate have been dictated by a (belated) translation into German of the influential book *Multidirectional Memory: Remembering the Holocaust in the Age of Decolonization* by Michael Rothberg (2009, German edition 2021), as well as the provocative formulation of the "German Catechism" by global historian of genocide, A. Dirk Moses (2021). In Moses's view, there have been several key points perpetuating German historical understanding of the Holocaust over the last few decades, which have represented it in terms of its uniqueness, including notions of a civilizational rupture, or anti-Semitism as a specifically German phenomenon that should not be confused with racism. Moses refuted those representations both on historical analytical and ethical grounds, claiming the Holocaust bore comparison to other instances of mass-scale violence in the nineteenth and twentieth centuries.

For three inter-related reasons, the above examples of recent intellectual debates devoted to key processes that shaped modernity – colonization and mass genocide – are suitable starting points for a chapter that aims to highlight the relevance of the concept of "collective memory" to historical sociology studies.

First, the public intellectuals taking part in these debates directly address collective memories. Michael Rothberg and other participants in the Holocaust comparability discussion argue that the Holocaust has enabled the articulation of other histories of mass atrocities rather than being a zero-sum game in competitive victimhood. Rothberg observed, while focusing on how the memory of the Holocaust was shaped and contributed to by the articulation of histories of other cases of extreme violence in various local settings, including in the Balkans, Eastern Europe, the Middle East, and North America, and, that in each of these cases, the allegedly universal memory of the Holocaust plays a different role and has different local meanings. Neither Rothberg nor his allies abstain from normative positions in on-going mnemonic conflicts over the shape of public memory, and yet they point to the intricacies of both history and memory (see also Rothberg 2019).

Similarly, Sarr and Savoy's report is significant not only because it claims that restitution of objects of colonial provenance is necessary, but because it charts the complexities of this process. In particular, it raises the issue of the transformation, translocation, and re-semanticization of cultural objects in time and space, and the fact that their original environments have altered or ceased to exist. In light of these changes, the report addresses the need to re-institutionalize artefacts in a new historical, cultural, social, and political setting. During this re-institutionalization process, objects might become elements of profound work on a new relational ethics. The goal is not only to "give back" objects, but rather to challenge gaps in knowledge about their provenance through "memory work … memorialization and … the work of writing or re-writing of history" (Sarr and Savoy 2018, 32). This work should be accompanied by the re-socialization of restituted objects in a variety of African cultural institutions from schools to newly funded museums. In effect, the authors of *The Restitution of African Cultural Heritage* make a normative argument stressing the need for the reinvention of African and European past(s) for the sake of the future. Viewed from a memory studies perspective, the participants in the debates in both France and Germany have taken on the role of "memory activists," a notion that will be addressed later in the chapter.

Second, both debates show that shifts in historical interpretation reflect wider social changes. Indeed, the public perception of museums with collections originating in colonial conquests has changed from them being regarded as institutions representing the supremacy of European knowledge, culture, and aesthetics to them being challenged as ethically dubious organizations symbolizing contested power structures and the problematic legacy of race science. Similarly, the view on the Holocaust as a unique event representing specifically German atrocities is currently being decentered and shifted toward a viewpoint presenting the Holocaust in relation to other forms of genocide and mass violence.

Finally, these two cases represent a larger epistemic shift in contemporary humanities and social sciences away from thinking in terms of traditional geopolitical categories and discrete nations and toward the examination of global convergences and connections. That shift has been clearly reflected for at least two decades in both the vast expansion of memory studies literature and the rapid development of robust memory-activist movements. The shift has contributed to new avenues of research that follow the trajectories of various agents, objects, ideas, and other phenomena beyond state-centered borders rather than attempting to tell the histories of nation states.

Thus, one of the key arguments of this chapter is that both memory activism and memory studies can enrich the way we have been used to thinking about historical sociology. At the same time, the chapter seeks to address a more general aim – to demonstrate the theoretical and empirical significance of the rise of multidisciplinary memory studies for historical sociology, with a particular focus on reinterpretations of the very concept of collective memory over the past century. The next section tries to relate memory studies and historical sociology as distinct but mutually significant fields of inquiry. It is followed by a section recapitulating the sociological premises of the theory of collective memory in its foundational moment in the interwar period. The chapter then moves on to the two sections that discuss two recent "waves" of memory studies concentrating on the consequences of rediscovering and reconceptualizing of the notion of "collective memory" for the research on nation states, power structures, media, generations, class, gender, and family, and paying attention to the transnational turn and the activist turn in more depth. Finally, the chapter concludes with some general considerations about relation of memory studies and historical sociology.

Situating Historical Sociology and Memory Studies: A Brief Overview

Memory studies and historical sociology developed in parallel and share some foci and questions, but explicit research encounters between them have been few even though it was the Durkheimian school in sociology that gave birth to the very concept of "collective memory." Defining and delineating both historical sociology and memory studies is difficult because in fact they both are internally only loosely connected by diverse theories, concepts, and methodologies from social sciences and humanities. To explore the relationship between them, let us initially focus on some of the main trends within each of this field; later, the discussion will move to some of their points of contact.

As far as historical sociology is concerned, its main promise has been to bring historical and temporal dimension to bear on sociological explanation. Although it has been drawing on the nineteenth and early twentieth century classics of sociology, such as Karl Marx or Max Weber, as a field of inquiry on its own right, historical sociology is relatively young, taking off in the second half of twentieth century as an answer to the domination of a presentist, survey type of sociology. At the time of its heyday, in the 1970s and 1980s, historical sociological inquiry revolved around two major categories – historical change and comparative historical analysis of nation states and other power structures in the *longue durée*. Importantly, for decades, historical sociology has reflected its European origins based on attempts to conceptualize the shift between "traditional" and "modern" societies, as well as the in-between process of modernization. It grounded its intellectual agenda in such ideas as progress as well as in critical dichotomies like "the centre" and "the peripheries" (Bhambra 2010; Sztompka 1993).

Today, this reflection has been either complemented or challenged by various branches of the humanities and their ambition to cover spaces and temporalities alternative to those established by older theories of modernization. In the story of historical sociology these developments have been sometimes captured under the rubric of the third wave and "cultural turn," although in fact they have reflected a far more profound epistemic change. Postcolonial studies have contributed to the shifting of our spatial imagination from "the West" and "the rest" to seeing the metropole and colony as single units in which flows of ideas, people, and objects take place. Furthermore, the way we think about historical sociology in postcolonial and decolonial studies has been significantly affected by attempts to: "provincialize Europe" (Chakrabarty 2000), question the Western discourses of (linear) time, space, and modernity by focusing on connectivities and subaltern subjectivities (Bhambra 2014), and "epistemically de-link" from the normative rhetoric of modernity (de Mignolo and Tlostanova 2012). Although focus on modernization and nation states (and their critical analysis) prevailed, in a broad sense historical sociology covers any sociological inquiry put in a historical perspective, including matters of the everyday and family life, art and culture, labor and economy, or politics and conflicts (see further, Bucholc and Mennell, this section of the handbook).

Memory studies have even more complex story of its origin to tell. Since Maurice Halbwachs developed the notion of collective memory in the interwar period, it has traveled far from its sociological home, embracing other fields of inquiry and gaining new momentum in the interdisciplinary memory boom that has proliferated in the social sciences and humanities from the 1980s until today (Olick and Robbins 1998; Sierp 2021). Although memory has always been a key topic for psychologists (keeping themselves rather distant from the transdisciplinary boom), several other disciplines have fostered research on memory, providing perspectives on and observations of its social foundations and functions.

Historians have explored many areas of collective memory, such as the evolution of mnemonic forms across traditional and modern societies (Yates 1966; Le Goff 1992; Connerton 2008), the role played by collective memory in the building and dismantling of empires and nation states (Anderson 1983, Hobsbawm and Ranger 1983; Nora 1984–1986), and connections between collective memory and the industrialization and deindustrialization processes (Berger 2019). Moreover, historians have contributed extensively to the literature on uses of the past in diverse organizations (Decker et al. 2020) and social movements (Berger et al. 2021) as well as to describing the memories of families, generations, and social groups and classes (Debouzy 1986; Slabáková 2022). Last but not least, they have examined contemporary conflicts in a historical perspective, with a particular focus on the remembrance and forgetting of the Holocaust and other forms of genocide and ethnic cleansing, decolonization, slavery, and the Stolen Generations. Indeed, in the field of memory, historians might have carried out Emile Durkheim's wish that the sole focus of their profession should be the implementation of a sociological program (Bellah 1959).

The ubiquity of memory research in history has been complemented by ongoing conceptual work in several other disciplines. Cultural studies have taken the lead in studying media of mnemonic transmission and proposing refinements to the initial Halbwachsian concept. For instance, a useful distinction between cultural memory (transmitted by media) and communicative memory (transmitted in social milieus) was introduced by Jan Assmann (2008b), and another one, between the canon (memory performing its functions) and the archive (the memory as a storage), was formulated by Aleida Assmann (2008a) to differentiate between mnemonic topics central to collective identities and others, currently dormant but with a potential to be retrieved into the canon. Moreover, cultural studies, by concepts such as "travelling memory" (Erll 2011), which focus on circulation, conflations, and conflicts of memory, significantly contributed to the disavowal of the methodological nationalism dominating memory studies in the 1980s and 1990s.

Finally, political and international relations studies have provided many neatly organized empirical comparative insights into the uses of the past in politics. Research topics include, for instance, the relations between memory and human rights (Levy and Sznaider 2006; David 2020), the role of memory in administration and policymaking (Dybris McQuaid and Gensburger 2019), and the role of memory in politics and transitional justice processes in post-authoritarian countries (Pettai and Pettai 2015; Bernhard and Kubik 2014). Legal studies have also explored the relations between memory and law by employing such concepts as "memory laws," a term denoting the legal governance of memory (Belavusau and Gliszczyńska-Grabias 2017).

In accounts of the history of the extremely rich body of scholarship within memory studies, there is a widespread belief that the field went through three waves, or phases (Erll 2011; Feindt et al. 2014). In this view, the interwar period represents memory studies' foundational moment. That was followed by a second wave in the 1980s marked by the "rediscovery" of Halbwachs' scholarship, which had been largely forgotten in post-war social science, and the current phase, which is marked by attempts to overcome the methodological nationalism of the 1980s and 1990s. Of course, like any approximation, this tripartite division has some flaws, not least of which is the manner in which it underplays the significance of some important works of the 1960s, such as those by Roger Bastide on diaspora memories (Grande 2021) or underestimates the reception and recurrence of Halbwachs's ideas in sociological peripheries, as occurred, for instance, in Poland in the 1960s (Kończal and Wawrzyniak 2018).

Therefore, not unlike the notion of waves in historical sociology (Riley 2006), the waves in memory studies should not be understood as one-directional and linear transitions. Conversely, they imply returns to original concepts while developing new ideas and research agendas. And yet, approximate though it is, the idea of waves captures well the main trends of interest in international memory studies scholarship. Halbwachs's works did indeed make a spectacular return at the end of the twentieth century, and since then, there have been many attempts to reformulate his ideas so as to increase their utility for research practices of various disciplines. The second and third waves made apparent "... the constitutive, rather than merely indicative, role of memory in social life" (Olick 2016, 38), while also emphasizing approaches that define "culture" as the real memory of a society (Assmann 2008a). Current research on politics of memory is "marked by an impressive degree of interdisciplinarity and by a wide range of international empirical objects [of study]" (Zubrzycki and Woźny 2020, 177). At the institutional level, the memory studies boom has resulted in a dedicated journal (*Memory Studies*, published since 2008) and an international academic organization (the Memory Studies Association, established in 2016), and the publication of important handbooks (Misztal 2003; Erll and Nünning 2010; Radstone and Schwarz 2010; Olick et al. 2011; Kattago 2014; Tota and Hagen 2017; Hutton 2016; Gutman and Wüstenberg, forthcoming) and significant book series (including titles published by Berghahn Books, Brill, de Gruyter, Palgrave and Routledge).

Where are then the intersection points between historical sociology and memory studies? In short, they lie in shared interests in history, time, and space. However, those shared interests require some clarification because they do not necessarily imply the same understanding and use of key concepts.

In historical sociology, historical perspective has been employed to study change in societies. Of course, history has been seen as a discipline with which historical sociology has engaged in more or less friendly epistemic debates concerning the use of evidence (with historians privileging arguments based on primary sources over secondary literature exploited by sociologists) and legitimate levels of generalizations (the Weberian method of ideal types often employed by sociologists is usually treated with suspicion by historians). In turn, memory studies have initially developed its conceptual apparatus by opposing "history" and "memory," believing history to be both a reservoir of past events developed in linear time and an objective science aiming at studying those events, and conversely, collective memory to be a combination of facts and fiction, taking place in nonlinear time and governed by emotions and affects, to be studied by sociologists (Halbwachs 1925). Later, for various reasons, not least under the influence of the narrative turn in humanities, memory scholars started to consider academic history as a specific form of collective memory (Assmann 1992). Moreover, those memory studies' scholars who are trained as historians have developed historical approaches to studying collective memories, including concepts such "history of the second degree" (Nora 1984–1986) or "mnemohistory" (Assmann 1998). Such approaches are of particular interest to historical sociology. With regard to time, unless temporalities become its specific research interests, historical sociology has been employing its linear understanding. Conversely, memory studies' main focus has been on reconstruction of various temporalities and their relations to the content of collective memories. Finally, one can see that both historical sociology and memory studies have undergone similar trajectories relating to conceptualizing space from a predominant stress on nation in the second wave to transnational spaces in the present research. And yet, the focus on space has been for years much more central to collective memory studies because of their core assumption that the processes of remembrance are spatially framed.

In what follows, without pretense to being exhaustive, this chapter tries to reconstruct the main foci of sociological perspectives in memory studies, along with other approaches whose arguments are closely linked to historical sociology. Overall, it argues that memory studies are perfectly positioned to add to both conventional and novel inquiries in historical sociology because the primary focus of the research is on the variety of temporalities and spatial references employed to think about history.

Collective Memory: Foundations

It was Émile Durkheim (1858–1917) who laid the foundations for the sociological perspective on memory. In a famous passage from *The Elementary Forms of Religious Life* (*Les Formes élémentaires de la vie religieuse* 1912), he described religious beliefs performed in and thereby strengthened by commemorative rituals incorporating affective and emotional elements as the binding forces of a social group.

> The mythology of a group is the system of beliefs common to this group. The traditions whose memory it perpetuates express the way in which society represents man and the world; it is a moral system and a cosmology as well as a history. So the rite serves and can serve only to sustain the vitality of these beliefs, to keep them from being effaced from memory and, in sum, to revivify the most essential elements of the collective consciousness. Through it, the group periodically renews the sentiment which it has of itself and of its unity; at the same time, individuals are strengthened in their social natures. (Durkheim 2008 [1912], 375)

In the passage quoted above, Durkheim was using the example of totemic societies. Given that Durkheim's key research interest, like that of many other European sociologists at the turn of the twentieth century, was forms of social integration in a modern world where traditional bonds were no longer effective, it could at first sight seem surprising that he and his school started their sociological inquiry by studying religious representations of premodern societies. And yet, Durkheim and his closest collaborators, Marcel Mauss and Henri Hubert, believed that they ought to be looking at indigenous and ancient societies because their alleged "lesser complexity" made it possible to gain insights into the very essence of general organization and universal features of human societies. Durkheimians also thought that representations from the sacred sphere stood for the most important values; societies worshipped the sacred because it presented an opportunity for them to represent themselves in it. One of the most influential concepts Durkheimians popularized while undertaking this research was that of "collective representations" standing for beliefs, values, and symbols shared by community members. In the light of Durkhemian School, collective representations, often conveyed through rituals, not only expressed social relations but also served as a tool for their regulation. That concept directly influenced the idea of "collective memory" formulated later by Maurice Halbwachs.

Durkheim also left an explicit legacy on how to think about uses of the past in modern societies. Although he viewed such societies as largely "anomic," that is, suffering from a crisis of values and standards of behavior, he believed that they would eventually establish their normative integrity by means of secular collective representations of the past. "[T]his state of incertitude and confused agitation cannot last forever," Durkheim wrote, "A day will come when our societies will know again those hours of creative effervescence, in the course of which new ideas arise and new formulae are found which serve for a while as a guide to humanity; and when these hours shall have been passed through once, men will spontaneously feel the need of reliving them from time to time in thought, that is to say, of keeping alive their memory by means of celebrations which regularly reproduce their fruits" (Durkheim 2008 [1912], 442–443).

Durkheim also hinted at some additional key characteristics of the social uses of the past: representations of the past are in fact malleable in time, they provide a sense of continuity and are maintained through social performativity. Two members of his school developed these ideas further: Maurice Halbwachs (1877–1945) and Stefan Czarnowski (1879–1937).

Maurice Halbwachs and His Legacy: "Social Frameworks of Memory" and "Collective Memory"

Maurice Halbwachs's approach to memory was both the sign of an intense intellectual struggle in which Durkheimian sociology tried to stake out a path for itself distinct from other scholarly disciplines (while also claiming its superiority over them) and the sign of a time preoccupied with memory. In the interwar period, Lev Vygotsky laid the foundations for the psychology of memory by advancing research on the cognitive functions of logical memory and, more generally, furthering a new approach toward the social foundations of the mind (Wertsch 1985). Another psychologist, Frederic Bartlett, who was influenced by Halbwachs, claimed that memories are anchored in cultural attitudes and personal habits rather than being direct representations of past events (Bartlett 1932). In turn, according to the German philosopher, Walter Benjamin, memory was "embodied perception and affective engagement" that does not so much reveal the past as serves as a mode of critical inquiry into the past (Simine 2013, 71). Finally, German art historian, Aby Warburg, mapped the "afterlife of antiquity" by showing how representations of Western antiquity had been reconstructed over the centuries (Johnson 2012). Warburg's unfinished *Mnemosyne Atlas* became the canonical reference in cultural memory studies. All these diverse approaches have one thing in common – they all reflected on social influences on human memory.

Maurice Halbwachs expanded on Durkheim's early concepts of "collective representations" by explicitly linking them to the vocabulary of "memory" and proposing his influential notions of "social frameworks of memory" and "collective memory." His insistence that memory is intrinsically social, and individuals cannot, therefore, remember independently of society, led him to engage in an argument with Sigmund Freud. He was also inspired by Henri Bergson, from whom he took his interest in the concept of time, but, unlike Bergson, he argued that time, when linked to memory, is socially constructed. He sharpened his sociological arguments in friendly debates on disciplinary perspectives on memory with, on the one hand, historians of the *Annales* school such as Marc Bloch and Lucien Febvre, and on the other, the psychologist Charles Blondel. Halbwachs maintained that an individual can recollect the past only within the social frameworks of language, space and time. Halbwachs developed this argument in a collection of essays published as *The Social Frameworks of Memory* (*Les cadres sociaux de la mémoire*, 1925), in which he argued that individuals can only remember within the context of the social groups to which they belong (family, religious circle, vocational groups, nation, social class, or others). Social groups *frame* both the content of their individual recollections (autobiographical memory) and the ways of representing the group's history (historical memory). Therefore, the content of individual memory depends on factors such as relevance and selection criteria, as well as the emotions and narrative patterns existing within each group. Halbwachs' parallel empirical research on urban life and the working class led him to underscore the relevance of the spatial and material environment within which social groups live, remember or recreate the past. In short, *The Social Frameworks of Memory* became a landmark text in sociology as it advanced the thesis that memories are socially, culturally, temporarily, and spatially contextualized. Halbwachs argued that group dynamics tend to erase representations that could divide individuals, and these dynamics also tend to adopt new values to emphasize (new) traditions that are more in accordance with the present needs. This approach, with its focus on individual memory as both an effect of group socialization and a key tool for maintaining social cohesion, inspired later social psychologists (Middleton and Brown 2011), oral historians (Abrams 2016), and sociologists working on biographical interviews in sociology (Wylegala 2019).

Halbwachs was, however, also preoccupied with advancing another, parallel concept, namely that of collective memory. In his posthumously published collection, *The Collective Memory* (*La mémoire collective* 1950), Halbwachs viewed collective memory as a container for group representations that could be studied independently of individual memories. Nonetheless, he foresaw a potential line of critical argument resting on the claim that adopting his notion of collective memory would be tantamount to accepting the reification some sort of group mind. Consequently, he was explicit in stressing that although not all individuals share the same recollections, the ways they express their autobiographical and historical memories depend on existing wider patterns of social determinants. Halbwachs took up that point as follows: "While the collective memory endures and draws strength from its base in a coherent body of people, it is individuals as group members who remember. While these remembrances are mutually supportive of each other and common to all, individual members still vary in the intensity with which they experience them. I would readily acknowledge that each memory is a viewpoint on the collective memory, that this viewpoint changes as my position changes, that this position changes as my relationship to other milieus changes. Therefore, it is not surprising that everyone does not draw on the same past of this common instrument. In accounting for that diversity, however, it is always necessary to revert to a combination of influences that are social in nature" (Halbwachs 1980 [1950], 48). The two concepts, social frameworks and collective memory, are actually two sides of the same coin, because, in practice, collective memory is not only shaped by interactions within a social group, it also, in turn, shapes individual memories by becoming one of the important "social frameworks" needed for their production.

Maurice Halbwachs's final contribution to memory studies was an empirical work titled *The Legendary Topography of the Holy Land* (*La Topographie légendaire des Évangiles en Terre sainte*, 1941), which showed how several generations of Christian pilgrims' views of the topography of the Holy Land changed over different historical periods. The Christian representations overlapped with Jewish and Islamic memory of the sacred place. The reconstruction of multiple layers of memory inscribed in the topography of the Holy Land helped Halbwachs to argue that each social group uses space as a framework to enclose and retrieve its remembrance and identity. *The Legendary Topography* was also a landmark piece of historical sociology because it underscored the temporal and spatial features of conflicting collective memories and demonstrated memory's malleability. It continues, to this day, to inspire scholarship on religion, as can be seen by the influential concept of "religion as a chain of memory," that is, the idea that mnemonic aspects of religious beliefs and practices make communities from individuals (Hervieu-Leger 2000; Bogumił and Yurchuk 2022).

Since the rediscovery of Halbwachs at the end of the twentieth century, there have been many attempts to reformulate his ideas in such a way as to increase their utility for the research practices of various disciplines. To give an example, the American cultural anthropologist James Wertsch proposed a reconceptualization of how collective memory works by advancing the concept of national "narrative templates" in his work on Russian history textbooks. Wertsch observed that various events in Russian history, such as the Mongol invasion, the Polish–Russian War (1609–1618), the Napoleonic Wars, or the Second World War, are told according to the same schema, which consists of: "(1) An 'initial situation' . . . in which the Russian people are living in a peaceful setting. (2) The initiation of trouble or aggression by an alien force, or agent, which leads to: (3) A time of crisis and great suffering, which is: (4) Overcome by a triumph over the alien force by the Russian people, acting heroically and alone" (Wertsch 2002, 93). In other words, groups possess shared narrative tools which their members use when writing histories. Wertsch also introduced the notion of "mnemonic standoff" to signify a situation in which individuals belonging to conflicted memory groups, such as victims and perpetrators, agree on certain contentious facts, for instance that killings took place, but still remain divided by the more durable narrative templates in which they frame those facts.

In historical sociology, Jeffrey K. Olick has persistently added to the terminology of collective memory, making connections with other sociological traditions, including those of Pierre Bourdieu, Norbert Elias, or Mikhail Bakhtin, and challenging the aforementioned popular criticism of Maurice Halbwachs's theory (and the Durkheimian school in general) for reifying collective memory. Among others, Olick introduces a distinction between "collected memories" and "collective memory" and addresses the gains and losses of the individualistic "collected" approach, which understands group memories as the sum of individual memories, so as not to risk committing the collective mind fallacy, such as in the case of sociological surveys, based on methodological individualism. Nevertheless, Olick favors the "collective" view, believing it more closely represents the interests of historical sociology, while at the same time underlining that the most common misunderstanding among researchers in his field is their tendency to view collective memory as "a thing, and not a process by, wrongly, comprehending it as something what societies have, and not what they do" (Olick et al. 2011, 45). In Olick's broad approach, collective memory is seen as a dynamic process composed of "variety of retrospective activities and products: collective representations (publicly available symbols, meanings, narratives and rituals), deep cultural structures (generative systems of rules or patterns for producing representations), social frameworks (groups and patterns of interaction), and culturally and socially framed individual memories" (Olick 1999). Viewed in this dynamic way, memory studies have significantly contributed to the central questions of historical sociology through the "temporalization of sociological concepts" and the "more processual theorization of modernity" (Olick et al. 2011, 42). Most of the recent research works on collective memory have indeed recognized the complexity of the processes in which memory is both transmitted and changed.

Stefan Czarnowski's Theory of Heroes and the Social Functions of the Past

While Halbwachs's ideas of "social frameworks of memory" and "collective memory" laid the foundation of memory studies and are still being debated today, another second-generation Durkheimian, Stefan Czarnowski, proposed a perspective on the development and functions of uses of the past that was of direct relevance to historical sociology. Czarnowski was a student of Durkheim's close collaborator, the archaeologist Henri Hubert, who contributed to the Durkheimian school with his ideas on the social construction of time. Czarnowski, whose works were published in French and Polish, but never translated into English, became the founder of a Polish strand of social constructivism in national studies. Internationally, his ideas circulated in various specialist contexts from religious to hero studies. But his significance for the memory studies has only been rediscovered recently (Schwartz 1996, 2000; Kończal and Wawrzyniak 2018; Vasilyev 2020).

Although Czarnowski never used the term "collective memory," he advanced a concept of historical heroes and heroines that accorded with the Durkheimian understanding of the role of collective representations. His notion of heroes, in which societies represent their core values and beliefs, was not only close to Durkheim's view of the social function of "totems," but also to some of Halbwachs's reflections on the role of collective memory. "The notion of hero is a function of several synthetic judgements, on the one hand, on how a group comprehends its existence as a community and, on the other hand, on how it emotionally relates to what constitutes its fundamental social value," Czarnowski claimed in his book *The Cult of Heroes and Its Social Condition: St. Patrick, National Hero of Ireland (Le culte des héros et ses conditions sociales: St. Patrick, héros national de L'Irelande* Czarnowski 1919, 234; *trans. JW*). In this work, he explored the case study of St Patrick's legend, arguing that, in the process of Ireland's Christianization, images of this fifth-century Christian missionary were gradually transformed, in response to changing values and political concerns, into representations depicting him as Ireland's national hero. In particular, Czarnowski argued that the cult of St Patrick was successfully established, because it connected old pagan beliefs with Christian mythology rather than replacing the former with the latter. Moreover, St Patrick's cult played an important role in the power struggle between two elite groups: the more traditional druids and more change-oriented filids (who weakened the druids' position in the Celtic clans, by turning to Christianity and promoting the worship of saints that incorporated the traits of Celtic heroes). The cult became useful in moments of danger and crisis like, for instance, when the Irish clergy chose to advance Patrick's cult when they were fighting for their autonomy against the Roman clergy; the Scandinavian invasions also contributed to St Patrick's elevation as a symbol of Irish unity. Czarnowski's book was a landmark piece of early historical sociology because it combined the theoretical concerns characteristic of the Durkheimian school with a multitude of historical details recovered in scrupulous work. In the main, it focused on change over centuries and illustrated well the historical dynamics of collective memory (Wawrzyniak 2019).

In his later work, Czarnowski (1938) developed his theory of culture encompassing reflection on the social construction of space, time and history. His concept of the "past in the present" was an equivalent to Halbwachs' notion of collective memory. By defining society as composed of "a small handful of the living and an immense mass of the dead" to underly his point on the role of the historic representations for embodying socially relevant ideas, values, and symbols, Czarnowski claimed that the representations of the past are strictly dependent on the present. Referring to historical, anthropological, and religious data, he explored diverse temporalities and different constructions of the past.

Czarnowski's ideas played a key role in forming a sociological reflection on time, space, and history in Polish sociology. To international readers he has been known mainly because of his previous work on heroes. While several generations of scholars have referred to it rather ritually, the most creative and unusual reception came from James Joyce, who was inspired by Czarnowski's ideas while writing his last book, *Finnegans Wake* (Joyce 1939; O'Dwyer 1980). One of this novel's protagonists has the capacity of disappearing in a state of dormancy from which he can be summoned back to reinvigorate society. The social willingness to call him back is realized through various rituals, including story-telling of legends of his life and death. In such a way Joyce connected the idea of the re-emergence and recollections of the past, inspired not only by the religious practices referenced in Czarnowski's *Cult* but also by the book's main claim on the socially bonding role of the rituals, performances, and invented histories.

From Collective Memory to Collective Forgetting

The second generation of Durkheimians, and Halbwachs and Czarnowski in particular, contributed to memory studies with concepts and research relevant to understanding how the past is being used in the present. While they demonstrated well the social functions of collective memory, the questions of what, why and how societies forget were only of secondary interest to them. However, that issue was addressed by many authors during the second and third wave of memory studies research. In his philosophical account of memory and forgetting, Paul Ricoeur (2004) pondered over the intrinsic relations between those two processes, paying attention, on the one hand, to acts of remembrance as struggles against forgetting, and on the other, to attempts at forgetting that involved memory. In the history and sociology fields, there is now a growing body of literature exploring silence as a "complex and rich social space that can operate as a vehicle of either memory or of forgetting" (Vinitzky-Seroussi and Teeger 2010). British historian, Jay Winter (2010), distinguishes different types and social constructions of silence that encompass the silence of those who cannot speak and those who choose not to speak, as well as the silence of groups with diverse dynamics, which is not the same as forgetting. Recent research by Holocaust scholar, Roma Sendyka (2016) focuses on identifying "non-sites of memory," namely, vernacular places that remain non-commemorated, yet they are familiar to local communities as spaces where killings took place, and they generate their affective reactions of shame, fear, and disgust.

From the perspective of historical sociology, Paul Connerton (2008) systematized reflection on collective forgetting by distinguishing seven types of such forgetting in terms of their social functions. Connerton proposed to speak of (i) "forgetting as repressive erasure," which occurs when a group's collective memory is subjected to physical or symbolic violence. Numerous instances of such forgetting can be gleaned from the struggles of ancient and modern regimes against the memories of their opponents. Connerton then introduces the concept of (ii) "prospective forgetting," which aims at restoring a group's cohesion after a great disruption such as a war, for instance, by underplaying the scale of the experienced traumatic events, as was the case in the immediate aftermath of the Second World War in Europe. More broadly, he argues for the existence of: (iii) "forgetting constitutive in the formation of a new identity," that is, forgetting that discards unworkable, undesirable, or unneeded elements of the past; (iv) "structural amnesia," which refers to those elements of the past that are overlooked due to a deficit of information at the expense of those that are considered more socially important; (v) "forgetting as annulment," which stems from the surfeit of information in print and digital societies; and (vi) "forgetting as planned obsolescence," when innovation erases old knowledge and skills. Finally, there is (vii) "forgetting as humiliated silence," which relates to situations when individuals and groups have no way of telling their stories because of a variety of reasons from social taboos to individual and collective traumas.

This brief overview of a few old and new canonical works in memory studies shows that if historical sociology's main focus is the mechanisms and structures of reproduction and change in time, then studies of collective memory and forgetting have significantly contributed to this line of inquiry. Almost a century has actually passed since the Annales historian, Marc Bloch (a friend, critic, and continuator of Halbwachs who wrote about the role of memory in feudal societies) noticed the importance of *The Social Frameworks of Memory*. He argued that Halbwachs' work "does us a great service that no one better than a historian – too often isolated by the necessities of work, lost in the details of erudition – could appreciate: he pushes us to reflect on the conditions of the historical development of humanity; indeed, what would this development look like without 'collective memory'?" (Bloch 2011 [1925]).

The Second Wave of Memory Studies: Nations, Power Struggles, Media

There have been various political and social reasons for the return of interest in the 1980s and 1990s in the concept of collective memory. Decolonization, the decay of communism, and the collapse of authoritarian regimes in Asia, South America, and Southern and Eastern Europe were followed by the establishment of various institutions dealing with transitional justice for perpetrators and victims of past crimes. The Second World War and the Holocaust, in particular, were revisited by the generation of 1968, who gained a position of public importance in the 1980s. More generally, the democratization of public life was coupled with a growing demand for the articulation of the memories of various groups long silenced historically, including women, people of color, and impoverished classes. The changing of carriers of memory from print to digital form has also played a role.

It is often assumed that the so-called second wave of memory studies was preoccupied with national frameworks of memory. While it certainly included groundbreaking scholarship at that issue (discussed below), the second wave was in fact much broader, encompassing also focus on media perpetuating cultural memory in the *longue durée*, as well as renewed interest in social frameworks of memory (family, gender, generations, class). Among them the idea of class memory has been critically examined in several oral history projects dealing with the legacies of Nazism and fascism (Niethammer and Plato 1987; Passerini 1987; Portelli 2017). Also, reconsideration of Karl Mannheim's (1952 [1928]) classical concept of generations opened new avenues of research on collective memory. For instance, a series of quantitative surveys conducted in several countries confirmed that memories of political events are dependent on age and that the early adulthood is the formative moment for "generational imprinting" of collective memories (Corning and Schuman 2015). Intergenerational transmission of memories within families has been another important subject of multidisciplinary research. Family involves, on the one hand, face-to-face interactions and emotional bonds, and on the other, it is a major institution of social structuration and socialization. In a family, various processes of remembering intersect with each other: individual memories, family stories, national collective memories (Welzer 2005). Marianne Hirsch (2012) coined the term "postmemory" to denote the collective memories of the generation of descendants of the Holocaust's survivors. They have been indirectly exposed to violence by means of (fragmented) narratives, (incomplete) images, and above all affective everyday practices in which they grew up. In effect they bear, collectively, the trauma of the others. Under specific circumstances, family memories can become politicized, as it happened, for instances, in the case of the Madres de Plaza de Mayo in Argentina, where women publicly protested against the violations of human rights of their relatives. More generally social frameworks of memory started to be reconsidered during the second wave and the interest in them has continued until the present moment. And yet it has been political structures-oriented research that made to the forefront of memory studies.

Collective Memory and Nation States

Many authors contributed to the debunking of uses of the past in nation states through the double lens of constructivist approach and modernization theory. Their studies show how states maintain their legitimacy by nurturing institutional memories, investing in history education, archives and museums, managing history policies, introducing memory laws, and supporting political commemorative performance. The constructivist perspective on nation states as products of nationalism originated in Ernest Renan's essay *What Is a Nation?* (1882), which argued that nations are based as much on what people jointly forget as what they remember. Renan's idea was developed a century later by Benedict Anderson in his work on "imagined communities" (Anderson 2006 [1983]). It was the emergence of print and wide circulation of books and newspapers, Anderson argued, that first made it possible for individuals to believe they shared common pasts with complete strangers. Moreover, Anderson identified several processes leading to the emergence of modern nations, challenging the idea that nationalism originated in Europe: as

Anderson pointed out, the earliest nationalist liberation movements were the children of "creole pioneers" in the North and South Americas. Also, in the colonial context of Asia and Africa, local elites were able, thanks to their travel and educational experiences, to realize their aspirations toward liberation and power by producing historical imaginaries of common roots. In addition to paying attention to linguistic nationalism by intellectual movements, as well as to the state-led nationalism as a response to those movements in large empires, Anderson emphasized the importance of monuments and museums representing common history and added two other key aspects of nationalism: the census, which counts the people, and the map, which symbolically demarcates their political space.

While, for Anderson, nations were product of modernization, the historians Eric Hobsbawm and Terence Ranger deconstructed "invented traditions" in a co-edited volume (1983), in which they stressed their fabrication and historical falsity. Despite the fact that they never quoted Halbwachs, they built an argument quite sociological in nature, as it viewed invented traditions as "a set of practices, normally governed by overtly or tacitly accepted rules and of a ritual or symbolic nature, which seek to inculcate certain values and norms of behaviour by repetition, which automatically implies continuity with the past" (Hobsbawm and Ranger 1983, 1). The volume, above all, presents a plethora of evidence that invented traditions served as tools for bolstering the legitimacy of elites and their power apparatuses in different parts of the world in the nineteenth century. This process was particularly evident in Africa where the traditions the British invented there took different forms to those in Europe or Asia. While in Europe the ruling classes were under a certain compulsion to negotiate their invented traditions with the lower classes, and in India collective representations were balanced against the local imperial traditions, in Africa invented pasts became a simple matter of command, control, and exploitation (Hobsbawm and Ranger 1983, 211).

The work of French historians under the guidance of Pierre Nora on *lieux de mémoire* (sites of memory) was another case in point of a historical study on nation states informed by sociological thinking (Nora 1984–1986). Not only did the authors identify sites, artifacts, and symbols at which collective memory crystallized in France (such as the Louvre, Marianne, July 14, or the struggle between the Gaullists and the Left), but they also meticulously traced their histories. The effect of such crystallization processes was what Nora called "the history of the second degree" of representations, their use and re-use, and entanglements with various present(s). In his introduction to over a 100 essays (divided into three main parts focusing, respectively, on *La République*, *La Nation* and *Les France*), Nora departed from Halbwachs by arguing that in late modern societies, collective memories had disappeared together with their natural environment of premodern *milieux de mémoire* (communities of memory). Also, national memories, the product of nineteenth-century nationalism, had lost their binding force. Therefore, Nora proposed a shift toward *lieux de mémoire* (sites of memory), which function as artificial substitutes for the no-longer-existent "natural" collective memory (Nora 1989). Since the publication of his several-volume study, the concept of *lieux de mémoire* has embarked on an international career and became a favored lens for studying national remembrance in many countries. However, the project has also been criticized for Nora's Eurocentrism and an approach that ignores both the French colonies and the French of diverse origins living in France (Tai 2001). Also, Nora's view that *milieux de mémoire* no longer exist has been profoundly questioned by works on family memory and local community memory (Bogumił and Głowacka-Grajper 2019).

Collective Memory and Power

Another key focus of memory studies scholarship has been on uses of the past in power struggles, and many case studies have contributed to the identification of the mnemonic mechanisms behind the securing of legitimacy of power in nation states. The remarkable shifts in Germany's methods of working through its Nazi past led to it becoming a particularly densely researched case study. In a full-fledged historical sociological analysis, Olick (2016) distinguished between three periods in the West German post-war memory politics by tracking changing constellations of collective memory, official memory, institutional structure, and political culture. In the first period, collective memory influenced the redesign of institutional structures and commemoration policy, which had a knock-on effect on political culture. In the second period (the 1960s), generational and institutional turns influenced changes in political culture, which resulted in further changes in collective memory. Finally, in the third period (the 1970s and 1980s), shifts in political culture toward neoconservatism brought changes in politics that led, in turn, to changes in official memory.

Scholars studying Eastern Europe and Asia in twentieth century have shown how future-oriented universalist revolutionary and working-class rhetoric has interplayed with national and religious local mythologies. The communist upheavals that uprooted the existing political and social order did not automatically result in changes within vernacular memories that continued to strive to commemorate the past and seek sources of identity and resistance in the past. The revolutionaries attempted to create a version of the official memory according to which they would be seen as the most legitimate holders of power. Such a task entailed gaining control over existing narratives of the past to stop them from falling into the hands of enemies of the revolution. This explains why newly appointed political authorities not only eliminated old representations of the past through "repressive erasure" (see Connerton's term introduced in the previous section of this chapter) but also creatively adapted existing images, symbols, and rituals (Wertsch 2002; Zaremba 2019).

Political memories of communist countries were additionally complicated by the experience of the Second World War. Focusing on how local authorities managed the past and present in the region of Vinnytsia in Central Ukraine, Amir Weiner (2002) analyzed how a myth of the Great Patriotic War combining revolutionary and national representations served as an essential social engineering instrument from the 1940s until the dissolution of the Soviet Union. In Soviet propaganda, the war was presented as a step toward communism taken by the masses. However, the war altered the primary focus of the revolutionary conception of the political community: class-based solidarity was displaced by the bonds that had resulted from blood spilled in the name of the motherland. The war also changed the archetypal hero, transforming the "revolutionary" into the "defender of the Soviet Union." Similarly, the ideal enemy changed from the "kulaks" and "enemies of the revolution" to "traitors" and "collaborators." Thus, the war myth transformed the basic categories and relations operating in Soviet society. According to Weiner, however, if the war had not genuinely affected social life, it would not have been possible to anchor the myth in society as an object of political manipulation. Once that process had taken place, the myth was not only sustained by communist institutions and activists, but also by informal structures. For that reason alone, attempts to use public memory of the war were much more effective at legitimizing the political system than appeals to the ideology of the revolution. The myth of the Great Patriotic War was so convincing that it out-survived Perestroika and still maintains its popularity today. Similarly, Mark Edele (2008) analyzed the rise of the veteran movement in the USSR from its beginnings in the immediate post-war period, when it was not officially welcomed, to the cult of the war period under Leonid Brezhnev's leadership, with the favorable climate it provided for transforming veterans into "a status group" benefiting from "scarce goods, services, and esteem" despite shortages in the communist economy. Edele's account ends with the veterans being established as "an institutionalized pillar of the political system." More generally, accounts like those from the Soviet Union and other case studies from other Eastern European countries such as Yugoslavia, East Germany, and Poland revealed the processes underlying negotiations between the proponents of narratives imposed by the state and various memory groups who strove, despite authoritarian constraints, for their version of history to be officially recognized.

Collective Memory, Media and the *Longue Durée*

The second wave of memory studies brought also landmark scholarship on the changes of forms of memory from a wider perspective. Memory scholars distinguish entire epochs based on the ways in which people remember the past and how the inventions of key media change them. The collective memory of oral societies crystallized in myths of origin and genealogy. Subsequently, the inventions of epigraphs and documents altered memorialization from hearing and repeating to seeing and rewriting (Le Goff 1992). Ancient civilizations brought libraries and archives, and, above all, calendars affecting different temporalities. Greek culture, which encompassed the complexities of both oral and written cultures, invented key mnemonic notions that for centuries influenced European cultures, societies and legal systems (Assmann 2011). The mnemotechnics, the art of memory, emerging in the Antiquity, later was connected with the Middle Ages or the Renaissance occult memory systems. As part of rhetoric, it was preserved until modernity, when mnemotechnics was abandoned due to the broad expansion of print and archiving systems (Yates 1966; Anderson 1983). The digital revolution replaced "collective memory" by "memory of the multitude" in two ways. Passive readers of print media were transformed into active users of digital media. And the relation of collective memory to place and space became problematic, because memory is transmitted, fragmented, and transformed in various digital channels and networks (Hoskins 2018; Mandolessi 2021).

One of the most influential ways of seeing continuity and change in collective memory in the longue durée was elaborated by German Egyptologist, Jan Assmann, whose "mnemohistory," viewed as a branch of historiography, is "concerned not with the past as such, but only with the past as it is remembered." The focus of mnemohistory is "cultural memory," which constitutes the storage of a given group's knowledge, which "exists in two modes: first in the mode of potentiality of the archive whose accumulated texts, images, and rules of conduct act as a total horizon, and second in the mode of actuality, whereby each contemporary context puts the objectivized meaning into its own perspective, giving it its own relevance" (Assmann 1995). In *Moses the Egyptian*, Assmann (1998) employs his concept of mnemohistory to study how ancient Egypt was remembered in Western culture from the Antiquity, through the Renaissance to Sigmund Freud's view on Moses. In particular, Assmann introduces the concept of the Mosaic Distinction related to representations of Moses (the differentiation between allegedly "true" monotheistic religions and "false" polytheistic religions) to illuminate how monotheistic cultures denied the Egyptians any part in their roots by condemning them as pagan idolaters. The Mosaic Distinction was paralleled by the concept of "the other" and became a source of religious violence in the world, most of which has been directed by monotheists against polytheistic religions. In this way Assmann connects some key themes that perpetuated European cultures encounters with the world from the Antiquity to the present day.

The Third Wave: Transnationalization and Memory Activism

Memory studies' interest in normative concerns has never been as acutely visible as it is now in what is its third wave, which is being shaped by the influence of postcolonial studies, transnational perspectives, and a global approach. The examples brought to the fore at the beginning of this chapter show that the current shift in humanities and social science has called into question some of the European identity tenets. In effect, the recent memory boom is posing significant challenges to the field of historical sociology that also have the potential to inform new research trends. I shall comment on two of these trends in memory studies: the transnational turn and the activist turn.

Transnational Turn

The first of these parallels the unprecedented transnationalization of research in the humanities and social science in general. Concepts such as "European memory"

(Pakier and Stråth 2010), "travelling memory" (Erll 2011), "entangled memories" (Feindt et al. 2014), "memory unbound" (Bond et al. 2016), and "transnational memory space" (Wüstenberg and Sierp 2020) currently proliferate in memory studies and look set to supplant the previous research stage based on methodological nationalism. This scholarship postulates that modern research should primarily concentrate on the travel, movement, and circulation of memory (De Cesari and Rigney 2014). The research foci are diverse, varying from everyday memories in the diaspora to memory lawmaking in international courts or the circulation of representations on the Internet.

One of the most influential is a discussion that relates to the extent to which globalization and the fragmentation of contemporary memories are taking place. Reflecting on the Holocaust and memory, Daniel Levy and Nathan Szneider (2006) argued that national memories were decaying in the global age. They labeled the new form of memory that had supplanted them "cosmopolitan memory." In their view, cosmopolitan memory gained in importance in the 1990s has been universal in adopting global values such as human rights and has revolved around the Americanized and mediatized vision of the Holocaust. Szneider and Levy's concept has met with mixed reception. Accepted by many, it has also been treated with skepticism: its diagnosis of the decay of nation states is premature and its hopes for the positive impact of cosmopolitan memory on global solidarity and humanistic care unrealistic. Anna Cento Bull and Hans Lauge Hansen (2016, 1–2) have argued that "cosmopolitan mode of remembering... has proved unable to prevent the rise of, and is being increasingly challenged by, new antagonistic collective memories constructed by populist neo-nationalist movements." In addition to "antagonistic" and "cosmopolitan" modes, they propose "agonistic" mode which implies reflexive, dialogic, multi-perspectivist framework of remembrance which does not however assume consensual outcome of the juxtaposing of different memories of the past, rather their effective coexistence.

How can historical sociology add to and benefit from the transnational turn in memory studies? A way to do it is to rethink how globalization has inspired memory studies to search for spatial alternatives to nation states and pay more attention to supranational connections and networks. In line with contemporary historiography, the "global" does not need to refer so much to the object of study but to following connections between the mnemonic agents, carriers, forms, events, sites, and practices across time and space. For instance, in a recent work, we introduced the notion of "regions of memory," while seeking to highlight processes that transcend national boundaries to illuminate lesser-known directions of memory movement in space Such regions are not defined by their spatial proximity, but rather by the effect they have for discourses about and representations of issues that bring them together topically, whether or not they are in close geographical proximity (Lewis et al. 2022). For instance, such spaces as Western Africa, Portugal, and Brazil are connected by the memories of land conquest, slavery and decolonization, or, East-Central Europe has been shaped by the layered and entangled legacies of the Holy Roman, the Habsburg, Ottoman, Romanov, Hohenzollern, Nazi, and Soviet Empires.

Jie-Hyun Lim and Eve Rosenhaft propose a notion of "mnemonic solidarity" in search of historicizing the dissonant pasts in global context, with a special consideration of the global South. They take for granted that global circulation of memories takes place "through processes of and practices of translation, cross-referencing, adaptive imagination, unilateral 're-purposing' and active dialogue, as well as competition" (Lim and Rosenhaft 2021, 4). And yet, they notice that this circulation involves two opposite directions of de-territorialization and re-territorialization in national contexts of universal mnemonic symbols such as the Holocaust. Finally, they postulate to study the regions of global South – Africa, Asia, and Latin America – on their own, and to elicit mnemonic voices of subaltern actors for global discourses.

Activist Turn

The current "activist turn" in memory studies is of interest to historical sociology because, thanks to its perspective "from below," it complements previously dominating focus on elites as agents of commemoration. The term has been variously used by different authors; however, the conceptualization particularly attractive to social science has been offered by Yifat Gutman and Jenny Wüstenberg, who define memory activism as strategic action that intentionally targets official memory to provoke or resist mnemonic change (Gutman and Wüstenberg 2021; Gutman and Wüstenberg, forthcoming). In this conceptualization, memory activists are civil society and social movements agents and, as such, even though they often act locally, they might draw on the global or regional mnemonic discourses. Memory activists come in various public roles: they can identify as public intellectuals, journalists, scholars, legal experts, religious leaders, artists, curators, associations of victims of violence or their families, and many others. Their aim is change of official memories encoded in public commemorative policies and law, educational practices, heritage site management, or media discourses and representations. To give illustrations of how memory activism works, one can point to how grassroots civic engagement in commemorating civilian victims of the Second World War has affected policy processes in the European Union (Lagrou 2009), or, bottom-up processes of remembering and forgetting of state orchestrated violence in military dictatorships in Latin America paved the way to truth-seeking committees and transitional justice measures in the region. The activism of Korean "comfort women" has contributed to global condemnation of war sexual violence. Memory activists have struggled against the institutional repression of varied events and processes like the Gulag, the Nakba, stolen generations, or child sexual abuse within the Catholic Church. In such cases, by raising the issues of public recognition of violence, memory activists bolstered the awareness of silenced pasts, put together the recollections of survivors, and supplied evidence for courts.

While memory activism is best known for its engagement against cases of mass atrocities, structural violence, and administrative racism, its potential for influencing historical change does not stop there. Among others, memory activists are important social actors trying to preserve the memory of environment, on the one hand, or an industrial past, on the other. For instance, in deindustrialization studies they are seen as agents who mobilize nostalgic resources to protect values and forms of social life that remain relevant in the present, but are endangered by political, economic, cultural, and technological change (Berger 2019). In short, memory activists enact the past to generate alternative processes of memory that challenge its dominant regimes, revisit historical circumstances, and current social relations to imagine new possibilities.

Although memory activism, as an analytical category, is a new addition to the scholarship and most of the research work done so far under its rubric has focused on the cases of challenging official narratives in the twentieth century, it can also become a handy tool to discuss cases from more distant historical perspective. As such it complements research on the bottom-up "counter-narratives" or "revisionist memories" by individuals, groups, and movements contesting official versions of the past. For instance, commemoration of the Persian Wars was the subject of contestations among the Greeks. In ancient Rome, L. Cornelius Sulla systematically destroyed monuments of his opponent C. Marius, opening the tradition of organized forgetting in public space (Stein-Hölkeskamp 2015). In the Middle Ages and early modernity, peasant revolts and religious movements arose from memories of injustice, and subsequently also became the subject of vernacular commemorations (Erdélyi 2015). As already discussed in the previous section, in the nineteenth century, memory activists contested on a mass scale the official memories of empire by reinventing other usable traditions for alternative political communities.

Overall, the relation between the activist turn and historical sociology can be approached by asking three interrelated questions. First, by inquiring how memory activists employ, contest, or change existing social, political, and cultural structures in the pursuit of their aims, and how memory activism institutionalizes and enforces new political, cultural, or social norms. On this point, Gutman and Wüstenberg (2021) provide a useful typology depended on the "mode" and "role" of memory activism. With regard to the "mode," memory activists can be antagonist "warriors" or agonistic "pluralists" in relation to other actors and institutional design of the state (see also Kubik and Bernhard 2014). "Warriors" respect only their own interpretation of history and in the struggle for this interpretation they might challenge the structure of the existing political institutions, while "pluralists" believe that there is enough space for various approaches to the past, making them permissive to others' positions as long as their fit the democratic "rules of the game." According to Gutman and Wüstenberg, both "warriors" and "pluralists" can operate in two different temporalities: "the past has ended" and "the past is on-going." In the former, commemorative activism is undertaken to assure that a condemned event will never repeat again. In the latter case, memory activists intervene in the course of events in an attempt to avoid violence or to follow it to create evidence for the sake of future commemoration in the post-conflict situation. The discussed typology also identifies roles performed by memory activists, including "victims," "resisters and heroes," "entangled agents," and "pragmatists." The experience of suffering is the essential impetus for victims' memory activism. In turn, resistors and heroes perform their agency as "moral participants in historical events" rather than as victims or implicated bystanders. Entangled agents do not necessarily have any direct links to the historical events but nevertheless feel obliged to commemorate them. Finally, pragmatists become activists due to more general commitments, for example, to democratic norms or to professional standards, and might also seek to function as mediators. Although Gutman and Wüstenberg's typology works best for the cases of twentieth-century memory activism related to mass violence, it can also inspire further historical reflection on the functions of memory activism in *longue durée*.

Second, one can ask how political, cultural, and social institutions sanction, shape, or prevent certain types of memory activism. For instance, forms of governance and memory policies affect possible forms of memory activism because they define the basic rules of commemoration and also open new prospects for commemorative actions (Dybris McQuaid and Gensburger 2019). Also, concrete design of heritage sector may offer both opportunities for and constraints on memory activism. In particular, museums and memorial sites provide spaces at which grassroots memory activism might start changes in wider institutional frameworks (Marcuse 2008; Sodaro 2018). For instance, Stefan Berger (2019) has argued that post-industrial memory activism can aid the building of "embedded capitalism," counteracting inequalities of neoliberalism, but if it is to become a socially productive and collectively shared sentiment, it needs to be supported by the heritage sector opened for the variety of "practical pasts" that are likely to empower communities.

Finally, one can examine what institutional legacies of the past are commemorated by memory activists and what agenda they entail for the future. As argued at the beginning of this chapter, the present decolonization discourse calls for the development of a new relational ethics between European and non-European countries. In and beyond museum discourse, memory activism revolving around racism and slavery in the Western World (Araujo 2020) or serfdom in Eastern Europe (Leszczyński 2020) are cases in point. Here, progressive activism clashes with right-wing movements, and re-emergent memories of the centuries-old structural violence against ethnic minorities and peasants have caused a "boomerang effect," as right-wing and populist groups strive to protect a positive image of the past. On both sides of the memory activism, collectively constructed nostalgia for "the golden age" serves as a means of activist mobilization. Nostalgic memory activism may be directed at reviving pasts lost in the colonization period, giving voice to subaltern groups, but it may also be an expression for heydays of empire, woven of patriotic stories, and national sentiments. Overall, both transnational and activist turns have provided the plethora of topics to rethink from a historical sociological perspective.

Conclusion

This chapter attempted to explain why collective memory studies matter so much for historical sociology by viewing the development of memory studies in three stages. The foundational moment for the study of collective memory can be traced to the publication of the seminal works of first- and second-generation sociologists such as

Emile Durkheim, Maurice Halbwachs, and Stefan Czarnowski, who shed light on why uses of the past are among the main factors binding social groups together. They also drew attention to basic assumptions that could be made about collective memory: it is built from elements representing social values, it is closely tied to emotions and affects, and, needless to add, it does not have to rely on factual accuracy.

The second phase was marked by a return of interest in collective memory in the late twentieth century. That phase was most notable for the laying of the unprecedented interdisciplinary conceptual groundwork required to facilitate the development of memory studies as a field of inquiry in its own right. From a historical sociological perspective, the most significant works from this phase reflected on interdependencies between nation states, nationalism, and the construction of the past. Authors such as Benedict Anderson, Ernst Gellner, and Eric Hobsbawm have described how the past is used for the purposes of the legitimation of power and in power struggles. A plethora of historical works have also appeared on how the usage of various media changes the organization of human memory.

Despite skeptics' predictions that memory studies would decline after the boom years of the 1980s and 1990s, the field became institutionalized, allowing it to enter its third phase, which has lasted ever since. There are also voices anticipating the fourth wave related to the memories of environment and Anthropocene (Craps 2018). There has been a growing number of research works published within and across disciplines and a rise in the relevance of memory studies for the heritage sector accompanied by demands from memory activist movements for changes to be made to public accounts of history. Supplemented by the findings of postcolonial studies, the third phase challenges the methodological nationalism and Eurocentrism of previous research. From a historical sociological perspective, it invites scholars, heritage professionals, and policymakers to rethink not only the issues of agency and memory, but also the spatial frameworks of global, regional, and local memory in the *longue durée*.

If historical sociology aims to bring history back into the fold of sociological inquiry, then the current memory phase could well be its closest ally. This chapter has argued that the reflections of collective memory studies can provide further insights into the most enduring concerns of historical sociology, notably, epochal social transformations, modernization processes, class formation and dissolutions, and the origins and decay of large state structures such as empires and nation states. However, memory studies, along with its conceptual toolkit created amid contemporary transnational inquiries and grassroots struggles over uses of the past, could also inspire a new wave of research within historical sociology. Not only do memory studies give a prominent role to "history of the second degree," to recall Pierre Nora's concept, but they can also aid critical reflection on older, traditionally Western-centric categories of historical inquiry. The alliances formed between scholars of collective memory and memory activists make the field sensitive to the diverse uses of the past around the world. Overall, memory studies occupy an important position at the forefront of today's humanities and social science and, therefore, deserve close attention of historical sociologists.

References

Abrams L (2016) Oral history theory. Routledge, London/New York
Anderson B (2006 [1983]) Imagined communities. Verso, London
Araujo AL (2020) Slavery in the age of memory. Engaging the past. Bloomsbury Academic, New York
Assmann J (1992) Das kulturelle Gedächtnis, Schrift, Erinnerung und Politische Identität in frühen Hochkulturen. Verlag C. H. Beck, Munich
Assmann J (1995) Collective memory and cultural identity (trans: Czaplicka J). New Ger Crit 65: 125–133. https://doi.org/10.2307/488538
Assmann J (1998) Moses the Egyptian. The memory of Egypt in Western monotheism. Harvard University Press, Cambridge, MA
Assmann A (2008a) Canon and archive. In: Erll A, Nünning A (eds) Cultural memory studies. An international and interdisciplinary handbook. De Gruyter, Berlin, pp 97–108
Assmann J (2008b) Communicative and cultural memory. In: Erl A. Nünning, A (eds), Cultural memory studies. An international and interdisciplinary handbook. De Gruyter, Berlin/New York, p. 109–118
Assmann A (2011) Cultural memory and Western civilization. Functions, media, archives. Cambridge University Press, Cambridge, UK
Bartlett FC (1932) Remembering. A study in experimental and social psychology. Cambridge University Press, Cambridge, UK
Belavusau U, Gliszczyńska-Grabias A (eds) (2017) Law and memory. Towards legal governance of history. Cambridge University Press, Cambridge, UK
Bellah RN (1959) Durkheim and history. Am Sociol Rev 24(4):447–461
Berger S (2019) Industrial heritage and the ambiguities of nostalgia for an industrial past in the Ruhr Valley, Germany. Labor Stud Work Class Hist Am 16(1):36–64
Berger S, Scalmer S, Wicke C (2021) Remembering social movements. Activism and memory. Routledge, London
Bhambra GK (2010) Historical sociology, international relations and connected histories. Camb Rev Int Aff 23(1):127–143. https://doi.org/10.1080/09557570903433639
Bhambra GK (2014) Connected sociologies. Bloomsbury, London
Bloch M (2011 [1925]) Mémoire collective, tradition et coutume. A propos d'un livre récent (trans: Silva JM). In: Olick JK, Vinitzky-Seroussi V, Levy D (eds) The collective memory reader. Oxford University Press, Oxford, UK, pp 151–155
Bogumił Z, Główacka-Grajper M (2019) Milieux de mémoire in late modernity. Local communities, religion and historical politics. Peter Lang, Frankfurt am Main
Bogumił Z, Yurchuk Y (2022) Memory and religion from a postsecular perspective. Routledge, London
Bond L, Craps S, Vermeulen P (eds) (2016) Tracing the dynamics of memory studies. Berghahn Books, New York
Bull A, Hansen HL (2016) On agonistic memory. Mem Stud 9(4):390–404. https://doi.org/10.1177/1750698015615935
Chakrabarty D (2000) Provincializing Europe. Princeton University Press, Princeton
Connerton P (2008) Seven types of forgetting. Mem Stud 1(1):59–71. https://doi.org/10.1177/1750698007083889
Corning A, Schuman H (2015) Generations and collective memory. University of Chicago Press, Chicago
Craps S (2018) Introduction – memory studies and the Anthropocene. A roundtable. Mem Stud 11(4):498–515. https://doi.org/10.1177/1750698017731068
Czarnowski S (1919) Le culte des héros et ses conditions sociales: Saint Patrick héros national de l'Irlande. Alcan, Paris
Czarnowski S (1938) Kultura. Biblioteka "Wiedzy i Życia", Warszawa

David L (2020) Past can't heal us. The dangers of mandating memory in the name of human rights. Cambridge University Press, Cambridge, UK

De Cesari C, Rigney A (eds) (2014) Transnational memory. Circulation, articulation, scales. De Gruyter, Berlin

de Mignolo W, Tlostanova M (2012) Learning to unlearn. Decolonial reflections from Eurasia and the Americas. Ohio State University Press, Columbus

Debouzy M (1986) In search of working-class memory. Some questions and a tentative assessment. Hist Anthropol 2(2):261–282. https://doi.org/10.1080/02757206.1986.9960769

Decker S, Hassard J, Rowlinson M (2020) Rethinking history and memory in organization studies. The case for historiographical reflexivity. Hum Relat. https://doi.org/10.1177/0018726720927443

Durkheim E (2008) The elementary forms of religious life (trans: Swain JW). Dover Publications, Inc., Mineola (Durkheim E (1912) Les Formes élémentaires de la vie religieuse: le système totémique en Australie. Alcan, Paris)

Dybris McQuaid S, Gensburger S (2019) Administrations of memory. Transcending the nation and bringing back the state in memory studies. Int J Polit Cult Soc 32:125–143. https://doi.org/10.1007/s10767-018-9300-3

Edele M (2008) Soviet veterans of World War II. A popular movement in an authoritarian society, 1941–1991. Oxford University Press, Oxford, UK

Erdélyi G (2015) Armed memory. Agency and peasant revolts in Central and Southern Europe (1450–1700). Vandenhoeck & Ruprecht, Gottingen

Erll A (2011) Travelling memory. Parallax 17(4):4–18. https://doi.org/10.1080/13534645.2011.605570

Erll A, Nünning A (eds) (2010) Cultural memory studies. An international and interdisciplinary handbook. De Gruyter, Berlin

Feindt G, Krawatzek F, Mehled D, Pestel F, Trimcev R (2014) Entangled memory. Toward a third wave in memory studies. Hist Theory 53(1):24–44. https://doi.org/10.1111/hith.10693

Grande T (2021) The other and memory in Roger Bastide. Trauma Mem 9(1):2–8. https://doi.org/10.12869/TM2021-1-01

Gutman Y, Wüstenberg J (2021) Challenging the meaning of the past from below. A typology for comparative research on memory activists. Mem Stud. https://doi.org/10.1177/17506980211044696

Gutman Y, Wüstenberg J (eds) (forthcoming) The Routledge handbook of memory activism. Routledge, London

Halbwachs M (1925) Les cadres sociaux de la mémoire. Alcan, Paris

Halbwachs M (1941) La Topographie légendaire des Évangiles en Terre sainte. Presses Universitaires de France, Paris

Halbwachs M (1980) The collective memory (trans: Ditter FJ, Ditter VY). Harper & Row, New York [La Mémoire collective. Presses Universitaires de France, Paris, 1950]

Herman A (2019) One year after the Sarr-Savoy report, France has lost its momentum in the restitution debate. The Art Newspaper, 12 November. Available at https://www.theartnewspaper.com/2019/11/12/one-year-after-the-sarr-savoy-report-france-has-lost-its-momentum-in-the-restitution-debate. Accessed 15 Oct 2021

Hervieu-Leger D (2000) Religion as a chain of memory (trans: Lee S). Wiley, London

Hirsch M (2012) The generation of postmemory. Writing and visual culture after the Holocaust. Columbia University Press, New York

Hobsbawm E, Ranger T (eds) (1983) The invention of tradition. Cambridge University Press, Cambridge, UK/New York

Hoskins A (2018) Digital memory studies. Media pasts in transition. Routledge, London

Hutton PH (2016) The memory phenomenon in contemporary historical writing. Palgrave, Cham. https://doi.org/10.1057/978-1-137-49466-5_2

Johnson CD (2012) Memory, metaphor, and Aby Warburg's atlas of images. Cornell University Press, Ithaca

Joyce J (1939) Finnegans Wake. Faber and Faber Limited, London
Kattago S (ed) (2014) The Ashgate research companion to memory studies. Routledge, London
Kończal K, Wawrzyniak J (2018) Provincializing memory studies. Polish approaches in the past and present. Mem Stud 11(4):391–404. https://doi.org/10.1177/1750698016688238
Kubik J, Bernhard M (eds) (2014) Twenty years after communism. The politics of memory and commemoration. Oxford University Press, Oxford, UK
Lagrou P (2009) The legacy of Nazi occupation: patriotic memory and national recovery in Western Europe, 1945–1965. Cambridge University Press, Cambridge, UK
Le Goff J (1992) History and memory (trans: Rendall S, Clarman E). Columbia University Press, New York
Leszczyński A (2020) Ludowa Historia Polski. WAB, Warszawa
Levy D, Sznaider N (2006) The Holocaust and memory in the global age. Temple University Press, Philadelphia
Lewis S, Olick J, Wawrzyniak J, Pakier M (eds) (2022) Regions of memory. Transnational formations. Palgrave, Cham
Lim J-H, Rosenhaft E (eds) (2021) Mnemonic solidarity. Global interventions. Palgrave, Cham
Mandolessi S (2021) Challenging the placeless imaginary in digital memories. The performation of place in the work of Forensic Architecture. Mem Stud 14(3):622–633. https://doi.org/10.1177/17506980211010922
Mannheim K (1952) Problems of generations. In: Mannheim K (ed) Essays on the sociology of knowledge (ed and trans: Kecksmeti P). Routledge and Kegan Paul, London (Mannheim K (1928) Das Problem der Generationem. Kölner Vierteljahrshefte für Soziologie 7:157–185, 309–330)
Marcuse H (2008) Legacies of Dachau. Cambridge University Press, Cambridge, UK
Middleton D, Brown SD (2011) Memory and space in the work of Maurice Halbwachs. In: Meusburger P, Heffernan M, Wunder E (eds) Cultural memories. Knowledge and space, vol 4. Springer, Dordrecht, pp 29–49
Misztal B (2003) Theories of social remembering. Open University Press, Maidenhead
Moses AD (2021) The German catechism. Geschichte der Gegenwart. Available at https://geschichtedergegenwart.ch/the-german-catechism/. Accessed 15 Oct 2021
Niethammer L, Plato A (eds) (1987) Wir kriegen jetzt andere Zeiten. Auf der Suche nach der Erfahrung des Volkdes in nachfaschistischen Ländern. Dietz, Berlin
Nora P (ed) (1984–1986) Les lieux de mémoire, vol 1–3. Gallimard, Paris (English translations appeared as Nora P, Kritzman D (eds) (1997–1998) Realms of memory. The construction of French past (trans: Goldhammer A), vol 1–2. Columbia University Press and Nora P, Jordan DP (eds) (2001) Rethinking France: les lieux de mémoire (trans: Trouille MS), vol 1–4. Chicago University Press, Chicago)
Nora P (1989) Between memory and history: les lieux de mémoire. Representations 26:7–24. https://doi.org/10.2307/2928520
O'Dwyer R (1980) Czarnowski and "Finnegans Wake". A study of the cult of the hero. James Joyce Q 17(3):281–291
Olick JK (1999) Collective memory. The two cultures. Sociol Theory 17(3):333–348
Olick JK (2016) The sins of the fathers. Germany, memory, method. University of Chicago Press, Chicago
Olick JK, Robbins J (1998) Social memory studies. From "collective memory" to the historical sociology of mnemonic practices. Annu Rev Sociol 24(1):105–140
Olick JK, Vinitzky-Seroussi V, Levy D (eds) (2011) The collective memory reader. Oxford University Press, Oxford, UK
Pakier M, Stråth B (eds) (2010) A European memory? Contested histories of politics of remembrance. Berghahn Books, New York
Passerini L (1987) Fascism in popular memory. The cultural experience of the Turin working class. Cambridge University Press, Cambridge, UK

Pettai EC, Pettai V (2015) Transitional and retrospective justice in the Baltic States. Cambridge University Press, New York
Portelli A (2017) Biography of an industrial town. Terni, Italy, 1831–2014. Palgrave, Cham
Radstone S, Schwarz B (eds) (2010) Memory. Histories, theories, debates. Fordham University Press, New York
Ricoeur P (2004) Memory, history, forgetting (trans: Blamey K, Pellaue D). Chicago University Press, Chicago
Riley D (2006) Waves of historical sociology. Int J Comp Sociol 47(5):379–386. https://doi.org/10.1177/0020715206068620
Rothberg M (2009) Multidirectional memory. Remembering the Holocaust in the age of decolonization. Stanford University Press, Stanford
Rothberg M (2019) Implicated subject. Beyond victims and perpetrators. Stanford University Press, Stanford
Sarr F, Savoy B (2018) The restitution of African cultural heritage. Toward a new relational ethics (trans: Burk DS). Available via http://restitutionreport2018.com/sarr_savoy_en.pdf. Accessed 15 Oct 2021
Schwartz B (1996) Introduction. The expanding past. Qual Sociol 19:275–282. https://doi.org/10.1007/BF02393272
Schwartz B (2000) Abraham Lincoln and the forge of national memory. Chicago University Press, Chicago
Sendyka R (2016) Sites that haunt. Affects and non-sites of memory. East Eur Polit Soc 30(4):687–702. https://doi.org/10.1177/0888325416658950
Sierp A (2021) Memory studies – development, debates and directions. In: Berek M et al (eds) Handbuch Sozialwissenschaftliche Gedächtnisforschung. Springer, Wiesbaden. https://doi.org/10.1007/978-3-658-26593-9_42-1
Simine SA (2013) Mediating memory in the museum. Palgrave Macmillan, Basingstoke
Slabáková R (ed) (2022) Family memory. Practices, transmissions and uses in a global perspective. Routledge, London
Sodaro A (2018) Exhibiting atrocity. Memorial museums and the politics of past violence. Rutgers University Press, New Brunswick
Stein-Hölkeskamp E (2015) Marius, Sulla, and the war over monumental memory and public space. In: Galinsky K (ed) Memory in ancient Rome and early Christianity. Oxford scholarship online. Oxford University Press, Oxford, UK. https://doi.org/10.1093/acprof:oso/9780198744764.003.0008
Sztompka P (1993) The sociology of social change. Wiley-Blackwell, London
Tai HTH (2001) Remembered realms. Pierre Nora and French national memory. Am Hist Rev 106(3):906–922. https://doi.org/10.2307/2692331
Tota AL, Hagen T (eds) (2017) Routledge international handbook of memory studies, 1st edn. Routledge, London
Vasilyev A (2020) Historical sociology of memory by Stefan Czarnowski. Higher School of Economics research paper no. WP BRP 195/HUM/2020. Available at SSRN: https://ssrn.com/abstract=3738134; https://doi.org/10.2139/ssrn.3738134. Accessed 15 Oct 2021
Vinitzky-Seroussi V, Teeger C (2010) Unpacking the unspoken. Silence in collective memory and forgetting. Soc Forces 88(3):1103–1122
Wawrzyniak J (2019) From Durkheim to Czarnowski. Sociological universalism and Polish politics in the interwar period. Contemp Eur Hist 28(2):172–187. https://doi.org/10.1017/S0960777318000516
Weiner A (2002) Making sense of war. The Second World War and the fate of the Bolshevik Revolution. Princeton University Press, Princeton
Welzer H (2005) Grandpa wasn't a Nazi: the Holocaust in German family remembrance. American Jewish Committee, New York
Wertsch J (1985) Vygotsky and the social formation of mind. Harvard University Press, Cambridge, MA

Wertsch J (2002) Voices of collective remembering. Cambridge University Press, Cambridge, UK

Winter J (2010) Thinking about silence. In: Ben-Ze'ev E, Ginio R, Winter J (eds) Shadows of war. A social history of silence in the twentieth century. Cambridge University Press, Cambridge, UK, pp 3–31

Wüstenberg J, Sierp A (2020) Agency in transnational memory politics. Berghahn Books, New York

Wylegala A (2019) Displaced memories. Remembering and forgetting in post-war Poland and Ukraine (trans: Lewis S). Peter Lang, Frankfurt

Yates FA (1966) Art of memory. Routledge, London

Zaremba M (2019) Communism – legitimacy – nationalism. Nationalist legitimization of the communist regime in Poland. Peter Lang, Frankfurt am Main

Zimmerer J (2011) Von Windhuk nach Auschwitz? Beiträge zum Verhältnis von Kolonialismus und Holocaust. LIT Verlag, Münster (English edition forthcoming, Routledge)

Zubrzycki G, Woźny A (2020) The comparative politics of collective memory. Annu Rev Sociol 46(1):175–194

Part VI

History of Sociology

Historiography and National Histories of Sociology: Methods and Methodologies

31

Fran Collyer

Contents

Introduction	808
Disciplines, Specialities, and Fields	809
The Specialities and Disciplines Bordering National Histories of Sociology	811
History and Historical Sociology	811
Histories of Ideas and the Sociology of Ideas	812
Intellectual Histories, Institutional Histories, and Sociology	815
Sociology of Knowledge	816
The History of Sociology and National Histories of Sociology	819
Histories of Sociology	819
National Histories of Sociology	819
Tensions Between Historical and Sociological Methodologies	820
Engagement with Sociology, and the Sociology of Institutions	823
Engagement with Sociology of Knowledge	825
Conclusion	827
References	829

Abstract

Histories of Sociology have become increasingly popular in recent decades, particularly national histories of Sociology, yet there have been few attempts to articulate a methodology and methods appropriate for the tasks at hand. This chapter examines the History of Sociology as a speciality of the discipline, defining its boundaries and tracing its own history. It investigates many national histories of sociology to ascertain their methods and methodologies, and reveals the extent to which these engage with existing methods and methodologies as found in the disciplines of Sociology and History, but also their related specialities, including the History of Ideas and History of Intellectuals, as well as Historical Sociology, the Sociology of Science, the Sociology of Ideas, the

F. Collyer (✉)
Sociology and Social Policy, University of Sydney, Sydney, NSW, Australia
e-mail: Fran.Collyer@sydney.edu.au

© The Author(s), under exclusive licence to Springer Nature Singapore Pte Ltd. 2022
D. McCallum (ed.), *The Palgrave Handbook of the History of Human Sciences*,
https://doi.org/10.1007/978-981-16-7255-2_64

Sociology of Intellectuals, the Sociology of Institutions, and the Sociology of Knowledge. This systematic review suggests the subspecialty has recently shifted from an emphasis on the Sociology of Ideas to a Sociology of Institutions, and begun to adopt, albeit implicitly, the principles of the Sociology of Knowledge in its increasing acknowledgement of the shaping of Sociology's terrain by the geo-socio-political context.

Keywords

Method · Methodology · History · Sociology · National histories · Sociology of Knowledge

Introduction

Sociology, as a worldwide practice of the twenty-first century, has developed into a very diverse and large discipline, with many speciality and subspecialty areas. With these realities in mind, this chapter turns the spotlight on the recent growth in interest in producing national histories of the discipline. This interest has been sparked in various ways, including the growing demand from governments and university administrators to justify Sociology (Dayé 2018: 523), but also through the efforts of individuals such as Sanja Magdalenic and Per Wisselgren, who, through the 2000s, founded and coordinated the Swedish national research network, and organized various events on the History of Sociology (including an International Sociological Association (ISA) RC08 conference session in 2008 at Umeå University); Sujata Patel's (2010) edited collection of *Diverse Sociological Traditions*, also with the ISA's support; and the efforts of John Holmwood and Stephen Turner with their series of national histories, *Sociology Transformed* (e.g., Turner 2014; Crothers 2018; Chen 2018).

This growth in national histories of Sociology raises a number of questions about the relationship between Sociology and History, and indeed about Sociology itself. Primary among these are questions about what sociologists mean by "history," and also by *national* histories, as the category "histories of Sociology" is, as shall be demonstrated in this chapter, very diverse. This diversity raises questions about how such histories are best approached, theoretically and methodologically. While sociologists *have* been engaged in reflection about the production of histories, the major methodological debate has been about how to understand *texts* – whether we should employ an historicist or presentist position (e.g., Seidman 1985) – rather than how to understand the social structures and social institutions that have produced these (Platt 2005: 29). As such, little has been said about how a history ought to be researched or analyzed, even though investigating a speciality's methods and methodologies is critically important.

Examination of the largely unexplored terrain of national histories has revealed the extent to which they draw their methodologies and methods from established disciplines and specialities, rather than offering something unique to the speciality.

Hence an essential first step is to define the nature of a discipline, a speciality, or a field. The chapter therefore begins with some definitions, noting considerable variation in the way key terms such as discipline, subdiscipline, speciality, and field, are used across literatures, disciplines, and countries. Once key terms are defined, the chapter proceeds with a discussion of the disciplines and specialities closely related to the History of Sociology, briefly setting out their key principles and illustrated with some of their major texts. This is an important step in understanding the history of any discipline, because modern disciplines developed in relation to one another, each defining their arenas of practice and cognitive spaces through formal and informal social and professional interaction, and in ongoing competition for limited resources (Collyer 2012a: 7). Moreover, while each discipline is in a continual process of creating "distinctions" as a means for survival (Bourdieu 1981), these disciplinary and speciality "spaces" are at the same time critical sites for the development of ideas and practices. Indeed concepts, theories, methodologies, and methods may originate in either a speciality area or the parent discipline. Grounded theory, the concepts of deviance and cultural lag, as well as theories of the professions, for instance, were birthed in the speciality of Medical Sociology (Collyer 2012a: 159).

The final section of the chapter provides an analysis of a systematic selection of national histories of Sociology, identifying their methods and methodologies, and showing how these have been drawn from its parent discipline, other disciplines, and other specialities. The chapter concludes with some reflections on national histories of Sociology and the political (in the broad sense of the term) nature of their methodologies.

Disciplines, Specialities, and Fields

Disciplines, subdisciplines, specialities, and fields are fundamental to the scientific division of labor. They proliferated during the twentieth century, and individual scholars now increasingly work across two or more categories, such as studies of the urban environment and race relations, or Sociology and Gender Studies. There is also considerable variation in both terminology and practice across and within countries, with marked disparity between disciplinary or speciality boundaries among otherwise similar societies. In many countries, disciplines function as both a cognitive division within an overall system of knowledge, and as an independent *organizational* unit within a university. In the United States, for instance, disciplines tend to be both knowledge structures and departmental units. The practice of combining disciplines into multidisciplinary organizational units in the face of severe funding cuts, as has occurred apace in British and Australian universities, complicates the picture however, as many of these are given discipline-neutral labels such as Social Inquiry or Schools of Social Science, though some units continue to contain discipline-based programs of study within the overall structure.

The term discipline is reserved in this chapter for formal, knowledge-producing units which serve as sources of professional and academic legitimation, identity, and recognition. Disciplines are also reproductive units, socializing new generations of scholars, and generally have strong social and professional networks, their own journals, and may, as found in the case of Sociology, share certain world views or motivations (e.g., a concern for the underprivileged or an interest in social reform). Thus disciplines are "forms of identification and affiliation, social as much as intellectual, psychic as much as political, ethical as much as methodological" (Mills 2008: 15).

The term specialities may, like disciplines, also refer to areas of intellectual endeavor and to the labels used for professional or academic recognition, but only in a few instances are they used as organizational labels. While the term "speciality" in the Natural Sciences and Medicine often refers to multidisciplinary areas of endeavor, in Sociology, the term usually signifies specific subjects or methodological areas of Sociology, albeit with possible connections to other specialities and disciplines. As such, Sociology is understood to have a growing number of speciality areas, and in contrast to the natural sciences, where specialization is viewed positively – as a mark of disciplinary progress and competence – sociologists often voice concerns about the extent of specialization, seeing it as a "fragmentation" of the discipline. Nevertheless, pragmatically speaking, as sociological knowledge expands, it becomes increasingly difficult to gain an understanding and competency across the whole discipline, and specialities have become strategically important to building a career and a presence in the national and/or international academic community.

Each speciality tends to gather, over time, its own traditions, major works, agendas, and problematics, and may even favor particular methods or methodologies. For some specialities, the ties binding practitioners to the parent discipline are tenuous, and the individuals working within them may adopt a dual identity (as is very common in the United States, see Collyer 2012a: 224) or even lead the individual to not use the professional identity of the parent discipline. Thus specialities are something of a heterogeneous category of knowledge units, with some tied very closely to the parent unit, while others are clearly constituted across two or more disciplines (a cross-disciplinary speciality). Where units lose all attachment to any parent disciplines, they are best discussed as a field (an area of knowledge with broad appeal, coalescing around a subject of interest without necessarily requiring a specific discipline's approach, method, methodology, or established tradition of key theories and theorists). Thus Medical Sociology and the Sociology of Health and Illness can be classified as either a speciality of Sociology or a cross-disciplinary speciality, and there can be significant tension between these two categories. Indeed a long-running debate within Medical Sociology has been about the difference between a sociology *in* medicine and a sociology *of* medicine, with the first signifying studies that apply sociological concepts or methods, but address the problems and concerns of Medicine; and the second referring to studies which owe their fealty primarily to the discipline of Sociology (Collyer 2012a: 103, 136–138). Other specialities such as Urban Sociology appear less problematic, but disciplinary borders are regularly transgressed, and only become problematic for the discipline in efforts to produce a programmatic statement about its legitimate field of operations.

The Specialities and Disciplines Bordering National Histories of Sociology

As prefigured in the introduction, histories of Sociology come in all shapes and sizes, and as a speciality area of Sociology, it has its own subspecialty, national histories of Sociology. This section examines the disciplines and specialities bordering the History of Sociology, describing their legitimate territories, approaches, methodologies, and methods.

History and Historical Sociology

Historical Sociology is one of the more clearly defined cross-disciplinary specialities of Sociology. It has only a small number of English language journals, primarily the *Journal of Historical Sociology*, and *InterDisciplines, Journal of History and Sociology*. Both are explicitly interdisciplinary, with the first having editors from Sociology, Anthropology, Geography, and History, and the second stating that it seeks to bring the two disciplines [History and Sociology] into a "productive relationship... making History more sensitive to questions of systematisation and theoretical reflection and by infusing Sociology with a sense of historicity" (Mense-Petermann et al. 2018). The speciality also has several "handbooks," such as the *Handbook of Historical Sociology* edited by Delanty and Isin (2003), as well as edited collections (e.g., Go and Lawson 2017), and even subspecialty collections (e.g., Hobden and Hobson 2001).

The *history* of this speciality has been documented (e.g., Tilly 2001; Wilson and Adams 2015), and many of its histories are chronologies of intellectual shifts, closely akin to the History of Ideas (as examined below), and offering typologies or phases across time. There have been debates within Historical Sociology over methods and methodologies, with a focus on differences between itself and the discipline of History. The histories of the two disciplines – History and the Social Sciences more broadly – followed entirely different paths during the institutionalization of the disciplines in the nineteenth and early twentieth centuries in Western Europe and North America. While Sociology and the other Social Sciences aligned themselves with the methods of the Natural Sciences, History was oriented toward the humanities, though it was for a long period dominated by an empiricist, objectivist, epistemology – particularly in the United States (Steinmetz 2005: 9). Both disciplines underwent changes in the late twentieth century, however, as the "cultural turn" brought greater similarities between the two. This is noted by Wilson and Adams (2015: 27), for instance, who suggest historical sociologists, like historians, "continue to grapple with the importance of sequence and causality in history," with the problems of specificity and generalizability, and with "the roles of context and culture in the explanation of social change" (Wilson and Adams 2015: 28). And yet it is the differences between the two which sometimes spark the most interest. These "demarcation" disputes surface over such matters as an apparent inattention to theory among historians, the lack of sufficient regard for the temporal dimensions of social life in Sociology (Aminzade 1992); as well as other differences in their area of focus (Tilly 2001: 6753); the methods they use (Wilson and Adams 2015: 29); and their approaches to explanation (Gorski 2013; Steinmetz 2005: 11). With regards methods, Wilson and Adams (2015: 29) propose that while the distinction between historians and historical sociologists has lessened in recent years, with the latter group making greater use of archival materials and the former employing more social theory, differences nevertheless remain with regards the explanation of change. With regards the latter, Gorski (2013) suggests

> ...most historians do not reflect much on what it means to explain something... In mainstream historical discourse, as in most non-scientific discourses, to explain an event or action simply means to identify the sequence of events that preceded it or a complex of motives that animated it. From this perspective, something counts as explained to the extent that it has been successfully inserted into a narrative sequence or an interpretive framework. (Gorski 2013: 357)

Sociologists, on the other hand, have a variety of ways to deal with historical analysis, such as (a) the identification of underlying processes as causal mechanisms, (b) interpretation rather than explanation, or (c) engage with a theory, "where change is located in material or symbolic aspects of the world, or in parts of it, such as in the value system or set of interests or the economy" (Gorski 2013: 357–358).

Historians have long emphasized the use of archives for their research. Sociologists, however, in seeking to write histories, have a particularly difficult problem with obtaining good data about Sociology. Many have discussed the inadequacies of the archives, the pressure on space for storage of records, overreliance on the memories of past sociologists or administrators (e.g., Platt 2005), a new dependence on web-sites which "disappear" without a trace (Husbands 2019), but also the individual way university-based sociologists work, with our own files and filing systems that are often discarded when we move offices or retire. Perhaps also, as Fleck (2015: 308) suggests, we try to avoid "bothering our peers" with surveys or questionnaires. In Australia, an otherwise well-organized, bureaucratic society, the central collection of statistics about Sociology staff and students is extremely poor, with numerous changes in collection categories and all of this dependent on information gathered from individual universities and their varying interest in collecting and sharing data with government. These problems are common also in Britain (Platt 2005). In part, a lack of records can be the result of disinterest, but knowledge and power are intimately entwined, and information is in many cases purposely not collected or withheld by management, administrators, and politicians. If writing history depends on the analysis of primary sources, as Halsey (2004: viii) suggests, then sociologists seeking to write their history are generally faced with a very difficult task.

Histories of Ideas and the Sociology of Ideas

Many early histories of Sociology (and some of the more recent ones) contain little analysis of the material elements of history –the institutions or institutional infrastructure essential to sustaining scholarly endeavor – and instead examine ideational components of the discipline. In this, they are akin to the History of Ideas, a speciality which Dayé suggests has more synergy with the Philosophy of Science, French Intellectual History or Historical Epistemology (and found in the work of Gaston Bachelard or Georges Canguilhem and others) than with History itself (Dayé 2018: 532). These specialities focus on the genesis of concepts, whether of Science, Social Science, or Medicine, and the pre-conditions of, or historical obstacles to, their emergence (Bourdieu 1998: 190). And here, "pre-conditions" and "obstacles" are not conceived in material terms (such as levels of research funding) but rather the kind of conceptual and metaphorical elements that serve as a foundation for building knowledge. Thus the History of Ideas has, at its core, "the vision that ideas beget ideas, either by differentiation or by refinement" (Fleck and Dayé 2015: 321). In this, it is quite opposed to the sociological approach, or at least to the basic principles of the Sociology of Knowledge, where there is a relation (of some assumed or specifiable kind) between ideas, practices, and social structures.

The speciality of the History of Ideas varies considerably between the Anglophone schools, and the French, German, and others, but has nevertheless traditionally been concerned with a limited set of ideas, generally the philosophical ideas of well-known thinkers such as Aristotle or Marx. There remain similarities across the schools, however, particularly in the way certain ideas are given a coherency, and then invested with an:

> irresistible force of logic... [and consequently] ideas are pictured as so many individual agents that "influence" subsequent thought and action in identifiable ways. Where the tracing of such discrete "influences" becomes difficult, or where broader and less articulate beliefs have to be accounted for, one imagines ideas being distorted and diluted as they "trickle down" from a surface of clearly stated propositions to a subsoil of incoherent but common opinion. (Ringer 1990: 277)

The History of Ideas is also conducted quite differently by professional historians of Ideas (e.g., Lovejoy 1936), and historians of the social sciences, and it differs again from the Sociology of Ideas. The latter two tend to focus not on single ideas, but "compositions of ideas, or "theories," as their unit of analysis" (Fleck and Dayé 2015: 321). Most sociologists would be familiar with this *sociological* orientation to ideas, which sets out a "sociological tradition" of patterns of thought over time (e.g., Parsons et al. 1961; Levine 1995). Works within the Sociology of Ideas, which is more correctly classified as a speciality of the Sociology of Knowledge (as discussed below), often follow a typical History of Ideas approach in their focus on intellectual change and the notion that "ideas beget ideas." Sociological approaches to the genesis of ideas nevertheless generally seek to avoid methodological individualism, which relies upon subjective, individual motivation to explain social phenomena (Ringer 1990: 277). Foucault, for example, regarded his approach as a "history of thought," distinguishing it from the History of Ideas by suggesting the latter focuses on the origins of a new concept and the other ideas that are part of its context, while the former:

> ...is the analysis of the way an unproblematic field of experience, or a set of practices which were accepted without question, which were familiar and out of discussion, becomes a problem, raises discussion and debate, incites new reactions, and induces a crisis in the previously silent behaviour, habits, practices, and institutions. (Foucault 2011)

Foucault's work thus stands as both an example of, and an exception to, the general approach to the Sociology of Ideas, which can be described, methodologically, as following either an internalist or externalist approach to historical changes in knowledge or ideas. Internalist approaches envisage knowledge developing according to an internal logic, and knowledge production occurs within narrowly defined spaces, for instance, within Abbott's (2001: 122, 126) "interactional units" (that is, the disciplines of American universities). In contrast, externalist approaches reduce developments to the general social and economic conditions of the period in question, such that the social characteristics of an individual or group (e.g., class or gender) shape or determine (depending on the theoretical preference followed by the author), the choices made by the individual, and the knowledge he or she produces.

Well-known examples of externalist readings include Mannheim's (1972) Sociology of Knowledge, which explains "the form and content of (nearly) all ideas by reference to non-ideational social factors, especially macro-level economic, political, ideological and other conditions 'external' to the realm of ideas itself" (Camic 2015). Examples of internalist forms of the Sociology of Ideas include Nisbet's (1966) historical survey of European Sociology and claims for five central "unit ideas"; Donald Levine's *Visions of the Sociological Tradition* (1995); Collins' (1985) thesis of the driving principle of intellectual production as an "intellectual law of small numbers"; and Abbott's (2001: 147) proposal of a "governing fractal dynamic" determining the social-scientific field and ensuring a "tiresome repetition" of debates.

The problem with the internalist approach, according to Camic (2015), is its ahistoricity. The present state of the disciplines (and perhaps its recent past) is taken to be the "uniform prototype" for all forms of knowledge production, and, in taking the disciplines as "given," and not examining how they produce knowledge differently across various periods and places, they are unable to adequately explain the development of new ideas:

> The result is an approach fundamentally at odds with a growing body of research on the history of intellectual fields, a literature that not only underscores the halting process by which various disciplines, ranging from Philosophy and Sociology, acquired institutional autonomy as academic units... but also... furnishes evidence for the intellectual openness, permeability, and porousness of these disciplines long after they attained relative institutional autonomy. (Camic 2015)

Sociology's past – and present – is clearly founded on disputes over the appropriate theoretical or methodological approach to a given subject matter, and no less so for the Sociology of Ideas. For Charles Camic and Neil Gross (2004), the Sociology of Ideas has not been well served by being a subspecialty of the larger, more-encompassing, Sociology of Knowledge. Camic and Gross see the Sociology of Knowledge as too concerned with macro-social factors, that is, the broad economic, political, and religious conditions of society, and how these shape the development of ideas. As a consequence, it has not given sufficient attention to the impact of institutions and their local configurations. Camic and Gross suggest an area of the Sociology of Knowledge be set aside for a *new Sociology of Ideas*, which would maintain a focus on the "social processes by which ideas grow, emerge, and change," while avoiding reductionism and preserving the details that account for the choices made by particular individuals (Camic and Gross 2004). Studies within the *new* Sociology of Ideas engage with varying theoretical traditions and methods, including archival work, qualitative interviews, ethnographic observation, and analyses from quantitative data. What *is* shared is a methodology – an approach to knowledge which investigates the processes of "making" knowledge from specific situations in time and place. In this, the *new* Sociology of Ideas shares some similarities to Institutional History, which has been popular in Social Studies of Science but not well represented in recent histories of the social and behavioral sciences (Fleck and Dayé 2015: 322).

Intellectual Histories, Institutional Histories, and Sociology

Histories of Sociology often do not distinguish between the discipline's *ideational* and its *institutional* development, proclaiming for instance, that "Sociology first began in Europe," without specifying the nature of the subject, "Sociology." This said, the majority of early studies focus on the ideational aspects of the discipline and have an affinity with either the Sociology of Ideas or the tradition of Intellectual History. Early forms of Intellectual History concentrated on the texts of a few "great thinkers" from the past, such as Locke or Hobbes, and while some contain a glimpse of the private life of scholars, this tends to be the province of autobiographical writing (Fleck and Dayé 2015: 320). From the 1970s, there was an expansion in the breadth of texts and type of thinkers that could be studied, but the speciality largely remains the "study of great books." Methodologically, Intellectual History, like the History of Ideas, aims to establish a pathway of ideas reaching from the present back through the decades. Many are explicitly presentist in approach, establishing connections between past and present ideas.

The *Sociology of Intellectuals* emerged in the 1920s and has tended to follow the methodological formulae of Intellectual History, providing accounts of intellectual progress and focusing on the "cognitive identity" of the discipline (e.g., Merton 1977: 5). This speciality has had a "chequered history," with periods when little work has been completed within the speciality proper, and others when it appears as a cohesive body of literature (Kurzman and Owens 2002: 63). During dormant periods, related work was carried out within other speciality areas such as the Sociology of Professions, or the Sociology of Knowledge, and sometimes such shifts have coincided with changing preferences for alternative identities among the intellectuals themselves (Kurzman and Owens 2002: 63). Thus the Sociology of Intellectuals has at times verged on being a speciality, but much of its work is claimed by other specialities. For example, Konrad and Szelényi's (1979) work belongs, quite rightly, to Intellectual History, but also Social Theory and the Sociology of Knowledge.

Institutional History offers an alternative to Intellectual History, documenting the formation of the bodies through which ideas are nurtured, developed, and transmitted: social networks, societies, university research centers, schools, departments, and so on. A *sociological focus* on institutional development offers an analysis of the processes whereby central ideas become embedded in social practices and structured into more resilient and permanent arrangements. Unlike intellectual histories, *institutional histories* can demonstrate actual contact or influence between specific social actors (or generations of actors), and divulge the social and political struggles through which disciplinary goals are furthered and bodies of formal knowledge protected from dissipation or misuse (Collyer 2012b: 117). Moreover, institutions – whether disciplines, universities, trade unions, or professional associations – have a formal structure which is deeply ingrained within a culture and organizes social action into "predictable and reliable patterns" (Streek and Thelen 2005: 13).

Conceiving of disciplines as institutions allows consideration of their similarities (and differences) across various national and geographic locations, and throws a spotlight on the political, economic, cultural, and societal factors which have shaped disciplinary formation. Less developed within the literature is conceptual analysis of the *process* of institutionalization. It is common to find institutionalization being discussed as an "end-point," that is, where a set of social practices or rules have "become institutionalised." Prior to institutionalization, participants in an intellectual community or network may have to rely on persuasion and personal worth to obtain cultural authority, and there is a high probability of bureaucratic, corporate, or religious intervention in the activities themselves. Through the processes of institutionalization, material supports and cultural resources become concentrated, formal mechanisms for communication and interaction are established, the legitimacy of a set of rules and sanctions is accepted, and there is increasing public or community recognition and support. The extent to which recent histories of Sociology have taken an institutional approach needs to be examined empirically, and the second part of this chapter will take some first steps in this process.

Sociology of Knowledge

A final speciality bordering the History of Sociology is the Sociology of Knowledge. Among sociologists, there is little consensus about whether the Sociology of Knowledge constitutes a speciality within the discipline or is a fundamental aspect of all sociological work. However, the evidence suggests the Sociology of Knowledge is one of the smaller and less understood specialities. While there are rarely conferences or journals entirely devoted to the Sociology of Knowledge, it has a set of adherents, some of whom work primarily in the speciality, has its own "founding" theorists, theoretical problematics, and its own, documented, history. Indeed most sociologists could identify some of the early proponents of the Sociology of

Knowledge – Scheler, Mannheim, Merton, Berger and Luckmann, Parsons, Coser, and Gouldner – although disputes over the precursors or earliest figures (e.g., Durkheim; Marx and Engels), and the more recent figures would be likely (e.g., Hartsock, Bourdieu). Evidently, the boundaries of the speciality are highly permeable, encroaching into the History and Philosophy of Science, the History of Ideas, the History of Intellectuals, and the Sociologies of Science, of Ideas, of Intellectuals, Institutions, and even the Sociology of Medicine (e.g., Fleck 1979 [1935]).

During some periods, the borders of the speciality have been quite rigid. Critics of Mannheim, including Merton, Parsons, and Lukács, argued that if a Sociology of Knowledge were to be developed, it must rid itself of its epistemological and philosophical elements (Meja and Stehr 1993: 66). Additionally, both Mannheim and Merton maintained Sociology could examine the content or substance of ideas only in certain fields, arguing that neither mathematics nor science could be investigated sociologically. Indeed, the Sociology of Knowledge and more specifically, the Sociology of Ideas, were, for a period, restricted to an investigation of some of the conditions which might "incline thinkers to certain topics, general worldviews, or stylistic conventions" (Camic and Gross 2004: 239). Significant shifts in these boundaries were brought about through developments in sociological studies of science, and the History and Philosophy of Science, particularly with Kuhn's (1979) [1962] publication on the study of scientific change; investigations into the discursive foundations of medical knowledge and practice (Foucault 1973; Fleck 1979); developments in Feminism in the light of Hartsock's (1983) standpoint theory (which challenged widespread notions about the objectivity of all knowledge, including scientific, medical, and social science knowledge); and encouragement for a "sociology of Sociology," with investigations into our own practices (e.g., Reynolds and Reynolds 1970). In addition, the work of Berger and Luckmann (1996) [1966] broke apart the boundaries between the study of organized forms of knowledge (i.e., disciplinary knowledge) and everyday knowledge, examining the (pre-theoretical) knowledge used by social actors in daily life, and perhaps, even more importantly, challenged the conventional view of ideas as determined by social reality by insisting that reality is itself a social construct. While this last proposition remains contentious, the understanding that *all forms of knowledge are socially mediated* became the dominant research hypothesis in Sociology, with the Sociology of Knowledge expanding into many new areas.

The current boundaries of the Sociology of Knowledge are still being articulated. Camic and Gross' proposal for a *new* Sociology of Ideas within the Sociology of Knowledge (discussed above) is a case in point. In addition, Swidler and Arditi proclaimed the birth of a *new* Sociology of Knowledge which is concerned with:

> how kinds of social organisation make whole orderings of knowledge possible, rather than focussing in the first instance on the differing social locations and interests of individuals or groups. It examines political and religious ideologies as well as science and everyday life, cultural and organisational discourses along with formal and informal types of knowledge. It also expands the field of study from an examination of the contents of knowledge to the forms and practices of knowing. (1994: 306)

The claim is that the *new* Sociology of Knowledge differs from previous versions, because it picks up on a shift, evident throughout the social sciences over the same period, away from materialism and social structure toward semiotic theories of communication, where it is maintained that human experience takes a conscious and communicable shape only through language, categories of thought, norms, and so on. Similar notions about the *new* Sociology of Knowledge are revealed in the large collection compiled by Meja and Stehr (2000), which differs from Swidler and Arditi's work only through its incorporation of at least two feminist challenges to conventional approaches to knowledge.

In the 20 years since the pronouncement of a *new* Sociology of Knowledge, work in the speciality has diversified further, and a "newer form" of the *new* Sociology of Knowledge is warranted. Despite the protestations of its proponents, studies in knowledge responding to the "cultural turn" and investigating semiotic theories of communication have never fully supplanted other theoretical and methodological inclinations. The speciality is, and has long been, a heterogeneous category, indeed perhaps more heterogeneous than previously recognized. Conceptual, methodological, and theoretical alternatives to semiotic and cultural readings of knowledge continue to make their mark within the Sociology of Knowledge, and these, in stark contrast, insist on the continuation of structural divisions in the production and circulation of knowledge. Examples can be readily found. For several decades, sociologists have been systematically examining the ongoing impacts of colonialism and imperialism on the way we make, or access, knowledge (e.g., Alatas 1972). The contemporary environment has also been reconceived as a knowledge-based society, showing how knowledge, and the way knowledge is treated, have restructured the economy, the labor market, the institutions, and the universities (e.g., Toffler 1980; Slaughter and Leslie 1997). These studies have often failed to be recognized under the umbrella of the Sociology of Knowledge, even while they focus in various ways on the social structuring of knowledge and its impacts, and take global inequality as a central feature of contemporary society (e.g., Connell 2007; Steinmetz 2005; Collyer et al. 2019).

The contemporary Sociology of Knowledge then is very much a heterogeneous speciality, employing a diversity of theoretical approaches, methods, and methodologies. Its studies range across the full methodological spectrum, some specifying the construction of objective variables and indicators to quantify and measure the extent or type of change associated with the use or production of knowledge, while others employ qualitative methodologies, emphasizing interpretation and the contextual nature of truth. It is thus apparent that the contemporary Sociology of Knowledge is theoretically and methodologically as diverse as its parent discipline, and that the growth of post-structuralism, postcolonial, Foucauldian, and Bourdieusian approaches have only augmented the range of possibilities for this speciality area.

The History of Sociology and National Histories of Sociology

Histories of Sociology

As a speciality, the History of Sociology sits within the broader field of the History of the Human, Social, or Behavioral Sciences, sharing many similar concerns, such as the development of knowledge institutions, the lives of important intellectuals, and the historical progression of ideas or advancements in social theory, but differing in its focus on one discipline. Dayé regards the History of Sociology as not yet fully consolidated as a speciality, as it "does not converge around a gravitational point – neither theoretically, nor methodologically, nor organisationally, in journals and associations" (Dayé 2018: 522). Moreover, the history of its own formation and character have not been as well articulated as other specialities such as Medical Sociology (see, for example, Collyer 2012a). However, it has had Chairs established, for example, in Russia (Zdravomyslova 2010: 144). And like other specialities, it has its own history and adherents, and formal groupings within professional associations, although these groups are not particularly large, and the speciality is yet to have its own journal. Nevertheless, as noted at the beginning of this chapter, the History of Sociology, and national histories of Sociology in particular, have recently seen more activity in the international arena, with a rising number of publications.

National Histories of Sociology

National histories of Sociology are a subspecialty of the History of Sociology and a fairly recent development. Like the speciality as a whole, the offerings here tend to be quite diverse theoretically and methodologically. The latter term, "methodology," used frequently in the Research Methods literature, generally refers to the underpinning philosophical or theoretical rationale for the selection of methods for an empirical study. It is, however, also used to cover the rationale for the selection of one's approach and procedures employed for *analysis*. Methodology thus embraces the reasoning behind the focus of the research – *what* is to be studied, counted as relevant or excluded from investigation. There are some methodologies which immediately spring to mind, and commonly used in Sociology, such as phenomenology, positivism, or ethnomethodology, but other theoretical frameworks can be utilized as methodologies: for example, theories of institutions or of scientific change. In this section, a collection of national histories of Sociology is examined, and the most notable feature revealed about this speciality is that the early national histories generally do not offer any description of the methods or methodologies employed in their studies, nor explicitly engage with social theory. More recent offerings are somewhat more likely to engage with theories, methods, and methodologies, though even here the interest is minimal, with often only implicit mentions in the text or footnotes. A second feature of this speciality is that where discussion is present, we find a broad diversity of approaches, with the studies borrowing methodologies, theoretical frameworks, guiding principles, and methods from the various disciplines and specialities discussed in this chapter. Such "borrowings" place the national histories in some danger from succumbing to the methodological problems that have long plagued these specialities, but equally introduce the possibility for new insights. These issues are examined in the next three sections.

Tensions Between Historical and Sociological Methodologies

Sociologists have long been critical of "Whig histories," where present circumstances are portrayed as a result of a grand procession of "great men and great deeds" – a problem endemic in the History of Ideas and the Sociology of Ideas. Cossu and Bortolini (2017), for example, in their study of Italian Sociology, reflect on previous efforts at telling their national history, suggesting claims about the institutionalization of Italian Sociology have been largely unsubstantiated, providing only a "socially legitimised self-image" which makes "a very long, difficult, agonistic process look like the outcome of the actions of a small cohort of great and brave men involved in a heroic journey, constantly fighting a hostile academic environment" (2017: 3). Cossu and Bortolini's (2017: 2) approach to overturning this fallacious image of Italian Sociology is to leave "the history" to the historians, and "do what we are supposed to do best: sociology." While they make extensive use of historical archives in search of data, they work "at the intersection of sociological and historical analyses. Time, context, and process are all important" (Cossu and Bortolini 2017: 4).

The tension between the differing methodologies of History and Sociology becomes particularly evident among national histories of the discipline. Sociologists are expected to be reflexive about the selection of data to substantiate causality and produce explanation. Yet for many years, there has been insufficient criticism of "sanitised" historical narratives which fail to take into account the less palatable aspects of the past, such as the effects of colonialism and imperialism, racism, and patriarchy (Alatas 1972; Connell 2007; Bhambra 2007). Such histories offer little acknowledgement of the part played by sociologists in other parts of the world, of the distortions introduced where we assume a single, dominant voice can speak for all, and where we ignore the role of geopolitics in defining the disciplinary landscape. Recent national histories are far more likely to pay attention to national differences in institutional and political structures (e.g., Hanafi 2010; Golenkova and Narbut 2015; Sooryamoorthy 2016), and it is not coincidental that these more reflexive studies come from scholars located in institutions outside the global metropole, where their experiences of marginality have shaped their approach to the subject matter.

The contrasting methodologies of History and Sociology and the problems of borrowing from other disciplines and specialities become particularly apparent where explanations are sought for periods of decline or growth. Wallace's (1992) history of Sociology at Columbia, for example, falls prey to the internalist/externalist dilemma that stalks the History of Ideas, following an internalist logic in explaining the resistance to the establishment of a department after 1891. Wallace relies heavily on the minutes of meetings from the archives to argue his case. His account suggests many of the developments (such as why Sociology flourished at one university but not at another) is explained by personal relationships between individuals (resentments, rivalries, etc.) and/or the personalities of the individuals themselves (holding unfavorable attitudes toward Jewish people, displays of poor behavior, bouts of poor health, etc.). Another example is Harley and Wickham's (2014) history of Australian Sociology, which largely explains the delayed formation of Sociology at the University of Melbourne as the result of alleged "poor performance" of the sociologists themselves:

> It is hard to imagine the attempt at the University of Melbourne going worse than it did. Through a combination of arrogance, overreach, and poor leadership, the discipline alienated many university authorities and those in other disciplines. (Harley and Wickham 2014: 9)

While it is important not to overlook the role of individuals in institutionalization, sociological analysis needs to look beyond individual failings for explanations of historical change. In the case of Australian Sociology, an externalist perspective would take into account other factors in the decades prior to 1959 that better explain the extended resistance to its institutionalization, such as the lack of state interest in dealing with social problems, the strength of other disciplines, including history and economics (Austin-Broos 2005: 268–269), the specific relations between capital and labor (Mitropoulos 2005), and the political agendas of the nation state (Collyer 2017). These may yield a different story of Australian Sociology and the enormous challenges it faced in a country with essentially no intellectual community, no scholarly tradition, and little government or university interest in supporting a critical discipline such as Sociology. A national history with an appropriate balance between an internalist and externalist view of the discipline is Masson and Schrecker's (2016) study of France. In that country, the discipline suffered through similarly difficult periods, with "internal" problems – such as the need for theoretical conformity or more sustained leadership – and "external" ones such as boundary difficulties with other disciplines, relations with Sociology in other countries, the devastation of war, and the availability of funds for contract-based research.

A number of the national histories of Sociology employ the methodological approach of *Historical Sociology* to provide a more appropriate balance between History and Sociology. Sooryamoorthy, for example, explicitly engages with Historical Sociology, stressing "the linearity of time and the progressive order of history" (2016: 3), and yet purposefully invokes a *sociological* approach to the subject matter in searching out the ways colonialism and apartheid have "embedded their perspectives and structures in knowledge production" (2016: 4). Tzeng's study of (2012) Sociology in Taiwan, Hong Kong, and Singapore also deals with "the tension between historical interest that pays attention to details with an aim to interpret the unique, and the sociological interest that works with theoretical categories with an aim to discover the typical" (Tzeng 2012: 83). Methodologically, Tzeng claims to work in "a constant dialectical journey between historical details and conceptual abstractions" (2012: 83–84), gathering historical evidence but using sociological methods and analytical concepts to complete the study. With regards methods, for example, Tzeng (2012: 85–86, 92) employs a multi-strategy design, collecting both qualitative and quantitative materials, analyzing documents from the archives (e.g., conference papers, class handouts, official publications from universities and governments, newspaper items), and mixing bibliometric analysis with 71 semi-structured interviews. And unlike traditional History, he organizes his study into three analytical, *sociological* categories: "the regional-geopolitical, the State-institutional, and the practitioner-level," each of which comprises a "structure-agent set," where the State is both "a collective agent in the structure of regional geopolitics," and a constituent of "the structural environment in which individual or collective sociologists work" (Tzeng 2012: 18).

An engagement between History and Sociology is similarly evident in national histories where there is a portrayal of events through several discrete, historical phases. Some favor the methodology of the History of Ideas, basing chronologies on intellectual shifts over time (e.g., Tilly 2001). However, this is not as common among recent *national* histories of Sociology, for although there are some examinations of change from one theoretical orientation to another (e.g., Ekerwald 2014), most prefer a focus on institutional rearrangements over time, drawing on empirical material such as the size of departments, staff and student enrolments, and perhaps state-discipline relationships to suggest periods of decline and growth (e.g., Larsson and Magdalenic 2015; Crothers 2018; Masson and Schrecker 2016; Turner 2014). In some countries, the varying fortunes of Sociology are more specifically articulated in relation to a set of political and ideological historical "drivers." Examples include Cordeiro and Neri's (2019) periodization of Brazilian Sociology into five phases built around the challenges of dictatorship and democracy; Sooryamoorthy's (2016: 3–4) description of South African Sociology as having passed through three marked phases: the colonial, the apartheid, and the democratic; and Chen's (2018) study of Chinese Sociology which identifies the changing influence of American Sociology versus that of the Chinese Communist Party.

One of the perennial problems of using historical "phases" to offer a narrative of the discipline is their susceptibility to critique: given the unlimited number of potential factors to be taken into account. Another is the way past accounts can rather easily become part of the discipline's self-image, discouraging fresh appraisals. As already noted above, this has been found to be the case for Cossu and Bortolini (2017) in Italy, but also for Turner (2014), who uncovers a similar "myth" about American Sociology as built around scientific and value-neutral sociology.

It appears from even this brief analysis of disciplinary histories, that opening the subspecialty of national histories of Sociology to a broad range of scholars, as has occurred in the recent decade, has lent added weight to arguments against assuming a universal pattern of development for Sociology across all societies, and encouraged a trend away from internalist histories. Of particular note is that the national histories from peripheral countries appear more likely to take account of the specific ideologies and policy proclivities of the state (e.g., Cordeiro and Neri 2019; Sooryamoorthy 2016; Chen 2018), as if the state and its administration in dominant countries such as the US are value-neutral, or alternatively, that these Sociology's have a greater capacity for autonomy, and can therefore have their histories told as an internalist narrative. We see this, for example, in Turner's history of American Sociology, which he admits to be internalist in emphasis (2014: 118).

This matter of internalism versus externalism coincides with the current debate about the possibilities for a global Sociology, of whether there is one or many, and whether a *national* Sociology is even possible. Archer (1991: 134), for instance, argues that we are increasingly interconnected through processes of globalization and little can now be understood in strictly local terms. Some, such as Beck (2000: 20), argue we must go beyond *methodological nationalism* and analyze Sociology beyond the "prism" of nation-states. These "distinct cultures" may well be composed of "fluid, interconnected, messy social processes" and their national boundaries may not adequately describe the social orders found within them, as Connell (2010: 44) suggests; but the evidence suggests an abundance of experiences of Sociology, with each "group" developing in unique geopolitical-institutional contexts. Moreover, it is quite clear that the internalism/externalism debate has an historiographic underside, for it is only the American and (some of the) European Sociologies where the histories are written *as if they developed* autonomously and independently of global events or developments in other nations. Only these Sociologies are presented as if they have been able to ignore or absorb developments elsewhere without apparent major disruption to their own patterns of practice. It is likely that this presentation of history reflects a difference in academic orientation across the globe, between those who practice *intraversion*, where scholars are inwardly focused and suspicious of knowledge from sources external to their own country, rather than *extraversion*, where they are oriented toward, and dependent on the institutions, concepts, and techniques of the metropole or institutions of the global North (Collyer et al. 2019: 10, 129). Evidence from comparative citation studies indicates the former – intraversion – is common in the US (the global North or metropole), while the latter – extraversion – is a characteristic of scholarly practice in peripheral countries (the global South). It appears practices of intraversion (particularly when combined with the positivism of much American Sociology, Steinmetz 2005: 41) and extraversion shape not just our citation practices but our historiographic writings, inhabiting the methodologies we use to narrate our national histories.

Engagement with Sociology, and the Sociology of Institutions

As previously noted, surprisingly few national histories of Sociology dialogue directly with sociological concepts and theories as a means to investigate the development of the discipline. This has been noted by others, including Platt (2005: 29), who is critical of the small amount of historical work undertaken in Sociology, and also the generally "unsociological" approach of the authors. There are, however, several recent national histories that fly in the face of this trend. Larsson and Magdalenic (2015: 3–4), for example, engage explicitly with three theoretical perspectives "as guiding principles" for their history of Swedish Sociology: boundary work in science, as a means to examine the disciplinary landscape and its tensions; gender, as it has effected individual's careers, but also the structure of the discipline; and the interdependent relationship between Sociology and the Swedish welfare state. A second example comes from Zdravomyslova (2010), who applies the sociological concepts of symbolic and administrative capital to identify the way various Russian Sociologists position themselves within the discipline. Other examples come from Crothers (2018: 2), who invokes Kuhn's conception of paradigms to discuss some of the methodological and conceptual shifts in New Zealand Sociology; Hanafi (2010), who uses Bourdieu's concept of field to examine the complex rules of the Palestinian research field; Patel (2010), and her concept of colonial modernity in the Indian setting; and Denis (2010) with the concepts of race and ethnicity in the Sociology of the Caribbean.

A more intensive engagement can be found between the national histories and the Sociology of Institutions, particularly among recent publications (e.g., So 2017). This trend must, in part, result from the influence of the *Sociology Transformed* series, where editorial guidance was given to examine "disciplines as national and global formations," and offer a "systematic discussion of the transformation" from the post-war period to the current day (source: series proposal, courtesy Stephen Turner). Most contributors interpreted this to mean an examination of the "context" within which Sociology developed, and to take an institutional approach, gathering evidence such as changes in research funding, student enrolments, staff numbers, or the formation of departments, journals, and professional associations.

Chen's (2018: 1) national History of Sociology in China is an example of an explicitly designed institutional approach. For Chen (2018: 5), this means examining the historical and institutional factors and charting "the material, symbolic and organisational resources, including funding supports, cultural and intellectual traditions, and networks, [and] associations." Chen notes that other studies of Chinese Sociology have been hampered by the professionalization of the discipline since the 1990s, "which has served to legitimise a positivist conception of knowledge accumulation as linear progress" (2018: 5). Taking an institutional perspective, on the contrary, encourages an "openness regarding the relative autonomy of disciplinary development vis-à-vis its material, social, and organisational basis" (Chen 2018: 5). Likewise, Cossu and Bortolini (2017) argue that an institutional approach helps avoid an excessive focus on the individual agency of a handful of "founders," and guards against "romanticising" the birth or delayed birth of the discipline. Instead it guides one's search for "something structural and more deeply embedded in the organisational context which prevented the full integration of Sociology into the academic system" (Cossu and Bortolini 2017: 5–6).

Chen's (2018) study suggests that an institutional approach does not necessarily preclude an investigation of the intellectual content of the discipline, nor its social relationships. And we find other "histories" which do precisely this, such as Scott's (2017: 44) history of British Sociology, which traces the intellectual trends and currents over time, but sits this within an institutional context (e.g., relations with, and changes to other disciplines), and also a political context (e.g., the expansion of the National Health Service). Zdravomyslova's (2010) study is another with a twin focus on institutional and intellectual development. Likewise, Pereyra's (2010) study of Argentinian Sociology explicitly combines an institutional and cognitive focus.

Despite the benefits of the institutional approach, this chapter reveals certain missing features of national histories. First, very little is said about the sociologists and the Sociology undertaken outside Sociology departments: the greater majority of individuals working in other areas of the universities, in inter- or multidisciplinary units, and those in industry, the community, and the public sector. Crothers' (2018) study mentions – though does not investigate – these groups, discussing the situation in terms of the "mainstream," that is, work performed within academic departments of Sociology, and "sociology" (without capitalization), where scholars and other intellectuals "are infected with a sociological perspective but practise their sociology beyond the confines of formal Sociology departments" (2018: 2–3). A rare exception to the trend is Azarya's (2010) study of Sociology, which explores the employment of sociologists by the armed forces and large companies in Israel.

Secondly, the focus of almost all national histories has been on departments of Sociology in the prestigious universities. Payne's (2019) study of sociologists working in the British "Polytechnics" – the previously vocational institutions that were reclassified as universities after 1992 – is therefore a welcome addition to the field. This is where the author estimates half of all British sociologists are employed, and where about half of all students are educated, and yet these institutions are virtually absent in other accounts of British Sociology (Payne 2019: 194–195).

And thirdly, very little appears about the shaping of Sociology by organizational arrangements within the institutions, and how these have changed over time. Yet, the mechanisms of appointment and promotion, the supervision of staff, career building and mentoring, collaboration and competition, and shifting levels of autonomy between departments and the management of universities are all fundamental features of the working lives of sociologists. Some historical, comparative, organizational studies – with good use made of organizational theory – are sorely needed to flesh out Sociology's past and present.

Engagement with Sociology of Knowledge

Very few national histories of Sociology engage directly with the Sociology of Knowledge, explicitly stating this to be their methodological or theoretical approach, and yet many develop their studies around questions central to the speciality. These may include the impact of social location, cultural conditions, forms of authority and power, institutional arrangements, and/or material conditions on sociological knowledge (the externalist approach); the manner in which knowledge is immanently produced but socially organized and informs practice (the internalist approach); or a third approach, where an attempt is made to overcome the internalist/externalist problematic prevalent within the Sociology of Ideas, by employing Bourdieusian or similar theories or methodologies (Ringer 1990: 275). The tendency to not identify oneself as working within the Sociology of Knowledge tradition is a well-known problem for the speciality, and probably will not be overcome until it loosens its ties to Philosophy or the cultural Sociology popularized in recent decades and develops a "new face" that takes into account the kind of critiques discussed in this chapter.

That said, there are national histories that explicitly engage with the Sociology of Knowledge, including Chen's (2018) study of Chinese Sociology. For Chen, the Sociology of Knowledge, as developed by Mannheim and Merton, "sought to illuminate the relationship between structural conditions and intellectual outcomes" (2018: 5), and he finds this and the institutional approach enable a focus on the way the "academic community and its intellectual production are being shaped by the state, universities, research institutes, professional associations and other agencies" (Chen 2018: 1). Likewise, Tzeng's (2012) history is explicitly framed within the Sociology of Knowledge and guided by its central problematic: the way human thought and knowledge is connected to the social context (2012: 60). The Sociology of Knowledge also provides theories for Tzeng's study. He adopts the functionalist approach set out by Merton to investigate the institutional foundations of Sociology, but is inspired by the phenomenological tradition of Berger and Luckmann in the examination of the "memories, intentions, ideas and rationalities of the actors." However, Tzeng found the conventional literature on the Sociology of Knowledge to be bereft of the geographical dimension of knowledge, particularly the "geography-bounded power relation between the former colonial powers where modern science was first invented and the former Asian colonies where such scholarship was introduced" (Tzeng 2012: 61). Tzeng therefore turns to the postcolonial literature, and writers such as Alatas (1972), Connell (2007), and Bhambra (2007). These, however, see Sociology as a Western project, and Tzeng in contrast, aims to show Sociology as having developed as a mix of both East and West (Tzeng 2012: 65–66).

A third national history explicitly adopting a Sociology of Knowledge methodology is Collyer's (2017) study of Australian Sociology. This seeks to "explain sociological knowledge within its political, economic and social structures," examining the discipline in a specific national context (Collyer 2017: 82). Addressing the question of whether there is, or can be, an *Australian* Sociology, Collyer proposes that even with the interconnectedness of all Sociologies:

> an axiom of the Sociology of Knowledge is that all knowledge is shaped by its political, institutional, cultural and economic context, and thus we should be able to, at the very least, identify some features of Australian Sociology that reflect its unique institutional, political and cultural context - if not also its location on the periphery and its particular relationship with the sociological metropole. (2017: 89)

Collyer goes on to identify the unique characteristics of Australian Sociology, including a sustained "brand" of critical Sociology, which strives "to relate social issues to power, public policy and social reconstruction" (Skrbis and Germov, in Collyer 2017: 90), an emphasis on both race and class, the strength of its feminism, its wide-ranging incorporation of other traditions, and its clear distinction from other disciplines such as psychology.

Many more national histories *implicitly* adopt the methodology of the Sociology of Knowledge. Hanafi (2010: 257), for instance, seeks to "analyse how the interface between structures of power within the Palestinian society and state, the international community and the market of research production influence themes of research and the relationship between donors and the NGOs." Hanafi demonstrates how *the making of knowledge* is shaped by the social context, concluding that "with much research taking place outside the university, and without its protection, researchers are open to attack by authorities and other groups, and consequently 'fail to be critical toward their own society'" (Hanafi 2010: 265).

Sooryamoorthy (2016: 3) is another who acknowledges a debt to Historical Sociology rather than the Sociology of Knowledge, arguing that the former provides concerns and methods to investigate social change and invigorate the discipline. He nevertheless implicitly adopts the principles of the Sociology of Knowledge, finding that various forms of political change have "embedded their perspectives and structures in knowledge production" (2016: 3–4), and "[t]hus different sociologies developed in the colonial period, apartheid times and in the democratic era" (Sooryamoorthy 2016: 4). Likewise, both Connell (2010) and Burawoy (2010) speak to the inequalities and hierarchies of the knowledge system within which Sociology sits, and the "global division of sociological labour that mirrors world political and economic power" (Burawoy 2010: 53). These kinds of analyses stretch the Sociology of Knowledge beyond its early concerns, asking new questions, such as: For whom is the knowledge is produced? How have the various Sociologies developed over time in the unequal geopolitical context? How is sociological knowledge made? What is the sociological division of labor and how does it differ from one location to the next? Such questions enable the possibility of greater progress to be made in understanding Sociology's past and present, and suggest there is much more to be gained by a continuation in interest in comparative, national studies of the various national sociologies.

Conclusion

The dearth of reflection on methods, methodologies, and theoretical frameworks in national histories of Sociology has been a surprising finding of this analysis. While few national studies have been written by individuals trained in history (Fleck and Dayé 2015: 319–320), one would have thought that their authors, almost all of whom are sociologists, would have paid more attention to sociological concepts and methods. Also surprising has been the extent to which there has been an adoption of the methods and methodologies of other disciplines and specialities, including, unfortunately, the well-known problems and weaknesses of these. For instance, despite ongoing critique of the Sociology of Ideas and the Sociology of Intellectuals for seeking to explain historical change or "progress" through a tale of "heroes" and prominent figures in the discipline, these approaches continue to appear in many histories of Sociology. It *is* true that the emphasis on the "heroes" of the discipline is most common in older histories (e.g., Diner 1975; Abrams 1968), yet Sociology's histories continue to maintain an emphasis on:

> the work of great men and - more recently - women, while the attention paid to ordinary sociologists and their routine practices, as well as to the social groups and institutions to which sociologists have belonged, has been very much less; we seem still to be anachronistically at the stage of studying the barons rather than the common people. (Platt 2005: 29)

Given the level of criticism about the Western canon and the extent to which women's Sociology has been ignored, it is nevertheless heartening to read national histories where attention has been paid to "re-inserting" "lost" actors and presenting revised stories of the past. Examples include Turner's (2014) *American Sociology*, which narrates women's involvement in the reform movement and the role of feminism in shaping Sociology's history and sustaining the discipline after the crises of the 1980s, and Calhoun et al. (2010), which incorporates du Bois and the women of Hull House into the history of American Sociology.

The revision of history to acknowledge the contributions of minorities, of women, and of scholars from the global South is not simply an exercise in political correctness. Being forced to examine missing actors reveals more about the "social mechanics" of the knowledge system. An excellent example of this is Ann Oakley's (2020) revised history of British Sociology, which traces the work of five early twentieth century, little known women associated with the London School of Economics and Political Science. The women conducted multiple empirical studies on the economics and sociology of (house)work, providing policy evidence that laid foundations for the welfare state, and extended the discipline's methodological "toolkit." Oakley's study demonstrates how systematic disinterest in the contributions of women in Sociology's history is tied into the hierarchical structure of sociological knowledge, where research method is less highly valued than sociological theory. The outcome has been a distortion in national histories of the discipline, as these have overwhelmingly focused on contributions to sociological theory over time, and largely ignored the history of methods, despite methodological innovation being central to the development of Sociology (Oakley 2020: 292–293). Oakley's study thus provides evidence of the need to critically investigate the "origin myths" of the discipline, as this reveals another aspect to the story of the "late" development of Sociology in Britain. If women's contributions are taken into account, it becomes clear that the "late" arrival of Sociology only describes the slow uptake of sociological work within the major institutions. Outside the universities, women were central to the production of sociological studies and innovations in method, but because they were not made welcome in the universities, and could not build a school, they could not transmit their knowledge inter-generationally. As a consequence, many of their methodological innovations were "discovered" many years later in the mainstream, as if they were new. Oakley (2020: 305) shows how the lack of attention to women's contributions has led to distortions in the historical record. Despite these few exceptions, we are nevertheless waiting for studies that systematically examine the contributions to Sociology outside the mainstream – in the community, in industry, in the corporations, and in the public sector – and from those who work the "hyphen" between disciplines and fields in multi- or interdisciplinary units.

It is evident that more attention must be paid to method and methodology. Admittedly, it is a difficult thing to ask of sociologists already burdened with demands. To undertake national histories, we need to be concerned about matters other sociologists might ignore – "doing" Sociology from both inside and outside; studying key individuals but not forgetting the majority of members upon whom the key people depend and are able to act; mapping the history of an area of practice and knowledge where the subject itself (Sociology) is disputed, heterogenous and with highly porous boundaries; and avoiding the methodological traps of other specialities and disciplines while adopting their more beneficial elements. Moreover, casting a light on our own identities and practices is particularly difficult when the very language, concepts, and methodologies employed by our subjects are the very tools we have at our disposal to make sense of what we see. And perhaps most difficult of all, being conscious, as sociologists, that we are making methodological, epistemological, ontological, and even political decisions at every step of the way, as we conduct our research, write, rewrite, and publish. Not an easy task!

References

Abbott A (2001) Chaos of disciplines. University of Chicago Press, Chicago
Abrams P (1968) The origins of British sociology: 1834–1914. University of Chicago Press, Chicago
Alatas SF (1972) The captive mind in development studies. Int Soc Sci J 34(1):9–25
Aminzade R (1992) Historical sociology and time. Sociol Methods Res 20(4):456–480
Archer M (1991) Presidential address. Int Sociol 6(2):131–147
Austin-Broos D (2005) Australian sociology and its historical environment. In: Germov J, McGee T (eds) Histories of Australian sociology. Melbourne University Press, Carlton, pp 245–266
Azarya V (2010) Academic excellence and social relevance: Israeli sociology in universities and beyond. In: Patel S (ed) The ISA handbook of diverse sociological traditions. Sage, Los Angeles, pp 246–256
Beck U (2000) What is globalisation? Polity, Cambridge
Berger PL, Luckmann T (1996) [1966] The social construction of reality. Anchor Books, Garden City
Bhambra G (2007) Rethinking modernity. Palgrave Macmillan
Bourdieu P (1981) The specificity of the scientific field. In: Lemert C (ed) French sociology. Columbia University Press, New York, pp 257–292
Bourdieu P (1998) George Canguilhem. Econ Soc 27(2–3):190–192
Burawoy M (2010) Forging global sociology from below. In: Patel S (ed) The ISA handbook of diverse sociological traditions. Sage, Los Angeles, pp 52–65
Calhoun C, Duster T, VanAntwerpen J (2010) The visions and divisions of American sociology. In: Patel S (ed) The ISA handbook of diverse sociological traditions. Sage, Los Angeles, pp 114–125
Camic C (2015) Das Verschwinden des 'Charakters'. In: Daye C, Moebius S (eds) Soziologiegeschichte. Suhrkamp (p. 310–337). [In English, The Eclipse of; Character' 2012 translation by author]

Camic C, Gross N (2004) The new sociology of ideas. In: Blau J (ed) Blackwell companion to sociology. Wiley Blackwell, pp 236–249

Chen H-F (2018) Chinese sociology. In: Holmwood J, Turner S (series eds) Sociology transformed. Palgrave Macmillan

Collins R (1985) The sociology of philosophies. Belknap Press of Harvard University Press, Cambridge

Collyer FM (2012a) Mapping the sociology of health and medicine. Palgrave Macmillan

Collyer FM (2012b) The birth of a speciality. Health Sociol Rev 21(1):116–130

Collyer FM (2017) From nation building to neo-liberalism. In: Korgen K (ed) The Cambridge handbook of sociology, vol 1. Cambridge University Press, pp 82–94

Collyer FM, Connell R, Maia J, Morrell R (2019) Knowledge and global power. Monash University Press

Connell RW (2007) Southern theory. Allen and Unwin, Crows Nest

Connell R (2010) Learning from each other. In: Patel S (ed) The ISA handbook of diverse sociological traditions. Sage, Los Angeles, pp 40–51

Cordeiro VD, Neri H (2019) Sociology in Brazil. In: Holmwood J, Turner S (series eds) Sociology transformed. Palgrave Macmillan

Cossu A, Bortolini M (2017) Italian sociology, 1945–2010. In Holmwood J, Turner S (series eds) Sociology transformed. Palgrave Macmillan

Crothers C (2018) Sociologies of New Zealand. In: Holmwood J, Turner S (series eds) Sociology transformed. Palgrave Macmillan

Dayé C (2018) A systematic view on the use of history for current debates in sociology. Am Sociol 49:520–547

Delanty G, Isin E (eds) (2003) Handbook of historical sociology. Sage

Denis A (2010) Ethnicity and race within sociology in the commonwealth Caribbean. In: Patel S (ed) The ISA handbook of diverse sociological traditions. Sage, Los Angeles, pp 292–300

Diner SJ (1975) Department and discipline. Minerva 13(4):514–553

Ekerwald H (2014) Svensk Sociologi – de många rösternas ämne [Swedish sociology]. In: Andersson G, Brante T, Edling C (eds) Det personliga är sociologiskt [The personal is sociological]. Liber AB, Stockholm, pp 79–94

Fleck L (1979) [1935] Genesis and development of a scientific fact. University of Chicago Press, Chicago

Fleck C (2015) The study of the history of sociology and neighbouring fields. Contemp Sociol: J Rev 44(3):305–314

Fleck C, Dayé C (2015) Methodology of the history of the social and behavioral sciences. In: Smelser NJ, Baltes PB (eds) International encyclopedia of the social and behavioral sciences, vol 15, 2nd edn. Elsevier, pp 319–325

Foucault M (1973) [1963] The birth of the clinic (trans: Sheridan-Smith AM). Pantheon, New York

Foucault M (2011) The Government of self and others (ed: Gros F, trans: Burchell G) Picador, New York

Go J, Lawson G (2017) Global historical sociology. Cambridge University Press, Cambridge

Golenkova Z, Narbut N (2015) History of sociological thought in Central and Eastern European countries. Serbian State Publisher of Textbooks, Belgrade

Gorski P (2013) Conclusion. Bourdieusian theory and historical analysis: maps, mechanisms, and methods. In: Gorski P (ed) Bourdieu and Historical Analysis Duke University Press: Durham and London, pp 327–366

Halsey AH (2004) A history of sociology in Britain. Oxford University Press, Oxford

Hanafi S (2010) Palestinian sociological production. In: Patel S (ed) The ISA handbook of diverse sociological traditions. Sage, Los Angeles, pp 257–267

Harley K, Wickham G (2014) Australian sociology. In: Holmwood J, Turner S (series editors) Sociology transformed. Palgrave Macmillan

Hartsock N (1983) The Feminist standpoint. In: Harding S, Hintikka MB (eds) Discovering reality. Synthese library, vol 161. Springer Netherlands, pp 283–310

Hobden S, Hobson J (2001) Historical sociology of international relations. Cambridge University Press, Cambridge

Husbands CT (2019) Sociology at the London School of Economics and Political Science, 1904–2015: sound and fury. Palgrave Macmillan

Konrad G, Szelényi I (1979) The intellectuals on the road to class power. Harvester Press, Brighton

Kuhn T (1979) [1962] The structure of scientific revolutions. University of Chicago Press, Chicago

Kurzman C, Owens L (2002) The sociology of intellectuals. Annu Rev Sociol 28:63–90

Larsson A, Magdalenic S (2015) Sociology in Sweden. In: Holmwood J, Turner S (series editors) Sociology transformed. Palgrave Macmillan

Levine D (1995) Visions of the sociological tradition. University of Chicago Press, Chicago

Lovejoy AO (1936) The great chain of being. Harvard University Press, Cambridge

Mannheim K (1972) [1936] Ideology and Utopia. Routledge and Kegan Paul, London

Masson P, Schrecker C (2016) Sociology in France after 1945. In: Holmwood J, Turner S (series editors) Sociology transformed. Palgrave Macmillan

Meja V, Stehr N (1993) The sociology of knowledge and the ethos of science. In: Leonard E, Strasser H, Westhues K (eds) In search of community. Fordham University Press, New York, pp 65–83

Meja V, Stehr N (eds) (2000) The sociology of knowledge. Edward Elgar, Cheltenham

Mense-Petermann U, Schlerka SM, Welskopp T (2018) Editorial. Interdisciplines 9(2):1–6

Merton RK (1977) The sociology of science. In: Merton RK, Gaston J (eds) The sociology of science in Europe. Southern Illinois Press, Carbondale

Mills D (2008) Difficult folk? Berghahn Books, New York

Mitropoulos A (2005) Discipline and labour. In: Germov J, McGee T (eds) Histories of Australian sociology. Melbourne University Press, Carlton, pp 101–121

Nisbet RA (1966) The sociological tradition. Heinemann, London

Oakley A (2020) Women, the early development of sociological research methods in Britain and the London School of Economics. Sociology 54(2):292–311

Parsons T, Shils E, Naegele KD, Pitts JR (1961) Theories of society. Free Press of Glencoe, New York

Patel S (2010) The ISA handbook of diverse sociological traditions. SAGE Publications Ltd,, London. https://doi.org/10.4135/9781446221396

Payne G (2019) Poor cousins. In: Panayotova P (ed) The history of sociology in Britain. Palgrave Macmillan, pp 191–220

Pereyra DE (2010) Dilemmas, challenges and uncertain boundaries of Argentinian sociology. In: Patel S (ed) The ISA handbook of diverse sociological traditions. Sage, Los Angeles, pp 212–222

Platt J (2005) What should be done about the history of British sociology? In: Halsey A, Runciman W (eds) British sociology seen from without and within. Oxford University Press, Oxford, pp 23–35

Reynolds L, Reynolds J (eds) (1970) The sociology of sociology. David McKay, New York

Ringer F (1990) The intellectual field, intellectual history, and the sociology of knowledge. Theory Soc 19:269–294

Scott J (2017) The development of sociology in Britain. In: Korgen K (ed) The Cambridge handbook of sociology, vol 1. Cambridge University Press, pp 37–49

Seidman S (1985) Classics and contemporaries. Hist Sociol 6:121–135

Slaughter S, Leslie L (1997) Academic capitalism. The Johns Hopkins Press, Baltimore

So AY (2017) Sociology in East Asia. In: Korgen K (ed) The Cambridge handbook of sociology, vol 1. Cambridge University Press, pp 50–62

Sooryamoorthy R (2016) Sociology in South Africa. In: Holmwood J, Turner S (series eds) Sociology transformed. Palgrave Macmillan

Steinmetz G (2005) Introduction. In: Steinmetz G (ed) The politics of method in the human sciences. Duke University Press, Durham/London, pp 1–56

Streek W, Thelen K (2005) Institutional change in advanced political economies. In: Streek W, Thelen K (eds) Beyond continuity. Oxford University Press, Oxford, pp 1–57

Swidler A, Arditi J (1994) The new sociology of knowledge. Annu Rev Sociol 20:305–329
Tilly C (2001) Historical sociology. In: Smelser NJ, Baltes PB (eds) International encyclopedia of the behavioral and social sciences. Elsevier, pp 6753–6757
Toffler A (1980) The third wave. William Morrow
Turner S (2014) American sociology. In: Holmwood J, Turner S (series eds) Sociology transformed. Palgrave Macmillan
Tzeng A (2012) Framing sociology in Taiwan, Hong Kong and Singapore. PhD thesis, Department of Sociology, University of Warwick. http://go.warwick.ac.uk/wrap/49816
Wallace RW (1992) Starting a department and getting it under way. Minerva 30(4):497–512
Wilson NH, Adams J (2015) Historical sociology. In: International encyclopedia of the social and behavioral sciences, vol 15, 2nd edn. Elsevier, pp 27–30
Zdravomyslova E (2010) What is Russian sociological tradition? In: Patel S (ed) The ISA handbook of diverse sociological traditions. Sage, Los Angeles, pp 140–151

The History of Sociology as Disciplinary Self-Reflexivity

32

George Steinmetz

Contents

Introduction	834
Part One: From the History of Science, to the History of Social Science, to the History of Sociology	836
The History of the History of Sociology	842
Part Two: *Why* Should One Write the History of Sociology?	846
Overcoming the Repression of Disciplinary Memory	846
The History of Social Science as a Guide to Scientific *Flourishing* (and Decay)	851
Understanding Social Science as a Determinant of Social Phenomena	852
The History of Sociology as Sociological Reflexivity	853
Conclusion	855
References	856

Abstract

This chapter addresses the question: how and why the history of sociology should be written. The chapter's first conclusion is that the history of sociology is an essential methodological component of sociology in general. In order to develop this argument, the chapter surveys the emergence of research on the history of sociology and the other human and social sciences, paying attention to the concepts, methods, theories, and justifications that have structured this literature. The key conceptual advance was the creation of a sociology of social science that attends closely to texts and contexts at differing distances from the immediate site of scientific production. The second section asks *why* sociologists should write the history of sociology, and argues that this work contributes to sociology in four main ways, (1) uncovering repressed elements of disciplinary memory and sources of contemporary scientific *doxa*; (2) shedding light on the conditions for the flourishing of knowledge (including sociological knowledge);

G. Steinmetz (✉)
Department of Sociology, University of Michigan, Ann Arbor, MI, USA
e-mail: geostein@umich.edu

© The Author(s), under exclusive licence to Springer Nature Singapore Pte Ltd. 2022
D. McCallum (ed.), *The Palgrave Handbook of the History of Human Sciences*,
https://doi.org/10.1007/978-981-16-7255-2_60

(3) examining the role of sociology itself as a determinant of non-sociological phenomena; and (4) working as under-laborers for scientific reflexivity, along the lines suggested by Pierre Bourdieu.

Keywords

History of science · History of social science · History of sociology · Reflexivity · Pierre Bourdieu

> The social history of social science, so long as it is also considered a science of the unconscious ... is one of the most powerful means of distancing oneself from ... the grip of an incorporated past which survives into the present. (Pierre Bourdieu, *A Lecture on the Lecture*)

Introduction

Sociologists rarely discuss the reasons for writing the history of their discipline in explicit terms (but see Peter 2001; Kruse 2001; Moebius 2015; Fleck and Dayé 2015). This is problematic, for several reasons. First, the subfield of the history of sociology seems particularly vulnerable to appearing narcissistic, whimsical, or antiscientific. One Nazi sociologist proclaimed that a science that takes itself as its own analytic object represents "the symptom of a profound sickness of an entire culture" and a "pathology of scientificity" (Eschmann 1934: 955). Such an odious figure may seem unworthy of our consideration, but Eschmann's words resonate with persistent objections from more reasonable corners. Some suggest that past forms of knowledge have been forgotten for good reasons that they have been defeated in tests of scientific strength. Hegel referred to Cicero's observation that "the whole of the history of Philosophy becomes a battlefield covered with the bones of the dead; it is a kingdom not merely formed of dead and lifeless individuals, but of refuted and spiritually dead systems, since each has killed and buried the other" (Hegel 1995, vol. 1: 17). Since the Ancient world, science has usually been seen "as an essentially unilinear, incremental process, dependent upon the steady accumulation and ordering of knowledge" (Baker 1975: 375).

A second set of challenges relates to demands to make social science immediately useful and "public," as entire university administrations in the USA have been taken over by calls to make scholarship "relevant." These arguments suggest that historical work on sociology amounts to a form of scientific navel-gazing. The graduate director in one top US sociology department recently issued a memo counseling students to avoid entering the professional "circle of doom" by writing on historical topics or on "a part of the world Americans know little about and care less about." This memo (in the author's possession) recommended further that students who insisted on writing on a country other than the USA "compare with the U.S.," since "Americans like that." The memo's advice was inaccurate in predicting graduate students' success on the academic job market, but it represented a powerful performative speech act, marginalizing graduate students working on non-parochial topics. The fact that this incident was not an outlier is revealed by an SSRC-funded survey of the top ten US sociology, political science, and economics departments, which found all of the chairs and graduate advisors strongly opposed to graduate students writing on foreign countries (Stevens, Miller-Idriss, and Shami, 2018). The history of sociology often focuses on historical eras and countries deemed irrelevant by this putative contemporary audience. Another taboo, according to this memo, is "me-search." The history of sociology probably also qualifies as "me-search," at least when it involves sociologists studying their own profession.

One possible response by historians of social science would be to fall back on doctrines of academic freedom or "the usefulness of useless knowledge" (Abraham Flexner). While these doctrines are valid (Steinmetz 2018), this paper argues that the history of sociology can have other use-values. It can explain why certain ideas operate as taken-for-granted *doxa* within current sociology, while other ideas are marginal. The history of sociology can contribute to understanding the conditions that allow science – including sociology – to flourish intellectually or to stagnate. The history of sociology sheds light on the causal role of sociological ideas as determinants of social processes and events. And the history of sociology is a key part of the reflexivity that many sociologists embrace and try to integrate into their work. In short, the history of sociology is, in fact, a useful activity. It is central to the sociological enterprise.

The most obvious contribution of research on sociology is self-reflexivity. Far from marginal, the sociology of sociology becomes "an indispensable instrument of the sociological method" in this respect (Bourdieu 1990: 178). In order to fulfill this promise, however, the history of sociology needs to be a science not only of the discipline's conscious but of its unconscious as well. The history of sociology is unique compared to other forms of intellectual and disciplinary history, insofar as its object – sociology – presents it with its own epistemology, ontology, and methodology.

The first section of this chapter surveys the emergence of research on the history of sociology and the other human and social sciences, in order to answer the question: how should one write the history of sociology? The chapter will pay special attention to the concepts, methods, theories, and justifications that have structured this literature. The key conceptual advance, it is argued here, was the creation of a sociology of social science that attends to both internal and external determinants and, indeed, that overcomes this distinction, in favor of a proliferation of contexts at differing distances from the immediate site of scientific production, weaving both highly proximate and widely ranging, environing scales into the umbrella term "context." This sociology of sociology pays attention to differing degrees of fluidity and closure between the scientific field and these environing other sites. Bourdieu's field theoretic approach helps to make sense of the sociologically *intermediate* levels that exist between the macro-epochal trends and social structures, on the one hand, and the immediate point of contact between scientists and their objects or writers and their texts. By theorizing this intermediate level of fields, sociology becomes central, once again, to the history of science. This is my sole disagreement with Daston's (2016) discussion of the changes in the history of science since Kuhn. When Daston writes that the history of science has bid "farewell to sociology and philosophy" because of their presentism and orientation toward general laws, she ignores the growing number of sociologists in this area who are committed to a "contexualized, practice-centered" approach and to ideas of contingent historicity.

The second section asks why sociologists should engage in the arduous work of writing the history of sociology. I will argue that historians of sociology contribute in at least the following four ways: (1) they uncover repressed elements of disciplinary memory and sources of contemporary scientific *doxa*; (2) they shed light on the conditions for the flourishing of knowledge (including sociological knowledge); (3) they examine the role of sociology itself as a determinant of non-sociological phenomena; and (4) they work as under-laborers for scientific reflexivity, along the lines suggested by Bourdieu.

Part One: From the History of Science, to the History of Social Science, to the History of Sociology

Until the twentieth century, historians of science pursued a fairly specific set of goals: to legitimate particular sciences and theories, celebrate particular geniuses, fight battles with scientific competitors, and contribute to improved scientific practice. Much of this took the form of *monumental* history, in Nietzsche's sense (Nietzsche 1997 [1874]). There was little interest in exploring the causes or contexts of scientific change, and even less interest in criticizing science. Starting in Ancient Greece, the history of science was triumphal and progressivist, taking the form of a grand narrative whose *dramatis personae* were scientific heroes (Daston 2001). Many histories of science written during the nineteenth century adopted the frameworks of French positivism or English Whig historiography, both of which were committed to the idea of inexorable scientific progress (McEvoy 1997). Ernst Mach made a slight correction to this approach in his book *The Science of Mechanics* (1883), arguing that the study of "the rejected and transient thoughts of the inquirers" could also be "very important and very instructive," because it "not only promotes the understanding of that which now is, but also brings new possibilities before us" (Mach 1883 [1919]: 254–255). Yet Mach's approach was still framed within a vision of unstoppable scientific progress. The English mathematician and philosopher Alfred North Whitehead wrote in 1917 that a "science which hesitates to *forget* its founders is lost" (Whitehead 1917: 115; my emphasis). Even the scientists celebrated in Vannevar Bush's *Science: The Endless Frontier*, which was "released shortly before the dropping of the atomic bomb," were described in "persistently individualistic" language as a "collectivity of heroic individuals, comparable in spirit to Daniel Boone or Davy Crockett, pushing back the frontiers" (Hollinger 1990: 900).

The earliest histories of the human and social sciences were guided by similar motives. The first example was Stanley's (1656) history of philosophical doctrines and "eminent" individuals. Although Stanley provided a bit of biographical information on the philosophers, he made no attempt to link philosophical ideas to their social contexts. For example, Stanley did not discuss the effects of slavery on Greek or Roman philosophy, except to note that Solon was more "profuse and delicate, and more luxurious in his verses than beseemes a philosopher" due to his "practicing Merchandise" (Stanley 1655: 30), that Xenophon received slaves as gifts (Stanley 1655, part 3: 104), that Anniceris paid a ransom to free Plato from slavery (Stanley 1655, part 4: 17), and that Bion and his family were sold into slavery (Stanley 1655, part 4: 22). Anticipating Mach, Johann Jakob Brückner argued in *Historia critica philosophiae* that the history of philosophy could "instruct men, what is to be avoided," and put them "upon their guard against repetition of attempts" (1791: 8). Brückner also suggested that the history of philosophy could elucidate the sources of present-day *doxa*, by leading to "the full discovery of the origin of many notions and practices, which have no other support than their antiquity" (Brückner and Enfield 1791: x). However, Brückner's approach was overshadowed in Germany by that of Kant, who understood the history of philosophy as a chronicle of the progress of reason itself. Kantians such as Wilhelm Tiedemann and Dietrich Tennemann continued writing the history of philosophy in this vein.

Hegel inaugurated an alternative strand that *rehistoricized* the sciences (Jaeschke 1993), including philosophy itself. Hegel lectured more frequently on the history of philosophy than any other topic. Clearly this was partly aimed at justifying Hegel's own system, and at legitimating philosophy in general. The history of philosophy paralleled world history itself, as "the medium of the development of the world spirit" (Jaeschke 1993: xvi). Hegel argued that each generation received all that each past generation had produced in science and in intellectual activity" as "an heirloom." All of this was then "changed, [as] the material worked upon" was "both enriched and preserved at the same time" in dialectical fashion (Hegel 1995, vol. 1: 3). Every past philosophy therefore "has been and still is necessary," and "thus, none have passed away, but all are affirmatively contained as elements in a whole" (Hegel 1995, vol. 1: 37). Philosophers needed to study "the essential ideas of philosophers in the past" because they "are all too prone to forget the origins and context of their own doctrines" (Beiser 1995: xxvii). In sum, "[t]he course of history does not show us the Becoming of things foreign to us, but the Becoming of ourselves and of our own knowledge" (Hegel 1995:, vol. 1: 4).

While this could be read as an alternative version of the doctrine of progress, Hegel broke with Kant in insisting that each philosophical school had to be understood on its own terms and "in the context of its own time, as the self-awareness of the ideals and values of its age" (Beiser 1995: xv, xxix): "[T]here is a definite Philosophy which arises among a people, and the definite character of the standpoint of thought is the same character which permeates all the other historical sides of the spirit of the people" (Hegel 1995, vol. 1: 53). One should not assume that one's own "principles are somehow natural, divine, eternal, or innate, when they are in fact only the product of a specific time and place, the self-awareness of the values and ideals of a specific culture" (Beiser 1995: xxvii). This did *not* mean, however, that "political history, forms of government, art and religion" were "related to Philosophy as its causes," but rather that they had the "same common root" as Philosophy, namely, the *Weltgeist* or spirit of the time (Hegel 1995, vol. 1: 54). Hegel's more historicist approach set a different tone than the Kantian historians, and it was emulated by historians of other disciplines.

Economics was the first social science to be subjected to historical analysis (Krauth 1978). In Germany, the "number of lectures on the history of economic thought" increased sharply in the wake of the 1848 Revolution (Lindenfeld 1997: 162). Karl Marx was an important participant in this historicization of economics, just as Hegel inaugurated a more thoroughly historical approach to philosophy. Marx developed his theory of capitalism by incorporating past economic theory, via a form of *immanent critique* of the field's tradition (Murray 1988). Marx also explored the resonances between economic theories and their social contexts (Marx 2000). He did not reduce economic doctrines to "dubious dealings with the bankers and brokers of the day," but tied them to the more abstract forces of commodity fetishism (Murray 1988: 98) and the value form.

Other disciplines were inspired by Hegel, Marx, and German historicism (Troeltsch 1922; Steinmetz 2020a) "to examine the past for its own sake, to see events in context, and to fathom the deeper motives for actions" (Beiser 1995: xi). The series *Geschichte der Wissenschaften in Deutschland* (*History of the Sciences in Germany*) published studies of the history of political science, legal science, and other nonnatural-science disciplines. These studies tended to follow the traditional approach by presenting the main doctrines, discoveries, and personalities, and boosting the legitimacy of new fields. However, a few authors pursued a more sociological approach. Wilhelm Roscher, one of the founders of the historical school of economics, explored the relations between economic thought and politics, arguing that Fichte's entire philosophy, including his economics, "bore the unmistakable imprint of the democratic-revolutionary era," i.e., the French Revolution (Roscher 1874: 639).

The history and sociology of science advanced between the 1920s and the 1960s. The most important intervention was the sociology of knowledge or *Wissenssoziologie*, developed by Max Scheler, Karl Mannheim, Georg Simmel, Ernst Grünwald, Paul Honigsheim, Wilhelm Jerusalem, Siegfried Marck, Alexander von Schelting, and Alfred Weber. Scheler presented himself as an anti-Marxist, but he borrowed "a number" of his "basic principles" from Marxism (Bukharin 1931: 17, note 13). According to Robert K. Merton, Scheler "flatly repudiates all forms of sociologism," but at the same time "indicates that different types of knowledge are bound up with particular forms of groups," including social classes (Merton 1949: 472). According to Scheler, "even some quite formal types of thought and valuation" (Bukharin 1931) took opposing values among the lower and upper social classes, including Pragmatism vs. Intellectualism, milieu-theoretical thought vs. nativist thought, and an optimistic view of future and pessimistic retrospection vs. a pessimistic view of the future and optimistic retrospection (Scheler 1926: 204–205). Scheler argued that modern science emerged from the combination of "two social strata that were originally separate": a group of educated upper classes involved in "free contemplation" and another "class of people who have rationally accumulated the experiences of work and craftsmanship" (Scheler 1925: 142). This thesis about the retardation of science due to Greek Philosophers' disdain for manual labor and experimentation had already been presented by F. J. Moore (1918: 2) in a book on the history of chemistry, and was repeated and developed by a number of subsequent theorists of science (e.g., Zilsel 2000 [1942]).

According to Mannheim, *Wissenssoziologie* sought to trace the connections between "the social position of given groups and their manner of interpreting the world." It was concerned with both the *contents* and the *forms* of knowledge. Mannheim credited Marx with "unmasking" the class interests behind ideologies and analyzing the reification of culture, but he went beyond Marx in attending to the role of "generations, status groups, sects, occupational groups, schools, etc.," as well as social classes and capitalism (Mannheim 1929: 276–277, 309–310). Merton argued that Mannheim and other *Wissenssoziologen* "neglected the analysis of the more firmly established disciplines" and exempted science "from existential determination" (Merton 1937: 494; 1949: 470), and others repeated this charge. Yet Merton clearly had not read Mannheim's Heidelberg habilitation thesis on Conservatism, sections of which were published in the *Archiv für Sozialwissenschaft und Sozialpolitik* (*Journal of Social Science and Social Policy*) in 1926 and in English translation in 1953. Mannheim argued that German conservative thought represented a submerged counter-current to modern rationalism that reemerged in reaction to the diffusion of the French Revolution and its Enlightenment ideas into Prussia, a "socially backward but culturally mature society" (1953: 88, 120). While many Conservative thinkers were characterized by their unstable social and economic status and were part of the "socially unattached intelligentsia," Mannheim also discussed established university professors such as Hegel, Savigny, and Ranke. Mannheim's view of "social unattached" intellectuals as "*ideologues* who can find arguments in favor of any political cause they may happen to serve" is more critical in his thesis on Conservatism (1953: 127) than in *Ideology and Utopia*, where that group is said to play "a decisive role, by virtue of its unique capabilities for openness and choice, in generating a synthesis out of incompatible ideologies and thus making possible an effective practical way out of crisis" (Kettler, Meja, and Stehr 1984: 79). Mannheim's thesis on Conservatism also refutes the claim by Camic and Gross (2002: 100) that he did not pay attention to the *contents* of knowledge or to the "*particular* men and women who produced them."

In his book *Man and Society in an Age of Reconstruction*, Mannheim (1935) analyzed the ways pragmatism, behaviorism, and official psychoanalysis were being deployed to influence social behavior in contemporary socio-political regimes. After 1940 Mannheim began to include the natural sciences and mathematics under the purview of the sociology of knowledge. Indeed, Mannheim was reproached by other critics precisely for arguing that "even scientific thought, and especially the social sciences," is "inescapably bound up with and 'corresponding' to the social position of the thinker" (von Schelting 1936: 665).

A further contribution was made by the *Commission on the History of Knowledge* at the Soviet Academy of Science (Graham 1985: 137). Nikolai Bukharin, Head of the Commission between 1930 and his execution in 1938, argued that both the natural and social sciences were "heavily mediated by social, economic, and political factors, and therefore cannot be separated from the society in which [they emerge]" (Bukharin 1931: 140). The Russian historian-scientist Boris Hessen (1931) argued that the main scientific problems tackled by physics in Newton's age were determined by the demands of production, warfare, and transportation. This did not mean that these external demands were translated directly into science, however. Scientific goals and theoretical concepts were developed by studying existing technology (Freudenthal and McLaughlin 2009:11). Hessen also emphasized the movement of elites into science and the merging of the mechanical and liberal arts. Ludwik Fleck, Robert Merton, and Thomas Kuhn went on to emphasize scientific communities, with differing accents.

Further advances toward a historical sociology of social science are attributable to the French historical school of epistemology ("*épistemologie française*"; Bitbol and Gayon 2015), whose leading figures were Gaston Bachelard, Alexandre Koyré, Georges Canguilhem, and Michel Foucault. Bachelard rejected notions of continuous progress, or a singular "scientific revolution," proposing instead the idea of continuous historical "ruptures" in scientific evolution. Errors and obstacles were therefore as important for the history of science as the successful discoveries. Deeply influenced by psychoanalysis, Bachelard also identified an array of blockages to scientific knowledge rooted in unconscious emotions, including scientists' anxieties about social status. According to Bachelard, the scientific fact cannot be accessed or assembled from empirical sense data, but has to be "conquered, constructed, and confirmed" via an epistemological break with spontaneous perceptions and common sense (Bourdieu et al. 1991 [1968]): 11). These ideas were crucial for Bourdieu's elaboration of a constructivist, reflexive approach to the history of science that foregrounded strategic struggle within scientific fields, in contrast to the pacified image of "scientific communities" associated with Merton and Kuhn.

The French historical school of epistemology also extended the history of science to include the social sciences, which the established disciplines had traditionally disdained as less "noble" (Fabiani 1989: 123–124). Canguilhem (1989) focused on psychology, the social science closest to philosophy in France. Foucault's historical studies of the human sciences also began with psychology (Foucault 1954). Foucault broadened his scope in *Les mots et les choses* (1966) to include linguistics, economics, and biology. Foucault's approach was also grounded in the new structuralist theories of discourse analysis, which opened up new ways of grouping, decoding, and eventually deconstructing scientific texts.

By the 1960s, then, there was already an impressive archive of methodological and theoretical resources in the history of the human sciences, reaching from Hegel to Foucault. Two additional interventions in the intervening period paved the way for current work in the history and sociology (or historical sociology) of sociology. The first encompasses Science and Technology Studies (STS), Sociology of Scientific Knowledge (SSK), and Actor-Network Theory (ANT), which are jointly referred to here as STS. The second set of contributions consists of discussions of writing history of sociology, discussed in the next section.

Many of the tenets of STS crystallize arguments developed by earlier authors. Like Marx and other Marxists, STS refuses to draw a sharp line between technology and science or to focus on "great scientific men." Like the sociologists of knowledge, STS pays attention to both contents and forms of knowledge. Like the French historical school of epistemology, STS understands scientific history as a discontinuous process, and urges analysts to pay attention to both defeated and victorious theories. Like Foucault and the Cambridge school of intellectual history, STS pays attention to discursive forms and semiotic filiations. The most innovative features of STS, perhaps, are its emphasis on *microlevel* practices in scientific laboratories and *networks* that exceed scientific disciplines and connect disparate actors and nonhuman actants in the production of scientific facts (Latour and Woolgar 1979).

The STS approach also has several weaknesses that make it ill-suited to serve as the sole foundation for a historical sociology of sociology. First, its focus on laboratories and its methodological "localism" does not translate well into the studies of the humanities and interpretive social sciences (cf. Gross and Fleming 2011). Second, its emphasis on heterogeneous networks underestimates the causal importance of logics of social closure around scientific fields (and social fields in general) (Bourdieu 2004). Third, its empiricist ontology is connected to a fundamentally affirmative and anti-critical stance (Latour 2004; compare Horkheimer 1972; Fassin forthcoming). Like positivism, STS seems to assume that "science normally is as it ought to be," which "helps to explain its studiously descriptive stance to the sciences it studies" (Fuller 2007: 20). Finally, STS rejects the idea of "conflict as a central feature of social life" and the premise that "the capacities of actors are often shaped by structures that exist prior to the entrance of particular actors on the scene" (Albert and Kleinman 2011: 265).

This chapter argues below that Bourdieusian field theory resolves most of the shortcomings of these approaches and provides the most fruitful approach to the history of social science, once three revisions are made. First, Bourdieu's approach *can* be articulated with the actor-network approach (see Collins 1998: 946, n. 4; Camic 2020, ch.2), or more generally with network approaches (Singh 2019), as long as the latter are separated from empiricist ontology and axiological neutrality (or anti-critical ethics). Second, Bourdieu's approach becomes more powerful once its analysis of semi-autonomous fields is combined with attention to historical-epochal processes that are not contained within any single field, such as war, revolution, crisis, the breakdown of regulatory paradigms, etc. (Steinmetz 2005). Third, Bourdieu's habitus-centric model of the individual subject becomes more realistic and adequate once it is grounded in psychoanalysis and a non-metaphorical theory of the unconscious (Steinmetz 2006, 2014). Before discussing in more detail this *neo-Bourdieusian* approach to the history of social science, however, the next section surveys the evolution of writing on the history of science and the more recent emergence of historical attention to the social sciences, leading to the a historiography of sociology.

The History of the History of Sociology

Sociologists have always shown some interest in the history of their discipline, and indeed, the first panels on "historical sociology" at the meetings of the American Sociological Society were actually focused on the history of sociology (Blumer 1933: 11). Yet this writing has typically completely unsociological and theoretically naïve forms. Much of it is a "pure Whig historiography which concentrated on a mere chronology of dogmas and theories and completely lacked all sociological elements of a disciplinary history" (Lepenies and Weingardt 1983: viii). It has often taken the form of a "tribal history" of the discipline playing "the role of guard and keeper of the canonical wisdom of more than [one] hundred years of great intellectual achievements" (Käsler 1999: 31; Idem., 2001: 10). Conventional histories of sociological theory include only rudimentary gestures toward contextualization, often along the lines of "national traditions."

Prior to the 1960s and 1970s almost all of these histories were oriented toward legitimating the new discipline or promoting specific approaches over others. Albion Small, founder of the University of Chicago sociology department, wrote about the pre-history of sociology just a few years after the discipline was introduced in the USA and European universities. Small's publications combine efforts to legitimate the new discipline with warnings about ignoring earlier work, lest one duplicate it. Small wrote:

> We cannot fully understand anything human unless, among other ways of apprehension, we understand it genetically, i.e., in its growth, in its evolution. Present problems of social interpretation are no exceptions to this rule. We cannot take a fully intelligent part in today's thinking unless we are intelligent about the antecedents which have prepared the way for present thinking (Small 1924: 404).

Small was happy to examine obsolete knowledge formations such as Cameralism, although he framed the latter as a stepping stone toward the more perfected approaches of present-day sociology (Small 1909). Durkheim's accounts of the history of French sociology and Pragmatist philosophy (Durkheim 1900, 1983) were surprisingly unsociological; the same was true of German-language studies such as Gumplowicz (1885: Ch. 1) and Stein (1897). Von Wiese (1926, 1960) was mainly concerned to promote his own branch of sociology, which is one that the discipline seems to have simply forgotten, rather than repressed. During the golden age of postwar American sociological neopositivism, a number of Columbia University sociology students wrote theses on the history of European sociology that sought to resurrect lost empirical traditions. This was part of a concerted effort to shift European sociology in a more "American" direction (Pollak 1979) and away from its supposedly overly theoretical orientation, a shift symbolized by frequent references to Fréderic Le Play and Charles Booth as alternative sociological founding fathers replacing Durkheim and Weber (e.g., Lerner 1959: 20–21).

The atheoretical fog hanging over the history of sociology began to lift once writers were exposed to Wissenssoziologie. This can be illustrated with three examples from very different contexts. The first example is the German theologian-sociologist Ernst Troeltsch, whose book on historicism traced the stark division between Anglo-French positivism and German historicism in the human sciences. Of course, French and Anglo-American sociologists tended to read this geographic difference as a marker of German sociology's immaturity, but the German historicists suggested a different interpretation: any "laws" in sociology were changeable and contingent (Troeltsch 1922: 475). German historicism had been strengthened by the "global catastrophe of the Great War," Troeltsch argued, while the positivist ideas were connected to the "ideals of the French revolution, or the English gentleman, or American democracy, or socialistic fraternity" (Troeltsch 1922: 10, 143).

The second example is Harry Barnes and Howard P. Becker, who were immersed in Weimar-era German sociology when they published the first volume of their encyclopedic *Social Thought from Lore to Science* in 1938. According to Barnes and Becker, their approach involved placing "all social theories in their full social and cultural contexts." They attributed this methodology to "Wissenssoziologie," which they intended to combine with a "historical ... mode of exposition" (Barnes and Becker 1938: ix). That said, Barnes and Becker did not systematically reconstruct the historical contexts of the various sociological movements. Barnes's *Introduction to the History of Sociology* (1948) was a more sustained attempt to combine textual and contextual analysis. Barnes was well suited for this task as one of the few American practitioners of a fully historical sociology during the middle decades of the twentieth century who was not a refugee from Nazi Germany, where the first wave of genuine historical sociology had emerged (Steinmetz 2010). In his introduction, Barnes returned to his previous claim that "the doctrines of any writer lose much of their significance if their relation to the prevailing social environment is not pointed out and the purposes of the work clearly indicated" (1948:3–4). Barnes went on to make a number of contextualist arguments, for example: the Stoics and Epicureans reflected the disorder of the Alexandrian empire (Ibid.: 10); social contract theory grew out of a society with parliamentary procedures, sound legal foundations, and commerce, which emphasized the "importance of contracts in the sphere of economic activities" (Ibid.: 29); the replacement of the cyclical view of history with the idea of progress after 1600 was due to the rise of science and rationalism (Ibid.: 38); comparative ethnology grew out of exploration and colonization (Ibid.: 70); and the transition from social philosophy to sociology reflected the rise of the nation state, the industrial revolution, and the discovery of new cultures which provided an external standpoint for criticizing one's own society (Ibid.: 46). Barnes pursued a similar approach when he turned to individual sociologists, analyzing the Austrian sociologist Ludwig Gumplowicz as "the classic example of the influence of a writer's social and political environment upon his theory":

> The almost unique cultural and ethnic diversity and the continual struggle of national groups and social classes in Austria-Hungary, as well as the control of political authority by a minority in both states of the Dual Monarchy, unquestionably colored, if not entirely determined, the main lines of his whole sociological system, based, as it was, upon the premises of ethnic diversity, group and class conflict, the political sovereignty of a ruling

minority, and the problems of national emancipation, cultural assimilation, and ethnic amalgamation" (Ibid.: 192).

The third example of a move toward a more theoretical history of sociology is Gurvitch and Moore's *Twentieth Century Sociology* (1945). The book's second half consisted of case studies of sociology in different countries and linguistic regions. The stark contrasts that emerge in this section call attention to the fact that sociology was precisely *not* like the natural sciences, which supposedly "know no country." Roger Bastide's chapter in *Twentieth Century Sociology* identified a tendency in Brazilian sociology toward theoretical syncretism, which he connected to the historical legacy of colonialism, slavery, and racial syncretism (Bastide 1945). Another chapter in Gurvitch and Moore's *Twentieth Century Sociology* traced the underdevelopment of British sociology to a lack: Britain was "the country of stability and gradual change," while sociology was "essentially the product of rapid social change and crisis" (Rumney 1945: 562). The various Russian sociological schools before and after the 1917 Revolution were motivated by "political history" and social "trends" (Laserson 1945). None of this added up to a full-fledged sociological approach, of course, but it was improvement.

The history of sociology began to emerge as a full-fledged subfield in the late 1960s in the USA, the 1970s in France, and the 1980s in West Germany. This shift is sometimes described as having stemmed from a generalized crisis of received traditions and postwar configurations in sociology and beyond. But this account does not explain the differences among the leading sites of sociological production. In the United States, a more sociological view of sociology was inspired to some extent by C. Wright Mill's *The Sociological Imagination* (1959) and Pitirim Sorokin's *Fads and Foibles in Modern Sociology and Related Sciences* (1956). Alvin Gouldner's important *Enter Plato* (1965) tried to assess the ways Hellenic society "gave rise to and shaped Plato's social theory" (Camic and Gross 2002: 102). Gouldner's *Coming Crisis of Western Sociology* (1970) analyzed the supposed "crisis" of the discipline of sociology and the role of "internal" factors (Parsonsian "positivism" and its dissolution) and external ones (as the global counter-cultural revolt) in producing this crisis. Here Gouldner introduced the idea that scientists' experientially based *background assumptions* shape the development of theory by making them "resonate" emotionally with certain ideas and not others. These assumptions do not "rest on evidence," but were *affectively* laden, because they were inculcated early in socialization (1970: 35). Indeed, Gouldner's *Coming Crisis* seemed to argue *performatively* that the history of sociology was also emerging as part of the critical crescendo he was analyzing.

In fact, Gouldner had no real emulators. Instead of flourishing, the history of sociology and the sociology of knowledge both "went into a steep decline in the late 1960s" (Camic and Gross 2002: 98). The *Journal of the History of Sociology* began an almost clandestine existence between 1978 and 1985 and then disappeared. Charles Camic (1997: 229), one of the few US sociologists specializing in disciplinary and intellectual history during this period, bemoaned the absence of interest among historical sociologists in the history of sociology. Even today, the history of sociology section of the American Sociological Association issues regular warnings that its membership is in danger of falling below the threshold required to keep sections going. There is still no dedicated journal for the history of sociology in the USA, although the international journal *Serendipities* (currently edited in Denmark) fills this gap. Indeed, the history of sociology is one of the rare subfields in sociology in which there are more fulltime specialists in countries such as Austria, Britain, Denmark, France, and Germany than in the USA.

In Germany, the history of sociology emerged at the end of the 1970s, and was initially concerned with recovering the repressed history of sociology in Nazi Germany and the work of the sociologists driven into exile by Hitler. At the time, former Nazis were disappearing from dominant positions in West German universities. A new generation emerged that was less focused on fighting these former Nazis politically, as in the 1960s, and more interested in recovering the details about sociology's past. Path-breaking studies of Nazi-era sociology were published by Klingemann (1996, 2009) and others (Schauer 2017). Most of this research was empirically meticulous but relatively untheorized. Two writers pointed out at the time that German sociology was only "starting to reconceptualize the traditional form of disciplinary history *sociologically* as a reflexive approach to its own past" (Hülsdunker and Schellhase 1986: 9). Eventually, however, some analysts began to reconnect their research into the history of sociology to the German interwar discussions of the sociology of knowledge (Endreß 2001). In the past two decades, a large, more theoretically grounded literature on the history of sociology has appeared in Germany and Austria, and the University of Graz has emerged as the center of this endeavor (see Moebius and Ploder 2017–2019).

This promising start can be connected to the French strand of writing the history of sociology, which is linked to Bourdieu's path-breaking efforts to create a sociological theory of practice linked to the concepts of habitus, symbolic capital, social field, and scientific reflexivity, which he defined as involving a historical reconstruction of scientific fields. There, the history of the social sciences, including sociology, has become a large subfield, organized around a number of journals, including *Anamnèse (Anamnesis), Bérose, Gradhiva, Revue d'histoire des sciences humaines (Journal of the History of the Human Sciences)*, and *Revue d'histoire et d'archives de l'anthropologie (Review of the History and Archives of Anthropology)*. One writer noticed already in 1997 that French sociologists "now seem to be infatuated" with the history of sociology (Hirschhorn 1997:5). It is not accidental that Endreß (2001) refers to Bourdieu as having made the most promising moves in the direction of connecting the history of sociology to the traditions of *Wissenssoziologie*, which had been interrupted by Nazism and migration to the USA and UK. Bourdieu's earliest field-theoretic work drew on Kurt Lewin's psychological field theory, which Bourdieu transformed into a novel theory of social structure (Steinmetz 2017b). Lewin's own epistemology was strongly grounded in the same German historicism that undergirded Mannheim's thinking (Mannheim 1924; Ash 1982).

Other proximate origins of Bourdieu's ideas can be traced to the French school of epistemology (Bachelard, Canguilhem, Foucault) and to Bourdieu's emphatic support for this work, which he understood as a key part of sociological reflexivity. This can be traced to his first-hand experience of coercive constraints on the production of social knowledge in Algeria, against the wider backdrop of postwar concerns with academic freedom in a country that had been occupied by Nazi Germany and the repressive Vichy regime (Steinmetz 2020b). Bourdieu's work became increasingly historical after 1970, when he began to participate in a community of historians and sociologists (Steinmetz 2011, 2017a). Although Bourdieu himself did not carry out sustained research on the history of sociology, his colleagues, students, and readers began to integrate concepts like field, habitus, and symbolic capital with detailed historical research in this area, relying on primary documents, archives, and interviews.

Part Two: *Why* Should One Write the History of Sociology?

Further discussion of Bourdieu's understanding of reflexivity will be deferred until the end of the next section, however. This section turns first to three additional arguments for the history of sociology. The history of sociology discloses contributions that have been repressed in the course of ongoing struggles for domination inside the discipline. Some of these forgotten elements are valuable in their own right. Others shed light on the sources of contemporary scientific *doxa*. Occasionally, the history of social science can illuminate the conditions for the flourishing of knowledge, including sociological knowledge. The history of sociology can examine the role of sociology as one determinant of non-sociological phenomena.

Overcoming the Repression of Disciplinary Memory

The sociological activity and artifacts that have been erased from core disciplinary memory are legion. There are many reasons for this disciplinary *amnesia*. Some past sociologists and theories are simply brushed aside as obsolete; others are actively resisted or even censored. Still others are repressed, in the psychoanalytic meaning of that term. Stoler (2011) rejects the psychoanalytic terms *amnesia* and *repression* in this context in favor of "aphasia," yet this term is just as problematic as any conceptual language that imports terms from the natural sciences into the social sciences. The word *aphasia* has a *linguistic* emphasis, whereas repressed material also takes nonlinguistic forms – embodied, emotional, visual, and material ones. The aphasia concept also elides the key *moral* dimension in discussions of forgetting (Ricoeur 2004, 412–456). Sociological work is repressed for various reasons. IT may offend political, aesthetic, moral, or epistemic sensibilities.

Psychoanalytic theory is concerned with symbolic material that is absent from the scene of consciousness but that remains active in the unconscious, producing symptoms, parapraxes, dreams, and other "compromise" formations. Recovered memories are rarely identical to the original repressed material, having undergone transformations via mechanisms like displacement and condensation, and due to changed interests and contexts of the present. Intellectual history serves as sociology's psychoanalyst, coaxing repressed material out of the collective disciplinary unconscious. Such recovered sociological material may seem valuable again, in changed conditions. Reconstructing the events leading to the original repression of memory may uncover some of the interests that underpin current disciplinary doxa. Of course, some writers have tried to cast epistemological, political, and ethical aspersions on the scientific practice of uncovering causal factors in accounting for empirical processes. In the sociology of knowledge, this gesture often relies on a translation of Mannheim's term enthüllen as "unmasking" (which is closer to *entmasken* in German) or as "unveiling" (*entschleiern* in German), rather than the more straightforward translation, "uncovering." Such translation games led Baehr (2019), for example, to repeat arguments made by Boltanski (2012) in characterizing the entire sociology of knowledge approach as "conspiratorial" and "paranoid," assimilating it to a surveillance operation. This recalls the attacks on Mannheim by American sociologists and Americanizing "young Turks" in postwar German sociology, who were opposed to Marxism and central European pessimism and who supported conventional views of scientific truth (de Gré 1941; Dahlke 1940; Scheuch 1990: 42).

One notable example of disciplinary repression involves historical sociology itself, along with the sociology of knowledge. The first two were linked endeavors during the Weimar Republic, and they were decimated by the rise of Nazism and the resulting forced emigration. After 1945, these traditions were subjected to a combination of "historical amnesia and epistemological critique" in Germany (Acham 1995: 291). The "Americanizing" generation of postwar West German sociologists experienced "massive pressure to distance themselves from German intellectual history," which was tied to the putative German *Sonderweg* or special path that was described as leading inexorably to Nazism (Klingemann 2009: 262; Steinmetz 1997). It was now "good form" to refer to oneself as a "sociologist in Germany" rather than a "German sociologist" and "to keep the older German sociology at arm's length" (Scheuch 1990; Tenbruck 1979: 79; Kruse 1998). By the same token, it was long taboo for German sociologists to examine the Nazi past in general or Nazism within the discipline (Käsler 2002: 166; Steinmetz 2017a: 489). Johannes Weyer (1986: 89) referred to this posture among postwar German sociologists as a posture of *not-coming-to-terms-with the past* (*Vergangenheits(nicht-)bewältigung*).

The rising generation of postwar sociologists, shaped by the anti-Weimar ideologies of the Nazi era and the current Americanizing wave, disavowed Weimar sociology along with the exiled historical sociologists who were still alive (mainly in the USA and UK). One young sociologist at the time, Dietrich Rueschemeyer, scoffed at Karl Mannheim's "historicist hypotheses" that could never be "operationalized in a form appropriate to industrial-style research." Rueschemeyer recommended replacing Mannheim and Scheler's approach with laboratory research on "small groups" – a fad in American social psychology at the time (Rueschemeyer 1958; Kuklick 1983: 291). Another young German sociologist, Heinz Maus, agreed that the intellectual path leading from Hegel, Romanticism, Dilthey, and Rickert to the historical sociology and epistemology of the Weimar era had been an intellectual dead end. Maus wrote on the history of sociology (1956, 1962) but in an unsociological way, revealing the scientific cost of sidelining *Wissenssoziologie*. As a student in Nazi Germany, Maus aligned himself with the dominant tendency in sociology (Klingemann 1996, 2009) by rejecting Mannheim and defending applied sociology (Maus 1939–1940); he continued to do so after the war (König and Maus 1962). Marxists joined in, accused *Wissenssoziologie* of value and political relativism and irrationalism (Lukács 1981). German-speaking sociologists only began disinterring their discipline's Nazi past in the 1980s, as discussed above. They also began to reconstruct the severed traditions of *Wissenssoziologie*. Although this particular approach has only been adopted by a few sociologists (Endreβ 2001), the recent *Handbuch Geschichte der deutschsprachigen Soziologie* (*Handbook of the History of German-Language Sociology*) embodies a radical shift (Moebius and Ploder 2017).

American sociology has also had a very weak disciplinary memory of historical sociology. Before 1933, there were just two historical sociologists in the USA, the aforementioned Becker and Barnes. Around three dozen historical sociologists became refugees from Nazi Germany starting in 1933, and the majority ended up in the USA. Most of them filled peripheral academic positions and were ignored by the other American sociologists. The main reason for this marginalization was the profound mismatch between the German *emigrés'* historicism and the positivism and scientism that dominated US sociology (Steinmetz 2005, 2010, 2020a). Most English language accounts of historical sociology ignore Becker and Barnes and the entire generational wave of German refugee historical sociologists, despite the fact that most of the latter ended up in the USA and Britain (e.g., Abrams 1982; Delanty and Isin 2003; Adams et al. 2005; Clemens 2007).

A related example of disciplinary repression involves the fate of Marxism in the Cold War USA, Nazi Germany, and West Germany. The practice of the history of science was decisively shaped by the Cold War. The second part of Merton's 1938 *Science, Technology and Society in Seventeenth-century England* dealt with the relations between science and technology. There Merton acknowledged his debt to Hessen's "provocative essay," writing "I follow closely the technical analysis of Professor B. Hessen" (Merton 1938: 501, n. 24). Merton's dissertation (1938) dedicated significantly less space to the "Weberian" hypothesis about Puritanism and science than the second hypothesis, derived from Hessen, but his dissertation is only remembered for first section (Freudenthal and McLaughlin 2009: 32). In 1949, Merton could still write that Marxism was "the storm-center of *Wissenssoziologie*" (Merton 1949: 462). Soon, however, Merton moved away from Marxism, noting in 1952 that investigating "the connections between sciences and society consititute [d] a subject matter which ha[d] become tarnished for academic sociologists who know that it is close to the heart of Marxist sociology" (Merton 1952: 15). Harvard historian of science James Conant discussed the "interconnection between science and society about which so much has been said in recent years by our Marxist friends [sic]" at a moment when such friendships were being shredded by McCarthyism (Conant 1947: 18). As an alternative to Marxism, Conant promoted local case studies of scientific research in order to demonstrate "the evolution of new conceptual schemes as a result of experimentation" (Conant 1947: 19). The Cold War was thus at the origin of a sociology of science that tended to restrict its gaze to the local scale and the laboratory, bracketing all wider contexts (Fuller 1994). Thomas Kuhn did discuss a variety of "external" factors and contexts in his 1957 book on the Copernican Revolution, including the beginning of the era of "voyages and explorations," the "Moslem invasion," and intellectual developments in the Reformation and Renaissance such as humanism, Neoplatonic philosophy, and the intensified testing of inherited Aristotelian ideas by scholasticism (Kuhn 1957: ch. 4). Kuhn's more famous *Structure of Social Revolutions* (1962) marked a step backward in this respect by shying away from contextualizing explanation by focusing on activities within scientific communities. As Stocking (1965: 214) noted at the time, *Structure* was "imperfectly historicist in its focus on the inner development of science to the deliberate neglect of external social, economic, and intellectual conditions." At the same time, Kuhn contributed powerfully to breaking the hold of positivism over the self-understanding and self-presentation of the natural sciences, by replacing the image of science "aiming for an ultimate theory of reality" with a less linear view of progress, and by showing that the sciences were like the humanities and social sciences in being "constituted as communities and traditions that periodically were subject to ideological strife" (Fuller 1994: 93, 82),

Marxism resurfaced in sociology in the late 1960s, and Marxists began to be hired in sociology departments, even in the United States. But mainstream sociologists' objections to Marxism quickly evolved to meet this new situation. Marxism was no longer decried as politically dangerous, but was now dismissed as boring, outmoded, and reductionist. As a result, there was little progress toward developing a neo-Marxist historiography of sociology, beyond Therborn's initial efforts (1976) to combine a "Marxism of Marxism" with a "sociology of sociology," and Gouldner's quasi-Marxist writings, discussed above. One of the few practicing historians of sociology in the USA at the time, Robert Alun Jones, did not even differentiate between historians of Marxism and Marxist historians of sociology (Jones 1983: 462). American sociologists unfamiliar with less "vulgar" forms or unorthodox of Marxism continued to be unable or unwilling to distinguish between a "reflection theory" of science and approaches that discuss more contingent connections between capitalism, politics, and science (e.g., Gouldner 1970; Steinmetz 2005). Neo-Marxist approaches – Frankfurt Critical Theory, Althusserian structuralism and Aleatory Materialism, and French and German Regulation Theory – barely registered among historians of social science. More recently, however, as the fear and loathing previously directed at Marxism has dissipated, a historiography of sociology inspired by newer versions of neo-Marxism has begun to seen the light of day, and researchers have started to reexamine the work of Marxists, including Marxist sociologists, in a more carefully contextualizing way (Gouarné 2013).

Another version of repressed disciplinary memory concerns *individual* sociologists. Ludwig Gumplowicz, whose relevance to the present global situation seems evident, is one example. Gumplowicz, a professor of "state sciences" in Graz, was the first German language author to use the words "sociology" and "sociological" (*Sociologie/soziologische*) in his book titles in the 1880s, although these terms had already been used in the titles of translated works by Herbert Spencer. Comte was being translated into German in the 1880s, when Gumplowicz's sociological texts were appearing, and Gumplowicz (1885) referred to Comte and Spencer, and to proto-sociologists such as Albert Schaeffle, Paul von Lilienfeld, and Adolf Bastian, but none of these writers used the noun *Sociologie* or the adjective *sociologisch* in any of their book titles prior to Gumplowicz. Gumplowicz was also was one of the first sociologists to focus on the role of war and violence in the formation of states and empires and shaping social processes more generally; the central significance of race, ethnicity, and "ethnocentrism" (his term); and the ubiquity of authoritarian forms of rule. He insisted on the lack of progress in social and political affairs, in a century obsessed with the idea of progress (Nisbet 1980). Gumplowicz understood race as a "social product, the result of social development," and not as something biological (Gumplowicz 1909: 196–197, note 1). Warfare was not determined by racial difference; instead, war imposed a racial format on struggles between warring groups (Gumplowicz 1883:194). All of this seems insightful and relevant to the contemporary world. Yet Gumplowicz's use of the word race and especially the title of his book, *The Race Struggle* (*Der Rassenkampf*), caused him to be seen falsely as a racist or even a proto-fascist (despite his Judaism).

A more surprising case of repression involves Max Weber and Emile Durkheim, at least during the two decades after WWII in Germany and France, respectively. Weber was the most frequently cited sociologist at the meetings of the German Sociological Association before 1933 (Käsler 1984: 36). He was not ignored in Nazi Germany, but the discipline's emphasis on applied, presentist, quantitative work made him less central (Klingemann 1996). The pervasive "Americanization" of (west) German sociology after 1945 and the turn to Soviet Marxism and a positivist, applied sociology in the East further reduced Weber's importance. The two-volume, 800-page handbook on sociology and Marxism in the GDR edited by Peter Christian Ludz (1972), a west German specialist on the GDR, includes numerous chapters on Marxism and industrial sociology and even one on game theory, but it has fewer than ten references to Weber. Lothar Peter (1991: 40) dates the first reference to Gramsci in an East German sociological publication to 1988, a year before the state's collapse. He argues that GDR sociology was divided between an abstract, dogmatic Marxism and a quantitative empiricism that sought to identify laws of social behavior (Peter 1991: 18–34). There was a parallel effort to dismiss Durkheim as outmoded in postwar France, at least at the "modernizing" pole of the reconstituting discipline. Gallup-trained survey sociologist and Vichy collaborator Jean Stoetzel was the leading voice in France of an "American inspired" sociology that sought to break with the Durkheimian legacy. Research teams were urged to produce standardized, commercialized, quantitative, applicable results (Blondiaux 1991).

In both cases, the repressed sociologists made a comeback, partly due to historical research. Weber was "reintroduced" to German sociologists at the meetings of the professional association in Heidelberg in 1964. Since then, Weber has been located at the very center of German sociology and of the history of sociology in Germany. Historical research on the *Max-Weber-Gesamtausgabe* has set impeccable standards for work in the history of the social sciences. In France, Philippe Besnard's (1983) work set similarly high standards for research on the Durkheimians, while Bourdieu and his school reestablished Durkheim as the main progenitor of French sociology (Wacquant 2001).

The History of Social Science as a Guide to Scientific *Flourishing* (and Decay)

The history of social science can also contribute to a realistic understanding of the conditions for the *flourishing* of knowledge, including social science. Barnes argued that the originality of intellectual life in ancient Greece was due to three factors located at different levels. The first was the localism that stemmed from the absence of a central, extensive state. This was also related to a high degree of "group self-consciousness which lay at the basis of those utopian or idealistic theories of society that appeared in the Republic of Plato and the Politics of Aristotle" (1948: 6). The second factor was freedom and liberty (for non-slaves) and the absence of state religion. The third, Barnes suggested, was the preference for deductive, a priori generalization.

In an analysis of interactions between historians and sociologists in France and Germany during the twentieth century, Steinmetz (2017a) identifies three axes of variation within interdisciplinary practices: (1) symmetry, or equal participation by the parties, versus asymmetry; (2) processual, dialogic, recursive, and open-ended interactions, versus an orientation toward discrete outputs; (3) autonomy of intellectual production within a given field or sphere, versus heteronomy, that is, its subsumption under the interests of external political or economic powers. This study determined that the more generative forms of interdisciplinarity occurred where interactions had the following features: (1) the two parties were equal in power, or symmetrical; (2) the interacting groups were motivated by autonomous intellectual problems, rather than by external compulsion, cajoling, or influence; (3) where interaction allowed a fusion of perspectives, rather than being organized around a division of labor between disciplines.

The opposite of scientific flourishing is scientific decay, or demise. Wagner and Wittrock (1991) present a model of the differential survival and development of three different approaches to social science: (1) comprehensive social science, which rejected "institutional tendencies towards both disciplinary segmentation and professional specialization," exemplified by the historical school of German economics and Max Weber's sociology; (2) formalized disciplinary discourse, exemplified by neoclassical economics, the legal theory of the state, and formal sociology, and (3) pragmatic specialization, as in the early American development of "separate organizations for economics, history, psychology, sociology" (Wagner and Wittrock 1991: 347). The first version suffered because it rejected standardized "skill requirements" and "codified set[s] of theorems and method" (Ibid.: 348). As for the formalized disciplines, they "hardly ever provided adequate analytical tools for an understanding of the immediate problems of contemporary society, and … soon proved unable to cover the analytical and informational needs of societal elites"

(Ibid.: 348). For example, the Great Depression emphatically illustrated the failure of neoclassical economics, which was "superseded or transformed by Keynesian, corporative, and technocratic-interventionist economic discourses." Yet it was these modernized, formalized disciplines that became dominant, especially after WWII, when the European social sciences were largely reoriented toward "the American model" (Ibid.: 333).

Understanding Social Science as a Determinant of Social Phenomena

A third *raison d'être* of the history of social science is related to the permeation of social reality by social science. This consubstantiality of social science and social reality takes different forms. First, social science represents a reflexive monitoring of society, according to Giddens (1985). Giddens also argues that "[t]he conceptual innovations and empirical discoveries of social scientists routinely 'disappear' back into the environment of events they describe, thereby in principle reconstituting it" and that "[m]odernity is inseparable from *the constitutive role of social science,* and reflection upon social life more generally, which routinely orders and reorders both the intimate and more impersonal aspects of the lives people lead" (1989: 251–252). This is one of the sorts of "looping effects" or feedback effects between knowledge and the realities it describes, discussed by Hacking (1999). The social sciences influence social policies, educational systems, foreign affairs, and other matters more deliberately. Historians of social science are thus able to shed light on the constructions of the world by social science. Even if one is interested in explaining social processes that ostensibly have nothing to do with intellectual or social scientific production, one may have to carry out intellectual or scientific histories in order to do any social scientific work at all. This is not the same as arguments about "information" or "knowledge society" – terms that suggest a revolutionarily new, "purified and sanitized version of free-market capitalism" (Fuller 2007: 84). Instead, this is another way in which the history of social science is an indispensable part of sociological methodology in general, one that all social researchers need to master. This is true even if "scientific discourse on society exists alongside other discourses" (Wagner and Wittrock 1991: 334), which may also shape the practices one is studying.

It is important to distinguish between the *intentional* application of social science to policy (or to the construction of social worlds), and the much broader and *unintentional* impact of social science on worlds beyond its borders (Steinmetz 2020b). Social science has been *deliberately* deployed to guide policies ranging from eugenics to social insurance, labor market policies, poverty relief, and schemes intended to "nudge" individuals toward preferred behaviors. Well-documented examples of deliberate social scientific policymaking include modernization theory and theories of "colonial development," both of which informed postwar projects of land reform, dam building, and the agrarian "Green Revolution" (Knöbl 2001; Cullather 2002; Steinmetz 2017c). But social science also shapes social practices in a myriad *unintentional* ways (Steinmetz 2004). Social processes and events are sometimes codetermined by social scientific discourses that ostensibly have little to do with them. Steinmetz tried to uncover some of the hidden causal chains explaining colonial policy in *The Devil's Handwriting*. He found that the ethnographic and civilizational accounts of European travelers, missionaries, and amateur ethnographers set the basic parameters of formal German colonial policies, even if this was not the conscious intent of those accounts, and even if they were generated years, sometimes decades or even centuries before colonial conquest. The history of social knowledge was a crucial component of the explanation of colonial genocide in German Namibia and of the partial breakdown of the rule of colonial difference in German Qingdao. This does not mean that such findings can be directly deployed in an effort to prevent genocide or encourage civilizational exchange in other times and places. There is no social scientific recipe for "genocide prevention." It does, however, point to the *some* conditions that have been conducive to these outcomes in the past, and therefore provides some guidance in the present.

The History of Sociology as Sociological Reflexivity

Finally, the history of social science can serve as an under-laborer for scientific reflexivity, perhaps along the lines suggested by Bourdieu. Social scientists have to hesitate before blindly adopting the instruments, theories, and concepts they find ready at hand. They need to reflect on what they are doing when they do social science, and which assumptions they unwillingly enact and which implicit understandings they may unwittingly reproduce. More positively, they need to consider how a reflexive practice can improve social science (and by extension, social life), contributing to flourishing within the scientific realm and in the social and civic extensions of the social sciences – and perhaps in social life more generally. This means that the history of sociology has to be a constitutive part of all forms of sociological reflexivity. The history of sociology should be included in all training programs in sociological methods (Bourdieu 1990: 178).

According to Bourdieu (following Durkheim and Bachelard), reflexivity involves a rupture with the analyst's own *spontaneous* or *common sense* categories. To accomplish this first step in reflexivity, the researcher needs to *objectify* the history of the social worlds and scientific categories in which they are situated and from which they emerged into the specific moment of their ongoing social investigation. Such reflexive practice does not involve confessional style or identitarian approaches taking the form of "I am writing as an X." Nor is reflexivity the same thing as psychoanalysis, although a more complete form of reflexive socioanalysis would require probing of the researcher's unconscious (Gingras 2004: 628). Bourdieu obliquely acknowledged this in his insistent and increasing use of psychoanalytic terminology and ideas (Steinmetz 2006) and in his quasi-psychoanalytic auto-analysis (Bourdieu 2007).

In fact, it is often quite misleading to equate a researcher's demographic characteristics outside the field of science and their sociologically pertinent characteristics within the field. Fields can transform the species and holdings of capital that people bring with them from "outside"; habitus is reworked by the logics of the scientific field; the balance of forces in the field determines which sorts of practices will be most valued at a given time. This means that a reflexive sociologist will have to consider the genesis and evolution of the field in which they are located, the space of positions at the moment of analysis, and the contents of those positions – the discipline's *illusio* and forms of *doxa* and *heterodoxy*. The sociologist will also need to situate the field in which they are active within the overarching fields of science, or social science, and within the metafield of power. The researcher will need to identify the specific *external* fields and institutions that have shaped the specific scientific practices under investigation and that may be impinging on them in the present (Bourdieu 2004). As a supplementary stage, the researcher can then situate themself within those historical spaces, perhaps in the form of a self-analysis. The end goal is to shed light on the *spontaneous* or *common sense* categories at the outset of research.

In a second reflexive move, the researcher needs to objectify the social actors and processes under investigation. This means that the sociologist cannot base their analysis on actors' self-descriptions, since there is no reason to believe that actors necessarily understand the underlying logic of their own actions (Bourdieu 1977: 1–30). Unlike structuralism, however, Bourdieu rejects the idea that social structures explain social action directly. Instead, the causal efficacy of social structures is mediated through the actor's habitus, which generates practices thorough processes of strategic, regulated improvisation. The objectification of practices using structural methods and qualitative or quantitative retranscription is therefore a first step in the process of objectification. In a final stage the analyst can return to the level of actors' subjectivities, whose meaning can now be better understood.

What form does this reflexive approach take in research in the history of social science? Bourdieu's mature theory of intellectual production argues that texts are defined by the interaction of an author's habitus with their positions in specific, relevant fields, and by the history of those fields, which explains the space of positions in a field at a given moment in time (Frangie 2009). This approach urges researchers to undertake a double historical socioanalysis of their own concepts and theories and the historical evolution and current dynamics of the fields in which they are working, as well as the concepts, theories, and positions in the fields they are studying and of the individuals who are the ultimate creators of social science.

Writing the historical sociology of modernization theory in the Cold War in the USA provides an illustration of this multilayered reconstruction of social constructions. One might begin such an investigation with the puzzling fact that there has been more historical research (at least in the USA) on modernization theory than most other forms of social science, including colonial developmentalism, which preceded and paralleled modernization (Steinmetz 2017c). One would then need to reconstruct the genesis of the fields in which this research was located within American academia, to understand the role modernization theory played and continues to play in the form of intellectual shadows, specters, and traces of concepts (e.g., "modernity") and theories ("alternative modernities"). The comparisons made by states and empires among themselves had lasting effects on sociology's ontological assumptions, including the very idea of "society" (Tenbruck 1992). Indeed, these practices of "lay comparativism" were translated unwittingly into comparative social science methods (Steinmetz (2021).

Asking about my own location might entail looking at both the particular disciplines that I inhabit and my trajectory through them, that is, the sociobiographical reasons for my interest in modernization theory and the genesis of my ideas. I would need to reconstruct the textual and practical artifacts of modernization theory – the books, conferences, publications, laboratories, and disciplinary and political homes such as US social science departments, foundations, branches of the US government, and think tanks, as well as sites of actual implementation of the theory domestically and overseas. Attention to the textual production of modernization theory would also require moving between social scientific sites and non-academic sites. I would then need to situate theory groups and individual theorists within these fields, and to ask whether there was a subfield of modernization theory and where it was located. I would need to explore the relational standing of modernization theory vis-à-vis the disciplines from which its scholars were drawn, asking whether it was a disciplinary subfield or a separate field. I might need to reconstruct the sociogenesis of the individuals who played dominant roles in the relevant fields, which would entail studying their social properties and life courses (family origins, education, career, private life, etc.). Finally, I would ask how modernization theorists conceived of the subjectivity of putatively "traditional" and "modern" communities, and how they understood these subjects' constructions of the subjectivity of others. Ideally this entire set of reconstructions would yield a picture of the reciprocal relations between modernization theory and its various contexts, including its relations with the contexts that it attempted to "modernize" (e.g., Cullather 2002).

In sum, the Bourdieusian approach to the historical sociology of social science combines contextual analysis, textual analysis, and analysis of individual social scientists and their practices (see also Moebius 2015). The relevant *contexts* range in scale from the broadest ones – the entire social space, an entire empire, a total historical event, etc. – through to the narrower field and institutions of scientific activity (examined both historically, from their genesis through to the present, and at the specific moment of scientific production), and on to the immediate context where a particular research project is carried out, a particular work created, or a particular idea generated.

Conclusion

The historical sociology of social science is thus a crucial part of all social scientific inquiry and methodology. It is an inherently interdisciplinary or transdisciplinary activity. Historical research on sociology touches on intellectual history and the sociology of ideas; the study of intellectuals and knowledge; the philosophy of science; subfields such as cultural sociology, political theory, and cultural anthropology; and entire disciplines in the humanities and social sciences. Every historian of sociology quickly discovers the shifting and variable boundaries between sociology and other disciplines and between sociology and non-disciplinary scientific spaces. The question arises almost inevitably whether it is even possible to speak of a sociological disciplinary field before the 1950s. The fact that one often cannot determine an author's disciplinary identity when reading current books on the history of social science is a sign of the interdisciplinary flourishing of this literature and also a sign of its institutional weakness. The history of sociology and social science is also an unusually international space, where seemingly unlikely locations emerge as centers, and where sociology's Anglophone monolingualism is still not taken for granted.

This interstitial and international identity can prove to be advantageous, especially in the early stages of the formation of a new field of knowledge. Without a clearly defined disciplinary home, it is more difficult for particular hegemonic groups to dominate and discipline this space. The pressures to align research practices with field-wide norms are stronger in disciplinary subfields than in interstitial knowledge spaces (Steinmetz 2017a). The absence of US hegemony in this subfield makes it easier for new voices to be heard (Connell 2007; Collyer, et al. 2019). This is a productive form of semi-marginality and hybrid identity that can be conducive to formulating novel intellectual agendas. Put differently, the very irrelevancy of the historical sociology of social science to the central competing agendas in sociology shields this activity from the leading mainstream epistemologies of *presentism, positivism, policy relevance, pragmatism, explanatory parsimony*, and *US-centrism*, all of which hover drone-like overhead in most corners of sociological activity.

References

Abrams P (1982) Historical sociology. Cornell University Press, Ithaca
Acham K (1995) Geschichte und Sozialtheorie. Zur Komplementarität kulturwissenschaftlicher Erkenntnisorientierungen [History and social theory. On the complementarity of cultural science knowledge orientations]. Karl Alber Verlag, Freiburg i. Br. [In German]
Adams J, Clemens E, Orloff A (2005) Introduction: social theory, modernity, and the three waves of historical sociology. In: Adams J, Clemens E, Orloff A (eds) Remaking modernity: politics, processes and history in sociology. Duke University Press, pp 1–72
Albert M, Kleinman DL (2011) Bringing Pierre Bourdieu to science and technology studies. Minerva 49:263–273
Ash MG (1982) The emergence of gestalt theory: experimental psychology in Germany 1890–1920. Unpublished PhD dissertation, Harvard University (History)
Baehr P (2019) The unmasking style in social theory. Routledge, London
Baker KM (1975) Condorcet, from natural philosophy to social mathematics. University of Chicago Press, Chicago
Barnes HE (1948) Introduction to the history of sociology. University of Chicago Press, Chicago
Barnes HE, Becker HP (1938) Social thought from lore to science, vol 1. D.C. Heath, Boston
Bastide R (1945) Sociology in Latin America. In: Gurvitch G, Moore WE (eds) Twentieth century sociology. Philosophical Library, New York, pp 632–633

Beiser FC (1995) Introduction. In: Hegel (ed) Lectures on the history of philosophy, vol 1. University of Nebraska Press, Lincoln, pp xi–xl
Besnard P (1983) The sociological domain. The Durkheimians and the founding of French sociology. Cambridge University Press, Cambridge
Blondiaux L (1991) Comment rompre avec Durkheim? Jean Stoetzel et la sociologie française de l'après-guerre (1945–1958) [How to break with Durkheim. Jean Stoetzel and French sociology after the war (1945–1958)]. Revue française de sociologie [French Journal of Sociology] 32(3): 411–441. [In French]
Blumer H (ed) (1933) Publication of the American sociological society, vol 27, no 1 February 1933, proceedings. American Sociological Society, Chicago
Boltanski L (2012) Mysteries and conspiracies: detective stories, spy novels and the making of modern societies. Polity Press, Cambridge
Bourdieu P (1977 [1972]) Outline of a theory of practice. Cambridge University Press, Cambridge
Bourdieu P (1990) A Lecture on the Lecture. In: Bourdieu P In Other Words. Essays Towards a Reflexive Sociology. Stanford University Press, Stanford, pp 177–198
Bourdieu P, Chamboredon J-C, Passeron J-C (1991 [1968]) The craft of sociology. Epistemological preliminaries. de Gruyter, New York
Bourdieu P (2004) Science of science and reflexivity. University of Chicago Press, Chicago
Bourdieu P (2007) Sketch for a self-analysis. Polity, Cambridge
Brückner JJ, Enfield W (1791) The history of philosophy: from the earliest times to the beginnings of the present century, vol 1. Printed for J. Johnson, London
Bukharin NI (1931) Theory and practice from the standpoint of dialectical materialism. In: Science at the cross roads. Papers presented to the International Congress of the History of Science and Technology, held in London from June 29th to July 3rd, 1931. Kniga, London, pp 1–23
Camic C (1997) Uneven development in the history of sociology. Schweizerische Zeitschrift fur Soziologie/Revue Suisse de sociologie 23:227–233
Camic C (2020) Veblen. Harvard University Press, Cambridge, MA
Camic C, Gross N (2002) Alvin Gouldner and the sociology of ideas: lessons from 'Enter Plato'. Sociol Q 43(1):97–110
Canguilhem G (1989 [1943]). The normal and the pathological. Zone Books, New York
Clemens ES (2007) Toward a historicized sociology: theorizing events, processes, and emergence. Annu Rev Sociol 33:527–549
Collins R (1998) The sociology of philosophies: a global theory of intellectual change. Belknap Press of Harvard University Press, Cambridge, MA
Collyer F, Connell R, Maia J, Morrell R (2019) Knowledge and global power: making new sciences in the south. Monash University Publishing, Clayton, Victoria, Australia
Conant JB (1947) On understanding science. An historical approach. Yale University Press, New Haven
Connell R (2007) Southern theory: the global dynamics of knowledge in social science. Polity, Cambridge
Cullather N (2002) Damming Afghanistan: modernization in a buffer state. J Am Hist 89(2):512–537
Dahlke O (1940) The sociology of knowledge. In: Barnes HE, Becker HP, Becker F (eds) Contemporary social theory. Appleton-Century Company Incorporated, New York, pp 64–89
Daston L (2001) The history of science. In: Smelser NJ, Baltes PB (eds) International encyclopedia of the social and behavioral sciences, vol 10. Pergamon, Oxford, pp 6842–6848
Daston L (2016) History of Science without Structure. In: Richards RJ, Daston L (eds) Kuhn's Structure of Scientific Revolutions at Fifty: Reflections on a Science Classic. University of Chicago Press, Chicago, pp 115–132
de Gré G (1941) The sociology of knowledge and the problem of truth. J Hist Ideas 2(1):110–115
Delanty G, Isin EF (eds) (2003) Handbook of historical sociology. SAGE, London
Durkheim É (1900) La sociologie en France au XIXe siècle [French sociology in the 19th century]. Revue bleue [Blue Journal] 12:609–613, 647–652. [In French]

Durkheim É (1983) Pragmatism and sociology. Cambridge University Press, Cambridge

Endreß M (2001) Zur Historizität Soziologischer Gegenstände und ihre Implikationen für eine wissenssoziologisch Konzeptualisierung von Soziologiegeschichte [On the historicity of sociolgical objects and its implications for a scientific conceptualization of the history of sociology]. Jahrbuch für Soziologiegeschichte 1997/1998 [Yearbook of the History of Sociology 1997/1998]:65–89. [In German]

Eschmann EW (1934) Die Stunde der Soziologie [The hour of sociology]. Die Tat [The Deed] 25(12):953–966. [In German]

Fabiani J-L (1989) Sociologie et histoire des idees. L'epistemologie et les sciences sociales [Sociology and the history of ideas. Epistemology and the social sciences]. In: Enjeux philosophiques des années 50 [Philosophical stakes of the 1950s]. Éditions du Centre Georges Pompidou, Paris, pp 115–130. [In French]

Fassin D (Forthcoming) That obscure object of post-humanism. Will it put an end to the social sciences and the humanities? In: Fassin D, Steinmetz G (eds) The social sciences through the looking-glass. Studies in the production of knowledge. Duke University Press, Durham

Fleck C, Dayé C (2015) Methodology of the history of the social and Behavioral sciences. In: Wright JD (ed) International encyclopedia of the social and behavioral sciences, vol 15, 2nd edn. Elsevier, Oxford, pp 319–325

Foucault M (1954) Maladie mentale et personnalité. Presses universitaires de France, Paris

Frangie S (2009) Bourdieu's reflexive politics. Socio-analysis, biography and self-creation. Eur J Soc Theory 12:213–229

Freudenthal G, McLaughlin P (eds) (2009) The social and economic roots of the scientific revolution: texts by Boris Hessen and Henryk Grossmann. Springer, Dordrecht

Fuller S (1994) Teaching Thomas Kuhn to teach the cold war vision of science. Contention 4(1):81–106

Fuller S (2007) The knowledge book: key concepts in philosophy, science and culture. Acumen, Stocksfield

Giddens A (1985) The nation-state and violence. Polity Press, Cambridge

Giddens A (1989) A reply to my critics. In: Held D, Thompson JB (eds) Social theory of modern societies: Anthony Giddens and his critics. Cambridge University Press, Cambridge, pp 249–301

Gingras Y (2004) Sociological reflexivity in action. Soc Stud Sci 40(4):619–631

Gouarné I (2013) L'introduction du marxisme en France: philosoviétisme et sciences humaines, 1920–1939 [The introduction of Marxism in France: Philosovietisme and the human sciences, 1920–1939]. Presses universitaires de Rennes, Rennes. [In French]

Gouldner A (1965) Enter Plato: classical Greece and the origins of social theory. Basic Books, New York

Gouldner A (1970) The coming crisis of western sociology. Basic Books, New York

Graham LR (1985) The socio-political roots of Boris Hessen: Soviet Marxism and the history of science. Soc Stud Sci 15(4):705–722

Gross N, Fleming C (2011) Academic conferences and the making of philosophical knowledge. In: Camic C, Gross N, Lamont M (eds) Social knowledge in the makin. University of Chicago Press, Chicago, pp 151–180

Gumplowicz L (1883) Der Rassenkampf: Sociologische Untersuchungen [The race struggle: sociological investigations]. Wagner'schen Univ.-Buchhandlung, Innsbruck. [In German]

Gumplowicz L (1885) Grundriss der Sociologie [Outline of sociology]. Manz'sche K. K. Hof-Verlags und Universitäts-Verlag, Wien. [In German]

Gumplowicz L (1909) Der Rassenkampf. Sociologische Untersuchungen [The race struggle: sociological investigations]. Wagner'sche Universitäts-Buchhandlung, Innsbruck. [In German]

Hacking I (1999) The social construction of what? Harvard University Press, Cambridge, MA

Hegel GWF (1995) Lectures on the history of philosophy. Greek philosophy to Plato. 3 vols. University of Nebraska Press, Lincoln/London

Hessen B (1931 [2009]) The Social and Economic Roots of Newton's Principia, In: Freudenthal G, McLaughlin (eds) The Social and Economic Roots of the Scientific Revolution: Texts by Boris Hessen and Henryk Grossmann. Springer, Dordrecht, pp 41–101

Hirschhorn M (1997) The place of the history of sociology in French sociology. Schweizerische Zeitschrift fur Soziologie 23(1):3–7

Hollinger DA (1990) Free Enterprise and free inquiry: the emergence of laissez-faire communitarianism in the ideology of science in the United States. New Lit Hist 21(4):897–919

Horkheimer M (1972) Critical Theory. Selected Essays. Herder and Herder, New York, 1972

Hülsdunker J, Schellhase R (1986) Zur Aktualität der Soziologiegeschichte [On the contemporaneity of the history of sociology]. In: Hülsdunker J, Schellhase R (eds) Soziologiegeschichte: Identität und Krisen einer "engagierten" Disziplin [History of sociology: identity and crises of an "engaged" discipline]. Duncker & Humblot, Berlin, pp 9–12. [In German]

Jaeschke W (1993) Einleitung [Introduction]. In: Hegel, Vorlesungen über die Geschichte der Philosophie [Lectures on the philosophy of history]. F. Meiner Verlag, Hamburg, pp vii–xl. [In German]

Jones RA (1983) The new history of sociology. Annu Rev Sociol 9:447–469

Käsler D (1984) Die frühe deutsche Soziologie 1900 bis 1934 und ihre Entstehungs-Milieus. Eine wissenschaftssoziologische Untersuchung [Early German sociology and the milieus of its creation. A study in the sociology of science]. Westdeutscher Verlag, Opladen. [In German]

Käsler D (1999) Klassiker der Soziologie [Classics of sociology] 2 vols. Beck, München. [In German]

Käsler D (2002) From republic of scholars to jamboree of academic sociologists: the German sociological society, 1909–1999. Int Sociol 17(2):159–177

Kettler D, Meja V, Stehr N (1984) Karl Mannheim and conservatism: the ancestry of historical thinking. Am Sociol Rev 49(1):71–85

Klingemann C (2009) Soziologie und Politik: Sozialwissenschaftliches Expertenwissen im Dritten Reich und in der frühen westdeutschen Nachkriegszeit [Sociology and politics. Social scientific expertise in the third Reich and in the early west German postwar period]. VS Verlag für Sozialwissenschaften Wiesbaden. [In German]

Klingemann C (1996) Soziologie im Dritten Reich [Sociology in the third Reich]. Nomos Verlagsgesellschaft, Baden-Baden. [In German]

Knöbl W (2001) Spielräume der Modernisierung. Das Ende der Eindeutigkeit [Modernization: the room for Manoeuvre. The end of unambiguity]. Velbrück, Weilerswist. [In German]

König R, Maus H (eds) (1962) Handbuch der empirischen Sozialforschung [Handbook of empirical social research]. F. Enke, Stuttgart. [In German]

Krauth W-H (1978) Disziplingeschichte als Form wissenschaftlicher Selbstreflexion – Das Beispiel der deutschen Nationalökonomie [The history of disciplines as a form of self-reflexivity. The example of German national economics]. Geschichte und Gesellschaft [History and Society] 4(4):498–519. [In German]

Kruse V (1998) Historische Soziologie als 'Geschichts- und Sozialphilosophie' – Zur Rezeption der Weimarer Soziologie in den fünfziger Jahren [Historical sociology as 'historical and social philosophy'. On the recepetion of Weimar sociology in the fifties]. In: Acham K, ed., Erkenntnisgewinne, Erkenntnisverluste. Kontinuitaten und Diskontinuitäten in den Wirtschafts-, Rechts- und Sozialwissenschaften zwischen den 20er und 50er Jahren [Knowledge gains, knowledge losses: discontinuities in the economic, legal, and social sciences between the 1920s and the 1950s]. F. Steiner, Stuttgart, pp 76–106. [In German]

Kruse V (2001) Wozu Soziologiegeschichte? Das Beispiel der deutschen historischen Soziologie [The history of sociology: What's it for?]. Jahrbuch für Soziologiegeschichte 1997/1998 [Yearbook of the History of Sociology 1997/1998]:105–114. [In German]

Kuhn TS (1957) The Copernican revolution: planetary astronomy in the development of western thought. Harvard University Press, Cambridge

Kuhn TS (1962) The structure of scientific revolutions. University of Chicago Press, Chicago

Kuklick H (1983) The sociology of knowledge: retrospect and prospect. Annu Rev Sociol 9:287–310
Laserson MM (1945) Russian sociology. In: Gurvitch G, Moore WE (eds) Twentieth century sociology. Philosophical Library, New York, pp 671–702
Latour B (2004) Why has critique run out of steam? From matters of fact to matters of concern. Crit Inq 30(2):225–248
Latour B, Woolgar S (1979) Laboratory life: the social contruction of scientific facts. Sage Publications, Beverly Hills
Lepenies W, Weingart P (1983) Introduction. In: Graham L, Lepenies W, Weingart P (eds) Functions and uses of disciplinary histories. D. Reidel, Dordrecht, pp ix–xx
Lerner D (1959) Social science: whence and whither. In: Lerner D (ed) The human meaning of the social sciences. Meridian Books, New York, pp 13–39
Lindenfeld DF (1997) The practical imagination: the German sciences of state in the nineteenth century. University of Chicago Press, Chicago
Ludz PC, ed. (1972) Soziologie und Marxismus in der Deutschen Demokratischen Republik [Sociology and Marxism in the German democratic republic]. 2 vols. Luchterhand, Neuwied. [In German]
Lukács G (1981 [1954]) The destruction of reason. Humanities Press, Atlantic Highlands
Mach E (1919 [1883]). The science of mechanics. A critical and historical account of its development. The Open Court Publishing Company, Chicago
Mannheim K (1929 [1959]) Ideology and utopia. An introduction to the sociology of knowledge. Harcourt, Brace, New York
Mannheim K. (1949 [1935]) Man and society in an age of reconstruction. Routledge & Kegan Paul, London
Mannheim K (1953 [1926]) Conservative thought. In: Paul Kecskemeti, ed., Karl Mannheim, essays on sociology and social psychology. Oxford University Press, New York, pp 74–164
Mannheim K (1924) Historismus [Historicism]. Archiv für Sozialwissenschaft und Sozialpolitik [Journal of Social Science and Social Policy] 52(1):1–60. [In German]
Marx K (2000) Theories of surplus value: Boks I, II, and III. Prometheus Books, Amherst
Maus H (1956) Geschichte der Soziologie [History of sociology]. In: Ziegenfuss W (ed) Handbuch der Soziologie [Handbook of sociology]. F. Enke, Stuttgart, pp 1–120. [In German]
Maus H (1962) A short history of sociology. Routledge & K. Paul, London
McEvoy JG (1997) Positivism, Whiggism, and the Chemical Revolution: A Study in the Historiography of Chemistry. History of Science 35:1–33
Merton RK (1937) The sociology of knowledge. Isis 27(3):493–503
Merton RK (1938) Science, technology and society in seventeenth century England. Osiris 4:360–632
Merton RK (1949) The sociology of knowledge. In: Merton (ed) Social theory and social structure. Free Press, Glencoe, pp 456–488
Merton RK (1952) Foreward. In: Barber B (ed) Science and the social order. Free Press, Glencoe, pp 7–20
Moebius S (2015) Methodologie soziologischer Ideengeschichte [Methodology of the sociological history of ideas]. In: Dayé C, Moebius S (eds) Soziologiegeschichte. Wege und Ziele [The history of sociology. Paths and goals], vol 2. Suhrkamp: Taschenbuch Wissenschaft, Berlin, pp 3–60. [In German]
Moebius S, Ploder A, eds. (2017–2019) Handbuch Geschichte der deutschsprachigen Soziologie [Handbook of the history of German-language sociology]. 3 vols. Springer VS, Wiesbaden. [In German]
Moore FJ (1918) A history of chemistry. McGraw-Hill, New York
Murray P (1988) Karl Marx as a historical materialist historian of political economy. Hist Polit Econ 20(1):95–105
Nietzsche F (1997 [1874]) On the uses and disadvantages of history for life. In: Nietzsche, Untimely meditations. Cambridge: Cambridge University Press, pp 57–124

Nisbet RA (1980) History of the idea of progress. Basic Books, New York
Peter L (2001) Warum und wie betreibt man Soziologiegeschichte? [Why and how do we do the history of sociology?] In: Jahrbuch für Soziologiegeschichte 1997/1998 [Jahrbuch für Soziologiegeschichte 1997/1998 [Yearbook of the History of Sociology 1997/1998]:9–64. [In German]
Peter L (1991) Dogma oder Wissenschaft? Marxistisch-leninistische Soziologie und staatssozialistisches System in der DDR [Dogma or science? Marxist-Leninist sociology and the state-socialist system in the GDR]. IMSF, Frankfurt am Main. [In German]
Pollak M (1979) Paul F. Lazarsfeld, fondateur d'une multinationale scientifique [Paul F. Lazarsfeld, founder of a scientific multinational]. Actes de la recherche en sciences sociales [Journal of Research in the Social Sciences] 25(1):45–59. [In French]
Ricœur P (2004) Memory, history, forgetting. University of Chicago Press, Chicago
Roscher W (1874) Geschichte der National-Oekonomik in Deutschland [History of national economics in Germany]. R. Oldenbourg, München. [In German]
Rueschemeyer D (1958) Probleme der Wissenssoziologie: Eine Kritik der Arbeiten Karl Mannheims und Max Schelers und eine Erweiterung der wissenssoziologischen Fragestellung, durchgefuehrt am Beispiel der Kleingruppenforschung [Probleme of the sociology of knowledge: a critique of the works of Karl Mannheim and Max Scheler and an extension of the sociological of knowledge problematic using the example of small group research]. Doctoral dissertation, Köln. [In German]
Rumney J (1945) British sociology. In: Gurvitch G, Moore WE (eds) Twentieth century sociology. Philosophical Library, New York, pp 562–585
Schauer A (2017) Soziologie in Deutschland zur Zeit des Nationalsozialismus [Sociology in Germany during the national socialist period]. In: Moebius S, Ploder A (eds) Handbuch Geschichte der deutschsprachigen Soziologie [Handbook of the history of German-language sociology], vol 1. Springer VS, Wiesbaden, pp 117–148. [In German]
Scheler M (1925) Wissenschaft und soziale Struktur [Science and social structure]. In: Verhandlungen des vierten Deutschen Soziologentages am 29. u. 30. Sept. 1924 in Heidelberg [Proceedings of the fourth German sociology meeting]. Mohr, Tübingen, pp 118–180. [In German]
Scheler M (1926) Die Wissensformen und die Gesellschaft [The forms of science and society]. Der Neue-Geist Verlag, Leipzig. [In German]
Scheuch E (1990) Von der deutschen Soziologie zur Soziologie in der Bundesrepublik Deutschland [From German sociology to sociology in the Federal Republic of Germany]. Österreichische Zeitschrift für Soziologie [Austrian Journal of Sociology] 15:30–50. [In German]
Singh S (2019) How should we study relational structure? Critically comparing the epistemological positions of social network analysis and field theory. Sociology 53(4):762–778
Small AW (1909) The cameralists, the pioneers of German social polity. The University of Chicago Press, Chicago
Small AW (1924) Origins of sociology. The University of Chicago Press, Chicago
Stanley T (1655) The history of philosophy, vol 1. Printed for Humphrey Moseley and Thomas Dring, London
Stein L (1897) Die sociale Frage im Lichte der Philosophie. Vorlesungen über Soziologie und ihre Geschichte [The social question in the light of philosophy. Lectures on sociology and its history]. F. Enke, Stuttgart. [In German]
Steinmetz G (1997) German exceptionalism and the origins of Nazism: the career of a concept. In: Kershaw I, Lewin M (eds) Stalinism and Nazism: dictatorships in comparison. Cambridge University Press, Cambridge, pp 251–284
Steinmetz G (2004) The uncontrollable afterlives of ethnography: lessons from German 'salvage colonialism' for a new age of empire. Ethnography 5(3):251–288
Steinmetz G (2005) Scientific authority and the transition to post-Fordism: the plausibility of positivism in American sociology since 1945. In: Steinmetz G (ed) The politics of method in the human sciences: positivism and its epistemological others. Duke University Press, Durham, pp 275–323

Steinmetz G (2006) Bourdieu's Disavowal of Lacan: Psychoanalytic theory and the concepts of "habitus" and 'symbolic sapital'. Constellations 13(4):445–464

Steinmetz G (2010) Ideas in Exile: Refugees from Nazi Germany and the failure to transplant historical sociology into the United States. Int J Polit Cult Soc 23(1):1–27

Steinmetz G (2011) Bourdieu, historicity, and historical sociology. Cult Sociol 5(1):45–66

Steinmetz G (2014) From Sociology to Socioanalysis : Rethinking Bourdieu's Concepts of Habitus, of modern Symbolic Capital, and Field along Psychoanalytic Lines. In Chancer L, Andrews, J (eds) The Unhappy Divorce of Sociology and Psychoanalysis. Diverse Perspectives on the Psychosocial. Palgrave Macmillan, London, 2014, pp 203–219

Steinmetz G (2017a) Field theory and interdisciplinary: relations between history and sociology in Germany and France during the twentieth century. Comp Stud Soc Hist 59(2):477–514

Steinmetz G (2017b) Field theory in bourdieusian sociology. Invited paper, plenary session on field theory, American Sociological Association, Montréal, August 17

Steinmetz G (2017c) Sociology and colonialism in the British and French empires, 1940s–1960s. J Modern Hist 89(3):601–648

Steinmetz G (2018) Scientific autonomy, academic freedom, and social research in the United States. Crit Hist Stud (fall):281–309

Steinmetz G (2020a) Historicism and positivism in sociology: from Weimar Germany to the contemporary United States. In: Paul H, van Veldhuizen A (eds) Historicism: a travelling concept. Bloomsbury, London, pp 57–95

Steinmetz G (2020b). Soziologie und Kolonialismus: Die Beziehung zwischen Wissen und Politik [Sociology and Colonialism: The Relations between Knowledge and Power]. Mittelweg 36, 29(3):17–36

Steinmetz G (2021) Komparative Soziologie, kritischer Realismus und Reflexivität [Comparative sociology, critical realism, and reflexivity]. Kölner Zeitschrift für Soziologie und Sozialpsychologie, Sonderheft, Praktiken des Vergleichens – Neue Formate der Klassifizierung, Bewertung und Vermessung [Cologne Journal of Sociology and Social Psychology. Special Issue, Practices of Comparisons – New Formats of Classification, Valuation, and Measurement]. Heintz B, Wobbe T, eds. [In German]

Stevens ML, Miller-Idriss C, Shami SK (2018) Seeing the world: how US universities make knowledge in a global era. Princeton University Press, Princeton

Stocking GW Jr (1965) On the limits of 'presentism' and 'historicism' in the historiography of the behavioral sciences. J Hist Behav Sci 1:211–218

Stoler AL (2011) Colonial aphasia: race and disabled histories in France. Publ Cult 23(1):121–156

Tenbruck FH (1979) Deutsche Soziologie im internationalen Kontext. Ihre Ideengeschichte und ihr Gesellschaftsbezug [German sociology in the international context: its history of ideas and relation to society]. In: Kölner Zeitschrift für Soziologie und Sozialpsychologie, Sonderheft 21, Deutsche Soziologie seit 1945. Entwicklungsrichtungen und Praxisbezug [Cologne Journal of Sociology and Social Psychology. Special Issue 21, German Sociology since 1945. Trajectories of Development and Practical Relevance]. Lüschen G, ed., pp 71–107. [In German]

Tenbruck FH (1992) Was war der Kulturvergleich, ehe es den Kulturvergleich gab? [What was cultural comparison, before there was cultural Comparitvism?]. In: Matthes J (ed) Zwischen den Kulturen? Die Sozialwissenschaften vor dem problem des Kulturvergleichs [Between the cultures? The social sciences faced with the problem of cultural comparison]. Göttingen, O. Schwartz, pp 13–36. [In German]

Therborn G (1976) Science, class and society: on the formation of sociology and historical materialism. NLB, London

Troeltsch E (1922) Der Historismus und seine Probleme [Historicism and its problems]. Mohr, Tubingen. [In German]

von Schelting A (1936) Review of *Ideologie und Utopie* by Karl Mannheim. Am Sociol Rev 1(4): 664–674

Wacquant L (2001) Durkheim and Bourdieu: the common plinth and its cracks. Sociol Rev 49(1): 105–119

Wagner P, Wittrock B (1991) States, institutions, and discourses: a comparative perspective on the structuration of the social sciences. In: Wagner PB, Wittrock B, Whitley R (eds) Discourses on society. The shaping of the social science disciplines. Kluwer Academic Publishers, Dordrecht, pp 331–358

Weyer J (1986) Soziologie – ein Phantomfach? Einige Konsequenzen der 1945 erfolgten Weichenstellungen für die Identität der heutigen Soziologie [Sociology – a phantom discipline? Some results of the foundational actions in 1945 for the identity of the contemporary discipline]. In: Hülsdünker J, Schellhase R, eds., Soziologiegeschichte. Identität und Krisen einer "engagierten" Disziplin [History of sociology: identity and crises of an "engaged" discipline]. Duncker & Humblot, Berlin, pp 87–103. [In German]

Whitehead AN (1917) The organisation of thought, educational and scientific. Williams and Norgate, London

Wiese LV (1926) Soziologie: Geschichte und Hauptprobleme [Sociology: History and Main Problems] W. de Gruyter, Berlin, 1926

Wiese LV (1960) Soziologie, Geschichte und Hauptprobleme [Sociology: History and Main Problems] 6th ed. W. de Gruyter, Berlin

Zilsel E (2000 [1942]) The social origins of modern science. Boston: Kluwer Academic Publishers

Locating the History of Sociology: Inequality, Exclusion, and Diversity

33

Wiebke Keim

Contents

Introduction: A Local Attempt to Locate the History of Sociology	866
The History of Sociology Is Embedded in a Broader History of European Modernity	873
The History of Sociology Is Embedded in Geopolitical Power Structures	875
The Discipline of Sociology Is Internally Divided into Centers and Peripheries	876
Mechanisms of Marginalization Have Been Historically Established Within the Discipline	879
Academic Libraries Provide Unequal Access to Literature from Different Places	880
Teaching Socializes New Generations of Sociologists into the Existing Biases Within the Discipline	882
The Center-Periphery Divide Has Epistemic Effects	883
Challenging the Established History of Sociology and the Status Quo: Counterhegemonic Potentials	884
Conclusion	886
References	886

Abstract

The history of sociology, until recently, has been the history of sociology in a few countries – France, Germany, Great Britain, the USA, and perhaps Italy. A review of a local library selection of history of sociology books illustrates this statement. Recent efforts have started to discuss why this is so. These deconstruct the self-sufficient narrative of sociology as being a reflection of an exclusive and self-contained European modernity characterized by moments such as the Enlightenment, the French or Industrial Revolutions. The argument is put forth that the discipline has remained strongly divided into centers and peripheries of knowledge production and circulation; that North Atlantic hegemony has been fostered by the introduction of bibliometric indicators based on ISI Web of Science; and that teaching and the socialization of new generations of sociologists

W. Keim (✉)
CNRS/SAGE (University of Strasbourg), Strasbourg, France
e-mail: wiebke.keim@misha.fr

© The Author(s), under exclusive licence to Springer Nature Singapore Pte Ltd. 2022
D. McCallum (ed.), *The Palgrave Handbook of the History of Human Sciences*,
https://doi.org/10.1007/978-981-16-7255-2_66

into the discipline have a decisive role in perpetuating unequal perceptions of knowledge. Thorough critiques of the limited scope of the history of sociology have been accompanied by attempts to highlight counterhegemonic potentials and to diversify its legacy.

Academic knowledge aspires to certain degrees of generality, even in the social sciences. This chapter deals with the fact that despite the epistemic pretension to produce generally valid bodies of knowledge, the places of production and circulation of knowledge as well as unequal relationships between those places have shaped the history of sociology. Furthermore, location has an impact on the way this history is written, as well as the way in which this history is perceived by scholars. This chapter outlines the extent to which and how the mechanisms of place have shaped the history of sociology, the epistemic effects they have produced, and the resistances which have emerged against the established perception of the discipline and the writing of its history.

Keywords

Places of knowledge production · Circulation of knowledge · Centers and peripheries · Global Souths · Canonization · Decentering the history of sociology

Introduction: A Local Attempt to Locate the History of Sociology

This attempt to locate the history of sociology shall begin from the author's own location. The closest sociological library is at the Sociology Institute of Freiburg University in Germany. Its collection shall provide us with a first impression of the localization of the history of sociology. Dedicated shelves display 14 books that promise a general history of the discipline: Among them one in four and one in three volumes, as well as one that appears twice as an updated and extended reedition: Wiese ([1926] 1971), Becker and Barnes ([1938] 1961), Barnes ([1948] 1961), Schoeck ([1952] 1964), Naumann (1958), Mills ([1960b] 1966), Aron ([1965] 1968), and Jonas (1976) (since volume one of the first edition was inaccessible, the following relies on this new edition that is identical with the first edition: Jonas [1968a–69]), Klages [1969], Käsler [1976b, 1978], Lepenies [1981a, b, c, d], Hauck [1984], Käsler [1999a, b], and Kneer and Moebius [2010]. Ten are by German authors, three by Anglophone authors (one in German translation), and one by a French author (in German translation). The first thing to catch one's attention is that all the books were either written or edited by men, that nearly all contributors to the edited volumes were also men, and that all those men were based in either Germany (in one case Germany and Austria), France, or the USA.

We can see where and how they locate the history of sociology in their books by searching through the "content"-pages. Again, *all* persons who appear with their names in the "Contents" are men, i.e., the history of the sociology was not only written but was also made exclusively by men. (On the reasons for the absence of women from the history of sociology, as well as selections of women to be included because of their contributions to the development of the discipline, see Honegger and Wobbe (1998) and Lengermann and Niebrugge-Brantley (1998). Thomas and Kukulan (2016) argue that in teaching classical sociology, the reception of female sociologists has started slowly). Altogether, the names of exactly 200 men appear on the content pages. Only 29 appear three times or more (M. Weber, Durkheim: 11; Comte: 10; Spencer, Simmel, Marx, Pareto: 8; Tönnies: 7; Mannheim: 6; von Wiese: 5; Michels, Sombart, A. Weber, Giddings: 4; Freyer, Sorel, Parsons, Homans, Merton, Lazarsfeld, Habermas, W.G. Sumner, Troeltsch, Tarde, Hobhouse, Hegel, Engels, Scheler, Mosca: 3) and none are black or colored. (On the exclusion of black sociologists, see the tireless work of Morrison on W.E.B. du Bois, e.g. Morris [2015]; on recovering Ibn Khaldun, see Dhaouadi [1990, pp. 319–320], Soyer and Gilbert [2012, p. 27], and Alatas [2014]). It comes as no surprise, given that we are in a German library, and that most books are in German, that most (15) were German sociologists (R. Michels could count as German and Italian), followed by six US-Americans (Lazarsfeld could count as US-Austrian), four French, two British, and two Italian. The selection of books therefore suggests that the history of sociology has happened basically in those five countries, since the men who are considered to be its main representatives were based there.

However, the early books, up to the end of the 1950s, do not speak about this localization. They do not question *where* sociology started. Instead, they are busy arguing *when* it started. Reference to place remains random: if at all, it is used as a mere encyclopedic ordering principle. The earliest book titled *History of sociology* by Leopold von Wiese ([1926] 1971) addresses the question of timing explicitly. For von Wiese, sociology started with the works of Comte and Spencer and this was a matter of social evolution: sociology emerges when "the national spiritual culture of a single people has reached a certain level of development" (Wiese [1926] 1971, pp. 14–15; translations of works in languages other than English are by the author), something that, he seems to assume, has happened only in four countries so far. His book then presents the history of the British, the USA, French, and German sociologies, but without reflection on the geopolitical implications of this localization.

Several authors separate a "prehistory" of sociology, one that is very long (e.g., 3000 years for Barnes [1948] 1961; 2500 years for Schoeck [1952] 1964), and often much more inclusive in terms of non-European traditions of social thinking, from the history of the discipline proper, which is mostly timed around Comte's successful suggestion of the discipline's current name, "sociology," and from that time is presented as an exclusively European endeavor (note, however, that the term "sociology" was not invented by Comte but by Sieyès: Guilhaumou 2006). The significance of other traditions of social ideas is not considered. For Barnes ([1948] 1961), who briefly mentions the situation in Spain, Novicow and Kovalevsky for Russia as well as Cornejo as the most successful representative of systematic sociology in Latin America, the relevant caesura separates predecessors from *systematic* sociology. For Schoeck ([1952] 1964), as for von Wiese, it is an evolutionist assumption which explains the European founding moment, in as far as the temporal horizon was delimited within the "occidental *Kulturkreis.*" There, the French Revolution, social and political crises caused the development of a "modern consciousness" (Schoeck [1952] 1964, p. 31). Naumann (1958) relates the emergence of sociology proper, defined, as opposed to former social thinking, by its distinctive method, directly to the Industrial Revolution. Interestingly, Naumann does ask the question of location:

> "If one looks at sociology as an academic discipline, one can easily confirm that until the middle of the 19th century, it was in the most proper sense of the term, homeless (...) For one century, sociology enjoyed a right to hospitality in many places" (Naumann, 1958, viii–avii). What he means, however, is not geographical location, but disciplinary and institutional anchoring: "Only a small part of those who accepted it [sociology] were based at university" (1958).

One of the early books on the shelf offers a slightly different perspective: Becker and Barnes' 1938 history of sociology ([1938] 1961). After presenting an extremely broad, global "pre-history" to sociology, the authors include, in the main part on the history of the discipline itself, developments in a wide variety of places – from Eastern European or Russian sociologies to India, China, and Japan. Becker mentions Gökalp as one of the first academic sociologists. He is also the only one who includes, for instance, Ibn Khaldun not only as a predecessor, but as a contributor to sociological conflict theory, or Novicov as a sociological critic of Darwinism. Despite this inclusiveness and diversification of locations, in this book as well, location appears as mainly a structuring principle in a very broad-based overview on global sociology. Becker, like Naumann, does not provide any reflection on place.

From the 1960s, as opposed to the rather encyclopedic accounts prior to that point, histories of sociology in the consulted library start to include sociological reflections on the history of the discipline. At the same time, the process of canonization is set into practice (see ▶ Chap. 36, "Social Theory and the History of Sociology" in the same volume). Mills is the first in the chronological row to speak about the "classics" of sociology as those whose "work represents the best that has been done by later 19th and earlier 20th-century sociologists, and remains directly relevant to the best work that is being done today" (Mills 1960c, p. 2). In this argumentative context, Mills includes a reflection which appears today as highly problematic, as outlined below: "In some part, I think, one's intellectual heritage is rather arbitrary, and must be. The order in which one comes upon various books, and the phase in one's development at which the exposure occurs - these are not entirely controllable. And, of course, not everyone is especially concerned with the origins or the conceptions he comes to use. Nonetheless, the conceptions of some men do continue to inform social inquiry and reflection" (Mills 1960c, p. 2).

We shall see in more detail below that the order in which one discovers sociological books, for instance during our undergraduate studies, is by no means random; there is today intensive debate, contestation, and negotiation regarding the origin of concepts that sociologists use; and the fact that "the conceptions of *some men* do continue to inform social inquiry and reflection" has become a problem.

Aron (1965), whose focus is on a few individual theorists (Montesquieu, Comte, Marx and de Tocqueville), and then turns to social thought after 1848, reflects, in his introduction, on the Cold War divide between the US and Russian sociologies, without further problematizing the geopolitics underlying this divide. Despite this globally observable divide, he still argues that the history of sociology started from a common moment, i.e., modernity and the singular capacity of modern society to be reflexive about social change.

Jonas (1968a–69, 1976) presents a very elaborate history of sociology in four volumes. He dates the emergence of sociology to a specific moment in European history, the Enlightenment. Jonas includes an argument around the significance of studying the history of the discipline: It serves not only as a convenient way into sociology theory, but also a means to locate current social theory within its own tradition. At this point, what happens to the non-North Atlantic traditions? Interestingly, after the main chapters on the core countries that we have defined above, Jonas includes a section on Latin American sociology (Jonas 1968b–69, p. 154 ff.) as well as a "Digression on other countries" (Jonas 1968b–69, pp. 158–163). France and Great Britain are identified as the two locations where sociology has a long tradition, because these two countries, as early as the eighteenth century, put forth the idea of autonomous laws inherent in society and made them an object of study. Germany, in turn, represents, for Jonas, the only country, on the grounds of its tradition of idealism, where there were productive intellectual grounds for an innovative reception of the French and British sociological debate. As opposed to these three exceptional situations, any other country in the world is marked by mere reception of sociology produced elsewhere. In Italy, the development of sociology only occurred once the ideas of Comte and Spencer had replaced the ancient Italian sociological endeavors that remained too limited within the tradition of antiquity, according to the author. While decolonization in Latin America provided fertile grounds for British and French reflections on social emancipation and some original thinking has occurred (e.g., Sarmiento Facundo, Mejia, de Hostos, Quesada, Colmo, Alvarez, and Cornejo), the ultimate judgment is that Latin American sociology remains within the stage of reception of theory produced in Europe.

Similarly, Klages (1969) assumes the history of sociology to be naturally European because of the specific historical context of emergence characterized by modern crises. He does reflect upon the point that sociology could be of particular importance for the countries of the Third World, not only as a means of self-understanding, but also as a vehicle for self-empowerment. However, in contrast to Aron, who tentatively indicated a sensitivity for the global geopolitical context, Klages fails to acknowledge the meaning of US hegemony when he states "countries that still have to struggle with the problems of a specific traditionalism, still considered as largely legitimate, rely, in their sociological theory-building, much more often than the big industrial nations on Parsons" (Klages 1969, pp. 13–14).

Käsler (1976b, 1978) is the first on the dedicated book shelf to codify the "classics" for German readers at the end of the 1970s. The lengthy "prehistories" that we frequently find up until the 1950s have by now disappeared. The narrative begins with Comte, and is followed by Marx, Spencer, Pareto, Tönnies, Simmel, and Durkheim in the first volume and features Mead, M. Weber, Scheler, Michels,

Geiger, Mannheim, and Schütz in the second. The will to insist on the German thinkers because the book is directed to a German audience is explicitly voiced. The interest in reading the classics, according to Käsler, is a practical one: finding one's "identity" as a sociologist in "being knowledgeable about the historical 'grounds' on which we stand" (Käsler 1976a, p. 11). While the construction of a "mausoleum" of past thinkers, a history of heroes or monumental figures has to be avoided, Käsler favors the use of the classics to discuss problems, theoretical or methodological, of relevance today. For this author, the specific characteristic of the "classics" is that they continue to be relevant beyond their historical context and can therefore still inform sociological work. Moreover, they set the standards by which current work is still measured. Käsler, in writing this, is very much aware there is no "born classic," but only "made classics" through the current needs and concerns of sociology (Käsler 1976a, p. 15). He is aware, for instance, that some of the German "classics" were turned into "classics" through their US-American reception. What all this means for the discipline outside the core countries is not reflected upon. We shall see below what the 1999 edition tells us about these questions.

Until this point, all works have structured their history of sociology chronologically, combined, in many cases, with a spatial ordering. This is a basic lexicon approach and does not contain any reflection on the meaning of geography. A few books also simply feature important representatives of the history of sociology. Lepenies (1981a, b, c, d), inspired by advances in science studies, is the first in the row to propose an alternative book structure. He focusses on institutions and includes a reflection on the construction of the history of the discipline through the scholarly community. On both accounts, however, the confinement to Europe remains unquestioned in his work. The issue of "national traditions" and "internationalisation" is raised, perhaps explicitly for the first time, reaching back to the failure of Worms' early attempts at internationalization, to the contemporary endeavors on behalf of the ISA. Truly international social science organizations – in the sense of representing a large variety of national associations and research communities – emerged after 1945 under the auspices of UNESCO (Heilbron et al. 2008). Alongside international associations in anthropology, political science, economics, legal science, and psychology, all created between 1948 and 1951, the International Sociological Association (ISA) was founded in 1949. Initially representing a small number of national associations from the core countries, it increasingly gained in international representation: associations from several Eastern European countries joined in the late 1950s, and throughout the 1960s and 1970s associate members from Latin America and Asia joined. Decolonization led to increasing membership from southern countries, and after 1989 the regions under former Soviet domination joined. However, this discussion remains again limited to transnational exchanges, in particular, to the migration of scholars, between three countries – France, Germany, and the USA.

Hauck (1984) may have offered a challenging intervention in the course of the history of sociology so far, since he explains in his foreword that his book directly relates to his teaching activity at the University of Jos in Nigeria. His aim was to systematize sociological theory by means of a critique of ideology. In relating the emergence of sociology to bourgeois society, Hauck produced a Marxist critique of evolutionism and modernization theory. Disappointingly, the Nigerian context of production does not seem to have left any impact on the author, since there is no reference to, or reflection on, the Nigerian teaching experience, apart from the mention in the foreword.

Käsler revisits his "Classics" in an updated edition more than 20 years later (Käsler 1999a, b). The selection of early figures, presented in volume I, remains exactly the same as in the first edition, although several chapters are written by different authors. There are three additions: Park, a chapter on the Durkheimians, Mauss and Halbwachs, and on Elias. The second volume adds a selection of more recent "classics": Lazarsfeld, Parsons, Adorno, a chapter on Freyer, Gehlen, and Schelsky, and also Aron, Homans, Merton, Mills, Goffman, Luhmann, Habermas, and Bourdieu. The criteria for this selection remain the same as in the first edition. The editor's introduction is virtually unchanged, but preceded by several new sections that, finally, explicitly address the problems inherent in the localization of the history of sociology as we have traced it to this point: Käsler uses the metaphor of "the house of sociology" to structure its history. He identifies Comte as its founding architect. However,

> [t]he concerted scholarly project of sociology, encompassing cultures and epochs, and initiated by scholars of the 19th century Occident, became a construction of the 20th century. Its creators were white-skinned Europeans who had grown up in the Judeo-Christian Occident during early modernity (Kaesler 1999, p. 11).

The way this reflection on the origins and their location is framed is telling of the way a renowned historian of sociology from a core country envisions what are, for him, new challenges, first and foremost of which is the ongoing internationalization of the discipline:

> International sociology represents today such a breadth that it is constantly necessary to self-critically *locate* anew its research questions, fields of study and methodology. If the sociological approaches of the early 20th century already did not formulate a disciplinary identity that reached consensus, the current appearance of international sociology is more and more diffuse. This lack of unity increasingly leads to a situation where sociologists from different camps can hardly communicate on their scholarly progress, since they seem to lack a common language – or maybe even a common object (Kaesler 1999, pp. 11–12, author's emphasis).

This is precisely why the study of the classics is necessary: "In order to bring it down to an (allegedly) simple denominator: scientific sociology lacks the authoritativeness/binding character [Verbindlichkeit] of a *canon*" (Kaesler 1999, pp. 11–12, emphasis in the text). Käsler's ambition, as he declares, is not to canonize the given set of classics, but rather engage the scholarly community in a negotiation around "what is important and good, what one should read and discuss," since it is only through collegial negotiation that the permanently necessary revisions, additions, and eliminations from the canon can be achieved. Nevertheless, the fear of a loss of disciplinary unity and identity, if not ultimately of control over the shape of the discipline, is more than evident in Käsler's account, when he confirms that:

> here, the contributions of sociological classics are of essential significance: They provide us with exemplary, standard texts, they represent the cultural memory of the discipline and in their historical succession they represent its life history (Kaesler 1999, p. 14).

Clearly, Käsler reacts at this point to the ongoing debates within the ISA at the time, namely to a series of articles in the ISA Journal, *International Sociology*, on the challenges of the internationalization or globalization of the discipline, where the indigenization debate initiated by Akiwowo, among others, generated debate. Contributions to this debate, provoking Archer's presidential speech "Sociology for one world: unity and diversity," were republished in the edited book *Globalisation, knowledge and society: readings from International Sociology* (see the presidential address Archer [1991] as well as the edited volume featuring the debate within the ISA journal *International Sociology*: Albrow and King [1990]). After the 1989 turning point, the voices from a multiplicity of places finally stirred up an international debate about the contours of the discipline. The limited role for the history and presence of places outside the core countries is clearly announced by Käsler:

> Of course, this house [of sociology] must not become a closed-up fortress, but has to always remain open and hospitable; but it will need to remain conscious of its intellectual core, otherwise it will fall into ruins (Kaesler 1999, p. 18).

Admitting other locations into the discipline clearly represents a threatening scenario.

The final book on the library shelf, by Kneer and Moebius (2010), offers a unique approach to the history of sociology: instead of starting from a fixed definition of the discipline, from a chronological order or from its representative figures, it builds a history from the discipline's defining controversies. Unfortunately, despite the general title of the book – "Sociological controversies. Contributions to a different history of the science of the social," it remains explicitly limited to German-language sociology.

To sum up, throughout the early decades following von Wiese's pioneering attempt, with the exception of the remarkable volume by Becker, histories of the discipline remain limited to four to five countries. They relegate the rest of the world's intellectual traditions – which they do include! – to its prehistory, with no consequences for the subsequent development of the discipline. "Place" does not matter, if at all, and appears merely as an encyclopedic structuring principle. Instead, the authors discuss the timing of sociology's emergence and date this through a narrative of European exclusiveness and exceptionalism.

From the 1950s, the "prehistory" and thus the majority world disappears almost completely. Historians proceed toward the codification of a canon and ascribe a specific, practical role to the study of the "classics": finding one's identity as a sociologist, and placing current theory into a long intellectual tradition. Place largely remains a structuring device and refers to four countries only. If at all, other places feature merely as places of reception of sociology produced in the core.

Lepenies (1981a, b, c, d), inspired by advances in the sociology of science, is the first to reorganize the history of sociology around specific themes, making it a social (and not merely intellectual) history. If he refers, for the first time, to "internationalisation," it is exclusively to exchanges between three of the core countries. At the end of the 1990s, Käsler's introduction to his second edition of the "Classics" (whose structure and contents remain otherwise nearly unchanged, if extended), realizes that place matters. Through the ISA, the existence and claims of sociologies outside the core countries has become visible – and threatening. The "classics" have a unifying role to play here.

The following sections present recent approaches that help us understand the observed role of place in the history of sociology. Of course, we assume that the place from which the history of sociology is told affects the way it is told and, in particular, the way it is located. The location and positioning of the historian of sociology has an impact on the shaping of the narrative. It therefore may be expected that a library in a German university primarily harbors books by German and North Atlantic scholars who tell the history of their discipline from a German and North Atlantic perspective. However, from ongoing debate among historians of sociology, and within the Research Committee 08 of the International Sociological Association, we know that the broad picture presented in this chapter would be similar in many libraries around the globe. The following sections will help explain the history of sociology as it is told by historians in a limited number of places, and why it remains focused on these places, thus becoming recognized as *the* history of sociology. There are several reasons for this.

The History of Sociology Is Embedded in a Broader History of European Modernity

The emergence of modern science has long been presented as deriving directly from the indigenous development of European modernity, reduced to a vision of the Renaissance, the Enlightenment, the French Revolution, and the Industrial Revolution. The significance of those developments for the rest of the world and the contribution of extra-European scholars to the achievements of modern science have long been omitted. This has changed only rather recently, through debates in global, connected, entangled (science) histories that have harshly criticized the Eurocentric limitations of prior history writing.

Attempts to write a more global history of the sciences, with a sensitivity to colonial or imperial science, have grown over the last decades (see, for example, Mac Leod 1982; Petitjean et al. 1992; Raj 2006; Todd 1993). Polanco (1990, 1992) has proposed a general model for the distinction of characteristic traits of central and peripheral science, rather than social science specifically. He can be distinguished as the author who most clearly conceptualizes this historical approach in theoretical terms. In analogy with Braudel's "world-economy," Polanco apprehends the development and global distribution of modern science since the sixteenth and seventeenth century as "world-science." Its historical center had been shifting within Europe, and more recently toward the USA. The concept of world-science highlights the spatial dimension and geographical location as an explanatory factor for the constitution of the sciences internationally. World-science is defined by Polanco as the largest expansion of a coherent and autonomous science system, divided into center, semi-periphery, and periphery, and thus exerts "domination effects." The formation of scientific communities on the periphery, as a result of the expansion of world-science, is characterized by its exogenous character.

Similar to Polanco's paralleling of world economy and world-science, Hountondji (1990, 1994b) draws a connection between historical subordination and the present situation of the sciences in the Global South. Drawing on dependency and world systems theory, he understands underdevelopment in the South as a consequence of their historical annexation from the world market and transposes this explanatory scheme to the domain of scientific development:

> In other words, we need to identify the specific, inevitable and structural shortcomings of scientific activity in Africa, and, perhaps, in the Third World as a whole. To such analysis, I wish to contribute the following hypotheses: scientific and technological activity, as practiced in Africa today, is just as 'extroverted', as externally oriented, as is economic activity; its shortcomings are, therefore, of the same nature. That is, they are not cognate or consubstantial with our systems of knowledge as such. On the contrary, they derive from the historical integration and subordination of these systems to the world system of knowledge and 'know-how', just as underdevelopment as a whole results, primarily, not from original backwardness, but from the integration of our subsistence economies into the world capitalist markets (Hountondji 1990, p. 7; see also, 1994a, p. 2).

Similar revisions have been proposed for the discipline of sociology given that it has been legitimated as the discipline of modernity *par excellence*. The history of sociology, until recently, was integrated into a broader, Eurocentric narrative of the development of European societies which single-handedly reached the peak of human development. The irresistible attraction of European modernity, to put it bluntly, then pulled the rest of the world in the same direction. Any other elements that would call for a more global, entangled history of the discipline were left to one side. The history of the discipline was written as one chapter in the broader history of European modernity. We could see this clearly in the attempts to date and contextualize the origins of the discipline in the selection of literature analyzed in the first section of this chapter. In the large majority of books, historians of sociology explain the emergence of the discipline as an offshoot of the internal and self-contained emergence of European modernity, including the development of "societal self-consciousness," which was to become sociology's task.

Underlying this narrative is the idea that Europe took the decisive step into modernity single-handedly. Connell's "Northern theory: the political geography of general social theory" (2006), integrated as the first part of her book "Southern theory" (2007), presents an alternative account of the development of sociological theory in historical perspective. It analyzes and reflects on the teaching of the canon in US-American study programs, an issue she had previously presented (Connell 1997). There she summarized:

> The familiar canon embodies an untenable foundation story of great men theorising European modernity. Sociology actually emerged from a broad cultural dynamic in which tensions of liberalism and empire were central. Global expansion and colonisation gave sociology its main conceptual frameworks and much of its data, key problems, and methods (Connell 1997, p. 1511).

Connell delves into the origins of sociology and closely scrutinizes the writings of a variety of early "classics," – a practice which has become increasingly rare, an adverse effect of "canonization." From Connell's reading of early sociological texts, the significance of the colonial project that European thinkers were justifying to some extent appears as crucial. This aspect of many historical writings has been excluded from the narrative. Many early writings contain reflections on societies outside Europe, social realities that had become known through colonial adventures and needed to be known and controlled to ensure colonial conquest. Those early social scientists often reasoned on extra-European societies within "grand ethnography" or evolutionist frameworks. Empirically, early sociology was often quite global in scope. Yet the direct impact of colonialism on European social theorizing is completely omitted in the series of books we have analyzed above.

Connell's works shed a different light on the location of "northern theory" within the history of sociology as well as on the process of canonization, and the omissions and distortions of large bodies of sociological literature it promoted. It can be a first element in rectifying the narrative given in the selection of literature analyzed in the first section. In it, Europe appeared as the place where modernity emerged, and therefore also as the place where sociology as a societal self-reflection on modernity is historically located. Because other world regions lag behind in social development and have entered modernity with delay, or have still not fully achieved the modern stage, they can, by implication, only copy Western European sociological production. By implication, they become secondary or subordinated places of knowledge production.

The History of Sociology Is Embedded in Geopolitical Power Structures

The way the history of sociology is written is not only embedded within a broader history of European modernity. Broader geopolitical power structures also affect the historical perception and representation of places outside of Europe. Global political, economic, and ideological power relations have impacted the ways in which the discipline has developed in different places, as well as the way those local developments are perceived by historians of sociology. Several scholars have tried to come to grips with the global constitution of the discipline of sociology, and in particular the unequal relationships between the identified core countries, or global center, on the one hand, and the manifold peripheries to it, on the other hand.

One of the first attempts to understand the global geopolitics of the discipline of sociology was presented in 1985 by Gareau. We had seen in our selection of literature that some authors commented on the global divide within the discipline with regard to the Cold War. Gareau's article "The multinational version of social science with emphasis upon the discipline of sociology" (1985) distinguishes three social scientific "blocks": Western social science in the USA and Western Europe, soviet Marxism-Leninism, and the peripheral social sciences of the South. Gareau empirically sustains his assumption that the three blocks communicate in hierarchical relationships. Basing himself on citation analysis and questionnaires, he comes to the following conclusion:

> The system has been exposed as one which features vertical relations; from the United States to the other Western social science powers and to most of the Third World; from the Soviet Union to the other socialist countries (except China) and to selected Third World countries; and from the middle range social science powers, mostly in Western Europe and elsewhere as well to their peripheries. The centres send out messages, but receive little inflow in return. The peripheries can have more than one centre, but it seems that they have a chief centre. A crucial feature of the system is the lack of communication among the peripheral states; their ties are with the centre. This is notably the case among the Third World states, although an exception – and this only to some extent – is Latin America (Gareau 1985, p. 107).

Gareau thus confirms the ethnocentric perspective of Western social science and the intellectual dependency and subordination of the Souths. Scholarly communication between the West and the South is unilateral and hierarchical. Gareau assumes a purely external determination of the observed intellectual hegemony: US-American social science is not that widely spread and recognized because of its "intrinsic value," but because of the political, economic, and cultural domination of the USA. Social scientific power corresponds to and relies on economic and political power, because the social sciences are part of a "knowledge industry."

The Discipline of Sociology Is Internally Divided into Centers and Peripheries

A later conceptualization of centers and peripheries within global sociology seeks to overcome the somewhat crude unilateral economic and geopolitical determinism inherent in Gareau's approach. Indeed, his perspective neglects the fact that conceptual, institutional, relational, and prestige factors *within* the social sciences cannot be exclusively reduced to the global economic and geopolitical situation. Other conceptual tools are required to handle the existing inequalities and divides between places and how they affect the production and circulation of social science knowledge.

One of the innovative aspects of the center-periphery approach at the time of its emergence was its conceptualization of the relationships between, and the reciprocal conditioning of, the global center and periphery. The three-dimensional model that has been developed within dependency theory for the global expansion of capitalism (Cardoso and Faletto [1969] 1978) can be transposed to the domain of the social sciences, in only partial analogy, for sure, as we are dealing here not with material goods but with knowledge. Three dimensions have thus to be distinguished for the sake of analytical clarity: a dimension of material and institutional infrastructures and internal organization, where we can distinguish developed from underdeveloped sociologies; a dimension related to the conditions of existence and reproduction, where we can distinguish autonomous and dependent sociologies; and a dimension of international position and recognition, where we can distinguish central and peripheral sociologies (Keim 2009).

First of all, the development of the social sciences requires an appropriate material, institutional and personal basis. Lack of the necessary material infrastructure, but in some cases also the suppression of academic freedom, seems to be one of major causes for the peripheral status of many southern places. A strongly developed sociology shows a high degree of institutionalization, with specialized centers for research and teaching, journals, and associations. Institutional development requires stable positions and sufficient funding opportunities for academic researchers, as well as a broader academic institutional framework and further infrastructure such as editing houses, a book market, information and communication technologies, well-equipped libraries, etc. These statements are congruent with Gareau's account. Furthermore, however, a developed sociology is characterized by its *internal* division of labor that covers and continuously develops all domains of sociological activity from empirical data collection and the realization of case studies at a low level of abstraction to conceptualization, methodology, and theory building. It therefore requires a functioning scholarly community, i.e., a "critical mass" of peers, that constantly communicates, cooperates, and critically discusses results, in a thematic as well as cognitive division of labor. The production of scholarly knowledge is always a collective endeavor. Furthermore, the scholarly community determines and maintains the requirements for accession and exclusion from the profession – curriculum development, teaching contents, examination, and certification. A developed sociology can thus be defined as a system of autonomous production of knowledge. Consequently, an underdeveloped sociology lacks one or several of the abovementioned characteristics. This first dimension, scientific development, is mainly determined by external factors such as the availability of funding, depending on national economies and government support in the first place, and scientific and higher education infrastructures. But the historically evolved hierarchies and inequalities in the production and circulation of social scientific knowledge remain intact even in countries with comparably strong local social sciences (e.g. the case of Japan, Koyano 1976; Lie 1996).

A second dimension of the center-periphery problem is often, but certainly not always, related to the state of development. At the level of this second dimension, referring to the conditions of existence of given sociologies, we can distinguish autonomous or dependent sociologies. An autonomous sociology has the capacity for self-reproduction and autonomous development at the level of its staff, institutions, and knowledge. Research results are communicated internally and can circulate more widely. On the contrary, dependent sociology requires a steady import of theories and concepts, teaching material, and research devices, and requires its staff to have academic degrees bestowed by the universities of the center. Moreover, it relies on a methodological-theoretical as well as personal basis to which it is unable to contribute. Autonomy is not to be confused with autarchy, in the sense that scholarly activity is in itself always internationally constituted. The difference lies in the fact that autonomous sociology benefits from international exchange and communication, whereas these are an essential requirement for dependent sociology. The problem of dependency has been aptly described by S.H. Alatas (1974, 2006b) and S.F. Alatas (2001, 2003, 2006a).

The third dimension, centrality and marginality, represents essentially an intrascientific problem referring to the position and function of given sociologies within the international arena. The terms centrality and marginality are used here to describe the relationship between different places of academic activity. Centrality refers to internationally visible sociologies that enjoy prestige in the international community and that are recognized as the core of the discipline. This applies to their institutions and scholarly authorities, teaching programs and degrees, as well as prestigious journals and editing houses. Their particular position confers on them the power of setting the dominant research agendas and teaching content, and fashionable methodological and theoretical approaches. Referring to a phenomenon of mutual recognition, definitions of marginal and central science are always somewhat tautological. Central science is often defined as the mainstream in the sense of the international bibliographic databases. However, these databases *are* the mainstream and they *set* the mainstream at the same time.

The hypothesis of such abstract, structural, global center-periphery models is that sociologies outside the North Atlantic core countries occupy a marginal position within the international arena. They lack international recognition, and not only are they largely ignored in the rest of the world, but ignorance of their sociology is not even considered a problem. Nonhegemonic sociologies rely on the institutions and scholarly production of the center, either because they have no local alternative – in this case marginality combines with underdevelopment and dependency – or because they remain oriented, despite local alternatives, to the more prestigious locations in the international field. Empirical analyses help to understand more detail about the impact of those center-periphery structures on global sociology as well as on the practices in different places (Keim 2008a, 2010a, b, 2011, 2017; Keim et al. 2014; as well as the two world reports on social sciences: International Social Science Council 1999, 2010). Center-periphery frameworks allow us to approach the hierarchies and divides between places of knowledge production in an analytical, systematic, and integrated way. The observed center-periphery structures within sociology at a global scale account for the way the history of sociology is narrated and for the visibility or invisibility of given places within that narrative.

Mechanisms of Marginalization Have Been Historically Established Within the Discipline

Recent scholarship has identified, empirically grounded, and analyzed, constitutive mechanisms that have historically evolved and consolidate the inequalities between locations of academic activity. The following briefly presents three examples of such mechanisms: bibliometric databases, the role of academic libraries, and teaching practices. Other such mechanisms would include, for instance, the disciplinary divides within the social sciences (Wallerstein et al. 1996), publication (Collyer 2016), as well as mobility and migration of students and scholars (Gérard 2013; Jeanpierre 2010; Karady 2002).

Bibliometric databases are not only used by scholars for literature research. They also play a crucial role in institutional or national science policy, research evaluation and promotion, as well as in the production of international university rankings. As such, they play a crucial role in producing and reproducing global divides between peripheral and central locations.

Gareau had already realized that the situation in Latin America was somewhat different from that in other southern world regions. In the meantime, Beigel (2010, 2013a, b) has produced detailed empirical analyses that confirm the particular realities of Latin American social sciences and their history in a long-term perspective. At the same time, the author sheds light on a powerful mechanism of production of centers and peripheries as well as of "segmented circuits": bibliometric databases and their use as evaluation indicators (Beigel and Salatino 2015).

In her work, Beigel shows that international circuits of publication, recognized by academic institutions in the region, have existed since the nineteenth century at the level of Latin America. They have been reinforced with the creation of regional bodies of research and teaching from the 1950s (e.g., Flacso and CLACSO). Today, Latin America has four publication circuits that allow for the construction of scholarly prestige and recognition within the region: first of all, the mainstream circuits with their impact measures and ranking of journals (ISI Web of Science [ISI-WoS], former Social Science Citation Index; SCOPUS). This mainstream circuit is internationally recognized, but remains limited in circulation in as far as access to those publications is expensive and therefore limited. It remains also limited in terms of indexation of high-quality academic journals from peripheral countries and in languages other than English. Second, the transnational open access circuits, such as Google Scholar or DOAJ; third, the regional open access circuits such as SciELO, Latindex, or Redalyc, all of which are highly valued within the Latin American social sciences and humanities. Finally, local circuits of nonindexed journals that are often published only in hard copy confer prestige to university-based academics. Latin America has also produced several national classification systems for journals (Núcleo Básico de Revistas CAICYT/CONICET in Argentina; Publindex in Colombia; QUALIS/CAPES in Brazil; and CONACYT in Mexico, see Beigel 2017, p. 851). Perhaps the most interesting observation is that regional publication circuits preexisted the introduction of the largely Anglophone mainstream with the rise of *ISI Web of Science* and that they have partly remained resilient to its introduction. In the course of its global imposition for the evaluation of scholarly production anywhere on the globe, this mainstream has marginalized the preexisting Latin American circuits.

Over the past 40 years, ISI Web of Science has monopolized the accumulation of scholarly prestige. It has penetrated national academic systems everywhere and been established as the single most important measure of "scientific excellence" (also see on the Arab regions, e.g., Hanafi 2011). While some scientific disciplines and geographic regions have benefited from this process, it has marginalized scholarly production in languages other than English, the social science and humanities disciplines and peripheral regions (Beigel and Salatino 2015, p. 13). The existence of Latin American circuits before the introduction of the global mainstream, however, has largely been ignored because the history of the social sciences has often used national boundaries to structure and limit (see above), whereas the Latin American case has been one of international communication from the outset. This internationalism, because it happened in a peripheral area, has not been perceived as such in the historiography of the disciplines. What needs to be taken into account is the pioneering role of Latin America in terms of alternative bibliometric databases, in particular in their open access version, that make the kind of research that Beigel has conducted possible. Beigel's research confirms that the bibliometric mainstream is not only an instrument to measure marginality (and not: productivity), but that it simultaneously produces marginality (Keim 2008b, p. 110 ff.). Highlighting the structuring impact of bibliometrics on the way different places of knowledge production are perceived is a crucial contribution to redressing the established narrative on the history of sociology as outlined in the first section of this chapter.

Academic Libraries Provide Unequal Access to Literature from Different Places

The academic library appears as a key factor in reproducing global divides between places. Empirical studies confirm, for instance, that it keeps important parts of globally published literature undiscoverable within Europe (Schmidt 2020). Indeed, as the first section of this chapter has shown, from a European perspective, published research results are unequally discoverable depending on their place of origin. Schmidt focuses on the hard structures that steer global scholarly communication, i.e., the (technical) systems of scholarly publishing and indexing, collections, and discovery systems. In line with the preceding section, Schmidt states that an important embodiment of those technical systems is the citation database, most prominently the Web of Science (WoS), as well as scientometrics more broadly. Academic libraries increasingly rely on scientometrics to constitute their collections. This consolidates the marginalization of literatures from peripheral places. Such "unjustified neglect" is a structural problem inherent in the institutional logics of libraries and of collection management.

In order to address the logics at play within European academic libraries, which is the location from where current research practices could be changed toward more democratic ways of doing research, Schmidt looks into academic library ethics and operations. At the institutional level, she does this through an analysis of collection policies, and at the actor level and professional values, through a survey on the self-perception of librarians. The key problem she identifies is that collection management is largely outsourced to commercial aggregators and vendors who base their strategies on mainstream scientometrics:

> Vendor-preselection products [...] decrease the libraries' agency in arranging records of knowledge, and support the maintenance of colonial power structures, since they seem to undermine the discoverability of small, 'local' or independent publishers' programmes (Schmidt 2020, p. 248).

This practice has led to an immense influence of commercial actors over the entire research information landscape.

This development combines with the professional ethos of librarians, "neutrality," that is largely apprehended as passive neutrality: the users' information needs guide the library' activities, the task is to provide access to what the user wants to access. But since the user cannot discover literature that is not indexed, this leads to a vicious circle. Schmidt instead advocates that the role of the library is "to counter existing biases through proactive acquisition of resources that confront dominant positions with whatever they tend to marginalise" (Schmidt 2020, p. 283).

Schmidt's book illustrates this point with a study on Southeast African scholarly production. African scholars are confronted with specific problems. They are caught between contradictory demands for internationalization on the one hand and local relevance on the other. The discussion about the current distinction between "international" and "local" journals, in the sense that international journals often count as more prestigious and central, is particularly interesting. It shows that different, incoherent definitions circulate and that it is actually unclear what makes a journal international or local. Furthermore, African scholars bump up against a generalization barrier because their contexts are framed as "cultural differences" with regard to general social theory. Therefore, they are also faced with "area studies incarceration" (Schmidt 2020, p. 147), i.e., their research is seen as being relevant first and foremost to African Studies, rather than to the social sciences' and humanities' core disciplines (on this, also see Wallerstein et al. 1996, as well as the empirical analysis of invited scholars in Keim 2010a).

"Decolonial scientometrics," similar to Beigel's approach mentioned above, confront mainstream scientometrics with alternative means of accounting for the scholarly production in peripheral places, combining university rankings, database indexing, researchers' CVs, and institutional records to construct an alternative database. Schmidt provides a thorough description of the African publication market and indexing landscape and complements the results for the region with an affiliation-based approach, zooming into one single institution, the University of Mauritius. The results reveal a substantial quantity of publications: Around 2000 journal titles exist in sub-Saharan Africa. We also find here concretely what has been outlined theoretically, e.g., the area studies incarceration in the form of an index of

African published literature exclusively included in the AfricaBib bibliography, managed by the African Studies Center Leiden Library, i.e., SSH research from Africa channeled into African Studies in Europe. The WoS coverage is insufficient, and with the ceasing of Africabib, the discoverability of African journals has worsened. An original insight generated through Schmidt's alternative citation analysis is that "local journals" are not limited to Southeast African authors, and that publishing locally is as important as publishing abroad: "These findings confirm that 'local' journals or publishers are, in fact, very 'international', not only in terms of authorship, but also in terms of where they are read and cited" (Schmidt 2020, p. 226). This means "local" publication does not automatically lead to a local audience, a result that contradicts authoritative studies like Mosbah-Natanson and Gingras (2014), who found that "Global South" scholars prefer to cite "'central' research." Schmidt demonstrates that this claim only holds when the underlying data is based on a "Global North" index. Overall, the poor coverage and limited discoverability of Africa-based research confirms the hegemonic bias in the global research system.

Teaching Socializes New Generations of Sociologists into the Existing Biases Within the Discipline

Let us get back to Mill's ([1960a] 1966) statement that the order in which one reads books and the moment in one's biography when one comes across certain books are overall random; that, furthermore, not everybody engages intensely with the origins of the concepts used. Recent research shows those assumptions to be fundamentally wrong, and particularly for those in peripheral places. A qualitative in-depth study, not on sociology but on political sciences in Argentina, demonstrates how Argentinian representatives of this neighboring discipline strategically use foreign knowledge to structure their careers (Rodríguez Medina 2013, 2014b). The author develops a conceptual framework based on approaches from science and technology studies to understand the role of knowledge circulation for peripheral scientific practice.

Indeed, the place of production of social science knowledge matters. It is particularly the unequal relationship between local and foreign knowledges in *teaching*, i.e., the hierarchy of taught knowledge that makes a huge difference in peripheral places in comparison with central places. It is because the European canon, for instance, is included in the very first semesters of the political science curriculum that Latin American students follow certain mobility patterns: The places that appear as central in the history of their disciplines are the most attractive ones for exchange programs, reinforcing what has been described as "academic dependency by choice." Rodríguez Medina's research confirms the same logic in an analysis of the reception of Luhmann among Latin American sociologists (Rodríguez Medina 2014a, c). Because of the prestige of authors imported from the center, a first generation could establish themselves as Luhmann experts in their home countries by canonizing Luhmann in the teaching curricula of their respective institutions. This paved the way for two subsequent generations of Luhmann specialists who built their careers on acquiring expertise in Luhmannian theory and, later, in adapting and applying it to their own contexts of research.

The innovative aspect of Rodríguez Medina's work is that he provides an analysis of mechanisms that are put into place in peripheral places and which reproduce the asymmetries between centers and peripheries locally. In departing from the hypothesis that knowledge always travels in materialized form, he approaches its circulation as a technology rather than, as historians of the discipline had done previously, as a decontextualized and detached intellectual exercise. His most critical conceptual intervention is certainly the concept of "subordinating object":

> My working hypothesis is that materialised knowledge coming from metropolitan centres – which I call subordinating objects – has the capacity (1) to organise curriculum and to canonise ideas, (2) to define conference themes, (3) to determine research agendas and (4) to regulate academic mobility. Through those capacities, subordinating objects attain a broader capacity to structure peripheral academic fields (Rodríguez Medina 2014b, p. 13).

Rodríguez Medina's results converge with those of others. Gérard, for instance, analyzes how student mobility transforms the disciplinary structuring of Mexican academia and creates hierarchies between researchers depending on their place of study (Gérard 2013; Gérard and Grediaga Kuri 2009). An analysis of the effects of the language of teaching on the reception and production of social science knowledge in Arabophone countries shows that language can have a similar effect as a subordinating object in peripheral places (Hanafi and Arvanitis 2014). Those insights rectify Mills' naïve assumption and shed a different light on the significance and functions of the established "canon." The location of sociological theory and traditions as they are taught worldwide has strong effects on the perception of different places of sociological knowledge production and continuously perpetuates the existing center-periphery divide.

The Center-Periphery Divide Has Epistemic Effects

There is no doubt that the scope for scholarly, international communication, including the global interconnectedness of social scientists, has increased considerably in recent decades. This interconnectedness, combined with social-scientific interest in globalization, has led to ongoing debates on the internationalization of the social sciences – something that Käsler (1999a) essentially perceived as a threat at the end of the 1990s. Optimistic voices, for example, within the International Sociological Association, talk confidently about the internationalization of their discipline, a favorite topic at world congresses. However, these developments have also led to fierce contest, and to resistance to the idea of a single, unified, and "truly global" sociology. Arguments against the vision of a globalized discipline have in turn provoked fears of the fragmentation of the discipline into localized, nationalized, or indigenized sociologies, as expressed by Käsler in the reedition of his "Classics," discussed above.

The implication is that the articulation between the commonly accepted and shared idea of the discipline and its local realizations is becoming increasingly problematic (Berthelot 1998). However, it is not paradoxical that the call for more local sociologies, often emerging from the Global South, appears at exactly the time of ever-increasing globalization. We need to take the dissident voices' backgrounds into account in order to understand that they come as no surprise. They are specific challenges to a North Atlantic domination that has to be resisted in order to develop an independent scholarly tradition, one that speaks from the context of origin.

Besides political challenges and resistance to North Atlantic domination, there is a fundamental epistemological problem. General social theory per se purports to produce universally valid statements, concepts, and theories. But this does not happen unless these statements have, first, engaged seriously with the sociological knowledge produced outside the identified core countries, and, second, have been adequately tested against empirical realities outside Western Europe and North America. This has hardly ever been done. The North Atlantic domination therefore leads to a strongly distorted form of universality. It is distorted because to date this claim of universality relies on both "radical exclusion" and "radical inclusion." These supposedly general theories do not take into account the experience of the majority of humanity, living in the global Souths. Nor do they recognize the social theories produced in the Souths. This can be called "radical exclusion." In turn, "radical inclusion" means that despite these multiple radical exclusions, general social theory is regarded as universally valid. The social realities in the southern hemisphere are thus subsumed, without further thought, under the claims produced in the North. This tendency, which only starts to be reflected upon, blurs the distinction between the universal and the particular, and the North Atlantic particular is thought to have universal validity. This is a fundamental epistemological problem for social science, i.e., for disciplines aiming at the formulation of generally valid claims about society (Keim 2008a).

Challenging the Established History of Sociology and the Status Quo: Counterhegemonic Potentials

In recent years, several attacks have been launched against the North Atlantic domination of the social sciences. These have included critiques of Eurocentrism (Amin 1988), the deconstruction of orientalism (Said [1978] 1994), attacks on anthropology and area studies (Mafeje 2001; Kwaschik 2018), calls for the provincialization of Europe (Chakrabarty 2000), the decolonization of sociology (Gutiérrez Rodríguez et al. 2010), and critiques of the coloniality of knowledge and epistemic hegemony (Bhambra 2010; Lander 2003; Mignolo 2004; Quijano 2000). At the same time, constructive approaches call for alternative discourses (Alatas SF 2006a), and have developed sociological concepts from knowledge contained in oral poetry (see the debate involving Akiwowo, Makinde and Lawuyi/Taiwo, in Albrow and King 1990; also Adésínà 2002; Sitas 2004).

The ISA has put considerable effort into representing more equally the diverse traditions within sociology (Burawoy et al. 2010; Cruz e Silva and Sitas 1996; Patel 2010). Alatas and Sinha, in "Sociological theory beyond the canon," suggest the need to push its boundaries, to include "other" voices and to read conventional "classics" and alternative ones together (2017). Connell (2007) presents a selection of "southern" sociological traditions, namely from Africa, the Muslim-dominated countries, Latin America, and India. Each chapter traces the specific regional theory development, highlighting discussions of local realities versus imported theoretical frameworks, and thus the struggles for intellectual emancipation from the dominant paradigms of Europe and North America. It seems important to note that "Southern theory" is not a call for fragmentation into localized, indigenized, or endogenized sociologies, but a powerful argument for a serious international debate on an equal footing, where the experiences and perspectives emerging from the Souths have to be fully acknowledged and might correct, complete, amplify, or supplement existing general theory, where necessary:

> It is helpful to think of social science not as a settled system of concepts, methods and findings, but as an interconnected set of intellectual projects that proceed from varied social starting points into an unpredictable future (Connell 2007, p. 228).

In this context of debate, "counter hegemonic currents" (Keim 2011) are understood as implicit challenges to North Atlantic domination. They include socially relevant, social science research and teaching, which has the potential to develop into theoretically relevant fields of knowledge production over time in the countries of the global Souths. An historical example is the emancipation of an entire continental community, Latin America, from the international mainstream through dependency theory, introducing a paradigm shift away from the then dominant, rather Eurocentric, modernization theory. The Latin American approaches have reached considerable prominence, if not even domination of the German field of development studies, for instance (Ruvituso 2020). Another example is the development of South African labor studies into an autonomous scholarly community, which has recently produced publications relevant to the field of labor studies (Keim 2017). The edited volume "Global knowledge production in the social sciences. Made in circulation" (Keim et al. 2014), deals with the issue of circulation between and across places that occupy unequal positions within the international scholarly community. There the authors argued:

> (t)he so-called 'rise of the South' in international relations and the global economy, accompanied by considerable increases in the research and higher education sectors in the more affluent emerging countries (China, India, Brazil, Turkey, etc.; see Royal Society 2011), is certainly one of the reasons why issues of international circulation and knowledge production are on the agenda today. Changes in circulation patterns within the social sciences as well as in perceptions thereof are also related to their institutionally fostered internationalisation, here in particular in sociology, through the creation of international scholarly associations, conferences and publications (Keim et al. 2014, p. 2).

Contributions to this edited volume argue that the ongoing debate on internationalization or globalization of the discipline could gain from having a closer look at the experience with alternative forms of cooperation, especially involving South-South networks. The abovementioned critiques of Eurocentrism and the coloniality of knowledge as well as the calls for alternatives often remain at a very abstract level. In the edited volume, Keim et al. sought to highlight, through engaging with colleagues from five continents, concrete examples of sociological research, of traveling theories and concepts, of precise studies discussed in the framework of international cooperation, i.e., to provide empirical, concrete examples. Ultimately, the book argues that social science knowledge as produced in circulation, i.e., through continuous international exchange, could indeed look different from established mainstream social science.

Conclusion

It appears the present double movement, in which the scholarly community becomes more internationalized while specific local claims also gain in status is not as paradoxical as it might appear. On the contrary, it seems this recent development has its foundations in the very history of the discipline, in the realities of its worldwide spread, and in the forms of its international constitution. Tensions between local and general sociologies could be regarded as a direct consequence of growing international communication. Increased international exchange and the gradual accession of peripheral sociologists to central fora confront scholars, who have to date regarded themselves as practising universally valid theory, with the problem of North Atlantic domination. However, the expected internationalization of the disciplines cannot be achieved on a more equal footing between North and South as long as this problem is not recognized and adequately discussed. Taking the social experience and theoretical production emerging from the global Souths seriously will enrich the disciplines and enable scholars to reflect upon the possibilities of generalizing their claims beyond the local context to a broader empirical basis. Departing from a more critical vision of the discipline's history and the role of different places in it, this remains the major task for the current and future generations of social scientists.

References

Adésìnà J (2002) Sociology and Yoruba studies: epistemic intervention or doing sociology in the 'vernacular'? Afr Sociol Rev 6(1):91–114

Alatas SH (1974) The captive mind and creative development. Int Soc Sci J XXVI(4):691–700

Alatas SF (2001) The study of the social sciences in developing countries: towards an adequate conceptualisation of relevance. Curr Sociol 49(2):1–28

Alatas SF (2003) Academic dependency and the global division of labour in the social sciences. Curr Sociol 51(6):599–613

Alatas SF (2006a) Alternative discourses in Asian social science. Responses to Eurocentrism. Sage Publications India, New Delhi u.a.O.

Alatas SH (2006b) The autonomous, the universal and the future of sociology. Curr Sociol 54(1): 7–23

Alatas SF (2014) Applying Ibn Khaldun. The recovery of a lost tradition in sociology. Routledge, London/New York

Alatas SF, Sinha V (2017) Sociological theory beyond the canon. Palgrave Macmillan UK, London

Albrow M, King E (eds) (1990) Globalization, knowledge and society: readings from International Sociology. Sage, London

Amin S (1988) L'eurocentrisme [Eurocentrism]. Anthropos, Paris

Archer MS (1991) Sociology for one world. Unity and diversity. Int Sociol 6(2):131–147

Aron R ([1965] 1968) Main currents in sociological thought 1. Penguin Books

Barnes HE (ed) ([1948] 1961) An introduction to the history of sociology. The University of Chicago Press, Chicago

Becker H, Barnes HE ([1938] 1961) Social thought from lore to science, 3rd edn. Dover, New York. [Online]

Beigel F (ed) (2010) Autonomía y dependencia académica. Universidad e investigación científica en un circuito periférico: Chile y Argentina (1950–1980) [Academic autonomy and dependency. The university and scientific research in a peripheral circuit: Chile and Argentina (1950–1980)]. Editorial Biblos, Buenos Aires

Beigel F (2013a) Centros y periferias en la circulación internacional del conocimiento [Centres and peripheries in the international circulation of knowledge]. Nueva Sociedad 245:110–123. [Online]. Available at http://ri.conicet.gov.ar/handle/11336/1232. Accessed 20 Oct 2018

Beigel F (2013b) The politics of academic autonomy in Latin America. Ashgate, Farnham

Beigel F (2017) Científicos periféricos, entre Ariel y Calibán. Saberes institucionales y circuitos de consagración en Argentina: las publicaciones de los investigadores del CONICET [Peripheral scientists, between Ariel and Calibán. Institutional knowledge and circuits of consecration in Argentina: the publications of CONICET researchers]. DADOS – revista de Ciências Sociais 60(3):825–865

Beigel F, Salatino JM (2015) Circuitos segmentados de consagración académica: las revistas de ciencias sociales y humanas en Argentina [Segmented circuits of academic consecration: social sciences and humanities journals in Argentina]. Información, cultura y sociedad 32:7–32. [Online]. Available at http://www.scielo.org.ar/scielo.php?script=sci_arttext&pid=S1851-17402015000100002

Berthelot J-M (1998) Les nouveaux défis épistémologiques de la sociologie [New epistemological challenges in sociology]. Sociologie et Sociétés XXX(1):1–16

Bhambra GK (2010) Sociology after postcolonialism: provincialized cosmopolitanisms and connected sociologies. In: Gutiérrez Rodríguez E, Boatcă M, Costa S (eds) Decolonizing European sociology: transdisciplinary approaches. Ashgate, Farnham/Burlington, pp 33–48. [Online]. Available at http://site.ebrary.com/lib/academiccompletetitles/home.action

Burawoy M, Chang M-K, Hsieh MF-Y (eds) (2010) Facing an unequal world: challenges for a global sociology. Institute of Sociology, Academia Sinica; Council of National Associations of the International Sociological Association, Academia Sinica, Taipei

Cardoso FH, Faletto E ([1969] 1978) Dépendance et Développement en Amérique latine [Dependency and development in Latin America]. PUF, Paris

Chakrabarty D (2000) Provincializing Europe: postcolonial thought and historical difference. Princeton University Press, Princeton

Collyer FM (2016) Global patterns in the publishing of academic knowledge. Global North, global South. Curr Sociol 66(1):56–73

Connell RW (1997) Why is classical theory classical? Am J Sociol 102(6):1511–1557

Connell R (2006) Northern theory: the political geography of general social theory. Theory Soc 35(2):237–264

Connell R (2007) Southern theory. The global dynamics of knowledge in social science. Polity Press, Cambridge

Cruz e Silva T, Sitas A (eds) (1996) Introduction – Southern African social science in the late 20th century: gathering voices: perspectives on the social sciences in Southern Africa. ISA Regional Conference for Southern Africa, Durban, ISA

Dhaouadi M (1990) Ibn Khaldun: the founding father of eastern sociology. Int Sociol 5(3):319–335

Gareau FH (1985) The multinational version of social science with emphasis upon the discipline of sociology. Current Sociology 33(3):1–165

Gérard E (2013) Dynamiques de formation internationale et production d'élites académiques au Mexique [The dynamics of international training and the production of academic elites in Mexico]. Revue d'Anthropologie des Connaissances 7, 1(1):317–344

Gérard E, Grediaga Kuri R (2009) Endogamia o exogamia cientifica ? La formacion en el extranjero, una fuerte influencia en las practicas y redes cientificas, en particular en las ciencias duras [Scientific endogamy or exogamy ? Training abroad, a strong influence on scientific practices and networks, in particular in the hard sciences]. In: Didou Aupetit S, Gérard E, Tuirán R (eds) Fuga de cerebros, movilidad académica y redes científicas: Perspectivas latinoamericanas [Brain drain, academic mobility and scientific networks: Latin American perspectives]. Instituto Politécnico Nacional (IPN), Centro de Investigación y de Estudios Avanzados (Cinvestav), México, pp 137–160

Guilhaumou J (2006) Sieyès et le non-dit de la sociologie: Du mot à la chose [Sieyès and the non-said in sociology: from words to things]. Revue d'Histoire des Sciences Humaines 15(2):117

Gutiérrez Rodríguez E, Boatcă M, Costa S (eds) (2010) Decolonizing European sociology: transdisciplinary approaches. Ashgate, Farnham/Burlington. [Online]. Available at http://site.ebrary.com/lib/academiccompletetitles/home.action

Hanafi S (2011) University systems in the Arab East. Publish globally and perish locally vs publish locally and perish globally. Curr Sociol 59(3):291–309

Hanafi S, Arvanitis R (2014) The marginalization of the Arab language in social science: structural constraints and dependency by choice. Curr Sociol 26(5):723–742

Hauck G (1984) Geschichte der soziologischen Theorie: Eine ideologiekritische Einführung [The history of sociological theories: an introduction from the perspective of critique of ideology]. Rowohlt, Reinbek bei Hamburg

Heilbron J, Guilhot N, Jeanpierre L (2008) Toward a transnational history of the social sciences. J Hist Behav Sci 44(2):146–160

Honegger C, Wobbe T (1998) Frauen in der Soziologie: Neun Portraits [Women in sociology: nine portraits]. C.H. Beck, München

Hountondji PJ (1990) Scientific dependence in Africa today. Res Afr Lit 21(3):5–15

Hountondji PJ (1994a) Démarginaliser [To demarginalise]. In: Hountondji PJ (ed) Les savoirs endogènes: pistes pour une recherche [Endogenous knowledge: pathways for research]. Codesria, Dakar, pp 1–37

Hountondji PJ (ed) (1994b) Les savoirs endogènes: pistes pour une recherche [Endogenous knowledge: pathways for research]. Codesria, Dakar

International Social Science Council (ed) (1999) World social science report 1999. Unesco, Paris

International Social Science Council (ed) (2010) World social science report 2010. Unesco, Paris

Jeanpierre L (2010) The international migration of social scientists. In: International Social Science Council (ed) World social science report 2010. Unesco, Paris, pp 118–121

Jonas F (1968a–69) Geschichte der Soziologie [History of sociology]. Rowohlt, Reinbek bei Hamburg. [Online]

Jonas F (1968b–69) III. Französische und italienische Soziologie. Mit Quellentexten [III. French and Italien sociology] Geschichte der Soziologie [History of sociology]. Rowohlt, Reinbek bei Hamburg

Jonas F (1976) Geschichte der Soziologie: Mit Quellentexten [History of sociology: with primary sources]. Rowohlt, Reinbeck bei Hamburg. [Online]

Kaesler D (1999) Was sind und zu welchem Ende studiert man die Klassiker der Soziologie? [What are and why does one study the classics of sociology?]. In: Käsler D (ed) Klassiker der Soziologie. Band 1: Von Auguste Comte bis Norbert Elias [Classics of sociology. Volume 1: from Auguste Comte to Norbert Elias]. Beck, München, pp 11–38

Karady V (2002) La migration internationale d'étudiants en Europe, 1890–1940 [The international migration of students in Europe, 1890–1940]. Actes de la Recherche en Sciences Sociales 145: 47–60

Käsler D (1976a) Einleitung [Introduction]. In: Käsler D (ed) Klassiker des soziologischen Denkens. Erster Band: Von Comte bis Durkheim [Classics of sociological thinking. Volume 1: from Comte to Durkheim]. Beck, München. [Online]. Available at http://www.dandelon.com/intelligentSEARCH.nsf/alldocs/A698B24903D32F85C1257306003C2AD7/

Käsler D (ed) (1976b) Klassiker des soziologischen Denkens. Erster Band: Von Comte bis Durkheim [Classics of sociological thinking. Volume 1: from Comte to Durkheim]. Beck, München. [Online]. Available at http://www.dandelon.com/intelligentSEARCH.nsf/alldocs/A698B24903D32F85C1257306003C2AD7/

Käsler D (ed) (1978) Klassiker des soziologischen Denkens. Zweiter Band: Von Weber bis Mannheim [Classics of sociological thinking. Volume 2: from Weber to Mannheim]. Beck, München

Käsler D (ed) (1999a) Klassiker der Soziologie. Band 1: Von Auguste Comte bis Norbert Elias [Classics of sociology. Volume 1: from Auguste Comte to Norbert Elias]. Beck, München

Käsler D (1999b) Klassiker der Soziologie. Band 2: Von Talcott Parsons bis Pierre Bourdieu [Classics of sociology. Volume 2: from Talcott Parsons to Pierre Bourdieu], 5th edn. Beck, München

Keim W (2008a) Distorted universality – internationalization and its implications for the epistemological foundations of the discipline. Can J Sociol 33(3):555–574

Keim W (2008b) Vermessene Disziplin: Zum konterhegemonialen Potential afrikanischer und lateinamerikanischer Soziologien [Blinkered discipline. On the counter-hegemonic potential of African and Latin American sociologies]. Transcript, Bielefeld

Keim W (2009) Social sciences internationally – the problem of marginalisation and its consequences for the discipline of sociology. Afr Sociol Rev 12(2):22–48

Keim W (2010a) Analyse des invitations de chercheurs étrangers par l'EHESS: Compétences reconnues et clivages Nord-Sud [An analysis of invitations of foreign researchers to the EHESS: acknowledged competencies and North-South-divides]. Cahiers de la Recherche sur l'Education et les Savoirs (9):33–52

Keim W (2010b) 'Aspects problématiques des relations internationales en sciences sociales: pour un modèle centre-périphérie' [Problematic aspects of the international relations within the social sciences: for a centre-periphery model]. Revue d'Anthropologie des Connaissances 4(3): 570–598

Keim W (2011) Counter hegemonic currents and internationalization of sociology. Theoretical reflections and one empirical example. Int Sociol 26(1):123–145

Keim W (2017) Universally comprehensible, arrogantly local. South African labour studies from the Apartheid era into the new millennium. Editions des Archives Contemporaines, Paris

Keim W, Çelik E, Ersche C, Wöhrer V (eds) (2014) Global knowledge production in the social sciences: made in circulation. Ashgate, Farnham/Burlington

Klages H (1969) Geschichte der Soziologie [History of sociology]. Juventa Verl, München. [Online]. Available at http://www.gbv.de/dms/hebis-mainz/toc/035165626.pdf

Kneer G, Moebius S (eds) (2010) Soziologische Kontroversen. Beiträge zu einer anderen Geschichte der Wissenschaft vom Sozialen [Sociological controversies. Contributions to another history of sociology]. Suhrkamp, Berlin

Koyano S (1976) Sociological studies in Japan – pre-war, post-war and contemporary stages. Curr Sociol 24(1):2–196

Kwaschik A (2018) Der Griff nach dem Weltwissen: Zur Genealogie von Area Studies im 19. und 20. Jahrhundert [The grip on world knowledge: on the genealogy of Area Studies in the 19th and 20th centuries]. Vandenhoeck & Ruprecht, Göttingen

Lander E (ed) (2003) La colonialidad del saber: eurocentrismo y ciencias sociales – perspectivas latinoamericanas [The coloniality of knowledge: eurocentrism and social sciences – Latin American perspectives]. CLACSO, Buenos Aires

Lengermann PM, Niebrugge-Brantley J (1998) The women founders. Sociology and social theory, 1830–1930. McGraw-Hill, Boston

Lepenies W (ed) (1981a) Geschichte der Soziologie. Studien zur kognitiven, sozialen und historischen Identität einer Disziplin. Band 1 [History of sociology. Studies on the cognitive, social and historical identity of a discipline. Volume 1]. Suhrkamp, Frankfurt (Main)

Lepenies W (ed) (1981b) Geschichte der Soziologie. Studien zur kognitiven, sozialen und historischen Identität einer Disziplin. Band 2 [History of sociology. Studies on the cognitive, social and historical identity of a discipline. Volume 2]. Suhrkamp, Frankfurt am Main

Lepenies W (ed) (1981c) Geschichte der Soziologie. Studien zur kognitiven, sozialen und historischen Identität einer Disziplin. Band 3 [History of sociology. Studies on the cognitive, social and historical identity of a discipline. Volume 3]. Suhrkamp, Frankfurt am Main

Lepenies W (ed) (1981d) Geschichte der Soziologie. Studien zur kognitiven, sozialen und historischen Identität einer Disziplin. Band 4 [History of sociology. Studies on the cognitive, social and historical identity of a discipline. Volume 4]. Suhrkamp, Frankfurt am Main

Lie J (1996) Sociology of contemporary Japan. Curr Sociol 44(1):1–66

Mac Leod R (1982) On visiting the moving metropolis: reflections on the architecture of imperial science. Hist Rec Aust Sci 5(3):1–16

Mafeje A (2001) Anthropology in post-independence Africa: end of an era and the problem of self-definition. [Online]. Available at http://multiworldindia.org/wp-content/uploads/2009/12/Social-ScientistsArchieMafeje-publication.pdf. Accessed Sept 2013

Mignolo W (2004) Colonialidad global, capitalismo y hegemonía epistémica [Global coloniality, capitalism and epistemic hegemony]. In: Sánchez Ramos I, Sosa Elízaga R (eds) América Latina: los desafíos del pensamiento crítico [Latin America: the challenges of critical thinking]. Siglo Veintiuno, México, pp 113–137

Mills CW ([1960a] 1966) Die Klassiker. Einleitung. In: Mills CW (ed) Klassik der Soziologie: eine polemische Auslese [Original title: images of man. The classic tradition in sociological thinking]. Fischer, Frankfurt am Main, pp 7–30

Mills CW (ed) ([1960b] 1966) Klassik der Soziologie: eine polemische Auslese [Original title: Images of man. The classic tradition in sociological thinking]. Fischer, Frankfurt am Main

Mills CW (1960c) Introduction: the classical tradition. In: Mills CW (ed) Images of man. The classical tradition in sociological thinking. George Braziller, New York, pp 1–17

Morris AD (2015) The scholar denied. W.E.B. Du Bois and the birth of modern sociology. University of California Press, Oakland

Mosbah-Natanson S, Gingras Y (2014) The globalization of social sciences?: Evidence from a quantitative analysis of 30 years of production, collaboration and citations in the social sciences (1980–2009). Curr Sociol 62(5):626–646

Naumann H (ed) (1958) Soziologie. Ausgewählte Texte zur Geschichte einer Wissenschaft [Sociology. Selected texts on the history of a discipline]. K. F. Koehler Verlag, Stuttgart

Patel S (2010) The ISA handbook of diverse sociological traditions. SAGE, Los Angeles

Petitjean P, Jami CJ, Moulin AM (eds) (1992) Science and empires – histoire comparative des échanges scientifiques – expansion européenne et développement scientifique des pays d'Asie, d'Afrique, d'Amérique et d'Océanie. Kluwer, Dordrecht

Polanco X (ed) (1990) Naissance et développement de la science-monde – production et reproduction des communautés scientifiques en Europe et en Amérique latine [Birth and development of world-science – the production and reproduction of scientific communities in Europe and Latin America]. Paris

Polanco X (1992) World-science: how is the history of world-science to be written?. In: Petitjean P, Jami CJ, Moulin AM (eds) Science and empires – histoire comparative des échanges scientifiques – expansion européenne et développement scientifique des pays d'Asie, d'Afrique, d'Amérique et d'Océanie [Science and empires – comparative history of scientific exchanges – European expansion and scientific development in the countries of Asia, Africa, America and Oceania]. Kluwer, Dordrecht, pp 225–242

Quijano A (2000) Coloniality of power and Eurocentrism in Latin America. Int Sociol 15(2): 215–232

Raj K (2006) Relocating modern science: circulation and the construction of scientific knowledge in South Asia and Europe; seventeenth to nineteenth centuries. Permanent Black, Delhi

Rodríguez Medina L (2013) Centers and peripheries in knowledge production. Routledge, New York

Rodríguez Medina L (2014a) Bounding Luhmann: the reception and circulation of Luhmann's theory in Hispanic America. In: Keim W, Çelik E, Ersche C, Wöhrer V (eds) Global knowledge production in the social sciences: made in circulation. Ashgate, Farnham/Burlington, pp 39–62

Rodríguez Medina L (2014b) Construyendo periferia: un microanálisis de objetos subordinantes como tecnologías epistémicas [Constructing periphery: a micro-analysis of subordinating objects as epistemic technologies]. Sociológica 29(83):9–46

Rodríguez Medina L (2014c) The Circulation of European Knowledge. Niklas Luhmann in the Hispanic Americas. Palgrave Macmillan, New York

Ruvituso CI (2020) From the South to the North: the circulation of Latin American dependency theories in the Federal Republic of Germany. Curr Sociol 68(1):22–40

Said EW ([1978] 1994) L'orientalisme – l'Orient créé par l'Occident [Orientalism]. Éd. du Seuil, Paris

Schmidt N (2020) The privilege to select. Global research system, European academic library collections, and decolonisation. Lund University, Faculties of Humanities and Theology, Lund

Schoeck H (ed) ([1952] 1964) Die Soziologie und die Gesellschaften. Problemsicht und Problemlösung vom Beginn bis zur Gegenwart [Sociology and societies. Problems and solutions from its beginnings into the present]. Karl Alber, Freiburg/München

Sitas A (2004) Voices that reason – theoretical parables. University of South Africa Press, Pretoria

Soyer M, Gilbert P (2012) Debating the origins of sociology: Ibn Khaldun as a founding father of sociology. Int J Sociol Res 5(1–2):13–30

The Royal Society (2011) Knowledge, networks and nations: Global scientific collaboration in the 21st century. The Royal Society, London. [Online]. Available at http://royalsociety.org/uploadedFiles/Royal_Society_Content/Influencing_Policy/Reports/2011-03-28-Knowledge-networks-nations.pdf. Accessed 1 Nov 2012

Thomas JE, Kukulan A (2016) Why don't I know about these women? The integration of early women sociologists in classical theory courses. Teach Sociol 32(3):252–263

Todd J (1993) Science in the periphery: an interpretation of Australian scientific and technological dependency and development prior to 1914. Ann Sci 50:33–36

von Wiese L ([1926] 1971) Geschichte der Soziologie [Soziologie – Geschichte und Hauptprobleme] [History of sociology (Sociology – history and key problems)], 9th edn. de Gruyter, Berlin

Wallerstein I et al (eds) (1996) Ouvrir les Sciences Sociales – Rapport de la Commission Gulbenkian pour la Restructuration des Sciences sociales [Open the social sciences. Report of the Gulbenkian Commission on the restructuring of the social sciences]. Descartes et Cie, Paris

Colonialism and Its Knowledges

34

Sujata Patel

Contents

Introduction	894
Part I: The Indigenous Approach	895
Part II: The Postcolonial and the Decolonial	900
Postcolonialism	900
Coloniality/Decoloniality	904
Conclusion	910
References	912

Abstract

This chapter offers a comparative historical analysis of three trends – the indigenous, the postcolonial, and decolonial – which have confronted the nineteenth century Western disciplinary field of sociology as a hegemonic field organized through the colonial grid. It maps the ontological-epistemic stances that these positions articulate to legitimize non-Western pathways to political modernity. It argues that distinct political contexts have organized the scholarship and research queries of these subaltern/non-hegemonic perspectives and analyzes these in terms of the two forms of colonialism: settler vs. non-settler colonialism. While highlighting some internal critiques that have informed these positions, it argues that these circuits of knowledge-making have created cognitive geographies which need to be taken into account to ensure non-hegemonic global social theory.

Keywords

Indigenous sociologies · Postcolonialism · Decoloniality · Power/Knowledge

S. Patel (✉)
Umea University, Umea, Sweden
e-mail: patel.sujata09@gmail.com

© The Author(s), under exclusive licence to Springer Nature Singapore Pte Ltd. 2022
D. McCallum (ed.), *The Palgrave Handbook of the History of Human Sciences*,
https://doi.org/10.1007/978-981-16-7255-2_68

Introduction

The past as a mirror to the present is an oft repeated phrase that explores what occurred earlier to grasp the contemporaneous. This statement is particularly apposite in examining the current sociological imagination which abounds with perspectives like the postcolonial, the decolonial, Eurocentrism, colonial modernity, Southern theory, and indigenous theories, together with concepts such as the captive mind, coloniality, colonial difference, extraversion, and subalternity. The past, in this case, is that of colonialism: a grid through which the politics of knowledge construction has been and is being debated today. It argues that postcolonial/decolonial is not one perspective – it includes many viewpoints and advocates differing conceptual frameworks on the politics of knowledge production. These have emerged to articulate divergent positions consequent to the impact of colonialism on these regions and have generated a wide range of knowledge projects on which noncolonial social sciences build. These conceptual frameworks can be broadly divided between those that have emerged in context with non-settler colonialism and settler colonialism (Patel 2021). This chapter limits its discussions to the scholarship of select scholars whose works are available in English language, thus linking this scholarship to their regions' historical and intellectual locations.

This chapter argues that the postcolonial/decolonial (used here as a generic concept) critically dissects dominant/hegemonic academic knowledge produced since the late nineteenth century in Europe through its university system and which have since defined the discipline of sociology (Heilbron 1995; Wallerstein et al. 1996). It also contends that these perspectives contain two organically interrelated parts: first, a methodology to study the "social" based on an ontological-epistemic viewpoint and second, a theory to assess alternate pathways toward modernities based on this ontology. While recognizing that the impact of the various postcolonial/decolonial perspectives has impacted the world in various ways, this chapter is limited to an assessment of three knowledge positions: the indigenous, the postcolonial (knowledge positions in non-settler colonialism) and the decolonial (knowledge position in settler colonialism). It makes an historical-comparative assessment of these three positions, which have been unevenly consolidated in cognitive geographical circuits and created territories and borders of debate and deliberation. The chapter suggests that initially, knowledge circuits originated in the early to mid-twentieth century within some Asian, African and Latin American nation-states post-Bandung (Nash 2003), and traces its travel from there to North America in the late twentieth century as postcolonial and decolonial perspectives. It also argues that contemporary globalization has not completely broken these cognitive circuits and hence the necessity of "learning from each other" (Connell 2010).

For each of the above-mentioned perspectives, this chapter traces existing scholarship on the following three questions: first, what theories in the field of sociology of knowledge/epistemology have scholars utilized to critique Western colonial assumptions? Second, what practices of knowledge-making – theories, methodologies, methods – have been extracted, utilized and re-designed to produce a sociological/social scientific analysis of their regions? Third, what has been the nature of the internal critiques subsequently generated?

The chapter is divided into three parts. The first part explores how the decolonization process of the mid-twentieth century in some nation-states of Asia and Africa charted out models to create alternate paths toward modernity. This section analyzes the emergence of proto theories on the politics of knowledge production. It traces how sociologists and anthropologists explored the ontological-epistemic by extracting from the culturist/philosophical ideas within the region/nation state in order to constitute the "social." Known as the indigenous/indigeneity perspective, this offers an analysis the scholarship of two of its exemplars, D.P. Mukerji (1894–1961) from India and Akiwowo Akinsola (1922–2014) from Nigeria. It also examines the scholarship of thinkers from these regions who extended and critiqued the indigenous above-mentioned positions and presented new concepts, such as the "captive mind" (Alatas 1972, 2006), academic dependency (Alatas 2003), endogeneity and extraversion (Hountondji 1995, 1997, 2009), and colonial modernity (Patel 2017, 2018, 2021).

Part Two shifts the discussion regarding colonialism and the social sciences to its debates in the Americas. It initially discusses the postcolonial position inscribed in Edward Said's *Orientalism* (1978) and that of the subaltern school on modern systems of knowledge. These assess the way the representations of the "other" are constituted in literature/language and the colonial archive. Next, it makes a critical evaluation of the ontological principles governing Anibal Quijano's (1928–2018) theory of the coloniality of power and indicates its historical-intellectual location. It elaborates the attributes of the decolonial position, its distinction from the postcolonial perspective, and notes the criticisms made by social scientists of the postcolonial and decolonial approaches while it affirms that the postcolonial/decolonial has emerged today as a global perspective. Part Three the conclusion maps the contributions made by the postcolonial/decolonial approaches in reconstituting global social theory.

Part I: The Indigenous Approach

For most of the newly independent nation-states, decolonization was not only about the transfer of political sovereignty to the "natives," but designing new pathways to free themselves from the economic and cultural dependencies generated by colonialism. At the Bandung conference in 1955, 29 Asian and African countries met to be constituted as a non-aligned political block that would design a new model of development outside the political influence of the first and second world's economic models of capitalism and communism (Nash 2003). These newly independent countries came together to develop and draw from new economic programs of import substituting industrialization that these nation-states were developing from earlier anti-colonial critiques. Critical to the project of economic autonomy was the investment by individual nation-states in intellectual infrastructure-institutions of teaching, research and publication which would support the growth of human resources and aid the development of autonomous social sciences and humanities. The Bandung initiative thus provided a prelude to, and a background from which to develop alternate social science models for growth and development and break knowledge circuits controlled by European and North American countries and create new ones. Over the next two decades, a series of conferences made possible (for example, in Cairo in 1957 that brought together communists from China and Russia and Marxist intellectuals from Algeria) an exchange of ideas and models. In 1973 the first Asian Conference on Teaching and Research in Social Sciences held in Shimla, India, finalized the project of creating an indigenous social science (Atal 1981).

The Shimla conference argued that colonialism had created academic colonialism and its tools, methods, and theories evoked intellectual servility, deference, and dependency to the West and the Western social sciences. It called for "self-rule" through the formation of indigenous social sciences. To "do" indigenous social science, meant the establishment of an institutional infrastructure that would aid the growth of conceptual and metatheoretical frameworks based on one's "own" culture and thereby evolve new scientific practices that could help social sciences to realize the nation's needs (Atal 1981). The conference defined four pathways to realize this goal: the constitution of social science concepts in local regional languages with the use of local resources; research by insiders ("natives"/citizens) rather than outsiders (non-citizens); determination of research priorities in terms of "national" priorities; and lastly, the formulation of new perspectives and paradigms for the social sciences in terms of local/national philosophical and cultural themes and intellectual legacies (Atal 1981).

In this context, it would be productive to note that the first intervention made in this emerging field of study was that of the Malay sociologist, Hussein Alatas, whose elaboration provided a definition for academic colonialism – he called it the "captive mind," which he defined as an "uncritical and imitative mind dominated by an external source, whose thinking is deflected from an independent perspective" (Alatas 1972: 692). Later, it was elaborated by Farid Alatas (2003) as academic dependency. Consequently, Hussein Alatas (2006) assessed how to conceive of an independent position which he called autonomous knowledges. He argued that there is no need to reject completely Western knowledges and fields of study. However, there is a need to extract those Western concepts and theories that can be of use within "native" societies. Some concepts and theories of Western knowledge, Hussein Alatas argues, are relevant because of their methodologies-these reflect on its own past and present and thus can be used for comparison across the world given their validity as scientific truths. However, he also contends that such concepts and theories can be extracted from the existing scholarship from other parts of the world-an example being the works of Ibn Khaldun (Alatas 2014). This orientation to indigenous studies – as the adaptation of Western thought to local contexts – can be seen today within Chinese sociology (Chen 2021, Xie 2020) and in Iranian sociology (Connell forthcoming). However, sociologists/anthropologists in some countries have invested their intellectual resources to formulate new perspectives for the social sciences drawn from their region's cultural/philosophical principles. Below, this chapter elaborates the indigenous perspectives of D.P. Mukerji from

India and of Akinsola Akiwowo from Nigeria where these experiments were most in evidence.

D.P. Mukerji's sociological queries followed those of European scholars in the mid- to late-nineteenth century. He wanted to construct a sociology to comprehend and examine the transition of India toward capitalism and modernity and thereby outline the current issues and problems affecting India's path toward modernity. His sociology recognized India as overburdened with poverty and backwardness, and that these processes were determined by colonial capitalist exploitation. He was constantly troubled by the question: how does one understand India's current economic problems, and what social science language does one need to construct to examine and evaluate them? Mukerji was focused on the problems of the contemporary: the plight of agricultural laborers, reconstitution of forms of non-free bondage and slavery, issues facing internal migrants and the contradictory impact of the processes of urbanization and industrialization in the context of colonialism. No wonder his sociology was perceived to be radical, a fact which he acknowledged when he identified himself by the term Marxisant (Joshi 1986; Madan 2007; Patel 2013).

But Mukerji was caught in another dilemma, one with which many Indian intellectuals of the day were fraught. He believed that the theories promulgated by European sociologists did not provide an understanding of Indian conditions. He was highly conscious of India's long civilizational history. The British, he argued, provided India with universal propositions based on their experience of the transition to capitalism. This model made the market the unifying element for organizing the country. Exporting this model to India, he contended, would be a disaster. Mukerji believed that in India it was its culture and its symbiosis that characterized its civilization over the *long duree*. This is best represented in terms of the following principles: that of acceptance, adaptability, accommodation, and assimilation. These cultural values were embodied in India's long history and needed to be retrieved to elaborate a sociological theory of, and for, modern India. They can be drawn from Sanskrit concepts such as *shantam* (harmony – that which sustains the universe amidst all its incessant changes), *shivam* (welfare – being the principle of coordination with the social environment), and *adavaitam* ("unity of unity" or synthesis) (Madan 2007, 2013).

Unlike Mukerji, who searched Hindu/Sanskrit texts to organize his sociology, Akiwowo's project was built on an excavation of tales, myths, and proverbs of the Yoruba – a group of about 47 million people inhabiting Nigeria, Benin, and Togo – to suggest:

> how ideas and notions contained in a type of African oral poetry can be extrapolated in the form of propositions for testing in future sociological theories in Africa or other world societies. (Akiwowo 1986:343)

Akiwowo argues that the concept of *asuwada* in Yoruba poems should be used as a key philosophical principle to organize a theory of sociation (Akiwowo 1986, 1999). *Asuwada* implies that although the unit of all social life is the individual, an individual as a "corporeal self needs fellowship of other individuals" (Makinde 1988: 62–3). Thus, community life based on common good is *sui generis* to the existence of the individual.

Akiwowo's and Makinde's elaboration of the *asuwada* concept to build an indigenous theory has led many to raise fundamental questions regarding the use of folk culture to construct a sociological theory. The queries range from the selection of the indigenous position to comprehend "colonial knowledges," given the variety of discussions within the African continent on this theme (Olaniyan 2000), to issues relating to the methodological. Questions have been asked why Akiwowo choose the *asuwada* principle over other, similar, indigenous concepts to define sociation. Additionally, scholars have queried the interpretations of the Yoruba poems, particularly given the similarities of these interpretations to Durkheimian functionalism. Adesina (2006, p. 5) argues that there may be differences among social scientists in the interpretation of these poems and thus they might express competing meanings of Yoruba poetry. In this case, which interpretation does one accept? What principles will allow us to debate and resolve these scientific issues? In these circumstances, what legitimacy does Akiwowo's *asuwada* sociology have?

Adesina (2006) suggests that all sociologies base themselves on particularities (including European and North American ones), and that the particularities elaborated in the Akiwowo project of indigenous sociology must meet with traditions of science. He asks: have we created methods to examine the truth of indigenous concepts and theories? Have we explored the reasons for its effectiveness? Why is it grounded in myth and magic? Can we dissociate it from these moorings and construct an endogenous science? To construct endogenous knowledge, it is important to move from "translation" to "formulation" (Adesina 2006: 9). This implies, according to Adesina, an engagement with modern science.

This argument reflects the distinction between the indigenous and the endogenous, as made by the philosopher Paulin Hountondji (1997, 2009) from Benin. Hountondji's distinction between endogenous (scientific) and indigenous (ideological) knowledge, gestures to the colonial constitution of the concept of indigenous. According to Hountondji, the concept of indigenous has emerged within the binary of West versus the East (Hountondji 1997). Western philosophy is perceived as universal while African philosophy is perceived as ethnic. More generally, Hountondji argues that the categorization of African philosophy as the "other" of Western knowledge systems precludes the development in Africa of its cultures of science in order to interrogate its own philosophical traditions and create thereby an internal dialogue with these kinds of knowledges. Hountondji thus defines indigenous as "epistemological sublimation from the socio-material experiences of African lived life" (Hountondji 1997:35).

Hountondji accepts that some Western ideas, concepts, and/or theories may have relevance to local contexts, others may not. Some kinds of knowledge may be adapted and assimilated; others may not be. However, he asks readers to query why African knowledges – local knowledges – did not develop new cultures of scientific traditions that can be accepted as "truths" or as science. For Hountondji, the answer lies in the concept of extraversion (externally produced knowledge). African knowledges, he argued are steeped in extraversion, and lack the autonomy to develop scientific practices to re-produce themselves. Hountondji identifies many attributes that defines extraversion, including the autonomy to produce and publish books and journals in independent publishing houses, to house these in libraries and archives, to critically apply research specializations, topics and questions together with an absence of philosophical location of concepts and its scientific understandings. Such scientific cultures, Hountondji argues, fuels and promotes academic tourist circuits with diasporic scholars circulating between the core and the periphery. The only alternative is to break the binary of the colonial/indigenous.

Was this also the problem with Indian sociological traditions and particularly of Mukerji's oeuvre? Patel (2013, 2017, 2020) has argued that it is important to assess how nationalism and colonialism were epistemically co-produced in order to comprehend the concept of the indigenous. Mukerji's sociology, she has argued was associated with traditional nationalist trends which searched for an alternate ontological language in the Hindu past, a past which was ironically described, explained, and conceptualized within the Orientalist project of the nineteenth century that used scriptures and literary sources tom prehend Hinduism. This project consorted with the interpretations given by Brahmin interlocutors to delineate the essentialist Hindu and thus Indian principles of the social (Dalmia 1996). Oriental and later Indological studies, the precursor of anthropology and sociology in India, legitimized the idea of India as an ancient Hindu civilization and searched for these civilizing traits in Hindu/Sanskrit texts. These were later translated by European Orientalists and then re-translated in regional languages. Nationalist sociology/anthropology was constituted through this prism as Indian intellectuals retrieved these interpretations in the late nineteenth and early twentieth century from within nationalist thought. These texts, Patel argues, carried not only a colonized gaze but also an Brahminic/upper caste male gaze. With nationalist sociologists like Mukerji not seeking their categories from within modernity but in India's Hindu past, the project of indigenous sociology in India has paradoxically reproduced the language of colonialism and arguably legitimized the notion of India as a traditional society (Patel 2006, 2013, 2017, 2021).

These criticisms apart, the search for the ontological through the indigenous/indigeneity project faded away from the mid-1970s onward. This trend coincided with the slow deterioration of the non-aligned movement and decline of government spending in universities (Onwuzuruigbo 2018). If it has a semblance of presence, it is in some contemporary social movements of Asian and African countries (Odora Hoppers 2002), in the discussions relating to the cognitive circuits that it had established on this theme. This drift has been reinforced due to the global developments in the 1980s and 1990s that has led not only to the dissolution of the Soviet bloc and its political and economic influence but also of communism as an ideology and practice and Marxism as a theoretical perspective. A unipolar world promoting capitalist globalization led by the USA/Europe has now emerged. Consequently, the world, which was divided a hundred years ago into the West-East axis has now re-grouped within the bipolar axis of the North-South (UNESCO 2010). There is once more a diffusion of ideas and scholarship, and a flow of human resources and research aid from the Global North. It is in this context we see the growth of scholarship within a South-South axis (Connell 2007; De Souza Santos 2014; Fiddian-Qasmiyeh and Daley 2018),

This chapter turns to map the discussion of similar ideas in the Americas and assess the continuities and breaks between these and the past.

Part II: The Postcolonial and the Decolonial

Postcolonialism

Postcolonialism grew as an academic project that deconstructed the hegemonic orientation of teaching, research and writing of English literature within mainstream American universities. Its key interlocutors were from the diasporic communities of west and south Asia. They brought to bear, in their teaching and study of mainstream English literature, the sensibilities and memories of the anti-colonial movements with the experiences of discrimination and prejudice faced by these communities in the USA, thereby querying the legitimacy given to the ideology of American exceptionalism by contemporary American scholarship (Sharpe 2000; Schwarz 2000). The entry of this diasporic community into departments of English literature in elite American universities was made feasible through changes introduced in immigration rules subsequent to the passage of a new Immigration Act in 1965. These changes abolished the quota system in favor of the Europeans and gave access to jobs for Asians, Africans, and Latin Americans in the USA.

However, postcolonialism, as a perspective which grew around Edward Said's text titled *Orientalism* (1978), was also an assault on the hegemonic ideology of American exceptionalism promoted then within American academia. These academics, dominated mainstream social science departments, included the political sociologist, S.M. Lipset (Sharpe 2000; Schwarz 2000). Exceptionalism as a theme was derived from the nineteenth century principle of "manifest destiny" promoted by settler Puritan communities. The settlers argued that the United States was destined to expand its dominion and spread democracy and capitalism across the entire North American continent-this belief subsequently justified the annexation of huge amounts of land in the Americas through military aggression or via legal means. No wonder, American exceptionalism as an academic project searched for unique features that determined America's rise as a global power. It found it in arguments within history: that the American nation did not carry its antecedents in feudalism, nor had a class of aristocracy, nor left and communist movements, and thus was an original nation. Exceptionalism argued that USA was the bearer of values such as freedom and liberty, individual responsibility, republicanism, representative democracy, and laissez-faire economics. The citizens of the USA, it was argued, were equal before the law and this made the USA morally superior to Europe (Shafer 1999).

Said's (1978) study, and that of others who established the postcolonial standpoint, questioned this rhetoric to show how the West has created and consumed an imaginary Orient to perpetuate its discursive power through literature and language. Said's description of Oriental studies/ Middle East studies portrayed it as a Western style of thought and an institution of power for exercising control over the Arabs and Islam. Said's work built on earlier perspectives that had argued that Oriental thought controls the consciousness of the "natives." This perspective built on the anti-imperialist Marxist and Communist approaches popular in the 1960s and the 1970s in the Arab world and had emerged subsequent to the war of independence in Algeria (Halliday 1993). These discussions had also found space in the various conferences held post Bandung in the Arab region, such as the first Afro Asian Solidarity conference in 1957 at Cairo, mentioned above. However, while borrowing from this idea, Said made a clean break from its Marxist genealogy, integrating Michel Foucault's structuralist/poststructuralist critique of disciplinary knowledge with an assessment of Orientalism. Using Foucault's concepts of power/knowledge and discourse, Said and his colleagues made a critique of images and representations as they related these with institutions that produced Orientalist knowledge.

This approach flowed through the book, *Orientalism,* and Said argued that Orientalism was not only a field of knowledge or a discipline (the first Chair in Arabic at Cambridge was established in 1643), or a set of institutions (by the mid-nineteenth century, Oriental Studies was a well-established academic discipline in most European countries), nor was it only a corporate institution which primarily studied oriental societies and their cultures within Western universities. Rather, he contended, Orientalism was a mode of thought based on a particular epistemology and ontology which established a division between the Orient and the Occident. Said used Foucault's concept of discourse to argue that it combined power with knowledge and thereby produced its objects for a discourse of power which is resistant to change and transformation because of its linguistic constitution. In Said's conceptual framework, there was no phenomena outside of language; for language is self-referential. Language/literature defines the character of Orientalism; it not only produces the Orient as an object of knowledge but also establishes its outcome in terms of the relations of power. Postcolonial studies, in this sense, is a radical methodology that questions both the past and the ongoing legacies of European colonialism in order to undo them by interrogating its epistemic authority with institutional power.

By the 1990s, postcolonialism was no longer restricted to the field of English literature; with the association of the subaltern studies scholars with this perspective, its ambit moved further to the discipline of history. The subaltern studies school emerged in the 1980s to provide an ontological-epistemic critique of nationalist elite Indian historiography (Guha 1983; Chatterjee, 1986; 1997). Introduced by Gayatri Spivak to the American academy in the late 1980s/early 90s, this scholarship reinvented itself as postcolonialist and created for itself a new avatar within the North American area studies departments and within mainstream academia. Henceforth its work was seen as being the "interrogation of the relationship between power and knowledge (hence of the archive itself and of history as a form of knowledge)" (Chakraborty 2000:15). It led to a shift of discussions from the search for the subaltern and a critique of elite nationalist historiography in colonial, nationalist and Marxist historiography to the discourses and texts that represent "the fabric of dominant structures and manifest... itself as power" (Prakash 1994:1482). If Said used literary texts to analyze the West's project of domination of the Orient, the subaltern scholars argued that the recovery of the subaltern subject was possible only by deconstructing the historical documents in the archive. In following this path, they made a break from their earlier concerns and this led to some Marxist historians disassociating themselves from the field of study (Sarkar 1997, 2002). Though these scholars assert that their search for the subaltern has not been abandoned, they also contend the subaltern and subalternity are organically connected and both can be extracted from the historical archive (Chakrabarty 2000).

Their position now fits with the postcolonialists, who have argued that "doing" post coloniality is "doing" politics against colonialism/imperialism. The oriental archive, they contend, lays bare the West's framing of the "other" and its politics. This perspective has led Said, his colleagues, and the subalterns, to confront their own fields of knowledge; in Said's case it is Oriental Studies/Middle Eastern Studies, and in the case of the subalterns, the discipline of history whose procedures they argue are enmeshed in the authority of the West. Thus, postcolonialism becomes not only a theory of knowledge but a "theoretical practice," a methodology that can transform knowledge from static disciplinary competence to activist intervention. It gives this privilege to scholars who can use it to expose unequal power, make it visible and be involved in its winding-up. Given that the politics of this scholarship is to end colonial knowledge, it can be argued that this idea is shared with the politics that has organized the mid-twentieth century project of the indigenous/indigeneity, mentioned above. In this sense, there is a continuity between the two projects. But many would disagree with this contention, and rather would assert that there is a definite break between the two. Contra postcolonialism, they would suggest, the indigenous perspective in both its variants – as an adaption of Western theories to local contexts and as a search for the ontological-epistemological – was about decentering the colonial/imperial cultures of learning and knowledge-making in order to constitute new political modernities. Additionally, the indigenous perspective promoted the quest for the ontological through an interrogation of the local cultural/philosophical traditions of the new nation-states; thus, built-in into this work is a sociological perspective that explores the cultural in relation to its geographies. The postcolonialist project is not about constituting and legitimizing new political modernities. Rather, its ontological position is to reject Western knowledge through its deconstruction of the literature/languages and documents in the archives. There is little to no engagement here with the relationship of processes and structures with literature, or with the way events, processes and systems engaged with the documents in the archive.

No wonder the postcolonial perspective has found moorings within North America and Europe academic circuits. However, in the last four decades, it has also faced stringent criticisms. It has also evoked enormous moral and political outrage – some have perceived it as a personal attack on Europe and on Europeans and on American literature. Critics have been candid of postcolonialism's dismissal of the many assumptions of social science knowledge production. This criticism has been threefold: First, postcolonial analysis has been considered too simplistic; it is argued that it naturalizes the various geographies constituting the Orient in a universal argument while generalizing the work of scholars, scholarship and their institutions as being marked completely by West's power-knowledge project. Consequently, it has had an impact on scholarship: every text is now perceived as a narrative of power rather than being one of the many constituting a corpus that has structured the history of ideas within its geographies of circulation. Secondly, postcolonialism's theoretical architecture is based on the affirmation of the epistemic difference between West vs the East and the use of the binary, we/they or I/other. This has led many to argue the West has created all the ills that organize the contemporary Orient(s). Given that colonialism accentuated the differences and inequalities already in place, critics suggest a need to fine tune theoretical interventions instead of naturalizing the binaries as universal axioms to comprehend knowledge politics. Third, following the above, the most important criticism relates to postcolonialism's anti-foundationalist position. If the "true" descriptions of the "real world" are rejected, how can we study it? This has led some sociologists to contend that the postcolonial approach is limited, prejudicial and too shallow to be of any value to contemporary social sciences grounded in evidence and empirical details (Turner 1989).

And yet, without the introduction of this approach, mainstream American academia embedded in various degrees of American exceptionalism, sometimes called ethnocentric provincialism, or European scholarship embedded in varieties of Occidentalism (Coronil 1996), would not have engaged with colonialism and reflected on how it has affected both American and European history and the history of its social science disciplines. Indeed, the study of Occidentalist assumptions in literature, literary criticism, historiography, and now sociology and social science theories, have led to the recovery of the work of scholars such as the sociologist W.E.B. du Bois (1868–1963) (Morris 2017), and an assessment of the way silences regarding racism, ethnicity, and indigeneity have organized the disciplines of sociology/social sciences in North America. Postcolonialism has made it possible to question the implied assumptions of the dominant discourses in the social sciences. Postcolonialism has also helped to interrogate categories and classifications systems, used by sociologists to reflect on their own discipline and comprehend how these are related to Eurocentric assumptions. This can be seen in the pioneering work of Connell (1997,2019), who queried the definition and meaning of "classical sociology," or that of Alatas and Sinha (2017), on how to redefine the sociological cannon in the context of its global practices which stretch beyond and before Europe's constitution of these. In addition, the publication of path breaking texts, such as those by Gurminder Bhambra (2013, 2014) and Julian Go (2016), have laid bare the practices of the discipline, deconstructing its power, and authority in defining what it is and should be. Consequently, contemporary scholars have found it easier to debate concepts such as internal colonialism, race and racism, ethnicity and minorities in their various regions. Lastly, postcolonialism, this author contends, is an important link in understanding decoloniality, which is discussed below.

Coloniality/Decoloniality

This chapter argues that with the enunciation by Anibal Quijano (2000) of the concept of coloniality, we have come a full circle in debates on colonialism and its knowledge. The concept of the coloniality/coloniality of power combines, in new and radical ways, the mid-twentieth century project of conceiving an ontological-epistemology and as well to establish the theoretical scaffolding for an alternate political modernity: this time to describe the frames of settler colonialism. Quijano integrates arguments of the world system approach drawn from Immanuel Wallerstein, with those from Latin American scholars such as Sergio Bagu and Gonzalez Casanova. He Quijano incorporates the above perspectives on settler colonialism and capitalism while simultaneously reusing Samir Amin's (1932–2018) concept of and theory elaborated in *Eurocentrism* (1989) in a new way. Amin, who extended Martin Bernal's argument on Eurocentrism elaborated in *Black Athena* (1987), defined it as a theory of world history, fine-tuned during the Renaissance, to assert Europe's uniqueness and superiority (Moghadam 1989). Amin states that it is not Greece and later Rome that were the cradle of European civilization, rather the civilizational roots of Europe are in the Orient – in Ancient Egypt and with the Phoenicians (the Levant). Amin thus challenged the contemporary economistic-oriented practices in Marxist historiography with his historical, sociological and philosophical analysis of the cultural and intellectual assumptions of European epistemologies.

Quijano follows a similar methodology to query Eurocentric assumptions regarding modernity, but his canvass is much larger than that of Amin. Following Bagu's contention that in Latin America "there was no servitude on a large scale, but slavery with multiple shades" (Biegal 2010:193); Quijano in his text *Coloniality of power, Eurocentrism and Latin America* (2000) maps the growth of the capitalist/colonial system and analyses the implications of the Iberian invasion of Latin America. His article argues that the slow consolidation of a trade circuit that linked Iberian countries with the Americas from the fifteenth century later integrated Africa into its trade routes with the beginning of slavery in the sixteenth century. The chapter shows how, over the next 250 years, a strong trade nexus developed between the Pacific and Atlantic Oceans and connected Asia and Europe around the colonial territory of "New Spain": a geographic space comprising not only the Americas, parts of Africa but also the Philippines. In this circuit, the agents of colonialism outside "America" had a limited and almost negligible role (Quijano 2000; Mignolo 2001). This analysis brings Quijano to contend that Latin America had been part of the world system since the late fifteenth century, and since that time was a settler colonized territory. Thus, the scaffolding that structured the theory of capitalism through Eurocentrism took place in this period, in the first phase, from the beginning of the sixteenth century to the eighteenth century.

For Quijano, Eurocentrism is the elaboration of a perspective on knowledge associated with colonial ethnocentrism and universal racial classification. It consists of two attributes: the constitution of the binary and a theory of linear history: "a peculiar dualist/evolutionist historical perspective" (Quijano 2000: 556). It helped to form the basis of the European scientific-technological development during the 18th/19th centuries, but was also imbricated in many other theories of universal history and culture. These premises also influenced the formation of the social sciences in the late nineteenth century, as Wallerstein et al. (1996; Wallerstein, 1997) have argued. Gradually, this Eurocentrism formed the contours of an ideology and of a rather diffuse common sense; it seduced the population it encountered and, in some periods, did its job through covert rather than overt oppression.

Quijano intertwines four main trends of coloniality as a system, all of which he delineates from the Latin American experience of settler colonialism – an economic process that extracted and transferred value through the control and subordination of forms of labor to capital – this in turn legitimized a social classification system around the category of "race," thereby creating a "racial division of labor." This system found justification through the institutions of the nation state and notions of democracy, and was thus presented and articulated through a Eurocentric theory of modernity, permeating individual and collective identities and constituting sociabilities (Quijano 2000: 544–545). In the above-mentioned chapter, we see Quijano mobilizing an historical-sociological approach to analyze a sociology of epistemology. More particularly, while assessing the interconnected processes of global and regional political economies and combining these with an analysis of social and political institutions, Quijano provides a sociological framework to examine colonial society while simultaneously offering an epistemological critique of contemporary sociological theories and an ontology with which to reconstitute it.

In Quijano's oeuvre, coloniality becomes an ontology: "what is, is the consequence of that which has been" (Gandarilla Salgado 2021:202), something that can be unraveled from the values and norms institutionalized in everyday life, within the family system and marriage alliances, within sexualities, in education, its pedagogies, and its philosophies. Contrary to contemporary Marxist orthodoxies, Quijano's theory of colonial capitalism argues that material changes and cognitive interventions occur concurrently. Thus, capitalism in its colonial form emerges when the control and subordination of slavery, servitude, and wage labor occur alongside one other. For Quijano, this model of work, related to land appropriation, resource extraction, and in-migration, was first institutionalized in the Americas and over time has become a global model. Thus, Quijano's methodology brings his observations and interpretations of the archival documentation with empirical evidence and actual events, and sieves these with objects and things that have organized experiences, while integrating these with the mechanisms, causes, power, and structures that have in turn produced these events. Obviously, a scientific practice is organizing Quijano's oeuvre. And yet it is important to mention, that here too, Quijano provides a note of caution. He argues that no methodology/method is autonomous from the ideologies of "the consequence of that which has been." Thus, methods and methodologies need to be deconstructed and located in the knowledge systems of its structuration, in order to comprehend "its purpose of inquiry into reality and for the production of its knowledge" (quoted in García-Bravo 2021:205). This implies that a reflexive assessment of the methodologies, its histories of use, and its philosophical origins need to be explored before these are re-used again.

In what way are Quijano's sociological practices linked with decoloniality? Led by Walter Mignolo, who argued that coloniality was the "dark side of modernity," the decoloniality group, who now have a formidable repertoire of publications, consider Quijano's theory of coloniality as a critical point of departure for outlining their position (Escobar 2007). Initiated in the early 1990s by Latin American scholars in the USA who initially came to be known as the postcolonial Latin American Studies Programme, the group argued that Eurocentrism emerged to clothe the organic linkages between coloniality/modernity (Castro-Gomez 1998; Coronil 2008). However, by 1998, the group realized it had little in common with the postcolonialists. The collective's understanding of Eurocentrism, Mignolo contended, was based on the Iberian colonization of the Americas (the first phase of modernity), while that of postcolonialism was formed through the British colonialism of the eighteenth or nineteenth century: which he argued, borrowing from Quijano, was the *second* phase of modernity (Mignolo 2007). Mignolo contends that because the postcolonialists focus on deconstructing the power/knowledge matrix within language/literature, and the archive through the methodology of poststructuralism, it cannot "de-link" from Eurocentrism. Postcolonialism, he argues, does not have the language to critique the colonizer from the episteme of the colonized, that is from an exterior position (Dussel 2000); it merely critiques modernity and thus remains internal to Eurocentrism. In addition, Grosfoguel (2007) contends that unlike the postcolonialists, the decolonial does not make a cultural argument. Decoloniality is an attempt to find an epistemic voice outside modernity with which to formulate new universals that have not inherited such totalitarian orientations (Tlostanova and Mignolo 2009: 131).

Drawing from Quijano, the decolonialists suggest that the world came to be interconnected in the circle of colonial/capitalist modernity in the sixteenth-century when the twin processes, that of the expulsion of Jews and Muslims from Spain and the elevation of Western Christianity to religious dominance, led to an early racial classification. The American continent became the first contact zone and battleground for the deployment of ideas of civilization, evangelization, empire, and racial difference, together with the subalternation of the knowledges of the colonized (Mignolo 2001). Mignolo proposes that Eurocentrism was born at this juncture, and thereby becomes the knowledge form of modernity/coloniality – an hegemonic representation and mode of knowing that claims universality for itself, and that relies on "a confusion between abstract universality and the concrete world hegemony derived from Europe's position as center" (Dussel 2000: 471; Quijano 2000: 549). Thus, decoloniality is in need of a new episteme that comprehends the historical processes necessary for the creation of an original set of concepts, and Mignolo (2017) has argued that it is among the indigenous groups (original inhabitants) within the regions of the Caribbean, Mesoameria, and the Andes one would find subaltern knowledge. In addition, the decolonialists draws on a variety of thinkers who have critiqued settler colonialism to create this new epistemology. These range from some in Latin America, such as Enrique Dussel, and others in Africa and the Caribbean, such as Frantz Fanon, Aime Cèsaire, and C.L.R. James whose concepts have been integrated with perspectives such as dependency theories, liberation theology, and ideas popularized by Latin America social movements (e.g., that of Zapatistas). Recently, some of the decolonialists have traced the legacy of decoloniality to the Bandung conference of 1955 (Maldonado-Torres et al. 2019).

The decolonial approach has developed a repertoire of concepts to present a distinctive position. These include Occidentalism-the formation of specific forms of racialized and gendered Western selves as the effect of Orientalist representations of the non-Western Other (see Rodríguez et al. 2010, on the way it organized European sociology); colonial difference (the epistemic division of modernity from coloniality and its use to create further divisions and differences in knowledge, which is different from the way it is used by Chatterjee 1986 see Patel, 2018); and imperial difference (the downgrading and hierarchization of European others, for example, the Ottomans, the Chinese or the Russians). Border epistemology, de-linking and pluriversality, as defined by Mignolo (2007), implies separating one's way of thinking from all forms of Enlightenment, its political and economic theories, including from Marxist theory where the alternatives of socialism and communism were promoted. De-linking, originally conceived by Amin (1987 and borrowed from Quijano) (Davis and Walsh 2020:11), implies both an epistemological and a political practice. It is presented as a radical project, which in Escobar's (2007) words is an "inquiry in the very borders of systems of thought" and which make possible "non-Eurocentric modes of thinking" (Escobar 2007: 180). For Mignolo, the "de-" of decoloniality helps to conceptualize it as "re-epistemic reconstitution, re-emergence, and resurgence" (Mignolo 2017).

I have argued that Quijano's methods differ from those used by the decolonial theorists. Quijano's gaze remained on the Latin American historical experience when elaborating a sociology of epistemology, while the decolonialists have increasingly used the methodologies of postmodernism, poststructuralism and deconstruction in their search for an alternative epistemic "voice" and in understanding of difference in narratives about indigeneity/racism/ethnicity. Rarely has there been an attempt to engage with the social science methods and methodologies used either by mainstream social sciences, new historical and sociological work based on quantitative methods or those used by subaltern/feminist works. (Although there are exceptions to this, see Coronil 2008). Also, in their work, there is a complete absence of political economy and discussion of economic development. Critics argue that though there are "gestures to the subregional and to the indigenous" there is very little "critical work that is supposed to undo or challenge the homogenizing work of colonialism and nation-building" (Salvatore 2010:344). Consequently, the decolonialists retain many of the criticisms that postcolonialists have faced as elaborated above, particularly its anti-foundationalism – a commitment to study the "real" world but not having the methods to do so.

Latin American historians and social scientists have been critical both of Quijano's concept of coloniality and the decolonial school in different ways (Bortoluci and Jansen 2013; Coronil 2008; Domingues 2009; Salvatore 2010). Some historians, while appreciative of Quijano's extensive canvass outlining the 500-year history of Latin America, have raised two queries. The first relates to the persistence of coloniality today: about whether it even has an ontological relevance given the sweep of capitalist modernity (Domingues 2009), and the genocide against, and the decimation of, the original inhabitants, the indigenous groups and it is important to note that indigenous has a different meaning in settler colonialism. Secondly, they have asked whether coloniality has had differential spatial impact in the Latin American continent. The ideology to rebuild after the destruction of the old settler colonialism has encouraged various legal interventions for land appropriation and new forms of incorporation, such as miscegenation. Historians have contended that empirical studies have affirmed that conquest/genocide of the "indigenous" groups (the original inhabitants of the land) and their displacement together with modernity, neo-colonial domination and backwardness have walked hand in hand in Latin America, creating sometimes mixed and sometimes uneven spatial processes (Salvatore 2010:339).

In addition, scholars who have accepted the idea that the fifteenth and sixteenth century encounters of the Spanish and the Portuguese should be considered colonial, have argued there is a very specific Latin American historical process consequent to the Iberian invasion into the region. While agreeing that Latin America is an important site for examining the representational and discursive practices that were developed for, and through, the operation of early European colonialism; they nevertheless argue for the need to make an assessment of the implications of Catholicism, the contrasting effects of a centralizing State juxtaposed by local bureaucracies, and an analysis of its peculiar racial ideologies across the region, before accepting the coloniality thesis (Bortoluci and Jansen 2013). In particular, social scientists in Latin America have raised fundamental questions regarding some aspects of the coloniality thesis. One of the major problems seems to be the lack of engagement with the various ways race and racialization have been organized across the Americas (Benzi 2021; Bortoluci and Jansen 2013). This begs the question of whether all forms of difference emerging from the Iberian colonial experience since the sixteenth century be categorized as racial and theorized as part of coloniality. More generally, scholars query whether the Iberian invasion of the Americas be assessed as a colonial encounter. Are the knowledge implications of the processes of settler colonialism, such as experienced by North, Central and Latin America and the Antipodes, similar to that of non-settled colonialism, such as experienced in countries of Africa, West and South Asia or those of Southeast Asia? (Veracini 2011). Given that the decolonial perspective is now used to examine forms of genocide in Rawanda, Armenia, Cambodia and the Palestine, this begs the question of whether settler colonialism has now emerged in a new avatar? (Wolfe 2006).

Today, Latin American scholarship (in Latin America) marks itself out against the decolonial position while emphasizing its genealogy within the specific processes that articulated dependency theory. It thereby has set itself in a distinct new cognitive circuit. Most Latin Americanists perceive their scholarship as being moored in the developments that took place in the region in the late 1940s and early 1950s, which they call the "Latin American" regional social science perspective when there was an attempt to chalk an alternate political modernity theory through the theories around dependency (Sorá and Blanco 2018). Thus, the ontological-epistemological position in this case is located in terms of the region. The geopolitical context for this intellectual development is similar to that experienced by the ex-colonial countries of Asia and Africa in the 50s, when the project of Nonalignment and Third-Worldism promoted alternate ways to perceive modernity. However, even if the context was similar, the content and implications of the Latin American intervention was, and is, distinct. The Latin American regional perspective makes a break from the earlier notion that Latin American unity is based on its shared history in the Iberian colonization and their identity within the common languages of Spanish and Portuguese. Instead, scholars argue that by the mid-twentieth century, a new way of thinking found expression in Latin America "in the face of the political and cultural domination of the United States and Western Europe" (Sorá and Blanco 2018:127). Four attributes defined this regional perspective and continue to define it: a concern with the present; an analysis of the contemporary processes based on an adaption of contemporary global sociological theories to Latin American contexts; an interdisciplinary perspective combining economy, sociology and political theory; and the use of existing social scientific methodologies and methods to assess these processes.

This Latin American regional perspective was conceptualized in the context of political developments in Cuba in the late 1950s and involved Latin American scholars from radical and left positions who developed a perspective that abandoned the "points of view of the great centres of world economy." It found support within the region from the first generation of scholars of all persuasions as they were liberated intellectually from the military regimes and involved themselves in the research and teaching of the social sciences in universities and research centers funded by the UN (e.g., the Economic Commission for Latin America and the Caribbean (CEPAL) established at Santiago de Chile in 1948), and private USA foundations and their own governments. Latin American governments set up a teaching institute, FLASCO (the Latin American School of Social Sciences) and a research institute CLAPCS (the Latin American Centre for Research in the Social Sciences). In addition, the establishment of the Latin American School of Sociology (ELAS), and the Latin American School of Political Science and Public Administration (ELACP), played important roles in creating a Latin American perspective. Since that time, a key role has been played by CLASCO (the Latin American Social Sciences Council), which has sourced funds from across the world to fund Latin American research (Sorá and Blanco 2018).

These developments were aided and supported by the setting up of professional associations such as ALAS (the Latin American Association of Sociology) together with publishing houses and journals. A critical moment emerged when from the late 1960s onward, for a decade or more, Santiago de Chile became the site for intellectual conversations across disciplines, perspectives, and institutions. It provided the intellectual infrastructure to frame the dependency school, which argued that development and underdevelopment were mutually reconstituted in the center (the USA) and periphery (Latin America) (Biegal 2010). The institutionalization of the dependency theories led to the further consolidation of regional intellectual circuits across Mexico, Brazil, Argentina and Chile, organizing interdisciplinary knowledge between the first and second generations and soon the third generation of intellectuals to further its spread across sub-regions. The recent growth of private universities, increased student enrolments and neoliberal initiatives (at the expense of radical politics), have led to a fragmentation in the above project (Sorá and Blanco 2018) with different national research groups emphasizing distinct research and theoretical foci and a significant number aligning with contemporary European theories (Bortoluci and Jansen 2013). And yet, most Latin American social scientists (and this includes some who advocate decoloniality, such as Coronil 2008), would agree that the ideas organized around dependency have influenced the subsequent growth and development of Latin American social sciences and retains its flavor even today.

Conclusion

This chapter has mapped the continuities and breaks that have occurred in the long history of ideas on the theme of "colonialism and its knowledges." The goal firstly, was to understand this history, to assess how this academic project, entangled in anti-colonial political movements, has been constituted to interrogate hegemonic knowledges and thereby develop new research agendas and conceptualize new ways of thinking, recast the old and create new methodologies and present new paradigms. The chapter also argues secondly, that to justify and legitimize these interventions, scholars have mapped out ontological-epistemic standpoints, three of which have been explored here: that of D.P. Mukerji, Akiwowo Akinsola and Anibal Quijano. The chapter has narrated the various twists and turns of the project as it progressed from its inception in the 1940s and 1950s as an indigenous/indigeneity perspective in Asia and Africa and as a dependency perspective in Latin America, to the subaltern/postcolonial, and on to the decolonial across Asia-Africa-Latin America to North America. In these journeys, this academic work found new legitimacy through the reconstitution of conventional methodologies, such as that of historical sociology and the constitution of versions of poststructuralist and deconstructive perspectives. Though the focus of this chapter has remained on this history and its distinct expressions across its articulation in various geographies, it also highlights the following themes that have organized these travels.

Substantively, this project has unraveled the need to distinguish between two kinds of colonialism – the first, that of non-settler colonialism and the second that of settler colonialism. Successful settler colonies "tame" "a variety of wildernesses, end up establishing independent nations," and repress, co-opt, and extinguish indigenous peoples and their alterities, and through this process manage ethnic diversity. Racialization is the key to this process. On the other hand, in non-settler colonial systems, a determination to exploit sustains a drive for the subordination of the colonized (Veracini 2011). It should be noted that the indigenous in the context of settler colonialism has a different genealogy than in the context of non-settler colonialism. While agreeing with Walter Mignolo's argument that contemporary social sciences have focused on the British colonialism that was experienced in the eighteenth/nineteenth century the chapter asserts that these two forms of colonialism have generated distinct knowledge systems and that the postcolonial and decolonial perspectives are separate and distinctive positions. This acknowledgement allows us to comprehend the way Eurocentrism was institutionalized in its two phases of modernity: from the sixteenth to eighteenth century and from the eighteenth or nineteenth to twentieth centuries. Colonialism was thus not an experience of one part of the world, rather it was an experience that engulfed the world and included the Americas, the African continent together with the Antipodes. This distinction presents us with a new way to understand colonialism and stretches its history backwards to the fifteenth and sixteenth centuries. This new definition of colonialism reframes ways to understand the world capitalist system and its cultures of modernity since the fifteenth/sixteenth century.

Additionally, the chapter has charted the processes that have institutionalized subaltern circuits of knowledge production to form alternate/non-hegemonic forms of thinking. These have been sustained because of their intimate connections with anticolonial/ antiimperialist perspectives together with ideas related to nationalism, left and radical viewpoints, political commitments of individuals and collectives of scholars; the politics of the newly independent nation-states; but also because of the growth of educational infrastructure that has organized the production and circulation of this knowledge: universities, research institutes, journals and books, publication houses, and professional associations. The chapter has mapped the political contexts that have helped to define new research queries and provide original perspectives; these becoming incubators for the growth of emerging scholarship. It has affirmed the thesis that new paradigms can emerge when subaltern perspectives encounter those formalized by normal social science. The chapter has elaborated how funding has been critical to sustain these knowledge systems and that without the support of governments, public and private social science research foundations and alternate intellectual networks, these would not have continued to operate. Despite these support structures, it is argued that this scholarship has remained in the margins, and is likely to do so until the mainstream itself changes.

We have also highlighted the tensions that these positions have generated with the demands of evidence and analysis, and more generally with the protocols of social science scholarship. Today, mainstream social science accepts that the social world is complex and heterogenous, contingent, and plastic, and thus cannot be interrogated through the principles formatted by the natural sciences. Given this, in recent times, there has been a more liberal understanding of the protocols demanded from the social sciences. However, there remains a schism in this literature between the post colonialists/decolonialists who promote an anti-foundational epistemology and scholars who insist that "real" evidence is necessary to make relevant and significant analytical arguments. This remains a critical tension in this project.

Consequent to the growth, in the last 80 to 100 years of these above-mentioned distinct and varied perspectives, we have seen a slow dismantling of the epistemological assumptions which have governed the Eurocentric epistemology of the discipline. No longer is modernity equated with the West – with the West being the center of modernity's geography. If this chapter has highlighted differences within the ex-colonial worlds and the global South regarding how colonialism should be perceived, we now recognize similar trends within Europe which have hegemonized the European others. This has raised the query of whether all forms of domination have been touched by colonialism. Consequently, a universal theory, a "one fits all" position, has become dysfunctional. No wonder most scholars now agree that it is important to write histories and to do sociologies in terms of differing scales and from epistemes that organize colonial/national margins. No wonder they also assert the necessity of interrogating the methods and methodologies of science: given the embeddedness of ways of seeing and knowing in power/knowledge dynamics which in turn are defined by dominant-subaltern circuits of colonial knowledge.

We live in an interconnected global world. From the fifteenth to the eighteenth and from the eighteenth to mid-twentieth century, colonialism established circuits of material and cognitive connections. Since that time, in the twenty-first century, we are integrated by many complex cognitive circuits within the overall divisions created by its colonial and nationalist histories, its geographies, and the unequal distribution of income, privilege, status, and power. These circuits have created differences and variations within the global world system consequent to the way events and processes have been organized and sociabilities constituted. Subsequently, it is imperative to understand these cognitive geographies and the system that reproduces them and to do research through these varying scales, bringing back local and regional scholarship as it connects the scholarship of organic intellectuals with formal knowledges. A decade back I argued for a need for diversity in knowledge traditions (Patel 2010). Today, more than ever, this is needed.

Acknowledgments The author acknowledges the contributions of the following: Fran Collyer for suggesting the title, Joao Maia for long conversations on this theme and Raewyn Connell for detailed comments.

References

Adesina J (2006) Sociology, endogeneity and challenge if transformation. Afr Soc Rev 10(2):133–150
Akiwowo AA (1986) Contributions to the sociology of knowledge from an African oral poetry. Int Sociol 1:343–358
Akiwowo AA (1999) Indigenous sociologies. Extending the Scope of the Argument. Int Soc 14: 115–138
Alatas SH (1972) The captive mind in development studies. Some neglected problems and the need for autonomous social science tradition in Asia. Int Soc Sci J 24(1):9–25
Alatas SF (2003) Academic dependency and the global division of labour in the social sciences. Curr Sociol 51(6):599–613
Alatas SH (2006) The autonomous, the universal and the future of sociology. Curr Sociol 54(7):7–23
Alatas SF (2014) Applying Ibn Khaldun: the recovery of a lost tradition in sociology. Routledge, London
Alatas F, Sinha V (2017) Sociological theory beyond the Cannon. Palgrave Macmillan, London
Albrow M, King E (1990) Globalization, knowledge and society: readings from international sociology. Sage, London
Amin S (1987) A note on the concept of delinking. Review X(3):435–444

Amin S (1989) Eurocentrism. Monthly Review Press, New York
Archer M et al (2016) What is critical realism? Perspective. A newsletter of ASA theory section. http://www.asatheory.org/current-newsletter-online/what-is-critical-realism. Downloaded on 25 Mar 2021
Atal Y (1981) The call for indenisation. Int Soc Sci J 33(1):189–197
Barlow TE (1999) Introduction: on 'colonial' modernity. In: Barlow TE (ed) Formations of colonial modernity in East Asia. Duke University Press, Durham, pp 1–20
Benzi D (2021) Coloniality of power, eurocentrism and Latin America from the perspective of macro-historical sociology in J.G. Gandrilla, M. García-Bravo and D. Benzi, Two decades of Aníbal Quijano's *Coloniality of Power, Eurocentrism and Latin America*. Contexto Internacional 43(1):212–218
Bernal M (1987) Black Athena Vol 1: the fabrication of ancient Greece 1785–1985. Rutgers University Press, New Brunswick
Bhambra GK (2013) The possibilities of, and for, global sociology: a postcolonial perspective. Polit Power Soc Theory 24:295–314
Bhambra GK (2014) Connected sociologies. Bloomsbury Academic, London
Biegal F (2010) Dependency analysis. The creation of new social theory in Latin America. In: Patel S (ed) ISA handbook of diverse sociological traditions. Sage, London, pp 189–200
Bortoluci JH, Jansen RS (2013) Toward a postcolonial sociology: the view from Latin America. Polit Power Soc Theory 24:199–229
Castro-Gomez S (1998) Latin American postcolonial theories. Peace Rev 10(1):27–33
Chakrabarty D (2000) Subaltern studies and postcolonial historiography. Nepalta.V fr South 1(1):9–32
Chatterjee P (1986) Nationalist thought and the colonial world: a derivative discourse? Zed Books for the United Nations University, London
Chatterjee P (1997) Our modernity. Sephis, CODESRIA, Dakar
Chen H (2021) Between north and south: Historicizing the indigenization discourse in Chinese sociology. J Hist Sociol 34(1):103–119
Connell RW (1997) Why is classical theory classical? Am J Sociol 27(6):1511–1557
Connell R (2007) Southern theory. The global dynamics of knowledge in social science. Polity, London
Connell R (2010) Learning from each other: sociology on a world scale. In: Patel S (ed) The ISA handbook on diverse sociological traditions. Sage, London, pp 40–52
Connell R (2019) Canons and colonies: the global trajectory of sociology. Estudos Históricos (Rio De Janeiro) 32(67):349–367
Connell R (forthcoming) Curriculum for revolution: Ali Shariati's practical plan and the radical politics of knowledge
Coronil F (1996) Beyond occidentalism: toward nonimperial geohistorical categories. Cult Anthropol 11(1):51–87
Coronil F (2008) Elephants in the Americas? Latin American postcolonial studies and global decolonisation. In: Morana M, Dussel E, Jauregui CA (eds) Coloniality at large. Latin America and the postcolonial debate. Duke University Press, Durham
Dalmia V (1996) Sanskrit scholars and Pandits of the old school: the Benares Sanskrit College and the constitution of authority in the late nineteenth century. J Indian Philos 24(4):321–337
Davis BP, Walsh J (2020) The politics of positionality: The difference between post-, anti-, and de-colonial methods. Cult Theory Crit. https://doi.org/10.1080/14735784.2020.1808801
De Souza Santos B (2014) Epistemologies of the south: justice against epistemicide. Paradigm Publishers, Boulder
Dirks N (2013) South Asian studies: Pasts and futures. South Asia Institute, Harvard University (video). https://mittalsouthasiainstitute.harvard.edu/event/south-asian-studies-pasts-and-futures/. Downloaded on 12 Dec 2020
Domingues JM (2009) Global modernization, 'coloniality' and a critical sociology for contemporary Latin America. Theory Cult Soc 26(1):112–133

Dussel E (1993) Eurocentrism and modernity (Introduction to the Frankfurt Lectures). boundary 2 20(3):65–76
Dussel E (2000) Europe, modernity and eurocentrism. Nepalta: Views from the South 1(3):465–478
Escobar A (2007) Worlds and knowledges otherwise. The Latin American coloniality/modernity research programme. Cult Stud 21(2–3):179–210
Fiddian-Qasmiyeh E, Daley P (2018) The Routledge handbook of south-south relations. Routledge, London
Gandarilla Salgado JG (2021) To realise the real mutation and not succumb to the procrustean bed in J.G. Gandrilla, M. García-Bravo and D. Benzi, Two decades of Aníbal Quijano's *Coloniality of Power, Eurocentrism and Latin America*. Contexto Internacional 43(1):199–205
García-Bravo M (2021) The interdisciplinary perspective in Aníbal Quijano in J.G. Gandrilla, M. García-Bravo and D. Benzi, Two Decades of Aníbal Quijano's *Coloniality of Power, Eurocentrism and Latin America*. Contexto Internacional 43(1):206–211
Go J (2013) Introduction: entangling postcoloniality and sociological thought. Polit Power Soc Theory 24:3–31
Go J (2016) Postcolonial thought and social theory. Oxford University Press, New York
Grosfoguel R (2007) The epistemic decolonial turn. Cult Stud 21(2–3):211–223
Guha R (1982) On some aspects of historiography of colonial India. In: Guha R (ed) Subaltern Studies, vol 1. Oxford University Press, Delhi, pp 1–9
Guha R (1983) Elementary aspects of peasant insurgency in colonial India. Oxford University Press, Delhi
Halliday F (1993) 'Orientalism' and its critics. Br J Middle East Stud 20(2):145–163
Heilbron J (1995) The rise of social theory. Polity Press, Cambridge
Hountondji P (1995) Producing knowledge in Africa today. The second Bashorun M. K. O. Abiola Distinguished Lecture. Afr Stud Rev 38(3):1–10
Hountondji P (1997) Introduction in endogenous knowledge. Research trails. CODESRIA, Dakar, pp 1–42
Hountondji P (2009) Knowledge on Africa. Knowledge by Africans. Two perspectives on African studies. RCCS Ann Rev 1:121–131
Joshi PC (1986) Founders of the Lucknow school and their legacy: Radhakamal Mukerjee and D. P. Mukerji. Econ Polit Wkly
Levi-Strauss C (1966) Anthropology: its achievements and future. Curr Anthropol 7(2):124–127
de Lissovoy N, Bailon R (2019) Coloniality. Key dimensions and critical implications, keywords in radical philosophy and education, common concepts for contemporary movements, vol 1. Brill, Leiden, pp 83–97
Madan TN (2007) Search for synthesis: the sociology of D.P. Mukerji. In: Uberoi P, Sundar N, Deshpande S (eds) Anthropology in the east: founders of Indian sociology and anthropology. Permanent Black, Delhi, pp 256–289
Madan TN (2013) Sociology at the University of Lucknow: The first half century (1921–1975) (Oxford in India readings in sociology and social anthropology)
Makinde M (1988) African philosophy, culture, and traditional medicine, Ohio University Center for International Studies Monographs in International Studies, Africa Series Number 53. Athens
Maldonado-Torres N et al (2019) Editorial introduction: Frantz Fanon, decoloniality and the spirit of Bandung. Bandung J Global South 6:153–161
Mignolo WD (2001) Coloniality at large. The western hemisphere and the colonial horizon of modernity. CR New Centen Rev 1(2, Fall):19–54
Mignolo WD (2007) Delinking. The rhetoric of modernity, the logic of coloniality and the grammar of de-coloniality. Cult Stud 21(2–3):44–514
Mignolo WD (2017) Interview, Walter Mignolo/part 2, key concepts. E-Interrelations. https://www.e-ir.info/2017/01/21/interview-walter-mignolopart-2-key-concepts/. Downloaded on 15 Mar 2021
Mignolo WD, Tlostanova MV (2006) Theorizing from the borders shifting to geo- and body-politics of knowledge. Eur J Soc Theory 9(2):205–221

Moghadam V (1989) Against eurocentrism and nativism: a review chapter on Samir Amin's eurocentrism and other texts. Social Democr 5(2):81–104

Morris A (2017) The scholar denied. W. E. B. Du Bois and the birth of modern sociology. University of California Press, Berkeley

Nash A (2003) Third Worldism. Afr Soc Rev Rev Afr Soc 7(1):94–116

Mukerji DP (1986) Some reflections. Econ Polit Wkly 21(33):1455–1469

Odora Hoppers (2002) Indigenous knowledge and the integration of knowledge systems. Towards a philosophy of articulation. New Africa Books, Claremont

Olaniyan T (2000) Africa: Varied colonial legacies in Henry Schwarz and Sangeeta Ray edited A companion to postcolonial studies, Oxford, Blackwell, pp 269–281

Onwuzuruigbo I (2018) Indigenising Eurocentric sociology: the 'captive mind' and five decades of sociology in Nigeria. Curr Soc Rev 66(6):831–848

Parry B (1997) Postcolonial. Conceptual category or chimera? In the politics of postcolonial criticism. Yearb Engl Stud 27:3–21

Patel S (2006) Beyond binaries. Towards self reflexive sociologies. Curr Soc 54(3):381–395

Patel (2010) Introduction. Diversities of sociological traditions. In: The ISA handbook of diverse sociological traditions. Sage, London, pp 1–18

Patel S (2013) Orientalist-eurocentric framing of sociology in India: a discussion on three twentieth-century sociologists. Polit Power Soc Theory 25:105–128

Patel S (2015) Colonial modernity and the Problematique of indigenous and indigeneity: South Asian and African experiences. In: Sabea H, Biegal F (eds) Academic dependency: the challenge of building autonomous social sciences in the south. EDIUNC-SEPHIS: Mendoza

Patel S (2017) Colonial modernity and methodological nationalism: The structuring of sociological traditions in India. Sociol Bull 66(2):125–144

Patel S (2021) Sociology's encounter with the decolonial: The problematique of indigenous vs that of coloniality, extraversion and colonial modernity. Curr Sociol 69(3):372–388. https://doi.org/10.1177/0011392120931143

Prakash G (1994) AHR forum. Subaltern studies as postcolonial criticism. Am Hist Rev 99(5):1475–1490

Quijano A (1993) Modernity, identity, and Utopia in Latin America. boundary 2 20(3):140–155

Quijano A (2000) Coloniality of power, eurocentrism, and Latin America. Nepalta. Views from South 1(3):533–580

Quijano A (2007) Coloniality and modernity/rationality. Cult Stud 21(2–3):168–178

Rodríguez EG, Boatcă M, Costa S (2010) Decolonizing European sociology different paths towards a pending project, transdisciplinary approaches. Ashgate, London/Fraham, Surrey

Said E (1978) Orientalism. Pantheon Books, New York

Salvatore RD (2010) The postcolonial in Latin America and the concept of coloniality: a Historian's point of view. AContracorrinte 8(1):332–348

Sarkar S (1997) The decline of the subaltern in subaltern studies. In: Sarkar S (ed) Writing social history. Oxford University Press, Delhi, pp 82–108

Sarkar S (2002) Postmodernism and the writing of history. In: Sarkar S (ed) Beyond nationalist frames. Relocating, postmodernism, hindutva, history. Permanent Black, Delhi, pp 154–195

Schwarz H (2000) Mission impossible: Introducing postcolonial studies in the US academy, in Henry Schwarz and Sangeeta Ray edited A companion to postcolonial studies, Oxford, Blackwell, pp 1–20

Shafer BE (1999) American exceptionalism. Annu Rev Polit Sci 2:44–263

Sharpe J (2000) Postcolonial studies in the house of US multiculturalism, in Henry Schwarz and Sangeeta Ray edited A companion to postcolonial studies, Oxford, Blackwell, pp 112–125

Sorá G, Blanco A (2018) Unity and fragmentation in the social sciences in Latin America. In: Heilbron J et al (eds) The social and human sciences in global power relations. Palgrave Macmillan, London, pp 127–152

Tlostanova M (2012) Postsocialist ≠ postcolonial? On post-soviet imaginary and global coloniality. J Postcolonial Writ 48(2):130–142

Tlostanova M, Mignolo W (2009) Global coloniality and the decolonial option. Kultur 6:130–147
Turner BS (1989) From orientalism to global sociology. Sociology 23(4):629–638
UNESCO (2010) World social science report: knowledge divides. UNESCO and ISSC, Paris
Veracini L (2011) Introducing settler colonial studies. Settler Colon Stud 1(1):1–12
Xie L (2020) Post-western sociologies: What and why? Chinese Journal of Sociology. J Chin Sociol 8(5). https://doi.org/10.1186/s40711-020-00141-8
Wallerstein I (1997) Eurocentrism and its avatars, the dilemmas of social science. New Left Rev I 226:21–39
Wallerstein I et al (1996) Open social sciences. Report of the Gulbenkian commission on the restructuring of the social sciences. Stanford University Press, Stanford
Wolfe P (2006) Settler colonialism and the elimination of the native. J Genocide Res 8(4):387–409

Knowledge Boundaries and the History of Sociology

35

Per Wisselgren

Contents

Introduction	918
Five Knowledge Boundary Concepts	919
Boundary-Work	919
Boundary Object	920
Boundary Organization	921
Trading Zone	921
Co-production	922
Conflict Versus Collaboration	922
Four History of Sociology Episodes	924
Establishing Boundary Objects in Early Pre-academic Sociology	924
Demarcating the Disciplinary Boundaries of Classical Sociology	926
Reformulating the Science-Policy Contract During the Interwar Period	927
Re-establishing Modern Postwar Sociology Historiographically	930
Conclusion: The History of Sociology as the History of Its Knowledge Boundaries	932
References	932

Abstract

This chapter draws on and introduces some of the more influential theoretical knowledge boundary concepts that have been developed in the intersecting areas of the sociology of knowledge, science and technology studies, and the history of science over previous decades. It is argued that together these concepts provide a rich analytical toolbox for a deepened understanding of how knowledge boundaries have, in different ways, shaped and reshaped the history of sociology. In the first part of the chapter, five of these concepts – boundary-work, boundary object, boundary organization, trading zone, and co-production – are introduced and discussed, while the second part provides empirical examples of their application in the form of four short, chronologically ordered, episodes from the history of

P. Wisselgren (✉)
Department of History of Science and Ideas, Uppsala University, Uppsala, Sweden
e-mail: per.wisselgren@idehist.uu.se

sociology. Although the boundary concepts differ in important respects, they also complement each other. The first two episodes show how knowledge boundaries have determined not only separation and exclusion, but also communication and inclusion in the early history of sociology. The third episode shows how the concepts can be combined to highlight various aspects and analytical levels of one and the same phenomenon, whereas the fourth and final episode shows how the disciplinary history of sociology itself has been used as a powerful tool for exclusion. These insights call for a historiographically reflexive approach to the history of sociology and its boundaries.

Keywords

Knowledge boundaries · History of sociology · Boundary-work · Boundary object · Boundary organization · Trading zone · Co-production

Introduction

To demarcate, transgress, and negotiate boundaries is a deeply human, social, and cultural activity, something that we are constantly doing or confronting in our everyday lives in order to understand and act in the world. Bounding a practice is also a way of defining what it is, of excluding outsiders and including insiders, of telling practitioners what behavior is appropriate within it, and of distributing values across its borders (Shapin 1992: 335). It happens on the small scale of social interaction, on the medium scale in local cultures within organizations, and on the large scale of nationalism and geopolitics (Tilly 2004: 213). In that sense, boundaries have a strong impact on social hierarchies and people's access to resources and are closely related to more general sociological questions concerning identity, power, and knowledge, as exemplified in a series of research overviews by Michèle Lamont and colleagues (Lamont 2001; Lamont and Molnár 2002; Lamont et al. 2015).

Within this broader research on social and symbolic boundaries, a dynamic strand of special relevance to the history of sociology and the human sciences has been concerned with the creation and spanning of knowledge boundaries and their societal relevance. Theoretically situated at the intersection of sociology of knowledge, science and technology studies (STS), and history of science, this research strand has developed a number of more specific knowledge boundary concepts and built up what can be described as a rich analytical toolbox over recent decades. Nevertheless, these theoretical boundary concepts have not been used with any great frequency or applied systematically to the history of sociology.

This chapter argues that several of these concepts offer appropriate heuristic devices and a potential for bringing analytical clarity to a number of central and recurrent issues in the history of sociology. These include questions about, for example, disciplinary formation, sociology's relation to both neighboring disciplines and extra-academic knowledge areas (such as social work, social policy and extra-academic social research), the relationship between different subdisciplinary areas and their internal hierarchies, sociology's more general societal role, as well as historiographical issues about canon-making and who to include and exclude from the history of sociology.

Another argument is that the general key question addressed about what is to be, and historically has been, regarded as authorized knowledge or "proper" sociology, should be posed as an historical and empirical question rather than predefined by presentist standards. In that sense, the chapter advocates a processual understanding of how the boundaries of sociology have been historically shaped and reshaped, demarcated and transgressed, stabilized and contested. This will encourage a contextualized sociology of knowledge approach which places the margins themselves and what is played out in the peripheral borderlands, at the center of our attention.

The chapter is structured in two parts. The first introduces a handful of the most well-established theoretical knowledge boundary concepts and discusses some of their similarities and differences, as well as some of their analytical qualities and implicit assumptions. Often these concepts are kept distinct and used one by one. An argument developed in this chapter is, however, that it is possible to regard them to a larger extent as complementary. In the second part of the chapter, the boundary concepts are applied and discussed in relation to four empirical formative episodes from the history of sociology, where different forms of knowledge boundary issues were at stake. The chapter concludes with a short summary and some reflections on the implications of understanding and studying the history of sociology as the history of its knowledge boundaries.

Five Knowledge Boundary Concepts

The reason the intersecting areas of sociology of knowledge, STS, and the history of science have been especially productive in developing knowledge boundary concepts is not very strange given that questions about the social aspects of knowledge-making, the relation between science and society, and the cultural contexts of science have been at the center of their focus since their inception and especially since the 1970s (Camic et al. 2011; Gieryn 1995; Porter and Ross 2003). More elaborated theoretical boundary concepts were, however, being formulated from the 1980s onwards. And during recent decades, the number of boundary concepts has increased steadily. Here only a handful of the most influential will be briefly introduced and summarized.

Boundary-Work

One of the earliest of the elaborated boundary concepts was *boundary-work*. Originally formulated by sociologist Thomas F. Gieryn in his 1983 *American Sociological Review* article "Boundary-work and the demarcation of science from non-science: Strains and interests in professional ideologies of scientists," the concept was subsequently developed in a longer chapter in the 1995 *Handbook of*

Science and Technology Studies (Gieryn 1995) and elaborated in full book-length format under the title *Cultural Boundaries of Science* (Gieryn 1999). If read alongside each other, it is possible to identify a slight change in the theoretical framing over time, from a sociology of the professions perspective in the first text, to a constructivist STS standpoint in the second, and a cultural "cartographic" approach in the book. Despite this, the analytical core idea is relatively distinct in all three texts with its focus on the professional interests of scientists and their strategic, rhetorical demarcations of knowledge boundaries, and their aim to establish their epistemic authority "downstream," where the distinction between authorized and less authorized knowledges is to be seen as an historically contingent and locally situated activity. In that sense, knowledge boundaries are historically changing, contextually variable and often ambiguous, internally inconsistent and sometimes disputed (Gieryn 1983: 792; 1995: 439–441; 1999: 4–5). More specifically, Gieryn discerns three forms of boundary-work strategies: *expulsion* (or *monopolization*), where competing rivals are excluded from within by defining them as outsiders with labels such as "pseudo" or "amateur"; *expansion* into domains claimed by other professions or occupations; and *protection of autonomy* from intrusion of external powers (Gieryn 1983: 791–792; 1995: 424; 1999: 15–18). Although the main focus in the 1983 article was restricted to the demarcation of science from nonscientific forms of knowledge, it was suggested that the concept can also be applied to ideological demarcations of disciplines, specializations, or theoretical orientations within science (Gieryn 1983: 792). Boundary-work has proved to be an immensely useful concept in illuminating the social aspects of scientific knowledge by its successful application in a wide range of studies. One example is Michael S. Evans, who has applied it with a particular focus on the role of the public in the early history of American sociology in terms of *audience boundary-work* (Evans 2009). Another example is David Hess's study of new religious movements in *Science in the New Age* (1993), which includes outsiders' boundary-work, where boundary negotiations appear as a mutual process rather as defined primarily from within.

Boundary Object

Another influential knowledge boundary concept is *boundary object*, which was originally formulated in an article by the sociologist, information theorist, and STS scholar Susan Leigh Star and the philosopher James R. Griesemer in *Social Studies of Science* 1989. Star and Griesemer are also interested in the borderlands of scientific practice, but from an actor-network and interactionist social worlds perspective. Their focus is on a distinctly different question, namely how communication across social worlds is possible. More specifically, the article deals empirically with the case of the University of California Berkeley Museum of Vertebrate Zoology during its formative years in the early twentieth century and how the various groups of amateurs, professionals, administrators, and others managed to find common grounds for cooperation. It was, in essence, the boundary objects that made this possible, that is, objects which are "both plastic enough to adapt to local needs and the constraints of the several parties employing them, yet robust enough to maintain a common identity across sites" (Star and Griesemer 1989: 393). Boundary objects can be simultaneously concrete and abstract, specific and general, conventional and customized and are often internally heterogenous. In this specific case, the museum itself as a *repository* of ordered piles of objects, abstracted *ideal types* in the form of maps, diagrams and atlases, *coincident boundaries* between different particular territories, and *standardized forms* like field notes or specified systems of rules were able to perform the function of boundary objects on different levels and make cooperation across the different social worlds possible in the absence of consensus (ibid: 410–411). The boundary object concept has, like boundary-work, been widely recognized and adopted and developed by scholars in many fields (Star 2010: 604). One similar and partly parallel concept is Joan Fujimura's *standardized packages*, which is described as a "gray box which combines several boundary objects [...] in ways which further restrict and define each" and embraces both collective work across divergent social worlds and fact stabilization (Fujimura 1992: 169). Together with Geoff Bowker, Star has also developed the concept of *boundary infrastructures* as "larger infrastructures of classification [...] that cross larger levels of scale than boundary objects" (Bowker and Star 1999: 589).

Boundary Organization

A third concept is *boundary organization*. This is similar to boundary object, but its focus is on the meeting of different social worlds within organizational contexts and it has boundary object as one of its definitional criteria. Coined by the political scientist David H. Guston, "boundary organization" is defined as institutions that mediate and stabilize the boundary between science and politics, involve participation of actors from different social worlds, provide space for boundary objects or standardized packages that make collaboration across these worlds possible, and include delegations of authority and integrity between principals and agents (Guston 2000: 6; Guston 2001: 400–401). Empirically, the concept has its origin in Guston's book *Between Politics and Science* (2000) which studies the roles of organizations such as the Office of Research Integrity and the Office of Technology Transfer in the history of research policy in twentieth-century USA. This theoretical concept has proved widely applicable, especially in studies on research and environmental policy studies, and has been developed and discussed with regard to *international boundary organizations* and questions about, for example, the increased level of complexity, contingency, and contestedness on the global political level (Miller 2001: 480; Wisselgren 2017: 150–151, 176).

Trading Zone

A fourth concept, which is also centered on the communicative and collaborative aspects of science and across scientific subcultures and groups of actors, is *trading zone*. Developed by the historian of science Peter Galison in his book *Image and Logic* (1997), the trading zone concept applies a more material and spatial approach, not necessarily focused on organizations but including knowledge sites such as laboratories, seminars, cafés, lounges, airports, and their structures. Empirically focused on the development of modern microphysics and the specific question about how the different scientific subcultures involved, including theorists, experimenters, and instrument-makers, were "intercalated" to each other without being homogenized, the trading zone provides "a social, material, and intellectual mortar binding together the disunified traditions" (Galison 1997: 803). Inspired by anthropological linguistics on pidgin languages to understand the exchange and collaborations between cultural subgroups, Galison argues that it is not about translation but about coordinating action and belief, an intercalated set of subcultures bound together through a complex of pidgins and creole (Galison 1997: 838). Galison's concept has in its turn inspired Helga Nowotny, Peter Scott, and Michael Gibbons in their attempt to develop a more general social theory on the relation between science and society with a focus on *transaction spheres* where "exchanges take place across disciplinary and institutional boundaries between science and society at large" (Nowotny et al. 2001: 144–147)

Co-production

A fifth and final concept, which is already well-established within STS-inspired research and which can be said to summarize this brief overview, is Sheila Jasanoff's *co-production*. Formulated in *States of Knowledge* (2004), the notion of co-production elaborates, like transaction spheres, on the general relationship between science and society. Based on the observations that science has never been isolated from society and that scientific knowledge and social order are always co-produced – as shown already by Steven Shapin and Simon Schaffer (1985) in their classic study on the controversy between Robert Boyle and Thomas Hobbes in the seventeenth century – the notion highlights "that the ways in which we know and represent the world [...] are inseparable from the ways in which we choose to live in it" (Jasanoff 2004: 2). Even if co-production should not be seen as a fully-fledged theory, but rather an integrative and interdisciplinary framework, the concept has proved useful for exploring the mutual and continually changing relationship between knowledge and social order (ibid: 3, 43).

Conflict Versus Collaboration

Together these five knowledge boundary concepts – boundary-work, boundary object, boundary organization, trading zone, and co-production – provide a rich analytical toolbox for studying the role of scientific boundaries, the social and cultural aspects of science, and the relation between science and society. The concepts also complement each other in the sense that some are concerned with analyses on a more concrete and actor-oriented level (boundary-work and boundary object), whereas others are more oriented towards the middle-range level (boundary organization and trading zone), and a third group on the more general social theoretical level (co-production, but also transaction spheres).

Evidently this list of knowledge boundary concepts could easily have been extended. For example, yet another category of theoretical boundary concepts that might merit inclusion is those focused on the roles and functions of individual actors, in their varying roles and capacities in guarding, transgressing, or negotiating different forms of knowledge boundaries as "gate keepers," "marginal people," "knowledge brokers," "hybrid persons," "go-betweens," "co-agents," "translators," "boundary spanners," etc. But for the purpose of this chapter – to argue for the potential of applying knowledge boundary concepts to the history of sociology by referring to a few examples – the five chosen concepts should suffice.

However, in order to understand their more specific original research questions, it is also important to be aware of the original empirical contexts of the concepts – ranging from a natural history museum in the early twentieth century, the historical development of microphysics, and the formation of US research policy – as well as differences in the theoretical approaches which led to their development: including professionalization theory, symbolic interactionism, anthropological linguistics, and organization theory. It is, for example, worth noting that boundary-work, especially in its original conceptualization, was based on a conflict-theoretical point of departure, whereas several of the latter concepts are more concerned with questions about collaboration, and in that sense may be described as more consensus-oriented. The point of these remarks is that it is of analytical importance to be critically aware of these contexts of origin and connected assumptions when *de*-contextualizing the concepts in order to avoid taking these latter assumptions for granted. Although it is probably not especially controversial to regard scientists' actions as strategically motivated in today's competitive research systems, there is a risk of anachronistically projecting that kind of intentionality by interpreting the historical actors as overly strategic and conflict-oriented – as well as, of course, the opposite. From an historical point of view, the suggested way forward is instead to reformulate these assumptions about the conflict-oriented aspects versus the collaborative aspects as open empirical questions.

Another point, which follows from the above, is that when posed as an empirical question, it will be clear that these concepts are not necessarily in opposition, but often complementary. Boundaries and borders are not only, as Lamont and Molnár aptly argue, "sites for the division of people into separate spheres and opposing identities and groups, but [also] sites for interaction between individuals from many backgrounds by hybridization, creolization and negotiation" (Lamont and Molnár 2002: 184). This will be demonstrated in the next section.

Four History of Sociology Episodes

In order to illustrate the usability and relevance of applying knowledge boundary concepts, this second part will exemplify how different forms of boundary disputes, negotiations, and collaborations have been central in the history of sociology. More specifically four short, chronologically ordered episodes from the history of sociology in Sweden will be highlighted. The empirical focus on the Swedish case is primarily motivated by the author's more detailed empirical knowledge about the four specific episodes. Another reason is that the Swedish case is less well-known than the development of sociology in, for instance, the USA, Britain, France, or Germany. A third argument, finally, is that although the Swedish case in some important respects – like any national case – is local and exceptional, the four episodes do after all, with an analytical focus on the conceptual applications rather than the historical singularities, mirror more general history of sociology patterns.

The selection of the four historical episodes is motivated by what Peter Wagner and others in more general terms have identified as the two "waves" of the institutionalization of sociology, where the first wave refers to the "classical era" around the turn of the twentieth century and the second to the re-establishment of modern sociology in the early post-WWII era (Wagner 2001: 8; Ross 2003: 208). The two additional episodes highlighted here are concerned with the preacademic era and the interwar period. More concretely, the four episodes focus on: first, the early preacademic social reform-oriented era of the nineteenth century; second, the institutionalization and academization of sociology around the turn of the century; third, the interwar period; and fourth, the re-institutionalization of academic sociology after the Second World War. What unites the four episodes is that they can all be said to highlight important formative moments in the history of sociology where a central and recurrent theme concerned the shaping and reshaping of its knowledge boundaries.

Establishing Boundary Objects in Early Pre-academic Sociology

Sociology was first established as an academic subject in the higher education system in Sweden in 1903. This is the year Gustaf Steffen became professor of economics and sociology at Gothenburg University College. But the academization of sociology did not, of course, emerge from nothing. Before establishing the specific boundaries of the new discipline and regulating its relationship with neighboring knowledge fields, it was even more fundamental to identify a shared and common knowledge object, a boundary object to gather around and articulate a need for. And in the Swedish case, as in most other national cases, this need was first articulated outside the university among a heterogenous group of socially engaged individuals with varying backgrounds but common concerns, as will be shown in this first episode. Here it is instructive to follow Steffen's own trajectory and background.

Gustaf Steffen (1864–1929) was originally trained as a natural scientist, in chemistry, and worked for several years as a journalist, first in Germany and then in England. What made him change track and become interested in sociology was his acutely rising awareness of the social problems of modern society that gave rise to the extensive discussions of the "social question" in the late nineteenth century. In this way, he became involved in the contemporary, broad, and middle-class based social reform movement, which united diverse groups of socially committed individuals, including intellectuals, women philanthropists, journalists, politicians, priests, doctors, scientists, and other groups who either came in contact with the social misery of the time or took part in the social debate for other reasons. In Steffen's case, this interest brought him into contact with the socialist movement and led him to study economics in Germany and then to the Fabian Society in England with its ideas about knowledge-based social policy reform (Wisselgren 2015). And within this broad international social reform movement, the need for a new science of modern society was frequently articulated, as shown in several other history of sociology studies (Turner and Turner 1990; Bulmer et al. 1991; Heilbron 2015).

At this time, the term "sociology" had a relatively vague and general meaning. In the Swedish context, the word had been used sporadically from the 1870s and became more frequent only from around 1880 with the translation of Herbert Spencer's *The study of sociology*. A group that played an important role in introducing the kind of popular sociology represented by Spencer and other amateur sociologists in Sweden as well as in other parts of the world, often in the context of the social question, consisted of authors and public intellectuals. As Raewyn Connell (1997) has shown, this genre of sociological literature was often published and distributed through the same channels and read and discussed by the same audience who also read the works of socially engaged realistic and naturalistic authors like Balzac, Zola, Turgenev, Dickens, and Eliot. It is also against this background that we should understand the perhaps surprising fact that Gustaf Steffen chose to present his very first thoughts in 1896 about sociology in the literary journal *Vintergatan*. When Wolf Lepenies (1988) locates the rise of sociology within the intermediary area between literature and science, with examples from France, Germany, and Britain, this pattern applies to Sweden as well.

More specifically, the broad range of groups involved in the promotion of sociology in Sweden in the late nineteenth century – including authors, philanthropists, politicians, and scientists – as well as its close connection to the social question, is illustrated by the private Lorén Foundation from which Steffen received repeated financial support and which enabled him to qualify for a chair in sociology. The Foundation, which had an explicit and twofold aim to promote the social sciences and contribute to the solution of the social question, was led by a board of five directors and one deputy, comprising two authors, one physician, two economists, and one mathematician. And the recipients of grants and stipends similarly mirrored the broad range of competencies within the social reform movement as a whole, including authors, women philanthropists, politicians, journalists, and natural scientists (Wisselgren 2015).

Summarized in terms of boundary concepts, this first episode from the preacademic phase in the history of sociology is primarily characterized by how the social question and the social reform movement – as plastic but robust boundary objects – provided common ground with relatively open and permeable knowledge boundaries for the diverse socially engaged groups in the early formation of sociology as a knowledge area.

Demarcating the Disciplinary Boundaries of Classical Sociology

Once Gustaf Steffen was formally appointed as Sweden's pioneering sociologist in 1903, his twofold task was to establish the epistemic authority of the new knowledge area by defining it in relation to neighboring disciplines and by proving its broader social and public relevance. As will be shown in this section, the success of Steffen's boundary-work was, however, ambiguous.

On the one hand, Steffen established himself as an authority and expert as Sweden's only professional sociologist for almost three decades. He wrote extensively and published a series of introductory and popular booklets on sociology as an academic subject in 1905–1912 and wrote entries on sociological topics for the most important domestic encyclopedia (*Nordisk familjebok*) as well as a number of books, including a theoretically grounded magnum opus in four volumes entitled *Sociologi: En allmän samhällslära* (*Sociology: A general theory of society*, 1910–1911). He participated as the sociological expert in major royal commissions together with other social science experts, corresponded with other classical sociologists such as Georg Simmel, and was generally acknowledged by the international sociological community. He also had a number of prominent students and was a well-known name to the general public debate. In that sense, Steffen established the epistemic authority of sociology as a new academic knowledge area, carving for it an academic and professional niche within the broader field of social knowledge and the new social sciences within the contemporary higher education landscape. In that sense, a boundary-work process of academization, specialization, and professionalization took place in which several of the "amateur" actors and groups, who had been actively involved in the preceding preacademic phase (such as authors, philanthropists, politicians but also natural scientists), were marginalized. In similar ways, the previously broad and vital public links to the social reform movement were partly disconnected (Wisselgren 2013; Evans 2009).

On the other hand, the boundary-work to be pursued was a challenging and ambiguous one, since Steffen's chair was divided between two subjects: economics and sociology. This challenge was not unique for Steffen, but shared by several of the classical generation of sociologists to which Steffen belonged. Born in the same year as Max Weber, Steffen corresponded with Georg Simmel, who was six years older and the same age as Émile Durkheim. Weber's first chair was in commercial law and his second in economics. Not until 1918 was Weber appointed to a chair in sociology, first in Vienna, and the year after in Munich, before his death in 1920. Simmel did not become professor until 1914 – in philosophy. And Durkheim's chair at the Sorbonne was shared between the subjects pedagogics and sociology. In Steffen's case, however, the challenge of representing economics and at the same time establishing sociology in the Swedish higher education system was burdensome and caused great problems. One reason was that economics was already established at the three other Swedish higher education institutions of the day, and since Gothenburg University College was an institutional newcomer, it had to adjust its formal graduation requirements to the standards laid down at the traditional universities. In practice this meant that Steffen, although he identified himself primarily as a sociologist, had to devote the major part of his teaching to economics, while the sociology courses were not introduced until the third cycle of studies (Wisselgren 1997). Wagner and Wittrock have in more general terms remarked upon the importance of the institutional setting and educational and professional practices for the cognitive development of any discipline, whereas Shapin has observed that "[p]ractices that have not succeeded in making their boundary-discourse stick are unlikely to be recognizable as distinct entities within the general stream of cultural life" (Wagner and Wittrock 1991: 332–341; Shapin 1992: 335). Although Steffen's courses in sociology were appreciated, he managed neither to secure the continuation of his sociological project by creating a more permanent school nor to acquire any disciples. It was partly for these reasons that sociology disappeared as a subject from the Swedish universities on his death in 1929 (Larsson and Wisselgren 2006).

Reformulating the Science-Policy Contract During the Interwar Period

After Steffen's death, it was to take 18 years before sociology was re-established as an academic subject in 1947. And since it has been recognized that Steffen's sociological production more or less ceased after 1918, the entire interwar period may appear as relatively "un-sociological." At the same time, it is generally acknowledged, especially among Swedish welfare state historians, that the interwar era stands out as foundational with regard to the early conceptualization of the "Swedish model" and its ideas about social engineering, including the notion that social reforms ought to be based on rigorous social research, where it is more or less taken-for-granted that Swedish social science was well-established and vital. This assumption is however problematized by the actual status of contemporary academic social science, which was generally restricted in terms of numbers of posts and its relatively low-intensive character compared to the situation after the Second World War – the important exception being economics, which prospered with the foundation of the Stockholm School of Economics and was both vital and dynamic.

To understand this interwar paradox – with the opposing images of the period as either dynamic or low-intensive with regard to the social sciences – we need to apply a broader and contextualized sociology of knowledge approach which takes into account the role of contemporary, extra-academic, policy-relevant, social research. In this research, the sociological perspective remained central, and it was an arena of important boundary-work processes between social science and social policy that had implications for the development of sociological and social scientific discourse. By doing so we will see, in this third episode, that social scientific knowledge was indeed of crucial importance during the interwar period, but also that these activities were primarily located outside, rather than within, academia. This will also give us reason to highlight the roles of Gunnar and Alva Myrdal as two of the strongest advocates of the sociological perspective during the interwar period and especially their thoughts about the boundary between science and politics.

There is much that could be – and has been – said about Alva and Gunnar Myrdal (for two examples emphasizing their sociological and social scientific contributions, see Jackson 1990; Ekerwald 2001). In this context, it should be enough as a general background to observe that they were both not only politicians but also trained as social scientists. Gunnar received his PhD in Economics from Stockholm University in 1927 and Alva had started a PhD project in psychology at Uppsala University when the couple received a Rockefeller stipend to visit the USA in 1929 – the same year as Steffen died. Their meeting with the contemporary dynamic, interdisciplinary, applied, and reform-oriented American social science and collaborative couples like Dorothy S. and W.I. Thomas had a decisive effect on their future careers. In a private letter, Alva explained this new orientation with great enthusiasm: "an economist and a social psychologist united in marriage and authorship makes naturally and easily a sociologist" (quoted from Lyon 2001: 220). From this moment on, a recurrent and central theme that emerged as a thread throughout both Gunnar and Alva Myrdal's writings would be the relationship between science and politics.

Here it is especially instructive to take a closer look at the book Gunnar wrote while in the USA in 1929, published in 1930 as *Vetenskap och politik i nationalekonomien* (translated into English under the title *The Political Element in the Development of Economic Theory* in 1953). Although the book was primarily concerned with economic theory, the basic ideas applied to social research more generally. The explicit aim of the book was to "explore the very boundary itself between politics and the science of economics" (Myrdal 1930: 10; own translation). Offered here is almost a textbook example of not only Gieryn's boundary-work in general but also the three more specific types of boundary-work strategies identified by Gieryn: *expulsion, expansion,* and *protection of autonomy* (Gieryn 1999: 15–18).

With regards the first, Myrdal's critique targeted the predominant contemporary neoclassical economics, which he regarded as value-laden and biased in promoting tacit ideological assumptions in the name of science. The task here, according to Myrdal, was to "demarcate the boundary between 'is' and 'ought'" in order to avoid confusion of "scientific facts with political ideas, theories and ideologies," or as he summarized it: "What is required is a methodological and consistently implemented theoretical boundary-work" (Myrdal 1930: 11–12; own translation). Essentially, the aim was to defend the scientific credibility of economics and social science.

At the same time, however, Myrdal was very keen to argue for the social and political relevance of economic and social research. It was not about restricting the social scientists' sphere of action, but on the contrary to *expand* it, to clarify not only the boundary between science and politics but also "the connections between the two realms" (ibid: 10; own translation). "This boundary-work is needed," Myrdal explained, "in order to really rationalize the political decisions" (ibid: 12; own translation). Hence, the solution was *not* that economists and social scientists should try to eliminate social and ideological elements from their work. On the contrary, Myrdal explained, they ought to explicate the value premises in order to make their results more transparent. In that way, policy-relevant social research which was both objective and practical would be possible, according to Myrdal, "because it never recommends anything except from an explicated and clarified interest and standpoint" (ibid: 285; own translation). In that sense, Myrdal's theory not only demarcated the boundary between science and politics, but also argued in favor of science-based political reforms.

The third and final part of Myrdal's boundary-work emphasized that the connection between science and politics should be one-directional. It was social science that should enrich political decision-making, not the other way around. Politicians on the other hand should stay away from social research, as he clarified in another text, where he emphasized the importance of "firmly rejecting every threat to our academic autonomy and especially of keeping political and other special interests outside academic administration" (Myrdal 1945: 69–70; own translation).

Consequently, Myrdal's theoretical boundary-work had a threefold aim. He wanted to ensure the scientific credibility of economic research by criticizing ideologically biased neoclassical theory, motivate its political relevance by expanding the sphere of social scientists into the political sphere, and protect the autonomy of science from any political intervention – all at the same time. In that sense, the passage between science and politics promoted by Myrdal can be metaphorically summarized in the form of a "cat flap," through which the social scientists could move relatively freely in both directions, whereas the politicians were strictly restricted to their own sphere (Wisselgren 2008a).

The practical implication of this influential discursive innovation, which reformulated the contract between science and politics, was that it theoretically legitimized the active participation and involvement of social scientists in policy-relevant contexts outside academia. Contextually, the importance of this innovation was amplified by its actual timing, with the Social Democratic Worker's Party entering upon its four decades of political power in 1932. Consequently, when Myrdal was invited to assist the new Social Democratic government with their economic policy in 1933, this was not regarded as a problem from Myrdal's point of view but indeed the opposite: an opportunity to turn his theoretical ideas into practical action. In a similar vein, the doors to the many royal commissions that were initiated in the 1930s and 1940s were left wide open for the inclusion of social scientists as experts, an opportunity that soon became an established and expected practice, laying the foundation for a strong domestic tradition of policy-relevant social research as well as science-based political reforms (Wisselgren 2008b; Fridjonsdottir 1991).

In this latter sense, we can see how the interwar episode not only allows us to analyze Gunnar Myrdal's theoretical ideas in terms of an unusually clear example of Gieryn's boundary-work strategies, but at the same time to apply other complementary boundary concepts. The very core idea in Myrdal's book can hence be described as a plastic and robust *boundary object* which made it possible for actors from different social worlds (social scientists and politicians) to establish common grounds for agreement, or as a new social contract regulating and stabilizing the relationship between science and politics in Guston's sense. Moreover, the royal commission can be interpreted either as a *boundary organization* (Guston), as a *trading zone* for the exchange of different forms of social scientific and political knowledge (Galison), or as a *transaction sphere* at the more general societal level (Nowotny et al.) where social scientific knowledge and social order were *co-produced* (Jasanoff). Whichever concept is preferred, the case offers a concrete example of Shapin and Schaffer's point "that the history of science occupies the same terrain as the history of politics" (Shapin and Schaffer 1985: 332).

Re-establishing Modern Postwar Sociology Historiographically

When sociology was re-established as an academic subject in Sweden after the Second World War, the disciplinary institutionalization process turned out to be much more successful. In 1947, Torgny T. Segerstedt (1908–1999) was appointed to the new chair in sociology at Uppsala University. In the following years, sociology was established at all other Swedish universities: Lund University in 1948 (chair in 1956), Stockholm University College in 1949 (chair in 1954), Gothenburg University in 1959, and Umeå University in 1965. The new discipline was consolidated with the foundation of the Scandinavian journal *Acta Sociologica* in 1955, the Nordic Sociological Association in 1960, the Swedish Sociological Association in 1961, and the domestic journal *Sociologisk forskning* in 1964 (Larsson and Magdalenić 2015). Although the disciplinary formation of sociology was part of a more general postwar expansion of the social sciences in Sweden and mirrored more general international trends during the second wave of institutionalization of modern sociology (Dalberg et al. 2019), international observers have commented upon both the rapidity of this development and the distinct profile of the new postwar sociology in Sweden as positivistic, quantitative, methodologically driven, and pragmatic with clear connections to social policy legislation (Reiss Jr. 1968).

The boundary-work pursued in designing the relatively homogenous profile of the new American-inspired, contemporary-oriented, specialized social scientific discipline under the self-consciously labeled "Uppsala School of Sociology" has been analyzed in more detail by Anna Larsson (Larsson 2001: 179–183; Larsson and Wisselgren 2006: 159; Larsson and Magdalenić 2015: 23–33). In this final episode, we will pay special attention to how internal disciplinary history filled an especially important role and function in the boundary-work of the new postwar sociologists. This will give us reason to reflect upon the more general function of disciplinary history and provide us with yet another pointer to enable us to understand why Gustaf Steffen has not been recognized in the traditional domestic narratives. One striking aspect is namely the way that postwar sociologists actively ignored the pioneering role of Steffen. On the contrary, when Torgny Segerstedt was asked to give his own personal retrospective account of the history of Swedish sociology, he introduced his chapter with the words: "It was in 1947 that the first independent professorship in sociology was established in this country. I became its first holder" (quoted from Larsson and Wisselgren 2006: 159). Later in the same text, Steffen is mentioned briefly but then in unfavorable terms as an outdated and speculative thinker influenced by French intuition philosophy. This ignorance of Steffen is revealing.

In 1945, in the crucial preparatory stages before establishment of the chair in Uppsala was being proposed by the Social Science Research Committee (*Socialvetenskapliga forskningskommittén*), which had been set up by the government in 1945 to investigate the state of Swedish social science and of which Segerstedt was a member, the deprecation of Steffen was similar. From his influential hybrid position, Segerstedt was offered an extraordinary opportunity to affect the research policy for the social sciences in a direction that favored him (Larsson 2001: 96–100; Larsson and Wisselgren 2006: 169). When the research committee presented its proposal to establish the chair in Uppsala, it was emphasized – in the sections written by Segerstedt himself – that it was a very specific type of sociology that was asked for. A rhetorical boundary-work manoeuver in this context was to distinguish between two types of sociology, characterized on the one hand as "the speculative form of sociology" represented by French and German scholars, and on the other, "a markedly empirical form," which uses quantitative methods and according to Segerstedt had "become most developed in the United States." The underlying normative direction implied in this seemingly harmless, dichotomized description, was then explicitly spelled out and underlined:

> It has to be strongly emphasized, what Swedish social scientific research needs is an empirical sociology, aimed at field surveys of modern society [...]. What is at stake is the avoidance of a speculative sociology, which on the basis of scarcely examined facts constructs far-reaching and fragile theses. (SOU 1946: 81; quoted from Larsson and Wisselgren 2006: 168)

As several historians and sociologists of science as well as more specific reception studies in the history of sociology have observed, disciplinary history often serves as a rhetorical vehicle in the bounding of collective identities (Lepenies and Weingart 1983; Platt 1996; Connell 1997). Usually, this normative function is enacted by extracting historical lines of ancestry where the present discipline takes a share of the accumulated prestigious heritage. What is remarkable in the case of the re-established Swedish postwar sociology is, however, the rather one-sidedly negative use of history. In the strategic bounding of a disciplinary identity, the pioneering role of Steffen was conceived as an historical encumbrance rather than a resource. The result was an effective, dichotomized foundation story which was rhetorically addressed to the lay public, political authorities, sponsors, and not least of all to the practicing sociologists. In this way, the discipline's history not only defined the collective memory of the past, but also glorified Segerstedt's own pioneering role and, by way of socialization and exclusion, structured the future. This narrative suited the cultural, political, and scientific setting in the years after the Second World War, in which the first generation of postwar sociologists contributed to the co-production of empirical studies and social policy proposals in the service of the emerging welfare state (Larsson and Wisselgren 2006: 173; Fridjonsdottir 1991: 254).

Conclusion: The History of Sociology as the History of Its Knowledge Boundaries

The disciplinary history of sociology can be written and narrated in many different ways and with varying objectives and readers in mind. This chapter has advocated an historical and sociological approach which makes use of the rich analytical toolbox of theoretical knowledge boundary concepts that have been developed in recent decades within the intersecting fields of the sociology of knowledge, science and technology studies, and the history of science.

Although the five boundary concepts introduced in the first part of the chapter – boundary-work, boundary object, boundary organization, trading zone, and co-production – differ in important respects with regard to their theoretical and empirical contexts of origin, their conflict- versus collaborative-oriented approaches, and their analytical levels, a general argument has been that they also complement each other. In the second part of the chapter, it was shown by applying the boundary concept empirically in the form of four short, chronologically ordered, episodes, how knowledge boundaries in different ways have shaped and reshaped the history of sociology in Sweden.

The first two episodes exemplified how knowledge boundaries have been conditions not only for separation and exclusion, but also for communication and inclusion during different phases of the early history of sociology. The third episode illustrated how the concepts can be combined analytically to highlight different aspects and levels of one and the same phenomenon, whereas the fourth and final episode showed how the internal disciplinary history of sociology in itself has been used as a powerful tool of exclusion in the efforts to re-establish a disciplinary identity strategically adjusted to the current postwar setting.

The disciplinary history of sociology is consequently not only an innocent attempt to commemorate what actually happened in the past. Like boundary-making in the broader social and symbolic sense, it can also be an activity that is closely related to more general questions about knowledge, identity, and power. A history of sociology that places the margins at the center and explores the historical shaping and reshaping of its boundaries is in no way a guarantee for avoidance of the most common historiographical problems. But it can be useful for developing a more reflexive understanding of the ways in which knowledge boundaries have been drawn and transgressed in the past as well as in the present.

References

Bowker GC, Star SL (1999) Sorting things out: classification and its consequences. MIT Press, Cambridge, MA

Bulmer M, Bales K, Sklar KK (eds) (1991) The social survey in historical perspective: 1880–1940. Cambridge University Press, Cambridge

Camic C, Gross N, Lamont M (2011) Introduction: the study of social knowledge making. In: Social knowledge in the making. University of Chicago Press, Chicago

Connell RW (1997) Why is classical theory classical? Am J Sociol 102(6):1511–1557

Dalberg T, Börjesson M, Broady D (2019) A reversed order: expansion and differentiation of social sciences and humanities in Sweden 1945–2015. In: Fleck C, Duller M, Karády V (eds) Shaping human science disciplines: institutional developments in Europe and beyond. Palgrave Macmillan, Cham, pp 247–287

Ekerwald H (2001) Alva Myrdal: making the private public. Acta Sociologica 43(4):343–352

Evans MS (2009) Defining the public, defining sociology: hybrid science-public relations and boundary-work in early American sociology. Public Underst Sci 18(1):5–22

Fridjonsdottir K (1991) Social science and the "Swedish model": sociology at the service of the welfare state. In: Wagner P, Wittrock B, Whitley R (eds) Discourses on society: the shaping of the social science disciplines. Kluwer, Dordrecht, pp 247–270

Fujimura JH (1992) Crafting science: standardized packages, boundary objects, and "translation". In: Pickering A (ed) Science as practice and culture. University of Chicago Press, Chicago, pp 168–211

Galison P (1997) Image and logic: a material culture of microphysics. University of Chicago Press, Chicago

Gieryn TF (1983) Boundary-work and the demarcation of science from non-science: strains and interests in professional ideologies of scientists. Am Sociol Rev 48(6):781–795

Gieryn TF (1995) Boundaries of science. In: Jasanoff S (ed) Handbook of science and technology studies. Sage, Thousand Oaks, pp 393–443

Gieryn TF (1999) Cultural boundaries of science: credibility on the line. University of Chicago Press, Chicago

Guston DH (2000) Between politics and science: assuring the integrity and productivity of research. Cambridge University Press, New York

Guston DH (2001) Boundary organizations in environmental policy and science: an introduction. Sci Technol Hum Values 26(4):399–408

Heilbron J (2015) French sociology. Cornell University Press, Ithaca

Hess DJ (1993) Science in the New Age: the paranormal, its defenders and debunkers, and American culture. University of Wisconsin Press, Madison

Jackson WA (1990) Gunnar Myrdal and America's conscience: social engineering and racial liberalism, 1938–1987. University of North Carolina Press, Chapel Hill

Jasanoff S (ed) (2004) States of knowledge: the co-production of science and the social order. Routledge, London

Lamont M (2001) Symbolic boundaries: overview. In: Smelser NJ, Baltes PP (eds) International encyclopedia of the social and behavioral sciences, vol 23. Elsevier, Amsterdam, pp 15341–15347

Lamont M, Molnár V (2002) The study of boundaries in the social sciences. Annu Rev Sociol 28:167–195

Lamont M, Pendergrass S, Pachucki M (2015) Symbolic boundaries. In: International encyclopedia of the social & behavioral sciences, vol 23, 2nd edn, pp 850–855

Larsson A (2001) Det moderna samhällets vetenskap: Om etableringen av sociologi i Sverige 1930–1955. Umeå universitet, Umeå

Larsson A, Magdalenić S (2015) Sociology in Sweden: a history. Palgrave Macmillan, Basingstoke

Larsson A, Wisselgren P (2006) The historiography of Swedish sociology and the bounding of disciplinary identity. J Hist Behav Sci 42(2):159–176

Lepenies W (1988) Between literature and science: the rise of sociology. Cambridge University Press, Cambridge

Lepenies W, Weingart P (1983) Introduction. In: Graham L, Lepenies W, Weingart P (eds) Functions and uses of disciplinary histories. Reidel, Dordrecht, pp 9–20

Lyon ES (2001) The Myrdals and the Thomases 1930–1940: the trials and tribulations of a cross-Atlantic research collaboration. In: Mucha J, Kaesler D, Winclawski W (eds) Mirrors and windows: essays in the history of sociology. Nicholas Copernicus University Press, pp 219–234

Miller C (2001) Hybrid management: boundary organizations, science policy, and environmental governance in the climate regime. Sci Technol Hum Values 26(4):478–500
Myrdal G (1930) Vetenskap och politik i nationalekonomien. Norstedt, Stockholm
Myrdal G (1945) Universitetsreform. Tiden, Stockholm
Nowotny H, Scott P, Gibbons M (2001) Re-thinking science: knowledge and the public in an age of uncertainty. Polity Press, Cambridge
Platt J (1996) A history of sociological research methods in America, 1920–1960. Cambridge University Press, Cambridge
Porter TM, Ross D (2003) Introduction: writing the history of the social science. In: Porter TM, Ross D (eds) The Cambridge history of science. Vol 7: The modern social sciences. Cambridge University Press, Cambridge, pp 1–10
Reiss AJ Jr (1968) The field. In: Sills DL (ed) International encyclopedia of the social sciences. Free Press, New York, pp 1–23
Ross D (2003) Changing contours of the social science disciplines. In: Porter TM, Ross D (eds) The Cambridge history of science. Vol. 7: The modern social sciences. Cambridge University Press, Cambridge, pp 205–237
Shapin S (1992) Discipline and bounding: the history and sociology of science as seen through the externalism-internalism debate. Hist Sci 30(4):333–369
Shapin S, Schaffer S (1985) Leviathan and the air-pump: Hobbes, Boyle, and the experimental life. Princeton University Press, Princeton
Star SL (2010) This is not a boundary object: reflections on the origin of a concept. Sci Technol Hum Values 35(5):601–617
Star SL, Griesemer JR (1989) Institutional ecology, "translations" and boundary objects: amateurs and professionals in Berkeley's Museum of vertebrate zoology, 1907–39. Soc Stud Sci 19(3): 387–420
Tilly C (2004) Social boundary mechanisms. Philos Soc Sci 34(2):211–236
Turner SP, Turner JH (1990) The impossible science: an institutional analysis of American sociology. Sage, Newbury Park
Wagner P (2001) A history and theory of the social sciences: not all that is solid melts into air. Sage, London
Wagner P, Wittrock B (1991) States, institutions and discourses. In: Wagner P, Wittrock B, Whitley R (eds) Discourses on society: the shaping of the social science disciplines. Kluwer, Dordrecht, pp 331–357
Wisselgren P (1997) Sociologin som inte blev av: Gustaf Steffen och tidig svensk socialvetenskap. Sociologisk Forskning 34(1–2):75–116
Wisselgren P (2008a) Vetenskap och/eller politik?: Om gränsteorier och utredningsväsendets vetenskapshistoria. In: Sundin B, Görandsdotter M (eds) Mångsysslare och gränsöverskridare: 13 uppsatser i idéhistoria. Historiska studier, Umeå, pp 103–119
Wisselgren P (2008b) Reforming the science-policy boundary: the Myrdals and the Swedish tradition of Governmental Commissions. In: Eliaeson S, Kalleberg R (eds) Academics as public intellectuals. Cambridge Scholars Publishing, Newcastle, pp 173–195
Wisselgren P (2013) "Not too many ladies, but too few gentlemen": on the gendered co-production of social science and its publics. In: Danell R, Larsson A, Wisselgren P (eds) Social science in context: historical, sociological, and global perspectives. Nordic Academic Press, Lund, pp 33–47
Wisselgren P (2015) The social scientific gaze: the social question and the rise of academic social science in Sweden. Routledge, London/New York
Wisselgren P (2017) From utopian one-worldism to geopolitical intergovernmentalism: UNESCO's Department of social sciences as an international boundary organization, 1946–1955. Serendipities: J Sociol Hist Soc Sci 2(2):148–182

Social Theory and the History of Sociology 36

Hon-Fai Chen

Contents

Introduction	936
Thought and Schools	937
Science and Tradition	938
History and Present	943
Rethinking the Canon	947
Decentering the West	952
Conclusion	957
References	957

Abstract

Instead of a mere chronological account of sociological ideas with biographical notes and ad hoc commentaries, the history of social theory aims to delineate how theory and theorizing have been profoundly shaped by changing historical contexts and social structures. Classical theorists such as Marx, Weber, and Durkheim are best understood as products of their times, while the definition of sociological classics is open to contestation and redefinition. Five major themes in the historical studies of sociological theory are discussed in this chapter, including thought and schools, science and tradition, history and the present, rethinking the canon, and decentering the West. In early studies, social theory was conceived as part of social thought, which consisted of a remarkable variety of theoretical schools and civilizational sources. Later on, a scientific approach to modern social theory and its development rose to prominence, against which the notion of a sociological tradition was set forth as a counterpoint. On the other hand, intellectual historians insisted on the importance of contextualizing social theory, and objected to the projection of contemporary theoretical debates onto classical works. In more recent studies, the idea of canon and its exclusivity has been

H.-F. Chen (✉)
Lingnan University, New Territories, Hong Kong
e-mail: honfaichen@ln.edu.hk

questioned. Marginalized and neglected theorists, including in particular women and black thinkers, were reappraised and reincorporated into the sociological canon. The Eurocentric bias of social theory, as a manifestation of the imperial origin and episteme of sociology, was also criticized. Notable attempts were made to utilize postcolonial thoughts for revisiting and reinventing social theory.

Keywords

Canon · Classical theory · Eurocentrism · History of social theory · Modernity · Scientism · Social thought · Sociological tradition

Introduction

This chapter will introduce some of the major works standing at the intersection of two distinct but closely related fields, namely, social theory and the history of sociology. An easy and straightforward way to delineate the intersection and delimit the scope of discussion would be to focus on the "history of sociological theory." This treatment is legitimate but deceptively simple, for it neglects the inherent tension between theoretical construction and historical interpretation. There is a kernel of truth in the common understanding that while theory consists of concepts and generalizations, history concerns primarily contexts and interpretations. As will be demonstrated, such a tension is not irreconcilable, but it does give rise to divergent approaches to the writing and rewriting of the history of sociological theory. Alongside this tension is the problem of defining "theory": should one include scientific propositions, moral thoughts, literary expressions, and/or philosophical traditions? Should one focus exclusively on the "great social thinkers" in Europe and North America, or cover those outside the West and others excluded from the sociological canon?

Notwithstanding these issues, the history of sociological theory could be taken to encompass all interpretive accounts that seek to situate social theorists and their works in the historical, institutional, and intellectual development of the sociological discipline. In this definition, the history of sociological theory is a subfield of the history of sociology, as the latter also covers the history of sociological research and institutions, the changing relationship of sociology to the other social sciences, and the sociology of idea and knowledge. Nevertheless, this definition could serve to exclude pure works in theoretical construction (see for example Stinchcombe 1968), textual exposition (Bierstadt 1981), or the analytical treatment of social theorists (Elster 1985).

Instead of examining the validity of specific historical accounts, the purpose of this chapter is to discern the multiple and changing ways in which the history of social theory and sociology has been written and rewritten. In other words, it aims to write a "history of histories" and on that basis identify the successive themes and major issues in the historical representations of sociological theory. Specifically, the following discussion will be structured around five key themes: (i) thought and schools; (ii) science and tradition; (iii) history and present; (iv) rethinking the canon; and (v) decentering the West. Instead of offering an exhaustive treatment of the subject, this chapter will focus on classical social theory, as it stands directly at the intersection of the theoretical and the historical.

Thought and Schools

A common tendency in early works of the history of sociology was to assimilate social theory to the ecumenical, time-honored corpus of social thought. This approach was epitomized by Howard P. Becker and Harry E. Barnes's (1938) *Social Thought from Lore to Science*, a three-volume treatise displaying an encyclopedic knowledge not only in modern sociological theory, but also social thought stemming from diverse historical and civilizational settings. For the authors, "social thought" covered not only sociology and social science, but also proverbial folklore and social philosophy insofar as these sought to address value-related questions such as "What is a good life? Why is it a good social life? Why is it good?" "How can it be safeguarded or attained?" (Becker and Barnes 1938: xxi). Judged by present-day standards, a remarkable feature of the treatise was its attempt to shed the positivistic and Eurocentric bias by situating pre-modern, non-Western social thought in its cultural and sociological context. For example, the entirety of the first volume was devoted to the social thought of preliterate peoples and ancient civilizations.

Nevertheless, the epistemological framework of the treatise was imbued with modernist assumptions. The authors introduced the ideal types of "sacred" and "secular" to contrast the isolated character of premodern and non-Western societies with the mobility and accessibility of modern Western society. These contrasts were reflected in the difference between modern sociology and its intellectual predecessors. However, this ideal-typical construction did not merely serve to celebrate European superiority. In the discussion of medieval social thought, a prime significance was assigned not to the Renaissance and Reformation, but rather to Europe's cultural contact with Islamic civilization, out of which the early modern theories of the nation-state and secular social thought were born. The immense contributions of Ibn Khaldun, an Islamic social thinker of the time, to the studies of history, geography, economy, the city, and the state were singled out for special attention (Becker and Barnes 1938: 266–79). While the development of modern sociology culminated in Comte, Marx and Spencer, the authors criticized the organic analogy and Social Darwinism in their works. A final point to note was the authors' worldwide survey of modern sociology, not only in the "social science centers" of Britain, France, Germany, Italy, and the United States, but also in Russia, Eastern Europe, the Balkans, Turkey, Spain, Portugal, Latin America, India, China, and Japan. National differences were articulated, and a great variety of sociological thinkers and "schools" in each of the national traditions was introduced.

The assimilation of social theory to social thought can be found in other early works. Emory Bogardus' *A History of Social Thought* was published in 1922 and saw four editions between 1940 and 1960. Interestingly, this introductory text was intended to shed light on contemporary social problems. Underlying the continuity of social thought from past to present was the human aspiration to progress, specifically the efforts to promote the welfare of fellow social members (Bogardus 1922/1928: 16–8). On top of abstract thinking, a more important function of social thought and social theory was to inculcate moral values, such as tolerance (ibid.: 649). Noteworthy here was the inclusion of two new chapters in the 1960 edition, which covered Howard Odum as the first sociologist of the American South, and Radhakamal Mukerjee as an Indian sociologist. The reason for this inclusion lay in the "cycle in the history of social thought," which flowed from the ancient East via Europe and America and back to the Orient. While in the 1928 edition only 18 pages were devoted to the social thought of China, India, and other ancient civilizations, it was later expanded to contain 54 pages and four separate chapters.

The common practice of mapping diverse "schools" of social thought and sociological theory was epitomized in *Contemporary Sociological Theories*, Pitrim Sorokin's monumental text in 1928. Sorokin aimed to distinguish the principal types or schools of sociological theory from the 1850s, and to test their propositions with quantitative evidence. A total of nine major schools were identified, each with its fundamental assumptions, branches, predecessors, and representatives. Sorokin's approach was deliberately anti-biographical, as the scientific worth of sociological theories was held to be independent of their authors. The treatment was also critical, as sociological theories were overcrowded with the "sterile flowers" of speculative theories and the "weeds" of prescriptive theories. Despite the existence of contradictory theoretical systems, there emerged a shared definition of sociology as the scientific study of the general characteristics of all classes of social phenomena, and their relationship with non-social phenomena (Sorokin 1928: 760–1). The number of sociological schools was trimmed in the companion volume *Sociological Theories of Today* (Sorokin 1966), as the early emphasis on geographical, biological, and psychological aspects of social phenomena gave way to integral theories of the social and cultural system. Sorokin's typological approach was later adopted by Don Martindale in *The Nature and Types of Sociological Theory*, with a more systematic classification of theories along various dimensions (Martindale 1960/1981: 622).

Insightful as it was, the end was sealed for the history of social thought when Barnes, a coauthor of *Social Thought from Lore to Science*, proclaimed that a major change had taken place in sociological theorizing in the 1940s. While the theoretical systems in early sociology focused on thinkers and schools, it was displaced by more specialized works (Barnes 1948). Since that time, the scientific status of sociology has replaced the moral relevance of social thought as the prominent theme in most historical accounts of the discipline.

Science and Tradition

The reorientation in the studies of sociological theory and its history can be traced back to 1937, when Talcott Parsons published *The Structure of Social Action* (hereafter SOSA). For admirers and critics alike, Parsons' *magnum opus* could be rightly regarded as a watershed in the history of modern social theory. From a disciplinary standpoint, the major contribution of SOSA was to delineate the "action frame of reference" as the analytical foundation of scientific sociology. But the SOSA could also be read as a peculiar interpretation of modern European intellectual history. This was evident in its widely quoted opening sentence, "Who now reads Spencer?" By this, Parsons intended to indicate the general movement of late nineteenth-century European social thought away from the utilitarianism, positivism, and evolutionism represented by Herbert Spencer. In Parsons' reading, modern social thinkers were engaged in a critical dialogue with utilitarianism and its problematic implications. Central to utilitarianism was the selection of means by individual actors in pursuit of subjective ends. While this "means-end schema" recognized the voluntaristic element of choice, it failed to address the problem of order that was pointedly formulated by Thomas Hobbes. To avoid the "war of all against all," most social thinkers were led to reduce the selection of means and ends either to determination by biological factors or the emanation of normative values. According to Parsons, such a theoretical dilemma found its partial solution in four European social thinkers, namely, Alfred Marshall, Vilfredo Pareto, Emile Durkheim, and Max Weber. Despite the diversity of biographical, historical, and intellectual backgrounds, there was a *convergence* among these thinkers toward a relational, systemic understanding of the actor, means, ends, conditions, and normative orientation as the irreducible elements of action or the "unit act." By setting forth the convergence thesis as a historical interpretation of modern social thought, Parsons' aim was to legitimize sociology as the science of normative values and social integration.

A critical discussion of SOSA will be offered in the subsequent sections. Here the focus will be put on works bearing the influence of Parsons' scientistic approach to the history of sociological theory. In terms of its intent and scope, Ronald Fletcher's (1971) *The Making of Sociology* could be taken as the British version of SOSA. The two-volume work sought to defend the scientific status of sociology by clarifying the core of theoretical ideas and its continuous development since the nineteenth century. Reminiscent of Parsons' convergence thesis, Fletcher aimed to demonstrate that there was a fundamental "agreement" over the nature and basic tenets of sociology, and that a conceptual "conspectus" had been laid down by its nineteenth-century founders and elaborated by twentieth-century theorists. But Fletcher was even more ambitious than Parsons in denying the existence of theoretical schools in the first place. Instead of initial divergence and eventual convergence, the history of modern sociological theory had been following a pattern of cumulative development from the very beginning (Fletcher 1971: 17). Fletcher also set out to cover the underrated works of British theorists, including John Stuart Mill, Edvard Westermarck, Leonard Hobhouse, and Morris Ginsberg. With these British contributions duly recognized, one might see a closer link between sociology and philosophy, but also with epistemology, moral thought, and evolutionary theory.

Specifically, Fletcher argued that the foundations of sociology were laid down by a small number of thinkers in the nineteenth century, including Comte, Mill, Spencer, and Marx, along with American sociologists like Lester Ward, William

Sumner, and Franklin Giddings (who were excluded in SOSA). These and other thinkers agreed that sociology was a science composed of "social statics" and "social dynamics." This basic agreement was developed as Tonnies, Westermarck, and Hobhouse revised the theories of evolution and social change. The analysis of the social system was extended to the objective study of social facts (Durkheim), the subjective understanding of social action (Weber), the psychological aspects of society (McDougall, Mead, Freud, Pareto, and Simmel), and the functional analysis of the cultural system (Malinowski and Radcliffe-Brown). In effect, Fletcher contended that the core of sociology had always been the structural-functional and comparative-evolutionary analysis of the social system. Contrary to Parsons, Spencer was never dead, and neither was there a breakdown of evolutionism and positivism.

The scientific ethos was no less evident in John Madge's widely acclaimed *The Origins of Scientific Sociology*. The aim of the historical essay was to demonstrate that "the discipline of sociology is at last growing up and is within reach of attaining the status of a science" (Madge 1962: 1). There was a shift from descriptive study and fragmented knowledge to a universalistic social science characterized by systematic theory and method. The history of sociology was represented in terms of landmark empirical studies and their methodological innovations. Durkheim's *Suicide* was a sociological classic by virtue of its application of statistical techniques and use of administrative records, whereas Thomas and Znaniecki's *The Polish Peasant* was pathbreaking in its use of personal documents in the life history approach. European theorists such as Max Weber were conspicuously absent in this account, as they did not contribute much to systematic empirical analysis. Theoretical ideas such as anomie and the four wishes were discussed, but the primary emphasis was given to research techniques. The history of scientific sociology was first and foremost the history of method, and only secondarily the history of theory.

But scientism did not exhaust the historical discourse even at its apex. A different approach was articulated in Robert Nisbet's *The Sociological Tradition*. In this celebrated work, Nisbet moved beyond the conventional emphasis on individual thinkers and sociological schools, focusing instead on the "unit-ideas" shared by different thinkers and schools and defining the scientific *and humanistic* concerns of classical sociology. According to Nisbet, five ideas were constitutive of the sociological tradition, namely, community, authority, status, the sacred, and alienation. These ideas were set forth in the nineteenth century in reaction to the notions of society, power, class, the secular, and progress, all of which embodied the Enlightenment vision of an individualistic, rationalistic, and anti-traditional social order. From the outset, sociological thought had been caught between modernism and conservatism. Sociology arose amidst the "Two Revolutions," that is, the processes of industrialization and democratization that gave birth to the modern world. On the other hand, sociology assimilated the conservative critique of liberalism, radicalism, and modernism in its longing for traditional values and medieval institutions. In this vein, the affinity of Comte, Le Play, Tonnies, and Durkheim with conservatism and medievalism was highlighted (Nisbet 1966: 15–6).

In this light, the sociological tradition and its unit-ideas can be taken as the modern expressions of conservative social philosophy (Nisbet 1966: 17). Tocqueville and Marx represented the liberal and radical poles of the sociological tradition, against which Weber, Durkheim, Simmel, and other theorists struggled to resolve the conflict between traditionalism and modernism. Apart from putting sociology in historical and ideological contexts, another major contribution of Nisbet was to bring to the fore its moral and aesthetic dimensions. Instead of addressing scientific puzzles or social problems, sociological concepts originated as moral ideas and artistic imaginations. Contrary to Madge, the lasting value of sociological classics such as *Suicide* did not reside in research techniques, but rather in their innovative theoretical ideas that were rooted in moral, religious, and artistic sources. While the treatment of scientific and social issues can be outdated, the images conveyed by Tonnies' *Gemeinschaft* and *Gesellschaft*, Weber's rationalization, Simmel's metropolis, and Durkheim's anomie continued to enliven our understanding of modern society (ibid.: 18–20).

The notion of a humanistic tradition of sociological thought received further treatment in later works. A central issue here was the unity and diversity of the sociological tradition. Randall Collins, for instance, proposed that there was a multiple but finite number of sociological traditions. While his original list had three traditions, namely, conflict theory, Durkheimianism, and micro-interactionism, the rational-utilitarian tradition was subsequently added (Collins 1994). A more sophisticated treatment was offered by Donald Levine's (1995) *Visions of the Sociological Tradition*. Unlike most conventional accounts of the history of sociology, Levine constructed a "dialogical" narrative that recognized the diversity of sociological traditions while engaging them in a common, ongoing dialogue. Seven national (along with ancient and international) traditions were identified, including Hellenic, British, French, German, Marxian, Italian, and American. Each tradition was defined by a distinctive moral vision centering upon the following questions: How can secular thought ground moral judgment? What was the source of human dispositions to act morally? And how were the facts of human experience and history to be explained? (Levine 1995: 100). Altogether, these traditions constituted a single, continuous line of moral and sociological thought, as they successively advanced competing visions of human nature and good society. Despite the impressive scope of his analysis, Levine's theoretical lineage remained Western.

Nisbet's emphasis on the ideological and aesthetic dimension was also taken up by other studies of sociological thought. An early example was from Irving Zeitlin (1968), who traced the origin of sociological theory to Enlightenment social thought in the eighteenth century. While Zeitlin recognized the conservative and romantic influences on Saint-Simon and Comte, a special place was reserved for Marx. As the "true heir" to the *philosophes*, Marx contributed to both the scientific study of human society and the critique of social institutions standing in the way of progress and reason. The debate with Marx's revolutionary-scientific sociology defined the overall direction of *fin-de-siècle* social theories, particularly Weber and Durkheim. Along the same line, Steven Seidman (1983) challenged the widespread conception that Durkheim and Weber were bourgeois sociologists at odds with Marxism. Rather, classical social theory represented a creative synthesis of European liberalism and revolutionary tradition. Seidman's reading was based on a reinterpretation of Enlightenment social thought. The "science of man" inaugurated by Montesquieu, Hume, and others represented an alternative strand of modern social thought, which served to counter the individualism and rationalism of social contract theory with an analysis of social institutions, moral values, and historical processes. This synthetic approach to the study of society was inherited by Marx, Weber, and Durkheim, who shared a commitment to individual freedom and democracy while resolving the crisis of European liberalism by incorporating the critical insights of the counter-Enlightenment and the egalitarian revolutionary tradition (Seidman 1983: 70).

On the aesthetic side, the definitive work was Wolf Lepenies' (1988) *Between Literature and Science: The Rise of Sociology*. Lepenies brought to light the contention between classical sociologists and men of letters for spiritual and moral leadership in the new culture of industrial society. In the nineteenth century, sociology struggled to keep a distance from literature. To legitimize itself as science, the discipline had to suppress the literary orientation in its early works. On the other hand, sociologists encountered a grave challenge from novelists such as Balzac and Zola, who claimed to practice sociology in a literary form adequate to the description (rather than explanation) of social life. While in France and Britain the strong presence of a literary tradition had provoked sociologists to build a grand system of positive science, in Germany the antithesis between literature and poetry had bred an asocial and anti-sociological propensity. The tension between feeling and reason found distorted expressions in the sociologists' zeal for the religion of humanity, Soviet communism, and Nazism. It was also manifested in the problematic role of women in the history of sociology, including the ambivalent relationships between Auguste Comte and Clotilde de Vaux, John Stuart Mill and Harriet Taylor, and Sidney and Beatrice Webb.

Taken together, Parsons and Nisbet signified two divergent approaches to the history of sociological theory. But the boundary between the scientistic and humanistic conceptions of theoretical development was not watertight. It was reflected in a compendium coedited by Nisbet himself and the Marxian sociologist Tom Bottomore. *A History of Sociological Analysis* (Bottomore and Nisbet 1979) was modeled upon Joseph Schumpeter's *History of Economic Analysis*, with a focus on the history of theoretical analysis rather than empirical research. It covered mainly the historical development of theoretical schools such as conservatism, Marxism, functionalism, and interactionism, with some discussion of methodological problems, conceptual issues, and historical trends. The contributors set out to chart the emergence and development of sociology as a theoretical and empirical science as distinct from prior social thought. Scientific progress had been achieved by Marx, Weber, and Durkheim, as they produced cumulative knowledge about the various aspects of the social structure and the social system. But while Nisbet seemed to follow the scientistic conception, he regarded the multiple paradigms in contemporary sociological theory as a step back from classical achievements. This view was consistent with his stress on the lasting value of the sociological tradition, a scientific *and* humanistic enterprise that had been abandoned but could be revived.

Another major work exemplifying the dual foci of scientific analysis and humanist value was Robert Friedrichs' (1970) *A Sociology of Sociology*. As an exercise in the history of sociology and the sociology of knowledge, Friedrichs's award-winning book aimed to examine whether sociology could become a unified scientific paradigm despite the persistent division of its theoretical schools. While sociologists failed to arrive at any substantive agreement over theoretical approaches, they shared certain self-images that had been prevalent during successive stages of disciplinary history. Instead of what Thomas Kuhn called the "normal" and "revolutionary" science, sociology had been oscillating between the "prophetic" and "priestly" modes of inquiry. Sociology in a prophetic mode was oriented toward the critique of social conditions from the standpoint of moral values. It found expression in the prominent themes of anomie, conformity, manipulation, contradiction, and alienation in sociological works prior to the 1950s. Sociology in a priestly mode, by contrast, was obsessed with the prediction and control of social phenomena with the application of research techniques. In embracing the doctrine of value neutrality, priestly sociologists like George Lundberg and Talcott Parsons gave up the role of social critic in favor of professional advisor. But a return to the prophetic mode began in the 1960s, as C. Wright Mills, Howard Becker, and Peter Berger, among others, called for a critical and humanistic sociology.

At another level, scientism and humanism shared the propensity to offer a schematic interpretation of disciplinary history. The development of sociological theory was conceived as following a single thread, be it a convergence toward analytical science or a crystallization of unit-ideas. This orientation stood in contrast to the approach of intellectual history, which aimed to provide in-depth case studies of specific thinkers in historical and social contexts.

History and Present

Among numerous attempts to write the intellectual history of sociological thinkers and their works, Lewis Coser's (1971) *Masters of Sociological Thought* was one of the founding texts. Coser began his work with an anecdote: an American student could not make sense of Weber's notion of value neutrality, owing to his ignorance of the political and intellectual context of Wilhelmine Germany. From this Coser stressed the importance of placing sociological theory in its sociohistorical and intellectual milieu: "...to understand the history of sociological theory more is required than a knowledge of formal propositions and theoretical structures... [That presupposes] some familiarity with the social and intellectual milieu in which these theories emerged" (Coser 1971: xvii). Specifically, Coser examined how theorists and their works were shaped by social origins (e.g., class background and generational experience); social positions (being central or marginal inside the intellectual circle); social networks (of friends and enemies); and audiences (academic or extra-academic, actual or imaginary). Simmel, for example, was an academic outsider, a gifted lecturer, and an essayist addressing the literary circle in Berlin. All the 12 thinkers covered in the text were male Europeans or Americans.

It would be impossible to introduce all major works in the intellectual history of sociological theory, but three can be singled out to illustrate the *biographical, institutional,* and *ideological* levels of analysis in this genre. Arthur Mitzman's (1969) *The Iron Cage* was a psychoanalytic reading of Weber's life and work. The key theme was Weber's generational revolt against his father and with it the "cultural superego" of his time. Instead of a cool advocate of value neutrality or a sober theorist of bureaucratization and rationalization; Weber was portrayed as a liberal bourgeoisie caught in a heroic struggle. Before 1897, Weber was resisting his authoritarian father. Such a sentiment was projected into Weber's study of the Junkers in East Germany and his support for the imperialist policy of Germany. After Weber's recovery from mental illness in 1902, there was a shift from ascetic rationalism to erotic mysticism and from religious to aristocratic charisma, as reflected in his comparative sociology of world religions and political sociology of legitimate domination. But Weber's ambivalence in facing the dilemma of freedom and modernity was never resolved. That was symptomatic of his contemporaries like Sombart and Michels, who saw themselves as the "epigones" of a reified world of bureaucratic and capitalist institutions created and imposed by their fathers.

Clark (1973) adopted an institutional perspective in explaining the success of Durkheim and his followers in inaugurating their program for sociology. The reason lay in the French university system, which was conducive to patronage networks among a small number of chair professors and students. This informal structure was favorable to the Durkheimians but not their intellectual competitors. While the social statisticians were anchored in governmental ministries, they constituted an isolated status group failing to promote quantitative techniques outside their circle. In contrast, the Institute of International Sociology founded by Rene Worms gathered a sizeable group of students and distinguished scholars. Yet it was too diverse to develop an integrated program of sociological theory and research. In further contrast to these two groups, Durkheim's career was boosted at crucial points by university administrators and government officials in the Third Republic. Instead of professional organizations, Durkheim's strategy of patronage building was to recruit followers from various disciplines to serve as collaborators in *Annee Sociologique*. Complementary to this strategy was Durkheim's definition of sociology, which was broad enough to cover a wide range of "social facts" in specialized fields such as law, education, linguistics, and religion. Ideological factors, such as the Dreyfus Affair, also played a role in enhancing the reputation and dominance of the Durkheimians.

Finally, Ross (1991) probed the ideological origins and features of the American social sciences, including economics, political science, and sociology. Above all, the American social sciences were marked by a peculiar tendency to reduce history to nature, that is, to conceive the social world as an unchanging, idealized liberal order governed by the laws of nature and amenable to technocratic control. Such ahistoricism was rooted in the national ideology of "American exceptionalism," the idea that America occupied an exceptional place in history and was exempted from poverty, class conflict, and the other problems of economic and political modernity (Ross 1991: 26). A reluctance to face historical uncertainty and change was embodied in the naturalistic and scientific approach that came to be established in the

American social sciences between 1870 and 1929. During the Gilded Age, there was a need to respond to the crisis brought by rapid industrialization and growing class conflict. But sociologists like Sumner and Ward either denied class warfare or resorted to "social forces" as the quasi-natural basis of social order. In the Progressive Era, a revised liberalism was proposed to justify social reform. Ross and Cooley set forth the notions of "social control" and "socialization" as the social-psychological means of containing social change. These notions fitted into a pluralist, behaviorist, and statistical model of the social world, whose "processes" and "cycles" were susceptible to "adjustment" and intervention.

While the intellectual history of sociological theory could be written in a variety of ways, some methodological issues were central to this approach. These issues were articulated by Charles Camic following his studies of Talcott Parsons' early works. Revisiting the SOSA at its 50th anniversary, Camic sought to put the text in its socio-intellectual context. This entailed an interdisciplinary struggle following the expansion of American higher education in the late nineteenth century. The struggle for recognition and autonomy between departments and disciplines, particularly between the established and emerging social sciences, defined Parsons' formative experience at Harvard in the 1920s and 1930s. Camic's argument was that Parsons intended SOSA to be a *charter*, that is, a quasi-official and public document spelling out the purposes and propositions of sociology, and defending its legitimate status against dominant players such as behavioral psychology and neo-classical economics. To this aim, SOSA served "at once as a treatise on scientific method, a defense of human voluntarism, a historical account of trends in Western social theory over three centuries, a statement of the analytical foundations of social theory, a study of the causes and solutions for the problem of social order – and, through it all, an attempt to classify the various sciences and specify their interrelations" (Camic 1989: 48). Parsons endeavored to accomplish all these tasks by resorting to the "great dichotomy" of conditional factors (heredity and environment) and normative factors (common values). All theoretical options were reduced to this binary framework. Instead of a disinterested inquiry in general theory, the insights, ambiguities, and limitations of SOSA could be understood only in light of the nature of its charter. The tendency to lift the SOSA from its original context not only led to a distortion of its meaning but also misled one to seek solutions for contemporary theoretical issues from a historically specific project.

From this and other studies, Camic was able to draw an important methodological lesson. In his introduction to an edited volume, Camic (1997) distinguished between two approaches to the studies of classical sociological theory. While *presentism* read a classical text by lifting it out of context and extracting its insights for present debates, *historicism* proceeded by reconstructing the biographical, institutional, and intellectual contexts in which the text was embedded (Camic 1997: 1–3). Like most contributors to the edited volume, Camic's perceived himself to be a historicist whose interest was not so much to arrive at authoritative interpretations than to recover theorists and works that have been marginalized, misremembered, or forgotten in the history of sociology. Methodologically put, for historicism, the use of studying sociological classics was to highlight the contingency of paradigm choices, and hence to explore alternative theoretical conceptions once available but eventually repressed (ibid.: 6). From a historicist perspective, presentism was best exemplified by ambitious attempts at theoretical synthesis, including Jeffrey Alexander's (1982) *Theoretical Logic in Sociology* and Jurgen Habermas' (1984) *The Theory of Communicative Action*. While differing from Parsons in their post-positivist and neo-Kantian orientations, these works adopted the grand vision and strategy of SOSA by assimilating and redeploying past ideas to address contemporary concerns.

But one should not readily conclude that presentism did not possess any intellectual worth. In the methodological reflections on his study of the liberal origin of European social theory, Seidman took to task the historicist assumption of a context-bound and discontinuous model of theoretical development. The problem with this assumption was its failure to differentiate the particular intentions of a classical author (e.g., Marx's critique of political economy), and the general problems addressed by a theoretical text (e.g., history, society, and human nature) that were relatively autonomous from specific contexts (Seidman 1983: 291). Seidman's proposal was to move beyond the dichotomy between historicism and presentism by underscoring the continuity of tradition in linking past and present. While Marx and Weber were bounded by their respective contexts, there was a continuous line of dialogue and argumentation running between them and through the sociological tradition.

More concrete examples can illustrate how historical contexts and contemporary concerns can be reconciled in the study of sociological theory. These can be found in interpretive works tackling the theme of *modernity*. Raymond Aron's masterpiece *Main Currents of Sociological Thought* was one such example. While introducing classical theorists, Aron could identify a common motif in their works, which was the comparative-historical analysis of the structure of modern society and its transformation. This was evident in Montesquieu's classification of laws, Comte's distinction between military and industrial societies, Tocqueville's contrast between aristocracy and democracy, and Marx's theory of capitalist and pre-capitalist formations. On the other hand, the responses of Comte, Tocqueville, and Marx to the 1848 Revolution were compared. Epitomizing the conservative, liberal, and socialist stances toward political modernity, these theorists advanced three general conceptions of the social and its relationship with the economic and the political (Aron 1965: 258–60). The mutual implication of history, modernity, and theory can also be found in Aron's treatment of Durkheim, Pareto, and Weber. Following their predecessors, these theorists adopted the approach of comparative-historical sociology to problematize the relationship between science and religion in modern society. At the same time, their theoretical formulations (social fact and morality, residues and deviations, rationalization and world religions) bore the stamp of their personalities, national traditions, and historical contexts.

An appropriate balance between historical contextualization and theoretical generalization can also be found in Anthony Giddens' (1971) *Capitalism and Modern Social Theory*. This landmark study was pivotal for establishing Marx's place in the *canon* of classical social theory alongside Durkheim and Weber. Based on a historical reading of their texts, Giddens demonstrated that there were significant parallels (but not convergence) between the three thinkers. Following Nisbet and Aron, Giddens regarded the birth of modern society in the Two Revolutions as the central problem of classical sociology. But a contextual understanding of Marx, Weber, and Durkheim would also consider the changing social and political structures of Britain, France, and Germany. Marx formulated his theory of class, capitalism, and revolution with cross-reference to the Industrial Revolution in Britain, the social and political upheaval in post-revolutionary France, and the economic and political backwardness of Germany. Weber set forth his political sociology of state bureaucracy and charismatic leadership at a time when Wilhelmine Germany became an industrial power and a unified nation-state. Durkheim envisioned sociology as the science of moral and social reconstruction, when the Third Republic in France was torn between liberal and reactionary forces. While Weber and Durkheim argued against historical materialism by highlighting the interplay of ideas and material interests as well as the relationship between collective representation and social structure, Marx in fact stood closer to them in upholding a dialectical (rather than mechanical) conception of consciousness and material conditions. Finally, all three theorists were concerned with division of labor and social change, though it was variously conceptualized in terms of class relation and alienation, social integration and anomie, and bureaucratic rationalization and the iron cage.

Rethinking the Canon

Yet the identification of modernity as the central problem of classical sociology and the corresponding canonization of Marx, Weber, and Durkheim as the core theorists of modernity were eventually put to challenge. The issue was raised by Raewyn Connell in her important essay "Why Is Classical Theory Classical?" Connell questioned not the inclusion or exclusion of particular theorists in the sociological canon, but rather the very idea of a founding moment and a classical era in sociology. This foundational myth served to legitimize the canon by defining sociology as the science of modern society, with a small number of theorists laying down its conceptual foundation in response to the great transformation of Europe in the late nineteenth century. But a survey of sociology textbooks and major studies in the history of sociology revealed that the canonical conception of theoretical development was in fact a late invention. Up to the First World War, the predominant theme of sociological studies and teaching had not been modernity but rather *global difference*, that is, the difference between the "civilized" West and its "primitive" others (Connell 1997: 1516–7).

According to Connell, the rise of sociology and the social sciences in the metropolitan centers, including in France, Britain, Germany, and the United States, was coterminous with the global expansion of European imperial powers. The connection between sociology and imperialism was revealed in the theories of evolution and progress, which were built on ethnographic data from the colonies. In terms of subject matter, nineteenth-century sociology was preoccupied with race, gender, and sexuality rather than class, alienation, and industrialization. Not all sociologists were racists, but the existence of racial difference and hierarchy was taken for granted. Contentious issues such as women's status and sexual mores occupied a central place in Comte's and Spencer's sociological treatises. The sociologists' use of comparative method embodied an "imperial gaze" that subsumed all social types under an evolutionary scheme. Finally, early sociologists were part of the political culture of empires, as they sought to shape public opinion by addressing the conflict between liberal bourgeois values and the violence of imperial expansion.

World War I signified the crisis of European imperialism and with it that of sociology. As "progress" became problematic, sociology lost its appeal in interwar Europe. The United States became the new center of sociology, with a shift of focus from global difference to social problems inside the metropole. It was reflected in the growth of urban sociology, the advance of statistical techniques, and the funded projects of social engineering in the 1920s and 1930s. The process of canonization took place in this context, as Parsons set out in SOSA to define a core of European thinkers as the founders of sociology. While Parsons' interpretation was subject to severe criticism, the existence and legitimacy of a sociological canon went undisputed. Rather, the discontent with Parsons lent force to canonization as it provided the impetus for the translations, commentaries, and applications of Durkheim, Weber, Simmel, and Marx, among others, since the 1960s. The canonical point of view was institutionalized through the pedagogy of the classics, as selected readings of sociological theory were adopted not only in America but disseminated throughout the world.

The next section will elaborate Connell's position against Eurocentrism. Here we will stay in the confines of Western sociology and consider some theoretical implications of debunking the canon. Camic, it can be recalled, problematized the contingency of theoretical choices by applying the historicist approach. One important choice was "predecessor selection," the process by which social theorists selected certain figures as the precursors of their own theoretical positions (Camic 1992: 422). In the case of Parsons, Camic sought to explain his identification of four European thinkers as the predecessors of sociology, and the *exclusion* of American institutionalists as a viable alternative. Parsons was familiar with the ideas of the institutionalists, in particular his Amherst teachers, Walton Hamilton and Clarence Ayres. In fact, the institutionalist critique of utilitarianism in terms of values and institutions should fit well with Parsons' theoretical position. That Parsons did not even mention the American institutionalists in SOSA was due to the negative reputation of institutional economics and the growing interest in the four European thinkers at Harvard in the 1920s and 1930s. In this context, Parsons was converted from his early exposure to the institutionalists toward the making of a more "credible" theoretical argument for his intellectual peers.

Baehr (2016) offered a different answer to the question of what makes classical theory classical. He would agree with Connell and Camic that there were no fixed or intrinsic criteria of classicality. But instead of reputation, Baehr focused on *reception*, the process through which a theoretical text was accorded with classical status. He spelt out four conditions of textual reception and classic formation. The first was "cultural resonance": the text must possess sufficient appeal to be considered significant or controversial. Examples were the reception of Simmel as a German source to legitimize American sociology, and of Durkheim's *The Rules of Sociological Method* as a widely cited but heavily criticized work. The second condition was "textual suppleness," in which the ambiguity and openness of a theorist's oeuvre provided the space for new and alternative readings. "Reader appropriation" referred to the selective adaptation of a text, idea or theorist to specific contexts, for instance the Americanization of Weber and his notion of the "iron cage." Finally, "social transmission and diffusion" referred to the promotion of a text, idea, or theorist via institutional platforms (as for Durkheim and Parsons) or individual agencies (as for Weber and Simmel) (Baehr 2016: 120–134). While the reception/formation of "classics" can be analyzed sociologically, Baehr contended that "founders" and "canons" were mythic notions. A discipline could not really be "founded," as it entailed the collective effort of an intellectual network (e.g., Durkheim and his circle in *Annie Sociogique*) and shadow group (e.g., the women behind Comte, Mill, and Weber). Sociological classics were also qualitatively different from religious canons, as they did not enjoy the definitive, integral, unalterable, and indisputable status of the latter (ibid.: 161–6).

Inasmuch as canonization was conceived as a social, contingent, and exclusionary process, the history of sociological thought was open to a reappraisal of forgotten and neglected theorists. Law and Lybeck (2015) noted the tendency of sociological theory to focus on "winners" such as Marx, Weber, and Durkheim at the expense of those "losers" in the struggle for intellectual recognition. It amounted to a "sociological amnesia" that impeded the creation of historically reflexive knowledge. To counteract this tendency, sociological theory should not only bring back the "failed" and "forgotten" sociologists in its history but also explain when, where, and why they came to be excluded. Some examples of sociological amnesia included Raymond Aron's circumscribed status as an interpreter of classical social theory rather than an original theorist, and the warm reception of Robert Bellah's *The Interpretation of Cultures* in contrast to Clifford Geertz's *Beyond Belief*. A hypothesis was set forth to explain these cases: the institutionalization of sociology as an academic discipline was at once conducive to the success of some theorists and the failure of others (Law and Lybeck 2015: 7–10). In this vein, a pioneering effort to reconsider neglected theorists, such as Karl Mannheim, Susanne Langer, and Alfred Schultz, can be found in a special issue in *Sociological Theory* (1994–95). Most recently, Conner et al. (2021) set out to revive another cluster of neglected theorists, including more obscure names such as John Stuart-Glennie, Annie Marion Maclean, and Gregory P. Stone. One may note that despite their variety, most of the forgotten and neglected theorists being discussed were Western thinkers.

Apart from individual thinkers, the critique of canonization and disciplinary exclusion was extended to specific fields. Collyer's (2010) discussion of health and medicine in the history of sociological theory is instructive here. Collyer in effect advanced Connell's critique of the "foundational myth" of sociology, by challenging the common assumption that its founders were preoccupied with modern industrial society but uninterested in matters of life, death, and illness. Instead of viewing medical sociology as the late development of a specialized field, Collyer showed that the problem of health, mortality, and disease figured prominently in nineteenth century sociological and public discourses. Founders such as Saint-Simon, Marx, Weber, and Durkheim treated health and related issues as social phenomena that could not be reduced to an individual problem of physical health. Collyer moved on to explain how sociology's early concern with health issues and call for public interventions were written out of disciplinary history. It had to do with the interrelated processes of medicalization and canonization. As the biomedical model attained professional dominance, sociology redefined health and illness primarily as a medical problem. This was evident in Parsons' theory of the medical profession and sick role in the 1950s, which reserved authority for biological science and medical treatment while limiting the sociological domain to the analysis of patient-doctor interaction.

If sociology has sought to canonize theorists pioneering the scientific study of modern society, a *postmodern* perspective has reinforced a critical and reflexive attitude toward the canon. Steven Seidman's (1994) *Contested Knowledge: Social Theory in the Postmodern Era* was a continuation of his study in the Enlightenment heritage of modern social theory. Instead of focusing on liberalism and the science of man, Seidman broadened his argument by underscoring the nature of social theory as scientific and *moral* inquiry. In their own historical periods, classical sociologists regarded themselves as public educators and moral advocates, holding out the promise "to help deliver humanity from oppression to freedom" (Seidman 1994: 1). Science not only served to establish the public authority of sociological knowledge but also furnished the means to public enlightenment. But there arose a tension between moral and scientific visions, when the founders formulated their social critiques while emphasizing the non-ideological nature of their scientific works. Contemporary theorists since Parsons have largely retreated from the role of public intellectual in their preoccupation with abstract conceptual and methodological issues. As sociological theory was scientized and canonized, it did not only exclude certain contributors but also became irrelevant to public debates. But the moral vision was being rearticulated with the dislodgment of the scientific canon. This was evident in the critical social science of C. Wright Mills, Jurgen Habermas, and Stuart Hall, followed by the deconstruction of science in post-structuralism, feminism, queer theory, and Afrocentrism. A postmodern canon of social theory should embrace moral perspectives such as the public philosophy of Robert Bellah, the interpretive sociology of Zygmunt Bauman, the standpoint theory of Dorothy Smith, and the historical social science of Immanuel Wallerstein.

Another laudable effort to revisit sociological theory from a postmodern perspective was Charles Lemert's short but insightful volume on classical social thinkers. For Lemert, the core of classical theory was not reason but rather *riddle*. Its primary task was to "think the unthinkable," that is, to make sense of the surprising and alien social experiences of modernity by revealing its uncertainties, contradictions, and dark side. A common problem of classical social theory was "Why, indeed, was the modern world neither as rational nor as progressive as its culture had promised?" (Lemert 2007: 120). From this, the classical theorists articulated five riddles of modernity: (i) why had the modern revolution not led to a better life for the masses; (ii) why did modernity's rational rules result in an unreasonable double-bind?; (iii) how, without religion, could industrial society overcome social conflict?; (iv) what if reason was unable to account for the unreasonable unconscious?; and (v) what would become of universal reason if social differences are real and intractable? (ibid.: 35–6). One way or another, these riddles centered on the *unreasonableness* of reason in modern society. While Marx, Weber, and Durkheim tackled the mysteries of modern economic, political, and social life, the psychoanalytic thought of Freud and the feminist thought of Gilman shed light on modernity's unthinkable dimensions. Without definitive solutions, the five riddles were open to reformulations, as could be found in the works of W.E.B. Dubois, Anna Julia Cooper, Georg Simmel, and Ferdinand de Saussure.

This section began with a discussion of Connell's view that the subject matter of early sociology was gender and racial differences rather than classes and cities. In more recent works in the history of sociological theory, the problematics of gender and race received its long-overdue scholarly attention. Aldon Morris's (2015) study of W.E.B. Dubois was a welcome addition to this scholarship. For Morris, Dubois was not only another neglected theorist but rather the true founder of scientific sociology in America. The conventional narrative, according to Morris, upheld the racist view that Black social scientists did not contribute much to the discipline, and canonical status was rightly reserved for White sociologists. But historical records show that Dubois and his "Atlanta School" had embarked upon systematic empirical research way ahead of the Chicago School. Growing up in a Black-hostile environment, Dubois bore the moral mission to refute racist prejudices and White privileges based on a meticulous collection and analysis of empirical data. With his students and colleagues, Dubois formulated and tested the hypothesis that sociological and economic factors were the root causes of racial inequalities. This position stood in contrast to Robert Park, whose theory of race-relation cycle was premised on an evolutionary conception of the Black people and their culture. In this vein, Morris disputed Park's founder status and intellectual integrity by documenting his unholy alliance with the conservative Black leader Booker T. Washington to suppress the mounting influence of Dubois and his Atlanta School. Dubois' scientific exchange and moral inspirations for Weber on the issue of race was another largely forgotten chapter in the history of sociology.

In line with Morris' treatment of Dubois, efforts were made to open the sociological canon to the "women founders." In their feminist reader in social theory and the history of sociology, Lengermann and Niebrugge (1998) identified 15 women who played an active role in the formation of sociology. Instead of being invisible, there was a significant presence of women social thinkers from 1830 to 1930. Harriet Martineau was the contemporary of Comte and Spencer, while Jane Addams, Charlotte Perkins Gilman, Anna Julia Cooper, Marianne Weber, Beatrice Webb, and the "Chicago Women's School" were of the same generation as Durkheim, Weber, Simmel, Mead, Thomas, and Park. These women thinkers maintained intellectual ties with each other and with their male counterparts. They produced representative works in sociology and published in professional journals. Despite their preeminence, the women founders were excluded from the sociological canon and erased from its historical record. Martineau, Cooper, Webb, and Marianne Weber were conceived by their male contemporaries as "others" lacking intellectual autonomy and authority. The exclusion of women was sealed as pure academic research was institutionalized and elevated above social reformism. With the change of the sociologists' work sites from social service agencies to state-funded universities, the women founders' concern with poverty, race, community, and voluntary group gave way to scientism and the doctrine of value neutrality.

Decentering the West

When Connell highlighted the theme of *global* difference in early sociology, she had in mind not only the substantive treatment of gender and race but also the Eurocentric bias in the works of classical theorists. This idea was elaborated in her subsequent work on "Southern theory." As Connell (2007) saw it, the ultimate purpose of opening the canon was to democratize social knowledge, which could not be accomplished without building a truly global sociology. This project entailed an extension of Connell's critique of the foundational myth and imperial origin of classical sociology. Contemporary social theorists in the North, including James Coleman, Anthony Giddens, and Pierre Bourdieu, tended to treat European experiences as universal while disregarding colonial experiences and perspectives. Instead of focusing exclusively on Western (including women and Black) thinkers, and perpetuating the hegemony of the social sciences in the metropoles, a proper task of social theory was to introduce and incorporate the subaltern perspectives of the global South.

To repudiate the assumptions of universalism and Eurocentrism in sociological theory, Connell proceeded to examine the life and work of neglected theorists from the South. A common theme of these theorists was the critique of European colonialism and the structural and intellectual dependence of former colonies. This included the search for an indigenous sociological and philosophical tradition among African theorists such as Akinsola Akiwowo and Paulin Hountondji; the defense of Islam as a modern, rational, and revolutionary tradition by Middle Eastern thinkers such as Al-Afghani and Ali Shariati; the dependency theory in Latin America, as formulated by Raul Prebisch and Fernando Cardoso among others; and the postcolonial studies in India, which were inaugurated by Ranajit Guha's subaltern studies and Ashis Nandy's theory of the colonial state of mind. In subsuming these works under the rubric of "Southern theory," Connell did not intend to impose a geopolitically bounded category, but rather initiate a critical reflection on the *relations* of power and exclusion between the metropoles and peripheries. The notion also served to obliterate the conventional division of labor between theory construction in the North and data collection in the South. As Connell (2007: viii) put it, Southern theory consisted of modern sociological texts for us to learn *from*, not just *about*.

Connell's efforts in criticizing Eurocentric social theory were not isolated. In the same year as Connell's *Southern Theory*, Gurminder Bhambra published her work on the same subject. According to Bhambra (2007), there were two paradigmatic assumptions about modernity, namely its rupture with the traditional past, as well as the difference between Europe and the rest of the world. Against this temporal-spatial framework, theories of modernity (and postmodernity, multiple modernities, global modernity, etc.) regarded Europe as the marker and leader of change, presuming that social structures originating from Europe would become universal. In this way, Eurocentrism was built into the epistemic framework of modern social science. In order to transcend this framework, Bhambra proposed to rewrite the history of sociology from a postcolonial perspective. Drawing upon Edward Said's critique of Orientalism, Bhambra aimed to show that the colonial encounter, the construction of the colonial gaze, and the silencing of the colonial other were constitutive of modern sociology and its politics of knowledge. Following the attempts of Scottish and French Enlightenment thinkers to distinguish stages of European cum universal history, classical social theorists viewed Western civilization and its relationship with other cultures in terms of rupture and difference, and conceptualized the "social" as the locus of historical progress and transformation. Instead of the modernist ethos, the history of sociology should attend to the changing relations between the colonizers and the colonized. By focusing on the interconnectedness of events and contexts in the metropoles and the peripheries, these "connected histories" could serve as a postcolonial approach pertinent to a critical understanding of the discipline.

A more recent initiative to learn from postcolonial studies is Julian Go's (2016) reflection and reconstruction of social theory. Go's project was to engage social theory and postcolonial thought in a mutual dialogue. Sociology originated from the "high imperialism" of the nineteenth century, when European empires were expanding and partitioning the non-Western world. For Go, social theory was not only born *in* empires; it was born *for* them. Sociology did not only have an imperial origin but also an imperial episteme (Go 2016: 4–5). Evolutionary schemes, racist remarks and an affirmative stance toward imperial expansion can be found in the works of the classical theorists. By contrast, postcolonial thought was born of anticolonial movements in Africa, Asia, and Latin America. It began in an era of high imperialism, and culminated in the wave of decolonization in the 1960s and the institutionalization of postcolonial studies in higher education in the 1980s. The foundation of postcolonial thought was laid by Frantz Fanon, Aimé Césaire, and W.E.B. Dubois, among others, who analyzed the economic, political, and cultural aspects of empire while envisioning a postcolonial world free of imperial power and domination. Despite its indifference to the colonial question, social theory needed to engage with postcolonial thought in order to address the problem of modernity in light of its entanglements with empire. The parochial concern with European and American modernity could be transcended by incorporating the postcolonial critique of the imperial episteme. In fact, the emphasis on power and social relations in postcolonial thought was not foreign to social theory; it could be found in Wallerstein's world-system theory, Bourdieu's field theory and Latour's actor-network theory. Insights from postcolonial thought could thus serve to reinforce the movement of social theory away from its implicit "metrocentrism."

The call for a postcolonial social science opened up new directions in the historical studies of sociology and social theory. The first concerned the *sociology of empire*. In his introduction to an edited volume, Steinmetz (2013) pointed out that sociologists have been studying empires ever since its inception. Four periods of theorizing and research on empire could be distinguished, which corresponded to successive phases in the global expansion of Western imperialism. The first phase was 1830–1890, when Africa was being partitioned by European powers. Founders of sociology like Comte, Tocqueville, and Marx gave their support and critiques of imperial expansion. The second phase, 1890–1918, witnessed the collapse of the Ottoman, Russian, and Austrian empires. Sociology was institutionalized in Europe and America, where sociologists theorized the empire, researched the colonies, and advised on colonial governance. While British sociologists advanced economic interpretations of imperialism, German sociologists favored political and military explanations. The third phase was 1918–1945, when European imperialism led to destructive warfare and anticolonial resistance. The period witnessed the proliferation of sociological theories, including the cyclical theories of empires/civilizations, non-economic theories of imperialism, theories of cultural contact and colonial transculturation, the interpretation of Nazism as a modern form of empire, and the idea of a *nomos* or "great space" for Europe. Finally, from 1945 onwards, the Cold War, national independence, and American new imperialism found expression in the hegemony of modernization theory, the countercurrent of Marxism, and the historical sociology of colonialism and empire. This hidden genealogy not only cast a new light on the history of sociology, but contained rich thoughts on the forms, developmental trajectories, determinants, and effects of empire (Steinmetz 2013: 43–6).

The second line of development was to write a *transnational history of sociology*. In various works, Sujata Patel has criticized the "methodological nationalism" of sociology, that is, the tendency to designate "society" or the nation-state as the object of sociological theory and research. Patel's position can be best illustrated with her historical account of sociology in India. In a recent paper, Patel (2021) distinguished three stages of development in Indian sociology. The first stage was the early 1930s, when Indian sociologists sought to indigenize the discipline by adopting a culturalist and Indological perspective. The second stage covered the 1960s and 1970s, when M.N. Srinivas' works became the dominant paradigm. While it represented a significant advance from the interpretation of ancient texts to the field investigation of rural society; Patel argued that the Srinivasian paradigm fails to transcend the epistemic framework of colonial modernity. In formulating the theory of Sanskritization and Westernization, Srinivas attributed the transformation of Indian social structure to the functional exigencies of modernization, and systematically neglected the transnational forces of imperialism, colonialism, and capitalism. His notion of dominant caste also served to exclude non-caste groups from the social and political projects of the modern Indian nation-state. These issues only became apparent in the third and current stage of Indian sociology, when colonial modernity was subject to critical reflection in subaltern and postcolonial studies.

By highlighting the necessity to transcend methodological nationalism and colonial modernity, Patel paved the ground for a transnational approach to the history of sociology and social theory. A notable attempt here was Stephane Dufoix's (2021) paper on the dialectics of the transnational and the local in the post-war development of sociology in Asia. While there was a proliferation of indigenization discourses in Asia during the 1960s and 1970s, the meaning of indigeneity should not be taken at face value. Despite its nationalist connotation, the "indigenous" was constructed in and through transnational academic networks. To illustrate this, Dufoix meticulously documented the history of three sociological associations, namely, the Division of Social Sciences of the United Nations Educational, Scientific and Cultural Organization (UNESCO), the International Sociological Association (ISA), and the International Social Science Council (ISSC). In addition to India, China, and Japan, sociologists from Philippines, Pakistan, and other Asian countries have utilized these institutional platforms to strengthen regional collaboration. Central to these inter-Asian and South-South initiatives was the urge to end the academic dependency of Asian social science, and to transcend the misplaced dichotomy between Western universalism and non-Western particularism. Instead of mutually isolated national traditions, alternative discourses such as indigenization, endogenous development, and decolonization should be conceived as interconnected efforts in the "world social science archipelago."

The third possibility for rethinking sociology and social theory entails the quest for a *new universalism*. In his groundbreaking work, Alatas (2006) purported to examine the historical past and current state of Asian social science, and on that basis proposes some possible ways to reclaim its autonomy and relevance vis-à-vis the "world social science powers" of Europe and America. By bringing together theoretical discourses from India, China, Korea, Malaysia, Singapore, Philippines, and Indonesia, Alatas compensated for the inadequate treatment of Asia in Connell's *Southern Theory*. These peripheries shared the experience of implanting Western social science into their local contexts. While Asian social science suffered similar problems (such as Eurocentrism and a lack of originality), there were notable contributions to the critique of Western hegemony and the construction of alternative discourses. The latter included attempts to indigenize and decolonize social science, along with sociological perspectives that were rooted in religious and national traditions. A common purpose of these alternative discourses was to decenter rather than denounce the West, as local experiences and intellectual resources were invoked along with the critical appropriation of Western ideas to build a global and universal social science (Alatas 2006: 82–3).

Following the lead of these alternative discourses, Alatas and Sinha (2017) set out to write a textbook on classical sociological theory without the conventional biases of Eurocentrism and Androcentrism. To reorient the teaching of classical theory, the authors sought to broaden and universalize the sociological canon. First, the historical contexts of classical sociology were broadened from European modernity to the colonial history of non-Western societies. It was followed by a critical re-reading (rather than cancellation) of Marx, Weber, and Durkheim, with a discussion of the Eurocentrism implicit in concepts such as the Asiatic mode of production. An introduction to non-Western thinkers was in order, covering Ibn Khaldun's historical sociology of state formation; Jose Rizal's theory of colonial society and subjectivity; Said Nursi's social theology; and Benoy Kumar Sarkar's materialist interpretation of the Hindu tradition. There was also a treatment of female thinkers, including Harriet Martineau, Florence Nightingale, and Pandita Ramabai. By juxtaposing Western and non-Western, male and female theorists, one could discern the universal themes of freedom and enslavement that defined the core of a global classical sociology.

The quest for a non-Eurocentric universalism was pursued along a different line by Xie Lizhong, a leading Chinese social theorist. For years, Xie had been adapting post-positivist and postmodern (though not postcolonial) social theory to the Chinese intellectual context. For Xie, the common target of these critical currents was the "given realism" of classical and modern social theory, that is, the presumption that social reality was objectively given and hence independent of social construction. In this vein, Xie (2021: 4–9) argued that indigenization discourses inadvertently upheld this modernist assumption, as privileged access to social reality was reserved for non-Western sociological traditions in much the same manner as Eurocentric social science. By contrast, postmodernism upheld a "discursive realism" for which social reality was constructed by social agents following discursive rules. These discursive rules and systems were plural in the sense that none could claim to represent the sole truth of social life. The proper task of sociology and social theory was to reconstruct the discursive systems against which multiple representations of social reality were constructed. As the expressions of peculiar social experiences, theories of capitalism, rationalism, and folk and urban societies were some discursive systems that could lay equal claim to universal validity.

While Xie recommended that sociologists should embrace pluralism, this should not be narrowly construed in geographical terms. Instead of pitting the distinctiveness of non-Western societies and cultures against the Western model, Xie suggested that the particular and universal features of *both* Western and non-Western societies should be incorporated in a given discursive and theoretical system. The critique of Western hegemony needed not give up the legitimate quest for universal social science, for otherwise one would be entrapped in untenable anti-epistemological projects. Xie's significance thus lies in the question he raises: how is it possible to counter Western hegemony and decenter social science, without leading to the relativization and fragmentation of knowledge? This concern was reflected in Xie's approach to "post-Western sociologies," a collaborative project of French and Chinese sociologists since 2010. In Xie's interpretation, "post-Western" should not be misconstrued as non-Western, de-Western or anti-Western. Inasmuch as pluralism was embraced, one must be careful not to relapse to the monist and realist perspective often implicit in the quests for intellectual hegemony *and* counter-hegemony. It could be achieved through the co-production of knowledge by Western and non-Western sociologists, who bring together their discursive systems for mutual illumination (Xie 2021: 3). Such a project should duly recognize the multiple forms of universalizable social knowledge, for which both Western and non-Western traditions could furnish useful insights and perspectives.

Conclusion

This chapter has reviewed some major works and approaches to the history of sociological theory. In the early studies, social theory was assimilated within the broader corpus of social and moral thought. While modern sociology has made scientific progress, a prominent theme is the variety of theoretical schools and civilizational sources. A major reorientation took place with the publication of Talcott Parsons' *The Structure of Social Action*, which inaugurated a scientistic interpretation of modern social theory. Yet far from monopolizing disciplinary discourses, scientific sociology soon provoked the reaction of humanistic scholars. For the latter, sociology embodied an intellectual, moral and aesthetic tradition with an unbroken continuity from the classical era to the present. Alongside these schematic interpretations, intellectual history aimed to examine particular theorists and works with reference to their biographical, institutional and ideological contexts. While historicism and presentism were posited as incompatible methodological options, modernity constituted the middle ground between theoretical generalization and historical contextualization.

More recent works in the history of sociology have re-thought the canon and decentered the West. The very idea of a canon has been questioned, as this served to exclude theorists, ideas, and paradigms on extra-intellectual grounds. Postmodern perspectives as well as women and Black thinkers have been reappraised to broaden the sociological imagination. A closely related target has been Eurocentrism, which was rooted in the historical linkage of sociology with European imperialism and colonialism. Postcolonial thought has been appropriated not only for the critique of Western hegemony, but also for opening new directions of theoretical development. These entail a historical understanding of the imperial entanglements of sociology, and the building of a global, transnational, and universalistic social science. At this point we have come full circle, as an inclusive treatment of Western and non-Western sociological thought is once again put on the agenda, this time with a keen awareness of the lingering effects of Eurocentrism.

References

Alatas SF (2006) Alternative discourses in Asian social sciences. Sage, New Delhi
Alatas SF, Sinha V (2017) Sociological theory beyond the canon. Palgrave Macmillan UK, London
Alexander J (1982) Theoretical logic in sociology. Routledge and Kegan Paul, London
Aron R (1965) Main currents of sociological thought. Weidenfeld and Nicolson, London

Baehr P (2016) Founders, classics, canons: modern disputes over the origins and appraisal of Sociology's heritage. Transaction Publishers, New Brunswick
Barnes HE (ed) (1948) An introduction to the history of sociology. University of Chicago Press, Chicago
Becker HP, Barnes HE (1938) Social thought from Lore to science. Dover Publications, New York
Bhambra G (2007) Rethinking modernity: postcolonialism and the sociological imagination. Palgrave Macmillan, Basingstoke
Bierstadt R (1981) American sociological theory: a critical history. Academic Press, New York
Bogardus E (1922/1928) A history of social thought. J.R. Miller, Los Angeles
Bottomore T, Nisbet R (eds) (1979) A history of sociological analysis. Heinemann, London
Camic C (1989) Structure after 50 years: the anatomy of a charter. Am J Sociol 95(1):38–107
Camic C (1992) Reputation and predecessor selection: parsons and the institutionalists. Am Sociol Rev 57:421–445
Camic C (ed) (1997) Reclaiming the sociological classics: the state of the scholarship. Blackwell Publishers, Malden, MA
Clark T (1973) Prophets and patrons: the French university and the emergence of the social sciences. Harvard University Press, Cambridge, MA
Collins R (1994) Four sociological traditions. Oxford University Press, New York
Collyer F (2010) Origins and canons: medicine and the history of sociology. Hist Hum Sci 23(2):86–108
Connell R (1997) Why is classical theory classical? Am J Sociol 102(6):1511–1557
Connell R (2007) Southern theory: social science and the global dynamics of knowledge. Polity, Cambridge/Malden
Conner C, Baxter NM, Dickens DR (eds) (2021) Forgotten founders and other neglected social theorists. Lexington Books, Rowman & Littlefield
Coser L (1971) Masters of sociological thought: ideas in historical and social context. Harcourt Brace Jovanovich, New York
Dufoix S (2021) Under Western eyes? Elements for a transnational and international history of sociology in Asia (1960s–1980s). J Hist Sociol 34:55–74
Elster J (1985) Making sense of Marx. Cambridge University Press, Cambridge, UK
Fletcher R (1971) The making of sociology: a study of sociological theory. Joseph, London
Friedrichs R (1970) A sociology of sociology. Free Press, New York
Giddens A (1971) Capitalism and modern social theory. Cambridge University Press, West Nyack
Go J (2016) Postcolonial thought and social theory. Oxford University Press, New York
Habermas J (1984) The theory of communicative action. Heinemann, London
Law A, Lybeck ER (eds) (2015) Sociological amnesia: cross-currents in disciplinary history. Ashgate, Farnham, Surrey
Lemert C (2007) Thinking the unthinkable: the riddles of classical social theories. Paradigm Publishers, Boulder
Lengermann PM, Niebrugge G (1998) The women founder: sociology and social theory 1830–1930. McGraw-Hill, Boston
Lepenies W (1988) Between literature and science: the rise of sociology. Cambridge University Press, Cambridge
Levine D (1995) Visions of the sociological tradition. University of Chicago Press, Chicago
Madge J (1962) The origins of scientific sociology. Tavistock Publications, London
Martindale D (1960/1981) The nature and types of sociological theory. Routledge & Paul, London
Mitzman A (1969) The iron cage: an historical interpretation of max weber. Grosset & Dunlap, New York
Morris A (2015) The Scholar Denied: W.E.B. Du Bois and the birth of modern sociology. University of California Press, Oakland
Nisbet R (1966) The sociological tradition. Basic Books, New York
Parsons T (1937) The structure of social action. Free Press, New York

Patel S (2021) Nationalist ideas and the colonial episteme: the antinomies structuring sociological traditions of India. J Hist Sociol 34:28–40

Ross D (1991) The origins of American social science. Cambridge University Press, Cambridge

Seidman S (1983) Liberalism and the origins of European social theory. Blackwell, Oxford

Seidman S (1994) Contested knowledge: social theory in the postmodern era. Blackwell, Cambridge, MA

Sorokin P (1928) Contemporary sociological theories. Harper & Brothers, New York

Sorokin P (1966) Sociological theories of today. Harper & Row, New York

Sociological Theory (1994–95) Special Issue on "Neglected Theorists," 12(3) and 13(1)

Steinmetz G (ed) (2013) Sociology and empire: the Imperial entanglements of a discipline. Duke University Press, Durham

Stinchcombe A (1968) Constructing social theories. Harcourt, Brace & World, New York

Xie L (2021) Post-Western sociologies: what and why? J Chinese Sociol 8(5):1–25

Zeitlin I (1968) Ideology and the development of sociological theory. Prentice-Hall, Englewood Cliffs

José Carlos Mariátegui and the Origins of the Latin American Sociology

37

Fernanda Beigel

Contents

Introduction	962
The Circulation of Mariátegui's Ideas in the Midst Between the Komintern and the Artistic Avant-Garde	963
A Founding Father for the Latin American Sociology and His Readings Within the Dependency Analysis School	965
Mariátegui's Contribution to the Development and Critique of Indigenism	968
Conclusions	971
References	972

Abstract

José Carlos Mariátegui was a prominent figure in the origins of the Latin American sociology, during the twentieth century. He created an indigenist perspective, a new methodology, and a rooted reading of Marxism. With his *Seven Essays to Interpret the Peruvian Reality* (1928), Mariátegui produced a significant move from the typical "national studies" towards the analysis of "the problem of the indigenous people," as well as a shift from positivism to Marxism. Moreover, this boosted the development of a conceptual paradigm that was later called the structural-historical method, the basis of dependency analysis. Eventually, in the midst of his contemporary social movements, this Peruvian essayist inaugurated the militant sociology that was to be developed across the region from the 1960s. This chapter suggests, accordingly, that Mariátegui should be considered as the founding father of this regional sociological tradition.

F. Beigel (✉)
Consejo Nacional de Investigaciones Científicas y Técnicas (CONICET), Mendoza, Argentina

CECIC- Universidad Nacional de Cuyo, Mendoza, Argentina
e-mail: mfbeigel@mendoza-conicet.gob.ar

Keywords

Mariátegui · Latin America · Indigenism · Sociological traditions

Introduction

José Carlos Mariátegui was a singular intellectual for his time, because in the mid-1920s, he affirmed that Marxism was a religion and, at the same time, declared his sympathies for the Soviet Union and Bolshevism, so he did not go unnoticed in the international communist movement. In fact, the figure of José Carlos Mariátegui (1894–1930) became a paradigm for his contemporaries and also constitutes a milestone for all those who begin to investigate the journey of Latin American Marxism. Although his work had significant weight in his time – whether to canonize it, vilify it, or manipulate it – it is not so easy to understand why it met with silence after his death; why it circulated again after the mid-twentieth century and; finally, why it assumed a new relevance role after the 1980s, particularly in the context of the recent indigenous emergence in Latin America.

The international circulation of Mariátegui's ideas effectively began while he was alive, towards the second half of the 1920s, when he published his famous magazine *Amauta* (1926–1930), and with the dissemination of some of his journalistic writings in political-cultural magazines, including *Repertorio Americano* (Costa Rica, 1919–1959), *Revista de Filosofía* (Buenos Aires, 1915–1929), and *Monde* (Paris, 1928–1935). With the death of Mariátegui and the institutionalization of the communist parties in Latin America, the circulation of his work, far from increasing, decreased significantly and only began to be read more profusely in the early 1980s when traditional Marxism entered a new phase. These ups and downs in the reception of Mariátegui's writings are linked to a kind of "sanitary fence" during the 1930s and lasted until the 1960s when they began to be read by dependency sociologists like Aníbal Quijano.

Well, why then was the sociological legacy of Mariátegui's so little acknowledged? Sociology in Latin America has a long history, emerging from three distinct paths of development of social knowledge: (a) the sociology chairs that were first established in 1882–1898 in various countries of the region; (b) social essays, which were often written by journalists; and (c) social reports and surveys produced by technicians working at public bureaus. The itinerary of each national sociology has been frequently explained through the opposition between "chair sociology" and "scientific sociology" – a founding myth largely based on Argentinian references. However, this national narrative has obscured the relevance of José Carlos Mariátegui, who was neither a "scientific sociologist" because of his adherence to Marxism, nor a "chair sociologist." An autodidact without formal education, Mariátegui was basically a journalist and essayist deeply involved in the social movements of his time.

This chapter argues that this Peruvian essayist was an important figure in the construction of a regional sociology during the twentieth century, which offered an endogenous perspective, its own methodology and a theoretical tradition. With Mariátegui's *Seven Essays to Interpret the Peruvian Reality* (1928) came a significant move from the more typical "national studies" towards the analysis of "the problem of the indigenous people," as well as a shift from positivism to Marxism. After a period of residence in Europe, Mariátegui returned to Perú and developed a conceptual paradigm that was later called the structural-historical method, the basis of dependency analysis. Eventually, writing in the midst of the social movements, this Peruvian essayist inaugurated the militant sociology that was to be developed across the region from the 1960s.

The Circulation of Mariátegui's Ideas in the Midst Between the Komintern and the Artistic Avant-Garde

Mariátegui was born in Perú in 1894 and died in 1930 at the early age of 36 years. His trajectory as a journalist was formed in his youth and became a more complex type of cultural practice known as "editorialism," as it was associated with the development of publishing companies and cultural journals. Mariátegui became a social essayist in his brief mature years as a part of his own "peruanization" and the rise of his interpretation of Marxism (Beigel 2003). His most important cultural project was the journal *Amauta* (1926–1930), which played a central role in the consolidation of indigenism in Perú and the artistic/political avant-garde in Latin America. Indigenism is the name of a cultural and political movement born by mid-nineteenth century, at first as a philanthropical perspective towards indigenous people. In the 1920, this movement evolved to nationalist and socialist trends both mainly developed by urban intellectuals or political leaders that did not belong to the indigenous communities. During the 1980s and 1990s, an indigenous uprising occurred in most countries and Indigenism was considered an external and outdated perspective to understand the more recent politics of indigeneity.

Around *Amauta*, Mariátegui created an intellectual network throughout the region and beyond, reaching its consolidation by 1929 with 12 publishing houses in Latin America, 1 in the United States, and 3 in Europe. Accordingly, during his life, the circulation of Mariátegui's ideas occurred within the cultural sphere and to a lesser degree in the contested environment of the international communist movement. From his very extended writings, only two books were published before his death: *La escena contemporánea* (Lima, Minerva, 1925) and *Siete Ensayos de Interpretación de la Realidad Peruana* (Lima, Biblioteca Amauta, 1928). His journal articles appeared in various media during the 1920s, but a complete compilation of his work came to light only in 1959 with the first edition of 20 volumes published by his sons.

The *Amauta* journal offers a fine explanation of Mariátegui's heterodox position vis-à-vis the international communist movement. This magazine acted as an agent for the famous magazine *Monde* (directed by Henri Barbusse), and promoted creative encounters between the new avant-garde which would, shortly after his death, fall out of grace with the arrival of "socialist realism." Between December

1929 and January 1930, various documents of the Communist International and other organizations linked to the Komintern, were published in *Amauta*, such as the Latin American Trade Union Confederation and the Anti-imperialist League (in which Mariátegui was a member). But the relationship of the Peruvian Socialist Party created by Mariátegui in 1928 with the Comintern was not an easy one, particularly because of the indigenous and nationalist position of the Peruvian group. The participation of Mariátegui's delegation in the Communist Conference held in June 1929 in Buenos Aires played a decisive role in this sense, where the problem of races was discussed, as we will see below. For this reason, during the last months of his life, there were no direct contacts between Mariátegui and the Comintern.

Mariátegui's projected exile to Buenos Aires, planned for March–April 1930, was managed by the editorial writer Samuel Glusberg, through his contacts within literary circles. Mariátegui's most important concern in his last weeks was to continue publishing *Amauta* and he believed he could do it from the Argentinian capital: but he would only transfer it from Lima if *Amauta* was permanently banned in Peru. It would be imprecise to affirm, however, that in this final journey, the Peruvian "prioritized" relations with the Buenos Aires cultural sphere rather than the leaders of the International. This last connection does not seem to have presented itself as a real and existing possibility. Mariátegui was never an officer at the Komintern, nor did he maintain personal ties with the Argentine Section or other Moscow representatives. In fact, the project of the trip to Buenos Aires had been planned for 3 years and came about only with a proposal from editorial writer and friend Samuel Glusberg. Mariátegui basically sought to develop his intellectual militancy and planned to publish his book *Defense of Marxism* on arrival. He was certainly mobilized by the pressure of police harassment and "isolation" that he confessed to having in Lima, but another vital need was much more urgent: in Buenos Aires, he could undergo a surgical operation and prolong his life.

Most of the intellectuals and groups with whom Mariátegui established a relationship during his life participated in the controversial process of interpreting his theoretical and political legacy. A few days after his death, some published articles portrayed him as an "abstract dogmatic," while his fellow party members declared in *Amauta* that he was a "revolutionary ideologue." Some considered Mariátegui an "aestheticizing" intellectual, distant from political action, and others characterized him as an "organizer of the Peruvian proletariat." There were also accusations of populism by the Russians, and the Peruvian replicas, who described Mariátegui as a "Marxist-Leninist-Stalinist." Thus, the first period of reception of Mariátegui's work, between 1930 and 1934, was marked by the internal conflicts of Latin American communism and the most recent debates in the Peruvian political field. The Argentine journal *Claridad* became a vehicle for a discussion centered on the break between Victor Raúl Haya de la Torre and Mariátegui; the creation of the Peruvian Socialist Party in light of the institutionalization process of the III International; the place of the Indian in the revolution; the relations between socialism; and anti-imperialism. Between the APRA and the Comintern, the editor of the *Amauta* was described with plentiful adjectives and conjugated in various verb tenses: creator of a "petit bourgeois Mariateguism"; founder of the Peruvian Communist Party; confusing populist; "nationalist" in his later years, but communist on the deathbed; among others. This response created a significant difficulty for the diffusion of Mariátegui's work, a difficulty that was equally useful for some communists (who saw in him a "liquidationist" tendency) and some Apristas (who considered him a theoretical threat).

The criticisms coming from the Komintern, somewhat ambiguous until 1931, opened a second combative phase against the direction Mariátegui had taken in Peruvian Marxism, and finally resulted in an open condemnation of "Mariateguismo" by 1934 within the Peruvian Communist Party. This "Mariateguismo" was nourished by Mariategui's own work, in which the Peruvian communists highlighted "great errors": the confusion between the national problem and the agrarian problem; the attribution to imperialism and capitalism of a progressive role in Peru; and "the replacement of revolutionary tactics and strategy by debate and open discussion." He was then judged both by his practice and by his theoretical propositions, after an alleged vindication of his figure. Moreover, he was accused of being an "insufficient" Marxist, from the "self-sufficiency" of Leninism, and it was also alleged that in Mariátegui's final years, including the final months of his career, he would have "fought his own Mariáteguism." Finally, the Peruvian communists declared that the position of the party "against Mariateguism is and must be one of implacable and irreconcilable combat" since it hindered the organic and ideological Bolshevization of its ranks (Arició 1978; Martínez de la Torre 1947).

Mariátegui's career ended, thus, exorcising the demons of communist politics, for his writings were clearly manipulated in the process of institutionalization of the sections of the International at this time. This rigidity of the movements began to break only with the death of Stalin in 1956. But other circumstances also came to improve the reading conditions of Mariátegui's work: the Cuban Revolution (1959) and the first edition of Mariátegui's complete works (1959). The former energized the Latin American intellectual field towards new readings of Marx and promoted new perspectives throughout the continent, thus developing new forms of political praxis. The (1959) edition of his works extended the reading of Mariátegui to ever wider circles. At the same time, the Lettere dal Carcere (Letters from the Prison) of Antonio Gramsci (1956) were published for the first time in Spanish, and with a very similar fate, both the Peruvian and the Italian began to be read by the nascent academic Marxism in Latin America (Beigel 2003).

A Founding Father for the Latin American Sociology and His Readings Within the Dependency Analysis School

In the traditional historiography of the origins of the Latin American sociology, there was a struggle between two ways of analyzing social phenomena: the empirical "scientific" sociology and the essayist sociology of "chairs" (Franco 1978). But in the more recent sociology of sociology, this opposition has been interpreted as a founding myth (Blanco 2005; Pereyra 2005). In the same drive and searching for the emergence of the Latin American sociology as a regional tradition, it is imperative to discuss the contribution of the Peruvian Jose Carlos Mariátegui (1894–1930), who was neither a "scientific sociologist" nor a "chair sociologist." An autodidact without formal education, Mariátegui never taught at any university other than the Popular University Gonzalez Prada, an informal space of training courses created for workers during the Peruvian University Reform in 1919. Nevertheless, Mariátegui's *Seven Essays* involve a rich and critical reading of Marxism and an intense engagement with locally produced knowledge on the history of Peru and Latin America. His dialogue with Francisco Garcia Calderon, who published *Le Perou Contemporain* in 1907 in Paris, anticipates the debate between the sociology of modernization and the structuralist perspective of the 1960s. Calderon was part of the Societe de Sociologie de Paris and participated actively in the journal *Revista de America*, published in France during the first decades of the twentieth century. According to Mejia Navarrete (2005), Mariátegui inaugurated "national studies" and the dualist interpretation of Peruvian society, opposing the development of a modern coast to a traditional and backward mountain range inhabited by indigenous people (Navarrete 2005: 305). A fourfold shift points to a comparison of Calderon and Mariátegui, or *Le Perou Contemporain* and *Seven Essays*. First, the move from the systemic "national study" towards the analysis of Peruvian reality linked to "the problem of the Indio": from positivism to Marxism. Second, the transition from dualism to structural heterogeneity. Third, the radical change from the academic writing exercised by Calderon, settled in Paris, to that of Mariátegui, from Lima, which engaged with social movements. Fourth, and finally, was the theoretical and practical move from Peru to Latin America as an historical process of observation and an intellectual community (Beigel 2019).

Dependency as a theoretical concept and the structural-historical method were developed analytically between the 1960s and 1970s, but were first sketched in the writings of Mariátegui. In his diagnosis of the local economic formation, he argued that the Peruvian problems were part of the historical continental process initiated by the colonial conquest. An incomplete independence had given birth to a republic that coexisted with precapitalist relations and servitude; accordingly, capitalism had evolved as an overlapping of ancient and new modes of production: (a) the locally powerful latifundismo (large estates with autonomous rules and production system); (b) the industrial bourgeoisie; (c) the foreign capital investing in mining and other national resources; and, finally, (d) the indigenous communities which subsisted in the mountain ranges with their ancient social and economic traditions. The first three, the dominant social groups, were articulated by the State. This was why Mariátegui argued that the liberal elite and "national" bourgeoisie were not dynamic actors for social change. Socialism and nationalism were, thus, complementary because "nation" was a project yet to be built (Mariátegui [1928] Mariátegui 1995).

During his short life, Mariátegui not only developed studies of social problems but dedicated a large part of his time to the creation of a regional circuit for the communication of indigenists and cultural groups throughout Latin America. Spearheaded by his journal *Amauta*, this network was a precursor to Latin American sociology as a locus and practice between the academy and politics, and in dialogue with local, regional, European traditions. Because of the fragmentary feature of this intellectual circuit – typical of a period prior to the institutionalization of sociology – his vision of structural heterogeneity, his original Latin-Americanization of Marxism, and his contribution to the definition of the "problem of the Indio" were not discussed until sociology and Marxism had a theoretical and empirical research encounter. All this was possible in the new conditions established by the new research institutes, the regional centers and faculties created after the Second World War. The Economic Commission for Latin America (ECLA), established in 1948, was a fundamental milestone in the development of socioeconomic knowledge in the region. The Commission systematized statistical information accumulated in public bureaus over previous decades, and stimulated national studies and the building of regional offices, as well as the technical training of the officials of the ministries of finance and planning. The Division of Social Studies, led by Jose Medina Echavarria, boosted the discussion of the social factors of development (see Consideraciones sociologicas sobre el Desarrollo Economico de America Latina-Sociological Considerations on Economic Development in Latin America, 1964). The Latin-Americanization of sociology was given further impetus by the creation of the Latin American Faculty of Social Sciences (1957) where the first graduate school of sociology started in 1958.

In previous studies, Beigel (2010) has explored the dynamic academic development that took place between 1964 and 1973, particularly in Chile, where many academic exiles arrived after the coup d'etat's in Argentina and Brazil. Santiago became the axis of this regional circuit, and it experienced an exceptional period of productivity in which new theories and concepts emerged. This contributed to the consolidation of indigenous sociological traditions, among them the theories of dependency and the debate on social marginality, crossed by interdisciplinary debates between economics and history. The Latin American Council for the Social Sciences (CLACSO), established in 1967, played a relevant role as a regional network, favoring the development of research groups and performing a determinant role in the regional circulation of sociological products. During those years, Chile became the main laboratory for an endogenous process of knowledge creation, in a context in which a political experience generated worldwide attention: the electoral victory of the Christian Democracy party in 1964 and soon followed by a democratic socialist alliance in 1973.

The dependency focus arose, therefore, in these academic circles as a theoretical problem with the intention to re-diagnose underdevelopment within a collective and interdisciplinary reflection. Dependency was outlined as an historical situation, occurring under certain national and international conditions, as the result of the global structure of underdevelopment. It was not conceived as an external imposition, but a relationship between industrialized and peripheral countries. The critique of developmental policies and economism led to questions on the: (a) rationality of the productive structure; (b) legitimacy principles of Latin American states; and (c) struggle for power. Accordingly, the social scientists working at the regional centers based in Santiago gave depth to the structuralist arguments on developmental policies. Most of them did this in discussion with ECLA, and pointing out its main limitations. Interestingly, Aníbal Quijano, who made a major contribution to the analysis of Peruvian class structure in the context of dependency, built his arguments through updating Mariátegui's *Seven Essays*, where he found a proposal of a structural-historical analysis of the relation between precapitalist and capitalist process of production and the labor force (Quijano 1977). One of Quijano's main interests was social marginality and its structural link with the expansion of capitalism, and he participated in the debate on the diagnosis of Latin American capitalism based on Mariátegui's legacy.

In addition to these reflections on the structuralist legacy, the readings of Mariátegui and other heterodox Marxists fed new contributions that added the final "stitches" to this theoretical and methodological focus. For example, a set of concepts that had been previously developed in the region, and which analyzed the historical relationship between social structures and political change, were outlined by Sergio Bagú in *Economía de la Sociedad Colonial* (Economy of Colonial Society), published in 1949. Bagú argued: "It wasn´t capitalism that appeared in America in the period we studied, but colonial capitalism. There was no servitude on a large scale, but slavery with multiple shades, hidden very often under complex and fallacious juridical formulas. Ibero-America was born to integrate the cycle of new-born capitalism and not to extend the agonizing feudalistic phase" (Bagú 1949:261). Bagú's project sought to create a unified history of capitalism in Latin America which would contribute to understanding its specific path far from the "semi-feudalistic" approach. Thus, in combination, the writings of José Carlos Mariátegui and colonial studies opened the path to an indigenous Marxist interpretation of dependent capitalism and enriched the development of the concept of structural heterogeneity.

Mariátegui's Contribution to the Development and Critique of Indigenism

The consolidation of Latin American sociology as a research field and regional space was indeed evidenced in the emergence of a local sociological tradition based on the historical-structural method and Marxism. A particular perspective combining class and nation was the basis of dependency analysis. But a previous tradition can be traced in the sociological studies of nation and class struggle: two social issues that were very actively developed in the first stirrings of Latin American Marxism. There are some homologies between the development of the concepts of class, race, and nation in Latin America and the sociological traditions in the United States and Europe, along with transnational dialogues, but also deep differences. Race had an erratic intervention in regional sociological research during the twentieth century, but conversely, a central role in the 1920s. Its appearance and later absence can be explained by the difficult relationship between Marxism and ethnicity, an issue that marks again Mariátegui's singular contribution to the birth of Latin American sociology.

In fact, the first sociological contribution to the analysis of race, nation, and class came with Mariátegui's *Thesis on the Problem of Race* (1929). His writings were, again, tightly linked to his praxis within the indigenist and workers' movements in his contemporary Peru. He argued that the Indian's problem was not ethnicity, but social and economic: related to land tenure. He believed indigenous people were the subject of the socialist revolution in Peru, given their history in the community and the fact that they formed the majority of the population. Noteworthy is his comprehension of the dominant anti-racial claim within the indigenous movement and the need to understand its roots. "When on the shoulders of the productive class weighs the harshest economic oppression added to the hatred and vilification of which it is a victim as a race, it is only a matter of clear and simple comprehension of the situation for this mass to rise up as one man to dump all forms of exploitation" (Mariátegui 1929).

His thesis was prepared for the communist conference held in Buenos Aires in June 1929, in a particular context. The official discourse existing in the Komintern was to preserve ethnic identity by conferring national autonomy on non-Russian communities that were part of the Union of Soviet Socialist Republics (USSR). On the other hand, the ideological and scientific struggle against biological racism prevented most Marxists from using the concept of race or stimulating its prevalence. During the conference, the Peruvian delegation presented Mariátegui's argument against considering the problem of the Indio as an ethnic-national issue, but instead as an economic problem, the result of the persistence of feudality. For the Peruvian delegates, the solution was Indo-American socialism and the revolution was a national task: to nationalize meant to decolonize. However, this did not prevent race from playing "a relevant role" in preparing for this revolution because only activists from the indigenous milieu, sharing a mindset and indigenous language, were able to achieve consent among their peers (Mariátegui 1929).

A fair assessment of Mariátegui's sociological contribution to the problem of race must be considered in light of two processes that mark his itinerary: (a) the context of the major political and cultural movement in which he took part, indigenism and (b) the shift that occurred along his praxis, from Eurocentric to rooted/creative Marxism. Concerning the first context, indigenism was a cultural and political movement existing from 1870 until 1970, basically concerned with indigenous redemption but featuring an externality from aboriginal communities. As a part of the socialist trend of indigenism, Mariátegui participated in the urban movement of his time that attempted to change the social conditions of indigenous people, postulating the project of Indo-American socialism. Race was, for him, a complementary issue, with class and nation being the main axes for defining "the problem of the Indio." However, and in contrast to other writings of the period, Mariátegui acknowledged the "exteriority" of indigenism. In one of his *Seven Essays*, he declared that indigenist literature did not offer a true image of the indigenes, because it was still a mestizo literature. "Precisely that is why it is called indigenist and not indigenous. The indigenous people´s literature is still to come and will arrive in its time. When the Indians themselves are able to produce it" (Mariátegui 1995 [1928]: 242).

Regarding the second process, in previous studies, Beigel (2003) has pointed out that the first display of Mariátegui's Marxism was in 1925 when he labelled his column "To Peruanize Peru," at the same time as he became a sociologist. His main concern was to build and understand the "primary problem" of his country. And he was already aware that this "Peruanization" had taken place within himself. When he returned to Peru in 1923, after 4 years in Europe, he was despised and considered a "Europeanizing" intellectual by nativists who rejected Marxism. By 1927, Mariátegui declared:

> "Regarding the confluence or alloy of indigenism and socialism, nobody attentive to contents and essence can be surprised. Socialism organizes and defines the claims of the masses, of the working class. And, in Peru, the masses – the working class – in four-fifths is composed of indians. Our Socialism wouldn't be Peruvian – it wouldn't even be Socialism – if it was not based on solidarity, firstly, with the claims of the Indians. In this attitude no opportunism is hidden. Neither artifice. If we give two minutes to a reflection of the meaning of socialism. This attitude is not fake, feigned or clever. It's just socialist" (Jose Carlos Mariátegui, 'Intermezzo polemico', Mundial, February 25, 1927).

When Mariátegui died, the Marxist tradition had a strong focus on class, while populism boosted the national perspective. Race, for its part, was subsumed by official indigenist policies, such as the celebration of miscegenation in Mexico and the conversion of Aztecas into a "national symbol." During the decades of 1940–1960, the state indigenisms assimilated the indigenous communities in the process of the consolidation of nation-states, while socialist indigenism was left to one side in the Marxist tradition. All this was expected to eventually change when the indigenous and black movements achieved their political and ideological autonomy in the struggle against the oligarchic foundations of societies and the prevalence of colonial relations – even in the Marxist tradition.

Actively part of this debate, but from a different perspective, Quijano (1992) transited from theories of dependency to coloniality, articulating class and race into a theory of domination. He argued that colonialism is an historical phenomenon that starts with the conquest of America, Asia, and Africa by the European powers and ends with decolonization revolutions developing from the beginning of the nineteenth century. Latin America was the first entity/historical identity of the current colonial-modern world system. It was constituted as the "Accidental Indies," the original space and beginning of a new pattern of power. It was the place of the first classification of the survivors of the colonial genocide as Indios – an "indigenization" expressed as a "racialization" that includes at the same time class and racial subalternity. Coloniality thus, has proven to be, in the last 500 years, more profound and lasting than colonialism (Quijano 2010). Quijano's "coloniality of power" reinforced the place of race-class as causal relations to explain inequalities, but tended to leave aside the national question typically central in the 1960s.

Conclusions

Mariátegui's contribution to Latin American Marxism has been certainly observed in the available literature (Aricó 1981; Terán 1986; Lowy 2007; Kohen 2010). His indigenist perspective, with its strengths and weaknesses, have also been discussed when his legacy came strikingly to life after the Zapotecan Subcommander Marcos named Mariátegui as the inspirer of their struggle in the Mexican jungle. Conversely, his role as the founder of the sociological basis for dependency analysis and the more entangled analysis of nation-race-class has generally gone unnoticed. Segato points out that until the beginning of the twenty-first century, it was rare in Latin America to find reports on the color of the poor, people in prison, or victims of police abuse. Race was mostly treated as a cultural feature but not part of the structural unequal condition of the population, even if it is a mark of subalternity and domination present since the Conquest (Segato 2010). The indigenist movement and state indigenism contributed greatly in this direction. As argued by Wade (2017), indigenous communities were analyzed as "peasants" and Afro-descendants were seen as "integrated" into an illusionary but persistent racial democracy. But the absence of Mariátegui in the sociological debate hardly responds to this circumstance.

As a result of the emergence of new indigenous/black movements by the mid-1990s, the "prosecution" of indigenism was fulfilled while research on race and gender entered increasingly into the regional sociological agenda. Eventually, it was after the collapse of communism that race came to the forefront. At first sacrificing the nation-class binomial, it was boosted by postcolonial studies and the belief that Latin-Americanism and dependency analysis were responsible for a now old perspective centered in the "nation-state" as a unit of analysis and "class" as a central principle of identity. The novelty was that the category of race became the main axis for the comprehension of domination and social inequality in the capitalist world system. These debates integrated a dialogue on eurocentrism that was already taking place in other spaces in the newly baptized "Global South." In fact, during these years, class was rephrased into the struggle between South and North, while the displacement of the category of "nation" was fulfilled. Race gained increasing interest in social research and has been a central issue for intersectional feminism, de-colonial perspectives, and studies on inequality.

After the victory of Evo Morales in the presidential elections of Bolivia (December 2005), a new type of government headed by an Aymaran leader changed the agenda of the social sciences. The "political instrument" that led Morales to power, the Movement for Socialism (MAS), was built in the midst of cocalero unionism and the massive demonstrations in defense of water and gas as natural (national) resources against transnational companies. These two forms of traditional struggle, historically embodied in syndicalism (class) and anti-foreigner movements (nation), arrived now with a sense of novelty. The growth of MAS and the social movements were based on the defense of traditional knowledge, communal justice, and indigene organizations.

In 2009, the Bolivian Constitution was completely reformed defining the country as pluri-national and recognizing all indigene nations and languages as official. In parallel, the Bolivian nation was affirmed, and a program of nationalizations performed. A broader, relevant change completed the new scenario: the resurrection of Latin-Americanism, starting with the regional initiatives of a pool of presidents including Hugo Chavez, Nestor Kirchner, Lula da Silva, and Evo Morales. In a research agenda with an already consolidated concern for race and class, nation came back on the scene, stirring up postcolonial theories that had based their reflections in the disappearance of nationalism as a collective identity.

Several critiques (Svampa 2016; Stefanoni 2017) were made to the developmentalist path used by Morales and the limits of de-patriarchalization, but undoubtedly a relevant morphological change took place in the Bolivian social milieu. To be an "Indio" was considered negative, and it was a symbol for subalternity in the history of the country. Currently, an Indianization of politics took place: to be an Indio meant having a form of relevant social capital impacting rapidly on the composition of the bureaucratic elites. Accordingly, the Bolivian experience, with its advances and backwardness, has been a decisive laboratory for a new perspective on inequality. And in the meantime, southern feminisms seem to be preparing the ground, or underground, for the massive emergence of a Latin American sociology of gender.

References

Aricó, (1978) José ARICÓ, Comp. Mariátegui y los orígenes del marxismo latinoamericano, 2 edición, Cuadernos de Pasado y Presente, N 60, México, Siglo XXI
Aricó, José (1981) La cola del diablo. Itinerario de Gramsci en América Latina, Buenos Aires, Puntosur
Bagú S (1949) Economía de la sociedad colonial. Ensayo de historia comparada de América Latina [Economy of Colonial Society. Essay on Comparative History of Latin America]. El Ateneo, Buenos Aires
Beigel F (2003) El itinerario y la brújula. El vanguardismo estético-político de José Carlos Mariátegui [The itinerary and the compass. Mariátegui and the political-aesthetical Avantgarde]. Biblos, Buenos Aires
Beigel F (2005) 'Una mirada sobre otra: el Gramsci que conoció Mariátegui' [A glance over another: what Gramsci did Mariátegui meet?]. Estudos de Sociologia, UNESP 10(18/19):23–49
Beigel F (2010) Dependency analysis: the creation of new social theory in Latin America. In: Patel S (ed) The ISA handbook of diverse sociological traditions. SAGE, London, pp 189–200
Blanco, Alejandro (2005) La Asociación Latinoamericana de Sociología: una historia de sus primeros congresos en Sociologías, Porto Alegre, año 7, número 14, pp. 22–49
Beigel F (2019) Latin American Sociology: A Centennial Regional Tradition, in Beigel, F. Ed. Key texts for Latin American Sociology. SAGE: London.
Franco, Rolando (1978) 25 años de Sociología Latinoamericana Revista Paraguaya de Sociología, Año 11, N30.
Gramsci, Antonio ([1956] 2001) Cuadernos de la cárcel. México, Ediciones ERA-Universidad Autónoma de Puebla
Löwy, M. (2007) Le marxisme en amérique latine de José Carlos Mariategui aux zapatistes du Chiapas. Actuel Marx Vol. 2 (n 42), p. 25 à 35

Martínez de la Torre, (1947) Apuntes para una interpretación marxista de Historia Social del Perú, Tomo II, Lima, Empresa Editora Peruana.

Mariátegui JC (1926) Presentación de Amauta [Presentation of Amauta]. Amauta 1:1

Mariátegui JC ([1928] 1995). Siete ensayos de interpretación de la realidad peruana [Seven interpretative essays on the peruvian reality]. Sociedad Editora Amauta, Lima.

Mariátegui JC (1927) 'Intermezzo polémico' [Controversial intermezzo]. Revista Mundial

Mariátegui, José Carlos (1929) 'El problema de la raza']The problema of race], Amauta, 25: 69–80.

Medina Echavarría J (1964) Consideraciones sociológicas sobre el desarrollo económico en América Latina [Social consideration on the economic development of Latin America]. Solar/Hachette, Buenos Aires

Mejía Navarrete, Julio (2005) El desarrollo de la sociología en el Perú. Notas introductorias, Sociologias, Porto Alegre, 7(14): 302–337

Pereyra, Diego (2005) La Asociación Latinoamericana de Sociología y su rol fundacional. Una historia sobre la organización institucional de la sociologóa en América Latina desde 1950 hasta 1960, Sociology: History, Theory and practices, Russian Society of Sociologists, Moscow-Glasgow, p. 155–173.

Quijano A (1992) "Raza, Etnia y Nación" [Race, ethnics and nation] in Mariátegui: Cuestiones Abiertas. In: José Carlos Mariátegui y Europa: El otro aspecto del descubrimiento. Ed. Amauta, Lima, pp 167–188

Quijano, Anibal (2010) "Bien Vivir para Redistribuir el poder. Los pueblos indígenas y su propuesta alternativa en tiempos de dominación global' [Well-living to redistribute power. Indigenous people and alternatives in times of global domination]. In Informe 2009–2010 Pobreza, desigualdad y desarrollo en el Perú. Oxfam, Lima.

Quijano, Aníbal (1977) Imperialismo y marginalidad en América Latina, Lima: Mosca Azul Editores.

Segato R (2010) 'Los cauces profundos de la raza latinoamericana: una relectura del mestizaje' [The deep roots of the Latin American race: re-reading miscegenation]. Crítica y Emancipación, Año II 3:11–44

Stefanoni P (2010) 'Bolivia después de las elecciones: ¿a dónde va el evismo?' [Bolivia after the elections, where is Evism leading to]. Nueva Sociedad 225

Svampa M (2016) Debates Latinoamericanos. Indianismo, Desarrollo, Dependencia, Populismo [Latin American debates. Indianism, development, Dependency, Populism]. EDHASA, Buenos Aires

Teránm O. (1986) Editorial Universidad Autónoma de Puebla, ICUAP.

Wade P (2017) Racism and race mixture in Latin America. Latin American Research Review 52(3): 477–485